Studies in Computational Intelligence

Volume 897

Series Editor

Janusz Kacprzyk, Polish Academy of Sciences, Warsaw, Poland

The series "Studies in Computational Intelligence" (SCI) publishes new developments and advances in the various areas of computational intelligence—quickly and with a high quality. The intent is to cover the theory, applications, and design methods of computational intelligence, as embedded in the fields of engineering, computer science, physics and life sciences, as well as the methodologies behind them. The series contains monographs, lecture notes and edited volumes in computational intelligence spanning the areas of neural networks, connectionist systems, genetic algorithms, evolutionary computation, artificial intelligence, cellular automata, self-organizing systems, soft computing, fuzzy systems, and hybrid intelligent systems. Of particular value to both the contributors and the readership are the short publication timeframe and the world-wide distribution, which enable both wide and rapid dissemination of research output.

The books of this series are submitted to indexing to Web of Science, EI-Compendex, DBLP, SCOPUS, Google Scholar and Springerlink.

More information about this series at http://www.springer.com/series/7092

Songsak Sriboonchitta ·
Vladik Kreinovich · Woraphon Yamaka
Editors

Behavioral Predictive Modeling in Economics

Springer

Editors
Songsak Sriboonchitta
Faculty of Economics
Chiang Mai University
Chiang Mai, Thailand

Vladik Kreinovich
Department of Computer Science
University of Texas at El Paso
El Paso, TX, USA

Woraphon Yamaka
Faculty of Economics, Center of Excellence
in Econometrics
Chiang Mai University
Chiang Mai, Thailand

ISSN 1860-949X ISSN 1860-9503 (electronic)
Studies in Computational Intelligence
ISBN 978-3-030-49727-9 ISBN 978-3-030-49728-6 (eBook)
https://doi.org/10.1007/978-3-030-49728-6

This Springer imprint is published by the registered company Springer Nature Switzerland AG
The registered company address is: Gewerbestrasse 11, 6330 Cham, Switzerland

Preface

One of the main objectives of economics—and, in particular, of econometrics, qualitative techniques for studying economic phenomena—is to predict consequences of different economic decisions and to come up with decisions that best benefit the economics. For this prediction, we need to know how people will behave under different circumstances. Traditional econometric techniques simplifyingly assumed—sometimes implicitly, sometimes explicitly—that each person always has all the available information and always has enough time and enough computing power to come up with the best possible strategy. In some cases, this assumption led to reasonable results, but in many other cases, it turned out to be too simplistic to make accurate predictions. In reality, people make decisions based on limited information and limited ability to process this information and, as a result, make decisions which are often far from optimal.

Nobel prizes have been awarded for discovering this sub-optimality of human behavior. This was not easy. But now comes an even more difficult part: how can we incorporate this often sub-optimal human behavior into our economic models? How can we come us with behavioral models that help us predict the consequences of economic decisions in situations where the current simplifying models cannot make good predictions? These are difficult tasks, this problem is far from being solved.

This volume contains the first results in this direction. Some of the papers describe newly developed behavioral models; some papers are more traditional—they prepare the ground for the future behavioral models; and some papers are foundational and theoretical; they will hopefully lead to successful models in a decade or so. Overall, papers from this volume present a good state-of-the-art slice of what can be done and what should be done in incorporating behavioral predictive models into econometrics.

We hope that this volume will help practitioners to become better knowledgable in behavioral techniques and help researchers to further develop this important research direction.

We want to thank all the authors for their contributions and all anonymous referees for their thorough analysis and helpful comments.

The publication of this volume was partly supported by the Faculty of Economics of the Chiang Mai University, Thailand. Our thanks to the leadership and staff of the Chiang Mai University for providing crucial support. Our special thanks to Prof. Hung T. Nguyen for his valuable advice and constant support.

We would also like to thank Prof. Janusz Kacprzyk (Series Editor) and Dr. Thomas Ditzinger (Senior Editor, Engineering/Applied Sciences) for their support and cooperation in this publication.

January 2020

Songsak Sriboonchitta
Vladik Kreinovich
Woraphon Yamaka

Contents

Theoretical Results

On Modeling of Uncertainty in Behavioral Economics

Hung T. Nguyen

Abstract This paper is about a closer look at mathematical analysis of various types of uncertainty, in natural science as well as in social sciences, for econometricians. The emphasis is upon the current effort to promote quantum probability as the most appropriate and reliable model for uncertainty in behavioral economics, both in human decision-making (e.g., in investment portfolio selection), and modeling of financial data, taking into account of human factor.

Keywords Bayesian probability · Belief functions · Bohmian mechanics · Choquet capacity · Fuzziness · Imprecise probabilities · Kolmogorov probability · Quantum entropy · Quantum mechanics · Quantum probability · Random sets

1 Introduction

It seems that history of science is repeated, and this time for social sciences. The first quarter of the 20th century was the time where quantum physics extended Newton and Einstein physics, from motions of big and very big things to extremely small things. The analogy is striking: with three Nobel Memorial Prizes in Economic Sciences in this first quarter of the 21st century *behavioral economics* surfaced as a new approach to economic studies, expanding traditional approach based upon standard notion of probability. Specifically, the analogy with physical laws (discovered from theoretical physics and validated by observed experiments) is the *modeling of the notion of uncertainty*.

A standard definition of behavioral economics is this. It uses the effects of psychological, cognitive and social factors on economic decisions of individuals and institutions and how those decisions vary from those implied by *classical theory* (based upon Kolmogorov stochastic analysis, e.g. [9, 14, 24, 25]).

It is all about *uncertainty!* As humans, we face uncertainty in almost all aspects of our life, and we make decisions in the face of it. As we will see, if uncertainty means

H. T. Nguyen (✉)
Department of Mathematical Sciences, New Mexico State University,
Las Cruces, NM 88003, USA
e-mail: hunguyen@nmsu.edu

Faculty of Economics, Chiang Mai University, Chiang Mai, Thailand

S. Sriboonchitta et al. (eds.), *Behavioral Predictive Modeling in Economics*,
Studies in Computational Intelligence 897,
https://doi.org/10.1007/978-3-030-49728-6_1

roughly that we do not know something, then there are various types of uncertainty, exemplified by randomness and imprecision (modeled by probability and fuzziness, respectively). Thus, when talking about uncertainty, it is essential to specify which type of uncertainty we are facing. In this paper, we focus on the "common" notion of uncertainty, namely uncertainty about the "occurrence of future events". We could call it "randomness" if we also distinguish two sub-types of randomness : intrinsic or not. This is exemplified by the distinction between "games of chance" and "games of strategies" (in microeconomics or else). The point is this. There is a fundamental difference between things happening by chance and those happening by free will. This is the distinction between natural and social sciences! As such, uncertainty in these two situations is different, and hence requires *different modeling.* For example, in (behavioral) finance, the portfolio chosen by an investor is based upon how she *feels* about the uncertainty involved. In other words, she has her own concept of uncertainty. Thus, if "theoretical" experts in uncertainty analysis are going to advise her how to behave, they must first correctly model her uncertainty! It is all about *how to model uncertainties?*

Clearly, as social sciences are much younger than natural sciences, approaches to the former are somewhat inspired from the latter. But like physics (where we cannot use Newton's mechanics to study mechanics of tiny particles), we should not use (standard) probability to, say, predict particles' motion, as Feynman [10] has quickly pointed out to probabilists and statisticians in 1951. The intrinsic nature of randomness in the particle's world is different than the randomness in games of chance whose uncertainty was modeled by probability since Blaise Pascal (1623–1662) and Pierre de Fermat (1601–1665). It requires another way to model *quantum uncertainty* (axioms and calculus).

Now, what is the "analogy" between particles' behavior (in "choosing" their paths of motion) and humans' free will, so that we could view quantum uncertainty as the uncertainty that humans actually face (and feel) when they make decisions?

Well, here is what the physicist Freeman Dyson ([5], p. 20) has to say:

"Atoms in the laboratory are weird stuff, behaving like active agents rather than inert substances. They make unpredictable choices between alternative possibilities according to the laws of quantum mechanics. It appears that mind, as manifested by the capacity to make choices, is to some extent inherent in every atom."

At the "beginning", there is probability as a quantitative model for (natural) uncertainty. What is probability? Well,

"After all, the theory of probability is only common sense reduced to calculation" and

"Probability is the most important concept in modern science, especially as nobody has the slightest notion what it means"

And when theorists advise people how to make decisions (using probability), here is what happened (Hawking [12], p. 47):

"Economics is also an effective theory, based on the notion of free will plus the assumption that people evaluate their possible alternative courses of action and choose the best. That effective theory is only moderately successful in predicting behavior because, as we all know, decisions are often not rational or are based on a defective analysis of the consequences of choice. That is why the world is in such a mess".

It is "interesting" to note that while developing the mathematical language for quantum mechanics in 1932 (including quantum probability), von Neumann ([27], the first version was in German, 1932; the reference is the English translation, published by Princeton University Press in 1955), moved on the inaugurate quantitative economics with his theory of games intended to model "economic behavior" in 1944 [28], in which, instead of using (his own) quantum probability to model *human uncertainty*, he used Kolomogorov probability (which was formulated in a general framework in 1933)! Now, it is our turn to do what von Neumann has missed! This is exemplified by recent works on *quantum games.*

But prior to the current research efforts in replacing Kolmogorov probability by quantum probability (for social sciences only, not for natural sciences, except quantum physics), what happened? Right after von Neumann's theory of decision-making (based on Kolmogorov probability) was taking place, Allais [1], followed by Ellsberg [8], pointed out that additivity property of probability is not "consistent" with the way humans evaluate chances. From which, and especially from the work of Kahneman and Tversky [13], worldwide efforts were focusing on extending probability theory to various non-additive *set-functions*, such as , *belief functions, imprecise probabilities, fuzzy measures*, intended to be models of uncertainty, in the hope they model better humans' view of uncertainty, and hence more realistic to use in decision-making. In closing his famous Text on the Theory of Choice [14] in 1088, Kreps has this to say:

"These data provide a continuing challenge to the theorist, a challenge to develop and adapt the standard models so that they are more descriptive of what we see. It will be interesting to see what will be in a course on choice theory in ten or twenty years time."

Now, over 30 years, what do we see as advances in a choice theory? Well, besides very recent efforts aiming at using quantum probability as the best model for humans' uncertainty for decision-making in all aspects of social activities (by analogy with intrinsic randomness of particles, and the complete success of quantum mechanics), all research works seem unsatisfactory. Why? Well, they forgot to ask psychlogists to help to validate their proposals! As Stephen Hawking [12] reminded us:

"In the case of people, since we cannot solve the equations that determine our behavior, we use the effective theory that people have free will. The study of our will, and of the behavior that arises from it, is the science of psychology".

What is the missing point in all research so far (in the efforts to propose non-additive models of uncertainty to capture human's factor in decision-making)? As experiments conducted by psychologists revealed, it is the *noncommutativity* of uncertainty measure. All current measures of uncertainty are commutative ones! As we will see, quantum probability is noncommutative.

This paper is organized as follows. In Sect. 2, we discuss all uncertainty measures in the literature. In Sect. 3, we elaborate on the notion of quantum probability. And, finally, for a tutorial paper such as this, we indicate how to use quantum probability in behavioral econometrics.

2 A Panorama of Classical Uncertainty Models

We start out by looking back at various quantitative models of uncertainty proposed so far in the literature, keeping, of course, quantum probability for the next section. These are commutative uncertainty measures. The purpose of this review is to take a closer look at how axioms of each proposed uncertainty measure have arised, and whether they were well founded, i.e., capturing the meaning of the uncertainty notion under considerations.

2.1 Probability

The oldest model is of course the quantitative model for the concept of chance. The idea that chance can be measured led to efforts to suggest quantitative measures of it. An excellent reading on "Great ideas about chance" is [7].

The state-of-the-art of the quantitative modeling of chance is the theory of probability as formulated by Andrey Kolmogorov in 1933. It is a mathematical theory, in the spirit of Hilbert's sixth problem, in which the theory is established by axioms. Probability is the quantitative model for the notion of chance. How to arrive at axioms for probability? Well, from games of chance, i.e., in uniform, finite cases, where the probability of an event to occur can be measured as the ratio of favorable cases over possible cases, to general situations, using measure theory in mathematical analysis, Kolmogorov arrived at the now well known axioms of a probability measure. Specifically, let $\mathscr{A}$ be a $\sigma-$field of subsets of a set Ω, representing events (say, of a random experiment whose possible outcomes are Ω), the probability measure is a set-function $P(.)$, defined on $\mathscr{A}$ with values in the unit interval $[0, 1]$, satisfying two axioms: $P(\Omega) = 1$, and countably additive (i.e., $P(\cup_{n\geq 1} A_n) = \sum_{n\geq 1} P(A_n)$, for finite or infinite countable n, and when the events A_n are pairwise disjoint). The triple $(\Omega, \mathscr{A}, P)$ is referred to as a probability space, describing a random phenomenon, from which a *probability calculus* follows. Kolmogorov probability theory is applied successfully to almost all areas of sciences, such as statistics and classical physics.

It should be noted that since events are subsets of the sample space, probability measures are *commutative* in the sense that $P(A \cap B) = P(B \cap A)$.

Remark
The concept of $\sigma-$ field (borrowed, of course, from measure theory) is considered, not only by technical reason (the power set 2^Ω of Ω is too big to define probability measures), but also by practical reasons. The similarity in quantum mechanics (see the next section) will make this appearent: "observables" in quantum mechanics must be self-adjoint operators.

Few more important things to note about probability. Since probability theory is based on the theories of functions and measures in mathematics, random variables are well defined, as well as their laws (distributions), where (Lebesgue) integral defines expected values (we will constrast these with the situation of uncertainty in quantum mechanics where another probability calculus is based on a new mathematics due to von Neumann [27]).

First, how probability calculus is applied to (human) decision-making? Testing of hypotheses is an example of decision-making. As stated clearly in [7] (p. 115):

"A cookbook frequentist hypothesis tester doesn't have to think. He calculates a p-value".

If so, where is the human's factor? It is well known by now that the "mechanical" use of p-values in hypothesis testing ("mechanical" means "no need of thinking"!) is "outdated", so that as far as statistical inference for applications is concerned, econometricians must turn to Bayesian testing, noting that Bayesian probability (Thomas Bayes, 1763) models also epistemic uncertainty. Before making a decision, the statistician needs to think a bit! That could be the choice of a (subjective) prior distribution. That is a human factor! Thus, in a sense, besides logical aspects, Bayesian testing (as a human decision-making process) captures some aspects of "behavioral decision-making".

Secondly, in the context of (human) decision-making, obviously the notion of *risk* plays an essential role. Since risk is caused by uncertainty, it seems reasonable that it can be measured (quantified) from the probability calculus. However, the semantic of risk should be examined carefully to propose an appropriate quantification of it in each given context. For example, if X is a (random loss) variable (say, in an investment), the notion of *variance* of X, which is appropriate to measure the error of the measurements, but not appropriate for "downside risk" in financial context, as the variance $V(X)$ is "symmetric". Without looking for another probability calculus reflecting human thinking, the state-of-the-art for "risk analysis", especially in financial econometrics, is this (for more details, see a text like [26]).

Let $\mathscr{X}$ be a class of random variables of interest. A (coherent) risk measure is a functional $\rho(.) : \mathscr{X} \to \mathbb{R}$ satisfying the following properties (axioms):

(i) $X \leq Y$ (a.s) $\Longrightarrow \rho(X) \leq \rho(Y)$
(ii) For $\lambda \geq 0$, $\rho(\lambda X) = \lambda\rho(X)$
(iii) For $a \in \mathbb{R}$, $\rho(X + a) = \rho(X) + a$
(iv) Sub-additivity: $\rho(X + Y) \leq \rho(X) + \rho(Y)$

Note that axioms (ii) and (iv) entail the convexcity of $\rho(.)$, i.e., for $\lambda \in [0, 1]$, we have $\rho[(\lambda X) + (1 - \lambda)Y] \leq \lambda\rho(X) + (1 - \lambda)\rho(Y)$, reflecting "diversification" in portfolio selection.

As such, $V(X)$, the value-at-risk $VaR_\alpha(X) = F_X^{-1}(\alpha)$ are not coherent risk measures. However, the tail value-at-risk $TVaR_\alpha(X)$ is, where, for $\alpha \in (0, 1)$,

$$TVaR_\alpha(X) = \frac{1}{1-\alpha}\int_\alpha^1 F_X^{-1}(t)dt$$

It is "interesting" to note that all risk measures can be defined in terms of another set-function called *(Choquet) capacity*. This is so since, unlike expected values (which is defined via Lebesgue measure), other risk measures are defined in terms of quantile functions (which exist regardless of whether the variable distributions are heavy-tailed or not) which turn out to be defined via another type of "integral". Indeed, if we let $g_\alpha : [0, 1] \to [0, 1]$, where $g_\alpha(x) = 1_{(1-\alpha,1]}(x)$, then

$$F_X^{-1}(\alpha) = \int_0^\infty g_\alpha(1 - F_X(t))dt = \int_0^\infty (g_\alpha \circ P)(X > t)dt = \int_0^\infty \nu(X > t)dt$$

where the set-function $\nu = g_\alpha \circ P$ (on $\mathscr{A}$) is a Choquet capacity which is not additive. Note however that ν is just a set-function (measuring capacities in Potential Theory) and not necessarily a measure of any type of "uncertainty" per se.

If we let $\nu = g_\alpha \circ P$ where $g_\alpha(x) = \min\{1, \frac{x}{1-\alpha}\}$, then $TVaR_\alpha(X) = \int_0^\infty \nu(X > t)dt$. Thus, risk measures based on probability have "Choquet integral representations". This suggests that general risk measures can be defined by capacities, via Choquet integrals, where by a capacity, we simply mean a set-function $\nu(.)$ such that $\nu(\varnothing) = 0$, $\nu(\Omega) = 1$ and monotone: $A \subseteq B \Longrightarrow \nu(A) \leq \nu(B)$, and the Choquet integral of X is

$$C_\nu(X) = \int_0^\infty \nu(X > t)dt + \int_{-\infty}^0 [\nu(X > t) - 1]dt$$

In summary, under the umbrella of probability calculus, Choquet capacities (and integrals), generalizing probability measures as (non-additive) set-functions, surface as "general risk measures", but not (yet) as uncertainty measures, recalling that risks came from uncertainties! But, if we just look at $C_\nu(X)$ as a risk measure defined from the capacity $\nu(.)$ directly, then it is possible to view $\nu(.)$ as the "uncertainty" generating it. The question is: What is the uncertainty that the capacity $\nu(.)$ is intended to provide a quantitative modeling?

2.2 Entropy

There is another concept of uncertainty on the state of a stochastic system, not capturing directly by probability, but based on it. For example, before tossing the (biased) coin, or performing an experiment on the system X, we are uncertain about its outcome. The uncertainty about the state 0 (resp. 1) is quantified by the probability p (resp. $1 - p$). But how about the uncertainty about its outcome which could be either 0 or 1? (when we do not know which value X will take, we say that we have uncertainty about its value). In other words, when we perform an experiment, X will take some value, i.e. X will be in some "state", but we are uncertain about which value it takes. We are talking about the *uncertainty on the state of a stochastic system*. This type of uncertainty is useful in many contexts where, say, comparing of stochastics systems is of concern.

Then the question is "How to *quantify* this type of uncertainty?", i.e., "How to "measure" it?". Well, for a random variable X, taking values in a finite set

$\{x_1, x_2, ..., x_m\}$ with (known) probability (mass) density function $f(x_i) = p_i, i = 1, 2, ..., m$, its global uncertainty should depend on $p_i's$, i.e., is some function $f_m(p_1, p_2, ..., p_m)$. To find it, we need to specify the axioms for such a measure. It is natural to require that $f_m(.)$ should satisfy the following axioms.

(i) For any m, $f_m(.)$ is symmetric (this axiom is natural since the numbering of the states x_i is arbitrary),
(ii) If we add the impossible event, say, $\varnothing$, to the states, then

$$f_{m+1}(p_1, p_2, ..., p_m, 0) = f_m(p_1, p_2, ..., p_m)$$

(iii) $f_m(.)$ is maximum for $p_1 = p_2 = ... = p_m = \frac{1}{m}$ (noting that the p_i are *rational numbers*)
(iv) If X takes $k = mn$ values, then

$$f_{mn}(r_{11}, ..., r_{1n}, r_{21}, ..., r_{2n}, ..., r_{m1}, ..., r_{mn})$$

$$= f_m(p_1, p_2, ..., p_m) + \sum_{j=1}^{m} p_j f_n(\frac{r_{j1}}{p_j}, ..., \frac{r_{jn}}{p_j})$$

where $r_{ji} = P(X = x_{ji})$, $j = 1, 2, ..., m$, $i = 1, 2, ..., n$, with x_{ji} denoting the states $(\sum_{j,i} r_{ji}r_{ji} = 1)$ and $p_j = \sum_{i=1}^{n} r_{ji}$.

It turns out that the solution is unique and given by

$$f_m(p_1, p_2, ..., p_m) = -\sum_j p_j \log p_j$$

called the *entropy* of X.

Entropy is a measure of a "global" uncertainty about a random variable (or stochastic system) X. It is a function of probability, but it is not about measuring the same type of uncertainty (i.e., uncertainty about the possible occurence of events). The entropy operator $H(.)$ operates on random variables (in fact on distribution of random variables) with values in $\mathbb{R}^+$ (for the time being, for finite random variables). Specifically, if we let $p = (p_1, p_2, ..., p_n)$ such that $p_i \in [0, 1]$, $\sum_{i=1}^{n} p_i = 1$ (i.e., p is a finite probability density), i.e., $p \in S_n$ (that simplex), then the domain of H is $\cup_{n=1}^{\infty} S_n$. Thus,

$$H(.) : \cup_{n=1}^{\infty} S_n \to [0, \infty)$$

If X has density p, we write $H(X) = H(p)$ which measures the amount of uncertainty on the state of X, and not the probability of getting some value!

Remark
While it is well known that the notion of entropy is useful in a variety of areas, such as physics, statistics (via E.T. Jaynes maximum entropy principle), it starts to enter social sciences as well, such as financial econometrics, since entropy can be used to

characterize the diversification in investment portfolios. We will elaborate on it in the next section where, for behavioral econometrics, its extention to *quantum entropy* will be mentioned.

2.3 Fuzzy Measures and Possibility

In 1965, Lotfi Zadeh realized that there are physical objects which cannot be described precisely in mathematical terms, but only in natural languages. And it is so because of their intrinsic imprecision. For example, "beautiful flowers", "heavy rain", "healthy people", etc. To use such imprecise "objects" in some scientific investigations, such as control engineering, it is necessary to describe them mathematically, i.e., creating a new mathematical language. This is a well known procedure in science, exemplified by Newton's calculus to describe "acceleration", and von Neumann's functional analysis to describe quantum events and quantum measurements. Upfront, the *fuzziness* of such objects, as Zadeh put it, is not a new type of uncertainty per se. Instead, it is a kind of ambiguity, imprecision, semantically. An outcome of a random experiment could by fuzzy. An event could be a fuzzy event.

Roughly speaking, a fuzzy set is a set with no sharply defined boundary. An ordinary (crisp) set is not fuzzy because we can tell precisely which elements are in it and which elements aren't. How to "describe" a fuzzy set in mathematical terms? Well, this is just an extention (generalization) problem in mathematics. Let A be a subset of Ω, we wish to extend is to a fuzzy subset of Ω. We cannot do it directly, so let's do it indirectly! just like we way we extend real numbers to complex numbers. By that we mean looking for an equivalent way to represent a subset, but this equivalent representation can be extended. Now a subset $A \subseteq \Omega$ is equivalent to its indicator (or membership) function $1_A(.) : \Omega \to [0, 1]$, since $x \in A \Longleftrightarrow 1_A(x) = 1$. Thus, a crisp set A is characterized by a function from Ω to the two-point set $\{0, 1\}$. If we extend memberships to partial memberships, i.e., extend $\{0, 1\}$ to the whole unit interval $[0, 1]$ to obtain membership functions for fuzzy sets. Thus, the mathematical definition of a fuzzy set (of Ω) is a function $\mu(.) : \Omega \to [0, 1]$. Given a fuzzy concept (in natural language) such as "healthy people" (in a population), the specification of its membership is subjective and clearly depends on several factors (contributing to the property of "healthy"). For all mathematics of fuzzy set theory and related "fuzzy" measures, see [17].

In 1974, Michio Sugeno considered *fuzzy measures* as extensions of Borel measures by essentially dropping additivity. The motivation is the problem of subjective evaluation of objects. Let Ω be the possible states of nature and $\omega_o \in \Omega$ is the true state but unknown to the observer. For a subset A of Ω, the degree to which A could contain ω_o is a number in $[0, 1]$, acting like a fuzzy membership. Such a degree is the value of a set function $\mu(.)$ defined on 2^{Ω}, called a *fuzzy measure* with properties just like a (Choquet) capacity. The distinction is in the meaning of the values of $\mu(A)$: For example, $\mu(A) = 0.7$ means that the observer's *degree of "belief"* that ω_o is in A is 0.7. A general fuzzy measure $\mu(.)$ is a set function such that $\mu(\varnothing) = 0$, and monotone increasing. In this sense, a fuzzy measure can be viewed as an uncertainty measure where uncertainty refers to as the ignorance of the true state of nature. This can be also viewed as the counterpart of set (interval) estimation in statistics, where probability of coverage plays the role of "degree of confidence".

In 1978, Zadeh considered *possibility theory* as a softening of probability [29] (events can be improbable but possible!) His theory of fuzzy sets provides the basis for quantifying the linguistic concept of possibility. By itself, possibility is another type of uncertainty. A possibility measure is a set function $Poss(.)$, defined on 2^{Ω} with values in $[0, 1]$, such that $Poss(\cup_{i\in I} A_i) = \sup_{i\in I} Poss(A_i)$ for any index set I. In particular, the membership function $f(.) : \Omega \to [0, 1]$ of a fuzzy subset of Ω acts as a *possibility distribution* (of a "fuzzy variable") generating the associated possibility measure $Poss(A) = \sup_{\omega\in A} f(\omega)$.

2.4 Belief Functions and Related Measures

When considering epistemic uncertainty, Bayesians interprete their subjective probabilities as degrees of belief. In 1976, Shafer [24] proposed a theory of *belief functions* as a "mathematical theory of evidence". The buzzword is *evidence.* An evidence does not contain all information needed to make a reliable decision or to carry out an analysis. A recent striking (important) example is the use of p-values to make decisions in testing problems without realizing that the knowledge of a p-value is only an evidence! As stated in [7] (p. 116–117):

"A Bayesian, looking for the probability that the effect is real given the evidence, would say that the p-value is only part of the story. There is the prior probability of a real effect, which may depend on the field".

"There is now a movement to go beyond mechanical use of p-values: the proper guidance is Bayesian".

The point is this. Frequentist testing uses only the p-values (an evidence) to reach decisions, whereas Bayesians use some available evidence *and* (subjective) prior information for that. If only evidence is available, then the inference should be based upon *another probability calculus.*

In a sense, Shafer's belief functions could be interpreted as an attempt to propose a "weak" probability calculus to handle (i.e., to make inference with) evidence (partial knowledge). It is about "partial information" and how to model uncertainty under it?

An evidence (e.g., a p-value) creates an uncertainty since it does not provide all information needed for the decider to decide. Shafer's belief function theory is an attempt to propose a modeling of this *type of uncertainty* (arising from an evidence) for decision analysis (extending the complete prior information of Bayesian statistics). How he modeled this type of uncertainty? He represented it quantitatively by a set function, called a belief function. To make things simple, let U be a finite set. A belief function on U is a function $F(.) : 2^U \to [0, 1]$ satisfying the axioms

(i) $F(\varnothing) = 0$
(ii) $F(U) = 1$
(iii) $F(.)$ is monotone of infinite order, i.e., for any $k \geq 2$, and $A_1, A_2, ..., A_k$, subsets of U, where $|I|$ denotes the cardinality of the set I, we have

$$F(\cup_{j=1}^{k} A_j) \geq \sum_{\varnothing \neq I \subseteq \{1,2,...,k\}} (-1)^{|I|+1} F(\cap_{i \in I} A_i)$$

Note that the axiom (iii) is simply a modification of Henri Poincare' 's equality for probability measure. It makes the set function $F(.)$ non additive. The value $F(A)$ is meant to be the "degree of belief" in A. Thus, beliefs are a weaker form of probabilities. They are so since the underlying evidence creates a partial knowledge situation (to specify a probability law).

You might be curious to figure out the rationale of the axiom (iii)! It came from a previous work of Dempster [6] on upper and lower probabilities induced by multi-valued mappings. Thus, taking beliefs functions as non additive uncertainty measures for the uncertainty created by evidence (partial knowledge), an associated *calculus* was developed for applications, such as decision-making, see e.g., [21].

But, as pointed out by Nguyen in 1978 [20] (see also [16]), a belief function $F(.)$ is a bona fide *distribution of a random set* S, defined on some $(\Omega, \mathcal{A}, P)$ taking values in 2^U, namely, $F(A) = P(S \subseteq A)$, so that, in fact, the analysis of belief functions is within the context of probability theory, where partial knowledge forces to consider random sets (set-valued random elements, such as in coarse data cases) rather than random variables (point-valued random elements). Note that, as a special case when S is a random variable X, the belief function $F(.)$ reduces to $F(A) = P(X \in A)$.

It should be noted that, if F is a belief function, then $F = \inf\{P : P \geq F\}$, where P's are probability measures on U. This is a special case of *imprecise probabilities* (say, in robust Bayesian statistics). For example, when U is the parameter space (of some statistical model), a robust Bayesian analysis refers to the consideration of a set of prior probability measures rather than just a single one. Specifically, let $\mathcal{P}$ be a set of priors, suitable for modeling possible prior information in a given problem. Inference will then be based upon the lower and upper envelops (as uncertainty measures):

$$L_-(A) = \inf_{P \in \mathcal{P}} \{P(A)\} \text{ and } L_+(A) = \sup_{P \in \mathcal{P}} \{P(A)\}$$

These are quantitative models for the uncertainty about the "correct" prior probability measure (which is known only to be in the set $\mathcal{P}$).

3 Entering Quantum Uncertainty

Upfront: All uncertainty measures in the previous section have one thing in common: they are commutative set functions, although except Kolmogorov probability, the others are non additive. They are commutative since they are set functions, i.e., mappings defined on sets (their structure is Boolean).

Here, let's talk first about another type of uncertainty, not in any social context, but in physics. The next section will connect it to social sciences.

It is well known that quantum mechanics is intrinsically random. Its randomness is the same as in games of chance, i.e., the probability of a quantum event (e.g., where an electron will land) has the same meaning as that of an ordinary event. However, even the meaning of probability is the same, it does not necessarily imply that its associated calculus (the laws to combine, manipulate probabilities of various events) is the same. How can it be? Well, do not forget that an uncertainty theory is based on its axioms, from which its associated calculus follows. For example, all calculi of probabilities follow from its axiom of (countably) additivity. Now, by nature, as Feynman pointed out [10] to statisticians, the calculus of probabilities in quantum mechanics does not obey the additivity axiom of Kolmogorov (similar to Ellsberg [8] from psychological experiments). Moreover, and this is important (and "consistent" with human's behavior), quantum probability is *noncommutative,* i.e., for two "quantum events" A and B, their "quantum" probability $Q(A\&B)$ could be different than $Q(B\&A)$. Of course that is possible if the connective "and" (&) is not commutative, which in turn, requires appropriate concept of "quantum events" which should be more general than sets.

In a tutorial exposition such as this, it suffices (following David Hilbert's advice "what is clear and easy to grasp attracts us, complications deter") to describe quantum probability as a simple *generalization* of Kolmogorov (commutative) probability theory to a noncommutative one, and in the finite case.

The simplest method for doing so consists of two steps. Just like the way Zadeh generalized crisp sets to fuzzy sets, first, we seek some equivalent representation of probability theory which can be extended. Second, we are precisely looking for a noncommutative extension. Thus, the strategy is looking for an equivalent representation which can be extended to a noncommutative theory.

Let $(\Omega, \mathscr{A}, P)$ be a finite probability space, say, with $\Omega = \{1, 2, ..., n\}$, and hence $\mathscr{A} = 2^{\Omega}$, and $P(A) = \sum_{j\in A} P(j)$, $A \subseteq \Omega$. A random variable X is a map from Ω to the real line $\mathbb{R}$. To extend this finite probability space to a noncommutative one, we proceed as follows.

Each random variable X can be represented equivalently as a $n \times n$ diagonal matrix $[X]$ with diagonal entries being $X(1), X(2), ..., X(n)$. Since a subset (event) A of Ω is equivalent to its indicator function $1_A(.) : \Omega \to \{0, 1\}$, i.e., the vector $(1_A(1), 1_A(2), ..., 1_A(n))'$ (transpose), so that, likewise, it can be represented by the $n \times n$ diagonal matrix $[A]$ whose diagonal terms are $1_A(1), 1_A(2), ..., 1_A(n)$. Finally, the probability density function $P(.) : \Omega \to [0, 1]$ is identified with the $n \times n$ diagonal matrix $[\rho]$ with diagonal entries being $P(1), P(2), ..., P(n)$.

With this identification (representation), we transform elements of the probability space $(\Omega, \mathscr{A}, P)$ into $n \times n$ (real) matrices which are linear operators on $\mathbb{R}^n$, noting that diagonal matrices are symmetric.

Now observe that the diagonal matrix $[A]$, representing an event, is a projector on the Hilbert space $\mathbb{R}^n$, and the probability density P is represented by the positive operator $[\rho]$ with unit trace.

Remark

(i) By a projector p we mean an orthogonal projection onto a closed subspace of the Hilbert space $\mathbb{R}^n$. p is a projector if $p = p^2$ and $p = p^*$ (transpose). An operator A is said to be positive if $< Ax, x > \geq 0$ for all $x \in \mathbb{R}^n$, and we denote it as $A \geq 0$. As such, it is necessarily symmetric. With the partial order $A \geq B$ if $A - B \geq 0$, projectors p are positive such that $0 \leq p \leq 1$ (identity operator on $\mathbb{R}^n$).

(ii) It is clear that the expected value of a random variable X, represented by $[X]$, under the density P, represented by $[\rho]$, is $tr([\rho][X])$, the trace of the matrix multiplication $[\rho][X]$. Thus, as we will see, trace operator will replace integral in the analysis.

(iii) The ordinary entropy of P, namely $H(P) = -\sum_{j=1}^{n} P(j) \log P(j)$, is the same as $-tr([\rho]\log[\rho])$ (quantum entropy) where $\log[\rho]$ is the diagonal matrix with diagonal terms being $\log P(1), \log P(2), ..., \log P(n)$.

With the above identification of $(\Omega, \mathscr{A}, P)$, we are now ready to extend Kolomogorov commutative probability to a noncommutative one, simply by extending the (commutative) sub- algebra of diagonal matrices to the *noncommutative* algebra of (arbitrary) symmetric matrices. Specifically, the sample space $\Omega = \{1, 2, ..., n\}$ is replaced by the (finitely dimensional) Hilbert space $\mathbb{R}^n$, the field of events $\mathscr{A}$ becomes the set of all projectors on $\mathbb{R}^n$, denote as $\mathscr{P}$ (a lattice, non boolean), and the probability P becomes a "density matrix" which is an arbitrary positive operator ρ (on $\mathbb{R}^n$) with unit trace. The triple $(\mathbb{R}^n, \mathscr{P}, \rho)$ is called a (finite dimensional) *quantum probability space.*

From this construction, we see that, when considering "quantum uncertainty", we arrive at a non boolean structure of events, and a noncommutative and non additive probability calculus.

Remark

(i) A general quantum probability space is of the form $(H, \mathscr{P}(H), \rho)$ where H is a complex, separable Hilbert space, $\mathscr{P}(H)$ is the set of projectors on H (quantum events), and ρ is a positive operator with unit trace. Quantum random variables, called observables, are self-adjoint operators.

(ii) While ρ plays in fact the role of a "probability density", it is a general representation of a "quantum probability measure" on $\mathscr{P}(H)$, for the dimension of H more than 2, as $tr(\rho p)$ for $p \in \mathscr{P}(H)$, which is the extension of $P_f(A) = E_f(1_A) = \int_A f(x)dx$ when the probability measure P on $(\Omega, \mathscr{A})$ has a probability density function f. This is due to Gleason's theorem, see, e.g., [23] for a proof. This fact justifies to take $(H, \mathscr{P}(H), \rho)$ as the general form of a quantum probability space.

(iii) In quantum mechanics, the density matrix ρ is obtained from the Schrodinger's wave equation which is the counterpart of Newton's law of classical mechanics.

Specifically, the random law governing the particle dynamics (with mass m in a potential $V(x)$) is a wave-like function $\psi(x, t)$, living on the complex, separable Hilbert space $L^2(\mathbb{R}^3, \mathscr{B}(\mathbb{R}^3), dx)$, which is the solution of the complex PDE

$$ih\frac{\partial\psi(x,t)}{\partial t} = -\frac{h^2}{2m}\Delta_x\psi(x,t) + V(x)\psi(x,t)$$

where Δ_x is the Laplacian (in $\mathbb{R}^3$), i is the complex unit $\sqrt{-1}$, and h is the Planck's constant, with the meaning that the wave function $\psi(x, t)$ is the "probability amplitude" of the position x at time t, i.e., the function $x \in \mathbb{R}^3 \to |\psi(x, t)|^2$ is the probability density function for the particle position at time t. Richard Feynman computed $\psi(x, t)$ using his "path integral" which is used in "quantum finance" [2]. This probability density allows physicists to compute, e.g., the probability that the particle will land in a neighborhood of a given position x.

Without going into details, we write $\psi(x, t) = \varphi(x)\eta(t)$ (separation of variables) with $\eta(t) = e^{-iEt/h}$, and using Fourier transform, we can choose $\varphi \in H$ with $||\varphi|| = 1$. Let φ_n be a (countable) orthonormal basis of H, we have $\varphi = \sum_n <\varphi_n, \varphi> \varphi_n = \sum_n c_n\varphi_n$, with $\sum_n |c_n|^2 = 1$. Then, $\rho = \sum_n c_n|\varphi_n >< \varphi_n|$ is a positive operator on H with

$$tr(\rho) = \sum_n <\varphi_n|\rho|\varphi_n> = \sum_n \int \varphi_n^*\rho\varphi_n = 1$$

Note that we use Dirac's notation, namely, for $\tau, \alpha, \beta \in H$, $|\alpha >< \beta|$ is the operator sending τ to $< \beta, \tau > \alpha = (\int \beta^*\tau dx)\alpha$. If A is a self-adjoint operator on H, then

$$tr(\rho A) = <\varphi|A|\varphi> = \sum_n c_n <\varphi_n|A|\varphi_n>$$

Thus, the "state" $\varphi \in H$ determines the density matrix ρ in $(H, \mathscr{P}(H), \rho)$. In other words, ρ is the density operator of the state φ.

(iv) Now as the "density matrix" ρ on a general quantum probability space plays the role of an ordinary probability density function f (whose ordinary entropy is $-\int f(x)\log f(x)dx$), its *quantum entropy* (as defined by von Neumann [27]) is $-tr(\rho\log\rho)$. As maximum (Kolmogorov) entropy principle provides equilibrium models in statistical mechanics or other stochastic systems, it also enters financial econometrics as the most diversified portfolio selection.

Let's specify von Neunman's quantum entropy in a simple case, e.g., when $H = \mathbb{C}^n$.

For a density matrix ρ (extension of a probability density function) on $(\mathbb{C}^n, \mathscr{P}(\mathbb{C}^n))$, $\rho\log\rho$ is a $n \times n$ self adjoint matrix (operator) which is defined as follows (by using spectral theorem). The spectral theorem says this. Since ρ is a self adjoint operator on $\mathbb{C}^n$, there exists an orthonomal basis of $\mathbb{C}^n$, $\{u_1, u_2, ..., u_n\}$ consisting of eigenvectors of ρ, with associated eigenvalues $\{\lambda_1, \lambda_2, ..., \lambda_n\}$ (the spectrum of ρ). If we let P_j be the projector onto the closed subspace spanned by u_j, then $\rho = \sum_{j=1}^n \lambda_j P_j$.

For $g(.) : \mathbb{R} \to \mathbb{R}$, $g(x) = x \log x$, the (self adjoint) operator $g(\rho) = \rho \log \rho$ is defined by $\sum_{j=1}^{n} g(\lambda_j) P_j$ whose trace is

$$tr(\rho \log \rho) = \sum_{j=1}^{n} <u_j, (\rho \log \rho) u_j> = \sum_{j=1}^{n} \lambda_j \log \lambda_j$$

so that the quantum entropy of ρ is $-tr(\rho \log \rho) = -\sum_{j=1}^{n} \lambda_j \log \lambda_j$ which depends only on the eigenvalues of ρ.

In summary, while *real analysis* is the language for Kolmogorov probability, *functional analysis* is the language of quantum probability, in which trace operator replaces integral.

4 An Uncertainty Model for Behavioral Economics

As social sciences are relatively young relative to natural sciences, and in view of successes of the latter, the approaches of the former were inspired from the latter (such as imitating laws of Newton's mechanics to model macroeconomic trading). It is often said that, however, there is a fundamental difference, namely the presence of uncertainty in topics of social sciences, such as economics, which makes things much more complicated, especially the impossibility of obtaining "laws" governing economics evolutions, since "all models are wrong". Three things to note. First, the notion of uncertainty in social sciences was not looked at carefully (without examining its sources). Second, econometricians seem to forget quantum mechanics which presents a (natural) science domain much closer to economic analysis, because of its intrinsic randomness. Third, and this is very important, there is a lack of help from psychologists (as Stephen Hawking have reminded us).

It is often said that (natural) science works since, after hypotheses and reasoning, a theory provides predictions of things (such as "the universe is expanding") which can be checked (by experiments, observations) whether these are correct or not. This is the power of science that provides reliability and allows us to trust its findings.

Again, as Stephen Hawking has pointed out, economic analyses based on standard probability, as a quantitative model of underlying uncertainty, are only "moderately" successful, perhaps this is so because of a defective modeling of the "real" uncertainty in decision-making and in the economic data we see. For example:

"A natural explanation of extreme irregularities in the evolution of prices in financial markets is provided by quantum effects" (S. and I. Segal, 1998).

Should we continue to use Ito stochastic calculus to handle Brownian motion or put it in the place where it belongs (i.e., instead, using *quantum stochastic calculus*, see e.g., [23])?

It all boils down to model the uncertainty in social sciences in some "appropriate" manner. Recall that, "at the beginning" when it was realized that "chance (in games of chance) can be measured", we measured chance in the same way that we used to measure geometric objects, such as length, area, volume, by using (Lebesgue) measure theory. An uncertainty measure is simply a measure with the restriction that the measure of the whole sampling space is 1. While keeping the (countably) additive axiom, we have an additive uncertainty measure (noting that, why rejecting the countable additivity axiom, the Bayesian school keeps the finite additivity property). When applying a such uncertainty measure (called probability) to social sciences, such as in (human) decision-making in microeconomics, exemplified by von Neumann's game theory, it does not work, see e.g., [8], in the sense that what we observe is not what the theory predicted. Of course, to be fair, von Neumann supposed that economic agents are *all* "rational".

Thus, if we think about physics, when a theory prediction does not match the observation, the theory is wrong! as exemplified by Einstein's mistake with his equations predicting that the universe is static, whereas by observations, or by Georges Lemaitre 's thinking, it is not so.

Starting with the help of psychologists, things look clearer. In making decisions, humans do not measure and manipulate uncertainty as theorists think. The field of behavioral economics is becoming somewhat "popular" now. But, what is more significant is the revealing of psychologists about other aspects of human perception, such as uncertainty needs not be an increasing function (e.g., in the so-called "Linda story"), the "order effect" (leading to the possibility that uncertainty is noncommutative), the "framing problem", the "bias" problem. These revelations require a completely different way of modeling the uncertainty in social sciences. All existing modifications of additive measures (to various types of nonadditive uncertainty measures, as reviewed in Sect. 2) are not quite appropriate since they do not capture the noncommutativity observed in psychological experiments. We need an uncertainty theory having, in addition, the noncommutativity property.

How to "come up" with such a theory? Well, there is an old saying: "There is nothing new under the sun"! We got it for free! It is precisely the uncertainty encountered a century ago in quantum mechanics, and it was quantified as quantum probability. Applying quantum probability to social sciences is precisely what is happening recently, exemplified by [4, 11], see also [15, 18, 19, 22].

Another important issue related to economic predictions from economic data (e.g., time series) is this. Facing uncertainty, say, in the fluctuations of stock returns (from observed data) we should ask "what is the random nature of the phenomenon?", rather than just apply some traditional approaches. Also, other relevant questions could be "Do factors effecting the fluctuations correlate with each other, or/and also exhibit some forms of interference?", "How to formulate the effect of an interference in economics?", "Can we always be able to model correlations by copulas?", "Is Kolmogorov probability calculus appropriate for such and such situation?".

As G. Box once famously said "All models are wrong, but some are useful", how could we figure out the "useful ones"? by cross validation, perhaps. Remember that quantum mechanics is successful because physicists can design measurement experiments to validate (test) their theories (not using P-values, of course!).

Now, recognizing that quantum probability is the best model we have so far for economic investigations is, of course, a crucial first step. As in usual statistics, the next important step is how to propose (quantum) statistic models from observed data, such as in financial econometrics, especially "useful" ones? where here by "useful models" we mean those which could produce reliable predictions.

Well, as in decision-making, the modeling of data should contain human factors, as fluctuations of stock returns were due also to human intervention! Then the "difficult" question is "How to incorporate human factors into a model?". Note that, perhaps because it is a difficult problem that traditional statisticians have tried to avoid it, by "argueing" that human factors have already subsumed into the random nature of the fluctuations. Could we do better than that?

Recently [5] revealed a clue: We can "see" the (explicit) human factor in modeling if we use Bohmian mechanics approach to quantum mechanics [3].

Recall that the standard interpretation of quantum mechanics is this. The dynamics of a particle with mass m is "described" by a wave function $\psi(x,t)$, where $x \in \mathbb{R}^3$ is the particle position at time t, which is the solution of the Schrodinger's equation (counterpart of Newton's law of motion of macroobjects):

$$ih\frac{\partial\psi(x,t)}{\partial t} = -\frac{h^2}{2m}\Delta_x\psi(x,t) + V(x)\psi(x,t)$$

As such, particles in motion do not have trajectories (in their phase space), or put it more specifically, their motion cannot be described (mathematically) by trajectories (because of the Heisenberg's uncertainty principle). The probability amplitude $|\psi(x,t)|^2$ is used to make probabilistic statements about the particle motion. There is another, not "official", interpretation of quantum mechanics in which it is possible to consider trajectories for particles, called Bohmian mechanics [3]. This mechanics formulation is suitable to use as models for, say, financial data (where time series data are like "trajectories" of moving objects). In fact, they could be considered as "useful models" in G. Box' sense since, as we will see, they display an extra term which can be used to represent the missing human factor.

Just for an exposition, here is the model offered by Bohmian mechanics, in one dimensional case ($x \in \mathbb{R}$).

The polar form of the complex-valued wave function $\psi(x,t)$ is $\psi(x,t) = R(x,t)\exp\{\frac{i}{h}S(x,t)\}$, with $R(x,t)$, $S(x,t)$ being real-valued. The Schrodinger's equation becomes

$$ih\frac{\partial}{\partial t}[R(x,t)\exp\{\frac{i}{h}S(x,t)\}]$$

$$= -\frac{h^2}{2m}\Delta_x[R(x,t)\exp\{\frac{i}{h}S(x,t)\}] + V(x)[R(x,t)\exp\{\frac{i}{h}S(x,t)\}]$$

Differentiating

$$ih\frac{\partial}{\partial t}[R(x,t)\exp\{\frac{i}{h}S(x,t)\}]$$

$$= -\frac{h^2}{2m}\frac{\partial^2}{\partial x^2}[R(x,t)\exp\{\frac{i}{h}S(x,t)\}] + V(x)[R(x,t)\exp\{\frac{i}{h}S(x,t)\}]$$

and identifying real and imaginary parts of both sides, we get, respectively

$$\frac{\partial S(x,t)}{\partial t} = -\frac{1}{2m}(\frac{\partial S(x,t)}{\partial x})^2 + V(x) - \frac{h^2}{2mR(x,t)}\frac{\partial^2 R(x,t)}{\partial x^2}$$

$$\frac{\partial R(x,t)}{\partial t} = -\frac{1}{2m}[R(x,t)\frac{\partial^2 S(x,t)}{\partial x^2} + 2\frac{\partial R(x,t)}{\partial x}\frac{\partial S(x,t)}{\partial x}]$$

Note that the equation for $\frac{\partial R(x,t)}{\partial t}$ gives rise to the dynamical equation for the probability density function $f_t(x) = |\psi(x,t)|^2 = R^2(x,t)$, namely

$$\frac{\partial R^2(x,t)}{\partial t} = 2R(x,t)\frac{\partial R(x,t)}{\partial t}$$

$$= 2R(x,t)\{-\frac{1}{2m}[R(x,t)\frac{\partial^2 S(x,t)}{\partial x^2} + 2\frac{\partial R(x,t)}{\partial x}\frac{\partial S(x,t)}{\partial x}]\}$$

$$= -\frac{1}{m}[R^2(x,t)\frac{\partial^2 S(x,t)}{\partial x^2} + 2R(x,t)\frac{\partial R(x,t)}{\partial x}\frac{\partial S(x,t)}{\partial x}]$$

$$= -\frac{1}{m}\frac{\partial}{\partial x}[R^2(x,t)\frac{\partial S(x,t)}{\partial x}]$$

On the other hand, the equation for $\frac{\partial S(x,t)}{\partial t}$ gives some analogy with classical mechanics in Hamiltonian formalism.

In Newtonian mechanics, the state of a moving object of mass m, at time t, is described as $(x, m\dot{x})$ (position $x(t)$, and momentum $p(t) = mv(t)$, with velocity $v(t) = \frac{dx}{dt} = \dot{x}(t)$). The Hamiltonian of the system is the sum of the kinetic energy and potential energy $V(x)$, namely,

$$H(x,p) = \frac{1}{2m}v^2 + V(x) = \frac{mp^2}{2} + V(x)$$

Hence, $\frac{\partial H(x,p)}{\partial p} = mp$, or $\dot{x}(t) = \frac{1}{m}\frac{\partial H(x,p)}{\partial p}$. If we look at

$$\frac{\partial S(x,t)}{\partial t} = -\frac{1}{2m}(\frac{\partial S(x,t)}{\partial x})^2 + V(x) - \frac{h^2}{2mR(x,t)}\frac{\partial^2 R(x,t)}{\partial x^2}$$

ignoring the term $\frac{h^2}{2mR(x,t)}\frac{\partial^2 R(x,t)}{\partial x^2}$ for the moment, i.e., the Hamiltonian $\frac{1}{2m}(\frac{\partial S(x,t)}{\partial x})^2 - V(x)$, then the velocity of this system is $v(t) = \frac{dx}{dt} = \frac{1}{m}\frac{\partial S(x,t)}{\partial x}$.

But the full equation has the term $Q(x,t) = \frac{h^2}{2mR(x,t)}\frac{\partial^2 R(x,t)}{\partial x^2}$ which we call it a "quantum potential", we follow Bohm to interprete it similarly, leading to the Bohm-Newton equation

$$m\frac{dv(t)}{dt} = m\frac{d^2x(t)}{dt^2} = -(\frac{\partial V(x,t)}{\partial x} - \frac{\partial Q(x,t)}{\partial x})$$

This equation provides the concept of "trajectory" for the "particle". Regardless of the debate in physics about this formalism of quantum mechanics, Bohmian mechanics is useful for economics! The quantum potential (field) $Q(x,t)$, giving rise to the "quantum force" $-\frac{\partial Q(x,t)}{\partial x}$, disturbing the "classical" dynamics, will play the role of "mental factor" (of economic agents) when we apply Bohmian formalism to economics.

It is possible then to apply the above Bohm-Newton equation to financial modeling incorporating human factor in it. With all economic quantities analogous to those in quantum mechanics, we proceed to solve the Schrodinger' s equation to obtain the (pilot) wave function $\psi(x,t)$ (representing expectation of traders in the market), where $x(t)$ is, say, the stock price at time t; from which we obtain the mental (quantum) potential $Q(x,t) = \frac{h^2}{2mR(x,t)}\frac{\partial^2 R(x,t)}{\partial x^2}$ producing the associated mental force $-\frac{\partial Q(x,t)}{\partial x}$. Next, solving the Bohm-Newton's equation to obtain the "trajectory" for $x(t)$. Of course, economic counterparts of quantities such as m (mass), h (the Planck constant) should be spelled out (e.g., number of shares, price scaling parameter, i.e., the unit in which we measure price change). The potential energy describes the interactions among traders (e.g., competition) together with external conditions (e.g., price of oil, weather, etc.) whereas the kinetic energy represents the efforts of economic agents to change prices.

References

1. Allais, M.: Le comportement de l'homme rationel devant le risk: critiques des postulats et axiomes de l'ecole americaine. Econometrica **21**(4), 503–546 (1953)
2. Baaquie, B.E.: Quantum Finance. Cambridge University Press, New York (2004)
3. Bohm, D.: Quantum Theory. Prentice Hall, Englewood Cliffs (1951)
4. Busemeyer, J.R., Bruza, P.D.: Quantum Models of Cognition and Decision. Cambridge University Press, Cambridge (2012)
5. Chopra, D., Kafatos, M.: You are the Universe: Discovering Your Cosmic Self and Why it Matters. Rider, London (2017)
6. Dempster, A.: Upper and lower probabilities induced by a multivalued mapping. Ann. Math. Stat. **38**, 325–339 (1967)
7. Diaconis, P., Skyrms, B.: Ten Great Ideas about Chance. Princeton University Press, Princeton (2018)
8. Ellsberg, D.: Risk, ambiguity, and the Savage axioms. Q. J. Econ. **75**(4), 643–669 (1961)
9. Etheridge, A.: A Course in Financial Calculus. Cambridge University Press, Cambridge (2002)
10. Feynman, R.: The concept of probability in quantum mechanics. In: Berkeley Symposium on Mathematical Statistics and Probability, pp. 533–541. University of California, Berkeley (1951)

11. Haven, E., Khrennikov, A.: Quantum Social Science. Cambridge University Press, Cambridge (2013)
12. Hawking, S., Mlodinow, L.: The Grand Design. Bantam Books, London (2010)
13. Kahneman, D., Tversky, A.: Prospect theory: an analysis of decision under risk. Econometrica **47**, 263–292 (1979)
14. Kreps, D.V.: Notes on the Theory of Choice. Westview Press, Boulder (1988)
15. Nguyen, H.T.: Toward improving models for decision making in economics. Asian J. Econ. Banking **3**(01), 1–19 (2019)
16. Nguyen, H.T.: An Introduction to Random Sets. Chapman and Hall/ CRC Press, Boca Raton (2006)
17. Nguyen, H.T., Walker, C., Walker, E.: A First Course in Fuzzy Logic, 4th edn. Chapman and Hall/ CRC Press, Boca Raton (2019)
18. Nguyen, H.T., Trung, N.D., Thach, N.N.: Beyond traditional probabilistic methods in economics. In: Kreinovich, V., et al. (eds.) Beyond Traditional Probabilistics Methods in Economics, pp. 3–21. Springer, Cham (2019)
19. Nguyen, H.T., Sriboonchitta, S., Thach, N.N.: On quantum probability calculus for modeling economic decisions. In: Kreinovich, V., Sriboonchitta, S. (eds.) Structural Changes and Their Econometric Modeling, pp. 18–34. Springer, Cham (2019)
20. Nguyen, H.T.: On random sets and belief functions. J. Math. Anal. Appl. **65**(3), 531–542 (1978)
21. Nguyen, H.T., Walker, A.E.: On decision making using belief functions. In: Yager, R., Kacprzyk, J., Pedrizzi, M. (eds.) Advances in the Dempster -Shafer Theory of Evidence, pp. 311–330. Wiley, New York (1994)
22. Nguyen, H.T., Thach, N.N.: A closer look at the modeling of economic data. In: Kreinovich, V., Thach, N.N., Trung, N.D., Van Thanh, D. (eds.) Beyond Traditional Probabilistic Methods in Economics, pp. 100–112. Springer, Cham (2019)
23. Parthasarathy, K.R.: An Introduction to Quantum Stochastic Calculus. Springer, Basel (1992)
24. Shafer, G.: A Mathematical Theory of Evidence. Princeton University Press, Princeton (1976)
25. Shiryaev, A.N.: Essentials of Stochastic Finance. World Scientific, New Jersey (1999)
26. Sriboonchitta, S., Wong, W.K., Dhompongsa, S., Nguyen, H.T.: Stochastic Dominance and Applications to Finance. Risk and Economics. Chapman and Hall/ CRC Press, Boca Raton (2010)
27. Von Neumann, J.: Mathematical Foundations of Quantum Mechanics. Princeton University Press, Princeton (1955)
28. Von Neumann, J., Morgenstern, O.: The Theory of Games and Economic Behavior. Princeton University Press, Princeton (1944)
29. Zadeh, L.A.: Fuzzy sets as a basis for a theory of possibility. J. Fuzzy Sets Syst. **1**, 3–28 (1978)

Using Machine Learning Methods to Support Causal Inference in Econometrics

Achim Ahrens, Christopher Aitken, and Mark E. Schaffer

Abstract We provide an introduction to the use of machine learning methods in econometrics and how these methods can be employed to assist in causal inference. We begin with an extended presentation of the lasso (least absolute shrinkage and selection operator) of Tibshirani [50]. We then discuss the 'Post-Double-Selection' (PDS) estimator of Belloni et al. [13, 19] and show how it uses the lasso to address the omitted confounders problem. The PDS methodology is particularly powerful for the case where the researcher has a high-dimensional set of potential control variables, and needs to strike a balance between using enough controls to eliminate the omitted variable bias but not so many as to induce overfitting. The last part of the paper discusses recent developments in the field that go beyond the PDS approach.

Keywords Causal inference · Lasso · Machine learning

1 Introduction

Over the last 40 years, economic research has evolved significantly, and the pace of this change shows no sign of abating. Particularly notable is the discipline's increasingly empirical orientation. That shift is itself a reflection of two other strands of

Invited paper for the International Conference of the Thailand Econometric Society, 'Behavioral Predictive Modeling in Econometrics', Chiang Mai University, Thailand, 8–10 January 2020. Our exposition of the 'rigorous lasso' here draws in part on our paper Ahrens et al. [1]. All errors are our own.

A. Ahrens
ETH Zürich, Zürich, Switzerland
e-mail: achim.ahrens@gess.ethz.ch

C. Aitken
Heriot-Watt University, Edinburgh, UK
e-mail: christopher.f.aitken@gmail.com

M. E. Schaffer (✉)
Heriot-Watt University, Edinburgh, UK
e-mail: m.e.schaffer@hw.ac.uk

S. Sriboonchitta et al. (eds.), *Behavioral Predictive Modeling in Economics*, Studies in Computational Intelligence 897,
https://doi.org/10.1007/978-3-030-49728-6_2

change: high-quality (administrative and private) data are now in plentiful supply and can be accessed for the purposes of research; and, methodologically, the profession has embraced 'credible' design-based research that exploits variation generated by quasi-experiments and randomised trials [4]. These advances have stimulated interactions with other disciplines, and based on bibliometric analysis, they appear to have increased the influence of economics (Angrist et al. [5]).

Economists have also started to incorporate into their toolkits methods developed by researchers who work at the nexus of statistics and computer science—the field of machine learning. These methods have famously been used successfully for difficult, diverse tasks such as facial recognition and accurate language translation. Many problems that we now face in the conduct of economic research are very close in spirit to those, and have been tackled with machine learning methods. For example, economic historians have gained a deeper understanding of intergenerational mobility by employing those algorithmic models to link individuals across datasets that lack clean identifiers and which are rife with measurement and transcription issues (Feigenbaum [28]). Similarly, in political economy, related methods have been used to quantify the 'partisanship' of congressional speeches (Gentzkow et al. [32]). Machine learning methods have even been used to demonstrate in detail the changing nature of economic research (Angrist et al. [5]).[1]

These applications demonstrate the power of machine learning methods for a particular type of problem. However, their applicability is not bounded to that domain. The models can also be employed to make the econometric techniques we rely on more credible and more robust, even when applied to design-based studies. This aspect of the frontier will be the focus of the present review.

1.1 What is Machine Learning?

'Machine learning' (ML) is a relatively new discipline,[2] and is newer still to many applied economists, so we begin with some definitions and by setting out how it relates to the familiar, more traditional field of econometrics. A caveat: as Humpty Dumpty told Alice in *Through the Looking Glass*, 'when I use a word, it means just what I choose it to mean'. Terminology in this area is still settling down.

Machine learning constructs algorithms that can learn from the data. **Statistical learning** is a branch of statistics that was born in response to machine learning. Statistical learning encompasses models which, naturally, are more statistical in nature, and importantly, it emphasises the principled assessment of uncertainty. The distinction between these fields is subtle and narrowing.

[1] Previous research on that topic had relied instead on anecdotal evidence, or restricted its attention to a subset of the literature that is not representative because it is not feasible to manually classify the full corpus of work (Hamermesh [36] and Backhouse and Cherrier [10]).

[2] That said, traditional methods such as k-means cluster analysis and ridge regression are often now associated with ML.

Both fields are conventionally divided into two areas: unsupervised and supervised learning problems. **Unsupervised learning problems** are defined by tasks for which there is no output variable, only inputs. A key objective is dimension reduction: in other words, we wish to reduce the complexity of the data. Some of these methods, such as principal component analysis (PCA) and cluster analysis, are already well-known to economists. Unsupervised machine learning can be used to generate inputs (features) for supervised learning (e.g., principal component regression), as we suggested previously. Whilst these techniques are important, we do not consider them here further. We refer interested readers to Gentzkow, Kelly and Taddy [33] or Athey and Imbens [8] for primers.

In **supervised learning problems**, the researcher has an outcome for each individual y_i and predictors $\boldsymbol{x}_i$. The objective of the researcher is to fit a model using the (training) data which can be used to accurately predict y_i (or classify it if y_i is categorical) using additional ('held out') data $\{\boldsymbol{x}_i\}_{i \in H}$, where $H = \{i : i > n\}$.

1.2 Econometrics and Machine Learning

We can think of applied econometrics – econometrics as practiced by economists – as consisting in large part of two different but related and overlapping activities: predictive inference and causal inference.

Prediction is often done by economists in the context of forecasting using time-series data. The typical forecasting question: how can we reliably forecast y_{t+s} (GDP growth, inflation, etc.) based on information available up to time t? 'Nowcasting' is a variation on forecasting, where the nowcast takes place at time t but y_t becomes available only in the future after a lag of s periods. Prediction of outcomes is also done in cross-section and spatial settings; see Bansak et al. [11] for an example (assignment of refugees across resettlement locations) and Mullainathan and Spiess [44] for a general discussion.

Causal inference is fundamentally different, or alternatively, is a very special form of predictive inference. The typical causal inference question: what is the predicted policy impact on y of a change in policy d, and how can we estimate that impact using data $\{(y_i, \boldsymbol{x}_i, d_i)\}_{i=1}^n$? (Note that we have separated out the 'treatment' variable d from the other covariates $\boldsymbol{x}$. We return to this point shortly.) The general framework used by applied economists allows for causal inference with respect to the treatment d. Usually the researcher specifies a model using theory and perhaps diagnostic or specification tests; the model is estimated using the full dataset; and parameter estimates and confidence intervals are obtained based either on large-sample asymptotic theory or small-sample theory.

Typical examples from labour economics would be the impact of changing the school-leaving age (and hence education levels) on wages, and the impact of minimum wages on employment levels. In this framework, the causal variable d is commonly thought of as a policy level, i.e., something that can be set or influenced by economic policymakers. The framework is, of course, used widely outside of economics as well, and much of the work currently being done in causal inference is cross-disciplinary. For example, it is now common for econometrics textbooks to explain causal inference in terms of the 'experimental ideal', often referred to as the 'gold standard' for causal inference: if researchers could conduct a randomised control trial (RCT) and set d to have different values in two random samples of subjects, what would be the (mean) difference in outcomes y between the two groups? And indeed, RCTs and field experiments are now part of the standard armoury of applied economists; applied development economics in particular has been revolutionised by this approach.

Curiously, although the distinction between predictive and causal inference is fairly straightforward to explain and fundamental to what applied economists do, it is not often treated clearly as such in textbooks at either the undergraduate or graduate level. The standard approach in econometrics has been to teach both using the same toolbox: specify a parametric model, show how it can be estimated using various methods (Least Squares, Instrumental Variables (IV), Generalized Methods of Moments (GMM), Maximum Likelihood, etc.), and discuss the conditions under which a causal parameter can be consistently estimated. Prediction and forecasting is typically covered separately in detail as part of time series analysis.

The connection between prediction in econometrics and machine learning is fairly obvious. Economists and econometricians who work in forecasting have taken great interest in machine learning methods, and are importing these techniques into their work. We do not discuss the work in this area here.

This paper, instead, looks at how machine learning methods can be used in estimating **causal** impacts, and this is where the distinction between $\boldsymbol{d}$ and $\boldsymbol{x}$ comes in. We focus in particular on the confounder or omitted variable bias problem. Omitted variable bias means that standard methods for estimating the treatment effect of d will yield coefficient estimates that are biased and inconsistent. The standard textbook remedy is to include 'controls' $\boldsymbol{x}$. The practical problem facing researchers is that often the choice of controls is very difficult, and in particular the set of potential controls may be **high-dimensional**. The standard framework for estimating causal effects assumes that both d and $\boldsymbol{x}$ are low-dimensional. If $\boldsymbol{x}$ is high-dimensional, the research has a problem: if all controls are inserted, overfitting means the model estimates will be badly biased; the researcher selects a small number of controls but they are the wrong ones, the model estimates will again be biased. Machine learning methods can address this problem, and in this paper we show how to employ one such method: the 'post-double-selection' (PDS) and related methods introduced by Belloni et al. [13, 19] that use a popular machine-learning estimator, the lasso or Least Absolute Shrinkage and Selection Operator [50].

The structure of the paper is as follows. In the next section we discuss in detail the lasso estimator and a particular version with a theory-driven penalty, the 'rigorous lasso'. We then discuss how this version of the lasso is used for causal inference in the PDS method of Belloni et al. [14]. We illustrate its use with what is literally a textbook example: the impact of a disamenity on housing prices, employed by Wooldridge in his widely-used undergraduate and graduate texbooks (Wooldridge [56, 57]) to illustrate the importance of including controls to address omitted variable bias. The last section briefly surveys the current literature and advances in this area.

The software used to implement the estimators used here is Stata and in particular the **lassopack** and **pdslasso** packages by Ahrens, Hansen, and Schaffer. See Ahrens et al. [1] for a detailed discussion of lassopack.

2 Sparsity and the Rigorous or Plug-In Lasso

2.1 High-Dimensional Data and Sparsity

The high-dimensional linear model is:

$$y_i = \boldsymbol{x}_i'\boldsymbol{\beta} + \varepsilon_i \tag{1}$$

We index observations by i and regressors by j. We have up to $p = \dim(\boldsymbol{\beta})$ potential regressors. p can be very large, potentially even larger than the number of observations n. For simplicity we assume that all variables have already been mean-centered and rescaled to have unit variance, i.e., $\frac{1}{n}\sum_i y_i = 0$ and $\frac{1}{n}\sum_i y_i^2 = 1$, and similarly for the predictors x_{ij}.

If we simply use OLS to estimate the model and p is large, the result is disaster: we overfit badly and classical hypothesis testing leads to many false positives. If $p > n$, OLS is not even identified.

How to proceed depends on what we believe the 'true model' is. Does the model include only a small number of regressors or a very large number with little individual contribution. In other words, is the model 'sparse' or 'dense'?

In this paper, we focus primarily on the 'sparse' case and specifically on an estimator that is particularly well-suited to the sparse setting, namely the *lasso* or 'Least Absolute Shrinkage and Selection Operator' introduced by Tibshirani [50]. One of the appealing features of the lasso is that it is both well-suited to the high-dimensional setting in terms of predictive performance, and at the same time the lasso solution is sparse, with most coefficient estimates set exactly to zero, thus facilitating model interpretation.

In the **exact sparsity** case of the p potential regressors, **only s regressors belong in the model**, where

$$s := \sum_{j=1}^{p} \mathbb{1}\{\beta_j \neq 0\} \ll n. \tag{2}$$

In other words, most of the true coefficients β_j are actually zero. The problem facing the researcher is that which are zeros and which are not is unknown.

We can also use the weaker assumption of **approximate sparsity**: some of the β_j coefficients are well-approximated by zero, and the approximation error is sufficiently 'small'. The discussion and methods we present in this paper typically carry over to the approximately sparse case, and for the most part we will use the term 'sparse' to refer to either setting.

The sparse high-dimensional model accommodates situations that are very familiar to researchers and that typically presented them with difficult problems where traditional statistical methods would perform badly. These include both settings where the number p of observed potential predictors is very large and the researcher does not know which ones to use, and settings where the number of observed variables is small but the number of potential predictors in the model is large because of interactions and other non-linearities, model uncertainty, temporal & spatial effects, etc.

2.2 *The Penalisation Approach and the Lasso*

There are various estimators available that can be used for regularisation in a high dimensional setting; the lasso is just one of these. The basic idea behind these estimators is *penalisation*: put a penalty or 'price' on the use of regressors in the objective function that the estimator minimizes.

One option is penalisation based on the number of predictors. This is the so-called ℓ_0 'norm'. For example, the estimator could minimize the residual sum of squares minus some 'price' λ for each nonzero coefficient:

$$\sum_{i=1}^{n} \left(y_i - \boldsymbol{x}_i'\boldsymbol{\beta}\right)^2 + \lambda \, \|\boldsymbol{\beta}\|_0 \tag{3}$$

where $\|\beta\|_0 = \sum_{j=1}^{p} 1\left\{\beta_j \neq 0\right\}$, i.e., the number of predictors. There is a cost to including many predictors, and minimisation of the objective function will include dropping predictors that contribute little to the fit. AIC and BIC are examples of this approach.

The problem with penalisation based on ℓ_0 'norm' is very simple: it is computationally infeasible if p is at all large (*NP*-hard). So we need another approach.

The lasso estimator minimizes the mean squared error subject to a penalty on the *absolute size* of coefficient estimates (i.e., using the ℓ_1 norm):

$$\hat{\boldsymbol{\beta}}_{\text{lasso}}(\lambda) = \arg\min \frac{1}{n} \sum_{i=1}^{n} \left(y_i - \boldsymbol{x}_i'\boldsymbol{\beta}\right)^2 + \frac{\lambda}{n} \sum_{j=1}^{p} \psi_j |\beta_j|. \tag{4}$$

The tuning parameter λ controls the overall penalty level and ψ_j are predictor-specific penalty loadings.

The intuition behind the lasso is straightforward: there is a cost to including predictors, the unit 'price' per regressor is λ, and we can reduce the value of the objective function by removing the ones that contribute little to the fit. The bigger the λ, the higher the 'price', and the more predictors are removed. The penalty loadings ψ_j introduce the additional flexibility of putting different prices on the different predictors x_{ij}. The natural base case for standardised predictors is to price them all equally, i.e., the individual penalty loadings $\psi_j = 1$ and they drop out of the problem (but we will see shortly that separate pricing for individual predictors is needed in some settings).

We can say 'remove' because in fact the effect of the penalisation with the ℓ_1 norm is that *the lasso sets the* $\hat{\beta}_j$*s for some variables to zero*. This is what makes the lasso so suitable to sparse problems: the estimator itself has a sparse solution.

In contrast to ℓ_0-'norm' penalisation, the lasso is computationally feasible: the path-wise coordinate descent ('shooting') algorithm allows for fast estimation.

It is also useful to compare the lasso to another commonly-used regularisation method, the ridge estimator. The ridge estimator uses the ℓ_2 norm:

$$\hat{\boldsymbol{\beta}}_{\text{Ridge}}(\lambda) = \arg\min \frac{1}{n} \sum_{i=1}^{n} \left(y_i - \boldsymbol{x}_i' \boldsymbol{\beta}\right)^2 + \frac{\lambda}{n} \sum_{j=1}^{p} \psi_j \beta_j^2. \tag{5}$$

The ridge estimator, like the lasso, is computationally feasible. But it typically does not have a sparse solution: all predictors will appear, and the predictors that contribute little will have small but nonzero coefficients.

The lasso, like other penalized regression methods, is subject to an attenuation bias. This bias can be addressed by post-estimation using OLS, i.e., re-estimate the model using the variables selected by the first-stage lasso [15]:

$$\hat{\boldsymbol{\beta}}_{\text{post}} = \arg\min \frac{1}{n} \sum_{i=1}^{n} \left(y_i - \boldsymbol{x}_i' \boldsymbol{\beta}\right)^2 \qquad \text{subject to} \qquad \beta_j = 0 \text{ if } \tilde{\beta}_j = 0, \tag{6}$$

where $\tilde{\beta}_j$ is the first-step lasso estimator such as the lasso. In other words, the first-step lasso is used exclusively as a model selection technique, and OLS is used to estimate the selected model. This estimator is sometimes referred to as the 'Post-lasso' [15].

In sum, the lasso yields sparse solutions. Thus, the lasso can be used for model selection. We have reduced a complex model selection problem into a one-dimensional problem. We 'only' need to choose the 'right' penalty level, i.e., λ. But what is the 'right' penalty?

2.3 *The Lasso: Choice of Penalty Level*

The penalisation approach allows us to simplify the model selection problem to a one-dimensional problem, namely the choice of the penalty level λ. In this section we discuss two approaches: (1) *cross-validation* and (2) *'rigorous'* or *plugin* penalisation. We focus in particular on the latter.

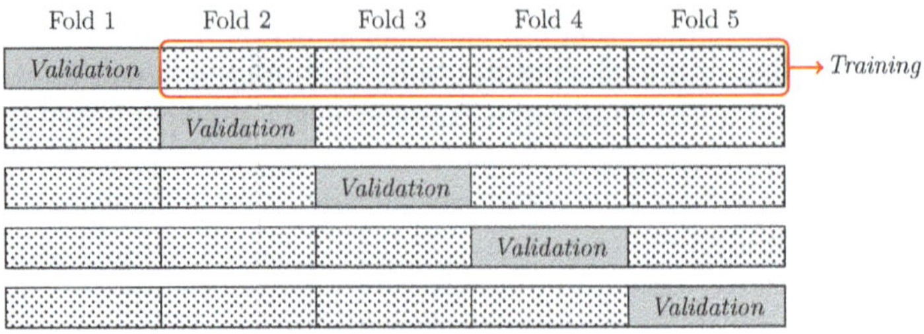

Fig. 1 This is K-fold cross-validation

The objective in *cross-validation* is to choose the lasso penalty parameter based on predictive performance. Typically, the dataset is repeatedly divided into a portion which is used to fit the model (the 'training' sample) and the remaining portion which is used to assess predictive performance (the 'validation' or 'holdout' sample), usually with mean squared prediction error (MSPE) as the criterion. Arlot and Celisse [6] survey the theory and practice of cross-validation; we briefly summarise here how it can be used to choose the lasso tuning parameter with independent data. In the case of independent data, common approaches are 'leave-one-out' (LOO) cross-validation and the more general 'K-fold' cross-validation.

In 'K-fold' cross-validation, the dataset is split into K portions or 'folds'; each fold is used once as the validation sample and the remainder are used to fit the model for some value of λ. For example, in 10-fold cross-validation (a common choice of K) the MSPE for the chosen λ is the MSPE across the 10 different folds when used for validation. LOO cross-validation is a special case where $K = 1$, i.e., every observation is used once as the validation sample while the remaining $n - 1$ observations are used to fit the model (Fig. 1).

Cross-validation is computationally intensive because of the need to repeatedly estimate the model and check its performance across different folds and across a grid of values for λ. Standardisation of data adds to the computational cost because it needs to be done afresh for each training sample; standardising the entire dataset once up-front would violate a key principle of cross-validation, which is that a training dataset cannot contain any information from the corresponding validation dataset. LOO is a partial exception because the MSPE has a closed-form solution for a chosen λ, but a grid search across λ and repeated standardisation are still needed.

Cross-validation with dependent data adds further complications because we need to be careful that the validation data are independent of the training data. For example, one approach used with time-series data is 1-step-ahead cross-validation (Hyndman et al. [38]), where the predictive performance is based on a training sample with observations through time t and the forecast for time $t + 1$. The main setting for this paper is independent data so we do not discuss the dependent case further.

2.4 The Rigorous or Plug-In Lasso

Bickel et al. [21] presented a theoretically-derived penalisation method for the lasso that assumed a known error variance. The method extended and feasible algorithms proposed in a series of papers by Belloni, Chernozhukov, Hansen, and coauthors to accommodate homoskedasticity with unknown variance, heteroskedasticity, non-Gaussian errors and clustered errors (e.g., [15, 17, 20, 23]). The approach is referred to in the literature as the 'rigorous' or 'plug-in' lasso. The rigorous lasso has several appealing features for our purposes. First, it has properties that enable it to be used straightforwardly in support of causal inference, the main topic of this paper. Second, it is theoretically and intuitively appealing, and a useful illustration of how theoretical approaches to high-dimensional problems can work. Lastly, it is computationally attractive compared to cross-validation, and hence of practical interest in its own right.

The rigorous lasso is consistent in terms of prediction and parameter estimation under assumptions about three important model characteristics:

- **Sparsity**
- **Restricted sparse eigenvalue condition**
- **The 'regularisation event'**

We consider each of these in turn.

We have already discussed *exact sparsity*: there is a large set of potentially relevant variables, but the true model contains only a small number of them. Exact sparsity is a strong assumption, and in fact it is stronger than is needed for the rigorous lasso. Instead, we assume *approximate sparsity*. Intuitively, some true coefficients may be non-zero but small enough in absolute size that the lasso performs well even if the corresponding predictors are not selected.

Belloni et al. [13] define the *approximate sparse model (ASM)*,

$$y_i = f(\boldsymbol{w}_i) + \varepsilon_i = \boldsymbol{x}_i'\boldsymbol{\beta}_0 + r_i + \varepsilon_i. \tag{7}$$

where ε_i are independently distributed, but possibly heteroskedastic and non-Gaussian errors. The elementary predictors $\boldsymbol{w}_i$ are linked to the dependent variable through the unknown and possibly non-linear function $f(\cdot)$. The objective is to approximate $f(\boldsymbol{w}_i)$ using the target parameter vector $\boldsymbol{\beta}_0$ and the transformations $\boldsymbol{x}_i := P(\boldsymbol{w}_i)$, where $P(\cdot)$ is a set of transformations. The vector of predictors $\boldsymbol{x}_i$ may be large relative to the sample size. In particular, the setup accommodates the case where a large number of transformations (polynomials, dummies, etc.) approximate $f(\boldsymbol{w}_i)$.

Approximate sparsity requires that $f(\boldsymbol{w}_i)$ can be approximated sufficiently well using only a small number of non-zero coefficients. Specifically, the target vector $\boldsymbol{\beta}_0$ and the sparsity index s need to satisfy

$$\|\boldsymbol{\beta}_0\|_0 := s \ll n \quad \text{with} \quad \frac{s^2 \log^2(p \vee n)}{n} \to 0 \tag{8}$$

and the resulting approximation error $r_i = f(\boldsymbol{w}_i) - \boldsymbol{x}_i'\boldsymbol{\beta}_0$ satisfied the bound

$$\sqrt{\frac{1}{n}\sum_{i=1}^{n} r_i^2} \leq C\sqrt{\frac{s}{n}}, \tag{9}$$

where C is a positive constant.

For example, consider the case where $f(\boldsymbol{w}_i)$ is linear with $f(\boldsymbol{w}_i) = \boldsymbol{x}_i'\boldsymbol{\beta}^\star$, but the true parameter vector $\boldsymbol{\beta}^\star$ is high-dimensional: $\|\boldsymbol{\beta}^\star\|_0 > n$. Approximate sparsity means we can still approximate $\boldsymbol{\beta}^\star$ using the sparse target vector $\boldsymbol{\beta}_0$ as long as $r_i = \boldsymbol{x}_i'(\boldsymbol{\beta}^\star - \boldsymbol{\beta}_0)$ is sufficiently small as specified in (9).

The *Restricted sparse eigenvalue condition* (RSEC) relates to the Gram matrix, $n^{-1}\boldsymbol{X}'\boldsymbol{X}$. The RSEC condition specifies that sub-matrices of the Gram matrix of size m are well-behaved (Belloni et al. [13]). Formally, the RSEC requires that the minimum sparse eigenvalues

$$\phi_{\min}(m) = \min_{1\leq\|\boldsymbol{\delta}\|_0\leq m} \frac{\boldsymbol{\delta}'\boldsymbol{X}'\boldsymbol{X}\boldsymbol{\delta}}{\|\boldsymbol{\delta}\|_2^2} \quad \text{and} \quad \phi_{\max}(m) = \max_{1\leq\|\boldsymbol{\delta}\|_0\leq m} \frac{\boldsymbol{\delta}'\boldsymbol{X}'\boldsymbol{X}\boldsymbol{\delta}}{\|\boldsymbol{\delta}\|_2^2}$$

are bounded away from zero and from above. The requirement that $\phi_{\min}(m)$ is positive means that all sub-matrices of size m have to be positive definite.[3]

The *regularisation event* is the third central condition required for the consistency of the rigorous lasso. Denote by $\boldsymbol{S} = \nabla\hat{Q}(\boldsymbol{\beta})$, the gradient of the objective function $\hat{Q}$ at the true value $\boldsymbol{\beta}$. $S_j = \frac{2}{n}\sum_{i=1}^{n} x_{ij}\varepsilon_i$ is the jth element of the score vector. The idea is to select the lasso penalty level(s) to control the estimation noise as summarised by the score vector. Specifically, the overall penalty level λ and the predictor-specific penalty loadings ψ_j are chosen so that the 'regularisation event'

$$\frac{\lambda}{n} \geq c \max_{1\leq j\leq p} \left|\psi_j^{-1} S_j\right| \tag{10}$$

occurs with high probability, where $c > 1$ is a constant slack parameter.

Denote by $\Lambda = \max_j |\psi_j^{-1} S_j|$ the maximal element of the score vector scaled by the predictor-specific penalty loadings ψ_j, and denote by $q_\Lambda(\cdot)$ the quantile function for Λ, i.e., the probability that Λ is at most a is $q_\Lambda(a)$. In the rigorous lasso, we choose the penalty parameters λ and ψ_j and confidence level γ so that

$$\frac{\lambda}{n} \geq c q_\Lambda(1-\gamma). \tag{11}$$

The intuition behind this approach is clear from a very simple example. Say that no predictors appear in the true model ($\beta_j = 0\ \forall\ j = 1, \ldots, p$). For the lasso to select no variables, the penalty parameters λ and ψ_j need to satisfy

[3] Bickel et al. [21] use instead the weaker *restricted eigenvalue condition (REC)*. The RSEC implies the REC and has the advantage of being sufficient for both the lasso and the post-lasso.

$\lambda \geq 2\max_j |\sum_i \psi_j^{-1} x_{ij} y_i|$.[4] Because none of the regressors appear in the true model, $y_i = \varepsilon_i$, and the requirement for the lasso to correctly identify the model without regressors is therefore $\lambda \geq 2\max_j |\sum_i \psi_j^{-1} x_{ij}\varepsilon_i|$. Since $x_{ij}\varepsilon_i$ is the score for observation i and predictor j, this is equivalent to requiring $\lambda \geq n\max_j |\psi_j^{-1} S_j|$, which is the regularisation event in (10). We want this event to occur with high probability of at least $(1-\gamma)$. We therefore choose values for λ and ψ_j such that $\frac{\lambda}{n} \geq q_\Lambda(1-\gamma)$. Since $q_\Lambda(\cdot)$ is a quantile function, by definition we will choose the correct model (no predictors) with probability of at least $(1-\gamma)$. This yields (11), the rule for choosing penalty parameters.[5]

The procedure for choosing λ is not yet feasible, because the quantile function $q_\Lambda(\cdot)$ for the maximal element of the score vector is unknown, as is the predictor-specific penalty loadings ψ_j. We discuss how these issues are addressed in practice in the next subsection.

If the sparsity and restricted sparse eigevalue assumptions ASM and RSEC are satisfied, if certain other technical conditions are satisfied,[6] and if λ and ψ_j are estimated as described below, then Belloni et al. [13] show the lasso and post-lasso obey:

$$\sqrt{\frac{1}{n}\sum_{i=1}^{n}\left(x_i'\hat{\boldsymbol{\beta}} - x_i'\boldsymbol{\beta}\right)^2} = O\left(\sqrt{\frac{s\log(p \vee n)}{n}}\right), \tag{12}$$

$$\|\hat{\boldsymbol{\beta}} - \boldsymbol{\beta}\|_1 = O\left(\sqrt{\frac{s^2\log(p \vee n)}{n}}\right), \tag{13}$$

$$\|\hat{\boldsymbol{\beta}}\|_0 = O(s) \tag{14}$$

Equation (12) provides an asymptotic bound for the prediction error. Equation (13) provides an asymptotic bound for the bias in the estimated $\hat{\boldsymbol{\beta}}$. Equation (14) provides a sparsity bound; the number of selected predictors in the estimated model does not diverge relative to the true model.

The 'oracle' estimator is the least squares estimator obtained if the s predictors in the model were actually known. This provides a useful theoretical benchmark for comparison. Here, if the s predictors in the model were known, the prediction error would converge at the oracle rate $\sqrt{s/n}$. Thus, the logarithmic term $\log(p \vee n)$ in (12)–(13) can be interpreted as the cost of not knowing the true model. For this reason, Belloni et al. [13] describe these rates of convergence as *near-oracle* rates.

For the case of the lasso with theory-driven regularisation, Belloni and Chernozhukov [15] have shown that post-estimation OLS, also referred to as post-lasso, achieves the same convergence rates as the lasso and can outperform the lasso in situations where consistent model selection is feasible (see also Belloni et al. [13]).

[4]See, for example, Hastie et al. ([35], Ch. 2).

[5]In this special case, the requirement of the slack is loosened and $c = 1$.

[6]These conditions relate to the use of the moderate deviation theory of self-normalized sums [39] that allows the extension of the theory to cover non-Gaussianity. See Belloni et al. [13].

The rigorous lasso has recently been shown to have certain appealing properties *vis-a-vis* the K-fold cross-validated lasso. The rates of convergence of the rigorous lasso are faster than those for the K-fold cross-validated lasso derived in Chetverikov et al. [24]. More importantly for our purposes – using the lasso to assist in causal inference – the sparsity bound for the K-fold cross-validated lasso derived in Chetverikov et al. [24] does not exclude situations where (14) fails badly, in the sense that the number of predictors selected via cross-validation is much larger than s. One of the implications is that cross-validation will select a penalty level λ that is 'too small' in the sense that the regularisation event (10) will no longer be guaranteed to occur with high probability. While it is possible to use the cross-validated lasso, and indeed other machine learning estimators, to address the confounder problem in causal inference, it should not be used in the basic 'post-double selection' framework we discuss below; other techniques are needed. We return to this point later.

2.5 Implementing the Rigorous Lasso

The quantile function $q_\Lambda(\cdot)$ for the maximal element of the score vector is unknown. The most common approach to addressing this is to use a theoretically-derived upper bound that guarantees that the regularisation event (10) holds asymptotically.[7] Specifically, Belloni et al. [13] show that

$$\mathrm{P}\left(\max_{1\le j\le p} c\left|S_j\right| \le \frac{\lambda\psi_j}{n}\right) \rightarrow 1 \text{ as } n \rightarrow \infty,\ \gamma \rightarrow 0 \tag{15}$$

if the penalty levels and loadings are set to

$$\lambda = 2c\sqrt{n}\Phi^{-1}(1-\gamma/(2p)) \qquad \psi_j = \sqrt{\frac{1}{n}\sum_i x_{ij}^2\varepsilon_i^2} \tag{16}$$

c is the slack parameter from above and $\gamma \rightarrow 0$ means the probability of the regularisation event converges towards 1. Common settings for c and γ, based on Monte Carlo studies are $c = 1.1$ and $\gamma = 0.1/\log(n)$, respectively.

The only remaining element is estimation of the ideal penalty loadings ψ_j. Belloni et al. [13], Belloni et al. [19] recommend an iterative procedure based on some initial set of residuals $\hat{\varepsilon}_{0,i}$. One choice is to use the d predictors that have the highest correlation with y_i and regress y_i on these using OLS; $d = 5$ is their suggestion. The residuals from this OLS regression can be used to obtain an initial set of penalty loadings $\hat{\psi}_j$ according to (16). These initial penalty loadings and the penalty level from (16) are used to obtain the lasso or post-lasso estimator $\hat{\boldsymbol{\beta}}$. This estimator is

[7]The alternative is to simulate the distribution of the score vector. This is known as the 'exact' or *X-dependent* approach. See Belloni and Chernozhukov [14] for details and Ahrens et al. [1] for a summary discussion and an implementation in Stata.

then used to obtain a updated set of residuals and penalty loadings according to (16), and then an updated lasso estimator. The procedure can be iterated further if desired.

The framework set out above requires only independence across observations; heteroskedasticity, a common issue facing empirical researchers, is automatically accommodated. The reason is that heteroskedasticity is captured in the penalty loadings for the score vector.[8] Intuitively, heteroskedasticity affects the probability that the term $\max_j |\sum_i x_{ij}\varepsilon_i|$ takes on extreme values, and this needs to be captured via the penalty loadings.

In the special case of homoskedasticity, the ideal penalisation in (16) simplifies:

$$\lambda = 2c\sigma\sqrt{n}\Phi^{-1}(1-\gamma/(2p)), \qquad \psi_j = 1. \tag{17}$$

This follows from the fact that we have standardised the predictors to have unit variance and hence homoskedasticity implies $E(x_{ij}^2\varepsilon_i^2) = \sigma^2 E(x_{ij}^2) = \sigma^2$. The iterative procedure above is used to obtain residuals to form an estimate $\hat{\sigma}^2$ of the error variance σ^2.

The rigorous lasso has also been extended to cover a special case of dependent data, namely panel data. In the cluster-lasso [20], arbitrary within-panel correlation is accommodated. We do not discuss it further here and refer the reader to Belloni et al. [20] for a presentation and Ahrens et al. [1] for a summary discussion and implementation in Stata.

3 Using Machine Learning to Assist Causal Inference: The Lasso and 'Post-Double-Selection'

3.1 *The Confounder Problem*

Programme evaluation is a fundamental part of the modern empirical social sciences. This is particularly true in economics, where the profession is frequently relied on to give sound advice about the structure and scope of large scale government policies. For theorists too, these studies are important, as they allow them to test the explanatory power of their models. As a result, it is imperative that these evaluations be as credible as possible.

In this literature, it is common to assume that unconfoundedness holds—in other words, conditioning on observable factors is sufficient to make the treatment assignment as good as random. For much of the literature's history, there was little substantive advice given to practitioners to help them satisfy this condition. Unfortunately, the issue is not a simple one.

---

[8]The formula in (16) for the penalty loading is familiar from the standard Eicker-Huber-White heteroskedasticty-robust covariance estimator.

Many of the most important questions posed in the empirical social sciences revolve around the causal effect of a programme or policy. However, estimating and drawing inferences about such effects is often difficult to achieve in a credible manner. If a researcher is fortunate, they may be able to conduct a randomised control trial (RCT), which would (in a number of important respects) simplify the task, since it allows one to calculate an unbiased estimate of the average effect of the treatment (that is, the policy, project or programme to which the 'units' were exposed). This is important, as *unbiasedness* imposes no requirement on the researcher to know anything specific about relevant covariates or confounders [26]. However, even in experimental setups it may not be possible for researchers to rely entirely on randomisation of treatment. In such circumstances the confounder problem reappears and researchers face the problem of choosing controls to address it.

In any case, in the discipline of economics researchers are often extremely limited in the extent to which they can employ randomised control trials [7]. RCTs are expensive, not always feasible and in some cases are unethical.

For instance, consider the evaluation of policy on minimum wages: it is hardly fair or politically feasible to randomly assign people or places to be subject to different minimum wage regulations. Studies evaluating the impact of minimum wage laws, as well as a myriad of other important policies, have often utilised other techniques which rely instead on observational data, where the assignment to treatment is not random. And because the assignment to treatment is not random, researchers are immediately confronted by the problem of omitted confounders, i.e., omitted variable bias.

Successfully addressing the omitted variable bias problem is challenging, even when confounders or proxies for them are observable. 'Just insert some controls' is not adequate advice.

- The dimensionality of the controls may be large, immediately posing a problem for the researcher if, as is often the case, they do not have a strong theoretical or other prior basis for reducing the number of controls.
- Include too many controls, and the estimated treatment effect will suffer from overfitting.
- Include too few controls, and it will suffer from bias.
- Typically the researcher also lacks information about whether interactions and/or polynomials would be required to adequately address the problem. A low-dimensional problem can easily become a high-dimensional problem this way.
- Classical hypothesis testing to reduce the number of controls are poorly suited to addressing the problem because of the resulting pre-test bias. False discovery control (multiple testing procedures) is problematic and rarely done.
- Perhaps most worrisome of all is the 'research degrees of freedom' [49] or 'garden of forking paths' [31] problem. Researchers may try many combinations of

controls, looking for statistical significance in their results, and then report only the results that 'work'.[9]

Recent work by various authors has shown how machine learning methods can be used to address this problem. In the next subsection, we look at one such method: the 'post-double-selection' (PDS) and related methods of Belloni, Chernozhukov, and Hansen [18]. These authors show that the 'rigorous lasso' can be used to select controls in a theory-driven, parsimonious and semi-automated way. 'Theory-driven' means that asymptotic properties of PDS estimators are known. 'Parsimonious' means that the controls selected can address the omitted variable bias problem and at the same time avoid gross overfitting. 'Semi-automated' means that researcher degrees of freedom in the selection of controls are reduced and hence 'p-hacking' is restrained.

3.2 The Lasso and Causal Inference

The main strength of the lasso is prediction rather than model selection. But the lasso's strength as a prediction technique can also be used to aid causal inference.

In the basic setup causal inference setup, we *already know* the causal variable of interest. No variable selection is needed for this. The lasso is used instead to **select controls** used in the estimation. These other variables are not themselves subject to causal inference. But using them means we can obtained improved causal inference for the variable we **are** interested in.

Why can we use the lasso to select controls even though the lasso is (in most scenarios) not model selection consistent? There are two ways to look at this:

- **Immunisation property:** The moderate model selection mistakes of the lasso do not affect the asymptotic distribution of the estimator of the low-dimensional parameters of interest [13, 19]. We can treat modelling the nuisance component of our structural model as a prediction problem.
- The **irrepresentable condition** states that the lasso will fail to distinguish between two variables (one in the active set, the other not) if they are highly correlated. These type of variable selection mistakes are not a problem if the aim is to control for confounding factors.

We note here in passing that the PDS lasso methodology can also be used to select instruments from a high-dimensional set in order to address endogeneity in instrumental variables (IV) estimation. We focus on the 'selection of controls' problem here; for discussion of the IV application, see Belloni et al. [13] for the development of the theory.

[9]These authors are careful to note that the problem readily arises when researchers make decisions contingent on their data analysis; no conscious attempt to deceive is needed. Deliberate falsification, sometimes called 'p-hacking', is special case and likely much rarer.

3.3 The Post-Double Selection (PDS) Estimator

Our model is

$$y_i = \underbrace{\alpha d_i}_{\text{aim}} + \underbrace{\beta_1 x_{i,1} + \ldots + \beta_p x_{i,p}}_{\text{nuisance}} + \varepsilon_i.$$

The causal variable of interest or "treatment" is d_i. The xs are the set of potential controls and not directly of interest. We want to obtain an estimate of the parameter α. The problem is the controls. We want to include controls because we are worried about omitted variable bias – the usual reason for including controls. But which ones do we use?

The naive approach does not work. If estimate the model using the rigorous lasso–but imposing that d_i is not subject to selection–and use the controls selected by the lasso, the estimated $\hat{\alpha}$ will be badly biased. The reason is that we might miss controls that have a strong predictive power for d_i, but only small effect on y_i. Similarly, if we only consider the regression of d_i against the controls, we might miss controls that have a strong predictive power for y_i, but only a moderately sized effect on d_i. See Belloni et al. [18].

Instead, we use the **Post-Double-Selection lasso**:

- Step 1: Use the lasso to estimate

$$y_i = \beta_1 x_{i,1} + \beta_2 x_{i,2} + \ldots + \beta_j x_{i,j} + \ldots + \beta_p x_{i,p} + \varepsilon_i,$$

  i.e., without d_i as a regressor. Denote the set of lasso-selected controls by A.
- Step 2: Use the lasso to estimate

$$d_i = \beta_1 x_{i,1} + \beta_2 x_{i,2} + \ldots + \beta_j x_{i,j} + \ldots + \beta_p x_{i,p} + \varepsilon_i,$$

  i.e., where the causal variable of interest is the dependent variable. Denote the set of lasso-selected controls by B.
- Step 3: Estimate using OLS

$$y_i = \alpha d_i + \mathbf{w}_i' \beta + \varepsilon_i$$

  where $\mathbf{w}_i = A \cup B$, i.e., the union of the selected controls from Steps 1 and 2.

The PDS method is easily extended to cover the case of multiple causal variables: simply repeat Step 2 for each causal variable and add the selected controls to the set B. We illustrate in the example of housing prices below.

Belloni et al. [13, 19] show that the estimate of α using the above methodology is consistent and asymptotically normal, under fairly general conditions. The restricted sparse eigenvalue is the same as in Sect. 2.4. There are further technical conditions relating to technical conditions; see (Belloni et al. [19], Condition SM). As above we need sparsity, but this approximate sparsity has to hold in both equations:

$$m(\boldsymbol{w}_i) = \boldsymbol{x}_i'\boldsymbol{\beta}_{m0} + r_{mi}, \qquad \|\boldsymbol{\beta}_{m0}\|_0 \leq s, \qquad \sqrt{\frac{1}{n}\sum_{i=1}^{n} r_{mi}^2} \leq C\sqrt{\frac{s}{n}} \tag{18}$$

$$g(\boldsymbol{w}_i) = \boldsymbol{x}_i'\boldsymbol{\beta}_{g0} + r_{gi}, \qquad \|\boldsymbol{\beta}_{g0}\|_0 \leq s, \qquad \sqrt{\frac{1}{n}\sum_{i=1}^{n} r_{gi}^2} \leq C\sqrt{\frac{s}{n}} \tag{19}$$

where $\frac{s^2 \log^2(p \vee n)}{n} \to 0$. C is constant like above. An important caveat is that justifies inference on the causal variable(s), but not on the selected high-dimensional controls.

The intuition behind the PDS lasso is that the procedure approximately orthogonalizes y_i and d_i with respect to the disturbance ε_i. A comparison with the Frisch-Waugh-Lovell (FWL) Theorem is perhaps helpful here.

Consider the simple regression model

$$y_i = x_{1,i}\beta_1 + x_{2,i}\beta_2 + u_i$$

and the researcher is interested in β_1. The OLS estimate $\hat{\beta}_1$ can be obtained simply by regression y_i against $x_{1,i}$ *and* $x_{2,i}$. Leaving $x_{2,i}$ out of the regression would cause the estimate $\hat{\beta}_1$ to suffer from omitted variable bias should the omitted $x_{2,i}$ be correlated with the included $x_{1,i}$.

The FWL Theorem states that the following procedure

1. Regress y_i against $x_{2,i}$ (call the residuals $\tilde{y}_i$)
2. Regress $x_{1,i}$ against $x_{2,i}$ (call the residuals $\tilde{x}_{1,i}$)
3. Regress $\tilde{y}_i$ against $\tilde{x}_{1,i}$

generates the *numerically* same estimate $\hat{\beta}_1$. By regressing against $x_{2,i}$, we derive orthogonalized versions of y_i and $x_{1,i}$ and, this way, account for the omitted variable bias.

The PDS methodology, in effect, finds a parsimonious set of controls that can be used to orthogonalize the outcome and treatment variables.

A closely related alternative to lasso PDS that is asymptotically equivalent is *Double-Orthogonalisation* (DO), proposed by Chernozhukov-Hansen-Spindler [23]. The PDS method is equivalent to FWL partialling-out all selected controls from both y_i and d_i. The DO method essentially partials out from y_i only the controls in set A (selected in Step 1, using the lasso with y_i on the LHS), and partials out from d_i only the controls in set B (selected in Step 2, using the lasso with d_i on the LHS). DO partialling-out can use either the lasso or Post-lasso coefficients. DO and PDS are asymptotically equivalent.

3.4 An Example: The Impact of a Disamenity on House Prices

The example we use is literally a textbook example, taken from Jeff Wooldridge's undergraduate and advanced graduate textbooks.[10] The original study is Kiel and McClain [41], who look at the impact of a new incinerator (local waste disposal) on the prices of houses near the incinerator. The location is North Andover, Massachusetts. Starting in 1979, rumours about building a new incinerator start to circulate. Construction of the incinerator starts in 1981. By 1981, information about the new incinerator should be reflected in house prices. We expect that houses that are closer to the site of the new incinerator should be negatively affected because of the perceived cost of the disamenity.

The example is used by Wooldridge to illustrate a 'differences-in-differences' estimation strategy: we compare the change in sales prices 1978–81 of houses near the incinerator vs the prices of houses far from the incinerator. We pool data for the years 1978 and 1981, and specify the model to estimate as

$$\log(rprice_i) = \beta_1 + \beta_2 \log(dist_i) + \beta_3 y81_i + \beta_4(\log(dist_i) \times y81_i) + \varepsilon_i \quad (20)$$

where $rprice_i$ is the sales price of house i in 1978 dollars, $dist_i$ is the distance in miles to the incinerator, and $y81_i$ is a dummy variable equal to 1 if the house sale took place in 1981 and equal to 0 if the house sale took place in 1978.

The omitted variables bias problem is obvious: we expect the choice of location of the incinerator to be related to the quality of housing nearby and to the nature of the location. It is likely the incinerator is built near low-quality housing and in an undesirable location. This will affect the estimate of the pure distance effect β_2 and possibly also the interaction effect β_3.

The potential set of controls includes variables that measure characteristics of the house itself (age, number of rooms, number of bathrooms, log house size in square feet, log land area in square feet), location characteristics (log distance to nearest interstate highway, log distance to the central business district). The textbook exposition cited above suggests using square terms of age and log(distance to highway) but not the central business district measures. Why higher-order terms should be included for some measures but not others, and why some measures should be included and not others, is difficult to justify or specify *a priori*.

In principle, including all possible levels and interactions of controls is appealing – in effect, a second-order Taylor approximation to an arbitrary control function – but in the traditional approach, where all these variables are included without penalisation, we run the risk of overfitting. The PDS methodology addresses this very easily; we include the full set of levels, squares and interactions (34 controls in all), and only those that contribute substantially to addressing the omitted variable bias are retained.

---

[10]Wooldridge [56], pp. 450-3, 474 and Wooldridge [57], pp. 153-4.

Table 1 Lasso-selected variables for Kiel-McClain example

Dependent variable:	Selected:
$\log(rprice)$	$baths \times \log(area)$, $\log(land) \times rooms$, $\log(land) \times \log(area)$, $log(area) \times rooms$, $\log(area) \times \log(CBD)$
$\log(distance)$	$\log(CBD)^2$
$\log(distance) \times y81$	$\log(CBD)^2$
Selected from:	Levels, squares and cross-products (34) of age, number of rooms, number of baths, log land area, log house area, log distance to interstate highway, log distance to central business district (CBD)
Unpenalised controls:	$y81$

Below we report the result of estimating with no controls at all, the PDS estimates, and the DO estimates using the lasso and post-lasso coefficients, using the Wooldridge textbook dataset of 321 observations from the Kiel-McClain study. Heteroskedasticity-robust penalty loadings are used in the PDS and DO estimations; heteroskedasticy-robust standard errors are reported for all four sets of results.

Table 1 reports the lasso-selected controls based on the separate estimations for $log(rprice)$, $log(distance)$ and $log(distance) \times y81$. The lasso estimation for the dependent variable in the structural equation $log(rprice)$ selects 5 out of the 34 possible penalized controls; all 5 are interactions. The lasso estimations for the explanatory variables in the structural equation, $log(distance)$ and $log(distance) \times y81$, both select only the square of the log of the distance to the central business district. In all these estimations, the dummy for 1981 $y81$ is always included; this is done by specifying that the variable has a zero penalty in the lasso estimations. An unpenalised intercept is also always included. The union of these selected predictors plus the unpenalised dummy $y81$ yields 7 controls for the PDS estimation. The selected predictors for the 3 separate lassos are separately partialled out from $log(rprice)$, $log(distance)$ and $log(distance) \times y81$ and then used in the DO estimations. Either the lasso or post-lasso OLS coefficients can be used for the partialling-out, yielding two different sets of DO estimation results.

Table 2 reports the no-controls, PDS, and DO estimation results. Coefficients and heteroskedasticity-robust standard errors are shown for all four estimations. The coefficient on the year dummy $y81$ is also shown, as is standard for a differences-in-differences estimation, but no standard error is displayed in the PDS estimation because the variable is treated as an unpenalised control rather than as a causal variable.[11] No coefficient on $y81$ is reported for the DO estimations as it is partialled-out along with the other controls.

Without controls, there is no incinerator effect but a strong positive distance effect. The impact of the inclusion of the controls is to make the distance effect much smaller and less precisely estimated; the incinerator effect remains small and becomes slightly

[11] To treat it as a causal variable and obtain a valid standard error, we would have to estimate an additional lasso regression with $y81$ as the dependent variable etc.

Table 2 Structural equation estimations for the Kiel-McClain example; heteroskedastic-robust standard errors in parentheses

Regressor	No controls	PDS	DO-lasso	DO-post-lasso
$\log(distance)$	0.317	0.060	0.003	0.046
	(0.038)	(0.065)	(0.057)	(0.062)
$\log(distance) \times y81$	0.048	0.022	0.041	0.017
	(0.077)	(0.050)	(0.047)	(0.048)
$y81$	−0.275	−0.075	(n.a.)	(n.a.)
	(0.762)	(n.a.)		

more precisely estimated. These results are similar to those in the examples and discussions in Wooldridge's textbooks, and illustrate the same point - the apparent distance effect is spurious and driven by omitted variable bias. But the controls are drawn from a more flexible functional form, selected in such a way that overfitting is avoided at the same time that omitted variable bias is addressed, and because the selection is automated it is relatively immune to suspicions of p-hacking.

3.5 *Caveats*

Probably the most important caveat to bear in mind when using the PDS methodology is the requirement that the confounder dimensionality is sufficiently sparse. The Kiel-McClain example is a good one in the sense that this assumption seems reasonable: the original authors set out the dimensions in which confounding could be an issue, provided proxies for these dimensions, and it is plausible that levels and interactions of these proxies is enough to approximate the problem.

In other applications, the sparsity assumption will be less plausible. For example, we might have employee data with codes for occupation, or sales data disaggregated by product code. It is natural to code characteristics using dummy variables, and it would be tempting to use this large set of dummies along with the PDS methodology to address the counfounder issue. But it is problematic to assume sparsity here, because it amounts to the assumption that most jobs or products are very similar and a few are very different.

4 Heterogeneous Treatment Effects

Up to this point, we have concerned ourselves with tools, still in development, that have made it possible to estimate causal effects using machine learning methods. The advantage of these techniques is that they allow us to address arbitrary and auxiliary assumptions which have the potential to weaken the validity of empirical work.

One of the most influential lines of current research in this area has focused on heterogeneity in treatment effects: it is unrealistic to assume that the effect of a policy, intervention or treatment (generally defined) does not vary for each individual to whom it was applied. The methods employed by researchers should be robust to that fact, and should explicitly target some element of the distribution of effects to ensure that the results can be reliably understood and interpreted.

Much of the modern literature on heterogeneous effects builds from a common framework, the Rubin Causal Model. Its roots lie with the foundational work conducted by Fisher [29, 30] and Neyman [45] on agricultural experiments, and it is now the dominant structure used in the analysis of causality in many fields, including statistics and econometrics.

The framework is simple: suppose we wish to evaluate the effect of a policy or decision, or something similar, on a number of individuals' outcomes, which we denote separately by $Y_i \in \mathbb{R}$. We record the treatment status of each individual with $W_i \in \{0, 1\}$, which takes the value 1 for individuals who were affected explicitly by the policy, and is 0 otherwise. The settings that this describes are intentionally and necessarily restricted; the treatment status of an individual is binary. We do not allow for varying treatment intensity.

Alongside this information, we have available for each individual additional data, contained in the p-dimensional column vector X_i, which we require to address confounding. This information can be (and often is) large in scale. There are few practical constraints on the nature of the data that we can include, but there is clear guidance about the inclusion of one particular type of variable: the vector of covariates must not contain any series which is directly affected by the treatment [55]. For instance, if one were interested in evaluating the effects of a past labour market programme, incorporating as a control participants' current employment status would be inappropriate.

Finally but perhaps most importantly, the realised outcome for each individual, Y_i, is assumed to be a function of their own 'potential' outcomes $(Y_i(0),\ Y_i(1))$, where $Y_i(0)$ represents the outcome that individual i would have experienced had they not received the treatment, and $Y_i(1)$ is defined analogously. When the treatment is assigned, we can only observe one of these values per person. The remaining half has become 'missing' by definition, which is the fundamental problem of causal inference. We can summarize this argument by writing $Y_i = W_i Y_i(1) + (1 - W_i) Y_i(0)$.

This framework is extremely flexible. The causal effect of the treatment is allowed to vary at the level of the individual, as $Y_i(1) - Y_i(0)$, and is effectively left unrestricted. As a result, it is typically the case that researchers target a summary of these effects. In econometrics, the targets (or 'estimands') most commonly selected are the average treatment effect (ATE),

$$\tau = \tau_{\text{ATE}} = E\left[Y_i(1) - Y_i(0)\right],$$

and the average treatment effect on the treated (ATET)

$$\tau_{\text{ATET}} = E\left[Y_i(1) - Y_i(0) \mid W_i = 1\right].$$

This is at least partly due to convenience, and some prominent researchers (e.g., the Nobelist econometrician James Heckman) have criticised others for not motivating their targets more carefully. However, as a result of advances at the intersection of machine learning and causal inference (and related programmes of research), this practice is beginning to change. Instead of simple summaries, it is now not uncommon for researchers to report the pattern of heterogeneity in effects. This is a topic we briefly return to in the conclusion.

The model is flexible in another important regard: it places no assumptions over the distribution of the potential outcomes. One could proceed to make such assumptions. However, it is not clear that one could justify such an assumption without a rigorous theoretical underpinning for it; and so it is likely that the imposition of such a condition would introduce a source of fragility into the model. As such we maintain the looser framework. Nevertheless we must impose a minimal degree of structure on the model to ensure that the estimands are identified. That structure is delivered by two assumptions.

The first, and most important, is **unconfoundedness**. Intuitively, it says that the assignment to treatment is 'as good as random', *conditional on the information contained within the covariates*. Formally, we express this as

$$W_i \perp\!\!\!\perp (Y_i(0),\ Y_i(1)) \mid X_i. \tag{21}$$

Notice the burden placed on the observed covariates: they need to be sufficiently comprehensive to ensure that this condition holds. But how does one know for a given set of circumstances that this assumption is credible? Earlier we discussed the difficult process of selecting controls, and it is important to emphasise that the same dilemma arises here. However, the solution that we presented is also appropriate: machine learning methods, which are designed for variable selection and high dimensional data, can naturally address this issue for us by shifting assumptions over the relevance (or otherwise) of covariates onto an assumption of sparsity, which encodes our prior belief that they are not all required—indeed *many* should be left out of the model.

Unfortunately, whilst machine learning methods allow us to remain agnostic about the inclusion and form of covariates, and consequently make unconfoundedness more palatable, their use is also likely to increase the frequency with which researchers encounter issues related to the second identifying assumption, **overlap**:

$$\kappa < P(W_i = 1 \mid X_i = x) < 1 - \kappa, \quad \kappa \in \left(0, \frac{1}{2}\right), \quad \forall x. \tag{22}$$

D'Amour et al. [25] demonstrate that there is an unfortunate link between the dimension of the covariates, p, and the overlap assumption. Strict overlap, which is the form presented above and the form required to ensure that the estimators we are interested in are $\sqrt{n}$-consistent, implies that the average imbalance in covariate means between the treated and control units converges to zero as the dimension p grows. This suggests that there are some instances in which it would be unsuitable to employ estimators that make use of machine learning methods without first pruning the set of covariates to be included. Trimming outliers may help to alleviate this problem, but D'Amour et al. [25] highlight that there are subtle issues with this fix.

To construct estimators in a setting such as this, one must turn to semiparametric theory. Semiparametric methods make it possible to leave unrestricted important components of the data generating process. This is particularly useful if we lack information about those components of the model, or in the event that those components involve nuisance functions that are not of direct interest. A semiparametric model is simply one that is indexed by an infinite-dimensional parameter [40]. That definition is quite wide: it encompasses everything from nonparametric models (which in no way restrict the set of possible probability distributions for the data), to regression models with parametric functional forms but errors that have an unspecified distribution. More specifically, one can view semiparametric models as possessing a finite-dimensional parameter of interest (which is parametric) and an infinite-dimensional nuisance parameter (which is the nonparametric component) (e.g. Begun et al. [12]). For example, in the context of causal inference it is often the case that researchers place assumptions on the treatment mechanism (in other words, they model it parametrically) but they leave the outcome model unrestricted. That reflects the relative strength of the information we can garner about the two models.

Many semiparametric models can be characterised and analysed through their influence function. An estimator $\hat{\Phi}$ for Φ is asymptotically linear if it can be approximated by a sample average, with i.i.d. data $D_i = (Y_i, X_i, W_i)$, in the following manner

$$(\hat{\Phi} - \Phi_0) = \frac{1}{n}\sum_{i=1}^{n} \phi_i (D_i) + o_p\left(\frac{1}{\sqrt{n}}\right), \tag{23}$$

where ϕ has mean zero and finite variance (that is, $E[\phi(D)] = 0$ and $V = E\left[\phi(D)\,\phi(D)'\right] < \infty$). Then by the Lindeberg-Levy central limit theorem, the estimator $\hat{\Phi}$ with influence function ϕ is asymptotically normal:

$$\sqrt{n}(\hat{\Phi} - \Phi_0) \rightsquigarrow N(0, V), \tag{24}$$

where $\rightsquigarrow$ denotes convergence in distribution. Thus, the estimator's asymptotic behaviour, up to a small degree of error, can be fully described by its influence function, which renders simple the construction of confidence regions and test statistics that have approximately correct coverage and size in large samples.

Equally, one can reverse this logic to create estimators given a candidate influence function. The end of this process is pleasingly simple: one solves the estimating equation formed from the sample analogue of the moment condition the influence function satisfies.[12]

$$\frac{1}{n}\sum_{i=1}^{n} \phi_i \left(D_i; \Phi, \hat{\eta}\right) = 0. \tag{25}$$

[12] Of course, this discussion raises a more fundamental question: given a specification like that above, how does one construct such a function? The answer is a little technical, unfortunately. Interested readers can find a detailed discussion in van der Vaart [52].

Note here the explicit dependence of the function on a nuisance parameter η and that the estimating equation is evaluated with $\hat{\eta}$, which is produced by another model.

Before we proceed it will be useful to mildly restrict and clarify the framework we are working with. It has two basic components: the first is an equation for the outcome, Y_i, as usual; the second describes the model for the assignment to treatment. Let $\mu(w, x) = E[Y_i(w) \mid X_i = x]$ for some $w \in \{0, 1\}$ and $e(x) = P(W_i = 1 \mid X_i = x)$, which is called the propensity score in the literature. Then

$$Y_i = \mu(W_i, X_i) + \varepsilon_i, \quad E[\varepsilon_i \mid X_i, W_i] = 0, \tag{26}$$

$$W_i = e(X_i) + \nu_i, \quad \text{and} \quad E[\nu_i \mid X_i] = 0. \tag{27}$$

The target we focus on is the ATE, which in this context is

$$\tau = E[\mu(1, X_i) - \mu(0, X_i)]. \tag{28}$$

Note that the confounders X_i are related to both the treatment *and* the outcome variables, and hence comparing the raw outcomes of the treated and control units would produce a biased estimate. The manner in which Y_i and W_i depend on X_i (through $\mu(W_i, X_i)$ and $e(X_i)$) is left unrestricted, and it accommodates general heterogeneity in the effect of the treatment. As described in (7), we tackle the problem of the unknown and complex form of the functions by using a linear combination of transformations of the covariates and the treatment indicator to approximate them. As we have said before, the number of terms required for this task may be considerable, so to obtain $\hat{\mu}(W_i, X_i)$ and $\hat{e}(X_i)$, we use methods from the machine learning literature, and particularly the lasso.

How do we link these estimates to τ? One approach is to construct an influence function which identifies the target parameter through its associated moment condition. The problem with this approach is that the nuisance parameters $\eta = (\mu, e)$ are estimated imperfectly. Again, as mentioned previously, machine learning methods address high-dimensionality through regularisation. That controls, and in fact substantially reduces, the variance of the predicted values produced by the model, but achieves that at the cost of the introduction of bias in its coefficient estimates. As a result, the influence function must be designed carefully to ensure that its associated moment condition is locally insensitive to the value of the nuisance parameter η around η_0. Formally, we want this orthogonality condition to hold:[13]

$$\partial_\eta E\big[\phi(D; \tau_0, \eta)\big]_{\eta=\eta_0} = 0.$$

It turns out that there is an interesting and useful connection between functions which possess this property and semiparametric efficiency. The efficient influence

[13] ∂_η is shorthand for $\partial/\partial\eta'$. This version of the condition is actually more stringent than required. A more general definition of it, based on Gateaux derivatives, can be found in Belloni et al. [16] and Chernozhukov et al. [22].

function is the (almost surely) unique function which satisfies the semiparametric efficiency bound (analogous to the Cramer-Rao lower bound) [47]. Aside from being efficient, it possesses a number of other interesting properties—one of which is orthogonality.[14,15] Given the target and the assumptions of unconfoundedness and overlap, its form (derived by Hahn [34]) is as follows:

$$\phi\left(y, w, x; \tau, \eta\right) = \frac{w\left(y - \mu\left(1, x\right)\right)}{e\left(x\right)} - \frac{\left(1 - w\right)\left(y - \mu\left(0, x\right)\right)}{1 - e\left(x\right)} + \left(\mu\left(1, x\right) - \mu\left(0, x\right)\right) - \tau. \tag{29}$$

We recover $\hat{\tau}$, as we explained, from the finite-sample analogue of the moment condition

$$\frac{1}{n}\sum_{i=1}^{n}\phi\left(Y_i, W_i, X_i; \hat{\tau}, \hat{\eta}\right) = 0, \tag{30}$$

The Double Machine Learning (DML) estimator of Chernozhukov et al. [22] extends this framework in one important direction. Machine learning methods have an inbuilt tendency to overfit the data. Regularisation attenuates that but, *generally speaking*, it does not completely address it. Across equations, when the true structural errors, ϵ_i and ν_i, are correlated with the estimation errors of the nuisance parameters, such as $\hat{e}\left(X_i\right) - e\left(X_i\right)$, poor performance can result. The dependence between these components, however, can be addressed fairly easily using sample splitting.[16]

Sample splitting begins with partitioning the data into two sets, an auxiliary fold and a main fold. With the auxiliary fold, one obtains $\hat{\mu}\left(W_i, X_i\right)$ and $\hat{e}\left(X_i\right)$, that is, estimates of the nuisance parameters. Then with the main fold, the treatment effect is estimated by plugging the results of the first step, $\hat{\tau}$, into the estimating equation defined by (30). This process carries a significant disadvantage: the final estimate is produced using only a portion of the data, and thus it is not efficient. The DML estimator makes use of an adapted form of the process called cross-fitting, where the procedure is repeated with the roles for the samples reversed. The resulting treatment effect estimates are averaged together, which ensures full efficiency. Finally, to guarantee that the performance of the estimator is not adversely affected by an unusual partition of the data, the procedure is repeated a 'large' number of times, say 100, and the median of the final set of values forms the estimate for the treatment effect. This procedure can be generalised to allow for splits of unequal size, which in finite samples may improve the performance of the estimator, as it allows more data to be used for the first stage, when the machine learning methods learn the structure of the nuisance functions.

This aspect of the estimator's design is crucial. In more conventional work, the bias induced by the data-adaptive nature of the machine learning methods would be

[14]The estimators that possess this property need not be semiparametrically efficient, but because they can be, we restrict our focus to those that are.

[15]Estimators based on the efficient influence function are also double-robust (Robins et al. [48]).

[16]Sample splitting was originally introduced by Angrist and Krueger [3] and Altonji and Segal [2] in the context of bias reduction of IV and GMM estimators.

analysed and controlled using empirical process theory. In particular, that approach would proceed by imposing constraints, called Donsker conditions, on the function class that contains the values of the nuisance estimator (Vaart and Wellner [51]). These conditions make it possible to conclude that terms responsible for the estimator's bias vanish asymptotically. However, the size—or complexity—of that function class must be suitably bounded for it to be Donsker.[17] Unfortunately, in the environment we are interested in, where the dimension of the covariates, p, is allowed to grow with the sample size, that requirement will not hold. If the complexity of the function class grows but at a rate that is sufficiently slow relative to n, it is possible to show that the estimator's bias, caused by overfitting, will tend to zero [22]. However, the assumptions required to demonstrate that result are typically restrictive in practice: for example, if one were to use the lasso to estimate the nuisance functions, the model would need to be very sparse for the bias to vanish.

With sample-splitting, those conditions become substantially weaker. The model is allowed to be more complex with that adaptation, and thus, the number of non-zero coefficients can be larger. Specifically, for the DML, if the outcome and propensity score models are estimated with the lasso, we require that $s_\mu s_e \ll n$, where s_μ and s_e denote their respective sparsity indices. Without the procedure (that is, relying solely on orthogonalisation), that condition is instead $s_\mu^2 + s_e^2 \ll n$ [22]. The former is clearly weaker than the latter, and embedded within the first condition is a useful trade-off. Say the model for the propensity score was expected to be very sparse, such that $s_e \ll \sqrt{n}$, perhaps because the assignment procedure is well-understood and is contingent on only a small number of factors (as it may be if one were looking at, for instance, a treatment prescribed by clinicians). In that case, the outcome model can be relatively dense and there is scope to include many covariates in the model. Thus, we can use external information to reason about the relative complexity of the assignment model and the process that determines the outcome, and then we can balance one against the other.

There are further, more general benefits to sample-splitting. The loosened requirements over the complexity of the model allow one to use, with little adjustment to the underlying theory, a large variety of machine learning methods for the estimation of η. Without the procedure, one would have to verify that those methods individually satisfied relevant complexity restrictions. The methods do not have to rely on approximate sparsity. Instead, they need only satisfy conditions on the quality of the approximations they provide for the nuisance functions. Crudely, for instance, we could require that both nuisance parameters are estimated at the $o_p\left(n^{-1/4}\right)$ rate, such that the product

$$E\left[\left(\hat{\mu}\left(w, X_i\right) - \mu\left(w, X_i\right)\right)^2\right]^{1/2} E\left[\left(\hat{e}\left(X_i\right) - e\left(X_i\right)\right)^2\right]^{1/2} = o_p\left(n^{-1/2}\right) \quad (31)$$

[17]There are a number of conditions that are intimately related to the size of the function class, as measured by its bracketing and covering numbers, which if satisfied are sufficient for it to be Donsker. See Vaart and Wellner [51] for a full statement.

is asymptotically negligible [8]. Notice that each estimator is allowed to converge at a rate that is substantially slower than that of a correctly specified parametric model.[18] Random forests, neural nets and L_2 boosting all satisfy this condition (see, e.g., Wager and Walther [54] and Luo and Spindler [43]), as do more traditional, flexible estimators, such as generalised additive models. Of course, ensembles of these are also permissible, and they are likely to perform at least as well as the best of the individual models.[19]

5 Related Developments

The pace of development in this field is rapid, so it would not be possible to comprehensively cover its related literature in the space we have. Instead, we will simply point to a number of new techniques, sampled from this body of work, which are tied to the methods we have discussed.

For the estimation of average treatment effects (and restricted versions thereof), Athey et al. [9] introduce the Approximate Residual Balancing (ARB) estimator. The method is split into two stages: first, the lasso is used to estimate the outcome model, which is assumed to be linear; then, the residuals from that are weighted and added back to the result from the first stage. Those weights are chosen to ensure that the distribution of the covariates for the treated and control units match closely (in-sample). The functional form restriction they impose allows them to obtain a tight guarantee on the model's finite-sample bias. Furthermore, the model is consistent even when the propensity score is very dense. In fact, their asymptotic results continue to hold when the propensity score cannot be consistently estimated. This requires strong sparsity of the outcome model ($s_\mu \ll \sqrt{n}$), though. Thus, it is a complement to the DML.

Ning et al. [46] developed a related method, which they call the High-Dimensional Covariate Balancing Propensity Score estimator, which is a modified version of the Horvitz and Thompson [37] method. Based on an adapted version of the lasso with a quasi-likelihood, they first estimate the coefficients in the propensity score model. Then, they use a weighted version of the lasso to estimate the outcome model for the treated and control groups separately. For the variables selected in the second step, they find calibrated new coefficients for the propensity score model using a quadratic programme which ensures the covariate distributions are balanced (as above). Those new coefficients are layered over (i.e., replace) the corresponding set from the original propensity score model, fit in the first step. Finally, with the modified model for $e(\hat{X}_i)$, the fitted probability of treatment is obtained, and that is used to weight the outcomes

[18]Note that this observation can be connected to the discussion in the paragraph above: if one of the models is estimated parametrically based on a relationship that is known to be true, the other model need only be consistent, since the product of the two rates would be $o_p\left(n^{-1/2}\right)$.

[19]Laan et al. [42] provide asymptotic justifications for weighted combinations of estimators, particularly those which use cross-validation to calculate the weights.

of the treated and control units. Aside from being $\sqrt{n}$-consistent, asymptotically normal and semiparametrically efficient (qualities it shares with the DML and ARB), it also possesses the sample-boundedness property: loosely, the estimated treatment effect is guaranteed to be reasonable because the estimate of each component of the target (e.g., $E[Y_i(1)]$) must lie within the range of outcomes observed in the data. Notably, the authors also incorporate sample-splitting and find that the level of sparsity required of the outcome and propensity score models matches that of the DML exactly.

Farrell et al. [27] focus on deep neural networks and develop the theory necessary to justify their use for causal inference. Specifically, they demonstrate that the most popular variant of the models at present, multilayer perceptrons with rectified linear activation functions, can be used to estimate the nuisance parameters as set out above at the required rate under appropriate smoothness conditions, and subsequently, one can conduct valid inference on treatment effects with the estimator.

Finally, Wager and Athey [53] developed new results for an adaptation of random forests (that they call causal forests) which show that they can consistently estimate conditional average treatment effects. That is, they can be used to estimate average treatment effects for a given value of x. Moreover, provided that the subsamples used for the estimation of each tree are large enough, the method is asymptotically normal and unbiased. Pointwise confidence intervals can thus be constructed: they develop results for an infinitesimal jackknife estimator for the variance of the forest which is also consistent. Results from a simulation study they conduct demonstrate the power of the theory, but they also highlight a number of features of the method which require further study to improve performance, including the splitting rules used for the trees.

There is a vast pool of work which extends these methods and others in important and relevant directions. The activity in this field and rapid progress made clearly indicate its importance, and we expect that the insights gleaned from this research will become increasingly important for applied economic research.

References

1. Ahrens, A., Hansen, C.B., Schaffer, M.E.: lassopack: model selection and prediction with regularized regression in Stata. The Stata J. **20**, 176–235 (2020)
2. Altonji, J.G., Segal, L.M.: Small-sample bias in GMM estimation of covariance structures. J. Bus. Econ. Stat. **14**, 353–366 (1996)
3. Angrist, J.D., Krueger, A.B.: Split-sample instrumental variables estimates of the return to schooling. J. Bus. Econ. Stat. **13**, 225–235 (1995)
4. Angrist, J.D., Pischke, J.-S.: The credibility revolution in empirical economics: how better research design is taking the con out of econometrics. J. Econ. Perspect. **24**, 3–30 (2010)
5. Angrist, J., Azoulay, P., Ellison, G., Hill, R., Lu, S.F.: Economic research evolves: fields and styles. Am. Econ. Rev. **107**, 293–297 (2017)
6. Arlot, S., Celisse, A.: A survey of cross-validation procedures for model selection. Statist. Surv. **4**, 40–79 (2010)

7. Athey, S., Imbens, G.W.: The state of applied econometrics: causality and policy evaluation. J. Econ. Perspect. **31**, 3–32 (2017)
8. Athey, S., Imbens, G.W.: Machine learning methods that economists should know about. Ann. Rev. Econ. **11**, 685–725 (2019)
9. Athey, S., Imbens, G.W., Wager, S.: Approximate residual balancing: debiased inference of average treatment effects in high dimensions. J. Roy. Stat. Soc.: Ser. B (Stat. Methodol.) **80**, 597–623 (2018)
10. Backhouse, R., Cherrier, B.: The age of the applied economist: the transformation of economics since the 1970s. Hist. Polit. Econ. **47** (2017)
11. Bansak, K., Ferwerda, J., Hainmueller, J., Dillon, A., Hangartner, D., Lawrence, D., Weinstein, J.: Improving refugee integration through data-driven algorithmic assignment. Science **359**, 325–329 (2018)
12. Begun, J.M., Hall, W.J., Huang, W.-M., Wellner, J.A.: Information and asymptotic efficiency in parametric-nonparametric models. Ann. Stat. **11**, 432–452 (1983)
13. Belloni, A., Chen, D., Chernozhukov, V., Hansen, C.: Sparse models and methods for optimal instruments with an application to eminent domain. Econometrica **80**, 2369–2429 (2012)
14. Belloni, A., Chernozhukov, V.: High dimensional sparse econometric models: an introduction. In: Alquier, P., Gautier, E., Stoltz, G. (eds.) Inverse Problems and High-Dimensional Estimation SE - 3. Lecture Notes in Statistics, pp. 121–156. Springer, Heidelberg (2011)
15. Belloni, A., Chernozhukov, V.: Least squares after model selection in high-dimensional sparse models. Bernoulli **19**, 521–547 (2013)
16. Belloni, A., Chernozhukov, V., Fernandez-Val, I., Hansen, C.: Program evaluation and causal inference with high-dimensional data. Econometrica **85**, 233–298 (2017)
17. Belloni, A., Chernozhukov, V., Hansen, C.: Inference for High-Dimensional Sparse Econometric Models (2011). http://arxiv.org/abs/1201.0220
18. Belloni, A., Chernozhukov, V., Hansen, C.: High-dimensional methods and inference on structural and treatment effects. J. Econ. Perspect. **28**, 29–50 (2014a)
19. Belloni, A., Chernozhukov, V., Hansen, C.: Inference on treatment effects after selection among high-dimensional controls. Rev. Econ. Stud. **81**, 608–650 (2014b)
20. Belloni, A., Chernozhukov, V., Hansen, C., Kozbur, D.: Inference in high dimensional panel models with an application to gun control. J. Bus. Econ. Stat. **34**, 590–605 (2016)
21. Bickel, P.J., Ritov, Y., Tsybakov, A.B.: Simultaneous analysis of lasso and dantzig selector. Ann. Stat. **37**, 1705–1732 (2009)
22. Chernozhukov, V., Chetverikov, D., Demirer, M., Duflo, E., Hansen, C., Newey, W., Robins, J.: Double/debiased machine learning for treatment and structural parameters. Econom. J. **21**, C1–C68 (2018)
23. Chernozhukov, V., Hansen, C., Spindler, M.: Post-selection and post-regularization inference in linear models with many controls and instruments. Am. Econ. Rev. **105**, 486–490 (2015)
24. Chetverikov, D., Liao, Z., Chernozhukov, V.: On cross-validated lasso in high dimensions. Annal. Stat. (Forthcoming)
25. D'Amour, A., Ding, P., Feller, A., Lei, L., Sekhon, J.: A Gaussian Process Framework for Overlap and Causal Effect Estimation with High-Dimensional Covariates, arXiv:1711.02582v3 [math.ST] (2019)
26. Deaton, A., Cartwright, N.: Understanding and misunderstanding randomized controlled trials. Soc. Sci. Med. **210**, 2–21 (2018)
27. Farrell, M.H., Liang, T., Misra, S.: Deep Neural Networks for Estimation and Inference (2019)
28. Feigenbaum, J.J.: Automated census record linking: a machine learning approach (2016). Working Paper
29. Fisher, R.A.: Statistical Methods for Research Workers, 5th edn. Oliver and Boyd Ltd., Edinburgh (1925)
30. Fisher, R.A.: The Design of Experiments, 8th edn. Hafner Publishing Company, New York (1935)

31. Gelman, A., Loken, E.: The garden of forking paths: why multiple comparisons can be a problem, even when there is no 'fishing expedition' or 'p-hacking' and the research hypothesis was posited ahead of time (2013). http://www.stat.columbia.edu/~gelman/research/unpublished/p_hacking.pdf
32. Gentzkow, M., Shapiro, J.M., Taddy, M.: Measuring group differences in high-dimensional choices: method and application to congressional speech. Econometrica **87**, 1307–1340 (2019)
33. Gentzkow, M., Kelly, B., Taddy, M.: Text as data. J. Econo. Lit. **57**, 535–574 (2019)
34. Hahn, J.: On the role of the propensity score in efficient semiparametric estimation of average treatment effects. Econometrica **66**, 315 (1998)
35. Hastie, T., Tibshirani, R., Wainwright, M.J.: Statistical Learning with Sparsity: The Lasso and Generalizations, Monographs on Statistics & Applied Probability. CRC Press, Taylor & Francis, Boca Raton (2015)
36. Hamermesh, D.S.: Six decades of top economics publishing: who and how? J. Econ. Lit. **51**, 162–172 (2013)
37. Horvitz, D.G., Thompson, D.J.: A generalization of sampling without replacement from a finite universe. J. Am. Stat. Assoc. **47**, 663 (1952)
38. Hyndman, R.J., Athanasopoulos, G.: Forecasting: Principles and Practice, 2 ed. (2018)
39. Jing, B.-Y., Shao, Q.-M., Wang, Q.: Self-normalized Cramér-type large deviations for independent random variables. Ann. Probab. **31**, 2167–2215 (2003)
40. Kennedy, E.H.: Semiparametric Theory and Empirical Processes in Causal Inference, arXiv:1510.04740v3 [math.ST] (2016)
41. Kiel, K., McClain, K.: House prices during siting decision stages: the case of an incinerator from rumor through operation. J. Environ. Econ. Manag. **28**, 241–255 (1995)
42. van der Laan, M.J., Dudoit, S., van der Vaart, A.W.: The cross-validated adaptive epsilon-net estimator. Stat. Decisions **24**, 373–395 (2006)
43. Luo, Y., Spindler, M.: High-Dimensional L2 Boosting: Rate of Convergence, arXiv:1602.08927v2 [stat.ML] (2019)
44. Mullainathan, S., Spiess, J.: Machine learning: an applied econometric approach. J. Econ. Perspect. **31**, 87–106 (2017)
45. Neyman, J.: On the Application of Probability Theory to Agricultural Experiments. Essay on Principles. Section 9. Translated by D. M. Dabrowska and T. P. Speed. Stat. Sci. **5**, 465–472 (1990)
46. Ning, Y., Peng, S., Imai, K.: Robust estimation of causal effects via a high-dimensional covariate balancing propensity score. Biometrika (2020)
47. Powell, J.: Estimation of Semiparametric Models. Elsevier Science B.V., Amsterdam (1994)
48. Robins, J.M., Rotnitzky, A., Zhao, L.P.: Estimation of regression coefficients when some regressors are not always observed. J. Am. Statis. Assoc. **89**, 846 (1994)
49. Simmons, J.P., Nelson, L.D., Simonsohn, U.: False-positive psychology: undisclosed flexibility in data collection and analysis allows presenting anything as significant. Psychol. Sci. **22**, 1359–1366 (2011)
50. Tibshirani, R.: Regression shrinkage and selection via the lasso. J. Roy. Stat. Soc.: Ser. B (Methodol.) **58**, 267–288 (1996)
51. van der Vaart, A.W., Wellner, J.A.: Weak Convergence and Empirical Processes. Springer, Heidelberg (1996)
52. van der Vaart, A.W.: Asymptotic Statistics, Cambridge University Press (1998)
53. Wager, S., Athey, S.: Estimation and inference of heterogeneous treatment effects using random forests. J. Am. Stat. Assoc. **113**, 1228–1242 (2018)
54. Wager, S., Walther, G.: Adaptive concentration of regression trees, with application to random forests, arXiv:1503.06388 [math.ST] (2019)
55. Wooldridge, J.M.: Violating ignorability of treatment by controlling for too many factors. Econom. Theory **21**, 1026–1028 (2005)
56. Wooldridge, J.M.: Introductory Econometrics: A Modern Approach, 4th edn. Cengage, Boston (2009)
57. Wooldridge, J.M.: Econometric Analysis of Cross Section and Panel Data, 2nd edn. MIT Press, Cambridge (2010)

Don't Test, Decide

William M. Briggs

Abstract There is no reason to use traditional hypothesis testing. If one's goal is to assess past performance of a model, then a simple measure or performance with no uncertainty will do. If one's goal is to ascertain cause, then because probability models can't identify cause, testing does not help. If one's goal is to decide in the face of uncertainty, then testing does not help. The last goal is to quantify uncertainty in predictions; no testing is needed and is instead unhelpful. Examples in model selection are given. Use predictive, not parametric, analysis.

Keywords Causation · P-values · Hypothesis testing · Model selection · Model validation · Predictive probability

1 Testing Is Dead

A vast amount of statistical practice is devoted to "testing", which is the art of deciding, regardless of consequence, that a parameter or parameters of a probability model does or does not take a certain value. Most testing is done with p-values. If the p-value is less than the magic number, the test has been passed; otherwise not. Everybody knows the magic number.

The foremost, and insurmountable, difficulty is that every use of a p-value contains a fallacy or is a mistake. This is proved several times over in [1–4]. Some have accepted that p-values should be abandoned, but, still desiring to test, they have moved to Bayes factors. See [5] about Bayes factors.

Bayes factors are at best only a modest improvement on model decision making. Unfortunately, their use retains several fundamental misconceptions shared by p-values about what parameters are and do. Many for instance believe that if a test has been passed, a cause has been proved. For example, suppose we have a regression that characterizes uncertainty in "amount of improvement" for some observable (say, in medicine). One parameter in the model represents presence of a new drug. If this parameter is positive and the model's parameter passes its test, frequentist or Bayes, almost everybody believes the new drug "works" *in a causal sense*. Meaning

W. M. Briggs (✉)
New York, USA
e-mail: matt@wmbriggs.com

S. Sriboonchitta et al. (eds.), *Behavioral Predictive Modeling in Economics*, Studies in Computational Intelligence 897,
https://doi.org/10.1007/978-3-030-49728-6_3

everybody thinks, regardless of any provisos they might issue, that the new drug really does *cause* an improvement above the old one.

This inference is certainly natural, just as it is certainly invalid. It is invalid even in those times in which the parameter really is associated with a causal effect. A conclusion in some logical arguments can be true even though the argument itself is invalid. Probability models can't prove cause, though they can assume it. See [6].

In any case, testing is not is what is wanted by the majority of users of statistical models. What they're after is quantifying uncertainty of *this* given *that*. In our example, patients wants to know what are the chances they will improve if they take the new drug. And that's not all. They will also likely want to know the cost, possibility of side effects, and so forth. Patients factor in all these things, along with the model's probability of improvement, into a decision whether to use the new drug or to go with the old, a decision which is personal to them—and *not* to the statistician creating the model. Patients likely won't do this "analysis" in a formal way, but what they *are* doing is deciding and not testing.

Testing is definite. Too definite. Testing under frequentism says a thing is certain—a parameter "is" equal to 0, say—with no measure of uncertainty attached. There are, in frequentist inference, measures of uncertainty of tests, but these are statements about tests different than the one in hand, about what happens to other tests at the limit. The limit is a time of no concern to anybody.

Here we argue that testing should be abandoned and replaced everywhere with deciding.

2 Model Uncertainty

The simplest case is one model. We have background evidence M which is known or accepted such that the uncertainty in the observable y is characterized by the probability model deduced from M. This is

$$\Pr(y|x\mathrm{DM}), \tag{1}$$

where the optional x are other measures probative of y, the optional D is n prior observations (y, x), and M the background information. In M are *all* relevant premises of the model, including premises about priors on the parameters of the model, if any. Finally, y is shorthand for $y \in s$ where s is some relevant set.

Suppose $\Pr(y|x\mathrm{DM}) = p$, then it is *true* the probability of y conditional on this model is p. That is, the probability has been deduced conditional on M. Whether M is true, or useful, is another matter entirely.

Next imagine two models are under contention, M_1 and M_2. Why there might be two, or more, is explored later. That there are two requires having some form of background knowledge B, from which these two probability models are deduced. B may be as simple as "I am considering M_1 and M_2."

In B is the implicit or explicit evidence from which we deduce the prior probability of the models. Explicit evidence might exist. It would have the form $Pr(\mathrm{M}_1|\mathrm{B}) = q$, from which $Pr(\mathrm{M}_2|\mathrm{B}) = 1 - q$. If no explicit evidence exists, and only the implicit evidence that these two (and no other) models are possible, then via the statistical syllogism we deduce $q = 1/2$.

Then, as is well known,

$$\Pr(y|x\mathrm{DB}) = \Pr(y|x\mathrm{DBM}_1)\Pr(\mathrm{M}_1|x\mathrm{DB}) + \Pr(y|x\mathrm{DBM}_2)\Pr(\mathrm{M}_2|x\mathrm{DB}). \quad (2)$$

This is expanded in the obvious way to $p > 2$ models. This humble formation is the solution to all old testing problems.

It is important to understand the relationship between any of the models. The first assumption is that any M_i is probative to $y \in s$ for at least some s. Leave aside difficult questions of infinite sets and so-called "improper" (read *not*) probabilities. Any actual y in use in finite (which is to say, all) settings, will actually take at most a finite number of values (even if y is *potentially* infinite). No measurement apparatus exists to probe any y with infinite precision, and no actual decision uses any but finite y.

Therefore, an implicit model which always exists for any actual y is that $\Pr(y = y_i|x\mathrm{D}) = 1/m$, where $y \in \{y_1, y_2, \ldots, y_m\}$, with $m < \infty$. These values represent the smallest possible actual measurements, or smallest decisionable measurements, of y. These will always actually exist in any real-life decision setting; even in those cases where y potentially belongs in some infinite set. This model says, knowing only that y can take values in a known set, the probability it takes any of these values, absent *any* other information, is uniform. This again uses the statistical syllogism for proof.

Thus, in order for an M_i to be probative of y, it must be that

$$\Pr(y \in s|x\mathrm{D}) \neq \Pr(y \in s|x\mathrm{DBM}_i) \quad (3)$$

for at least some $s \in \{y_1, y_2, \ldots, y_m\}$, where s is any collection of y_k. If there is instead equality, then M_i adds nothing to our knowledge of the uncertainty of y and is not therefore a proper model *of* y.

The same holds true in comparing any two models. As long as, given B, no M_i can be deduced from any other M_j, and each M_i is probative of y, then each model adds something to the knowledge of the uncertainty of y. Whether this addition is useful or good or cost efficient are, of course, separate questions. All we require is that each model is separately probative in the sense of (3) and that

$$\Pr(y \in s|x\mathrm{DBM}_i) \neq \Pr(y \in s|x\mathrm{DBM}_j) \quad (4)$$

for at least some s. If there is equality at all s, then $\mathrm{M}_i \equiv \mathrm{M}_j$.

It is crucial to understand that if the probabilities for two models only differ at $s = \{y_k\}$ for a singular k, then each model is still adding different information—conditional on D, B, and x, of course.

Another clarification. Suppose B indicates M_1 is a normal model with a conjugate prior on the parameters, and that M_2 is also a normal model with a Jeffrey's prior on the parameters. These are *two separate models* in our terminology because they will give different probabilities to various s. Models are *different* when they are not logically equivalent and when they are probative to y—conditional on D and B (and x, when present).

It is true one of these models might be better with regard to the predictions, and the *decisions conditional on those predictions*. But if that is not known in advance (in B), then there is no way to know that one should prefer one model over the other.

This finally brings us to the difference between testing and deciding and their relationship to uncertainty.

3 Testing Versus Deciding

Suppose we have the situation just mentioned, two normal models with different priors for the observable y. We'll assume these models are probative of y; they are obviously logically different, and practically different for small n. At large n the difference in priors vanishes.

A frequentist would not consider these models, because in frequentist theory all parameters are fixed and ontologically exist (presumably in some Platonic realm), but a Bayesian might work with these models, and might think to "test" between them. What possible reasons are there to test in this case?

First, what is being tested? It could be which model fits D, the past data, better. But because it is always possible to find a model which fits past data perfectly, this cannot be a general goal. In any case, if this is the goal—perhaps there was a competition—then all we have to do is look to see which model fit better. And then we are done. There is no testing in any statistical sense, other than to say which model fit best. There is *no* uncertainty here: one is better *tout court*.

The second and only other possibility is to pick the model which is most likely to fit future data better.

Fit still needs to be explained. There are many measures of model fit, but only one that counts. This is that which is aligned with a decision the model user is going to make. A model that fits well in the context of one decision might fit poorly in the context of another. Some kind of proper score is therefore needed which mimics the consequences of the decision. This is a function of the probabilistic predictions and the eventual observable. Proper scores are discussed in [1]. It is the user of the model, and not the statistician, who should choose which definition of "fit" fits.

There is a sense that one of these models might do better at fitting, which is to say predicting, future observables. This is the decision problem. One, or one subset of models, perhaps because of cost or other considerations, must be chosen from the set of possible models.

There is also the sense that if one does not know, or know with sufficient assurance, which model is best at predictions, or that decisions among models do not have to be made, that the full uncertainty across models should be incorporated into decisions.

The two possibilities are handled next.

3.1 Put Your Best Fit Forward

Now it is easy to calculate the so-called posterior of every model; i.e.

$$\Pr(\mathrm{M}_i|\mathrm{DB}) = \frac{\Pr(y|x\mathrm{BM}_i)\Pr(\mathrm{M}_i|x\mathrm{B})}{\sum_i \Pr(y|x\mathrm{BM}_i)\Pr(\mathrm{M}_i|x\mathrm{B})}. \quad (5)$$

Recall D = (y, x), the previous observations.

This can be of interest in the following sense. The background information B supplies the models, M_i. There are only three possibilities for B. (1) B specifies a strictly causative M, i.e. it has identified all the causes of y in conjunction with x. There is then no reason to have any rival models. (2) From B we *deduce* a probability model, as in the case of a die throw, for instance. Again, there is no need of a rival model, for the exact probability of y (possibly given x) has been deduced. (3) B specifies only *ad hoc* models, usually chosen by custom and experimentation. There is no sense then that any M_i specifies is the true model, though one may be best at either fitting D or in predicting future y.

The model posteriors thus represent how well *ad hoc* models fit past data. But since our interest is prediction, we want to know how good each model will make predictions, if we are going to pick just one model (or possibly some subset of models). Thus "best" has to be defined by the decisions to be made with it.

Predicting which model will be best runs like this: (1) Fit each model M_i and form (1), i.e. the predictive form of y. (2) Use $\Pr(y|x\mathrm{DM}_i) = p_i$ to probabilistically predict y. (3) Compute $S(\mathrm{D}(y, x), p_i)$, a proper score reusing the previous data D; S_i may be a vector of scores of length n, or a single number. (4) The probability of seeing scores "like" S_i in new predictions (of size n) is the posterior probability of M_i. This "like" has to be strengthened considerably in future work.

The probabilities and the costs, or rather consequences, of each S_i are then used in a standard decision setting. For example, it could be in the case of two models $\Pr(\mathrm{M}_1|\mathrm{DB}) > \Pr(\mathrm{M}_2|\mathrm{DB})$ but that

$$\Pr(\mathrm{M}_1|\mathrm{DB})S(\mathrm{D}(y, x), p_1) < \Pr(\mathrm{M}_2|\mathrm{DB})S(\mathrm{D}(y, x), p_2)$$

M_2 is picked if higher S are better (for this decision). Note carefully that the costs of the model, and all other elements related to the decision to be made, are incorporated in S.

3.2 *Be Smooth*

There is not much to be said about using all uncertainty across all models, except to reiterate that all known or accepted uncertainties should be used in making predictions and in ascertaining model performance. That is, use this:

$$\Pr(y|x\text{DB}) = \sum_i \Pr(y|x\text{DBM}_i)\Pr(\text{M}_i|x\text{DB}). \tag{6}$$

In other words, don't test: use all available information. The differences in the models are of no consequence, or they do not matter to any decisions that will be made. Do not pick the "best" model—use all of them! The weights between models are specified in B, as before. The posterior predictive distribution smooths over all models.

In particular, there is no reason in the world to engage in the usual practice of model selection, if model selection is used to discover causes or "links". Links are causes but said in other words. Modelers speak of links when the modeler knows that probability models cannot discover causes, but still wants to say his measure is a cause; hence, "link". There are innumerable headlines like this: "Eating bananas is *linked to* irascibility." People who want to be irascible then begin eating bananas, and vice versa, believing in the causal power of the "link".

Model selection often happens like this. A modeler is using a regression model (which are ubiquitous), and begins with a selection of measures x_i, for $i = 1 \dots p$. Now what is often missed about modeling, or rather not emphasized, is that at this point the modeler has already made an infinite number of "tests" by not including every possible x there is. And he did this not using any information from *inside* the model. He did *no* testing, but made decisions. This is a direct acknowledgement that statistical *testing is not needed.*

There must have been some reason for the set of x chosen, some suspicion (which belongs formally in B) that these x are related in the causal chain of y in some way. There is therefore, based on that reason alone, no reason to toss any of these x away. Unless it is for reasons of costs and the like. But that situation (decision) has already been covered in the previous sub-section.

It may also be that there are too many x, that $p > n$, for instance, and the modeler does not want to use methods where this can be handled. It may be, too, that some of the x were included for dubious reasons. It could be, that is to say, the modeler wants to try $\text{M}_1 = \text{M}(x_1, x_2, \dots, x_p)$ and $\text{M}_2 = \text{M}(x_1, x_2, \dots, x_{-j}, \dots, x_p)$, i.e. a model with and without x_j.

Absent cost and the like there just is no reason to test, i.e. to decide between M_1 and M_2. Let B decide the weighting, the initial q_i, use all uncertainty and don't test. Unless q_i is high for a dubiously entered x_j, the model with it will not get great weight in the posterior prediction calculation. Use both models and compute (6).

Not that I recommend it, but this is the solution to the all-subsets regression "problem". For interest, an example is given below.

4 Giving It a Try

4.1 Deciding

This is a somewhat contrived example, using appendicitis data from [7]. There are $n = 443$ patients on which the presence of appendicitis is measured, along with age, sex, white blood count, and the results from each of ten medical examinations, which are either positive or negative. In reality, these are simple costless tests, such as examining for nausea or right lower quadrant pain. For our purposes we will assume that we can only "afford" one of these tests and have to pick one—perhaps because of monetary or time costs.

The models are logistic regressions, each with age, sex, and white blood count and one of the ten tests. A uniform distribution over models is used for our B ($q_i = 1/10$). The scoring-decision function is to assume a posterior predictive probability of appendicitis greater than 1/3 indicates the presence of appendicitis, such that an action along those lines is taken, such as an ultrasound or exploratory laparotomy. Predictions which are accurate incur no costs. False positives receive a penality of 1, and false negatives, which are medically more dangerous, receive a penality of 2. All tests are thought to cost the same amount (in time or expenses).

Figure 1 shows the results. This is the model posterior, computed by (5) by the score of each model. The number of the test is also given. It is clear the model with the highest posterior probability, test 6, does not have the lowest score.

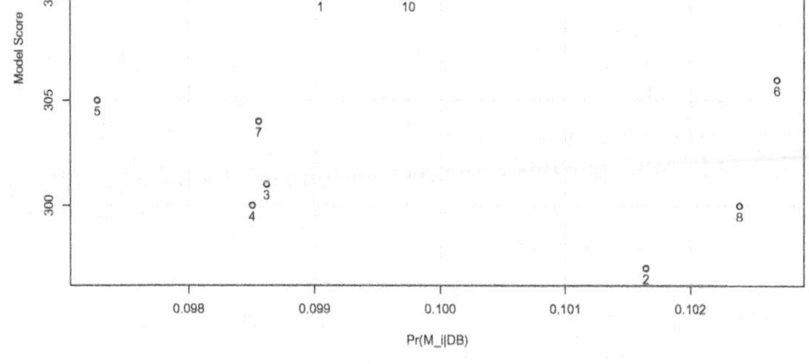

Fig. 1 This is the model posterior, computed by (5), plotted against the score of each model. See the text for model score calculation. The number of the test is also given.

Picking which model to implement requires some form of decision analysis. Here we computed a sum total score, but the distribution of scores could have been used instead. Something along those lines is illustrated in the next section. What's really needed is a *second* modeling of scores. These scores are all in-sample based on the posterior predictive distributions of the observable, and as such they are predictions of what future scores would look like—given an individual model is used.

The lesson is clear: model posterior probabilities do not translate directly into a preference ordering. That depends on the score and decisions to be made. This example, as said, is contrived, but the steps taken would be exactly the same in real situations.

4.2 Full Uncertainty

Data made available from [8] is used to predict levels of prostate specific antigen (PSA) > 10, which are levels commonly taken to indicate prostate cancer or prostate difficulties. Available as probative measures are log cancer volume, log prostate weight, age, log of benign prostatic hyperplasia levels, seminal vesicle invasions, log of capsular penetration, and Gleason score. The data is of size $n = 97$, with 67 of the observations used to fit models, and 30 set aside to assess model performance. The `rstanarm` package version 2.18.2 with default priors in R version 3.5.2 was used for all calculations.

It is not known which combination of these measures best predict, in a logistic regression, PSA > 10. Some attempt at an all-subset regression might be attempted. There are seven potential measures, giving $2^7 = 128$ different possible models. It was decided, in our B, to give each of these 128 models identical initial weight. We will not decide between any of them, and will instead use all of them, as in equation (6) to make posterior probability predictions. We will test the performance of the "grand" model on the 30 hold outs.

As a matter of interest in how measures relate to predictions, the posterior prediction of PSA > 10 as a function of Age is given in Fig. 2.

The red dots are the model (6), i.e. the grand model integrated across all 128 possible submodels. The predictions for those models are also given as small black dots for comparison. The probabilities for high PSA for 40 year olds differs only little from 80 year olds. Whether this difference in important depends on the decisions that will be made by the model. It is not up to the statistician to decide what these decisions will be, or there import.

The Brier score, which is the square of the difference between the observable and its predicted probability, was used to assess model performance. The scores for all models, including the grand model, are given in Fig. 3.

The asterisks are the mean Brier score in the hold out data; thus, these are genuine out-of-sample performances. The range of Brier scores for the 30 data points are given as vertical lines. The red dashed line is the grand model. The blue line is the "best" model according to minimum mean Brier score.

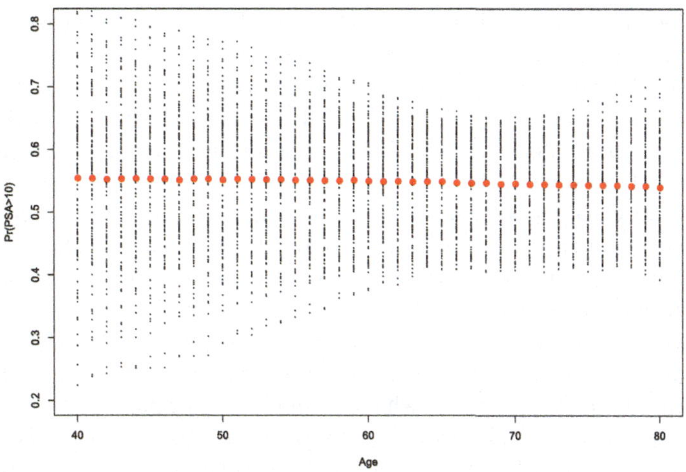

Fig. 2 The red dots are the model (6), i.e. the grand posterior predictive model integrated across all 128 possible submodels. The predictions for those models are also given as small black dots for comparison.

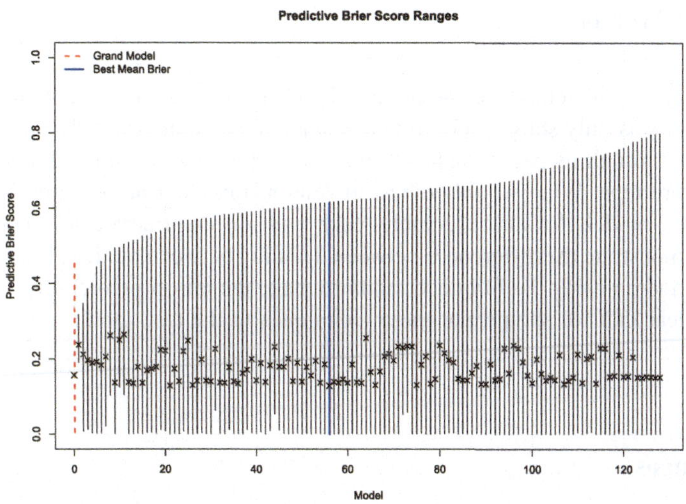

Fig. 3 Brier scores for all models, including the grand model. The asterisks are the mean Brier score in the hold out data. The range of Brier scores for the 30 data points are given as vertical lines. The red dashed line is the grand model. The blue line is the "best" model according to minimum mean Brier score.

Some points are interesting. The model with the least range, but higher than average mean score, was the so-called null model; i.e. a logistic regression with only an intercept. All models must be judged relative to this, since it says, in effect, every man has the same fixed chance of a PSA > 10. If a more "sophisticated" model cannot beat this, it should not be used. Which we can see is the case for many models.

The mean Brier score of the grand model is not much higher than the best model. The grand model has the advantage of a shorter range of Brier scores, which is to say, a lower maximum (worse) score. Indeed, only 10 models have smaller ranges, but of those none has a lower mean score. And 5 of those models have higher minimum scores, which is worrisome.

Now whether the mean Brier score is truly best, or whether the model with a low mean score but also with low maximum is best, attributes belonging to the grand model, depends on the decision to which the model will be put.

The grand model uses all measures, at various weights, as it were. The lowest-mean-score model used log cancer volume, log prostate weight, age, and Gleason score. We saw above that age was not especially predictive, though.

5 Last Words

The strategy of searching or computing all subset of linear models is not recommended. It was only shown here to put the model averaging (or rather integrating) procedure through its paces, and to provide a ready-made set of comparison models for a problem. What would be best to demonstrate the true potential is to have actual rival models for an observable, each with different advocates. This would nicely demonstrate the differences between deciding between models, based on predictions, and on using the full uncertainty to make the fairest predictions.

It is clear that we should put these techniques to the test. Don't test: decide to decide.

References

1. Briggs, W.M.: Uncertainty: The Soul of Probability, Modeling & Statistics. Springer, New York (2016)
2. Briggs, W.M.: Everything wrong with p-values under one roof. In: Kreinovich, V., Thach, N., Trung, N., Thanh, D. (eds.) Beyond Traditional Probabilistic Methods in Economics, pp. 22–44. Springer, New York (2019)
3. Briggs, W.M., Nguyen, H.T., Trafimow, D.: The replacement for hypothesis testing. In: Kreinovich, V., Sriboonchitta, S. (eds.) Structural Changes and Their Econometric Modeling, pp. 3–17. Springer, New York (2019)
4. Briggs, W.M.: Asian J. Bus. Econ. **1**, 37 (2019)
5. Briggs, W.M., Nguyen, H.T.: Asian Journal of Business and Economics **1** (2019, accepted)
6. Briggs, W.M.: arxiv.org/abs/1507.07244 (2015)

7. Birkhahn, R., Briggs, W.M., Datillo, P., Deusen, S.V., Gaeta, T.: Am. J. Surg. **191**(4), 497 (2006)
8. Hastie, T., Tibshirani, R., Friedman, J.: The Elements of Statistical Learning. Springer, New York (2009)

Weight Determination Model for Social Networks in a Trust-Enhanced Recommender System

Mei Cai, Yiming Wang, Zaiwu Gong, and Guo Wei

Abstract Social networks (SNs) among users are proven to be beneficial in enhancing the quality of recommender systems. But "information cocoons" in an SN make recommendations homogeneity and vision narrow. We constructed a recommender network in which the trust relationship is derived from a social network viewpoint and found that the trust relation is measured more appropriate by group interactions in the SN than by using individual interactions. Through analyzing characters of fuzzy measures, we weigh non-additive trust relationships between groups in the trust network. Further, a weight determination model for the trust network in recommender systems, on the basis of the knowledge of trust values, is developed for obtaining fuzzy measures for the Choquet integral. The novelty of this model is that it applies parameters necessary to describe the trust complementary and redundancy between experts and Choquet integral and it also utilizes fuzzy measures to deduce the effect of trust complementary and redancy in recommendations. The model can help remove deviations caused by interactions, and improve the accuracy and objectivity of recommendations.

Keywords Social networks (SNs) · Non-additive fuzzy measure · The trust complementary and redundancy · Trust-enhanced recommender systems

---

M. Cai (✉) · Y. Wang · Z. Gong
School of Management Science and Engineering, Nanjing University of Information Science and Technology, Nanjing 210044, Jiangsu, China
e-mail: caimei@nuist.edu.cn

Y. Wang
e-mail: 954480696@qq.com

Z. Gong
e-mail: zwgong26@163.com

G. Wei
Department of Mathematics and Computer Science, University of North Carolina at Pembroke, Pembroke, NC 28372, USA
e-mail: guo.wei@uncp.edu

S. Sriboonchitta et al. (eds.), *Behavioral Predictive Modeling in Economics*, Studies in Computational Intelligence 897,
https://doi.org/10.1007/978-3-030-49728-6_4

1 Introduction

According to the Six Degrees of Separation, all people are six, or fewer, social connections away from each other i.e., anyone is connected to a stranger through a maximum of five people. Today, with the popularization and development of the internet, social networks (SNs) demonstrate this theory once again in a meaningful way through constructing "stranger ties", namely relationships with strangers from which more information among people will be obtained. SNs have expanded to a substantial segment of internet users of various ages. Here, an SN (see e.g., [1, 2]) refers to a social structure under which individuals are linked together to share their ideas, interests and values. Adopting an SN, users can reduce their social and economic costs when finding how people present or hide their true ideas or purposes and get connected more conveniently with others. However, the current instruments and theories cannot explain how such information flows between the participants [3].

R. Brown, a British anthropologist, first proposed the concept of social networking, which focuses on how culture provides for the behavior of internal members of bounded groups [1, 2] (such as tribes, villages, etc.). Peng et al. [3] put forward his point of view. According to Brown, social networks refer to a relatively stable system, mainly composed of individuals with certain social relationships.

When people are making purchases, the factors they need to consider are mainly two aspects. On the one hand, they will listen to the evaluations of the producers and the evaluations conveyed by the sellers. On the other hand, they will also refer to acquaintances and opinion leaders [4, 5]. People often ask their friends and acquaintances for recommendations and they also like to share their opinions and favorites such as books, movies, music with others in social networks like Facebook.

Online social networks provide platforms for users with similar interests easily communicate with each other. The company believes that it is necessary to make a specific scientific forecast for the cold start of new users. The government also needs to use existing social networks as the main platform for the government to improve services and deliver information. Because the basic filtering methods in recommender systems (RSs), such as collaborative filtering and content-based filtering, require the historic rating of each user. However, such information is not always available for new users of cold-start problem. Social network-based approaches have been shown useful in reducing the problems with cold start users. By using the Trust-enhanced Recommender Systems (Trust-enhanced RSs) [6] one can find a path from a new user to get a personalized recommendation, you need to collect the reviewer's evaluation information and filter it to effectively retain the information that can be trusted. There are two ways to obtain Trust-enhanced RSs. One is obtaining the explicit feedback. For example, conduct a questionnaire survey and ask participants to rate network members about trust relationships. The other way is to effectively combine the above factors through a public social networking platform such as facebook or twitter, and then evaluate. The importance among members is evaluated based on interpersonal trust and is represented in Trust-enhanced RSs [7–9].

Chen et al. [5] introduced a new system model based on a trustworthy social network that can be used to obtain valuable information on social networks. Victor et al. [6] putting forward their own ideas, they proposed a trust model, which is to extract effective information in bilateral relations. The model saves the source of valuable information, so that users can conduct interactive analysis. Cross [7] developed a method that extracts some effective social factors from the information in the social network to generate a scientific method for group recommendation. Quijano-Sanchez et al. [8] studied a new model, the population model, which mainly studies the interactions between members in social networks to extract the potential impacts of their interactive information to make a scientific prediction of a group's perception of a project. Capuano et al. [9] established a new model based on group decision making (GDM), which can make scientific improvements in social impact.

The application of online SNs support methodology for recommender systems may face some unexpected influences. In our paper, we want to discuss the "information cocoons" proposed by Cass Sunstein, a leading American Constitutional lawyer [10]. Sunstein proposed that the recommendation under the SN algorithm creates a more unreal "mimetic environment" for the SN's user: "the information homogeneity is serious, the vision is narrow, and users cannot access more heterogeneous information. The users are stuck in the "information cocoon room".

When looking at the Tremolo or today's headlines in China, or looking at a certain aspect of the content, one will continue to receive the same type of content in the future, and if you are not interested in it, such content will no longer appear in front of you. The existing recommendation methods make our contact circle smaller and smaller, and it is difficult for new things to come in. So ours vision is always limited to a very narrow area.

In information dissemination, the public's own information needs are not omni-directional. The public only pays attention to the things they choose and the areas that make them happy. Over time, they will shackle themselves in a cocoon house. Sinhaand Swearingen [11] compared recommendations made by RSs to recommendations made by the user's friends, and found that these two kinds recommendations cannot supply satisfactory recommended items. The former were often "new" and "unexpected", while the latter mostly served as reminders of previously identified interests. Their research results further illustrated the influence of information cocoon effect in recommendation systems.

Take the 2016 US presidential election as an example. Professors, College students, financiers in the eastern United States (Washington, New York, Boston), and actors, Internet and technologists in the western United States (Los Angeles, San Francisco, Seattle) generally believe that Hillary Clinton is the winner. Farmers in the Central Plains and industrial workers in the Great Lakes generally believe that Trump is the winner. Hillary's supporters, who had already prepared the celebrations and items to celebrate her victory, waited for the results of the vote. Professors and students watched live television in classrooms, waiting for the last carnival. As a result, Trump won a big victory, and Hillary's supporters were shocked. Because the members of their social networks are like-minded people, and the information they usually receive came from the recommendation of their friends, they have no chance to receive information that disagrees with them. Such a recommendation system gives them the illusion that the world is like them. Our social network has become a huge echo room. This echo room exists in many aspects of our lives.

This phenomenon exists in recommender systems in a long run. Users do not get satisfactory services. A reasonable recommendation method becomes more and more important to change this phenomenon. We think the trust complementary and redundancy, where the former means more unfamiliar information will strengthen the trust, while latter means more familiar information will unstrengthen the trust, in trust relationships among SNs are the cause of "information cocoon".

Interactions between groups, each has three or more variables, often produce synergistic and redundant interactions [12]. Murofushi and Sugeno [13] considered the workstations in a workshop as an example to explain synergy and redundancy effects. Bottero et al. [14] took into account the synergy and redundancy effects among pairs of criteria in Multiple criteria decision making. Grabisch [15] described the role of the players in the cooperation project through the concept of super-additive and sub-additive, and concentrated on the opinions and problems, and drew conclusions through experts or voting.

Recently, some most common applications of fuzzy measures, laying in multi-criteria decision-making (MCDM), when interacting criteria are not appropriately considered, can cause some bias effects in evaluation [16], like echo room effect. Specifically, the problem centers at the hypothesis, stating that the given multiple criteria are theoretically independent, which however is often incorrect in practice [23].

The incorporation of fuzzy measures and Choquet integral provides an approach to handle criteria interaction. Choquet integrals [17, 18] based on non-additive measures have wide applications. Lourenzutti et al. [19] proposed a Choquet based integral where he considered groups of decision makers (DMs) with all their different opinions and criteria interaction. Bottero et al. [14] applied the Choquet integral to assess the activities for real-world problems. Some extensions of Choquet integral in the MCDM field have been proposed [20].

Therefore, in this paper, we will investigate the recommendation problems in a Trust-enhanced recommender system considering complex interactions among members in SNs. The decision process on internet possesses two main characteristics: non-availability of the importance of experts and the non-additivity of the importance of features. In our study, recommendation problems are viewed as a special group decision-making problem, and we utilize Choquet integral to obtain a more effective result. The recommendations are personalized and the effects from the trust complementary and redundancy will be reduced.

The remainder of this article will make the following arrangements. In Sect. 2, we present a brief review of Trust models and the trust propagation. In Sect. 3, we propose a trust-enhanced recommender network to describe the trust relationships, especially interactions among individual experts. Weight determination for experts based on trust relationships is carried out in Sect. 4. In addition, according to the aggregation operator, Choquet integral is applied to complete the recommendation process. Section 5 illustrates the model with a numerical example. In Sect. 6, the study limitation is discussed and the conclusion is drawn.

2 Trust Models and Trust Propagation

In broad terms, trust models are defined as a way to deal with problems such as social trust [21], cloud service [22, 23], P2P [24, 25], recommender systems [26]. As for recommender systems and e-commerce, uncertainties and risks in online transactions make it difficult for consumers to assess the credibility of unfamiliar entities on the Internet. Based on the user interaction behavior to measure the credibility of the seller objectively, building a trust computing model with the strong anti-attack ability and true trust has become a research hotspot in the industry. At present, the application of trust models in social networks mainly focuses on the calculation of trust relationships.

Trust models refer to a scientific approach to effective trust assessment of individuals or the whole, including the trust propagation and the trust aggregation. We assume that:

- A user is much more likely to believe statements from a trusted acquaintance than from a stranger.
- Trust is transitive, irreversible and monotone non-decreasing.

The trust value is mainly used to detect the trust level of the comparison agent and other agents. Trust values can be used to rank in the recommendation system; for example, for the same information, priority can be given to more rogue sources of information. Therefore, we need to introduce the trust value, and compare and save the information source of the trust value.

Definition 1 [6]. A trust value (t, d) is an element of $[0, 1]^2$ where t is called the trust degree, and d the distrust degree.

A trust score space $\mathcal{BL} = ([0, 1]^2, \leq_t, \leq_k, \neg)$ consists of the set $[0, 1]^2$ of trust scores (t_i, d_i), a trust ordering $\leq_t$, a knowledge ordering $\leq_k$, and a negation $\neg$ defined by

$$(t_1, d_1) \leq_t (t_2, d_2) \text{ iff } t_1 \leq t_1 \text{ and } d_1 \geq d_2;$$

$$(t_1, d_1) \leq_k (t_2, d_2) \text{ iff } t_1 \leq t_1 \text{ and } d_1 \leq d_2;$$

$$\neg(t_1, d_1) = (d_1, t_1)$$

The basis of this approach is that intuition deserves precision and scores on intuitionistic fuzzy sets, which can be used to assess trustworthiness and to define the concept of knowledge deficit.

Definition 2 [27]. The trust score and knowledge deficit associated to an orthopair of trust/distrust values (t, d) are

$$TS(t, d) = t - d \tag{1}$$

$$KD(t, d) = |1 - t - d| \tag{2}$$

Victor et al. [6] explained that orthopairs of trust/distrust values (t, d) satisfying $KD(t, d) = 0$, have perfect knowledge (i.e., complete trust state), while all others will have a deficit in knowledge. The combination of both trust score and knowledge deficit can be used to propose order relations for the set of orthopairs of trust/distrust values.

Victor et al. [6] proposed the concept of a propagation operator, suggesting the screening of expert information through a trusted third party (TTP). Through the description of Fig. 1, it can be concluded that in the three experts E_1, E_2 and E_3, there is no clear indication of trust value or untrusted value between expert E_1 and E_3, although an indirect orthopair of trust/distrust values between expert E_1 and E_3 can be obtained by propagating the corresponding values of the path via expert E_2.

The propagation operator was defined in [6]:

$$P((t_{0i}, d_{0i}), (t_{im}, d_{im})) = (\mathcal{T}(t_{0i}, t_{im}), \mathcal{T}(t_{0i}, d_{im})) \tag{3}$$

with $\mathcal{T}$ being a t-norm.

This propagation operator reflects the expected of trust behavior in a simple social network in Fig. 1.

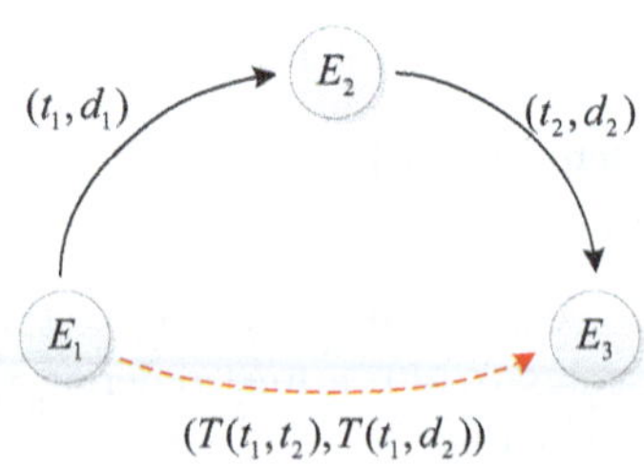

Fig. 1. Trust propagation of orthopairs of trust/distrust values between E_1 and E_3 via E_2

3 Trust-Enhanced Recommender Networks

3.1 The Trust Complementary and Redundancy in a Trust Network

SNs are built on an open sharing, like-mindedness and mutual respect. The issue of trust has become an important part of social networking and actual social interaction [28]. In the construction process of most models, the social factors are not considered, and the trust problem of interpersonal interaction has not been considered [9]. The foundation of the construction of SNs is ideally a social trust network. All Trust-enhanced RSs allow users to evaluate projects they have already experienced and to present their own personal insights [6]. On this issue, it has been suggested that people are more willing to trust the people around them than anonymous users [11]. Take Filmtrust as an example. Sinha and Swearingen [11] built a trust-aware network to extract trust relationships among users from feedback. Trust-enhanced Networks are an appropriate tool to extract importance information when recommendations are given based on social networks.

We construct a special trust network for our problem, where there are only two possible states between any two experts: having relations or no relations. The relation is presented as a trust value (t_{ij}, d_{ij}), where t_{ij} means the degree of E_i trusts E_j and d_{ij} means the degree of E_i distrusts E_j. No relation means is reflected by no edge connected these two experts. We use the important index $TS(t_{ij}, d_{ij})$ in Definition 2 to describe the relationships $\succ_t$ and $\succ_d$ in Trust-enhanced Recommender Network:

We define $\succ_t$ as follows:

$TS(t_{ij}, d_{ij}) > TS(t_{im}, d_{im}) > 0 \Rightarrow expert\ E_j \succ_t expert\ E_m$ according to expert E_i, which means E_i trusts E_j more than E_m;
$TS(t_{ij}, d_{ij}) < TS(t_{im}, d_{im}) < 0 \Rightarrow expert\ E_j \succ_d\ expert\ E_m$ according to expert E_i, which means E_i distrusts E_j more than E_m.

Our trust model is based on an important conclusion that trust is propagative, but not transitive [28]. Transitivity implies propagation, but the reverse is not necessarily true. This conclusion is quite different from previous researches, such as [29, 30]. An SN reflects a person's personality needs. A person has many personalities, so there are many criteria for selecting friends. Because different types of friends can meet his different personality needs. So friends of friends are not necessarily friends. That is to say, your two friends are not necessarily to be friends. The trust propagation operator is used to obtain trust values $P((t_{ij}, d_{ij}), (t_{jk}, d_{jk}), \cdots, (t_{lm}, d_{lm}))$, if there is a path $i \rightarrow j \rightarrow \cdots l \rightarrow m$ from expert E_i to E_m. We can often meet the situation that there is more than one path from expert E_i to E_m. So we can use $P(\cdot)$ to get another trust value different from the previous one. If we admit the assumption trust relation is transitive, then the trust values calculated from different paths should be the same. According to our intuition, the trust relationship should not change with different third parts between these two persons. To resolve this contradiction, some methods are provided to make trust values consistent [11, 12]. But we think it is not suitable to make trust values consistent. In fact, the trust value from E_i to E_m is fixed and will not change when somebody enters or leaves the group. We can apply the propagation operator to calculate the trust value from E_i to E_m because there is an edge between E_i to E_m. If this edge does not exist, the application of the propagation operator is not appropriate. The trust value calculated by the propagation operator is a group trust value involve all third part of the path, not the direct trust value from E_i to E_m. So the trust values of different paths are different.

We use μ to measure the trust degree[1] of the set $\{E_i, E_m\}$, E_i and E_m be disjoint subsets of X, there are two results:

1. $\mu(E_i \cup E_m) < \mu(E_i) + \mu(E_m)$, shows the set $\{E_i, E_m\}$ get less trust from E_o than separate, which are called the trust redundancy. Since in this situation, $TS(P((t_{ij}, d_{ij}), (t_{jk}, d_{jk}), \cdots, (t_{lm}, d_{lm}))) < 0$, we can conclude the third parts in the path show E_i trusts E_j and their influence to E_o should be deduced in order to deduce the effect of echo room because the preferences of E_i and E_j confirm each other;
2. $\mu(E_i \cup E_m) > \mu(E_i) + \mu(E_m)$, shows the set $\{E_i, E_m\}$ get more trust from E_o than separate, which are called the trust complementary. Since in this situation, $TS(P((t_{ij}, d_{ij}), (t_{jk}, d_{jk}), \cdots, (t_{lm}, d_{lm}))) < 0$, we can conclude the third parts in the path show E_i distrusts E_j and their influence to E_o should be increased in order to supply more choices to E_0 because the preferences of E_i and E_j are mutual exclusive.

We explain in detail. Compared to classic additive measures, the trust relations of experts is substituted by monotonicity.

$(t_{im}, d_{im}) = (1, 0)$ means E_m is perfectly trusted by E_i. $P((t_{0i}, d_{0i}), (t_{im}, d_{im})) = (\mathcal{T}(t_{0i}, t_{im}), \mathcal{T}(t_{0i}, d_{im})) = (t_{0i}, d_{0i})$ means the biggest trust score of E_m is equally to E_i, or trust score calculated by propagation operators is no bigger than $E_i's$. That means E_i and E_m satisfy two quite same personality needs of E_0. We can conclude their interest and hobby must be similar. In other words, if we want to present E_0's personality, E_i is enough to show one aspect. In fact, this two friends should reduce their impact as a whole. Considering the third parts' influence, the trust is diminishing when taking them as a group, so we conclude

$$\mu(E_i \cup E_m) < \mu(E_i) + \mu(E_m)$$

$(t_{im}, d_{im}) = (0, 1)$ means E_m is perfectly distrusted by E_i, we can get $P((t_{0i}, d_{0i}), (t_{im}, d_{im})) = (\mathcal{T}(t_{0i}, t_{im}), \mathcal{T}(t_{0i}, d_{im})) = (0, t_{0i})$. $TS(P((t_{0i}, d_{0i}), (t_{ij}, d_{ij}))) = -t_{0i}$ means through third parts, E_m should be distrusted by E_0. We can conclude E_i and E_m's interest and hobby must be different. That shows E_0 has two quite different aspects reflected by E_i and E_m. The more E_i distrusts by E_m, the more different these

[1]In previous studies, trust degree of the set $\{E_i, E_m\}$ is 0 or 1. 0 means completely distrust, while 1 means completely trust. We design the trust degree is a number in interval [0, 1]. So the measure of trust degree is a fuzzy measure.

two aspects are. In fact, if we want to know E_0' personality from the full aspects, this two experts should increase their impact as a whole. These two different people can become E_0's friends at the same time which just shows that there are two different needs in E_0's personality. Considering the third parts' influence, the distrust is increasing when taking them as a group, so we conclude

$$\mu(E_i \cup E_m) > \mu(E_i) + \mu(E_m)$$

In the recommendation process, we need to aggregate every recommenders' evaluation. Since E_i and E_m have distrust relations, the aggregation result will cause a mutual strengthening effect and an item liked by two experts distrusted each other will be strongly recommended. Adversely, an item will not be strongly recommended. If we want to avoid echo room effect and give more pertinent recommendations, we need to reduce the effects of the trust complementary and redundancy. The aim of next section is to construct a recommendation method and provide a good personalized recommendation.

3.2 Construction of a Trust-Enhanced Recommender Network

Social networks are described as connected graphs where nodes represent entities and edge their interdependencies. We establish a trust-enhanced recommender network simplified from a social network. Suppose that a graph $G = \{N, L\}$ is a social network structure consisting of a set of nodes N and a set of edges L. We extract sub-network called a Trust-enhanced Recommender network from the SN. The SN can be simplified to a small network where all elements in the network have influences on a special person who we want to give recommendation advice. This special person is named central node in the graph. A Trust-enhanced Recommender network is defined as follows:

Definition 3. Suppose $TN \subset G$ is a weighted, directed graph which is around a central point $E_o \in N$. TN is called a Trust-enhanced Recommender network composed by a set of nodes E_i and a set of edges L_{ij}, the nodes can present the state of the individual, and the trust relationship between the nodes can be presented through the edge. A directed graph edge from node E_i to node E_j represents the trust relationship from node E_i to node E_j and is given a pair of value (t_{ij}, d_{ij}). Node $E_o \in N$ (central node) is a special one in the network, and it presents the person who needs to give recommendation advice. Other nodes called experts are trusted by him and can give recommendations. We delete the node E_i which have no direct connection with node E_o or $TS(t_{0i}, t_{0i}) < 0$. But if there are third parts in the path from node E_i to node E_m, we retain the nodes in the path. TN is a special SN for E_0. Fig. 2 is an example of an SN and Fig. 3 is an example of a Trust-enhanced Recommender Network based on Fig. 2.

Fig. 2 An example of SN

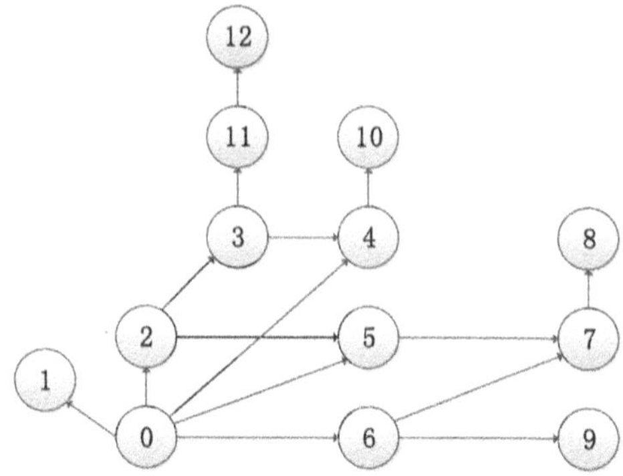

Fig. 3 An example of a Trust-enhanced Recommender Network based on Fig. 2

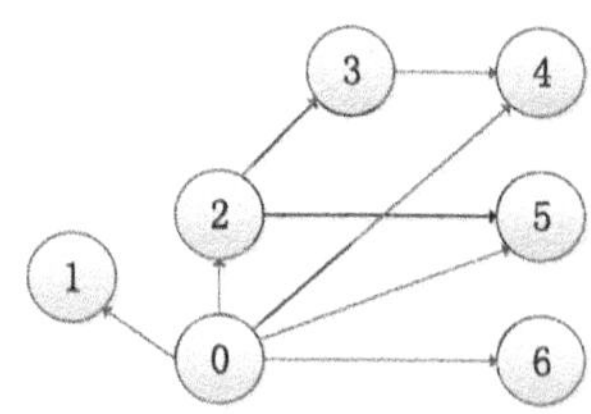

We construct such a Trust-enhanced Recommender network which is based on the following assumption.

(1) If E_0 shows directly distrust to E_i or no knowledge about E_i, we will not consider the recommendation form E_i. So in our Trust-enhanced Recommender Network, we delete all nodes without direct edges connecting with the central node or $TS(t_{0i}, t_{0i}) < 0$.
(2) We believe that in the recommendation process, the recommended results are also not transitive (just as the trust relationship does not transfer). So in the first hypothesis we removed the nodes that are not directly related to the central node. However, if a node itself does not connect directly with the central node, but it connects indirectly to the central node through a path connecting to the central node, it is retained (see Fig. 3). These nodes form a chain of trust.
(3) Assuming E_i and E_j are interdependent, their influence on the central node are not additive. Through analyzing the interdependences among the nodes in the trust chains from E_i to E_j, we can know how much the degree of the trust complementary and redundancy in the path from E_i to E_j.

The concept of Trust-enhanced Recommender network is proposed for group recommendation. Trust scores can serve as a judgement in the group decision making. The one with a higher trust score is assumed to have more influence. The weights of three experts E_j, E_k and E_l are obtained according to the trust score from E_o to them. If we do not consider the trust complementary and redundancy among the network, we can obtain the weight as follows:

$$w(j) = TS(oj)/(TS(oj) + TS(ok) + TS(ol)) \quad (4)$$

$$w(k) = TS(ok)/(TS(oj) + TS(ok) + TS(ol)) \quad (5)$$

$$w(l) = TS(ol)/(TS(oj) + TS(ok) + TS(ol)) \quad (6)$$

The aggregation result is not fully satisfactory. The reason is that too much importance is given to E_j and E_k, which results in a redundant sense.

Because E_j trust E_k, a product which is recommended by E_j will also be recommended by E_k. The recommended result overestimates the effects of E_j and E_k, and underestimates the effect of E_l. We think the interaction among nodes causes this phenomenon.

There are two kinds of nodes in $N \backslash E_o$: nodes that have direct connections to the center node and nodes that connect to the central node indirectly. Each node of the second type should not be considered individually but should be considered in a set containing itself and other nodes in the paths connecting them (refer to Fig. 3). This is an interaction set. This recommendation problem is a group decision-making problem where the decision environment is complex and quite different from traditional decision making problems. If in the recommender system one ignores the interaction among SNs, then the recommendation result different from SN's true preference will be obtained. Recommended products are gradually confined to a small circle which dominated by several individuals familiar with each other, and other voices are increasingly difficult to be heard. In a relatively closed "cocoon house", opinions in a small group will be repeated constantly, which makes people everywhere in a relatively closed environment.

4 Weight Determination for Experts in a Trust-Enhanced Recommender Network

We now turn to a model that can be applied for calculating the weights of experts based on a Trust-enhanced Recommender Network.

4.1 Measurement of Interdependence Among Nodes in a Trust-Enhanced Recommender Network

According to Definition 3, edge from node E_i to node E_j is defined as a pair of value (t_{ij}, d_{ij}). We quantify the interaction index from E_i to E_j and denote it as $I_{ij} = TS(t_{ij}, d_{ij}) \neq 0$. Since the edge can connect two nodes in the graph, we think interaction only exists between two nodes. So for any group T with more than three nodes, the interaction index $I(T) = 0$. The interaction index is defined as

$$I(T) = \begin{cases} TS(t_{ij}, d_{ij}), & when|T| = 2, T = \{E_i, E_j\} \\ 0, & when|T| > 2 \end{cases} \tag{7}$$

We apply a discrete fuzzy measure μ on N to describe coefficient of trust relationship form E_0 which is a monotonic set function. Given a set T where any nodes at least connect with E_o, $\mu(T)$ presents the trust E_o to the set.

$$\mu(T) = \sum_{E_j \in N} I_j \tag{8}$$

$$I_j = \sum_{T \subset N \backslash E_j} \frac{(n-t-1)!t!}{n!} \times [\mu(T \cup E_j) - \mu(T)] \tag{9}$$

where $t = |T|, n = |N|$.

I_j is similar to the Shapley value [31] and it shows that the expert interaction. I_j is different from $\mu(j)$.

Thus the interaction index between E_i an E_j has another representation, presented as the sum over all combinations $T \subseteq N \backslash \{E_i, E_j\}$

$$I_{ij} = \sum_{T \subset N \backslash ij} \frac{(n-t-2)!t!}{(n-1)!} \times \mu(T \cup \{E_i, E_j\}) - \mu(T \cup E_i) - \mu(T \cup E_j) + \mu(T)$$

where $t = |T|, n = |N|$.

According to function (7)–(9), we conclude that the interdependences in trust network can be view as a $2-order$ additive fuzzy measure. We get the following functions:

$$\mu(i) = TS(t_{oi}, d_{oi}) \tag{10}$$

$$\mu(i \cup j) = TS(t_{oi}, d_{oi}) + TS(t_{oj}, d_{oj}) - TS(t_{ij}, d_{ij}) \tag{11}$$

Fuzzy measure μ can fully present the trust relation in a Trust-enhanced Recommender Network.

4.2 Weight Determination Model for Experts

The determination of fuzzy capacity on coalitions (i.e. groups, subsets) has proven difficult because the capacities are defined over sets, not only on individual elements, and all such subsets of elements need to be identified. We thus assume that we have at our disposal we are unable to tell which element is more important than another one, and what interdependences are among them. We suppose that the

distrust values. The status and
determined by the degree of
ly promote the integration of

hts of groups of experts in an

, 1] on N is a monotonic set

th the recommendation result,

is given a higher capacity, and
capacity;
with $TS(t_{ij}, d_{ij})$. The bigger the

n between fuzzy capacity and
measures [15], f is a linear

(12)

(13)

from:

$(TS(t_{ij}, d_{ij}))x_i x_j)$ (14)

nce between E_i and E_j. then
not edge connecting E_i and E_j.
dundancy between E_i and E_j.
E_i and E_j has trust relationship.
plementary between E_i and E_j.
ans E_i and E_j has distrust

Situation (i) is in fact a situation whe
nodes. We think these two nodes are inde
two nodes are dependent. We need to ap
trust redundancy (reflecting a mutual weak
the trust complementary (reflecting a
(iii) between DMs. When we want to ut
recommend products to someone, we ne
situations. Our methodology takes into ac
expert's interdependences in terms of the
represent the interdependence between
[16, 20]. In this way, we acknowledge th
on himself but also on his interdependen

$$I_i' = f\left\{ \sum_{T \subseteq N \setminus E_i} \frac{(n-t-1)!}{n!} \right.$$

For a given Mobius representation μ,
resentation I' is given by [15]

$$I_i' = \mu'(0i) + \frac{1}{2} \sum_{E_j \in N}$$

From above functions we can conclude

Proof. $\sum_{E_i \in N} I_i' = \sum_{E_i \in N} \mu'(0i) + \sum_{E_j \in N \setminus E_i} \mu'($
According to Definition 4, $\mu'(N) =$

$\sum_{E_i \in N} I_i' = 1.$

I_i' is suitable to be the importance of an
After above analysis, we construct We
The model for identifying fuzzy capacities
Model (1)

$$maxmin$$

subject to

$|\mu'(i) - k \times TS(t_{oi}, d_{oi})| < \varepsilon$ ranking of
$E_i \in N$ *there is*

$|a(ij) - k \times TS(t_{ij}, d_{ij})| < \varepsilon$

relative importance of experts is consistent with trust/distrust values. The status and importance of experts in the trust network can be determined by the degree of interdependence between nodes, which can positively promote the integration of individual opinions.

We use the fuzzy capacity to characterize the weights of groups of experts in an SN:

Definition 4. A discrete fuzzy capacity $\mu' : 2^n \rightarrow [0, 1]$ on N is a monotonic set function ($n = |N|$) which satisfies the following:

(1) $\mu'(\emptyset) = 0$ and $\mu'(N) = 1$ (boundary conditions),
(2) $0 \leq \mu'(T) \leq 1$,
(3) $S \subseteq T, \mu'(S) \leq \mu'(T)$ (monotonicity condition).

Assumption

(1) Anyone without direct trust has no influence with the recommendation result, so his capacity is 0.
(2) Anyone with direct trust with higher trust score is given a higher capacity, and the one with equal trust score is given an equal capacity;
(3) Interaction index I'_{ij} of two nodes is consistent with $TS(t_{ij}, d_{ij})$. The bigger the $TS(t_{ij}, d_{ij})$, the bigger I'_{ij} is.

With above assumption, we construct a connection between fuzzy capacity and measure of interdependence. For 2-additive fuzzy measures [15], f is a linear function of μ, we define:

$$\mu'(i) = f(\mu(i)) \tag{12}$$

$$I'_{ij} = f(I_{ij}) \tag{13}$$

For any $K \subset X$ such that $|K| \geq 2$, we can deduce from:

$$I'(K) = f(\sum_{i=1}^{n}(TS(t_{oi}, d_{oi}))x_i + \sum_{\{i,j\}\subset X}(TS(t_{ij}, d_{ij}))x_i x_j) \tag{14}$$

with $x_i = 1$ if $i \subset K$, $x_i = 0$, otherwise.

(i) I'_{ij} is zero when there is no interdependence between E_i and E_j. then $\mu'(i \cup j) = \mu'(i) + \mu'(j)$. That means there is not edge connecting E_i and E_j.
(ii) I'_{ij} is negative when there exists the trust redundancy between E_i and E_j. $\mu'(i \cup j) < \mu'(i) + \mu'(j)$. This situation means E_i and E_j has trust relationship.
(iii) I'_{ij} is positive when there exists the trust complementary between E_i and E_j. $\mu'(i \cup j) > \mu'(i) + \mu'(j)$. This situation means E_i and E_j has distrust relationship.

Situation (i) is in fact a situation where there is no path connecting these two nodes. We think these two nodes are independent. Situation (ii) and (iii) show these two nodes are dependent. We need to apply parameters necessary to describe the trust redundancy (reflecting a mutual weakening effect, situation (ii) and to describe the trust complementary (reflecting a mutual strengthening effect, situation (iii) between DMs. When we want to utilize the recommendation information to recommend products to someone, we need to reduce the impact from above two situations. Our methodology takes into account the above situations by modeling expert's interdependences in terms of the Shapley index I_{ij} [31], which is used to represent the interdependence between criteria in MCDM by researchers, e.g. [16, 20]. In this way, we acknowledge that the importance of E_i not only depends on himself but also on his interdependences with other experts.

$$I_i' = f\left\{ \sum_{T \subseteq N \setminus E_i} \frac{(n-t-1)!t!}{n!} [\mu(T \cup \{E_i\}) - \mu(T)] \right\} \tag{15}$$

For a given Mobius representation μ, the corresponding interdependence representation I' is given by [15]

$$I_i' = \mu'(0i) + \frac{1}{2} \sum_{E_j \in N \setminus E_i} \mu'(ij), E_i \in N \tag{16}$$

From above functions we can conclude $\sum_{E_i \in N} I_i' = 1$.

Proof. $\sum_{E_i \in N} I_i' = \sum_{E_i \in N} \mu'(0i) + \sum_{E_j \in N \setminus E_i} \mu'(ij)$

According to Definition 4, $\mu'(N) = 1 \Rightarrow \sum_{E_i \in N} \mu'(i) + \sum_{E_j \in N \setminus E_i} \mu'(ij) = 1$. So $\sum_{E_i \in N} I_i' = 1$.

I_i' is suitable to be the importance of an element E_i.

After above analysis, we construct Weights determination model for experts. The model for identifying fuzzy capacities (weights) is given at below:

Model (1)

$$maxminze(\varepsilon) \tag{17}$$

subject to

$$\begin{aligned} &|\mu'(i) - k \times TS(t_{oi}, d_{oi})| < \varepsilon \quad \text{ranking of experts (weights on inletons)}, \\ &E_i \in N \quad \textit{there is direct path between } E_o \textit{ and } E_i \end{aligned} \tag{18}$$

$$|a(ij) - k \times TS(t_{ij}, d_{ij})| < \varepsilon \quad \text{ranking of pairs of DMs}$$

$$(E_i, E_j \textit{ belong to } \text{interdependence set}) \tag{19}$$

$$\left.\begin{array}{l} \sum\limits_{E_i \in N} a(i) + \sum\limits_{\{E_i, E_j\} \subseteq N} a(ij) = 1 \\ a(i) \geq 0 \forall E_i \in N \\ a(i) + \sum\limits_{E_j \in T} a(ij) \geq 0 \\ \forall E_i \in N, T \subset N \backslash E_i \end{array}\right\} \text{boundary and monotonicity condition} \tag{20}$$

$$\left.\begin{array}{l} a(i) = \mu'(i) \\ a(ij) = \mu'(i) + \mu'(j) - \mu'(i \cup j) \end{array}\right\} \text{definition of Möbius transform} \tag{21}$$

In this Model, we assume f and $TS(t_{ij}, d_{ij})$ hold a linear relation, i.e., f can be written as a linear function $k \times TS(t_{ij}, d_{ij})$, which satisfies the consistency in assumption. The threshold δ can be tuned as wished. The process of determining weights for DMs in an SN is showed in Fig. 4.

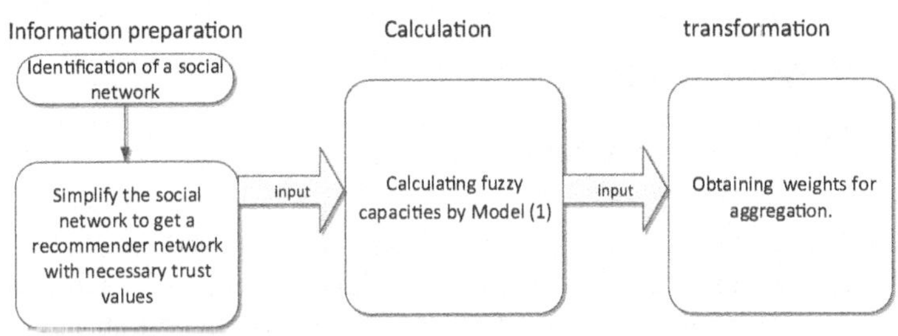

Fig. 4 The process of Weights determination

4.3 The Recommendation Based on a Trust-Enhanced Recommender Network

The recommendation problem based on a Trust-enhanced Recommender Network is the following:

Let $P = \{P_1, P_2, \cdots, P_m\}$ be a set of items needed to be recommended. There is a set of experts $E = \{E_1, E_2, \cdots, E_n\}$ giving evaluation information about P. The set of experts has direct trust relationships with a special user and the evaluation set is $X = \{x_1, x_2, \cdots, x_m\}$. RSs provide a utility function $g : X \times P \rightarrow R$ aggregating evaluations from a Trust-enhanced Recommender Network. The central issue of RSs is to select the utility function g which is appropriate to predict the special user's preference according to the Trust-enhanced Recommender Network. Through Model (1), we obtain fuzzy capacity.

Definition 5 [17]. The discrete Choquet integral of f on X with respective to μ can be defined as follows:

$$\int_X f d\mu = C_\mu(f) = \sum_{i=1}^{n} [f(x_{(i)}) - f(x_{(i-1)})]\mu(A_{(i)}) \tag{22}$$

where $(\cdot)$ indicates a permutation on X such that $0 \le f(x_{(1)}) \le f(x_{(2)}) \le \cdots \le f(x_{(n)}) \le 1$ and $A_{(i)} = \{x_i, x_{i+1}, \cdots, x_n\}$, $A_{(n+1)} = \emptyset$.

The Choquet integral can be expressed using M�o bius transform of μ in a very instructive way when the measure is 2-additive

$$C_\mu(x) = \sum_{i \in N} a(i)x_i + \sum_{a(ij) > 0} a(ij)(x_i \wedge x_j) + \sum_{a(ij) < 0} a(ij)(x_i \vee x_j) \tag{23}$$

When the capacity is additive, $C_\mu(x)$ can be seen as a weighted sum. The global evaluation is calculated by the Choquet integral.

5 Illustrative Examples

5.1 An Numerical Example

A person wants to buy a dust coat, with four types of dust coats, i.e., to select from $P = \{\mathrm{A, B, C, D}\}$. In order to reach more users and more items, a Trust-enhanced RS gives recommendations through analyzing friends' preferences about the certain items. The process of the recommendation is as follows:

Step 1:

Construct a Trust-enhanced Recommender Network through visualizing information collected and calculated from the platform of a Web-based social network.

(i) the trust information collection
(ii) the trust value assessment
(iii) visualization of trust relationships
(iv) construction of a Trust-enhanced Recommender Network (see Fig. 4)
(v) according to the set of experts $E = \{E_1, E_2, \cdots, E_6\}$, we collect the evaluations about $P = \{A, B, C, D\}$ (see Table 1).

Table 1 Preference matrices for items

	e_1	e_2	e_4	e_5	e_6
A	9	6	6	9	9
B	8	6	6	8	8
C	7	8	8	6	7
D	5	7	7	9	5

Table 2 Trust values of a trust-enhanced recommender network

(t_{ij}, d_{ij})	1	2	3	4	5	6
0	(0.60, 0.20)	(0.80, 0.30)		(0.90, 0.10)	(0.70, 0.05)	(0.80, 0.20)
2			(1.00, 0.10)	(0.80, 0.20)	(0.70, 0.50)	
3				(0.80, 0.20)		

Table 3 The fuzzy capacities

a_1	a_2	a_4	a_5	a_6	a_{24}	a_{25}
0.364	0.181	0.138	0.252	0.455	−0.138	−0.253

Table 4 Global evaluation

	A	B	C	D
$C_\mu(x)$	7.462	6.448	7.358	6.714

Step 2

Weights determination for experts in a Trust-enhanced Recommender Network

(i) the interdependence calculation

Calculate $TS(t_{0i}, d_{0i})$, $TS(t_{0j}, d_{0j})$ if E_i and E_j directly connect to E_o. If there is a path between E_i and E_j (no direct connection, we obtain trust scores $TS(P(i \to j))$ by the propagation operator (Eq. 1). The trust values (t_{ij}, d_{ij}) are in Table 2.

(ii) construct Weight determination model

Their fuzzy capacities are shown in Table 3.

Step 3

Global evaluation calculating

Applying Choquet integral Eq. (23) to get the global evaluation (Table 4).

5.2 Comparison and Analysis

In the above, there are also many methods of the trust propagation between every experts. In the following, we will use the weighted average calculation to derive the distribution of trust values and untrusted values in multiple trust paths to determine the trust value.

The trust values are shown in Table 5 (Table 6).

Our method concludes A $\succ$ C $\succ$ D $\succ$ B, while the example for comparison concludes C $\succ$ A $\sim$ D $\succ$ B. Reasons for this difference include:

1. 1. In a Trust-enhanced Recommender Network, $\{e_2, e_4, e_5\}$ is a group with complex trust relation. Because of trust redundancy, we do weaken weights of them and strength weights of others. Weighted averaging does not consider this effect, so weights of e_1 and e_5 are in fact weaken.
2. In the example, $a_{24} = -0.138$, which is also the value of $-a_4$. This means e_4's influence can be absolutely absorption. Victor et al. [6] defined "Knowledge absorption" to describe this phenomenon. $a_{25} = -0.253$, which means knowledge is partially absorbed. e_2 as a special node in the Trust-enhanced Recommender Network should be payed special attention.

Table 5 Trust values of a trust-enhanced recommender network

(t_{ij}, d_{ij})	1	2	3	4	5	6
0	(0.60, 0.20)	(0.80, 0.30)		(0.90, 0.10)	(0.70, 0.05)	(0.80, 0.20)
TTD				(0.80, 0.20)	(0.70, 0.50)	

Table 6 Weights of experts in a trust-enhanced recommender network

w_1	w_2	w_4	w_5	w_6
0.136	0.169	0.271	0.220	0.203

Table 7 Global evaluations

	A	B	C	D
Weighted averaging	7.203	6.966	7.373	7.203

Table 8 Preference matrices for items

	e_1	e_2	e_4	e_5	e_6
A	9	6	6	9	9
B	6	9	9	6	6
C	6	9	9	6	6
D	6	9	9	6	6

Table 9 Weights of experts in two situations

	a_1	a_2	a_4	a_5	a_6	a_{24}	a_{25}
Situation1	0.3	0.3	0.2	0.3	0.4	−0.2	−0.3
Situation2	0.2	0.2	0.1	0.2	0.3	0	0

Table 10 Global preferences of the two situations

	A	B	C	D
Situation 1	4.5	3	3	4.5
Situation 2	7.5	7.5	7.5	7.5

In order to give a good comparison, an extreme example is given. Table 8 is the preference matrix for items. Table 9 shows the weights of experts.

In Situation 1, we consider the interdependency and extract corresponding information to measure these relations. Situation 2 does not consider the interdependency or miss corresponding information. So we apply Choquet integral to aggregate preferences in Situation 1 and weighted averaging operator to aggregate preferences in Situation 2. The global preferences are shown in Table 10.

From Table 10, we find there is no preference among these four items in Situation 2. But preferences can be distinguished in Situation 1. Because even $\{e_2, e_4, e_5\}$ has the equal importance to $\{e_1, e_6\}$, "Knowledge absorption" makes preference of $\{e_1, e_6\}$ more important than preference of $\{e_2, e_4, e_5\}$. $\{e_1, e_6\}$ gives high evaluations to A and D and these two items get high global evaluations.

6 Conclusion

In this study a method to solve the weight determination problem in SNs is developed, for helping recommend new items for users using information form the trust network. Reorganizing the trust relationship among elements in a SN allows us to design a fuzzy capacity to reflect the non-additive nature of the problem.

There are two main innovations. Firstly, we rediscover the Trust Complementary and Redundancy in a trust network. In our model, we assume trust relations can be propagative but not transitive. From the aspect of the Trust Complementary and Redundancy, we give reasonable explanation of our assumption. Secondly, we apply non-additive measures to describe trust relations and transform these fuzzy measures to weights of nodes in recommendation. The established weight determination model can be used to manipulate the non-additivity between elements. Hence, the presented method creates a framework for innovation.

There are still some limitations in our paper. We apply existing propagation operator in our Trust-enhanced Recommend Network to obtain trust relationships. We need to modify this propagation operator to improve the information efficiency of the recommendation in the future work. Although $\mathcal{T}$-norm in this operator is easier, some information is lost.

In another aspect, we extract the Trust-enhanced Recommend Network from an SN. That process is omitted and we get the result. But it will be a hard job to complete. This process will be investigated in the future research.

Compliance with Ethical Standards

Acknowledgements This work was supported by National Natural Science Foundation of China (NSFC) (71871121, 71401078, 71503134), Top-notch Academic Programs Project of Jiangsu High Education Institutions, and HRSA, US Department of Health & Human Services (No. H49MC0068).

Conflict of Interest This section is to certify that we have no potential conflict of interest.

This article does not contain any studies with human participants or animals performed by any of the authors.

References

1. Yu, Y., et al.: The impact of social and conventional media on firm equity value: a sentiment analysis approach. Decis. Support Syst. **55**(4), 919–926 (2013)
2. Verbraken, T., et al.: Predicting online channel acceptance with social network data. Decis. Support Syst. **63**, 104–114 (2014)
3. Peng, S., et al.: Influence analysis in social networks: a survey. J. Netw. Comput. Appl. **106**, 17–32 (2018)
4. Chen, R.-C., et al.: Merging domain ontologies based on the WordNet system and fuzzy formal concept analysis techniques. Appl. Soft Comput. **11**(2), 1908–1923 (2011)
5. Cheng, L.-C., Wang, H.-A.: A fuzzy recommender system based on the integration of subjective preferences and objective information. Appl. Soft Comput. **18**, 290–301 (2014)
6. Victor, P., et al.: Gradual trust and distrust in recommender systems. Fuzzy Sets Syst. **160**(10), 1367–1382 (2009)
7. Levin, D.Z., Cross, R.: The strength of weak ties you can trust: the mediating role of trust in effective knowledge transfer. Manag. Sci. **50**(11), 1477–1490 (2004)
8. Quijano-Sanchez, L., et al.: Social factors in group recommender systems. ACM Trans. Intell. Syst. Technol. **4**(1), 1–30 (2013)
9. Capuano, N., et al.: Fuzzy group decision making for influence-aware recommendations. Comput. Hum. Behav. **101**, 371–379 (2019)
10. Sunstein, C.R.: Infotopia. Law Press China, Beijing (2008)

11. Sinha, R., Swearingen, K.: Comparing recommendations made by online systems and friends. In: DELOS Workshop: Personalisation and Recommender Systems in Digital Libraries (2001)
12. Timme, N., et al.: Synergy, redundancy, and multivariate information measures: an experimentalist's perspective. J. Comput. Neurosci. **36**(2), 119–140 (2014)
13. Murofushi, T., Sugeno, M.: An interpretation of fuzzy measures and the Choquet integral as an integral with respect to a fuzzy measure. Fuzzy Sets Syst. **29**(2), 201–227 (1989)
14. Bottero, M., et al.: On the Choquet multiple criteria preference aggregation model: Theoretical and practical insights from a real-world application. Eur. J. Oper. Res. **271**(1), 120–140 (2018)
15. Grabisch, M.: K-order additive discrete fuzzy measures and their representation. Fuzzy Sets Syst. **92**(2), 167–189 (1997)
16. Marichal, J.-L., Roubens, M.: Determination of weights of interacting criteria from a reference set. Eur. J. Oper. Res. **124**(3), 641–650 (2000)
17. Choquet, G.: Theory of capacities. Annales de l'institut Fourier **5**, 131–295 (1954)
18. Horanská, Ľ., Šipošová, A.: A generalization of the discrete Choquet and Sugeno integrals based on a fusion function. Inf. Sci. **451–452**, 83–99 (2018)
19. Lourenzutti, R., et al.: Choquet based TOPSIS and TODIM for dynamic and heterogeneous decision making with criteria interaction. Inf. Sci. **408**, 41–69 (2017)
20. Grabisch, M.: The application of fuzzy integrals in multicriteria decision making. Eur. J. Oper. Res. **89**(3), 445–456 (1996)
21. Nepal, S., et al.: Strust: a trust model for social networks. In: 2011 IEEE 10th International Conference on Trust, Security and Privacy in Computing and Communications, pp. 841–846. IEEE (2011)
22. Pawar, P.S., et al.: Trust model for optimized cloud services. In: IFIP International Conference on Trust Management, pp. 97–112. Springer, Heidelberg (2012)
23. Li, X., Du, J.: Adaptive and attribute-based trust model for service-level agreement guarantee in cloud computing. IET Inf. Secur. **7**(1), 39–50 (2013)
24. Can, A.B., Bhargava, B.: Sort: a self-organizing trust model for peer-to-peer systems. IEEE Trans. Dependable Secure Comput. **10**(1), 14–27 (2012)
25. Chen, D., et al.: A trust model for online peer-to-peer lending: a lender's perspective. Inf. Technol. Manag. **15**(4), 239–254 (2014)
26. Shaikh, R., Sasikumar, M.: Trust model for measuring security strength of cloud computing service. Procedia Comput. Sci. **45**, 380–389 (2015)
27. Wu, J., et al.: Trust based consensus model for social network in an incomplete linguistic information context. Appl. Soft Comput. **35**, 827–839 (2015)
28. Sherchan, W., et al.: A survey of trust in social networks. ACM Comput. Surv. **45**(4), 1–33 (2013)
29. Wu, J., et al.: Uninorm trust propagation and aggregation methods for group decision making in social network with four tuple information. Knowl.-Based Syst. **96**, 29–39 (2016)
30. Wu, J., et al.: A visual interaction consensus model for social network group decision making with trust propagation. Knowl.-Based Syst. **122**, 39–50 (2017)
31. Grabisch, M., Labreuche, C.: Fuzzy measures and integrals in MCDA. Multiple Criteria Decision Analysis: State of the Art Surveys, New York, NY (2016)

Survey-Based Forecasting: To Average or Not to Average

Kayan Cheng, Naijing Huang, and Zhentao Shi

Abstract Forecasting inflation rate is of tremendous importance for firms, consumers, as well as monetary policy makers. Besides macroeconomic indicators, professional surveys deliver experts' expectation and perception of the future movements of the price level. This research studies survey-based inflation forecast in an extended recent sample covering the Great Recession and its aftermath. Traditional methods extract the central tendency in mean or median and use it as a predictor in a simple linear model. Among the three widely cited surveys, we confirm the superior forecasting capability of the Survey of Professional Forecasters (SPF). While each survey consists of many individual experts, we utilize machine learning methods to aggregate the individual information. In addition to the off-the-shelf machine leaning algorithms such as the Lasso, the random forest and the gradient boosting machine (GBM), we tailor the standard Lasso by differentiating the penalty level according to an expert's experience, in order to handle for participants' frequent entries and exits in surveys. The tailored Lasso delivers strong empirical results in the SPF and beats all other methods except for the overall best performer, GBM. Combining forecasts of the tailored Lasso model and GBM further achieves the most accurate inflation forecast in both the SPF and the Livingston Survey, which beyonds the reach of a single machine learning algorithm. We conclude that combination of machine learning forecasts is a useful technique to predict inflation, and averaging should be exercised in a new generation of algorithms capable of digesting disaggregated information.

Keywords Forecast combination · Inflation · Information aggregation · Lasso · Professional forecasters

K. Cheng · Z. Shi (✉)
Department of Economics, The Chinese University of Hong Kong, 928 Esther Lee Building, Shatin, New Territories, Hong Kong SAR, China
e-mail: zhentao.shi@cuhk.edu.hk

K. Cheng
e-mail: kychengkayan@gmail.com

N. Huang
School of Economics, Central University of Finance and Economics, Beijing, China
e-mail: huang.naijing@gmail.com

S. Sriboonchitta et al. (eds.), *Behavioral Predictive Modeling in Economics*, Studies in Computational Intelligence 897,
https://doi.org/10.1007/978-3-030-49728-6_5

1 Introduction

Individual firms' price setting processes add up to generate an economy's price rigidity and consequently inflation transition. Understanding the ensemble from the micro level into the macro level is essential for monetary policies. Among the central issues in macroeconomics is the nature of short-run inflation dynamics. Despite decades of intensive investigation, this topic remains to be one of the most fiercely debated and few definitive answers have surfaced. At stake, among other things, is the understanding of business cycles and the appropriate conduct of monetary policies. Expected future inflation rate is one of the most crucial factors that influence economic agents' decision-makings. Households and firms demand accurate inflation forecast to negotiate salary with employees and to make optimal consumption and investment decisions across periods. Real values of wages, real interest rates and real exchange rates would be adjusted during inflation in the absence of corresponding nominal movements. Central banks call for reliable projection of future inflation to monitor the status of the economy and to implement fiscal and monetary policies.

To improve the quality of inflation forecast, enduring effort has been made by economists in identifying new forecast models and revising the existing ones. Four major types of inflation forecasting models have been developed. (i) Time series models based on historical information, including the classical univariate time series models like the autoregressive integrated moving average (ARIMA) model, and the nonlinear or time-varying univariate models, such as the random walk inflation forecasting model (Atkeson et al. [2]), the unobserved components stochastic volatility (UC-SV) model (Stock and Watson [20]), to name a few. (ii) The Phillips curve models accommodate real economic activities. Commonly used activity variables include the unemployment rate, the output gap, the output growth, and perhaps their conjunctions with other variables. Although both the backward-looking Phillips curves and the New Keynesian Phillips curves have been developed, the latter appears infrequently (and only recently) in the inflation forecasting literature and there is little evidence to buttress its performance. (iii) The third family computes inflation forecasts from explicit or implicit inflationary expectations of others. These forecasts subsume regressions based on implicit expectations derived from asset prices. Examples include forecasts extracted from the term structure of nominal Treasury debt, which embodies future inflation expectations according to the Fisher relation, and forecasts extracted from the Treasury inflation-Protected Securities (TIPS) yield curve. (iv) Survey-based measurement is filtered to help with forecast, for example Grant and Thomas [11], Mehra [13] and Ang et al. [1] among others.

Individual forecasts from independent sources are valuable to central banks, and surveys help to record these inflation expectations absent from other macroeconomic variable. For example, surveys about future inflation reflect people's perceptions of coming price movements, which are essential elements in the price and wage setting processes, and therefore can heavily influence the actual inflation out-turns. Rich information contained in the survey data make it a valuable predictor for future inflation, and its value has been heeded in the literatures. Ang et al. [1] conducted a comprehensive comparison over 39 forecasting models derived from the above four categories and demonstrated the superiority of survey results in forecasting inflation. The superior forecasting capacity of survey data on inflation rate was confirmed by multiple studies (e.g. Ang et al. [1]; Carlson [7]; Mehra [13]; Thomas [22]). In this paper, we focus on the usage of survey data on predicting inflation. We seek the best out-of-sample performance by aggregating information from professional forecasters.

Traditional forecasting methods are typically linear time series models wherein the estimators are justified by asymptotics under a class of data generating processes. However, in these classical models, the number of variables must be considerably fewer than the sample size in order to allow large-sample theory to kick in. This reduces the amount of data that can enter the models for tightening up the prediction. Machine learning techniques are not bound by this constraint and allow for a much larger pool of variables. In this paper, we extract information from many individual forecasters via different machine learning methods to improve the forecasting performance. We consider the most popular ones, including the Lasso, the random forest, and the gradient boosting machine (GBM). While most of the applied researches employing machine learning borrow off-the-shelf methods, we propose a tailored Lasso model to penalize forecasts according to experience. It is helpful if we tailor machine learning methods for the economic context. As economists enter and exit the survey at different periods of time, the individual response data are in form of a sparse matrix. The sparse nature of the data could be handled with Lasso.

In terms of the empirical performances of these competitive methods, we find that the machine learning methods are generally stronger than the classical mean or median models. The tailored Lasso model improves the standard Lasso but is still inferior to GBM. It implies that in this application of inflation rate forecast, GBM smoothly handles the potential nonlinearities that the Lasso based methods miss. Given that the tailored Lasso and GBM tackle the same forecast problem from different angles, it is natural to consider a forecast combination of the two. It turns out that a simple combination by the constrained least squares in SPF offers the overall best forecast of the inflation rate.

The empirical results deliver a yes-and-no response to the question "to average or not to average" in the title of this paper. A short answer is that we should average in a new way. The classical average by the mean or median is inferior to the new generation of machine learning algorithms. However, faced with the economic indicators whose generating processes is difficult to capture and may subject to evolution, the new way of averaging that combines machine learning methods to attack different aspects of the complexity offers optimism.

This paper is organized as follow. Section 2 introduces the machine learning methods we used for forecasting inflation; Sect. 3 describes what data and how data is employed in this study. Section 4 shows the out-of-sample performance results of models in Sect. 2. Section 5 concludes.

2 Methodology: Machine Learning Models Using Surveys from Individual Forecasters

Forecasts of inflation, no matter made by Federal Reserve officials or private sector economists, is informative for changing the stance of monetary policy. The surveys to professional forecasters has been proved successful in out-of-sample prediction. Even in the combined forecast, the data consistently place the highest weights on the survey forecasts while little weight was given on other forecasting methods (Ang et al. [1]). In this paper, we focus on using survey data to predict inflation, and try to extract information from individual professional forecasters. Throughout this paper, the accuracy of the out-of-sample forecasts is measured by the root-mean-square error (RMSE)

$$RMSE = \sqrt{T^{-1}\sum_{t=1}^{T}(\text{Predicted}_t - \text{Realized}_t)^2}.$$

The prediction will be exercised via the following ways.

2.1 *Machine Learning Forecasting Methods*

This section describes the machine learning methods that we invoke in the forecast of inflation. All of our estimates share the basic objective of minimizing the root-mean-square prediction error (RMSE). The difference lies in the regularization, such as adding parameterization penalties and robustification against outliers. These modifications are designed to avoid overfitting and improve models' out-of-sample predictive performance.

2.1.1 The Standard Lasso Model

The least-square shrinkage and selection operator (Lasso, Tibshirani [23]) is the most intensively studied statistical method in the past 15 years. The Lasso is also popular in econometrics given its rich range of extensions (Caner [5], Belloni et al. [3], Shi [17], Su et al. [21], Lee et al. [12], to name a few). Lasso penalizes the L_1-norm of the coefficients, in addition to the typical sum of squared residuals for the ordinary least squares. In our context,

$$\min_{\beta_i}\left\{\sum_{t=1}^{T}\left(\pi_{t-4,t} - \sum_{i=1}^{N}\beta_i\pi^f_{t-4,t,i}\right)^2 + \lambda\sum_{i=1}^{N}|\beta_i|\right\}$$

where $\pi^f_{t-4,t,i}$ is individual i's predicted annual inflation rate for quarter t, and $\pi_{t-4,t}$ is the realization of its corresponding target. When $\pi^f_{t-4,t,i}$ is missing, it is replaced by the simple average of observed predictions at time t. Intuitively, during the periods that only a few of the forecasters are active, the mean forecast provides natural smoothing to reduce the volatility of individual forecasters.

2.1.2 Tailored Lasso Model

Professional forecasters frequently enter and exit surveys since most of the surveys were launched decades ago (Capistrán and Timmermann [6]; Diebold and Shin [8], Figs. 1 and 2). To accommodate this peculiarity of the survey data and boost the performance of the standard Lasso model, we proposed the *tailored Lasso* model, which is tailored to differentiate the penalty level for forecasters with diverse experience:

$$\min_{\beta_i} \left\{ \sum_{t=1}^{T} \left(\pi_{t-4,t} - \sum_{i=1}^{N} \beta_i \pi^f_{t-4,t,i} \right)^2 + \lambda_1 \sum_{i=1}^{N} \exp(\lambda_2 / S_{it}) \, |\beta_i| \right\} \quad \text{s.t. } 0 \le \beta_i \le 1$$

where $S_{i,t}$ is the accumulated number of predictions that individual i has made up to the time t. This optimization problem involves two tuning parameters, λ_1 and λ_2. In this study, we fix $\lambda_2 = \ln(10)$ for the SPF and $\lambda_2 = \ln(5)$ for the Livingston Survey. On the other hand, λ_1 is selected through series of cross validation. Details of the cross validation procedure are presented in the Appendix.

The model penalizes more heavily on an individual who has exercised fewer predictions. Intuitively, such a forecaster is less experienced and his forecast tends to be unreliable and less trustworthy. At the extreme case where $S_{i,t} = 0$, the penalty term $\lambda_1 \exp(\lambda_2 / S_{it}) \, |\beta_i|$ automatically rules out this person.

2.1.3 The Random Forest and Gradient Boosting

Lasso and tailored Lasso capture individual predictor's impact on the future inflation, but they do not account for the interactions among individuals. It is possible to include multivariate function of individual predictors. However, without prior assumptions about the specification of interactions, the model suffers from excessive computational burden. The regression tree is a popular alternative to form data-driven interactions among predictors. Different from linear models, tree models are fully nonparametric. A regression tree can approximate potentially severe nonlinearities. However, trees are prone to overfit and therefore must be heavily regularized. In this paper, we consider two "ensemble" tree based machine learning methods that combine experts from many different trees into a single forecast, which are the random forest (Breiman [4]) and the gradient boosting (Friedman [9]). In econometrics, Wager and Athey [25] developed treatment effect inference for random forecast, and

Fig. 1 Entry and Exit of Experts for the SPF (1982Q3–2018Q3) • *Notes* The y-axis marked the 241 IDs and the x-axis marked the time period from 1982Q3 (leftmost) to 2018Q3 (rightmost). A cross on the figure indicates an instance when the expert made a forecast in the SPF

Shi [16], Phillips and Shi [15], and Shi and Huang [18] extended boosting in various settings.

3 Data

The focus of this paper is to predict inflation rates through survey data. Besides the actual annual inflation rates, we also gathered inflation forecasts from the three most commonly quoted surveys: (i) the Survey of Professional Forecasters, (ii) the Livingston Survey, and (iii) the Survey of Consumers by the University of Michigan.[1] We followed Ang et al. [1] to construct the following variables.

[1]Note that sample periods mentioned in this section would be defined as the periods of data which we retrieved from corresponding sources. However, since data from surveys are denoting one-year-ahead forecasts of the inflation rates, to provide clearer interpretation, we define sample periods as the data's forecasting periods in implementation. That is, the sample periods of surveys is lagged by one year in all other sections.

Fig. 2 Entry and Exit of Experts for the Livingston Survey (1952Q2–2018Q2) • *Notes* The y-axis marked the 427 IDs and the x-axis marked the time period from 1982Q3 (leftmost) to 2018Q3 (rightmost). A cross on the figure indicates an instance when the expert made a forecast in the Livingston Survey

3.1 Historical Annual Inflation Rates

The historical annual inflation rate is the macroeconomic indicator we target in our forecasting models. In this study, we employed the seasonally adjusted *Consumer Price Index for All Urban Consumers: All Items* (CPIAUCSL) to calculate the historical annual inflation rate from 1982Q4 to 2018Q3. CPIAUCSL is monthly data published by the U.S Bureau of Labor Statistics and we obtain the index from December 1981 to September 2018 from the database of the Federal Reserve Bank of St. Louis.[2] Following the below stated definition of the annual inflation rate, only monthly index levels for March, June, September and December are used. The annual inflation rate at quarter t is defined as $\pi_{t-4,t} = \ln(CPI_t/CPI_{t-4})$, where CPI_t is the CPIAUCSL level at the last month of quarter t.

[2]Available at FRED St. Louis: https://fred.stlouisfed.org/series/CPIAUCSL/.

3.2 Surveys

The SPF provides economists' expected percentage changes in the CPI levels on a quarterly basis starting from 1981Q3.[3] We use variables "CPI3", "CPI4", "CPI5" and "CPI6" from SPF, which represent forecast for the annual CPI inflation rate for 1 to 4 quarters later respectively. Based on the above 4 variables, the annual inflation forecast from the SPF at quarter t is constructed as the simple average,

$$\pi_{t,t-4}^{fs} = \frac{1}{4}\left(CPI3_{t-4} + CPI4_{t-4} + CPI5_{t-4} + CPI6_{t-4}\right).$$

The sample period of SPF is from 1981Q4 to 2017Q3, and 3 separate datasets of mean responses, median responses and individual responses would be used. For the individual response, 241 respondents who made at least one forecast from 1981Q3 (The earliest avaliable data for the SPF) to 2017Q3 are included in the analysis.

In contrast to the SPF, the Livingston Survey records economists' expectation on future CPI levels instead of future CPI movements on a half-yearly basis.[4] We build the Livingston Survey forecast based on its variables "Base Period" and "Forecast12Month", representing the actual CPI level at the base period when the survey is issued and respondents' expected CPI level in 12 months. Annual inflation forecast from the Livingston Survey at quarter t is calculated as,

$$\pi_{t,t-4}^{fl} = \frac{12}{14}\ln\left(\frac{\text{Forecast 12 Month}_{t-4}}{\text{BasePeriod}_{t-4}}\right).$$

Forecasts are adjusted by a factor of $\frac{12}{14}$ by considering the observation of Carlson [7] that forecasters participated in the Livingston Survey usually make forecast for 14 months periods instead of 12 months. Such adjustment is employed by Ang et al. [1], Mehra [13] and Thomas [22]. Sample period of the Livingston Survey is from 1981Q4 to 2017Q2. Separate datasets for mean responses, median responses and individual responses are used. 427 respondents who participated at least once from 1951Q2 to 2017Q2 are included in the analysis.

As shown obviously in Fig. 3, the performance of the Livingston Survey is relatively poor and unstable before 1952Q2. It motivates Ang et al. [1] to select 1951Q2 as the start of their sample period and form the forecasting period from 1952Q2. Taking this observation into account, we therefore only consider forecasters who have participated in the above mentioned periods.

While the two surveys mentioned before are polled among professionals, the Survey of Consumers by the University of Michigan reports American households'

[3]Available at FRED Philadelphia https://www.philadelphiafed.org/research-and-data/real-time-center/survey-of-professional-forecasters/data-files/cpi/.

[4]Available at FRED Philadelphia https://www.philadelphiafed.org/research-and-data/real-time-center/livingston-survey/historical-data/.

expected percentage change in price levels in 1 year on monthly basis.[5] As individual data is not provided by the survey, only the mean and median responses ("$px1_{mean}$", "$px1_{med}$") are included in our study and the full sample period is from 1981Q4 to 2017Q3. Annual inflation forecast from the Survey of Consumers at quarter t is denoted as $\pi^{fm}_{t,t-4} = px1_{t,t-4}$.

3.3 Data Overview

Figures 3, 4, 5 and 6 overview the surveys' performances. Figures 3 and 5 plot the mean and median forecasts of distant surveys together with the historical inflation rates from 1948Q1 to 2018Q3. Figures 4 and 6 illustrate the forecasting errors (subtracting the actual inflation rates from surveys' mean and median forecasts) together with the actual inflation rates for the period of 1979Q1 to 2018Q3, where two of the three surveys have been established.

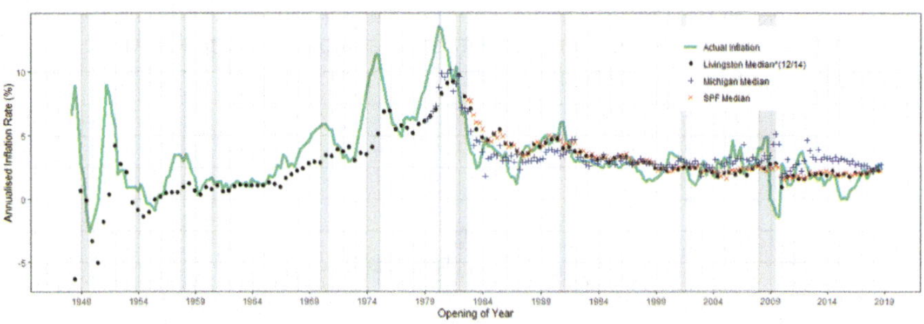

Fig. 3 Median forecasts of surveys and the actual inflation rate (1948 Q1–2018Q3)

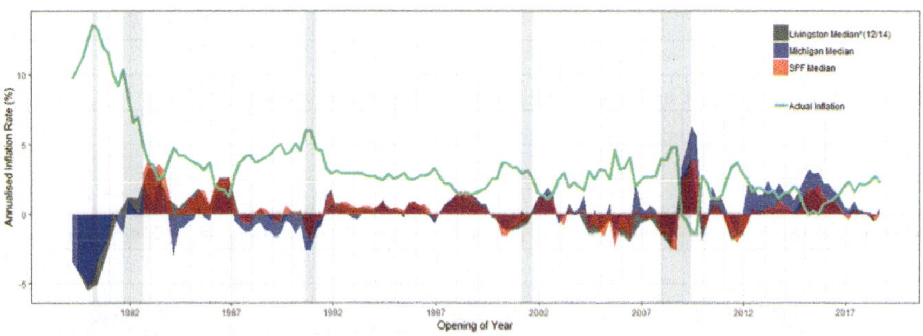

Fig. 4 Difference between the median forecasts of surveys and the actual inflation rate (1979 Q1–2018Q3)

[5]The Regents of the University of Michigan. (2018). Expected Change in Prices During the Next Year. Available at https://data.sca.isr.umich.edu/data-archive/mine.php.

3.4 Overall Performance of the Surveys

We observe from Figs. 3 and 5 that all surveys track the inflation rate well in most of the time, although the performance deteriorates when the actual inflation is highly volatile, which is unsurprising. Surveys conducted by professional economists, namely the SPF and the Livingston Survey, are more stable and conservative, while the Survey of Consumers is more fluctuated. The Survey of Consumers are lagged behind the actual data around 2005 to 2010, and two separate peaks are witnessed right after the peaks of the actual inflation rate. It might demonstrate that households form their expectations mainly base on past observations.

Figures 4 and 6 show that the performances of the SPF and the Livingston Survey are more comparable whereas that of the Survey of Consumers is less satisfactory. The forecasting errors are huge for the Survey of Consumers' mean forecasts and the situation has been worsening after the 2008 financial crisis. For the SPF and the Livingston Survey, no significant difference is observed with regard to the surveys' mean and median forecasts. The comparisons between their predictabilities will be discussed in a later section, where we confirm that both the SPF and the Livingston Survey perform better with their mean forecast while the Survey of Consumers wins with its median forecast. Such comparison extends Ang et al. [1] which only discussed the median forecast.

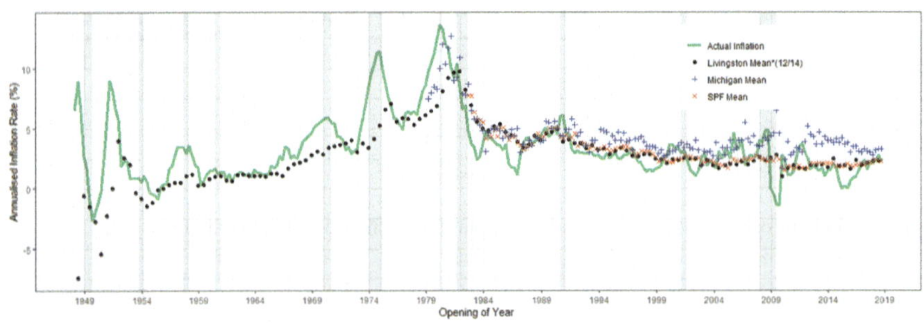

Fig. 5 Mean forecasts of surveys and the actual inflation rate (1948 Q1–2018Q3)

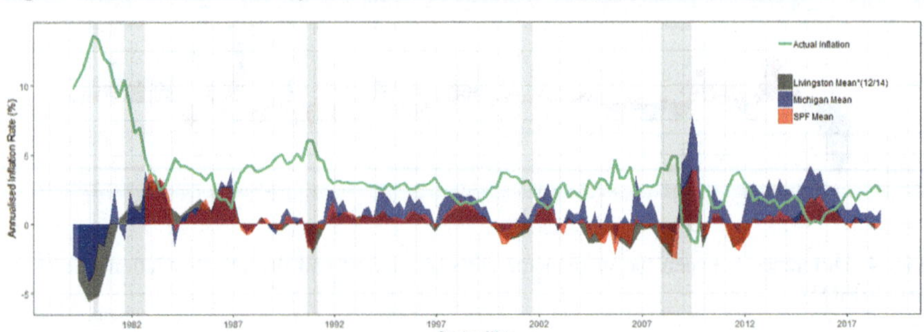

Fig. 6 Difference between the mean forecasts of surveys and the actual inflation rate (1979 Q1–2018Q3)

4 Empirical Results

In this section, we compare the forecasting performance of the mean or median models along with the machine learning methods introduced in Sect. 2.

It is well-known that the long economic time series may be contaminated by structural breaks. Instead of using the longest time series, our full sample period is constructed between 1981Q4 and 2017Q3, where 1981Q4 is the earliest period which information of all of the three surveys is available. An alternative sample period from 1984Q4 to 2017Q3 is used to skip the period where a plunge of inflation rate was recorded for robustness check.

To mitigate the impact of potential structural breaks, a set of rolling window out-of-sample forecasts are conducted with fixed window length of 10 years and 1 quarter. In the full sample period (1982Q4 to 2018Q3), the out-of-sample period begins at 1993Q1 (the SPF) and 1993Q2 (the Livingston Survey). In the alternative sample period, the out-of-sample period starts at 1996Q1 (the SPF) and 1996Q2 (the Livingston Survey).

All machine learning methods require tuning parameters in practice. The tuning parameters are determined in a data-driven manner by time series cross validation that takes into account the temporal dependence. The choices of tuning hyperparameters and the implementation of cross validation are detailed in the Appendix.

4.1 *Linear Models for Mean or Median Predictor*

Ang et al. [1] consider three survey data-based forecasting models:

- Naive form: $\pi_{t-4,t} = \pi^{fi}_{t-4,t}$;
- Bias adjusted form 1: $\pi_{t-4,t} = \alpha_1 + \beta_1 \pi^{fi}_{t-4,t} + \varepsilon_{t-4,t}$;
- Bias adjusted form 2: $\pi_{t-4,t} = \alpha_1 + \alpha_2 D_t + \beta_1 \pi^{fi}_{t-4,t} + \beta_2 D_t \pi^{fi}_{t-4,t} + \varepsilon_{t-4,t}$ where D_t is a dummy variable which equals to 1 when the actual inflation rate at quarter t exceeds its past 24-month moving average, i.e. $D_t = \mathbf{1}\left\{\pi_{t-4,t} - \frac{1}{8}\sum_{j=0}^{7} \pi_{t-j-4,t-j} > 0\right\}$.

We apply the three specifications to the mean and median forecasts of the SPF ($\pi^{fi}_{t-4,t} = \pi^{fs}_{t-4,t}$), the Livingston Survey ($\pi^{fi}_{t-4,t} = \pi^{fl}_{t-4,t}$) and the Survey of Consumers ($\pi^{fi}_{t-4,t} = \pi^{fm}_{t-4,t}$). Variables $\pi^{fs}_{t-4,t}$, $\pi^{fl}_{t-4,t}$ and $\pi^{fm}_{t-4,t}$ follow the definitions stated in Sect. 3.

Table 1 reports the RMSE results across 6 models in our full sample period and the alternative sample period. For the SPF and the Livingston Survey, both sample periods suggest that the mean forecast is slightly better than their median counterparts, regardless of the specifications. In all cases except for those of the Survey of Consumers, mean forecasts produce a lower RMSE. For the Survey of Consumers, less reliable forecasts made by non-professionals may act as extreme values to undermine the simple average.

Table 1 RMSE results for models using mean or median forecasts

Forecast by		Naive from		Bias adjusted1		Bias adjusted 2	
		Mean	Median	Mean	Median	Mean	Median
Full sample period	SPF	1.1612	1.1676	1.1848	1.1883	1.1903	1.1939
(1982Q4–2018Q3)	Livingston	1.1757	1.1834	1.1957	1.2052	1.2989	1.3364
	Consumers	2.0513	1.5584	1.1926	1.1795	1.1978	1.1919
Alt. sample period	SPF	1.2134	1.2224	1.2470	1.2535	1.2573	1.2632
(1985Q4–2018Q3)	Livingston	1.2321	1.2433	1.2524	1.2660	1.3518	1.3962
	Consumers	2.1087	1.6482	1.2248	1.2154	1.2352	1.2397

• *Notes* The full sample period's out-of-sample evaluation begins at 1993Q1 (SPF, Survey of Consumers) and 1993Q2 (Livingston Survey). The alternative sample period's out-of-sample evaluation begins at 1996Q1 (SPF and Survey of Consumers) and 1996Q2 (Livingston Survey)

We confirm the superiority of the naive model with mean forecasts from the SPF. Forecasters' mean responses in the SPF yield the most accurate predictions for future inflation rates in both samples. Our result also offers a clearer picture on the performances across surveys. The SPF is preferred to the Livingston Survey and the Survey of Consumers in forecasting inflation. The Survey of Consumers, which non-professional households have participated, is the least favoured one according to our analysis. This pattern is also agreed across the two subsamples. This is in sharp contrast to Ang et al. [1], where the SPF and the Livingston Survey are winners in two separate sample periods. Published before the Great Recession, their mixed results reflect the situation during the "Great Moderation".

Finally, selection of models matters across different surveys. In Table 1, the naive form of model stays robust for the SPF and the Livingston Survey, and obtains the lowest RMSE in both sample periods. When we switch to the Survey of Consumers, however, the bias adjusted form 1 is the best while the naive form is the worst. The bias adjustment for the Survey of Consumers' median forecasts makes them comparable to those of the other two surveys, and even beats the best performing model of the Livingston Survey in the alternative sample period. During the alternative sample period, the Survey of Consumers achieves an RMSE of 1.2154 with the Bias adjusted form 1 using its median forecasts. The best performing model of the Livingston Survey attains an RMSE of 1.2321 under the naive form of model with its mean data.

4.2 Machine Learning for Individual Predictors

Out-of-sample RMSEs of machine learning methods based on individual information are reported in Table 2. Again, the performance of the SPF is better than that of the Livingston Survey. Comparing the results in Table 1 and Table 2, all methods employing individual forecasters' predictions from the SPF outperform the best

Table 2 RMSE results for models using forecasts from individual forecasters

		Lasso	Tailored lasso	Random forest	GBM
Full sample period	SPF	1.0908	1.0900	1.1303	1.0095
(1982Q4–2018Q3)	Livingston	1.2008	1.1931	1.2313	1.1439
Alt. sample period	SPF	1.1548	1.1539	1.2070	1.0339
(1985Q4–2018Q3)	Livingston	1.2699	1.2632	1.2599	1.2336

•*Notes* The time window is the same as that in Table 1. Out-of-sample forecasts are computed by the rolling window method with fixed window length: 10 years and 1 quarter

performing model using mean or median. Given that the ordinary Lasso model is already strong enough to outperform SPF's naive mean forecasts, the tailored Lasso model improves its RMSE further and beats all other methods except for the all time winner—GBM. Performance of the tailored Lasso model is also recognized for its much lower computational cost when comparing to GBM.

Across the 4 machine learning methods using forecasts from individual forecasters, GBM is the winner in all cases. For the Livingston Survey, GBM is also the only model that can beat the survey's naive mean forecast in the full sample period, while such outperformance is not recognized in the alternative sample period. In general, machine learning methods are very competitive in forecasting because of their flexibility along with data-driven regularization.

4.3 Combining Forecasts of the Tailored Lasso Model and the Gradient boosting Machine

Lasso is a generalization of linear models, and the tailored Lasso adapts to the context of survey data to combat the missing variables. Both the random forecast and GBM are based on regression trees to handle potential interactions and nonlinearities. They are techniques focusing on different aspects of the forecast. It has been well documented that forecast combination can achieve better accuracy than individual forecasts (e.g., Granger and Ramanathan [10]; Nowotarski et al. [14]; Stock and Watson [19]). Given that the tailored Lasso and GBM are the best performers in their respective categories, can we combine them to further enhance the forecasting accuracy?

From our previous computations, forecasts of both models are ready from 1993Q1 (the SPF) and 1993Q2 (the Livingston Survey) for the full sample period, while from 1996Q1 (The SPF) and 1996Q2 (The Livingston Survey) for the alternative sample period. We combine these sets of forecasts by simple averaging and by the Ordinary Least Squares (OLS) and the Constrained Least Squares (CLS) approaches based on expanding windows of window length 5 years and 2 quarters (SPF) and 5 years and

Table 3 RMSE results for models' forecasts combinations

		Simple averaging combination	OLS combination	CLS combination
Full sample period	SPF	1.1280	1.1450	1.1138
(1982Q4–2018Q3)	Livingston	1.2446	1.2806	1.2569
Alt. sample period	SPF	1.1485	1.1524	1.1220
(1985Q4–2018Q3)	Livingston	1.3099	1.3542	1.3340

•*Notes* For the full sample period, its out-of-sample period begins at 1998Q3 (SPF) and 1998Q4 (Livingston Survey). For the alternative sample period, its out-of-sample period begins at 2001Q3 (SPF) and 2001Q4 (Livingston Survey. For the forecasts combination, out-of-sample forecasts are computed by the expanding window method with initial window length 5 years and 2 quarters (SPF); 5 years and 1 quarter (Livingston Survey)

1 quarter (Livingston Survey). Given the forecasts from the tailored Lasso model ($\pi_{t-4,t}^{TLasso}$) and GBM ($\pi_{t-4,t}^{GBM}$), we combine the two methods as follows.

- Simple Averaging Combination: $\hat{\pi}_{t-4,t}^{SA} = \left(\hat{\pi}_{t-4,t}^{TLasso} + \hat{\pi}_{t-4,t}^{GBM}\right)/2$.
- OLS Combination: $\hat{\pi}_{t-4,t}^{OLS} = \alpha + \beta_1 \hat{\pi}_{t-4,t}^{TLasso} + \beta_2 \hat{\pi}_{t-4,t}^{GBM}$, where the coefficients $(\alpha, \beta_1, \beta_2)$ are estimated by OLS.
- CLS Combination: $\hat{\pi}_{t-4,t}^{CLS} = \beta_1 \hat{\pi}_{t-4,t}^{TLasso} + \beta_2 \hat{\pi}_{t-4,t}^{GBM}$ where the coefficients (α, β_1, β_2) are estimated minimizing the standard criterion function of OLS while subject to the constraints $\beta_1 + \beta_2 = 1$ and $\beta_1 \geq 0, \beta_2 \geq 0$

To evaluate the combinations' performances, we also present the results of all mentioned models with an adjusted out-of-sample period. For the full sample period, the out-of-sample period now begins at 1998Q3 (SPF) and 1998Q4 (Livingston Survey). For the alternative sample period, its out-of-sample period begins at: 2001Q3 (SPF) and 2001Q4 (Livingston Survey). RMSE results of the above 3 combinations are listed in Table 3.

The combination built by OLS is the worst among the three. Although OLS combination (Granger and Ramanathan [10]) enables us to correct for biasness through its intercept term, Nowotarski et al. [14] points out such unbiasedness comes at the expense of a poorer performance for highly correlated individual forecasts.

To alleviate the shortcomings of OLS combination, we combine the two machine learning methods through CLS advocated by Nowotarski et al. [14]. The CLS is a variant of OLS combinations with additional constraints that no intercept term is imposed and the coefficients have to be summed up to 1. From the tables, CLS in SPF obtains the lowest RMSE which can even beat GBM. In addition to the significant improvement in the full sample periods, a tiny drop of RMSE is recorded in the alternative sample periods, from 1.1221 (GBM) to 1.1220 (CLS).

However, such relative merit is not repeated in the Livingston Survey. One possible reason would be the unstable nature of its experts and thus the unstable relative performances of the models. According to Table 4, for the Livingston Survey, the tailored Lasso model outperforms GBM in the alternative sample period while it loses in the other instances. As a result, instead of adopting time-varying models that

Table 4 RMSE results for models using forecasts from individual forecasters (Out-of-sample periods adjusted to evaluate forecasts combinations)

		Best forecast combination	Lasso	Tailored lasso	Random forest	GBM
Full sample period	SPF	1.1138	1.2016	1.2004	1.2411	1.1150
(1982Q4–2018Q3)	Livingston	1.2446	1.3077	1.2959	1.2535	1.2494
Alt. sample period	SPF	1.1221	1.2441	1.2387	1.3014	1.1221
(1985Q4–2018Q3)	Livingston	1.3099	1.3656	1.3353	1.2953	1.3397

•*Notes* Out-of-sample forecasts are computed by the rolling window method with fixed window length: 10 years and 1 quarter. The time window is the same as in Table 3

Table 5 RMSE results for models using mean or median forecasts (Out-of-sample periods adjusted to evaluate forecasts combinations)

Forecast by		Naive from		Bias adjusted 1		Bias adjusted 2	
		Mean	Median	Mean	Median	Mean	Median
Full sample period	SPF	1.2407	1.2528	1.3008	1.3089	1.3115	1.3186
(1982Q4–2018Q3)	Livingston	1.2726	1.2828	1.3048	1.3211	1.4083	1.4580
Alt. sample period	SPF	1.2869	1.3002	1.3239	1.3300	1.3337	1.3393
(1985Q4–2018Q3)	Livingston	1.3349	1.3463	1.3433	1.3499	1.4475	1.4867

•*Notes* The time window is the same as in Table 3

depend on past data, we try the time-invariant simple averaging combinations. Such neat and simple combination method has documented success (Stock and Watson [19]). Surprisingly, the simple averaging combination provides the best performing forecasts for the Livingston Survey across the 3 combinations, and across all existing models in Table 4 and Table 5. Although for the SPF, this simple averaging combination fails to outperform GBM, it is still the best across all other methods.

With CLS approach for the SPF and the simple averaging method for the Livingston Survey, combining forecasts from the tailored Lasso and GBM yields the best results for surveys. Using CLS forecasts combination with the SPF thus offers the most accurate inflation forecasts in both sample periods, which should be considered as a valuable tool to inform decision makers.

5 Conclusion

This research conducts a comprehensive forecasting performance comparison on inflation of different models using survey data, including classic mean or median models and various machine learning models. For the commonly used mean or median forecasting model, the SPF achieves the most accurate forecast across the two sample periods we consider. Besides, mean forecasts of surveys performs consistently better than their median forecasts for professional surveys. Although the naive form of model is robust for the SPF and the Livingston Survey, for surveys incorporating respondents that are non-professional households, a model with bias adjustment could bring better forecasts. We have demonstrated that the bias adjusted form has allowed the Survey of Consumers to beat the Livingston Survey's best performing forecasts. However, in general machine learning methods outperform the classic mean or median forecasting models. In addition to the standard Lasso model, we proposed the tailored Lasso to adapt to the missing individual responses in the surveys. When we apply the new method on the best performing survey, the SPF, the model performs satisfactorily and is powerful in beating all models except for GBM. We further combine the forecasts of the tailored Lasso model and GBM to obtain the most accurate inflation forecasts across all methods.

Acknowledgements Zhentao Shi acknowledges financial support from the Research Grants Council (RGC) No. 14500118.

Appendix: Cross Validation and Tuning of the Hyperparameters

Hyperparameters in models of Sect. 2.1 were tuned within sets of designed grids through a series of nested cross-validation procedure. The nested cross-validation are described in Fig. 7 while the tuning grids of the hyperparameters are displayed in Table 6.

The nested cross-validation is conducted by both the inner loops and the outer loops, as shown Fig. 7. Given that the outer loop is built by a series of rolling windows to train for optimal parameters, each of the training set is then divided into training subsets for tuning hyperparameters. Optimal model is then selected according to the RMSE. In this study, lengths of training subsets were set to be 80% of the training

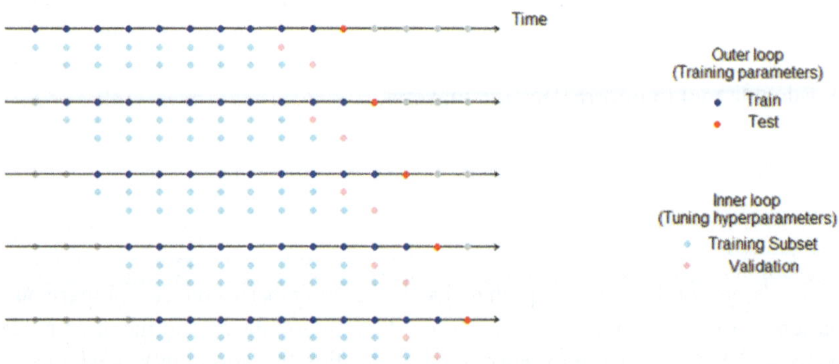

Fig. 7 The Nested Cross Validation Procedure • *Notes* This graph is for illustration purpose only and the number of data points involved are different from the real data

Table 6 Tuning Grids of Hyperparameters in models of Sect. 2.1

		SPF	Livingston
Lasso	λ	$\exp^{-x}$, $x = \{1, 1.5, \ldots, 9.5, 10\}$	
Tailored Lasso	λ_1	$\ln(10)$	$\ln(5)$
	λ_2	$\exp^{-x}$, $x = \{1, 1.5, \ldots, 9.5, 10\}$	
Random forest	No. of trees	500	
	No. of candidate features to split at each node	$\{5, 10, 15, 20\}$	
	Min. node size	$\{30, 40, 50\}$	
GBM	Max. depth of tree	$\{1, 2, 3, 4, 5\}$	
	No. of trees	$\{100, 200, \ldots, 400, 500\}$	
	Shrinkage	$\{0.01, 0.10\}$	
	Min. no. of observations in the leaf nodes	5	3

length. According to Varma and Simon [24] (2006), this nested cross-validation approach estimates the true error with nearly zero bias.

Table 6 lists the grids of the tuning hyperparameters on which we evaluate the machine learning methods in Sect. 2.1. For Lasso, we took exponential for the lambda values λ and λ_2, in order to be compatible with the standard way to compare the forecasting errors with values of $\ln(\lambda)$, instead of λ.

For our tailored Lasso, the adaptive penalty level λ_1 is designed to be $\ln(x)$. Based on our design, the penalty factor of the model, $\lambda_1 \exp(\lambda_2/s_{it})$, is more sensitive to an increase in the accumulated number of forecasts s_{it} when s_{it} is at a relatively low level. In other words, while forecasters who have accumulated 2 forecasts are penalized much lighter than those who have made merely 1 forecast, the difference between the penalties imposed on forecasters who have accumulated 20 forecasts and 21 forecasts are much smaller. It is intuitive that professional skills improve with experience in a decreasing rate. On the other hand, as SPF consist of quarterly data and the Livingston Survey is of half-yearly data, we set $\lambda_1 = \ln(10)$ for the SPF while $\lambda_1 = \ln(5)$ for the Livingston Survey. Optimal design of the adaptive factor λ_1 requires further tuning in the future.

Furthermore, since more trees is in general beneficial to the performance of the random forest algorithm, we skipped its tuning procedure and take the default value of 500. On the other hand, to avoid potential overfitting caused by applying an excessive number of trees on GBM algorithm, we tuned for the number of trees with values from 100 to 500 for a fair comparison between GBM and the random forest.

Since data on individual response is not available for the Survey of Consumers, models in Sect. 2.1 would be applied on the datasets of the Survey of Profession Forecasters and the Livingston Survey, but not the Survey of Consumers. Besides, parameters in models of Sect. 2.1 were tuned through a series of nested cross-validation procedure. This nested cross-validation approach could estimate the true error with nearly zero bias (Varma and Simon [24]).

References

1. Ang, A., Bekaert, G., Wei, M.: Do macro variables, asset markets, or surveys forecast inflation better? J. Monetary Econ. **54**, 1163–1212 (2007)
2. Atkeson, A., Ohanian, L.E., et al.: Are phillips curves useful for forecasting inflation? Fed. Reserve Bank Minneap. Q. Rev. **25**, 2–11 (2001)
3. Belloni, A., Chen, D., Chernozhukov, V., Hansen, C.: Sparse models and methods for optimal instruments with an application to eminent domain. Econometrica **80**, 2369–2429 (2012)
4. Breiman, L.: Random forests. Mach. Learn. **45**, 5–32 (2001)
5. Caner, M.: Lasso-type GMM estimator. Econom. Theory **25**, 270–290 (2009)
6. Capistrán, C., Timmermann, A.: Forecast combination with entry and exit of experts. J. Bus. Econ. Stat. **27**, 428–440 (2009)
7. Carlson, J.A.: A study of price forecasts. In: Annals of Economic and Social Measurement, vol. 6, no. 1, pp. 27–56. NBER (1977)
8. Diebold, F.X., Shin, M.: Machine learning for regularized survey forecast combination: partially-egalitarian lasso and its derivatives. Int. J. Forecast. **35**(4), 1679–1691 (2018)
9. Friedman, J.H.: Greedy function approximation: a gradient boosting machine. Ann. Stat. **29**, 1189–1232 (2001)
10. Granger, C.W., Ramanathan, R.: Improved methods of combining forecasts. J. Forecast. **3**, 197–204 (1984)
11. Grant, A.P., Thomas, L.B.: Inflationary expectations and rationality revisited. Econom. Lett. **62**, 331–338 (1999)
12. Lee, J.H., Shi, Z., Gao, Z.: On lasso for predictive regression. arXiv:1810.03140 (2018)
13. Mehra, Y.P.: Survey measures of expected inflation: revisiting the issues of predictive content and rationality. FRB Richmond Econ. Q. **88**, 17–36 (2002)
14. Nowotarski, J., Raviv, E., Trück, S., Weron, R.: An empirical comparison of alternative schemes for combining electricity spot price forecasts. Energy Econ. **46**, 395–412 (2014)
15. Phillips, P.C., Shi, Z.: Boosting the Hodrick-Prescott filter. arXiv:1905.00175 (2019)
16. Shi, Z.: Econometric estimation with high-dimensional moment equalities. J. Econom. **195**, 104–119 (2016)
17. Shi, Z.: Estimation of sparse structural parameters with many endogenous variables. Econom. Rev. **35**, 1582–1608 (2016)
18. Shi, Z., Huang, J.: Forward-selected panel data approach for program evaluation. arXiv:1908.05894 (2019)
19. Stock, J.H., Watson, M.W.: Combination forecasts of output growth in a seven-country data set. J. Forecast. **23**, 405–430 (2004)
20. Stock, J.H., Watson, M.W.: Why has us inflation become harder to forecast? J. Money Credit Bank. **39**, 3–33 (2007)
21. Su, L., Shi, Z., Phillips, P.C.: Identifying latent structures in panel data. Econometrica **84**, 2215–2264 (2016)
22. Thomas, L.B.: Survey measures of expected us inflation. J. Econ. Perspect. **13**, 125–144 (1999)
23. Tibshirani, R.: Regression shrinkage and selection via the lasso. J. Roy. Stat. Soc.: Ser. B (Methodol.) **58**, 267–288 (1996)
24. Varma, S., Simon, R.: Bias in error estimation when using cross-validation for model selection. BMC Bioinform. **7**, 91 (2006)
25. Wager, S., Athey, S.: Estimation and inference of heterogeneous treatment effects using random forests. J. Am. Stat. Assoc. **113**, 1228–1242 (2018)

PD-Implied Ratings via Referencing a Credit Rating/Scoring Pool's Default Experience

Jin-Chuan Duan and Shuping Li

Abstract Letter-based credit ratings are deeply rooted in commercial and regulatory practices, which in a way impedes a wider and quicker adoption of scientifically rigorous and operationally superior and granular credit risk measures such as probability of default (PD). This paper reverse-engineers a mapping methodology converting PDs to letter ratings by referencing the realized default rates of different rating categories experienced by a commercial rating agency such as standard & Poor's or Moody's.

1 Introduction

Credit rating as an organized business service can be traced back to early 1900s with pioneers such as Moody's and the forerunners of the Standard & Poor's. With Fitch joining later in 1924, we have the commonly named "the Big Three" credit rating agencies that dominate the global rating landscape today. Consumer credit ratings are also common with arguably the FICO credit score introduced in 1989 being the best known. In-house credit scoring systems also abound, and literally every lending institution (banks, finance companies, and modern P2P lending platforms) has some sort of credit scoring system in place, because being able to separate good from bad borrowers is vital to the long-run survival of any lending institution.

J.-C. Duan
Business School, Risk Management Institute and Department of Economics, National University of Singapore, Singapore, Singapore
e-mail: bizdjc@nus.edu.sg

S. Li (✉)
Risk Management Institute, National University of Singapore, Singapore, Singapore
e-mail: rmils@nus.edu.sg

S. Sriboonchitta et al. (eds.), *Behavioral Predictive Modeling in Economics*, Studies in Computational Intelligence 897,
https://doi.org/10.1007/978-3-030-49728-6_6

Credit rating/scoring is an act of classifying obligors in a market segment (corporates, sovereigns/supranationals, regional governments, or consumers) into some pre-defined risk categories; for example, Standard & Poor's (S&P) slots the obligors in its corporate rating pool into 21 rating categories prior to default, i.e., AAA, AA+, AA, AA-, $\cdots$, C. In general, the model involved can either be a result of a summary quantitative score complemented with qualitative judgement, for example S&P ratings, or a pure quantitative score like FICO. Credit rating/scoring is often criticized for its lack of granularity, meaning not refined enough to differentiate credit risks of borrowers. Typical rating/scoring models are not probabilistically based through a formal statistical analysis. We contend that the two issues are intertwined with the latter being the source of problem. Were a credit risk assessment system based on a statistical model, one would have obtained probabilities of default (PDs) which are naturally granular.

Furthermore, the quality of a probabilistic credit rating model lends itself to a rigorous examination by an out-of-sample performance study. This point can be better appreciated by comparing the Altman [1] Z-score with a logistic regression model. The Altman Z-score is based on a classification tool known as the linear discriminant analysis where several financial ratios are applied to separate firms in a balanced sample consisting of equal numbers of bankrupt vs. survived firms. The Z-score provides granular relative credit quality rankings on firms, but no PD is attached to a particular score. In short, the Altman Z-score is a credit rating/scoring model with granularity but not probabilistic. To make a credit assessment system probabilistic, one may, for example, adopt the logistic regression with binary outcomes (default/bankruptcy vs. survival) on the same set of financial ratios to generate PDs. Abandoning a balanced sample is a must, because defaults are rare and a balanced sample will greatly overstate default occurrences; for example, US public firms has experienced roughly a 0.2% annual default rate on average over a long time span. In short, deploying a balanced sample instead of its original natural state would lead to a highly distorted PDs even if one deploys the logistic regression. There are other conceptual and implementation issues of importance concerning the implementation of the logistic regression or any probabilistic model, and among them the natural data dependency over time for any obligor stands out as an obvious one.[1]

The objective of this paper is to develop a generic technique that can suitably turn any granular PD model into a credit rating/scoring system. The need for such reverse engineering is rather obvious in terms of business applications. The long tradition of credit rating practice has developed a deeply entrenched management infrastructure (business conventions, regulatory regimes and reference knowledge) around it. A credit rating of, say, S&P BBB- and above is known as an investment-grade obligor meeting certain regulatory and/or fiduciary requirements. Merely providing a PD value, regardless of its granularity and scientific quality, simply will not meet usage requirements under many circumstances. In short, a PD system critically needs a rating-equivalent interpretation for its outputs in order to facilitate its business and

[1]Please see Shumway [5]. Dealing with corporate termination for reasons other than default is another important aspect, and for this, readers are referred to Duffie et al. [2] and Duan et al. [4].

regulatory adoption. The logical basis for this reverse engineering is to assign PD implied rating (PDiR) by referencing the realized default experience of a rating system on a rating pool that is deemed relevant.

Technique-wise, we utilize data-cloning sequential Monte Carlo optimization to determine the suitable boundary PD levels for each of many rating categories. As to the data, we take the PDs generated by the Credit Research Initiative (CRI) at the National University of Singapore for their quality and global coverage. The CRI-PDs are a product of the forward intensity model of Duan et al. [4] and are available on a daily frequency for all exchange-traded firms globally (over 70,000 firms in 133 economies at the time of this writing).[2] We will provide the PDiR methodology in this paper and provide the specific results of referencing the S&P global rating pool. For example, a firm having its 1-year PD in the range between 0 and 0.9 basis points (bps) can be viewed as an AAA rated firm by benchmarking against the S&P global rating pool's historical average realized 1-year default rates.

2 Methodology

Table 1 (columns 1 and 2) is an example of the 20-year average realized 1-year default rates for different rating cohorts provided in the 2018 S&P report for its global rating pool.[3] Mapping the CRI 1-year PD to the respective PDiR requires one to define the (upper or lower) bound for all rating cohorts. Obtaining the boundary values to set the PDiR requires us to first smooth the average realized 1-year default rates (ADRs) for different rating cohorts in a rating pool, and then to search for the proper boundary values to separate PDs while matching the expected 1-year PD for each of all rating cohorts to its corresponding cohort-specific smoothed ADR. The expectation conditional on a rating category must be taken with reference to the empirical distribution of the CRI 1-year PDs which are taken once a year at the year-end and averaged over the same period over which the ADRs are computed.

In the following, we provide the technical details for the PDiR mapping method.

2.1 The Smoothed ADRs

First, note that AAA and AA+ rating cohorts typically lack default occurrence for the S&P or Moody's rating pool. Furthermore, the reported default rate for the lowest S&P credit quality category, i.e., CCC/C, actually consists of CCC+, CCC, CCC-, CC and C, but does not offer default rates for individual rating subcategories. The boundary values for these subcategories could not possibly be determined without first extrapolating/interpolating ADRs for these subcategories. A sensible approach

[2] See CRI Technical Report [7].

[3] See S&P's [6] 2018 Annual Global Corporate Default and Rating Transition Study.

is to perform a linear regression of the logit-transformed ADRs on the cohorts with meaningful values, and through which to predict ADRs for those without. That is,

$$\text{logit}(ADR_p) = \log\left(\frac{ADR_p}{1 - ADR_p}\right) \tag{1}$$

where p represents some rating cohort and ADR_p is the average one-year realized default rate for that cohort.

Table 1 One-year PD to PDiR by referencing the S&P global rating pool

Rating category		Observed S&P average default rate (bps)	Smoothed S&P average default rate (bps)	CRI PD lower bound (bps)	CRI PD upper bound (bps)
AAA		0.00	0.29	0.00	0.90
AA+		0.00	1.06	0.90	1.22
AA		2.15	1.64	1.22	2.05
AA−		3.85	2.53	2.05	3.08
A+		5.95	3.91	3.08	4.89
A		7.95	6.06	4.89	7.60
A−		7.30	9.35	7.60	11.25
BBB+		9.75	14.46	11.25	18.51
BBB		16.15	22.35	18.51	26.83
BBB−		25.85	34.52	26.83	44.44
BB+		31.25	53.29	44.44	65.14
BB		48.45	82.18	65.14	103.34
BB−		100.55	126.53	103.34	158.14
B+		185.75	94.36	158.14	245.03
B		419.05	297.44	245.03	371.75
B−		753.50	452.67	371.75	573.38
CCC/C	CCC+	2723.65	683.22	573.38	831.61
	CCC		1018.64	831.61	1232.25
	CCC−		1492.37	1232.25	1895.82
	CC		2134.05	1895.82	2367.01
	C		3935.64	2367.01	10,000

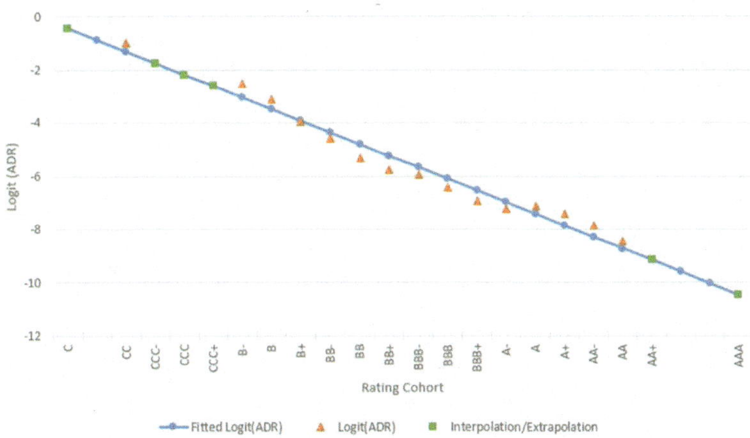

Fig. 1 Logit(ADR) and rating categories

Figure 1 displays the linear regression result on the S&P global rating pool, which provides a fairly accurate description of the relationship between the logit-transformed ADRs and the rating categories, which starts with the value of 1 and increase by 1 for each rating notch improvement except that the spacing between AAA and AA+ is set to three ticks and between CC and C is set to two ticks. Note also that we have assumed the S&P reported ADR for the combined category of CCC/C is representable by the CC cohort. Recall that we are only able to obtain the realized default rate for this combined category. The design is motivated by the fact that without extra spacing, the PDiR would have led to a far greater number of AAA or C firms as compared to the S&P rating practice. The smoothed ADRs are those fitted logit-transformed values after being transformed back to probabilities.

2.2 *The Upper PD Boundaries Defining the PDiR Classes*

The PDiR is designed by matching as closely as possible the average model values with those smoothed ADRs over all rating categories. Specifically, we aim to ensure the average 1-year probability of default (APD) in each rating cohort, based on the empirical distribution of the 1-year PD in the CRI universe of exchange-listed firms, to be close to its smoothed ADR counterpart. In a technical sense, we need to first construct the empirical distribution and then find a set of upper bounds to consistently slot PD values into different rating categories.

The empirical distribution is constructed with year-end snapshots of the CRI-PDs for the active firms in each of the previous 20 years. The final empirical distribution is the average of the 20 year-end distributions with the number of firms ranging from 22,219 in December 1998 to 35,689 in December 2017. In total, over half-

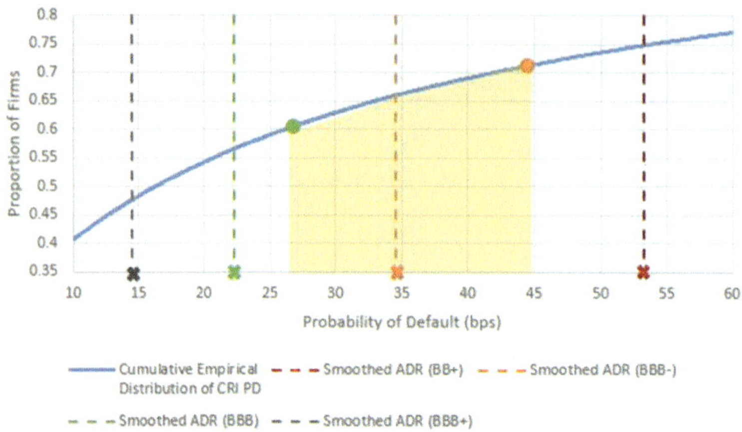

Fig. 2 Empirical distribution of CRI 1-year PD, boundary values, and smoothed ADR

a-million PD values are deployed in characterizing this empirical distribution. To define this empirical distribution, each of the year-end distributions shares the same 3,000 variably-spaced grid points used to provide sensible spacing in PDs. Specifically, lower PD values are given more refined spacing; for example, the spacing is determined with an increment of $2e^{-5}$ for PD less than 0.01 bps, and $1e^{-3}$ for PD between 100 bps and 1,000 bps. Figure 2 presents a segment (between 10 and 60 bps) of this empirical distribution constructed from the universe of over half-a-million CRI 1-year PDs for all exchange-traded firms globally.

Assuming a set of PD upper bounds are in place, the APD for a rating category is a conditional expected value of PD based on the empirical distribution, meaning that the PD is constrained to fall between the upper and lower boundary values for that rating category. The conditional expected value for a rating category p with lower and upper bounds B_{p-1} and B_p equals

$$APD_p = \frac{\int_{B_{p-1}}^{B_p} x dF(x)}{F(B_p) - F(B_{p-1})} \tag{2}$$

where $F(x)$ be the cumulative empirical distribution. The lower bound is naturally the upper bound of the adjacent and better rated category. Since the values of these upper bounds are unknown, we need to devise an optimization target and utilize a numerical scheme to find the values that solve the optimization problem.

Since PD only take values between 0 and 10,000 bps, the lower bound for the best credit-quality category and the upper bound for the worst category are naturally set. The remaining boundary values are selected by minimizing the sum of squared relative differences between the APD and the smoothed ADR for all rating categories, where the smoothed ADR is used as the base for computing the relative difference. Since there are 21 rating categories, 20 unknown upper bounds,

$\theta = \{B_1, B_2, \ldots, B_{20}\}$, are to be solved together. Note that each B_p is constrained between two adjacent smoothed ADRs. Mathematically, these 20 boundary values influencing APDs are chosen to minimize the following objective function:

$$L(\theta) = \sum_{p \in \{AAA, AA+, \ldots, C\}} \left(\frac{APD_p - \widehat{ADR_p}}{\widehat{ADR_p}} \right)^2 \tag{3}$$

where $\widehat{ADR_p}$ represents the smoothed ADR for rating category p.

The shaded area in Fig. 2 illustrates the idea where the range of PDs is for the BBB- category. The goal is to find boundary values so that the average PD confined to this range is as close as possible to the smoothed ADR for the BBB- category along with other categories. Note that the above objective function is not smooth with respect to θ because the empirical distribution is not smooth. Usual gradient-based optimization algorithms cannot be directly applied to solve this problem. We thus deploy the density-tempered sequential Monte Carlo (SMC) method with data cloning by Duan, Fulop, and Hsieh [3] to obtain the optimal solution. Our problem is technically simpler because our model has no latency or Monte Carlo error in assessing $L(\theta)$.

2.3 *The Data-Cloning Density-Tempered SMC Algorithm*

The general idea underpinning the algorithm is to first turn the optimization problem into a sampling problem where by a simple transformation of $L(\theta)$, such as $f(\theta) \propto \exp[-L(\theta)]$, makes $f(\theta)$ a density function up to a norming constant. If one is able to sample θ under $f(\theta)$ without having to know the norming constant, the sample (particle cloud in the language of SMC) will provide a Monte Carlo solution to the original minimization problem with the solution being the θ in the sample corresponding to the highest density value.

Good sampling in one step is generally infeasible, and thus sampling must be carried out sequentially. Let $\theta^{(n)}$ denote the particle cloud at the n-th intermediate tempering step, and $I(\theta)$ stand for the density of the initialization sampler used to generate the initial particle cloud. Similar to importance sampling, $I(\theta)$ is arbitrary in principle except that its support must contain the support of the target, i.e., $f(\theta)$. Like importance sampling, a missing norming constant will not matter because self-normalization into probabilities will remove that norming constant anyway. Sequential sampling targets a series of self-adaptive intermediate density functions, and the terminal state produces $f(\theta)$. The n-th intermediate target density is by design a tempered value of $f(\theta)$ with γ_n (values between 0 and 1) as the tempering parameter in the following expression:

$$f_n(\theta^{(n)}) \propto \left(\frac{\exp[-L(\theta^{(n)})]}{I(\theta^{(n)})} \right)^{\gamma_n} \times I(\theta^{(n)}) \tag{4}$$

Clearly, setting $\gamma_0 = 0$ in the above produces the initialization density, $I(\theta)$, and $\gamma_{n^*} = 1$ yields the target density, $f(\theta)$. The sequential sampling scheme is to build a self-adaptive bridge taking n from 0 to n^*.

The data-cloning density-tempered sequential Monte Carlo algorithm comprises five steps as described below.

- **Step 1: Initialization**
Draw an initial particle cloud, say, a set of 1,000 particles denoted by $\theta^{(0)}$, with each dimension of θ from a truncated normal distribution (each B_p is bounded by its two adjacent smoothed ADRs) and independent of other dimensions. Specifically, for each parameter B_p, mean is set to be the average of the two adjacent smoothed ADRs and standard deviation is set as $1/6$ of the distance between the two adjacent smoothed ADRs.
- **Step 2: Reweighting and resampling**
At the beginning of each tempering step n, perform reweighting to set the stage for advancing to the next γ_n in the self-adaptive sequence. Let $w^{(n)}$ denote the weights vector corresponding to the 1,000 particles and "$\cdot$" be the element-by-element multiplication. Thus,

$$w^{(n)} = w^{(n-1)} \cdot \left(\frac{\exp[-L(\theta^{(n-1)})]}{I(\theta^{(n-1)})} \right)^{\gamma_n - \gamma_{n-1}} \tag{5}$$

where the next γ_n will be chosen to maintain a minimum effective sample size (ESS), say, 50% of the intended sample size, which turns out to be 500 in our case. Note that the ESS is defined as $\frac{(\sum_{i=1}^{1000} w_i)^2}{\sum_{i=1}^{1000} w_i^2}$. It should be theoretically clear that such a γ_n always exists if $w^{(n-1)}$ exceeds the threshold ESS. Practically, γ_n advances quite fast toward 1.
Resample $\theta^{(n-1)}$ according to the weights, $w^{(n)}$, to obtain an equally-weighted sample, $\theta^{(n)}$, which will be true regardless of whether $\theta^{(n-1)}$ is equally-weighted. Whenever resampling has been performed, $w^{(n)}$ must be reset to the vector of 1's to reflect the fact that sample is already equally-weighted.
- **Step 3: Support boosting**
To avoid particle impoverishment, i.e., shrinkage of the empirical support, we can apply the Metropolis-Hastings (MH) move to boost the empirical support, which is conducted as follows:

  - Propose $\theta_i^* \sim \mathcal{Q}\left(\cdot \,\middle|\theta^{(n)} \right), i = 1, \ldots, 1000$. In this implementation we are using a truncated normal kernel with means and standard deviations derived from $\theta^{(n)}$, and each dimension of $\theta^{(n)}$ is independently sampled. In principle, correlations across different dimensions of $\theta^{(n)}$ can be factored into the proposal sampler. However, imposing independence in this case does not really impede the performance of the MH move in terms of the acceptance rate defined next.
  - Compute the MH acceptance rate, α_i, for each of the 1,000 particles:

$$\alpha_i = \min\left(1, \frac{f_n(\theta_i^*)\mathcal{Q}(\theta_i^{(n)} \mid \theta^*)}{f_n(\theta_i^{(n)})\mathcal{Q}(\theta_i^* \mid \theta^{(n)})}\right). \tag{6}$$

– With probability α_i, set $\theta_i^{(n)} = \theta_i^*$, otherwise keep the old particle.

Repeatly perform the above MH move until two conditions are met. First, the ESS reaches at least 90% of the original sample size or 900 in our case, and second, the accumulative acceptance rate reaches 100%.

- **Step 4: Repeat Steps 2 and 3 until reaching** $\gamma = 1$
- **Step 5: Data Cloning**
  To increase precision, we deploy data-cloning steps as in Duan, Fulop, and Hsieh [3]. Specifically, Steps 1–4 are re-run with the target density powered up to 2^k in round k; that is, $f(\theta)^{2^k}$. Cloning can be likened to repeatedly using the same data sample while treating them as if they were independently obtained. In round k, a new set of particles will be generated using the means and standard deviations derived from the set of 1,000 particles obtained in round $k-1$. To make the procedure more robust, we deploy slightly enlarged standard deviations in this re-initialization. Cloning is considered satisfactorily completed if two consecutive rounds yielding no significant improvement to the maximum $f(\theta)$ value.

2.4 *Map 1-Year PD to PDiR*

After the upper (or lower) bounds are obtained, it is straightforward to map individual 1-year PDs to implied ratings. However, the PD of a firm may oscillate across a particular upper (or lower) bound, causing frequent changes to implied ratings and making them unattractive in applications. We thus remove by time smoothing such rating changes due to oscillation across a boundary. Specifically, the PDiR deployed by the CRI is the result of first computing a two-week (10 business days) moving average PD, and then mapping the moving average against the upper and lower bounds defining different rating categories.

3 Application

The PDiR offers a qualitative measure of a firm's credit quality through referencing the historical observed default rates of some rating pool. Table 1 presents the upper and lower PD bounds defining different rating categories that were obtained with our methodology by referencing the S&P global corporate rating pool. With the PDiR in place, one is better able to make sense of the PDs produced by a default prediction model, and in this case, the CRI 1-year PDs. The PDiR offer a kind of credit risk assessment in line with the business convention. For example, the PDiR

Fig. 3 PDiR and 1-year PD of NII Holdings Inc.

offers a convenient way of identifying speculative-grade corporates by the standard definition of being below BBB-. In short, the PDiR has the best of the two worlds, providing intuitive business meaning to the results of a scientifically rigorous default prediction model.

One can better appreciate the PDiR by observing its time series vis-a-vis the PD time series. Figure 3 plots the PDiRs along with the PDs for NII Holdings, an American telecommunication company, from January 2, 2012 to the business day immediately preceding its Chap. 11 bankruptcy filing on September 15, 2014. It is evident in this figure of its credit quality deterioration over the period. Its 1-year PD already rose above 1,000 bps toward the end of 2013, and the PDiR dropped to CCC and lower. Although either way reveals NII Holdings' credit problem, a CCC or lower rating may trigger a more intuitive response/action within the established risk management infrastructure of typical financial institutions.

4 Conclusion

We illustrate the PDiR methodology using the NUS-CRI 1-year PD along with the historical default experience of the S&P global corporate rating pool. The PDiR construction can of course reference different corporate rating pools; for example, referencing Moody's. Our methodology is generic and need not be confined to corporate ratings. For example, one has constructed a new PD model for consumer credits and would like to benchmark it against the existing scoring system. The PDiR methodology can generate the PD implied credit scores, and with which operational continuity can be ensured without having to fundamentally alter the existing management infrastructure (credit approval, credit limits, etc) that was built over the years around credit scores.

References

1. Altman, E.: Financial ratios, discriminant analysis and the prediction of corporate Bankruptcy. J. Finan. **23**(4), 589–609 (1968)
2. Duffie, D. and Saita, L. and Wang, K. (2007). Multi-period Corporate Default Prediction with Stochastic Covariates. Journal of Financial Economics, 83
3. Duan, J.C., Fulop, A., Hsieh, Y.W.: Data-cloning SMC^2: a global optimizer for maximum likelihood estimation of latent variable models. Comput. Statist. Data Anal. **143**, 106841 (2020)
4. Duan, J.C., Sun, J., Wang, T.: Multiperiod corporate default prediction–a forward intensity approach. J. Economet. **170**(1), 191–209 (2012)
5. Shumway, T.: Forecasting bankruptcy more accurately: a simple hazard model. J. Bus. **74**, 101–124 (2001)
6. Standard & Poor's: 2018 Annual Global Corporate Default and Rating Transition Study, 9 April 2018. https://www.spratings.com/documents/20184/774196/2018AnnualGlobalCorporateDefaultAndRatingTransitionStudy.pdf
7. NUS-CRI Staff (2020). NUS-RMI Credit Research Initiative Technical Report Version: 2020 Update 1. The Credit Research Initiative at the National University of Singapore. (http://rmicri.org)

Tail Risk Measures and Portfolio Selection

Young C. Joo and Sung Y. Park

Abstract Since Markowitz [13] propose the mean-variance efficient portfolio selection method it has been one of the frequently used approach to the portfolio optimization problem. However, as we know, this approach has critical draw backs such as unstable assets weights and poor forecasting performance due to the estimation error. In this study, we propose an improved portfolio selection rules using various distortion functions. Our approach can make up for the pessimism of economic agents which is important for decision making. We illustrate the procedure by four well-known datasets. We also evaluate the performance of proposed and many other portfolio strategies to compare the in- and out-of-sample value at risk, conditional value at risk and Sharpe ratio. Empirical studies show that the proposed portfolio strategy outperforms many other strategies for most of evaluation measures.

1 Introduction

When many practitioners and researchers considering a risky investment, one of the main goals is finding a portfolio that minimizes risk and maximizes returns. Especially, once we have set the target return, reducing the risk problem is the primary issue. Since Markowits [13] introduced mean-variance (MV) portfolio selection problem, many studies have tried to find appropriate risk measures. As we know, classical MV portfolio use moment-based risk measures such as variance and/or semivariance. These risk measures have some advantages over others that they are very simple to calculate and have well-known characteristics. However, it also has critical drawbacks such as difficulty describing risks of tail events. Most of all, some

Y. C. Joo
Shandong University, 27 Shanda Nanlu, Jinan, China
e-mail: jooyc@sdu.edu.cn

S. Y. Park (✉)
Chung-Ang University, 84 Heukseok-ro, Dongjak-gu, Seoul, Korea
e-mail: sungpark@cau.ac.kr

S. Sriboonchitta et al. (eds.), *Behavioral Predictive Modeling in Economics*, Studies in Computational Intelligence 897,
https://doi.org/10.1007/978-3-030-49728-6_7

times a portfolio has a larger variance than other portfolios even if it is a better portfolio than others (Copeland and Weston [5]). These drawbacks lead to a desire for better and more appropriate risk measures. To characterize the behavior of the risk measure, Artzner et al. [3] suggested four properties of coherent risk measure: monotonicity, subadditivity, linearly homogeneity, and translation invariance. After they proposed a set of properties for the coherent risk measure, many researchers began to study such measure.

Value at Risk (Jorion [11]; Duffie and Pan [8]) is one of the commonly used risk measures. Value at Risk (VaR_α) has an advantage of expressing losses for an amount of money. However, it does not consider the losses when tail events occur and, moreover, is not a coherent risk measure as it fails to satisfy subadditivity. To overcome the drawbacks of VaR_α, Conditional Value at Risk (CVaR_α) has been introduced (Artzner et al. [3]; Rockafellar and Uryasev [14]; Acerbi and Tasche [2]; Bassett et al. [4]). However, CVaR_α also has a limitation in that it only measures risk at certain levels selected by the investors.

To classify risk assessment more systematically, many studies developed risk measures that can be used for the problem of asset allocation (Wang et al. [18]; Artzner et al. [3]; Bassett et al. [4]; De Giorgi [6]). These studies discussed economic agents' preferences towards risk and extended the usual portfolio selection strategy by focusing on distortion risk measures instead of moment-based risk measures. Distortion risk measures can be obtained from the expected utility with a tilted probability measure. One of the great advantages of distortion risk measures is its well-reflected implications in terms of behavior.

Currently, there have only been a few studies considering risk measures in the asset allocation problem. Bassett et al. [4] address the problem of the Choquet expected utility theory on risk assessment and describe its application to portfolio choice. They provide a general approach to pessimistic portfolio allocation. The pessimistic portfolio selection strategy can easily implement any distortion function, satisfying all the properties needing to be coherent. In this study, we use well-known distortion risk measures to generate general pessimistic preference. Moreover, we focus on the portfolio optimization problem empirically.

This study proposes a new approach to the portfolio selection problem by incorporating distortion risk measures. We also evaluate several portfolio selection strategies and show that the performances of the proposed portfolios are better in terms of many evaluation measures: value at risk, conditional value at risk and Sharpe ratio. To evaluate various portfolio strategies, we compare the in- and out-of-sample performance of the portfolios for well-known empirical datasets.

Our study makes several contributions to the portfolio selection under measuring risk. First, its main contribution is showing the relationship between level of risk aversion and risk measures. A distortion function is constructed with a distortion parameter, λ, which represents the level of risk aversion or pessimism. As the λ increases, the magnitude of curvatures of the distortion function increases. We use the non-decreasing and concave distortion functions to develop the portfolio so that riskier events receive more weight than less risky events.

Second, we show the effect of increasing out-of-sample performances of the portfolio rule. Many researchers have pointed out that the conventional MV portfolio strategy shows poor performances due to estimation error. Even though the portfolios demonstrate positive in-sample performances, their out-of-sample performances are poor. However, we show the proposed portfolio strategy outperforms other existing strategies in most in- and out-of-sample cases for empirical datasets. Intuitively speaking, the proposed portfolio strategy can reduce estimation error.

Third contribution of our study is providing a simple way to solve the robust portfolio optimization problem. From the ambiguity perspective, the proposed portfolio optimization strategy is one solution for the robust optimization problem. Basically, the computed values of the risks depend on the distribution of the considered dataset. Many studies assume the loss distribution is known or it is estimated using nonparametric methods (Ghaoui et al. [10]; Delage and Ye [7]; Wozabal [20]). Therefore, a primary benefit of the proposed approach is that it does not require estimating the loss distribution. Portfolio selection based on pessimistic strategy is a robustified method. Moreover, outcomes lie on the same line with solving the minimax problem naturally. Hence, portfolio decisions from distortion risk measures and pessimistic strategies can minimize the risk as well as maximize the return of the portfolio. From the given results, we propose that our portfolio strategy has been shown to not only reduce risk, but also increase return.

The remainder of this study is organized as follows. In Sect. 2, we define the risk and measures of risk that satisfy the four axioms of coherence. In Sect. 3, we present a description of quantile regression methods for measuring distortion risks and show the technical presentation for conditional quantile regression in terms of distortion risk measures. In Sect. 4, we provide empirical applications of portfolio optimization for risk measures, under restrictions, to compare in- and out-of-sample performances. Finally, Sect. 5 offers a conclusion to this paper.

2 Properties of Risk Measures

2.1 Coherence

A risk measure is a function of real-valued random variables, $X \in \mathcal{X}$ on $(\Omega, \mathcal{A})$, representing an uncertainty; $\varrho : \mathcal{X} \to \mathbb{R}$. One of the main characteristics of the risk measure is that an uncertain return can be interpreted as a functional value. In particular, Artzner et al. [3] suggest properties for coherent risk measures that provide the behavior of risk measures:

- Monotonicity: $X, Y \in \mathcal{X}$, with $X \leq Y \Rightarrow \varrho(X) \geq \varrho(Y)$.
- Subadditivity: $X, Y, X + Y \in \mathcal{X} \Rightarrow \varrho(X + Y) \leq \varrho(X) + \varrho(Y)$.
- Linearly homogeneity: For all $\lambda \geq 0$ and $X \in \mathcal{X}$, $\varrho(\lambda X) = \lambda \varrho(X)$.
- Translation invariance: For all $\lambda \in \mathbb{R}$ and $X \in \mathcal{X}$, $\varrho(\lambda + X) = \varrho(X) + \lambda$.

When these four conditions are satisfied, a risk measure is referred to as coherent. Monotonicity refers to a random variable Y having better value than X, wherein the risk measure has the same order. Subadditivity implies that diversification does not increase risk. In other words, a diversified portfolio should reduce risk. Linearly homogeneity means that the risk measure does not change even when the unit of loss or return changes. Finally, translation invariance implies that if a constant, λ, is added to a risk, it is the same as adding a constant, λ, to the risk measure.

There has been much discussion about which risk measure is most relevant in portfolio selection. One of the most commonly used measures of risk is variance of a portfolio's return. Variance has some advantages, such as being simple to understand and calculate. However, it is not systematic. This type of risk measure does not satisfy monotonicity, and thus is not a coherent risk measure.

Recently, a frequently used risk measure has been classical VaR_α. For given random variables X, the VaR_α at the confidence level $\alpha \in (0, 1)$ can be written as

$$VaR_\alpha(X) = F_X^{-1}(\alpha) = \inf\{x \in \mathbb{R} | F_X(x) > \alpha\},$$

where $F_X(\cdot)$ denotes the cumulative probability function of a given random variable X, so We can express VaR_α as the quantile function of X, $F_X^{-1}(\cdot)$. VaR_α can be expressed as a α quantile of money losses. Despite these benefits, VaR_α does not satisfy subadditivity, and therefore is not a coherent risk measure.

The most popular coherent distortion risk measure is CVaR_α (Rockafellar and Uryasev [14]).

$$CVaR_\alpha(X) = -\frac{1}{\alpha}\int_0^\alpha F_X^{-1}(t)dt.$$

CVaR_α can measure the risk under low probability. In addition, it is a non-decreasing and concave coherent risk measure. To quantify the risk in a variety of circumstances, a number of risk measures were proposed: distortion risk measures (Wang [17]), convex risk measures (Föllmer and Shied [9]; Ruszczynski and Shapiro [15]), and spectral risk measures (Acerbi and Simonetti [1]). In this study, We develop a portfolio selection rule using various distortion risk measures.

2.2 Distortion Risk Measures

Applications of distortion risk measures to portfolio optimization started with dual theory (Yaari [21]). Yaari developed a dual theory concerning the behavior toward risk. The idea of probability and risks carries out replacement of the wealth utility function with a probability distortion function. On that basis, distortion risk measures transform the probability distribution in order to measure the risk considering all levels of probabilities.

Let g be a distortion function: g is non-decreasing and concave, g maps from $[0, 1]$ to $[0, 1]$ and $g(0) = 0$ and $g(1) = 1$. Distortion risk measures re-weigh the probabilities of expected utilities and create an inflated probability of the worst outcome. Wirch and Hardy [19] have shown that for positive loses, if the distortion function is concave, then a distortion risk measure is coherent.

It is of great importance to choose the distortion function because different risk measures originate from the chosen distortion function. Distortion risk measures developed by Yaari [21] measure risk by reflecting a distortion function, g, on the cumulative distribution function, $F_X(x)$. Applying the Choquet integral, one can write a distortion risk measure ϱ_g

$$\varrho_g(X) = -\int_{-\infty}^{\infty} x dg(F_X(x)) \tag{1}$$

$$= -\int_{-\infty}^{0} [1 - g(\bar{F}_X(x))]dx + \int_{0}^{\infty} g(\bar{F}_X(x))dx, \tag{2}$$

where $\bar{F}_X(x) = 1 - F_X(x)$.

The quantile based risk measure is one of the most commonly used distortion risk measures. For example, the distortion function of $\text{VaR}_\alpha(X)$ can be shown as $g_\alpha(u) = \mathbb{1}(1-\alpha, 1](u)$ by which a Choquet integral can be expressed

$$\text{VaR}_\alpha(X) = \int_{-\infty}^{0} [1 - g(P(F_X(x) \geq 1 - \alpha))]dx + \int_{0}^{\infty} g(P(F_X(x) < 1 - \alpha))dx = F_X^{-1}(1-\alpha).$$

$$g_{VaR}(u) = \begin{cases} 0, & \text{if } 0 \leq u < 1 - \alpha \\ 1, & \text{if } 1 - \alpha \leq u \leq 1. \end{cases}$$

The distortion function defining the CVaR_α can be denoted by $g_\alpha(u) = min \{u/(1-\alpha), 1\}$:

$$g_{CVaR}(u) = \begin{cases} \frac{u}{1-\alpha}, & \text{if } 0 \leq u < 1 - \alpha \\ 1, & \text{if } 1 - \alpha \leq u \leq 1. \end{cases}$$

Both distortion functions, $g_{VaR}(\cdot)$ and $g_{CVaR}(\cdot)$, are non-decreasing. However, $g_{VaR}(\cdot)$ is not a concave function, and therefore, it is not a coherent risk measure. By contrast, since $g_{CVaR}(\cdot)$ is continuous and concave, it is coherent.

One notable aspect of the distortion functions of VaR_α and CVaR_α is that they do not consider risks above the α level. However, the following distortion functions assign weights to all levels of probabilities. A simple example of distortion function is a one-parameter family (One), hereafter

$$g_{One,\lambda}(u) = 1 - (1-u)^\lambda,$$

where λ, $-\infty \leq \lambda \leq \infty$ and $\lambda \geq 1$, is a measure of risk aversion. In the process, $g_{One,\lambda}(\cdot)$ becomes more distorted as $\lambda \to \infty$.

The power distortion function (Power), hereafter has a similar functional form with the one-parameter family

$$g_{Power,\lambda}(u) = (u)^{\lambda}, \quad 0 \leq \lambda \leq 1.$$

The main difference with One and Power is the shape of functions as λ changes. These two functions distort in different directions. When λ increases, One has a more distorted shape but Power has a less distorted shape. Wang [17] suggested a distortion risk measure of a parametric family known as Wang's transform (Wang), hereafter

$$g_{Wang,\lambda}(u) = \Phi[\Phi^{-1}(u) + \lambda], \quad u \in [0, 1],$$

where Φ denotes a standard normal distribution, and λ reflects the risk premium. Wang equally applies to assets and losses with λ. Especially, Wang is the major distortion function unifying economic, financial, and actuarial pricing theories. It shows how the mean of distorted distribution function can be used as the distortion risk measure. The shape of the distortion functions is determined by a risk premium parameter λ. When λ get a higher (lower) value, the distortion function becomes more (less) distorted in shape.

3 Portfolio Selection Strategies

We will illustrate the extension of general pessimistic portfolio allocation and coherent distortion risk measures. The aim of Bassett et al. [4], who coined pessimistic portfolio allocation, is to apply Choquet expected utility theory to the portfolio choice problem. In contrast to conventional portfolio optimization problems solving least squares, they consider quantile regression.

Given a return series x_{it} for an asset i and time t, $i = 1, \cdots, N$, $t = 1, \cdots, T$ on X, and benchmark, Y, we can say $Y = X'\pi$ is a portfolio of assets with a weight vector π. To minimize risk subject to a constraint on the mean return, they use α-risk. Specifically, the measure of risk in this literature is the α-risk, $\varrho_{g_\alpha}(X)$, of a random sample X, and is the negative Choquet g_α expected return. The α-risk can be formulated as

$$\hat{\rho}_{g_\alpha} = (n\alpha) \min_{\xi \in \mathbb{R}} \sum_{i=1}^{n} \rho_\alpha(x_i - \xi) - \hat{\mu}_n, \tag{3}$$

where $\hat{\mu}_n$ is an estimator of $E(X) = \mu$. These equations can be expressed as

$$\min_{\pi} \varrho_{g_\alpha}(X'\pi). \tag{4}$$

$$s.t. \quad \mu(X'\pi) = \mu_0 \quad and \quad \mathbf{1}'\pi = 1.$$

Taking the first asset as numeraire, the objective function of the linear quantile regression problem can be written as

$$\min_{(\beta,\xi)\in\mathbb{R}^p} \sum_{i=1}^{n} \rho_\alpha(x_{i1} - \sum_{j=2}^{p}(x_{i1} - x_{ij})\beta_j - \xi), \tag{5}$$

$$s.t. \quad \bar{x}'\pi(\beta) = \mu_0,$$

where ξ denotes α a sample quantile of the considered portfolio returns distribution, μ_0 is the target return and $\pi(\beta) = (1 - \Sigma_{j=2}^{p}\beta_j, \beta')'$. (5) is a type of linear quantile regression theorized by Koenker and Bassett [12]. Following the (5), they solve a linear function estimating α a conditional quantile function. The estimators $\hat{\xi}$ and $\hat{\beta}$ are the α sample quantile of the chosen portfolio's returns distribution and weights of the portfolio, respectively. They estimate mean-variance portfolio before finding the empirical α-risk given μ_0. We denote quantile regression portfolio optimization without a target return, μ_0, as CVaR, and with a target return, μ_0, as CVaR.R.

Consider the combination of distortion risk measures and the quantile regression problem. When distortion risk measures are all positive, they consist in weighted averages with the quantiles of the returns distribution. The weighted averages of quantile risks can be expressed as

$$\varrho_g(X) = \sum_{k=1}^{m} g_k \varrho_{g_{\alpha k}}(X), \tag{6}$$

$$\min_{(\beta,\xi)\in\mathbb{R}^{p+m}} \sum_{k=1}^{m}\sum_{i=1}^{n} g_k \rho_{\alpha k}(x_{i1} - \sum_{j=2}^{p}(x_{i1} - x_{ij})\beta_j - \xi_k), \tag{7}$$

$$s.t. \quad \bar{x}'\pi(\beta) = \mu_0,$$

where g_k denotes positive weights: $k = 1, \cdots, m$ and $\mathbf{1}'g_k = 1$. Note that general quantile risks are fundamental building blocks more so than any coherent distortion risk measure represented by a piecewise linear concave function (Bassett et al. [4]). Equation (7) provides an efficient way to optimally combine distortion risk measures with quantile regression problems. This approach is easy to carry out; since the distortion parameters are all positive, the $\rho_{\alpha k}$ function rescales the argument under the weights, g_k. In order to see the relationship between risk aversion parameter λ and performances of the portfolio, we use five levels of λ, $\lambda = (0.1, 0.25, 0.5, 0.75, 0.9)$ for Power and Wang, and $\lambda = (1.1, 5.5, 15, 35, 50)$ for One.

For a portfolio selection based on distortion risk measures, we use three types of distortion functions: one-parameter family (One), Power distortion function (Power), and Wang's transformation (Wang). In addition, we consider the case of a given target return constraint, μ_0, and denote each portfolio as One.R, Power.R, and Wang.R (Table 1).

Table 1 List of portfolio optimization models

Model	Abbreviation
From the previous studies	
Equally weighted	Naive
Mean variance	MV
Minimum variance	MinV
Ledoit and Wolf's (2003) shrinkage	LW
Minimize $\text{CVaR}_{0.05}$	MinCVaR
Minimize $\text{CVaR}_{0.05}$ with target return	MinCVaR.R
Developed in this study	
Wang's transformation	Wang
Wang's transformation with target return	Wang.R
One parameter family	One
One parameter family with target return	One.R
Power distortion function	Power
Power distortion function with target return	Power.R

Notes This table lists the portfolio optimization strategies we use. The upper six list is portfolio selection strategies from the previous studies and the lower six list is portfolio selection strategies developed in this study. Note that we get the portfolio target return from the equally weighted portfolio given in-sample period

To compare the performance of each portfolio selection strategy, we consider the most common ones. These portfolios assume investors estimate asset return, μ, and covariance matrix, Σ, using historical return data. First, the equally weighted portfolio (Naive) strategy involves holding equal weight $w = 1/N$ in each of the N risky assets. This strategy does not need any optimization or estimation.

Markowitz's [13] mean-variance (MV) efficient portfolio is one of the most common formulations of the portfolio selection problem. In the MV optimal portfolio, the investor optimizes the tradeoff between the mean and variance of portfolio returns. The unbounded MV optimization problem, $\max_w w'\mu - w'\Sigma w,\ w'\mathbf{1} = 1$, is the quadratic utility function of a rational investor (Scherer 2002).

Minimum variance portfolios (MinV) have been used to exclude means in the optimizer. Since the minimum variance portfolio strategy does not require an expected return in the optimization process, it can be written as $\min_w w'\Sigma w,\ w'\mathbf{1} = 1$. If the investor ignores expected returns, they can replace the estimated return vector with a vector of one.

The fourth strategy is the shrinkage type portfolio selection rule. Ledoit and Wolf (2003) suggest shrinkage covariance matrix estimator (Σ_{LW}). They consider the sample covariance matrix Σ_S and a highly structured estimator Σ_T, then estimate $\hat{\Sigma}_{LW}$ by shrinking Σ_S toward Σ_T. Ledoit and Wolf's shrinkage estimator can be estimated as $\hat{\Sigma}_{LW} = \delta\Sigma_T + (1-\delta)\Sigma_S$, where δ is the shrinkage constant, $\delta \in [0, 1]$. Ledoit and Wolf's shrinkage type portfolio can be written as $\min_w w'\hat{\Sigma}_{LW} w,\ w'\mathbf{1} = 1$, and we denote this shrinkage type portfolio by LW.

Table 2 List of datasets

Dataset	N	T	Range	Abbreviation
6 portfolios sorted by size and book-to-market	6	648	1964. 01–2017. 12	6 BM
10 U.S. industry portfolios	10	648	1964. 01–2017. 12	10 IP
25 portfolios sorted by size and book-to-market	25	648	1964. 01–2017. 12	25 BM
48 U.S. industry portfolios	48	648	1964. 01–2017. 12	48 IP

Notes This table lists the datasets we use. We obtained the all datasets from Kenneth French's homepage. We used the one month U.S. T-Bill to compute excess returns. In columns 2 and 3, N and T denote the number of risky assets and observations in each data set, respectively. All datasets have the same time period spanned, January 1964–December 2017. The abbreviations are given in the last column

4 Empirical Analysis

4.1 Empirical Datasets and Evaluation Measures

To compare the in- and out-of-sample performance of portfolio selection strategies, we use four excess monthly return datasets, 6 and 25 portfolios sorted by size and book-to-market (6BM and 25BM) and 10 and 48 U.S. industry portfolios (10IP and 48IP). We use the one month U.S. T-Bill to compute excess returns. The datasets cover the same time period from January 1964 to December 2017 (Table 2).[1]

At first, to evaluate the performance of each strategy, we calculate the in- and out-of-sample Sharpe ratio (SR). In order to compare the results, we use a rolling window scheme. We consider two window lengths, $W = 120$ and 240. For the in-sample cases, evaluation measures depend on the estimated parameters over the chosen window. The average of in-sample estimated SR is given

$$\mathrm{SR}_{in} = \frac{1}{(T-W)} \sum_{t=W}^{T} \frac{\hat{w}_t' \hat{\mu}_t}{(\hat{w}_t' \hat{\Sigma}_t \hat{w}_t)^{1/2}}, \tag{8}$$

where $\hat{\mu}_t$, $\hat{\Sigma}_t$, and $\hat{w}_t$ denote the estimated mean vector, covariance matrix, and the portfolio weight vector at time t for the window $[t-W+1, t]$, respectively. The out-of-sample portfolio returns of the resulting portfolio are based on each asset's next period return. Following a rolling window scheme, the out-of-sample portfolio return at time $t+1$ is calculated by $\hat{r}_{t+1} = \hat{w}_t' \hat{\mu}_{t+1}$, where $\hat{\mu}_{t+1}$ denotes the return at time $t+1$. The out-of-sample SR can be written as

[1] All datasets are obtained from Kenneth French's homepage.

$$\tilde{m} = \frac{1}{(T-W)} \sum_{i=W}^{T} \hat{r}_{i,t+1}, \tag{9}$$

$$\tilde{\sigma}^2 = \frac{1}{(T-W-1)} \sum_{i=W}^{T} (\hat{r}_{i,t+1} - \tilde{m})^2, \tag{10}$$

$$\mathrm{SR}_{out} = \frac{\tilde{m}}{\tilde{\sigma}}, \tag{11}$$

To compare the risk of each portfolio, we compute the in- and out-of-sample $\mathrm{VaR}_{0.05}$ and $\mathrm{CVaR}_{0.05}$,

$$\mathrm{VaR}_{in} = \frac{1}{(T-W)} \sum_{t=W}^{T} VaR_{0.05}(\hat{r}_t), \tag{12}$$

$$\mathrm{VaR}_{out} = VaR_{0.05}(\tilde{m}), \tag{13}$$

$$\mathrm{CVaR}_{in} = \frac{1}{(T-W)} \sum_{t=W}^{T} CVaR_{0.05}(\hat{r}_t), \tag{14}$$

$$\mathrm{CVaR}_{out} = CVaR_{0.05}(\tilde{m}), \tag{15}$$

4.2 VaR and CVaR

Table 3 reports in- and out-of-sample $\mathrm{VaR}_{0.05}$ for $W = 120$. Four columns on the left side are in-sample and on the right side are out-of-sample results. The top six rows give the results from existing models compared to those of my portfolio optimization model, presented in the following rows. First of all, we find that the pattern of the performances of the portfolios move in the same direction as the level of distortion.

Overall, Naive has higher $\mathrm{VaR}_{0.05}$ than other portfolios in both in- and out-of-sample cases. One3 outperforms all other portfolios in every in-sample case. Note that using distortion risk measures reduces the $\mathrm{VaR}_{0.05}$ in both cases. $\mathrm{VaR}_{0.05}$ from the $W = 240$ reported in Table 4. The main differences with the $W = 120$ case are: first, increasing the number of observations reduces the $\mathrm{VaR}_{0.05}$ for all portfolios; and second, no more One3 outperforms all other portfolios. Specifically, from the in-sample data, Power3 has the lowest $\mathrm{VaR}_0.05$ in 6BM and 10IP, and One3 and One.R3 have the lowest $\mathrm{VaR}_{0.05}$ in 25BM and 48IP, respectively. However, for the out-of-sample cases, my model outperforms other portfolios in 6BM and 25BM. As shown in the results, distortion risk measures reduce $\mathrm{VaR}_{0.05}$ regardless of window length and target return restrictions. In order to see the risk more specifically, we also look at the $\mathrm{CVaR}_{0.05}$.

Tables 5 and 6 give the $\mathrm{CVaR}_{0.05}$ of portfolio returns for $W = 120$ and 240. From the in-sample results of $W = 120$, we find that target return restriction increases the $\mathrm{CVaR}_{0.05}$ in most cases. However, distortion risk measures reduce the risks for

Table 3 Value at risk total ($\alpha = 0.05$, W = 120)

	In-sample				Out-of-sample			
	6BM	10IP	25BM	48IP	6BM	10IP	25BM	48IP
Naive	0.0698	0.0777	0.0715	0.0701	0.0715	0.0771	0.0737	0.0651
MV	0.0517	0.0428	0.0392	0.0317	0.0607	0.0524	0.0544	0.0638
MinV	0.0461	0.0426	0.0340	0.0324	0.0512	0.0555	0.0560	0.0627
LW	0.0541	0.0452	0.0507	0.0361	0.0658	0.0561	0.0672	**0.0539**
MinCVaR	0.0493	0.0459	0.0311	0.0214	0.0616	0.0621	0.0656	0.0759
MinCVaR.R	0.0542	0.0440	0.0362	0.0224	0.0634	0.0555	0.0571	0.0696
Wang1	0.0478	0.0438	0.0388	0.0428	0.0514	0.0582	0.0533	0.0660
Wang2	0.0472	0.0434	0.0373	0.0404	0.0510	0.0556	**0.0530**	0.0645
Wang3	0.0468	0.0430	0.0360	0.0373	0.0523	0.0540	0.0556	0.0644
Wang4	0.0465	0.0424	0.0349	0.0343	0.0532	0.0530	0.0573	0.0660
Wang5	0.0463	0.0422	0.0342	0.0326	0.0518	0.0526	0.0581	0.0656
One1	0.0491	0.0457	0.0423	0.0491	0.0521	0.0608	0.0573	0.0718
One2	0.0457	0.0422	0.0314	0.0272	0.0487	0.0529	0.0554	0.0734
One3	**0.0447**	**0.0400**	**0.0246**	**0.0206**	0.0541	0.0548	0.0550	0.0751
One4	0.0475	0.0437	0.0307	0.0214	0.0594	0.0578	0.0656	0.0759
One5	0.0506	0.0473	0.0312	0.0214	0.0606	0.0603	0.0652	0.0759
Power1	0.0452	0.0421	0.0309	0.0259	**0.0492**	0.0543	0.0561	0.0683
Power2	0.0454	0.0422	0.0312	0.0272	0.0494	0.0535	0.0551	0.0673
Power3	0.0458	0.0422	0.0326	0.0290	0.0503	0.0526	0.0556	0.0681
Power4	0.0462	0.0424	0.0339	0.0317	0.0510	0.0532	0.0579	0.0651
Power5	0.0465	0.0427	0.0346	0.0344	0.0518	0.0537	0.0562	0.0639
Wang.R1	0.0593	0.0488	0.0516	0.0458	0.0670	0.0582	0.0659	0.0664
Wang.R2	0.0553	0.0452	0.0469	0.0417	0.0624	0.0533	0.0617	0.0676
Wang.R3	0.0525	0.0434	0.0427	0.0374	0.0600	0.0545	0.0580	0.0669
Wang.R4	0.0518	0.0429	0.0400	0.0336	0.0587	0.0540	0.0580	0.0640
Wang.R5	0.0516	0.0428	0.0391	0.0319	0.0595	0.0529	0.0561	0.0653
One.R1	0.0588	0.0493	0.0514	0.0464	0.0666	0.0603	0.0687	0.0693
One.R2	0.0518	0.0429	0.0376	0.0275	0.0572	**0.0517**	0.0546	0.0719
One.R3	0.0508	**0.0400**	0.0313	0.0199	0.0575	0.0550	0.0542	0.0693
One.R4	0.0526	0.0429	0.0328	0.0210	0.0609	0.0567	0.0602	0.0688
One.R5	0.0553	0.0463	0.0337	0.0212	0.0618	0.0590	0.0588	0.0699
Power.R1	0.0521	0.0428	0.0377	0.0291	0.0612	0.0542	0.0586	0.0636
Power.R2	0.0524	0.0424	0.0394	0.0298	0.0597	0.0555	0.0586	0.0641
Power.R3	0.0540	0.0434	0.0424	0.0327	0.0643	0.0553	0.0602	0.0638
Power.R4	0.0576	0.0467	0.0483	0.0383	0.0662	0.0581	0.0630	0.0608
Power.R5	0.0606	0.0509	0.0532	0.0450	0.0670	0.0585	0.0716	0.0689

Notes This table reports in- and out-of-sample $\mathrm{VaR}_{0.05}$ of each portfolio selection rules for 120 window length. The numbers in bold indicate the best results in the portfolio strategies. The numbers following the abbreviations are the level of distortion and .R denotes the portfolio with a target return restriction, respectively

Table 4 Value at risk total ($\alpha = 0.05$, $W = 240$)

	In-sample				Out-of-sample			
	6BM	10IP	25BM	48IP	6BM	10IP	25BM	48IP
Naive	0.0689	0.0760	0.0700	0.0671	0.0712	0.0742	0.0720	0.0633
MV	0.0524	0.0460	0.0431	0.0391	0.0581	0.0457	0.0505	0.0595
MinV	0.0465	0.0445	0.0371	0.0398	0.0540	**0.0454**	0.0451	0.0601
LW	0.0524	0.0447	0.0416	0.0409	0.0646	0.0490	0.0553	**0.0535**
MinCVaR	0.0503	0.0465	0.0397	0.0349	0.0598	0.0527	0.0595	0.0735
MinCVaR.R	0.0554	0.0465	0.0414	0.0348	0.0635	0.0530	0.0573	0.0654
Wang1	0.0488	0.0446	0.0402	0.0457	0.0522	0.0479	0.0483	0.0588
Wang2	0.0472	0.0445	0.0393	0.0436	0.0517	0.0485	0.0478	0.0582
Wang3	0.0467	0.0445	0.0383	0.0419	0.0501	0.0483	0.0479	0.0578
Wang4	0.0466	0.0444	0.0377	0.0405	0.0491	0.0478	0.0458	0.0585
Wang5	0.0465	0.0443	0.0373	0.0397	0.0514	0.0469	**0.0450**	0.0592
One1	0.0493	0.0447	0.0429	0.0493	0.0536	0.0479	0.0481	0.0626
One2	0.0468	0.0439	0.0362	0.0376	0.0518	0.0468	0.0452	0.0592
One3	0.0468	0.0429	**0.0317**	0.0309	0.0591	0.0466	0.0513	0.0677
One4	0.0490	0.0442	0.0364	0.0345	0.0598	0.0478	0.0545	0.0737
One5	0.0509	0.0470	0.0441	0.0353	0.0593	0.0526	0.0601	0.0735
Power1	0.0469	0.0442	0.0353	0.0364	0.0551	0.0456	0.0461	0.0587
Power2	0.0467	0.0443	0.0356	0.0369	0.0525	0.0464	0.0461	0.0589
Power3	**0.0464**	**0.0441**	0.0364	0.0382	0.0512	0.0462	**0.0450**	0.0605
Power4	0.0480	0.0442	0.0377	0.0399	0.0549	0.0460	0.0452	0.0593
Power5	0.0499	0.0450	0.0383	0.0410	**0.0487**	0.0482	0.0455	0.0598
Wang.R1	0.0588	0.0503	0.0545	0.0486	0.0539	0.0502	0.0658	0.0645
Wang.R2	0.0562	0.0472	0.0488	0.0451	0.0552	0.0464	0.0575	0.0611
Wang.R3	0.0536	0.0463	0.0449	0.0420	0.0529	0.0475	0.0537	0.0594
Wang.R4	0.0523	0.0463	0.0432	0.0402	0.0519	0.0486	0.0511	0.0587
Wang.R5	0.0521	0.0462	0.0425	0.0392	0.0548	0.0466	0.0502	0.0597
One.R1	0.0585	0.0503	0.0541	0.0499	0.0535	0.0541	0.0590	0.0646
One.R2	0.0525	0.0463	0.0414	0.0378	0.0614	0.0491	0.0497	0.0611
One.R3	0.0523	0.0448	0.0379	**0.0303**	0.0607	0.0498	0.0548	0.0650
One.R4	0.0548	0.0459	0.0400	0.0339	0.0646	0.0518	0.0590	0.0661
One.R5	0.0580	0.0484	0.0422	0.0347	0.0674	0.0533	0.0629	0.0655
Power.R1	0.0538	0.0462	0.0432	0.0370	0.0603	0.0495	0.0526	0.0612
Power.R2	0.0534	0.0459	0.0434	0.0368	0.0603	0.0491	0.0510	0.0616
Power.R3	0.0546	0.0459	0.0451	0.0389	0.0567	0.0510	0.0565	0.0609
Power.R4	0.0575	0.0485	0.0508	0.0427	0.0553	0.0519	0.0601	0.0609
Power.R5	0.0603	0.0522	0.0564	0.0481	0.0529	0.0539	0.0648	0.0621

Notes This table reports in- and out-of-sample $VaR_{0.05}$ of each portfolio selection rules for 240 window length. The numbers in bold indicate the best results in the portfolio strategies. The numbers following the abbreviations are the level of distortion and .R denotes the portfolio with a target return restriction, respectively

25BM and 48IP. In the 6BM and 10IP cases, MinCVaR has the lowest in-sample $\text{CVaR}_{0.05}$. The $\text{CVaR}_{0.05}$ of out-of-sample shows different results. From the 6BM and 25BM, One3 and Power5 have the lowest $\text{CVaR}_{0.05}$ but from the 10IP and 48IP, MV and LW show the best results. In the in-sample case of $W = 240$, MinCVaR has lower $\text{CVaR}_{0.05}$ than other portfolios for 6BM, 10IP, and 25BM, but in the 48IP, One.R5 outperforms in other portfolios. For the out-of-sample cases, my portfolio selection rules have lower $\text{CVaR}_{0.05}$ in 6BM, 10IP, and 25BM. However, in the 48IP, LW has the lowest value of $\text{CVaR}_{0.05}$. Note that even if a portfolio outperforms all other portfolios in the in-sample, it does not correlate to a better outcome in the out-of-sample. For instance, when the level of distortion increases, most of the $\text{CVaR}_{0.05}$ decreases for in-sample cases, but this trend disappears in out-of-sample cases.

From Figs. 1 and 2, we can see changes in the trend of the in-sample portfolio return's $\text{CVaR}_{0.05}$ density as changing the level of distortion. From the first row, figures present the density of in-sample $\text{CVaR}_{0.05}$ of portfolio returns for each dataset. Each column is composed of Wang, One, and Power. The solid black lines indicate the results of the lowest level of distortion for all figures. We can see a common trend in Wang and One that the $\text{CVaR}_{0.05}$ densities move toward the left side as the distortion level increases, while the densities from Power move toward the right. The reason for this movement of densities in opposite directions is due to the difference in distortion functions. As the level of the distortion of a function increases, the higher risk assigns higher weights in the Wang and One models, but gives lower weights from Power. Therefore, as the distortion level, λ, increases, the densities of $\text{CVaR}_{0.05}$ from Wang and One move to the left side and Power moves to the right side.

4.3 Sharpe Ratio

Before taking a look at the Sharpe ratio, it is useful to see the densities of portfolio returns. Figures 3 and 4 present in- and out-of-sample portfolio return densities with 120 window length. Overall, figures show that the in-sample portfolio returns are increasing as the level of distortion increases but there are more substantial changes of $\text{CVaR}_{0.05}$ This pattern is getting clearer as observations increase, but it disappears in out-of-sample cases. However, from the figures, we find that the pessimistic portfolio improves the performances on both tail part of the portfolio return densities.

Tables 7 and 8 present the SR for $W = 120$ and 240, respectively. From the in-sample cases with $W = 120$, we find that Power.R1 outperforms other portfolios in 6BM, 10IP, and 25BM and Wang1 outperforms others in 48IP. Unfortunately, this improvement for the in-sample does not lead to the out-of-sample cases. In these, the SR of Power1 shows the highest value for 6BM and 25BM, while for the 10IP, Wang.R5 has the highest SR. However, for the 48IP, Naive shows the best performance. In Table 8 shows the effect of relatively long estimation windows, $W = 240$. The overall SR increases as the number of observations increases. In the case of in-sample, Naive and MV has the highest SR for 6BM and 25BM, respectively. For 10IP, Power.R1 outperforms other portfolios and for 48IP, LW shows the best

Table 5 Conditional value at risk total ($\alpha = 0.05$, W = 120)

	In-sample				Out-of-sample			
	6BM	10IP	25BM	48IP	6BM	10IP	25BM	48IP
Naive	2.2429	2.4964	2.3190	2.2074	2.1927	2.4193	2.2631	2.1509
MV	1.5688	1.2562	1.1358	0.8575	1.7161	**1.4837**	1.6464	1.7668
MinV	1.4135	1.2636	1.0234	0.8662	1.5638	1.5336	1.5841	1.7942
LW	1.5442	1.2797	1.3597	0.9542	1.7605	1.4929	1.7054	**1.5528**
MinCVaR	**1.1848**	**1.0190**	0.6240	0.4289	1.6005	1.7009	1.9749	2.1386
MinCVaR.R	1.5369	1.1932	0.9620	0.4978	1.7798	1.5254	1.7312	1.9715
Wang1	1.5647	1.3878	1.2312	1.2770	1.6540	1.6062	1.5706	1.9069
Wang2	1.5482	1.3682	1.2032	1.2176	1.6399	1.5866	1.5790	1.8766
Wang3	1.5246	1.3448	1.1641	1.1340	1.6207	1.5588	1.5708	1.8367
Wang4	1.4987	1.3241	1.1208	1.0503	1.6015	1.5467	1.5679	1.8176
Wang5	1.4817	1.3110	1.0938	0.9980	1.5876	1.5416	1.5695	1.8172
One1	1.6017	1.4318	1.3302	1.4440	1.6793	1.6286	1.6182	2.0004
One2	1.4346	1.2770	1.0245	0.8955	1.5383	1.5320	1.5599	1.9490
One3	1.2982	1.1441	0.7604	0.4463	**1.5240**	1.5488	1.7153	2.0843
One4	1.2038	1.0373	**0.6224**	0.4289	1.5663	1.6078	1.9326	2.1386
One5	1.1995	1.0245	0.6238	0.4289	1.6001	1.7660	1.9715	2.1386
Power1	1.3660	1.2158	0.9098	0.6495	1.5316	1.5228	1.6052	1.8580
Power2	1.3966	1.2408	0.9576	0.7376	1.5417	1.5293	1.5932	1.8605
Power3	1.4396	1.2759	1.0239	0.8661	1.5630	1.5392	1.5776	1.8445
Power4	1.4748	1.3068	1.0805	0.9764	1.5847	1.5511	1.5617	1.8138
Power5	1.5063	1.3348	1.1122	1.0431	1.6106	1.5805	**1.5486**	1.7953
Wang.R1	1.8173	1.4566	1.5093	1.3079	1.8640	1.6348	1.8325	1.8913
Wang.R2	1.7426	1.3909	1.4062	1.2150	1.8510	1.5426	1.7307	1.8564
Wang.R3	1.6574	1.3380	1.2920	1.1109	1.7877	1.5415	1.6263	1.8081
Wang.R4	1.6188	1.3063	1.2147	0.9946	1.7554	1.5073	1.6461	1.7449
Wang.R5	1.6013	1.2897	1.1802	0.9379	1.7359	1.5070	1.6305	1.7734
One.R1	1.8085	1.4738	1.5133	1.3356	1.8938	1.6679	1.8042	1.8893
One.R2	1.5742	1.2667	1.1489	0.8699	1.7076	1.5037	1.5798	1.8479
One.R3	1.4799	1.1400	0.9108	0.4707	1.6124	1.5001	1.6251	1.9318
One.R4	1.4046	1.0476	0.7851	**0.4278**	1.6108	1.5433	1.7548	1.9270
One.R5	1.3985	1.0387	0.7696	0.4284	1.6623	1.5562	1.7627	1.9365
Power.R1	1.5130	1.1739	0.9860	0.6398	1.6747	1.4990	1.6844	1.7829
Power.R2	1.5697	1.2215	1.0734	0.7222	1.7053	1.5121	1.6935	1.7707
Power.R3	1.6673	1.3035	1.2347	0.8982	1.7859	1.5479	1.7113	1.7914
Power.R4	1.7810	1.3947	1.4141	1.0913	1.8463	1.6327	1.8289	1.8000
Power.R5	1.8496	1.4936	1.5381	1.2696	1.9101	1.6914	1.8777	1.9007

Notes This table reports in- and out-of-sample $CVaR_{0.05}$ of each portfolio selection rules for 120 window length. The numbers in bold indicate the best results in the portfolio strategies. The numbers following the abbreviations are the level of distortion and .R denotes the portfolio with a target return restriction, respectively

Table 6 Conditional value at risk total ($\alpha = 0.05$, W = 240)

	In-sample				Out-of-sample			
	6BM	10IP	25BM	48IP	6BM	10IP	25BM	48IP
Naive	2.2402	2.4723	2.3205	2.2001	2.1886	2.4079	2.2572	2.1228
MV	1.6354	1.3490	1.3103	1.0887	1.7804	1.3603	1.6309	1.5965
MinV	1.4665	1.3211	1.1748	1.1089	1.5848	1.4348	1.4575	1.5935
LW	1.5655	1.3298	1.2579	1.1291	1.8711	1.4562	1.6023	**1.5051**
MinCVaR	**1.3324**	**1.1397**	**0.9051**	0.7118	1.7417	1.5163	1.9323	2.1311
MinCVaR.R	1.6048	1.2658	1.1513	0.7626	1.9115	1.4880	1.7037	1.9288
Wang1	1.5920	1.4195	1.2985	1.3449	1.6206	1.5137	1.4709	1.6298
Wang2	1.5624	1.4096	1.2815	1.3114	1.6100	1.4972	1.4677	1.6222
Wang3	1.5424	1.3928	1.2570	1.2653	1.6021	1.4847	1.4572	1.6042
Wang4	1.5227	1.3747	1.2326	1.2235	1.5904	1.4720	1.4502	1.5865
Wang5	1.5103	1.3631	1.2174	1.1969	1.5867	1.4636	1.4452	1.5788
One1	1.6157	1.4449	1.3536	1.4312	1.6294	1.5283	1.5179	1.6561
One2	1.4742	1.3277	1.1745	1.1359	**1.5662**	1.4390	**1.4174**	1.6315
One3	1.4087	1.2410	1.0514	0.8551	1.6143	1.3707	1.4782	1.8457
One4	1.3555	1.1622	0.9139	0.7133	1.7184	1.4301	1.8196	2.1165
One5	1.3470	1.1490	0.9204	0.7104	1.7666	1.5153	1.9661	2.1537
Power1	1.4330	1.2907	1.1195	1.0024	1.5775	1.4099	1.4447	1.6584
Power2	1.4500	1.3088	1.1458	1.0522	1.5735	1.4255	1.4400	1.6279
Power3	1.4746	1.3347	1.1823	1.1237	1.5718	1.4494	1.4418	1.6042
Power4	1.5293	1.3569	1.2145	1.1845	1.6034	1.4690	1.4451	1.5916
Power5	1.5962	1.3786	1.2341	1.2246	1.5771	1.5051	1.4461	1.5954
Wang.R1	1.8502	1.5220	1.6101	1.3776	1.8771	1.5064	1.9821	1.7046
Wang.R2	1.7708	1.4292	1.4898	1.3049	1.8393	1.4224	1.7636	1.6609
Wang.R3	1.6969	1.3992	1.3992	1.2354	1.7607	**1.3570**	1.6570	1.6292
Wang.R4	1.6683	1.3851	1.3548	1.1817	1.7636	1.3691	1.6141	1.5981
Wang.R5	1.6558	1.3768	1.3373	1.1517	1.7843	1.3728	1.5936	1.5813
One.R1	1.8505	1.5350	1.6064	1.4218	1.8872	1.4913	1.8780	1.7500
One.R2	1.6307	1.3574	1.3136	1.1162	1.7562	1.4259	1.6041	1.5687
One.R3	1.5824	1.2772	1.1773	0.8262	1.7994	1.3905	1.6581	1.8352
One.R4	1.5542	1.2068	1.0773	0.7053	1.8697	1.4823	1.7908	1.8743
One.R5	1.5533	1.1997	1.0604	**0.7051**	1.9486	1.5527	1.9147	1.9637
Power.R1	1.5912	1.2907	1.2040	0.9215	1.8161	1.3960	1.7125	1.7086
Power.R2	1.6267	1.3211	1.2610	0.9961	1.7874	1.3912	1.7134	1.6458
Power.R3	1.6996	1.3560	1.3621	1.0943	1.8038	1.4331	1.7657	1.5954
Power.R4	1.8102	1.4444	1.5138	1.2125	1.8369	1.4779	1.8847	1.6172
Power.R5	1.8674	1.5724	1.6505	1.3480	1.8262	1.5708	1.9596	1.7656

Notes This table reports in- and out-of-sample $CVaR_{0.05}$ of each portfolio selection rules for 240 window length. The numbers in bold indicate the best results in the portfolio strategies. The numbers following the abbreviations are the level of distortion and .R denotes the portfolio with a target return restriction, respectively

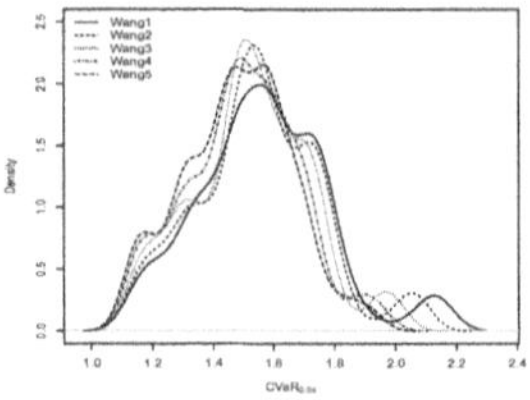

(a) Wang-6BM

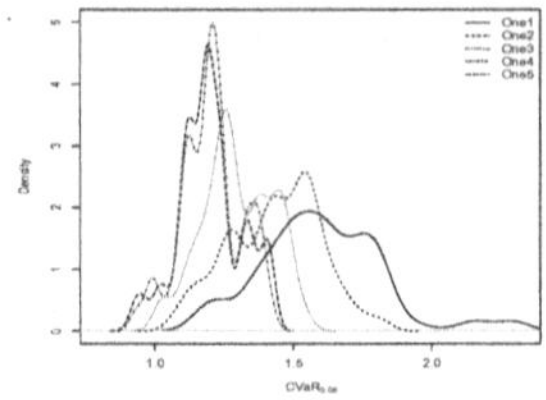

(b) One-6BM

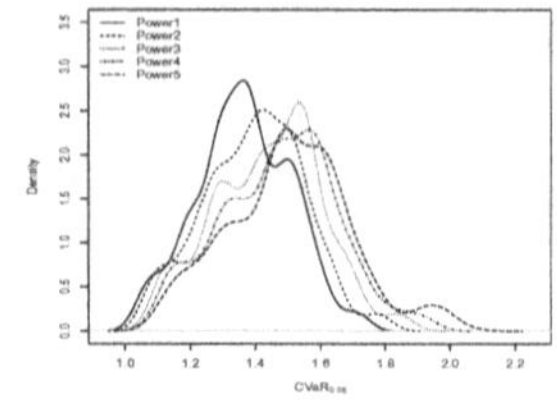

(c) Power-6BM

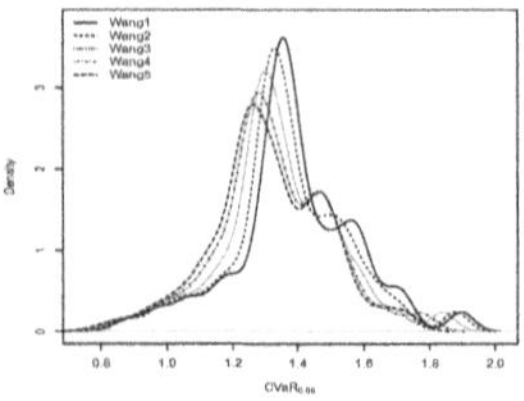

(d) Wang-10IP

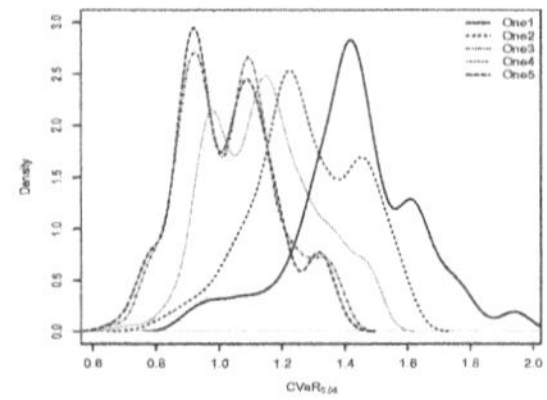

(e) One-10IP

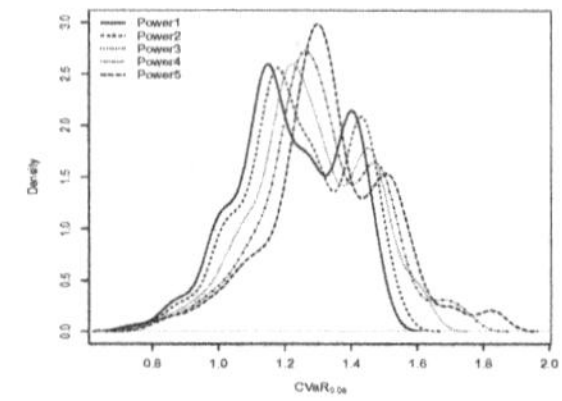

(f) Power-10IP

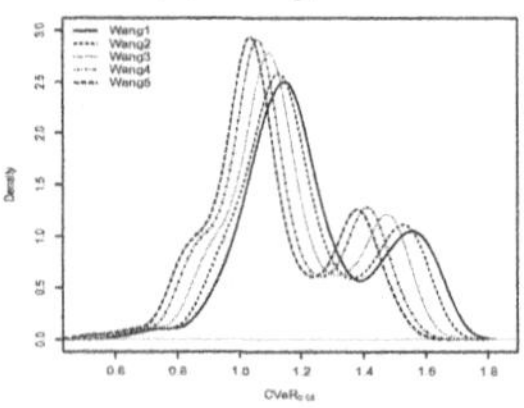

(g) Wang-25BM

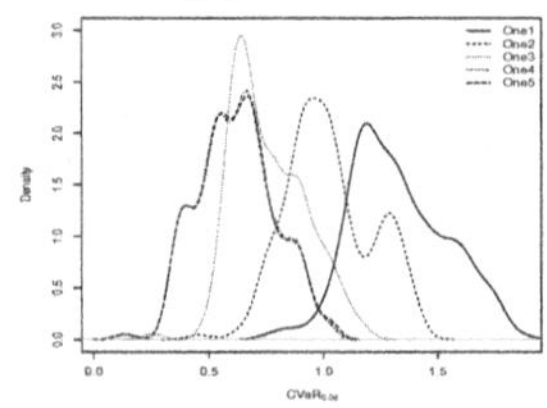

(h) One-25BM

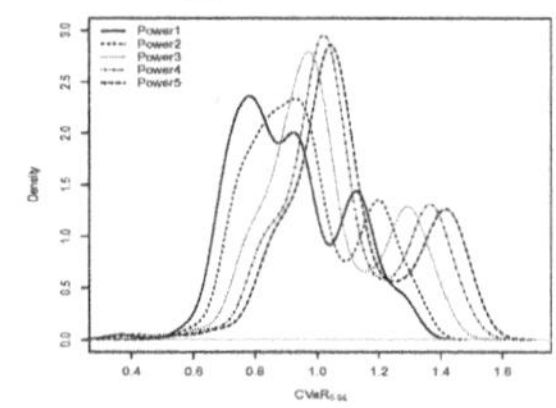

(i) Power-25BM

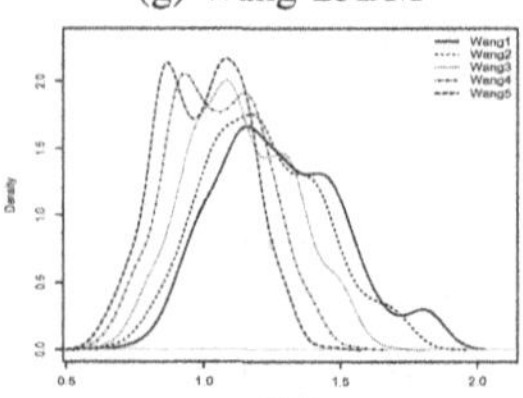

(j) Wang-48IP

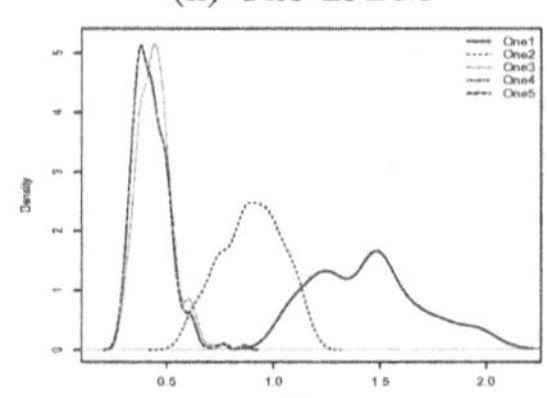

(k) One-48IP

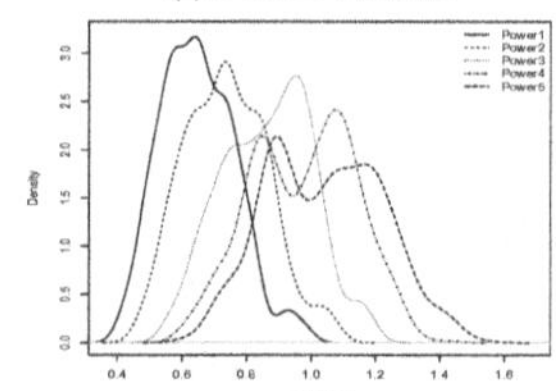

(l) Power-48IP

Fig. 1 Density of in-sample conditional value at risk ($\alpha = 0.05$, W = 120) *Notes* These figures show the in-sample $\text{CVaR}_{0.05}$ of the portfolio return based on the distortion risk measures for each dataset with 120 window length. From the first column, each column is drawn from Wang's transformation, one parameter family and power function, respectively. The numbers following the abbreviations indicate the level of distortion

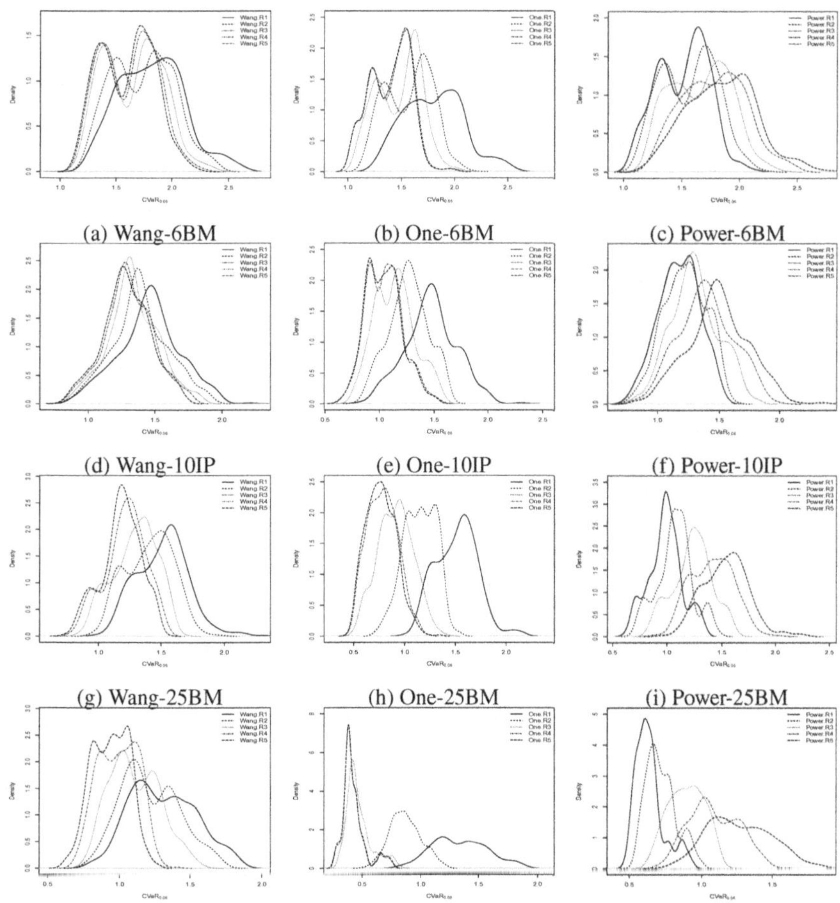

(a) Wang-6BM (b) One-6BM (c) Power-6BM

(d) Wang-10IP (e) One-10IP (f) Power-10IP

(g) Wang-25BM (h) One-25BM (i) Power-25BM

(j) Wang-48IP (k) One-48IP (l) Power-48IP

Fig. 2 Density of in-sample conditional value at risk with target return ($\alpha = 0.05$, $W = 120$) *Notes* These figures show the in-sample $CVaR_{0.05}$ of the portfolio return with a target return restriction based on the distortion risk measures for each dataset with 120 window length. From the first column, each column is drawn from Wang's transformation, one parameter family and power function, respectively. The numbers following the abbreviations indicate the level of distortion

Table 7 Sharpe ratio total (W = 120)

	In-sample				Out-of-sample			
	6BM	10IP	25BM	48IP	6BM	10IP	25BM	48IP
Naive	2.4387	2.6492	2.5726	1.1170	0.1982	0.1626	0.1662	**0.1502**
MV	2.4387	2.6492	2.5726	1.1170	0.2020	0.1873	0.2237	0.0893
MinV	1.7006	1.8733	1.5185	1.5413	0.2561	0.1684	0.2816	0.0802
LW	1.3185	1.7049	1.4256	1.6853	0.1688	0.1764	0.2029	0.1228
MinCVaR	1.8505	1.2722	1.4981	0.8470	0.2468	0.1133	0.2192	0.0394
MinCVaR.R	2.4388	2.6494	2.5818	1.1185	0.1715	0.1738	0.2273	0.0741
Wang1	1.6720	1.8223	1.4235	**1.7358**	0.2490	0.1593	0.2693	0.0938
Wang2	1.7036	1.8677	1.4443	1.7240	0.2498	0.1658	0.2698	0.0866
Wang3	1.7407	1.8884	1.4793	1.6555	0.2527	0.1710	0.2760	0.0815
Wang4	1.7755	1.8767	1.5171	1.5653	0.2556	0.1737	0.2794	0.0771
Wang5	1.7923	1.8612	1.5374	1.4987	0.2579	0.1747	0.2826	0.0738
One1	1.5905	1.8426	1.3695	1.6371	0.2442	0.1512	0.2558	0.0912
One2	1.8449	1.7428	1.6459	1.1538	0.2636	0.1799	0.2877	0.0673
One3	1.8845	1.5026	1.8630	0.8885	0.2618	0.1530	0.2725	0.0430
One4	1.8733	1.3262	1.5262	0.8470	0.2500	0.1325	0.2319	0.0394
One5	1.8998	1.3152	1.4975	0.8470	0.2465	0.1145	0.2216	0.0394
Power1	1.8339	1.7624	1.5851	1.2168	**0.2655**	0.1680	**0.2907**	0.0673
Power2	1.8264	1.7813	1.5790	1.2556	0.2642	0.1705	0.2897	0.0712
Power3	1.8079	1.8054	1.5646	1.3248	0.2619	0.1725	0.2879	0.0748
Power4	1.7725	1.8129	1.5517	1.4099	0.2583	0.1729	0.2877	0.0769
Power5	1.7169	1.7582	1.5478	1.5214	0.2562	0.1660	0.2873	0.0840
Wang.R1	2.4396	2.6499	2.5855	1.1176	0.1597	0.1315	0.1976	0.0883
Wang.R2	2.4391	2.6500	2.5870	1.1178	0.1684	0.1648	0.2058	0.0986
Wang.R3	2.4398	2.6498	2.5886	1.1178	0.1866	0.1740	0.2138	0.0955
Wang.R4	2.4405	2.6499	2.5882	1.1181	0.1923	0.1794	0.2191	0.0980
Wang.R5	2.4400	2.6500	2.5855	1.1182	0.2000	**0.1886**	0.2188	0.0885
One.R1	2.4393	2.6502	2.5854	1.1177	0.1673	0.1322	0.1881	0.0800
One.R2	2.4389	2.6493	2.5805	1.1174	0.2035	0.1872	0.2297	0.0845
One.R3	2.4399	2.6500	2.5839	1.1176	0.2059	0.1704	0.2446	0.0662
One.R4	2.4401	2.6508	2.5849	1.1176	0.1994	0.1499	0.2367	0.0667
One.R5	2.4400	2.6510	2.5869	1.1175	0.1955	0.1453	0.2381	0.0696
Power.R1	**2.4434**	**2.6526**	**2.5977**	1.1197	0.1937	0.1731	0.2299	0.0944
Power.R2	2.4412	2.6509	2.5895	1.1184	0.1898	0.1690	0.2342	0.0982
Power.R3	2.4410	2.6503	2.5881	1.1183	0.1860	0.1740	0.2231	0.1060
Power.R4	2.4394	2.6501	2.5886	1.1188	0.1645	0.1309	0.1962	0.1133
Power.R5	2.4395	2.6502	2.5866	1.1190	0.1519	0.1174	0.1732	0.0760

Notes This table reports in- and out-of-sample Sharpe ratio of each portfolio selection rules for 120 window length. The numbers in bold indicate the best results in the portfolio strategies. The numbers following the abbreviations are the level of distortion and .R denotes the portfolio with a target return restriction, respectively

Table 8 Sharp ratio total (W = 240)

	In-sample				Out-of-sample			
	6BM	10IP	25BM	48IP	6BM	10IP	25BM	48IP
Naive	**6.0118**	6.4113	**6.3816**	2.4450	0.1608	0.1382	0.1607	0.1614
MV	**6.0118**	6.4113	**6.3816**	2.4450	0.2384	0.2510	0.2692	0.1498
MinV	3.4808	4.5997	2.6986	3.9688	0.3012	0.2293	0.3500	0.1476
LW	3.0885	4.4179	3.1070	**4.7442**	0.2327	0.2286	0.2995	**0.1912**
MinCVaR	2.9250	2.9014	2.6997	2.4483	0.2744	0.1915	0.3145	0.0996
MinCVaR.R	6.0113	6.4116	6.3663	2.4455	0.2050	0.2255	0.2616	0.1043
Wang1	3.1030	4.6465	2.6324	4.3342	0.3058	0.2204	0.3340	0.1341
Wang2	3.3613	4.7578	2.6846	4.4010	0.3059	0.2230	0.3373	0.1402
Wang3	3.5653	4.6982	2.7683	4.1870	0.3044	0.2250	0.3424	0.1471
Wang4	3.6175	4.6273	2.8237	3.9935	0.3028	0.2262	0.3474	0.1520
Wang5	3.6269	4.5763	2.8509	3.9200	0.3023	0.2262	0.3509	0.1545
One1	3.1648	4.5606	2.5748	4.3481	0.3143	0.2115	0.3274	0.1273
One2	3.5990	4.2595	2.9349	3.2924	0.2995	0.2250	0.3628	0.1575
One3	3.5945	3.7979	3.0225	2.7708	0.2873	0.2228	0.3597	0.1363
One4	3.0594	3.2079	2.7742	2.4509	0.2747	0.2030	0.3432	0.1028
One5	2.8126	2.8330	2.7053	2.4387	0.2700	0.1902	0.2993	0.1025
Power1	3.5391	4.2673	2.9293	3.6957	0.2969	0.2293	**0.3651**	0.1484
Power2	3.5403	4.3461	2.9140	3.7360	0.2986	0.2282	0.3633	0.1521
Power3	3.4675	4.4662	2.8752	3.8634	0.3018	0.2267	0.3578	0.1551
Power4	3.0892	4.5764	2.8048	4.0475	0.3038	0.2258	0.3516	0.1513
Power5	2.9221	4.5015	2.7230	4.0010	**0.3384**	0.2217	0.3469	0.1465
Wang.R1	6.0077	6.4132	6.3514	2.4453	0.2532	0.2165	0.2130	0.1296
Wang.R2	6.0074	6.4123	6.3475	2.4455	0.2478	0.2362	0.2328	0.1341
Wang.R3	6.0093	6.4118	6.3522	2.4454	0.2459	0.2463	0.2623	0.1413
Wang.R4	6.0098	6.4117	6.3537	2.4453	0.2415	**0.2521**	0.2659	0.1518
Wang.R5	6.0099	6.4116	6.3545	2.4453	0.2410	0.2483	0.2705	0.1536
One.R1	6.0083	6.4126	6.3525	2.4453	0.2547	0.2154	0.2210	0.1313
One.R2	6.0114	6.4114	6.3747	2.4450	0.2398	0.2414	0.2757	0.1598
One.R3	6.0104	6.4118	6.3648	2.4453	0.2232	0.2466	0.2585	0.1378
One.R4	6.0080	6.4123	6.3530	2.4455	0.2073	0.2281	0.2514	0.1105
One.R5	6.0082	6.4124	6.3502	2.4455	0.1964	0.2038	0.2487	0.1060
Power.R1	6.0046	**6.4136**	6.3074	2.4461	0.2253	0.2437	0.2724	0.1459
Power.R2	6.0078	6.4126	6.3354	2.4457	0.2431	0.2469	0.2635	0.1446
Power.R3	6.0072	6.4121	6.3482	2.4456	0.2552	0.2319	0.2441	0.1500
Power.R4	6.0083	6.4126	6.3418	2.4455	0.2546	0.2320	0.2152	0.1441
Power.R5	6.0080	6.4128	6.3496	2.4458	0.2502	0.1852	0.2050	0.1240

Notes This table reports in- and out-of-sample Sharpe ratio of each portfolio selection rules for 240 window length. The numbers in bold indicate the best results in the portfolio strategies. The numbers following the abbreviations are the level of distortion and .R denotes the portfolio with a target return restriction, respectively

SR. Unlike these in-sample results, my portfolio strategies outperform most other portfolios in out-of-sample cases too. Specifically, Power1 and Power5 have the highest SR for 6BM and 25BM, and Wang.R4 has the highest SR for 48IP. In general, the portfolio strategies using distortion risk measures outperform other portfolio rules in out-of-sample cases.

5 Concluding Remarks

This study proposes a new asset allocation strategy using pessimistic portfolio selection rules with various distortion risk measures. The proposed method incorporates pessimistic preferences in which the distortion parameter, λ, can be interpreted as the degree of pessimistic preference of the portfolio selection strategy. In addition, calibrating the level of distortion, the effect of distortion parameter change on the optimal portfolio is studied. From empirical applications, we find that portfolio returns and the corresponding risk measures change systematically over the level of λ. Moreover, the risk for the portfolio decreases as the level of the distortion increases.

To compare portfolio performances using the rolling window method, we use three performance evaluation measures: $\text{VaR}_{0.05}$, $\text{CVaR}_{0.05}$ and Sharpe ratio. Using empirical datasets, we find that portfolio selection rules with distortion risk measures decrease portfolio risk significantly. Moreover, in terms of SR, proposed portfolio selection rules outperform other portfolio strategies in most out-of-sample cases.

The empirical findings given in this paper have important implications. First, none of the portfolio strategies is dominate for all of evaluation measures. However, the proposed pessimistic portfolio strategy based on distortion risk measures can reduce risks, as well as estimation errors. Second, the proposed pessimistic portfolios tend to have lower risks.

In this study, we do not select the portfolio weights which minimized the forecasted $\text{CVaR}_{0.05}$ but choose weights based on the in-sample $\text{CVaR}_{0.05}$. It would be useful to select portfolio weights based on the forecasted $\text{CVaR}_{0.05}$. Moreover, it would yield more precise optimal portfolio weights if the cross-validation method were used to select the best portfolio. This is a topic we would like to pursue in my future research.

Acknowledgment This research was supported by the Chung-Ang University research grant in 2020.

References

1. Acerbi, C., Simonetti, P.: Portfolio optimization with spectral measures of risk. Rot S Man (2002)
2. Acerbi, C., Tasche, D.: Expected shortfall: a natural coherent alternative to value at risk. Econ. Notes **31**, 379–388 (2002)

3. Artzner, P., Delbaen, F., Eber, J.M., Heath, D.: Coherent measures of risk. Math. Financ. **9**, 203–228 (1999)
4. Bassett, G.W., Koenker, R., Kordas, G.: Pessimistic portfolio allocation and choquet expected utility. J. Financ. Econ. **2**, 477–492 (2004)
5. Copeland, T.E., Weston, J.F.: Financial Theory and Corporate Policy. Pearson Addison Wesley, Boston (1998)
6. DeGiorgi, E.: Reward risk portfolio selection and dtochastic dominance. J. Bank. Financ. **29**, 895–926 (2005)
7. Delage, E., Ye, Y.: Distributionally robust optimization under moment uncertainty with application to data-driven problems. Oper. Res. **58**, 595–612 (2010)
8. Duffie, D., Pan, J.: An overview of value at risk. J. Deriv. **4**, 7–49 (1997)
9. Föllmer, H., Schied, A.: Stochastic Finance: An Introduction in Discrete Time. Walter de Gruyter, Berlin (2004)
10. Ghaoui, L.E., Oks, M., Oustry, F.: Worst-case value-at-pisk and robust portfolio optimization: a conic programming approach. Oper. Res. **51**, 543–556 (2003)
11. Jorion, P.: Value at Risk: The New Benchmark for Managing Financial Risk. Irwin, Chicago (1997)
12. Koenker, R., Bassett, G.: Regression quantiles. Econometrica **46**, 33–50 (1978)
13. Markowitz, H.: Portfolio selection. J. Financ. **7**, 77–91 (1952)
14. Rockafellar, R.T., Uryasev, S.: Optimization of conditional value-at-risk. J. Risk **2**, 21–41 (2000)
15. Ruszczynski, A., Shapiro, A.: Optimization of convex risk functions. Math. Oper. Res. **31**, 433–452 (2006)
16. Scherer, B.: Portfolio resampling: review and critique. Financ. Anal. J. **58**, 98–109 (2002)
17. Wang, S.S.: A class of distortion operators for pricing financial and insurance risks. J. Risk Insur. **67**, 15–36 (2000)
18. Wang, S.S., Young, V.R., Panjer, H.H.: Axiomatic characterization of insurance prices. Insur. Math. Econ. **21**, 173–183 (1997)
19. Wirch, J.L., Hardy, M.: Distortion risk measures: coherence and stochastic dominance. Insur. Math. Econ. **32**, 168 (2003)
20. Wozabal, D.: Robustifying convex risk measures for linear portfolios: a nonparametric approach. Oper. Res. **62**, 1302–1315 (2014)
21. Yaari, M.E.: The dual theory of choice under risk. Econometrica **55**, 95–115 (1987)

Why Beta Priors: Invariance-Based Explanation

Olga Kosheleva, Vladik Kreinovich, and Kittawit Autchariyapanitkul

Abstract In the Bayesian approach, to describe a prior distribution on the set [0, 1] of all possible probability values, typically, a Beta distribution is used. The fact that there have been many successful applications of this idea seems to indicate that there must be a fundamental reason for selecting this particular family of distributions. In this paper, we show that the selection of this family can indeed be explained if we make reasonable invariance requirements.

1 Formulation of the Problem

In the Bayesian approach (see, e.g., [2, 4]), when we do not know the probability $p \in [0, 1]$ of some event, it is usually recommended to use a Beta prior distribution for this probability, i.e., a distribution for which the probability density function $\rho(x)$ has the form

$$\rho(x) = c \cdot x^{\alpha-1} \cdot (1-x)^{\beta-1},$$

where α and β are appropriate constants and c is a normalizing constant—guaranteeing that

$$\int_0^1 \rho(x)\, dx = 1.$$

There have been numerous successful application of the use of the Beta distribution in the Bayesian approach. How can we explain this success? Why not use some other family of distributions located on the interval [0, 1]?

In this paper, we provide a natural explanation for these empirical successes.

O. Kosheleva · V. Kreinovich (✉)
University of Texas at El Paso, El Paso, TX 79968, USA
e-mail: vladik@utep.edu

O. Kosheleva
e-mail: olgak@utep.edu

K. Autchariyapanitkul
Faculty of Economics, Maejo University, Chiang Mai, Thailand
e-mail: kittar3@hotmail.com

S. Sriboonchitta et al. (eds.), *Behavioral Predictive Modeling in Economics*, Studies in Computational Intelligence 897,
https://doi.org/10.1007/978-3-030-49728-6_8

Comment. The need for such an explanation is especially important now, when the statistician community is replacing the traditional p-value techniques with more reliable hypothesis testing methods (see, e.g., [3, 7]), methods such as the Minimum Bayesian Factor (MBF) method which is based on using a specific class of Beta priors

$$\rho(x) = c \cdot x^{\alpha-1}$$

that correspond to $\beta = 1$; see, e.g., [5].

2 Analysis of the Problem and the Main Result

Main Idea. We want to find a natural prior distribution on the interval $[0, 1]$, a distribution that describes, crudely speaking, how frequently different probability values p appear. In determining this distribution, a natural idea to take into account is that, in practice, all probabilities are, in effect, conditional probabilities: we start with some class, and in this class, we find the corresponding frequencies.

From this viewpoint, we can start with the original probabilities and with their prior distribution, or we can impose additional conditions and consider the resulting conditional probabilities. For example, in medical data processing, we may consider the probability that a patient with a certain disease recovers after taking the corresponding medicine. We can consider this original probability—or, alternatively, we can consider the conditional probability that a patient will recover—e.g., under the condition that the patient is at least 18 years old.

We can impose many such conditions, and, since we are looking for a universal prior, a prior that would describe all possible situations, it makes sense to consider priors for which, after such a restriction, we will get the exact same prior for the corresponding conditional probability.

Let Us Describe this Main Idea in Precise Terms. In general, the conditional probability $P(A \mid B)$ has the form

$$P(A \mid B) = \frac{P(A \,\&\, B)}{P(B)}.$$

Crudely speaking, this means that when we transition from the original probabilities to the new conditional ones, we limit ourselves to the original probabilities which do not exceed some value $p_0 = P(B)$, and we divide each original probability by p_0.

In these terms, the above requirement takes the following form: for each $p_0 \in (0, 1)$, if we limit ourselves to the interval $[0, p_0]$, then the ratios p/p_0 should have the same distribution as the original one.

Definition 1. *We say that a probability distribution with probability density $\rho(x)$ on the interval $[0, 1]$ is* invariant *if for each $p_0 \in (0, 1)$, the ratio x/p_0 (restricted to the values $x \le p_0$) has the same distribution, i.e., if*

$$\rho(x/p_0 : x \le p_0) = \rho(x).$$

Proposition 1. *A probability distribution is invariant if and only if it has a form*

$$\rho(x) = c \cdot x^a$$

for some c and a.

Proof. The conditional probability density has the form

$$\rho(x/p_0 : x \le p_0) = C(p_0) \cdot \rho(x/p_0),$$

for an appropriate constant C depending on p_0. Thus, the invariance condition has the form

$$C(p_0) \cdot \rho(x/p_0) = \rho(x).$$

By moving the term $C(p_0)$ to the right-hand side and denoting $\lambda \stackrel{\text{def}}{=} 1/p_0$ (so that $p_0 = 1/\lambda$), we conclude that

$$\rho(\lambda \cdot x) = c(\lambda) \cdot \rho(x), \tag{1}$$

where we denoted $c(\lambda) \stackrel{\text{def}}{=} 1/C(1/\lambda)$.

The probability density function is an integrable function—its integral is equal to 1. It is known (see, e.g., [1]) that every integrable solution of the functional equation (1) has the form

$$\rho(x) = c \cdot x^a$$

for some c and a. The proposition is thus proven.

Comment. It is worth mentioning that namely these distributions—corresponding to $\beta = 1$—are used in the Bayesian approach to hypothesis testing [5, 6].

How to Get a General Prior Distribution. The above proposition describes the case when we have a single distribution corresponding to a single piece of prior information. In practice, we may have many different pieces of information. Some of these pieces are about the probability p of the corresponding event E, some may be about the probability $p' = 1 - p$ of the opposite event $\neg E$.

According to Proposition 1, each piece of information about p can be described by the probability density

$$c_i \cdot x^{a_i},$$

for some c_i and a_i. Similarly, each piece of information about $p' = 1 - p$ can be described by the probability density

$$c'_j \cdot x^{a'_j}$$

for some c'_j and a'_j. In terms of the original probability $p = 1 - p'$, this probability density has the form

$$c'_j \cdot (1 - x)^{a'_j}.$$

Since all these piece of information are independent, a reasonable idea is to multiply these probability density functions. After multiplication, we get a distribution of the type

$$c \cdot x^a \cdot (a - x)^{a'},$$

where $a = \sum_i a_i$ and $a' = \sum_j a'_j$. This is exactly the Beta distribution—for $\alpha = a + 1$ and $\beta = a' + 1$.

Thus, we have indeed justified the use of Beta priors.

Acknowledgments This work was supported by the Institute of Geodesy, Leibniz University of Hannover. It was also supported in part by the US National Science Foundation grants 1623190 (A Model of Change for Preparing a New Generation for Professional Practice in Computer Science) and HRD-1242122 (Cyber-ShARE Center of Excellence).
This paper was written when V. Kreinovich was visiting Leibniz University of Hannover.

References

1. Aczel, J., Dhombres, J.: Functional Equations in Several Variables. Cambridge University Press, Cambridge (1989)
2. Gelman, A., Carlin, J.B., Stern, H.S., Dunson, D.B., Vehtari, A., Rubin, D.B.: Bayesian Data Analysis. Chapman & Hall/CRC, Boca Raton (2013)
3. Gelman, A., Robert, C.P.: The statistical crises in science. Am. Sci. **102**(6), 460–465 (2014)
4. Kock, K.R.: Introduction to Bayesian Statistics. Springer, Heidelberg (2007)
5. Nguyen, H.T.: How to test without p-values. Thail. Stat. **17**(2), i–x (2019)
6. Page, R., Satake, E.: Beyond p-values and hypothesis testing: using the minimum Bayes factor to teach statistical inference in undergraduate introductory statistics courses. J. Educ. Learn. **6**(4), 254–266 (2017)
7. Wasserstein, R.L., Lazar, N.A.: The ASA's statement on p-values: context, process, and purpose. Am. Stat. **70**(2), 129–133 (2016)

Ranking-Based Voting Revisited: Maximum Entropy Approach Leads to Borda Count (and Its Versions)

Olga Kosheleva, Vladik Kreinovich, and Guo Wei

Abstract In many practical situations, we need to make a group decision that takes into account preferences of all the participants. Ideally, we should elicit, from each participant, a full information about his/her preferences, but such elicitation is usually too time-consuming to be practical. Instead, we only elicit, from each participant, his/her ranking of different alternatives. One of the semi-heuristic methods for decision making under such information is Borda count, when for each alternative and each participant, we count how many alternatives are worse, and then select the alternatives for which the sum of these numbers is the largest. In this paper, we explain the empirical success of the Borda count technique by showing that this method naturally follows from the maximum entropy approach—a natural approach to decision making under uncertainty.

1 Formulation of the Problem

Need for Voting and Group Decision Making. In many real-life situations, we need to make a decision that affects many people. Ideally, when making this decision, we should take into account the preferences of all the affected people. This group decision making situation is also known as *voting*.

What Information Can be Used for Voting: From the Simplest Majority Voting to the Most Comprehensive Situations. The simplest—and most widely used—type of voting is when each person selects one of the possible alternatives. After this selection, all we know is how many people voted for each alternative.

O. Kosheleva · V. Kreinovich (✉)
University of Texas at El Paso, El Paso, TX 79968, USA
e-mail: vladik@utep.edu

O. Kosheleva
e-mail: olgak@utep.edu

G. Wei
Department of Mathematics and Computer Science, University of North Carolina at Pembroke, PO Box 1510, Pembroke, NC 28372, USA
e-mail: guo.wei@uncp.edu

S. Sriboonchitta et al. (eds.), *Behavioral Predictive Modeling in Economics*, Studies in Computational Intelligence 897,
https://doi.org/10.1007/978-3-030-49728-6_9

Clearly, the more people vote for a certain alternative, the better is this alternative for the community as a whole. Thus, if this is all the information we have, and we do not plan to extract any additional information from the participants, then a natural idea is to select the alternative that gathered the largest number of votes. (Another idea is to keep only the alternatives with the largest number of votes and vote again.)

In this scheme, for each person, we only take into account one piece of information: which alternative is preferable to this person. To make more adequate decision, it is desirable to use more information about people's preferences. An ideal case is when we use full information about people's preferences; we will discuss this case in the following text. This is ideal but this requires too much elicitation and is, thus, not used in practice.

An intermediate stage—when we use more information than in the simple majority voting—is when we ask the participants to rank all the alternatives, and use these rankings to make a decision.

Ranking-Based Voting: A Brief Reminder. The famous result by a Nobelist Kenneth Arrow shows that it is not possible to have a ranking-based voting scheme that would satisfy all reasonable fairness-related properties [14, 16, 17]. So what can we do?

One of the schemes used in such voting is the *Borda count* (see, e.g., [16, 17], when for each participant i and for each alternative A_j, we count the number b_{ij} of alternatives that the i-th participant ranked lower than A_j. Then, for each alternative A_j, we add up the numbers corresponding to different participants, and we select the alternatives with the largest value of the corresponding sum $\sum_{i=1}^{n} b_{ij}$.

Why Borda Count? Borda count is often successfully used in practice. However, the fact that there are several other alternative schemes prompts a natural question: why namely Borda count and why not one of these other schemes?

In this paper, we provide an explanation for the success of Borda count: namely, we show that the Borda count (and its versions) naturally follow from the maximum entropy approach—a known way for making decisions under uncertainty.

2 What if We Have Complete Information About the Preferences: Reminder

How to Describe Individual Preferences. In order to describe what should we do when only know the rankings, let us first recall what decision we should make when we have full information about the preferences. To describe this, we need to recall how to describe these preferences.

In decision theory (see, e.g., [4, 8, 13–15]), a user's preferences are described by using the notion of *utility*. To define this notion, we need to select two extreme alternatives:

- a very bad alternative A_- which is worse than anything that we will actually encounter, and
- a very good alternative A_+ which is better than anything that we will actually encounter.

For each number p from the interval $[0, 1]$, we can then form a lottery $L(p)$ in which:

- we get A_+ with probability p and
- we get A_- with the remaining probability $1 - p$.

Then:

- For $p = 0$, the lottery $L(p)$ coincides with the very bad alternative A_- and is, thus, worse than any of the alternatives A that we encounter:

$$L(0) = A_- < A.$$

- For $p = 1$, the lottery $L(p)$ coincides with the very good alternative A_+ and is, thus, better than any of the alternatives A that we encounter:

$$A < L(1) = A_+.$$

Clearly, the larger p, the better the lottery. Thus, there exists a threshold p_0 such that:

- for $p < p_0$, we have $A(p) < A$, and
- for $p > p_0$, we have $A < A(p)$.

This threshold is known as the *utility* of the alternative A; it is usually denoted by $u(A)$.

In particular, according to this definition:

- the very bad alternative A_- has utility 0, while
- the very good alternative A_+ has utility 1.

To fully describe people's preferences, we need to elicit, from each person i, this person's utility $u_i(A_j)$ of all possible alternatives A_j.

Utility is Defined Modulo Linear Transformations. The numerical value of utility depends on the selection of values A_- and A_+.

One can show that if we use a different pair of alternatives (A'_-, A'_+), then the resulting new utility values $u'(A)$ are related to the original values $u(A)$ by a linear dependence: $u'(A) = k + \ell \cdot u(A)$ for some k and $\ell > 0$.

Utility-Based Decision Making Under Probabilistic Uncertainty. In many practical situations, we do not know the exact consequences of different actions. For each action, we may have different consequences $c_1, \ldots, c_m$, with different utilities $u(c_1), \ldots, u(c_m)$. We can also usually estimate the probabilities $p_1, \ldots, p_m$ of different consequences. What is the utility of this action?

This action is equivalent to selecting c_i with probability p_i. By definition of utility, each consequence c_i is, its turn, equivalent to a lottery in which we get A_+

attains the largest possible value. This, in its turn, is equivalent to maximizing the product $\prod_{i=1}^{n}(b_{ij}+1)$, or, alternatively, to maximizing the sum $\sum_{i=1}^{n}\ln(b_{ij}+1)$.

Acknowledgments This work was supported in part by the National Science Foundation grants 1623190 (A Model of Change for Preparing a New Generation for Professional Practice in Computer Science) and HRD-1242122 (Cyber-ShARE Center of Excellence).

References

1. Ahsanullah, M., Nevzorov, V.B., Shakil, M.: An Introduction to Order Statistics. Atlantis Press, Paris (2013)
2. Arnold, B.C., Balakrishnan, N., Nagaraja, H.N.: A First Course in Order Statistics, Society of Industrial and Applied Mathematics (SIAM). Pennsylvania, Philadelphia (2008)
3. David, H.A., Nagaraja, H.N.: Order Statistics. Wiley, New York (2003)
4. Fishburn, P.C.: Utility Theory for Decision Making. Wiley, New York (1969)
5. Jaimes, A., Tweedie, C., Magoc, T., Kreinovich, V., Ceberio, M.: Selecting the best location for a meteorological tower: a case study of multi-objective constraint optimization. In: Ceberio, M., Kreinovich, V. (eds.) Constraint Programming and Decision Making, pp. 61–66. Springer, Berlin (2014)
6. Jaynes, E.T., Bretthorst, G.L.: Probability Theory: The Logic of Science. Cambridge University Press, Cambridge (2003)
7. Kosheleva, O., Kreinovich, V., Lorkowski, J., Osegueda, M.: How to transform partial order between degrees into numerical values. In: Proceedings of the 2016 IEEE International Conferences on Systems, Man, and Cybernetics SMC 2016, Budapest, Hungary, 9–12 October 2016
8. Kreinovich, V.: Decision making under interval uncertainty (and beyond). In: Guo, P., Pedrycz, W. (eds.) Human-Centric Decision-Making Models for Social Sciences, pp. 163–193. Springer, Heidelberg (2014)
9. Lorkowski, J., Kreinovich, V.: Interval and symmetry approaches to uncertainty – pioneered by Wiener – help explain seemingly irrational human behavior: a case study. In: Proceedings of the 2014 Annual Conference of the North American Fuzzy Information Processing Society NAFIPS 2014, Boston, Massachusetts, 24–26 June 2014
10. Lorkowski, J., Kreinovich, V.: Likert-type fuzzy uncertainty from a traditional decision making viewpoint: how symmetry helps explain human decision making (including seemingly irrational behavior). Appl. Comput. Math. **13**(3), 275–298 (2014)
11. Lorkowski, J., Kreinovich, V.: Granularity helps explain seemingly irrational features of human decision making. In: Pedrycz, W., Chen, S.-M. (eds.) Granular Computing and Decision-Making: Interactive and Iterative Approaches, pp. 1–31. Springer, Cham (2015)
12. Lorkowski, J., Kreinovich, V.: Fuzzy logic ideas can help in explaining Kahneman and Tversky's empirical decision weights. In: Zadeh, L., et al. (eds.) Recent Developments and New Direction in Soft-Computing Foundations and Applications, pp. 89–98. Springer, Cham (2016)
13. Luce, R.D., Raiffa, R.: Games and Decisions: Introduction and Critical Survey. Dover, New York (1989)
14. Nguyen, H.T., Kosheleva, O., Kreinovich, V.: Decision making beyond Arrow's 'impossibility theorem', with the analysis of effects of collusion and mutual attraction. Int. J. Intell. Syst. **24**(1), 27–47 (2009)
15. Raiffa, H.: Decision Analysis. McGraw-Hill, Columbus (1997)
16. Saari, D.G.: Chaotic Electrions!. American Mathematical Society, Providence (2001)
17. Saari, D.G.: Disposing Dictators, Demystifying Voting Paradoxes: Social Choice Analysis. Cambridge University Press, New York (2008)

Monitoring Change Points by a Sign Change Method

Bo-Ru Lai and Hsiuying Wang

Abstract Signal detection is an important problem in monitoring or other related fields. Wang [16] proposed a sign change method which based on the difference of the traditional chart statistic and the JS-type chart statistic to detect out-of-control signals. Since this approach is not necessary to construct control limits to detect a signal, it can supply an efficient way for signal detection. We design and conduct a simulation study to compare the performance of the sign change method with those based on the traditional charts in this study. From the simulation result, we show that the sign change method can detect out of control signals more efficiently based on the MC1 or the EMWMA control chart statistics.

Keywords Hotelling-chart · MCUSUM · MEWMA · James-Stein estimator

1 Introduction

While we want to detect shifts in a manufacturing process with multivariate characteristic, multivariate control chart is a practical tool. The first multivariate control chart discussed in this study is the Hotelling-T^2 control chart Hotelling [6], which detects mean shift just based on the most recent observations, leading to insensitive detection of small and moderate mean shifts. The second is the multivariate CUSUM control chart Crosier [2], that detects shifts by either reducing every multivariate observation to a scalar or constituting a CUSUM vector from the observations. In addition, Pignatiello and Runger [11] established an improved CUSUM control chart, MC1, and proved that it performs better than several other CUSUM control charts. The last is the MEWMA control chart, proposed by Lowry et al. [9], which utilizes all data information from the previous to the last observations to set control limits that has the advantage of smaller average run length to detect small mean shifts in the process mean.

B.-R. Lai · H. Wang (✉)
Institute of Statistics, National Chiao Tung University, Hsinchu, Taiwan
e-mail: wang@stat.nctu.edu.tw

S. Sriboonchitta et al. (eds.), *Behavioral Predictive Modeling in Economics*, Studies in Computational Intelligence 897,
https://doi.org/10.1007/978-3-030-49728-6_10

As we use control charts, there are two important processes, Phase I and Phase II. In Phase I, a set of historical data are gathered to estimate the parameter and construct control limits. In Phase II, the data are collected orderly over time to check whether the parameter has altered from the Phase I estimator. Most of the control chart schemes concentrate on supplying modified methods for Phase II monitoring. Nevertheless, the study discussed in the phase I is not as deep as in the Phase II in the literature. Usually before Phase II monitoring, one must use Phase I study to estimate the in-control process parameter to construct the control limits. Hence, whether the Phase I estimation is accurate becomes a crucial step for Phase II study.

In order to monitor a univariate characteristic which is assumed from a normal distribution, the number of parameter requested to estimate is usually less than or equal to 2. For instance, there exist only two parameters, mean and variance rto estimate in a univariate normal process. Thus, the sample mean and sample variance are optimal estimators under the squared error loss in this case.

However, as the number of parameters requested to estimate is greater than 2 such as the case for a normal process with multivariate characteristic, it is well-known that the traditional estimators are not optimal under the squared error loss. For instance, the sample mean is inadmissible for estimating the multivariate normal mean under the squared error loss as the dimension is greater than 2 (Berger 1985). As a result of the inadmissibility characteristic of the sample mean, the Phase I parameter estimation for monitoring multivariate process could be improved.

It is well-known that the shrinkage estimator, such as the James-Stein estimator, has the smaller mean squared error than the traditional estimator. Wang, Huwang and Yu [17] proposed a method to construct multivariate control charts based on the James-Stein estimator. Hence, in this paper, we improve the Phase I estimation by replacing the traditional estimator with the shrinkage estimator. Because one of the famous shrinkage estimators is the James-Stein estimator, we propose modified multivariate control charts on the basis of the James-Stein estimator in the Phase I study and show significant improvements of the modified charts compared with the conventional charts. For executing this, several famous charts on the basis of the traditional estimators and the James-Stein estimators are studied in the Phase II monitoring.

Wang [16] proposed a sign change method which depends on the difference of the two statistic values based on the conventional chart and the modified chart to detect a mean shift signal. Because this method is not essential to set control limits to detect a signal, it can offer a convenient way for signal detection.

In this study, we mainly use the simulation study to evaluate the performance of the sign change method.

2 JS-type Control Charts

We first review the history of the shrinkage estimator in the literatures and then discuss the modified control charts by improving the control charts based on the shrinkage estimators.

2.1 *Shrinkage Estimation*

Let $X_1, ..., X_n$ be a random sample from a p-dimensional multivariate normal distribution $N(\mu, \Sigma)$ that has mean vector μ and $p \times p$ positive definite matrix Σ. In order to discuss two different cases of Σ, suppose that Σ is the identity matrix in this subsection. In the next subsection, suppose that Σ is any covariance matrix. The sample mean $\bar{X} = \sum_{i=1}^{n} X_i/n$ is the traditional estimator for μ. Stein (1956) has showed that to estimate μ, under the squared error loss, $\bar{X}$ is not admissible for any $p \geq 3$.

James and Stein [7] proposed a modified estimator named James-Stein estimator which has a smaller mean squared error than $\bar{X}$. Many studies for the shrinkage estimators are referred to (Strawderman and Cohen [13]; Efron and Morris [5]; Draper and Van Nostrand [4]; Casella 1980; Wang 1999, [14]).

In this study, we mainly utilize the James-Stein estimator to construct modified control charts. The James-Stein estimator is denoted as

$$\bar{X}_0^{JS} = (1 - \frac{p-2}{n(\bar{X}-\nu)'\Sigma^{-1}(\bar{X}-\nu)}) \cdot (\bar{X}-\nu) + \nu, \tag{1}$$

where ν is a fixed vector that the $\bar{X}_0^{JS}$ shrinks to.

That is, there is an estimator $\bar{X}_0^{JS}$ such that the mean squared error (MSE), $E[(\bar{X}_0^{JS}-\mu)'\Sigma^{-1}(\bar{X}_0^{JS}-\mu)]$, of $\bar{X}_0^{JS}$ is smaller than or equal to the mean squared error (MSE), $E[(\bar{X}-\mu)'\Sigma^{-1}(\bar{X}-\mu)]$, of traditional mean $\bar{X}$ for all μ and the strict inequality holds for some μ while $p \geq 3$.

Lehmann and Casella [8] proposed that ν can be any p-dimensional vector. A modified James-Stein estimator is denoted as

$$\bar{X}^{JS} = (1 - \frac{p-2}{n(X-\nu)'\Sigma^{-1}(\bar{X}-\nu)})^{+} \cdot (\bar{X}-\nu) + \nu, \tag{2}$$

which is proved to have smaller MSE than the above James-Stein estimator (1). Here the notation x^+ is defined to be

$$x^+ = \begin{cases} x \text{ if } x > 0, \\ 0 \text{ otherwise.} \end{cases}$$

It is worth noting that in this paper, we mainly utilize the form (2) as the James-Stein estimator for constructing better control charts.

Note that a confidence set on the basis of the shrinkage estimator is preferable to that on the basis of the traditional estimator, leading to the potential of the James-Stein estimator for constructing better control charts.

Generally, a $1-\alpha$ confidence set for μ on the basis of the sample mean $\bar{X}$ is

$$C = \{\mu : (\bar{X} - \mu)'\Sigma^{-1}(\bar{X} - \mu) \leq c/n\},$$

where c is the $1-\alpha$ cutoff point of a chi-square distribution with p degrees of freedom. The confidence set C has a coverage probability $Pr(C)$ equals $1-\alpha$.

Nevertheless, the confidence set on the basis of the James-Stein estimator is

$$C^{JS} = \{\mu : (\bar{X}^{JS} - \mu)'\Sigma^{-1}(\bar{X}^{JS} - \mu) \leq c/n\} \tag{3}$$

C^{JS} has been proved to have a higher coverage probability than $1-\alpha$ analytically and numerically. Hence,

$$Pr(C^{JS}) \geq Pr(C), \tag{4}$$

for c satisfying some conditions and the strict inequality holds for some μ while $p \geq 3$ (Brown 1966; Joshi 1967; Hwang and Casella 1982; DasGupta et al. 1995).

Because the confidence sets C^{JS} and C have the same volume, characteristic (4) demonstrates that the confidence set C^{JS} could contain a larger proportion of the population than the confidence set C with the same capacity, which results in better performance of the confidence set C^{JS}. Utilizing (4), in the Sect. 2.2, we introduce several control charts in the Phase II study to monitor the mean vector μ by the James-Stein estimator (2) used in the Phase I estimation.

Note that even though v in (2) can be chosen to be any vector, the performance of the James-Stein estimator relies on v. If we do not have any preference of choosing v, we can choose the value of v to be the zero vector or close to the sample mean.

2.2 Modified Control Charts

We first review some well-known control charts for monitoring μ and then provide improved forms of these charts. For convenience, we only take the case of subgroup size 1 (i.e., individual observation) into account in our study. Nevertheless, the outcomes could be easily generated to the case which the subgroup size is greater than 1. In the following, the notation $\bar{X}$ represents the sample mean of the in-control observations in the Phase I study and $X_i \sim N(\mu, \Sigma), i = 1, 2, \cdots$, represents the observation of subgroup i in the Phase II monitoring, where Σ is unknown. In real utilizations, the covariance matrix Σ is often unknown and the sample covariance matrix S is used to substitute for Σ in calculation.

1. The T^2 chart
   The first control chart discussed is the Hotelling-T^2 control chart Hotelling [6]. While we know the covariance matrix, the monitoring statistic for the sample i is

$$T_i = (X_i - \bar{X})' \Sigma^{-1} (X_i - \bar{X}). \tag{5}$$

   The T^2 chart gives an out-of-control signal if the monitoring statistic

$$T_i > c_1, \tag{6}$$

   where c_1 is the constant to achieve the particular average run length (ARL_o) (Tracy et al. 1992).
2. The MC1 chart
   The MC1 control chart is proposed by Pignatiello and Runger [11] and has been showed that it can modify the multivariate CUSUM control charts Crosier [3]. The monitoring statistic for the MC1 chart is

$$MC1_i = max\{\|C_i\| - kn_i, 0\},$$

   where $k > 0$,

$$C_i = \sum_{l=i-n_i+1}^{i} (X_l - \bar{X}), \tag{7}$$

$$n_i = \{ \begin{matrix} 1 & \text{if } MC1_{i-1} \leq 0, \\ n_{i-1} + 1 & \text{if } MC1_{i-1} > 0, \end{matrix}$$

$$\|C_i\| = \sqrt{C_i' \Sigma^{-1} C_i}$$

   and

$$MC1_0 = 0.$$

   The MC1 control chart gives an out-of-control signal if the monitoring statistic

$$MC1_i > c_2, \tag{8}$$

   where c_2 is the particular constant to achieve a desired ARL_0. The parameter k is selected to be half of the distance between μ^* and μ_0 by Pignatiello and Runger [11], where μ_0 is the in-control mean value and μ^* denotes a particular out-of-control mean value.

Because the improved control charts are obtained by using the James-Stein estimator to substitute for the sample mean in the original charts. These improved control charts are named as the JS-type control charts.

3 Sign Change Method and Simulation

In this section, we first introduce an approach which is proposed in Wang [16]. This approach is based on the conventional control chart statistic and JS-type control chart statistic to derive a sign change method and then we conduct a simulation study for comparing this method to the existing methods.

3.1 Method

In the conventional control chart method, we have to set control limits and then we can detect the mean shift signal based on the control limits. Nevertheless, except the case that we know what distribution the monitoring statistics follow, for most control charts, the derivation of control limits depends on enormous numerical calculation or simulation. If we can rely on a method which is not essential to set control limits for signal detection, then it can offer us a more convenient way to detect an out-of-control signal.

Wang [16] proposed a sign change method which based on comparing two statistics to detect a sign. The method is illustrated as follows.

Let W_i and W_i^{JS} denote the traditional control chart statistic and the corresponding JS-type control chart statistic. Let $D_i = W_i - W_i^{JS}$. Wang [16] used the change point of the sign of D_i to detect out-of-control signals. This approach has been named as the sign change method in Wang [16].

3.2 Simulation Study

We conduct a simulation study to compare the performance of the traditional control chart and the sign change method. Suppose that $\mu = \mu_0$ as the manufacture process is in control and $\mu = \mu^*$ while the manufacture process is out-of-control. Since the covariance matrix is usually unknown in real situation, we mainly take the unknown covariance matrix case into consideration in this study. In our simulation, let the vector v in the (11) be the zero vector $\mathbf{0}$. That is, the James-Stein estimator (11) shrinkage to the zero point $\mathbf{0}$. Besides, we assume the covariance matrix is unknown so we utilize the sample covariance matrix to be the covariance matrix estimation.

First, we give an example in a simulation study. We randomly generate 30 samples with dimension 5 and the data is shown in the Table 16. Let the first 14 sample be generated from the process $\mu = \mu_0$ and the other 16 samples are generated from an out of control situation $\mu = \mu^*$. Note that the 15th sample is the first sample from the out-of-control process. The first 10 sample is assumed to be in the in-control process and are used for parameter estimation in phase I, and the last 20 sample are used to detect an out-of-control signal in Phase II. From the sign change method, we have D_i, i = 1, ..., 30 for the Hotelling-T^2, MC1 and EWMA chart, respectively. The signs of the D_i values, which are the differences between the conventional control chart statistic and JS-type control chart statistic are shown as below

1. The sign for the D_i based on T^2 chart and the JS-T^2 chart
   $+, -, +, +, -, -, +, -, -, +, +, -, -, -, +, -, -, -, -, -, -, +, -, -, -, -, -, -, -, -$
2. The sign for the D_i based on MC1 chart and the JS- MC1 chart
   $+, +, +, +, +, +, +, 0, -, +, +, +, +, +, +, +, +, -, -, -, -, -, -, -, -, -, -, -, -, -$
3. The sign for the D_i based on MEWMA chart and the JS- MEWMA chart
   $+, -, -, -, -, -, -, -, -, +, +, +, +, +, -, -, -, -, -, -, -, -, -, -, -, -, -, -, -, -$

The outcomes from the methods based on the MC1 and MEWMA control charts show the sign change method can detect the change point efficiently. The T^2 chart detects an out-of-control signal at a wrong point which does not have any mean shift. The data display that the sign for the MEWMA chart obviously changes in the 15th sample(the 5th sample in phase II), there is an significant shift exactly at the point. In addition, The data show that the sign for the MC1 chart obviously changes in the 18th sample (the 8th sample in phase II), we can find that the 18th sample is significantly smaller in x_2 and x_3 (Table 11). We can detect the out-of-control signal by the sign change method easily from the MC1 and MEWMA control charts. However, in most application, the change may not always be such evident for us in monitoring.

In the simulation study, we repeat 1000 times. We calculate the counts which the change point is detected in the 15th to the 20th sample and for various values of $d = ||\mu^* - \mu_0||$ by the sign change method and traditional control chart.

To have an objective comparison, we first calculate the count of the sign change method while the process is in control and then construct the control limits of every chart based on a required sign change count. The counts are calculated for various out-of-control cases. If we found a larger count number in an out-of-control condition, we can say the sign change method has better property than another because it accurately detects the out-of-control signal. The comparison of the Hotelling-T^2, MC1 and EWMA control chart and the corresponding sign change method are presented in the following tables, mainly setting the $\mu = \mu_0$ from zero vector to large vector in Phase I, which displays the counts for the sign change method and the traditional charts for different values of d and for various dimensions.

3.2.1 Comparion for Hotelling-T² Chart

Tables 1, 2 and 3 provides the performances for the SC-T² (sign change method based on T² chart) and Hotelling-T² charts. The results show that the Hotelling-T² charts are always better than the sign change method. The counts of SC-T² get smaller even d becomes bigger. The reason may be that the T² chart is not sensitive to small mean shift. In this section, the notation $\underline{v}$ denotes the vector $v(1, ..., 1)'_{1\times p}$.

Table 1 The counts (accurate signal) of signals of the SC-T² and Hotelling-T² chart when $\mu_0 = \underline{0}$ occurring at the 15th–20th observation

d	$p = 3$		$p = 5$		$p = 7$	
	SC-T²	T²	SC-T²	T²	SC-T²	T²
0	96.031	95.423	100.677	99.91	99.006	98.154
0.5	93.567	97.256	98.747	102.266	97.9	101.571
1.0	87.803	101.933	94.69	107.527	95.071	105.852
1.5	81.438	104.824	90.139	111.81	91.019	112.109
2.0	76.205	105.952	85.738	114.942	87.088	117.667
2.5	71.907	106.078	81.735	116.154	83.819	121.765
3.0	69.333	106.126	78.628	116.917	80.473	123.254
c_1		3.922		10.38		26.552

Table 2 The counts (accurate signal) of signals of the SC-T² and Hotelling-T² chart when $\mu_0 = \underline{0.05}$ occurring at the 15th–20th observation

	$p = 3$		$p = 5$		$p = 7$	
0	95.191	94.632	100.212	99.486	98.251	97.406
0.5	93.828	96.156	99.507	101.712	98.903	99.206
1.0	89.492	100.539	97.617	106.596	96.59	104.46
1.5	84.151	103.693	93.459	111.194	94.457	110.162
2.0	79.686	104.784	90.32	114.286	90.118	114.802
2.5	76.062	104.914	86.771	116.021	86.964	119.265
3.0	73.727	105.489	84.194	115.96	84.898	120.597
c_1		3.90		10.362		26.317

Table 3 The counts (accurate signal) of signals of the SC-T^2 and Hotelling-T^2 chart when $\mu_0 = \underline{0.1}$ occurring at the 15th–20th observation

d	$p = 3$		$p = 5$		$p = 7$	
	SC-T^2	T^2	SC-T^2	T^2	SC-T^2	T^2
0	94.541	93.954	98.16	97.424	97.173	96.329
0.5	94.226	95.088	99.225	98.987	98.322	98.639
1.0	90.618	99.275	98.792	102.989	96.726	103.092
1.5	86.704	102.606	95.694	107.933	94.894	109.336
2.0	82.174	103.733	92.565	111.129	92.017	113.85
2.5	79.115	103.845	90.331	112.184	90.166	118.116
3.0	76.822	104.096	87.378	112.976	88.058	119.052
c_1		3.886		10.253		26.193

3.2.2 Comparion for MC1 Chart

Tables 4, 5 and 6 give us information about the performances for the SC-MC1 (sign change method based on MC1 chart) and MC1 control charts. The constant k is selected to equal 0.5. The outcomes display that the sign change method for MC1 chart is generally better than the MC1 chart while the dimension p and μ_0 become larger and larger.

Table 4 The counts (accurate signal) of signals of the SC-MC1 and MC1 chart when $\mu_0 = \underline{0}$ occurring at the 15th–20th observation

d	$p = 3$		$p = 5$		$p = 7$	
	SC-MC1	MC1	SC-MC1	MC1	SC-MC1	MC1
0	99.726	99.804	96.206	95.876	79.091	78.559
0.5	100.157	102.271	96.327	96.683	80.328	78.936
1.0	100.755	105.471	98.775	99.915	84.688	81.512
1.5	102.757	106.654	103.481	102.677	91.896	83.534
2.0	105.633	107.918	110.173	103.601	100.191	85.386
2.5	109.808	107.207	116.734	104.226	108.017	85.87
3.0	112.39	107.773	122.517	103.82	115.302	85.556
c_2		1.332		2.56		4.823

Table 5 The counts (accurate signal) of signals of the SC-MC1 and MC1 chart when $\mu_0 = \underline{0.05}$ occurring at the 15th–20th observation

d	$p = 3$		$p = 5$		$p = 7$	
	SC-MC1	MC1	SC-MC1	MC1	SC-MC1	MC1
0	100.562	100.54	96.173	95.81	79.562	78.948
0.5	100.942	102.771	96.372	96.806	80.953	79.701
1.0	102.348	105.563	99.306	99.412	86.511	82.082
1.5	104.868	107.351	106.67	102.549	94.64	84.318
2.0	110.694	108.072	114.682	103.526	104.829	85.545
2.5	114.974	108.53	121.967	103.68	114.387	86.124
3.0	118.7	107.996	129.212	103.944	122.824	86.622
c_2		1.335		2.557		4.833

Table 6 The counts (accurate signal) of signals of the SC-MC1 and MC1 chart when $\mu_0 = \underline{0.1}$ occurring at the 15th–20th observation

d	$p = 3$		$p = 5$		$p = 7$	
	SC-MC1	MC1	SC-MC1	MC1	SC-MC1	MC1
0	100.96	100.997	97.481	97.085	80.996	80.399
0.5	101.851	102.109	97.524	98.524	82.239	80.959
1.0	104.411	105.993	101.316	102.031	88.946	83.903
1.5	109.026	107.862	109.851	104.207	99.129	85.782
2.0	114.95	108.252	119.792	105.535	111.377	87.925
2.5	120.717	107.986	129.57	105.67	122.387	88.519
3.0	124.881	108.565	137.557	105.423	131.295	88.651
c_2		1.336		2.573		4.866

3.2.3 Comparion for MEWMA Chart

Tables 7, 8 and 9 show the performances for the SC- MEWMA (sign change method based on MEWMA chart) and MEWMA control charts. The constant λ is selected to be 0.2. The results displays that the counts of the sign change method is always better than the MEWMA control chart no matter what the dimension p and μ_0 are.

The following Tables 10, 11 and 12 shows the error detection for calculating the counts which we detect the change point in the 11th to the 14th sample. We only consider the case of $\mu_0 = \underline{0.05}$ for the Hotelling-T^2 chart, the MC1 chart and the MEWMA chart.

Table 7 The counts (accurate signal) of signals of the SC-MEWMA and MEWMA chart when $\mu_0 = \underline{0}$ and $\lambda = 0.2$ occurring at the 15th–20th observation

d	$p = 3$		$p = 5$		$p = 7$	
	SC-type	MEWMA	SC-type	MEWMA	SC-type	MEWMA
0	127.824	127.441	102.462	101.799	92.456	91.631
0.5	132.146	131.284	105.955	102.771	94.643	92.772
1.0	141.815	135.26	112.746	105.894	100.042	94.58
1.5	149.541	138.101	120.061	107.117	106.084	96.869
2.0	154.761	138.929	126.015	107.634	111.153	98.157
2.5	157.954	138.846	129.575	107.391	115.639	98.674
3.0	159.961	138.712	133.428	107.993	118.457	97.597
c_3		2.492		5.896		13.994

Table 8 The counts (accurate signal) of signals of the SC-MEWMA and MEWMA chart when $\mu_0 = \underline{0.05}$ and $\lambda = 0.2$ occurring at the 15th–20th observation

d	$p = 3$		$p = 5$		$p = 7$	
	SC-type	MEWMA	SC-type	MEWMA	SC-type	MEWMA
0	128.832	128.485	103.73	103.058	94.65	93.845
0.5	138.563	131.826	111.319	104.534	99.424	95.251
1.0	150.733	137.632	121.988	107.405	108.21	97.724
1.5	161.048	140	131.077	108.095	114.989	99.569
2.0	166.017	139.897	138.459	108.894	121.738	100.197
2.5	169.13	139.834	142.418	108.362	125.633	100.049
3.0	171.093	140.19	145.95	109.119	129.595	99.958
c_3		2.506		5.931		14.13

Table 9 The counts (accurate signal) of signals of the SC-MEWMA and MEWMA chart when $\mu_0 = \underline{0.1}$ and $\lambda = 0.2$ occurring at the 15th–20th observation

d	$p = 3$		$p = 5$		$p = 7$	
	SC-type	MEWMA	SC-type	MEWMA	SC-type	MEWMA
0	131.951	131.581	108.012	107.33	98.109	97.301
0.5	146.072	134.811	118.258	108.407	105.914	98.936
1.0	160.557	141.115	131.088	111.136	115.811	100.987
1.5	170.315	143.441	142.263	113.144	125.144	103.467
2.0	176.662	143.468	149.655	113.859	131.601	103.603
2.5	180.56	143.643	154.358	114.269	136.224	104.667
3.0	182.213	144.111	157.372	114.499	139.862	104.91
c_3		2.54		6.033		14.367

Table 10 The counts (error signal) of signals of the SC-T^2 and Hotelling-T^2 chart when $\mu_0 = \underline{0.05}$ occurring at the 11th–14th observation

	$p = 3$		$p = 5$		$p = 7$	
0	891.92	894.355	883.996	883.162	886.68	876.313
0.5	892.355	895.017	884.417	883.578	886.031	876.913
1.0	892.283	894.834	884.041	883.442	886.481	876.418
1.5	892.281	894.684	884.368	883.399	885.935	876.69
2.0	892.117	894.842	884.03	883.506	887.421	876.996
2.5	892.304	895.017	884.669	883.22	887.377	876.266
3.0	891.929	894.501	884.22	883.829	886.94	877.261
c_1		3.90		10.362		26.317

Table 11 The counts (error signal) of signals of the SC-MC1 and MC1 chart when $\mu_0 = \underline{0.05}$ occurring at the 11th–14th observation

d	$p = 3$		$p = 5$		$p = 7$	
	SC-MC1	MC1	SC-MC1	MC1	SC-MC1	MC1
0	743.22	890.971	677.252	895.477	664.907	912.915
0.5	743.121	891.065	676.694	896.199	664.323	913.466
1.0	743.45	891.824	677.781	896.735	665.401	913.483
1.5	743.228	891.948	676.267	895.913	665.165	913.455
2.0	742.47	891.835	676.977	896.04	664.637	913.694
2.5	742.558	891.455	677.862	896.222	664.615	913.633
3.0	742.723	892.004	677.243	896.044	664.978	913.323
c_2		1.335		2.557		4.833

Table 12 The counts (error signal) of signals of the SC-MEWMA and MEWMA chart when $\mu_0 = \underline{0.05}$ and $\lambda = 0.2$ occurring at the 11th–14th observation

d	$p = 3$		$p = 5$		$p = 7$	
	SC-type	MEWMA	SC-type	MEWMA	SC-type	MEWMA
0	662.126	859.88	702.644	890.435	727.779	899.621
0.5	661.881	859.73	702.865	890.463	726.736	899.494
1.0	662.793	859.147	701.614	890.327	726.338	899.358
1.5	662.046	859.366	701.451	891.299	727.251	899.257
2.0	662.851	860.037	701.823	891.013	726.306	899.483
2.5	662.425	860.163	701.687	891.628	727.709	899.903
3.0	662.003	859.81	701.311	890.881	726.162	900.039
c_3		2.506		5.931		14.13

3.2.4 Known Covariance Case

In addition to the unknown covariance matrix case, we also conduct a simulation study with known covariance matrix in this subsection. We only consider the case of $\mu_0 = 0.05$ for the Hotelling-T^2 chart, the MC1 chart and the MEWMA chart. Tables 13, 14 and 15 display that there are a little difference in the counts and significant change in the control limits.

Table 13 The counts (accurate signal) of signals of the SC-T^2 and Hotelling-T^2 chart when $\mu_0 = \underline{0.05}$ and known Σ occurring at the 15th–20th observation. The counts (accurate signal) of signals of the SC-T^2 and Hotelling-T^2 chart when $\mu_0 = \underline{0.05}$ and known Σ occurring at the 15th–20th observation

	$p = 3$		$p = 5$		$p = 7$	
0	99.788	99.46	110.512	110.057	117.886	117.339
0.5	99.517	99.96	112.472	111.371	121.363	117.842
1.0	95.474	101.378	110.46	113.328	123.065	120.062
1.5	90.283	101.893	108.373	113.705	121.702	121.814
2.0	85.973	102.186	105.418	114.796	119.759	123.09
2.5	81.965	102.762	102.702	114.285	118.478	122.974
3.0	80.145	102.345	100.646	114.952	116.175	122.603
c_1		2.99		5.454		7.889

Table 14 The counts (accurate signal) of signals of the SC-MC1 and MC1 chart when $\mu_0 = \underline{0.05}$ and known Σ occurring at the 15th–20th observation

d	$p = 3$		$p = 5$		$p = 7$	
	SC-MC1	MC1	SC-MC1	MC1	SC-MC1	MC1
0	106.475	106.871	116.108	116.215	115.085	114.992
0.5	105.398	107.67	114.11	117.466	112.073	116.412
1.0	103.591	110.846	112.057	120.018	110.146	120.283
1.5	103.817	111.774	113.356	122.907	112.451	123.754
2.0	107.014	112.061	119.095	123.384	118.738	125.856
2.5	110.885	111.301	124.851	123.284	126.88	126.059
3.0	114.554	111.349	131.187	122.913	134.922	125.13
c_2		1.087		1.703		2.31

Table 15 The counts (accurate signal) of signals of the SC-MEWMA and MEWMA chart when $\mu_0 = \underline{0.05}$, known Σ and $\lambda = 0.2$ occurring at the 15th–20th observation

d	$p = 3$		$p = 5$		$p = 7$	
	SC-type	MEWMA	SC-type	MEWMA	SC-type	MEWMA
0	128.746	128.652	102.188	101.764	81.052	80.485
0.5	139.817	130.015	108.988	102.849	86.507	80.356
1.0	155.116	132.622	120.327	103.901	94.792	80.655
1.5	167.842	133.68	132.917	104.017	104.716	81.306
2.0	174.258	134.733	142.606	104.77	115.328	81.669
2.5	178.165	133.485	149.283	104.171	123.134	81.113
3.0	181.165	133.608	154.435	104.161	129.853	81.181
c_3		1.853		3.249		4.60

4 Real Data Example

We give a data example by using the oil prices of four different types. The data set is composed of prices for Unleaded Gasoline 92, Unleaded Gasoline 95, Unleaded Gasoline 98, and Premium Diesel. The data are the mean of the price for every month in the island of Taiwan. There are 25 samples with 4 variables in the data set from November, 2009 to November, 2011. The data are presented in Table 16. Suppose that the covariance matrix is unknown. We utilize the first 10 samples as the data in Phase I for estimating the in-control parameters. By using the sign change method in the previous section, we calculate the $D_i, i = 1, ..., 25$ for the MEWMA chart and MC1 chart. The signs of the D_i values, which are the differences between $MC1_i$ of $MC1_i^{JS}$ are

$$-, -, +, +, +, +, +, +, +, -, -, -, -, +, +, +, +, +, +, +, +, +, +, +, +$$

The signs of the D_i values, which are the differences between the MEWMA control chart and JS-type MEWMA control chart are

$$-, -, -, +, 0, -, +, +, 0, -, -, -, -, +, +, +, +, +, +, +, +, +, +, +, +$$

For the statistics calculated in Phase II, the D_i for the MC1 chart and the MEWMA chart changes sign in the 4th sample in the Phase II study (the 14th sample in the 25 sample). While contrasting with the real data, we found that the oil prices increase for all different types from 13th to 14th sample and become larger continually. It is reasonable to believe there is a small mean shift at the 14th sample.

By applying a sign change method to the real data, we lead to the same result for both charts. It reveals the robustness of this method for detecting the out-of-control signal.

Table 16 Oil price data

Month	92	95	98	*PD*
2009/11	29.83	30.53	32.03	27.39
2009/12	29.54	30.24	31.75	27.21
2010/01	29.49	30.19	31.69	27.12
2010/02	28.18	28.88	30.38	25.79
2010/03	28.44	29.13	30.65	26.09
2010/04	29.34	30.03	31.55	27.03
2010/05	29.16	29.86	31.37	26.86
2010/06	28.54	29.23	30.73	26.13
2010/07	28.57	29.27	30.78	26.16
2010/08	28.74	29.44	30.95	26.33
2010/09	28.58	29.28	30.79	26.17
2010/10	29.24	29.94	31.45	26.94
2010/11	29.78	30.48	31.99	27.52
2010/12	30.33	31.03	32.53	28.13
2011/01	30.79	31.5	33	28.61
2011/02	30.98	31.68	33.19	28.81
2011/03	31.92	32.62	34.13	29.83
2011/04	32.2	32.9	34.41	30.19
2011/05	31.67	32.37	33.88	29.65
2011/06	31.38	32.08	33.58	29.37
2011/07	31.21	31.91	33.41	29.23
2011/08	30.9	31.6	33.1	28.94
2011/09	31.07	31.77	33.26	29.07
2011/10	30.89	31.59	33.09	28.95
2011/11	31	31.7	33.2	29.15

5 Conclusion and Further Study

Wang [16] proposed a sign change method which based on the difference of the conventional chart statistic and JS-type chart statistic to detect signs. In this paper, we mainly concentrate on the simulation study to compare the performance of sign change method with those based on conventional control charts. While considering different $\mu = \mu_0$, the simulation result displays that the sign change method performs better than the conventional charts. Furthermore, in the real data example, we can detect the shift of oil prices by the sign change method. James-Stein estimator has the advantage that it can gather information from other dimensions to estimate one dimensional estimator. With the above-mentioned property of the James-Stein estimator, the sign change method has been a more convenient and efficient tool for detecting out-of-control signal based on the MC1 or EWMA control chart statistics.

References

1. ASTV Manager Online, 1 January 2013. Manager. http://www.manager.co.th/Home/ViewNews.aspx?NewsID=9550000157897
2. Chatterjee, S., Qiu, P.: Distribution-free cumulative sum control charts using bootstrap-based control limits. Ann. Appl. Stat. **3**, 349–369 (2009)
3. Crosier, R.B.: Multivariate generalizations of cumulative sum quality-control schemes. Technometrics **30**, 291–303 (1988)
4. Draper, N.R., Van Nostrand, R.C.: Ridge regression and James-Stein estimation: review and comments. Technometrics **21**, 451–466 (1979)
5. Efron, B., Morris, C.: Empirical bayes on vector observations—an extension of Stein's method. Biometrika **59**, 335–347 (1972)
6. Hotelling, H.: Multivariate quality control. In: Eisenhart, Hastay, Wallis (eds.) Techniques of Statistical Analysis. McGraw-Hill, New York (1947)
7. James, W., Stein, C.: Estimation with quadratic loss. In: Proceedings of the Fourth Berkeley Symposium on Mathematical Statistics and Probability, vol. 1, pp. 361–379 (1961)
8. Lehmann, E.L., Casella, G.: Theory of Point Estimation, 2nd edn. Springer, New York (1998)
9. Lowry, C.A., Woodall, W.H., Champ, C.W., Rigdon, S.E.: A multivariate exponentially weighted moving average control chart. Technometrics **34**, 46–53 (1992)
10. Montgomery, D.C.: Statistical Quality Control, 6th edn. Wiley, Hoboken (2009)
11. Pignatiello, J.J., Runger, G.C.: Comparisons of multivariate CUSUM charts. J. Qual. Technol. **22**, 173–186 (1990)
12. Stein, C.: Inadmissibility of the usual estimator for the mean of a multivariate distribution. In: Proceedings of the Third Berkeley Symposium on Mathematical Statistics and Probability, vol. 1, pp. 197–206 (1956)
13. Strawderman, W.E., Cohen, A.: Admissibility of estimators of the mean vector of a multivariate normal distribution with quadratic loss. Ann. Math. Stat. **42**, 270–296 (1971)
14. Wang, H.: Improved confidence estimators for the multivariate normal confidence set. Statistica Sinica **10**, 659–664 (2000)
15. Wang, H.: Comparison of p control charts for low defective rate. Comput. Stat. Data Anal. **53**, 4210–4220 (2009)
16. Wang, H.: A sign change method to detect out-of-control signals. Technical report (2019)
17. Wang, H., Huwang, L., Yu, J.H.: Multivariate control charts based on the James-Stein estimator. Eur. J. Oper. Res. **246**, 119–127 (2015)

Find Trade Patterns in China's Stock Markets Using Data Mining Agents

Baokun Li, Ziwei Ma, and Tonghui Wang

Abstract In this paper we develop an autonomous agent-based market microstructure simulation with two kinds of agents using different trade strategies. Both types of agent follow different rules to trade, with decisions made on data mining algorithms and real data from China's stock markets and NASDAQ in the US. The time range for this study covers last six years, the most volatile period in last 30 years in China's markets. The generalizations show that, during the collapse, the agent traders change their trading strategies more uniformly and frequently than usual days. At the relative low turning points, agent traders either do not sell, or sell and buy more actively than the previous day. Also before publication of some news, agents on related listed companies could show a special pattern lasting for many days.

Keywords Market collapse · Insider trade · Data mining · Agent

1 Introduction

Trade patterns have been concerned and studied by traders, researchers, and regulators for a long time [3]. There are many papers dealing with this topic using various means. Among the methods, agent based modeling and data mining are two special ones that do not use fixed criterion to make decisions. It has been a long time that

---

B. Li (✉) · Z. Ma · T. Wang
Department of Mathematical Sciences, New Mexico University, Las Cruces, USA
e-mail: ziweima@nmsu.edu

T. Wang
e-mail: twang@nmsu.edu

Faculty of Mathematical Sciences and Civil Engineering, Beijing Institute of Technology, Zhuhai, China
e-mail: 18103@bitzh.edu.cn

S. Sriboonchitta et al. (eds.), *Behavioral Predictive Modeling in Economics*, Studies in Computational Intelligence 897,
https://doi.org/10.1007/978-3-030-49728-6_11

scholars used agents to study both experiment market and real stock market [1, 2, 6]. And data mining methods have been around in the research field for stock market for about 20 years [4, 5]. Instead of making decisions directly from data mining models as in most papers in this field, we construct agents who's actions are influenced by data mining results.

A data mining agent is a computer program designed to generate specific type of data, along with identifying patterns among those data types. A data mining agent can save valuable employee time, as it avoids the need to assign data monitoring roles to specific employees. These agents are typically used to detect trends in data, alerting organizations to paradigm shifts so effective strategies can be implemented to either take advantage of or minimize the damage from alterations in trends.

A data mining agent could be considered as a very limited type of virtual employee. In effect, this agent is nothing more than an employee tasked with trading stock shares based on inflow real time data and dynamic data mining decisions.

For example, an agent is programmed to monitor stock prices for a specific company, at each time spot, several data mining models are generated to predict the price trend based on the renewed trade data, the agent makes decisions to buy, or sell, or keep current status with some predetermined strategy.

In this way, a data mining agent acts to save valuable employee time, as it is no longer necessary to assign these elementary monitoring roles to specific employees. This frees up man hours in the organization, allowing employees to divert their attention elsewhere until the data mining agents alert them that something in the system is actually worth observing. Without the use of these agents, individual employees would have to observe and record changes in the surveyed systems on a daily basis.

Additionally, data mining agents can be used to sift through database records, retrieving specific requested information that would otherwise prove tedious or difficult for a human to retrieve. For example, a data mining agent can easily and tirelessly sift through millions of records to find something as tedious as "All sales exceeding 50 dollars from January 1st, 2001 to May 24th, 2019." Whereas a human could become tired and make mistakes during a particularly long and boring search, an agent will never fail to retrieve its stated objective.

Although useful, data mining agents have their limitations. With the current state of artificial intelligence technology, it is difficult for a data mining device to detect hidden or complex patterns more effectively than a skilled human. Thus, while these agents have their place in rote or constricted observations with specifically defined parameters, they are not as suitable for highly detailed patterns or those necessitating a touch of human intuition. In this paper, an agent is programmed to monitor stock prices for a specific company, at each time spot, several data mining models are generated to predict the price trend based on the renewed trade data, the agent makes decisions to buy, or sell, or keep current status with some predetermined strategy. With the assumption that these agents do not affect the real movements of the market. The results or patterns generated by the agents are considered as an auxiliary measure of market.

2 Data and Preparement

In this paper covers the results of our agents during last few years, starting from early 2014 to the August 2019. We collect minute data everyday for all listed companies in China's stock markets after closure of the China's stock markets, including prices of open, high, low and close which are useful to make candle graphs for minutes, five minutes, 15 min and hour intervals. Human technique traders use these graphs and trade amounts to make decisions. By mimicing their experiences, we transform these graphs and trade amounts into values that could be put into data mining models. Here we use five minute candle sticks and trade amounts for consequtive three days, we use these values as input, and the dependent variable is a binary which is one if the close price in tomorrow is higher than the close price today. At each time spot, we set up three kinds of models: classification tree, neural network, and logistic regression (Table 1).

When setting up the models dynamically, the time interval is not a day, but five minutes. This interval is properiate for stocks traded in China's stock markets, since all stocks are traded frequently and in large amounts everyday. But when using the method in US and other stock markets, intervals should be set larger.

The trading strategies are not unique. We follow three traditional strategies, Momentum trading, Value investing, and Wyckoff method(a strategy called Volume Spread Analysis). Momentum Trading is a strategy in the investment world where traders will only concern themselves with the "hot picks" of the day, the high percentage and volume movers of that trading session. Traders in this methodology will look for assets that move a significant percentage on a given day, in high volume.

Table 1 Variables transformed from transaction data in previous days

Variable	Description	Formular
x_1	Spread of previous day	$100(c_{t-1} - o_{t-1})/o_{t-1}$
x_2	Relative position of close price	$(c_{t-1} - l_{t-1})/(h_{t-1} - l_{t-1})$
x_3	Spread of previous second day	$100(c_{t-2} - o_{t-2})/o_{t-2}$
x_4	Relative position of close price in the previous second day	$(c_{t-2} - l_{t-2})/(h_{t-2} - l_{t-2})$
x_5	Spread of previous third day	$100(c_{t-3} - o_{t-3})/o_{t-3}$
x_6	Relative position of close price in the previous third day	$(c_{t-3} - l_{t-3})/(h_{t-3} - l_{t-3})$
x_7	Rate of previous ten days	$100(c_{t-1} - c_{t-11})/c_{t-11}$
x_8	Rate of previous twenty days	$100(c_{t-1} - c_{t-21})/c_{t-21}$
x_9	Standardised trade volume of previous day	$(v_{t-1})/const$
x_{10}	Standardised trade volume of previous second day	$(v_{t-1})/const$
x_{11}	Relative trade volume of previous day	$(v_{t-1} * 2)/(v_{t-2} + v_{t-3})$

This trading strategy is not a long-term or buy-hold type of strategy. Instead, traders will hold their long or short positions for only as long the stock is moving quickly.

Value Investing, on the other hand, is definitely more of a medium to long-term prospect. Value investors are specifically looking for undervalued stocks, those assets that traders believe are trading below their intrinsic values. Investors who employ this strategy believe strongly that the market overreacts to both good and bad news.

The Wyckoff Method use both price and volume to derive supply and demand, effort and results, cause and effect, then make decisions according to these measures. For example, if the market rises with contracting spread and volume, the market is not showing demand. Without demand, it is not likely to continue rising. Conversely, if the market falls with decreasing spread and volume, the market is not interested in selling. Thus, it is not likely that the market will continue to fall.

3 Buying Strategy and Selling Strategy, Patterns in China's Stock Market

In this paper, the agents mimic the traders with different strategies (or combined strategies), an agent makes decision using model predictions and these strategies, i.e., when the model predictions coincide with some strategy or strategies, the agent buy, keep, sell, or no action. Following are the strategies for our two kinds of agents (Table 2).

Each day, the benefit earned by such an agent is called daily power, or simply, power. The power calculated with the buying strategy is called buying power, the power calculated with the selling strategy is called selling power. We calculate buying power and selling power for each stock, and adding all buying powers and selling powers as the buying power and selling power for the whole market. Please note that

Table 2 Buying and selling strategies

Action	Buying strategy	Selling strategy
Buy	Three consecutive downs and three predictions are all downward	Price up and at least two predictions are upwards
Sell	Loses money in the last period	Loses money in the last period
Keep	Wins money in the last period	Wins money in the last period
Sell short	High price, last period down, and three predictions are all upward,	High price, last period down, and three Predictions are all downward
Keep	Wins money in the last period	Wins money in the last period
Buy to cover	Loses money in the last period	Loses money in the last period
Idle	No position, and no chance	No position, and no chance

the whole market contains all stocks in Shanghai Stock Exchange Market and all stocks in Shenzhen Stock Exchange Market.

According to long time observation, we generalize the following relations between the powers and price movements. Basically,

(1) If the price(or index) goes up, the buying power is larger than the selling power. (2) usually, in consecutive two transaction days, the buying power and selling power go to opposite directions. In other words, if one power goes up, then the other will go down.

If the two relations are violated, then something uncommon may have happened.

4 Buying Power and Selling Power in Crazy Bull 2015

There are many explanations about Bullish and Bearish periods in stock markets, like wave theory, demand and supply theory, Herd Behavior theory. Using our buying power and selling power method, We got a different explanation for the market movements during the Crazy bullish period and the despert bearish periods.

China's stock market index, we choose the SSE Composite Index(SCI) as the representation of the whole stocks in China's stock markets, i.e., SSE and Shenzhen SE. because the SCI and Shenzhen SE Composite Index(SSCI) always have the same trends during the last 20 years. In the Fig. 1, the top panel contains the candles for the SCI variations day by day; the middle panel contains corresponding volume for each day; the bottom panel contains corresponding buying power and selling power for each day. Note: In the top and middle panel, red color means index going upward, green color means index going downward. In the bottom panel the red zigzag line is for the buying power, the blue zigzag line is for selling power. The coloring is the opposite to the coloring convention in US and other financial markets.

In the first half of 2015, China's stock market entered a bullish period, in majority of days, the buying power is above the selling power, and the average difference

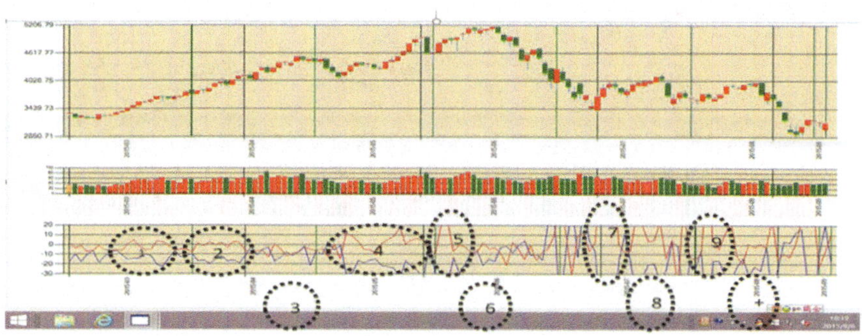

Fig. 1 China's 2015 roller coaster market, and day to day buying and selling power

is getting larger wave after wave, see the parts circled 1,2,4,5,7,9. If the average difference gets smaller in the new wave, index diving happens, see the parts circled 3,6,8,10(+). In the market collapse period from mid-June to early July, the buying and selling power pattern is not simply the opposite of the pattern in the rising parts. The selling power vibrates more vehemently than that in usual trade days.

4.1 Collapsing Pattern

Usually stock markets do not collapse frequently, but for the last few years in China, in the high-speed development process, due to De-leveraging in early 2018 and the kicking off of the China-US trade war in June 2018, China's stock markets collapsed twice. In Fig. 2, in the parts of two circles, It happens again that the selling power vibrates more vehemently than that in usual trade days.

4.2 Bottom Signals

Traders composition in China's stock markets may be different from that in advanced countries. The private investors are used to trading their stock shares themselves, and they account for more than 90% of the traders in markets. They make decisions mainly from news or social networks and media. So China's stock markets are more volatile than markets in developed countries, Consequently collapses and bouncing back at bottoms occurred more frequently.

So far there are two kinds of bottom signals. In the first kind, after the market index falling down for many days, the buying power and selling power both go up, which represents a phenomena that part of the traders are selling vehemently, at

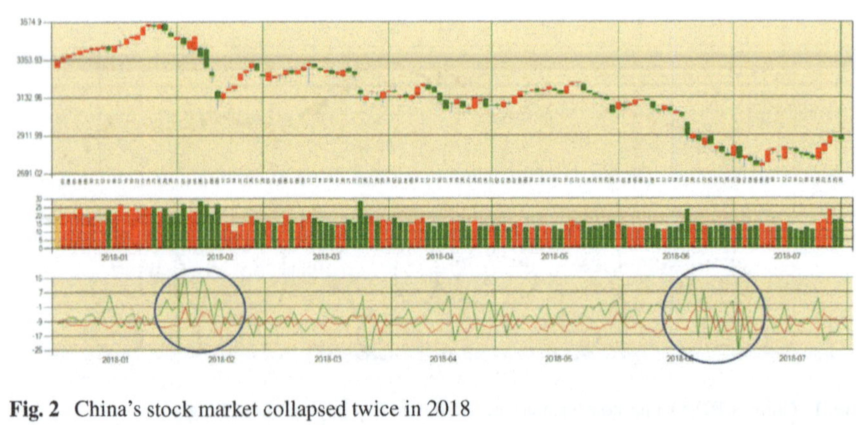

Fig. 2 China's stock market collapsed twice in 2018

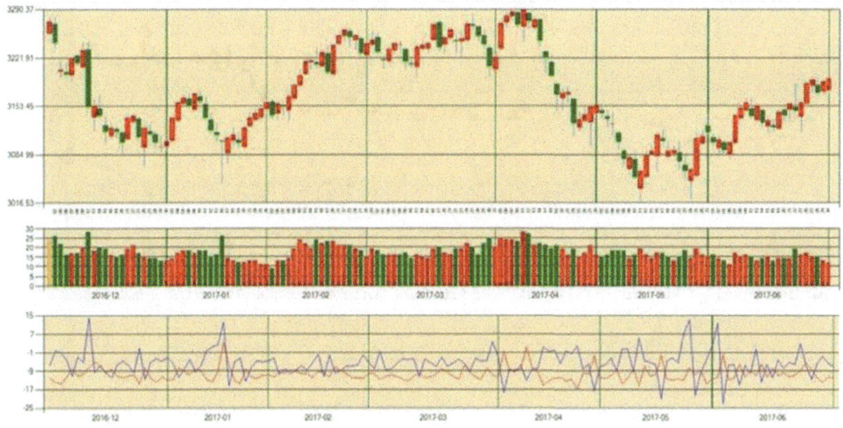

Fig. 3 Relative lowest points in China's stock market in 2017

the same time, part of the traders are buying actively. In the second kind, after the market index falling down for many days, the selling power became very low, close to historical low, which represents a phonomena that the traders are very reluctant to sell stock shares.

In above Fig. 1, in mid-August. 2015 , the two bottom index candlesticks look like inverted hammers, the corresponding buying power and selling power both go upward. This pattern means that the both buying transaction and selling transaction are active.

In Fig. 3, in mid-Jan, end of June, and early July 2017, the selling power line changes from high position to very low position the next day, in another words, traders sells stocks in crazy, but next day, they keep stock shares as treasure stones.

4.3 Special Pattern Before News Release

If a individual stock's buying and selling powers show some very uncommon patterns, it is possible there are insider transactions happened. For example, in Fig. 4, BZUN a company listed in NASDAQ, partially owned by Jack Ma. In Jan 10, 2016, Jack Ma, the Chairman of Alibaba in China, met with USA president Donald Trump to afford a platform to sell US products to China to mitigate the trade tension between the US and China. In the week before the meeting, the buying power is higher than the selling power (the green line is above the red line). After the meeting, the stock price soars in the next several months.

In Fig. 4, in the two circled periods, buying power is higher than the selling power (the green line is above the red line) as the stock price gets down or does not change much.

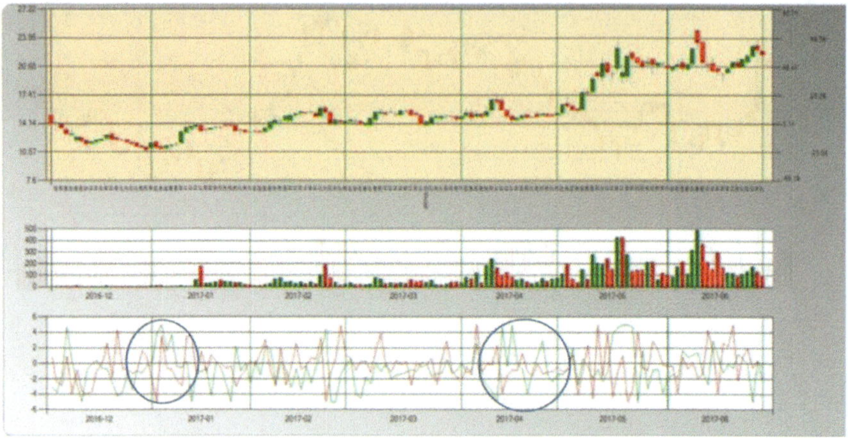

Fig. 4 Special pattern before President Trump's meeting with Jack Ma

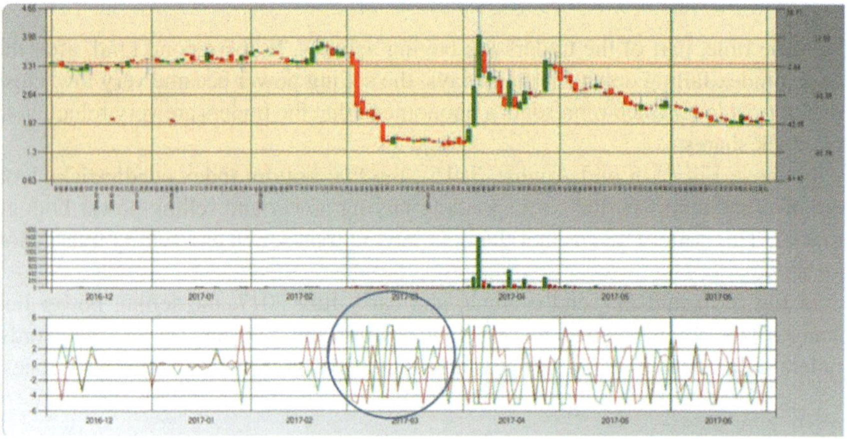

Fig. 5 Special pattern before proclamation of Xiong An New Area

For another example, in Fig. 5, CALI, also a chinese company listed in NASDAQ, its location is close to Xiong An New Area, a state-level new area in the Baoding area of Hebei, China. Xiong An is established in April 2017. Because the establishment means a lot to listed companies like CALI, many universities and state administrations will move to the New Area in a couple of years, companies around the New Area listed both in China's and US stock markets are favored by traders in both markets.

In Fig. 5, in the first 20 days in March, 2016, buying power is higher than the selling power (the green line is above the red line) as the stock price gets down and then stays at bottom. After the proclamation of the news at the April first, CALI's price doubled in three days.

Please note that the special pattern in BZUN and CALI works only if good news follows and possible insider trading exists. For most cases with such patterns in NASDAQ or in China's markets, usually either good or bad news follows after such pattern appears. In China's stock markets, the majority of cases are that good news follows; while in NASDAQ, good news and bad news both appears. The reason for the difference is that trade rules are different in the two countries. In China, trades can only buy long, but in the US, traders can trade in both directions to make money.

5 Conclusion and Further Study

This study mainly focuses on finding trade patterns in China's stock markets using agents simulated with the data mining algorithms. On one hand, China's stock markets experienced many collapses and we have chance to generalize trade patterns about special time spots; On the other hand, China's financial regulators changed many regulations during last several years, and these changes result in changes in trader's habit, therefore some rules generalized have to be abandoned. Because we only tried the method on NASDAQ in the first half of 2016, NASDAQ did not experience a roller coaster period at the time. In the future, we will try the agent method on US markets for a long time to know whether the buying and selling powers show the same pattern as in China's markets in some special turning points. We are also interested in finding the strategies investors could use to make money in real markets.

References

1. Kluger, B., McBride, M.: Intraday trading patterns in an intelligent autonomous agent-based stock market. J. Econ. Behav. Organ. **79**(3), 226–245 (2011)
2. Chany, N., LeBaronz, B., et al.: Agent-based models of financial markets: a comparison with experimental markets. SSRN Electron. J., August 2000
3. Tamersoy, A., et al.: Large scale insider trading analysis: patterns and discoveries. IEEE/ACM ASONAM, August 2013
4. Kannan, K., Sekar, P., et al.: Financial stock market forecast using data mining techniques. In: Proceedings of the International MultiConference of Engineers and Computer Scientists 2010 Vol I, IMECS 2010, Hong Kong, March 2010
5. Chun, S.-H., Kim, S.: Data mining for financial prediction and trading: application to single and multiple markets. Expert Syst. Appl. **26**(2), 131–139 (2004)
6. Kim, M., Kim, M.: Group-wise herding behavior in financial markets: an agent-based modeling approach. PLoS ONE **9**(4), e93661 (2014)

The Decomposition of Quadratic Forms Under Matrix Variate Skew-Normal Distribution

Ziwei Ma, Tonghui Wang, Baokun Li, Xiaonan Zhu, and Yuede Ma

Abstract In this paper, several properties of the noncentral skew Wishart distribution are studied and two results of decomposition properties are established. In general, a random matrix, which follows noncentral skew Wishart distribution with the degrees of freedom $k > 1$, can be decomposed into the sum of two independent random matrices, one having the noncentral skew Wishart distribution and another having the noncentral Wishart distribution. For illustration of these results, the multivariate one-way classification model with skew-normal error is considered as an application.

1 Introduction

A p-dimensional random vector $\mathbf{Z}$ has a multivariate skew-normal distribution with the shape parameter α, denoted by $\mathbf{Z} \sim SN_p(\alpha)$, if its probability density function (PDF) given by

$$f(\mathbf{z}; \alpha) = 2\phi_p(\mathbf{z})\Phi(\alpha'\mathbf{z}), \qquad \mathbf{z} \in \Re^p, \tag{1}$$

Z. Ma · T. Wang (✉)
Department of Mathematical Sciences, New Mexico State University, Las Cruces, USA
e-mail: twang@nmsu.edu

Z. Ma
e-mail: ziweima@nmsu.edu

B. Li
School of Mathematical Science and Civil Engineering, Beijing Institute of Technology, Zhuhai Campus, Beijing, China
e-mail: bali@swufe.edu.cn

X. Zhu
Department of Mathematics, University of North Alabama, Florence, USA
e-mail: xzhu7@una.edu

Y. Ma
School of Science, Xi'an Technological University, Xi'an, China
e-mail: mayuede0000@163.com

S. Sriboonchitta et al. (eds.), *Behavioral Predictive Modeling in Economics*, Studies in Computational Intelligence 897,
https://doi.org/10.1007/978-3-030-49728-6_12

where $\alpha \in \Re^p$, $\phi_p(\cdot)$ is the PDF of standard p-dimensional normal distribution and $\Phi(\cdot)$ is the cumulative distribution function (CDF) of standard univariate normal distribution, respectively, by Azzalini and Dalla Valle [3]. The location parameter $\mu \in \Re^p$ and the scale parameter, non negative definite $p \times p$ matrix Σ can be introduced through affine transformation, say $\mathbf{Y} = \mu + \Sigma^{1/2}\mathbf{Z}$, denoted as $\mathbf{Y} \sim SN_p(\mu, \Sigma, \alpha)$. Since then there are extensive studies on the class of multivariate skew-normal distribution in both theoretical development and realistic applications (see, Azzalini and Capitanio [2], Genton et al. [8], Gupta and Huang [11], Gupta and Chang [9], Gupta et al. [10], Gupta et al. [12], Wang et al. [18] and references therein). These contributions derive more detailed features, like higher order moments, moment generating functions (MGF), stochastic representations, the distribution of quadratic forms, extend to more general classes, and explore applications of multivariate skew-normal distribution. Reader are referred to two monographs, Genton [7] and Azzalini [4] for a comprehensive and updated development related to univariate and multivariate skew-normal distributions.

The distribution of quadratic forms plays an important role in statistical inference. Under the multivariate skew-normal setting, many researchers studied the distribution of quadratic forms. Genton et al. [8] and Gupta and Huang [11] discussed the distribution of quadratic forms when the location parameter equals zero. For the case when location parameter is not zero, Wang et al. [18] introduced the noncentral skew chi-square distribution, and derived a version of Cochran's theorem. Ye and Wang [20] and Ye et al. [22] defined noncentral skew F-distribution, the ratio of the quadratic forms of skew-normal vectors which is extended to noncentral closed skew F-distribution by Zhu et al. [25], and applied to linear mixed model with skew-normal random effects and the variance components model with skew-normal random errors. Recently, Ma et al. [17] discussed the decomposition properties of noncentral skew chi-square distribution.

Random matrix is of an excellent tool for modelling three-way data set, such as multivariate repeated measurements or longitudinal/sequential data. Therefore, Chen and Gupta [5] initially introduced matrix variate skew-normal distribution by extending the PDF from multivariate case. After that, there are more generalization of matrix variate skew-normal distribution studied by Harrar and Gupta [13], Akdemir and Gupta [1] and Young et al. [23]. Recently, Gallaugher and McNicholas [6] introduced three classes matrix variate skewed distribution using mixture construction which can be viewed as generalization of matrix variate skew-normal distribution. However, to our best knowledge, the literature of the distribution of quadratic forms under matrix variate skew-normal distribution is scarce. In Ye et al. [21], the noncentral skew Wishart distribution is introduced which leads to results in constructing confidence regions and hypothesis testing on location parameter under multivariate skew-normal setting by Ma et al. [15, 16]. In this study, we dig in the distribution of matrix variate quadratic forms under matrix variate skew-normal setting.

The remainder of the paper is organized as follows. In Sect. 2, a brief introduction on matrix variate skew-normal distribution. The noncentral skew Wishart distribution is introduced in Sect. 3. Followed by Sect. 4, some properties of noncentral skew

Wishart distribution are presented. The main results of decomposition of noncentral skew Wishart distribution are given in Sect. 5.

2 The Matrix Variate Skew-Normal Distribution

Let $M_{k\times p}$ be the set of all $k \times p$ matrices over the real field $\Re$ and $\Re^k = M_{k\times 1}$. For any $B \in M_{k\times p}$, we will use B', B^+ and $r(B)$ to denote the transpose, the Moore-Penrose inverse and the rank of B, respectively. For nonnegative definite $C \in M_{n\times n}$ with $r(C) = r < n$, let $C^{-1/2}$ be the symmetric matrix such that $C^{-1/2}C^{-1/2} = C^+$. For $B \in M_{k\times s}$ and $C \in M_{s\times p}$ and $D \in M_{p\times q}$, we use $B \otimes C$ to denote the Kronecker product of B and C, $\text{Vec}(BCD) = (B \otimes D')\text{Vec}(C)$, and $\langle B, C\rangle = \text{tr}\left(B'C\right)$ to denote the trace inner product. Also, for any full column rank matrix $A \in M_{k\times p}$, we use $P_A = A(A'A)^{-1}A'$ to represent the projection matrix, and let I_k denote the identity matrix with order k and $\mathbf{1}_k = (1, \ldots, 1)' \in \Re^k$.

Definition 2.1. *Let $Z \sim M_{k\times p}$ be a random matrix, it is said to have standard matrix variate skew-normal distribution if its PDF is given by*

$$f(Z;\Gamma) = 2\phi_{k\times p}(Z)\,\Phi\left(tr\left(\Gamma'Z\right)\right), \qquad Z \in M_{k\times p},$$

where $\phi_{k\times p}(Z) = (2\pi)^{-kp/2}\, etr\left(-Z'Z\right)$ and $\Phi(\cdot)$ is the CDF of univariate standard normal distribution, denoted as $Z \sim SN_{k\times p}(\Gamma)$.

We can obtain the moment generating function (MGF), expectation and convariance of Z as follows.

Proposition 2.1. *Let $Z \sim SN_{k\times p}(\Gamma)$ and $\Delta = \frac{\Gamma}{[1+tr(\Gamma'\Gamma)]^{1/2}}$. Then*
(i) The MGF of Z is

$$M_Z(T) = 2etr\left(T'T/2\right)\Phi\left(etr\left(\Delta'T\right)\right), \qquad T \in M_{k\times p}.$$

(ii) The expectation and convariance matrix are

$$E(Z) = \sqrt{\frac{2}{\pi}}\Delta, \qquad Cov(\mathbf{z}) = I_{kp} - \frac{2}{\pi}Vec(\Delta)\,Vec(\Delta)',$$

where $\mathbf{z} = Vec(Z)$.

Proof. We know that $Z \sim SN_{k\times p}(\Gamma)$ is equivalent to $\mathbf{z} \sim SN_{kp}\left(0, I_{kp}, \text{Vce}(\Gamma)\right)$ and the MGF of $\mathbf{z}$ is

$$M_{\mathbf{z}}(\mathbf{t}) = E\left(\exp\left(\mathbf{z}'\mathbf{t}\right)\right) = 2\exp\left(\mathbf{t}'\mathbf{t}/2\right)\Phi\left\{\frac{\text{Vec}(\Gamma)'\mathbf{t}}{\left(1+\text{Vec}(\Gamma)'\,\text{Vec}(\Gamma)\right)^{1/2}}\right\}, \qquad \mathbf{t} \in \Re^{kp}. \quad (2)$$

For $T \in M_{k\times p}$, let $\mathbf{t} = \mathrm{Vec}(T)$ in Eq. (2), we have

$$M_Z(T) = E\left(\mathrm{tr}\left(T'Z\right)\right) = 2\mathrm{etr}\left(T'T/2\right)\Phi\left\{\frac{\mathrm{tr}\left(\Gamma'T\right)}{\left(1+\mathrm{tr}(\Gamma'\Gamma)\right)^{1/2}}\right\},$$

which gives the result in (i). Similarly, from the facts

$$E(\mathbf{z}) = \sqrt{\frac{2}{\pi}}\mathrm{Vec}(\Delta) \qquad \text{and} \qquad Cov(\mathbf{z}) = I_{kp} - \frac{2}{\pi}\mathrm{Vec}(\Delta)\,\mathrm{Vec}(\Delta)',$$

we obtain the desired results immediately. □

Let $M \in M_{n\times q}$, $A \in M_{k\times n}$ and $B \in M_{p\times q}$. Consider the random matrix $Y = M + A'ZB$ with $Z \sim SN_{k\times p}(\Gamma)$. We define the distribution of Y as a generalized version of matrix variate skew-normal distribution which can be easily expressed on the basis of the distribution of Z.

Definition 2.2. *Let $Z \sim SN_{k\times p}(\Gamma)$. The distribution of $Y = M + A'ZB$ is said to have matrix variate skew-normal distribution with the location parameter $M \in M_{n\times q}$, the scale matrices $A \in M_{k\times n}$ and $B \in M_{p\times q}$, and the shape parameter $\Gamma \in M_{k\times p}$, denoted as $Y \sim \mathscr{SN}_{n\times q}(M, A, B, \Gamma)$.*

We obtain some properties of the distribution of Y as follows.

Proposition 2.2. *Suppose that $Y \sim \mathscr{SN}_{n\times q}(M, A, B, \Gamma)$. Let $\Delta = \Gamma/\left(1 + tr\left(\Gamma'\Gamma\right)\right)^{1/2}$. Then*
(a) The MGF of Y is given by

$$M_Y(T) = 2etr\left(T'M + \frac{B'BT'A'AT}{2}\right)\Phi\left\{\frac{tr\left(\Gamma'ATB'\right)}{\left(1+tr\left(\Gamma'\Gamma\right)\right)^{1/2}}\right\}, \qquad T \in M_{n\times q}. \tag{3}$$

(b) The expectation and covariance matrix of Y are

$$E(Y) = M + \sqrt{\frac{2}{\pi}}A'\Delta B, \qquad Cov(\mathbf{y}) = \left(A'\otimes B'\right)\left(I_{kp} - \frac{2}{\pi}Vec(\Gamma)\,Vec(\Gamma)'\right)(A\otimes B),$$

where $\mathbf{y} = Vec(Y)$.
(c) The PDF of Y, if it exists, is given by

$$f(Y; M, A, B, \Gamma) = 2\phi_{n\times q}(Y; M, V, \Sigma)\,\Phi\left(tr\left(\Gamma_1'V^{-1/2}(Y-M)\,\Sigma^{-1/2}\right)\right),$$

where $V = A'A$, $\Sigma = B'B$, and $\Gamma_1 = \frac{V^{-1/2}A'\Gamma B\Sigma^{-1/2}}{\left[1+Vec(\Gamma)'\left(I_{kp}-P_A\otimes P_B\right)Vec(\Gamma)\right]^{1/2}}$.

Proof. For (a), by definition of the MGF and Proposition 2.1, we obtain that the MGF of Y is

$$M_Y(T) = E\left(\operatorname{etr}(T'Y)\right) = \operatorname{etr}(T'M)\, E\left(\operatorname{etr}\left((ATB')'Z\right)\right)$$
$$= 2\operatorname{etr}\left(T'M + \frac{T'A'ATB'B}{2}\right)\Phi\left(\operatorname{tr}\left(\Delta' ATB'\right)\right).$$

For (b), the desired results can be directly computed by using the properties of expectation and covariance operators and Proposition 2.1. For (c), we suppose A and B are of full column rank matrices, then V and Σ are positive definite. Let $Y_1 = M + V^{1/2} Z_1 \Sigma^{1/2}$ with $Z_1 \sim SN_{k\times p}(\Gamma_1)$ where

$$\Gamma_1 = \frac{V^{-1/2} A' \Gamma B \Sigma^{-1/2}}{\left[1 + \operatorname{Vec}(\Gamma)'\left(I_{kp} - P_A \otimes P_B\right)\operatorname{Vec}(\Gamma)\right]^{1/2}}.$$

By (a), the MGF of Y_1 is

$$M_{Y_1}(T) = 2\operatorname{etr}\left(T'M + \Sigma T' V T/2\right)\Phi\left\{\operatorname{tr}\left(\Delta_1' V^{1/2} T \Sigma^{1/2}\right)\right\}, \tag{4}$$

where $\Delta_1 = \Gamma_1 / \left[1 + \operatorname{tr}\left(\Gamma_1'\Gamma_1\right)\right]^{1/2}$. Substituting Γ_1 back in equation (4), it is clear that (3) and (4) are equivalent, i.e. $M_Y(T) = M_{Y_1}(T)$. Thus by the uniqueness of MGF, Y and Y_1 have the same PDF. By (i) of Proposition (2.1), the PDF of Y_1 is

$$f(Y_1; M, A, B, \Gamma_1) = 2\phi(Y_1; M, V, \Sigma)\,\Phi\left\{\operatorname{tr}\left(\Gamma_1' V^{-1/2}(Y_1 - M)\Sigma^{-1/2}\right)\right\}.$$

Remark 2.1. *The Definition 2.2 is an extension from the definition of matrix variate skew-normal distribution given by Ye et al. [21] with the general scale parameter B and shape parameter Γ, where*

$$SN_{k\times p}(M, V\otimes\Sigma, \gamma\otimes\alpha') = \mathcal{SN}_{k\times p}(M, V^{1/2}, \Sigma^{1/2}, \gamma\otimes\alpha'),$$

and they are equivalent when the PDF of Y exists. So Proposition 2.2 presents an extended results from related results in Ye et al. [21]. When V and Σ are singular, we can extend our results without difficulty by using the relevant results on singular multivariate skew-normal distribution in Li et al. [14].

We consider the properties of matrix variate skew-normal distribution under linear transformation which will be useful to study the distribution of its quadratic forms in Sect. 3.

Theorem 2.1. *Let $Y \sim SN_{k\times p}(M, V\otimes\Sigma, \Gamma)$ and let $X = C'YD$ for matrices $C \in M_{k\times s}$ and $D \in M_{p\times t}$. Then, the MGF of X is given by*

$$M_X(T) = 2etr\left(DT'C'M + \frac{\Sigma DT'C'VCTD'}{2}\right)\Phi\left\{\frac{tr(\Gamma' ACTD'\Sigma^{1/2})}{[1 + tr(\Gamma'\Gamma)]^{1/2}}\right\}, \tag{5}$$

for $T \in M_{s\times t}$.

Proof. The proofs are similar to that of Proposition 2.1 given in Ye et al. [21]. □

Corollary 2.1. *Let* $Y \sim SN_{k\times p}(M, V \otimes \Sigma, \Gamma)$, *and* Q *be an* $k \times k$ *orthogonal matrix, then*

$$Q'Y \sim SN_{k\times p}(Q'M, Q'VQ \otimes \Sigma, Q'\Gamma). \tag{6}$$

Proof. Directly apply the Theorem 2.1 to obtain the desired result. □

3 Noncentral Skew Wishart Distribution

Definition 3.1. *Let* $Y \sim SN_{k\times p}(M, I_k \otimes \Sigma, \Gamma)$. *The distribution of* $Y'Y$ *is defined as the noncentral skew Wishart distribution with degrees of freedom* k, *the scale matrix* Σ, *the noncentral parameter matrix* $\Lambda = M'M$, *and the skewness parameter matrices* $\Delta_1 = \Gamma'M$, *and* $\Delta_2 = \Gamma'\Gamma$, *denoted by* $Y'Y \sim SW_p(k, \Sigma, \Lambda, \Delta_1, \Delta_2)$.

Remark 3.1. *The Definition 3.1 is an extension of Definition given by Ye et al. [21] with a general skewness parameter.*

The following lemma will be used in calculating the MGF of the quadratic form.

Lemma 3.1. *Let* $\mathbf{U} \sim N_n(\mathbf{0}, \Sigma)$. *For any scalar* $p \in \Re$ *and* $\mathbf{q} \in \Re^n$, *we have*

$$E\left(\Phi\left(p + \mathbf{q}'\mathbf{U}\right)\right) = \Phi\left\{\frac{p}{\left(1 + \mathbf{q}'\Sigma\mathbf{q}\right)^{1/2}}\right\}.$$

The proof of Lemma 3.1 is given in Zacks [24].

Proposition 3.1. *Let* $Z \sim SN_{k\times p}\left(0, I_{kp}, \Gamma\right)$, $Y = M + A'Z\Sigma^{1/2}$, *and* $Q = Y'WY$ *with symmetric* $W \in M_{n\times n}$. *Then the MGF of* Q *is*

$$M_Q(T) = \frac{2\exp\left\{\langle T, M'WM\rangle + 2\left\langle M, \left[\left(WA' \otimes T\Sigma^{1/2}\right)\left(I_{kp} - 2\Psi\right)^{-1}\left(AW \otimes \Sigma^{1/2}T\right)\right](M)\right\rangle\right\}}{\left|I_{kp} - 2\Psi\right|^{1/2}}$$
$$\times \Phi\left\{\frac{\left\langle \Gamma, \left[\left(I_{kp} - 2\Psi\right)^{-1}\left(A \otimes \Sigma^{1/2}\right)\right](L)\right\rangle}{\left[1 + Vec(\Gamma)'\left(I_{kp} - 2\Psi\right)^{-1} Vec(\Gamma)\right]^{1/2}}\right\},$$

for symmetric $T \in M_{p\times p}$ *such that* $\rho(\Psi) < 1/2$, *where* $\Psi = (AWA)' \otimes \left(\Sigma^{1/2}T\Sigma^{1/2}\right)$ *and* $L = 2WMT$.

Proof. Note that $Y = M + A'Z\Sigma^{1/2} = M + \left(A' \otimes \Sigma^{1/2}\right)(Z)$, then we have

$$\begin{aligned} Q = Y'WY &= \left[M + \left(A' \otimes \Sigma^{1/2}\right)(Z)\right]' W \left[M + \left(A' \otimes \Sigma^{1/2}\right)(Z)\right] \\ &= M'WM + M'W\left[\left(A' \otimes \Sigma^{1/2}\right)(Z)\right] + \left[\left(A' \otimes \Sigma^{1/2}\right)(Z)\right]' WM \\ &\quad + \left[\left(A' \otimes \Sigma^{1/2}\right)(Z)\right]' W \left[\left(A' \otimes \Sigma^{1/2}\right)(Z)\right], \end{aligned}$$

and

$$\begin{aligned}\langle T, Q\rangle &= \langle T, M'WM\rangle + \left\langle T, M'W\left[\left(A'\otimes \Sigma^{1/2}\right)(Z)\right] + \left[\left(A'\otimes \Sigma^{1/2}\right)(Z)\right]' WM\right\rangle \\ &\quad + \left\langle T, \left[\left(A'\otimes \Sigma^{1/2}\right)(Z)\right]' W\left[\left(A'\otimes \Sigma^{1/2}\right)(Z)\right]\right\rangle \\ &= \langle T, M'WM\rangle + \left\langle Z, \left(A\otimes \Sigma^{1/2}\right)(L)\right\rangle + \langle Z, \Psi(Z)\rangle,\end{aligned}$$

where L and Ψ are given above. By Definition 2.1, the MGF of Q is given by

$$\begin{aligned}M_Q(T) &= \int \exp\left\{\langle T, M'WM\rangle + \left\langle Z, \left(A\otimes \Sigma^{1/2}\right)(L)\right\rangle + \langle Z, \Psi(Z)\rangle\right\} f(Z)\,dZ \\ &= \frac{2\exp\{\langle T, M'WM\rangle\}}{(2\pi)^{kp/2}} \int \exp\left\{-\frac{1}{2}\mathrm{Vec}(Z)'\left(I_{kp}-2\Psi\right)\mathrm{Vec}(Z)\right. \\ &\quad \left. + \mathrm{Vec}(Z)'\,\mathrm{Vec}\left(\left(A\otimes \Sigma^{1/2}\right)(L)\right)\right\} \Phi\left\{\mathrm{Vec}(\Gamma)'\mathrm{Vec}(Z)\right\} dZ \\ &= \frac{2\exp\left\{\langle T, M'WM\rangle + 2\left\langle M, \left[\left(WA'\otimes T\Sigma^{1/2}\right)\left(I_{kp}-2\Psi\right)^{-1}\left(AW\otimes \Sigma^{1/2}T\right)\right](M)\right\rangle\right\}}{(2\pi)^{kp/2}} \\ &\quad \times \int \exp\left\{-\frac{1}{2}H'\left(I_{kp}-2\Psi\right)H\right\} \Phi\left\{\mathrm{Vec}(\Gamma)'\mathrm{Vec}(Z)\right\} dZ\end{aligned}$$

where $H = \mathrm{Vec}(Z) - \left(I_{kp}-2\Psi\right)^{-1}\mathrm{Vec}\left(\left(A\otimes \Sigma^{1/2}\right)(L)\right)$. Note that

$$\mathrm{Vec}(\Gamma)'\mathrm{Vec}(Z) = \mathrm{Vec}(\Gamma)'H + \left\langle \mathrm{Vec}(\Gamma), \left[\left(I_{kp}-2\Psi\right)^{-1}\left(A\otimes \Sigma^{1/2}\right)\right](L)\right\rangle.$$

By Lemma 3.1, we obtain

$$\begin{aligned}M_Q(T) &= \frac{2\exp\left\{\langle T, M'WM\rangle + 2\left\langle M, \left[\left(WA'\otimes T\Sigma^{1/2}\right)\left(I_{kp}-2\Psi\right)^{-1}\left(AW\otimes \Sigma^{1/2}T\right)\right](M)\right\rangle\right\}}{\left|I_{kp}-2\Psi\right|^{1/2}} \\ &\quad \times E\left\{\Phi\left\{\mathrm{Vec}(\Gamma)'H + \left\langle \mathrm{Vec}(\Gamma), \left[\left(I_{kp}-2\Psi\right)^{-1}\left(A\otimes \Sigma^{1/2}\right)\right](L)\right\rangle\right\}\right\} \\ &= \frac{2\exp\left\{\langle T, M'WM\rangle + 2\left\langle M, \left[\left(WA'\otimes T\Sigma^{1/2}\right)\left(I_{kp}-2\Psi\right)^{-1}\left(AW\otimes \Sigma^{1/2}T\right)\right](M)\right\rangle\right\}}{\left|I_{kp}-2\Psi\right|^{1/2}} \\ &\quad \times \Phi\left\{\frac{\left\langle \mathrm{Vec}(\Gamma), \left[\left(I_{kp}-2\Psi\right)^{-1}\left(A\otimes \Sigma^{1/2}\right)\right](L)\right\rangle}{\left[1+\mathrm{Vec}(\Gamma)'\left(I_{kp}-2\Psi\right)^{-1}\mathrm{Vec}(\Gamma)\right]^{1/2}}\right\}.\end{aligned}$$

□

By Proposition 3.1, we obtain the MGF of $X'X \sim SW_p(k, \Sigma, \Lambda, \Delta_1, \Delta_2)$ where $X = M + Z\Sigma^{1/2}$.

Corollary 3.1. *Let $X = M + Z\Sigma^{1/2}$ and $Z \sim SN_{k\times p}\left(0, I_{kp}, \Gamma\right)$ is given by*

$$M_{X'X}(T) = \frac{2\exp\left\{\langle T, \Lambda\rangle + 2\left\langle \Lambda, T\Sigma^{1/2}\left(I_p - 2\Sigma^{1/2}T\Sigma^{1/2}\right)^{-1}\Sigma^{1/2}T\right\rangle\right\}}{\left|I_p - 2\Sigma^{1/2}T\Sigma^{1/2}\right|^{k/2}} \times \Phi\left\{\frac{2\left\langle T, \Sigma^{1/2}\left(I_p - 2\Sigma^{1/2}T\Sigma^{1/2}\right)^{-1}\Delta_1\right\rangle}{\left[1 + \left\langle \Delta_2, \left(I_p - 2\Sigma^{1/2}T\Sigma^{1/2}\right)^{-1}\right\rangle\right]^{1/2}}\right\}, \tag{7}$$

for symmetric $T \in M_{p\times p}$ such that $\rho(T\Sigma) < 1/2$.

Remark 3.2. *Note that when the skewness parameter matrix $\Delta_1 = 0$, then the MGF of $X'X$ is reduced to*

$$M_{X'X}(T) = \frac{2\exp\left\{\langle T, \Lambda\rangle + 2\left\langle \Lambda, T\Sigma^{1/2}\left(I_p - 2\Sigma^{1/2}T\Sigma^{1/2}\right)^{-1}\Sigma^{1/2}T\right\rangle\right\}}{\left|I_p - 2\Sigma^{1/2}T\Sigma^{1/2}\right|^{k/2}}$$

which is free of Δ_2, denoted by $X'X \sim W_p(k, \Sigma, \Lambda)$.

Lemma 3.2. *Let $A \in M_{k\times n}$, $M \in M_{n\times p}$, $\Lambda \in M_{p\times p}$, $m \le k$ be a positive integer, $\Sigma \in M_{p\times p}$ be positive definite, and $W \in M_{n\times n}$ be nonnegative definite with rank m. If*

$$\frac{\exp\left\{\left\langle T, M'TM\right\rangle + 2\left\langle M, \left[\left(WA'\otimes T\Sigma^{1/2}\right)\left(I_{kp} - 2\Psi\right)^{-1}\left(AW\otimes\Sigma^{1/2}T\right)\right](M)\right\rangle\right\}}{\left|I_p - 2\Psi\right|^{1/2}} = \frac{\exp\left\{\langle T, \Lambda\rangle + 2\left\langle \Lambda, T\Sigma^{1/2}\left(I_p - 2\Sigma^{1/2}T\Sigma^{1/2}\right)^{-1}\Sigma^{1/2}T\right\rangle\right\}}{\left|I_p - 2\Sigma^{1/2}T\Sigma^{1/2}\right|^{m/2}},$$

for symmetric $T \in M_{p\times p}$ such that $\max\{\rho(\Psi), \rho(T\Sigma)\} < 1/2$, *where $\Psi = (AWA')\otimes\left(\Sigma^{1/2}T\Sigma^{1/2}\right)$, then*

(i) AWA' is idempotent of rand m and
(ii) $\Lambda = M'WM = M'WVWM = M'WVWVWM$

with $V = A'A$.

The proof of Lemma 3.2 is similiar to that of Corollary 2.3.2 given in Wong et al. [19].

Theorem 3.1. *Let $Z \sim SN_{k\times p}\left(0, I_{kp}, \Gamma\right)$, $Y = M + A'Z\Sigma^{1/2}$, and $Q = Y'WY$ with nonnegative definite $W \in M_{n\times n}$. Then the necessary and sufficient conditions under which $Q \sim SW_p(m, \Sigma, \Lambda, \Delta_1, \Delta_2)$, for some $\Delta_1 \in M_{p\times p}$ including $\Delta_1 = 0$, are:*

(i) AWA' is idempotent of rank m,
(ii) $\Lambda = M'WM = M'WVWM = M'WVWVWM$,
(iii) $\Delta_1 = \Gamma AWM/d$, and
(iv) $\Delta_2 = \Gamma' P_1 P_1' \Gamma/d^2$,

where $V = A'A$, $d = \sqrt{1 + tr(\Gamma' P_2 P_2' \Gamma)}$ *and* $P = (P_1, P_2)$ *is an orthogonal matrix in* $M_{k\times k}$ *such that*

$$AWA' = P \begin{pmatrix} I_m & 0 \\ 0 & 0 \end{pmatrix} P' = P_1 P_1'. \tag{8}$$

Proof. On one hand, we assume that (i)–(iv) hold. By (i), there exist an orthogonal matrix $P \in M_{k\times k}$ such that equation (8) holds. Then for nonnegative definite W, we get

$$P_1 P_1' AW = AW. \tag{9}$$

Let $K = P_1' AWM$ and $X = K + P_1' Z \Sigma^{1/2}$, where $Z \sim SN_{k\times p}(0, I_{kp}, \Gamma)$. By Proposition 2.2, we have $X \sim SN_{m\times p}(K, I_m \otimes \Sigma, \Gamma_1)$, where $\Gamma_1 = P_1'\Gamma/d$ where $d = \sqrt{1 + \mathrm{tr}(\Gamma' P_2 P_2' \Gamma)}$. Note that Λ, Δ_1 and Δ_2 chosen here are equivalent to (ii)-(iv), respectively. Thus it suffices to show that $Q \overset{d}{=} X'X$, which means that Q and $X'X$ have the same distribution.

By equation (8) and (9), we have

$$\begin{aligned}
Q &= \left(M + A'Z\Sigma^{1/2}\right)' W \left(M + A'Z\Sigma^{1/2}\right) \\
&= M'WM + \Sigma^{1/2} Z' AWM + M'WA'Z\Sigma^{1/2} + \Sigma^{1/2} Z' AWA' Z \Sigma^{1/2} \\
&= K'K + \Sigma^{1/2} Z' P_1 K + K' P_1' Z \Sigma^{1/2} + \Sigma^{1/2} Z' P_1 P_1' Z \Sigma^{1/2} \\
&= \left(K + P_1' Z \Sigma^{1/2}\right)' \left(K + P_1' Z \Sigma^{1/2}\right) \overset{d}{=} X'X.
\end{aligned}$$

Thus we obtain the desired result.

On the other hand, consider $Q \sim SW_p(m, \Sigma, \Lambda, \Delta_1, \Delta_2)$. Let $X \sim SN_{m\times p}(K, I_m \otimes \Sigma, \Gamma_1)$, $\Lambda = K'K$, $\Delta_1 = \Gamma_1' K$ and $\Delta_2 - \Gamma_1' \Gamma$. By Definition 3.1, $Q \overset{d}{=} X'X$ so that $M_Q(T) = M_{X'X}(T)$. Note that for the case when $\Delta_1 = 0$, the distribution of Q is the noncentral Wishart distribution. By Proposition 3.1 and Corollary 3.1, we obtain

$$\begin{aligned}
&\frac{\exp\left\{\langle T, M'WM\rangle + 2\left\langle M, \left[\left(WA' \otimes T\Sigma^{/2}\right)\left(I_{kp} - 2\Psi\right)^{-1}\left(AW \otimes \Sigma^{1/2} T\right)\right](M)\right\rangle\right\}}{\left|I_{kp} - 2\Psi\right|^{1/2}} \\
&= \frac{\exp\left\{\langle T, \Lambda\rangle + 2\left\langle \Lambda, T\Sigma^{1/2}\left(I_p - 2\Sigma^{1/2} T \Sigma^{1/2}\right)^{-1} \Sigma^{1/2} T\right\rangle\right\}}{\left|I_p - 2\Sigma^{1/2} T \Sigma^{1/2}\right|^{m/2}}
\end{aligned}$$

and

$$\Phi\left\{\frac{\left\langle \Gamma_1, \left[\left(I_{kp} - 2\Psi\right)^{-1}\left(A \otimes \Sigma^{1/2}\right)\right](L)\right\rangle}{\left[1 + (\mathrm{Vec}\,(\Gamma))'\left(I_{kp} - 2\Psi\right)^{-1} \mathrm{Vec}\,(\Gamma)\right]^{1/2}}\right\} = \Phi\left(\frac{2\left\langle T, \Sigma^{1/2}\left(I_p - 2\Sigma^{1/2} T \Sigma^{1/2}\right)^{-1} \Delta_1\right\rangle}{\left[1 + \left\langle \Delta_2, \left(I_p - 2\Sigma^{1/2} T \Sigma^{1/2}\right)^{-1}\right\rangle\right]^{1/2}}\right). \tag{10}$$

Let $\Omega = I_p - 2\Sigma^{1/2} T \Sigma^{1/2}$, then $(I_{kp} - 2\Psi)^{-1} = (I_k - AWA') \otimes I_p + (AWA') \otimes \Omega^{-1}$ and equation (10) is reduced to

$$\frac{\langle \Gamma, AWMT\Sigma^{1/2}\Omega^{-1}\rangle}{\left[d^2+\Gamma' P_1 P_1'\Gamma\right]^{1/2}} = \frac{\langle T, \Sigma^{1/2}\Omega^{-1}\Delta_1\rangle}{\left[1+\langle \Delta_2, \Omega^{-1}\rangle\right]^{1/2}}. \tag{11}$$

Note that $\Omega^{-1} = \sum_{i=0}^{\infty}\left(2\Sigma^{1/2}T\Sigma^{1/2}\right)^i$. Let $\Omega^{-1} = I_p$ in equation (11), then we have

$$\langle \Sigma^{1/2}T, \Delta_1\rangle = \sqrt{\frac{1+\langle \Delta_2, I_p\rangle}{1+\mathrm{tr}\,(\Gamma'\Gamma)}}\langle \Sigma^{1/2}T, \Gamma' AWM\rangle,$$

so that

$$\Delta_1 = \sqrt{\frac{1+\langle \Delta_2, I_p\rangle}{1+\mathrm{tr}\,(\Gamma'\Gamma)}}\Gamma' AWM \tag{12}$$

Let $\Omega^{-1} = 2\Sigma^{1/2}T\Sigma^{1/2}$ and substitute Δ_1 back to equation (11), then we obtain

$$\frac{1+\langle \Delta_2, I_p\rangle}{1+\langle \Delta_2, 2\Sigma^{1/2}T\Sigma^{1/2}\rangle} = \frac{1+\mathrm{tr}(\Gamma'\Gamma)}{d^2+2\mathrm{tr}\left(\Gamma' P_1 P_1'\Gamma\Sigma^{1/2}T\Sigma^{1/2}\right)}. \tag{13}$$

Further, let $T = 0$ in equation (13), we have

$$1+\langle \Delta_2, I_p\rangle = \left(1+\mathrm{tr}\left(\Gamma'\Gamma\right)\right)/d^2. \tag{14}$$

By equation (12)–(14), we have

$$\Delta_1 = \Gamma' AWM/d, \qquad \text{and} \qquad \Delta_2 = \Gamma' P_1 P_1'\Gamma/d^2. \qquad \square$$

The following result is a direct consequence of Theorem 3.1.

Corollary 3.2. *Let $Z \sim SN_{k\times p}\left(0, I_{kp}, \Gamma\right)$, $Y = M + A'Z\Sigma^{1/2}$ and W be nonnegative definite in $M_{n\times n}$ with rank m. Then $Q = Y'WY \sim SW_p\left(m, \Sigma, \Lambda, \Delta_1, \Delta_2\right)$ if and only if for some $\Delta_1 \in M_{p\times p}$ including $\Delta_1 = 0$:*

(i) $W = WVW$,
(ii) $\Lambda = M'WM$,
(iii) $\Delta_1 = \Gamma' AWM/d$, *and*
(iv) $\Delta_2 = \Gamma'\Gamma/d^2$,

where $V = A'A$, $d = \sqrt{1+\Gamma' P_2 P_2'\Gamma}$ and $P = (P_1, P_2)$ is an orthogonal matrix in $M_{k\times k}$ such that

$$AWA' = P\begin{pmatrix} I_m & 0 \\ 0 & 0 \end{pmatrix}P' = P_1 P_1'.$$

Example 3.1. *Consider the multivariate one-way classification with skew-normal random errors given by*

$$Y = M + \left(I_a \otimes \mathbf{1}_b\right)\mathscr{E}_1 + \mathscr{E}_0$$

where $n = ab$, Y is a $n \times p$ random matrix, $M = \mathbf{1}_n \mathbf{v}'$ with a fixed effect $\mathbf{v} \in \Re^p$, $\mathscr{E}_1 \sim N_{a\times p}(0, I_a \otimes \Psi_1)$, $\mathscr{E}_0 \sim SN_{n\times p}(0, I_n \otimes \Psi_0, \Gamma)$, with $\Gamma = \mathbf{1}_n \otimes \alpha'$, and $\mathscr{E}_1$ and $\mathscr{E}_0$ are independent. Let $\Sigma_1 = b\Psi_1 + \Psi_0$, $\Sigma_0 = \Psi_0$, and $Q = Y'Y$. Then under the null hypothesis that $\Sigma_1 = \Sigma_0 = \Sigma$ (i.e. $\Psi_1 = 0$), we obtain that $Q \sim SW_p(n, \Psi_0, \Lambda, \Delta_1, \Delta_2)$ where $\Lambda = n\mathbf{v}\mathbf{v}'$, $\Delta_1 = \Gamma'\mathbf{1}_n\mathbf{v}'$, and $\Delta_2 = \Gamma'\Gamma$.

4 The Properties of Noncentral Skew Wishart Distribution

The Sum of Skew Wishart and Wishart Matrices

Proposition 4.1. *If $U_1 \sim SW_p(k_1, \Sigma, \Lambda_1, \Delta_1, \Delta_2)$ and $U_2 \sim W_p(k_2, \Sigma, \Lambda_2)$ are independently distributed, then*

$$U = U_1 + U_2 \tag{15}$$

is distributed according to $SW_p(k_1 + k_2, \Sigma, \Lambda_1 + \Lambda_2, \Delta_1, \Delta_2)$.

Proof. Consider the MGF of U_1 and U_2 given by equation (7) and (8) with corresponding parameters. Since U_1 and U_2 are independent, the MGF of U equals the product of MGF of U_1 and U_2. Then the desired result follows immediately. □

The distribution of $D'UD$

Let $U \sim SW_p(k, \Sigma, \Lambda, \Delta_1, \Delta_2)$ and $D \in M_{p\times s}$ with $\Lambda = M'M$, $\Delta_1 = \Gamma'M$ and $\Delta_2 = \Gamma'\Gamma$ for some $M \in M_{k\times p}$, $\Gamma \in M_{k\times p}$, the following result give the distribution of $D'UD$.

Theorem 4.1. *Let $W = D'UD$ with $U \sim SW_p(k, \Sigma, \Lambda, \Delta_1, \Delta_2)$ and $D \in M_{p\times s}$. Then*

$$W \sim SW_s\left(k, \Sigma^*, \Lambda^*, \Delta_1^*, \Delta_2^*\right)$$

where $\Sigma^* = D'\Sigma D$, $\lambda^* = M^{*'}M^*$, $\Delta_1^* = \Gamma'M^*$ and $\Delta_2^* = \Gamma'\Gamma$ with $M^* = MD$, $\Gamma^* = c_0\Gamma\Sigma^{1/2}D\Sigma^{*1/2}$ and $c_0 = (1 + Vec(\Gamma)(I_p - P_{\Sigma^{1/2}D})Vec(\Gamma))^{-1/2}$.

Proof. By the definition, $U \stackrel{d}{=} Y'Y$ where $Y \sim SN_{k\times p}(M, I_k \otimes \Sigma, \Gamma)$. Therefore, let $X = YD$, $W \stackrel{d}{=} X'X$ where $X \sim SN_{k\times s}(M^*, I_k \otimes \Sigma^*, \Gamma^*)$ by applying Theorem 2.1. Thus the desired results follows immediately by the definition 3.1. □

Based on Theorem 4.1, we obtain the marginal distributions of U which can be written as the form $D'UD$ for some $D \in M_{p\times s}$.

Corollary 4.1. *Let $U \sim SW_p(k, \Sigma, \Lambda, \Delta_1, \Delta_2)$ be partitioned into q and $p-q$ rows and columns, say*

$$U = \begin{pmatrix} U_{11} & U_{12} \\ U_{21} & U_{22} \end{pmatrix}.$$

Then U_{11} follows non central skew Wishart distribution with the parameters given by substituting $D = \begin{pmatrix} I_p \\ 0 \end{pmatrix}$.

Proof. It is straightforward to apply Theorem 4.1 with $D = \begin{pmatrix} I_p \\ 0 \end{pmatrix}$.

5 The Decomposition of Noncentral Skew Wishart Distributions

At first, we present a fundamental result on the decomposition of noncentral skew Wishart distribution.

Theorem 5.1. *Let $U \sim SW_p(k, \Sigma, \Lambda, \Delta_1, \Delta_2)$ for $k \geq 2$, then U can be partitioned into the sum of two independent random matrices U_1 and U_2, $U = U_1 + U_2$, where $U_1 \sim SW_p(s, \Sigma, \Lambda^*, \Delta_1^*, \Delta_2^*)$ and $U_2 \sim W_p(k-1, \Sigma, \Lambda - \Lambda^*)$ where $s = r(\Delta_2)$ if and only if*

1. $\Lambda^* = \Delta_1' \Delta_1 / tr(\Delta_2)$;
2. $\Delta_i^* = \Delta_i$ *for* $i = 1, 2$.

Proof. "If part" is trivial since the MGF of U can be written as the product of the MGF's of U_1 and U_2. By the uniqueness of MGF, U and $U_1 + U_2$ follow the same distribution. For the proof of "only if" part, let two $k \times p$ matrices M and Γ satisfy the following conditions

(a) $M'M = \Lambda$;
(b) $\Gamma'\Gamma = \Delta_2$;
(c) $\Gamma'M = \Delta_1$.

Furthermore, we pick an orthogonal matrix $Q = (Q_1, Q_2)$ such that $Q_2'\Gamma = 0$. Then, let $Y_1 \sim SN_{s\times p}(Q_1'M, I_s \otimes \Sigma, Q_1'\Gamma)$ and $Y_2 \sim N_{(k-s)\times p}(Q_2'M, I_{k-s} \otimes \Sigma)$ be independently distributed. Consider the joint distribution

$$Y = \begin{pmatrix} Y_1' \\ Y_2 \end{pmatrix} \sim SN_{k\times p}\left(Q'M, I_k \otimes \Sigma, Q'\Gamma\right). \tag{16}$$

On one hand, the direct computation shows $Y'Y \sim SW_p(k, \Sigma, \lambda, \delta_1, \delta_2)$. On the other hand, it is clear that $Y'Y = Y_1Y_1' + Y_2'Y_2$, and $Y_1Y_1' \sim SW_p(s, \Sigma, \Lambda^*, \Delta_1^*, \Delta_2^*)$ and $Y_2'Y_2 \sim W_p(k-s, \Sigma, \Lambda^{**})$ with

$$\Lambda^* = M'Q_1Q_1'M,\ \Delta_1^* = \Gamma'Q_1Q_1'M,\ \Delta_2^* = \Gamma'Q_1Q_1'\Gamma, \quad \text{and} \quad \Lambda^{**} = M'Q_2Q_2'M.$$

It is clear that $\Lambda = \Lambda^* + \Lambda^{**}$ since $QQ' = Q_1Q_1' + Q_2Q_2' = I_k$. Also, we have $\Delta_1^* = \Delta_1$ and $\Delta_2^* = \Delta_2$ since $Q_2'\Gamma = 0$. □

Remark 5.1. *Note that the decomposition of a noncentral Wishart distributed matrix is depending on the rank of Δ_2 and the decomposition is just for the degree of freedom and noncentral parameter Λ which is similar with the decomposition of noncentral skew chi-square distribution.*

For the definition of noncentral skew Wishart distribution defined by Ye et al. [21], we have the following corollary.

Corollary 5.1. *Let $U \sim SW_p(k, \Sigma, \Lambda, \Delta_1, \Delta_2)$ with $\Lambda = M'M$, $\Delta_1 = \alpha \mathbf{1}_k' M$ and $\Delta_2 = k\alpha\alpha'$ for $\alpha \in \Re^p$. Then U is able to be decomposed into the independent sum of U_1 and U_2 with $U_1 \sim SW_p(1, \Sigma, \Lambda_*, \Delta_1, \Delta_2)$ and $U_2 \sim W_p(k-1, \Sigma, \Lambda - \Lambda^*)$.*

Example 5.1. *In Example 3.1, let $W_1 = \bar{J}_{ab}$, $W_2 = I_a \otimes \bar{J}_b - \bar{J}_{ab}$, $W_3 = I_a \otimes \left(I_b - \bar{J}_b\right)$ and $Q_i = Y'W_iY$ for $i = 1, 2, 3$. It is clear that*

$$Q = Y'Y = \sum_{i=1}^{3} Q_i.$$

From Example , we know $Q \sim SW_p\left(n, \Psi_0, \Lambda, \Delta_1, \Delta_2\right)$. We are interested in two questions:

(a) What are the distributions of these three components Q_1, $Q_2$and Q_3?
(b) Are these three components independently distributed?

For question (a), to find the distributions of these three components, we apply Corollary 3.2, we obtain the following

(i) $Q_1 \sim SW_p\left(1, \Psi_0, \Lambda_1, \Delta_{11}, \Delta_{12}\right)$ where $\Lambda_1 = n\mathbf{v}\mathbf{v}'$, $\Delta_{11} = n\alpha\gamma'$, and $\Delta_{12} = n\alpha\alpha'$;
(ii) $Q_2 \sim SW_p\left(a-1, \Psi_0, \Lambda_2, \Delta_{21}, \Delta_{22}\right)$ where $\Lambda_2 = 0$, $\Delta_{21} = 0$, and $\Delta_{22} = 0$; and
(iii) $Q_3 \sim SW_p\left(a\left(b-1\right), \Psi_0, \Lambda_3, \Delta_{31}, \Delta_{32}\right)$ where $\Lambda_3 = 0$, $\Delta_{31} = 0$, and $\Delta_{32} = 0$;

For question (b), we can apply Corollary 5.1 to show Q_1, Q_2 and Q_3 are independently distributed which will be a useful results to proceed analysis for this model.

References

1. Akdemir, D., Gupta, A.K.: A matrix variate skew distribution. Eur. J. Pure Appl. Math. **3**(2), 128–140 (2010)
2. Azzalini, A., Capitanio, A.: Statistical applications of the multivariate skew normal distribution. J. Roy. Stat. Soc. Ser. B (Stat. Methodol.) **61**(3), 579–602 (1999)

3. Azzalini, A., Dalla-Valle, A.: The multivariate skew-normal distribution. Biometrika **83**(4), 715–726 (1996)
4. Azzalini, A.: The Skew-Normal and Related Families. Cambridge University Press, Cambridge (2013)
5. Chen, J.T., Gupta, A.K.: Matrix variate skew normal distributions. Statistics **39**(3), 247–253 (2005)
6. Gallaugher, M.P., McNicholas, P.D.: Three skewed matrix variate distributions. Stat. Probab. Lett. **145**, 103–109 (2019)
7. Genton, M.G.: Skew-Elliptical Distributions and Their Applications: A Journey Beyond Normality. CRC Press, Boca Raton (2004)
8. Genton, M.G., He, L., Liu, X.: Moments of skew-normal random vectors and their quadratic forms. Stat. Probab. Lett. **51**(4), 319–325 (2001)
9. Gupta, A.K., Chang, F.C.: Multivariate skew-symmetric distributions. Appl. Math. Lett. **16**(5), 643–646 (2003)
10. Gupta, A.K., Gonzalez-Faras, G., Dominguez-Molina, J.A.: A multivariate skew normal distribution. J. Multivar. Anal. **89**(1), 181–190 (2004)
11. Gupta, A.K., Huang, W.-J.: Quadratic forms in skew normal variates. J. Math. Anal. Appl. **273**(2), 558–564 (2002)
12. Gupta, A.K., Nguyen, T.T., Sanqui, J.A.T.: Characterization of the skew-normal distribution. Ann. Inst. Stat. Math. **56**(2), 351–360 (2004b)
13. Harrar, S.W., Gupta, A.K.: On matrix variate skew-normal distributions. Statistics **42**(2), 179–194 (2008)
14. Li, B., Tian, W., Wang, T.: Remarks for the singular multivariate skew-normal distribution and its quadratic forms. Stat. Probab. Lett. **137**, 105–112 (2018)
15. Ma, Z., Chen, Y.-J., Wang, T. : Inferences on location parameter in multivariate skew-normal family with unknown scale parameter (2019)
16. Ma, Z., Chen, Y.-J., Wang, T., Peng, W.: The inference on the location parameters under multivariate skew normal settings. In: International Econometric Conference of Vietnam, pp. 146-162. Springer, Cham (2019)
17. Ma, Z., Tian, W., Li, B., Wang, T.: The decomposition of quadratic forms under skew normal settings. In International conference of the Thailand econometrics society, pp. 222-232. Springer, Cham (2018)
18. Wang, T., Li, B., Gupta, A.K.: Distribution of quadratic forms under skew normal settings. J. Multivar. Anal. **100**(3), 533–545 (2009)
19. Wong, C.S., Masaro, J., Wang, T.: Multivariate versions of Cochran's theorems. J. Multivar. Anal. **39**(1), 154–174 (1991)
20. Ye, R., Wang, T.: Inferences in linear mixed models with skew-normal random effects. Acta Mathematica Sinica, English Series **31**(4), 576–594 (2015)
21. Ye, R., Wang, T., Gupta, A.K.: Distribution of matrix quadratic forms under skew-normal settings. J. Multivar. Anal. **131**, 229–239 (2014)
22. Ye, R., Wang, T., Sukparungsee, S., Gupta, A.K.: Tests in variance components models under skew-normal settings. Metrika **78**(7), 885–904 (2015)
23. Young, P.D., Patrick, J.D., Ramey, J.A., Young, D.M.: An alternative matrix skew-normal random matrix and some properties. Sankhya A **82**(1), 28–49 (2019)
24. Zacks, S.: Parametric Statistical Inference: Basic Theory and Modern Approaches. Elsevier, Amsterdam (1981)
25. Zhu, X., Li, B., Wang, T., Gupta, A.K.: Sampling distributions of skew normal populations associated with closed skew normal distributions. Random Oper. Stochast. Equ. **27**(2), 75–87 (2019)

How to Gauge a Combination of Uncertainties of Different Type: General Foundations

Ingo Neumann, Vladik Kreinovich, and Thach Ngoc Nguyen

Abstract In many practical situations, for some components of the uncertainty (e.g., of the measurement error) we know the corresponding probability distribution, while for other components, we know only upper bound on the corresponding values. To decide which of the algorithms or techniques leads to less uncertainty, we need to be able to gauge the combined uncertainty by a single numerical value—so that we can select the algorithm for which this values is the best. There exist several techniques for gauging the combination of interval and probabilistic uncertainty. In this paper, we consider the problem of gauging the combination of different types of uncertainty from the general fundamental viewpoint. As a result, we develop a general formula for such gauging—a formula whose particular cases include the currently used techniques.

1 Formulation of the Problem

Need to Gauge Uncertainty. Measurements are never absolutely accurate, the measurement result $\widetilde{x}$ is, in general, different from the actual (unknown) value x of the corresponding quantity. To understand how accurate is the measurement, we need to gauge the corresponding uncertainty, i.e., to provide a number describing the

I. Neumann
Geodetic Institute, Leibniz University of Hannover, Nienburger Str. 1, 30167 Hannover, Germany
e-mail: neumann@gih.uni-hannover.de

V. Kreinovich (✉)
University of Texas at El Paso, El Paso, TX 79968, USA
e-mail: vladik@utep.edu

T. N. Nguyen
Banking University of Ho Chi Minh City, 56 Hoang Dieu 2, Quan Thu Duc, Thu Duc, Ho Chi Minh City, Vietnam
e-mail: Thachnn@buh.edu.vn

S. Sriboonchitta et al. (eds.), *Behavioral Predictive Modeling in Economics*, Studies in Computational Intelligence 897,
https://doi.org/10.1007/978-3-030-49728-6_13

corresponding measurement error $\Delta x \stackrel{\text{def}}{=} \widetilde{x} - x$. For different types of uncertainty, it is natural to use different characteristics. For example:

- For probabilistic uncertainty, when we know the probability distribution of the corresponding measurement error, a natural measure of deviation is the standard deviation σ.
- For interval uncertainty, we only know the upper bound Δ on the absolute value of the measurement error; this upper bound is an appropriate measure of uncertainty.

It is reasonable to select a characteristic that is described in the same unit as the measured quantity itself. In this case, if we change the measuring unit to the one which is λ times smaller, then:

- not only all numerical value x should multiply by λ ($x \to x' = \lambda \cdot x$), but also
- the corresponding characteristic of uncertainty should change the same way:

$$u \to u' = \lambda \cdot c.$$

Similarly, if we simply change the sign of the quantity—which, for many quantities like coordinate or charge, does not change its physical sense—then the corresponding characteristic of uncertainty should not change: $u' = u$.

In general, when we go from x to $x' = c \cdot x$, then the corresponding characteristic of uncertainty should change as $u' = |c| \cdot u$.

Need to Combine Uncertainty and to Gauge the Combined Uncertainty. The measurement error often consists of several components:

$$\Delta x = \Delta x_1 + \ldots + \Delta x_k.$$

For each of these components Δx_i, we usually know the corresponding characteristic of uncertainty u_i. Based on these characteristics, we need to estimate the characteristic u of the overall uncertainty Δx.

A similar problem occurs when we process data, i.e., when, based on the measurement results $\widetilde{x}_i$, we compute the value of some auxiliary quantity y depending on x in known way, as $y = f(x_1, \ldots, x_n)$, for some algorithm $f(x_1, \ldots, x_n)$. To estimate y, we use the measurement results $\widetilde{x}_i$ and thus, come up with an estimate $\widetilde{y} = f(\widetilde{x}_1, \ldots, \widetilde{x}_n)$. We need to estimate the resulting approximation error

$$\Delta y = \widetilde{y} - y = f(\widetilde{x}_1, \ldots, \widetilde{x}_n) - f(x_1, \ldots, x_n)$$

$$= f(\widetilde{x}_1, \ldots, \widetilde{x}_n) - f(\widetilde{x}_1 - \Delta x_1, \ldots, \widetilde{x}_n - \Delta x_n).$$

Measurements are usually reasonably accurate, so the measurement errors Δx_i are small, and thus, we can safely ignore terms which are quadratic or of higher order in terms of Δx_i and consider only the linear terms. Then, $\Delta y = \sum\limits_{i=1}^{k} c_i \cdot \Delta x_i$, where c_i

is the value of the corresponding partial derivative $c_i \stackrel{\text{def}}{=} \frac{\partial f}{\partial x_i}$ computed as the point $(\widetilde{x}_1, \ldots, \widetilde{x}_n)$. Once we know the uncertainty characteristics u_i of each measurement error Δx_i, we can find the uncertainty characteristics $U_i = |c_i| \cdot u_i$ of each term $X_i = c_i \cdot \Delta x_i$. Based on these characteristics, we need to estimate the uncertainty characteristic of the sum $\Delta y = X_1 + \ldots + X_n$.

The need to characterize the joint uncertainty by a single number comes from the desire to select a single less uncertain option. For example, in the traditional Markowitz's portfolio allocation problem (see, e.g., [3]), when we have full information about all the probabilities, the objective is to find, among all portfolios with the given value of expected rate of return, the one with the smallest possible standard deviation (which, in this case, corresponds to the smallest possible risk). In many practical situations, we know probabilities only with some uncertainty. As a result, for each portfolio, in addition to the random uncertainty, we have an additional uncertainty caused by the fact that we only have partial knowledge about the corresponding probabilities.

- If we minimize the random component, we risk missing a huge interval component.
- If we minimize the interval component, we risk missing a huge random component.

It is more adequate to minimize the appropriate combination of both uncertainties, this will make sure that none of the components become too large.

How Uncertainty is Combined and Gauged Now. In the case of probabilistic uncertainty, if we know the standard deviations σ_i of each component Δx_i and we have no information about their correlation, a natural idea is to assume that the error components are independent. The same conclusion can be made if we use the Maximum Entropy approach, which recommend to select, among all possible joint distributions, the one with the largest possible value of entropy [2]. It is well known that the variance of the sum of several independent random variables is equal to the sum of their variances, so for the variance σ^2 of the sum Δx we have $\sigma^2 = \sigma_1^2 + \ldots + \sigma_n^2$ and $\sigma = \sqrt{\sigma_1^2 + \ldots + \sigma_n^2}$.

On the other hand, if we know that for each i, the component Δx_i can take any value from the interval $[-\Delta_i, \Delta_i]$, then the largest possible value Δ of the sum is attained when each of the components Δx_i attains its largest possible value Δ_i, so we have $\Delta = \Delta_1 + \ldots + \Delta_n$.

In these two cases, we have two different formulas for combining uncertainty: if we know the uncertainty characteristics u_i of the components, then the uncertainty characteristic u of the sum is equal:

- in the first case, to $u = \sqrt{u_1^2 + \ldots + u_n^2}$ and
- in the second case to $u = u_1 + \ldots + u_n$.

What is a General Case? We are looking for the binary combination operation $u * u'$ which has the following properties:

- $u * 0 = u$, meaning that adding 0 should not change anything, including the accuracy;
- the sum does not depend on the order in which we add the components, so the result of combination should also not depend on the order in which we combine the components; so, we should have $u * u' = u' * u$ (commutativity) and

$$u * (u' * u'') = (u * u') * u'' \text{ (associativity);}$$

- monotonicity: if we replace one of the components with a less accurate one (with larger u), the result cannot become more accurate: if $u_1 \le u_2$ and $u'_1 \le u'_2$, then we have $u_1 * u'_1 \le u_2 * u'_2$.

It turns out (see, e.g., [1]) that under these conditions and under the above-described scale-invariance, every combination operation has:

- either the from $u * u' = (u^p + (u')^p)^{1/p}$ for some $p > 0$,
- or the form $u * u' = \max(u, u')$ (that corresponds to the limit case $p \to \infty$).

Remaining Problem: What if We Combine Uncertainties of Different Type? In many cases, we have different information about the uncertainty of different components. For example, the measurement error is often represented as the sum of a systematic error (the mean value) and the remaining part which is known as a random error. About the random error component, we usually know the standard deviation, so it can be a viewed as a probabilistic uncertainty; see, e.g., [6]. However, about the systematic error component, we only know the error bound—so it is the case of interval uncertainty. How should we gauge the result of combining uncertainties of different type?

How the Combination of Uncertainties of Different Type is Gauged Now? There are several ways to gauge the combination of probabilistic and interval uncertainty.

The first way takes into account that, in practice, probability distributions are often either Gaussian (normal) or close to Gaussian [4, 5]. This empirical fact is easy to explain: in many cases, the measurement error is a result of a large number of independent small factors, and it is known that the distribution of the sum of the large number of small independent random variables is close to Gaussian (and tends to Gaussian when the number of components tends to ∞)—this fact is known as the *Central Limit Theorem* (see, e.g., [7]).

Strictly speaking, a normally distributed random variable with 0 mean can take arbitrarily large values—since its probability density function $\rho(x)$ remains positive for all values x. However, from the practical viewpoint, the probabilities of very large values are so small that, for all practical purposes, such values can be safely ignored. Thus, in practice, we assume that all the values of a normal random variable with 0 mean and standard deviation σ are located in an interval

$$[-k_0 \cdot \sigma, k_0 \cdot \sigma],$$

where k_0 depends on how small the probability we can ignore; usually, people take k_0 equal to 2 (corresponding to 5%), 3 (0.1%) and 6 (10^{-6}%).

So, a random error component with standard deviation σ implies that this component lies in the interval $[-k_0 \cdot \sigma, k_0 \cdot \sigma]$. So, all we have to do to combine it with the interval uncertainty $[-\Delta, \Delta]$ is to combine the two intervals, and get

$$\Delta + k_0 \cdot \sigma.$$

Another frequently used approach is based on the Maximum Entropy idea, according to which, if we do not know the exact distribution, then, out of all possible probability distributions, we should select the one whose entropy is the largest; see, e.g., [2]. For example, if all we know that the systematic error is located on the interval $[-\Delta, \Delta]$, then, out of all possible probability distributions on this interval, we should select the distribution whose entropy is the largest—which turns out to be the uniform distribution on this interval. One can easily find that for the uniform distribution on the interval $[-\Delta, \Delta]$, the standard deviation is equal to $\dfrac{\Delta}{\sqrt{3}}$. Thus, to combine it with the random error component with known standard deviation σ, it is sufficient to use the general formula for combining standard deviations, and get

$$\sqrt{\frac{\Delta^2}{2} + \sigma^2}.$$

Need for a General Approach—and What We Do in this Paper. So what is the general formula? This is a problem to which, in this paper, we provide an answer.

2 Definitions and the Main Result

Let us assume that $\mathcal{T}$ is the set of possible types of uncertainty with T elements. For simplicity, let us enumerate the types, i.e., let us identify $\mathcal{T}$ with the set

$$\{1, 2, \ldots, T\}.$$

By combining uncertainties from some subset $S \subseteq \mathcal{T}$, we get, in effect, a new type of uncertainty. Thus, we have, in effect, as many types of uncertainty as there are nonempty subsets $S \subseteq \mathcal{T}$. Since uncertainties can be of different type, in order to properly combine them, we need to know the type. Thus, an uncertainty is described not just by a number, but also by a type.

Definition 1. *Let a finite set $\mathcal{T}$ be given. By an* uncertainty, *we mean a pair (u, S), where:*

- u *is a non-negative real number and*
- S *is a non-empty subset of the set* $\mathscr{T}$.

Definition 2. *Let a finite set* $\mathscr{T}$ *be given. By a* combination operation, *we mean a binary operation* $*$ *that maps two uncertainties* (u, S) *and* (u', S') *into a new uncertainty* $(u'', S \cup S')$ *and that has the following properties:*

- *the operation* $*$ *is commutative and associative;*
- *the operation* $*$ *is monotonic in the following sense: if* $u_1 \le u_2$ *and* $u_1' \le u_2'$, *then we have* $u_1'' \le u_2''$, *where* $(u_1, S) * (u_1', S') = (u_1'', S \cup S')$ *and*

$$(u_2, S) * (u_2', S') = (u_2'', S \cup S');$$

- *scale-invariance: for every* $\lambda > 0$, *if* $(u, S) * (u', S') = (u'', S \cup S')$, *then*

$$(\lambda \cdot u, S) * (\lambda \cdot u', S') = (\lambda \cdot u'', S \cup S');$$

- *zero-property: for each set* S, *we have* $(u, S) * (0, S) = (u, S)$; *and*
- *non-zero property: if* $u > 0$ *and* $(u, S) * (u', S') = (u'', S \cup S')$, *then* $u'' > 0$.

Proposition 1. *For each combination operation, there exist positive values* $c_1, \ldots, c_T$ *such that:*

- *either* $(u_1, \{1\}) * (u_2, \{2\}) * \ldots * (u_T, \{T\}) = ((c_1^p \cdot u_1^p + \ldots + c_T^p \cdot u_T^p)^{1/p}, \mathscr{T})$ *for all* u_i
- *or* $(u_1, \{1\}) * (u_2, \{2\}) * \ldots * (u_T, \{T\}) = (\max(c_1 \cdot u_1, \ldots, c_T \cdot u_T), \mathscr{T})$ *for all* u_i.

Proof. Due to the zero property, we have

$$(u, \mathscr{T}) \stackrel{\text{def}}{=} (u_1, \{1\}) * (u_2, \{2\}) * \ldots * (u_T, \{T\})$$

$$= (u_1, \{1\}) * (0, \{1\}) * \ldots * (0, \{1\}) * \ldots * (u_T, \{T\}) * (0, \{T\}) * \ldots * (0, \{T\}),$$

where each term $(0, \{t\})$ is repeated T times. Due to associativity and commutativity, we can rearrange the terms and get

$$(u, \mathscr{T}) = (u_1'(u_1), \mathscr{T}) * \ldots * (u_T'(u_T), \mathscr{T}),$$

where we denoted

$$(u_t'(u_t), \mathscr{T}) \stackrel{\text{def}}{=} (u_t, \{t\}) * (0, \{1\}) * \ldots * (0, \{T\}).$$

Due to non-zero property, if $u_t = 1$, then $u_t'(1) \neq 0$. Let us denote $c_t \stackrel{\text{def}}{=} u_t'(1) > 0$. Then, due to scale-invariance, we have $u_t'(u_t) = c_t \cdot u_t$ and thus,

$$(u, \mathcal{T}) = (c_1 \cdot u_1, \mathcal{T}) * \ldots * (c_T \cdot u_T, \mathcal{T}).$$

For the values of type $\mathcal{T}$, we get the usual properties of the combination operation from [1], so we conclude that for uncertainties of this type, we have either $(u, \mathcal{T}) * (u', \mathcal{T}) = ((u^p + (u')^p)^{1/p}, \mathcal{T})$ or $(u, \mathcal{T}) * (u', \mathcal{T}) = (\max(u, u'), \mathcal{T})$. In both cases, we get exactly the formulas from the proposition, The proposition is thus proven.

Comment. As expected, both existing methods for combining uncertainty are particular cases of this general approach—corresponding to $p = 1$ and $p = 2$.

Acknowledgments This work was supported by the Institute of Geodesy, Leibniz University of Hannover. It was also supported in part by the US National Science Foundation grants 1623190 (A Model of Change for Preparing a New Generation for Professional Practice in Computer Science) and HRD-1242122 (Cyber-ShARE Center of Excellence).

This paper was written when V. Kreinovich was visiting Leibniz University of Hannover.

References

1. Autchariyapanitkul, K., Kosheleva, O., Kreinovich, V., Sriboonchitta, S.: Quantum econometrics: how to explain its quantitative successes and how the resulting formulas are related to scale invariance, entropy, and fuzziness. In: Huynh, V.-N., Inuiguchi, M., Tran, D.-H., Denoeux, Th. (eds.) Proceedings of the International Symposium on Integrated Uncertainty in Knowledge Modelling and Decision Making IUKM 2018, Hanoi, Vietnam, 13–15 March 2018
2. Jaynes, E.T., Bretthorst, G.L.: Probability Theory: The Logic of Science. Cambridge University Press, Cambridge (2003)
3. Markowitz, H.M.: Portfolio selection. J. Financ. **7**(1), 77–91 (1952)
4. Novitskii, P.V., Zograph, I.A.: Estimating the Measurement Errors. Energoatomizdat, Leningrad (1991). (in Russian)
5. Orlov, A.I.: How often are the observations normal? Ind. Lab. **57**(7), 770–772 (1991)
6. Rabinovich, S.G.: Measurement Errors and Uncertainties: Theory and Practice. Springer, New York (2005)
7. Sheskin, D.J.: Handbook of Parametric and Nonparametric Statistical Procedures. Chapman and Hall/CRC, Boca Raton (2011)

Conditional Dependence Among Oil, Gold and U.S. Dollar Exchange Rates: A Copula-GARCH Approach

Zheng Wei, Zijing Zhang, HongKun Zhang, and Tonghui Wang

Abstract This paper investigates the dependence structure among nominal crude oil (WTI), gold, and specific U.S. dollar against four major currencies (Euro, British Pound, Japanese Yen and Canadian Dollar) on a daily basis over the last decade. In order to capture the tail dependence between commodity market and USD exchange rates, we apply both bivariate zero tail and tail copulas, combined with the AR-GARCH marginal distribution for gold, oil and exchange rates daily returns. The primary findings are as follows. Firstly, based on the concordance and correlation coefficient, we find that there is a positive correlation between gold and crude oil prices, and a negative dependence between gold and currencies as well as oil and currencies. Secondly, the crude oil price can be viewed as a short term indicator in the exchange rates movement; the crude oil price also can be viewed as a short term descend indicator of gold price, while the gold price is an short term rise indicator of oil price. Thirdly, small degree of conditional extreme tail dependence for all considered pairs are observed. Our results provide useful information in portfolio diversification, asset allocation and risk management for investors and researchers.

Z. Wei
Department of Mathematics and Statistics, University of Maine, Orono, ME 04469-5752, USA
e-mail: zheng.wei@maine.edu

Z. Zhang · H. Zhang
Department of Mathematics and Statistics, University of Massachusetts Amherst, Amherst, MA 01003, USA
e-mail: zhangzijingapply@gmail.com

H. Zhang
e-mail: hongkunz@gmail.com

T. Wang (✉)
Department of Mathematical Sciences, New Mexico State University, Las Cruces, NM 88003-8001, USA
e-mail: twang@nmsu.edu

S. Sriboonchitta et al. (eds.), *Behavioral Predictive Modeling in Economics*, Studies in Computational Intelligence 897,
https://doi.org/10.1007/978-3-030-49728-6_14

1 Introduction

As a financial indicator, gold is classed as one of the most important commodities and one of the most stable monetary asset. As a multifaceted metal through the centuries, it has common ground with money in that it acts as a unit of value, a store of wealth, medium of exchange and a hedging instrument. Therefore, gold has always been used as a hedge against inflation, deflation and currency devaluation. Gold also plays an important role with significant portfolio diversification properties. An abundance of research point to the benefits of including gold holdings that leads to a more balanced portfolio (Johnson and Soenen [11]; Ciner [6]; Shafiee and Topal [15]).

Since the international gold and foreign exchange markets are both dominated by the U.S. dollar, the relationship between gold and U.S. exchange rates have received much attention, especially after the international financial crisis. Moreover, the price of oil, another one of the most important commodities, is also dominated in U.S. dollar. The importance of crude oil in global economy will continue during this century as a unique raw material responsible for power generation and lots of derivatives production. Hence, due to its effect on world economic growth and energy costs, the behavior of crude oil price has attracted considerable attention. Also, the oil price and inflation rate are two main macroeconomic variables that influence the gold market.

The above motivations demonstrate the importance in measuring and capturing the stylized facts exhibited in the oil price, gold price and U.S. dollar exchange rates, as well as the relationship among them. In this paper, we focus on investigating both the conditional dependence and the extreme comovement of gold, crude oil and U.S. dollar exchange rates on each other using a copula-GARCH approach. The analysis of our study is not merely for risk management and market trading issues, but also for the better regulation of foreign exchange markets.

In recent years, a number of methods have been employed to explore the relationship between gold prices or oil prices with US dollar exchange rate. Sjaastad and Scacciavillani [14] identified the effect of major currency exchange rates on the prices of gold. A variation in any exchange rate will result in an immediate adjustment in the prices of gold. The power of such phenomenon is also suggested by Capie et al. [5] where assessed the role of gold as a hedge against the dollar and concluded that the negative relationship was found between gold prices and the sterling-dollar, yen-dollar exchange rates. Recently, Sari et al. [16] examined the co-movement and information transmission among precious metals, oil price, and dollar-euro exchange rate. Joy [12] applied the dynamic conditional correlations model on 23 years weekly data for 16 major dollar-paired exchange rates and find that the gold has behaved as a hedge against dollar. For the theory on oil prices and dollar exchange rates, Krugman [10], Golub [9], and Rogoff [13] identified the important relation between the oil prices and the exchange rate movements. Using various data set on oil prices and dollar exchange rates over different time period, the extensive evidences on the co-movement between two variables can also be found in literature, see Amano et al. [3], Akram [1], Basher [4], Wu et al. [27], and Aloui et al. [2]. To offer a comparative view, we summarize the key findings of major studies in the related literature in Table 1.

Table 1 Previous research on the interactions among gold prices, oil prices and exchange rates

Studies	Purposes	Data	Methodology	Primary findings
Recent literature on modeling gold prices and USD exchange rates				
Sari et al. [16]	This study examines the co-movements among the prices of metals, oil price, and the exchange rate	Daily data (1999–2007)	The forecast error variance decomposition on impulse response functions	The evidence of a weak long-run equilibrium relationship but strong feedbacks in the short run
Pukthuant hong and Roll [18]	The paper investigated relationship between Dollar, Euro, Pound, and Yen	Daily data (1971–2009)	GARCH	The gold price expressed in a currency can be associated with weakness in that currency and vice versa
Joy [12]	This paper addresses a practical investment question if the gold act as a hedge against the US dollar	Weekly data (1986–2008)	Multivariate GARCH	The gold has acted, increasingly, as an effective hedge against currency risk associated with the US dollar
Yang and Hamori (2014)	The paper investigates the dynamic dependence structure between specific currencies(GBP, EUR, JPY) and gold prices	Daily data (2012–2013)	Copula - GARCH	Lower and upper conditional dependences between currencies and gold were weaker during the financial turmoil period than normal period
Recent literature on modeling oil prices and USD exchange rates				
Akram [1]	The author investigates the contribution of a decline in real interest rates and the US dollar to higher commodity prices	Quarterly data (1990–2007)	Structural VAR model	A fall in the value of the US dollar leads to drive up commodity prices, including crude oil price

(*continued*)

Table 1 (*continued*)

Studies	Purposes	Data	Methodology	Primary findings
Wu et al. [27]	The authors examine the economic value of comovement between WTI oil price and U.S. dollar index futures	Weekly data (1990–2009)	Copula - GARCH	The dependence structure between oil and exchange rate returns becomes negative and decreases continuously after 2003
Basher et al. [4]	The authors study the dynamic link between oil prices, exchange rates and emerging market stock prices	Monthly data (1988–2008)	Structural VAR model	Positive shocks to oil prices tend to depress emerging market stock prices and the trade-weighted US dollar index in the short run
Aloui et al. [2]	The authors study the conditional dependence structure between crude oil prices and U.S. dollar exchange rates	Daily data (2000–2011)	Copula - GARCH	The rise in the price of oil is found to be associated with the depreciation of the dollar

In this study, we use a Copula - GARCH model to capture the conditional volatility and dependence structures of gold, crude oil and USD exchange rates on each other. To appropriately investigate the behavior of considered assets, AR-GARCH models have been chosen to describe and measure the conditional mean and conditional volatility of returns. The advantage of our method is that we standardize the return series by filtering out the influence of the conditional mean and the volatility using AR-GARCH models; then we apply the copula approach to analyze the tail dependence for the standardized residues. The conditional dependence and tail dependence analysis are based on copula approach with proper marginal distributions. The reason to apply copula based approach to our data is that copulas allow for better flexibility in joint distributions than multivariate normal and Student-t distributions. In addition, copulas not only capture linear dependence as correlation, but also describe nonlinear dependence of different financial markets. Moreover, since copulas present rich patterns of tail dependence, it helps us to examine changes in the dependence structure during a financial crisis period.

The data we used are daily log returns of gold price, Brent and WTI prices, and specific exchange rates which including U.S. dollar against four major currencies (Euro, British Pound, Japanese Yen and Canadian Dollar) from March 1, 2006 to March 18, 2016. Since Brent is the reference for about two-thirds of the oil traded around the world, and WTI the dominant benchmark for oil consumed in the United

States, daily prices of Brent and WTI are used in this study to represent crude oil market. To investigate the dynamic of conditional dependence among gold, oil and U.S. dollar exchange rates, we first select the most appropriate marginal for each time series asset returns among four types of marginal models. Then, we apply copula models (elliptical and Archimedean copulas) on the standardized residuals to describe the conditional dependence structure between all considered pairs. We select Gaussian copula, Student-t copula, Clayton, Gumbel, BB7 copulas and their rotated versions copulas to compare and contrast with the conditional correlation.

The remainder of the article is organized as follows. Section 2 presents the Copula - GARCH methodology used for this study. In Sect. 3, we describe the data, give and discuss empirical results. Summarization and conclusion in Sect. 4.

2 Methodology

2.1 Marginal Distributions

The complexity of modeling financial time series is mainly due to the existence of stylized facts. After investigating daily log returns of gold value, Brent and WTI prices, and each of the four U.S. dollar exchange rates, the following three properties are concerned in this study. First one is that the price variations generally displays small autocorrelations while the corresponding squared returns or absolute returns are generally strongly autocorrelated. The second is leptokurtosis, which means financial time series tendency to have distributions that exhibit fat tails and excess peakedness at the mean. The third is the volatility clustering that large absolute returns are expected to follow large absolute returns and small absolute returns are expected to follow small absolute returns.

To capture these stylized facts, we use the autoregressive moving average model $ARMA(p, q)$ to quantify the conditional mean and the univariate generalized autoregressive conditional heteroscedasticity model GARCH(1,1) to capture the conditional variance. This modeling approach is advantageous in that it offers the possibility to separately model the margins and association structure of different variables.

Let $\{r_t\}_{t=T-n+1}^{T}$ be the time series representing the daily log return on a financial asset price. Here we fixed a constant memory n so that at the end of day T our data consist of the last n daily log returns $\{r_{T-n+1}, \cdots, r_{T-1}, r_T\}$. Assume the dynamics of $\{r_t\}_{t=T-n+1}^{T}$ be a realization from an $ARMA(p, q)$-GARCH(1,1) process, which are given by

$$\begin{aligned} r_t &= \mu_t + \sigma_t z_t \\ \mu_t &= \mu + \textstyle\sum_{i=1}^{p} \phi_i r_{t-i} + \sum_{j=1}^{q} \theta_j \epsilon_{t-j} + \epsilon_j \\ \sigma_t^2 &= \omega + \alpha (r_{t-1} - \mu_{t-1})^2 + \beta \sigma_{t-1}^2, \end{aligned} \tag{1}$$

where the innovations z_t are white noise process with zero mean, unit variance, and marginal distribution function F; $\omega > 0$, $\alpha > 0$, and $\beta > 0$. The conditional mean

$\mu_t = \mathbb{E}(r_t|\mathscr{F}_{t-1})$, and the conditional volatility $\sigma_t^2 = \text{Var}(r_t|\mathscr{F}_{t-1})$ are measurable with respect to $\mathscr{F}_{t-1}$ which is the σ-algebra generated by information about the return process available up to time $t-1$.

The traditional GARCH model assumes a normal distribution for the innovations z_t. However, to capture the leptokurtosis properties for considered return series, we consider various marginal distributions for z_t, which includes normal, skewed normal, Student-t and skewed Student-t distributions. For each considered return series, we specify the marginal distribution by comparing with Akaike information criterion (AIC) under different assumptions of innovation marginal distributions.

2.2 Copula Function

Recently, the study of copula functions have been a popular phenomenon in constructing joint distribution functions and modeling statistical dependence in real multivariate data. Copulas have been applied to many areas including finance [26], actuarial science [19], medical research [20], econometrics [21], environmental science [22], just to name a few. Copulas provide flexible representations of the multivariate distribution by allowing for the dependence structure of the variables of interest to be modeled separately from the marginal structure. We here briefly review the multivariate copulas. For the general copula theory, see [23–25].

A bivariate copula is a joint cumulative distribution function (CDF) on $[0, 1]^2$ with standard uniform marginal distributions. More precisely, a bivariate **copula** (or 2-**copula**) is a function $C : [0, 1]^2 \mapsto [0, 1]$ satisfying following properties:

(i) $C(u, 0) = C(0, v) = 0$, for $u, v \in [0, 1]$,
(ii) $C(u, 1) = u$, $C(1, v) = v$, for $u, v \in [0, 1]$, and
(iii) For any $u \leq u'$, $v \leq v'$, $C(u', v') - C(u, v) - C(u', v) + C(u, v) \geq 0$.

Let $(X_1, X_2)^T$ be a 2-dimensional random vector with CDF denoted as $H(x_1, x_2)$, and marginal CDF's $F_1(x_1)$, $F_2(x_2)$. Sklar's theorem [17] states that if the marginals of $(X_1, X_2)^T$ are continuous, then there exist a unique copula C such that

$$H(x_1, x_2) = C(F_1(x_1), F_2(x_2)).$$

2.3 Copula Models of Conditional Dependence Structure

In this paper, we consider two families of copulas: elliptical copulas (Gaussian copula and Student-t copula) and Archimedean copulas (Clayton, Gumbel, and BB7 copulas). These copula models allow us to study the conditional dependence structure and to evaluate the degree of tail dependence.

The normal and the Student-t copulas are constructed based on the elliptically contoured distribution such as multivariate Gaussian or Student-t distributions, respectively. Consider random variables X_1 and X_2 with standard bivariate normal distribution:

$$H_\rho(x_1, x_2) = \int_{-\infty}^{x_1}\int_{-\infty}^{x_2} \frac{1}{2\pi\sqrt{1-\rho^2}} \exp\left(-\frac{t^2+s^2-2\rho st}{2(1-\rho^2)}\right) dtds,$$

where ρ is the Pearson correlation between X_1 and X_2. The marginal distributions of X_1 and X_2 follow standard normal distributions $N(0, 1)$ with distribution function Φ. Then, the **Gaussian copula** is defined by

$$C_G(u, v) = H_\rho(\Phi^{-1}(u), \Phi^{-1}(v)),$$

where $\rho \in (-1, 1)$ is the correlation coefficient, and if $\rho = 0$ the Gaussian copula is reduced to be independent copula.

For random variables X_1 and X_2 with standard bivariate Student t distribution,

$$H_t(x_1, x_2; \rho, \nu) = \int_{-\infty}^{x_1}\int_{-\infty}^{x_2} \frac{1}{2\pi\sqrt{1-\rho^2}} \left(1+\frac{t^2+s^2-2\rho st}{2(1-\rho^2)}\right)^{-\frac{\nu+2}{2}} dtds$$

we let T_ν denote the standard univariate Student t distribution function with degree freedom ν for the marginals X_1 and X_2 . Then the **Student-t copula** is defined by

$$C_t(u, v) = H_t(T_\nu^{-1}(u), T_\nu^{-1}(v); \rho, \nu), \tag{2}$$

where $\rho \in (-1, 1)$ and $\nu > 0$. The Gaussian copula is symmetric and has no tail dependence while the Student t copula is also symmetric and can capture extreme dependence between variables. The trivariate Gaussian copula and t copula can be defined in similar fashion. Both the trivariate Gaussian copula and t copula associated with the random variables X_1, X_2 and X_3 has a correlation matrix, inherited from the elliptical distributions, and t-copula has one more parameter, the degrees of freedom (df). The correlation matrix in elliptical copulas determines the dependence structure.

Archimedean copula family, a very popular family of parametric copula, contains the most widely used copulas like, Ali-Mikhail-Haq, Clayton, Frank, Gumbel, and Joe as the nest models [7]. The bivariate Archimedean copula is defined as

$$C(u_1, u_2) = \phi^{[-1]}(\phi(u_1) + \phi(u_2)), \tag{3}$$

where $\phi : [0, 1] \to [0, \infty]$ is a continuous strictly decreasing convex function such that $\phi(1) = 0$ and $\phi^{[-1]}$ is the pseudo-inverse of ϕ, i.e.,

$$\phi^{[-1]}(t) = \begin{cases} \phi^{-1}(t) & \text{if } 0 \le t \le \phi(0) \\ 0 & \text{if } \phi(0) \le t \le \infty. \end{cases}$$

The convex function ϕ is called the generator function of the copula C. If $\phi(t) = \frac{1}{\theta}(t^{-\theta} - 1), \theta > 0$, then C defined in Eq. (3) is the Clayton copula. If we set $\phi(t) = (-\log t)^{\theta}, \theta \ge 1$, C defined in Eq. (3) is the Gumbel copula. Furthermore, C defined in Eq. (3) is called the BB7 copula when $\phi(t) = (1 - (1 - t)^{\theta})^{-\delta}, \theta \ge 1$, and $\delta > 0$.

One limitation for the Clayton copula, Gumbel copula, and the BB7 copula is that they only allow the positive association. And in this paper, we employ the rotated Clayton, Gumbel, and BB7 copulas to model the negative dependence among variables. Note that the t-copula defined in Eq. (2), the Clayton, Gumbel, the BB7 copula are able to capture the tail dependence. Furthermore, the Clayton, Gumbel, the BB7 copula are asymmetric copulas which can be utilized in modeling the asymmetric dependence and asymmetric tail dependence among variables during bear and bull markets.

For an absolutely continuous copula C, the **copula density** is defined to be

$$c(u, v) = \frac{\partial^2 C(u, v)}{\partial u \partial v}. \tag{4}$$

2.4 *Estimation of Copulas*

In the copula literature there are several commonly used estimation methods, for instance, the maximum likelihood (ML) estimation, the inference functions for margins (IFM) [24], and the maximum pseudo-likelihood (MPL) estimation [8]. The ML and IFM methods require the specification of parametric models for the marginals. In contrast, the advantage of MPL method is that it uses the rank-based estimators for the marginals, thus it is robust against misspecification of the marginal models. In this paper we take the advantage of the MPL method to estimate the proposed class of copulas, as it is not influenced by the choice of the marginal distributions.

Given a sample of m observations $(x_{11}, x_{21}), \ldots, (x_{1m}, x_{2m})$ from a random vector $(X_1, X_2)^T$, and let $C(u, v)$ be the associated copula. We first compute the normalized ranks or the rescaled empirical distributions for the variable X_1 and X_2, which are defined as: $u_i = \frac{r_i}{m+1}, v_i = \frac{s_i}{m+1}$, for $i = 1, \ldots, m$, where r_i and s_i are the rank of x_{1i} and x_{2i} among m data points from X_1, X_2, respectively. The pseudo log-likelihood function for the parameters in the copula is

$$\ell^p(\theta) = \log \prod_{i=1}^{n} c(u_i, v_i) = \sum_{i=1}^{n} \log c(u_i, v_i) \tag{5}$$

where $c(u, v)$ is the copula density of $C(u, v)$ in Eq. (4). We can obtain the maximum pseudo-likelihood estimators (MPLE) for the parameters by maximizing Eq. (5) with respect to θ.

3 Data and Empirical Results

3.1 Data Description and Stochastic Properties

To study the dynamical correlations, risk contagion and portfolio risks among gold price, oil prices, and exchange rates, we select the daily gold price in London bullion market quoted in U.S. dollars per gram, daily closing oil prices in US dollars per barrel of West Texas Intermediate (WTI), and five U.S. dollar (USD) exchange rates over the period from March 1, 2006 to March 18, 2016. As for the exchange rates, we employ the data come from the amount of USD per unit of each of the four major currencies in international trade: Euro (EUR), British Pound Sterling (GBP), Japanese Yen (JPY) and Canadian Dollar (CAD). The data used in this study are all taken from the database of *Quandl*. Figure 1 provides the time series plots of daily spot oil prices, gold prices, as well as USD exchange rates.

To develop an accurate track record of asset performance, the initial price data are transformed into daily log-returns. Let p_t denotes the asset price on day t, then the corresponding daily percentage change is defined by

$$r_t = 100 \log \frac{p_t}{p_{t-1}} = 100(\log p_t - \log p_{t-1}).$$

We show the time paths of considered daily log returns in Fig. 2. According to Fig. 2, we observe that there are more isolated pronounced peaks than one would expect

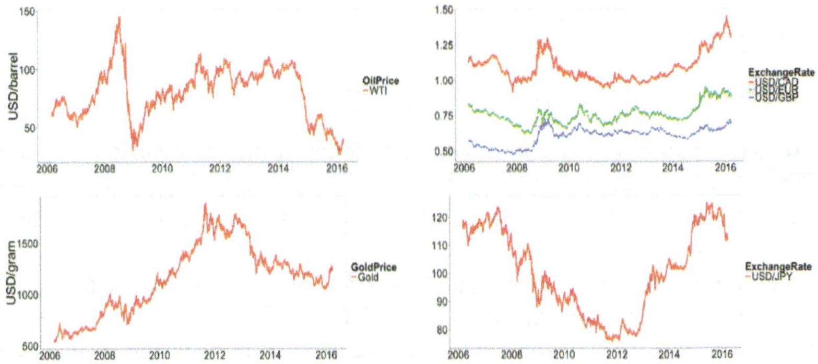

Fig. 1 Time series plots of oil prices (upper left), gold value (lower left) and USD exchange rates(upper and lower right) from 2006-3-1 to 2016-3-18.

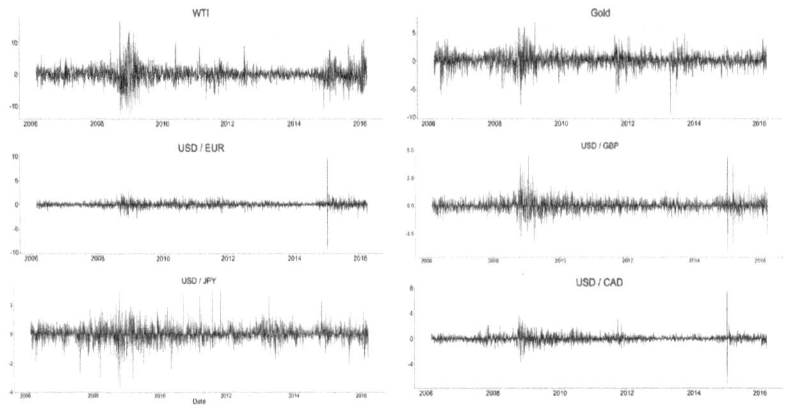

Fig. 2 Daily returns on crude oil, gold and USD exchange rates from 2006-3-1 to 2016-3-18.

Table 2 Descriptive statistics and stochastic properties of return series from 2006-3-1 to 2016-3-18.

	WTI	Gold	USD/EUR	USD/GBP	USD/JPY	USD/CAD
Panel A: Summary statistics						
Obs.	2532	2623	3661	3661	3661	3661
Min.	−12.83	−9.6	−9.4	−4.23	−3.68	−7.21
Max.	16.41	6.84	9.84	4.6	3.06	7.39
Mean	−0.02	0.03	0	0.01	0	0
Std. dev	2.47	1.26	0.59	0.49	0.48	0.52
Skew	0.14	−0.38	0.46	0.64	−0.19	0.51
Kurtosis	4.68	4.75	67.54	13.04	6.33	39.54
Panel B: Ljung-Box Q-statistics						
L-B $Q(5)$	361.64	272.48	177.14	18.879	42.461	51.517
L-B $Q(10)$	400.63	323.79	180.92	32.153	51.185	57.071
L-B $Q_2(5)$	870.18	1188.2	2896.2	696.64	108.88	2394.5
L-B $Q_2(10)$	1058.8	1703.5	2896.3	837.62	183.89	2400.3
S-W W	0.27775	0.15051	0.76637	0.8637	0.9114	0.79191

from the Gaussian series. Besides that, the high instability and volatility clustering behavior are also noticed in all return series. Those return series also exhibit two important price shocks, one is around the 2008 global financil crisis, the other one ranges from 2015 until recently.

Table 2 reports the descriptive statistics and distributional characteristics of all return series. As can be seen in Panel A of Table 2, the mean of all returns are quite small. As expected, the standard deviation of crude oil returns are larger than that of gold since oil is traded more heavily and actively than gold. Meanwhile, comparing with the standard normal distribution with skewness 0 and kurtosis 3, we confirm that all returns are lightly skewed and exhibit excess kurtosis. To check the autocorrelation of those returns, the Ljung-Box statistics is applied for returns at lag 5 and 10, i.e. Q(5) and Q(10), and squared returns $Q_2(5)$ and $Q_2(10)$. The Ljung-Box statistics for both return series and squared return series confirm that all sample returns have strong autocorrelation.

Based on above statistical analysis, we discover that all return series exhibit stationary, non-normally distributed, autocorrelated, and volatility clustering properties, which supports our choice of using the ARMA-GARCH based approach to analyze the conditional mean and conditional volatility for all returns.

3.2 Marginal Distribution Specifications and Parameter Estimations

In order to filter out the autocorrelation of the considered return series, the ARMA model is used in this paper. According to the censored orders of autocorrelation and partial autocorrelation function graphs, an AR(1) model is singled out through numerous trials. Examing the result of the Ljung-Box test for the residual series of the AR(1) model, it can be seen in Table 4 that almost all of the autocorrelation coefficients fall within the given confidence interval as well as their squred values. We hence conclude that the conditional mean of all the considered return series can be well fitted by the AR(1) model.

Whereas the considered return series has significant volatility clustering, the ARCH LM test is carried out for the residual series of the AR(1) model above. The result indicates that GARCH model needs to be adopted since there is high-order ARCH effect. According to the requirements that the AIC value should be relatively small, and model coefficients must be significant and positive, the GARCH(1,1) model is the best when comparison are made among GARCH(1,1), GARCH(1,2), GARCH(2,1) and GARCH(2,2) models. Because of the fat tail of the return, we consider different distributions, including normal, skewed normal, Student-t, and skewed Student-t distributions, for the innovation term z_t. The most appropriate distribution for z_t are chosen based on the information criteria AIC. As reported in Table 3, the return series of WTI and Gold can be adequately modeled by GARCH(1,1) model with skewed Student-t distribution; while for the return series of all USD exchange rates, GARCH(1,1) model with Student-t distribution is the most appropriated marginal distribution.

Thereafter, we apply the AR(1)-GARCH(1,1) model based on correspondingly specified innovation distribution to model the marginal distributions of considered return series. Table 4 summarizes the marginal distribution estimation results as well as diagnostic of the residuals. In Panel A of Table 4, μ and ϕ are respectively estimates of a constant and an autoregressive coefficient in the conditional mean equation; ω, α, and β are the coefficients of the conditional variance equation (see Eq. (1)); while γ is the degree of freedom as well as $skew$ represents the skewness parameter of the innovation distributions. We note that for all the return series, the conditional

Table 3 AIC of GARCH(1,1) model with different innovation distributions for modeling the conditional heteroscedasticity.

	WTI	Gold	USD/EUR	USD/GBP	USD/JPY	USD/CAD
norm	9.565341	8.909433	1.462971	1.153586	1.243578	1.191535
snorm	9.352697	8.493306	1.453081	1.149500	1.243839	1.174589
std	6.561036	5.407278	**1.178011**	**0.9509337**	**1.020435**	**0.8782591**
sstd	**6.559500**	**5.405993**	1.178552	0.9514799	1.020910	0.8786730

Table 4 Maximum likelihood estimation result of parameters in AR(1)-GARCH(1,1) models for each return series and the descriptive statistics of standardized residual series.

	WTI	Gold	USD/EUR	USD/GBP	USD/JPY	USD/CAD
Panel A: Quasi-maximum likelihood estimation of AR(1)-GARCH(1,1) models for returns						
Mean equation						
μ	0.018770	0.034694	−0.0074404	−0.0048215	0.0030268	−0.00065804
ϕ	−0.033442	−0.018618	−0.0115148	0.0434363	0.0615313	0.06941182
Variance equation						
ω	0.024691	0.011855	0.0021825	0.0050164	0.0092501	0.00324397
α	0.064767	0.035795	0.0587397	0.1400125	0.2373281	0.10712361
β	0.932880	0.958417	0.9762747	0.9770070	0.9564173	0.96291750
γ	8.542901	4.346358	2.2506521	2.0852898	2.0987041	2.20365290
$skew$	0.939485	0.969979	–	–	–	–
Panel B: Ljung-Box Q statsitics of standardized residuals						
L-B $Q(20)$	11.56275	27.15204	20.46524	20.37437	38.88932	23.32544
L-B $Q_2(20)$	21.88527	5.109431	28.26894	152.7406	26.18052	11.86768

variance term β with values above 0.93, which indicates that conditional variance is majorally past dependent and thus highly persistent over time. Moreover, all the degrees of freedom term γ are statistically significant with positive values, with relatively high value for the oil returns. Panel B of Table 4 reports Ljung-Box $Q(20)$- and $Q_2(20)$-statistics to justify the empirical results of the specified marginal distribution models. According to Panel B, except the USD/GBP and USD/JPY exchange rates, it indicates that no autocorrelation up to lag 20 for standardized residuals and squared standardized residuals for all the rest return series. Moreover, these results for the ACFs of the standardized residuals and squared standardized residuals confirms that the standardized residuals are not autocorrelated, which support our model specifications. Thereafter, instead of using raw returns, we use standardized residuals obtained from the GARCH fit to copula estimation.

Then, the copula functions are estimated based on the Pseudo data through the MPL method as described in Sect. 2.3. We consider the standardized residuals obtained from GARCH models and transform them into uniform variates. Moreover, we check the rank correlation coefficients for the dependence between the gold prices

Table 5 Correlation estimates of the Kendall's τ and the Spearman's ρ between oil prices, gold price and exchange rates.

	Overall sample (March 1, 2006–March 18, 2016)		Crisis period (July 1, 2008–June 30, 2009)	
	Kenall's τ	Spearman's ρ	Kenall's τ	Spearman's ρ
Gold-WTI	0.122	0.177	0.11	0.161
Gold-USD/EUR	−0.23	−0.333	−0.221	−0.328
Gold-USD/GBP	−0.182	−0.265	−0.029	−0.034
Gold-USD/JPY	−0.08	−0.119	−0.058	−0.086
Gold-USD/CAD	−0.181	−0.263	−0.368	−0.526
WTI-USD/EUR	−0.19	−0.274	−0.256	−0.358
WTI-USD/GBP	−0.164	−0.236	−0.231	−0.325
WTI-USD/JPY	0.061	0.09	0.24	0.341
WTI-USD/CAD	−0.261	−0.375	−0.315	−0.442

and oil prices, gold prices and exchange rates, as well as oil prices and exchanges rates, respectively. Table 5 reports the Kendall's τ and Spearman's ρ statistics for both the overall sample period and crisis period. We select the crisis period from July 1st, 2008 to June 30th, 2009 because the key trigger event for the global financial crisis on summer 2008, and the real GDP rebound modestly to 1.8 percent growth in 2009 according to the U.S. quarterly national GDP reports [28].

The monotone property of Kendall's τ and Spearman's ρ indicates the negative association relationship for all pairs between gold and exchange rates, as well as oil and exchange rates except oil to USD/JPY. And we observe the positive Kendall's τ and Spearman's ρ for gold and oil prices as excepted. In overall period, the constant correlations for the gold price and oil prices are positive and range from 0.12 to 0.25, while the constant correlation between the gold price and USD exchange rates are all negative and range from −0.058 to −0.526. Moreover, during the crisis period, the association between the gold price and oil prices, as well as the gold price and the USD/CAD are higher than that of the overall period; while the association between the gold price and the rest currencies are smaller than that of the overall period. It implies that the gold price are more deviated from the oil prices rather than the currencies. However, by comparing the correlations for oil prices with others in overall period and crisis period, we found that both of the comovement between oil prices and gold price, as well as currencies are substantially higher during the crisis period. It indicates that the oil prices are deviated not only by gold price but also the USD exchange rates. Specifically, we conclude that the association between USD/CAD exchange rate with both gold and oil are significant high during the crisis period.

To investigate furthur on the dynamical coorelation between all considered pairs, we provide the dynamical curves which display the changes of association measure (the rolling Kendall's tau) in Fig. 3. The figures are constructed by the following steps: (1) we start to compute the Kendall's tau by using the standardized residuals including

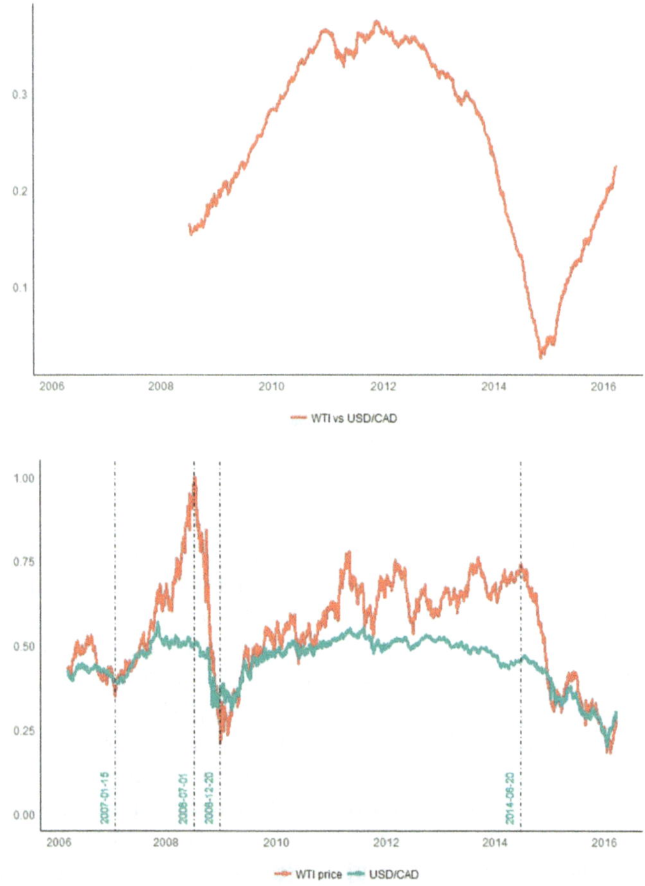

Fig. 3 Plot of negative dynamic Kendall's tau (the rolling Kendall's tau) between WTI price and USD/CDA exchange rate (top panel), normalized WTI price and negative normalized USD/CDA exchange rate chart (bottom panel).

the data between the period March 1st, 2006 and July 1st, 2008; (2)Kendall's tau is then calculated by shifting one data point at a time until the time window reaches up to March 18th, 2016. From the top panel of Fig. 3, we can see that the association between WTI price and USD/CAD exchange rates are peaked between the year 2010 and 2011(end of crisis period). The bottom panel of Fig. 3 indicates the normalized WTI price and negative normalized USD/CAD exchange rate. Note that we utilize the negative USD/CAD exchange rates because the negative association between oil prices and exchange rates as shown in Table 5. Notice that a rise or fall of WTI price at the time Jan 1st, 2007, July 1st, 2008, Dec 20th, 2008, June 20th, 2014, and Jan 10th, 2016 were followed by similar motion in the USD/CAD exchange rates. This indicates that crude oil (WTI price) is a good short-term indicator in the move in asset prices (USD/CDA exchange rates).

4 Conclusion

This paper investigates the dependence structure among gold, nominal crude oil and major U.S. dollar exchange rates from March 1, 2006 to March 18, 2016. Based on a copula-GARCH approach, we examine the conditional dependence structure and the extreme comovement on returns between paris of gold and oil, gold and currencies, as well as oil and currencies. We first apply the AR(1)-GARCH(1,1) model based on different innovation distributions to model the margins. The adoption of this filtering step is motivated by the stylized facts of our financial returns including non-normal distributed, autocorrelation of squared returns and volatility clustering. Then, different copula models are fitted to standardized residuals from the best fitted marginal models. The comparison results of various copula models show that the Student-t copula outperforms other copulas for fitting the conditional dependence structure of all considered pairs.

Empirical results show that (i) each of the analyzed series of gold, oil and currencies returns can be adequately described with the proposed AR(1)-GARCH(1,1) model based on either Student-t or skewed Student-t innovation distributions; (ii) there are positive dependence between gold and oil, negative dependence between gold and currencies, as well as oil and currencies, as indicated by the Kendall's τ and Spearman's ρ concordance, and the correlation coefficient; (iii) there is a small degree of conditional dependence in the extreme tail of all considered pairs; (iv) furthermore, we found that the crude oil price was a good short-term indicator in the move in asset prices like exchange rates. The crude oil price was a short term descend indicator of gold price, and the gold price was an short term rise indicator of oil price. The above findings lead us to conclude that the U.S. dollar depreciation was a key factor in driving up the crude oil price and gold price, while gold market and oil market are positively associated.

Besides the applied contribution, our paper have three main contributions for investors. First, the results of the study provide useful information for investors in asset allocation and portfolio diversification. Second, we show that gold has served as a hedge against fluctuation in the U.S. dollar exchange rates. Moreover, the appreciation of the U.S. dollar are found to coincide with a decrease in crude oil prices. Third, taking into account the extreme comovement between different assets, investors can improve the accuracy of market risk forecasts.

Acknowledgements H.K.Z. was partially supported by NSF grant DMS-1151762 and by a grant from the Simons Foundation (337646, HZ).

References

1. Akram, Q.F.: Commodity prices, interest rates and the dollar. Energy Econ. **31**(6), 838–851 (2009)
2. Aloui, R., Aissa, M.S.B., Nguyen, D.K.: Conditional dependence structure between oil prices and exchange rates: a copula-GARCH approach. J. Int. Money Financ. **32**, 719–738 (2013)
3. Amano, R.A., Van Norden, S.: Oil prices and the rise and fall of the US real exchange rate. J. Int. Money Financ. **17**(2), 299–316 (1998)
4. Basher, S.A., Haug, A.A., Sadorsky, P.: Oil prices, exchange rates and emerging stock markets. Energy Econ. **34**(1), 227–240 (2012)
5. Capie, F., Mills, T.C., Wood, G.: Gold as a hedge against the dollar. J. Int. Financ. Mark. Inst. Money **15**(4), 343–352 (2005)
6. Ciner, C.: On the long run relationship between gold and silver prices A note. Glob. Financ. J. **12**(2), 299–303 (2001)
7. Genest, C., Rivest, L.P.: Statistical inference procedures for bivariate Archimedean copulas. J. Am. Stat. Assoc. **88**(423), 1034–1043 (1993)
8. Genest, C., Rémillard, B., Beaudoin, D.: Goodness-of-fit tests for copulas: a review and a power study. Insur. Math. Econ. **44**(2), 199–213 (2009)
9. Golub, S.S.: Oil prices and exchange rates. Econ. J. **93**(371), 576–593 (1983)
10. Krugman, P.: Oil shocks and exchange rate dynamics. In: Exchange Rates and International Macroeconomics, pp. 259-284. University of Chicago Press (1983)
11. Johnson, R., Soenen, L.A.: Gold as an investment asset: perspectives from different countries. J. Invest. **6**(3), 94–99 (1997)
12. Joy, M.: Gold and the US dollar: hedge or haven? Financ. Res. Lett. **8**(3), 120–131 (2011)
13. Rogoff, K.: Oil, productivity, government spending and the real yen-dollar exchange rate (No. 91-06). Federal Reserve Bank of San Francisco (1991)
14. Sjaastad, L.A., Scacciavillani, F.: The price of gold and the exchange rate. J. Int. Money Financ. **15**(6), 879–897 (1996)
15. Shafiee, S., Topal, E.: An overview of global gold market and gold price forecasting. Resour. Policy **35**(3), 178–189 (2010)
16. Sari, R., Hammoudeh, S., Soytas, U.: Dynamics of oil price, precious metal prices, and exchange rate. Energy Econ. **32**(2), 351–362 (2010)
17. Sklar, A.: Fonctions de ŕepartition *à n* dimensions et leurs marges. Publ. Inst. Statist. Univ. Paris **8**, 229–231 (1959)
18. Pukthuanthong, K., Roll, R.: Gold and the dollar (and the Euro, Pound, and Yen). J. Bank. Financ. **35**(8), 2070–2083 (2011)
19. Frees, E.W., Valdez, E.A.: Understanding relationships using copulas. North Am. Actuar. J. **2**(1), 1–25 (1998)
20. de Leon, A.R., Wu, B.: Copula-based regression models for a bivariate mixed discrete and continuous outcome. Stat. Med. **30**(2), 175–185 (2011)
21. Patton, A.J.: Modelling asymmetric exchange rate dependence*. Int. Econ. Rev. **47**(2), 527–556 (2006)
22. Zhang, L., Singh, V.P.: Bivariate rainfall frequency distributions using Archimedean copulas. J. Hydrol. **332**(1), 93–109 (2007)
23. Nelsen, R.B.: An Introduction to Copulas, 2nd edn. Springer, New york (2006)
24. Joe, H.: Dependence Modeling with Copulas. CRC Press, Boca Raton (2014)
25. Wei, Z., Wang, T., Nguyen, P.A.: Multivariate dependence concepts through copulas. Int. J. Approx. Reason. **65**(2015), 24–33 (2015)
26. Umberto, C., Elisa, L., Walter, V.: Copula Methods in Finance. Wiley, Hoboken (2004). xvi+293
27. Wu, C.C., Chung, H., Chang, Y.H.: The economic value of co-movement between oil price and exchange rate using copula-based GARCH models. Energy Econ. **34**(1), 270–282 (2012)
28. Financial Crisis Inquiry Commission, & United States. Financial Crisis Inquiry Commission: The financial crisis inquiry report: Final report of the national commission on the causes of the financial and economic crisis in the United States. Public Affairs (2011)

Extremal Properties and Tail Asymptotic of Alpha-Skew-Normal Distribution

Weizhong Tian, Huihui Li, and Rui Huang

Abstract Extreme value theory has emerged as one of the most important statistical disciplines for the applied sciences. In this paper, the extremal properties of univariate alpha-skew-normal distribution was discussed. In addition, asymptotic tail dependence coefficients of the bivariate alpha-skew-normal distribution are investigated.

1 Introduction

Extreme value theory (EVT) has emerged as one of the most important statistical disciplines for the applied sciences. The distinguishing feature of EVT is to quantify the stochastic behavior of a process at unusually large or small levels. Specifically, EVT usually requires estimation of the probability of events that are more extreme than any other that has been previously observed. The Fisher–Tippett theorem [14] plays a key role in extreme value theory. It states that after suitable normalization, if the maximum (or minimum) of n independent and identically distributed random variables converges weakly to Q as $n \to \infty$, then Q is one of the following three families of extreme value distributions (EVDs), which are described by their cumulative density function (cdfs),

---

W. Tian (✉)
Department of Mathematical Sciences, Eastern New Mexico University, Portales, NM 88130, USA
e-mail: weizhong.tian@enmu.edu

H. Li
School of Science, Xi'an University of Technology, Xi'an 710048, Shaanxi, China
e-mail: 15834396228@163.com

R. Huang
Department of Mathematics and Statistics, Washington State University, Pullman, WA 99163, USA
e-mail: rui.huang2@wsu.edu

S. Sriboonchitta et al. (eds.), *Behavioral Predictive Modeling in Economics*, Studies in Computational Intelligence 897,
https://doi.org/10.1007/978-3-030-49728-6_15

(i) Gumbel distribution, $\Lambda(x) = \exp\{-e^{-x}\}$, for $-\infty < x < \infty$,
(ii) Frechet distribution, $\Phi_a(x) = \exp\{-x^{-a}\}$, for $x > 0$ and $a > 0$,
(iii) Weibull distribution, $\Psi_a(x) = \exp\{-(-x)^a\}$, for $x \leq 0$ and $a > 0$.

In other words, for each n, let $X_1, X_2, \cdots, X_n$ be independent and identically distributed (i.i.d) univariate random variables, then

$$\lim_{n\to\infty}\left(\frac{M_n - b_n}{a_n} \leq x\right) = G(x),$$

where $M_n = Max(X_1, X_2, \cdots, X_n)$ or $M_n = Min(X_1, X_2, \cdots, X_n), a_n, b_n$ are normalising constants, and $G(x)$ is a member of the EVDs. For details, see Embrechts et al. [13], Kotz and Nadarajah [18], and Coles [10].

Mills [21] introduced the inequality and ratio about the standard normal distribution, which are given as follows,

$$x^{-1}(1 + x^{-2})^{-1}\phi(x) < 1 - \Phi(x) < x^{-1}\phi(x), \quad \text{for} \quad x > 0$$
$$\frac{1 - \Phi(x)}{\phi(x)} \approx x^{-1}, \quad \text{for} \quad x \to \infty,$$

where $\phi(\cdot)$ and $\Phi(\cdot)$ are the probability density function (pdf) and cdf of the standard normal distribution, respectively.

Based on Mills inequality and ratio, Resnick [23] discussed the properties of extreme value based on normal distribution. The exact uniform convergence rate of the asymmetric normal distribution of the maximum and minimum to its extreme value limit was investigated by Chen and Huang [9]. After the class of skew normal distributions was introduced by Azzalini [4, 5], it has been received increasing interests, Tian et al. [25] discussed the distortion risk measures under skew normal setting and later on, Tian et al. [26] introduced the class of multivariate extended skew normal distribution, for details to see Azzalini [6]. Specifically, there are some contributions concern of the extremal properties for this new family. Liao et al. [20] derived the normalization constants a_n, b_n to the Gumbel extreme value distribution from the skew normal samples. Asymptotic behaviors of the extremes of the skew- distribution are studied by Peng et al. [22]. And recently, Beranger et al. [8] discussed the extremal properties of the extended skew normal distribution, which was introduced by Arellano-Valle and Genton [3].

It was well known that the concept of tail dependence was a useful tool to describe the dependence between extremal data in finance, which was proposed by Ane and Kharoubi [1]. The common measure of tail dependence was given by the so-called tail dependence coefficient and the upper tail dependence coefficient, λ_U, of $\mathbf{X}$ was defined by Sibuya [24] as follows,

$$\lambda_U = \lim_{u\to 1^-} P(F_1(X_1) \geq u | F_2(X_2) \geq u),$$

where $\mathbf{X} = (X_1, X_2)^{'}$ is a bivariate random vector with marginal distributions F_1 and F_2, respectively. After that, there were a lot of findings of tail dependence in asymmetric distributions, which have a wild application in finical filed. Bortot [7] studied the tail dependence of the bivariate skew-Normal and skew-t distributions. Fung and Seneta examined [16] the tail dependence for two different types of skew-t distributions.

In this paper, we will use the alpha-skew-normal distribution, which was proposed by Elal-Olivero [12]. A random variable Z is said to have an *alpha-skew-normal distribution* with skewness $\alpha \in \Re$, if its probability density function takes the following form,

$$f(z) = \frac{\left[(1-\alpha z)^2 + 1\right]}{2+\alpha^2}\phi(z), \tag{1}$$

which is denoted by $z \sim ASN(\alpha)$. Our contributions concerns the derivation of the extremal properties of the univariate alpha skew normal distribution and the tail asymptotic for the bivariate alpha-skew-normal distribution.

The remainder of this paper is set out as follows, the extremal properties of universe alpha-skew-normal distribution are discussed in Sect. 2. In Sect. 3, the coefficients of lower and upper tail dependence for the bivariate alpha-skew-normal distribution are derived. Some conclusions are given in Sect. 4.

2 Extremal Properties of the Univariate Alpha-Skew-Normal Distribution

In order to get the normalising constants of extreme value distribution for the alpha-skew-normal distribution, we need to investigate Mill's inequality and ratio for the alpha-skew-normal distribution.

Theorem 1. *Let $F_\alpha(x)$ and $f_\alpha(x)$ denote the cdf and pdf of $ASN(\alpha)$, respectively. For all $x > 0$, we have*

$$L_\alpha(x) < \frac{1-F_\alpha(x)}{f_\alpha(x)} < U_\alpha(x),$$

where the lower bound $L_\alpha(x)$ and the upper bound $U_\alpha(x)$ are given as follows,
(i) for $\alpha > 0$,

$$L_\alpha(x) = \frac{1}{x}\left[1 + \frac{2\alpha^2}{(1-\alpha x)^2+1}\frac{1}{1+x^2}\right]\left(1+\frac{1}{x^2}\right)^{-1},$$

$$U_\alpha(x) = \frac{1}{x}\left[1 + \frac{2\alpha^2}{(1-\alpha x)^2+1}\right],$$

(ii) for $\alpha < 0$

$$L_\alpha(x) = \frac{1}{x}\left[1 + \frac{4\alpha}{[(1-\alpha x)^2+1]x^3} + \frac{2\alpha^2 x}{[(1-\alpha x)^2+1](1+x^{-2})}\right]\left(1+\frac{1}{x^2}\right)^{-1},$$

$$U_\alpha(x) = \frac{1}{x}\left[1 - \frac{1}{x}\frac{2\alpha}{(1-\alpha x)^2+1} - \frac{2\alpha^2}{(1-\alpha x)^2+1}\right].$$

Theorem 2. *Let* $X \sim ASN(\alpha)$, *for* $\alpha \in \Re$, *as* $x \to \infty$, *we have*

$$\frac{1-F_\alpha(x)}{f_\alpha(x)} \approx \frac{1}{x}. \tag{2}$$

Proof: The result can be obtain from Theorem 1. □

Theorem 3. *Suppose* $X \sim ASN(\alpha)$, *for* $\alpha \in \Re$ *and* $x \to \infty$, *we have*

$$1 - F_\alpha(x) \approx c(x)\exp\left\{-\int_1^x \frac{g(t)}{f(t)}dt\right\},$$

where

$$c(x) \to \frac{\alpha^2}{(2+\alpha^2)\sqrt{2\pi e}}, \quad as \quad x \to \infty,$$
$$f(x) = \frac{1}{x}, \quad and$$
$$g(x) = 1 - \frac{1}{x^2} \to 1, \quad as \quad x \to \infty.$$

Proof: According to Theorem 2, as $x \to \infty$, we know that

$$\begin{aligned} 1 - F_\alpha(x) &\approx \frac{2+\alpha^2x^2-2\alpha x}{[(2+\alpha^2)\sqrt{2\pi}]x}e^{-\frac{x^2}{2}} \\ &= \frac{2/x^2+\alpha^2-2\alpha/x}{(2+\alpha^2)\sqrt{2\pi}}xe^{-\frac{x^2}{2}} \\ &= \frac{2/x^2+\alpha^2-2\alpha/x}{(2+\alpha^2)\sqrt{2\pi}}e^{\ln x-\frac{x^2}{2}}. \end{aligned}$$

Since $e^{\ln x - \frac{x^2}{2}} = \exp\left\{-\int_1^x \frac{t^2-1}{t}dt + \frac{1}{2}\right\}$, therefore, we get

$$c(x) = \frac{2/x^2 + \alpha^2 - 2\alpha/x}{(2+\alpha^2)\sqrt{2\pi}},$$
$$f(x) = \frac{1}{x}, \quad \text{and}$$
$$g(x) = 1 - \frac{1}{x^2}.$$

□

Theorem 4. *Let $X_1, \cdots, X_n$ be i.i.d random variable with $X_i \sim ASN(\alpha)$ for $i = 1, \cdots, n$. Define $M_n = Max(X_1, \cdots, X_n)$, then we have*

$$\lim_{n\to\infty} P(M_n \le \alpha_n x + \beta_n) = \Lambda(x),$$

where the normalising constants α_n and β_n are given as follows,
(i) for $\alpha > 0$

$$\alpha_n = (2\ln n)^{-\frac{1}{2}},$$
$$\beta_n = (2\ln n)^{\frac{1}{2}} + \frac{2\ln\alpha - \ln(2+\alpha^2) - \ln\sqrt{2\pi}}{(2\ln n)^{\frac{1}{2}}} + \frac{\ln 2 + \ln\ln n}{2(2\ln n)^{\frac{1}{2}}}.$$

(ii) for $\alpha < 0$

$$\alpha_n = (2\ln n)^{-\frac{1}{2}},$$
$$\beta_n = (2\ln n)^{\frac{1}{2}} + \frac{2\ln(-\alpha) - \ln(2+\alpha^2) - \ln\sqrt{2\pi}}{(2\ln n)^{\frac{1}{2}}} + \frac{\ln 2 + \ln\ln n}{2(2\ln n)^{\frac{1}{2}}}.$$

In the following, we discuss the converge rates of the distribution of M_n.

Theorem 5. *For the normalising constants α_n and β_n given in Theorem 4, we have*

$$F_\alpha(\alpha_n x + \beta_n) - \Lambda(x) \approx \Lambda(x)\left[1 + \frac{\alpha}{8\pi\sqrt{2\ln n}} - \frac{\alpha^2\left(1 + \frac{\ln 2 + \ln n}{2(2\ln n)}\right)}{8\pi\sqrt{2}}\right], \quad as \quad n \to \infty. \tag{3}$$

Proof: For $\alpha > 0$, let $U_n = \alpha_n(x) + \beta_n$, where α_n and β_n are given in Theorem 4, as $n \to \infty$, we have

$$\tau_n = n(1 - F_\alpha(U_n)) \to \frac{n}{(2+\alpha^2)\sqrt{2\pi}}\left(\frac{2}{U_n} - 2\alpha + \alpha^2 U_n\right)\exp\left\{-\frac{U_n^2}{2}\right\}.$$

U_n and U_n^2 can be obtained in Theorem 3, and also $U_n^{-1} = (2\ln n)^{-\frac{1}{2}} + o\left(\frac{\ln\ln n}{(2\ln n)^{\frac{1}{2}}}\right)$, therefore,

$$\tau_n(x) = \exp\{-x\}\left[-\frac{\alpha}{8\pi\sqrt{2\ln n}} - \frac{\alpha^2\left(1+\frac{\ln 2+\ln n}{2(2\ln n)}\right)}{8\pi\sqrt{2}}\right] + o\left(\frac{\ln\ln n}{(2\ln n)^{\frac{1}{2}}}\right),$$

for $\tau(x) = \exp\{-x\}$, we have

$$\tau(x) - \tau_n(x) = \exp\{-x\}\left[1 + \frac{\alpha}{8\pi\sqrt{2\ln n}} - \frac{\alpha^2\left(1+\frac{\ln 2+\ln n}{2(2\ln n)}\right)}{8\pi\sqrt{2}}\right].$$

Thus, by using Theorem 2.4.2 in Leadbetter et al. [19], the result is obtained. □

3 Tail Asymptotic for the Bivariate Alpha Skew Normal Distribution

The bivariate extension comes directly from the univariate alpha-skew-normal distribution and is defined as follows. A random variable $\mathbf{X}$ is said to have a *bivariate alpha-skew-normal distribution* with skewness $\boldsymbol{\alpha} \in \Re^2$, if its probability density function takes the following form,

$$f(\mathbf{x}) = \left[(1 + \boldsymbol{\alpha}'\mathbf{x})^2 + 1\right] K^{-1}\phi_2(\mathbf{x}), \tag{4}$$

where $\phi_2(\cdot)$ is the probability density function of a bivariate normal distribution $N_2(\mathbf{0}, \Sigma)$, $\boldsymbol{\alpha} \in \Re^2$, and $K = 2 + \boldsymbol{\alpha}'\Sigma\boldsymbol{\alpha}$. We denoted it by $\mathbf{X} \sim BASN(\mathbf{0}, \Sigma, \boldsymbol{\alpha})$.

In this section, we discuss the coefficients of the lower and upper tail dependence of $\mathbf{X} \sim BASN(\mathbf{0}, \Sigma, \boldsymbol{\alpha})$, which are defined as

$$\lambda_L = \lim_{u\to 0^+} P(X_1 \le F_1^{-1}(u)|X_2 \le F_2^{-1}(u)),$$
$$\lambda_U = \lim_{u\to 1^-} P(X_1 \ge F_1^{-1}(u)|X_2 \ge F_2^{-1}(u)),$$

where F_1^{-1} and F_2^{-1} are the marginal inverse distribution functions for X_1 and X_2, respectively. We know that, if λ_L (λ_U) exist and is positive, $\mathbf{X}$ is said to have asymptotic lower (upper) tail dependence. In the following, we consider the tail asymptotic dependence for the $\mathbf{X} \sim BASN(\mathbf{0}, \Sigma, \boldsymbol{\alpha})$, where $\Sigma = \begin{pmatrix} 1 & \rho \\ \rho & 1 \end{pmatrix}$ and $\boldsymbol{\alpha} = (\alpha_1, \alpha_2)'$.

Theorem 6. *Let $\mathbf{X} \sim BASN(\mathbf{0}, \Sigma, \boldsymbol{\alpha})$, where $\boldsymbol{\alpha} = (\alpha_1, \alpha_2)'$, for $z \to -\infty$, we have*

$$P(X_1 \le z) = F_1(z) \approx \frac{(\alpha_1 + \alpha_2\rho)^2}{K\sqrt{2\pi}}|z|e^{-\frac{z^2}{2}},$$
$$P(X_2 \le z) = F_2(z) \approx \frac{(\alpha_2 + \alpha_1\rho)^2}{K\sqrt{2\pi}}|z|e^{-\frac{z^2}{2}},$$

where $K = 2 + \alpha_1^2 + \alpha_2^2 + 2\rho\alpha_1\alpha_2$.

Proof: According to the Proposition 5 in Ara and Louzada [2], we obtain the marginal probability density functions of X_1 and X_2, respectively,

$$f(x_1) = K^{-1}\left[(1-\alpha_1 x_1)^2 + 1 + (1-\rho^2)\alpha_2^2 + (\alpha_2\rho x_1)^2 + 2\alpha_2\rho x_1(\alpha_1 x_1 - 1)\right]\phi(x_1), \quad \text{and}$$
$$f(x_2) = K^{-1}\left[(1-\alpha_2 x_2)^2 + 1 + (1-\rho^2)\alpha_1^2 + (\alpha_1\rho x_2)^2 + 2\alpha_1\rho x_2(\alpha_2 x_2 - 1)\right]\phi(x_2),$$

where $K = 2 + \alpha_1^2 + \alpha_2^2 + 2\rho\alpha_1\alpha_2$.
Therefore,

$$P(X_1 \le z) = K^{-1}\int_{-\infty}^{z}\left[(1-\alpha_1 x_1)^2 + 1 + (1-\rho^2)\alpha_2^2 + (\alpha_2\rho x_1)^2 + 2\alpha_2\rho x_1(\alpha_1 x_1 - 1)\right]\phi(x_1)dx_1,$$
$$= \Phi(z) + \frac{2\alpha_1 + 2\alpha_2\rho}{K\sqrt{2\pi}}e^{-\frac{z^2}{2}} - \frac{(\alpha_1+\alpha_2\rho)^2}{K\sqrt{2\pi}}ze^{-\frac{z^2}{2}}.$$

According to Feller [15],

$$\Phi(z) \approx \frac{1}{\sqrt{2\pi}}|z|^{-1}e^{-\frac{z^2}{2}}, \quad as \quad z \to -\infty. \tag{5}$$

Thus, as $z \to -\infty$,

$$F_1(z) \approx \frac{(\alpha_1+\alpha_2\rho)^2}{K\sqrt{2\pi}}|z|e^{-\frac{z^2}{2}}.$$

Similarly for $P(X_2 \le z)$. □

Theorem 7. *Let* $\mathbf{X} = (X_1, X_2)' \sim BASN(\mathbf{0}, \Sigma, \boldsymbol{\alpha})$, *where* $\boldsymbol{\alpha} = (\alpha_1, \alpha_2)'$, *then, as* $u \to 0^+$,

$$F_1^{-1}(u) \approx -\sqrt{-\ln\left[\frac{4K\pi u^2}{(\alpha_1+\rho\alpha_2)^4}\ln\left[\frac{K\sqrt{2\pi}u}{(\alpha_1+\rho\alpha_2)^2}\right]\right]}, \quad and$$
$$F_2^{-1}(u) \approx -\sqrt{-\ln\left[\frac{4K\pi u^2}{(\alpha_2+\rho\alpha_1)^4}\ln\left[\frac{K\sqrt{2\pi}u}{(\alpha_2+\rho\alpha_1)^2}\right]\right]},$$

where $K = 2 + \alpha_1^2 + \alpha_2^2 + 2\rho\alpha_1\alpha_2$ *and* $F_i^{-1}(\cdot)$ *is the inverse probability marginal function of* X_i, *for* $i = 1, 2$.

Proof: We work on $i = 1$ first, and it would be the same arguments for $i = 2$. Let $z(u) = F_1^{-1}(u)$, then equation (3.1) can be expressed into the following form,

$$u = a|z(u)|^b e^{-c|z(u)|^d},$$

where $a = \frac{(\alpha_1+\rho\alpha_2)^2}{\sqrt{2\pi}K}$, $b = 1$, $c = \frac{1}{2}$ and $d = 2$. According to Corless et al. [11], and Theorem 1 in Fung and Seneta [17], we obtain

$$
\begin{aligned}
z(u) &= -\left\{ \ln\left(\frac{\left(\frac{(\alpha_1+\rho\alpha_2)^2}{K\sqrt{2\pi}u}\right)^2}{2\ln\left(\frac{(\alpha_1+\rho\alpha_2)^2}{K\sqrt{2\pi}u}\right)} \right) + O\left(\frac{\ln\ln\left(\frac{(\alpha_1+\rho\alpha_2)^2}{K\sqrt{2\pi}u}\right)^2}{2\ln\left(\frac{(\alpha_1+\rho\alpha_2)^2}{K\sqrt{2\pi}u}\right)} \right) \right\}^{\frac{1}{2}} \\
&= -\left\{ -\ln\left[-\frac{4K\pi u^2}{(\alpha_1+\rho\alpha_2)^4} \ln\left(\frac{K\sqrt{2\pi}u}{(\alpha_1+\rho\alpha_2)^2} \right) \right] + O\left(\frac{\ln\ln\left(\frac{(\alpha_1+\rho\alpha_2)^2}{K\sqrt{2\pi}u}\right)^2}{2\ln\left(\frac{(\alpha_1+\rho\alpha_2)^2}{K\sqrt{2\pi}u}\right)} \right) \right\}^{\frac{1}{2}} \\
&\approx -\sqrt{-\ln(-4K\pi u^2 \ln(K\sqrt{2\pi}u))} \\
&\approx -\sqrt{-\ln\left(\frac{2K\sqrt{2\pi}u^2}{(\alpha_1+\rho\alpha_2)^2} \right)} \\
&\approx -\sqrt{-2\ln u}.
\end{aligned}
$$

□

Theorem 8. *Let* $\mathbf{X} = (X_1, X_2)' \sim BASN(\mathbf{0}, \Sigma, \boldsymbol{\alpha})$, *where* $\boldsymbol{\alpha} = (\alpha_1, \alpha_2)'$, *then as* $u \to 0^+$,

$$
\lambda_L(u) \approx 2^{\frac{2-\rho^2}{2(1-\rho^2)}} (K\sqrt{2\pi})^{\frac{\rho^2}{2(1-\rho^2)}} \frac{(\alpha_2+\rho\alpha_1)^{2(1-\rho^2)}\sqrt{1-\rho^2}}{|\alpha_1+\alpha_2\rho|} u^{\frac{1}{1-\rho^2}}.
$$

Corollary 1. *Let* $\mathbf{X} = (X_1, X_2)' \sim BASN(\mathbf{0}, \Sigma, \boldsymbol{\alpha})$, *where* $\boldsymbol{\alpha} = (\alpha_1, \alpha_2)'$, *then as* $u \to 0^+$,

$$
\lambda_U(u) \approx 2^{\frac{2-\rho^2}{2(1-\rho^2)}} (K\sqrt{2\pi})^{\frac{\rho^2}{2(1-\rho^2)}} \frac{(\alpha_2+\rho\alpha_1)^{2(1-\rho^2)}\sqrt{1-\rho^2}}{|\alpha_1+\alpha_2\rho|} (1-u)^{\frac{1}{1-\rho^2}}.
$$

Proof: Since $-\mathbf{X} \sim BASN(\mathbf{0}, \Sigma, -\boldsymbol{\alpha})$, the result can be obtained from $\lambda_U^X(u) = \lambda_L^{-X}(1-u)$. □

4 Conclusions

In this paper, we studied the extremal properties based on the univariate alpha skew normal distribution, as well as we presented the coefficients of lower and upper tail dependence and studied the tail asymptotic for the bivariate alpha skew normal distribution. Further studying are possible for the alpha skew normal distribution, such as, inferences in linear mixed models, extremal properties of multivariate case and matrix variate alpha skew normal distribution.

Acknowledgements We would like to express our gratitude to professor Vladik Kreinovich and professor Hung T. Nguyen. We also thank anonymous referees for their valuable comments and suggestions that help to improve this article significantly.

Appendix

Proof of Theorem 1: According to the Eq. (1),

$$\frac{1-F_\alpha(x)}{f_\alpha(x)} = \frac{\int_x^\infty \left[(1-\alpha z)^2+1\right]\phi(z)dz}{\left[(1-\alpha x)^2+1\right]\phi(x)} = \frac{\int_x^\infty \left[(1-\alpha z)^2+1\right]e^{-\frac{z^2}{2}}dz}{\left[(1-\alpha x)^2+1\right]e^{-\frac{x^2}{2}}}.$$

For $x > 0$,

$$\frac{1}{x^2}\int_x^\infty \left[(1-\alpha z)^2+1\right]e^{-\frac{z^2}{2}}dz > \int_x^\infty \frac{\left[(1-\alpha z)^2+1\right]}{z^2}e^{-\frac{z^2}{2}}dz$$
$$= \frac{\left[(1-\alpha x)^2+1\right]e^{-\frac{x^2}{2}}}{x} - \int_x^\infty \left[\frac{2\alpha(1-\alpha z)}{z}e^{-\frac{z^2}{2}} + \left[(1-\alpha z)^2+1\right]e^{-\frac{z^2}{2}}\right]dz.$$

Thus,

$$(1+\frac{1}{x^2})\int_x^\infty \left[(1-\alpha z)^2+1\right]e^{-\frac{z^2}{2}}dz > \frac{\left[(1-\alpha x)^2+1\right]}{x}e^{-\frac{x^2}{2}} - 2\alpha\int_x^\infty \frac{1-\alpha z}{z}e^{-\frac{z^2}{2}}dz$$
$$= \frac{\left[(1-\alpha x)^2+1\right]}{x}e^{-\frac{x^2}{2}} + 2\sqrt{2\pi}\alpha^2\Phi(-x) - 2\alpha\int_x^\infty \frac{e^{-\frac{z^2}{2}}}{z}dz. \tag{6}$$

For $\alpha > 0$, we have

$$\left(1+\frac{1}{x^2}\right)\int_x^\infty \left[(1-\alpha z)^2+1\right]e^{-\frac{z^2}{2}}dz > \frac{\left[(1-\alpha x)^2+1\right]}{x}e^{-\frac{x^2}{2}} + 2\sqrt{2\pi}\alpha^2\Phi(-x),$$

$$\left(1+\frac{1}{x^2}\right)\frac{\int_x^\infty \left[(1-\alpha z)^2+1\right]e^{-\frac{z^2}{2}}dz}{\left[(1-\alpha x)^2+1\right]e^{-\frac{x^2}{2}}} > \frac{1}{x} + \frac{2\sqrt{2\pi}\alpha^2\Phi(-x)}{\left[(1-\alpha x)^2+1\right]e^{-\frac{x^2}{2}}},$$

According to $\frac{\sqrt{2\pi}\Phi(-x)}{e^{-\frac{x^2}{2}}} = \frac{1-\Phi(x)}{\phi(x)} > x^{-1}(1+x^{-2})^{-1}$, we get

$$\frac{\int_x^\infty \left[(1-\alpha z)^2+1\right]e^{-\frac{z^2}{2}}dz}{\left[(1-\alpha x)^2+1\right]e^{-\frac{x^2}{2}}} > \left[\frac{1}{x}+\frac{2\alpha^2}{[(1-\alpha x)^2+1]}\left(x^{-1}(1+x^{-2})^{-1}\right)\right]\left(1+\frac{1}{x^2}\right)^{-1}.$$

By Eq. (6), we know

$$\begin{aligned}\int_x^\infty \left[(1-\alpha z)^2+1\right]e^{-\frac{z^2}{2}}dz &< \frac{\left[(1-\alpha x)^2+1\right]}{x}e^{-\frac{x^2}{2}} - 2\alpha\int_x^\infty \frac{e^{-\frac{z^2}{2}}}{z}dz + 2\alpha^2\int_x^\infty e^{-\frac{z^2}{2}}dz \\ &< \frac{\left[(1-\alpha x)^2+1\right]}{x}e^{-\frac{x^2}{2}}\left[1-2\alpha\frac{x\int_x^\infty \frac{e^{-\frac{z^2}{2}}}{z}dz}{[(1-\alpha x)^2+1]e^{-\frac{x^2}{2}}}+\frac{2\alpha^2 x\int_x^\infty e^{-\frac{z^2}{2}}dz}{[(1-\alpha x)^2+1]e^{-\frac{x^2}{2}}}\right] \\ &< \frac{\left[(1-\alpha x)^2+1\right]}{x}e^{-\frac{x^2}{2}}\left[1+\frac{2\alpha^2 x\int_x^\infty e^{-\frac{z^2}{2}}dz}{[(1-\alpha x)^2+1]e^{-\frac{x^2}{2}}}\right].\end{aligned}$$

According to Mills inequality,

$$\frac{\int_x^\infty [(1-\alpha z)^2+1]e^{-\frac{z^2}{2}}dz}{[(1-\alpha x)^2+1]e^{-\frac{x^2}{2}}} < \frac{1}{x}\left[1+\frac{2\alpha^2}{(1-\alpha x)^2+1}\right].$$

Therefore,

$$\left[\frac{1}{x}+\frac{2\alpha^2}{[(1-\alpha x)^2+1]}\left(x^{-1}(1+x^{-2})^{-1}\right)\right]\left(1+\frac{1}{x^2}\right)^{-1} < \frac{1-F_\alpha(x)}{f_\alpha(x)} < \frac{1}{x}\left[1+\frac{2\alpha^2}{(1-\alpha x)^2+1}\right].$$

In the following, we consider the case $\alpha < 0$. By Eq. (6) and Mill's inequality, we have

$$\begin{aligned}&\left(1+\frac{1}{x^2}\right)\int_x^\infty [(1-\alpha z)^2+1]e^{-\frac{z^2}{2}}dz \\ &> \frac{[(1-\alpha x)^2+1]}{x}e^{-\frac{x^2}{2}} - 2\alpha\int_x^\infty \frac{e^{-\frac{z^2}{2}}}{z}dz + 2\alpha^2\int_x^\infty e^{-\frac{z^2}{2}}dz \\ &> \frac{[(1-\alpha x)^2+1]}{x}e^{-\frac{x^2}{2}} - 2\alpha\left[\frac{e^{-\frac{x^2}{2}}}{x^2}-2\int_x^\infty z^{-3}e^{-\frac{z^2}{2}}dz\right] + 2\alpha^2\int_x^\infty e^{-\frac{z^2}{2}}dz \\ &> \frac{[(1-\alpha x)^2+1]}{x}e^{-\frac{x^2}{2}}\left[1-\frac{2\alpha}{[(1-\alpha x)^2+1]x}+\frac{4\alpha x^{-2}\int_x^\infty e^{-\frac{z^2}{2}}dz}{[(1-\alpha x)^2+1]e^{-\frac{x^2}{2}}}+\frac{2\alpha^2 x\int_x^\infty e^{-\frac{z^2}{2}}dz}{[(1-\alpha x)^2+1]e^{-\frac{x^2}{2}}}\right] \\ &> \frac{[(1-\alpha x)^2+1]}{x}e^{-\frac{x^2}{2}}\left[1+\frac{4\alpha}{[(1-\alpha x)^2+1]x^3}+\frac{2\alpha^2}{[(1-\alpha x)^2+1]}\frac{x}{1+x^{-2}}\right].\end{aligned}$$

Thus,

$$\frac{\int_x^{\infty}\left[(1-\alpha z)^2+1\right]e^{-\frac{z^2}{2}}dz}{\left[(1-\alpha x)^2+1\right]e^{-\frac{x^2}{2}}} > \frac{1}{x}\left[1+\frac{4\alpha}{\left[(1-\alpha x)^2+1\right]x^3}+\frac{2\alpha^2}{\left[(1-\alpha x)^2+1\right]}\frac{x}{1+x^{-2}}\right]\left(1+\frac{1}{x^2}\right)^{-1}.$$

In the other hand,

$$\begin{aligned}
x\int_x^{\infty}\left[(1-\alpha z)^2+1\right]e^{-\frac{z^2}{2}}dz &< \int_x^{\infty} z\left[(1-\alpha z)^2+1\right]e^{-\frac{z^2}{2}}dz\\
&= e^{-\frac{x^2}{2}}\left[(1-\alpha x)^2+1\right]-2\alpha\int_x^{\infty}(1-\alpha z)e^{-\frac{z^2}{2}}dz\\
&= e^{-\frac{x^2}{2}}\left[(1-\alpha x)^2+1\right]-2\alpha\int_x^{\infty}e^{-\frac{z^2}{2}}dz-2\alpha^2 e^{-\frac{x^2}{2}}\\
&= e^{-\frac{x^2}{2}}\left[(1-\alpha x)^2+1\right]\left[1-\frac{2\alpha\int_x^{\infty}e^{-\frac{z^2}{2}}dz}{e^{-\frac{x^2}{2}}\left[(1-\alpha x)^2+1\right]}-\frac{2\alpha^2}{(1-\alpha x)^2+1}\right]\\
&< e^{-\frac{x^2}{2}}\left[(1-\alpha x)^2+1\right]\left[1-\frac{2\alpha}{[(1-\alpha x)^2+1]x}-\frac{2\alpha^2}{(1-\alpha x)^2+1}\right].
\end{aligned}$$

Therefore,

$$\begin{aligned}
&\frac{1}{x}\left[1+\frac{4\alpha}{\left[(1-\alpha x)^2+1\right]x^3}+\frac{2\alpha^2}{\left[(1-\alpha x)^2+1\right]}\frac{x}{1+x^{-2}}\right]\left(1+\frac{1}{x^2}\right)^{-1}\\
&< \frac{1-F_\alpha(x)}{f_\alpha(x)} < \frac{1}{x}\left[1-\frac{2\alpha}{[(1-\alpha x)^2+1]x}-\frac{2\alpha^2}{(1-\alpha x)^2+1}\right]. \qquad \square
\end{aligned}$$

Proof of Theorem 4**:** For $\alpha > 0$, by Theorem 3, we know that $F_\alpha(x)$ belongs to the family of Gumbel extreme value distribution, see Leadbetter et al. [19]. Therefore by Proposition 1.1 in Resnick [23], the normalising constant $\alpha_n > 0$ and $\beta_n \in \Re$ can be determined by

$$1-F_\alpha(\beta_n)=\frac{1}{n}, \quad \text{and} \quad \alpha_n = f_\alpha(\beta_n).$$

Thus, there exist $U_n = U_n(x)$ such that

$$n(1-F_\alpha(U_n(x))) = e^{-x}.$$

By Theorem 2, as $n \to \infty$, we have

$$\frac{e^x n f_\alpha(U_n)}{U_n} \to 1, \quad \text{and}$$

$$\frac{n\left[(1-\alpha x)^2+1\right]\exp\left\{x-\frac{U_n^2}{2}\right\}}{U_n(2+\alpha^2)\sqrt{2\pi}} \to 1.$$

Taking logarithms for both sides, as $n \to \infty$,

$$\ln n - \ln U_n + \ln\left[(1-\alpha U_n)^2 + 1\right] - \ln(2+\alpha^2) - \ln(\sqrt{2\pi}) + x - \frac{U_n}{2} \to 0.$$

Since $\ln\left[(1-\alpha U_n)^2 + 1\right] \to 2\ln(\alpha U_n)$ as $U_n \to \infty$, so we have,

$$\ln n + \ln U_n + 2\ln\alpha - \ln(2+\alpha^2) - \ln(\sqrt{2\pi}) + x - \frac{U_n^2}{2} \to 0, as \quad n \to \infty. \tag{7}$$

Therefore,

$$\frac{U_n^2}{2} \to \ln n, \quad \text{as} \quad n \to \infty,$$

and as $n \to \infty$, we have

$$\ln U_n = \frac{1}{2}(\ln 2 + \ln\ln n) + o(1).$$

Plug it into Eq. (7), as $n \to \infty$, we have

$$\ln n + 2\ln\alpha + \frac{1}{2}(\ln 2 + \ln\ln n) - \ln(2+\alpha^2) - \ln(\sqrt{2\pi}) + x - \frac{U_n^2}{2} + o(1) \to 0,$$

and,

$$U_n^2 = 2\ln n\left[1 + \frac{x}{\ln n} + \frac{2\ln\alpha - \ln(2+\alpha^2) - \ln(\sqrt{2\pi})}{\ln n} + \frac{\ln 2 + \ln\ln n}{2\ln n} + o\left(\frac{1}{2\ln n}\right)\right]. \tag{8}$$

Therefore, as $n \to \infty$, we obtain

$$\begin{aligned} U_n &= (2\ln n)^{\frac{1}{2}}\left[1 + \frac{x}{2\ln n} + \frac{2\ln\alpha - \ln(2+\alpha^2) - \ln(\sqrt{2\pi})}{2\ln n} + \frac{\ln 2 + \ln\ln n}{4\ln n} + o\left(\frac{1}{2\ln n}\right)\right] \\ &= \frac{x}{(2\ln n)^{\frac{1}{2}}} + (2\ln n)^{\frac{1}{2}} + \frac{2\ln\alpha - \ln(2+\alpha^2) - \ln(\sqrt{2\pi})}{(2\ln n)^{\frac{1}{2}}} + \frac{\ln 2 + \ln\ln n}{2(2\ln n)^{\frac{1}{2}}} + o\left(\frac{1}{(2\ln n)^{\frac{1}{2}}}\right) \\ &= \alpha_n(x) + \beta_n. \end{aligned}$$

Therefore, α_n and β_n are obtained. For $\alpha < 0$, the proof will be similar. □

Proof of Theorem 8**:** According to the coefficient of lower tail dependence, for $z_1 = F_1^{-1}(u)$ and $z_2 = F_1^{-1}(u)$, as $u \to 0^+$, we have

$$\begin{aligned}
&P(X_1 \le z_1, X_2 \le z_2)\\
&= \int_{-\infty}^{z_1}\int_{-\infty}^{z_2} \frac{(2 - 2\alpha_1 x_1 + \alpha_1^2 x_1^2) + (\alpha_2^2 x_2^2 + 2\alpha_1\alpha_2 x_1 x_2 - 2\alpha_2 x_2)}{K2\pi(1-\rho^2)^{1/2}}\\
&\times \exp\left\{-\frac{(x_2-\rho x_1)^2}{2(1-\rho^2)} - \frac{x_1^2}{2}\right\} dx_2 dx_1\\
&= \int_{-\infty}^{z_1}\int_{-\infty}^{z_2} \frac{2 - 2\alpha_1 x_1 + \alpha_1^2 x_1^2}{K2\pi(1-\rho^2)^{1/2}} \exp\left\{-\frac{(x_2-\rho x_1)^2}{2(1-\rho^2)} - \frac{x_1^2}{2}\right\} dx_2 dx_1\\
&+ \int_{-\infty}^{z_1}\int_{-\infty}^{z_2} \frac{\alpha_2^2 x_2^2 + 2\alpha_1\alpha_2 x_1 x_2 - 2\alpha_2 x_2}{K2\pi(1-\rho^2)^{1/2}} \exp\left\{-\frac{(x_2-\rho x_1)^2}{2(1-\rho^2)} - \frac{x_1^2}{2}\right\} dx_2 dx_1.
\end{aligned}$$

According to Eq. (5), as $z_2 \to -\infty$, we have,

$$\begin{aligned}
&\int_{-\infty}^{z_1}\int_{-\infty}^{z_2} \frac{1}{K2\pi(1-\rho^2)^{1/2}}(2 - 2\alpha_1 x_1 + \alpha_1^2 x_1^2) \exp\left\{-\frac{(x_2-\rho x_1)^2}{2(1-\rho^2)} - \frac{x_1^2}{2}\right\} dx_2 dx_1\\
&\approx K^{-1}\left[(2+\alpha_1^2)\Phi(z_1) + \left(\frac{2\alpha_1 - \alpha_1^2 z_1}{\sqrt{2\pi}} \exp\{-\frac{z_1^2}{2}\}\right)\right]\left[\frac{1}{\sqrt{2\pi}}\left|\frac{z_2}{(1-\rho^2)^{1/2}}\right|^{-1} \exp\left\{-\frac{z_2^2}{1-\rho^2}\right\}\right].
\end{aligned}$$

As $z_1 \to -\infty$,

$$\begin{aligned}
&\int_{-\infty}^{z_1}\int_{-\infty}^{z_2} \frac{1}{K2\pi(1-\rho^2)^{1/2}}\left(2 - 2\alpha_1 x_1 + \alpha_1^2 x_1^2\right) \exp\left\{-\frac{(x_2-\rho x_1)^2}{2(1-\rho^2)} - \frac{x_1^2}{2}\right\} dx_2 dx_1\\
&\approx \frac{\alpha_1^2}{2K\pi}\left|\frac{z_2}{(1-\rho^2)^{1/2}}\right|^{-1}|z_1| \exp\left\{-\frac{1}{2}\left(z_1^2 + \frac{z_2^2}{1-\rho^2}\right)\right\}.
\end{aligned}$$

Similarly, we obtain

$$\begin{aligned}
&\int_{-\infty}^{z_1}\int_{-\infty}^{z_2} \frac{1}{K2\pi(1-\rho^2)^{1/2}}(\alpha_2^2 x_2^2 + 2\alpha_1\alpha_2 x_1 x_2 - 2\alpha_2 x_2) \exp\left\{-\frac{(x_2-\rho x_1)^2}{2(1-\rho^2)} - \frac{x_1^2}{2}\right\} dx_2 dx_1\\
&\approx \frac{\alpha_2^2}{2K\pi}\left|\frac{z_1}{(1-\rho^2)^{1/2}}\right|^{-1}|z_2| \exp\left\{-\frac{1}{2}\left(z_1^2 + \frac{z_2^2}{1-\rho^2}\right)\right\}.
\end{aligned}$$

Therefore, as $z_1 \to -\infty$ and $z_2 \to -\infty$,

$$P(X_1 \le z_1, X_2 \le z_2) \approx \frac{(1-\rho^2)^{1/2}}{2K\pi}(\alpha_1^2|z_2|^{-1}|z_1| + \alpha_2^2|z_2||z_1|^{-1}) \exp\left\{-\frac{1}{2}\left(\frac{z_2^2}{1-\rho^2} + z_1^2\right)\right\}. \quad (9)$$

By Theorem 7, we knew that, as $u \to 0^+$, $z_1 \approx -\sqrt{-\ln\left(\frac{2K\sqrt{2\pi}u^2}{(\alpha_1+\rho\alpha_2)^2}\right)}$ and $z_2 \approx -\sqrt{-\ln(\frac{2K\sqrt{2\pi}u^2}{(\alpha_2+\rho\alpha_1)^2})}$, thus, plug in Eq. (9), we have

$$
\begin{aligned}
P(X_1 \le z_1, X_2 \le z_2) \approx& \frac{(1-\rho^2)^{\frac{1}{2}}}{K\sqrt{2\pi}} \exp\left\{\frac{1}{2}\left[\frac{\ln(\frac{2K\sqrt{2\pi}u^2}{(\alpha_1+\rho\alpha_2)^2})}{1-\rho^2}\right] + \frac{1}{2}\ln\left(\frac{2K\sqrt{2\pi}u^2}{(\alpha_1+\rho\alpha_2)^2}\right)\right\} \\
&\times \left\{\alpha_1^2\sqrt{\frac{\ln(\frac{2K\sqrt{2\pi}u^2}{(\alpha_1+\rho\alpha_2)^2})}{\ln(\frac{2K\sqrt{2\pi}u^2}{(\alpha_2+\rho\alpha_1)^2})}} + \alpha_2^2\sqrt{\frac{\ln\left(\frac{2K\sqrt{2\pi}u^2}{(\alpha_2+\rho\alpha_1)^2}\right)}{\ln\left(\frac{2K\sqrt{2\pi}u^2}{(\alpha_1+\rho\alpha_2)^2}\right)}}\right\} \\
\approx& \frac{(1-\rho^2)^{\frac{1}{2}}}{K\sqrt{2\pi}}\left[\frac{2K\sqrt{2\pi}}{(\alpha_2+\rho\alpha_1)^2}\right]^{\frac{1}{2(1-\rho^2)}}\left[\frac{2K\sqrt{2\pi}}{(\alpha_1+\rho\alpha_2)^2}\right]^{\frac{1}{2}} u^{\frac{2-\rho^2}{1-\rho^2}} \\
\approx& 2^{\frac{2-\rho^2}{2(1-\rho^2)}}(K\sqrt{2\pi})^{\frac{\rho^2}{2(1-\rho^2)}}\frac{(\alpha_2+\rho\alpha_1)^{2(1-\rho^2)}\sqrt{1-\rho^2}}{|\alpha_1+\alpha_2\rho|}u^{\frac{2-\rho^2}{1-\rho^2}}.
\end{aligned}
$$

Therefore, according to $\lambda_L(u) = P(X_1 \le z_1 | X_2 \le z_2) = \dfrac{P(X_1 \le z_1, X_2 \le z_2)}{u}$, the result is obtained. □

References

1. Ane, T., Kharoubi, C.: Dependence structure and risk measure. J. Bus. **76**(3), 411–438 (2003)
2. Ara, A., Louzada, F.: The Multivariate Alpha Skew Gaussian Distribution. Bulletin of the Brazilian Mathematical Society, New Series, pp. 1–21 (2019)
3. Arellano-Valle, R.B., Genton, M.G.: An invariance property of quadratic forms in random vectors with a selection distribution, with application to sample variogram and covariogram estimators. Ann. Inst. Stat. Math. **62**(2), 363–381 (2010)
4. Azzalini, A.: A class of distributions which includes the normal ones. Scand. J. Stat. **12**, 171–178 (1985)
5. Azzalini, A., Dalla Valle, A.: The multivariate skew normal distribution. Biometrika **83**(4), 715–726 (1996)
6. Azzalini, A.: The skew-normal and related families, vol. 3. Cambridge University Press, Cambridge (2013)
7. Bortot, P.: Tail dependence in bivariate skew-normal and skew-t distributions (2010). www2.stat.unibo.it/bortot/ricerca/paper-sn-2.pdf
8. Beranger, B., Padoan, S.A., Xu, Y., Sisson, S.A.: Extremal properties of the univariate extended skew-normal distribution, Part A. Stat. Probab. Lett. **147**, 73–82 (2019)
9. Chen, S., Huang, J.: Rates of convergence of extreme for asymmetric normal distribution. Stat. Probab. Lett. **84**, 158–168 (2014)
10. Coles, S.: An Introduction to Statistical Modeling of Extreme Values. Springer, New York (2001)
11. Corless, R.M., Gonnet, G.H., Hare, D.E., Jeffrey, D.J., Knuth, D.E.: On the Lambert W function. Adv. Comput. Math. **5**(1), 329–359 (1996)
12. Elal-Olivero, D.: Alpha-skew-normal distribution. Proyecciones (Antofagasta) **29**(3), 224–240 (2010)
13. Embrechts, P., Kluppelberg, C., Mikosch, T.: Modelling Extremal Events: For Insurance and Finance. Springer, New York (1997)
14. Fisher, R.A., Tippett, L.H.C.: Limiting forms of the frequency distribution of the largest or smallest member of a sample. Proc. Cambridge Philosophical Soc. **24**, 180–190 (1928)
15. Feller, W.: An introduction to probability theory and its applications (1957)

16. Fung, T., Seneta, E.: Tail dependence for two skew T distributions. Stat. Probab. Lett. **80**(9–10), 784–791 (2010)
17. Fung, T., Seneta, E.: Tail asymptotics for the bivariate skew normal. J. Multivariate Anal. **144**, 129–138 (2016)
18. Kotz, S., Nadarajah, S.: Extreme Value Distributions: Theory and Applications. Imperial College Press, U.K (2000)
19. Leadbetter, M.R., Lindgren, G., Rootzén, H.: Extremes and Related Properties of Random Sequences and Processes. Springer, New York (1983)
20. Liao, X., Peng, Z., Nadarajah, S., Wang, X.: Rates of convergence of extremes from skew-normal samples. Stat. Probab. Lett. **84**, 40–47 (2014)
21. Mills, J.P.: Table of the ratio: area to bounding ordinate, for any portion of normal curve. Biometrika **18**(3–4), 395–400 (1926)
22. Peng, Z., Li, C., Nadarajah, S.: Extremal properties of the skew-T distribution. Stat. Probab. Lett. **112**, 10–19 (2016)
23. Resnick, S.I.: Extreme Values, Regular Variation and Point Processes. Springer, New York (2013)
24. Sibuya, M.: Bivariate extreme statistics, I. Ann. Inst. Stat. Math. **11**(2), 195–210 (1959)
25. Tian, W., Wang, T., Hu, L., Tran, H.D.: Distortion Risk Measures Under Skew Normal Settings. In: Econometrics of Risk, pp. 135–148. Springer, Cham (2015)
26. Tian, W., Wang, C., Wu, M., Wang, T.: The multivariate extended skew normal distribution and its quadratic forms. In: Causal Inference in Econometrics, pp. 153–169. Springer, Cham (2016)

Practical Applications

Strategy, Culture, Human Resource, IT Capability, Digital Transformation and Firm Performance–Evidence from Vietnamese Enterprises

Nguyen Van Thuy

Abstract Digital transformation is the application of technology to all aspects of the business. If this process is effective, it will completely transform the business operation, then it will increase the business efficiency. The study measures the factors affecting digital transformation and the impact of digital transformation on innovation and firm performance. Using quantitative methods based on data of 180 Vietnamese enterprises with digital transformation, the results show that there are four factors that influence Digital Transformation: IT Capability, Digital Business Strategy, Human resource capability, Organizational Culture. Among these four factors, the research findings also confirm that digital transformation has a direct impact on innovation and firm performance. Based on these findings, some specific policy implications will be proposed in order to make firm performance more effective.

1 Introduction

Digital Transformation has gained great research interests in both academia and practice. There are many concepts of digital transformation. According to Hess [13], *"Digital transformation is concerned with the changes digital technologies can bring about in a company's business model, which result in changed products or organizational structures or in the automation of processes. These changes can be observed in the rising demand for Internet-based media, which has led to changes of entire business models"*. Another concept of digital transformation from Gartner, digital transformation is the use of digital technologies to change business models, create new opportunities, revenue and value. According to Microsoft, digital transformation is a rethinking of how organizations gather people, data, and processes to create new values. *This research uses the concept of digital transformation in enterprises as the process of changing from the traditional model to digital businesses by*

N. V. Thuy (✉)
Banking Academy, Hanoi, Vietnam
e-mail: thuynv@hvnh.edu.vn

S. Sriboonchitta et al. (eds.), *Behavioral Predictive Modeling in Economics*, Studies in Computational Intelligence 897,
https://doi.org/10.1007/978-3-030-49728-6_16

applying new technologies such as big data (Big Data), Internet of Things (IoT), Cloud computing ... change the way of management, leadership, work processes, corporate culture to create new opportunities and values. The wave of digital transformation has been promoting very strongly in organizations and business in many countries around the world. Successful digital transformation can bring organizations closer to customers and help create new values for business development by creating new business opportunities and new strategies (Berman) [3]. Digital transformation has helped businesses in improving production and profit growth. It not only supports business, but also promotes growth and it is a source of competitive advantage. Nowadays, digital transformation (DT) is an inevitable trend because of the rapid change in customers' demands due to the level of technological changes and market competition. This trend not only creates opportunities for countries and organization to move forward but also poses a risk of being left behind in the 4.0 industries. However, digital transformation in businesses is a difficult, complex process with low success rate. The fact that only 11% of surveyed enterprises succeed in DT (Forrester's 2016) has posed a big question of what makes some businesses digitize successfully and become prosperous while many other businesses are left behind in the digital transformation race for both researchers and business executive. For transition economy like Vietnam now a days, technology development is currently at an early stage. Vietnamese businesses have initially invested in technology to carry out digital transformation but the level of success is not high. At present, there have not been many specific studies on digital transformation, the impact of digital transformation on innovation and operational efficiency. This study is conducted in the context of Vietnamese businesses, in order to identify and evaluate the influence of factors affecting the successful digital transformation. At the same time, this research also assesses the impact of digital transformation on innovation and performance in businesses at firm level.

2 Model and Research Hypotheses

2.1 Critical Success Factors in Digital Transformation

Digitalization has fundamentally changed the business, human resource as well as social models. Therefore, digital transformation is inevitable. Businesses must cope with challenges and take advantage of opportunities with an appropriate strategy. Many factors influencing the success of digital transformation in businesses such as *strategy, human resource, culture, and Information Technology (IT) capability* have been pointed out by many previous studies. Kane [16] found out that it is the strategy, not technology, that leads the digital transformation process. Strategy plays an important role in the digital transformation process of enterprises. Strategy to lead a digital transformation process towards success as vision, scope, goals, roadmap of implementation. Some other studies have shown that *corporate culture* considerably affects the success of digital transformation. According to the findings of Hartl and Hess [12], businesses with a high degree of openness (openness to new ideas, willingness to change, customer centered) will promote the willingness to accept, implement, change thinking, thereby helping businesses master the digital transformation process successfully. *Human resources* also have the impact on the success of the digital transformation process. According to Horlacher [14], the most important influencing factor in the personnel group is chief digital officer, who directs and manages the process of digital transformation. Then, factors affecting digital transformation in the company are ability, skills and resistance (Petrikina et al.) [21]. Additionally, Piccinini [22] showed the importance of attracting, recruiting, and keeping people with new talents and the ability to proficiently integrate digital technology with business know-how. Technology also plays an important role in digital transformation. Technology creates new business opportunities and new strategies. It not only supports business, but also promotes growth and is a source of competitive advantage. Technology in DT is *IT capability*. IT capability included IT Infrastructure capability, IT business spanning capability, and IT proactive stance. IT capability is the basic platform for digital transformation. The study of Nwankpa and Roumani [19] affirmed the impact of IT capability to the success of digital transformation in businesses. By overview, the main Critical Success Factors (CSFs) are *IT Capability, Digital Business Strategy, Human resource capability, Organizational Culture*. In the context of Vietnamese enterprises in the early stages of digital transformation. According to the report *"Readiness for the Future of Production Report 2018"* published by the World Economic Forum (WEF) in January 2018 [17], Vietnam is not in the group of countries ready for the future production. In particular, some indexes are weakly evaluated such as *"Technology and innovation index"* ranked 90/100; The index *"Human capital"* is ranked 70/100. Component indicators such as *"Firm-level technology absorption", "Impact of ICTs on new services and products", "Ability to Innovate"* ranked 78/100, 70/100 and 77/100. There will be many factors that affect successful digital transformation'firm in Vietnam that need to be tested. Therefore, it can be hypothesized:

H_1: *IT capability of businesses has a positive relationship with digital transformation.*

H_2: *Corporate culture has a positive relationship with digital transformation.*

H_3: *Enterprise human resources have a positive relationship with digital transformation.*

H_4: *Enterprise strategy has a positive relationship with digital transformation.*

2.2 The Relationship Between Digital Transformation and Innovation in Firms

According to Daft [10], *innovation* is the creation and discovery of new ideas, practices, processes, products or services. In the context of increasingly competitive

business, innovation is recognized as a key determinant for businesses to create sustainable values and competitive advantages (Wang and Wang) [25]. Innovation can be classified into two levels: improvements and new directions (Verganti) [24]. Díaz-Chao [11] shows that businesses which have implemented digital transformation can introduce new practices and innovation initiatives in their business operations. This relationship in the context of Vietnamese enterprises will be verified through the hypothesis:
H_5: *Digital transformation has a positive effect on innovation in businesses*

2.3 The Relationship Between Digital Transformation and Firm Performance

Firm performance is a general quality indicator that involves many different factors, and it reflects the level of input usage of the business. The firm performance is often expressed through characteristic indicators such as profit, growth and market value (Cho and Pucik) [9]. In other words, firm performance is a measure of how businesses can meet their goals and objectives compared to their competitors (Cao and Zhang) [8]. When digital transformation is successful at higher levels, businesses can achieve the goal of improvement in providing products and services to customers through the enhancement in the ability to customize products or services to each customer, thereby improving customer satisfaction and reducing selling costs. That equals increased business efficiency (Brynjolfsson and Hitt), (Nwankpa and Roumani) [6, 19]. Therefore, the hypothesis is:
H_6: *Digital transformation has a positive effect on firm performance*

2.4 The Relationship Between Innovation and Firm Performance

The impact of innovation on firm performance has been shown by many studies. Innovation improves firm performance, adds potential value and brings invisible resources to businesses (Wang and Wang) [25]. The more creative businesses are the more responsive to customer needs and can develop more possibilities that lead to better performance (Calantone) [7]. Therefore, the hypothesis is:
H_7: *Innovation has a positive effect on firm performance*
The research model is proposed in Fig. 1:

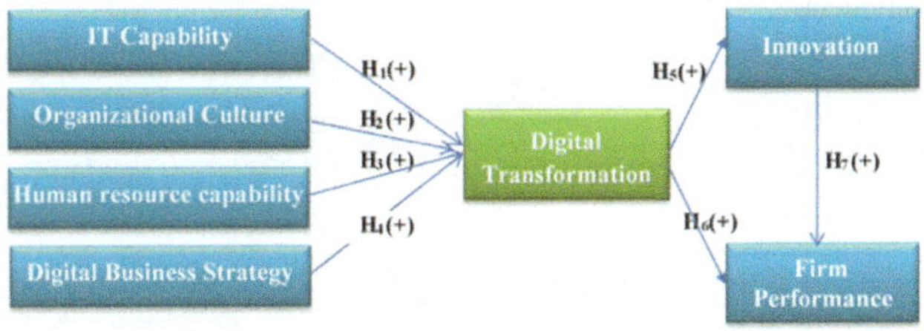

Fig. 1 Proposed research model

3 Data and Methodology

3.1 Data Measurement

In this study, these concepts including IT Capability, Digital Business Strategy, Human resource capability, Organizational Culture, Digital Transformation, Innovation, Firm Performance will be used as variables in research model. All the scales of these variables are inherited from previous studies and adapted to the context of Vietnam. 5-point Likert scale is used to evaluate these above variables, where 1 is completely disagree and 5 is completely agree. The scale of the variable *IT Capability (IT)* is inherited from the scales of Bharadwaj [5], Nwankpa and Roumani [19] which include 03 observed sub-variables. The scale of the variable *Digital Business Strategy (DS)* is inherited from the scales of Bharadwaj et al. [4]which include 04 sub-variables. The scale of the variable *Human resource capability (HR)* consists of 04 observed sub-variables is inherited from the findings of the study by Kane (2015) [16]. The scale of the variable *Organizational Culture (OC)* is inherited from the research findings of Hartl and Hess [12] which including 05 scales. The scale of the variable *Digital Transformation (DT)* consists of 03 observed variables inherited from the research findings of Aral and Weill (2007) [2]. The scale of the variable Innovation (IN) consists of 02 observed variables inherited from the research findings of (Hsu & Sabherwal, 2012) [15]. The scale of the variable *Firm Performance (FP)* variable is inherited from the research findings of Nwankpa and Roumani [19]. The measurement scale and references are shown in Table 1.

The survey questionnaire was designed based on the observed variables measured in the model. In addition, the survey also had other questions such as business size, business lines, ...

Table 1 Measurements

Variable	Scale
IT Capability Bharadwaj (2000), Nwankpa and Roumani (2016)	Relative to other firms in your industry, please evaluate your firm's IT Capability in the following area on a scale of 1 - 5 (1 = poorer than most; 5 = superior to most).
	The enterprise always invests in the latest technology infrastructure (IT_1)
	The enterprise always integrates business strategic planning and IT planning; (IT_2)
	The enterprise constantly seek new ways to enhance the effectiveness of IT use (IT_3)
Digital Business Strategy Bharadwaj, Omar A Pavlou, Paul A Venkatraman N, (2013)	Relative to other firms in your industry, please evaluate your firm's Digital Business Strategy in the following area on a scale of 1 - 5 (1 = poorer than most; 5 = superior to most).
	The enterprise clearly defined scope of digital business strategy (DS_1)
	The enterprise clearly defined scale of digital business strategy (DS_2)
	The enterprise clearly defined speed of decision making (DS_3)
	The enterprise clearly defined sources of value creation and capture (DS_4)
Human resource capability Osmundsen (2018)	Relative to other firms in your industry, please evaluate your firm's Human resource capability in the following area on a scale of 1 - 5 (1 = poorer than most; 5 = superior to most).
	The enterprise has a chief digital officer in digital transformation process (HR_1)
	Employees' firm fully participate in the digital transformation process (HR_2)
	The degree of employee interest in the digital transformation process (HR_3)
	The enterprise has policies to attract people with new talent and the ability to integrate digital technology expertise with business know-how (HR_4)
Organizational Culture Hess, Thomas (2017)	Relative to other firms in your industry, please evaluate your firm's culture in the following area on a scale of 1 - 5 (1 = poorer than most; 5 = superior to most).
	The enterprise has an openness towards change (OC_1)
	All activities of the enterprise are customer-centered (OC_2)
	The enterprise is constantly improving (OC_3)
	The enterprise is always ready to adapt to change (OC_4)
	The enterprise is always encourage the acquisition of new skills and knowledge (OC_5)
Digital Transformation Aral and Weill (2007)	Relative to other firms in your industry, please identify the degree to which your company uses digital technologies on a 1to 5 scale (1 = strongly disagree, 5 = strongly agree)
	Our firm is driving new business processes built on technologies such as big data, analytics, cloud, mobile and social media platform (DT_1)
	Our firm is integrating digital technologies such as social media, big data, analytics, cloud and mobile technologies to drive change (DT_2)
	Our business operations is shifting toward making use of digital technologies such as big data, analytics, cloud, mobile and social media platform (DT_3)
Innovation Hsu and Sabherwal (2012)	Please identify the degree to you agree with the statement on a scale of 1 to 5 (1 = strongly disagree, 5 = strongly agree).
	Our firm develops and produces new products or services continually (IN_1)
	Our firm gives priority to making efforts to increase the quality of products or services (IN_2)
Firm Performance Nwankpa and Roumani (2016)	Relative to other direct competitors, indicate how well your firm performed during the last 3 years on a scale of 1 to 5 (1 = poorer than most; 5 = superior to most).
	Profitability is high (FP_1)
	Customer retention rate is high (FP_2)
	Return on Investment is high (FP_3)
	Sales growth is high (FP_4)

3.2 Data Collection

The sample of the survey are Vietnamese firms which have been implementing digital transformation from January 2019 to March 2019 by direct survey. The subjects of these firms' surveys are chief digital officers or project leaders of digital transformation. The sample results include are 180 valid forms included in the analysis. Sample characteristics are shown in Table 2.

3.3 Methodology

After having the data, SPSS 20 & AMOS 20 were used to test the hypothesized relationships in the research model as well as evaluate the reliability of measurement scale based on Cronbach Alpha reliability coefficients, EFA and CFA, SEM Bayesian.

Table 2 Sample characteristics

Classification		(%) Respondent
Industry	Retail Trade	11
	Information Technology	22
	Finance and Insurance	17
	Entertainment	15
	Manufacturing	11
	E-commerce	18
	Others	6
Firm's size (Number of employees)	Less than 100	58
	100-200	24
	200-500	12
	500-1000	5
	Above 1000	1
Digital Transformation maturity	Beginner	38
	Intermediate Level	42
	Advanced	15
	Expert	5

4 Results

4.1 Reliability of Measurement Scales

A reliability test of scales is performed with Cronbach's Alpha reliability coefficient and 02 scales—DS_4 and OC_5 are eliminated because their item-total correlations are less than 0.3. Removing these 2 scales to test the reliability of the remaining scales with the Cronbach's Alpha reliability coefficient, all observed variables are found to have item-total correlations greater than 0.3 and the Cronbach's Alpha reliability coefficient of all factors is greater than 0.6 so the scales of the components DS, OC, HR, IT, DT, IN, FP are all accepted and included in the next factor analysis. Detailed results assessing measurement scale by Cronbach Alpha reliability for the 2th time is shown in Table 3.

Table 3 Results assessing measurement scale by Cronbach Alpha reliability

ID	Variable	Number of observed variables	Cronbach's Alpha
1	IT Capability	3	0,860
2	Digital Business Strategy	3	0,870
3	Human resource capability	4	0,830
4	Organizational Culture	4	0,861
5	Digital Transformation	3	0,789
6	Innovation	2	0,693
7	Firm Performance	4	0,819

Table 4 KMO and Bartlett's test

Kaiser-Meyer-Olkin Measure of Sampling Adequacy.		0.787
Bartlett's Test of Sphericity	Approx. Chi-Square	2067.430
	df	276
	Sig.	0.000

Table 5 Total variance explained

Component	Initial Eigenvalues			Extraction Sums of Squared Loadings		
	Total	% of Variance	Cumulative %	Total	% of Variance	Cumulative %
1	5.554	24.149	24.149	5.554	24.149	24.149
2	2.698	11.732	35.882	2.698	11.732	35.882
3	2.269	9.865	45.746	2.269	9.865	45.746
4	2.062	8.967	54.713	2.062	8.967	54.713
5	1.800	7.828	62.541	1.800	7.828	62.541
6	1.324	5.755	68.296	1.324	5.755	68.296
7	1.079	4.692	72.987	1.079	4.692	72.987
8	.733	3.187	76.174			
9	.634	2.755	78.929			

4.2 Exploratory Factor Analysis

The exploratory factor analysis (EFA) is used to reassess the degree of convergence of observed variables by components. Research on the implementation of KMO test and Bartlett's test in factor analysis shows that KMO coefficient = 0.787 > 0.5, Bartlett's test value is significant (Sig. =0.000 < 0.05) indicates that an EFA factor analysis is appropriate (Table 4).

Indicators all have factor loadings *"Factor loading"* > 0.5. At Eigenvalues greater than 1 and with factor extraction used as Principal Axis Factoring (PAF) with Varimax orthogonal rotation, factor analysis extracted 7 factors from 24 observed variables and with the extracted variance at 72% (greater than 50%), which is satisfactory (Tables 5 and 6).

4.3 Confirmatory Factor Analysis (CFA) and Structural Equation Model (SEM)

Confirmatory Factor Analysis (CFA): Based on the result of EFA: 4 factors (14 observed variables) affecting successful Digital Transformation (3 observed variables), Digital Transformation impacts on Firm Performance (4 observed variables), Digital Transformation impacts on Innovation (2 observed variables), Innovation impacts on Firm Performance, study Confirmatory factor analysis (CFA) the first time by AMOS 20 software. Result of the CFA has all observed variables's weight are allowed standard (>= 0.5) so that scale achieves convergence value (Anderson and Gerbing)[1]. The Result of CFA is shown in the following Fig. 2 and Table 7:

Table 6 Result of explore factor analysis EFA

Rotated Component Matrix[a]

	Component						
	1	2	3	4	5	6	7
OC2	.871						
OC3	.845						
OC4	.799						
OC1	.739						
HR4		.841					
HR1		.841					
HR3		.785					
HR2		.714					
FP1			.853				
FP2			.808				
FP4			.804				
FP3			.689				
DS3				.894			
DS1				.882			
DS2				.799			
IT3					.892		
IT1					.878		
IT2					.845		
DT2						.777	
DT1						.770	
DT3						.710	
IN1							.839
IN2							.814

Extraction Method: Principal Component Analysis.
Rotation Method: Varimax with Kaiser Normalization.
a. Rotation converged in 6 iterations.

SEM Structural Equation Model: SEM analysis was performed on AMOS 20 software of which the results are shown in Fig. 3.

The SEM results (Fig. 2) show that the weights of the observed variables all reached the standard level allowed (>= 0.5) and statistically significant with the p-values all equal to 0.000. Thus, it is possible to conclude that the observed variables used to measure the component variables of the scale achieved convergent validity. SEM shows that the model has 223 degrees of freedom, chi-squared test statistic = 274.933 with p-value = 0,000 < 0.05; Chi-square/df = 1.233 which satisfies the <3 requirement and indicators show that the model is compatible with market data (CFI = 0.971; TLI = 0.967; GFI = 0.885 and RMSEA = 0.036 < 0.08). The components of the variables DS, OC, HR, IT, DT, IN, FP have no correlation between the errors of the observed variables, so they all achieve the unidirectional property. In addition, the study examined the discriminant value of concepts in the model. The results show that the concepts in the model achieve discriminant validity.

4.4 Hypothesis Testing

In order to test the hypotheses of the research model, the research conducted tests by using Bayesian SEM. In Bayesian SEM, all estimated coefficients are reported in terms of posterior distribution. Parameters significance testing can be conducted by using 95% confidence interval which is the lower limit percentiles of 2.5% and the upper limit percentiles of 97.5% of the posterior distribution (Ntzoufras) [18]. The significance of a parameter depends on whether or not a zero value lies in a confidence interval (Sari, Wardhani and Astutik) [23]. If a confidence interval does not contain a zero value, the parameter is significant. Otherwise, if the zero value is included

Fig. 2 CFA Model

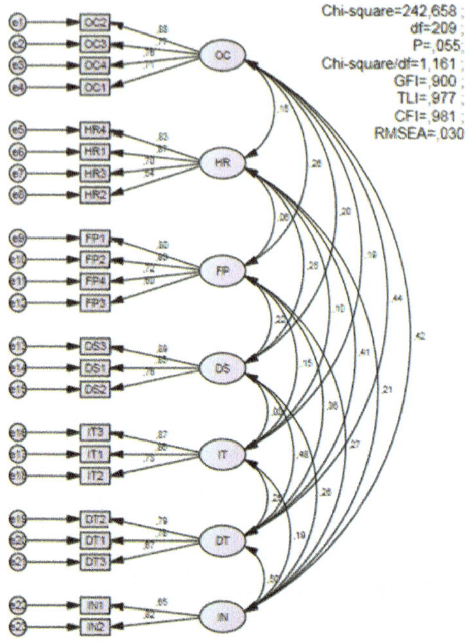

Table 7 Synthetic reliability and extracted variance

	Factor	Estimate	S.E.	C.R.	P
OC	Organizational Culture	0.649	0.092	7.047	***
HR	Human resource capability	0.716	0.112	6.398	***
FP	Firm Performance	0.519	0.088	5.878	***
DS	Digital Business Strategy	1.06	0.147	7.209	***
IT	IT Capability	0.707	0.105	6.741	***
DT	Digital Transformation	0.442	0.077	5.757	***
IN	Innovation	0.405	0.119	3.398	***

in the confidence interval, the parameter is not significant or has no effects on the dependent variable. The parameter significance testing can be carried out on both the measurement model and the structural model. In the measurement model, a significant indicator means it is acceptable to measure the corresponding latent construct. In the structural model, the parameter significance testing can determine whether there is significant influence between the latent constructs. The regression weights and the associated 95% confidence intervals between the measurement indicators and corresponding latent constructs and between the latent constructs are shown in Tables 8 and 9, respectively. From the results presented in Table 9, the factor loadings are all significant at the 5% level. While, from the results of the structural model as shown in Table 9, it is known that of the seven parameters tested, six parameters were recognized as significant. It can be concluded that hypotheses 1, 2, 3, 4, 5 and 6 were supported, while hypothesis 7 was not supported. Hypothesis 7, stating that Innovation has a significant effect on Firm Performance, is not supported

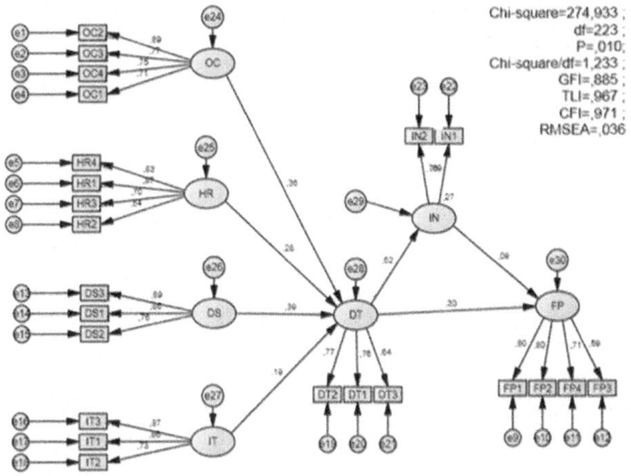

Fig. 3 Analysis results structural equation model

by the regression weight of IN on FP with a confidence interval from −0.114 to 0.365. For Hypothesis 1, which addressed that IT Capability has a significant effect on Digital Transformation, the confidence interval for the regression parameter is from 0.043 to 0.265, significantly larger than zero. Hypothesis 2, which asserted that Organizational Culture has a significant effect on Digital Transformation, has a confidence interval between 0.144 and 0.383, a parameter significantly greater than zero. Support was found for Hypothesis 3, stating that Human resource capability can significantly affect Digital Transformation, with the confidence interval range from 0.098 to 0.363. Hypothesis 4, which asserted that Digital Business Strategy has a significant effect on Digital Transformation, has a confidence interval between 0.143 and 0.337, a parameter significantly greater than zero. Support was found for Hypothesis 5, stating that Digital Transformation can significantly affect Innovation, with the confidence interval range from 0.355 to 0.748. Finally, Hypothesis 6 was supported, which said that Digital Transformation has a significant effect on Firm Performance, with a confidence interval between 0.09 and 0.572. So, these six hypotheses are supported. The final research model is shown in Fig. 4 and the standardized weights SEM are shown in Table 10.

Table 8 Bayesian SEM measurement model results

Path	Mean	S.D.	95% Lower bound	95% Upper bound
OC3<--OC	0.835	0.078	0.686	1
OC4<--OC	0.891	0.091	0.723	1.068
OC1<--OC	0.772	0.073	0.623	0.909
HR1<--HR	1.056	0.093	0.894	1.256
HR3<--HR	0.81	0.097	0.651	1.085
HR2<--HR	0.553	0.091	0.388	0.76
FP2<--FP	0.981	0.099	0.801	1.175
FP4<--FP	0.833	0.089	0.67	1.014
FP3<--FP	0.668	0.09	0.488	0.847
DS1<--DS	0.904	0.069	0.779	1.048
DS2<--DS	0.669	0.06	0.551	0.78
IT1<--IT	1.032	0.094	0.877	1.245
IT2<--IT	0.795	0.085	0.645	0.988
DT1<--DT	1.051	0.093	0.887	1.246
DT3<--DT	0.733	0.078	0.596	0.899
IN2<--IN	1.039	0.161	0.741	1.354

Table 9 Bayesian SEM results hypotheses testing

Path	Mean	S.D.	95% Lower bound	95% Upper bound
DT<--HR	0.216	0.063	0.098	0.363
DT<--DS	0.229	0.048	0.143	0.337
DT<--IT	0.153	0.058	0.043	0.265
DT<--OC	0.268	0.061	0.144	0.383
IN<--DT	0.542	0.1	0.355	0.748
FP<--DT	0.328	0.128	0.09	0.572
FP<--IN	0.124	0.122	-0.114	0.365

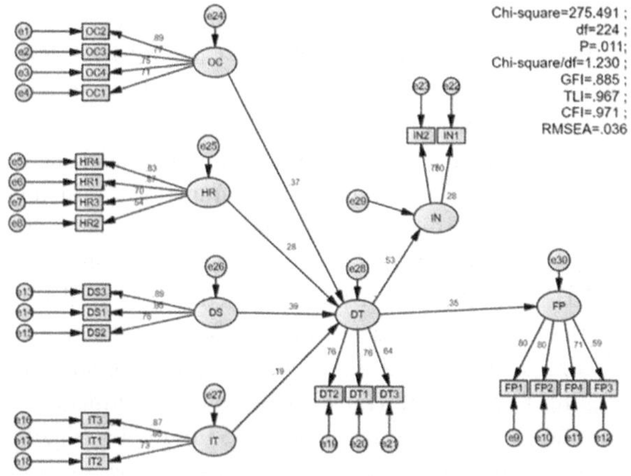

Fig. 4 The final research model

Table 10 The standardized weights SEM

Path			Estimate
DS	--->	DT	0.387
OC	--->	DT	0.366
HR	--->	DT	0.280
IT	--->	DT	0.188
DT	--->	FP	0.351
DT	--->	IN	0.528

5 Discussion

This study explores the factors affecting successful digital transformation and tested the relationship between digital transformation, and innovation and business efficiency. IT Capability, Digital Business Strategy, Human resource capability, Organizational Culture all positively affect Digital Transformation. Digital Transformation positively affect Innovation and Firm Performance. Innovation also positively affects Firm Performance. This result is consistent with published studies (Osmundsen, Nwankpa and Roumani, Kane and Díaz-Chao) [11, 16, 19, 20].

Research results show that the 04 factors including *Digital Business Strategy, Organizational Culture, Human resource capability, IT Capability* affect *Digital Transformation* in descending levels of impact. Among four factors, *Digital Business Strategy* has the greatest positive impact on Digital Transformation at coefficient of 0.387 (Table 10). *Organizational Culture* has the second greatest positive impact on Digital Transformation at 0.366 standard units (Table 10). This result is consistent with the research findings of Kane, Hartl and Hess [12, 16]. For businesses, if strategy is like the "Seed" then culture can be considered the "Land". If the "Land" is not favorable, no matter how hard one tries, the "Seed" cannot germinate and grow, and vice versa. In Digital Transformation, if these two factors are not combined well in businesses will lead to disorientation and the digital transformation process of businesses will inevitably fail.

Human resource capability is the third factor affecting digital transformation at coefficient of 0.280. And an interesting result is that *IT Capability* is the factor least affecting digital transformation at coefficient of 0.188 (Table 10). These results are consistent with the research findings of Horlacher, Petrikina et al. Piccinini, Nwankpa and Roumani and Kane [14, 16, 19, 22]. Everything is based on technology foundation. No matter what product or service is created, technology should be applied. The beginning of the technology era with new rules has reset the definitions of demand and market for all industries. In the digital business model, technology creates new opportunities and brings value to the business vision for each business segment. Technology creates a competitive advantage through which business practice and management will have a clearer view of the market and the current capabilities of the company. However, digital transformation must be executed in not only the employees' mindsets but also, in the way they approach and solve problems daily and hourly. At a business whose employees do not have digital thinking, and culture does not foster innovation, modern technology cannot create sustainability and development for the business.

The research findings also show that successful Digital Transformation has a significant impact on Innovation at coefficient of 0.528 (Table 10). This result is consistent with the research findings of Díaz-Chao [11], Nwankpa and Roumani [19]. Successful digital transformation helps businesses promote innovation and creation of products and services. Digitizing businesses helps business leaders analyze and measure markets and customers, and compete better. Based on the information, businesses can create better products and services than the current ones. The research findings show that successful Digital Transformation has an immense impact on Firm Performance at coefficient of 0.351. This result is consistent with the research findings of Nwankpa and Roumani [19], Brynjolfsson and Hitt [6]. Digital transformation helps businesses achieve many encouraging results such as: Cost reduction; Customer strategy improvement; Operating system consolidation; Better analysis; More focus on potential customers; New products/services; Accurate market segmentation; Global customer experience; Increase in agility and innovation. These research findings are empirical evidences demonstrating the benefits of digital transformation on firm performance.

6 Conclusions

These research findings are empirical evidence showing the strong impact of many factors on digital transformation success in businesses including *Digital Business Strategy, Organizational Culture, Human resource capability, IT Capability*. Among all factors, strategic and cultural factors have greater impacts than technology capability. The study also measured the effect of digital transformation on corporate performance and innovation and the effect of innovation on firm performance.

Based on the research results, some solutions should be recommend to help Vietnamese enterprises convert successfully to contribute to promoting innovation and improving operational efficiency as follows:

Firstly, Vietnamese enterprises need to be well aware and appreciate the role of digital transformation in the current industrial revolution 4.0. Shifting numbers is not only an indispensable trend, in today's digital era, which is an opportunity for businesses to surpass the industrial revolution 4.0, but also a risk of being left behind and being left further away for those are not related to it . Digital transformation is not simply a higher level of application and development of information and communication technology (ICT), but a strategy for digital data and technology to transform of model, process, product as well as output of production and business processes in the enterprise.

Secondly, strategy is a key factor determining the success of the digital transformation process of the business. Traditional strategy is no longer appropriate. Instead, an effective digital technology strategy creates initial experiences and integrates those experiences into the strategic process. In addition, high-ranking leaders of enterprises need to change their thinking about the digital transformation of enterprises, and also orient digital strategy thinking to create new opportunities, new business strategies to promote growth and creation that, in turn, create competitive advantage for businesses.

Thirdly, digital transformation is not only an activity to optimize workflows, increase productivity, generate profits but also be the cultural foundation of a business. And that culture must be built and practiced within a long and persistent strategy. In Vietnam, the coordination between strategy and culture in general and in digital transformation in particular has many difficult problems. In digital transformation, businesses need to build an effective cultural strategy according to these steps: (1) The enterprises should identify the "Objectives—Vision—Core values" of the business in digital transformation and put them into practice. The strategy is built with clear scope and objectives, supported by leaders—visionaries, who always promote and build a variable and innovative corporate culture; (2) They should recruit employees who are capable of complementing each other. Upon corporate culture established to promote innovation, risk tolerance, businesses need to focus on recruiting, attracting and retaining talents with the deepest commitment to digital progress (deeply committed to digital progress); (3) they should use tools to measure and evaluate results. When the strategy is established, the business needs to review how the established goals are performed to evaluate the results and effectiveness. Digital transformation is not only an activity to optimize workflows, increase productivity, create profits but above all, it must be the cultural foundation of a business. And that culture must be built and practiced within a long and persistent strategy.

Fourthly, in order to transform successful, it is also necessary to implemented supporting policies from the government to have a synchronous, connected and efficient digital transformation. These solutions on transforming the digital economy, developing digital infrastructure, developing a digital work force, building a legal environment will ensure safety and reliability for enterprises to carry out digital conversion.

Despite the extreme effort in doing research, it still has many obvious limitations. The biggest and most obvious limitation is that the sample size is quite small due to the small number of businesses involved in digital transformation and surveyed. In the future, when the number of businesses implementing digital transformation successfully increases, the next research will focus on developing in the direction of: (1) Research on different types of businesses. (2) Additional research on factors affecting digital transformation in businesses.

Acknowledgments This work was supported by the CDOs of Vietnamese enterprises. We also acknowledge the partial support of the Vietnam Association of Small and Medium Enterprises during the survey.

References

1. Anderson, J.C., Gerbing, D.W.: Structural equation modeling in practice: a review and recommended two-step approach. Psychol. Bull. **103**, 411 (1988)
2. Aral, S., Weill, P.: IT assets, organizational capabilities, and firm performance: how resource allocations and organizational differences explain performance variation. Organ. Sci. **18**, 763–780 (2007)
3. Berman, S.J.: Digital transformation: opportunities to create new business models. Strategy Leadersh. **40**, 16–24 (2012)
4. Bharadwaj, A., El Sawy, O.A., Pavlou, P.A., Venkatraman, N.: Digital business strategy: toward a next generation of insights. Manage. Inf. Syst. Q. **37**, 471–482 (2013)
5. Bharadwaj, A.S.: A resource-based perspective on information technology capability and firm performance: an empirical investigation. Manage. Inf. Syst. Q. **24**, 169–196 (2000)

6. Brynjolfsson, E., Hitt, L.M.: Beyond computation: information technology, organizational transformation and business performance. J. Econ. Perspect. **14**, 23–48 (2000)
7. Calantone, R.J., Cavusgil, S.T., Zhao, Y.: Learning orientation, firm innovation capability, and firm performance. Ind. Mark. Manage. **31**, 515–524 (2002)
8. Cao, M., Zhang, Q.: Supply chain collaboration: impact on collaborative advantage and firm performance. J. Oper. Manage. **29**, 163–180 (2011)
9. Cho, H.J., Pucik, V.: Relationship between innovativeness, quality, growth, profitability, and market value. Strateg. Manage. J. **26**, 555–575 (2005)
10. Daft, R.L.: A dual-core model of organizational innovation. Acad. Manage. J. **21**, 193–210 (1978)
11. Díaz-Chao, Á., Sainz-González, J., Torrent-Sellens, J.: ICT, innovation, and firm productivity: new evidence from small local firms. J. Bus. Res. **68**, 1439–1444 (2015)
12. Hartl, E., Hess, T.: The role of cultural values for digital transformation: insights from a Delphi Study (2017)
13. Hess, T., Matt, C., Benlian, A., Wiesböck, F.: Options for formulating a digital transformation strategy. Manage. Inf. Syst. Q. Executive **15**, 123–129 (2016)
14. Horlacher, A., Klarner, P., Hess, T., Thomas, T.: Crossing boundaries: Organization design parameters surrounding CDOs and their digital transformation activities (2016)
15. Hsu, I.C., Sabherwal, R.: Relationship between intellectual capital and knowledge management: an empirical investigation. Decis. Sci. **43**, 489–524 (2012)
16. Kane, G.C., Palmer, D., Phillips, A.N., Kiron, D., Buckley, N.: Strategy, not technology, drives digital transformation. MIT Sloan Manage. Rev. **14**, 1–25 (2015). Deloitte University Press
17. Martin, C., Samans, R., Leurent, H., Betti, F., Drzeniek-Hanouz, M., Geiger, T.: Readiness for the Future of Production Report 2018, World Economic Forum (2018)
18. Ntzoufras, I.: Bayesian Modeling Using WinBUGS. Wiley, Hoboken (2011)
19. Nwankpa, J.K., Roumani, Y.: A firm performance perspective, Yaman, IT capability and digital transformation (2016)
20. Osmundsen, K., Iden, J., Bygstad, B.: Digital Transformation: Drivers, Success Factors and Implications, In: Mediterranean Conference on Information Systems (MCIS), vol. 37 (2018)
21. Petrikina, J., Krieger, M., Schirmer, I., Stoeckler, N., Saxe, S., Baldauf, U.: Improving the readiness for change-Addressing information concerns of internal stakeholders in the smartPORT Hamburg (2017)
22. Piccinini, E., Hanelt, A., Gregory, R., Kolbe, L.: Transforming industrial business: the impact of digital transformation on automotive organizations (2015)
23. Sari, D.K., Wardhani, N.W.S., Astutik, S.: Parameter estimation of structural equation modeling using bayesian approach. CAUCHY **4**, 86–94 (2016)
24. Verganti, R.: The innovative power of criticism. Harvard Bus. Rev. **94**, 18 (2016)
25. Wang, Z., Wang, N.: Knowledge sharing, innovation and firm performance. Expert Syst. Appl. **39**, 8899–8908 (2012)

Do the Macao's Pillar Industries Have an Impact on Inbound Tourism?

Bing Yang, Jianxu Liu, and Songsak Sriboonchitta

Abstract The purpose of this article is to select 12 countries/regions (China, Hong Kong, Taiwan, Australia, Indonesia, Japan, Malaysia, New Zealand, the Philippines, South Korea, Singapore, and Thailand) for quarterly data (2008Q1–2018Q4) analysis using panel–ARDL model with Bayesian MCMC method to estimate the relationship between Macao's pillar industries and inbound tourism. The results show that there is a long–term relationship between word–of–mouth effect and Macao's tourism demand. Tourism price has a significant negative impact on the entry of tourists. Seasonality has a significant negative impact in the first quarter. It is recommended to launch the Spring Festival family travel products to attract family members to travel together. The gaming industry has a positive effect, and the real estate industry has a negative impact on the development of tourism.

1 Introduction

Macao is a region with developed economy and high degree of internationalization. It is also the city with the highest per capita GDP in China. Pillar industry appears in many articles, it is one which contributes a large proportion of GDP/GNP. Macao's pillar industries include export processing, real estate construction, tourism, gambling, and financial services. Macao's main income has been from gaming and tourism. The number of inbound tourists, which has continued to be on an increasing trend (Fig. 1), has always been the standard to measure the current situation of its

B. Yang · S. Sriboonchitta
Faculty of Economics, Chiang Mai University, Chiang Mai, Thailand
e-mail: yuanbing310@163.com

S. Sriboonchitta
e-mail: songsakecon@gmail.com

J. Liu (✉)
Faculty of Economics, Shandong University of Finance and Economics, Jinan, China
e-mail: liujianxu1984@163.com

S. Sriboonchitta et al. (eds.), *Behavioral Predictive Modeling in Economics*, Studies in Computational Intelligence 897,
https://doi.org/10.1007/978-3-030-49728-6_17

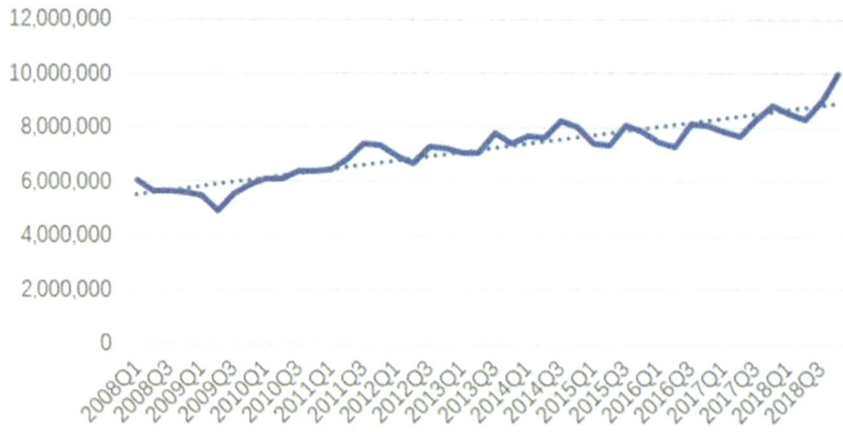

Fig. 1 The number of inbound tourist arrivals from 2008Q1 to 2018Q4

tourism industry. Since 2004, Macao's economy has grown strongly and prices have soared. There has been a phenomenon of speculation in real estate. According to DSEC (Government of Macao Special Administrative Region Statistics and Census Service), the GDP share in 2017 of the gaming industry was 49.1%, that of the real estate industry was 10.6%, while that of tourism reached 14.2%. The World Tourism Organization (UNWTO) recently released a new report, "Tourism Highlights 2018", containing the latest data showing that Macao's international tourism receipts reached 35.6 billion US dollars, entitling Macao to rank sixth in the world tourism. The growth of Macao's tourism industry is mainly driven by mainland tourists (Fig. 2). While gaming is an important part of Macao's development, the territory has a growing number of regular leisure tourists, the report said. With the inbound tourist growth rates of roughly 15% in 2010 and about 10% in 2018, it is of great significance to analyze the influence of other pillar industries on Macao inbound tourism.

Tourism is one of the key areas of human global activity and an "effective economic power tool" (Mishra [1]). Macao's tourism industry is driven by the gaming industry, while the real estate industry is driven by both tourism and gaming industry. Since 2006, Macao has surpassed Las Vegas in gambling revenue for the first time, becoming the first "gambling city" in the world, and its "gambling city" image is deeply rooted in people's mind. A large amount of foreign investment continues to pour into Macao, and the gaming industry and tourism industry continue to flourish, driving the all-round development of Macao's economy. According to stamp duty in 2006, the total revenue from selling residential units in Macao is 19.09 billion MOP (Macao Pataca), out of which the revenue from selling to the non-residents is 11.1 billion MOP, indicating that the main investment in real estate industry is foreign investment.

Among many scholars who studied on both the gaming industry and the tourism industry, Leiper [2] is the first to analyze the relationship between the gaming industry

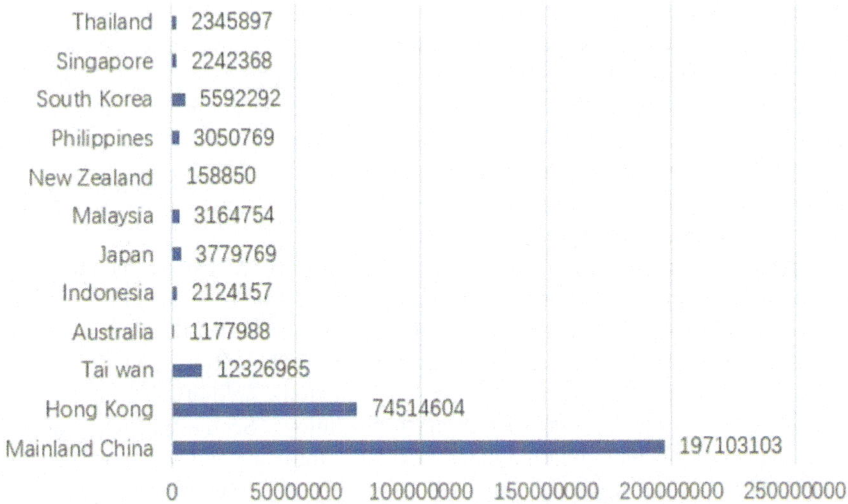

Fig. 2 Tourist Arrivals to Macao From Mainland China, Hong Kong, Taiwan, Australia, Indonesia, Japan, Malaysia, New Zealand, the Philippines, South Korea, Singapore and Thailand (2008Q1–2018Q4)

and tourism in general. He noted that gaming industry is an important part of the economy as most gambling tourists come from foreign countries, such as from the United Kingdom to Las Vegas, or from France to Monte Carlo; but Macao is a destination for both foreign gaming tourists and tourists from China's origins because gambling is illegal in China. The Asian gaming market is a thriving business, however. Most scholars who studied the advantages and the disadvantages of the development of the gaming industry and tourism often emphasized the negative impacts on the whole society, such as in the economic, social, cultural, and environmental aspects. After having studied the social, economic and environmental impacts brought about by the development of Macao's gaming industry, Wan [3, 4] provided policy advice to the tourism industry related to Macao's gaming industry. Gu et al. [5] pointed out that casino tax has great potential, and discussed the impact of casino tax on tourism. Gambling is no longer the glory of the past, Lu [6] said, it is considered to be 'Achilles heel' of Macao's economy. Israel and Mehrez [7] distinguished "gaming" and "gambling" into "gaming"—which is the term for the legal form of the particular activity, and "gambling"–which characterizes mostly the illegal one.

A large number of published studies in the literature on tourism analysis focus on econometric and time series modeling techniques. Lim [8] argued that disposable income should be used as a measure of income in tourism demand model. Song and Witt [9] pointed out that gross domestic product (GDP) can be used in empirical research as income variable. Witt and Witt [10] used the exchange rate between destination and origin and the individual CPI variables to calculate tourist price. Syriopoulos [11], Seddighi and Shearing [12], Kim and Song [13] found that autoregressive distributed lag (ARDL) dynamic model should be used in tourism demand study and forecast, and also found the word of mouth effect on tourism demand. Few researchers use the Bayesian model. Wong, Song, and Chon [14] developed a Bayesian VAR (BVAR) model by using Bayesian prior on model estimates. They found that BVAR showed significantly better predictive performance than non-Bayesian counterparts. Assaf et al. [15] introduced Bayesian global vector autoregressive estimation techniques (BGVAR) for modelling and forecasting the demand for international travel within Southeast Asia.

This study aims to expand the literature on the impacts of Macao's pillar industries on inbound tourism with two main contributions. First, it is the first ever to study the long-term relationship among the number of inbound tourists, income, prices, gaming industry, and real estate industry. Second, dynamic panel-ARDL model estimated by the Bayesian method is applied for the first time in the field of tourism.

This paper after the introduction is organized as follows. Section 2 provides details of the methodology, including the theoretical basis for using and selecting data and models. Section 3 discusses the empirical results. Finally, Sect. 4 includes the conclusion and discussion.

2 Methodology

2.1 Data and Variable Selection

Tourism demand has significant seasonality with obvious high and low (Witt and Witt [16]). Therefore, it is necessary to use quarterly data.

This study covers the period of quarterly data from 2008Q1 (quarter 1) to 2018Q4 (quarter 4). Obtained from the DSEC are the data on tourist arrivals, total revenue of gaming industry, total revenue of selling house, GDP, CPI, and the exchange rate of Macao. The data on GDP, CPI, and the exchange rates of 12 countries/regions obtained from a data-service provider Wind Info.

2.2 Model Specification

Many studies have shown that the number of tourist arrival is an important indicator for determining tourism demand. Therefore, we use the number of tourist arrivals as the measure of tourism demand. For the tourism price index which is an important factor for visitor's decision, we adopt the relative price which is the tourism price of the destination country divided by the tourism price of the source country/region (Lim and McAleer [17]). The demand model is simply:

$$TA_{i,t} = AY_{i,t}^{\beta_{01}} P_{i,t}^{\beta_{02}} H_{i,t}^{\beta_{03}} G_{i,t}^{\beta_{04}} e_{i,t}, \tag{1}$$

where
$TA_{i,t}$ is the number of tourists from country/region i to Macao at time t;
i = {Mainland China, Hong Kong, Taiwan, Australia, Indonesia, Japan, Malaysia, New Zealand, the Philippines, South Korea, Singapore and Thailand};
$Y_{i,t}$ is income per capita in country/region i at time t;
$P_{i,t}$ is tourism price;
$H_{i,t}$ is the total revenue from selling houses to non-residents (from country/region i) in Macao at time t;
$G_{i,t}$ is the total revenue of the gaming industry in Macao at time t;
$e_{i,t}$ is the error term.

The tourism price equation is as follows:

$$P_{i,t} = \frac{CPI_{i,t} \setminus EX_{i,t}}{CPI_{mc,t} \setminus EX_{mc,t}}, \quad (2)$$

where CPI is the consumer price index, EX is the exchange rate.

Gaming and housing activities are closely related to the visitors' income. Therefore, for each of the 12 countries/regions, a weight will be given to calculate the expenditure of tourists from that origin for gaming and house purchase in Macao.

The weight function is defined as:

$$w_{i,t} = \frac{Y_{i,t}TA_{i,t}}{\sum Y_{i,t}TA_{i,t}} \quad (3)$$

By taking logarithm on both sides of the Eq. (1), we can get the linear function for ordinary least squares (OLS) regression which is easier to estimate. (Croes and Sr [18]; Song and Witt [9]; Vanegas and Croes [19]). The following function is defined:

$$\begin{aligned} lnTA_{i,t} &= \alpha_0 + \beta_{01} lnY_{i,t} + \beta_{02} lnP_{i,t} + \beta_{03} lnH_{i,t} + \beta_{04} lnG_{i,t} \\ &\quad + \beta_{05} DQ_2 + \beta_{06} DQ_3 + \beta_{07} DQ_4 + u_{i,t}, \end{aligned} \quad (4)$$

where $\alpha_0 = lnA$ is drift component; $u_{i,t} = lne_{i,t}$. β_{01} is income elasticity, β_{02} is tourism price elasticity, β_{03} is housing revenue elasticity, and β_{04} is gaming revenue elasticity. DQ_2, DQ_3 and DQ_4 are the seasonal dummy variables for the second quarter, third quarter and fourth quarter. It is used to capture the seasonality of tourism demand.

Equation (4) is a static model, in which we will use the dynamic panel model (panel-ARDL model) to estimate the tourist's decision-making process. The panel-ARDL model initially takes the following form:

$$lnTA_{i,t} = \alpha_0 + \sum_{j=1}^{p_1} \alpha_j lnTA_{i,t-j} + \sum_{j=0}^{p_2} \beta_j lnY_{i,t-j} + \sum_{j=0}^{p_3} \gamma_j lnP_{i,t-j}$$
$$+ \sum_{j=0}^{p_4} \delta_j lnH_{i,t-j} + \sum_{j=0}^{p_5} \varphi_j lnG_{i,t-j} + \phi_1 DQ_2 + \phi_2 DQ_3 + \phi_3 DQ_4 + \varepsilon_{i,t}, \quad (5)$$

where p_1 to p_5 are the lag numbers decided by the Akaike information criterion (AIC) and the Schwarz information criterion (SIC). α_j, β_j, γ_j, δ_j, φ_j are the coefficients to represent the long-term dynamic of the model.

This Panel-ARDL model will be estimated based on Bayesian statistical theory with Markov Chain Monte Carlo (MCMC) method with a prior probability distribution. The parameters of Panel-ARDL will be estimated on the posterior probability distribution. This is to say this method can be implemented using a posterior distribution function of model parameters from the prior distribution function of the sample data. Bayes' theorem can be written in the following form:

$$f(\theta \mid y_t) = \frac{f(y_t \mid \theta) f(\theta)}{f(y_t)}, \quad (6)$$

where $f(y_t \mid \theta) f(\theta)$ is the likelihood function, θ is the vector of parameters, $f(\theta)$ is the prior probability and the $f(\theta \mid y_t)$ is marginal probability. The marginal probability $f(\theta \mid y_t)$ is based on the likelihood function $f(y_t \mid \theta) f(\theta)$ and the observed data y_t. When the closed-form solution to a function is not available, the Bayesian parameter estimation method can be used [20]. And it is required to apply the sampling based approaches, particularly the MCMC method and Gibbs sampling.

Assuming the equation is shown as follows:

$$Y_{it} = \alpha + \beta X_{it} + \varepsilon_{it}, \quad (7)$$

where α accounts for individual effect and it has normal distribution.

Then we can get the following:

$$P(y \mid \alpha, \widetilde{\beta}, h) = \prod_{i=1}^{n} \frac{h^{\frac{T}{2}}}{2\prod^{\frac{T}{2}}} \left\{ exp\left[-\frac{h}{2} (Y_i - \alpha_i - \widetilde{X}_t \widetilde{\beta})' (Y_i - \alpha_i - \widetilde{X}_t \widetilde{\beta}) \right] \right\} \quad (8)$$

The non-hierarchical prior assumption for the coefficients β^* and the variance of ε are shown below:

$$\beta^* \sim N(\beta^*, V), h \sim G(s^{-2}, v) \quad (9)$$

where β^* is normal distribution, h is Gamma distribution. h is the variance of ε.

The likelihood function for the random coefficients β_i is as below:

$$P(y \mid \beta, h) = \prod_{i=1}^{n} \frac{h^{\frac{T}{2}}}{2\prod^{\frac{T}{2}}} \left\{ exp\left[-\frac{h}{2} (Y_i - \widetilde{X}_t \widetilde{\beta})' (Y_i - \widetilde{X}_t \widetilde{\beta}) \right] \right\} \quad (10)$$

The hierarchical priors and their variance are shown as follows:

$$\beta_i \sim N\left(\mu_\beta, V_\beta\right), h \sim G\left(s^{-2}, v\right) \tag{11}$$

where β_i is normal distribution, h is Gamma distribution. h is the variance of ε.

3 Empirical Results

This section mainly introduces the test results of panel-ARDL model on Bayesian MCMC method.

The long-term dynamic relationship results of 12 countries/regions visiting Macao are shown in Table 1. According to the length of data, combined with the AIC and BIC, the variables of tourist arrival, income level and tourism price lag are optimal in the first order.

For the long-term tourists coming to Macao, one lag period has a significant effect. The results show that the "word-of-mouth" effect plays an important role in tourism demand. When tourists are willing to travel, most of them will let their family and friends to listen to their opinions (word-of-mouth) and then choose the destination. Income level coefficient has no statistical significance.

The Panel-ARDL model captures the significant negative impact of current period tourism price on inbound tourists to Macao. Higher tourism prices will lead to a drop in the number of tourists. When the tourism price increases by 1%, the number of tourists will decrease 0.4502%. The lower tourism price will attract more tourists, and more tourists will have more consumption. The elasticities of income and price sit between 0 and 1, which is more sensitive in current period than other period, Dees [21] also reported in other studies.

The coefficient of gaming industry is significant and positive, indicating that the gaming industry has a positive impact on tourism, and the gaming industry is actively promoting the development of tourism. After more than 100 years of historical changes, Macao's gaming industry has gradually formed a unique and diversified development framework, which plays a crucial role in Macao's economy and tourism. Among the tourists who come to Macao, most people choose to come to Macao for vacation and gambling in casinos. In 2018, the gaming industry accounted for 25.78% of Macao's total government revenue. The gaming industry has become a pillar industry of Macao's tourism. Thus, the fortunes of Macao's tourism industry depend on the fortunes of the gaming industry.

Macao's real estate industry and tourism are negatively correlated. The development of the real estate industry has not promoted the development of tourism. Since Macao's return to China, the real estate industry has developed rapidly and housing price has soared. According to the law of development, Macao's real estate industry has entered a period of adjustment. Land prices in Macao are a fifth of those in Hong Kong, but they are almost as expensive. Macao has a property bubble after introducing a series of policies to attract real estate industry investment.

Table 1 Posterior estimates of the parameters

Variables	Mean	SD	2.5%	50%	97.5%
$lnTA(-1)$	0.7601	0.0211	0.7183	0.7605	0.8011
lnY	−0.2488	0.1518	−0.5504	−0.2480	0.0481
$lnY(-1)$	−0.0008	0.1397	−0.2765	−0.0009	0.2771
lnP	−0.4502	0.1415	−0.7315	−0.4516	−0.1659
$lnP(-1)$	0.4549	0.1412	0.1714	0.4563	0.7360
lnG	0.2585	0.0218	0.2158	0.2583	0.3018
lnH	−0.0139	0.0071	−0.0275	−0.0139	−0.0002
DQ_2	0.0212	0.0154	−0.0091	0.0212	0.0512
DQ_3	0.0176	0.0155	−0.0129	0.0176	0.0483
DQ_4	0.0193	0.0153	−0.0108	0.0193	0.0493
$Constant$	0.2539	0.0765	0.1052	0.2546	0.4031
σ^2	0.0138	0.0009	0.0121	0.0121	0.0157

Source Authors' estimation; AIC (Akaike Information Criterion) used in lag length selection criteria for the ARDL specification. All of the statistics estimated by R Gui

For example, if mainland residents invest 1 million MOP to buy real estate in Macao and put 500,000 MOP in time deposits, the whole family can obtain permanent resident status in Macao. However, our model estimates that the real estate industry did not promote the development of tourism from 2008 to 2018. Real estate tourism is more of a domestic, short-haul, short-term tourism. Because of the characteristics of real estate development, people usually have a purposeful visit when choosing real estate, and the long-term dependence on real estate projects, including related supporting facilities, design features of living environment, etc.

Seasonality result indicated the impact of the first quarter is significant and negative. The number of tourists from Mainland China, Hong Kong and Taiwan accounts for about 80% of the total number of tourists to Macao. The three regions had two very important festivals which are the New Year and the Spring Festival in the first quarter. The two important festivals are the days of family reunion, exploring relatives and friends, and rarely traveling outbound. This phenomenon is consistent with the decline in the number of tourists in the first quarter.

Bayesian model comparison is a method of model selection based on Bayes factors which is the ratio of marginal likelihoods between two models. Numerator models are given as rows, and denominator models as columns. The models under consideration are statistical models [22]. The aim of the Bayes factor is to quantify the support for a model over another, regardless of whether these models are correct [23]. The log of the Bayes factor is 468 indicating very decisive support for model 1 and the results are shown in Table 2.

All the variables' trace and density distribution are shown in Fig. 2. As model1 corresponds to the decisive support, we will test change point detection for model1. The expected change point is 356. The most probable change point is detected at

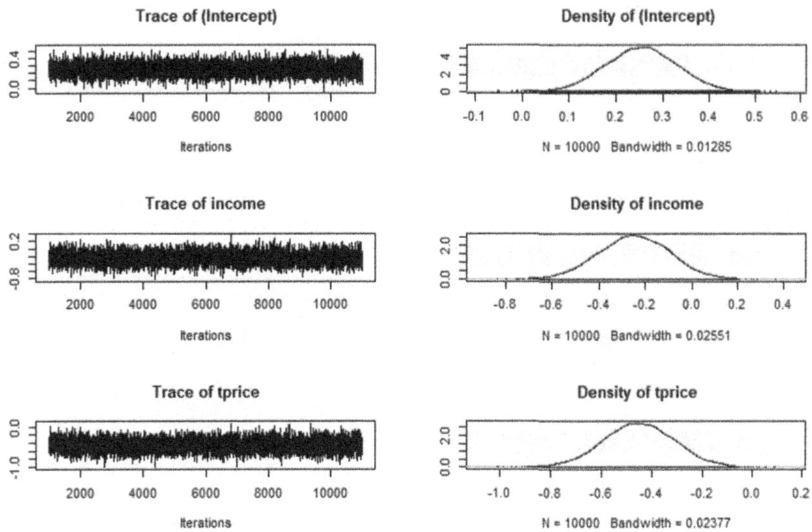

Fig. 3 Trace and density distribution of variables

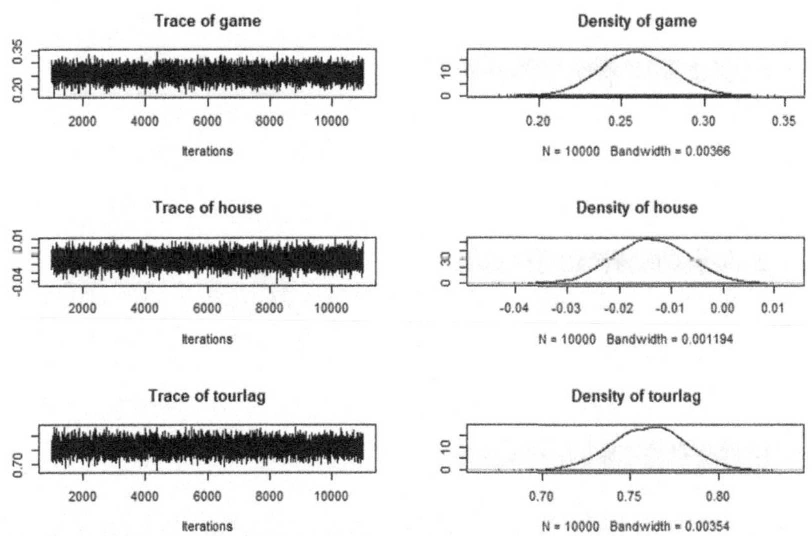

Fig. 3 (continued)

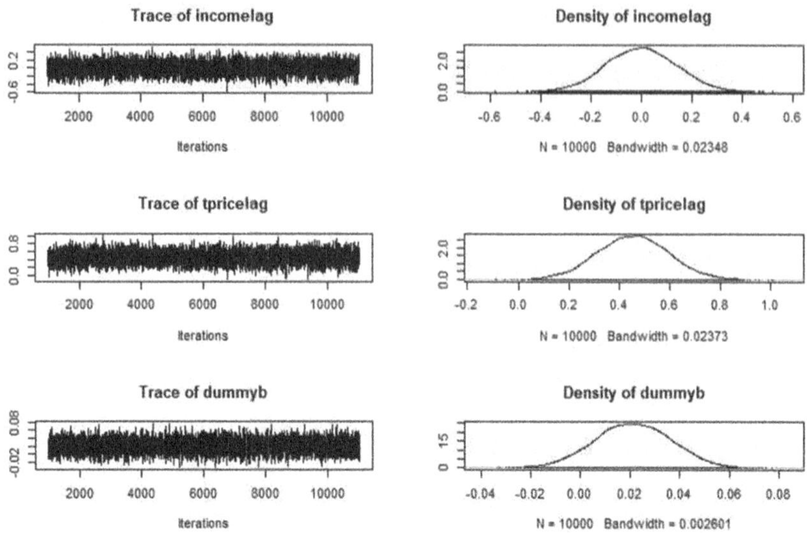

Fig. 3 (continued)

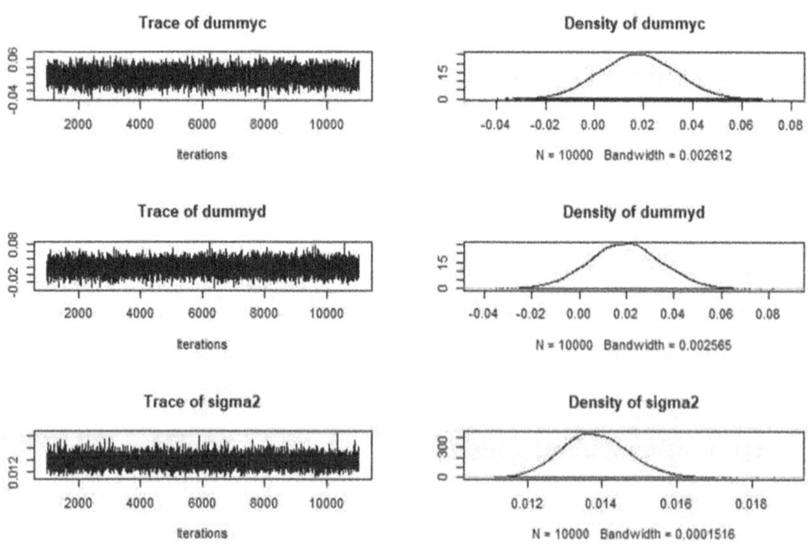

Fig. 3 (continued)

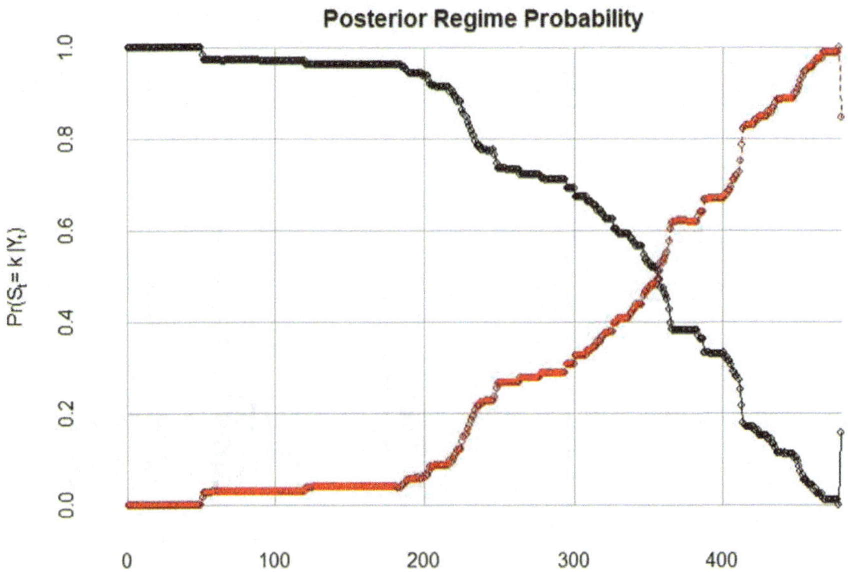

Fig. 4 Posterior regime probability

Table 2 The matrix of the natural log Bayes Factors

	Model0	Model1	Model2	Model3	Model4	Model5
Model0	0.0	−467.9	−388.7	−314.0	−279.84	−283.16
Model1	468	0.0	79.2	153.9	188.05	184.73
Model2	389	−79.2	0.0	74.7	108.86	108.86
Model3	314	−153.9	−74.7	0.0	34.12	30.80
Model4	280	−188.1	−108.9	−34.1	0.0	−3.32
Model5	283	−184.7	−105.5	−30.8	3.32	0.0

Source Authors' estimation. All of the statistics estimated by R Gui

the fourth quarter of 2008 in the Philippines. Figure 4 displays posterior regime probability, and Fig. 5 displays posterior density of regime change probabilities.

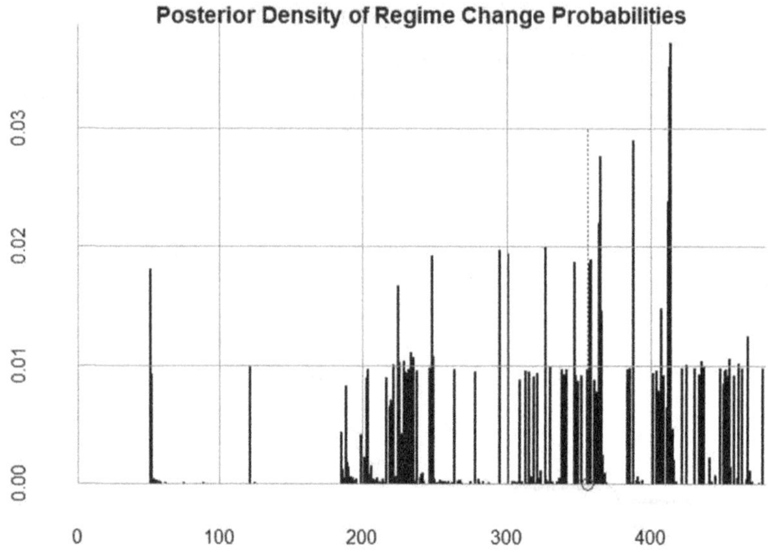

Fig. 5 Posterior density of regime change probabilities

4 Conclusion

This study uses the panel-ARDL model with Bayesian MCMC method to analyze the long-term effects of Macao's pillar industries on Macao's tourism. We find that the word-of-mouth effect plays an important role in the long-term relationship of tourism decision making. The tourism industry of Macao should pay more attention to the services for the tourists who have entered the territory for many times and develop more new tourism projects to increase the attraction. The current tourism price has a negative impact on the number of inbound tourists. Macao's gaming industry promotes the development of tourism, while the development of real estate industry has a negative impact on the tourism.

The long-term empirical results of this study provide important policy enlightenment for Macao. Firstly, tourism price has a significant negative impact on inbound tourists. It is recommended to offer low-cost air and hotel travel packages to reduce travel cost and increase their willingness to travel. Secondly, the impact of the first quarter is significant and negative. It is suggested that the travel agency launch suitable campaigns for the Spring Festival family travel, family travel, and New Year travel abroad projects. Finally, Macao's gaming industry has boosted tourism. The development of the gaming industry cannot be separated from the gambling policy of surrounding countries/regions. One condition of the no-listing is that Hong Kong does not have casinos. If casinos are opened in Hong Kong, the gaming industry of Macao would certainly be hit across the board. The development of gaming industry will further promote the stability and development of Macao's tourism industry.

Macao's real estate industry has put a damper on tourism. In further study, we can pay attention to the sustainable development of Macao tourism.

References

1. Mishra, P., Himanshu, R., Mohapatra, S.: Causality between tourism and economic growth: empirical evidence from India. Eur. J. Soc. Sci. **18**(4), 518–527 (2011)
2. Leiper, N.: Tourism and gambling. Geo J. **19**, 269–275 (1989)
3. Wan, Y.K.P., Li, X.C., Kong, W.H.: Social impacts of casino gaming in Macao: a qualitative analysis. Tourism **59**, 63–82 (2011)
4. Wan, Y.K.P.: The social, economic and environmental impacts of casino gaming in Macao: the community leader perspective. J. Sustain. Tour. **20**, 737–755 (2012)
5. Gu, X., Tam, P.S., Lei, C.K., Chang, X.: The economics of taxation in casino tourism with cross-border market power. Rev. Dev. Econ. **20**, 113–125 (2016)
6. Lu, P.: Study on the moderate diversification of industrial structure of Macao. Am. J. Ind. Bus. Manag. **6**, 176–187 (2016)
7. Israeli, A., Mehrez, A.: From illegal gambling to legal gaming: casinos in Israel. Tour. Manag. **21**, 281–291 (2000)
8. Lim, C.: Review of international tourism demand models. Ann. Tour. Res. **24**, 835–849 (1997)
9. Song, H., Witt, S.F.: Tourism Demand Modelling and Forecasting: Modern Econometric Approaches. Pergamon, Oxford (2000)
10. Witt, S.F., Witt, C.A.: Modeling and Forecasting Demand in Tourism. Academic Press, London (1992)
11. Syriopoulos, T.: A dynamic model of demand for Mediterranean tourism. Int. Rev. Appl. Econ. **9**, 318–336 (1995)
12. Seddighi, H.R., Shearing, D.F.: The demand for tourism in North East England with special reference to Northumbria: an empirical analysis. Tour. Manag. **18**, 499–511 (1997)
13. Kim, S., Song, H.: Analysis of tourism demand in South Korea: a cointegration and error correction approach. Tour. Anal. **3**, 25–41 (1998)
14. Wong, K.K., Song, H., Chon, K.S.: Bayesian models for tourism demand forecasting. Tour. Manag. **27**(5), 773–780 (2006)
15. Assaf, A.G., Li, G., Song, H., Tsionas, M.G.: Modeling and forecasting regional tourism demand using the Bayesian global vector autoregressive (BGVAR) model. J. Travel Res. (2018)
16. Witt, S.F., Wit, C.A.: Forecasting tourism demand: a review of empirical research. Int. J. Forecast. **11**(3), 447–475 (1995)
17. Lim, C., McAleer, M.: Co-integration analysis of quarterly tourism demand by Hong Kong and Singapore for Australia. Appl. Econ. **33**(12), 1599–1619 (2001)
18. Croes, R.R., Sr, M.V.: An econometric study of tourist arrivals in Aruba and its implications. Tour. Manag. **26**, 879–890 (2005)
19. Vanegas, S., Croes, R.R.: Evaluation of demand : US tourists to Aruba. Ann. Tour. Res. **27**(4), 946–963 (2000)
20. Gelfand, A.E., Smith, A.F.M.: Sampling-based approaches to calculating marginal densities. J. Am. Stat. Assoc. **85**(401), 946–963 (1990)
21. Dees, S., Mauro, F.D., Pesaran, M.H., Smith, L.V.: Exploring the international linkages of the euro area: a global VAR analysis. J. Appl. Econom. **22**(1), 1–38 (2007)
22. Morey, R.D., Romeijn, J.-W., Rouder, J.N.: The philosophy of Bayes factors and the quantification of statistical evidence. J. Math. Psychol. **72**, 6–18 (2016)
23. Ly, A., Verhagen, J., Wagenmakers, E.-J.: Harold Jeffreys's default Bayes factor hypothesis tests: explanation, extension, and application in psychology. J. Math. Psychol. **72**, 19–32 (2016)

Impact of Investment Structure by Economic Sectors and Other Factors on Economic Growth: Evidence from Vietnam with SGMM Estimation and Bayes Factor Approach

Huong Thi Thanh Tran and Hang Thu Pham

Abstract Investment and sectoral economic structure are two of the important factors affecting economic growth, especially in developing countries. However, there are few studies in Vietnam assessing the impact of investment structure by economic sectors on economic growth. Moreover, in foreign studies on this issue, most quantitative methods and models have not addressed the non-homogeneous and endogenous problems in the research model. Different from previous studies, a combination of System Generalized Method of Moments (SGMM) and Bayes Factor approach will be used to evaluate the impact of investment structure by economic sectors and other factors on GDP of Vietnam. Results of two-step SGMM estimation combined with the Minimum Bayes's Factor approach with data sets of 51 provinces of Vietnam in the period of 2013–2017 show that the proportion of investment of industry sector (RIC_IC) as well as service sector (RIC_SV), labor productivity (LP), control index of public service delivery (Pa6) and provincial competitiveness index (PCI) have positive impacts on Vietnam's GDP growth in the period 2013–2017. In order to promote Vietnam's economic growth, it is necessary to continue applying policies to attract investment in industry and service sectors with special emphasis on the industry sector, encouraging the provision of public services and strengthening provincial competitiveness to boost high economic growth.

Keywords Investment structure · Economic sectors · Factors · Economic growth · SGMM method · Bayes factor

H. T. T. Tran (✉)
Department of Statistics, Faculty of Accounting and Auditing, Banking Academy of Vietnam, 12 Chua Boc Street, Dong Da District, Hanoi, Vietnam
e-mail: huongttt76@hvnh.edu.vn

H. T. Pham
Faculty of Economics, Banking Academy of Vietnam, 12 Chua Boc Street, Dong Da District, Hanoi, Vietnam

S. Sriboonchitta et al. (eds.), *Behavioral Predictive Modeling in Economics*, Studies in Computational Intelligence 897,
https://doi.org/10.1007/978-3-030-49728-6_18

Table 1 Investment structure by economic sectors of Vietnam period 2013–2017 (current price)

	Total investment (billion VND)	Agriculture sector		Industry sector		Service sector	
		Investment (billion VND)	Percentage (%)	Investment (billion VND)	Percentage (%)	Investment (billion VND)	Percentage (%)
2013	1094542	63658	5.82	478967	43.76	551917	50.42
2014	1220704	61524	5.04	582753	47.74	576427	47.22
2015	1366478	76523	5.60	647983	47.42	641972	46.98
2016	1487638	87473	5.88	685057	46.05	715108	48.07
2017	1668601	100116	6.00	762551	45.70	805934	48.30

Source The authors' calculation from the Statistical Yearbook of Vietnam, GSO

1 A Summary of Investment Structure by Economic Sectors of Vietnam in the Period of 2013–2017

Looking at the figures in Table 1 and Chart 1 (Fig. 1), it can be seen that the proportion of investment in agriculture sector was very low compared with in industry and service sectors. The proportion of investment in industry sector tended to decrease slightly while the percentage of investment in service sector increased slightly.

Total investment in the period of 2013 to 2017 increased significantly (about 52.5%), of which the strongest increase was invested in industry sector; then to the service sector and the lowest increase is the agriculture sector. This makes the proportion of investment in the agriculture sector increase slightly in this period but still accounts for a negligible part of the investment structure (accounting for 6% in 2017). The proportion of investment of the industry sector is the biggest increase (2017 compared to 2013 increased by 1.94% points); meanwhile, the proportion of investment in the service sector decreased more than before (2017 compared with 2013 decreased by 2.12% points).

Chart 1 (Fig. 1) clearly shows the orientation of Vietnam's industrialization and modernization in recent years. However, for Vietnam, the number of people relying heavily on agricultural production is still large. Hence, investing in this sector should be carefully considered from the perspective of comprehensive development. Investment throughout indirect effects also plays an extremely important role. For example, investment in technology development for agricultural production; in education and training and in transport development, ... Due to the actual data of Vietnam, it is difficult to extract indirect investment for the agriculture sector. This is a major limitation in analyzing the structure of investment by economic sectors.

Although the agriculture sector still plays an important role in economic activities in Vietnam, the signs of the decreasing proportion of investment in the service sector in parallel with increasing trend of agriculture sector might be not a positive sign. In the context that the economy is industrializing and labor shifting among sectors, there is also a tendency of agricultural labor to gradually shift to the remaining

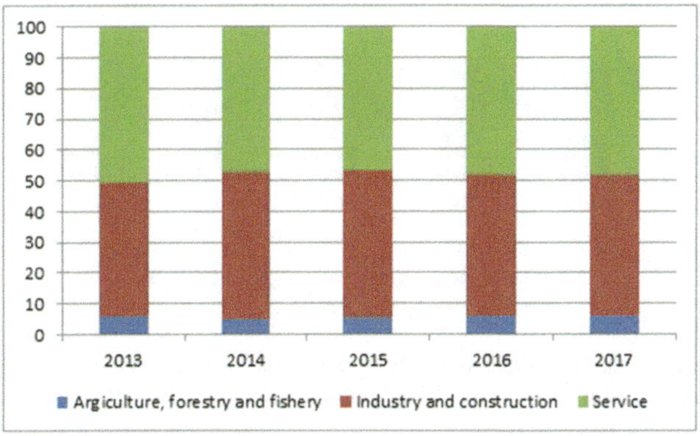

Chart 1 Structure of investment by economic sectors of Vietnam in the period of 2013–2017. *Source* The authors' calculation from the Statistical Yearbook of Vietnam, GSO

two areas. The increase in investment in both absolute and relative proportions in agriculture sector is not a driving force for positive economic growth. However, the answer to this fact also depends on the strategy of using investment in the agriculture sector following the trend of industrialization and depending on the situation in the efficiency of using investment in this area.

For Vietnam as well as other developing countries, economic growth is still one of the most important goals, because developing countries must grow faster to combat the risk of lagging far behind is economic compared to other countries. Is the growth of investment for the sectors closely related to the results and achievements of economic growth? Table 2 describes the growth rate of investment and GDP by economic sector in Vietnam during 2013–2017, partly showing the correlation between investment and GDP growth.

The data sheet shows that, in general, the growth rate of investment increases corresponding to the increase in GDP growth. For sectors, only the growth rate of investment for agriculture sector is negative, but VA growth in 2014 is still higher than the previous year. Also in 2014, the growth rate of investment in the industry sector increased sharply, but on the contrary, VA growth in this sector was quite stable, not

Table 2 Growth rate of investment and GDP by economic sectors of Vietnam in the period 2013–2017 (at constant prices in 2010)

	Total		Agriculture sector		Industry sector		Service sector	
	Investment (billion VND)	GDP	Investment (billion VND)	VA	Investment (billion VND)	VA	Investment (billion VND)	VA
2013	7.31	5.42	20.67	2.63	7.49	5.08	5.80	6.72
2014	9.80	5.98	−4.80	3.44	19.63	6.42	2.97	6.16
2015	9.06	6.68	22.43	2.41	8.01	9.64	8.70	6.33
2016	9.84	6.21	13.90	1.36	7.13	7.57	12.06	6.98
2017	10.76	6.81	12.92	2.90	9.93	8.00	11.29	7.44

Source The authors' calculation from the Statistical Yearbook of Vietnam, GSO

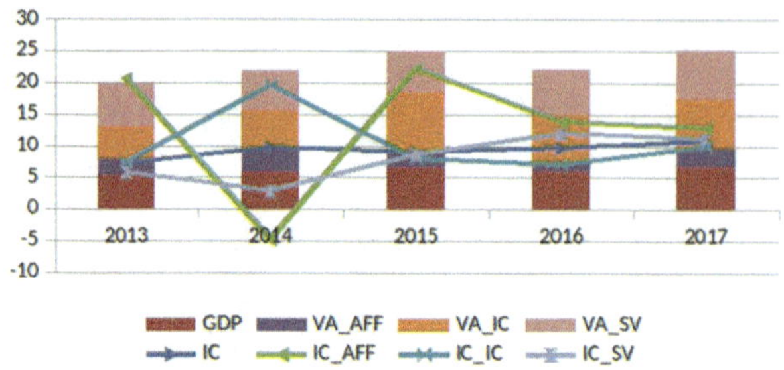

Chart 2 Growth rate of investment and GDP by economic sectors of Vietnam in the period 2013–2017 (at constant prices in 2010). *Source* The authors' calculation from the Statistical Yearbook of Vietnam, GSO

Table 3 Capital productivity by economic sectors of Vietnam in the period of 2013–2017 (at constant prices in 2010) Unit: times

	Total	Agriculture sector	Industry sector	Service sector
2013	2.92	8.58	2.21	2.22
2014	2.82	9.32	1.96	2.29
2015	2.75	7.80	1.99	2.24
2016	2.66	6.94	2.00	2.14
2017	2.57	6.32	1.97	2.06

Source The authors' calculation from the Statistical Yearbook of Vietnam, GSO

a sudden phenomenon such as investment. Similar to the agriculture sector, in 2014, the growth rate of investment in this area decreased significantly, but VA growth in the service sector only decreased slightly compared to 2013. Chart 2 (Fig. 2) illustrates more clearly by price in comparison with 2010.

Data in Table 2 and Chart 2 (Fig. 2) show that, except for 2014, the growth rate of investment of the whole economy as well as economic groups is always higher than the growth rate of GDP. GDP growth is mainly based on increasing investment. If the investment is increased, the economy will have a higher growth rate. How to reduce investment but still have positive growth and increase with high level is an important issue for a country where the accumulated domestic investment from the economy is still quite low. That is, the threshold of increasing investment is always a big question for Vietnam in this period. Besides, the data in Table 2 also shows that the agriculture sector has the highest growth rate of investment but the smallest growth rate of GDP. Industry sector has the lowest investment growth rate but the highest growth rate of GDP.

From the data in Table 1 and in Table 3, the agriculture sector has the lowest proportion of investment (much lower than that of industry and service sector) but the highest use of investment. In the research period, the agriculture sector only accounted for 5.82% to 6% of the investment, but the capital productivity in agriculture increased the most in 3 sectors and in 2017 was 6.32 times—this is no small achievement of the agriculture sector. In addition, the corresponding industry and service sectors only have increased capital productivity in less than 2 times compared to the previous year. Thus, the use of capital in agriculture and forestry and fishery seems to bring more positive results relative to the current status of using capital in industry and services.

Practical figures on capital structure by economic sectors in Vietnam in the period 2013–2017 are posing many problems for the Government in terms of policies and capital efficiency. In order to promote economic growth, Vietnam should continue to focus on investment in the industry and service sectors as the capital productivity of these two groups is lower than that of agriculture, forestry and fishery as well as capital productivity of the entire economy.

2 Research Overview

2.1 Studies on Economic Sectoral Structure with Economic Growth

The structure of the economic sector affecting growth is a long-standing economic and theoretical issue. This correlation has been documented in many classical studies both in developed and developing countries.

The first theory that addresses the issue of economic structure is the dual-sector theory of Lewis [22]. This theory pointed out the relationship between agriculture and industry in the process of growth. This theory divided the economy into two areas: agriculture and industry sector and solved the relationship between the two regions in the process of growth. With the assumption that the industry sector was full of labor, the model showed that economic growth was associated with the ability to accumulate, invest and expand the production of the industry sector because of the stagnant agriculture sector with many redundant workers. With the view that industrial development and increased investment in the industry sector will be the biggest drivers of economic growth, this model was fully oriented towards goals and solutions to increase investment and increase capital accumulation of industrial sector. In addition, the dual–sector model did not mention the service area in its study due to limitations of historical circumstances at the time of the study conducted. Similarly, the theory of shifting industry structure affecting economic growth from the viewpoint of the New Classical economic school also drew conclusions about the role of sectors in economic growth which was similar to Lewis. The fundamental difference was that, with the Neoclassical view when the production function evaluated science and technology, the economy should allocate additional investment to the agriculture sector (but with a lower proportion) to ensure an advantage for the industry sector in the process of economic growth and development.

Rostow's "take off" theory showed the relationship between economic restructuring and growth through the theory of development stages. According to Rostow, the development process of each country was divided into 5 stages including: traditional society, preparation for takeoff, takeoff, post-industrial and high consumer society. Through five stages of Rostow's study, the industry structure shifted clearly, from the traditional social stage, economic activities were mainly agricultural activities, gradually the agricultural sector decreased to billions. Importantly, until stage 5, the proportion of the agricultural sector was too small, almost disappearing from the industry structure. In contrast, the industry and service sector gradually increased in proportion and played a key role in the process of industry restructuring. From the third stage, the industry sector grew very strong and accounted for the largest proportion, until the fourth stage, the industrial sector still played a role in guiding the process of industry restructuring. However, when the society changed to stage 5, the industry sector tended to slow down, the proportion tended to decrease, replacing the service area. In the highest stage of social change, the service sector would play the most important role and accounted for the largest proportion in the sector structure. Rostow's model had emphasized the role of investment factors affecting overall economic growth in stages 3 and 4. However, this model had not clearly indicated the structure of investment capital in how each sector changes as the economy grows and grows higher.

The balanced development theory of Nurkse [29] and Rostenstein (1943) considered that it must be equally developed in all economic sectors to transfer economic structure translation quickly. This theory was suitable for developing countries to implement inward industrialization or import substitution. However, in fact, this theory had revealed certain disadvantages: the development of an economy with a balanced and completed structure had pushed economies to a closed situation, which could not take advantage of the benefits from the external environment. This development trend went against the trend of globalization. At the same time, this strategy went beyond the resources of most poor countries to be able to fulfill the expected structural objectives, due to the lack of capital and other resources to promote comprehensive development of industries. The basic point of view of Nurke and Rostenstein will led to an investment strategy to be implemented in all sectors to ensure the balance for the purpose of economic growth.

A separate study for countries where "Monsoon Asia" had the advantage of growth and development according to the agriculture sector, Oshima [31] introduced a new theory of the relationship between industry and agriculture in the process of restructuring from an economy dominated by agriculture to industry under specific conditions of monsoon Asian countries. In Oshima's view, excessing labor would not be frequent in monsoon Asian countries with agricultural development advantages. According to his research, Oshima said that for countries with agricultural development advantages, economic growth would be associated with the development of the agriculture sector first. Hence, in order to promote economic growth and development, this group of countries needed to invest in deep development. According to the process of restructuring the industry of the Oshima model, the issue of investment was mentioned in all three stages, however, the study did not show an overview of the investment structure in all three zones.

Recently, the theory of Syrquin [40] described the process of economic restructuring in the modern period according to three phases: primary production, industrialization and developed economy. Basically, the content of Syrquin's theory also gave quite similar points to Rostow's research. Through 3 stages, the agriculture sector would tend to occupy a large proportion in the first stage. Then, agriculture would gradually reduce the proportion. The industry structure would gradually shift from primary production to processing areas. Finally, when the economy entered a development stage, the service sector would play the most important role in GDP and in the labor structure. At the same time, aggregate factor productivity contributed the most in economic growth. This study also confirmed that investment and increasing investment were one of the factors that play an important role in economic growth. However, the role of investment factors in each economic sector was not clearly recorded.

Empirically, many studies assessed the role of sectoral structure in general economic growth. Some case studies led to interesting and relatively separate conclusions for the following countries:

- A group of quantitative studies confirmed that the leading role of the industry sector would lead to high economic growth, typically, the study of van Ark [44] with research data from 8 Western European economies. Fonfria et al. [11] studied the "rewards of restructuring" for Spain's manufacturing industry, resulting in the impact of industry structure on economic growth. The study of Lanjouw (2001) gave the same conclusion. Rodrick (2011), Rodrik (2014) in the study of unconditional convergence production of 100 countries in recent decades had shown that, for underdeveloped countries with low income, policies to restructure was a prerequisite for promoting growth. This study also affirmed that developing countries that want to promote rapid economic growth to catch up with other countries should shift their structure toward high-proportioned industry sector
- The second research trend suggested that the process of promoting agricultural growth in the economic structure would cause high economic growth. In a recent study on productivity growth in Asia, van Ark and Timmer [43] also reaffirmed the contribution of economic structure shifting to the growth of low-developed economies (the research on 7 asian countries in the period 1963–1996). Gyfason and Zoega [15] had also quantified the effects of sectoral restructuring on the growth of countries around the world by considering the effects of changes in the share of GDP in agriculture and the transition of migrant labor from rural to urban areas with World Bank data set for 86 countries (without Vietnam). Similar results were drawn from Matsuyama (2009); Gollin et al. [13]; Ngai and Pissarides [27]; Dollar and Kraay [9]. However, the role of the agricultural sector in the economic structure would changes along with the economic development periods. The richer the country was, the lower agriculture growth contributed (Christiaensen and Demery (2010); Ligon and Sadoulet [23]).
- The third research trend focused on the role of the service sector in the economic structure. Hansen and Prescott [16]; Vollrath [45] asserted that the shift towards service growth and some non-agricultural sectors will encourage economic growth.

accurately measurement of the probability of the hypothesis occurring without regard to the probability of the selected sample. Secondly, most of the logic in constructing hypotheses is valid for binary logic, and less valuable for multivalued logic. Meanwhile, most economic events occur in polynomial logic, so p-value will be less accurate. Lastly, the conclusion that "when p-value is small, the hypothesis H0 is rejected" is not entirely theoretical or practical. One evidence is visible about this disadvantage of p-value when studying the Lindley's paradox (Nguyen [28]). One of the remarkable methodological recommendations in recent times is the Bayesian approach to overcome the above disadvantages of quantitative methods using p-value. Many studies have shown and theoretically proved that according to the Bayes approach, the estimated results will be more reliable (Goodman [14], Nguyen [28]).

In summary, most of the theory of economic structure and economic restructuring considered the restructuring of investment in the period of industrialization. Besides, the theories also show that the forms of economic restructuring of the countries are very different and diversified and it is difficult to find a common pattern for all countries. Therefore, understanding the process of economic restructuring in each country for each stage of development is a complicated issue. This purpose requires researchers to approach in many different ways. Studies on the impact of economic restructuring on Vietnam's economic growth have been just only approached by output structure (structure in GDP) or structure by ownership. Very few studies have evaluated the impact of structural shifting on inputs such as investment by economic sectoral structure. Moreover, in terms of methods, almost all studies only use index methods, shift share analysis, VAR or VECM for time-series data. Besides, some studies used 2SLS to estimate the impact of investment on economic growth, however, these methods have not solved the heterogeneity endogeneity problems in the research model. Moreover, with recent methodological approaches, models based on the results of p-value are being criticized for lack of reliability. Therefore, our study will focus on assessing the impact of sectoral investment structure on economic growth in Vietnam for the period of 2013–2017. The study also aims to approach both old and new methods in order to increase the accuracy in evaluating the estimated results. It is expected that with theoretical and practical research efforts as well as a new quantitative approach, this research can contribute more to a comprehensive assessment of the role of investment with economic growth in developing countries.

3 Research Methods and Data Sources

3.1 Data Source

Data sources for calculating the indicators of gross domestic product (GDP), labor productivity (LP), the proportion of investment of industry sector (RIC_IC), the proportion of investment of the service sector (RIC_SV) are collected from the Statistical Yearbook of 51 provinces and cities in Vietnam. This survey is yearly conducted by General Statistic Office of Vietnam.

Corruption control index (Pa4), public administrative procedure control index (Pa5), public service ability control index (Pa6) are collected from The Viet Nam Provincial Governance and Public Administration Performance Index (PAPI) yearly survey which are conducted by many official institutions. It is a joint collaboration between the Centre for Community Support Development Studies (CECODES) under the Viet Nam Union of Science and Technology Associations (VUSTA), and the United Nations Development Programme (UNDP) in Viet Nam since 2009, with the close partnership and support of the Centre for Theory Work of the Viet Nam Fatherland Front from 2009 to 2010, the Front Review from 2010 to 2012, the Commission for People's Petitions under the National Assembly Steering Committee in 2012, and the Centre for Research and Training of the Viet Nam Fatherland Front–VFF–CRT from 2013. The Real–Time Analytics joined the implementation consortium in 2015 to provide technical platforms for real–time PAPI data collection and fieldwork monitoring. Since 2012, the Ho Chi Minh National Political Academy (HCMA) has been partnering with UNDP Viet Nam in conducting action-based research using PAPI data and providing PAPI findings to political leaders and executives through high–level training programmes.

Provincial Competitiveness Index (PCI) was first introduced in 2005, the Provincial Competitiveness Index (PCI) on Vietnam's business environment conducts an annual business survey, assessment and ranking of the economic governance quality of provincial authorities in creating a favorable business environment for development of the private sector. This survey is mainly conducted by the Vietnam Chamber of Commerce and Industry (VCCI) which is a national organization that assembles and represents the business community, employers, and business associations of all economic sectors in Vietnam.

All variables are summarized in Table 4.

3.2 Methodology

In order to assess the impact of the restructuring of investment and factors on Vietnam's economic growth, the equation form will be considered as follow:

$$GDP_{ij} = \alpha_0 + \alpha_1 LP_{ij} + \alpha_2 RIC\_IC_{ij} + \alpha_3 RIC\_SV_{ij} + \alpha_4 Pa4_{ij} + \alpha_5 Pa5_{ij} + \alpha_6 Pa6_{ij} + \alpha_7 PCI_{ij} + u_{ij}$$

Where:

- GDP_{ij}: Logarithm of GDP of province i in year j is calculated according to constant prices in 2010
- LP_{ij}: Logarithm of labor productivity of province i year j is calculated according to constant prices in 2010
- RIC_IC_{ij}: Proportion of implemented investment of industry–construction sector to the total implemented investment capital of the province i year j
- RIC_SV_{ij}: Proportion of implemented investment capital of the service sector to the total implemented investment capital of the province i year j

Table 4 Measurement of variables used in the research model

Variables	Unit	Measurement
Gross domestic product (GDP)	Billions VND	Gross domestic product at the constant price in 2010
Labor productivity (LP)	Million VND/person	GDP/Labor
Proportion of investment capital of industry–construction sector (RIC_IC)	%	Investment of industry–construction/total investment
Proportion of investment capital of service sector (RIC_SV)	%	Investment of service sector/Total investment
Pa4		The Papi4 is control corruption index and comprises 4 sub-indices, including: Limits on Public Sector Corruption, Limits on Corruption in service, Equity in employment, Willingness to fight corruption
Pa5		The Papi5 is Public Administrative Procedures and comprises 4 sub-indices, including: Certification procedures, Construction permits, Land procedures, Personal procedures
Pa6		The Papi6 is Public service delivery and comprises 4 sub-indices, including: Health, Education, Infrastructure, Law and order
PCI (Provincial Competitiveness Index)		The overall PCI comprises 10 sub-indices, reflecting economic governance areas that affect private sector development. A province that is considered to perform well on the PCI is the one that has: 1) low entry costs for business start-up; 2) easy access to land and security of business premises; 3) a transparent business environment and equitable business information; 4) minimal informal charges; 5) has limited time requirements for bureaucratic procedures and inspections; 6) limit crowding out of private activity from policy biases toward state, foreign, or connected firms; 7) proactive and creative provincial leadership in solving problems for enterprises; 8) developed and high-quality business support services; sound labor training policies; and 9) fair and effective legal procedures for dispute resolution

Source The authors' synthesis from the Statistical Yearbook of Vietnam, GSO

- $Pa4_{ij}$: Control corruption index of province i year j
- $Pa5_{ij}$: Public Administrative Procedures of province i year j
- $Pa6_{ij}$: Public service delivery of province i year j
- PCI: Provincial Competitiveness Index of province i year j

According to previous theory and empirical researches, it is expected that labor productivity (LP), the proportion of investment of industry and construction sector (RIC_IC) and the proportion of investment capital of the service sector (RIC_SV), $Pa4$, $Pa5$, $Pa6$ and PCI will impact positively on GDP.

3.3 Process of Estimation and Verification

There are three common methods used: Pooled ordinary least squares (POLS), fixed effect model (FEM), and random effect model (REM). The best model is chosen through the following tests:

- Breusch-Pagan test to choose between POLS or REM, FEM models. The hypothesis pair for the testing platform is:
  - H_0: No random effects exist ($\delta 2u = 0$)
  - H_1: random effects
    If the $p - value < 2 < 0.0000$, then reject the hypothesis H0, which means that the model has a random impact then we should use the REM instead of POLS.
- Testing Hausman to select FEM or REM. The hypothesis pair for the testing platform is:
  - H_0: There is no significant difference between the estimated values of the two models FEM and REM
  - H_1: There are significant differences between the estimated values of the two models FEM and REM
    If the p–value of the test $\chi^2 < 0.0000$, then reject the H_0 hypothesis, it is not recommended to use the REM but rather use the FEM.

However, one disadvantage of panel data with a large number of observable individuals in a short time series often results in changes of variance of residual and it is very difficult to overcome this problem. In addition, if there is an endogenous variable problem in the model, then REM and FEM estimates are no longer effective. To solve the problem in this study, the model's defects were tested and SGMM (System Generalized Method of Moments) should be used to analyze the impact of investment in economic sector and factors to GDP.

Then, to retest the obtained model, it is supposed that the results should be tested again by Bayesian approach. According to Goodman [14], Minimum Bayes's factor (MBF) should be calculated to compare with results with p-value from SGMM. Goodman [14] addressed the objectivity of MBF as follows:

The minimum Bayes's factor is unique function of the data that is at least as objective as the P value. Process for translating between Fisher's P-value and the MBF is summarized as below:

- **Step 1**: Since p–value is calculated based on the observed value, one must formulate MBF in exactly the same way. It is composed of two probability written as: $P(Evidence|H_0)$ and $P(Evidence|H_\alpha)$
- **Step 2**: Given the fact that a lesser P value means less support for H0, one must again formulate the MBF in the same manner. This mean, when calculate the MBF, one must place $P(Evidence|H_0)$ in the numerator, while $P(Evidence|H_A)$ is placed in the denominator. Therefore:

$$MBF = \frac{P(Evidence|H_0)}{P(Evidence|H_\alpha)}$$

- **Step 3**: Covert the MBF into a gaussian approximation form (Standard Normal Distribution form) for calculation purposes. The formula is:

$$MBF = \frac{P(Evidence|H_0)}{P(Evidence|H_\alpha)} = e^{-z^2/2}$$

where z represents the corresponding z–value from the null effect.

4 Estimated Results of the Model

4.1 Impact of Sectoral Investment and Factors to Vietnam's GDP Growth in the Period 2013–2017

In order to assess the impact of sectoral investment restructuring and factors on Vietnam's GDP growth in the 2013–2017 period, we used the panel data regression model with the data of 51 provinces and cities of Vietnam (Table 5).

Then, the results of estimating the effect of restructuring investment by economic sectors and factors to Vietnam' GDP growth as follows:

Breusch–Pagan test results show that the model has a random impact, then we should use the FEM or REM. Hausman test results proves that the fixed effect model (FEM) outperforms the overall OLS model and the random effect model (REM) in expressing the impact of investment by economic sectors to Vietnam's GDP growth (shown in Tables 6, 7).

Wald and Wooldridge test results show that the FEM exists in the phenomenon of variance of residual changing and autocorrelation. To solve this problem, FEM is used with robust standard error adjustment for the results of column 5 of Table 6. However, the FEM model with robust option has not resolved the endogenous problem. Hence, GMM (Generalized Method of Moments) method should be the most appropriate

Table 5 Statistics of variables in the model

Variable	Obs	Mean	Std. Dev.	Min	Max
GDP1	255	10.44551	.9051995	8.457878	13.35209
LP1	255	3.989425	.7032476	2.966398	7.844062
RIC_IC	255	42.61066	19.74453	4.849439	86.9109
RIC_SV	255	49.24076	18.68558	9.683965	88.14148
Pa4	254	6.014055	.7145991	0	7.51
Pa5	254	6.948661	.5375502	0	7.79
Pa6	254	6.981181	.567059	0	8.03
PCI	254	59.14069	3.450018	50.32	70.6944

Source The authors' calculation

Table 6 Effect of restructuring investment by economic sectors and factors to Vietnam's GDP growth

	OLS	RE	FE	FE-Robust	System GMM
GDP_{t-1}					.33694 (.03460)***
LP	.8663 (.0542)***	.66880 (.03304)***	0.6367 (.0329)***	.6367 (.2406)***	.65137 (.0346)***
RIC_IC	.01514 (.00415)***	.0035 (.00402)	−.00338 (.00438)	−.00338 (.0021)	.006 (.00118)***
RIC_SV	.0127 (.00426)***	.00475 (.00369)	−.00008 (.00391)	−.00008 (.00149)	.00837 (.00168)***
Pa4	.00475 (.05899)	.02702 (.0309)	.0371 (.0304)	.0371 (.0326)	−.0027 (.01084)
Pa5	−.14127 (.0919)	−.09617 (.04481)**	−.09039 (.04381)**	−.09039 (.05438)	−.19703 (.02692)***
Pa6	.1272 (.09104)	.09394 (.04691)**	.0806 (.0460)*	.0806 (.03481)**	.16322 (.02663)***
PCI	.0486 (.01048)***	.01245 (.00589)**	.0067 (.0058)	.00671 (.01046)	.0328 (.00574)***
Constant	2.9094 (.71201)***	6.5078 (.53696)***	7.4992 (.5644)***	7.4992 (.41362)***	

Standard error in parentheses; *Significant at the 10% level;
**Significant at the 5% level; ***Significant at the 1% level.
Source The authors' calculation

method. The selection of GMM not only allows solving the endogenous problem in the research model but also has advantages over the instrument variable model (IV) because it is easier to select instrument variables by exogenous variables. Therefore, through GMM method, we should introduce many instrument variables to easily approach the condition of an over-identification of the estimator. Furthermore, the estimating from Arellano and Bond [1] research is suitable for short-table data with short time series. In this study, the time series was only five years in 51 provinces and cities, so the GMM method introduced by Arellano and Bond [1] was chosen.

Table 7 Testing results of model's errors

	OLS	RE	FE	FE-Robust	System GMM
Observations	255	255	255	255	204
Adj R-squared	0.6898	0.6663	0.6246	0.6246	
F test that all = 0	78.16		56.63	47.88	
Pro>F	0.0000		0.0000	0.0000	
Wald (chi2)		450.73			2.93e+06
Prob>Chi2		0.0000			0.0000
Breusch and Pagan test		319.01			
Pro>Chibar2		0.0000			
F test that all u_i = 0			34.32		
Pro>F			0.0000		
Hausman test			20124.74		
Prob>Chi2			0.0000		
Wald test for heteroscedasticity			4.0e+05		
Prob>Chi2			0.0000		
Wooldridge test for autocorrelation			0.051 0.8224		
AR(1)					z = −1.39 Pr > z = 0.166
AR(2)					z=0.15 Pr > z=0.882
Hansen-test					chi2(58) = 37.73 Prob >chi2 = 0.982
Sagan-test					chi2(59) = 129.08 Prob >chi2 = 0.000

Source The authors' calculation

Implementing two-step SGMM (System–Generalized Method of Moments) with endogenous GDP variables, labor productivity (LP), investment ratio of industry sector (RIC_IC), and service investment ratio (RIC_SV), administrative management efficiency (as measured by Pa4, Pa5, Pa6) obtained the estimated results at column 6 Table 6. Besides, Arellano and Bond [1] have proposed two key tests to check the Over-identification of GMM model. The first test is the Sargan test or the Hansen test for the over-identification of the model. The second test used is the Arellano-Bond test to test for autocorrelation. Hansen test on the validity of the model gives the value P = 0.928 > 0.05 and the Arellano–Bond (AR2) test for the second-order chain correlation for P = 0.882 > 0.05 (column 6 of Table 7) shows the model has over-identification and there is no second-order chain correlation. Therefore all the results in the GMM model are significant.

After using the SGMM to estimate the results of the impact of factors on economic growth, the Bayesian factor method was used to verify the results. Inheriting the

Table 8 Testing results with Minimum Bayes Factor (MBF)

	P value (Z Score)	Minimum Bayes factor	Decrease in probability of the null hypothesis, %		Strength of evidence
		From	To no less than		
LP	0.000 (3.29)	0.0045 (1/224)	90 50 25	3.9 0.4 0.1	Strong to very strong
RIC_IC	0.000 (3.29)	0.0045 (1/224)	90 50 25	3.9 0.4 0.1	Strong to very strong
RIC_SV	0.000 (3.29)	0.0045 (1/224)	90 50 25	3.9 0.4 0.1	Strong to very strong
Pa4	0.803 (3.85)	0.7143 (1/1.4)	90 50 25	86.5 41.67 19.23	Weak
Pa5	0.000 (3.29)	0.0045 (1/224)	90 50 25	3.9 0.4 0.1	Strong to very strong
Pa6	0.000 (3.29)	0.0045 (1/224)	90 50 25	3.9 0.4 0.1	Strong to very strong
PCI	0.000 (3.29)	0.0045 (1/224)	90 50 25	3.9 0.4 0.1	Strong to very strong

Calculations were performed as follows:
A probability (Prob) of 90% is equivalent to an odds of 9, calculated as Prob/(1 – Prob)
Posterior odds = Bayes factor × prior odds; thus (1/224) × 9 = 0.0045
Probability = odds/(1 + odds); thus 0.0045/1.0045
Source The authors' calculation

Bayesian factor calculation method from Goodman [14], the results obtained are as follows (Table 8):

The above table gives us a conclusion that is completely similar to the results estimated from the SGMM method. With the variables of labor productivity (LP), the rate of investment in the industrial and agricultural sector (RIC_IC and RIC_SV respectively), administrative management efficiency (represented as variables Pa5, Pa6) and Provincial competitiveness index (PCI) are important and significant factors affecting economic growth. Only the Pa4, which represents the provincial corruption control, has no meaning to the outcome of GDP growth. Minimum Bayes factor values of LP, RIC_IC, RIC_SV, Pa5, Pa6 and PCI variables are all larger than the corresponding P–value. This shows that according to Bayes factor test, these factors have significant effects on GDP growth at provincial level.

4.2 Discussion

First of all, the proportion of investment of the industry and service sectors have great significance to the provincial GDP growth. The data in Table 6 shows that: the rate of investment for industry sector increases by 1%, then the provincial GDP growth increases by 0.6% respectively. In addition, the proportion of investment in the service sector increases to 1%, which will stimulate GDP growth by 0.8%, respectively. This result is quite similar to the reality that is occurring in the Vietnamese economy about recent 10 years. The proportion of the service sector increased rapidly and remained stable, while the proportion of the industry sector showed signs of abnormal change. This results in a contribution to the service sector's tendency to be more stable and higher than that of the industry sector. This is not a good sign. The underlying reason is that Vietnam is in the industrialization phase, thus, it is essential that the industry sector should be prioritized and promoted both in terms of growth rate and the leading role in contributing to economic growth. In fact, the implemented investment and the growth rate of industry sector were completely unstable. Hence, the key position for promoting economic growth is the service sector. Given the economic context of a low-income country like Vietnam, the value generated from the service sector may come from the informal sector in the economy. This shows signs of uncertainty in the process of industrialization and the efficiency of investment in the industry sector is quiet low. Investment impact changes the proportion contributing to GDP of sectors: This is an inevitable consequence of investment. The more investment a sector absorbs, the more likely it is to contribute to GDP. Determining which investment allocated for the most is critical of the nation's development. But the experiences of many countries have shown that the inevitable path one country grow fast with high speed is to increase investment to create development in the industry and service sector. Therefore, in order to achieve expected goals, Vietnam cannot be outside the rule of development. In order to promote economic growth, Vietnam needs to maintain the orientation of investment restructuring towards increasing the proportion of investment in the industry sector (RIC_IC) with higher effectiveness, and service sector (RIC_SV).

Increasing labor productivity (LP) is one of important factors contributing to improve the competitiveness of an enterprise as well as an economy. The estimation results show that, if labor productivity increased by 1%, GDP would increase by 0.65% respectively. In particular, in the context of Vietnam, we should build an effective growth model with high labor productivity, quality and competitiveness. Vietnam also needs to mobilize, allocate and effectively use credit resources and market mechanism. In order to gradually shift from low to high growth model, Vietnam needs to exploit and maximize domestic resources and effectively use external resources.

The third group of indicators show the ability of provincial administrative management, including indicators to control corruption, improve administrative procedures and the ability to provide public services. Administrative management capacity is measured through a survey of local residents on the management activities of the provincial government. The estimation results show quite a lot of noticeable issues and need to be further researched on this impact. First of all, the provincial ability to control corruption is thought to have no effect on its growth. In addition, improving administrative procedures for socio–economic activities of the population seems to have an adverse effect on the province's GDP growth. Finally, the only provincial government improving the ability to provide public services will have a positive impact on that provincial GDP growth. Specifically, if the people's satisfaction with the provision of public services by the provincial government increases by 1 point, the GDP will tend to increase by 16%. The above results can be explained by a number of practical socio–economic issues as follows: Government control of local corruption may not be effectively implemented. With the anti-corruption goals set by the Government, the people may feel more satisfied, but in fact, the corruption control has not brought economic benefits to the locality when the money is not available to be reimbursed to the state budget. On the other hand, improving the administrative procedures for the population may not be the key point to create economic growth. In some provinces, when rapid industrialization occurs, residents of rural areas will be compensated for the loss of agricultural land, or residents of the vicinity of urban areas will sell their residential land to earn income. These activities sometimes lead to an increase in unemployment but not generating economic growth. The final index of public service delivery plays an active role in promoting economic growth. The underlying reason is that these public services have a direct impact on people' health and provide essential infrastructure for economic activities. As a result, the more actively the provincial government improves its ability to provide public services, the more economic growth it will encourage. The final factor studied in the model is an indicator of provincial competitiveness (PCI). The results showed that if PCI increased by 1 point, it will create incentives for GDP growth to increase by 3.2%, respectively. Each province's competitiveness is a sign that the province has created a favorable business environment and encouraged private businesses or not. A province with a large PCI index will be more attractive to attract capital investment, especially foreign direct investment. Moreover, with a transparent investment environment, favorable mechanisms for businesses will encourage the investment process to be more efficient, greatly contributing to the overall economic growth of the province. The attraction of investment is difficult but using it to bring high efficiency is an important issue. In order to promote economic growth in the coming period of time, in parallel with promoting labor productivity growth (LP), Vietnam needs to focus on increase the proportion of investment of industry and service sectors (RIC_IC and RIC_SV). Improving public services (Pa6) as well as the competitiveness of each province (PCI) to contribute to the rapid and sustainable economic growth of Vietnam.

5 Conclusion

The research results have shown the proportion of investment of the construction-industry and service sectors (RIC_IC and RIC_SV), labor productivity (LP), provincial delivery public service (Pa6) and provincial competitiveness (PCI) have a positive impact on Vietnam's GDP growth. Quantitative results show that the implementation of restructuring capital investment towards increasing the proportion of investment capital in the two sectors, industry-construction and service sector has brought positive impacts on the economy and economic growth model, such as strongly increasing GDP growth rate, increasing investment efficiency and contributing to economic model transformation. Therefore, to promote Vietnam's economic growth in the coming period of time, it is necessary to:

1. Focus on restructuring and modernizing economic sectors to improve productivity and added value of each sector in particular and the whole economy in general. We should continue to restructure the economy and focus on restructuring the production and service sectors. We also should restructure the economy towards increasing the proportion of processing industry, reducing the proportion of mining industry and improving the competitiveness of industry–construction sector. Besides, Vietnam needs to maintain solutions to boost labor productivity, internal capital productivity, increase technology content and proportion of domestic value in products. On the other hand, the Government should focus on investing in a number of foundational industries, having competitive advantages and strategic implications for fast and sustainable growth associated with environmental protection. In order to attach importance to investing in industrial production of components and component assemblies, Vietnam needs to promote a number of deep-participating and effective goods items into production networks and value chains and global distribution as well as create conditions for businesses to propose investment projects to restructure the economy.
2. We should restructure the service sector in the direction of improving service quality, focus on investing in facilities and developing a variety of products, especially products with competitive capacity. We also need to maintain the growth rate of service sectors higher than GDP growth rate by focusing on investment to develop a number of advantageous service industries. One important thing is that we should encourage enterprises to invest in developing tourism, attach importance to developing forest, sea and tourism.
3. For the last element, Vietnam need to improve labor productivity as well as capital productivity of the entire economy by implementing solutions to improve the quality of human resources, strengthen scientific and technological potentials. Specifically:

   (i) In addition to focusing on training high quality human resources, it is necessary to focus on improving the quality of vocational training, providing sufficient skilled labor sources to meet the requirements of economic development, adapting to the requirements of associations. Then, it is import to take advantage of the period of golden population structure and increase people's ability to grasp job opportunities. There are specific mechanisms and policies to select and use talents.

(ii) Strengthening scientific and technological potentials and building a national innovation system, promoting the creative capacities of all individuals, businesses and organizations. Research and promulgate breakthrough mechanisms to encourage scientific research and technical improvement, apply new science and technology to production to increase labor productivity and increase the value of products. Then, we should encourage and create favorable conditions for enterprises of all economic sectors to research and develop science and technology. Supporting import source technology, high technology and strict control of technology import and improving the effectiveness of operation of funds for scientific and technological development are also necessary. At last, Vietnam need to build a number of modern scientific and technological research and development institutes, develop innovative innovation centers and technological nurseries.

4. Strengthen administrative management capacity and especially the ability to provide public services of local governments.
5. Increasing provincial competitiveness to create a favorable business environment for investors, especially foreign investment.

References

1. Arellano, M.B.: Some tests of specification for panel data: Monte Carlo evidence and an application to employment equations. Rev. Econ. Stud. **58**, 277–297 (1991)
2. Aschauer, D.A.: Is public expenditure productive? J. Monet. Econ. **23**, 177–200 (1989)
3. Aubyn, M., Afonso, A.: Macroeconomic rates of return of public and private investment: crowding-in and crowding-out effects. European Central Bank Working Paper Series, 03/2008 (2008)
4. Beddies, C.: Investment, Capital Accumulation, and Growth Some Evidence from the Gambia 1964-98. IMF Working Paper No. 99/117 (1999)
5. Christiaensen, L., Demery, L., Kuhl, J.: The (Evolving) Role of Agriculture in Poverty Reduction. Working Paper No. 2010/36, UNU-WIDER (2010)
6. Coutinho, R., Gallo, G.: Do Public and Private Investment Stand in Each Other's Way. WDR Background Paper, World Bank, October 1991
7. Crowder, W., Himarios, D.: Balanced growth and public capital. an empirical analysis. Appl. Econ. **29**(8), 1045–1053 (1997)
8. Cullison, W.: Public Investment and Economic Growth. Federal Reserve Bank of Richmond Economic Quarterly **79**(4), 19–33 (1993)
9. Dollar, D., Kraay, A.: Growth is good for the poor? J. Econ. Growth **7**(3), 195–225 (2002)
10. Fabricant, S.: Employment in manufacturing 1899–1939. NBER, New York (1942)
11. Fonfria, A., Alvarez, I.: Structural change and performance in Spanish manufacturing, Some evidence on the structure bonus hypothesis and explanatory. Universidad Complutense de Madrid, Mimeo, Madrid (2005)
12. Ghali, K.H.: Public investment and private capital formation in a vector error correction model of growth. Appl. Econ. **30**(6), 837–844 (1998)

13. Gollin, D., Parente, S., Rogerson, R.: The food problem and the evolution of international income levels. J. Monet. Econ. **54**(4), 1230–1255 (2007)
14. Goodman, S.N.: Toward evidence-based medical statistics. 2: Bayes factor, Ann. Intern. Med. **130**(12), 1005–1013 (1999)
15. Gyfason, G.Z.: The road from Agriculture. CESifo Venice Summer InStitute Workshop on Institution and Growth, 24–25 July 2004 (2004)
16. Hansen, G., Prescott, E.: Malthus to Solow. Am. Econ. Rev. **92**(4), 1205–1217 (2002)
17. Huong, T.T.T., Huyen, T.H.: An investigation into the impacts of FDI, domestic investment capital, human resources and trained workers on economic growth in Vietnam. Studies in Computational Intelligence, vol. 760, pp. 940–951. Springer (2018)
18. Khan, M., Semlali, A.: Financial development and Economic growth: An Overview. IMF Working Paper, WP/00/209 (1999)
19. Khan, M.: Government investment and economic growth in the developing world. Pakistan Dev. Rev. **35**(4 Part I), 419–439 (1996)
20. Khan, M.S., Kumar, M.S.: Public and private investment and the growth process in developing countries. Oxford Bull. Econ. Stat. **59**(1), 69–88 (1997)
21. Lanjouw, J., Lanjouw, P.: The rural non-farm sector: issues and evidence from developing-countries. J. Inter. Assoc. Agric. Econ. **26**(1), 1–23 (2001)
22. Lewis, W.A.: Economic development with unlimited supplies of labor. Manchester School Econ. Soc. Study **22**, 139–191 (1954)
23. Ligon, S., Sadoulet, E.: Estimating the Effects of Aggregate Agricultural Growth on the Distribution of Expenditures. World Development Report 2008 (2008)
24. Mallick, S.K.: Determinants of long term growth in India: a keynesian approach. Progress Dev. Stud. **2**(4), 306–324 (2002)
25. Matsuyama, K.: Structural change in an interdependent world: a global view of manufacturing decline. J. Eur. Econ. Assoc. **7**(2–3), 478–486 (2009)
26. Munell, A.: Why has productivity growth declined? Productivity and public investment. New England Econ. Rev. 3–22, January/February 1990
27. Ngai, L., Pissarides, C.: Structural change in a multisector model of growth. Am. Econ. Rev. **97**(1), 429–443 (2007)
28. Nguyen, H.T.: How to test without P-value. Thail. Stat. **17**(2), i–x (2019)
29. Nurkse, R.: Problems of Capital Formulation in Underdeveloped Countries. Oxford University Press, New York (1961)
30. Odedokun, M.O.: Relative effects of public versus private investment spending on economic efficiency and growth in developing countries. Appl. Econ. **29**(10), 1325–1336 (1997)
31. Oshima, H.T.: Economic Growth in Monsoon Asia: A Comparative Survey. University of Tokyo Press, Tokyo (1987)
32. Pereira, A.: Public capital formation and private investment: what crowds in what? Public Finance Rev. **29**(1), 3–25 (2001a)
33. Rabnawaz, A., Jafar, R.S.: Impact of public investment on economic growth. South Asia J. Multidisc. Stud. (SAJMS) **1**(8), 62–7 (2015)
34. Ramirez, M.D., Nazmi, N.: Public investment and economic growth in Latin America: an empirical test. Rev. Dev. Econ. **7**(1), 115–126 (2003)
35. Rodrick, D.: The Future of Economic Convergence. NBER Working Papers 17400, National Bureau of Economic Research, Inc (2011)
36. Rodrick, D.: An African Growth Miracle?. NBER Working Papers 20188, National Bureau of Economic Research, Inc (2014)
37. Rostenstein, P.N.: Problems of Industrialisation of Estern and South-Eastern Europe. Econ. J. **53**(210/211), 202–211 (1943)
38. Rostow, W.W.: The stages of Growth: A Non - Communist Manifesto. Cambridge Press, Cambridge (1960)
39. Sturm, J., Kuper, G.H., De Haan, J.: Modelling Government Investment and Economic Growth on a Macro Level: A Review. University of Groningen, CCSO Series No. 29, Department of Economics (1996)

40. Syrquin, M.: Pattenrns of structural change. In: Chenery, H., Srinivasan, T.N. (eds.) Handbook of Development Economics, vol. 1, pp. 205–248. North Holland, Amsterdam (1988)
41. Tran, T.T.H.: Statistical study of Vietnam's Economic Structure. Labor Publishing House, Vietnam (2017)
42. Tran, T.T.H.: Mot so van de ve co cau va chuyen dich co cau von dau tu theo nganh kinh te cua Viet Nam giai doan 2005–2013. Tap chi Con so va Su kien, No.5/2015, 39–41 (2015)
43. van Ark, B., Timmer, M.: Asia's Productivity Performance and Potential: The Contribution of Sectors and Structural Change (2003). http://ggdc.nl/databases/10sector/2007/papers/asiapaper4.pdf
44. van Ark, B.: Sectoral Growth Accounting and Structural Change in Postwar Europe. Research Memorandum GD-23, GGDC, Groningen (1995)
45. Wooldrige, J.M.: Econometric Analysis of Cross Section and Panel Data. The MIT Press, Cambridge (2002)
46. Tran, T.T.H.: Nghien cuu thong ke co cau kinh te Vietnam. Labor Publishing House, Vietnam (2017)

Anomaly Detection for Online Visiting Traffic as a Real-Estate Indicator: The Case of HomeBuyer

Arcchaporn Choukuljaratsiri, Nat Lertwongkhanakool, Pipop Thienprapasith, Naruemon Pratanwanich, and Ekapol Chuangsuwanich

Abstract Real-estate development involves a large amount of cash flow, yet the overall process takes a long period of time to complete, leading to a high risk of on-going demand change, competitor, and society. Since many buyers these days preview properties from online sources, the total user amount viewing each project and the market segment can imply a current trend for purchasing demand. Instead of monitoring a user logging file manually every day, we develop an auto-alarm system to detect anomaly events. In particular, we apply a seasonal auto-regressive integrated moving average (SARIMA) model to the number of user views varying in a seasonal manner from week to week. We then use Bollinger Bands, a widely used statistical indicator, to draw a boundary for any incident sporadically deviating from the expected. This system can alarm the real-estate developers whether their target customer interest is still staying on their properties or moving towards different areas or competitors so that they could abruptly adjust their strategy in time.

Keywords Anomaly detection · Online user log · Market metric · Time-series data analysis

---

A. Choukuljaratsiri (✉) · N. Lertwongkhanakool · P. Thienprapasith
Home Dot Tech Co., Ltd., Bangkok, Thailand
e-mail: arcchaporn@homedottech.com

N. Lertwongkhanakool
e-mail: nat@homedottech.com

P. Thienprapasith
e-mail: pipop@homedottech.com

N. Pratanwanich
Department of Mathematics and Computer Science, Faculty of Science, Chulalongkorn University, Bangkok, Thailand
e-mail: naruemon.p@chula.ac.th

E. Chuangsuwanich
Department of Computer Engineering, Faculty of Engineering, Chulalongkorn University, Bangkok, Thailand
e-mail: ekapol.c@chula.ac.th

S. Sriboonchitta et al. (eds.), *Behavioral Predictive Modeling in Economics*, Studies in Computational Intelligence 897,
https://doi.org/10.1007/978-3-030-49728-6_19

1 Introduction

"Asking the market what is happening is always a better approach than telling the market what to do", said Bollinger [5]

Internet-based data provide meaningful insights into market conditions of the economy. Nuarpear Warn, et al. (2016) studied Thai labor market behavior at the macro level from online job-boards and resume postings [19]. Apart from reflecting government policy effectiveness, the number of visits could be a useful non-financial metric for real-estate developers to evaluate their marketing strategy. An abrupt change in the total number of online visitors on each project site could imply customer movement such as increasing brand recognition [18, 27].

Developed by Home Buyer Group, "home.co.th" is one of the top Thai real-estate websites serving 10 million users per year. An example of the real-estate project detail and mortgage information pages are depicted in Fig. 1. User behaviors within the site including clicks, time spent for each web page, access time, to name but a few have been logged in the database. The number of users visiting on property pages are frequently monitored through a visualization tool upon request in the time frame of interest. When an unusual event in the number of users is spotted, further exploration is to be carried out for any related explanation regarding the market insights.

From Fig. 2, for instance, we noticed the declining number of visitors to our website after a new loan to value (LTV) regulation which restricts the amount of money borrowed the buyers for their second mortgage was announced by the Thai government in October 2018 [26]. LTVs play a pivotal role as a higher ratio implicitly suggests an increase in the repayment burden of a mortgage and a decrease in the initial liquidity constraint on potential home buyers [14]. The effect of the newly launched LTV regulation was moreover emphasized when we observed the number of visitors looking for mortgage information has drastically increased since then (see Fig. 2). This could demonstrate the greater concern over the loan approval.

Since it is laborious to monitor the number of people visiting online every day visually, we, Home Dot Tech—a subsidiary firm of Home Buyer, has developed the automatic alarm algorithm using anomaly detection, the process of finding patterns in data that do not conform to the expected behavior [22].

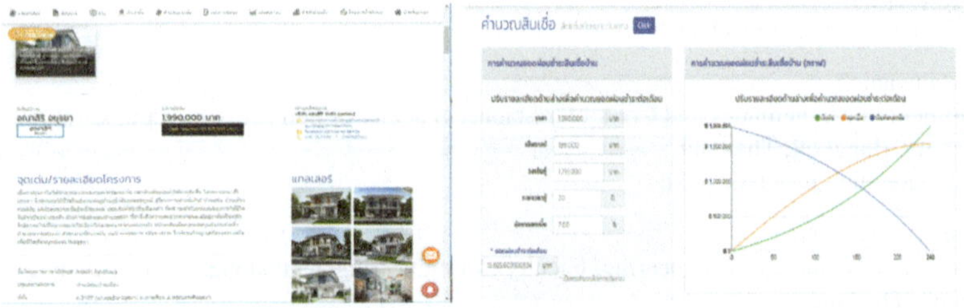

Fig. 1 Project detail page (left) and mortgage calculation page (right) at http://home.co.th

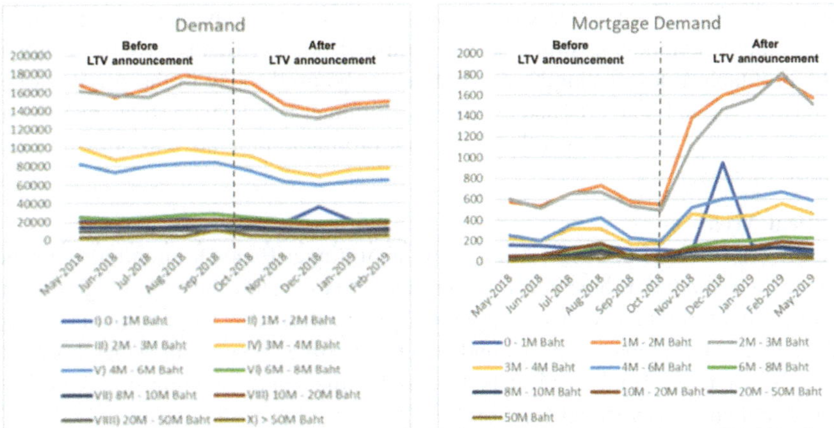

Fig. 2 (left) Total number of user views on the Bangkok property information pages categorized by price levels. (right) Total number of user views on the mortgage calculation page for Bangkok properties categorized by price levels (source: Home Dot Tech)

2 Data

We collected the data from the user log file which has been tracking the historical visits at our site (http://home.co.th). The data set used in this paper was the number of unique users who had visited the mortgage information page daily. As a pilot study, we focused on those users who had viewed the property projects located nearby any Skytrain or subway station in Bangkok and Metropolitan region from June 1st, 2018 to November 29th, 2018.

3 Related Works

Anomaly Detection
Anomaly detection refers to the process of finding patterns in data that do not conform to expected behavior. These non-conforming patterns are often referred to as anomalies, outliers, discordant, observations, etc. Anomaly detections find extensive use in a wide variety of applications such as networking monitoring, healthcare, smart devices, smart cities, IOT, fraud detection, cloud. In network security, anomaly detection is regarded as one of the powerful mechanisms to identify and detect any possible threat [15].

Bollinger Bands, one of the famous anomaly detection technique, were introduced by market investor John Bollinger in 1992 as a way to trigger buy and sell signals on shares by comparing current prices to the medium-term moving average. For a daily time series, Bollinger Bands are traditionally defined as two standard deviations either side of a 20-day moving average (i.e. a "window size" of 20 days and a "bandwidth" of 2 standard deviations). Although very simple, Bollinger Bands have been shown to be profitable rules of thumb and continue to be used today in the stock market [20]. However, in Lento [16] had tested BB (Bollinger Bands) with stock investment information, and it had shown that BB is not profitable in relative to buy-and-hold trading strategy.

Medical Area

In the work of Pagel in 2015 [20], they used Bollinger Bands to monitor real-time patients' amount in UK hospitals in order to identify the start and end of high patient load period during the winter season which is variant by year. This attempt is to help the hospital better prepare capacity and planning in time. They claimed that this work is the first time that Bollinger Bands was used in the health care industry, regarding of its simplicity yet practical solution.

Business Area

Many application areas face challenges from big data properties, including, network monitoring and detection, geo-spatial data, vehicular traffic, market prediction, and business forecasting. Currently, the vast majority of business analytic efforts are spent on descriptive-analytic and predictive analytic with typical methodologies including data mining, machine learning, artificial intelligence, and simulation.

Alshura et al. [2] has done literature reviews of publications related to big data in marketing since 2010–2018 and found that big data utilization toward the marketing domain is still in the early stage. Somehow, big data can benefit marketers in understanding the changing customers' tastes and preferences, creating a new strategy, and know more about customer attitude. Furthermore, it can enhance CRM (Customer Relationship Management) system, such that what is appropriate product and service, or how to advertise in the market. However, they concluded that data itself could not lead to better marketing, but it is insight derived from, suitable decision, and action, that makes all the difference. They also found that reliance too much on big data often yield to promotion on short-term sales, which erode long-term marketing.

According to Booth and Jansen [6], one of the main corporate website and relational marketing strategies (RMS) is content and media provision, aiming to draw more visitors and immersing them within the site. In Schäfer's work [24], they developed web mining-based methodology to utilize clickstream data as metrics to evaluate the relational marketing strategies. The visitor activities were analyzed in a different aspect and weighted them by a business expert according to their relevance for each strategies. For example, commerce RMS focus on buy/download transaction as an evaluation metric. On the other hand, content/media RMS focuses on long session duration, while support/service RMS focuses on finding (low page depth).

Shevales et al. [25] proposed a portal of business for a web application, using 4 models which are admin, user, 3rd party, and supplier, where every business can do marketing and manage data mining. The concept of data mining consists of anomaly detection, associations, classification, regression, rule learning, summarization, and clustering. They concluded that marketing is self-motivated and bendable, change with the change in needs and preferences of the customers and market forces, like opposition government policies and marketing circumstances.

Financial Area

Branch [11] had demonstrated the use of a pivot table as a simple technique to test the stock market anomaly by using the average changing percentage of S&P500 market price per month. Their result shows that the January effect, a financially conventional description of the US stock market anomaly occurred in January generally effected by monetary policy, which has disappeared since the 21st century.

A recent important discovery is that search engine traffic (i.e., the number of requests submitted by users to search engines on the www) can be used to anticipate the dynamics of social phenomena. Successful examples include unemployment levels, cars, and home sales. Recent works applied this approach to stock prices and market sentiment. Bordino's study [7] showed that daily trading volumes of stocks traded in NASDAQ-100 are correlated with daily volumes of queries related to the same stock.

Real-Estate Area

The study by Bork [8] explores the historical record of US house sentiment, which had been surveyed from public opinion about house buying conditions since 1975. The question is in such a way that if the current time is a good time to buy housing for investment or not. This work concluded that the house sentiment index derived from survey results using PLS (partial least square) could yield a better result to national price prediction, by using DMA (dynamic model averaging), than the conventional factors such as mortgage loan outstanding.

Alexander [1] has examined the relationship between US commercial real-estate search volume and market monthly transaction. They found a strong correlation between search volume indices (SVI) from google trend and transaction amount, with a variant time-lag period. Such models based on google trend alone outperform the baseline models which apply macro-economics factors in all cases.

Similarly, Ralf [21] study the sentiment trends of online search query data with the UK housing market transaction volume, and confirm it as a suitable indicator.

Public Area

As for the governance domain, Coglianese [13] has proposed a systematical framework to evaluate the performance of regulations and regulatory policies. The schema of a model, include behavioral change and intermediate outcome steps, etc. One of the evaluation indicators is "Impact/Effectiveness", that is, how much would each regulatory option change the targeted behavior? Three indicators that could be used as measurement are activities, behaviors, and outcomes. Using measures of behavior may in some cases facilitate stronger inferences about a regulation's immediate effects. Evaluations can be drawn on proxies, or measures that correlate with the ideal but not causally linked to them. The example includes hospital admissions by patients with relevant symptoms as a proxy for negative health effects.

Agricultural Area
In an attempt to solve the inequality in India from the fluctuated agricultural price, Madaan et al. [17] have proposed the model to detect anomaly of potato and onion price, to identify incidents of stock hoarding by the traders. The model is based on the food price forecast, which was trained from historical time-series price data by ARIMA (Auto-Regressive Integrated Moving Average) method. Subsequently, they applied obtained news articles about hoarding related incidents together with machine learning to treat each hoarding period as a positive set.

4 Methodology

We have developed an anomaly detection system to monitor and make an alarm automatically when the number of people browsing the site was higher or lower than expected.

SARIMA. A seasonal autoregressive integrated moving average model. The forecast of SARIMA is based on an autoregressive process (AR) which is factored by its own past values up to p orders and a moving average process (MA) which is informed by previously predicted errors. On top of that, it also accounts for seasonal patterns which repeat every s-period of time up to P, Q orders for both AR and MA processes. When trend is present, the SARIMA model can corrects by differencing the values in the order of d and D in both non-seasonal and seasonal contexts respectively.

The model is controlled by seven parameters $\{p, d, q, P, D, Q, s\}$ where its short-hand notation is written as SARIMA $(p, d, q)(P, D, Q)s$ and its formal form is give by Eq. (1) [3, 12, 23],

$$\Phi_P(B^s)\phi(B)\Delta_s^D \Delta^d y_t = \Theta_Q(B^s)\Theta(B)\epsilon_t, \tag{1}$$

where

$\phi(B) = 1 - \phi_1 B - \phi_2 B - \ldots - \phi_p B^p$; Non-seasonal AR

$\Phi_P(B^s) = 1 - \Phi_1 B - \Phi_2 B - \ldots - \Phi_P B^{Ps}$; Seasonal AR

$\theta(B) = 1 + \theta_1 B + \Theta_2 B + \ldots + \theta_q B^q$; Non-seasonal MA

$\Theta_Q(B^s) = 1 + \Theta_1 B + \Theta_2 B + \ldots + \Theta_Q B^{Qs}$; Seasonal MA

$\Delta^d = (1 - B)^d$; Non-seasonal differencing

$\Delta_s^D = (1 - B^s)^D$; Seasonal differencing

$B^n y_t = x_{t-n}$; Backshift operator.

In this paper, we applied SARIMA to predict the total number of visitors by taking the weekly visiting patterns into account. To estimate the expected number of visits daily, we trained the model on the historical observations up to the day of being estimated, using the repeated changing period (seasonal order) set to 7 days. Since we assumed that the total viewers followed a linear trend, we set the AR and differencing orders (p, d, P, D) to 1 and the moving average orders (q, Q) to 0.

On each day, if the difference between the estimated and the actual total visitors, known as a prediction error, is high, it means that the current number of prospective customers is extraordinarily far from expectation. This could potentially trigger the marketing team or any stakeholders to pay more attention to what is happening in the market so that they can take action in time. However, market segments that differ in locations, property types, and prices usually have different volatility in the number of customers. Thus, the prediction error threshold should not be fixed overtime or to all market segments. Instead, for each market segment, we consider the weekly visiting behaviors to determine a suitable threshold for a particular period of time. This can be achieved by setting up Bollinger Bands, the technique we applied after fitting the SARIMA model.

Bollinger Bands

By definition, Bollinger Bands consists of an upper and lower bound enveloping a moving average (MA) within a given period. A moving average is a series of averages of historical data points within a given window that is rolled over time in a data set.

$$UpperBound = MA + K * SD$$

$$LowerBound = MA - K * SD$$

The upper bound is at K times of the standard deviation above the moving average and the lower bound is at the K times and N-period standard deviation below the moving average. The efficiency depends on its parameters namely the window size to calculate moving average and standard deviation (SD), and the multiple magnitudes of the standard deviation (K) [4].

To conform with the weekly seasonal order of SARIMA model, the moving average of the prediction errors was calculated from the previous 7 days. We put the upper and lower bounds computed by 1.75 times of the standard deviation above and below the 7-day moving average respectively. The prediction errors within this band reflect the fluctuation of total users from expected in a weekly period. In other words, we employed this band as a boundary limit for a normal situation. Once an error jumps outside of this band, the system will give an alarm automatically.

5 Results

To validate our system, we split the last one-third of the data set covering the time that the LTV regulation was launched as a test set. Of all 70 days, 8 days were marked as true alarms due to abrupt changes in slope, compared to the previous data pattern.

It is noted that the alarms were marked after the first 7 days in the test set since the calculation of Bollinger Bands were based on the 7-day moving average. For each day in the test set, we estimated the number of visits using the SARIMA model which had been trained by the past observations up to the day of prediction. Figure 3 demonstrates the expected number of visits of those 70 days from Sep, 21st to Nov 29th, 2018. The results emphasize that the visiting behavior was repeated every 7 days, forming the weekly pattern. To justify whether a prediction error has deviated from the expected, we applied three different bandwidths of Bollinger Bands by varying the multiples (K) of the standard deviation to be 1.5, 1.75, and 2. Figure 4 draws the results of the 7-day moving average of the prediction errors and their upper and lower bounds when using K equal to 1.75. The points hitting or outside of the bands were considered alarms. With 1.75SD bands, only one out of five alarms were falsely triggered.

We mark an alarm made by the system to be a true positive when it was within 5 days after the supposed-to-alarm day, otherwise a false positive. The performance of each bandwidth was evaluated using precision and recall as computed in Equation below.

$$Precision = \frac{TP}{TP + FP};$$

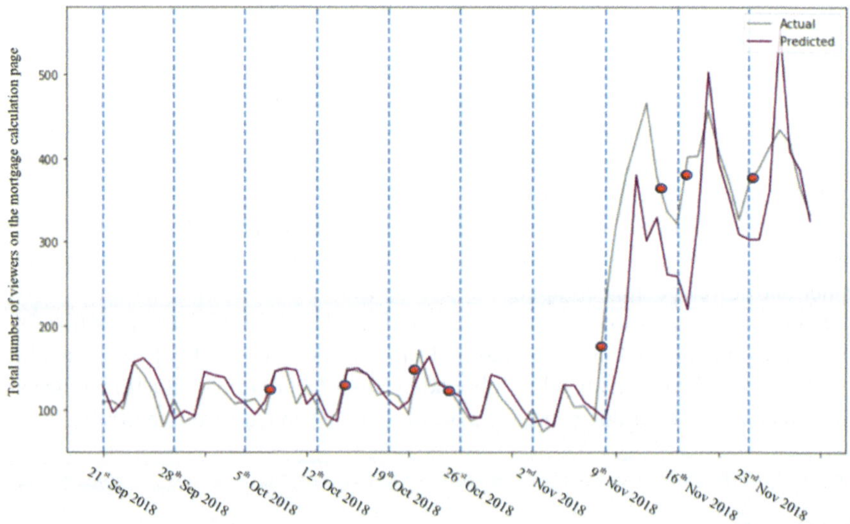

Fig. 3 Actual and predicted number of total mortgage viewers (Source: Home Dot Tech). The gray and purple lines represent the actual and the predicted number of viewers on the mortgage calculation page, respectively. The blue dashed vertical lines divide the time-series data into 7-day time slots of the repeated weekly pattern. The red dots are the days supposed to be abnormal

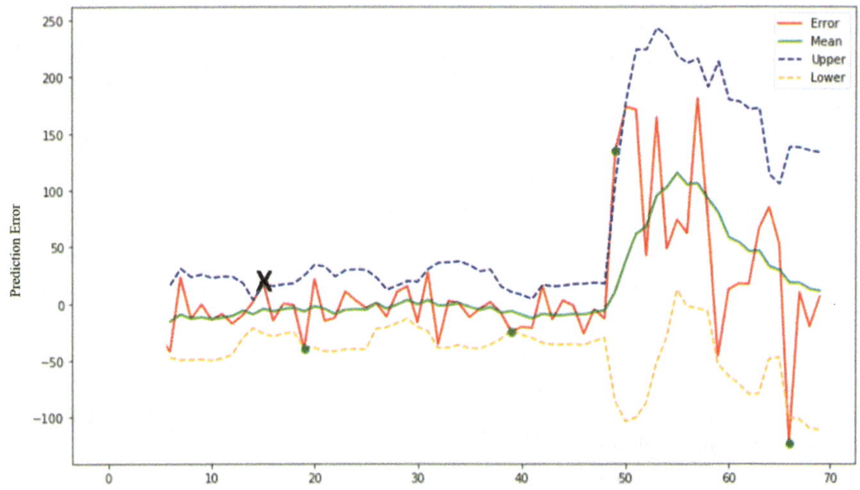

Fig. 4 Envelope of Bollinger Bands on the prediction errors from the SARIMA model. The green line is the 7-day moving average on the prediction errors drawn with the red line. The upper and lower bounds were calculated using 1.75 SD deviated from the moving average. The green dots and black cross represent true and false alarms

Table 1 Experiment results of different Bollinger Bands. *(MA: Moving Average, SD: Standard Deviation)*

Band size	Precision	Recall
$MA \pm 1.5SD$	0.50	0.63
$MA \pm 1.75SD$	0.80	0.50
$MA \pm 2SD$	1.00	0.33

$$Recall = \frac{TP}{P};$$

where TP, FP, and P denote the number of true positives, false positives, and all positive labels, respectively. The result is shown in Table 1. We can see that the larger the K value, the more precise the model. However, with the larger K, fewer alarms were reported. In practice, we can use all three levels together to represent the severity of the abnormality and let the users make judgments and interpretations themselves.

6 Case Study of Market Metrics on home.co.th

As another example, we employed our system to scan all 3,780 segments, categorised by 62 provinces, 6 property types, and 10 price levels, for anomaly detection on 12th May 2019. An alarm was classified as a low, medium, and high degree of abnormality if it was deviated from 1.5SD (low), 1.75SD (medium) and 2SD (high) respectively. In addition, we also reported the alarm type as increasing if the error was identified

Table 2 Anomaly detection on 12th May 2019

Region	HouseType	PriceLevel	Anomaly level	Direction	Suspicious Project ID	Viewer Change/Day
Samut Sakhon	Datached_House	3M-4M Baht	High (2.0SD)	Increasing	**11004**	8.00
					5932	5.5
					8353	5.0
Chon Buri	Shophouse	2M-3M Baht	High (2.0SD)	Increasing	7045	4.5
					7380	2.0
					8982	2.0

above the upper bound or decreasing if it was found below the lower bound. As a result, 26 segments were reported as anomalies in which 12, 8, and 6 of them had low, medium, and high levels of abnormality respectively.

The top two segments with the highest level of anomalies are reported in Table 2. We further examined whether any particular projects were the main influence contributing to the severe deviation from the expected. To do this, we computed the last three days' total user change by each project, and select the top three in each segment. The outstanding increase volume is on Project ID 11004, which we found that a promotional event was going to be arranged in the following two weeks, explaining one of the reasons why more buyer attention being drawn on this project higher, pulling the total number of visits in this region upwards sharply in a short time.

7 Conclusion

We have developed an alarming system that automatically provides an alarm if the number of online visits deviated from expected. In particular, we trained the SARIMA model to estimate the expected number of visits and applied the three levels of Bollinger Bands on the prediction errors to classify the anomalies with low, medium, and high severity. The larger the band, the higher the degree of abnormality. Our results show that major anomalies were able to be detected and explained by an approaching policy or market campaign. This metric can be used as an evaluation tool of policy effects on prospective national demand. On the other hand, such anomaly severity could also be an effective market indicator to evaluate if the launching marketing strategy has an impact on the attention of buyers. However, a few false alarms were reported. Tuning the Bollinger Band size is required to filter out uninteresting alarms, depending on the data volatility. In the future, alternative methods such as artificial neural networks can be studied to improve the estimation of the expected. Moreover, when more anomalies are labeled and available, the band sizes may be automatically adjusted.

Acknowledgements This work was supported by The Joint Research between Home Dot Tech Co., Ltd. and Faculty of Engineering, Chulalongkorn University.

References

1. Alexander Dietzel, M., Braun, N., Schäfers, W.: Sentiment-based commercial real estate forecasting with Google search volume data. J. Prop. Invest. Financ. **32**(6), 540–569 (2014)
2. Alshura, M.S., Zabadi, A., Abughazaleh, M.: Big data in marketing arena. Big opportunity, big challenge, and research trends: an integrated view. Manag. Econ. Rev. **3**(1), 75–84 (2018)
3. Apichit, H., Krerk, P.: SARIMA based network bandwidth anomaly detection. In: The Ninth International Joint Conference on Computer Science and Software Engineering (JCSSE) (2012)
4. Audrius, K., Ugnius, M.: Calibration of Bollinger Bands parameters for trading strategy development in the Baltic stock market. Inzinerine Ekonomika-Eng. Econ. (2010)
5. Bollinger, J.: Using Bollinger Bands. Stocks Commod. **10**(2), 47–51 (1992)
6. Booth, D., Jansen, B.J.: A review of methodologies for analyzing websites. In: Web Technologies: Concepts, Methodologies, Tools, and Applications, pp.145–166. IGI Global (2010)
7. Bordino, I., Battiston, S., Caldarelli, G., Cristelli, M., Ukkonen, A., Weber, I.: Web search queries can predict stock market volumes. PloS One **7**(7), e40014 (2012)
8. Bork, L., Møller, S.V., Pedersen, T.Q.: A new index of housing sentiment. Manag. Sci. **66**, 1563–1583 (2019)
9. Box, G.E., Jenkins, G.M., Reinsel, G.C., Ljung, G.M.: Time Series Analysis: Forecasting and Control. Wiley, Hoboken (2015)
10. Box-Steffensmeier, J.M., Freeman, J.R., Hitt, M.P., Pevehouse, J.C.: Time Series Analysis for the Social Sciences, p. 195. Cambridge University Press, Cambridge (2014)
11. Branch, B., Orland, W., Ma, A.: Using pivot table to test market anomaly. Int. J. Bus. Account. Finance **13**(1), 14–25 (2019)
12. Chang, X., Gao, M., Wang, Y., Hou, X.: Seasonal autoregressive integrated moving average model for precipitation time series. J. Math. Stat. **8**, 500–505 (2012)
13. Coglianese, C.: Measuring regulatory performance evaluating the impact of regulation and regulatory policy (2012)
14. Danilo, L., Valerio, V.: With (more than) a little help from my bank. Loan-to-value ratios and access to mortgages in Ital. Questioni di economia e finanza (occasional papers) (2016)
15. Habeeb, R.A.A., Nasaruddin, F., Gani, A., Hashem, I.A.T., Ahmed, E., Imran, M.: Real-time big data processing for anomaly detection: a survey. Int. J. Inf. Manag. (2018)
16. Lento, C., Gradojevic, N., Wright, C.S.: Investment information content in Bollinger Bands? Appl. Financ. Econ. Lett. **3**(4), 263–267 (2007)
17. Madaan, L., Sharma, A., Khandelwal, P., Goel, S., Singla, P., Seth, A.: Price forecasting & anomaly detection for agricultural commodities in India (2019)
18. Marek, S., Lucie, S., Frantisek, M.: Marketing effectiveness by way of metrics. Econ. Manag. **16**, 1323–1328 (2011)
19. Nuarpear, W.L., Voraprapa, N., Paphatsorn, S.: Labour market insights: The power of internet-based data (2016)
20. Pagel, C., Ramnarayan, P., Ray, S., Peters, M.J.: A novel method to identify the start and end of the winter surge in demand for pediatric intensive care in real time. Pediatr. Crit. Care Med. **16**(9), 821–827 (2015)
21. Ralf, H., Manuel, K.: GECO's weather forecast for the U.K. housing market: to what extent can we rely on Google econometrics? J. Real Estate Res. **36**(2), 253–281 (2014)
22. Riyaz, A.A.H., Fariza, N., Abdullah, G., Ibrahim, A.T.H., Ejaz, A., Imran, M.: Real-time big data processing for anomaly detection: a survey. Int. J. Inf. Manag. **45**, 289–307 (2018)
23. Saz, G.: The efficacy of SARIMA models for forecasting Inflation rates in developing countries: the case for Turkey. Int. Res. J. Financ. Econ. **62**, 111–142 (2011)
24. Schäfer, K., Günther, O.: Evaluating the effectiveness of relational marketing strategies in corporate website performance. In: WSBI, pp. 35–49 (2011)
25. Shevale Bhakti, R., Shaikh Ruksar, N., Pawar Bhagyshri, S., Jadhav, D.K.: Web based business marketing by using data mining concept (2018)
26. Somruedi, B., Kanana, K.: 30% down payment for 3rd mortgage (2018). https://www.bangkokpost.com/business/1573326/30-down-payment-for-3rd-mortgage
27. Tim, A.: Marketing metrics. Bus. Strat. Rev. (2000)

A Bayesian Analysis of the Determinants of China's Overseas Contracted Projects in Countries Along the Belt and Road Initiative

Mengjiao Wang, Jianxu Liu, and Songsak Sriboonchitta

Abstract The Belt and Road Initiative (BRI), also known as the New Silk Road, was adopted by the Chinese government in 2013 to enhance regional connectivity and cooperation on a transcontinental scale connecting China and East Asia to Europe and Africa. This paper investigates the main determinants of China's Overseas Contracted Projects (OCPs) in 45 Countries along the two principal axes of BRI—the Silk Road Economic Belt (the Belt) and the 21st Century Maritime Silk Road (the Road) by Bayesian linear regression models. We further divide the whole sample into the higher-income and lower-income subgroups, and the before-BRI and after-BRI subgroups, to compare the different determinants of China's OCPs in countries of different income levels and during different periods. The Findings indicate that: (1) In general, China's OCPs have resource-driven and labour-driven motivations while the backward infrastructure and the political instability of host countries are the hindrance of their development; (2) The abundant resources of higher-income countries are the important element to attract China's OCPs, while the main determinants of China's OCPs in lower-income countries are their development potential and labour resources; (3) The positive correlations of China's OCPs with local economic growth rate and infrastructure development turn to be significant after the BRI.

1 Introduction

China's Belt and Road Initiative (BRI), also known as One Belt and One Road (OBOR) Initiative, was launched in 2013 by China's President Xi Jinping to promote the construction of the Silk Road Economic Belt and the 21st-Century Maritime Silk

M. Wang · S. Sriboonchitta
Faculty of Economics, Chiang Mai University, Chiang Mai, Thailand
e-mail: mengjiaow1991@163.com

S. Sriboonchitta
e-mail: songsakecon@gmail.com

J. Liu (✉)
Faculty of Economics, Shandong University of Finance and Economics, Jinan 250000, China
e-mail: liujianxu1984@163.com

S. Sriboonchitta et al. (eds.), *Behavioral Predictive Modeling in Economics*, Studies in Computational Intelligence 897,
https://doi.org/10.1007/978-3-030-49728-6_20

Road with primary targets of enhancing policy co-ordination, strengthening facilities connectivity, facilitating unimpeded trade, deepening financial integration, and building people-to-people bonds among participating countries [12]. By the end of 2015, 65 countries[1] have been identified as participating in the BRI [3]. The Initiative has been achieving remarkable results in international economic cooperation and infrastructure development.

Infrastructure development is a crucial element of the BRI and has played a fundamental role in fostering regional cooperation and development, especially at the early stage of the Initiative [8]. However, most of the countries along the Belt and Road are developing countries with inadequate investment in infrastructure. The backward infrastructure of these countries is a crucial challenge for China to encounter in the OBOR Initiative [13].

China's Overseas Contracted Projects (OCPs) refer to the overseas projects undertaken by Chinese contractors (project contracting companies) through the bidding process. As a basic form of investment cooperation, OCPs have become a major pathway of China to help BRI countries to establish efficient infrastructure and transportation systems. According to the statistics from the Ministry of Commerce of China, the total turnover (revenues) of China's OCP reached 1138.29 billion yuan in 2017 with 7.5% annual growth rate and the value of the newly signed contracts was 1791.12 billion yuan with 10.7% annual growth rate. The improvement of infrastructure will encourage more investment and economic cooperation between China and BRI partner countries because it is also aligned with the development goals of the host countries, thus producing a win-win result that benefits all countries involved [1].

However, Overseas Contracted Project (OCP) is usually considered a "high-risk business" largely due to the lack of adequate overseas environmental information [15]. The complexities of investment climate and market characteristics of BRI partner countries are the challenges for China to guarantee the smooth cooperation process of BRI. Identifying the main determinants of China's contracted projects in BRI partner countries has great significance. First and foremost, a proper assessment of the challenges in destination countries is crucial to help China's contractors to avoid environmental risks and guarantee the sustainable development of overseas projects; Meanwhile, it helps to verify the underlying relationship between China's OCPs and the development indicators of host countries, so that the potential contributions of the Initiative to its partner countries in terms of economic growth and infrastructure improvement could be identified.

Foreign direct investment (FDI) and OCP are two major forms of investment cooperation between China and countries along the Belt and Road. Many studies focus on the analysis of the determinants, influencing factors, or risk management of FDI, but much less attention is paid to OCP. Zhi [15] indicates that there are two sources influencing contracted projects overseas. One is the environmental impacts which are also called external influences and the other is internal impacts which consist of the uncertainties in the project itself. This paper mainly focuses on the first

[1]The Belt and Road Initiative is an open and inclusive concept, the Government of China has never provided an official list of approved BRI participants.

source to examine the correlations between China's OCPs and the environmental factors and market characteristics of host countries.

In terms of the external determinants of contracted projects overseas, Lu et al. [9] mention that materials, labor, and machinery may be the major direct costs of China's contracted projects in BRI partner countries while the long distance and transportation costs are also challenges for China's contracted projects. Chang et al. [2] identify the influences of six macro factors on international construction projects, including economic performance, political stability, international environment, legal and regulatory frameworks, social safety and attitude towards foreigners. Particularly, the political risk of host countries is an issue of crucial importance for Chinese construction enterprises. Niagara and Datche [11] indicate that the economic, social, political, and physical environment, and the level of technological advance of host countries are all attributes used to measure external factors.

This paper uses the data of China's contracted projects in 45 BRI partner countries[2] to investigate the determinants of China's OCPs by Bayesian linear regression models. The contributions of this paper to related literature are threefold. Firstly, this paper compensates for the lack of studies on the influencing factors of China's OCPs in BRI partner countries. Secondly, it further compares the different influencing factors of China's OCPs in countries of different income levels, and the differences over different periods. It contributes to providing more target suggestions for China's contractors in different areas and helps to find out the changing characteristics after the Initiative was launched. Thirdly, the Bayesian linear regression model helps to avoid the problem with P value [6], and provides the results from a different perspective compared with the often-used frequentist statistical methods.

The remainder of this paper is organized as follows. Section 2 presents the descriptive statistics of the data. The research method is provided in Sect. 3. Section 4 shows empirical results and Sect. 5 concludes.

2 Data Description

This paper uses the annual turnover of China's contracted projects in BRI countries as the explained variable. The annual turnover in this study refers to the business revenue from contracted projects during the reference period. Meanwhile, the value of newly signed contracts of China's contracted projects in BRI countries is also used as the explained variable to conduct the robustness test. The value of newly signed contracts implies the total amount of contracts signed by Chinese contracting enterprises before the end of each reference period.

[2]The 45 BRI countries selected in this paper include Mongolia, Russia, Poland, Czech Republic, Hungary, Romania, Bulgaria, Ukraine, Belarus, Turkey, Iran, Iraq, United Arab Emirates, Saudi Arabia, Qatar, Kuwait, Oman, Yemen, Jordan, Israel, Georgia, Azerbaijan, Egypt, Kazakhstan, Kyrgyz Republic, Tajikistan, Uzbekistan, Turkmenistan, Vietnam, Lao PDR, Cambodia, Thailand, Malaysia, Singapore, Indonesia, Brunei, the Philippines, Myanmar, Timor-Leste, India, Pakistan, Bangladesh, Afghanistan, Nepal, Sri Lanka.

Table 1 China's overseas constructed projects in BRI partner countries: annual turnover and value of newly signed contracts (USD 10,000 dollars) from 2005 to 2017

Groups	Annual turnover		Value of newly signed contracts	
	Obs	Mean	Obs	Mean
Total	585	104236.8	480	144650.2
Higher-income group	286	95965.8	236	123607.5
Lower-income group	299	112148.2	244	165003.0
Before-BRI group	360	70486.0	300	108713.8
After-BRI group	225	158238.1	180	204544.3

Table 1 presents the annual turnover and the value of newly signed contracts of China's OCPs from 2005 to 2017 in the 45 countries along the Belt and Road. We divide the 45 countries into two groups, the higher-income group and the lower-income group, according to the country classifications by income level (2008) from World Bank.[3] In addition, the whole sample is roughly divided into two periods: the before-BRI period (2005–2012) and the after-BRI period (2013–2017).

Based on related literature (see details in Corkin [4], Duan et al. [5] and Luo et al. [10]), 11 explanatory variables are selected to investigate the determinants of China's OCPs in countries along the Belt and Road. These variables can be broadly divided into two categories: First is the environmental factors of host countries, including the level of economic development ($lngdppc$), economic growth ($gdpgr$), trade openness ($open$), natural resource availability ($resources$), labour force ($labour$), infrastructural level ($tele$ and $lnair$), and the political stability (pol) of the host countries which are all proved to be the key determinants of overseas investment. Second is the bilateral cooperation indicators of China and host countries: the exchange rate ($lnexr$), export from China ($lnexport$) and distance between the two countries ($lndist$). These indicators reflect the market characteristics of China and host countries, and are crucial to explaining the location choice and decision making of China's contractors. The proxy indicators of variables and sources of data are shown in Table 2.

[3]The World Bank classifies the world's economies into four income groups—high, upper-middle, lower-middle, and low based on Gross National Income (GNI) per capita (current US$). For the 45 BRI partner countries, the high and upper-middle income countries classified by the World Bank are assigned into the higher-income group, whereas the lower-middle and low classes are assigned into the lower-income group. There are 22 countries in the higher-income group: Russia, Poland, Czech, Hungary, Romania, Bulgaria, Belarus, Turkey, Iran, United Arab Emirates, Saudi Arabia, Qatar, Kuwait, Oman, Jordan, Israel, Azerbaijan, Kazakhstan, Thailand, Malaysia, Singapore, Brunei. The other 23 countries are in the lower-income group.

Table 2 Descriptions of variables and sources of data

Variables	Proxy for	Sources
lnturnover	Logarithm of the annual turnover of China's OCPs (USD 10,000 dollars)	Wind Economic Database
lncamount	Logarithm of value of the newly signed contracts (USD 10,000 dollars)	Wind Economic Database
lngdppc	Logarithm of GDP per capita (current US$)	World Bank
gdpgr	GDP growth (annual %)	World Bank
resources	Total natural resources rents (% of GDP)	World Bank
labour	Labor force participation rate (% of total population ages 15–64)	World Bank
tele	Fixed telephone subscriptions (per 100 people)	World Bank
lnair	Logarithm of air transport, freight (million ton-km)	World Bank
open	Trade (% of GDP)	World Bank
pol	Political Stability and Absence of Violence/Terrorism (range from −2.5 to 2.5)	World Bank
lnexr	Logarithm of exchange rate (currencies/CNY)	World Bank
lnexport	Logarithm of China's export value to host country (USD 10,000 dollars)	National Bureau of Statistics of China
lndist	Logarithm of the distance of capital cities between China and host country	CEPII Database

3 Method

3.1 *Basics of Bayesian Inference*

This paper uses Bayesian linear regression models to estimate the determinants of China's OCPs in countries along the Belt and Road.

Different from the frequentist statistics, Bayesian models assume all parameters to be random. The Bayesian inference is conducted based on the so-called Bayes' theorem:

$$p(B \mid A) = \frac{p(A \mid B)\, p(B)}{p(A)}, \tag{1}$$

where A and B are random vectors.

In the Bayesian framework, the unknown parameter vector θ is assumed to be random with prior distributions $P(\theta) = \pi(\theta)$ given data vector y. Let $f(y_i \mid \theta)$ donate the probability density function of y_i given θ, the likelihood function of a linear regression model is

$$L(\theta; y) = f(\theta; y) = \prod_{i=1}^{n} f(y_i \mid \theta). \quad (2)$$

By applying Bayes' theorem of formula (1), we can get the posterior distribution of θ, which helps to tell how θ is distributed given the observed data:

$$p(\theta \mid y) = \frac{p(y \mid \theta)\, p(\theta)}{p(y)} = \frac{f(y; \theta)\, \pi(\theta)}{m(y)}, \quad (3)$$

where $m(y)$ implies the marginal distribution of y which is defined by

$$m(y) = \int f(y; \theta)\, \pi(\theta)\, d(\theta), \quad (4)$$

which is not dependent on the parameter vector θ, so that formula (3) can be reduced to

$$p(\theta \mid y) \propto L(\theta; y)\, \pi(\theta). \quad (5)$$

The regression model can be simply represented by

$$Y_{it} = \alpha + \beta X_{it} + \varepsilon_{it}. \quad (6)$$

The Bayesian regression of this study is run Stata 15.0 software. Normal priors are used for regression coefficients β and inverse-gamma prior is used for the variance of ε_{it}.

3.2 Markov Chain Monte Carlo Methods (MCMC)

Metropolis—Hastings (MH) algorithm is described by Hastings [7] as a more general version of the Metropolis algorithm. The algorithm can be summarized as in the following steps:

A. Set up the initial values for $\theta_* = q(\cdot \mid \theta_{t-1})$;
B. Compute the acceptance probability (Bayes factor) and submit it to a "committee" to judge according to the rule:

$$r(\theta_* \mid \theta_{t-1}) = min\left\{\frac{p(\theta_* \mid y)\, q(\theta_{t-1} \mid \theta_*)}{p(\theta_{t-1} \mid y)\, q(\theta_* \mid \theta_{t-1})}, 1\right\}; \tag{7}$$

C. Sample u from the uniform distribution $u(0, 1)$;
D. Set $\theta_t = \theta_*$ if $u < r$, otherwise $\theta_t = \theta_{t-1}$;
E. Repeat step B to D for $j = 1, 2, ..., N$ to obtain $\theta_1, \theta_2, ..., \theta_N$.

The adaptive random-walk MH algorithm is used for sampling from a posterior distribution in this paper. The form of updates is based on a symmetric random walk proposal distribution Z_t, so that

$$\theta_* = \theta_{t-1} + Z_t, \tag{8}$$

and

$$Z_t \sim N\left(0, \rho^2 \Sigma\right), \tag{9}$$

where ρ is the scalar controlling the scale of random jumps for generating updates, and Σ is the covariance matrix.

4 Empirical Results

This paper firstly investigates the determinants of China's OCPs for the whole sample of the 45 BRI partner countries. Then, we conduct regressions on the higher-income group and lower-income group countries separately to investigate the differences for the influencing factors of China's OCPs in countries with different levels of income. We further compare the differences of different periods, before and after the BRI was launched. The regression results are reported in Table 3. The posterior mean values are used as the point estimators of Bayesian estimation.

4.1 Regression Results for the Whole Sample

From model (1), the coefficient of natural resource availability ($resources$) is 0.034 with a confidence interval of [0.027, 0.042]. The assumption of the Bayesian model that all parameters are random implies that in this study the mean of the posterior distribution for the parameters of $resources$ is 0.034, and there is 95% probability that the parameters of $resources$ fall between 0.027 and 0.042. We can infer that the turnover of China's OCPs is positively related to the natural resource availability of host countries. As Chinese contractors in BRI countries have resource seeking motivations, the inadequacy domestic resources should be a driving factor for China to develop contracted projects in BRI partner countries.

Table 3 Regression results of the determinants of China's OCP in BRI partner countries

	(1) Whole sample	(2) Higher-income	(3) Lower-income	(4) Before BRI	(5) After BRI
lngdppc	0.015 [−0.045, 0.061]	−0.011 [−0.237, 0.217]	**0.441** **[0.325, 0.556]**	**−0.220** **[−0.350, −0.098]**	−0.163 [−0.332, 0.013]
gdpgr	0.016 [−0.007, 0.038]	−0.002 [−0.039, 0.033]	0.025 [−0.002, 0.052]	0.026 [−0.000, 0.053]	**0.083** **[0.030, 0.137]**
resources	**0.034** **[0.027, 0.042]**	**0.048** **[0.036, 0.061]**	−0.011 [−0.023, 0.000]	**0.042** **[0.032, 0.052]**	**0.043** **[0.028, 0.058]**
labour	**0.011** **[0.004, 0.019]**	0.008 [−0.007, 0.022]	**−0.022** **[−0.031, −0.014]**	0.009 [−0.001, 0.019]	0.011 [−0.003, 0.026]
tele	**−0.018** **[−0.027, −0.007]**	0.007 [−0.011, 0.027]	**−0.021** **[−0.039, −0.004]**	**−0.019** **[−0.033, −0.005]**	**0.017** **[0.001, 0.033]**
lnair	0.024 [−0.010, 0.060]	0.017 [−0.066, 0.098]	**−0.112** **[−0.153, −0.069]**	**0.073** **[0.020, 0.123]**	0.051 [−0.012, 0.111]
open	0.001 [−0.001, 0.003]	**0.003** **[9.89e−06, 0.007]**	**−0.011** **[−0.014, −0.008]**	**0.002** **[0.000, 0.004]**	0.000 [−0.003, 0.003]
pol	**−0.287** **[−0.376, −0.199]**	**−0.443** **[−0.640, −0.225]**	**0.156** **[0.088, 0.227]**	**−0.235** **[−0.313, −0.158]**	**−0.147** **[−0.240, −0.057]**
lnexr	**0.091** **[0.063, 0.117]**	−0.051 [−0.146, 0.043]	**0.226** **[0.201, 0.248]**	**0.062** **[0.014, 0.109]**	**0.067** **[0.000, 0.131]**
lnexport	**0.730** **[0.694, 0.767]**	**0.766** **[0.669, 0.872]**	**0.669** **[0.600, 0.740]**	**0.766** **[0.678, 0.855]**	**0.582** **[0.481, 0.681]**
lndist	**−0.647** **[−0.706, −0.589]**	**−1.391** **[−1.579, −1.194]**	−0.020 [−0.130, 0.093]	**−0.485** **[−0.564, −0.410]**	**−0.659** **[−0.747, −0.574]**
cons	5.095 [5.039, 5.148]	10.498 [10.365, 10.633]	1.570 [1.485, 1.650]	4.706 [4.606, 4.820]	8.341 [8.116, 8.566]
sigma2	1.406 [1.244, 1.585]	1.718 [1.454, 2.022]	0.542 [0.456, 0.650]	1.246 [1.062, 1.461]	1.189 [0.977, 1.441]

Note: Posterior mean is reported as estimator, 95% equal-tailed credible intervals in parentheses

Similarly, the annual turnover of China's OCPs has a positive relationship with the labor resource (*labour*) of host countries. China's contracted projects in BRI partner countries mainly focus on the fields of electric power engineering, transport infrastructure and general construction with huge labor demand. Thus the labor resource of host countries is one of the factors that influence the location choice of China's contracted projects.

The turnover of China's contracted projects is negatively related to the fixed telephone subscriptions (*tele*) which is one of the indicators of infrastructure conditions of BRI partner countries. This reflects the challenges that China's contractors are facing from the backward infrastructure of host countries. Nevertheless, no meaningful conclusion could be drawn from air transport (*lnair*), another indicator of infrastructure level of host countries, since the credible interval of its coefficient contains zero.

The political stability (*pol*) of host countries has a negative impact on China's contracted projects in BRI countries. The political instability of some BRI countries is one of the major challenges for Chinese contractors. Our finding corresponds to Zhang et al. [14] who suggest that many countries along the Belt and Road have long been struggling with terrorist attacks, crimes and corruption, thus political risks pose important challenges for infrastructure projects and transnational investment.

The turnover of China's OCPs is positively related to the exchange rates ($lnexr$) of Chinese Yuan Renminbi (CNY) to currencies of BRI partner countries. The turnover of China's contracted projects in host countries tends to increase when CNY appreciates, since the appreciation of CNY helps reduce the operating and labor cost of Chinese contractors.

The significantly positive value of the coefficient for China's export to host countries ($lnexport$) indicates the positive correlation between China's OCPs and export. Export is an indicator of the closeness of trade and investment relationships between China and BRI countries, Chinese contractors are likely to choose countries with more export flows from China. On the contrary, China's OCPs could also help promote the export from China to BRI partner countries. From the comments by the Ministry of Commerce of China, the increase of contracted projects remarkably promotes exports of China, since the overseas projects have a huge demand for equipment and materials.

Geographical distance ($lndist$) between China and host countries has a negative influence on the development of China's OCPs because the greater geographical distance leads to the higher transportation cost for China's contractors.

For the other variables ($lngdppc$, $gdpgr$ and $open$), since their credible intervals all contain zero, we can not draw conclusions that there are significant correlations between these variables and the turnover of China's OCPs.

It can be summarized that the turnover of China's OCPs as a whole tends to be higher in countries with richer natural and labour resources, while the backward infrastructure and political instability of host countries have negative impacts on China's contracted projects. The appreciation of CNY against currencies of BRI partner countries and export from China to host countries both contribute to higher revenues of China's contracted projects, while the geographical distance obstructs the development of China's OCPs as it increases the transportation cost.

4.2 Different Determinants of China's OCP in Countries with Different Income Levels

There are several obvious differences in terms of the influencing factors of China's OCPs in countries of different income levels through the comparison of model (2) and model (3).

First of all, the posterior mean of the coefficients for GDP growth rate ($gdpgr$) is positive and significantly different from zero for the lower-income group, but it is not significant for the higher-income group. Higher-income countries have ideal economic foundation, but economic potential is not a major attraction factor for China's overseas contractors. In contrast, many lower-income countries are in the period of accelerated development with a huge demand for infrastructure construction which is a main factor that attracts China's investment and contracted projects.

In the second place, the coefficient of variable *resources* is positive and significant for the higher-income group, but not significant for the lower-income group. In fact, many higher-income countries along the Belt and Road are resource-rich countries, such as the United Arab Emirates, Saudi Arabia, Qatar, and Kuwait. These countries have rich oil and gas resources. Because most China's contracted projects focus on energy and petrochemical related industries, such resources availability in these countries becomes a main motivation for Chinese contractors to undertake external contracted projects.

Thirdly, the labor resource in lower-income countries has negative relationship with the turnover of China's OCPs. This implies the abundant labor supply in host countries inhibits the development of China's contracted projects. Similarly, both of the two indicators of infrastructure (*tele* and *lnair*) are also proved to have negative impacts on China's contracted projects in lower-income countries. In reality, the backward infrastructure is the main obstruction for lower-income countries to achieve faster development. Moreover, infrastructure construction needs enormous labor input and provides the opportunities for labor services from China. Therefore, the lower-income countries with huge demand for infrastructure development and inadequate labor force attract more contracted projects from China.

At last, the turnover of China's OCPs is negatively related to the political stability (*pol*) of higher-income countries, while positively related to the political stability of lower-income countries. Many BRI countries with low-income level face the problems of political instability which are also risks and challenges for Chinese international contractors. As for the higher-income countries with the more stable political environment, the political instability is not a major concern for China's contractors, instead, they may focus more on the development potential, infrastructure, and resources availability of host countries even though the political environment of these countries may be comparatively less stable than other countries in the same group.

4.3 Different Determinants of China's OCP in Different Periods

There are two major differences in the determinants of China's OCPs between different periods from the comparisons of model (4) and model (5).

First of all, in the after-BRI period, the coefficient of GDP growth rate (*gdpgr*) is positive and significantly different from zero compared with the before-BRI period. It shows the significant positive relationship between the turnover of China's OCPs and the economic growth rate of the BRI partner countries after the Initiative was proposed which is in line with the "mutual benefits" purpose of the Initiative. China's contracted projects in the BRI countries contribute to improving infrastructure and creating employment, thus accelerating the local economic growth while the rapid economic growth of host countries also provides more benefits for Chinese contractors.

Table 4 Robustness test of the determinants of China's OCP in BRI partner countries

	(6) Whole sample	(7) Higher-income	(8) Lower-income	(9) Before BRI	(10) After BRI
lngdppc	**−0.170** **[−0.272, −0.042]**	**−0.386** **[−0.608, −0.151]**	**0.338** **[0.215, 0.452]**	**−0.380** **[−0.465, −0.265]**	**−0.398** **[−0.633, −0.092]**
gdpgr	**0.045** **[0.013, 0.076]**	0.031 [−0.010, 0.070]	0.019 [−0.028, 0.058]	**0.046** **[0.011, 0.086]**	**0.109** **[0.047, 0.175]**
resources	**0.045** **[0.035, 0.055]**	**0.066** **[0.053, 0.079]**	−0.004 [−0.019, 0.014]	**0.054** **[0.041, 0.066]**	**0.055** **[0.037, 0.073]**
labour	**0.013** **[0.002, 0.025]**	0.013 [−0.008, 0.033]	**−0.021** **[−0.037, −0.004]**	**0.018** **[0.001, 0.032]**	0.014 [−0.002, 0.035]
tele	−0.009 [−0.021, 0.001]	0.016 [−0.005, 0.038]	−0.020 [−0.046, 0.005]	−0.012 [−0.030, 0.004]	**0.031** **[0.012, 0.052]**
lnair	0.023 [−0.027, 0.073]	0.017 [−0.080, 0.119]	**−0.131** **[−0.193, −0.070]**	**0.073** **[0.022, 0.128]**	0.000 [−0.081, 0.078]
open	0.001 [−0.002, 0.003]	0.002 [−0.001, 0.006]	**−0.015** **[−0.019, −0.010]**	0.002 [−0.001, 0.005]	−0.002 [−0.006, 0.002]
pol	**−0.161** **[−0.250, −0.067]**	−0.089 [−0.332, 0.169]	**0.325** **[0.109, 0.578]**	**−0.146** **[−0.217, −0.078]**	−0.003 [−0.232, 0.204]
lnexr	**0.114** **[0.066, 0.163]**	−0.015 [−0.0781, 0.052]	**0.258** **[0.189, 0.329]**	**0.089** **[0.030, 0.155]**	0.066 [−0.0146, 0.145]
lnexport	**0.818** **[0.747, 0.910]**	**0.976** **[0.904, 1.051]**	**0.683** **[0.623, 0.740]**	**0.836** **[0.689, 0.981]**	**0.820** **[0.709, 0.950]**
lndist	**−0.665** **[−0.788 −0.538]**	**−1.175** **[−1.300, −1.040]**	**0.137** **[−0.098, 0.397]**	**−0.511** **[−0.658, −0.391]**	**−0.616** **[−1.007, −0.274]**
cons	5.338 [5.086, 5.572]	8.588 [8.290, 8.908]	1.373 [1.291, 1.448]	4.626 [4.151, 5.069]	6.698 [6.159, 7.222]
sigma2	1.945 [1.698, 2.224]	2.417 [2.008, 2.933]	0.872 [0.717, 1.064]	1.919 [1.614, 2.286]	1.630 [1.307, 2.047]

Note: Posterior mean is reported as estimator, 95% equal-tailed credible intervals in parentheses

In addition, it is not surprising that comparing with the before-BRI period, the coefficient of *tele* becomes positive and significant. With the advance of the BRI, the infrastructure conditions of host countries are improving, so that in the after-BRI period, infrastructure is no longer an obstacle for the development of China's OCPs.

4.4 Robustness Test

The newly signed contract value of China's contracted projects in BRI countries is used as the explained variable to conduct the robustness test. Table 4 provides the results of the robustness test.

Model (6) reports the Bayesian linear regression results for the whole sample, while Models (7) and (8) show the results of two groups with different income levels. Meanwhile, Models (9) and (10) present the results from the sub-samples of different periods.

The signs of the coefficients for most explanatory variables are consistent with the results from model (1) to (5), except for the significance and size for a few coefficients which change in a small scope from the evidence credible intervals. In model (6), the coefficients of resource availability ($resources$), labour force ($labour$), exchange rate ($lnexr$), and export ($lnexport$) are still positive and significant, whereas the coefficients of political stability (pol) and distance ($lndist$) are still negative and significant. The conclusions from models (7) and (8) are unanimous with models (2) and (3), suggesting that natural resource availability has a significantly positive impact on China's OCPs in higher-income countries, while the main determinants of China's OCPs in lower-income countries are GDP growth rate and labor resource. The results of models (9) and (10) are also consistent with those of models (4) and (5).

The results of the robustness test further help to verify the credibility of our conclusions.

5 Conclusions

This paper investigates the determinants of China's contracted projects in 45 countries along the Belt and Road by Bayesian linear regression models, and the following major conclusions are obtained:

First, China's contracted projects in BRI countries in general have resource-driven and labor-driven motivations. The appreciation of RMB and China's export help promote the development of China's contracted projects in countries along the Belt and Road. However, Chinese contractors face the challenges from the backward infrastructure and the political instability of host countries, while geographical distance also has certain negative effects on the progress of China's OCPs.

Second, factors influencing China's contracted projects are different in countries of different income levels. It is obvious that China's contractors are attracted to the abundant natural resources of the higher-income countries along the Belt and Road, while the main determinants of China's OCPs in lower-income countries are their development potential (GDP growth rate) and labor resources.

Third, comparing with the former period, the correlation between China's OCPs in BRI countries and the speed of local economic development turns to be significantly positive after the Initiative was launched, so is the correlation between China's OCPs and infrastructure of host countries.

These conclusions help provide suggestions to the further development of China's contracted projects in countries along the Belt and Road. For one thing, Chinese contractors should pay attention to the adverse impacts of risk factors in host countries, such as the backward infrastructure and the political instability. Next, countries with different income levels have different market characteristics and investment climate, thus the strategic decision-making of Chinese contractors should be based on different local conditions. China's overseas investment and projects should keep the mutual benefit and win-win strategies, to guarantee the successful and sustainable implementation of the Belt and Road Initiatives.

References

1. Cai, P.: Understanding China's belt and road initiative. Lowy Institute For International Policy (2007)
2. Chang, T., Deng, X., Hwang, B. G., Zhao, X.: Political Risk Paths in International Construction Projects: Case Study from Chinese Construction Enterprises. Advances in Civil Engineering (2018)
3. Chin, H., He, W.: The Belt and Road Initiative: 65 countries and beyond. Fung Business Intelligence Center, Hong Kong (2016)
4. Corkin, L.: Chinese construction companies in Angola: a local linkages perspective. Resour. Policy **37**(4), 475–483 (2012)
5. Duan, F., Ji, Q., Liu, B.Y., Fan, Y.: Energy investment risk assessment for nations along China's Belt & Road Initiative. J. Clean. Prod. **170**, 535–547 (2018)
6. Gelman, A., Loken, E.: The statistical crisis in science. The best writing on mathematics (2015)
7. Hastings, W.K.: Monte Carlo sampling methods using Markov chains and their applications. Biometrika **57**(1), 97–109 (1970)
8. Huang, Y.: Understanding China's Belt & Road initiative: motivation, framework and assessment. China Econ. Rev. **40**, 314–321 (2016)
9. Lu, W., Liu, A.M., Rowlinson, S., Poon, S.W.: Sharpening competitive edge through procurement innovation: perspectives from Chinese international construction companies. J. Constr. Eng. Manag. **139**(3), 347–351 (2012)
10. Luo, C., Chai, Q., Chen, H.: "Going global" and FDI inflows in China: "One Belt & One Road" initiative as a quasi-natural experiment. World Econ. **42**(6), 1654–1672 (2019)
11. Niagara, P., Datche, E.: Factors affecting the performance of construction projects: a survey of construction projects in the coastal region of Kenya. Int. J. Sci. Res. Publ. **5**(10), 1–43 (2015)
12. Shao, Z.Z., Ma, Z.J., Sheu, J.B., Gao, H.O.: Evaluation of large-scale transnational high-speed railway construction priority in the belt and road region. Transp. Res. Part E Logist. Transp. Rev. **117**, 40–57 (2018)
13. Yii, K.J., Bee, K.Y., Cheam, W.Y., Chong, Y.L., Lee, C.M.: Is transportation infrastructure important to the One Belt One Road (OBOR) Initiative? Empirical evidence from the selected Asian countries. Sustainability **10**(11), 4131 (2018)
14. Zhang, C., Xiao, C., Liu, H.: Spatial big data analysis of political risks along the belt and road. Sustainability **11**(8), 2216 (2019)
15. Zhi, H.: Risk management for overseas construction projects. Int. J. Project Manage. **13**(4), 231–237 (1995)

Herding Behavior from Loss Aversion Effect in the Stock Exchange of Thailand

Kunsuda Nimanussornkul and Chaiwat Nimanussornkul

Abstract This paper investigates the effect of loss aversion behavior, by adapting the loss aversion utility function of Berkelaar et al. [3], and taking into account the US-China trade war, on the herding behavior in the Stock Exchange of Thailand. It used the daily data from 01/04/2010 to 05/31/2019 for estimation using Ordinary Least Squares and Quantile Regression. The results did not report the herding behavior in the Stock Exchange of Thailand in both upward and downward trends of the market. The testing of the effect of loss aversion bias on herding behavior revealed the loss aversion to have effect on the herding behavior in some industries when the returns are high. Furthermore, the US-China trade war has a bearing herding behavior in some industries.

1 Introduction

The efficient market hypothesis (EMH) has become an area of interest for fund and portfolio managers. The EMH assumes that a financial market is efficient if the prices reflect all the available information and the investors or fund and portfolio managers have rational expectations about the evolution of future prices. Therefore, all stocks in the financial market are correctly priced [10] and investors can adjust the portfolio optimization of stocks accordingly as suggested by some previous studies e.g. Autchariyapanitkul et al. [1], Zhu et al. [24], Sirisrisakulchai et al. [20], and Autchariyapanitkul et al. [2]. However, in reality and from other research findings as well as contending hypotheses there are irrational investment behaviors that have caused abnormal returns and losses in the financial market. Herding is one of such

K. Nimanussornkul (✉) · C. Nimanussornkul
Center of Excellence in Econometrics, Faculty of Economics,
Chiang Mai University, Chiang Mai 50200, Thailand
e-mail: kunsudan@gmail.com

C. Nimanussornkul
e-mail: Chaiwatnim@yahoo.com

S. Sriboonchitta et al. (eds.), *Behavioral Predictive Modeling in Economics*, Studies in Computational Intelligence 897,
https://doi.org/10.1007/978-3-030-49728-6_21

irrational behaviors, which challenge the validity and reliability of the EMH [4]. The herding behavior is an investment behavior in a group of investors who trade in the same direction over some time but they do not make investment decisions based on their rational expectation analysis. The reasons behind the herding behavior are many. The discussion below suggests the logic that the group of investors is better informed than individual investors. Therefore, the herding behavior may lead the stock prices away from their fundamental values, hence causing the financial market to become inefficient [4], and de-stabilized [21].

The methodologies for empirical investigation of the herding behavior can be classified into two main paths: the first path is the use of micro-data or primary data and the second path is the use of aggregate price or market data. The first path introduced by Lakonishok et al. [16] is a straightforward measurement of whether there is a tendency of fund or portfolio managers to buy (sell) on the same side of the market and if it is so, then we can conclude that there is a herding behavior in the market. The second path proposed by Christie and Huang [8] is to measure investors, herding from the market consensus by investigating the cross-sectional standard deviation (CSSD) of the observed asset return from the cross-sectional average return in the portfolio. However the cross-sectional standard deviation (CSSD) has a problem about the outliers making Chang et al. [5] try to improve it by means of the cross-sectional absolute deviation (CSAD) [21].

Many previous studies on herding behavior use CSAD. The results revealed herding behavior in some countries and some situations of uptrend and downtrend of the markets, as well as the impact of the crisis on herding behavior in some countries. For example, Chiang and Zheng [6] found herding behavior in advanced stock markets and Asian market but not in Latin American and the US markets. Moreover, the positive correlations between the stock return dispersions in Asian markets were also found (except Malaysia). Bui et al. [4] found the herding behavior in the rising market in Indonesia, the Philippines, Malaysia, and Vietnam and found the herding behavior in the downtrend market only in Malaysia. Putra et al. [18] found the herding behavior in Indonesia and Singapore, and their results showed no evidence of herding behavior during the global financial crisis in Indonesia. Previous studies which adopted the CSAD for reducing the multicollinearity problem were carried out by Yao et al. [23] and Filip et al. [10]. Meanwhile, Chen et al. [7] used the Quantile Regression instead of the Ordinary Least Squares and found herding behavior in the rising market more than in the downtrend market and no evidence of herding spillover effect from the US to China market.

In Thailand, there are studies on herding behavior such as those by Kulvanich and Boonvorachote [15], Padungsaksawasdi and Pooissarakit [17], and Rattanasri and Vichitthamaros [19] which used the cross-sectional absolute deviation (CSAD) and estimated by Ordinary Least Squares. Kulvanich and Boonvorachote [15] and Rattanasri and Vichitthamaros [19] analyzed the herding behavior in each sector in the Stock Exchange of Thailand (SET) and they found the herding behavior in some sectors while Padungsaksawasdi and Pooissarakit [17] analyzed both markets, the Stock Exchange of Thailand (SET) and the Market for Alternative Investment (MAI), and the results showed no existence of herding behavior.

As the hearding behavior of the investors in the financial market can arise from diverse reasons, the loss aversion tendency can be one as well. The loss aversion behavior is manifested when the investors are distinctly more sensitive to losses than to gains [3]. Previous papers studying the effect of loss aversion on herding behavior include Decamps and Lovo [9] and Kendall [14] that pursued laboratory experiment with the standard sequential trading model of Glusten and Milgron [11]. They found the different risk types of traders to affect the herding behavior in the stock market but did not find the effect of loss aversion on the herding behavior.

Therefore, this paper would like to investigate the effect of the loss aversion bias on herding behavior, in each industry in the Stock Exchange of Thailand employing the loss aversion utility function proposed Berkelaar et al. [3] in a Quantile Regression at the presence of the US-China trade war. The rest of this paper is organized as follows. Section 2 presents the methodology used in this study that are the herding measure, the loss aversion, and the Quantile Regression. The data description and empirical results are shown in Sect. 3. Section 4 is the conclusion.

2 Methodology

Tests of Herding. In this section, we briefly describe the two common methods for detecting the herding behavior: the CSSD using cross-sectional standard deviation of stock returns proposed by Christie and Huang [8], and the CSAD proposed by Chang et al. [5] using the cross-sectional absolute deviation.

The cross-sectional standard deviation (CSSD) method proposed by Christie and Huang [8] as follows:

$$CSSD_t = \sqrt{\frac{\sum_{i=1}^{N}(r_{i,t} - r_{m,t})^2}{N-1}} \tag{1}$$

Where $r_{i,t}$ is the observed stock return of firm i in the industrial groups at time t, $r_{m,t}$ is the equal-weight average of stock of N returns in the industrial groups at time t, and N is the number of firms in the industrial groups. However, this method has an outliers problem. So Chang et al. [5] proposed the cross-sectional absolute deviation (CSAD) as follows:

$$CSAD_t = \frac{1}{N}\sum_{i=1}^{N}|r_{i,t} - r_{m,t}| \tag{2}$$

Chang et al. [5] explained that the rational asset pricing model predicts the relationship between the CSAD and the absolute value of market returns to be positive and linear, and no herding behavior exists in the market. Then Chang et al. [5] adopted the following regression model for testing the herding behavior:

$$CSAD_t = \alpha_0 + \alpha_1|r_{m,t}| + \alpha_2 r_{m,t}^2 + \varepsilon_t \tag{3}$$

If no herding behavior exists in the market and the rational asset pricing model holds, the regression of this equation should show a linearity model, implyies the $\alpha_2 = 0$. On the other hand, a non-linear model with a statistically significantly negative α_2 implies the existence of herding behavior in the market. Yao et al. [23] and Filip et al. [10] tried to reduce the multicollinearity problem by substituting the variable $r_{m,t}$ to be the change between the variable $r_{m,t}$ and the arithmetic mean of the variable $r_{m,t}$ ($\bar{r}_{m,t}$) and hence enlarging a lag term of the dependent variable ($CSAD_{t-1}$) in the model in order to increase the power of the model. Therefore, this paper applies the model following Yao et al. [23] and Filip et al. [10], as shown below:

$$CSAD_t = \alpha_0 + \alpha_1 |r_{m,t}| + \alpha_2 (r_{m,t} - \bar{r}_{m,t})^2 + \alpha_3 CSAD_{t-1} + \varepsilon_t \tag{4}$$

Where the null hypothesis (H_0) is $\alpha_2 = 0$ implying no herding behavior exists in the market, and the rational pricing model holds, while the alternative hypothesis (H_1) is $\alpha_2 < 0$ implying the existence of herding behavior.

Testing Asymmetry in Herding Behavior. We test the existence of the herding behavior of investors in the uptrend and downtrend markets as follows:

$$CSAD_t^{UP} = \alpha_0^{UP} + \alpha_1^{UP} |r_{m,t}^{UP}| + \alpha_2^{UP} (r_{m,t}^{UP} - \bar{r}_{m,t}^{UP})^2 + \alpha_3^{UP} CSAD_{t-1}^{UP} + \varepsilon_t^{UP} \quad \text{if} r_{m,t} > 0 \tag{5}$$

$$\begin{aligned} CSAD_t^{DOWN} = \alpha_0^{DOWN} + \alpha_1^{DOWN} |r_{m,t}^{DOWN}| + \alpha_2^{DOWN} (r_{m,t}^{DOWN} - \bar{r}_{m,t}^{DOWN})^2 \\ + \alpha_3^{DOWN} CSAD_{t-1}^{DOWN} + \varepsilon_t^{DOWN} \quad \text{if} r_{m,t} \leq 0 \end{aligned} \tag{6}$$

Where $r_{m,t}^{UP}, \bar{r}_{m,t}^{UP}, r_{m,t}^{DOWN}, \bar{r}_{m,t}^{DOWN}$ represent the market return and the arithmetic mean of market return, both for an upward and downward trend in the market. The alternative hypothesis (H_1) is $\alpha_2 < 0$, implying the existence of herding behavior.

Testing the Effect of the Loss Aversion on Herding. The loss aversion utility function proposed by Berkelaar et al. [3], which was modified from the experiments of Kahneman and Tversky [13], is expressed by:

$$U(W) = \begin{cases} -A(\theta - W)^{\gamma_1} & \text{for} \quad W \leq \theta \\ +B(W - \theta)^{\gamma_2} & \text{for} \quad W > \theta \end{cases} \tag{7}$$

Where $U(W)$ is the utility from wealth, θ is the reference point, $A > B$ has to hold for loss aversion, and $0 < \gamma_1 \leq 1, 0 < \gamma_2 < 1$. The outcomes from the experiments of Kahneman and Tversky [13] showed $A = 2.25$, $B = 1$, and $\gamma_1 = \gamma_2 = 0.88$. In this paper, we apply the Eq. (7) to be the equation below:

$$U(W) = \begin{cases} -A(0 - r_j)^{\gamma_1} & \text{for} \quad r_j \leq 0 \\ +B(r_j - 0)^{\gamma_2} & \text{for} \quad r_j > 0 \end{cases} \tag{8}$$

Table 1 Interpret from Minimum Bayes Factor

Bayes Factor (MBF)	Strength of evidence for H_1
1 to 1/3	Weak
1/3 to 1/10	Moderate
1/10 to 1/30	Substantial
1/30 to 1/100	Strong
1/100 to 1/300	Very strong
<1/300	Decisive

Remark: Adapted from Held and Ott [12]

Where $U(r_j)$ is the utility from return, r_j is the return of the industry j at time t, the reference point is zero (this means the return of the industry is zero). We use $A, B, \gamma_1,$ and γ_2 in the same way as Kahneman and Tversky [13]. The effect of loss aversion on the herding behavior is analyzed as follows:

$$CSAD_t = \alpha_0 + \alpha_1 |r_{m,t}| + \alpha_2 (r_{m,t} - \bar{r}_{m,t})^2 + \alpha_3 CSAD_{t-1} + \alpha_4 Loss \times (r_{m,t} - \bar{r}_{m,t})^2 + \varepsilon_t \quad (9)$$

Where $Loss$ is the utility from return $U(r_j)$ in Eq. (8). The alternative hypothesis (H_1) is $\alpha_4 < 0$, implying the impact of the loss aversion on the herding behavior.

Testing the Effect of the US-China Trade War on Herding. The analysis of the impact of the US-China trade war, which started on July 6, 2018 [22], on herding behavior by using Eq. (10):

$$CSAD_t = \alpha_0 + \alpha_1 |r_{m,t}| + \alpha_2 (r_{m,t} - \bar{r}_{m,t})^2 + \alpha_3 CSAD_{t-1} + \alpha_5 Crisis \times (r_{m,t} - \bar{r}_{m,t})^2 + \varepsilon_t \quad (10)$$

Where $Crisis$ is a dummy variable, which takes the value 1 for the period 07/06/2018–05/31/2019, and $zero$ otherwise. The alternative hypothesis (H_1) is $\alpha_5 < 0$, implying the impact of the US-China trade war on herding behavior.

Estimation and Statistics. The paper estimates the Eqs. (4), (9) and (10) by using Ordinary Least Squares (OLS) and Quantile Regression, but Eqs. (5) and (6) estimated by using only OLS. Then we apply the Minimum Bayes Factor (MBF) approach using $-eplogp$ calibration, which p is p-value. The MBF is interpreted follow Table 1.

Data. This paper used the daily data of the Stock Exchange of Thailand (SET) and eight industries in the Stock Exchange of Thailand (SET), namely, Agro & Food Industry, Consumer Products, Financial, Industrial, Property & Construction, Resources, Services, and Technology from 01/04/2010 to 05/31/2019. The stock return is calculated as $r_t = 100[(P_t - P_{t-1})/P_{t-1}]$, where P_t denotes the stock price.

Table 2 shows a statistical summary of the data used in this paper. The number of observations is 2,298. By comparison of the mean values of CSAD, the Resources industry (CSAD6) has the highest mean value, whereas Industrial industry (CSAD4) has the highest standard deviation. It means the Industrial industry (CSAD4) has an unusual cross-sectional variations due to shocks or news.

3 Empirical Results

Tests of Herding. The paper estimated Eq. (4) using the Newey-West consistent estimator (1987) in Ordinary Least Squares (OLS) and Quantile Regression. The results of herding behavior test are shown in Table 3. The negative coefficient of α_2 and the Minimum Bayes factor (MBF) of α_2 corresponding to the very strong evidence supporting the alternative hypothesis imply the existence of herding behavior but only at a lower quantile of CSAD2 or Consumer Products industry. For other industries, the MBF of α_2 provides weak evidence for alternative hypothesis while the MBF of α_1 provides decisive evidence for alternative hypothesis (except a lower quantile of CSAD1), which imply no herding behavior exists in the rest of industries and the rational pricing model holds. This finding is similar to the study of Padungsaksawasdi and Pooissarakit [17] which concluded that there was no herding behavior in SET. These results are contradictory to the studies of Kulvanich and Boonvorachote [15] and Rattanasri and Vichitthamaros [19] because their studies are at the sector-level (the subset of the industry), whereas the present study is at the industry-level.

Testing Asymmetry in Herding Behavior. The paper tested asymmetry in herding behavior by estimating Eqs. (5) and (6) using the Newey-West consistent estimator (1987) in Ordinary Least Squares (OLS). Tables 4 and 5 provided the test results of asymmetry in herding behavior in an upward and downward market. The Minimum Bayes factor (MBF) of α_2 presenting weak evidence for alternative hypothesis and the

Table 2 Data description

	CSAD1	CSAD2	CSAD3	CSAD4	CSAD5	CSAD6	CSAD7	CSAD8	CSADall
Mean	1.3309	1.3415	1.2446	1.4634	1.4764	1.5482	1.4419	1.5300	1.4281
Median	1.2297	1.2297	1.1563	1.3819	1.3966	1.2903	1.3739	1.4342	1.3568
Maximum	26.9391	4.8518	4.1883	27.7331	3.9339	10.3247	4.0953	5.8432	7.5216
Minimum	0.4262	0.2545	0.4896	0.5311	0.6365	0.2943	0.6282	0.5372	0.6966
Std. Dev.	0.7077	0.5571	0.4258	0.7200	0.4413	0.8664	0.3940	0.4889	0.3707
Skewness	21.1291	1.5497	1.7328	22.1943	1.2456	2.2367	1.5512	1.3811	3.7017
Kurtosis	748.3143	7.1779	8.3775	782.6442	5.6290	11.2356	7.7297	7.4974	43.1631
Observations	2,298	2,298	2,298	2,298	2,298	2,298	2,298	2,298	2,298

Remark: CSAD1 is CSAD of Agro & Food Industry, CSAD2 is CSAD of Consumer Products,
CSAD3 is CSAD of Financial, CSAD4 is CSAD of Industrial,
CSAD5 is CSAD of Property & Construction, CSAD6 is CSAD of Resources,
CSAD7 is CSAD of Service, CSAD8 is CSAD of Technology, and
CSADall is CSAD of the whole of industry groups

Table 3 Results from Testing of Herding Behavior

	Intercept (α_0)	$\|r_m\|$ (α_1)	$(r_m - \bar{r}_m)^2$ (α_2)	$CSAD_{t-1}$ (α_3)	$Adj.R^2$	Interpret
CSAD1						
OLS	0.8805	0.4015	0.0280	0.1522	0.7354	No Herding
MBF	0.0000	0.0000	0.0000	0.7348		
0.1	0.6036	0.2877	0.0264	0.1362	0.1711	No Herding
MBF	0.0000	0.9716	1.0000	0.0581		
0.5	0.6899	0.3427	0.0308	0.2765	0.2560	No Herding
MBF	0.0000	0.0000	0.0000	0.0000		
0.9	0.6557	0.5805	0.0204	0.5878	0.3825	No Herding
MBF	0.0000	0.0000	0.0000	0.0000		
CSAD2						
OLS	0.6398	0.7679	−0.0222	0.2794	0.4930	No Herding
MBF	0.0000	0.0000	0.9286	0.0000		
0.1	0.3943	0.7297	−0.0266	0.1529	0.2332	Herding
MBF	0.0000	0.0000	0.0033	0.0000		
0.5	0.6117	0.7181	−0.0018	0.2667	0.2823	No Herding
MBF	0.0000	0.0000	1.0000	0.0000		
0.9	0.9126	0.7241	0.0194	0.4416	0.3282	No Herding
MBF	0.0000	0.0000	1.0000	0.0000		
CSAD3						
OLS	0.5733	0.4213	−0.0102	0.3523	0.5313	No Herding
MBF	0.0000	0.0000	1.0000	0.0000		
0.1	0.4373	0.3774	−0.0133	0.2358	0.2384	No Herding
MBF	0.0000	0.0000	1.0000	0.0000		
0.5	0.5263	0.3975	−0.0014	0.3621	0.3021	No Herding
MBF	0.0000	0.0000	1.0000	0.0000		
0.9	0.7602	0.3681	0.0417	0.4884	0.3736	No Herding
MBF	0.0000	0.0000	0.1924	0.0000		
CSAD4						
OLS	1.0303	0.4159	0.0265	0.1187	0.7839	No Herding
MBF	0.0000	0.0000	0.0000	0.6219		
0.1	0.7254	0.4050	0.0137	0.0915	0.2220	No Herding
MBF	0.0000	0.1078	1.0000	0.7355		
0.5	0.7620	0.3499	0.0290	0.3085	0.3060	No Herding
MBF	0.0000	0.0000	0.0000	0.0000		
0.9	0.9262	0.3615	0.0278	0.4759	0.4113	No Herding
MBF	0.0000	0.0000	0.0000	0.0000		
CSAD5						
OLS	0.7035	0.3668	−0.0022	0.3646	0.5651	No Herding
MBF	0.0000	0.0000	1.0000	0.0000		
0.1	0.5470	0.4032	−0.0073	0.2377	0.2862	No Herding
MBF	0.0000	0.0000	1.0000	0.0000		

(*continued*)

Table 3 (*continued*)

	Intercept (α_0)	$\|r_m\|$ (α_1)	$(r_m - \bar{r}_m)^2$ (α_2)	$CSAD_{t-1}$ (α_3)	$Adj.R^2$	Interpret
0.5	0.6355	0.3483	0.0010	0.3899	0.3302	No Herding
MBF	0.0000	0.0000	1.0000	0.0000		
0.9	0.8498	0.4104	−0.0087	0.5072	0.3886	No Herding
MBF	0.0000	0.0000	1.0000	0.0000		
CSAD6						
OLS	0.7501	0.5950	0.0182	0.1750	0.5684	No Herding
MBF	0.0000	0.0000	1.0000	0.0000		
0.1	0.5038	0.4544	−0.0046	0.0797	0.2518	No Herding
MBF	0.0000	0.0000	1.0000	0.0000		
0.5	0.6739	0.4004	0.0648	0.2177	0.3280	No Herding
MBF	0.0000	0.0000	0.3134	0.0000		
0.9	0.7966	0.7398	0.0497	0.4583	0.4378	No Herding
MBF	0.0000	0.0000	0.0000	0.0000		
CSAD7						
OLS	0.7180	0.3617	0.0088	0.3655	0.5386	No Herding
MBF	0.0000	0.0000	0.9455	0.0000		
0.1	0.5692	0.3769	−0.0043	0.2568	0.2426	No Herding
MBF	0.0000	0.0000	1.0000	0.0000		
0.5	0.7020	0.3616	0.0081	0.3577	0.2928	No Herding
MBF	0.0000	0.0000	0.3903	0.0000		
0.9	0.8497	0.3603	0.0152	0.5108	0.3720	No Herding
MBF	0.0000	0.0000	0.6700	0.0000		
CSAD8						
OLS	0.8098	0.3180	−0.0053	0.3175	0.3649	No Herding
MBF	0.0000	0.0000	1.0000	0.0000		
0.1	0.6132	0.2501	0.0083	0.1865	0.1663	No Herding
MBF	0.0000	0.0000	0.7744	0.0000		
0.5	0.7684	0.3236	−0.0049	0.3034	0.2002	No Herding
MBF	0.0000	0.0000	1.0000	0.0000		
0.9	1.0839	0.3792	−0.0170	0.4416	0.2342	No Herding
MBF	0.0000	0.0000	1.0000	0.0000		
CSADall						
OLS	0.7318	0.4012	0.0019	0.3460	0.6019	No Herding
MBF	0.0000	0.0000	1.0000	0.0000		
0.1	0.6420	0.3821	−0.0033	0.2650	0.3302	No Herding
MBF	0.0000	0.0000	1.0000	0.0000		
0.5	0.5877	0.3709	0.0018	0.4441	0.4173	No Herding
MBF	0.0000	0.0000	1.0000	0.0000		
0.9	0.5207	0.3391	0.0200	0.6637	0.5011	No Herding
MBF	0.0000	0.0000	0.4772	0.0000		

Remark: MBF is Minimum Bayes Factor

Table 4 Results of asymmetry in herding behavior in upward market

	Intercept (α_0)	$\lvert r_m \rvert$ (α_1)	$(r_m - \bar{r}_m)^2$ (α_2)	$CSAD_{t-1}$ (α_3)	$Adj.R^2$	Interpret
CSAD1	0.9304	0.5319	0.0228	0.0701	0.8316	No Herding
MBF	0.0000	0.0000	0.0000	0.9685		
CSAD2	0.6669	0.7986	−0.0210	0.2501	0.4765	No Herding
MBF	0.0000	0.0000	1.0000	0.0000		
CSAD3	0.6428	0.4604	0.0084	0.2819	0.5176	No Herding
MBF	0.0000	0.0000	1.0000	0.0000		
CSAD4	1.0370	0.5514	0.0214	0.0708	0.8530	No Herding
MBF	0.0000	0.0000	0.0000	0.7798		
CSAD5	0.7224	0.4108	0.0052	0.3365	0.5354	No Herding
MBF	0.0000	0.0000	1.0000	0.0000		
CSAD6	0.7670	0.5860	0.0517	0.1599	0.6158	No Herding
MBF	0.0000	0.0000	0.1322	0.0000		
CSAD7	0.7559	0.4139	0.0116	0.3215	0.4826	No Herding
MBF	0.0000	0.0000	0.4073	0.0000		
CSAD8	0.9123	0.3775	−0.0009	0.2168	0.3516	No Herding
MBF	0.0000	0.0000	1.0000	0.0000		
CSADall	0.8011	0.4477	0.0221	0.2791	0.5081	No Herding
MBF	0.0000	0.0000	0.9895	0.0111		

Remark: MBF is Minimum Bayes Factor.

MBF of α_1 presenting decisive evidence for alternative hypothesis in the both of an upward and downward trend in every industry indicate that no herding behavior exists in both upward and downward trends of the market, and the rational pricing model holds. This finding is similar to the study of Padungsaksawasdi and Pooissarakit [17] which revealed no herd behavior in both upward and downward trends of the market. However, the results are different from the studies at the sector-level of Kulvanich and Boonvorachote [15] and Rattanasri and Vichitthamaros [19].

Testing the Effect of the Loss Aversion on Herding. The impact of the loss aversion on herding behavior was tested Eq. (9) using the Newey-West consistent estimator (1987) in Ordinary Least Squares (OLS) and Quantile Regression. The estimation results are reported in Table 6. The negative coefficient α_4 and its MBF corresponding to substantial evidence supporting the alternative hypothesis imply the impact of loss aversion to herding behavior in the Financial industry (CSAD3), Industrial industry (CSAD4), the Resources industry (CSAD6), Technology industry (CSAD8) and the

Table 5 Results of asymmetry in herding behavior in downward market

	Intercept (α_0)	$\|r_m\|$ (α_1)	$(r_m - \bar{r}_m)^2$ (α_2)	$CSAD_{t-1}$ (α_3)	$Adj.R^2$	Interpret
CSAD1	0.6021	0.3676	−0.0024	0.3876	0.4291	No Herding
MBF	0.0000	0.0000	1.0000	0.0000		
CSAD2	0.7143	0.7690	−0.0194	0.2217	0.4660	No Herding
MBF	0.0000	0.0000	0.9946	0.0000		
CSAD3	0.6291	0.3386	−0.0008	0.3281	0.4953	No Herding
MBF	0.0000	0.0000	1.0000	0.0000		
CSAD4	0.8018	0.4104	−0.0052	0.2915	0.5112	No Herding
MBF	0.0000	0.0000	1.0000	0.0000		
CSAD5	0.8307	0.2975	0.0117	0.2928	0.5524	No Herding
MBF	0.0000	0.0000	0.6596	0.0000		
CSAD6	0.7075	0.6318	−0.0155	0.1939	0.5366	No Herding
MBF	0.0000	0.0000	1.0000	0.0000		
CSAD7	0.7990	0.3047	0.0229	0.3229	0.5448	No Herding
MBF	0.0000	0.0000	0.4719	0.0000		
CSAD8	0.8668	0.1991	0.0116	0.3332	0.3302	No Herding
MBF	0.0000	0.0000	0.9138	0.0000		
CSADall	0.7415	0.3356	0.0092	0.3564	0.7048	No Herding
MBF	0.0000	0.0000	0.9782	0.0000		

Remark: MBF is Minimum Bayes Factor.

Whole of industry (CSADall), especial at higher quantile. The results indicated that herding behavior is more likely to occur at higher quantile if the investors have loss aversion.

Testing the Effect of the US-China Trade War on Herding. For testing the effect of the US-China trade war on herding behavior, the Newey-West consistent estimator (1987) was used in Ordinary Least Squares (OLS) and Quantile Regression to estimate Eq. (10) the results are presented in Table 7. The negative coefficient α_5 and its MBF presenting to substantial evidence for the alternative hypothesis imply the effect of US-China trade war to herding behavior in the Financial industry (CSAD3), Property and Construction industry (CSAD5), the Resources industry (CSAD6) and Service industry (CSAD7). Meanwhile, only the MBF of α_5 in the Whole of industry (CSADall) corresponding to the moderate evidence. The results indicate that herding behavior is occurring in some industries when the trade war between the US and China is happening.

Table 6 Results of testing the effect of the loss aversion on herding

	Intercept (α_0)	$\|r_m\|$ (α_1)	$(r_m - \bar{r}_m)^2$ (α_2)	$CSAD_{t-1}$ (α_3)	$Loss$ (α_4)	$Adj.R^2$	Interpret
CSAD1							
OLS	0.8439	0.5293	−0.0416	0.1532	0.0039	0.7373	No Loss
MBF	0.0000	0.0000	0.3784	0.7303	0.0134		
0.1	0.5757	0.4188	−0.0433	0.1295	0.0044	0.1768	No Loss
MBF	0.0000	0.0000	0.8406	0.0913	0.1629		
0.5	0.6696	0.4438	−0.0343	0.2733	0.0037	0.2570	No Loss
MBF	0.0000	0.0000	0.0134	0.0000	0.0000		
0.9	0.6765	0.5231	0.0629	0.5823	−0.0024	0.3823	No Loss
MBF	0.0000	0.0111	1.0000	0.0000	1.0000		
CSAD2							
OLS	0.6512	0.7111	0.0309	0.2783	−0.0063	0.4938	No Loss
MBF	0.0000	0.0000	1.0000	0.0000	0.9567		
0.1	0.3999	0.7002	0.0264	0.1513	−0.0106	0.2332	No Loss
MBF	0.0000	0.3812	1.0000	0.0000	1.0000		
0.5	0.6376	0.5976	0.1146	0.2613	−0.0127	0.2843	No Loss
MBF	0.0000	0.0000	0.9858	0.0000	0.8226		
0.9	0.9236	0.6577	0.1223	0.4336	−0.0152	0.3286	No Loss
MBF	0.0000	0.0086	1.0000	0.0000	1.0000		
CSAD3							
OLS	0.5917	0.3622	0.0375	0.3477	−0.0052	0.5347	No Loss
MBF	0.0000	0.0000	1.0000	0.0000	0.8450		
0.1	0.4366	0.3797	−0.0150	0.2361	0.0002	0.2380	No Loss
MBF	0.0000	0.2336	1.0000	0.0000	1.0000		
0.5	0.5562	0.3235	0.0630	0.3488	−0.0072	0.3042	No Loss
MBF	0.0000	0.0000	0.9854	0.0000	0.7535		
0.9	0.7848	0.1933	0.1926	0.4883	−0.0166	0.3812	Loss
MBF	0.0000	0.4841	0.1078	0.0000	0.0320		
CSAD4							
OLS	1.0684	0.2849	0.0892	0.1195	−0.0034	0.7872	No Loss
MBF	0.0000	0.3093	0.9585	0.6193	1.0000		
0.1	0.6946	0.5034	−0.0445	0.0917	0.0040	0.2302	No Loss
MBF	0.0000	0.0000	0.4073	0.7254	0.0156		
0.5	0.7506	0.4361	−0.0169	0.2969	0.0025	0.3073	No Loss
MBF	0.0000	0.0000	1.0000	0.0000	0.7283		
0.9	0.9706	0.1433	0.2161	0.4744	−0.0106	0.4148	Loss
MBF	0.0000	0.0006	0.0000	0.0000	0.0000		
CSAD5							
OLS	0.7063	0.3578	0.0043	0.3641	−0.0006	0.5649	No Loss
MBF	0.0000	0.0000	1.0000	0.0000	1.0000		
0.1	0.5528	0.3972	−0.0001	0.2335	−0.0014	0.2863	No Loss
MBF	0.0000	0.0000	1.0000	0.0000	0.9200		

(continued)

Table 6 (*continued*)

	Intercept (α_0)	$\|r_m\|$ (α_1)	$(r_m - \bar{r}_m)^2$ (α_2)	$CSAD_{t-1}$ (α_3)	$Loss$ (α_4)	$Adj.R^2$	Interpret
0.5	0.6382	0.3384	0.0058	0.3911	−0.0003	0.3300	No Loss
MBF	0.0000	0.0000	1.0000	0.0000	1.0000		
0.9	0.8523	0.3513	0.0409	0.5141	−0.0045	0.3888	No Loss
MBF	0.0000	0.0021	1.0000	0.0000	1.0000		
CSAD6							
OLS	0.8033	0.4286	0.1298	0.1679	−0.0114	0.5847	Loss
MBF	0.0000	0.0000	0.0033	0.0000	0.0000		
0.1	0.5135	0.4063	0.0346	0.0802	−0.0037	0.2566	No Loss
MBF	0.0000	0.0000	0.2542	0.0000	0.8559		
0.5	0.7264	0.2284	0.2054	0.2062	−0.0168	0.3406	Loss
MBF	0.0000	0.0581	0.0000	0.0000	0.0000		
0.9	0.8412	0.6002	0.1671	0.4413	−0.0128	0.4391	No Loss
MBF	0.0000	0.0375	0.7338	0.0000	0.8723		
CSAD7							
OLS	0.7198	0.3557	0.0142	0.3650	−0.0006	0.5384	No Loss
MBF	0.0000	0.0000	0.9955	0.0000	1.0000		
0.1	0.5695	0.3778	−0.0046	0.2564	0.0000	0.2422	No Loss
MBF	0.0000	0.0086	1.0000	0.0000	1.0000		
0.5	0.7074	0.3550	0.0122	0.3549	−0.0003	0.2925	No Loss
MBF	0.0000	0.0000	1.0000	0.0000	1.0000		
0.9	0.8856	0.2663	0.0856	0.4993	−0.0062	0.3725	No Loss
MBF	0.0000	0.6712	1.0000	0.0000	1.0000		
CSAD8							
OLS	0.8304	0.2610	0.0323	0.3142	−0.0034	0.3674	No Loss
MBF	0.0000	0.0000	0.6350	0.0000	0.1807		
0.1	0.6219	0.2205	0.0232	0.1884	−0.0014	0.1671	Loss
MBF	0.0000	0.0000	0.0033	0.0000	0.0997		
0.5	0.8007	0.2177	0.0655	0.3011	−0.0059	0.2031	Loss
MBF	0.0000	0.0000	0.2868	0.0000	0.0738		
0.9	1.1405	0.2006	0.0857	0.4404	−0.0072	0.2384	Loss
MBF	0.0000	0.0783	0.0241	0.0000	0.0000		
CSADall							
OLS	0.7464	0.3526	0.0447	0.3425	−0.0046	0.6044	No Loss
MBF	0.0000	0.0000	0.9983	0.0000	0.9948		
0.1	0.6284	0.4010	−0.0190	0.2736	0.0015	0.3306	No Loss
MBF	0.0000	0.0000	0.1663	0.0000	0.3044		
0.5	0.5887	0.3679	0.0054	0.4437	−0.0007	0.4171	No Loss
MBF	0.0000	0.0000	1.0000	0.0000	1.0000		
0.9	0.5366	0.2351	0.1318	0.6643	−0.0123	0.5023	Loss
MBF	0.0000	0.0111	0.1956	0.0000	0.0941		

Remark: MBF is Minimum Bayes Factor.

Table 7 Results of testing the effect of the US-China trade war on herding

	Intercept (α_0)	$\lvert r_m \rvert$ (α_1)	$(r_m - \bar{r}_m)^2$ (α_2)	$CSAD_{t-1}$ (α_3)	$Crisis$ (α_5)	$Adj.R^2$	Interpret
CSAD1							
OLS	0.8812	0.4049	0.0278	0.1514	−0.0406	0.7355	No Effect
MBF	0.0000	0.0000	0.0000	0.7386	1.0000		
0.1	0.6053	0.2898	0.0260	0.1346	−0.0210	0.1710	No Effect
MBF	0.0000	0.9854	1.0000	0.0661	1.0000		
0.5	0.6920	0.3481	0.0306	0.2749	−0.0510	0.2562	No Effect
MBF	0.0000	0.0000	0.0000	0.0000	0.7587		
0.9	0.6557	0.5805	0.0204	0.5878	−0.0219	0.3822	No Effect
MBF	0.0000	0.0000	0.0000	0.0000	1.0000		
CSAD2							
OLS	0.6406	0.7718	−0.0231	0.2787	−0.0688	0.4929	No Effect
MBF	0.0000	0.0000	0.8972	0.0000	1.0000		
0.1	0.3949	0.7417	−0.0288	0.1519	−0.1636	0.2339	No Effect
MBF	0.0000	0.0000	0.0000	0.0000	1.0000		
0.5	0.6117	0.7193	−0.0020	0.2659	0.0949	0.2821	No Effect
MBF	0.0000	0.0000	1.0000	0.0000	1.0000		
0.9	0.9126	0.7268	0.0214	0.4415	−0.0905	0.3280	No Effect
MBF	0.0000	0.0000	1.0000	0.0000	0.9939		
CSAD3							
OLS	0.5761	0.4278	−0.0115	0.3494	−0.0704	0.5319	No Effect
MBF	0.0000	0.0000	1.0000	0.0000	0.3028		
0.1	0.4408	0.3853	−0.0156	0.2327	−0.1048	0.2399	No Effect
MBF	0.0000	0.0000	1.0000	0.0000	0.5551		
0.5	0.5294	0.4039	−0.0024	0.3586	−0.0660	0.3030	Effect
MBF	0.0000	0.0000	1.0000	0.0000	0.0856		
0.9	0.7517	0.3564	0.0443	0.4962	0.0225	0.3733	No Effect
MBF	0.0000	0.0000	0.1443	0.0000	1.0000		
CSAD4							
OLS	1.0305	0.4165	0.0265	0.1187	−0.0154	0.7838	No Effect
MBF	0.0000	0.0000	0.0000	1.0000	1.0000		
0.1	0.7250	0.4050	0.0137	0.0919	−0.0302	0.2217	No Effect
MBF	0.0000	0.1309	1.0000	0.6619	1.0000		
0.5	0.7643	0.3516	0.0289	0.3069	−0.0075	0.3057	No Effect
MBF	0.0000	0.0000	0.0000	0.0000	1.0000		
0.9	0.9248	0.3573	0.0280	0.4775	0.0351	0.4113	No Effect
MBF	0.0000	0.0000	0.0000	0.0000	1.0000		
CSAD5							
OLS	0.7076	0.3734	−0.0034	0.3613	−0.0759	0.5659	Effect
MBF	0.0000	0.0000	1.0000	0.0000	0.0411		
0.1	0.5442	0.4063	−0.0081	0.2397	−0.0259	0.2865	No Effect
MBF	0.0000	0.0000	1.0000	0.0000	0.9140		

(continued)

Table 7 (*continued*)

	Intercept (α_0)	$\lvert r_m \rvert$ (α_1)	$(r_m - \bar{r}_m)^2$ (α_2)	$CSAD_{t-1}$ (α_3)	$Crisis$ (α_5)	$Adj.R^2$	Interpret
0.5	0.6447	0.3546	−0.0002	0.3831	−0.0648	0.3307	No Effect
MBF	0.0000	0.0000	1.0000	0.0000	0.3347		
0.9	0.8451	0.4074	−0.0081	0.5126	−0.1133	0.3901	Effect
MBF	0.0000	0.0000	1.0000	0.0000	0.0000		
CSAD6							
OLS	0.7603	0.6137	0.0145	0.1677	−0.2218	0.5720	Effect
MBF	0.0000	0.0000	1.0000	0.0000	0.0000		
0.1	0.5059	0.4699	−0.0076	0.0790	−0.1177	0.2536	No Effect
MBF	0.0000	0.0000	1.0000	0.0000	0.5726		
0.5	0.6763	0.4283	0.0601	0.2105	−0.1079	0.3301	Effect
MBF	0.0000	0.0000	0.4410	0.0000	0.0241		
0.9	0.8154	0.7477	0.0485	0.4481	−0.2919	0.4422	Effect
MBF	0.0000	0.0000	0.0000	0.0000	0.0000		
CSAD7							
OLS	0.7198	0.3679	0.0076	0.3636	−0.0570	0.5391	No Effect
MBF	0.0000	0.0000	0.9957	0.0000	0.9135		
0.1	0.5726	0.3974	−0.0079	0.2504	−0.1151	0.2443	Effect
MBF	0.0000	0.0000	1.0000	0.0000	0.0394		
0.5	0.7059	0.3640	0.0078	0.3548	−0.0493	0.2931	No Effect
MBF	0.0000	0.0000	0.4466	0.0000	0.9903		
0.9	0.8497	0.3603	0.0152	0.5108	−0.0025	0.3717	No Effect
MBF	0.0000	0.0000	0.7106	0.0000	1.0000		
CSAD8							
OLS	0.8098	0.3125	−0.0045	0.3185	0.0234	0.3651	No Effect
MBF	0.0000	0.0000	1.0000	0.0000	0.8506		
0.1	0.6158	0.2462	0.0089	0.1867	0.0287	0.1673	No Effect
MBF	0.0000	0.0000	0.6715	0.0000	0.6358		
0.5	0.7630	0.3131	−0.0035	0.3098	0.0372	0.2008	No Effect
MBF	0.0000	0.0000	1.0000	0.0000	0.5153		
0.9	1.0818	0.3783	−0.0182	0.4435	0.0299	0.2341	No Effect
MBF	0.0000	0.0000	1.0000	0.0000	0.5927		
CSADall							
OLS	0.7358	0.4083	0.0005	0.3425	−0.0828	0.6028	Effect
MBF	0.0000	0.0000	1.0000	0.0000	0.2860		
0.1	0.6437	0.3824	−0.0036	0.2657	−0.0670	0.3316	Effect
MBF	0.0000	0.0000	1.0000	0.0000	0.1618		
0.5	0.5914	0.3709	0.0018	0.4420	−0.0648	0.4173	No Effect
MBF	0.0000	0.0000	1.0000	0.0000	0.8252		
0.9	0.5293	0.3457	0.0185	0.6574	−0.0275	0.5015	Effect
MBF	0.0000	0.0000	0.6159	0.0000	0.3077		

Remark: MBF is Minimum Bayes Factor.

4 Conclusions

This paper tested the herding behavior by applying the models of Yao et al. [23] and Filip et al. [10]. It performed the testing in both uptrend and downtrend of the market. Specifically, it analyzed the effects of loss aversion and the US-China trade war on the herding behavior in eight industries in the Stock Exchange of Thailand, namely, Agro & Food Industry, Consumer Products, Financial, Industrial, Property & Construction, Resources, Services, and Technology using Ordinary Least Squares and Quantile Regression. For the statistical test, the Minimum Bayes factor (MBF) was employed. The daily data from 01/04/2010 to 05/31/2019 of the industrial groups in the Stock Exchange of Thailand (SET) were used.

The results found no herding behavior in all industries except the Consumer Products industry at the lower quantile (Q = 0.1), and thus the rational pricing model holds. Moreover, no herding was evident in both upward and downward trends of the market. This finding is similar to the study of Padungsaksawasdi and Pooissarakit [17], which concluded that no herding behavior exists in SET. For the effect of loss aversion on herding behavior, the paper found the herding behavior is more likely to occur in some industry at higher quantile (Q=0.9) where the investors have loss aversion. Furthermore, the effect of the trade war between the US and China on the herding behavior is more likely to occur in Financial, Property and Construction, Resource, and Service industries.

Acknowledgements This paper has been partially supported by Center of Excellence in Econometrics, Faculty of Economics, Chiang Mai University.

References

1. Autchariyapanitkul, K., Piamsuwannakit, S., Chanaim, S., Sriboonchitta, S.: Optimizing stock returns portfolio using the dependence structure between capital asset pricing models: a vine copula-based approach. Stud. Comput. Intell. **622**, 319–331 (2016)
2. Autchariyapanitkul, K., Sriboonchitta, S., Chanaim, S.: Portfolio optimization of stock returns in high-dimensions: a copula-based approach. Thai J. Math. 11–23 (2014)
3. Berkelaar, A.B., Kouwenberg, R., Post, T.: Optimal portfolio choice under loss aversion. Rev. Econ. Stat. **86**, 973–987 (2004)
4. Bui, N.D., Nguyen, L.T.B., Nguyen, N.T.T.: Herd behavior in Southeast Asian stock markets - an empirical investigation. Acta Oeconomica **65**, 413–429 (2015)
5. Chang, E.C., Cheng, J.W., Khorana, A.: An examination of herd behavior in equity markets: an international perspective. J. Bank. Finance **24**, 1651–1679 (2000)
6. Chang, T., Zheng, D.: An empirical analysis of herd behavior in global stock markets. J. Bank. Finance **34**, 1911–1921 (2010)
7. Chen, Y.C., Wu, H.C., Huang, J.J.: Herd behavior and rational expectations: a test of China's market using quantile regression. Int. J. Econ. Financ. Issues **7**, 649–663 (2017)
8. Christie, W.G., Huang, R.D.: Following the pied piper: do individual returns herd around the market? Financ. Anal. J. **51**, 31–37 (1995)
9. Decamps, J.P., Lovo, S.: A note on risk aversion and herd behavior in financial markets. Geneva Risk Insur. **31**, 35–42 (2006)

10. Filip, A., Pochea, M., Pece, A.: The herding behaviour of investors in the CEE stocks markets. Procedia Econ. Finance **32**, 307–315 (2015)
11. Glosten, L., Milgrom, P.: Bid, ask, and transaction prices in a specialist market with heterogeneously informed traders. J. Financ. Econ. **14**, 71–100 (1985)
12. Held, L., Ott, M.: On p-values and Bayes factors. Ann. Rev. Stat. Appl. **5**, 393–419 (2018)
13. Kahneman, D., Tversky, A.: Prospect theory: an analysis of decision under risk. Econometrica **47**, 263–291 (1979)
14. Kendall, C.: Herding and Contrarianism: A Matter of Preference? USC Marshall School of Business (2018)
15. Kulvanich, J., Boonvorachote, T.: Herding behavior analysis in the Stock Exchange of Thailand. Kasetsart J. **34**, 43–59 (2013)
16. Lakonishok, L., Shleifer, A., Vishny, R.W.: The impact of institutional trading on stock prices. J. Financ. Econ. **32**, 23–43 (1992)
17. Padungsaksawasdi, C., Pooissarakit, I.: Herd behavior in the Stock Exchanges in Thailand. Chulalongkorn Bus. Rev. **37**, 51–80 (2015)
18. Putra, A.A., Rizkianto, E., Chalid, D.A.: The analysis of herding behavior in Indonesia and Singapore Stock Market. Adv. Econ. Bus. Manag. Res. **36**, 197–206 (2017)
19. Rattanasri, W., Vichitthamaros, P.: The study of herding behavior in the Stock Exchange of Thailand during 2010–2015. WMS J. Manag. Walaklak Univ. **7**, 9–20 (2018)
20. Sirisrisakulchai, J., Autchariyapanitkul, K., Harnpornchai, N., Sriboonchitta, S.: Portfolio optimization of financial returns using fuzzy approach with NSGA-II algorithm. J. Adv. Comput. Intell. Intell. Inf. **19**(5), 619–623 (2015)
21. Spyrou, S.: Herding in financial markets: a review of the literature. Rev. Behav. Financ. **5**, 175–194 (2013)
22. Wong, D., Koty, A.C.: The US-China trade war: a timeline. China Briefing (2019). DIALOG. http://www.china-briefing.com/news/the-us-china-trade-war-a-timeline/. Accessed 11 July 2019
23. Yao, J., Ma, C., He, W.P.: Investor herding behaviour of Chinese stock market. Int. Rev. Econ. Financ. **29**, 12–29 (2014)
24. Zhu, K., Yamaka, W., Sriboonchitta, S.: Multi-asset portfolio returns: a Markov switching copula-based approach. Thai J. Math. **14**(Special Issue applied mathematics), 183–200 (2016)

Technical Efficiency and Spatial Econometric Model: Application to Rice Production of Thailand

Thunyawadee Sucharidtham and Satawat Wannapan

Abstract Bayesian statistics and the MCMC simulation applied to the spatial panel data is presented in this paper focusing on rice farming in Thailand. Rice is chosen because it can be grown in all geographic regions of the country as the main staple and for domestic market and export. The rice sector has been crucial in all facets of the Thai society and rice farmers have been regarded as the backbone of the country. However, there are some obstacles curbing the improvement of rice production and rice farming systems. This paper is conducted employing the panel time-series data to investigate the spatial effects on the productivity of rice farming in Thailand. Yearly panel time-series data obtained from 76 provinces of Thailand is used for this study which includes three sections: (1) The overview of rice production in Thailand, (2) Technical efficiency reports for regional rice farming areas, and (3) Spatial panel econometric investigation to arrive at some significant findings for making effective policy recommendations. The empirical results from this study can be useful for those with interest in political benefit and contribute to the body of knowledge necessary for further research in agricultural economics.

Keywords DEA · Bootstrapping method · Bayesian approach · Spatial econometrics · Panel data · Thai rice products

1 Background of the Research

Spatial dependency has considerably been overlooked in computational economic models. This approach has been recently employed to regionally measure the

---

T. Sucharidtham (✉)
Administration and Liberal of Art, Rajamangala University of Technology Lanna, Chiang Mai, Thailand
e-mail: thunyawadee@gmail.com

S. Wannapan
Faculty of Economics, Chiang Mai University, Chiang Mai, Thailand
e-mail: lionz1988@gmail.com

S. Sriboonchitta et al. (eds.), *Behavioral Predictive Modeling in Economics*, Studies in Computational Intelligence 897,
https://doi.org/10.1007/978-3-030-49728-6_22

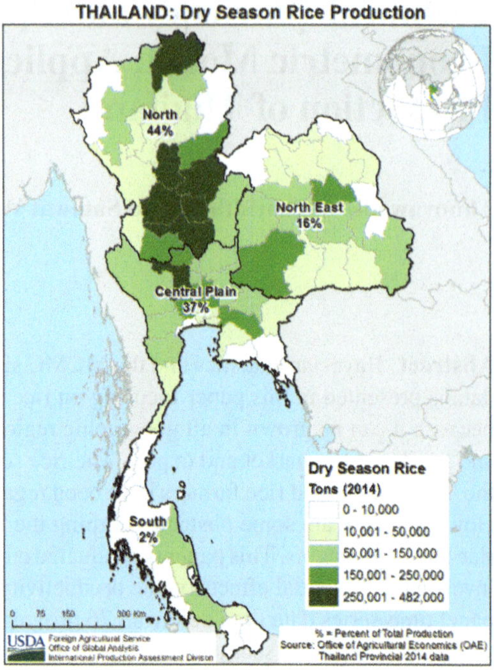

Fig. 1 South East Asia - rice production maps in dry season of Thailand

productivity of many goods, especially agricultural products. As noted by Pede et al. [22], spatial econometrics was first attempted in regional science for studies primarily on spatial problems and solutions for urban or local centers and regions according to Anselin [3] and [2] who subsequently addressed the signification of neighbors influences on economic decisions because neighborhood interactions can change individuals perspectives, information sets, preferences, and behavioral consequences. In agriculture, neighborhood interactions have primarily been pinpointed for investigating the drivers of technology adoption in agricultural production [11, 19, 22].

In Thailand today, rice still remains the dominant crops grown in different regions across the country despite the different environmental conditions and socio-economic settings. The interesting question is which area should be emphasized for economic reasons. Consider Fig. 1, the spatial map of dry season rice production in Thailand during 2014 crop year confirms rice can be grown and harvested throughout Thailand despite the threats of hot and dry climate. The shade of color levels can graphically explain that rice is separately produced in all regions of Thailand. Thus, this is why the spatial econometric model is crucially needed.

One thing that is inevitably necessary is the technical efficiency (TE) or the ability of a farmer to achieve the optimal output feasible, given a set of inputs. TE is frequently measured in terms of the ratio of the observed output to the frontier output, given the corresponding level of inputs used by the farm [22]. Technically,

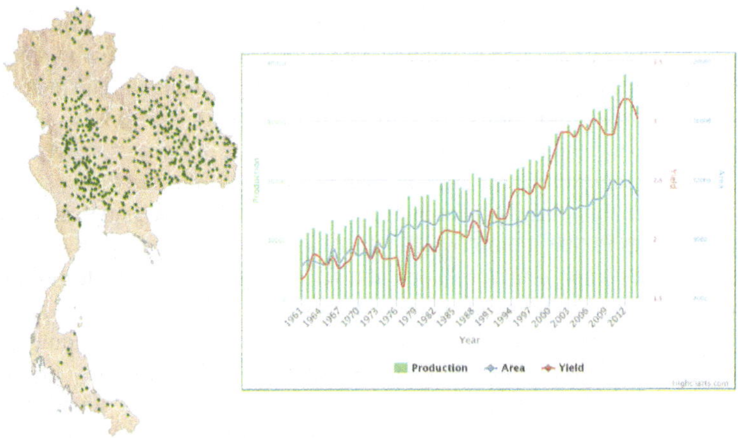

Fig. 2 Rice production map, and productivity trends during 1961 to 2012

the efficiency scores are possibly driven by several indicators such as technology adoption, capital inputs, and labors, even non-renewable and renewable resources. In academic fields, TE has been broadly mentioned in many researches, especially in the vigorous literature on TE in agriculture by such scholars as Gebregziabher et al. [9], Rejesus et al. [23], Shantha et al. [24], and Balde et al. [4]. However, the different contribution of our paper is not the typical form of the production frontier. The data envelopment analysis (DEA) following the paper of Simar and Wilson [27]; [21] is employed to investigate the efficiency of rice production in Thailand. With a bootstrapping non-parametric model, the bootstrapped DEA scores are technically consistent and serve as a valuable tool for implementing statistical inference on DEA [21].

Because of the limitation of the conventional efficiency score, following the paper of Pede et al. [22], spatial analysis or regional considerations in econometrics are crucial to capture how individuals probably behave similarly. Basically, the hypotheses regarding similar behavior of people in the same cohort are: (I) endogenous effects as manifested by the propensity of an individual to act in some way which varies with group behavior, (II) exogenous effects as manifested by the propensity of an individual to act in some way which varies with the exogenous characteristics of the group, and (III) correlated effects as reflected by individuals in the same group acting similarly because they face the same institutional environments and share similar individual characteristics. Agriculturally, neighboring producers are technically assumed to face similar climatic situations (conditions). This can be literally seen in the academic works of Conley and Udry [6], Ward and Pede [28], and Ostrom [20].

Data used in this paper is covered by a panel time-series analysis. As seen in Fig. 2, the green spots in the map (left-hand side) implies areas that rice is grown and cropped. Obviously, it seems that rice productions from Thailand are continuously efficient (the right-hand side graph). In this paper, collected samples were observed as the panel data from 76 provinces of the country. The time-series range of data covers during 2009 to 2017, extending updated observations. The data was officially provided by Office of Agricultural Economics, Bangkok, Thailand. The factors using in the paper are rice productions (tons), planted areas (Rai),[1] fertilizer usages (tons), farmer families, and the measurement of rainfall (mm.).[2] Analyses were performed following the spatial panel estimation from the outline in Millo and Piras [18]; Pede et al. [22]. However, this study is different from previous studies on spatial dependencies among TE scores in that it employs panel time-series data in every 76 provinces of Thailand with the capital factor (lands and fertilizers), labor variable (the number of farmer families), and uncontrollable exogenous variable (rainfalls). Additionally, with alternative statistics, the Bayesian approach is employed for inferencing research results. Hence, this paper is proposed to investigate three objects. The former is to generally survey in descriptive information for explaining sensible situations of rice production in Thailand. The second section is to econometrically capture the trends of technical efficiencies of rice productions in Thailand. Lastly, third is the spatial econometrics for the panel time-series data of rice production by monitoring regional areas in Thailand.

2 Theory and Methodology

2.1 *Data Review*

The general review of collected data was presented in Table 1. Rice farming productivity was set to be the dependent variable in the spatial panel time series estimation. The rice data was also mentioned in the spatial analysis by Suphannachart [25]. However, the panel used data in this paper was the extension which contained over-9 samples collection consecutive years.

Consider to the independent variables following Suphannachart [25] and Suphannachart and Warr [26], technical efficiency scores (TE?unbiased) and biased correctors (Biased) are the factors estimated by the boot- strapping DEA theory. These two specific variables were included in the spatial panel estimation to investigate the efficiency of Thai rice products. Next, the spatial factor (ρ) is the special one which is modified by accumulating the regional logistic measurements. This variable implies spatial effects are related to the efficiency of rice production in Thailand.

Other independent variables called planted areas (ln_land), fertilizer usages (ln ferti), rice farming families (ln_fam), and rainfalls (ln_water) are included to be the factors affecting rice productivity. These variables are observed by following the agricultural economics paper conducted by Suphannachart [25].

[1] A rai is a unit of area which equals to 1,600 square meters (16 ares, 0.16 hectares, 0.3954 acres)

[2] The standard instrument for the measurement of rainfall is the 203 mm (8 in.) rain gauge. This is essentially a circular funnel with a diameter of 203 mm which collects the rain into a graduated and calibrated cylinder. The measuring cylinder can record up to 25 mm of precipitation

Table 1 Data reviews and definitions

Variables	Definition	Data source
Dependent variable		
ln_rice	The panel samples of Thai rice productivity collected from 76 provinces during 2009 to 2017. The data was transformed into the natural log form.	Office of Agricultural Economics, Bangkok, Thailand
Independent variable		
TE_unbiased	The technical efficiency scores with bias clearance provided by the bootstrapping DEA estimation. The TE value was between 0 and 1	
Biased	The biased scores estimated by the bootstrapping DEA estimation	
ρ	The special factor indicating the effect from logistic connections in each province of Thailand. This indicator was modified for testing the regional effect inside the spatial panel estimation	GISTHAI, Department of Geology, Faculty of Science, Chulalongkorn University
ln_land	The panel data describing the number of crop areas in each province in Thailand during 2009 to 2017. The unit was counted as Rai which equaled 1,600 square meters. The data was transformed into the natural log form	Office of Agricultural Economics, Bangkok, Thailand
ln_ferti	The panel data explaining the number of fertilizer usages in the regional monitor in Thailand during 2009 to 2017. The unit was measured as tons. The data was transformed into the log form	Office of Agricultural Economics, Bangkok, Thailand
ln_fam	The panel data expressing the number of rice farmer households in each province in Thailand during 2009 to 2017. The data was modified into the natural log form	Office of Agricultural Economics, Bangkok, Thailand
ln_water	The panel data describing the measurement of rainfalls in regional areas in Thailand during 2009 to 2017. The data was already into the log form	Office of Agricultural Economics, Bangkok, Thailand

2.2 Production Technology and Technical Efficiency

The calculation of technical efficiency starts with a description of the structure of production technology. It refers to productivity with the definition of feasible sets for inputs and outputs (technology set). As found in Fare and Primont [8]; Kumbhakar and Lovell [10]; Abatania et al. [1], the technology set (T) is expressed by

$$T = \left[(x, y) : x \in \mathrm{R}_{+}^{N}, y \in \mathrm{R}_{+}^{M}, x \; can \;\; produce \;\; y\right], \tag{1}$$

where x indicates an $N \times K$ input matrix of non-negative real numbers and y denotes a non-negative $N \times M$ output matrix. The technology set can also be defined in terms of output sets and input sets. Considering into the input requirement set below, it is expressed as,

$$L(y) = [x \;\; can \;\; produce \;\; y] = \{x : (x, y) \in T\}, \tag{2}$$

where $L(y)$ is the set of inputs, which consists of all input vectors, x can produce a given output vector y. On the other hand, the output and input sets imply production possibility frontiers against which the technical efficiency performance of production activities can be measured. The distance of a production activity from the frontier is a measure of efficiency. In this case, the input orientation is used to operate the output farmer production. As the clarification by Abatania et al. [1], the input-oriented distance function is defined as follows:

$$D_i(y, x) = max\{\lambda : x|\lambda \in L(y)\}, \tag{3}$$

where $\lambda : x$ is the distance from a producer to the boundary of production possibilities. An input distance function adopts an input-conserving approach to the measurement of the distance from a producer to the boundary of production possibilities, which provides the maximum amount by a producers input vector [10].

For the boundary of production possibilities, it was sensible to state that DEA as an analysis to evaluate the technical efficiency is potentially suitable for multi-output technologies in a practical sense. Also, Bootstrapping methods have been developed for estimating the bias and correcting the efficiency estimates. Following Simar and Wilson [27]; Chaitip et al. [5], the panel bootstrapping method is conducted to explain statistical exactness of panel data envelopment analysis. The bootstrapping method was firstly proposed by Efron [7]. This method is a powerful tool for improving statistical estimations, especially for the data is not the normal distribution. Fundamentally, The panel DEA is the extension vision of DEA analysis and its TE_{it} can be expressed as,

$$TE_{it} = 0Q_{it}/0P_{it}, \tag{4}$$

where TE_{it} is the technical efficiency of each firm ($i = 1, ..., n$; the numeral identity of firms, $t = 1, ..., n$; the number of time varying). $0Q_{it}/0P_{it}$ is the ratio of technical efficiency belongs to a firm (i) at time (t). The section of efficiency calculations is the application to use the bootstrapping DEA by technically insert into the formula (3). The scores of efficient ratios are ranged between zero and one. The model can be derived as follows:

$$\bar{D}_{1\,bit}^{\,*} = \sum_{it}^{n} \left[\frac{y_{bit}^{n}}{n}, \frac{x_{bit}^{n}}{n} \right], \quad where\ \ D_i(y, x) = max\{\lambda : x|\lambda \in L(y)\}. \tag{5}$$

$\bar{D}_{1\,bit}^{\,*}$ is the TE_{it} belongs to rice productions produced from each area in Thailand. The sets of TE_{it} are provided by the bootstrapping data generation. $\hat{\emptyset}_{it(Lower*)} < \theta_{it} < \hat{\emptyset}_{it(Upper*)}$ is the confident interval of a panel bootstrapping method, where $\hat{\emptyset}_{it(Lower*)}$ is the low boundary of the bootstrapping mean and $\hat{\emptyset}_{it(Upper*)}$ is the high boundary of the bootstrapping mean.

2.3 *Bayesian Spatial Panel Econometrics for the Efficiency Score and Factors Affecting to Thai Rice Productions*

Interestingly, Bayesian inference is still computationally feasible for more complex spatial-temporal models with the advances in computing technology and Markov chain Monte Carlo (MCMC) algorithms showed that there was currently ample opportunity to resolve these computational difficulties in spatial panel data analyses and consider more complex and thus more realistic spatial panel models. Bayesian inference was novel for the analysis of spatial panel data [15, 29].

As found in many researches working on the Bayesian approach in spatial panel analyses, LeSage [14] took a Bayesian perspective on model comparison for cross-sectional and static panel data models considerably simplified the task of selecting an appropriate model. The spatial Durbin model (SDM) and spatial Durbin error model (SDEM) were compared and used to involve local spillovers. LeSage and Chih [16] extended the heterogeneous coefficients spatial autoregressive panel analysis by applying the Bayesian model for seeking a capable of products estimations of spill-in and spill-out effects for each region in the sample. Additionally, LeSage and his team [16] did apply the Bayesian approach in the spatial panel model to investigate retail fuel price rivalries, which are six categories of fuel brands, Aral, Jet, Esso, Shell, Total, and a category for independents. Moreover, Leorato and Mezzetti [13] applied the hierarchical Bayesian model for spatial panel data. The purpose of the paper was to analyze spatially dependent panel data by means of a separable covariance matrix. Recently, Li et al. [17] drawn a spatial lag panel smoothing transition regression (SLPSTR) model by considering spatial correlation of dependent variable across the panel smooth transition regression model. Thus, this section is about the general Bayesian hierarchical model for spatial areal unit data.

2.3.1 Data Likelihood

The model is fundamentally started with the data likelihood. The data region S is partitioned into n non-overlapping areal units $S = (S_1, ..., S_n)$, which are related to a corresponding set of responses $Y = (Y_1, ..., Y_n)^T$ which indicate rice products, and a vector of known offsets $O = (O_1, ..., O_n)^T$ which refer to efficient scores. The spatial pattern in the response is modeled by a matrix of covariates $X = \left(x_1^T, ..., x_n^T\right)^T$ and a set of random effects $\phi = \left(x_1^T, ..., x_n^T\right)$ which are inserted into any spatial autocorrelation that remains in the data after the covariate effects have been accounted for. The vector of covariates for areal unit S_k are symbolized by $x_k^T = \left(1, x_{k1}, ..., x_{kp}\right)$, the first element corresponds to an intercept term. The general model is an extension of a generalized linear model which is expressed as [12]

$$Y_k|\mu_k \sim f\left(y_k|\mu_k, \nu^2\right), \quad for\ k = 1, ...n, \tag{6}$$

$$g(\mu_k) = x_k^T\beta + \phi_k + O_k. \tag{7}$$

The responses Y_k can be the binomial, Gaussian or Poisson families of distributions $f\left(y_k|\mu_k, \nu^2\right)$. The expected value of Y_k is denoted by $E(Y_k) = \mu_k$, while ν^2 is an additional scale parameter that is required if the Gaussian family is employed. The expected values of the responses are related to the linear predictor via an invertible link function $g(.)$. The vector of regression parameters are indicated by $\beta\ \left(\beta_0, ..., \beta_p\right)$, non-linear covariate effects can be incorporated into the above model by including natural cubic spline or polynomial basis functions in X.

2.3.2 Prior Distributions

Considering into the starting point of the model, an independent Gaussian prior is specified for each regression parameter β_j, which contains $\beta_j \sim N(m_j, \nu_j)$ for j = 0, ..., p. The scale parameter υ^2 in the formula (6) for the Gaussian likelihood is assigned a uniform prior distribution, which is $\nu^2 \sim U(0, M_\nu)$, where the diffuse specification $M_\nu = 1$ is the default value. Hence, the simplest being the independent prior is denoted by $\theta_k \sim N(0, \sigma^2)$ and $\sigma^2 \sim U(0, M_\sigma)$, where θ_k is a substance for ϕ_k in the data likelihood in the formula (6). The variance parameter is assigned a uniform prior on the interval $(0, M_\sigma)$, where as before the default value is $M_\nu = 1$. This specification is appropriate if the covariates included in the formula (6) have removed all of the spatial structure in the response, leaving the random effects to account for the possible effects of over-dispersion (for binomial and Poisson models) [12].

3 Empirical Results

3.1 Descriptive Information for Observed Spatial Data

Descriptive statistics of the collected variables were inserted into the efficiency testing were presented in Table 2. In total 684 observations were the panel sets of received data from the Office of Agricultural Economics (OAE), Bangkok, Thailand. Obviously, data was not simply distributed due to the probability significance of Jarque-Bera testing. The maximum points and minimum points are considerably different. In other words, there was a huge gap of residual terms and fluctuated effects on the efficiency model. The data collected during 2009–2017 was the yearly time-series trends of factors affecting Thai rice production, including planted areas, fertilizer usages, the number of farming families, and rainfalls.

In terms of graphical presentation, as seen in the Appendix A, information in Figs. 3, 4, 5, 6, and 7 corresponded to each input factor (as seen in Fig. 2 in the Appendix A) such as agricultural land usages for rice growing, fertilizer utilization, and farming families. More specifically, the X axis represents regional zones of 76 provinces in Thailand. Number one to 17 are Northern provinces, 18 to 36 are Northeastern provinces, 37 to 47 Central regional provinces, 48 to 55 Eastern provinces, 56 to 62 Western provinces, and 63 to 76 are southern provinces of Thailand. Interestingly, the figures sensibly show that rice production was mostly produced in the Northern and the Central area of the country, and the Northeastern region was the area that intensively used in lands, fertilizers, and the highest number of rice farming families. On the other hands, South of Thailand was the area that rice is not economically grown.

Table 2 Descriptive information for factors affecting to rice production during 2009 to 2017

	Rice production (tons)	Crop areas (Rai)	Fertilizer usage (tons)	Rice farmer household (units)	Dailyprecipitation (mm.)
Mean	388250.1	800546.9	23047.4	49020.6	1665.9
Median	224092.5	405950.5	14794.5	27029.0	1474.5
Maximum	4314831.0	4314831.0	112045.0	243328.0	5559.4
Minimum	31.0	63.0	1.0	7.0	654.2
Std. dev.	492639.3	951749.7	26686.3	55342.1	733.4
Skewness	3.0	1.6	1.5	1.5	1.9
Kurtosis	17.7	4.8	4.5	4.3	7.5
Jarque-Bera	7165.6	369.4	319.0	293.4	989.6
Probability	0.00000	0.00000	0.00000	0.00000	0.0000
Observations	684	684	684	684	684

In the case of natural resource crucially needed to produce rice, rainfall statistics is graphically shown in Fig. 7 (seen in the Appendix A). Considering to details of the figure, yearly raining data is opposite to rice planting areas. The most highest frequencies of rainfalls are displayed in the east and south areas of the country. In particular, South of Thailand has the highest quantity of rains per year, but rice is not the main agricultural product of it. For this issue, it is undeniable that water management would be the critical problem for rice productions in Thailand.

3.2 The Empirical Results of Bootstrapping Panel Data Envelopment Analysis

In the case of efficiencies of rice producing, Figs. 8, 9, and 10 graphically present the scatter plots of technical efficiencies (TE) for Thai rice production (seen in the Appendix A). Considering into the details of Fig. 8, with the configuration of biases to the efficiency computation, the result empirically shows that all of regional areas in Thailand has low rates (scores) of efficiency in rice production since the difference of zoning management and rainfalls in each region. In terms of details in Fig. 9 (seen in the Appendix A), with the clearance of biases to the efficiency computation, the scatter graph displays a variety of efficient levels. However, the interesting point is the technical efficiency of rice producing in the north-east region. Although this area has the highest numbers of planting lands and rice farming families, the efficiency is still the lowest score when comparing with other regions, which is under 0.5 mostly. This critical issue also continuously appears in the scatter plot of bootstrapping efficiencies of rice productions with modified bias corrections (seen the details of Fig. 10 in the Appendix A). The result confirms that the inefficiency of rice producing in the north-east region is mostly under 0.4 approximately because of intensive labor inputs and the lack of technological adaptation.

3.3 Bayesian Spatial Panel Econometrics for the Efficiency Scores and Factors Affecting Thai Rice Production

According to the Bayesian spatial panel estimations and the powerful tool by MCMC simulations (10,000 simulated samples), the results obtained from the bootstrapping panel data envelopment analysis (Bootstrap DEA) in Sect. 3.2 are used to be the effect which relates to the fluctuation of Thai rice production. The empirical results presented in Table 3 indicate the strong conclusion that the spatial analysis can provide a better estimation compared with the non-spatial model. In other words, the comparison of the deviance information criterion (DIC) presented in the table strongly supported that parametric estimations by the spatial model are more precise than the counterpart.

Table 3 The empirical result of the relationship between Thai rice production and technical efficient computations by Bayesian spatial panel estimations

	Non-Spatial Model (MCSE)	Spatial Model (MCSE)
Intercept	10.9220 (0.0037)	10.4385 (0.0039)
TE_unbiased	1.9389 (0.0068)	2.2408 (0.0067)
Biased	−3.6111 (0.2660)	−2.3899 (0.1173)
ρ		0.2010 (0.0003)
DIC	4083.822	1988.331
Log marginal likelihood	−2113.3349	−1061.6926
MCMC iterations	12,500	12,500
Burn-in	2,500	2,500
MCMC sample size	10,000	10,000
Observations	684	684

Note: '*' is the Monte Carlo Standard Error (MCSE)

Table 4 The empirical result of factors affecting to Thai rice production by Bayesian random-effect panel estimations

	Bayesian mean (MCSE*)
Intercept	1.1996 (1.2137)
Lands (log form)	0.6429 (0.0122)
Fertilizer (log form)	0.4043 (0.0069)
Family (log form)	−0.0623 (0.0069)
Rainfall (log form)	−0.0611 (0.0193)
ρ	0.0382 (0.0011)
DIC	1330.943
Log marginal likelihood	−679.45367
MCMC iterations	12,500
Burn-in	2,500
MCMC sample size	10,000
Observations	684

Note: '*' is the Monte Carlo Standard Error (MCSE)

According to parametric details in the table, the spatial factor (ρ) indicates a positive effect on the rice production (0.2010). It is sensible to state that regional effects, especially logistic measurements, were the crucial element for rice production forecasting. Also, the technical efficiency (TE_unbiased) is positive on the production (2.2408). Conversely, the biased correction (Biased) which implies the miscalculation of the DEA provided the negative effect on the estimation (−2.3899).

Considering deeply into the factors affecting Thai rice production presented in Table 4, the results are the random-parametric means provided by the combination between Bayesian statistics and MCMC simulations. Planted areas and fertilizer usages are the factors positively drive the increment of rice production in every region, 1.4737 and 0.6502, respectively. On the other hand, the increasing number of rice farming families and rainfalls are the negative sign for the production.

4 Conclusion and Policy

The successful application of the Bayesian statistics and the MCMC simulation using the spatial panel data is presented in this paper. With the flexible condition and random-parametric outcomes which efficiently fit for random observations, the empirical results can be stated that they are more sensible. From this, it is a good opportunity to use the Bayesian approach in agricultural economics research, especially the spatial estimation and prediction of rice production.

Rice is the main agricultural product that is favorably grown in Thailand. It is the foodstuff that is preferably chosen in the main dish on dining tables, and it is also recorded as a main agricultural exporting product. To make a general view of rice production cannot be the best answer for implementing policies. This paper is conducted to employ panel time-series data for spatially estimating factors affecting the productivity of rice farming in Thailand, and the result confirms that the estimation of rice production should include a spatial factor for improving the reliability of parametric outcomes. Thus, 684 observations are collected as yearly panel information of seventy-six provinces during 2009–2017. The purpose of the research is divided into three sections. The first is to describe general information of rice farming situations around the country. With the bootstrapping panel data envelopment analysis (Bootstrap DEA), the second aim is to econometrically investigate efficiency of rice production in each of the 76 provinces for technically explaining the critical problem in the regional scale of rice planting areas. Lastly, the third section is to spatially estimate the panel time-series data for seeking a solution to suitably recommend agricultural policies for regional rice farming.

The general view of rice production in Thailand can report problems that essentially need to make a further research. In other words, the descriptive data indicates rice farming is mostly produced in the central-to-north belt of the country, but rainfalls are mainly recorded high in the central-to-south area of the country. This reflects the problem of water management for rice production. Another obstruction that is predominant is the lack of technology adoption in rice farming. This is reported from the empirical result by calculating the bootstrapping panel data envelopment analysis (Bootstrap DEA). It is obvious that intensive labor inputs in the northeastern region, which is the largest agricultural area in Thailand and has the highest number of rice farming families, has the lowest score of rice production efficiency. Thus, zoning management should be practically mentioned.

As a result, although the problems raised in this paper are the difficult task which cannot be solved in a short-run period, spatial panel econometrics employed in the third section can provide some useful hints to seek the success of policy launching. The empirical spatial results confirm that mass products of rice are not always produced by using a plenty of planting lands. Intensive zoning management is inevitably needed to find the most suitable places that have the capacity to do both in-season rice yields and off-season paddy fields. Intensively agricultural settlements should be realistically activated in the right place and right time. Since rice settlements are deeply depended on the number of skilled farming labors, the increment of high work-rate labors is very important. Skilled farming labors should be regionally encouraged by a state-of-art educational system. Additionally, technological adoptions should be practically implemented in some parts. Fertilizer usages are the topic that is chronic. Contact farming and self-reliance farming must be politically balanced. Also, water management is one of top difficult problems in Thailand. The technical way to move mass of water from the south area to settlement places for rice farming can potentially be the long-term solution to sustainably improve rice productions in the upcoming future.

Appendix A

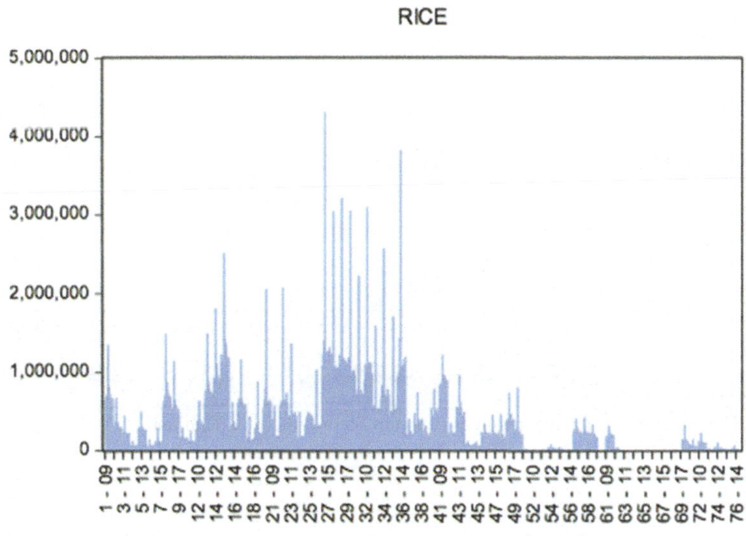

Fig. 3 The panel bar of rice productions (tons) in 76 provinces of Thailand

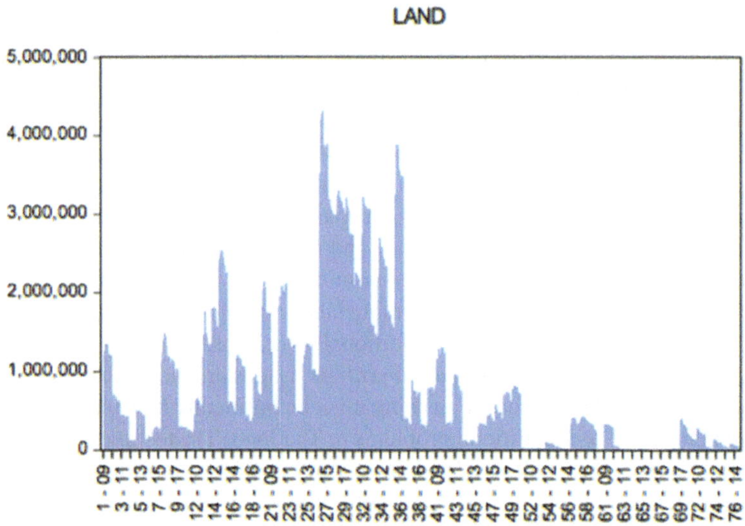

Fig. 4 The panel bar of rice planting areas (Rai) in 76 provinces of Thailand

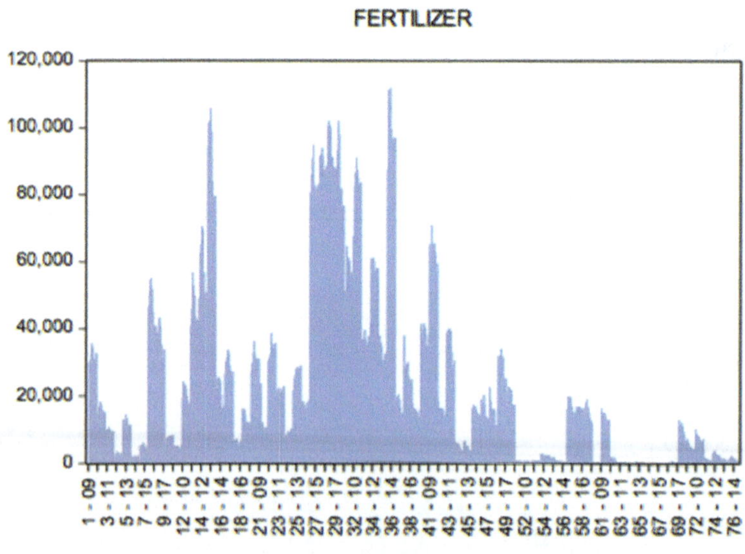

Fig. 5 The panel bar of fertilizer usages (tons) in 76 provinces of Thailand

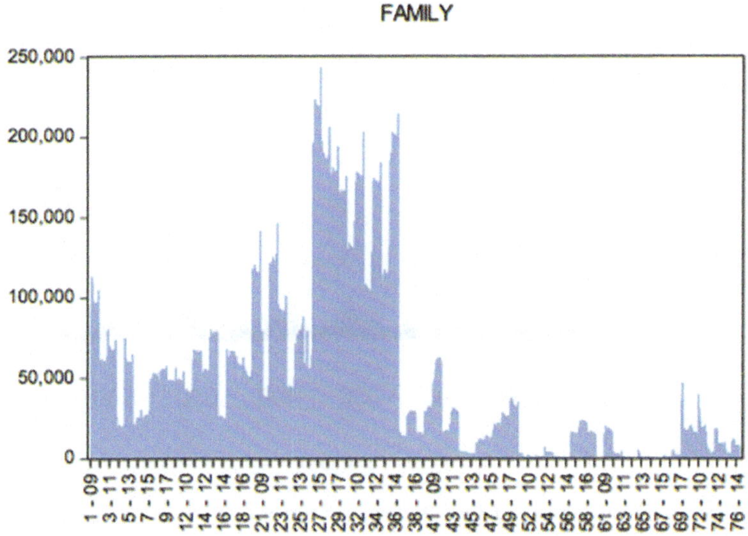

Fig. 6 The panel bar of rice farmer families (units) in 76 provinces of Thailand

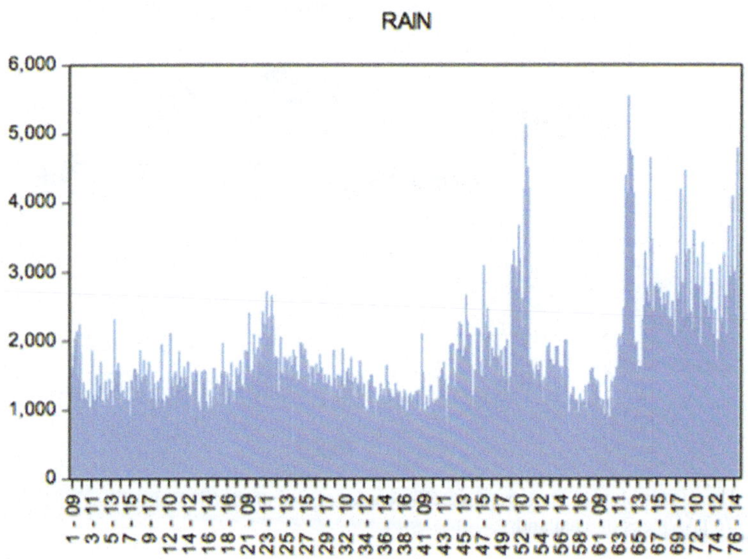

Fig. 7 The panel bar of yearly rainfall measurement (mm.) in 76 provinces of Thailand

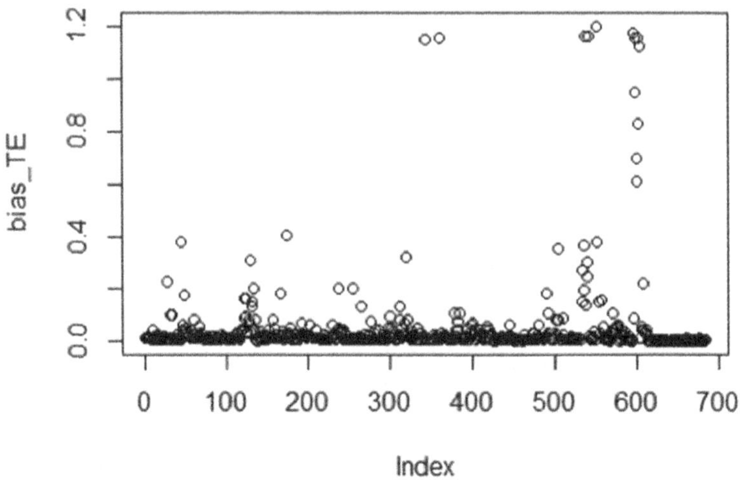

Fig. 8 The scatter plots of bootstrapping efficiencies of rice productions assuming bias corrections in 76 provinces of Thailand

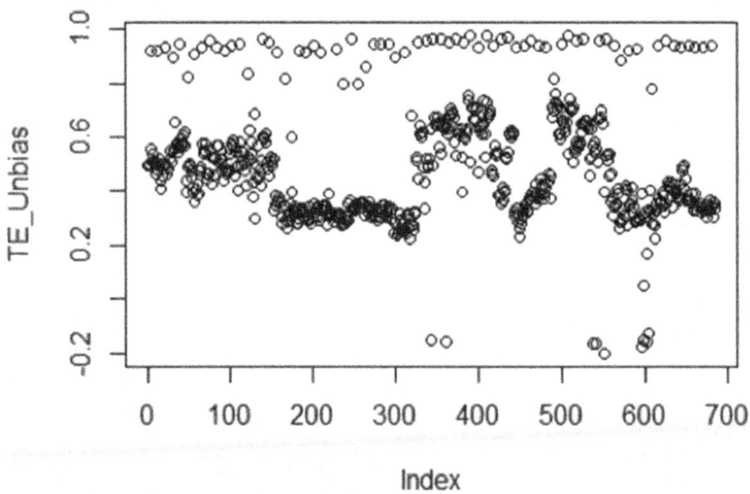

Fig. 9 The scatter plots of bootstrapping efficiencies of rice productions assuming no bias corrections in 76 provinces of Thailand

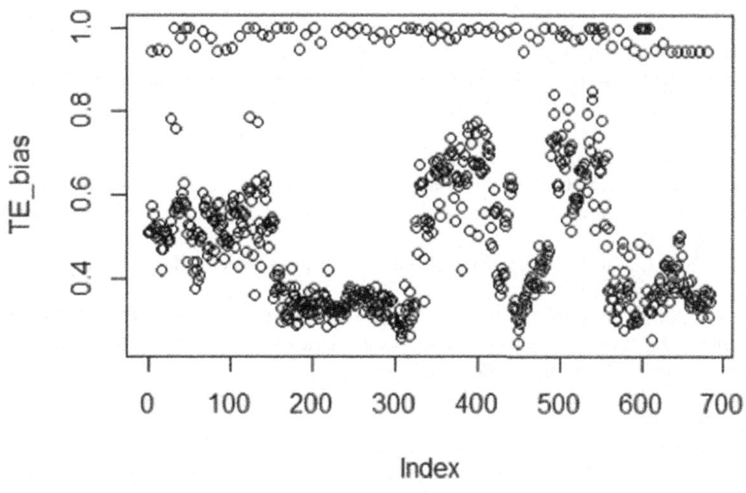

Fig. 10 The scatter plots of bootstrapping efficiencies of rice productions with modified bias corrections in 76 provinces of Thailand

References

1. Abatania, L.N., Hailu, A., Mugera, A.W.: Analysis of farm household technical efficiency in Northern Ghana using bootstrap DEA. Paper presented at the 56th Annual Conference of the Australian Agricultural and Resource Economics Society, The Esplanade Hotel, Fremantle WA, 7–10 February 2012 (2012)
2. Anselin, L.: Spatial externalities. Int. Reg. Sci. Rev. **26**(3), 147–152 (2003)
3. Anselin, L.: Spatial econometrics: methods and models. In: Studies in Operational Regional Science, vol. 4, Springer, Dordrecht (1988). https://doi.org/10.1007/978-94-015-7799-1
4. Balde, B.S., et al.: An analysis of technical efficiency of mangrove rice production in the Guinean Coastal Area. J. Agric. Sci. **6**(8), 179–196 (2014)
5. Chaitip, P., Chaiboonsri, C., Inluang, F.: The production of Thailands sugarcane: using panel data envelopment analysis (panel DEA) based decision on bootstrapping method. Procedia Econ. Financ. **14**, 120–127 (2014)
6. Conley, T.G., Udry, C.R.: Learning about a new technology: pineapple in Ghana. Am. Econ. Rev. **100**, 35–69 (2010)
7. Efron, B., Tibshirani, R.: Improvements on cross-validation: the.632+ bootstrap method. J. Am. Stat. Assoc. **92**(438), 548–560 (1997)
8. Fare, R., Primont, D.: Multi-Output Production and Duality: Theory and Applications. Kluwer Academic Publishers, Dordrecht (1995)
9. Gebregziabher, G., Namara, R.E., Holden, S.: Technical efficiency of irrigated and Rain-Fed smallholder agriculture in Tigray, Ethiopia: a comparative stochastic frontier production function analysis. Q. J. Int. Agric. **51**(3), 203–226 (2012)
10. Kumbhakar, S., Lovell, C.: Stochastic Frontier Analysis. Cambridge University Press, Cambridge (2000)
11. Langyintuo, A.S., Mekuria, M.: Assessing the influence of neighborhood effects on the adoption of improved agricultural technologies in developing agriculture. Afr. J. Agric. Resource Econ. **2**(2), 151–169 (2008)
12. Lee, C.: CARBayes: an R package for Bayesian spatial modeling with conditional autoregressive priors. J. Stat. Softw. **55**(13), 1–23 (2013)

13. Leorato, S., Mezzetti, M.: Spatial panel data model with error dependence: a Bayesian separable covariance approach. Bayesian Anal. **11**(4), 1035–1069 (2016)
14. LeSage, J.P.: Software for Bayesian spatial model comparison. Rgion et Dveloppement **40**, 11–23 (2014)
15. LeSage, J.P.: Spatial econometric panel data model specification: a Bayesian approach. SSRN (2013). https://ssrn.com/abstract=2339124
16. LeSage, J.P., Colin, V., Chih, Y.Y.: A Bayesian heterogeneous coefficients spatial autoregressive panel data model of retail fuel price rivalry. Ruhr Economic Papers, No. 617 (2016). ISBN 978-3-86788-717-5
17. Li, K., Fang, L., Lu, T.: Bayesian panel smooth transition model with spatial correlation. PLoS ONE **14**(3), 1–12 (2019)
18. Millo, G., Piras, G.: splm: Spatial panel data models in R. J. Stat. Soft. **47**(1), 1–38 (2012)
19. Maertens, A., Barrett, C.B.: Measuring social networks effects on agricultural technology adoption. Am. J. Agric. Econ. **95**(2), 353–359 (2012)
20. Ostrom, E.: Collective action and the evolution of social norms. J. Econ. Perspect. **14**(3), 137–158 (2000)
21. Tziogkidis, P.: Bootstrap DEA and hypothesis testing. Cardiff Economics Working Papers, No. E2012/18, Cardiff University, Cardiff Business School, Cardiff (2012)
22. Pede, V.O., McKinley, J., Singbo, A., Kajisa, K.: Spatial dependency of technical efficiency in rice farming: the case of bohol. Selected Paper prepared for presentation at the 2015 Agricultural & Applied Economics Association and Western Agricultural Economics Association Annual Meeting, San Francisco, CA, 26–28 July 2015 (2015)
23. Rejesus, R.M., Martin, A.M., Gypmantasiri, P.: Meta-impact assessment of the irrigated rice research consortium. IRRI Special Report, Los Banos, Philippines (2013)
24. Shantha, A., Asan Ali, B.G.H., Bandara, R.A.G.: Technical efficiency of paddy farming under major irrigation conditions in the dry-zone of Sri Lanka: a parametric approach. Aust. J. Basic Appl. Sci. **7**(6), 104–112 (2013)
25. Suphannachart, W.: Spatial analysis of research-productivity nexus: a case of Thai rice sector. In: 2017 International Conference on Applied Economics (ICOAE). Springer Proceedings in Business and Economics (2017). ISBN 978-3-319-70055-7
26. Suphannachart, W., Warr, P.: Research and productivity in Thai agriculture. Aust. J. Agric. Resource Econ. **55**, 35–52 (2011)
27. Simar, L., Wilson, W.P.: Sensitivity analysis of efficiency scores: how to bootstrap in nonparametric frontier models. Manag. Sci. **44**(1), 49–61 (1998)
28. Ward, P.S., Pede, V.O.: Capturing social network effects in technology adoption: the spatial diffusion of hybrid rice in Bangladesh. Aust. J. Agric. Resource Econ. **58**, 1–17 (2014)
29. Zheng, Y., Zhu, J., Li, D.: Analyzing spatial panel data of cigarette demand: a Bayesian hierarchical modeling approach. J. Data Sci. **6**, 467–489 (2008)

Artificial Neural Network with Histogram Data Time Series Forecasting: A Least Squares Approach Based on Wasserstein Distance

Pichayakone Rakpho, Woraphon Yamaka, and Kongliang Zhu

Abstract This paper aims to predict the histogram time series, and we use the high-frequency data with 5-min to construct the Histogram data for each day. In this paper, we apply the Artificial Neural Network (ANN) to Autoregressive (AR) structure and introduce the AR—ANN model to forecast this histogram time series. However, the conventional approach to estimate this histogram time series may not give reliable forecasting. Thus, we deal with this type of data considering the approach of Irpino-Verde [8]. The Least squares-based Wasserstein distance is employed to measure the loss function of AR—ANN model for histogram-valued data. We apply our model to forecast the US. Stock returns and examine the performance of the model with other competing models, namely AR—ANN based Billard-Diday [3] and AR models, using the root of the mean square error. The empirical results demonstrate that the AR—ANN model based Irpino-Verde approach performs better than other models.

Keywords Artificial Neural Network · Autoregressive model · Wasserstein distance · Histogram data

1 Introduction

Researchers and professionals have long investigated forecasting stock markets. Various types of models generally used to forecast the behavior of time series, such as Autoregressive (AR) model and Moving Average (MA) model, ARMA, nonlinear

P. Rakpho · W. Yamaka (✉)
Center of Excellence in Econometrics, Faculty of Economics,
Chiang Mai University, Chiang Mai, Thailand
e-mail: woraphon.econ@gmail.com

P. Rakpho
e-mail: pichayakone@gmail.com

K. Zhu
Faculty of Business Administration and Accountancy, Khon Kaen University, Khon Kaen, Thailand

S. Sriboonchitta et al. (eds.), *Behavioral Predictive Modeling in Economics*,
Studies in Computational Intelligence 897,
https://doi.org/10.1007/978-3-030-49728-6_23

AR, and nonlinear MA, etc. These models attempt to construct the time series based on a historical of the data to forecast future observations. Although these techniques are statistically powerful, they have low success rates when applied to forecast stock markets. Thus, the benefits of these models are limited in the general forecasting problem Maciel and Ballini [11]. Recently, artificial neural networks (ANNs) have been increasing consideration in forecasting theory as it has several advantageous features: (1) it provides a flexible way to investigate the nonlinear relationships among variables, (2) it has a high performance in pattern recognition and detecting complex nonlinear relationships among a large number of variables [15]. Thus, the ANN is considered as a tool for prediction the stock market. Also, we will construct the ANN structure based on the AR process; thus, the AR—ANN is introduced.

In the forecasting literature, most of the historical data is low frequencies, which may lead to unreliable forecasting. In many practices, the time series are usually collected in the form of a single value. They are almost the only type of time series that is usually considered. It was evident that the close daily price might not reflect the real behavior of the data set (See [2], [12]). Moreover, Arroyo and Maté [1] mentioned that this tradition data type does not faithfully describe the hold phenomena where a set of realizations of the observed variable is available for each time point. Thus, they suggest a Symbolic data like histogram-value data. This type of data is new and has been receiving more and more attention because they can summarize vast sets of data [8].

As the histogram data allows us to keep the variability in the data description in the intraday. So, we can preserve all the information of the data and gain more information about the prices. González-Rivera and Arroyo [6] suggested that if the volatility and distribution in the intraday might bring wrong inferences and results. Thus, the methodological approaches for analyzing this type of data is essential as it must guarantee that the information is preserved.

Many statistic methods have been proposed to make a reference to histogram-valued data. Billard and Diday [3] proposed a mid-point of the histogram. In other words, it refers to the average of the histogram-valued data. Irpino and Romano [9] suggested finding the optimal histogram representation of large data sets based on the average and center of the quantile function of the histogram. Dias and Brito [5] proposed the symmetric quantile distributions, and the representative of the histogram can be obtained from the explicative one considering the respective quantile functions. In the estimation point of views, Irpino and Verde [10] compared the performance of the least squares estimation based on the mid-point data of Billard and Diday [3], based on the quantile functions of Dias and Brito [4] and based on the quantile functions with the Wasserstein distance of Irpino and Romano [9], that is used to measure the loss function or the distance between the averages of the distributions and centered quantile distributions. They found that the technique of Irpino and Romano [9] perform better than the other two methods.

In order to forecast the stock returns, we thus consider the histogram-valued data of the stock market returns and apply the AR—ANN model to forecast this type of data. To the best of our knowledge, the histogram-valued data forecasting using neural networks is limited. Some of them are Arroyo and Maté [1], Arroyo, González-Rivera, Maté [2], Rakpho, Yamaka and Tansuchat [12]. However, none of them use the AR—ANN model to forecast the histogram-valued data. This paper thus attempts to fill the gap of applying this model to predict the stock returns. In addition, we also develop the estimation of this model by applying the Wasserstein distance of Irpino and Romano [9] to construct the loss function.

The remainder of the paper is organized as follows. In Sect. 2 we briefly explain the methodology employed. The next Sect. 3, the data and the descriptive statistics are presented. Section 4 analyses the empirical estimation results present goodness of fit indices are proposed, and Sect. 5 presents the conclusions.

2 Methodology

2.1 Histogram Data

Prior to introduce the forecasting model, AR—ANN. Let we briefly review that structure of the histogram-valued data which is a particular case of Numerical Probabilistic Symbolic Data (NPSD). Let $Y_t = \{y(1), ..., y(T)\}$ be a histogram-valued variable at time t, the histogram-valued variables can be expressed as follows:

$$H_{y(t)} = \left\{\left[\underline{I}_{y(t)1}, \overline{I}_{y(t)1}\right], \omega_{t1}; \left[\underline{I}_{y(t)2}, \overline{I}_{y(t)2}\right], \omega_{t2}; ... \left[\underline{I}_{y(t)n_t}, \overline{I}_{y(t)n_t}\right], \omega_{tn_t}\right\}, \quad (1)$$

where $\underline{I}_{y(t)i}$ and $\overline{I}_{y(t)i}$ denote the lower and upper boundary of the interval $i \in \{1, 2, ..., n_t\}$ in the histogram. n_t is the number of subintervals for the in the histogram t. Note that $\underline{I}_{y(t)i} \leq \overline{I}_{y(t)i}$ and $\underline{I}_{y(t)i+1} \geq \overline{I}_{y(t)i}$. ω_{it} is a nonnegative frequency associated to the subinterval $\left[\underline{I}_{y(t)i}, \overline{I}_{y(t)i}\right]$ and $\sum_{i=1}^{n_t} \omega_{it} = 1$. Then, according to Dias and Brito [4] we can transform the histogram $y(t)$ using the empirical quantile value, $\Omega^{-1}_{y(t)}$

$$\Omega^{-1}_{y(t)} = \begin{cases} \underline{I}_{y(j)1} + \frac{k}{w_{t1}} a_{y(t)1} & if \quad 0 \leq k < w_{t1} \\ \underline{I}_{y(t)2} + \frac{k - w_{t1}}{w_{t2} - w_{t1}} a_{y(t)2} & if \ w_{t1} \leq k < w_{t2}) \\ \vdots & \\ \underline{I}_{y(t)n_t} + \frac{k - w_{tn_{t-1}}}{1 - w_{tn_{t-1}}} a_{y(t)n_t} & if \ w_{tn_{j-1}} \leq k \leq 1 \end{cases}, \quad (2)$$

where $a_{y(t)i} = \overline{I}_{y(t)i} - \underline{I}_{y(t)i}$ with $i \in \{1, ..., n_t\}$ and $w_{t0} = 0$ and $w_{tl} = \sum_{h=1}^{n_t} \omega_{th}$. Note that when we work with histogram-valued variables, the frequency associated to the subinterval ω_{ih} and the number of subintervals in the histograms n_t may be different; the subintervals of histograms.

2.2 Autoregressive Models for Histogram-Valued Variables

An autoregressive (AR) model predicts future behavior based on past behavior. The process is basically a linear regression of the data in the current series against one or more past values in the same series. AR(p) model is an autoregressive model where specific lagged values of are used as predictor variables. Lags are where results from one time period affect following periods. The notation AR(p) indicates an autoregressive model of order p. The AR(p) model is defined as

$$\Omega^{-1}_{y(t)} = c + \sum_{i=1}^{p} \varphi_i \Omega^{-1}_{y(t-i)} + \varepsilon_t, \tag{3}$$

where c is constant term, $\Omega^{-1}_{y(t-i)}$ is the quantile function of the histogram-valued data at time $t - i$, ε_t is white noise and $\varphi_{i,\ldots,}\varphi_p$ are the parameters of the model.

2.3 Autoregressive Artificial Neural Network (AR(p)—ANN(j))

ANN is the data-driven flexible models that can approximate a large class of non-linear problems to any desired level of accuracy. The multilayer feed-forward back propagation method is used to find the optimal weight in the model [16]. In this study, we propose to fit the ANN(j) model for histogram data under the AR(p) specification. The model consists of three layers, namely input, Hidden, and output layers. Each layer is connected by the weight parameters. The network architecture for prediction our histogram-valued data is illustrated in Fig. 1.

In AR(p)—ANN(j) forecasting problem, the number of lag of inputs y_{t-i} and number of neurons (neuron size) $\sum_j$ in the hidden layer is flexible whereas the output layer has only one neuron y_t. Thus, we can evaluate the model as

$$\Omega^{-1}_{y(t)} = f\left(\alpha_0 + \sum_{j=1}^{q} \alpha_j g\left(p_{0j} + \sum_{i=1}^{p} p_{ij} \Omega^{-1}_{y(t-i)}\right)\right), \tag{4}$$

where p_{ij} is the weight between j hidden neuron and input layer, α_{ij} is the weight between j hidden neuron and output layer. v_0 and u_0 is the bias of input and output layer, respectively. g and f are the input and output activation functions, respectively. As there are many types of activation functions, in this study, we conduct the Tanh, Sigmoid, Linear and Exponential activation function.

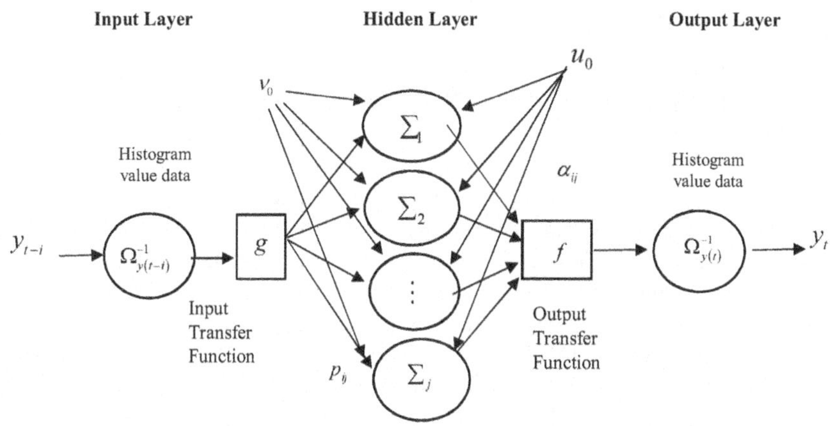

Fig. 1 The architecture of three-layer artificial neural network (ANN)

2.4 *Wasserstein Distance-Based Artificial Neural Network*

In this study, we apply the ℓ_2 Wasserstein distance of Verde and Irpino [13] to construct the loss function of the AR—ANN model. The ℓ_2 Wasserstein distance is used to measure the distance between two probability functions, which are the observed the probability function $h_1(y)$ and estimated probability function $h_2(y)$. Thus, in this study, the calculate the loss function or the distance between quantile functions of the observed y_t and estimated $\widehat{y}_t$, say $\Omega^{-1}_{y(t)}$ and $\hat{\Omega}^{-1}_{y(t)}$, respectively. Thus, we can find our loss function as

$$SSE\,(\alpha, p) = \sum_{i=1}^{n} d_w^2 \left(\Omega^{-1}_{y(t)}, \left[f \left(\alpha_0 + \sum_{j=1}^{q} \alpha_j g \left(p_{0j} + \sum_{i=1}^{p} p_{ij} \Omega^{-1}_{y(t-i)} \right) \right) \right] \right) \tag{5}$$

$$SSE\,(\alpha, p) = \sum_{t=1}^{T} \int_0^1 {\Omega^{-1}_{e(t)}}^2 dt \tag{6}$$

2.5 *The Goodness of Fit Evaluation*

In this study, we present the criteria that can be used for evaluating the goodness of fit of ANN model for histogram-valued data. The root mean square error based on the Wasserstein distance is expressed as

$$RMSE_w = \sqrt{\frac{SSE}{T}}. \tag{7}$$

3 Data Specification

To evaluate the performance of our AR—ANN for histogram valued data. We construct the model using the stock returns form the NASDAQ market of the United States (US). We collect high-frequency data with 5-minute of Blue-Chip stocks in US consist of Qualcomm Inc $(QCOM)$, Citigroup Inc. $(CITI)$ and AT&T Inc. (AT). The sample period from January 2017–April 2019, all data are collected from Bloomberg database. The movement of the high-frequency data with 5-minute of are plotted in Fig. 2, with each return data have 45,148 observations. In Table 1 presents the descriptive statistics and the result shows that in all cases, the mean is close to 0 $QCOM$ $(5.67E-06)$, $CITI$ $(2.74E-06)$ and AT $(-7.62E-06)$. The result of Jarque–Bera Normality test indicates that distributions are not normally distributed. Minimum Bayes factor (MBF) values indicate that the returns are decisive not

Table 1 Descriptive statistics

	$QCOM$	$CITI$	AT
Mean	5.67E-06	2.74E-06	−7.62E-06
Median	0.0000	0.0000	0.0000
Maximum	0.1566	0.0328	0.0409
Minimum	−0.1009	−0.0444	−0.0518
Std. Dev.	0.0021	0.0016	0.0014
Skewness	6.9582	0.4807	−2.7384
Kurtosis	868.1954	52.0174	156.3617
Jarque-Bera	1.41E+09	4521641	44301102
MBF-JB	0.0000	0.0000	0.0000
Unit root test	−51.1858	−215.3278	−214.166
MBF-Unit root	0.0000	0.0000	0.0000
Observations	45,148	45,148	45,148

Note MBF denotes Minimum Bayes factor which computed by e^{plogp}, where p is $p-$ value (see Vovk [14] and Held and Ott [7])

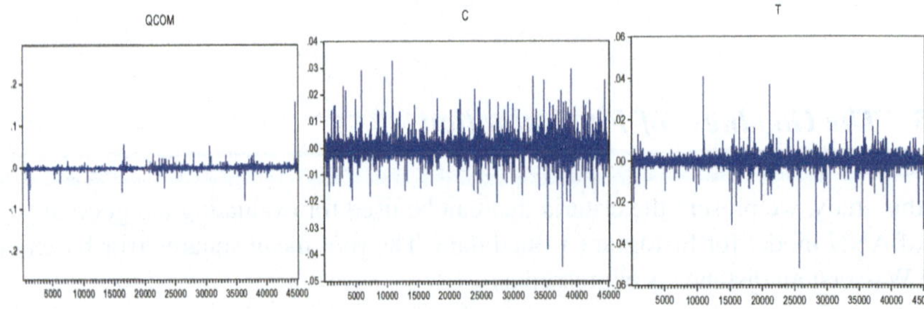

Fig. 2 High-frequency data with 5 min of Qualcomm Inc (QCOM), Citigroup Inc. (C) and AT&T Inc. (T) the sample period from January 2017–April 2019

normally distributed. Unit root test is also conducted to investigate the stationary of the data and the result shows that all returns are stationary with decisive evidence. Consequently, these variables can be used in our forecasting problem. Note that the interpretation of MBF is referred to as Vovk [14].

4 Estimated Results

We make out-sample forecasts to evaluate models' predictive abilities. To highlight predictive ability of the AR—ANN model for Histogram-valued data based on the Irpino-Verde [8] approach, we consider two competing models, say the Histogram-valued data based on the Billard-Diday [3] and the conventional AR model presented in Sect. 2.2. In addition, we also concern about the lag structure, the number of neuron and the activation functions of our models, thus lag 1, 2 and 3; the neuron sizes 1, 2, 3 and 4 together with four activation functions (Tanh, Sigmoid, Linear and Exponential) are also examined and compared in this empirical study. Thus, there are 96 AR(p−ANN(j)) specifications and 3 AR(p) specifications, totally 99 model specifications are estimated and compared in this study for each stock returns. In order to compare the three models, we computed the goodness of fit measures presented in 2.3. Each time series is divided into two subsets: in-sample and out-of-sample (test-samples). The in-sample contained approximately 80% of the data and rest was used as out-of-sample 20%. We computed the goodness of fit measures to consider best models presented in Tables 2, 3 and 4. In Tables 2, 3 and 4, we present the comparison results for out-sample forecast of $QCOM$, $CITI$, and AT, respectively. We can see that AR—ANN based Irpino-Verde approach applied to all the stock returns has good prediction capability and, as can be seen in the low rate of compared with than Billard-Diday approach as well as the conventional AR model, this model is capable of learning from the data and can process good results for forecasting. It can be seen that the AR(1)—ANN(2) model based Irpino-Verde approach with tanh activation function is perform the best specification for $QCOM$ stock return. Likewise, it can be verified that the AR(2)—ANN(1) model based Irpino-Verde approach with tanh activation function gives better performance than other specifications for $CITI$ stock returns. Finally, in the case of AT, AR—ANN, AR(3)—ANN(4) model based Irpino-Verde approach with expo activation function outperformed the other model specifications. According these finding, it can be clearly observed that AR—ANN model based Irpino-Verde approach has higher histogram prediction accuracy in terms of than AR—ANN model based Billard-Diday approach and conventional AR model.

Table 2 The goodness of fit indices of Qualcomm Inc (QCOM) Histogram data

Qualcomm Inc (QCOM)

Lag1

Billard-Diday model

Goodness of fit	size=1				size=2				size=3				size=4				AR model
	tanh	sigmoid	linear	exp	tanh	sigmoid	linear	exp	tanh	sigmoid	linear	exp	tanh	sigmoid	linear	exp	
$RMSE_W$	0.00124	0.00127	0.00128	0.00126	0.00127	0.0013	0.00125	0.00124	0.00131	0.00126	0.00125	0.00126	0.00122	0.00127	0.0012	0.00128	0.00122

Irpino-Verde model

Goodness of fit	size=1				size=2				size=3				size=4				AR model
	tanh	sigmoid	linear	exp	tanh	sigmoid	linear	exp	tanh	sigmoid	linear	exp	tanh	sigmoid	linear	exp	
$RMSE_W$	0.00103	0.00145	0.00107	0.0009	0.00082	0.00089	0.00105	0.00143	0.00096	0.00105	0.00111	0.00088	0.00095	0.00095	0.00098	0.00106	0.00117

Lag2

Billard-Diday model

Goodness of fit	size=1				size=2				size=3				size=4				AR model
	tanh	sigmoid	linear	exp	tanh	sigmoid	linear	exp	tanh	sigmoid	linear	exp	tanh	sigmoid	linear	exp	
$RMSE_W$	0.00124	0.00125	0.00129	0.00129	0.00124	0.0013	0.0012	0.00124	0.00129	0.00128	0.00128	0.00122	0.00125	0.00131	0.00124	0.00127	0.0012

Irpino-Verde model

Goodness of fit	size=1				size=2				size=3				size=4				AR model
	tanh	sigmoid	linear	exp	tanh	sigmoid	linear	exp	tanh	sigmoid	linear	exp	tanh	sigmoid	linear	exp	
$RMSE_W$	0.00089	0.00095	0.00086	0.00125	0.00093	0.00101	0.00107	0.00118	0.00092	0.00096	0.00085	0.00089	0.00098	0.00097	0.00096	0.00092	0.00098

Lag3

Billard-Diday model

Goodness of fit	size=1				size=2				size=3				size=4				AR model
	tanh	sigmoid	linear	exp	tanh	sigmoid	linear	exp	tanh	sigmoid	linear	exp	tanh	sigmoid	linear	exp	
$RMSE_W$	0.00121	0.00125	0.00125	0.00129	0.00127	0.00121	0.00117	0.00123	0.00129	0.00118	0.00125	0.00122	0.00128	0.0012	0.00123	0.00121	0.00135

Irpino-Verde model

Goodness of fit	size=1				size=2				size=3				size=4				AR model
	tanh	sigmoid	linear	exp	tanh	sigmoid	linear	exp	tanh	sigmoid	linear	exp	tanh	sigmoid	linear	exp	
$RMSE_W$	0.00121	0.00125	0.00138	0.00131	0.00129	0.00126	0.00124	0.00131	0.00133	0.00121	0.00119	0.00123	0.00123	0.00128	0.00123	0.00129	0.00091

Table 3 The goodness of fit indices of Citigroup Inc. (C) Histogram data

Citigroup Inc. (C)

Lag1

Billard-Diday model

Goodness of fit	size=1				size=2				size=3				size=4				AR model
	tanh	sigmoid	linear	exp	tanh	sigmoid	linear	exp	tanh	sigmoid	linear	exp	tanh	sigmoid	linear	exp	
$RMSE_W$	0.000704	0.000697	0.000715	0.000706	0.000694	0.000708	0.0007	0.00073	0.000692	0.000711	0.000708	0.000703	0.000764	0.000713	0.000686	0.000695	0.000698

Irpino-Verde model

Goodness of fit	size=1				size=2				size=3				size=4				AR model
	tanh	sigmoid	linear	exp	tanh	sigmoid	linear	exp	tanh	sigmoid	linear	exp	tanh	sigmoid	linear	exp	
$RMSE_W$	0.000691	0.000702	0.000694	0.000736	0.000691	0.000704	0.000713	0.0007	0.000684	0.000699	0.000706	0.00072	0.000707	0.000734	0.000703	0.00071	0.000697

Lag2

Billard-Diday model

Goodness of fit	size=1				size=2				size=3				size=4				AR model
	tanh	sigmoid	linear	exp	tanh	sigmoid	linear	exp	tanh	sigmoid	linear	exp	tanh	sigmoid	linear	exp	
$RMSE_W$	0.000693	0.000704	0.000697	0.000705	0.000729	0.000718	0.000704	0.000704	0.000745	0.000702	0.000695	0.000709	0.000737	0.000719	0.000702	0.000705	0.0007

Irpino-Verde model

Goodness of fit	size=1				size=2				size=3				size=4				AR model
	tanh	sigmoid	linear	exp	tanh	sigmoid	linear	exp	tanh	sigmoid	linear	exp	tanh	sigmoid	linear	exp	
$RMSE_W$	0.00069	0.000703	0.000705	0.000704	0.000725	0.000697	0.000722	0.000705	0.000694	0.0007	0.000735	0.000697	0.000713	0.000706	0.0007	0.000693	0.000699

Lag3

Billard-Diday model

Goodness of fit	size=1				size=2				size=3				size=4				AR model
	tanh	sigmoid	linear	exp	tanh	sigmoid	linear	exp	tanh	sigmoid	linear	exp	tanh	sigmoid	linear	exp	
$RMSE_W$	0.000697	0.000712	0.000706	0.000707	0.000715	0.000699	0.000707	0.0007	0.000762	0.000702	0.000696	0.000705	0.000762	0.000702	0.000707	0.000704	0.000704

Irpino-Verde model

Goodness of fit	size=1				size=2				size=3				size=4				AR model
	tanh	sigmoid	linear	exp	tanh	sigmoid	linear	exp	tanh	sigmoid	linear	exp	tanh	sigmoid	linear	exp	
$RMSE_W$	0.000694	0.000694	0.000705	0.000713	0.000698	0.000698	0.000697	0.000702	0.000704	0.000699	0.000739	0.000698	0.000701	0.000693	0.000705	0.000707	0.000696

Table 4 The goodness of fit indices of AT&T Inc. (T) Histogram data

AT&T Inc. (T)																	
Lag1																	
Billard-Diday model																	
Goodness of fit	size=1				size=2				size=3				size=4				AR model
	tanh	sigmoid	linear	exp	tanh	sigmoid	linear	exp	tanh	sigmoid	linear	exp	tanh	sigmoid	linear	exp	
$RMSE_W$	0.000758	0.000764	0.000767	0.000794	0.00075	0.000768	0.000753	0.000761	0.000748	0.000788	0.000748	0.00075	0.000756	0.000772	0.000757	0.000813	0.000752
Irpino-Verde model																	
Goodness of fit	size=1				size=2				size=3				size=4				AR model
	tanh	sigmoid	linear	exp	tanh	sigmoid	linear	exp	tanh	sigmoid	linear	exp	tanh	sigmoid	linear	exp	
$RMSE_W$	0.000756	0.000748	0.000749	0.000806	0.000748	0.000757	0.00075	0.00076	0.00075	0.000749	0.000741	0.000759	0.000762	0.000784	0.000761	0.000754	0.00073
Lag2																	
Billard-Diday model																	
Goodness of fit	size=1				size=2				size=3				size=4				AR model
	tanh	sigmoid	linear	exp	tanh	sigmoid	linear	exp	tanh	sigmoid	linear	exp	tanh	sigmoid	linear	exp	
$RMSE_W$	0.00082	0.00077	0.00073	0.00075	0.00079	0.00076	0.00078	0.00075	0.00074	0.00078	0.00076	0.00075	0.00076	0.00077	0.00073	0.00077	0.00076
Irpino-Verde model																	
Goodness of fit	size=1				size=2				size=3				size=4				AR model
	tanh	sigmoid	linear	exp	tanh	sigmoid	linear	exp	tanh	sigmoid	linear	exp	tanh	sigmoid	linear	exp	
$RMSE_W$	0.000748	0.000766	0.000755	0.000752	0.000768	0.000769	0.00074	0.000753	0.000759	0.000748	0.000749	0.000753	0.000784	0.000744	0.00075	0.000742	0.000748
Lag3																	
Billard-Diday model																	
Goodness of fit	size=1				size=2				size=3				size=4				AR model
	tanh	sigmoid	linear	exp	tanh	sigmoid	linear	exp	tanh	sigmoid	linear	exp	tanh	sigmoid	linear	exp	
$RMSE_W$	0.000758	0.000818	0.000787	0.000729	0.00075	0.000804	0.000761	0.000742	0.000797	0.000806	0.00077	0.000757	0.000763	0.000766	0.000757	0.000739	0.00074
Irpino-Verde model																	
Goodness of fit	size=1				size=2				size=3				size=4				AR model
	tanh	sigmoid	linear	exp	tanh	sigmoid	linear	exp	tanh	sigmoid	linear	exp	tanh	sigmoid	linear	exp	
$RMSE_W$	0.000739	0.000757	0.000793	0.000753	0.000747	0.00075	0.000759	0.000773	0.000779	0.000747	0.000738	0.00075	0.000775	0.000778	0.00075	0.00072	0.000754

5 Conclusion

In investment decision making, stock price prediction is an essential activity for both financial firms and private investors. Therefore, the observed historical data is used to forecast future data. However, most of the forecasting research normally used the single-valued data each day, which may not represent the full information of the data. To account for the variability or uncertainty of the stock return within the day, this study considers this histogram-valued data which is a particular kind of data studied in Symbolic Data Analysis. Each entity under analysis corresponds to a distribution that may be represented by a histogram or by a quantile function. In the forecasting model point of view, one of the most used methods for forecasting is the Autoregressive (AR) model. Nonetheless, the errors in prediction using this approach are often quite high. Hence, the recent researches attempted to improve the forecasting models employing a variety of techniques. In this paper, we extend the field of expert systems, forecasting, and model by applying an Artificial Neural Network (ANN) to AR model generating an AR—ANN. Besides, we also applied the ordinary least squares based on Wasserstein distance to obtain the optimal weight parameters between the neurons and input and output layers.

The AR—ANN model, is applied to forecast the Blue-Chip stocks in the US the sample period from January 2017–April 2019, all data are collected from Bloomberg database. We deal with the histogram data in AR—ANN using the approach of Irpino and Verde [8]. The performance of the model is examined by comparing with the AR(1)—ANN based on Billard-Diday [3] approach as well as the conventional AR model. We conduct the $(RMSE_W)$ as the goodness-of-fit evaluation in this comparison. We note that the predicted and observed quantile functions for each unit are used in the $RMSE_W$ measure.

The results show an overall improvement in forecasting using the AR—ANN method as compared to an AR model, it is found that the AR—ANN model fits better than the AR model in term of $RMSE_W$. In addition, our study also confirms a higher performance of the Irpino and Verde approach over the Billard-Diday approach in dealing with the histogram-valued data forecasting.

References

1. Arroyo, J., Maté, C.: Forecasting histogram time series with k-nearest neighbours methods. Int. J. Forecast. **25**(1), 192–207 (2009)
2. Arroyo, J., González-Rivera, G., Maté, C.: Forecasting with interval and histogram data. Some financial applications. Handbook of empirical economics and finance, pp. 247–280 (2010)
3. Billard, L., Diday, E.: Symbolic Data Analysis: Conceptual Statistics and Data Mining. Wiley, Hoboken (2006)
4. Dias, S., Brito, P.: A new linear regression model for histogram-valued variables. In: 58th ISI World Statistics Congress, Dublin, Ireland (2011). http://isi2011.congressplanner.eu/pdfs/950662.pdf

5. Dias, S., Brito, P.: Linear regression model with histogram–valued variables. Statistical Analysis and Data Mining: The ASA Data Science Journal, 8(2), 75–113 (2015). Dueker, M.J.: Markov switching in GARCH processes and mean-reverting stock-market volatility. Journal of Business and Economic Statistics, 15(1), 26–34 (1997)
6. González-Rivera, G., Arroyo, J.: Time series modeling of histogram-valued data: the daily histogram time series of S&P500 intradaily returns. Int. J. Forecast. **28**(1), 20–33 (2012)
7. Held, L., Ott, M.: How the maximal evidence of p-values against point null hypotheses depends on sample size. Am. Statist. **70**(4), 335–341 (2016)
8. Irpino, A., Verde, R.: A new Wasserstein based distance for the hierarchical clustering of histogram symbolic data. In: Data Science and Classification, pp. 185–192. Springer, Heidelberg (2006)
9. Irpino, A., Romano, E.: Optimal histogram representation of large data sets: Fisher vs piecewise linear approximation. In: Noirhomme-Fraiture, M., Venturini, G. (Eds.) EGC, vol. RNTI-E-9 of Revue des Nouvelles Technologies de l'Information, Cepadu es-Editions, pp. 99–110 (2007)
10. Irpino, A., Verde, R.: Linear regression for numeric symbolic variables: a least squares approach based on Wasserstein Distance. Adv. Data Anal. Classif. **9**(1), 81–106 (2015)
11. Maciel, L.S., Ballini, R.: Neural networks applied to stock market forecasting: an empirical analysis. J. Brazilian Neural Netw. Soc. **8**(1), 3–22 (2010)
12. Rakpho, P., Yamaka, W., Tansuchat, R.: Risk valuation of precious metal returns by histogram valued time series. In: International Conference of the Thailand Econometrics Society, pp. 549–562. Springer, Cham (2018)
13. Verde, R., Irpino, A.: Dynamic clustering of histogram data: using the right metric. In Selected contributions in data analysis and classification, pp. 123–134. Springer, Heidelberg (2007)
14. Vovk, V.G.: A logic of probability, with application to the foundations of statistics. J. Roy. Stat. Soc.: Ser. B (Methodol.) **55**(2), 317–341 (1993)
15. Xu, Q., Zhuo, X., Jiang, C., Liu, Y.: An artificial neural network for mixed frequency data. Expert Syst. Appl. **118**, 127–139 (2019)
16. Zhang, G.P.: Time series forecasting using a hybrid ARIMA and neural network model. Neurocomputing **50**, 159–175 (2003)

The Determinants of Planned Retirement Age of Informal Worker in Chiang Mai Province, Thailand

Pimonpun Boonyasana and Warattaya Chinnakum

Abstract This study examines the determinants of planned retirement age of informal workers in Chiang Mai Province, Thailand. The least absolute shrinkage and selection operator estimator regression (LASSO) and the ordinary least squares estimator regression (OLS) are employed to determine variables important for planned retirement age equation. The computational cost of this method is we need to take into account 2 times fewer variables compared with standard approaches in the simple regression literature. A cross-sectional study is conducted among informal workers in Chiang Mai Province, Thailand. A total of 400 informal workers are enrolled. A number of important participants, demographic characteristics, financial variables and health factors are considered in the planned retirement age equation. The study compares the results of two methods, namely the LASSO and the OLS. According to the minimum value of The out-of-sample Root Mean Squared Error, the LASSO performs better than the OLS. The results suggest that age of respondent, number of children, own accommodation, sex of respondent, and marital status (single) have positive impact on planned retirement age. In contrast, cost of bone disease treatment, daily expenses, caretaker after retirement, chronic disease, education level, cost of treatment for diseases, income after retirement, financial burden after retirement, and savings group deposits have negative impact on planned retirement age.

Keywords Planned retirement age · LASSO · Informal workers

1 Introduction

The informal sector is important in most developing countries, particularly in Thailand. Although there is no unique definition of the informal economy, it typically refers to economic transactions that are not explicitly observed or recognized.

P. Boonyasana (✉) · W. Chinnakum
Centre of Excellence in Econometrics, Faculty of Economics,
Chiang Mai University, Chiang Mai, Thailand
e-mail: pui.econ@gmail.com

W. Chinnakum
e-mail: warattaya_chin@hotmail.co.th

S. Sriboonchitta et al. (eds.), *Behavioral Predictive Modeling in Economics*,
Studies in Computational Intelligence 897,
https://doi.org/10.1007/978-3-030-49728-6_24

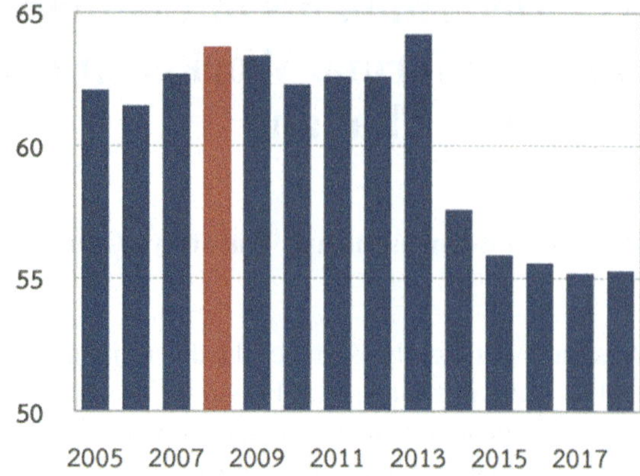

Fig. 1 The Informal Employment Survey in 2005–2017 (% of Total Employment)

In Thailand, the National Statistical Office (NSO) has initiated the Informal Employment Annual Survey, with the first one conducted in 2005. This annual survey serves as the main data source for estimation of the size of the informal sector. The survey is usually conducted during the third quarter of each year simultaneously with the labor force survey. The definition of informal employment used by the [10] covers employed persons who are not protected and not eligible for social security provision. [10] revealed that in 2017 about 21.2 million workers, or 55.3% of the total employment (Fig. 1), were engaged in the informal sector, with proportion of male and female workers not much different, i.e., 11.5 million or 55.2% and 9.5 million or 44.8%, respectively. When comparing the results for 2005–2017, it was found that the proportion of informal employment remained stable throughout the period. However, MacKellar [6] reported that the Thai pension system covers only about 10.5 million people or around 30% of the labor force, leaving out the two-thirds of Thailand's 36 million labor force that are estimated to be in the informal sector.

People's planned retirement age has had a huge impact on many economic and social issues, especially the economic welfare and the social security system. The accurately estimated people's planned retirement age is important for the social security reserve. According to the data from the Thailand's National Statistical Office [10] in 2018 which are shown in Fig. 2, people age 60 and over are made up of 18.2% of the total labor force. The data from the population estimates and projections conducted by the National Economic and Social Development Board [9] indicate that Thailand will become a complete aged society[1] in 2021, when one out of five Thais will be age 60 years or older. It is further projected that Thailand will become a super aged society[2] in ten years later, in 2031, when 28% of the population will be elderly. Thus the impact

[1]Complete aged society denotes a society in which the proportion of population age 60 years or more is more than 20% of the total population (or where the population age 65 years or more is over 14% of the total).

[2]Super aged society denotes a society in which the proportion of population age 60 years or more is more than 28% of the total population (or where the population age 65 years or more is over 20% of the total).

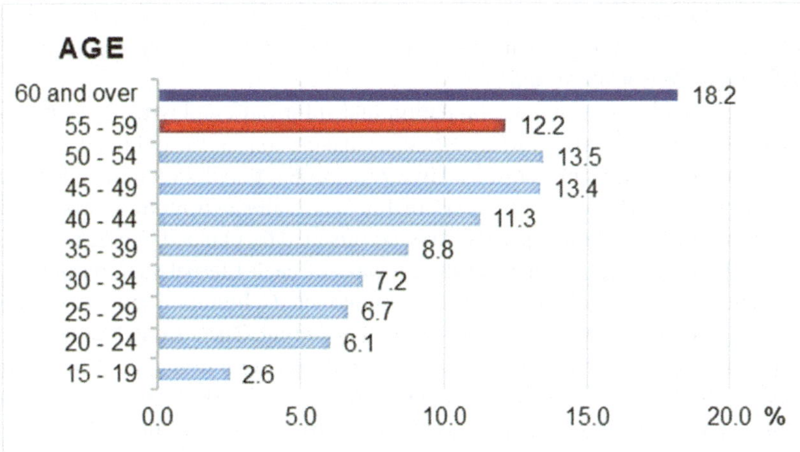

Fig. 2 Age profile of the labour force of Thailand in 2018

of the increasing number of senior citizens, on the country's labor demographics and the poverty among senior citizens must be monitored and prepared for.

A survey report on older persons in Thailand in 2007–2017 by the Foundation of Thai Gerontology Research and Development Institute [4] points out that the top three sources of income for the elderly in 2007 were children (52%), employment (29%), and spouse (6%) (Table 1). In 2011 and 2017, the top two sources of income were children and employment, but the proportion had declined. Instead, the third most common source of income became the elderly welfare subsidy, and that showed sign of increasing importance.

As for the working condition of the Thai elderly, 4,361,331 persons or 36.95% of the total aged population are employed. The labor-force participation rates of Thai elderly persons are quite high and higher than the world average (30.2%). Totally 3,847.6 thousand persons or 88.22% of total aged employment are informal employment.[3]

Table 1 Percent of source of income of elderly, 2007–2017

Source of income	2007	2011	2017
Children	52.3	40.1	34.9
Employment	28.9	35.1	30.9
Spouse	6.1	11.4	19.9
Saving	2.9	2.6	2.3

Source Foundation of Thai Gerontology Research and Development Institute [4]

[3]Informal workers consists of those working in the informal sector or households, excluding regular workers with social security benets provided by the employers and the workers in the formal sector without any employment and social security benets provided by the employers [8].

Chiang Mai is the second-largest province of Thailand. Its workforce consists mostly of informal workers totaling 624,616 people or 60.18%, whereas 413,365 or 39.82% are in the formal system.

The high incidence of aging population and informal employment are a major challenge for the Thai government and policymakers to provide public expenditure on pensions and health care services. The ageing of the population may reduce economic growth but, if those extra years of life were allocated more to work rather than to retirement, economic growth may be raised. The planned age of retirement is obviously the key determinant of the division of each life between participation and dependency. However, Scharn et al. [11] conducted a literature search to identify relevant articles with determinants on retirement timing. Their result found that the determinants of retirement timing were classified into eight domains: demographic factors, health factors, social factors, social participation, work characteristics, financial factors, retirement preferences, and macro effects. In total, they identified 49 determinants, ranging from one (social, and retirement preferences) to 21 determinants (work characteristics) per domain. In our study, to reduce the absolute values of regression parameter estimates and improve prediction accuracy, we employed Least Absolute Shrinkage and Selection Operator (LASSO) that was proposed by Tibshirani [12]. The objective of this paper is to analyze the determinants of planned retirement age of informal workers in Chiang Mai Province, Thailand.

The remainder of the paper is organized as follows. Section 2 contains a brief literature review. In Sect. 3 we present a rigorous description of the methodology used in the analysis. Section 4 describes the data and presents the estimated models and empirical results. Finally, Sect. 5 discusses the findings and draws conclusions.

2 Literature Review

The existing literature on the retirement planning issues is wide ranging both in terms of the different techniques employed and in terms of the determinants. There have been many studies on the planned retirement age. For example, Montalto et al. [7] analyzed the planned retirement age of respondents in USA. They included financial, demographic and perception variables, and found 13 of the 28 independent variables to be significant. The significant determinants included non-investment income, financial and non-financial assets, IRA/Keogh plans (these are US individual and self-employed retirement plans), defined-benefit superannuation, less-skilled occupations, lower education levels, age, race, and life expectancy. Van Erp et al. [13] studied non-financial determinants of retirement in the Netherlands. They concluded that social norms, default options, and reference-dependent utility are likely explanations for the observed individual propensity to retire at standard retirement ages. De Souza et al. [2] used country comparable census data for 1970–2010 to analyse the labor force participation trends and health of male workers aged 50 years and over in for 17 Latin American countries. Their result revealed, for instance, older men aged 60–64 have very similar health patterns compared to those aged 55–59, but very different patterns of labor force participation. The result indicated that the same health status translates into lower labor force participation today than in the past and a potential to raise retirement ages for an extended working life period, given improvements in health over the recent decades.

Most of them used traditional estimation or ordinary least squares (OLS) estimation and several determinants provided in the literature. However, the Least Absolute Shrinkage and Selection Operator (LASSO) estimation has not yet been investigated in the planned retirement age literature.

3 Research Methodology

3.1 Least Absolute Shrinkage and Selection Operator (LASSO)

Prior to our describing the Least Absolute Shrinkage and Selection Operator (LASSO) estimation in the regression framework, let us briefly explain the concept of the conventional Ordinary Least Squares (OLS) estimation. Considering the linear regression model:

$$\begin{aligned} Y_i &= X_i'\beta + \varepsilon_i \\ &= \beta_0 + x_{i1}\beta_1 + \ldots + x_{ik}\beta_k + \varepsilon_i \end{aligned} \tag{1}$$

where Y_i and X_i are the response and explanatory variables of respondent i. $\beta = (\beta_0, \beta_1, \ldots, \beta_k)$ is the estimated parameter and k is the number of explanatory variables. ε_i is referred to as the error whereas the distribution is normal. The parameters of this model are estimated by loss function minimization. Thus, we can obtain the optimal parameters by using OLS.

$$\left(\widehat{\beta}_{OLS}\right) = \arg\min_{\beta} \sum_{i=1}^{N} \left[Y_i - \beta_0 - \sum_{j=1}^{k} x_{ij}\beta_j \right]^2 \tag{2}$$

Then, we turn our attention to LASSO which is the extension of the OLS. In this estimation, the loss function of OLS is minimized with the additional constraint, so-called lasso penalty, which is $|\beta| \leq t$ and t is the upper bound for the sum of the coefficients. Thus, this is an estimation method that involves penalizing the absolute size of the regression coefficients [12]. In other words, LASSO minimizes the mean squared error subject to a penalty on the absolute size of coefficient estimates.

$$\left(\widehat{\beta}_{LASSO}\right) = \arg\min_{\beta} \sum_{t=1}^{N} \left[Y_i - \beta_0 - \sum_{j=1}^{k} x_{ij}\beta_j \right]^2 \tag{3}$$

$$\text{subject to} \|\beta\|_1 \leq t$$

where $\|\beta\|_1 \equiv \sum_{j=1}^{P} |\beta_j|$ is the L1 norm and $t \geq 0$ is a tuning parameter.

The LASSO problem can be equivalently written in Lagrangian form as:

$$\left(\widehat{\beta}_{LASSO}\right) = \arg\min_{\beta} \left(\frac{1}{2} \sum_{t=1}^{N} \left[Y_i - \beta_0 - \sum_{j=1}^{k} x_{ij}\beta_j \right]^2 + \lambda \|\beta\|_1 \right) \tag{4}$$

As is common practice, λ is allowed to vary and a value for it is chosen through cross validation. The larger the value of λ, the greater the amount of shrinkage. The relation between λ and the upper bound t is a reverse relationship. Indeed as t becomes infinity, the problem becomes an ordinary least squares and λ becomes 0. Viceversa as t becomes 0, all coefficients shrink to 0 and λ goes to infinity [1]. In this research paper, we are going to use LASSO for its variable selection property. When we minimize the optimization problem some coefficients are shrank When we minimize the optimization problem some coefficients are shrank to zero, i.e. $\hat{\beta}_j(\lambda) = 0$, for some values of j (depending on the value of the parameter λ). In this way the features with coefficient equal to zero are excluded from the model. Tibshirani [12] motivated the LASSO with two major advantages over OLS. First, due to the nature of the lasso penalty, the LASSO could force parameter estimates to be zero. In this sense, this estimator can remove irrelevant independent variables out from the regression model. Secondly, LASSO can outperform OLS in terms of prediction accuracy due to the bias-variance trade off.

4 Data and Results

4.1 Data

The study employs the quantitative method to obtain the relevant primary data. Sampling was drawn from informal workers which age between 55 and over and work in Chiang Mai Province, Thailand, where there was a projected informal workers of 854,630 person in 2016 as against the 300,000 persons in 2000 [10]. A total of 400 questionnaires were randomly distributed. The data were collected by survey. Respondents were asked when they planned (or had) retirement, and the reasons which may (or did) influence their decisions to retire. They were also asked about their demographic characteristics (age, marital status, education level, etc.) and financial information (income, expenditure and the use of financial institutions).

In this study, the LASSO (Least Absolute Shrinkage and Selection Operator), presented by Tibshirani [12], regression algorithm is employed to implement the variable selection for these feature variables.

The dependent variable is the planned retirement age of the respondent. There are 36 independent variables in the planned retirement age equation. The independent variables include respondent's demographic characteristics, financial variables, variables capturing access to resources, and health factors. Table 2 gives a list of all covariates used and a description of each variable. Means for continuous variables and percentages for dichotomous variables also are presented in Table 2. The LASSO and OLS equation can be represented by

$$\begin{aligned} Planned_i =& \beta_0 + \beta_1 AGE_i + \beta_2 SEX_i + \beta_3 Single_i + \beta_4 Married_i + \beta_5 EDU_i + \beta_6 HH_i \\ &+ \beta_7 Child_i + \beta_8 Grandchild_i + \beta_9 Income_i + \beta_{10} HHinc_i + \beta_{11} DE_i \\ &+ \beta_{12} SE_i + \beta_{13} ME_i + \beta_{14} TF_i + \beta_{15} Dep_i + \beta_{16} VFund_i + \beta_{17} Coop_i \\ &+ \beta_{18} Saveg_i + \beta_{19} Cash_i + \beta_{20} House_i + \beta_{21} Car_i + \beta_{22} Moto_i \\ &+ \beta_{23} LifeInsurance_i + \beta_{24} IncomeRet_i + \beta_{25} LoadRet_i + \beta_{26} Depen_i \\ &+ \beta_{27} Insure60_i + \beta_{28} HOME_i + \beta_{29} Disea_i + \beta_{30} Hypertension_i \\ &+ \beta_{31} Eye_i + \beta_{32} Bone_i + \beta_{33} Disbles_i + \beta_{34} Heart_i + \beta_{35} Deafness_i \\ &+ \beta_{36} Paresis_i + \varepsilon_i \end{aligned} \tag{5}$$

where $i = 1, 2, ..., \mathrm{N}$ represents sample. β are vectors of unknown parameters to be estimated.

4.2 Results

Table 3 presents the regression model of informal workers. The contents in Table 3 include the LASSO regression ($\hat{\beta}_{LASSO}$), OLS regression ($\hat{\beta}_{OLS}$), and stepwise regression ($\hat{\beta}_{Stepwise}$) coefficient. The out-of-sample Root Mean Squared Error (RMSE) is a standard commonly used for selecting statistical model. A good model is the one that has minimum out-of-sample RMSE among all the other models. According to Table 3, LASSO regression can perform better than OLS regression and stepwise regression.

When considering the LASSO regression, the following 19 variables, out of the total 36 variables, are important to explain the planned retirement age, listed in alphabetical order of variable abbreviation, where we also show the sign of the corresponding coefficient in parenthesis: AGE_i (Age of respondent, positive), $Bone_i$ (Cost of bone disease treatment, negative), Car_i (Car installments, positive), $Child_i$ (Number of children, positive), DE_i (Daily expenses, negative), $Depen_i$ (Caretaker after retirement, negative), $Disea_i$ (Chronic disease, negative), $Disbles_i$ (Cost of treating diabetes, positive), EDU_i (Education level, negative), $HOME_i$ (Own accommodation, positive), $Hypertension_i$ (Cost of treatment for blood pressure, negative), $IncomeRet_i$ (Income after retirement, negative), $LoadRet_i$ (Financial burden after

Table 2 Bigrams frequency.

Variable	Description	Mean/Percent
$Planned_i$ (dependent variable)		65.9700
Intercept		
Demographic factors		
AGE_i	Age of respondent	53.9725
SEX_i	Sex of respondent (reference category: female)	23.0000%
$Single_i$	Marital status: Single (reference category: single)	65.0000%
$Married_i$	Marital status: Married (reference category: married)	8.2500%
EDU_i	Education level	3.1300
Financial variable/access to resource		
HH_i	Household size	3.3965
$Child_i$	Number of children	0.9425
$Grandchild_i$	Number of grandchildren	0.3283
$Income_i$	Income of respondent (baht per month)	14,778.5875
$HHinc_i$	Household income (baht per month)	34,265.7500
DE_i	Daily expenses (baht per month)	4,674.8250
SE_i	Social expenses (baht per month)	452.6250
ME_i	Medical expenses (baht per month)	720.9250
TF_i	Child tuition fee (baht per semester)	2,946.7500
Dep_i	Deposit (baht)	59,811.5000
$VFund_i$	Village fund deposit (baht)	4,366.0000
$Coop_i$	Cooperative deposits (baht)	721.2500
$Saveg_i$	Savings group deposits (baht)	423.7500
$Cash_i$	Cash (baht)	3,395.0000
$House_i$	House installments (baht per month)	899.7500
Car_i	Car installments (baht per month)	1,133.8000
$Moto_i$	Motorcycle installments (baht per month)	577.1400
$LifeInsurance_i$	Life insurance (baht per year)	3,443.9750
$IncomeRet_i$	Income after retirement	9.0000%
$LoadRet_i$	Financial burden after retirement	28.5000%

(continued)

Table 2 (*continued*)

Variable	Description	Mean/Percent
$Depen_i$	Caretaker after retirement	89.2500%
$Insure60_i$	Health insurance after 60 years of age	87.7500%
$HOME_i$	Own accommodation	80.2500%
Health factors		
$Disea_i$	Chronic disease (reference category: have disease)	42.0000%
$Hypertension_i$	Cost of treatment for blood pressure (baht per year)	452.6000
Eye_i	Cost of the treatment of eye diseases (baht per year)	192.5500
$Bone_i$	Cost of bone disease treatment (baht per year)	276.4250
$Disbles_i$	Cost of treating diabetes (baht per year)	651.6750
$Heart_i$	Cost of treatment for heart disease (baht per year)	858.1500
$Deafness_i$	Cost of treating hard of hearing (baht per year)	12.5000
$Paresis_i$	Cost of treatment of Paralysis/Paralysis (baht per year)	12.5000

retirement, negative), ME_i (Medical expenses, positive), $Saveg_i$ (Savings group deposits, negative), SE_i (Social expenses, positive), SEX_i (Sex of respondent, positive), $Single_i$ (Marital status: Single, positive), and TF_i (Child tuition fee, positive).

1. **Demographic factors**
   Demographic factors (including AGE_i, SEX_i, $Single_i$, EDU_i) are generally significant. The result of age of respondent (AGE_i) implies that when age of respondent (AGE_i) increases by 1 year then the planned retirement age will increase by 0.2526 years. Holding all other things constant, females have planned retirement age 0.2480 higher than males and single person have planned retirement age 0.0147 higher than otherwise. The result of education level (EDU_i) implies that when education level (EDU_i) increases by 1 level then the planned retirement age will decrease by 0.1884 years.
2. **Financial variables and access to resources**
   The results of number of children ($Child_i$), daily expenses (DE_i), social expenses (SE_i), medical expenses (ME_i), child tuition fee (TF_i), and car installments (Car_i) indicate the positive impact of six variables on planned retirement age. When number of children ($Child_i$) increases by 1 person then the planned retirement age will increase by 0.3886 years. When daily expenses (DE_i) increases

by 1 baht then the planned retirement age will increase by 0.0003 year. When social expenses (SE_i) increases by 1 baht then the planned retirement age will increase by 0.0001 year. When medical expenses (ME_i) increases by 1 baht then the planned retirement age will increase by 0.0001 year and when child tuition fee (TF_i) increases by 1 baht then the planned retirement age will increase by 0.0001 year. When car installments (Car_i) increases by 1 baht then the planned retirement age will increase by 0.0001 year. The result of savings group deposits ($Saveg_i$) indicates its negative impact on planned retirement age, implying that when savings group deposits ($Saveg_i$) increases by 1 baht then the planned retirement age will decrease by 0.0002 year. Holding other things equal, those who will earn income after retirement ($IncomeRet_i$) have planned retirement age 1.3651 years lower than otherwise and those who get financial burden after retirement ($LoadRet_i$) have planned retirement age 0.0079 year lower than otherwise. The results also show that those who have caretaker after retirement ($Depen_i$) will have planned retirement age 1.6213 years lower than otherwise. Persons who have their own accommodation ($HOME_i$) will have planned retirement age 0.6431 years higher than otherwise.

3. **Health factors**
   Poor health factor has a significant impact on the planned retirement ages. Holding other things equal, those who have chronic disease have planned retirement age 0.1282 year lower than those in good health status. The results of cost of treatment for blood pressure ($Hypertension_i$) and cost of bone disease treatment ($Bone_i$) imply that when cost of treatment for blood pressure ($Hypertension_i$) increases by 1 baht then the planned retirement age will decrease by 0.0004 year and when cost of bone disease treatment ($Bone_i$) increases by 1 baht then the planned retirement age will decrease by 0.0004 year. The result of cost of treating diabetes ($Disbles_i$) indicates the positive impact on planned retirement age. When cost of treating diabetes ($Disbles_i$) increases by 1 baht then the planned retirement age will increase by 0.0001 year.

Table 3 LASSO, OLS and Stepwise regression of planned retirement age

Variable	$\hat{\beta}_{LASSO}$	$\hat{\beta}_{OLSPost-LASSO}$	$\hat{\beta}_{OLS}$	$\hat{\beta}_{Stepwise}$
$Planned_i$ (dependent variable)				
Intercept	54.5347	50.3700	51.3400	47.1590
Demographic factors				
AGE_i	0.2526	0.3215	0.3113	0.3707
SEX_i	0.2480	0.8653	0.8340	
$Single_i$	0.0147	1.3870	1.3680	
$Married_i$			−0.1296	
EDU_i	−0.1884	−0.3804	−0.4633	

(*continued*)

Table 3 (*continued*)

Variable	$\hat{\beta}_{LASSO}$	$\hat{\beta}_{OLSPost-LASSO}$	$\hat{\beta}_{OLS}$	$\hat{\beta}_{Stepwise}$
Financial variable/access to resource				
HH_i			−0.1116	
$Child_i$	0.3886	0.7420	0.8897	
$Grandchild_i$			−0.0661	
$Income_i$			0.0001	
$HHincome_i$			0.0001	0.0001
DE_i	0.0003	0.0004	0.0004	0.0005
SE_i	0.0001	0.0004	0.0004	
ME_i	0.0001	−0.0002	−0.0003	
TF_i	0.0001	0.0001	0.0001	
Dep_i			0.0001	
$VFund_i$			−0.0001	
$Coop_i$			0.0001	
$Saveg_i$	−0.0002	−0.0003	−0.0003	
$Cash_i$			0.0001	
$House_i$			0.0001	
Car_i	0.0001	0.0001	0.0001	
$Moto_i$			0.0003	
$LifeInsurance_i$			0.0001	
$IncomeRet_i$	−1.3651	−1.7680	−1.8100	
$LoadRet_i$	−0.0079	−0.7875	−0.7054	
$Depen_i$	−1.6213	−2.2050	−1.9850	
$Insure60_i$			−0.0944	
$HOME_i$	0.6431	1.0680	1.0760	
Health factors				
$Disea_i$	−0.1282	−0.2415	−0.2588	
$Hypertension_i$	−0.0004	−0.0004	−0.0004	−0.0008
Eye_i			0.0002	
$Bone_i$	−0.0004	−0.0007	−0.0009	−0.0007
$Disbles_i$	0.0001	0.0002	0.0002	
$Heart_i$			0.0001	
$Deafness_i$			0.0016	
$Paresis_i$			−0.0003	
Selected predictors	19	19	36	5
in-sample RMSE	5.0703	4.9128	4.8817	5.2698
out-of-sample RMSE	4.9751	5.4906	5.7703	5.4253

Note The blank entires correspond to variables omitted. $\beta_{OLSPost-LASSO}$ is defined by OLS applied to the model selected by LASSO

5 Conclusion

This study aims to measure the determinants of planned retirement age of informal workers in Chiang Mai Province, Thailand. A total of 400 questionnaires were randomly distributed. The data were collected by survey. The Least Absolute Shrinkage and Selection Operator (LASSO) estimator proposed by Tibshirani [12] was employed. There are several pieces of research that used ordinary least squares (OLS) and generalized method of moments (GMM) estimation. However, the LASSO estimation has not yet been investigated in the planned retirement age analysis. The primary advantage of the LASSO estimation is that it can provide a very good prediction accuracy, because shrinking and removing the coefficients can reduce variance without a substantial increase of the bias. This is especially useful when we have a small number of observation and a large number of features [3].

Thirty six variables, including demographic factors, financial variables or access to resources, and health factors included as independent variables in the model, were studied. The results showed that, the estimation using LASSO estimator performed better than OLS estimator and stepwise estimator. The results also indicate that only 19 variables, namely age of respondent, cost of bone disease treatment, number of children, daily expenses, caretaker after retirement, chronic disease, education level, own accommodation, cost of treatment for blood pressure, income after retirement, financial burden after retirement, savings group deposits, sex of respondent, and marital status: single, are significant determinant of retirements planning.

Planned retirement age increases when individuals are old, female, single and less-educated, but is decreased by such factors as household size, earning income after retirement or having caretaker after retirement.

The results of this research also contradict some of the previous studies. Previous works (for example, Honig [5]) indicated that retirement decisions of single female did not differ substantially from those of single male. In contrast, this research provides strong evidence for differences in planning for retirement between the single male and female.

Acknowledgments This work was supported by the Center of Excellence in Econometrics, Faculty of Economics, Chiang Mai University, Thailand. We also acknowledge the partial support from Chiang Mai University for the project "Preparation for Elder Life of Informal Workers in Chiang Mai Province".

References

1. Buhlmann, P., van de Geer, S.: Statistics for High-Dimensional Data: Methods. Theory and Applications. Springer, Heidelberg (2011)
2. De Souza, L.R., Queiroz, B.L., Skirbekk, V.F.: Trends in health and retirement in Latin America: are older workers healthy enough to extend their working lives? J. Econ. Ageing **13**, 72–83 (2019)

3. Fonti, V.: Feature Selection using LASSO. Research Paper in Business Analytics (2017). https://beta.vu.nl/nl/Images/werkstuk-fonti_tcm235-836234.pdf
4. Foundation of Thai Gerontology Research and Development Institute. http://www.dop.go.th/download/knowledge/th1552463947-147_1.pdf
5. Honig, M.D.: Retirement expectations: differences by race, ethnicity, and gender. Gerontol. **36**, 373–382 (1996)
6. MacKellar, L.: Pension systems for the informal sector in Asia. SP Discussion Paper No. 0903
7. Montalto, C.P., Yuh, Y., Hanna, S.: Determinants of planned retirement age. Financ. Serv. Rev. **9**, 1–15 (2000)
8. National Commission for Enterprises in the Unorganized Sector: The Challenge of Employment in India – An Informal Economy Perspective (2009)
9. National Economic and Social Development Board and the United Nations Population Fund: Impact of Demographic Change in Thailand (2011). https://thailand.unfpa.org/sites/default/files/pub-pdf/demographic%20eng.pdf. Accessed July 2019
10. National Statistical Office: The Informal Employment Survey 2018. http://www.nso.go.th/sites/2014en/Pages/survey/Social/Labour/The-Informal-Employment-Survey.aspx. Accessed July 2019
11. Scharn, M., Sewdas, R., Boot, C.R.L., Huisman, M., Lindeboom, M., van der Beek, A.J.: Domains and determinants of retirementtiming: a systematic review of longitudinal studies. BMC Public Health **18**, 1083 (2018)
12. Tibshirani, R.: Regression shrinkage and selection via the Lasso. J. R. Stat. Soc. Ser. B (Methodol.) **58**(1), 267–288 (1996)
13. van Erp, F., Vermeer, N., van Vuuren, D.: Non-financial determinants of retirement. CPB Discussion Paper (2013)

Portfolios Optimization Under Regime Switching Model: Evidences in the American Bonds and Other Financial Assets

Bing Yang, Payap Tarkhamtham, Pongsutti Phuensan, and Kongliang Zhu

Abstract In this study, we aim to investigate the high dimension portfolio optimization by using Markov Switching Copula-based GJR-GARCH model. The proposed model is flexible and can capture the dependence structure that changes over time. This model is applied to 8 times series, including DJIA, FTSE, COMEX Gold, US Dollar Index, Crude Oil, and US Bonds (one-month, 2-year, and 5-year. In order to construct a portfolio, first we use GJR-GARCH to capture the volatility of each asset. Then, the Markov Switching copula is used to measure the dependence across assets. Finally, the results from MS-Copula is used to construct portfolios and Value at Risk and Expected Shortfall are used for optimal portfolio selection.

1 Introduction

In a statistical model, if the variations of an error term show such an irregular pattern as heterostkedasticity, they tend to cluster for a period of time. In order to estimate volatility clustering in financial markets, Engle [6], an economist and the 2003 winner of the Nobel Memorial Prize for Economics, developed the autoregressive conditional heteroscedasticity (ARCH) model as the pioneer in 1982. After that, the ARCH model has been further extended into many subsequent related time series

B. Yang · P. Tarkhamtham
Faculty of Economics, Chiang Mai University, Chiang Mai, Thailand
e-mail: yuanbing310@163.com

P. Tarkhamtham
e-mail: payap.tar@gmail.com

P. Phuensan · K. Zhu (✉)
Faculty of Business Administration and Accountancy, Khon Kaen University, Khon Kaen, Thailand
e-mail: 258zkl@gmail.com

P. Phuensan
e-mail: pongphu@kku.ac.th

S. Sriboonchitta et al. (eds.), *Behavioral Predictive Modeling in Economics*, Studies in Computational Intelligence 897,
https://doi.org/10.1007/978-3-030-49728-6_25

financial modellings. Bollerslev [4] proposed the generalized autoregressive conditional heteroscedasticity (GARCH) model which has become very popular. Even GARCH model can help describe financial markets where volatility may change, so far many GARCH based studies have worked on stock and commodity portfolio returns using the conventional procedure where the correlation coefficient is usually used to calculate the interdependence of returns of two or more financial assets. However, correlation only can work well with the dependence between random variables in the linear regression, where the distributions are normal. The linearity and normality assumptions are not practical if we need to measure two or more variables that are correlated but not normally distributed.

Financial time series often exhibit the phenomenon of asymmetry in conditional mean and variance, with fat-tailed distributions which need to be analyzed generally with the use of Copula, see, Patton [18]. Copula-based multivariate model has been applied to financial time series to specify the correlation. For applications of Copula-based multivariate models to make portfolio decisions, see Patton [17], Hong et al. [13], Garcia and Tsafack [7] and Christoffersen et al. [5], and Kaas et al. [14] on Value-at-Risk. Since the copula-based GARCH model can be utilized to investigate asymmetric or tail dependence structure (see Patton [18], and Wu et al. [22]), then Autchariyapanitkul et al. [2] and Ayusuk and Sriboonchitta [3] investigated multivariate t-copula and Vine copula combined with GARCH model to interpret portfolio risk structure for multi-dimensional asset allocation issue. But most scholars still worked on the strong assumption of no dynamic behavior of economic and financial variables. Financial crises or abrupt changes in government policy will cause dramatic breaks in many economic time series, therefore, we need to employ various time series models to analyze nonlinear dynamic patterns such as volatility clustering, asymmetry and amplitude dependence.

Although copula models are quite successful, they are unable to capture more complex structure of the dependence. In the past three decades, the Markov Switching model, developed by Hamilton [11], has been extended to Copula models. Zhu et al. [23] took the advantage the Markov Switching model and propose the Markov Switching Copula-based GARCH models to characterize regime switching, volatility clustering time series in multi-asset portfolio returns. The model has been proved to be successful in capturing the different states of dependence. Therefore, this paper will apply this model to measure the dependence of American bonds and other financial assets. Also, we extend the model to quantify the level of financial risk within the American bonds-financial asset portfolio. Finally, this paper will obtain the optimal weight for the portfolios which varies with the VaR and ES in the market upturn and the market downturn.

The rest of this study is organized as follows. Section 2 presents the methodology for all models. In Sect. 3 we describe the dataset of financial assets. Section 4 discusses our empirical results and Sect. 5 concludes.

2 Methodology

2.1 ARMA(p,q)-GJR-GARCH(1,1)

To model the marginal distribution of each random variable, we employ a univariate ARMA(p,q)-GJR-GARCH(m,n) specification that can be described as

$$r_t = \phi_0 + \sum_{i=1}^{p} \phi_i r_{t-1} + \sum_{j=1}^{q} \theta_j \varepsilon_{t-i} + \varepsilon_t \tag{1}$$

$$\varepsilon_t = \sigma_t z_t \tag{2}$$

$$\sigma_t^2 = \omega + (\alpha + \gamma I_{t-1})\varepsilon_{t-1}^2 + \beta\sigma_{t-1}^2 \tag{3}$$

$$I_{t-1} = \begin{cases} 0, & \text{if } \varepsilon_{t-1} \geq 0 \\ 1, & \text{if } \varepsilon_{t-1} < 0 \end{cases}$$

where Eq. 1 is the conditional mean equation and Eq. 2 is variance equation. ε_t is the residual term, which consists of standard variance σ_t and standardized residual z_t which is proposed to have a Gaussian distribution, a Student-t distribution, a generalized error distribution (GED), a skewed GED and a skewed-t distribution. The best-fit ARMA(p,q)-GJR-GARCH(m,n) will give the standardized residuals to be transformed into a uniform distribution in (0, 1).

2.2 Copula

Consider the multivariate case with n random variables, given n variables $x_1, \cdots, x_n$ with marginal distribution $F_1(x_1), \cdots, F_n(x_n)$, Sklar's theorem [20] introduced a linkage between distributions of $x_1, \cdots, x_n$ and bind their margins using copula function. That is $H(x_1, \cdots, x_n) = C(F_1(x_1), \cdots, F_n(x_n)) = C(u_1, \cdots, u_n)$ where the marginal distributions $u_1, \cdots, u_n$ are uniform in the [0, 1] interval. If margins $F_1(x_1), \cdots, F_n(x_n)$ are continuous distribution functions, then there is a unique copula function C but if $F_1(x_1), \cdots, F_n(x_n)$ are discrete then C is not unique. For multivariate case, the copula density c is obtained by

$$c(F_1(x_1), \cdots, F_n(x_n)) = \frac{h\left(F_1^{(-1)}(u_1), \cdots, F_n^{(-1)}(u_n)\right)}{\prod_{i=1}^{n} f_i\left(F_i^{(-1)}(u_i)\right)} \tag{4}$$

where
h = the density function associated with H
f_i = the density function of each marginal distribution
c = the copula density

There are two main classes of copula functions, namely Elliptical and Archimedean. However, this study will focus on the Elliptical class. Elliptical copula function is in quadratic form of the copula function which has symmetrical tail dependence. The dependence structure, related to this function, is the Pearson's correlation which has the value of its parameter in the $[-1, 1]$ interval. The copula functions in Elliptical class are the Gaussian and the Student-t copulas.

1. Gaussian copula
The Gaussian, or Normal, copula is a linear correlation matrix with symmetric function because the upper and the lower tail dependences are equal, and so it has no tail dependence in this function. In the multivariate case, let Φ () be standard normal cumulative distribution, thus Gaussian copula density can be written as

$$f_{(n)}(R) = \frac{1}{\left|R^{1/2}\right|} exp\left\{\frac{-1}{2}\gamma\left(R^{-1} - I\right)\gamma'\right\}\left(\prod_{i=1}^{n} exp\left\{\frac{-1}{2}\gamma_i^2\right\}\right)^{-1} \quad (5)$$

where $\gamma_i = \Phi^{-1}(F_i(x_i)) = \Phi^{-1}(u_i)$, $u_i = F_i(x_i; \theta_i)$, R is correlation matrix, and I is the identity matrix.
2. Student-t copula
The Student-t copula has a linear correlation coefficient and has symmetrical tail dependence. However, it can capture some tail dependence. Thus, the multivariate student-t copula density can be written as

$$f_{(t)}(X) = \frac{\Gamma\left[(v+n)/2\right]|R|^{\frac{-1}{2}}}{\sqrt{v^n \pi^n (v/2)}}\left\{1 + \frac{R^{-1}(x-\mu)'(x-\mu)}{v}\right\}^{-\frac{v+n}{2}} * \prod_{i=1}^{n}\left\{1 + \frac{(x-\mu)'(x-\mu)}{v}\right\}^{\frac{v+n}{2}} \quad (6)$$

where, v is degree of freedom parameter and Γ is gamma distribution.

2.3 Value at Risk with Copula

Value at Risk (VaR) and conditioned Value at Risk or Expected Shortfall (ES) has been widely used to measure risk since the 1990s. The VaR of portfolio can be written as

$$VaR_\alpha = inf\left\{l \in R : P(L > l) \leq 1 - \alpha\right\} \quad (7)$$

where, α is a confidence level with a value [0, 1] which presents the probability of Loss L to exceed l but not larger than $1-\alpha$. While an alternative method, ES, is the extension of the VaR approach to remedy two conceptual problems of VaR (Halulu [10]). Firstly, VaR measures only percentiles of profit-loss distribution with difficulty to control for non-normal distribution. Secondly, VaR is not sub-additive. ES can be written as

$$ES_{\alpha} = E\left(L \mid L > VaR_{\alpha}\right) \tag{8}$$

To find the optimal portfolios, Rockafellar and Uryasev [19] introduced the portfolio optimization by calculating VaR and extend VaR to optimize ES. The approach focuses on the minimizing of ES to obtain the optimal weight of a large number of instruments. In other words, we can write the problem as in the following

The objective function is to

$$Minimize ES_{\alpha} = E\left(L \mid L > inf\left\{l \in R : P\left(L < l\right) \leq 1-\alpha\right\}\right) \tag{9}$$

Subject to

$$R_p = \sum_{i=1}^{n} (w_i r_i)$$

$$\sum_{i=1}^{n} (w_i) = 1$$

$$0 \leq w_i \leq 1, \quad i = 1, 2, \cdots, n$$

where R_p is an expected return of the portfolios, w_i is a vector of weight portfolio, and r_i is the return of each instrument.

2.4 Regime Switching Copula

In general, financial time series exhibit different behavior and lead to different dependencies over time; for this reason, the dependence structure of the variables may be determined by a hidden Markov chain with two regimes or more. Therefore, it is reasonable to extend the copula to Markov Switching (Hamilton [11]) and obtain Markov Switching copula. Thus, the model becomes more flexible since it allows the dependence copula parameter $\left(R^{S_t}\right)$ to be governed by an unobserved variable at time t (S_t). Let S_t be the state variable, which is assumed to have two states ($k = 2$), namely high dependence regime and low dependence regime. The joint distribution of $x_1, \cdots, x_n$ conditional on S_t, is defined as

$$\left(x_{1,t}, \cdots, x_{n,t} \mid S_t = i\right) \sim c_t^{S_t}\left(u_{1,t}, \cdots, u_{n,t} \mid \theta_{c,t}^{S_t}, R_{c,t}^{S_t}\right) \quad i = 1, 2 \tag{10}$$

The unobservable regime (S_t) is governed by the first order Markov chain, which is characterized by the following transition probabilities (P):

$$P_{ij} = Pr\left(S_{t+1} = j \mid S_t = i\right) \ and \sum_{j=1}^{k=2} p_{ij} = 1 \quad i, j = 1, 2 \tag{11}$$

where $\left(P_{ij}\right)$ is the probability of switching from regime i to regime j, and these transition probabilities can be formed in a transition matrix P, as follows:

$$P = \begin{bmatrix} P_{11} & P_{12} = 1 - P_{11} \\ P_{21} = 1 - P_{22} & P_{22} \end{bmatrix} \tag{12}$$

Following, Song [21], the Gaussian copula density function from Eq. (2) can be rewritten in the likelihood function form as

$$L_{(G)}\left(u_1, \cdots, u_n \mid \theta_1, \cdots, \theta_n, R\right) = \frac{1}{\left|R^{1/2}\right|} \\ * \prod_{i}^{T} \left(exp\left\{ \frac{-1}{2} \gamma \left(R^{-1} - I\right) \gamma' \right\} \prod_{j=1}^{n} f_i\left(x_{ij}; \theta_j\right) \right) \tag{13}$$

where $f_i\left(x_{ij}; \theta_j\right)$ is the density function obtained from the ARMA-GARCH step and we assume this function to be fixed. Similarly, the Student-t copula density function from Eq. (3) can be rewritten in the likelihood function form as

$$L_{(t)}\left(u_1, \cdots, u_n \mid \theta_1, \cdots, \theta_n, R, v\right) \\ = \prod_{i=1}^{T} \left(\frac{\frac{\Gamma[(v+n)/2]|R|^{-1/2}}{\sqrt{v^n \pi^n} \Gamma(v/2)}}{\left\{1 + \frac{R^{-1}(x-\mu)'(x-\mu)}{v}\right\}^{-\frac{v+n}{2}}} \prod_{j=1}^{n} f_i\left(x_{ij}; \theta_j, v\right) \right) \tag{14}$$

In this study, the method of Kim's filtering algorithm (Kim and Nelson [16]) is conducted to filter the state variable (S_t) and let $L_{(t)}\left(u_1, \cdots, u_n \mid \theta_1, \cdots, \theta_n, R, v\right)$ and $L_{(G)}\left(u_1, \cdots, u_n \mid \theta_1, \cdots, \theta_n, R\right)$ be an respectively, thus we can write the log likelihood of two-regime Markov Switching copula as

$$Gaussian \quad logL_N\left(\theta_{N,S_t}, R_{N,S_t}, P\right) \\ = \sum_{S_t=1}^{2} logL(N) \, Pr\left[S_t \mid \theta_{N,S_{t-1}}, R_{N,S_{t-1}}, P\right] \\ Student-t \quad logL_T\left(\theta_{T,S_t}, R_{T,S_t}, P\right) \\ = \sum_{S_t=1}^{2} logL(T) \, Pr\left[S_t \mid \theta_{T,S_{t-1}}, R_{T,T_{t-1}}, P\right] \tag{15}$$

To evaluate the log-likelihood in Eq. (15), we need to calculate the weight $Pr\left[S_t \mid \theta_{N,S_{t-1}}, R_{N,S_{t-1}}\right]$ and $Pr\left[S_t \mid \theta_{T,S_{t-1}}, R_{T,T_{t-1}}\right]$ for $S_t = 1, 2$ because the estimation of the Markov Switching copula needs inferences on the probabilities of S_t.

$$Pr\left(S_t = 1 \mid w_t\right) = \frac{logL\left(\theta_{S_t=1}, R_{S_t=1}, P\right) Pr\left(S_t = 1 \mid w_{t-1}\right)}{\sum_{S_t=1}^{2} logL\left(\theta_{S_t}, R_{S_t}, P\right) Pr\left[S_t = S_t \mid w_{t-1}\right]} \tag{16}$$

$$Pr\left(S_t = 2 \mid w_t\right) = 1 - Pr\left(S_t = 1 \mid w_t\right) \tag{17}$$

where w is all the information set of the model.

3 Dataset and Estimation

In this study, we use the data set including Dow Jones Industrial Average Index (DJIA), FTSE 100 Index (FTSE), COMEX Gold Price (COMEX), US Dollar Index (USD), Crude oil price (OIL), United States 1-Month Bond Yield (ONEM), United States 2-Year Bond Yield (TWOY), and United States 5-Year Bond Yield (FIVEY). The data set covers the period January 2002 to June 2018, totally 3924 observations. The data are collected from Thomson Reuter. Then, we transform the data into log return by $r_t = log(p_t/p_{t-1})$, where r_t is the rate of return at time t and p_t is the price at time t and similarly p_{t-1} is the price at time $t-1$.

Table 1 shows the descriptive statistics of the rates of return for each asset. We can see that almost all assets have the positive and close to zero average rate of return except ONEM, which is negative and close to zero. The result reveals excess kurtosis and positive skewness for OIL, ONEM, TWOY, and negative skewness for DJIA, FTSE, COMEX, USD, FIVEY. Moreover, all variables are not normally distributed, according to the results of the Jarque-Bera test. The Minimum Bayes Factor (MBF) for all variables are equal to zero which means the decisive evidence against the null hypothesis.

In order to estimate the Markov switching Copula, first we use the ARMA-GARCH process to get the standardized residuals and then transform them into uniform distribution [0, 1]. Then, comparison is made between the two Copula families which are Gaussian and Student-t Copula. In order to choose the best model, the log-likelihood is used as the criterion.

Then, we simulate 10,000 replications of the portfolio returns for each regime from Monte Carlo Simulation. Then we multiply the inverse of the marginal distribution with the random variable to obtain ε_{it}. And the return of each variable can be

Table 1 Descriptive statistics

	DJIA	FTSE	COMEX	USD	OIL	ONEM	TWOY	FIVEY
Mean	0.0001	0.0000	0.0002	0.0000	0.0002	−0.0003	0.0000	0.0000
Median	0.0002	0.0002	0.0002	0.0000	0.0005	0.0000	0.0000	0.0000
Maximum	0.0448	0.0483	0.0297	0.0109	0.0713	1.2919	0.1290	0.0718
Minimum	−0.0356	−0.0402	−0.0417	−0.0118	−0.0567	−1.1454	−0.1114	−0.0980
Std.Dev.	0.0049	0.0052	0.0050	0.0023	0.0104	0.1378	0.0190	0.0137
Skewness	−0.1651	−0.1035	−0.3988	−0.0019	0.0340	0.3826	0.1187	-0.0106
Kurtosis	12.1422	11.8319	7.8230	4.7254	7.1883	17.5153	7.5226	6.5978
Jarque − Bera	13683.22	12760.40	3907.26	486.73	2868.89	34544.30	3353.42	2116.42
	(0.0000)	(0.0000)	(0.0000)	(0.0000)	(0.0000)	(0.0000)	(0.0000)	(0.0000)

Note The value in parentheses () is Minimum Bayes Factor (MBF), which can be computed by $-e^p logp$, where p is p-value. The MBF can be interpreted as follows: MBF between 1—1/3 is considered weak evidence for H_1, 1/3—1/10 moderate evidence, 1/10—1/30 substantial evidence, 1/30—1/100 strong evidence, 1/100—1/300 very strong evidence, and <1/300 decisive evidence supporting H_1. ([8, 12], and [9])

computed by the following formula $r_{it} = \widetilde{u_{it}} + \sqrt{h_{it}}\varepsilon_{it}$. Where $\widetilde{u_{it}}$ is the simulated mean from ARMA. To compute the portfolio return in each regime, we specify an equally weighted portfolio return, that is,

$$PR_t = 0.125DJIA_t + 0.125FTSE_t + 0.125COMEX_t + 0.125USD_t \\ + 0.125OIL_t + 0.125ONEM_t + 0.125TWOY_t + 0.125FIVEY_t$$

4 Empirical Result

4.1 GARCH Results

In this study, we used ARMA-GJR-GARCH process to capture volatility and estimate the marginals. We selected the optimal lag and marginal distribution assumption for ARMA(p,q)-GJR-GARCH(1,1) by using AIC and found that all returns are ARMA(1,1)-GJR-GARCH(1,1). Then, we also used the AIC to select various margins assumption and found that the margins of all variables have student-t distribution.

Table 2 shows the parameters of each ARCH and GARCH. The estimated ARCH effects are equal to 0.1526, 0.1547, 0.0541, 0.0312, 0.0876, 0.2706, 0.0428, and 0.0282. The values of the GARCH coefficient are 0.9224, 0.9280, 0.9561, 0.9629, 0.9506, 0.9948, 0.9417, and 0.9503. We then used Ljung-box test and ARCH-LM test to ensure that residuals are independent and identically distributed (i.i.d) and no autocorrelation exists. The results of Ljung-box test show that all variables present weak evidence against null hypothesis. The results of ARCH LM test shows that DJIA, USD, OIL, ONEM, and TWOY provide weak evidence against null hypothesis. Then, COMEX and FIVEY correspond to moderate evidence against null hypothesis and FTSE provides substantial evidence against null hypothesis. We can conclude that residuals are i.i.d and have no autocorrelation. Then, we adjust residuals to be standardized residuals by divided by variance.

4.2 Model Selection

In this section, we transform the standardized residuals from the previous process into uniform distribution [0, 1]. Then, we compare four copula functions, namely Gaussian, MS Gaussian, Student-t and MS Student-t copulas. The AIC, BIC and Loglik are used to select the best fit model from our proposed models. In Table 3, the results show that MS-copula with Student-t function presents a lower AIC, BIC, and

Table 2 Estimates of GARCH model parameters

	DJIA	FTSE	COMEX	USD	OIL	ONEM	TWOY	FIVEY
C	0.0003	0.0003	0.0002	−0.0002	0.0004	−0.00002	−0.0004	−0.0002
	(0.0000)	(0.0000)	(0.0001)	(0.0000)	(0.0001)	(0.0001)	(0.0001)	(−0.0002)
$AR1$	−0.1029	0.8955	−0.1432	−0.0806	−0.3642	0.8338	0.2381	0.2174
	(0.2060)	(0.0202)	(0.6358)	(0.7366)	(1.1901)	(0.0172)	(0.1202)	(0.1557)
$MA1$	0.0443	−0.9142	0.1627	0.0602	0.3254	−0.8642	−0.3281	−0.2832
	(0.2065)	(0.0183)	(0.6363)	(0.7278)	(1.2089)	(0.0154)	(0.1164)	(0.1527)
$OMEGA$	0.0000	0.0000	0.0000	0.0000	0.0000	−0.0233	0.0000	0.0000
	(0.0000)	(0.0000)	(0.0000)	(0.0000)	(0.0000)	(0.0039)	(0.0000)	(0.0000)
$ARCH$	0.1526	0.1547	0.0541	0.0312	0.0876	0.2706	0.0428	0.0282
	(0.0035)	(0.0083)	(0.0090)	(0.0046)	(0.0021)	(0.0499)	(0.0095)	(0.0044)
$GARCH$	0.9224	0.9251	0.9561	0.9629	0.9506	0.9948	0.9417	0.9503
	(0.0062)	(0.0088)	(0.0042)	(0.0016)	(0.0044)	(0.0001)	(0.0097)	(0.0063)
$GAMMAL$	−0.1519	−0.1618	−0.0320	0.0038	−0.0787	0.5673	0.0290	0.0410
	(0.0123)	(0.0091)	(0.0113)	(0.0080)	(0.0098)	(0.0763)	(0.0102)	(0.0099)
$SHAPE$	8.4125	11.6326	5.2996	10.3263	9.8913	2.1000	6.5990	9.3002
	(0.9761)	(1.6439)	(0.8170)	(1.5842)	(1.2855)	(0.0267)	(0.5820)	(1.1100)
$Ljung-Box$	0.0975	2.126	0.8826	0.0152	0.5153	2.0000	0.2193	0.3907
	[0.5769]	[0.7606]	[0.9984]	[0.4168]	[0.9627]	[0.7671]	[0.7770]	[0.9128]
$ARCH-LM$	0.0782	0.0014	0.0115	0.5032	0.7179	0.0760	0.0217	0.0168
	[0.5274]	[0.0822]	[0.2222]	[0.9590]	[0.9970]	[0.5211]	[0.8829]	[0.2655]

Note The value in parentheses () is Minimum Bayes Factor (MBF), which can be computed by $-e^{p} log p$, where p is p-value. The MBF can be interpreted as follows: MBF between 1—1/3 is considered weak evidence for H_1, 1/3—1/10 moderate evidence, 1/10—1/30 substantial evidence, 1/30—1/100 strong evidence, 1/100—1/300 very strong evidence, and <1/300 decisive evidence supporting H_1. ([8, 12], and [9])

Table 3 Model selection

	AIC	BIC	Loglik
Gaussiancopula	−8365.7600	−8190.0640	4210.8800
MSGaussiancopula	−9465.5960	−9101.6540	4790.7980
studenttcopula	−9091.8170	−8909.8460	4574.9090
MSstudenttcopula	−9911.5890	−9535.0970	5015.7940

higher log-likelihood (Loglik) than others. Therefore, this paper adopts MS-copula with Student-t function to be inferenced in this study.

4.3 Markov Switching Student-T Copula

Table 4 provides the multivariate Student-t copula parameters of Markov switching regime. These solutions can be employed to construct efficient portfolio and produce optimal solutions for best expected returns with minimum loss, which will be presented in the last section.

From Table 4, we can see that almost all the dependence parameters in regime 1 are lower than regime 2. Thus, regime 1 can be interpreted as the low dependence regime or market upturn, while regime 2 as the high dependence regime or market downturn. From their study, Karimalis and Nomikos [15] found that the dependence between assets during market upturns is less than the market downturn. Then, we can confirm that regime 1 is the market upturn regime, while regime 2 is the market downturn regime.

We observe that the dependence parameters of all pairs are either negative or positive. The dependence parameters of ONEM in regime 1 are positive with DJIA, FTSE, USD, and OIL, while negative with COMEX, TWOY, and FIVEY. In regime 2, they are positive for FESE, COMEX, USD, and OIL, while negative for DJIA, TWOY, and FIVEY. Then, the dependence parameters of TWOY in regime 1 are negative except for COMEX and FIVEY. Likewise, in regime 2 the dependence parameters of TWOY are negative for DJIA, FTSE, OIL AND ONEM, while positive for COMEX, USD, and FIVEY. For the FIVEY, the dependence parameters in regime 1 are negative for DJIA, FTSE, USD, and OIL, while positive for COMEX. In regime 2, the dependence parameters are negative for DJIA, FTSE, and OIL, while positive for COMEX and USD.

Considering DJIA, the dependence parameters in regime 1 are positive for FTSE, USD, and OIL, except for COMEX. In regime 2, the dependence parameters are positive for FTSE, COMEX, and OIL, except for USD. Then, the dependence parameters of FTSE in regime 1, they are positive for USD and OIL, while they are negative for COMEX. In regime 2, they are positive for COMEX and OIL, while they are negative for USD.

Table 4 Empirical copula parameters

*Regime*1

	DJIA	FTSE	COMEX	USD	OIL	ONEM	TWOY	FIVEY
DJIA	1	0.4913	−0.1111	0.1129	0.0335	0.0236	−0.2917	−0.3011
		(0.0149)	(0.0212)	(0.0210)	(0.0217)	(0.0216)	(0.0201)	(0.0198)
FTSE	0.4913	1	−0.0668	0.1658	0.1076	0.0481	−0.1958	−0.2086
	(0.0149)		(0.0215)	(0.0207)	(0.0210)	(0.0216)	(0.0204)	(0.0203)
COMEX	−0.1111	−0.0668	1	−0.3940	0.1134	−0.0306	0.1588	0.1625
	(0.0212)	(0.0215)		(0.0173)	(0.0212)	(0.0221)	(0.0212)	(0.0212)
USD	0.1129	0.1658	−0.3940	1	−0.1253	0.0509	−0.3406	−0.3394
	(0.0210)	(0.0207)	(0.0173)		(0.0207)	(0.0217)	(0.0192)	(0.0192)
OIL	0.0335	0.1076	0.1134	−0.1253	1	0.0125	−0.0328	−0.0461
	(0.0217)	(0.0210)	(0.0212)	(0.0207)		(0.0220)	(0.0219)	(0.0218)
ONEM	0.0236	0.0481	−0.0306	0.0509	0.0125	1	−0.1119	−0.0967
	(0.0216)	(0.0216)	(0.0221)	(0.0217)	(0.0220)		(0.0223)	(0.0223)
TWOY	−0.2917	−0.1958	0.1588	−0.3406	−0.0328	−0.1119	1	0.9297
	(0.0201)	(0.0204)	(0.0212)	(0.0192)	(0.0219)	(0.0223)		(0.0037)
FIVEY	−0.3011	−0.2086	0.1625	−0.3394	−0.0461	−0.0967	0.9297	1
	(0.0198)	(0.0203)	(0.0212)	(0.0192)	(0.0218)	(0.0223)	(0.0037)	

*Regime*2

	DJIA	FTSE	COMEX	USD	OIL	ONEM	TWOY	FIVEY
DJIA	1	0.6489	0.1268	−0.3928	0.4458	−0.0047	−0.2837	−0.4066
		(0.0171)	(0.0290)	(0.0262)	(0.0254)	(0.0288)	(0.0275)	(0.0258)
FTSE	0.6489	1	0.1934	−0.3421	0.4545	0.0632	−0.2161	−0.3191
	(0.0171)		(0.0289)	(0.0278)	(0.0260)	(0.0286)	(0.0263)	(0.0261)
COMEX	0.1268	0.1934	1	−0.2806	0.2416	0.0353	0.0284	0.0261
	(0.0290)	(0.0289)		(0.0267)	(0.0282)	(0.0285)	(0.0282)	(0.0291)
USD	−0.3928	−0.3421	−0.2806	1	−0.4285	0.0015	0.0490	0.1243
	(0.0262)	(0.0278)	(0.0267)		(0.0244)	(0.0284)	(0.0304)	(0.0318)
OIL	0.4458	0.4545	0.2416	−0.4285	1	0.0455	−0.1870	−0.2831
	(0.0254)	(0.0260)	(0.0282)	(0.0244)		(0.0292)	(0.0278)	(0.0276)
ONEM	−0.0047	0.0632	0.0353	0.0015	0.0455	1	−0.0739	−0.0517
	(0.0288)	(0.0286)	(0.0285)	(0.0284)	(0.0292)		(0.0281)	(0.0289)
TWOY	−0.2837	−0.2161	0.0284	0.0490	−0.1870	−0.0739	1	0.7328
	(0.0275)	(0.0263)	(0.0282)	(0.0304)	(0.0278)	(0.0281)		(0.0151)
FIVEY	−0.4066	−0.3191	0.0261	0.1243	−0.2831	−0.0517	0.7328	1
	(0.0258)	(0.0261)	(0.0291)	(0.0318)	(0.0276)	(0.0289)	(0.0151)	

	Regime 1	Duration	Regime 2	Duration				
Regime 1	0.9788	13.5170	0.0212	0.2928				
Regime 2	0.0400	0.5902	0.9600	14.1649				

Then, considering the dependence parameters of COMEX, there is negative relationship with USD while positive with OIL for both regimes. Moreover, the dependence parameter among USD and OIL are negative in both regimes. As we mentioned, the presence of positive dependence among assets gives us some economic inference that these prices are moving in the same direction, and vice versa.

Moreover, the probability matrix of these financial assets are showed in Table 4 as well. The $Pr(S_t = 1)$ is 97.88% and $Pr(S_t = 2)$ is 96.00% while the probabilities of switching between these two regimes are less than 5%. The probabilities explained that low dependence regime and high dependence regime are persistent due to the high values of the probabilities. In addition, the duration of staying persistent is short for both regimes, with the duration equal to 13.5170 days for the low dependence regime and 14.1649 days for the high dependence regime.

4.4 Regime Probabilities

In this section, we analyze the evolution of the regime probabilities at each period. As we mentioned before, regime 1 is a low dependence regime or market upturn while regime 2 is a high dependence regime or market downturn. From Fig. 1 which presents the regime probabilities at each time point for low and high dependence regime, we found the interesting period (red box plot) from 2006 to 2009 which corresponds to market downturn. In fact, this period corresponds to the global financial crisis which created a large negative effect to the global economy.

4.5 Value at Risk and Expected Shortfall Estimation

In this section, we estimate the expected values of VaR and ES on an equally weighted portfolio based on the Markov switching Student-t copula.

Table 5 shows the estimation results of VaR and ES at 1%, 5%, and 10% probabilities. For regime 1 or market upturn, the estimated VaR values are 2.1861%, 0.9780%, and 0.6386%, respectively, while the estimated ES values are, respectively, 2.8388%, 1.5408%, and 1.1387%. The estimated VaR values indicate there are 1%, 5%, and 10% that this portfolio's return will fall 2.1861%, 0.9780%, and 0.6386%. By ES measure, there are 1%, 5%, and 10% probabilities that this portfolio's return will fall 2.8388%, 1.5408%, and 1.1387%. For regime 2 or market downturn, the estimated VaR shows at 1%, 5%, and 10% probabilities, this portfolio's return will fall 1.9105%, 0.9955%, and 0.6376%, respectively; while the estimated ES shows at 1%, 5%, and 10% probability, then this portfolio's return will fall 2.5916%, 1.5811%, and 1.1139%, respectively.

Table 6 shows the results of the optimal portfolio. We selected the optimum portfolio based on Conditional Sharpe ratio which was proposed by Agarwal and Naik [1]. They replace the standard deviation in the Sharpe ratio by an alternative risk mea-

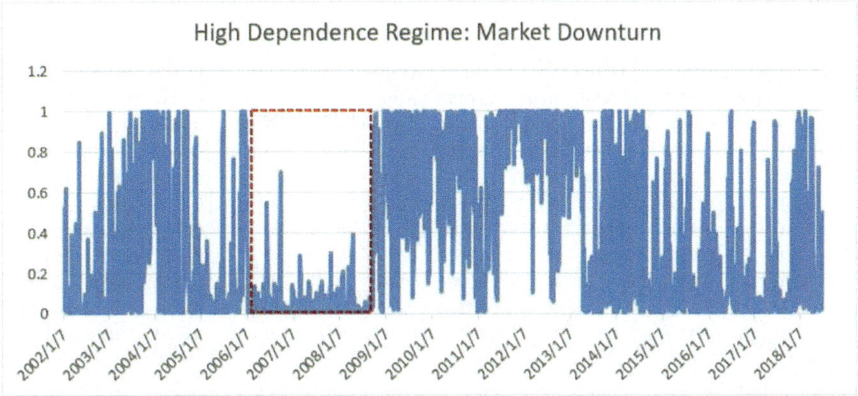

Fig. 1 Filtered probabilities of Market downturn regime

Table 5 Risk measurement

*Regime*1		
	Var%	ES%
1%	−2.1861	−2.8388
5%	−0.9780	−1.5408
10%	−0.6386	−1.1387
*Regime*2		
	Var%	ES%
1%	−1.9105	−2.5916
5%	−0.9955	−1.5811
10%	−0.6376	−1.1139

sure (Expected Shortfall). We then assume the risk-free rate equal to zero. So, we can compute the Conditional Sharpe ratio by $SR = er/ES$ where ES is the expected return and is the Expected Shortfall.

We found that in regime 1 or market upturn the optimal portfolio is portfolio number 7 with the largest Sharpe ratio around 0.0905. Likewise, in regime 2 or market downturn the optimal portfolio is portfolio number 5 of which its Sharpe ratio is about 0.0891. Therefore, for a risk taking investor who wants a high returns, he can allocate his investment to FTSE 33.98%, OIL 46.24%, and ONEM 19.78% in order to get the highest return at 0.0568% with 0.6353% risk. On the other hand, a risk averse investor can invest in DJI 6%, FTSE 3.13%, COMEX 17.79%, USD 64.52%, OIL 1.93%, ONEM 1.63%, TWOY 4.79%, and FIVEY 0.21% to face with the lowest risk (0.1437%).

Table 6 Optimal portfolio's weight

*Regime*1

Port 1	DJIA	FTSE	COMEX	USD	OIL	ONEM	TWOY	FIVEY	Risk%	Return%	Sharpe
1	0.0600	0.0313	0.1779	0.6452	0.0193	0.0163	0.0479	0.0021	0.1437	0	0
2	0	0.1195	0.1751	0.5727	0.0784	0.0477	0	0.0066	0.1815	0.0083	0.0456
3	0	0.1910	0.1490	0.4447	0.1363	0.0786	0	0.0004	0.2469	0.0176	0.0714
4	0	0.2638	0.1168	0.3151	0.1945	0.1098	0	0	0.3261	0.0267	0.0819
5	0	0.3367	0.0842	0.1855	0.2526	0.1410	0	0	0.4119	0.0357	0.0868
6	0	0.4095	0.0516	0.0559	0.3108	0.1722	0	0	0.5012	0.0448	0.0894
7	0	0.4440	0	0	0.3627	0.1933	0	0	0.5682	0.0514	0.0905
8	0	0.4093	0	0	0.3960	0.1947	0	0	0.5884	0.0532	0.0904
9	0	0.3745	0	0	0.4292	0.1963	0	0	0.6108	0.0550	0.0901
10	0	0.3398	0	0	0.4624	0.1978	0	0	0.6353	0.0568	0.0894

*Regime*2

Port 2	DJIA	FTSE	COMEX	USD	OIL	ONEM	TWOY	FIVEY	Risk%	Return%	Sharpe
1	0.0752	0.0630	0.1150	0.6718	0.0307	0.0126	0.0310	0.0007	0.1431	0	0
2	0	0.1495	0.0863	0.6379	0.0710	0.0553	0	0	0.1748	0.0117	0.0669
3	0	0.2127	0.0370	0.5501	0.1036	0.0966	0	0	0.2287	0.0189	0.0828
4	0	0.2739	0	0.4541	0.1346	0.1374	0	0	0.2957	0.0260	0.0879
5	0	0.3291	0	0.3337	0.1608	0.1764	0	0	0.3645	0.0325	0.0891
6	0	0.3843	0	0.2132	0.1871	0.2154	0	0	0.4373	0.0390	0.0891
7	0	0.4394	0	0.0927	0.2135	0.2544	0	0	0.5125	0.0455	0.0887
8	0	0.4743	0	0	0.2394	0.2863	0	0	0.5742	0.0507	0.0883
9	0	0.4412	0	0	0.2646	0.2942	0	0	0.5877	0.0517	0.0880
10	0	0.4082	0	0	0.2897	0.3021	0	0	0.6026	0.0528	0.0876

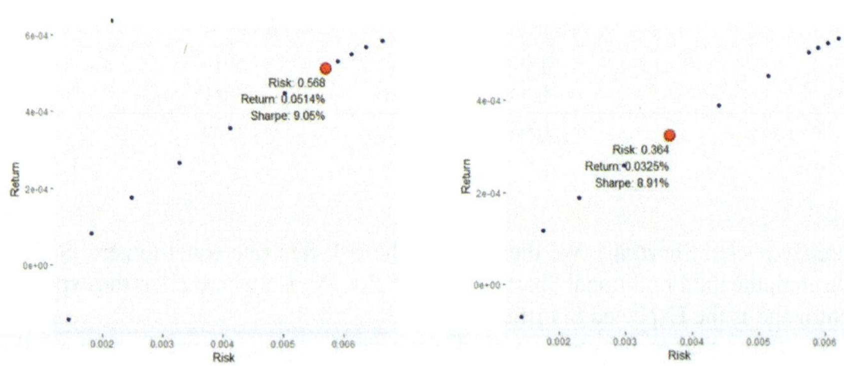

(a) Regime 1 (b) Regime 2

Fig. 2 Efficient frontier for two regimes

Figure 2 demonstrates the efficiency frontier for high dependence regime and low dependence regime, including the 10 portfolios in the Table 6. This study provides the optimal weighted investment for these financial assets in the market upturn and market downturn as well.

5 Conclusion

This study uses the Markov Switching Copula-based GARCH model to characterize regime switching on volatility clustering time series of multi-asset portfolio returns. We focus on the United States' bonds and other financial assets for the period 2002:01-2018:06. This study found that MS-copula with Student-t function presents lower AIC than Gaussian copula and no regime switching function. Therefore, this study applies MS-copula with Student-t function to be inferenced in the study. Then, we can conclude that the high dependence regime is a market downturn and the low dependence regime is a market upturn. Finally, the results of Value at Risk (VaR) and Expected Shortfall (ES) illustrate that the investors will face similar risk for both market upturn and market downturn. We also obtained the optimal weight for the portfolios which vary with ES in the market upturn and market downturn.

References

1. Agarwal, V., Naik, N.Y.: Risks and portfolio decisions involving hedge funds. Rev. Finan. Stud. **17**(1), 63–98 (2004)
2. Autchariyapanitkul, K., Chanaim, S., Sriboonchitta, S.: Portfolio optimization of stock returns in high-dimensions: A copula-based approach. Thai J. Math. 11–23 (2014)
3. Ayusuk, A., Sriboonchitta, S.: Risk analysis in Asian emerging markets using canonical vine copula and extreme value theory. Thai J. Math. 59–72 (2014)
4. Bollerslev, T.: A conditionally heteroskedastic time series model for speculative prices and rates of return. Rev. Econom. Statist. **69**, 542–547 (1987)
5. Christoffersen, P., Errunza, V., Jacobs, K., Langlois, H.: Is the potential for international diversification disappearing? A dynamic copula approach. Rev. Finan. Stud. **25**(12), 3711–3751 (2012)
6. Engle, R.F.: Autoregressive conditional heteroscedasticity with estimates of the variance of UK inflation. Econometrica **50**, 987–1007 (1982)
7. Garcia, R., Tsafack, G.: Dependence structure and extreme comovements in international equity and bond markets. J. Bank. Financ. **35**, 1954–1970 (2011)
8. Goodman, S.N.: Toward evidence-based medical statistics. 1: the P value fallacy. Ann. Internal Med. **130**(12), 995–1004 (2018)
9. Goodman, S.N.: Toward evidence-based medical statistics. 2: the Bayes factor. Ann. Internal Med. **130**(12), 1005–1013 (2018)
10. Halulu, S.: Quantifying the risk of portfolios containing stocks and commodities. Doctoral dissertation, Bogaziçi University (2012)
11. Hamilton, J.D.: A new approach to the economic analysis of nonstationary time series and the business cycle. Econometrica: J. Econometric Soc. **57**(2), 357–384 (1989)
12. Held, L., Ott, M.: On p-values and Bayes factors. Ann. Rev. Stat. Appl. **5**, 393–419 (2018)

13. Hong, Y., Tu, J., Zhou, G.: Asymmetries in stock returns: statistical tests and economic evaluation. Rev. Finan. Stud. **20**, 1547–1581 (2007)
14. Kaas, R., Laeven, R.J.A., Nelsen, R.B.: Worst VaR scenarios with given marginals and measures of association. Math. Econ. **44**(2), 146–158 (2009)
15. Karimalis, E.N., Nomikos, N.K.: Measuring systemic risk in the European banking sector: a Copula CoVaR approach. Eur. J. Finan. **24**(11), 944–975 (2018)
16. Kim, C.-J., Nelson, C.: State-Space Models with Regime Switching: Classical and Gibbs Sampling Approaches with Applications. MIT press, Cambridge (1999)
17. Patton, A.J.: On the out-of-sample importance of skewness and asymmetric dependence for asset allocation. J. Financ. Econ **2**, 130–168 (2004)
18. Patton, A.J.: Modelling asymmetric exchange rate dependence. Int. Econ. Rev. **47**(2), 527–556 (2006)
19. Rockafellar, R.T., Uryasev, S.: Conditional value-at-risk for general loss distributions. J. Banking Finan. **26**(7), 1443–1471 (1959)
20. Sklar, A.: Fonctions de repartition an dimensions et leurs marges. Publications de l'Institut de Statistique de L'Universit de Paris **8**, 229–231 (1959)
21. Song, P., Fan, Y., Kalbfleisch, J.D.: Maximization by parts in likelihood inference. J. Amer. Statist. Assoc. **100**, 1145–1158 (2005)
22. Wu, C.C., Chung, H., Chang, Y.H.: The economic value of co-movement between oil price and exchange rate using copula-based GARCH models. Energy Econ. **34**(1), 270–282 (2012)
23. Zhu, K., Yamaka, W., Sriboonchitta, S.: Multi-asset portfolio returns: a Markov switching copula-based approach. Thai J. Math. 183–200 (2016)

Impact of Economic Policy Uncertainty on the Stock Exchange of Thailand: Evidence from the Industry-Level Stock Returns in Thailand

Siriluk Punwong, Nachatchapong Kaewsompong, and Roengchai Tansuchat

Abstract This study quantifies the economic policy uncertainty of Thailand and investigates the impact of this uncertainty on agriculture and food (agro-food); and technology stock sectors. A two-step procedure was employed in this study. First, we measure the economic policy uncertainty during January 2004–December 2018. We collect the frequency of internet searches and use the method of Baker et al. (2016) to compute the economic policy uncertainty index for Thailand. We verify our index based on R-Squared and the result shows that R-Squared between the SET index and our EPU is 0.7006. This indicates a high performance of our EPU in explaining the SET index. Second, we apply the DCC-GARCH model to present the relationship between Thai EPU index and two stock sectors. Our results reveal that the mean value of the dynamic conditional correlation between the EPU index and agro-food sector stock return is −0.0890, while that of the EPU index and technology sector stock return pair is −0.0488. Thus, the EPU did produce a negative impact on the two stock sectors under investigation.

Keywords Economic policy uncertainty · Stock Exchange of Thailand · Industry-level stock returns · Thailand

S. Punwong
Faculty of Economics, Chiang Mai University, Chiang Mai 50200, Thailand
e-mail: siriluk.punwong@gmail.com

N. Kaewsompong (✉) · R. Tansuchat
Center of Excellence in Econometrics, Faculty of Economics, Chiang Mai University, Chiang Mai 50200, Thailand
e-mail: nachat_flysky@hotmail.com

R. Tansuchat
e-mail: roengchaitan@gmail.com

S. Sriboonchitta et al. (eds.), *Behavioral Predictive Modeling in Economics*, Studies in Computational Intelligence 897,
https://doi.org/10.1007/978-3-030-49728-6_26

1 Introduction

Numerous studies revealed that the policy-related economic uncertainty had a significant impact on financial markets [2, 17]. Most of the literature focused on the impact of policy shocks on the stock market since it plays a vital role in financial markets [3, 8, 11, 18, 24, 25]. Economic policy uncertainty has been found to have a negative effect on stock market in many countries such as Australia [8], U.S. [4], China [30], Hong Kong, Malaysia, South Korea [6], and so on. This uncertainty has brought some cost to the financial markets, especially stock market [13, 14]. When the level of uncertainty increases, a higher stock market volatility is presented. Following the advent of the global financial crisis in 2008, governments of many countries have avoided the economic recessions in their countries by increasing the level of intervention, adjusting or making a new economic policy in the form of monetary, fiscal, and regulatory policy [2]. Those policies promulgated by the government often caused uncertainty in the stock market.

For the reasons mentioned above, it is interesting to examine the effect of economic policy uncertainty on Thai stock market. In the past, Thai stock market, the Stock Exchange of Thailand (SET), has faced many uncertainty events such as Black May, Tom Yam Kung crisis, the United Front for Democracy against Dictatorship (UDD) assembly, and Thailand's rice-pledging scheme's failure. The SET index responded with a large decline at the presence of economic policy uncertainties. Thailand's domestic economy and stock market have mutable according to policy change and political actions. For instance, Vithessonthia and Techarongrojwong [28] found that the monetary policy announcements in changing interest rates have a significant effect on stock prices. The study by Khositkulporn et al. [19] noted that stock market investors or indirect investors are more sensitive to political events with any violence. Due to the SET and banking industrial equity have increased volatility after the uncertain political events such as coup, unrest and violence occurred. Furthermore, the Thai economy has been significantly negative as a result of increased political uncertainty, particularly impact on private investment in the short run and economic growth in the long run [23]. Although numerous studies have examined the effect of policy change and political actions in the modern Thai economic history on the economy and SET, only one published study that constructed an index to use a proxy for uncertainty. Luangaram and Sethapramote [23] measured political uncertainty or calculated political uncertainty index (PUI) for Thailand for using analyze quantifying the economic impacts. The PUI for Thailand based on the frequency of Thai-newspaper articles containing the keyword that involved with political uncertainty. This index can well reflect the uncertainty of Thai politics when compared with worldwide governance indicators. Also, it can be used in time with data consistency. However, political uncertainty is one part of uncertainty in the economy while both policy change and political actions in the modern Thai economic history seem to have caused a significant effect on SET. So far, no published study has been found on the impact of policy-related economic uncertainty on the stock market in the case of Thailand. This is due to the economic policy uncertainty measurement is not proposed and computed yet. Thus, it is worth measuring the economic policy uncertainty in Thailand and investigating its effect on the SET.

By doing so, we take two procedures. First, we measure the economic policy uncertainty in Thailand using the index based on the frequency of internet searches as EPU index can immediately reflect the action of investors as well as the emotion of the people in the country. Second, we further apply our proposed index to investigate the impact of economic policy uncertainty on SET index.

In the literature related to the measurement of economic policy uncertainty, there are many studies that introduce various approaches for measuring the uncertainties. For example, Baker et al. [5] developed the method to get a new economic policy uncertainty (EPU) index as a proxy for movements in policy-related economic uncertainty over time. The acquisition of Baker index is based on newspaper coverage frequency involved with EPU. Then, it was introduced into the adjustment process database for indexing by standardizing each paper's EPU rate to unit standard deviation, next average the standardized EPU rates across papers by month and then multiplicatively rescale the resulting series to a mean of 100. The method of Baker et al. [5] for calculating EPU index used in many countries such as China, Japan, India, Singapore, Europe, and so on. Although, in each country set in the different keywords and using local language, the results have shown the same direction that the EPU index spiked sharply around the period of the major economic or political shocks. Because of Baker index to be a useful proxy for EPU, there are many scholars used this EPU index to understand the effect between policy-related economic uncertainty and corporate investment [29], equity market uncertainties [1], stock returns [21], stock markets [22], prudential regulation, bank lending [16] and so on. Most recently, Tobback et al. [27] suggested using text mining approach in the articles and newspaper related to EPU. Although both studies could provide a reliable EPU which reflects the uncertainty of the country, the trend of people interested in reading the newspaper has been on a rapid decline because of the transition of technology to the digital world [26]. Furthermore, the information from newspaper cannot indicate investors behaviors. Some news related to EPU on the newspaper may not reflect the decision of investors in the stock markets. Therefore, in this study, we turn our attention to collecting the data from another source, say internet. The study of Castelnuovo and Tran [9] advocated that the internet searches is another source of the data which could reflect the economic policy uncertainty. As mentioned in Liemieux and Peterson [20], when searching for a product, consumers respond to greater product uncertainty by increasing search. Consumers always focus on maximizing search utility; search continues as long as the marginal benefit of doing so exceeds the marginal cost. According to this approach, the internet searches can be applied to quantify the behavior of the people nowadays. In addition, we can limit the scope of the data by using the concept of text mining. Thus, some selected keywords are considered in this study.

In our study, the economic policy uncertainty in Thailand is computed from the search behavior of people. We first collect the frequency of internet searches in the form of Google trends and then apply the method of Baker et al. [5] to build the Thai EPU index. Next, we investigate the effect of EPU index on two largest market capitalization sectors in SET market consisting of agriculture and food and technology sectors. In the methodological point of view, we use the Dynamic Conditional Correlation GARCH (DCC-GARCH) model introduced by Engle [12] for examining the dynamic correlation between EPU and stock industry. This model has been successfully applied to measure the dynamic correlation in many empirical studies (see, [7, 10]). We thus expect that this model can detect the possible changes in correlation overtime corresponding to the dynamic behavior of the investors in the financial market as well as the searching behavior of the people.

The rest of this paper consists of four sections. Section 2 presents the methodology used. Section 3 specifies the corresponding data. Section 4 presents and discusses the empirical results. Section 5 is the conclusion.

2 Methodology

2.1 Measuring Economic Policy Uncertainty for the Stock Exchange of Thailand

We first aim to construct the economic policy uncertainty (EPU) index for the Stock Exchange of Thailand (SET). This index can be viewed as a proxy for the policy-related economic uncertainty existing in the Thai stock market. The construction of the index is based on the assumption that the behavior of internet users that seek for online information when they are uncertain about some occurrence. In other words, this assumption is that the high uncertainty can be sourced from the intensifying search frequency. Fortunately, the frequency of internet searches is already reported by Google trends. Hence, we consider the searching behavior based on the google searches. Then, we use the method of Baker et al. [5] to calculate the EPU index for Thailand. The procedure of construction the index can be summarized as follows.

First is the identification of the possible keywords. The search terms were subjectively selected from the Bank of Thailand's monetary policy statements by referring to words that are connected to "economic policy uncertainty existing in Thailand". There are 58 keywords for generating the EPU index of the SET such as, "Securities Exchange", "Bank loan", "Bankruptcy", "Consumer confidence", "Default", "Monetary policy", "Oil price", "Stock market", "Tax reform", etc. We then searched these keywords on Google trends to get the volume of searching by internet users. The data of Google trends derived from each raw data point $R_{ij,t}$ - which is, the frequency of a word i in a group of searched words j in a month t divided by total searches T in the same month. This can be written in formula as $X_{ij,t} = R_{ij,t}/T_t$, where $X_{ij,t}$ is the re-scaling to range between 0 and 100 of a word i in a group of searched words j in a month t.

Second, the EPU index of the SET was then obtained by the method of Baker et al. [5]. The method as mentioned previously uses the frequency of data following five steps by first determining that X_{ij} is the scaled frequency counts for keyword set

$i = 1, 2, ..., 58$ in month t, and let T_1 and T_2 denote the time intervals : (i) Compute the times-series variance, ${\sigma_i}^2$, in the interval T_1 for each keyword set i. (ii) Standardize X_{it} by dividing through by the standard deviation σ_i for all t. This operation yields for each keyword set a series Y_{it} with unit standard deviation in the interval T_1. (iii) Compute the mean over keyword of Y_{it} in each month to obtain the series Z_t. (iv) Compute M, the mean value of Z_t in the interval T_2. (v) Multiply Z_t by $(100/M)$ for all t to obtain the normalized EPU time-series index.

2.2 Dynamic Conditional Correlation GARCH (DCC-GARCH) Model

In order to investigate the co-movements between economic policy uncertainty (EPU) and industry-level stock returns, we used the dynamic conditional correlation (DCC) model of Engle [12] to measure the correlation over time. The DCC model is used to model the volatility dynamic and conditional correlations of EPU and industry-level stock return volatilities. Thus, the triradiate DCC-GARCH (1,1) model is considered in this study. The model is constructed as follows:

$$r_t = \mu + e_t, where\ e_t | \Omega_{t-1} \tilde{} F(0, H_t)\, and\, t = 1, ..., T \quad (1)$$

where $r_t = (r_{1t}, r_{2t}, r_{3t})'$ is a vector of two stock sector returns and EPU index growth. This can be written in formula as $r_t = \ln(p_{i,t}/p_{i,t-1})$, $p_{i,t}$ denotes the industry sector i^{th} index and the EPU index growth at time t; $\mu = (\mu_1, \mu_2, \mu_3)'$ denotes a vector of the mean coefficients; $e_t = (e_{1t}, e_{2t}, e_{3t})'$ is a vector of the error terms; Ω_{t-1} is the information set up to time $t-1$. $F(.)$ denotes a trivariate distribution. H_t denotes a 3×3 conditional variance-covariance matrix. Note that the e_t vector has the following conditional variance-covariance matrix:

$$e_t = \sqrt{H_t z_t} \quad (2)$$

$$H_t = D_t R_t D_t, \quad (3)$$

where z_t is a 3×1 vector of iid errors such that $E[z_t] = 0$ and $E[z_t z_t^T] = I$, D_t is a diagonal matrix with the time-varying standard deviations on the diagonal which is obtained from univariate GARCH(p,q) model and R_t is the conditional correlation matrix, and D_t is a diagonal matrix with time-varying standard deviations on the diagonal.

$$D_t = diag(\sqrt{h_{11,t}}, \sqrt{h_{22,t}}, \sqrt{h_{33,t}}) \quad (4)$$

The elements in the diagonal matrix D_t are standard deviations following the univariate GARCH (1,1) process.

$$D_t = \begin{bmatrix} h_{11,t} & 0 & 0 \\ 0 & h_{22,t} & 0 \\ 0 & 0 & h_{33,t} \end{bmatrix} \tag{5}$$

where

$$h_{it} = \omega_i + \alpha_{ip} e_{i,t-1}^2 + \beta_i h_{i,t-1} \tag{6}$$

and the correlation matrix R_t is,

$$R_t = \begin{bmatrix} 1 & \rho_{12,t} & \rho_{13,t} \\ \rho_{12,t} & 1 & \rho_{23,t} \\ \rho_{13,t} & \rho_{23,t} & 1 \end{bmatrix} \tag{7}$$

Since the covariance matrix H_t is by definition positive definite, R_t should be positive definite. Moreover, all the elements in the correlation matrix must be equal to or less than one. To meet the requirements, the correlation matrix R_t can be decomposed into:

$$R_t = Q_t^{*-1} Q_t Q_t^{*-1} \tag{8}$$

where,

$$Q_t = (1 - a - b)\overline{Q} + a e_{t-1} e_{t-1}^T + b Q_{t-1} \tag{9}$$

$$Q_t^* = \begin{bmatrix} 1 & \frac{q_{12,t}}{\sqrt{q_{11,t} q_{22,t} q_{33,t}}} & \frac{q_{13,t}}{\sqrt{q_{11,t} q_{22,t} q_{33,t}}} \\ \frac{q_{12,t}}{\sqrt{q_{11,t} q_{22,t} q_{33,t}}} & 1 & \frac{q_{23,t}}{\sqrt{q_{11,t} q_{22,t} q_{33,t}}} \\ \frac{q_{13,t}}{\sqrt{q_{11,t} q_{22,t} q_{33,t}}} & \frac{q_{23,t}}{\sqrt{q_{11,t} q_{22,t} q_{33,t}}} & 1 \end{bmatrix} \tag{10}$$

where, Q_t is conditional correlation matrix with time varying at time t. Q_t^* is a 3×3 matrix containing the square root of the diagonal element of Q_t. The conditional correlation among the EPU index and the industry-level stock return volatilities in R_t is given by $\rho_{12,t}$, $\rho_{13,t}$, $\rho_{23,t}$.

3 Data Selection and Description Statistics

We consider the data from various keywords from Google trends. The keywords are chosen based on the Bank of Thailand's monetary policy statement (see, Table 1). Our monthly Google trend data is collected from www.google.com covering the period from January 2004 to December 2018, while the monthly agro-food sector stock index and technology sector stock index are obtained from the Thomson Reuters

Table 1 The keywords used for constructing the EPU index

Keywords	Agriculture industry, Asset, Bank loan, Bank of Thailand, Bankruptcy, Bond, Budget, Business outlook, Cabinet, Commodity, Consumer price index, Consumer confidence, Consumer spending, Credit rating, Crisis, Debt, Debt ceiling, Deflation, Derivative, Drought, Default, Exchange rate, Federal Reserve System, Fiscal policy, Foreign direct investment, Flood, Fuel price, Gas price, Government debt, Gross domestic product, Home price, Home sales, Inflation interest rate, Investment management, Legislation, Loan, Minimum wage, Monetary policy, Oil price, Oil supply, Policy, Protest, Real Estate, Recession, Reform, Regulation, Stock market, Securities and exchange commission, Share price, Stock price, Stock exchange, SET, Tax reform, Thai economy, Wage, World economy

Table 2 Descriptive statistics of the agro-food sector stock returns and the technology sector stock returns

Returns	Mean	Max	Min	Std.	Skewness	Kurtosis
Agro-Food	0.0078	0.1258	−0.2630	0.0523	−0.9148	3.7733
Technology	0.0021	0.1377	−0.2498	0.0608	−0.5262	1.5771

database. Note that all google trends are scoped only in Thailand, thus the frequency of internet searching is computed from domestic searching.

Table 2 presents the descriptive statistics of the agro-food sector stock returns and the technology sector stock returns. In this table, the skewness of both industry stock returns is less than zero, indicating that the distribution of the data series is skewed to the left. We also found that the technology sector stock return shows a higher kurtosis when compared to agro-food sector stock return. We can conclude that the distributions for the two data series do not follow the normal distribution.

4 Empirical Analytical Results and Discussion

4.1 The Economic Policy Uncertainty for the Stock Exchange of Thailand

The economic policy uncertainty (EPU) index obtained from the procedures explained in Sect. 2.1 is plotted in Fig. 1. We then examine the reliability of our index using

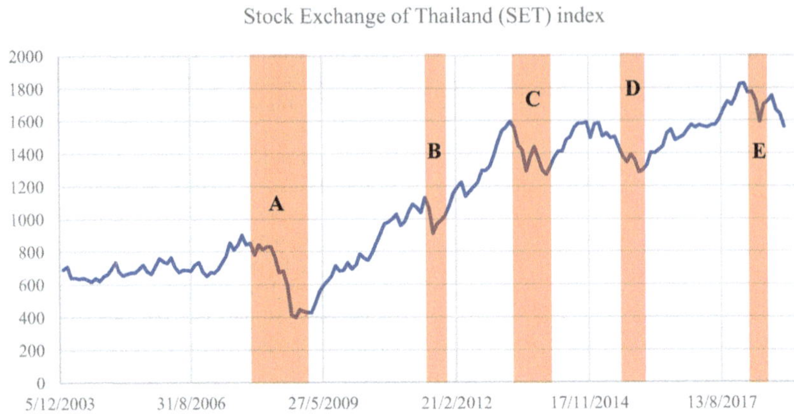

Fig. 1 The Stock Exchange of Thailand (SET) index

visual analysis. The index is compared with the Stock Exchange of Thailand (SET) index along our sample periods. Figures 1 and 2 plot the EPU index and the SET index respectively. Table 3 refers to major uncertainty-related events corresponding to the peak of EPU index. The EPU index peaked during the global financial crisis and the seizure of Suvarnabhumi International Airport by people's alliance for democracy in 2008 (shaded phase A) which resulted in the decrease of investor confidence. Furthermore, this index had a sharp increase during the European sovereign debt crisis (shaded phase B), the change of the US monetary policies, Thai government dissolving its parliament, and the corruption in the Thai rice pledge project (shaded phase C), the Ratchaprasong bombing (shaded phase D), trade war, increase in interest rate, and oil price (shaded phase E). We compare the movement of EPU index and stock index with a special concern to the major uncertainty-related events (A, B, C, D, and E). We can observe that the stock index plummeted corresponding to the five peaks of EPU index. This indicates that our EPU index is reasonable and reliable for represent Thailand economic policy uncertainty.

In addition, we use the R-Squared (R^2) for describing the relationship between the SET index and our EPU index. The calculated R^2 is 0.7006, implying that approximately 70% of SET index can be explained by our index. When we considered the strength of the relationship between our EPU index and the SET index, the result shows a negative relationship with the value −0.8370.

4.2 Dynamic Conditional Correlation-GARCH Result

After we obtained the EPU index, we then examine the impact the EPU on the industry-level stock returns in SET market. We apply DCC-GARCH model to find the correlation among the volatilities of EPU index, agro-food sector, and

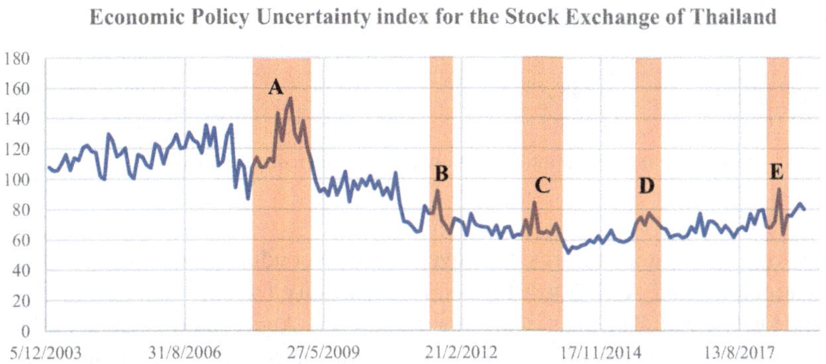

Fig. 2 The Economic policy uncertainty of Thailand

Table 3 Lists of the major uncertainty-related events that occurred when the EPU index is peaked

Letter	Date	Event
A	Jan-08	Suffer the consequences from subprime crisis
	Sep-08	The collapse of Lehman Brothers
	Nov-08	The seizure of Suvarnabhumi International Airport by People's Alliance for Democracy
B	Sep-11	The European sovereign debt crisis
C	May-13	The Federal Reserve's program to reduce holding the bonds
	Dec 13	Thai prime minister (PM) Yingluck dissolved parliament
	Jan-14	Corruption in the rice pledge project
D	Aug-15	The horrific Ratchaprasong bomb
E	Jun-18	The effects of the trade war may shrink the Thai export market
		The upward trend of interest rates for both Federal Reserve System (the Fed) and the Bank of Thailand (BOT)
		The world crude oil prices continued to decline due to falling demand and increasing crude oil prices

Table 4 Augment Dickey-Fuller test and ARCH tests

	EPU	Agro-Food	Technology
ADF (Constant)	−0.3487***	−8.1937***	−9.9466***
ARCH (12) Test	12.1835***	45.0196***	19.4502***

Note: *** Decisive evidence supporting stationarity and heteroscedasticity

technology sector stock returns. However, we must conduct a unit root test to ensure the stationarity and ARCH test for heteroscedasticity for our data, before examining the DCC-GARCH model. Consider Table 4, the Augmented Dickey–Fuller test (ADF) shows that agro - food sector stock returns, technology sector stock returns, and EPU index are stationary. For the heteroscedasticity, ARCH test confirms that there is heteroscedasticity in our series [15].

According to the results of the ADF and ARCH tests, we can further employ a DCC-GARCH model. The estimated parameters of DCC-GARCH (1,1) model are shown in the Table 5. The estimated coefficients for three GARCH equations corresponding to the return of agro-food sector (Agro_r), the return of technology sector stock (Tech_r), and EPU growth (EPU_g) are presented in Rows 2–10. These coefficients are estimated from the GARCH part. The statistical inference of this study is based on the Minimum Bayes factor (MBF). Thus, if the MBF is less than 0.33, it indicates a strong evidence supporting the ARCH(α) and GARCH(β) effects in the volatility. We also observe that the summation of ARCH(α) and GARCH(β) for all series are close to 1, indicating that there is a high volatility in our series. In addition, Table 5 provides estimated DCC parameters, a and b. The MBF of parameter a is 0.6692, but the MBF of parameter b is 0.0000, indicating a weak evidence of parameter a but strong evidence of parameter b in the DCC part.

We then considered the dynamic conditional correlations between Thai EPU and agro-food sector stock return, and between Thai EPU and technology sector stock return. Table 6 provides the descriptive statistics of the two dynamic conditional correlation pairs. We can see that EPU has a greater impact on the agro-food sector (mean value −0.0890) than on the technology sector (mean value −0.0488). Furthermore, the mean correlations between EPU and two stock sectors are all negative, indicating that the increase in the uncertainty has a negative effect on the industry-level stock market of Thailand.

Finally, the dynamic correlations between Thai EPU-agro & food sector stock return and Thai EPU-technology sector stock return are illustrated in Figs. 3 and 4, respectively. According to Figs. 3 and 4, we can observe a large fluctuation as it exhibited a wide range of the movement of correlation. We can observe a high fluctuation during the global financial crisis and the protests against government (shaded phase A), the European sovereign debt crisis (shaded phase B), the change of the US monetary policies, dissolving of the Thai parliament, and the corruption in the rice pledge project (shaded phase C), the horrific Ratchaprasong bombing (shaded phase D), trade war, increase in interest rate, and oil price (shaded phase E). These five phases are corresponding to the peak of EPU index in Fig. 2. During the European

Table 5 Estimated parameter results

Parameter	Estimator	Std. error	t value	MBF
$[Agro\_r].\omega_1$	3.6825	2.4546	1.5002	0.7357
$[Agro\_r].\alpha_1$	0.1826	0.1919	0.9513	0.9976
$[Agro\_r].\beta_1$	0.7010	0.1299	5.3955	0.0000
$[Tech\_r].\omega_2$	0.0236	0.5268	0.0449	0.0954
$[Tech\_r].\alpha_2$	0.0000	0.0149	0.0001	1.0000
$[Tech\_r].\beta_2$	0.9990	0.0010	1043.6	0.0000
$[EPU\_g].\omega_3$	0.0844	0.7266	0.1162	0.2392
$[EPU\_g].\alpha_3$	0.0000	0.0098	0.0000	0.0000
$[EPU\_g].\beta_3$	0.9990	0.0009	1148.1	0.0000
a	0.0696	0.0446	1.5614	0.6922
b	0.0000	0.2317	0.0000	0.0000

Table 6 Descriptive statistics of the dynamic conditional correlations between EPU and agro-food sector stock return; and EPU and technology sector stock return

Variable	Mean	Max	Min	Std.	Skewness	Kurtosis
EPU—Agro and Food	−0.0890	−0.0451	−0.1326	0.0112	0.1547	3.5798
EPU—Technology	−0.0488	−0.0228	−0.0733	0.0059	0.3954	4.1912

sovereign debt crisis (shaded phase B), both agro-food and technology sectors in the SET had a sharp increase in their dynamic conditional correlations. Their values had reached the maximum correlation value at −0.0451, and −0.0228, respectively. Note that the maximum correlation values here indicate weak correlation. On the contrary, the lowest values are −0.1326, and −0.0733, respectively. Apparently, the correlation between EPU and agro-food; and that between EPU and technology are quite high during the financial crisis and the protests against government (shaded phase A). However, during the global financial crisis, the protests against government (shaded phase A), and the horrific Ratchaprasong bombing (shaded phase D) seem to play an important role in the correlation between EPU and agro-food sector; and that between EPU and technology sector as the correlation is quite high. It can indicate that both stock sectors are more sensitive to the global financial crisis, the protests against government, and the horrific Ratchaprasong bombing than the other uncertain events.

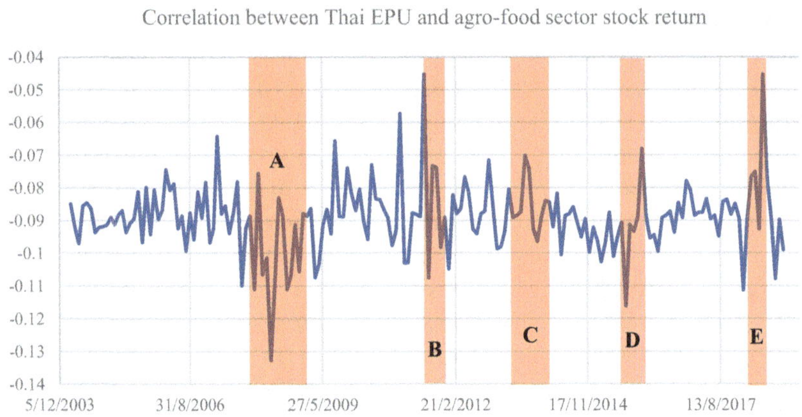

Fig. 3 DCC between Thai EPU and agro - food sector stock returns

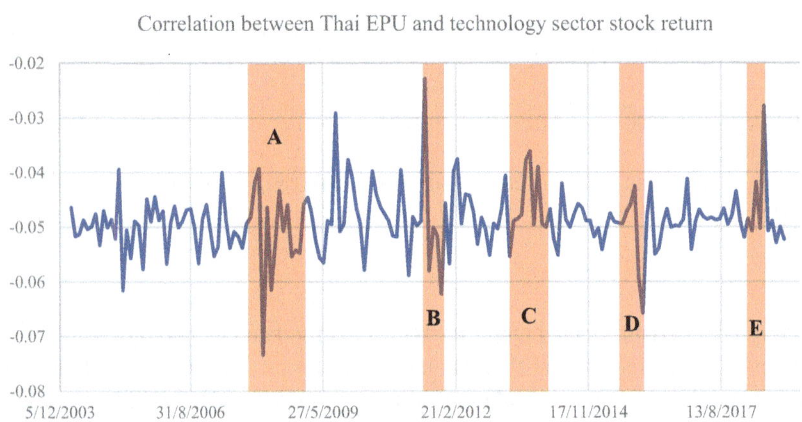

Fig. 4 DCC between Thai EPU and technology sector stock returns

5 Conclusion and Policy Implications

This paper focuses on the development of Thai economic policy uncertainty (EPU) index and the application of this index to investigate its impact on the agro-food and technology stock sectors in Thailand. Specifically, we constructed the index as a proxy for economic policy uncertainty in Thailand. Our index is calculated from the frequency of internet searches reported by Google trends (focusing only in Thailand). We then apply the method of Baker et al. (2016) to build Thai economic policy uncertainty index. After that, we apply the DCC-GARCH model to present the relationship between Thai EPU and agro-food sector; and between Thai EPU and technology sector. We consider these two sectors as they are the two largest markets in terms of capitalization in the Thai stock market.

Our results show that the dynamic conditional correlations between EPU and two stock sectors are all negative along our sample period. This indicates that an increase in the Thai economic policy uncertainty may lead to the decrease in the returns in both sectors. It is obvious that the global financial crisis, the protests against government, and the horrific Ratchaprasong bombing seem to play an important role in the correlation between EPU and agro-food; and between EPU and technology stock sectors as the correlation is quite high. This indicates that both stock sectors are more sensitive to the global financial crisis, the protests against government, and the horrific Ratchaprasong bombing than the other uncertain events. In addition, we also found that the impact of Thai EPU on the agro-food sector is greater than that on the technology sector.

Acknowledgements The authors are grateful to the Center of Excellence in Econometrics for the financial support for this study, we also appreciate Dr. Laxmi Worachai toward strengthening the quality of the paper.

References

1. Ajmi, A.N., Aye, G.C., Balcilar, M., El Montasser, G., Gupta, R.: Causality between US economic policy and equity market uncertainties: evidence from linear and nonlinear tests. J. Appl. Econ. **18**(2), 225–246 (2015)
2. Andersen, T.G., Bollerslev, T., Diebold, F.X.: Roughing it up: including jump components in the measurement, modeling and forecasting of return volatility. Rev. Econ. Stat. **89**, 701–720 (2007)
3. Antonakakis, N., Chatziantoniou, I., Filis, G.: Dynamic co-movements of stock market returns, implied volatility and policy uncertainty. Econ. Lett. **120**, 87–92 (2013)
4. Baker, S.R., Bloom, N., Davis, S.J.: Measuring Economic Policy Uncertainty. Stanford University Working paper (2013)
5. Baker, S.R., Bloom, N., Davis, S.J.: Measuring economic policy uncertainty. Q. J. Econ. **131**, 1593–1636 (2016)
6. Balcilar, M., Gupta, R., Kim, W.J., Kyei, C.: The role of economic policy uncertainties in predicting stock returns and their volatility for Hong Kong, Malaysia and South Korea. Int. Rev. Econ. Finance **59**, 150–163 (2019)
7. Billio, M., Caporin, M., Gobbo, M.: Flexible dynamic conditional correlation multivariate GARCH models for asset allocation. Appl. Econ. Lett. **2**(2), 123–130 (2006)
8. Brogaard, J., Dtzel, A.: The asset-pricing implications of government economic policy uncertainty. Manage. Sci. **61**(1), 3–18 (2015)
9. Castelnuovo, E., Tran, D.T.: Google It Up! A Google Trends-based Uncertainty Index for the United States and Australia, MPRA Paper No. 82297(2017)
10. Chodchuangnirun, B., Yamaka, W., Khiewngamdee, C.: A regime switching for dynamic conditional correlation and GARCH; application to agricultural commodity price and market risks. In: International Symposium on Integrated Uncertainty in Knowledge Modelling and Decision Making, pp. 289–301 (2018)
11. Dzielinski, M.: Measuring economic uncertainty and its impact on the stock market. Finance Res. Lett. **9**, 167–175 (2012)
12. Engle, R.: Dynamic conditional correlation: a simple class of multivariate generalized autoregressive conditional heteroskedasticity models. J. Bus. Econ. Stat. **20**(3), 339–350 (2002)
13. Fernández-Villaverde, J., Guerrón-Quintana, P., Kuester, K., Rubio-Ramírez, J.: Fiscal volatility shocks and economic activity. Am. Econ. Rev. **105**(11), 3352–3384 (2011)

14. Gilchrist, S., Sim, J.W., Zakrajsek, E.: Uncertainty, Financial Friction and Investment Dynamics. NBER Working Papers 20038 (2010)
15. Goodman, S.N.: Toward evidence-based medical statistic: the Bayes factor. Ann. Intern. Med. **130**(12), 1005–1021 (1999b)
16. Hu, S., Gongb, D.: Economic policy uncertainty, prudential regulation and bank lending. Finance Res. Lett. **29**, 373–378 (2019)
17. Johnson, T.C., Lee, J.: On the systematic volatility of unpriced earnings. J. Finance Econ. **114**, 84–104 (2013)
18. Karnizova, L., Li, J.: Economic policy uncertainty, financial markets and probability of US recession. Econ. Lett. **125**, 261–265 (2014)
19. Khositkulporn, P., Suthakawatin, P., Yanuphorm, J.: The effect of Thailand's political uncertainty on volatility of stock market exchange, banking industrial equity and business sentiment. J. Eng. Appl. **12**(2), 290–300 (2017)
20. Liemieux, J., Peterson, R.: Purchase decline as a moderator of the effects of price uncertainty on search behavior. J. Econ. Psychol. **32**, 33–44 (2011)
21. Li, X.L., Balcilar, M., Gupta, R., Chang, T.: The causal relationship between economic policy uncertainty and stock returns in China and India: evidence from a bootstrap rolling window approach. J. Emerg. Mark. Finance **52**(3), 674–689 (2016)
22. Li, X.M., Peng, L.: US economic policy uncertainty and co-movements between Chinese and US stock markets. Econ. Model. **61**, 27–39 (2017)
23. Luangaram, P., Sethapramote, Y.: Economic impacts of Political Uncertainty in Thailand. PIER Discussion Papers 86. Bangkok: Puey Ungphakorn Institute for Economic Research (2018)
24. Ozoguz, A.: Good times or bad times? investors' uncertainty and stock returns. Rev. Finance Stud. **22**(11), 4377–442 (2008)
25. Pastor, L., Veronesi, P.: Uncertainty about government policy and stock prices. J. Finance **67**, 1219–1264 (2012)
26. Peiser, W.: Cohort replacement and the downward trend in newspaper readership. Newsp. Res. J. **21**(2), 11–23 (2000)
27. Tobback, E., Naudts, H., Daelemans, W., Fortuny, E., Martens, D.: Belgian economic policy uncertainty: improvement through text mining. Int. J. Forecast. **34**, 355–365 (2018)
28. Vithessonthia, C., Techarongrojwong, Y.: Do monetary policy announcements affect stock prices in emerging market countries? the case of Thailand. J. Multinational Financial Manag. **23**(5), 446–469 (2013)
29. Wang, Y., Chen, C., Huang, S.: Economic policy uncertainty and corporate investment: evidence from China. Pac. Basin Finance J. **26**, 227–243 (2014)
30. Xiong, X., Bian, Y., Shen, D.: The time-varying correlation between policy uncertainty and stock returns: evidence from China. Physica A **499**, 413–419 (2018)

Analyzing the Relationship Among Aging Society, Investment in Artificial Intelligence and Economic Growth

Kantika Khanthawithoon, Paravee Maneejuk, and Woraphon Yamaka

Abstract Artificial Intelligence (AI) is performing an increasingly significant role in everyday life as it has the potential to increase the efficiency of productivity and boost the economy. While the technology world is advancing, the mortality rate and the fertility rate of people are declining. These demographic trends have urged many countries to pay a concern to the apparent problems associated with the aging society like the shrinking proportion of the working-age population, which in turn can bring about the decline in productivity and the national economic slow-down. Hence, we aim to investigate the relationship among investment in AI, Aging society, and Economic growth in three advanced countries, namely China, Japan, and the U.S. using the Copula-based simultaneous equations model. The results obtained from the model can conform to our hypothesis only in the case of China and Japan. However, the result of the U.S., is deviated from the hypothesis but challenging.

Keywords Aging society · Artificial intelligence · Copula-based simultaneous equations

1 Introduction

Economic growth is one of the measures of the well-being of the economy. The country with high economic growth usually has a better standard of living. Generally, high economic growth is a consequence of great productivity. According to

K. Khanthawithoon (✉)
Faculty of Economics, Chiang Mai University, Chiang Mai 50200, Thailand
e-mail: kanthika1996@gmail.com

P. Maneejuk · W. Yamaka
Center of Excellence in Econometrics, Faculty of Economics,
Chiang Mai University, Chiang Mai 50200, Thailand
e-mail: mparavee@gmail.com

W. Yamaka
e-mail: woraphon.econ@gmail.com

S. Sriboonchitta et al. (eds.), *Behavioral Predictive Modeling in Economics*,
Studies in Computational Intelligence 897,
https://doi.org/10.1007/978-3-030-49728-6_27

the simple production function, the economic output is almost produced by capital (K) and labor (L). When the capability of the input factors increases, the GDP also increases. However, nowadays, the world is progressing faster as a result of technological progress. Even though, high economic growth and development can improve health care services, lower the mortality rate, and enable people to have a longer life than before, there is a decline in fertility rate among baby-boomers and Gen X. These situations are driving a nation into an aging society, which is followed by a labor shortage problem. Intuitively and factually, problems emerging from an aging society have included a shortage of economically active manpower, a decrease in the productivity, reduction in the government spending on education, transport network, infrastructure, and other facilities to spend more on cost and pension for the elderly. Obviously, the problem is on the labor side. With a shrinking labor force, the burden of production is on capital and technology. Then, are there any labor substitutable production inputs in the aging society? This study considers technology as a way out of labor shortage problem because technology can improve the productivity of all factors of production, i.e., capital and labor. Hence, this study focuses on an aging society, investment in AI, and economic growth in order to respond to the current situation. Since all these three factors are tending to affect each other in the same period of time, we can characterize the circumstances to see an overview through two scenarios as follow.

Scenario 1: Artificial Intelligence and Economic Growth

The last decades witnessed the rapidly increased adoption and use of advanced technologies as AI, which plays an increasingly significant role in everyday life. The AI market has a potential in its path supported by the amount of investment which increases every year. Hence, AI has a potential effect on the decrease in the work hour,

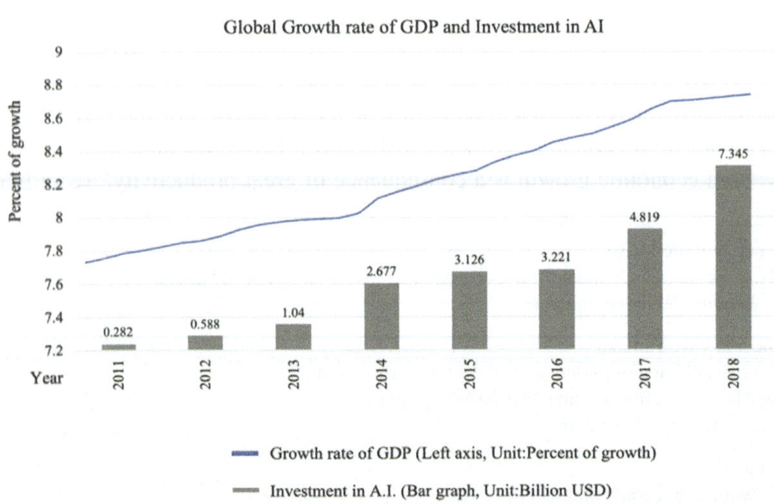

Fig. 1 A plot of global growth rate of GDP (percentage) and investment in AI (Billion USD)

more efficiency; it helps people do more work, in less time, with fewer resources, facile to organize and plan the scale of work with efficient outcomes. From all these things, AI can boost the economy with higher productivity. The critical point is that AI can increase the productivity of both capital and labor, thereby affect economic growth increases. Additionally, the growth rate of GDP is rising overtimes, which is in the same direction with investment in AI. Additionally, Fig. 1 shows the growth rate of GDP that has risen overtimes from 2011 to 2018, which is in the same direction with investment in AI. This notion is supported by Aghion and Jones [2] who investigated how AI affects economic growth and the proportion of income production, which are capital and labor. They found that AI can boost economic growth either permanently or temporarily. Nordhaus [10] emphasized that AI has the potential to increase productivity and wideness to the extent of human labor ability. Particularly the countries will have a higher readiness in technology. Szczepański [16] studied AI and economic impact in 12 developed countries, where generating economic growth more than 0.5% of the world economic output or double annual global economic growth rates. He believed that AI could drive the growth through three essential ways, which are an increase in labor productivity, create a new virtual workforce to solving the problems and self-learning, and the benefit from diffusion of innovation.

Scenario 2: Artificial Intelligence and Aging Society

In this decade, the circumstance of the aging society has been concerned in many countries. The older population is increasing, and the growth rate of the older population also rises. Meanwhile, many countries experience the situation of decreasing fertility rate, which decreases the number of working-age people and also the output productivity. Even though the population in this world is getting older, technology is also growing up. In the last decades, the rapid increase in the adoption of advanced technologies was found in many sectors. AI can replace humans in some production units, implying that it can be a solution to the labor shortage problem in the aging society while increasing the economic growth at the same time. Figure 2 shows that the number of older populations aged above 64, the growth rate of GDP and investment in AI increase together from 2011 to 2018. Acemoglu and Restrepo [1] investigated the relationship between an aging society and AI and found that the relationship was positive. This finding is similar to the study of Furman and Kennedy [4] who suggest that AI has a potential to improve a production productivity and may be a way of dealing with labor shortage from the aging society. Schneider et al. [14] also investigated a relationship between AI (and robotics) and shrinking labor force in Japan and suggested that including AI in a production process could lead to labor disruptions. They explained that the coming age of automation technology and AI was a sign of replacing labor by high technology in a non-manufacturing sector. This is because Japan has forced by shrinking population and decline in the labor force from the aging society. Thus, many industries tend to invest more in advanced technology such as robots and AI in order to prevent themselves from the labor shortage in the future. These evidences can support the existence of a positive relationship between AI and aging society.

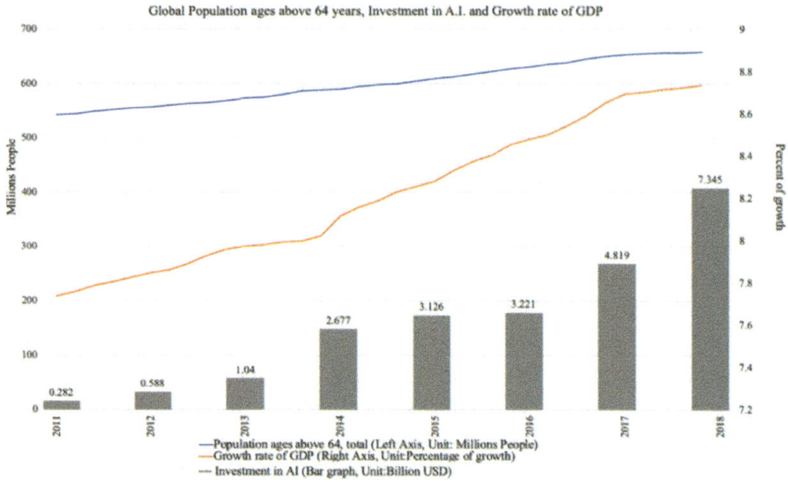

Fig. 2 A plot of global population ages above 64 years (Millions of people), investment in AI (Billion USD) and growth rate of GDP (percentage)

These two scenarios have been proved separately by many studies. However, this study believes that these two scenarios may happen at the same time and probably affect each other. First, AI affects economic growth of a country and second, an aging society can have an impact on AI. Therefore, this study intends to investigate a simultaneous link between these scenarios. We employ a Copula-based simultaneous equations model, developed by Choktaworn et al. [3] to combine the two scenarios into a system. In addition, this study considers China, Japan, and the U.S. as the sample countries because they have very outstanding characteristics of an aging society and AI use. They are also known as the leader in AI, due to their massive investment in AI and convenient environment provided by the government. Also, they have a population above 64 years old, over 7% of the total population, which is over the standard criteria of the United Nations to be defined as an aging society. The results of this study will provide new and interesting insights about the relationship among three variables, consisting of economic growth, AI investment, and aging society, and also show how the results are different across the U.S., China, and Japan.

After the introduction, Sect. 2 provides the methodology involving the simultaneous equations model. Section 3 presents the Copula-based simultaneous equations model. Section 4 deals with data description and empirical model application and discussion on the results. Conclusions are given in Sect. 5.

2 Methodology

This empirical study estimates the relationships using the copula-based simultaneous equations model, and then compare the estimated results to the copula-based simultaneous equations model. The comparison is conducted to investigate whether the copula-based simultaneous equations model has a better performance than the simultaneous equations model.

2.1 Review of Simultaneous Equations Model

Simultaneous equations is a system equation representing a set of relationships among variables as the endogenous variable in one equation affects another equation. For more information, refer to Greene [6], Ramanathan [13], Sirohi [17], and Zellner [19]. The correlation between explanatory variables and errors term is the causation of this simultaneous system. It can be defined in general form as

$$y_i = [Y, X]\begin{bmatrix} T_i \\ \beta_i \end{bmatrix} + \varepsilon_i i = 1, ..., M. \tag{1}$$

Suppose we have M equations in the system, y_i is a $T \times 1$ vector of dependent or endogenous variables, X is a $T \times K$ matrix of K exogenous variables, Y is $T \times M$ matrix of endogenous variables. β_i and T_i are $K \times 1$ vector of unknown parameters to be estimated in the vector of error terms in the equation i^{th} is $T \times 1$ vector. So that the extension form of simultaneous equations can be written as

$$\begin{bmatrix} y_1 \\ y_2 \\ \vdots \\ y_M \end{bmatrix} = \begin{bmatrix} T_{11}y_2 + \cdots + T_{1M}y_{M-1} + \beta_{11}X_{1i} + \cdots \beta_{1K}X_{Ki} \\ T_{21}y_1 + \cdots + T_{2M}y_{M-1} + \beta_{12}X_{1i} + \cdots \beta_{2K}X_{Ki} \\ \vdots \\ T_{M1}y_1 + \cdots + T_{MM}y_{M-1} + \beta_{1M}X_{Mi} + \cdots \beta_{MK}X_{Ki} \end{bmatrix} + \begin{bmatrix} \varepsilon_1 \\ \varepsilon_2 \\ \vdots \\ \varepsilon_3 \end{bmatrix}, \tag{2}$$

where we can define the coefficient of the endogenous variable to be zero.

The simultaneous equations model, it is assumed that the error term has no correlation between observations but there are correlations across equations.

$$E\left[\varepsilon_{it}\ \varepsilon_{jt}\middle| X\right] = 0_{ij}, ; t \neq s, i \neq j, \tag{3}$$

where i and j represent the number of each equation, then t and s represent the observations at the time. Nevertheless, we certainly accept for contemporaneous correlation which can be written as

$$E\left[\varepsilon_{it}\ \varepsilon_{jt}\middle| X\right] = \sigma_{ij}, \tag{4}$$

This can imply that the simultaneous equations model accepts non-zero covariance between the error terms of different equations. Generally, the error terms are assumed to have a normal distribution, $\varepsilon_t = (\varepsilon_{1t}, ..., \varepsilon_{Mt}) \sim N(0, \sum)$, where $\sum$ is a variance-covariance matrix for M equations, then

$$E(\varepsilon_t \varepsilon'_t) = \Omega = \Sigma \otimes I_T, \quad (5)$$

where I is a T - dimensional identity matrix and $\otimes$ represents the matrix Kronecker product.

3 Copula-Based Simultaneous Equations Model Estimation

Generally, the estimation of the simultaneous equations assumes a multivariate normal distribution, which may not be flexible and reliable to join the equations. Thus, this study employs the Copula-based simultaneous equations model, whose performance has been proved by various studies. This model is more flexibility to accommodate the different marginal distributions across different equations.

In the statistics, it is the joint distribution of random variables which is marginally uniformly distributed. Sklar [15] proposed Sklar's Theorem that can explain the linkage between the marginal distributions. Copula has two different classes which are Elliptical Copulas and Archimedean Copula. The Elliptical Copulas that we employ in this study are Gaussian and Student's t. The Archimedean family comprises Clayton, Gumbel, Frank, Joe, Rotated of Clayton 90, and 270° and Rotated Gumbel 90 and 270°. For more details, refer to Nelsen [9], Hofert [7], Nagler [8], Pastpipatkul et al. [11], and Stulajter [18].

3.1 Model Estimation

The estimation of Copula-based simultaneous equations model is to follow these steps: Ordinarily, the first step of the copula-based model is using a maximum likelihood method to accomplish the initial values. Second, use the chain rule to construct the simultaneous equations model written as

$$\begin{aligned} \tfrac{\partial^2}{\partial u_1 \partial u_2} F(u_1 u_2) &= \tfrac{\partial^2}{\partial u_1 \partial u_2} c(u_1 u_2) \\ &= f_1(x_1) f_2(x_2) c(u_1, u_2), \end{aligned} \quad (6)$$

In Eq. (6), both of u_1 and u_2 are the marginal distributions which can be either Gaussian or Student's t distribution. $f_1(x_1)$ and $f_2(x_2)$ are density functions of GDP and AI equations, respectively. The density function of Copulas is defined by $c(u_1, u_2)$. This investigation uses some of Copula families, for instance Gaussian,

Student-t, Clayton, Gumbel, Frank, Joe and the rotated Copulas to construct the joint distribution function of a bivariate random variable. Therefore, using the logarithm to transform Eq. (6) in order to obtain

$$\ln L = \sum_{i=1}^{T} (\ln l(\theta_1 | X_{GDP}) + \ln l(\theta_2 | X_{AI}) + \ln c(u_1, u_2)) \quad (7)$$

where in $l(\theta_1 | X_{GDP})$ and $l(\theta_2 | X_{AI})$ both are the logarithms of the likelihood function of two equations. In this study, we consider normal and student's t likelihood functions for GDP and AI equations.

The last term of Eq. (7), $\ln c(F_1, (u_1), F_2(u_2))$ implies the bivariate Copula density, which is assumed for Gaussian, Student-t, Clayton, Gumbel, Frank, Joe and the rotated Copulas. Finally, we apply the maximum likelihood estimation to maximize the simultaneous equations Copula likelihood function in Eq. (7) to accomplish the final estimation of the Copula-based model. Thus, compare the result of each Copula families and select an appropriate type suitable for the data by using the Akaike information criterion (AIC), then we obtain the best-fit Copula and parameter for the model.

4 Empirical Study: CHINA, JAPAN and the U.S

In this section, we apply the model with the real data set, in order to analyze the relationship among investment in AI, the population aged 65 and above (% of total population) and real GDP growth rate. To obtain the performance of each variable and observe the data through the real circumstance in China, Japan, and the U.S., this section starts with describing the data used in this study. Then we select the best-fit copula family, next estimating the result and discussion.

Data

The variables are comprising of the real GDP growth (GDP_t^i)retrieved from CEIC database, investment in AI (AI_t^i) retrieved from CrunchBase, population aged 65 and above (% of total population) (AGE_t^i) and the other control variables, which are consumption(CS_t^i), government expenditure (GE_t^i), the export of goods and services (EX_t^i), labor force (LF_t^i), high technology export (HT_t^i), research and development (RD_t^i), and, expenditure E-Government development index (EG_t^i) retrieved from World Bank database. This study uses the quarterly data set of China, Japan, and the U.S., covering Q1/2006 to Q4/2018. Then, we reconstruct the data to the growth rate for stationary.

The Model Specification

The simultaneous equations model comprised of two main equations, which are GDP and AI equations with a variable specified of each country. From Eqs. (8) and (9) i represent the country, namely China, Japan, and the U.S., we note that the model is estimated country-by-country.

$$GDP_t^i = \alpha_0 + \alpha_1 AI_t^i + \alpha_2 CS_t^i + \alpha_3 GE_t^i + \alpha_4 EX_t^i + \alpha_5 LF_t^i + \varepsilon_{1t} \quad (8)$$

$$AI_t^i = \beta_0 + \beta_1 GDP_t^i + \beta_2 AGE_t^i + \beta_3 HT_t^i + \beta_4 RD_t^i + \beta_5 EG_t^i + \varepsilon_{2t} \quad (9)$$

4.1 Selecting a Copulas

Before employing a Copula function to join simultaneous model equations, we have to select the best Copula families for the model by using AIC, and the results for each country showed in Tables 1, 2, and 3. This empirical study defines Normal and Student's t distributions for the error terms of equations. Then we can frame the possible outcomes in the form of four types (Marginal distribution of GDP equation/Marginal distribution of AI equation)
which are:

Type 1: Normal/Normal distributions
Type 2: Normal/Student's t distributions
Type 3: Student's t/Normal distributions
Type 4: Student's t/Student's t distributions

From Table 1, the best model for China is the Student's t based simultaneous equations model, where the marginal distribution of both equations is normal. From Table 2, Rotated Clayton 90° based simultaneous equations model is selected for

Table 1 AIC criteria for model selection for China

Copulas	Marginal			
	Type 1	Type 2	Type 3	Type 4
Gaussian	734.3796	702.6651	804.0724	663.1138
Student's t	**341.0000**	636.3181	593.8166	639.1532
Frank	616.5379	831.2355	1599.1224	3147.0556
Clayton	487.1226	2223.6193	1847.7048	2816.1838
Rotated Clayton 90°	1161.4415	2601.7024	2742.7126	557.7367
Rotated Clayton 270°	580.2847	675.2283	836.5315	523.2956
Joe	845.2034	684.8361	850.5369	2558.2754
Gumbel	1701.3275	419.6284	1937.1114	521.5863
Rotated Gumbel 90°	963.6706	837.9199	1277.0172	485.1298
Rotated Gumbel 270°	524.9896	1929.3643	2288.7493	2367.5243

Table 2 AIC criteria for model selection for Japan

Copulas	Marginal			
	Type 1	Type 2	Type 3	Type 4
Gaussian	354.7119	444.8665	433.7217	377.5326
Student's t	353.9206	433.7426	427.8859	354.1647
Frank	429.9231	345.3367	351.3033	370.3238
Clayton	416.1274	382.9208	350.1734	290.6548
Rotated Clayton 90°	**273.9876**	347.0446	339.1668	384.1404
Rotated Clayton 270°	423.7231	469.6821	429.1769	381.7453
Joe	444.1792	477.0285	565.7185	390.7958
Gumbel	387.2892	411.0815	396.5572	365.6861
Rotated Gumbel 90°	473.0959	443.2131	569.3524	310.2365
Rotated Gumbel 270°	313.0043	557.8311	621.7321	306.7125

Table 3 AIC criteria for model selection for the U.S.

Copulas	Marginal			
	Type 1	Type 2	Type 3	Type 4
Gaussian	701.8016	628.3947	717.6529	656.5936
Student's t	**407.1315**	708.8253	452.8829	646.4018
Frank	645.7132	661.8572	1301.6963	662.7751
Clayton	1344.1077	1238.5992	659.7378	653.0749
Rotated Clayton 90°	1310.7685	1231.0489	645.5962	648.9792
Rotated Clayton 270°	1545.4954	1478.0064	655.4912	666.5778
Joe	1623.4976	1487.4465	655.1898	664.3008
Gumbel	660.3088	1777.3237	1822.8574	662.3775
Rotated Gumbel 90°	1031.0829	682.9816	916.9884	680.7387
Rotated Gumbel 270°	654.6859	956.9589	670.8195	701.244

Japan and the marginal distribution of both equations is normal. From Table 3, the appropriate model for the U.S. is Student's t based simultaneous equations model, where both equations have marginal distribution as normal. Hence, we can notice that our case study countries have the same marginal distribution as normal but the difference in the type of copula.

4.2 Estimating Parameters of GDP and AI Equations

In this section, we represent the estimated results of the copula-based model compare the simultaneous equations model, and for China, Japan, and the U.S. In order to make statistic inference, this study uses the Minimum Bayes' Factor (MBF) for hypothesis testing. According to Goodman [5] , the interpretation of MBF between 1/3 to 1 is anecdotal evidence for the alternative hypothesis, 1/10 to 1/3 is moderate evidence, 1/30 to 1/10 is substantial evidence, 1/100 to 1/30 is strong evidence, 1/300 to 1/100 is very strong evidence and lower than 1/300 is extreme evidence for null hypothesis. Then, this study presents the result in two different forms, equation form and table form containing the MBF values.

4.2.1 Estimates of China

This study estimated the parameters from the copula-based model and compare the results with the conventional model, that is the simultaneous equations model. The AIC value of the copula-based model is less than the AIC of the conventional model. This implies that the copula-based model performs better than the conventional model. Also, as a result of the simultaneous equations model, we found that the GDP equation relies on the investment in AI and consumption. Differently, in copula-based simultaneous equations model, GDP equation indicates the positive relationship between investment in AI and real GDP growth rate or for every additional investment in AI, the real GDP growth rate is expected to increase by 0.788%. The economic growth also depends on consumption, government expenditure, the export of goods and services, and labor force respectively. In the AI equation, the estimated results from the copula-based simultaneous equation model indicate that every additional unit of real GDP growth rate can contribute investment in AI by 0.645%. Also, an increase in the older population can create investment in AI positively at 6.557%, while the AI equation also response to a change in high technology export, and research and development expenditure. The estimated results correspond to the characteristic of China's economy in the present-day. According to the National Bureau of Statistics of China, the statistics in 2018 announced that it is experiencing a significant drop in both its birth and marriage rates most likely due to its one-child policy, giving rise to aging society problem. In the meantime, the government of China has supported the AI ecosystem through the leader of AI in each sector, where the government is the center in order to reach the world's leading level by 2030 and considers AI is one way to boost the economy. Obviously, AI will be more advantageous in the country lacking labor, because in the future AI will be a smarter and cheaper technology. This can make AI more productive in an aging society. This can be one of the reasons why the magnitude of the older population is more than the other main variables.

As shown in Table 4, the copula-based simultaneous equation model has a better achievement than the simultaneous equations model in terms of AIC value. In addition, the MBFs of AI^{CN} and LF^{CN} in the GDP equation suggest decisive evidence. We can clarify that these two variables have MBF value at 0.0033 or accept the alternative hypothesis and reject the null hypothesis and other variables are inconclusive.

Table 4 Estimation of Copula-based simultaneous equations model for GDP and AI of CHINA

	The simultaneous equations model (Conventional model)				The Copula-based Simultaneous equations model			
GDP equation	Coeff.	Std. err.	T-value	MBF	Coeff.	Std. err.	T-value	MBF
Intercept	−2.2398	3.3483	−0.6690	1.0000	4.6046	7.4863	0.6150	1.0000
AI^{CN}	0.0341	0.0501	0.6810	1.0000	0.7885	0.0103	−7.0745	0.0000
CS^{CN}	3.4175	1.7113	1.9970	0.4134	0.9360	29.4291	0.0318	1.0000
GE^{CN}	−0.5915	1.6201	−0.3650	1.0000	1.3101	29.5444	−0.0443	1.0000
EX^{CN}	−0.7595	0.5829	−1.3030	0.8723	0.1376	29.9886	0.0045	1.0000
LF^{CN}	−30.6120	53.8211	−0.5690	1.0000	72.2712	0.1350	535.2034	0.0000
AI equation	Coeff.	Std. err.	T-value	MBF	Coeff.	Std. err.	T-value	MBF
Intercept	−54.0418	38.5741	−1.4010	0.8130	−48.4487	115.6716	−0.4188	1.0000
GDP^{CN}	0.2684	0.3945	0.6800	1.0000	0.6452	339.5823	0.0019	1.0000
AGE^{CN}	2.6765	9.1739	0.2920	1.0000	6.5571	0.0303	−1.7394	0.5625
HT^{CN}	−0.0441	1.0771	−0.0410	1.0000	0.1480	46.6193	−0.0031	1.0000
RD^{CN}	101.2338	119.3206	0.8480	1.0000	19.7141	252.4837	0.0780	1.0000
EG^{CN}	29.5467	20.5627	1.4370	0.7897	−33.9704	72.2508	−0.4701	1.0000
AIC	860.303				341.000			

Note CN corresponds to China

4.2.2 Estimates of Japan

This study found that the copula-based model provides the AIC value, which is less than the simultaneous equations model. Thus, we can indicate that the copula-based model has a better performance than the simultaneous equations model. In Table 5, the copula-based simultaneous equations model represents the strong positive relationship between investment in AI and real GDP growth rate or for every additional investment in AI, the real GDP growth rate is expected to increase by 8.634% at a higher degree than other main variables. Including consumption and labor force influence to the real GDP growth rate. In the AI equation, the estimated result from the copula-based simultaneous equations model specifies that every additional real GDP growth rate can expect investment in AI to increase for 1.161%, also increase in the older population can create investment in AI positively at 4.509%. According to the results, investment in AI and real GDP growth rate can generate each other positively with the different levels and involved in the endogenous growth theory, when the technology improves, then the growth rates in the economy will grow up while the number of older populations can generate investment in AI. Raj and Seamans [12] believed that AI and robotics could substitute or complement for human labor in manufacturing and services establishments, which are reasonable for the aging society that is lack of human labor. However, in the case of Japan, the government tends to support the development of AI rather than robotics. This is because AI has higher efficiency and lower cost than a robot. It also attempts to construct the cluster of AI by supporting private companies. Statistics show that investment in AI, older population and economic growth increase in the same direction in Japan. In the last few decades, Japan drove the economy with robotics, but now the government sup-

Table 5 Estimation of Copula-based simultaneous equations model for GDP and AI of JAPAN

	The simultaneous equations model (Conventional model)				The Copula-based Simultaneous equations model			
GDP equation	Coeff.	Std. err.	T-value	MBF	Coeff.	Std. err.	T-value	MBF
Intercept	0.3312	0.2591	1.2780	0.8790	−1.5852	0.1669	−9.4974	0.0000
AI^{JP}	−0.0129	0.0151	−0.8560	1.0000	8.6346	0.0014	60.3651	0.0000
CS^{JP}	−0.0230	0.0685	−0.3360	1.0000	3.0482	0.0017	0.1734	1.0000
GE^{JP}	−0.0943	0.0877	−1.0750	0.9713	−8.2474	0.0019	−0.4328	1.0000
EX^{JP}	0.0692	0.0588	1.1770	0.9314	−1.6336	0.0028	−0.5805	1.0000
LF^{JP}	−0.0946	1.5283	−0.0620	1.0000	4.1076	0.0447	0.9188	0.9997
AI equation	Coeff.	Std. err.	T-value	MBF	Coeff.	Std. err.	T-value	MBF
Intercept	22.2264	42.2487	0.5260	1.0000	1.2409	1.9824	10.8699	0.0000
GDP^{JP}	−0.7886	1.3941	−0.5660	1.0000	1.1610	0.1749	66.3739	0.0000
AGE^{JP}	−1.0481	0.8748	−1.1980	0.9215	4.5094	0.0304	−1.4828	0.7467
HT^{JP}	1.4528	0.5674	2.5610	0.1354	1.8816	0.0370	0.5083	1.0000
RD^{JP}	7.2759	6.2161	1.1700	0.9346	−3.7892	0.2052	−1.8463	0.4880
EG^{JP}	−7.6746	18.3754	−0.4180	1.0000	−1.3640	0.6463	−2.1103	0.3245
AIC	580.2840				273.9876			

Note JP corresponds to Japan

ports AI in order to obtain higher efficiency. As Japan is facing problems associated with the aging society, the government supports more AI ecosystems to apply and augment the technology to improve the quality of life and drive the economy at the same time.

In Table 5, reports that the copula-based simultaneous equations model carry out the more significant outcome than the simultaneous equations model, according to the lower AIC value. Furthermore, MBF value, where AI^{JP} in GDP equation and GDP^{JP} AI equation are considered decisive evidence, as the MBF value at 0.00333 or accept the alternative hypothesis and reject the null hypothesis. Then, we can confirm that the real GDP growth rate and investment in AI affect each other at an extreme level.

4.2.3 Estimates of the U.S.

According to the AIC value of the copula-based simultaneous equations model is lower than the simultaneous equations model, we can explain that the copula-based simultaneous equations model has greater performance than the simultaneous equations model. As a result, in the GDP equation, the copula-based simultaneous equations model shows that every additional investment in AI goes hand in hand with 7.810% decrease in real GDP growth rate. In the AI equation, the copula-based simultaneous equations model describes that every additional of the real GDP growth rate can expect investment in AI to an increase for 5.5%. Also, an increase in the older population can generate investment in AI positively by 3.46%. It can be seen that the real GDP growth rate and the number of older populations can positively create

Table 6 Estimation of Copula-based simultaneous equations model for GDP and AI of the U.S.

	The simultaneous equations model (Conventional model)				The Copula-based simultaneous equations model			
GDP equation	Coeff.	Std. err.	T-value	MBF	Coeff.	Std. err.	T-value	MBF
Intercept	0.7599	0.2122	3.5800	0.0088	7.9868	1.4243	5.6075	0.0000
AI^{US}	0.0003	0.0009	0.3330	1.0000	−7.8107	2.2325	−3.4986	0.0184
CS^{US}	0.7065	0.3181	2.2210	0.2672	7.6398	0.0082	92.8948	0.0000
GE^{US}	−0.1804	0.0529	−3.4050	0.0151	9.2354	9.1813	1.0059	0.9908
EX^{US}	0.0311	0.0534	0.5820	1.0000	2.3532	1.5286	1.5394	0.7206
LF^{US}	−2.0608	0.9150	−2.2520	0.2524	2.0551	3.8414	−5.3499	0.0000
AI equation	Coeff.	Std. err.	T-value	MBF	Coeff.	Std. err.	T-value	MBF
Intercept	−619.7570	1391.8780	−0.4450	1.0000	−6.1929	0.1610	−3845.4621	0.0000
GDP^{US}	5.1900	21.1520	0.2450	1.0000	5.5009	0.2009	27.3797	0.0000
AGE^{US}	−7.0100	18.4220	−0.3810	1.0000	3.4673	2.2706	−1.5270	0.0715
HT^{US}	−2.4080	2.9530	−0.8150	1.0000	−2.5891	0.1871	−13.8343	0.0000
RD^{US}	−16.7800	109.2900	−0.1540	1.0000	−1.5516	0.4395	−35.2989	0.0000
EG^{US}	869.5760	1143.3410	0.7610	1.0000	8.6998	0.1404	6195.3317	0.0000
AIC	710.0349				407.1315			

Note US corresponds to the U.S

investment in AI. Even though advanced technology especially AI, is considered a crucial engine for economic growth in the future, it is enabled to drive the U.S. economy during this time. Conversely, the result also shows that investment in AI will not generate economic growth in the U.S. which is different from our assumption. There is a possibility that at this time, the U.S. is facing the economic crisis then affects directly to economic growth. Additionally, the investment in AI in the U.S. mostly occurs in private companies and specifically in the IT sector. Thus, private AI investment seems not to have a significant impact on the overall economy and unable to generate economic growth in the U.S.

As shown in Table 6, according to the AIC values, the copula-based model has a lower AIC value when compared to the simultaneous equations model. Thus, we can explain that the copula-based model generates more accuracy than the conventional model. Moreover, the MBF criteria of the copula-based model, AI^{US} has very strong evidence, which is the MBF value at 0.0333 or accepts the alternative hypothesis, $CS^{US}$and LF^{US} have decisive evidence and in AI equation, GDP^{US},HT^{US}, $RD^{US}$and EG^{US}have decisive evidence with MBF value at 0.0033 or accepts the alternative hypothesis. AGE^{US} is with substantial evidence at 0.1 or accept the alternative hypothesis.

To conclude the result of this case study, we cannot deny that technology has become a part of our life. We have adopted many technologies across the sector to reach higher efficiency. Currently, we are facing an aging society, which is a huge problem for the economy, because the population is getting old, then the capacity of the production sector can slow down. Since this study investigates two scenarios as a system, the result shows that in China and Japan relationship are positive with an increase in investment in AI, older population and economic growth, which correspond to our assumption. Conversely, the U.S. increase in investment in AI is present with a decrease in economic growth. That an increase in the older population can create investment in AI happens in all cases of study, implying that the countries having more older population tend to adopt more AI in their country. Then, if we compare the magnitude of the result, we found that in GDP equation, when investment in AI increases Japan obtains the highest impact of an increase in real GDP growth rate then followed by China and the U.S. In AI equation, the countries that can increase real GDP growth rate and can create the most degree of investment in AI is the U.S. then followed by Japan and China. In the part of the adoption of AI in an aging society, when an increase in the older population can generate investment in AI is effectively in China, then followed by Japan and the U.S. The MBF criteria of AI in Japan and China are decisive evidence and, in the U.S. is very strong evidence. GDP in Japan and the U.S. are decisive evidence and in China is anecdotal evidence. AGE in the U.S. is substantial evidence and in China and Japan are anecdotal evidence. Then, we see the overview of MBF values, and some variables obtain less satisfactory results due to the limitation of investment in AI data. More observations are needed to obtain the consequence in the long term and three main variables also depend on other factors beyond the study.

5 Conclusion

This study aims to analyze the relationship among aging society, investment in AI and economic growth. Then, we employ the Copula-based simultaneous equations model to construct the system of equations in order to estimate AI and economic growth equations. In addition, we also compare the performance of the copula-based simultaneous equations model with the simultaneous equations model. Thus, we found that the copula-based simultaneous equations model that we use provide better performance than the simultaneous equations model. According to the copula-based model has a lower AIC value than the simultaneous equations model. In this study, we investigate these relationships, particularly in highly advanced technology countries namely, China, Japan, and the U.S.

The result obtained from the copula-based model reveals that China and Japan, being an aging society, can go hand in hand with investment in AI and economic growth, which corresponds to our expectations. We found some difference in the case of the U.S. the investment in AI cannot create economic growth. According to AI and Economic growth scenario; we found that in China and Japan, investment in AI can create economic growth and economic growth can create investment in AI as well. Differently, in the U.S. investment in AI cannot create economic growth, but economic growth can create investment in AI. Japan has the highest magnitude of investment in AI, and the U.S. has the highest magnitude of economic growth. The MBF values of AI suggest decisive evidence for China and Japan and very strong evidence for the U.S. The MBF values of GDP indicate decisive evidence in Japan and the U.S. According to AI and Aging society scenario; we found that in China,

Japan and the U.S., an aging society, can lead to investment in AI. China has the highest magnitude of being an aging society followed by Japan and the U.S. The MBF value of AGE suggests substantial evidence for the U.S.

Acknowledgements The authors are grateful to the Center of Excellence in Econometrics for the financial support for this study, we also appreciate Dr.Laxmi Worachai toward strengthening the quality of the paper.

References

1. Acemoglu, D., Restrepo, P.: Secular stagnation? the effect of aging on economic growth in the age of automation. Am. Econ. Rev. **107**(5), 174–79 (2017)
2. Aghion, P., Jones, B.F., Jones, C.I.: Artificial intelligence and economic growth (No. w23928). National Bureau of Economic Research (2017)
3. Choktaworn, K., Maneejuk, P., Yamaka, W.: An empirical examination of maximum entropy in copula-based simultaneous equations model. Thai J. Math. **13**, 243–256 (2017)
4. Furman, J., Seamans, R.: AI and the economy. Innov. Policy Econ. **19**(1), 161–191 (2019)
5. Goodman, S.N.: Toward evidence-based medical statistic. 2: the Bayes factor. Ann. Inter. Med. **130**(12), 1005–1021 (1999b)
6. Greene, W.H.: Simultaneous-Equations Models. Econometric Analysis International Edition, pp. 378–383. Pearson Education, London (2003)
7. Hofert, M., Pham, D.: Densities of nested Archimedean copulas. J. Multivariate Anal. **118**, 37–52 (2013)
8. Nagler, T., Vatter, T.: Solving estimating equations with copulas. arXiv preprint arXiv:1801.10576 (2018)
9. Nelsen, R.: An Introduction to Copulas, vol. 272, 2nd edn. Springer, New York (2006)
10. Nordhaus, W.D.: Are we approaching an economic singularity? Information technology and the future of economic growth (No. w21547). National Bureau of Economic Research (2015)
11. Pastpipatkul, P., Maneejuk, P., Sriboonchitta, S.: Do Copulas improve an efficiency of seemingly unrelated regression model? Int. J. Intell. Technol. Appl. Stat. **9**(2), 105–122 (2016)
12. Raj, M., Seamans, R.: AI, Labor, productivity, and the need for firm-level datat. In: The Economics of Artificial Intelligence (2018)
13. Ramanathan, R.: Simultaneous Equations model, Introductory Econometrics with Application, pp. 542–544. Roche, Basel (2002)
14. Schneider, T., Hong, G., Le, A.: Land of the rising ROBOTS. In: International Monetary Fund, Publications Finance & Development, vol. 55, no. 2, pp. 28–31 (2018)
15. Sklar, A.: Fonctionsder épartition'an dimensions et leurs marges, vol. 8, pp. 229–231. Publications de l'Institut de Statistique de l'Universit é de Paris (1959)
16. Szczepanski, M.: Economic impacts of artificial intelligence (AI), European Parliamentary Research Service (PE 637.967) (2019)
17. Sirohi, S., Choudhary, B.: Simultaneous equations: theory and applications in agriculture. In: Analytical Techniques for Decision Making in Agriculture, pp. 69–79 (2013)
18. Stulajter, F.: Introduction to copula functions and their application in portfolio and risk management. In: 4th International Conference Risk Management and Modeling, VŠB-TU Ostrava, Faculty of Economics, Department of Finance (2008)
19. Zellner, A., Palm, F.: Time series analysis and simultaneous equation econometric models. J. Econometrics **2**(1), 17–54 (1974)

Sustainable Entrepreneurship on Thailand's SMEs

Chalerm Jaitang, Paravee Maneejuk, and Pitchaya Boonsrirat

Abstract This study aims to investigate whether small and medium-sized enterprises (SMEs) in Thailand have the propensity to catching up with the global development agenda, sustainable entrepreneurship. This study collects the data from 231 entrepreneurs in the service sector in Thailand and designs the questionnaire according to the concept of sustainable entrepreneurship. Then, in the data analysis process, we introduce the logistic regression with the shrinkage method alternatively to the classical logistic regression model. A brief idea of this method is to involve penalizing size of the coefficients in the likelihood function, in which three penalty functions namely ridge regression, the lasso, and the elastic net, are considered in this study. Then, the performance of the models is evaluated through prediction accuracy and AUC. The results show that the Elastic Net slightly outperforms the other models. Also, the results reveal that the characteristics of enterprises and entrepreneurs, and the opinions on sustainable entrepreneurship aspects have the essential roles and positive effects on the propensity of entrepreneurs' interest in sustainable entrepreneurship.

Keywords Sustainable economics · Shrinkage method · Penalty function · Lasso regression · Ridge regression · Elastic net

1 Introduction

The report of the Office of Small and Medium Enterprise Promotion (OSMEP) indicates that the amount of small and medium enterprises (SMEs) in Thailand were 3,046,793 enterprises, which represented a volume growth of 1.30% and accounted for 99.78% of all enterprises in 2017. However, within the structure of the number of SMEs, there were 3,028,495 small operators which represented 99.40% of the SMEs. Thai SMEs also contributed to the GDP amounted 6,551,718 million Bath

---

C. Jaitang · P. Boonsrirat
Faculty of Economics, Prince of Songkla University, Songkhla 90110, Thailand
e-mail: chalermjaitang@gmail.com

P. Maneejuk (✉)
Center of Excellence in Econometrics, Faculty of Economics,
Chiang Mai University, Chiang Mai 50200, Thailand
e-mail: mparavee@gmail.com

S. Sriboonchitta et al. (eds.), *Behavioral Predictive Modeling in Economics*,
Studies in Computational Intelligence 897,
https://doi.org/10.1007/978-3-030-49728-6_28

with the growth rate of 5.1%, while Thailand economic (its GDP) expanded only 3.9% in 2017. Small and medium enterprises generated the value of 42.4% of the overall GDP in 2017: Small Enterprises (SE) generated 30.0% of GDP, and Medium Enterprises (ME) generated 12.4% of GDP. These statistics show the importance of Thailand SMEs as the economic mechanism and engine.

Based on the small and medium enterprises contribute to the Thai economy, the services sector played the dominant role by SME GDP in 2017, accounted for 40.9% of SME GDP. The next two key players are trade and repairs sector, and the manufacturing sector, which contributed favorably towards on SME GDP at 29.9 and 22.9% respectively. Within the sub-sector, the top three business categories with highest SME GDPs were real estate, hotels & restaurants, and transport & communication that all of them were categorized in service sector contributed to the overall GDP for the categories amounted to 96.5, 93.0, and 37.2% respectively. (Office of Small and Medium Enterprise Promotion [12]).

At present, OSMEP is the primary institution in supporting the operation of Thailand SMEs. It formulates and implements the five-year Promotional Master Plan that serves as a roadmap for SMEs and set up strategic plans for SMEs development. The First SME Promotional Master Plan was launched in 2002, and the current is the Fourth SME Promotional Master Plan (2017–2021). There are several government agencies responsible for SME promotion in the form of measures, projects and services provided to SMEs in various aspects, including the financial support, development and enhancement their competitiveness, and marketing and investment in oversea. Furthermore, the period of the current 4th Promotional Master Plan also aligns with the current 12th National Economic and Social Development Plan (2017–2021), which is responsible under the Office of National Economic and Social Development Board (NESDB) of Thailand. Also, the central theme of the current National Economic and Social Development Plan is based on sustainable development which the United Nations defined as development that meets the needs of the present without compromising the ability of future generations, and the three core elements of sustainable development are economic growth, social inclusion, and environmental protection.

Sustainable entrepreneurship, a spin-off concept of the sustainable development, is defined as the continuing commitment by businesses to behave ethically and contribute to economic development while improving the quality of life of the workforce, their families, the local and global community as well as future generations. Crals and Vereeck [2] explained that the gains of sustainable entrepreneurship included a positive reputation, motivation on employees, and business risk control. The business that focuses on sustainable entrepreneurship is said to deal with the 3 P's: People, Planet, and Profit. The first aspect, "people" is about the behavior of companies in social and ethical issues. It mainly deals with the issues of human rights, especially on how the business treats its employees and whether it creates social cohesion. The second aspect, "planet" is about the business practices of the company that involve the impacts left on natural resources and environment, also the sustainable remedy chosen to mitigate the adverse impacts. Finally, the third aspect, "profit", which is the very essence of a business enterprise, does not relate solely to the purely financial results of an enterprise. Profit is also about the use and allocation of value-added for employment, investments in machines, infrastructure and sponsoring, and distribution (e.g., labor participation).

A four force model of how motivators and barriers affect sustainable performance, but the barriers (e.g. lack of finance, lack of expertise, lack of information) are more pronounced than the motivators (e.g., customer demands, reduce wastes, improved working environment) for SMEs in the United States [11]. Wei-Loon, Roaimah and Juan Rizal [14] studied the factors influencing the propensity to sustainable entrepreneurship of small and medium enterprises in Malaysia, and identified that perceived feasibility, sustainable attitude and perceived desirability positively and significantly influenced such propensity. While social norm was not playing a significant role since most of the entrepreneurs were owned by the sole proprietor; thus, they usually make decisions by themselves. Unfortunately, it was contradicted with Meek, Pacheco and York [10] who identified that social norm was also an influential factor of an individual's intention towards environmental entrepreneurial actions. Thus, sustainable practice or commitment to their business, especially SMEs depends on business characteristics, owner perceptions, and attitudes, entrepreneurs' intention, social norms.

Sustainable entrepreneurship is approached and applied mostly by large companies. In this study, we will investigate whether the SMEs in Thailand have the propensity to catching up with the global development agenda, sustainable entrepreneurship. This study uses the questionnaires collected from 231 entrepreneurs in the service sector, which has the most significant contribution to Thai SME GDP. The questionnaires were designed to answers the question by using the enterprises' characteristics and entrepreneurs' opinion on five aspect checklists (corporate governance, risk management, environmental management, labor and human rights, and society management) which covered the concept of sustainable entrepreneurship from SET of Thailand. Section 2 provides data descriptions of independent variables used in the study, which are collected from entrepreneurs. Section 3 presents the methodology. The empirical finding and discussion are presented in Sect. 4. Finally, Sect. 5 concludes a summary of the methodology and the main findings.

2 Data Descriptions and Variables

This section presents the variables. The model applied the crucial 14 independent variables which were corrected by questionnaires, from 231 small and medium enterprises in the service sector, consisting of their characteristics of enterprises, characteristics of entrepreneurs, entrepreneurs' experiences with the office of small and medium enterprise promotion, and entrepreneurs' opinion scores on the concept of sustainable entrepreneurship. There are five aspect checklists about the concept of sustainable entrepreneurship: corporate governance, risk management, environmental management, labor, and human rights, and society management, which measure how much entrepreneurs agree with the concept in each aspect. The definitions and its classification of all independent variables as well as their labels are illustrated in Table 1 While the dependent variable is binary choices that one represents enterprises interested to operate the business following the sustainable entrepreneur concept, and zero represents enterprises which were not interested in the sustainable entrepreneur concept on their businesses. The number of enterprises was classified into two groups consisting of 184 enterprises (79.65%) interested in the sustainable entrepreneur and 47 enterprises (20.35%) not interested in the sustainable concept for applying on their businesses. The 231 observations (enterprises) were corrected in 2018, covering Chiang Mai, Khon Kean, Song Khla, and Bangkok that represent the different regions in Thailand.

3 Research Methodology

This study deals with a binary classification problem in which the response variable, attention of sustainable entrepreneurship, has only two outcomes. The entrepreneur that is interested in sustainable entrepreneurship will be coded as 1 and 0 if he is not interested. Consequently, we employ the most frequently used model, logistic regression, to perform the study. However, instead of using the classical logistic regression, this study considers the logistic regression with the shrinkage method, which involves penalizing size of the coefficients. The idea of the penalized logistic regression will be described in this section. Additionally, this study considers three penalty functions, namely ridge regression, the lasso, and the elastic net, which will also be explained in this section.

3.1 Logistic Function

Logistic function with shrinkage method is also known as the penalized logistic regression. The model is similar to the classical logistic regression but incorporates a penalty term in the log-likelihood function to shrink the coefficient estimates to zero. Before we discuss the penalty function, we shall spend the first part of this section on a review of the concept and functional form of the classical logistic regression.

In the simple logistic regression, the dependent variable is a binary choice taking values 0 or 1. To explain the concept of model mathematically, we suppose that we have n individuals or enterprises participating in our survey. The term $Y$is n-dimensional vector of response variables taking values 0 or 1. Let the number of explanatory variables equal to k. The term X stands for $n \times k$ a matrix of explanatory variables. Thus, for the i^{th}individual we denote $X_i = (x_{i1}, \ldots, x_{ik})$. Then the conditional probability that $P\,(Y_{i=1} = 1)$ is denoted p_i, which is given by

$$P\,(Y_{i=1} = 1\,|X_i) = p_i = \frac{e^{X_i\beta}}{1 + e^{X_i\beta}}, \tag{1}$$

Table 1 Data descriptions

Variable	Label	Description
Status	X1,i	0 = sole proprietor, 1 = juristic person
Age	X2,i	Age of enterprise (years)
Employees	X3,i	Number of employees, including part-time employees
Size	X4,i	0 = small enterprise, 1 = medium enterprise
Loan	X5,i	0 = not loan, 1 = loan from financial institutions or their relatives
Education	X6,i	1 = Junior high school and lower, 2 = Senior high school, 3 = Diploma/High Vocational Certificate, 4 = Bachelor Degree, 5 = higher Bachelor Degree
Consult OSMEP	X7,i	1 = had the experience consulting about business with the Office of Small and Medium Enterprise Promotion, 0 = have not
Consult OSS	X8,i	1 = had the experience consulting about business with SME one stop service, 0 = have not
Known	X9,i	1 = known about Promotional Master Plan for SME development, 0 = do not
Corporate governance	X10,i	The average opinion scores of corporate governance aspect: owner and manager in 1 person, communicating to members and staffs, encouraging members to follow business ethics, measuring and monitoring them (1 very disagree, 2 disagree, 3 fair, 4 agree, 5 very agree)
Risk management	X11,i	The average opinion scores of risk management aspect: efficiency and standard, analyzing business risks (covering economic, society and environment), sharing business risk information to member, continuing a business plan (1 very disagree, 2 disagree, 3 fair, 4 agree, 5 very agree)
Environmental management	X12,i	The average opinion scores of environmental management aspect: setting the quantity goals on environment management, planning the energy reduction and resources reduction, measuring the plans, communicating business goals to stakeholders (1 very disagree, 2 disagree, 3 fair, 4 agree, 5 very agree)
Labor and human rights	X13,i	The average opinion scores of labor and human rights aspect: protecting human rights abuses, planning to develop employees potential, assessment employees performance, supporting employees health, recording illness & injury hours to improve workplace (1 very disagree, 2 disagree, 3 fair, 4 agree, 5 very agree)
Society management	X14,i	The average opinion scores of society management aspect: evaluating customers satisfaction to improve goods and service, transparent communicating to customers, having activities or projects related to community and assessment them (1 very disagree, 2 disagree, 3 fair, 4 agree, 5 very agree)

Note: i represents the enterprise.

where β is a vector of unknown parameters or the regression coefficients. Then, the maximum likelihood estimation (MLE) is used to estimate the regression coefficients. We maximize the likelihood function to find the estimator that corresponds to the observed data. Thus, the likelihood function is given by (Cule, Vineis, and De Iorio [4]).

$$L(\beta, Y_i) = \prod_{i=1}^{n} p_i^{Y_i} (1 - p_i)^{1-Y_i}. \tag{2}$$

Then, we take the logarithm of Eq. (2) so obtain the log-likelihood function:

$$LL(\beta, Y_i) = \sum_{i=1}^{n} \{Y_i \log(p_i) + (1 - Y_i) \log(1 - p_i)\}. \tag{3}$$

We can obtain the value of coefficients by differentiating Eq. (3) with respect to β and setting the derivative to zero.

3.2 Shrinkage Method

According to the Shrinkage method, the size of the coefficients is penalized. We incorporate a penalty function in the log-likelihood function, Eq. (3),

$$LL_{penalty}(\beta, \lambda, Y) = -LL(\beta, Y) + \lambda P(\beta) \tag{4}$$

The idea of the penalized log-likelihood function, $LL_{penalty}(\beta, \lambda, Y)$, is to add the penalty function, $P(\beta)$, into the conventional log-likelihood function (Eq. (3)). The penalty function can be viewed as the additional constraint of the model. It is a shrinkage penalty that exhibits a size constraint on the coefficients [5]. However, the penalty function is adjusted by the regularized parameter, whose value is greater than or equal to zero, $\lambda \geq 0$. The idea of the regularized parameter is to control the bias-variance trade-off, which will be discussed later in Sect. 2.4. After obtaining the penalized log-likelihood function, we minimize this function to find the coefficient estimates. In general, the coefficients can be poorly determined and having a high variance when the model contains a lot of correlated variables. A large negative coefficient can cancel a large positive coefficient if the corresponding variables are correlated. The penalty imposes a size constraint on the coefficient to allay the cancellation [5]. Next, we should discuss a form of the penalty function. As mentioned earlier, this study considers three penalty functions, namely Ridge, Lasso, and the elastic net, which can be described as follows.

3.2.1 Ridge Regression

Ridge regression was firstly introduced by Hoerl and Kennard [7] as an alternative method to the ordinary least squares regression when there exist correlated explanatory variables. Later, the ridge regression is extended to the logistic regression [8], where the penalized log-likelihood function is written as

$$LL_{penalty}^{Ridge}(\beta, \lambda, Y_i) = \sum_{i=1}^{n} \left\{ Y_i \cdot \log(p_i) + (1 - Y_i) \cdot \log(1 - p_i) - \lambda \sum_{j=1}^{k} \beta_j^2 \right\}. \tag{5}$$

The log-likelihood function is maximized subject to a constraint on the sum of the squared values of the regression coefficients. Generally, ridge regression includes all predictors but with smaller coefficients [5]. In Eq. (5) it can be seen that when $\lambda = 0$, the penalized log-likelihood function equals the log-likelihood function without a penalty. To estimate the unknown parameter β, we minimize the penalized log-likelihood function Eq. (5) with respect to β, thus

$$\widehat{\beta}_{Ridge}(\lambda) = \arg\min_{\beta} \left\{ -LL_{penalty}^{Ridge}(\beta, \lambda, Y_i) \right\}. \tag{6}$$

As a remark, ridge regression does not perform variable selection. It shrinks the coefficient estimates toward zero but is refrained from putting them exactly to zero [7]. As a result, ridge regression is restricted to include all predictor variables in the model. Thus, when the number of predictor variables, k, is very large, it is difficult to interpret the estimated results. Although it does not result in a sparse model, ridge regression has shown to achieve high prediction accuracy when the predictor variables are highly correlated [7]. Since high correlation is frequent in high dimensional data, ridge regression is regularly used [1].

3.2.2 Least Absolute Shrinkage and Selection Operator: The LASSO

Lasso standing for the least absolute shrinkage and selection operator was originally introduced by Tibshirani [13]. The idea of the lasso penalty is to maximize the log-likelihood function subject to a constraint on the sum of the absolute values of the regression coefficients. According to the constraint, the variable that is unlikely to have a significant impact on the dependent variable, whose coefficient will be set exactly equal to zero. Moreover, if there is a group of highly correlated predictors, the lasso tends to randomly select one predictor of that group and neglect the remaining predictors [16]. That means lasso perform a variable selection by shrinking some coefficients and setting them to be zero. With that in mind, the lasso penalized log-likelihood function is written by,

$$LL_{penalty}^{Lasso}(\beta, \lambda, Y_i) = \sum_{i=1}^{n} \left\{ Y_i \cdot \log(p_i) + (1 - Y_i) \cdot \log(1 - p_i) - \lambda \sum_{j=1}^{k} |\beta_j| \right\}. \tag{7}$$

One obvious point we can see from Eq. (7) when comparing it with the log-likelihood of the ridge regression, Eq. (5), is the constraint -or how they deal with the regression coefficients. The ridge maximizes the log-likelihood subject to the squared β-value, as shown in Eq. (5), while the lasso considers the absolute β-value. Then, we minimize the penalized log-likelihood function Eq. (7) with respect to β,

$$\widehat{\beta}_{Lasso}(\lambda) = \arg\min_{\beta} \left\{ LL_{penalty}^{Lasso}(\beta, \lambda, Y_i) \right\}. \tag{8}$$

Besides, the lasso is restricted to select a maximum of n predictor variables in the model [1]. Consequently, when there exists a high dimensional problem -more predictors than observations, $k \gg n$, it follows that no more than n predictor variables can be included in the model [16]. There could be more than n coefficients that are non-zero, but the predictors are restricted to enter the model [15].

3.2.3 The Elastic Net

Later Zou and Hastle [16] proposed the elastic net to remove the limitation on the number of selected variables as appear in the lasso. Additionally, it promotes a grouping effect [16]. A method exhibits the grouping effect if the coefficients of a group of highly correlated variables are nearly equal. The elastic net combines the penalty terms of ridge regression and the lasso and gives them a weight denoted by α. The parameter α can be viewed as a regulator. When α is close to zero, we obtain the ridge penalty, and if it is close to one, we obtain the lasso penalty. Therefore, the penalized log-likelihood function for the elastic net is,

$$LL_{penalty}^{Elastic\,net}(\beta, \lambda, Y_i) = \sum_{i=1}^{n} \left\{ Y_i \cdot \log(p_i) + (1 - Y_i) \cdot \log(1 - p_i) - \lambda \sum_{j=1}^{k} (1 - \alpha)\beta_j^2 + \alpha |\beta_j| \right\} \tag{9}$$

This linear combination of the lasso and ridge penalty term was suggested as a penalty function by Friedman et al. [6]. The solution to Eq. (9) is retrieved by using coordinate descent [16]. Similar to the Ridge and LASSO estimation, we minimize the penalized log-likelihood function Eq. (9) with respect to β,

$$\hat{\beta}_{\text{Elastic net}}(\lambda) = \underset{\beta}{\operatorname{argmin}} \left\{ LL_{penalty}^{Elastic\,net}(\beta, \lambda, Y_i) \right\} \tag{10}$$

In contrast to the lasso, the elastic net can select all predictor variables even though $k \gg n$. Furthermore, it removes the constraint that was caused by grouped correlated predictors that were mentioned earlier for the lasso. Conclusively, the elastic net encourages a grouping effect that is to eliminate the trivial predictors but to include all groups of correlated predictors [15]. We note that, in this estimation, we divide the data into a training set and testing set, then we find the parameter α based on the training set only and then check the prediction accuracy on the testing set.

3.3 Cross-validation

This section provides a general description of cross-validation and describes how cross-validation is used to find the regularization parameter, λ, and the fitted model. Cross-validation is a validation method for accessing the performance of a predictive model. It is widely used for estimating prediction error. It sets aside one part of the data to fit the model and different part to test the model [5]. Consider a data set with n observations and K-fold cross-validation. The n observations are split into K subsets equally. We use one subset, says the K^{th}subset as a validation set while the other subsets are considered combined training sets to fit the model. The model fitted to the training data is then used for prediction on the validation set. This process is repeated for every subset.

In this study, cross-validation is used to select the regularization parameter λ. According to Extended Bayesian Information Criteria (EBIC) as suggested by Chen and Chen [3], the K-fold cross-validation estimate of the EBIC can be written as,

$$EBIC^{CV}(\lambda,\alpha) = -2LL_{penalty}\left(\widehat{\beta}_K(\lambda,\alpha)\right) + v(\lambda,\alpha)\log(n) + 2\tau v\log(k), \tag{11}$$

where $0 \leq \tau \leq 1$ is the additional parameter for controlling the sizes of the additional penalty. $\left(\widehat{\beta}_K(\lambda,\alpha)\right)$ is the maximum likelihood estimator of $\beta(\lambda,\alpha)$ at given λ and α, which is based on all the data except the observations in the K^{th}subset. The term $v(\lambda,\alpha)$ is a degree of freedom at a given λ and α. We then select the optimal λ and αsuch that the value of $EBIC^{CV}(\lambda,\alpha)$ is smallest. We note that this procedure is repeated for each fold so that every data point is used for validation once.

4 Empirical Results

4.1 Elastic NET Comparison

In Fig. 1 the prediction accuracy and AUC for Elastic Net are plotted against α (mixing percentage). As demonstrated by the red line, $\alpha = 0.1, 0.2, 0.4, 0.9$ achieve the highest prediction accuracy. However, as demonstrated by the blue line, $\alpha = 0.9$ achieve the highest AUC. According to these two measurements, we may conclude that the mixing percentage in Elastic Net is 0.9 indicating that the penalty term of Elastic Net depends greatly on Lasso penalty.

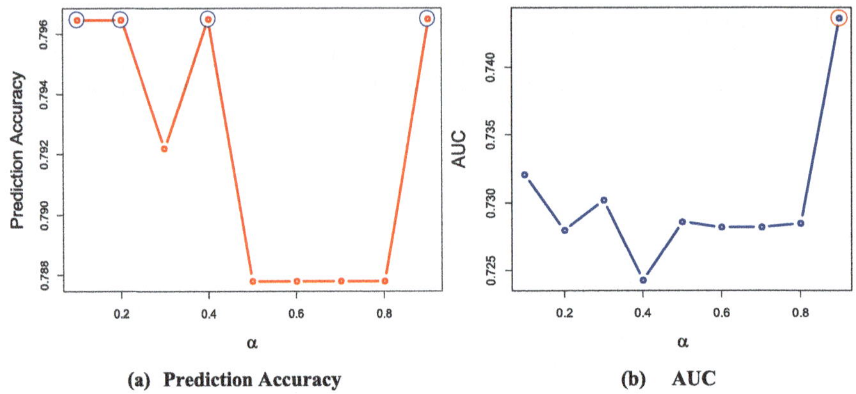

Fig. 1 Prediction accuracy (a) and AUC (b) versus (mixing percentage) for the Elastic net

4.2 Empirical Finding and Discussion

Table 2 shows the estimated results of the MLE-LASSO, MLE-Ridge, Elastic Net, and MLE without penalty term (Classical Logit). Following the study of Maneejuk, Yamaka, and Nachaingmai [9], this study evaluates the performance of the models using prediction accuracy and AUC. As presented in Table 2, Elastic Net can perform slightly better than the other models, in term of the prediction accuracy and AUC. Estimated results show that the Elastic Net drops 5 out of 14 independent variables form the empirical model. The coefficients of the variables X1, X2, X8, X9, and X12 are set precisely equal to zero, meaning that they are unlikely to have a significant impact on the dependent variable. Besides, the results of Elastic Net consistent with the results of the Lasso as the variables are also neglected in the Lasso model, including X7

Then, we calculate the marginal effects of the Elastic Net and present the results in Table 3. According to the empirical results indicate that some characteristics of enterprises, characteristics of entrepreneurs, and their attitudes on sustainable entrepreneurship concept have significant effects on the propensity to operate the business as sustainable entrepreneurship. We found that nine variables of Elastic Net model affect, and only five independent variables (enterprise's status (X1), enterprise's age (X2), consulting with SME one-stop service (X8), entrepreneurs known about Promotional Master Plan for SME development (X9), and average opinion scores of environmental management aspect (X12)) are not significant in the chosen model consisting with the result of the classical logit model.

The characteristics of enterprises which consist of the number of employees, the size of enterprises, and the loan of enterprises, have favorable effects on the propensity of entrepreneurs' interests in sustainable entrepreneurship. That means the higher number of employees (X3) will increase the propensity on the entrepreneurs' interests. The size of business (X4) has the highest marginal effect on entrepreneurs'

Table 2 Estimation results

	Lasso	Ridge	Elastic net $\alpha = 0.9$	Classical logit
Intercept	−3.9047	−1.3213	−4.3823	−4.9220*
X1		0.0537		−0.0238
X2		0.0006		−0.0047
X3	0.0075	0.0024	0.0092	0.0124
X4	0.017	0.2003	0.6431	14.521
X5	0.3264	0.1056	0.484	0.6803*
X6	0.5721	0.1664	0.8675	1.1580*
X7		0.0313	−0.4079	−0.9044*
X8		0.081		0.0054
X9		0.1154		0.157
X10	0.4661	0.1459	0.6474	0.7939*
X11	0.2006	0.0742	0.2437	0.2795
X12		0.033		−0.2483
X13	0.1039	0.2663	1.1346	16.391
X14		0.0659	0.0954	0.2137
Prediction accuracy	0.7922	0.7965	0.7965	0.7952
AUC	0.7297	0.7229	0.7484	0.7483

Note: * denotes strong evidence supports the effect of the independent variable on the dependent variable.

Table 3 Marginal effects of the elastic net $\alpha = 0.9$

Selected variable	dF/dx
X3	0.0013
X4	0.1692
X5	0.0689
X6	0.1421
X7	−0.1025
X10	0.0971
X11	0.0362
X13	0.1391
X14	0.0206

interests to sustainable entrepreneurship. If the enterprise is medium, it will increase the propensity to operate business as sustainable entrepreneurship comparing with the small enterprise. Moreover, the study found that the enterprises used to loan from financial institutions or their relatives (X5) have a higher probability than small enterprises. Almost of enterprises have not used to loan because they had the financial access problem, confirming the four force framework of motivators and barriers in Natarajan and Wyrick [11]. This result (X5: loan) also significantly consistent with the classical logit model. The secure financial access relaxed the enterprise to manage and adjust the businesses they want to be.

The characteristics of entrepreneurs, entrepreneurs' education level (X6), positively affects on entrepreneurs' interests in sustainable entrepreneurship. The higher entrepreneurs' education level is beneficial for individuals' perceive on planet and people such as consumers and their employees who can encourage the enterprises in the long run. While the experience to consult with the office of small and medium promotion (X7) negatively affects on entrepreneurs' interests in sustainable entrepreneurship. Individuals may be tired in procedures even though they know the advantages of consulting with the government office. Both results of X6 and X7 are consistent with classical logit results. However, the effect of entrepreneurs' education level (X6) and the experience to consult with the office of small and medium promotion (X7) may change (not consistent) in case the manufacture sector and trade sector.

The entrepreneurs' attitudes on five aspects of sustainable entrepreneurship concept show whether entrepreneurs agree or not incorporate governance concept, risk management concept, environmental management concept, labor and human right concept, and society management concept. The 4 out of 5 aspects have a positive effect on the propensity to operate business as sustainable entrepreneurship. First and foremost, the higher opinion scores of labor and human rights will increase the propensity on sustainable entrepreneurship practice in business. The opinion scores of corporate governance aspect, risk management aspect, and society management aspect are also beneficial on sustainable entrepreneurship practice respectively that correspond with two elements of the sustainable entrepreneurship: profit and people. The exciting finding is that the opinion scores on environmental management are not significant for entrepreneurs' interests doing business as sustainable entrepreneurship that contradicted with the one element of sustainable entrepreneurship concept, planet element, as some measurements on environmental management were reliable and contrast with the entrepreneurs' opinion. The average opinion scores on environmental management are 4.18, which is less than the other three aspects like risk management, labor and human right, and social management. However, the opinion scores on environmental management are slightly higher than the corporate governance aspect. This study shows that the propensity to operate enterprise as sustainable entrepreneurship in the service sector will increase and be higher if the entrepreneurs have perceived on corporate governance, risk management, labor, and human right, and society management covering at least 2 out of 3 elements of sustainable entrepreneurship concept. It does not mean that Thai entrepreneurs do not care about the environmental issue, but they have to adjust themselves and their business, and the government should do it as a national campaign for sustainability in the long run.

5 Conclusion

Sustainable entrepreneurship is defined as the continuing commitment by enterprises to behave ethically and contribute to economic development while improving the quality of life of the workforce, their families, the local and global community as well as future generations, and therefore the sustainable entrepreneurship concerns on enterprises' profit, enterprises' stakeholders and environment. This study thus aims to investigate whether the SMEs in Thailand have the propensity to catching up with the global development agenda called sustainable entrepreneurship. We use a questionnaire to collect the data from 231 entrepreneurs in the service sector, which has the most significant contribution to Thai SME GDP. The questionnaires were designed to answers the question by using the enterprises' characteristics and entrepreneurs' opinion on corporate governance, risk management, environmental management, labor and human rights, and society management, which cover the concept of sustainable entrepreneurship.

Consequently, this study deals with a binary classification problem in which the response variable, attention of sustainable entrepreneurship, has only two outcomes: interested in sustainable entrepreneurship (coded as 1) and not interested in it (coded as 0). To analyze the data under the concept of binary choices, we introduce the logistic regression with the shrinkage method, which involves penalizing size of the coefficients. This study considers three penalty functions, namely ridge regression, the lasso, and the elastic net. The performance of the models is evaluated through prediction accuracy and AUC. It is found that the Elastic Net performs better than the other models.

The estimated results from the Elastic Net show that the characteristics of enterprises and entrepreneurs, and the opinions on sustainable entrepreneurship aspects have the crucial roles and positive effects on the propensity of entrepreneurs' interest in sustainable entrepreneurship. The medium enterprises have a higher tendency to interest and apply the concept of sustainable entrepreneurship in their business than the small enterprises, especially the individuals who can loan from the financial institutions or relatives. The enterprises naturally concern on cost and profits to survive in the market. The entrepreneurs' opinion scores on the concept of sustainable entrepreneurship have an impact on the propensity to operate business as sustainable entrepreneurship concept, mainly labor and human rights aspect which represents the businesses concern on stakeholders (people) such as employees, consumers, and their partners. However, the government should encourage and support the small and medium enterprises to operate business following the way as sustainable entrepreneurship and to campaign about the energy and green business since some enterprises lack knowledge and expertise to consult. The accessible financial ability and perceive on businesses' stakeholders should be extraordinarily done by Thai government because those will be beneficial on sustainable entrepreneurship.

Acknowledgements The authors are grateful to the Center of Excellence in Econometrics for the financial support for this study, we also appreciate Dr. Laxmi Worachai toward strengthening the quality of the paper.

References

1. Algamal, Z.Y., Lee, M.H.: Applying penalized binary logistic regression with correlation based elastic net for variables selection. J. Mod. Appl. Stat. Meth. **14**(1), 168–179 (2015)
2. Carls, E., Vereeck, L.: Sustainable entrepreneurship in SMEs: theory and practice. In: Proceedings in 3rd Global Conference in Environmental Justice and Global Citizenship, Copenhagen (2004)
3. Chen, J., Chen, Z.: Extended Bayesian information criteria for model selection with large model spaces. Biometrika **95**(3), 759–771 (2008)
4. Cule, E., Vineis, P., De Iorio, M.: Significance testing in ridge regression for genetic data. BMC Bioinform. **12**(1), 372 (2011)
5. Friedman, J., Hastie, T., Tibshirani, R.: The Elements of Statistical Learning. Data Mining, Inference, and Prediction. Springer Series in Statistics, 2nd edn. Springer, New York (2008)
6. Friedman, J.H., Hastie, T., Tibshirani, R.: Regularization paths for generalized linear models via coordinate descent. J. Stat. Softw. **33**(1), 1 (2010)
7. Hoerl, A.E., Kennard, R.W.: Ridge regression: biased estimation for nonorthogonal problems. Technometrics **12**(1), 55–67 (1970)
8. Le Cessie, S., Van Houwelingen, J.C.: Ridge estimators in logistic regression. J. R. Stat. Soc. Ser. C (Appl. Stat.) **41**(1), 191–202 (1992)
9. Maneejuk, P., Yamaka, W., Nachaingmai, D.: Bayesian analysis of the logistic kink regression model using metropolis-hastings sampling. In: International Econometric Conference of Vietnam, pp. 1073–1083. Springer, Cham (2019)
10. Meek, W., Pacheco, D., York, J.: The impact of social norms on entrepreneurial action: evidence from the environmental entrepreneurship context. J. Bus. Ventur. **25**(2), 493–509 (2010)
11. Natarajan, S.G., Wyrick, A.D.: Framework for implementing sustainable practices in SMEs in the United States. In: Proceedings of the World Congress on Engineering. World Congress on Engineering, London (2011)
12. The Office of SMEs Promotion: SME yearly report in 2018. The Office of SMEs Promotion, Bangkok (2018)
13. Tibshirani, R.: Regression shrinkage and selection via the lasso. J. R. Stat. Soc. Ser. B (Methodol.) **58**(1), 267–288 (1996)
14. Wei-Loon, K., Roaimah, O., Juan Rizal, S.: Factors influencing propensity to sustainable entrepreneurship of SMEs in malaysia. Proc. Soc. Behav. Sci. **172**, 570–577 (2015)
15. Zhou, D.X.: On grouping e_ect of elastic net. Stat. Probab. Lett. **83**(9), 2108–2112 (2013)
16. Zou, H., Hastie, T.: Regularization and variable selection via the elastic net. J. R. Stat. Soc. Ser. B (Stat. Methodol.) **67**(2), 301–320 (2005)

Impact of Economic Policy Uncertainty on Thailand Macroeconomic Variables

Kanwara Ponlaem, Nachatchapong Kaewsompong, Paravee Maneejuk, and Jirakom Sirisrisakulchai

Abstract This study aims to investigate the impact of the international economic policy uncertainty (U.S., China, and Japan) on the Thai macroeconomy by using Global Vector Autoregressive (GVAR) model. The model is used to investigate the dynamic relationship across the country. The Generalized impulse response function is also used to examine the response of the macroeconomic indicators against the EPU shock. The results confirm that the economic policy uncertainty can transmit from one country to another country through the trading relationship. In the case of Thailand, the one standard error positive shock of the economic policy uncertainty (EPU) in major economies lead to a decline in Thai economic activities. We find that the foreign EPUs contribute a large impact on Thai export and import.

Keywords Economic policy uncertainty · Macroeconomic variables · GVAR model

1 Introduction

The global economic and financial crisis in 2008 caused the great contraction in economic activities and increased in macroeconomic volatility, many people concern

K. Ponlaem
Faculty of Economics, Chiang Mai University, Chiang Mai 50200, Thailand
e-mail: kanwara.pl@gmail.com

N. Kaewsompong (✉) · P. Maneejuk · J. Sirisrisakulchai
Center of Excellence in Econometrics, Faculty of Economics, Chiang Mai University, Chiang Mai 50200, Thailand
e-mail: Nachatchapong.kaew@cmu.ac.th

P. Maneejuk
e-mail: mparavee@gmail.com

J. Sirisrisakulchai
e-mail: Jirakom.s@cmu.ac.th

S. Sriboonchitta et al. (eds.), *Behavioral Predictive Modeling in Economics*, Studies in Computational Intelligence 897,
https://doi.org/10.1007/978-3-030-49728-6_29

about an economic uncertainty existing around the world. Baker et al. [1] developed a new index of economic policy uncertainty (EPU) based on newspaper coverage frequency and employed a VAR model to investigate the effect of the U.S. index on their macroeconomic variables. Their results showed the increase in the EPU index of the U.S. causes a decline in their GDP, industrial productions, and employment. However, the economic uncertainty not only affects their economy but also spillover effect on other countries that have a relationship with them. As shown in many previous studies, such as Istiak and Serletis [15], they found that the EPU of G7 countries response to real output in their countries. Fontaine et al. [11] and Fontaine et al. [12] showed that the EPU contagion from China is important for all countries, especially, economies like the U.S., EA, South Korea, and Russia. These countries are reacted to the uncertainty shocks and resulting in falling of industrial production, inflation, and export as well as increasing in unemployment. Furthermore, the study of Colombo [7] and Kido [18] found the U.S. EPU had contributed the spillover effect on a foreign exchange market, industrial production, and price in the Euro area, China, and Japan.

Thailand's economy is highly depending on trade since the value of export and import has increased every year and the values are accounted for about 60% of GDP [23]. The most important countries for Thailand trading are China, the U.S., and Japan as they have a percentage share to a total average value more than 50% of all trading countries [3]. As the economy of Thailand had an interrelationship with other countries through trading, they would have to face uncertainty or unexpected event from abroad especially the trading partner countries that can affect the Thai economy. For example, the global economic crisis started in the U.S. in 2008 affected to their economy and spread to other countries including Thailand. This crisis caused the depreciation of currency and a declined the trade and thereby leading to decrease in output and further high unemployment. In addition, China and the U.S. are engaged in a trade war as each country continues to dispute tariffs placed on goods traded between them [24]. The war has an indirect effect on other countries, for instance, Thailand as a supplier of such intermediates and raw materials provide to China, lost more than 0.1% of total GDP. The impacts of the crisis and the trade war are expected to create the economic uncertainty which is further transfer to Thailand's economy through an interdependence between them. Therefore, it is important to investigate the effect of the economic policy uncertainty (EPU) from the trading partner countries on Thailand macroeconomic variables.

To study the interrelation between the economic policy uncertainty and the macroeconomics variable using Generalized Impulse Response function (GIRF) method. To quantify this function, we employ the Global Vector Autoregressive (GVAR) model, which is originally suggested by Pesaran et al. [25] and further developed by Dees et al. [8]. The GVAR model is a global model that associates individual country-specific models in which domestic variables are related to country-specific foreign variables in a consistent manner, which are constructed from the domestic variables to match the relative importance of the rest of the world for the country under consideration. The model allows us to solve the curse of dimensionality problem on modeling in the interconnections across the country by combining time series, panel data, and factor analysis techniques [6]. The existing literature contained several studies employed the GVAR model to investigated the macroeconomic spillover effect include Bernd and Khan [5], Risquete and Hernandez [27], Samargandi and Kutan [28] and Feldkircher and Korhonen [10], they studied the spillover effect of public debt, fiscal policy, private credit, and macroeconomic shock from China on economic growth and found that the analysis of the macroeconomic effect from one country has a significant impact on another country.

The article is structured as follows. Section 2 interprets the GVAR model in a theoretical form, from the structure of the model to its solutions. Section 3 provides a specification and estimation of the GVAR model. Section 4 represents the impulse response of the macroeconomic variables to the international transmission of EPU shocks. Finally, we present the conclusion of the study in Sect. 5.

2 A Concept of Global Vector Autoregressive (GVAR)

The Global Vector Autoregressive (GVAR) model is introduced by Pesaran et al. [26] and further developed in Dees et al. [8]. The model is a precise way to analyze interdependence across economies. This model consists of two steps. First, separating estimation of country-specific VAR with exogenous variable (VARX) models for each country, that be enhanced by appropriately weighted averages of corresponding weakly exogenous foreign variables. Second, combining the country-specific VARX into the global model.

Let N be the number of countries in the model indices by $i = 0, 1, 2, \dots, N$ where country $i = 0$ serves as the reference country and $t = 1, 2, \dots, T$ denoting time. Formally, consider the country-specific VARX (1, 1) models for country i (with no constant and no time trends for simplicity) are given by:

$$x_{it} = \Phi_{i1} x_{i,t-1} + \Lambda_{i0} x_{it}^{*} + \Lambda_{i1} x_{i,t-1}^{*} + u_{it} \tag{1}$$

where x_{it} is a $k_i \times 1$ vector of domestic variables, x_{it}^{*} is a $k_{it}^{*} \times 1$ vector of the foreign variables, $\Phi_{i,1}$ is $k_i \times k_i$ vector of coefficient of domestic variables, $\Lambda_{i,0}$ and $\Lambda_{i,1}$ are $k_i \times k_i^{*}$ vector of coefficient of foreign variables, and u_{it} is a $k_i \times 1$ vector of serially uncorrelated and cross-sectionally weakly dependent process.

Foreign-specific variables are computed by the weighted averages of the corresponding domestic variables of all countries, with the weights also being country-specific; that is:

$$x_{it}^{*} = \sum\nolimits_{j=0}^{N} w_{ij} x_{jt} \tag{2}$$

where w_{it}, $j = 0, 1, \dots, N$, are a set of weights such that $w_{ii} = 0$ and $\sum_{j=0}^{N} w_{ij} = 1$. The weights are predetermined and are meant to capture the importance of country j for the i th economy.

In order to allow for long-run co-integrating relations within and between regions, the country-specific VARX model of Eq. (1) can be estimated in error-correction form or VECM model:

$$\Delta x_{it} = \alpha_i \left(\beta'_{ix} x_{i,t-1} + \beta'_{ix^*} x^*_{i,t-1}\right) + \Lambda_{i0} \Delta x^*_{it} \tag{3}$$

where $ECM_{i,t-1} = \beta'_{ix} x_{i,t-1} + \beta'_{ix^*} x^*_{i,t-1}$ is the vector of long-run cointegrating relation, also known as error-correlation terms (VECM).

In a second step, define a $(K \times 1)$ vector x_t of all domestic variables as $x_t = \left(x'_{0t}, x'_{1t}, \ldots, x'_{Nt}\right)'$. Note here that not all countries need to have the same set of variables so defining a selection matrix that picks up country-specific variables from the global vector of endogenous variables, and re-written x_{it} as $x_{it} = S_i x_t$, where S_i is an appropriate $(k_i \times K)$ selection matrix. Then, defining a $\left(k^*_i \times 1\right)$ vector x^*_{it} of foreign variables as $x^*_{it} = W_i x_i$, where W_i is an appropriate $\left(k^*_i \times K\right)$ link matrix by collecting the individual trade weights w_{ij} in each country. Using the link matrix (W_i) and selection matrix (S_i) in the country-specific VARX model. From Eq. (1) can be rewritten as:

$$S_i x_t = \Phi_{i1} S_i x_{t-1} + \Lambda_{i0} W_i x_t + \Lambda_{i1} W_i x_{t-1} + u_{it} \tag{4}$$

re-arrange:

$$(S_i - \Lambda_{i0} W_i)\, x_t = (\Phi_{i1} S_i + \Lambda_{i1} W_i)\, x_{t-1} + u_{it} \tag{5}$$

which yields:

$$G_i x_i = H_i x_{t-1} + u_{it} \tag{6}$$

where G_i and H_i are $(k_i \times K)$ matrices.

and these individual models are then stacked to yield the model for x_t given by:

$$G x_t = H x_{t-1} + u_t \tag{7}$$

where

$$G = \begin{pmatrix} G_0 \\ G_1 \\ \vdots \\ G_N \end{pmatrix}, H = \begin{pmatrix} H_0 \\ H_1 \\ \vdots \\ H_N \end{pmatrix}, u_t = \begin{pmatrix} u_{0t} \\ u_{1t} \\ \vdots \\ u_{Nt} \end{pmatrix} \tag{8}$$

finally, the GVAR model is obtained after pre-multiplying Eq. (7) by G^{-1}:

$$x_t = G^{-1} H x_{t-1} + G^{-1} u_t \tag{9}$$

3 GVAR Specific Procedures

We estimate the GVAR model for four countries including the U.S., China, Japan, and Thailand, using quarterly data from 2004Q1 to 2018Q4. To capture the effects of the economic policy uncertainty on the macroeconomic variables, we selected real GDP (gdp) and unemployment rate ($unemp$) that focus on output and labor input in the production process. The important macroeconomics variables for Thailand's economy are export (ex) and import (im) value, which is accounted for 65% of Thai GDP. We selected the real effective exchange rate ($reer$) to illustrate on the financial market. To removes seasonal patterns, we use the seasonal adjustment census X-13 (SA) method and transformed into natural logarithms in all data. For each country-specific model, we include domestic variables, $x_{it} = \{ex_{it}, im_{it}, gdp_{it}, unemp_{it}, reer_{it}\}$, and foreign variables, $x^*_{it} = \left\{ex^*_{it}, im^*_{it}, gdp^*_{it}, unemp^*_{it}, reer^*_{it}\right\}$, as well as the EPU index, which consists of EPU^{US}_t, EPU^{CH}_t, and EPU^{JP}_t. Note that EPU index can be viewed as the exogenous (X) variable.

Specifically, our VARX(p, q) can be constructed as follows:

$$x_{it} = \alpha + A_{i1}x_{i,t-1} + \ldots + A_{ip}x_{i,t-p} + B_{i0}x^*_{it} + B_{i1}x^*_{i,t-1} + \ldots + B_{iq}x^*_{i,t-q} + C_{i0}EPU^h_t + u_{it} \tag{10}$$

where x_{it} is a $k_i \times 1$ vector of the domestic variables, x^*_{it} is a $k^*_i \times 1$ vector of the foreign variables, EPU^h_t is a $s \times 1$ vector of the EPU index for country $h \in$ U.S., China, Japan, A_i is a $k_i \times k_i$ vector of a coefficients associated to lagged domestic variables, B_{i0} and B_{i1} are a $k_i \times k^*_i$ vector of a coefficient related to contemporaneous and lagged foreign variables, C_{i0} is a $k_i \times s$ vector of a coefficients related to the global variable, α_0 is a constant term, and u_{it} is the serially uncorrelated and cross-sectionally weakly dependent process.

3.1 *Trade and Aggregation Weight*

To establish the foreign variables, we set $x^*_{it} = \sum_{j=0}^{N} w_{ij}x_{jt}$, where $w_{ii} = 0$ and $\sum_{j=0}^{N} w_{ij} = 1$. In this study, we construct the weights by international trades. In Table 1, we represent the fixed trade weights which are calculated from the average quarterly shares of exports and imports over the period 2005Q1–2017Q4 and the 4×4 matrix of weights based on trade flows among our four countries

Table 1 Trade weighted

	U.S.	China	Japan	Thailand
U.S.	0.0000	0.5603	0.3841	0.2440
China	0.6757	0.0000	0.5206	0.3634
Japan	0.2756	0.3654	0.0000	0.3926
Thailand	0.0487	0.0743	0.0953	0.0000

Note Trade weighted are computed as average quarterly shares of imports and exports, shown in columns by country, so that each column, but not a row, sums to one

3.2 Specification and Estimation of the Country-Specific VARX Models

Before estimating the individual country-specific VARX model, we must achieve a unit root test to ensure the data are stationary. The ADF unit root test results show the variables are integrated of order one (I(1)) in all cases. The model contain ex_{it}, im_{it}, gdp_{it}, $unemp_{it}$, and $reer_{it}$ as endogenous variables and ex^*_{it}, im^*_{it}, gdp^*_{it}, $unemp^*_{it}$, $reer^*_{it}$, EPU^{US}_t, EPU^{CH}_t, and EPU^{JP}_t as an exogenous variables. When the country-specific VARX are formulated, we then determine the rank of their cointegrating space. We select the order of the individual country-specific VARX (p_i, q_i) models, where p_i indicates the number of lags of the domestic variables and q_i indicates the number of lags of the foreign variables. The selected lag order of the country-specific VARX model is shown in Table 2. Then, we estimate each country-specific VARX model in its vector error correcting form using the Johansen's reduced-rank procedure. The number of cointegrating relations for each country-specific VARX model is selected according to the trace statistic and presented in the last column of Table 2. Cointegrating relations can be interpreted as long-run relationships among variables. We can observe that there exists a Cointegration in the model, thus we can conclude that our VARX has a long run relationship.

Table 2 VARX order and number of cointegrating relations in the country-specific models

Country	VARX (p_i, q_i)		#Cointegrating
	p_i	q_i	Relations
U.S.	2	1	3
China	2	1	4
Japan	2	1	3
Thailand	2	1	1

Note the lag selection is based on Akaike Information Criteria (AIC)

Table 3 Contemporaneous effects of the foreign variables on their domestic counterpart

Country	Domestic variables (x_{it})				
	ex	*im*	*gdp*	*unemp*	*reer*
U.S.	0.09	0.40*	0.02*	0.68	0.40
	[1.94]	[4.63]	[3.61]	[1.79]	[2.40]
China	0.32	0.98*	−2.39	−0.01	−0.26
	[1.07]	[5.03]	[−1.47]	[−0.27]	[−2.48]
Japan	0.20	0.73*	0.00	0.18	−1.18*
	[0.81]	[4.97]	[0.14]	[1.64]	[−4.38]
Thailand	1.00*	0.16	0.01	−0.32	−0.24
	[5.09]	[0.54]	[0.14]	[−0.46]	[−1.68]

Note denote decisive evidence based on the MBF test (Goodman 1999). White's heteroskedastic-robust t-ratios are given in the square brackets

3.3 Contemporaneous Effects of Foreign Variables on Their Domestic Counterparts

We present the coefficients together with the robust t-ratios computed by White's heteroskedasticity-consistent variance estimators in Table 3. Most of the coefficients reveal the positive international linkages between the domestic and foreign variables. The export value are increased in 1% of the foreign export leads to an increase of 1% in the Thai export, which implies a relatively strong relationship between the domestic and foreign export value. Each country's import value seems to share the same positive tendency when the foreign import value increase. This indicates that there are strong positive relations between the domestic import and foreign import. The impact of real GDP is moderate positive in all countries, except for China. Concerning the labor market, the impact of foreign unemployment rate quite different. We observe a positive effect of foreign unemployment on the local unemployment in U.S. and Japan, while there are negative for Thailand and China. Another interesting feature is the negative relationship between the domestic REER and the foreign REER.

4 Impulse Responses Discussions

To assess the short-term effects of three economic policy uncertainty (EPU) shocks on the Thai macroeconomy. The GIRF of Koop et al. [19] is estimated based on VARX. We examine the implications of a one standard error of EPU shock from the U.S., China, and Japan to Thai macroeconomics variables. Most of existing empirical works on the study of the impact of economic policy uncertainty (EPU) are a focus on the developed countries, while little attention is paid to the developing economies (e.g., [2, 4, 7, 14, 17]).

4.1 The Responses of Thai Macroeconomic Variables to the U.S. EPU Shock

Figure 1 shows the GIRFs of a one standard error of the U.S. economic policy uncertainty (EPU) shock. The responses of export and real GDP of four economies are quite similar. We can observe that the response of Chinese real GDP to the US shock increase and it is not converging to the equilibrium. We expect that as China is a major economy and it is a high relationship between China and the U.S, thus the shock of the US could contribute a large impact on the Chinese economy. The response of Japanese Yen also show that great response and it does not converge to the equilibrium. These results are confirmed by the study of Han et al. [13], they investigated the impact of the U.S. EPU on other economies (including China and Japan) and they found that the economic uncertainty from the U.S. causes the Chinese and Japanese economy slowdown, especially, a declining in net export. In the cases of other macroeconomic variables, we also found that the response of import, export, and unemployment show a positive response in the first quarter then it begins to decrease in the second quarter.

Considering on the Thai macroeconomy perspective, the study of Luangaram et al. [22] found the political uncertainty index of Thailand has a significant impact on Thailand economic growth, but they only focus on the political uncertainty. Thailand is a small open economy, which is the economy open to international trade cause the growth of the economy depends on the economic policy uncertainty from the trading partner. So, our study has to investigate the impact of the economic uncertainty from the important trading partner on Thailand's macroeconomy and our results show the economy is highly depending on the U.S. in term of import and export. We can observe a large response of these two variables to the shock of EPU from the U.S. When we consider in the long run, the responses of all variables in Thailand are not recovering to the equilibrium. So, the Thai government should intervene in economic activities to reduce the impact of the U.S. EPU.

4.2 The Responses of Thai Macroeconomic Variables to Chinese EPU Shock

The GIRFs of a one standard error shock to the Chinese EPU are plotted in Fig. 2. The consequences of the shock of Chinese EPU on the macroeconomic variables are mostly negative in all the analyzed countries. There are only the response of unemployment and REER that are positive to the shock of Chinese EPU. These two variables seem not to recovery to the equilibrium. Following Fontaine et al. [11], Fontaine et al. [12], Jiang et al. [16], they are also found the Chinese EPU rapidly effect to the exchange rate, unemployment, and export in other countries, especially in the U.S.

In the case of Thailand, the response of export, import, real GDP and Thai currency are negative against the shock of Chinese EPU, which confirms that the changing economic policy in China has a significant impact to the Thai economy and international trade. Conversely, the unemployment of Thailand shows a negative response to the shock. We then compare the impact of the Chinese EPU, we find that the Thai import sector shows the largest negative response to the Chinese EPU, while unemployment shows the greatest positive response to the Chinese EPU. The results are agreeable with the study of Kotcharin et al. [20], they demonstrate that a firm, which is borrowing a capital to investment influenced by China's EPU, especially in Thai shipping industries have a higher level of debt when the international trade is declined and Dungey et al. [9] also found the Chinese uncertainty shocks have a more negative impact on Thailand's domestic output.

4.3 The Responses of Thai Macroeconomic Variables to Japanese EPU Shock

Many studies are mainly focused on the affected to the developed country [15, 21] and there are not the studies of the Japanese EPU shocks to Thailand's economy. In this paper, it is essential to investigate because Japan is a major trading partner country with Thailand. Figure 3 displays the GIRFs associated with a one standard error shock to the Japanese EPU. The Japanese EPU has a negative impact on the export in all analyzed economies in the first quarter and the response reverse to the positive after quarters 3–4. We also find that the responses of the macroeconomic variables do not converge to their equilibrium after the shock occurrence.

The response of Thailand's GDP has a weak negative response after the existence of the Japanese EPU and tend to stabilize in the long run, meaning that Thailand economy seems not to sensitive with the shock of the Japanese economy. The other two indicators including the unemployment rate and real effective exchange rate (REER) seems to diverge from the equilibrium. Considering Thailand's international trade variables, the responses of import and export variable are initially decreasing about 3–4 quarters. There are show a positive response to the shock of Japanese EPU. We also find that the response of Thai unemployment expresses the greatest positive response to the shock of Japanese EPU.

Finally, the responses of Thai macroeconomic variables against three EPU shocks are compared. The results are plotted in Fig. 4. The EPU shock of these three countries leads to the negative response of Thailand's macroeconomic variables. We observe that Thai real GDP and unemployment show the largest response to the shock of Chinese EPU when compared to the other EPUs. Japanese EPU plays an important role in the Thai currency, export sector, unemployment in the long run, as their response to the shock of Japanese are greater than other shocks. Japan is the largest foreign investment in Thailand as they have a Japan - Thailand Economic Partnership Agreement (JTEPA) agreement that consists of trade in goods, rules of origin of products, trade in services, and investment.

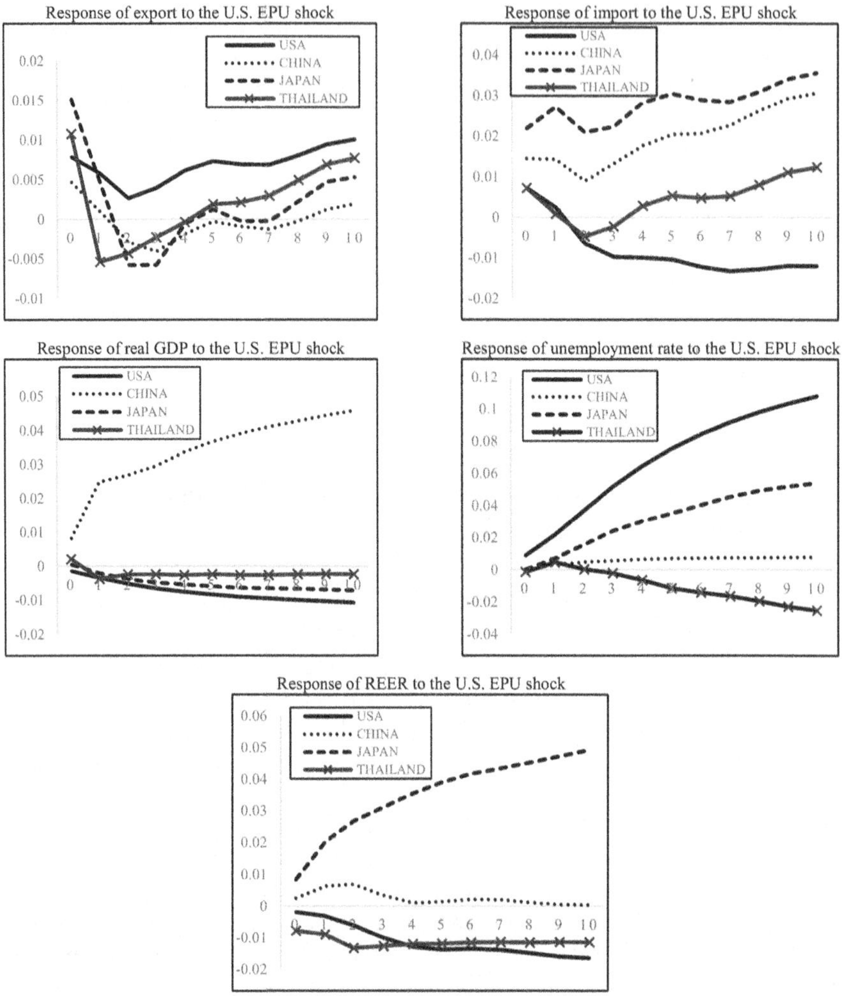

Fig. 1 Generalized impulse responses of the macroeconomic variables from all economies to the U.S. EPU shock

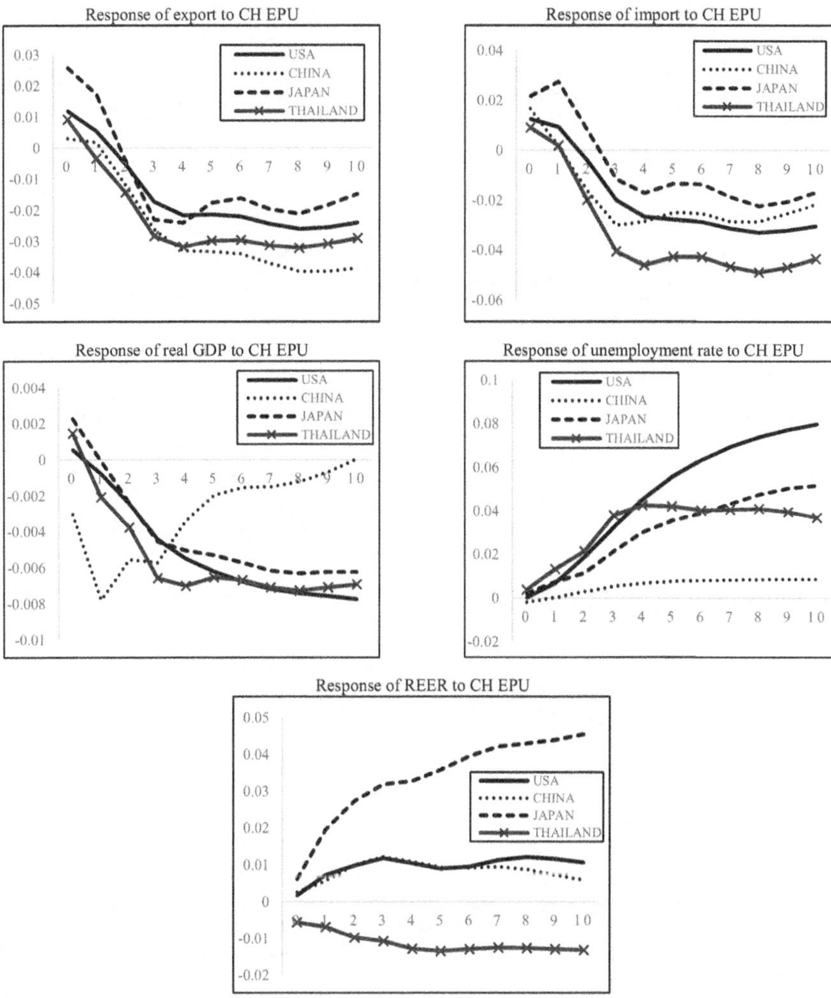

Fig. 2 Generalized impulse responses of the macroeconomic variables from all economies to the Chinese EPU shock

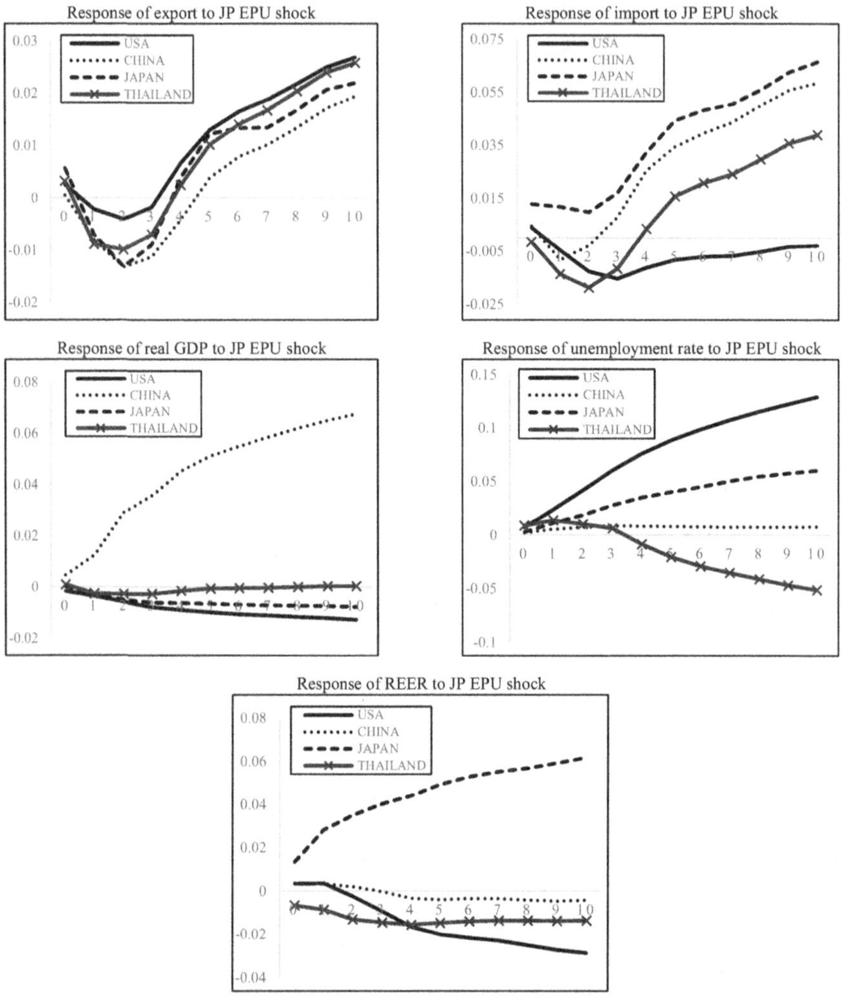

Fig. 3 Generalized impulse responses of the macroeconomic variables from all economies to the Japanese EPU shock

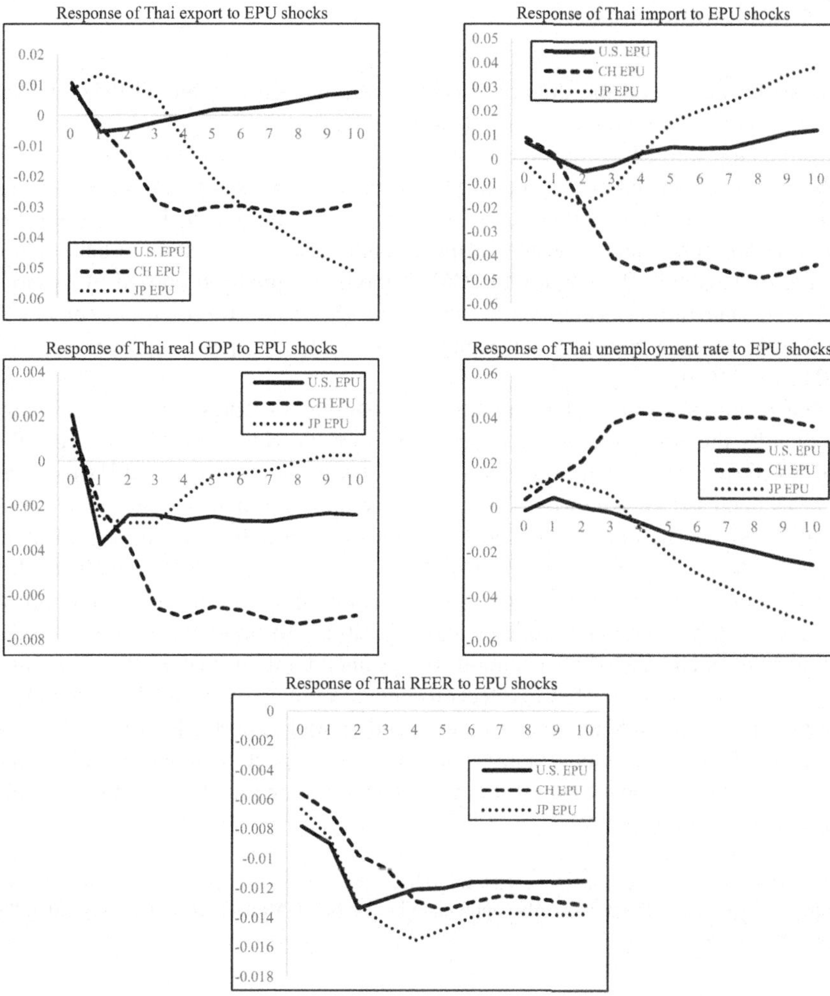

Fig. 4 Summarizing the Generalized impulse responses of the Thai macroeconomic variables to the U.S, Chinese and Japanese EPU shocks

5 Conclusion

This study focuses on the impact of economic policy uncertainty (EPU) on Thai macroeconomic variables. Our main objective attempts to investigate which countries show the largest of EPU impact to Thai macroeconomy. The EPU is measured form indexes that introduced by Baker et al. [1] and the EPU indexes are considered in this study concern the three countries, i.e., the U.S., China, and Japan. The EPU of these countries are important to consider as these countries are the most important trading partner of Thailand. We employ the GVAR model originally proposed by Pesaran et al. [26] and further developed in Dees et al. [8]. Then, we apply this model to investigate our four countries consisting the U.S., China, Japan, and Thailand from 2004Q1 to 2018Q4.

Our results show that Thailand's macroeconomic variables have been strongly affected by the economic policy uncertainty from the U.S., China, and Japan. We find that the responses of macroeconomic variables are affected by the EPU shocks, as there are decrease when they face economic uncertainty, particularly, the great impact on the Thai international trade and seem not to recover back to the equilibrium. So, after the existence of the EPU shock from U.S., Japan, and China, The Thai government must be concern about the economic policy uncertainty from abroad and pick suitable economic policies and regulations to boost the economy to the equilibrium and managing the economic policy uncertainty from the U.S., China, and Japan by impose a fiscal policy to effect on demand aggregate through the government taxation and then influence to consumer spending and import. Moreover, the Bank of Thailand must also impose a monetary policy to impact the money supply in the economy, which influences interest rates and the inflation rate. It also impacts on net exports and employment.

Acknowledgements The authors are grateful to the Center of Excellence in Econometrics for the financial support for this study, we also appreciate Dr. Worapon Yamaka toward strengthening the quality of the paper.

References

1. Baker, S.R., Bloom, N., Davis, S.J.: Measuring economic policy uncertainty. Q. J. Econ. **131**(4), 1593–1636 (2016)
2. Balli, F., Uddin, G.S., Mdassar, H., Yoon, S.: Cross-country determinants of economic policy uncertainty spillovers. Econ. Lett. **156**, 179–183 (2017)
3. Bank of Thailand: The Monetary Policy Report (2018). https://www.bot.or.th/Thai/PressAndSpeeches/Press/News2562/n0162e.pdf. Accessed 2 Jan 2019
4. Belke, A., Dubova, I., Osowski, T.: Policy uncertainty and international financial markets: the case of Brexit. Appl. Econ. **50**, 3752–3770 (2018)
5. Bernd, K., Khan, N.S.: Spillover effects of debt and growth in the euro area: evidence from a GVAR model. Int. Rev. Econ. Finance **49**, 102–111 (2017)
6. Cesa-Bianchi, A., Pesaran, H.M., Rebucci, A.: Uncertainty and economic activity: a global perspective. IDB Working Paper (2014)

7. Colombo, V.: Economic policy uncertainty in the US: does it matter for the Euro area? Econ. Lett. **121**, 39–42 (2013)
8. Dees, S., Mauro, F.D., Pesaran, M.H., Smith, L.V.: Exploring the international linkages of the Euro area: a global VAR analysis. J. Appl. Econ. **22**, 1–38 (2007)
9. Dungey, M., Vehbi, T.: The influences of international output shocks from the US and China on ASEAN economies. J. Asian Econ. **39**, 59–71 (2015)
10. Feldkircher, M., Korhonen, I.: The rise of China and its implications for emerging markets: evidence from a GVAR model. SSRN Electron. J. **20**, 61–89 (2012)
11. Fontaine, I., Didier, L., Razafindravaosolonirina, J.: Foreign policy uncertainty shocks and US macroeconomic activity: evidence from China. Econ. Lett. **155**, 121–125 (2017)
12. Fontaine, I., Razafindravaosolonirina, J., Didier, L.: Chinese policy uncertainty shocks and the world macroeconomy: evidence from STVAR. China Econ. Rev. **51**, 1–19 (2018)
13. Han, L., Qi, M., Yin, L.: Macroeconomic policy uncertainty shocks on the Chinese economy: a GVAR analysis. Appl. Econ. **48**(51), 4907–4921 (2016)
14. Huang, Z., Tong, C., Qiu, H., Shen, Y.: The spillover of macroeconomic uncertainty between the U.S. and China. Econ. Lett. **171**, 123–127 (2018)
15. Istiak, K., Serletis, A.: Economic policy uncertainty and real output: evidence from G7 countries. Appl. Econ. **50**, 4222–4233 (2018)
16. Jiang, Y., Zhu, Z., Tian, G., Nie, H.: Determinants of within and cross-country economic policy uncertainty spillovers: evidence from US and China. Finance Res. Lett. **31**, 1–10 (2019)
17. Junttila, J., Vataja, J.: Economic policy uncertainty effects for forecasting future real economic activity. Econ. Syst. **42**, 569–583 (2018)
18. Kido, Y.: The transmission of US economic policy uncertainty shocks to Asian and global financial markets. N. Am.n J. Econ. Finance **46**, 222–231 (2018)
19. Koop, G., Pesaran, M.H., Potter, S.M.: Impulse response analysis in nonlinear multivariate models. J. Econ. **74**(1), 119–147 (1996)
20. Kotcharin, S., Maneenop, S.: The Giant's pull: how macroeconomic conditions in China explain leverage decisions in Thailand's shipping SMEs. Asian J. Shipp. Logist. **34**(4), 337–344 (2018)
21. Liming, C., Ziqing, D., Zhihao, H.: Impact of economic policy uncertainty on exchange rate volatility of China. Finance Res. Lett. **31**, 1–5 (2019)
22. Luangaram, P., Sethapramote, Y.: Economic impacts of political uncertainty in Thailand. Puey Ungphakorn Inst. Econ. Res. **86**, 1–35 (2018)
23. Maneejuk, P., Yamaka, W., Sriboonchitta, S.: Analysis of global competitiveness using copula-based stochastic frontier kink model. Robust. Econ., 543–559 (2017)
24. Massimiliano, C.: The impact of the US-China trade war on East Asia (2018). https://voxeu.org/article/impact-us-china-trade-war-east-asia. Accessed 27 Dec 2018
25. Pesaran, H.H., Shin, Y.: Generalized impulse response analysis in linear multivariate models. Econ. Lett. **58**(1), 17–29 (1998)
26. Pesaran, H.M., Schuermann, T., Weiner, M.S.: Modeling regional interdependencies using a global error-correcting macroeconometric model. J. Bus. Econ. Stat. **22**, 129–162 (2004)
27. Risquete, R., Hernandez, R.: Macroeconomic effects of fiscal policy in the European Union: a GVAR model. Empir. Econ. **48**(4), 1587–1617 (2015)
28. Samargandi, N., Kutan, A.M.: Private credit spillovers and economic growth: evidence from BRICS countries. J. Int. Financ. Mark. Inst. Money **44**, 56–84 (2016)

Peter Hochmeier

Die Bedeutung von Spagyrik und Alchymie in der Naturheilkunde

Den Gutmütigen gewidmet

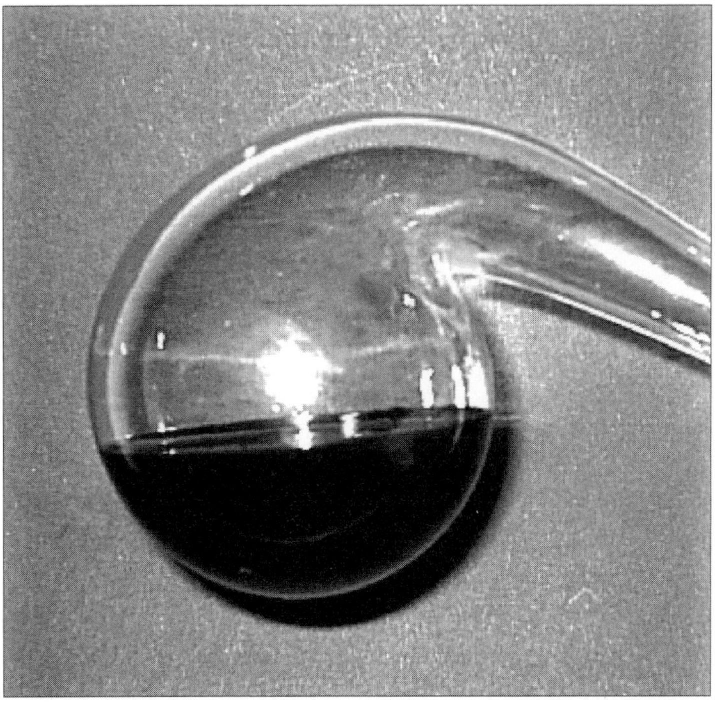

Drum, so lern *Alchymiam, die* sonst *Spagyria* heißt.
Die lernt das Falsche scheiden vom Gerechten.
(Paracelsus)

INHALT

II

Vorwort

Es ist mitunter der nachdenkliche Gesichtszug, das Anheben der Schultern, das Was-soll-ich-dazu-sagen-Ausstrecken der Daumen, das Erweitern des Augenrunds durch Hochziehen der Augenbrauen, das Einziehen der Luft mit gespanntem Nasenrücken und all diese sonderlichen Phänomene zu betrachten, welche auftreten, wenn der versierte Spagyriker oder Alchymist vom unbedarften Laien mit der Frage bedroht wird: „Was ist denn eigentlich Spagyrik?" Oder noch unverschämter: „Was ist Alchymie?"

Eine bedenksame Pause, die ersten obligaten „Naja's", ein höflich bemühtes Lächeln, wissend und doch hilflos, geben dem Kundigen Zeit, sich davon ein Bild zu machen, was den Eindringling wohl zu seiner Frage verleitet hat. Dieser tritt einen Schritt zurück, selbst die Heftigkeit seiner Neugierde bemerkend, und die Erwartung einer kurzen, einfachen und sogleich verständlichen Antwort scheint fortgeblasen wie Staub vom klärenden Wind. Er empfindet eine geheimnisvolle Atmosphäre, was ihn auf eine endlich bevorstehende Enthüllung hoffen lässt.

Es folgt eigentlich nicht gleich eine Enthüllung, sondern es breitet sich ein zarter, wundersamer Duft aus, welcher zweifellos dem Chymischen Gärtlein entströmt. Und dieser aromatische, in die tiefsten Schichten dringende Hauch gewährt dem Fragenden ein Hineinschnuppern und erweckt unmerklich ein Empfinden, dass dort noch weit herrlichere Düfte verborgen wären. Da er nun zu jenen Gutmütigen gehört, denen es weder an Geist, noch an Herz oder Verstand mangelt, erscheint ihm die darauffolgende Antwort wie eine Berührung, die dann in seinem Innersten noch lange nachwirkt und zu einer ungewissen Überzeugung heranwächst, die sich am Besten mit den Worten „dacht' ich's mir doch, dass es das gibt" beschreiben lässt.

Wie immer und mehr oder weniger ausführlich nun der Kundige formuliert – sein Besucher wird zweifellos und in jedem Falle Worte hören wie: „... altes Wissen ... Heilkunst ... Pflanzen ... Metalle ... Planeten ... Prinzipien ... Elemente ... Paracelsus ... scheiden ... destillieren ... Veredelung ..." und einige weitere Specifica mehr. Und er wird den Besucher überrascht, angetan und sehr nachdenklich entlassen.

‚Dieser Nachhall ... Als wäre es ein Erinnern. Da geht es um Dinge, die ich verstehe und doch wieder nicht verstehe. Ich verstehe sie, weil sie zusammenschwingen mit dem Wissen meines Herzens. Ich rieche da Wahrheit. Sollte es tatsächlich eine Art geben, die Natur zu begreifen, veredelnd, heilsam einzuwirken, und der Mensch wusste seit Urzeiten damit umzugehen? Obwohl ich so fühlte, hielt ich es letztendlich für einen naiven Wunschtraum.'

Die Kunst zu lösen und zu verbinden

Das dem Griechischen entstammende Wort ‚Spagyrik' bedeutet: Teilen, Trennen, Scheiden, Herausnehmen und Verbinden, Wiedervereinigen, Einsammeln, Zusammenfügen. Dem entspricht das lateinische *solve et coagula.*

Es bezeichnet die direkteste und eingewurzelte Art des Menschen, sich der Natur und ihren ‚Dingen' – nach Verstehen suchend – zu nähern. Dabei deutet der zweite Teil, das Wiedervereinigen, auf jene hohe Qualität hin, die in der Natur in vollem Umfang alleine dem Menschen zu Eigen ist. Dieser hat die Veranlagung ein ‚natürliches Ding' auseinanderzunehmen und es danach wieder so zusammenzusetzen, dass dem ‚Ding', welches in jedem Falle ein mehr oder weniger lebendiges ist, kein Schaden daraus entsteht. Die Dinge einfach zu öffnen, zerreißen, zerteilen, einen Teil davon behalten und den Rest fortzuwerfen – das ist zwar alltäglich bei Menschen und Tieren jeder Art und hat auch meist seine guten Gründe (was täte man sonst mit Bananen?) – aber die Idee der Wiedervereinigung geht noch einen wesentlichen Schritt weiter.

Zum einen liegt dieser ‚spagyrischen' Haltung des Menschen ein Bedürfnis, eine Sehnsucht zugrunde, eine Art ‚gesunder Neugierde', etwas Verborgenes zu enthüllen, sichtbar, erfahrbar und wenn möglich verständlich zu machen.

Während er dieser Neugierde folgt, fühlt der Mensch: das darf er tun – die Natur akzeptiert seine Kompetenz. Er darf die Dinge der Natur erforschen und nutzen. Das ist keine Selbstverständlichkeit, sondern eine besondere Qualität im Verhältnis der Natur zum Menschen. Deshalb öffnet er die Schale der Nuss, legt die Wurzeln aus dem Erdreich frei oder bricht das Kupfer aus dem Stollen. Die besondere menschliche Qualität liegt aber nun darin, dass er all das in der Haltung macht, die Ganzheit bewahren zu wollen, weil er begreift, dass alles Lebendige ein Ganzes ist.

Der – im alchymistischen Sinne – Naturkundige erkennt die Nussschale als ähnliche Ausdrucksform einer Kräftekonstellation wie seine eigene Schädeldecke und die Frucht darin als sein Gehirn. Er weiß, dass hier eine besondere Harmonie zwischen Mond und Saturn besteht – zwischen dem Diffusen, Grenzenlosen und dem Begrenzenden, Fixen. Er begreift das Kupfer als eine Ausdrucksform der proportionierenden, harmonisierenden Naturkraft, welche auch seine Nieren dominiert und als Venus am Nachthimmel sichtbar ist. Er achtet die gesamte Natur als einen mütterlichen Urgrund, aus dem sein eigener Körper gebildet ist, und von dessen Zustand die Qualität seines eigenen Lebens in dieser Welt abhängt. Mit tief empfundener Dankbarkeit nimmt er aus der Fülle dieser Schatztruhe, und, wenn nötig, wird er dazu beitragen, dadurch entstandene Wunden oder Missverhältnisse sofort wieder zu korrigieren.

Verliert er diese Haltung, dann verliert er aus Sicht der Natur nicht weniger als den Rang des Menschen und gleicht eher einem einsamen Raubtier oder einem dumpfen Parasiten.

Noch heute gibt es unter den Resten alter Kulturvölker Heilkundige, die sich mit einer kleinen Gabe an die Erde revanchieren, nachdem sie eine Heilpflanze ausgegraben haben, und es gibt Weise, die sich beim Baum liebevoll für den gespendeten Schatten bedanken. Wenn man zuzeiten von ‚menschlicher Kultur' sprach, dann meinte man damit auch, dass die Äcker, die Wälder, die Stollen, die Flüsse mit dementsprechenden Respekt behandelt und für Teile eines großen Zusammenhanges gehalten wurden. Wie sich der Mensch zur irdischen Natur verhält, wie er sich in ihr bewegt, aus ihr nimmt, wie er sie pflegt und was er ihr gibt – das bildet die Signaturen des Grades der ‚Kultur' und bildet zudem die Basis für ein womöglich zu erlangendes Wissen über die ‚Dinge der Natur' und eine Grundlage für die Heilkunst.

Soweit es das *Solve*, das Spao, das ‚aus der Natur nehmen', das Auseinandernehmen, Teilen und Öffnen betrifft, geschieht es bei entsprechendem ‚Hintergrund' respektvoll und aufmerksam, ohne willkürliches Vergeuden, Beschädigen oder Verlieren. Jeder einzelne Teil der Natur steht in wesentlichem Zusammenhang mit allen anderen Teilen. Ein Trachten des spagyrisch wirkenden Menschen ist es, Teil und Zusammenhang zu begreifen. Die Blüten oder Früchte des Rosenstrauches können zur Freude und Zierde, zur Nahrung oder Medizin verwendet werden, die Teile der Eiche zum Gerben und Färben, als Bau- oder Brennholz, zum Getränk, zur Medizin und zu anderem.

Und so wie jeder Mensch nach Art und Zunft das Seinige der Natur entnimmt, sucht auch der Heilkundige darin und verfährt damit auf seine Weise. Er muss aber tiefer suchen als manch Anderer, denn sein Handwerk ist eine große Bürde und seine Kunst von höchster Schönheit. Den Alten galten jene Weisen oder ‚Philosophen' als ‚heilkundig', die weit ins Wesen der Dinge vordrangen und jeden Schritt dorthin zusammenhängend mit allen anderen Schritten erkannten. Das lernten sie von solchen, die ‚kundig' waren, und sie lernten in der Schule des Lebens und der Natur. In der Sudhoff'schen Gesamtausgabe der paracelsischen Schriften findet man die, vermutlich während eines Vortrags des *Paracelsus* in Basel vermerkte Notitz eines Studenten:

„Ein Spagyriker ist, wer aus seiner Erfahrung
ein gewisses Etwas hat."

Das *Coagula*, das *Ageiros*, das Zusammenfügen, ist nach dem *Solve* der wesentlich menschentypische Schritt dieses ,rechten Umgangs mit den Dingen der Natur' *(Paracelsus)*. Dieser Schritt zeugt von Bemühen oder Verbundenheit, aber auch von Natur-Erfahrung und vom bewussten Vorhaben, der Natur die ,Kunst des Menschen' hinzuzufügen – auf veredelnde oder zumindest stimmige Weise. Das Wiedervereinigen beweist der Natur, dass hier ein Mensch am Werk ist. Damit ist der vielleicht wesentlichste Aspekt von Spagyrik und Alchymie angedeutet.

> Die Natur gibt nichts an den Tag, das vollendet sei, sondern der Mensch muss es vollenden. Diese Vollendung heißt *Alchymia*.
> Ein Alchymist ist ein Bäcker, indem er Brot bäckt, ein Rebmann,
> indem er den Wein macht, ein Weber, indem er das Tuch macht."
>
> *Paracelsus*: >Paragranum< 1530

Im weitesten Sinne ist damit jede Art von Kunst gemeint, Baukunst, Malerei, Schmiedekunst oder das kunstvolle Aufschließen von Gartengemüse und sein Wiedervereinigen auf dem Teller. Ja, es ist sogar das, wenn ein Kind eine handvoll Kiesel aufklaubt und danach irgendwo wieder zu einem Muster legt.

Soweit es aber jene besondere Kunst betrifft, die auch die ,hermetische', die ,hohe' und die ,königliche' genannt wird, ist darin eine ganz besondere Absicht enthalten, folgend der Erkenntnis, dass der Mensch in der Natur heilend und veredelnd wirken kann. Das ,kann' ist milde ausgedrückt – in Wahrheit ,muss' der Mensch der Natur helfen, denn sie hätte sonst keinen ähnlichen, derart kompetenten Freund.

Auf die Frage, ,wann' der Mensch wirkend in die Natur dieser Welt trat, antworten die ältesten Überlieferungen mit: Beides war zugleich da. Der Mensch und die Natur. Es war von Anfang an die ,Idee' der Erde, ein Planet der Menschen zu sein. Hermetisch ausgedrückt: Der Mensch ist die ,Quintessenz' der Welt. Der Begriff Quintessenz bedeutet nicht alleine ,Idee', sondern in der alchymistischen Praxis ist sie das ,merkuriale Salz'. Das ist mit dem Hinweis: „Ihr seid das Salz der Erde" gemeint. ,Wahre Menschen' sind bereits ein Ausdruck jener Quintessenz und wirken dementsprechend elixirisch, dh. veredelnd auf die Natur sowie auf andere, im Menschwerdungsprozess befindlichen ,Menschen' ein. Obgleich dies eigentlich ein Thema der Spiritualität ist, muss dennoch bemerkt werden, dass in diesem Umstand auch ein Grund der Heilkunst liegt, weil Krankheit und Heilung, genauso wie Leid und Erlösung, untrennbar mit Menschwerdung verbunden sind.

Auch wenn – den ältesten Überlieferungen zufolge – Gestalt und Körper des Menschen sowie seine Möglichkeiten, in der Welt zu leben, während Jahrmillionen nicht immer die selben waren, wie wir sie heute kennen, hat er doch stets wesentlich zur Entwicklung der Erde beigetragen, indem er diese Erde als einen erweiterten Menschenkörper begriff, den es in grauer Vorzeit aus der feinstofflichen Ideenwelt in die grobstoffliche Sphäre zu manifestieren galt. So besteht schon immer eine enge Verwandtschaft zwischen dem Menschenkörper und dem Erdenkörper, die unter anderem den Signaturkundigen zu der Bemerkung „wie innen so außen" veranlasst und zum Vergleichen der körperäußeren und -inneren Formen anleitet.

Diese Verwandtschaft besteht in Körper und Geist und erstreckt sich weit über die Erdensphäre hinaus in den Makrokosmos der Planeten und Gestirne. Das, was sich zum Beispiel im Geist des Menschen als zielstrebiges Wollen ausdrückt, bildet das Angesicht, wärmt Magen und Blut und wirkt im Mark der Knochen.

Es ist ein Aspekt des Feuer-Elements, der mit einem weiten Spektrum an Naturkräften korrespondiert und sich in zahllosen Formen, Gestalten und Situationen ausdrückt.

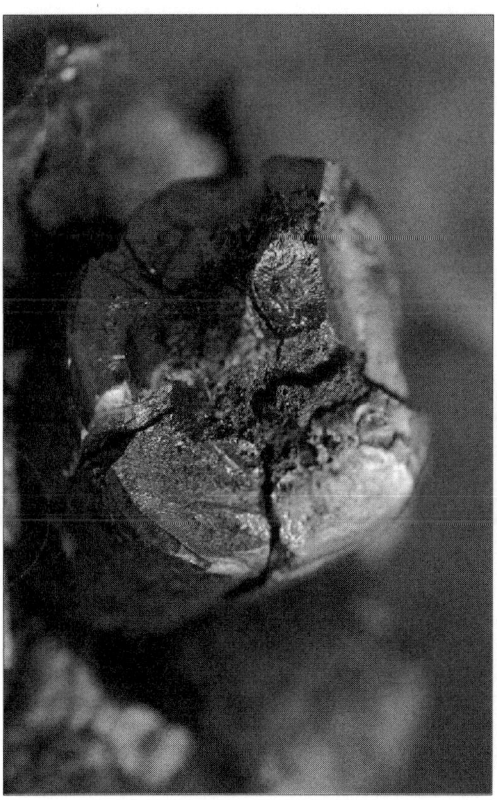

Feurige Kräfte versiegeln das Erdendasein von Körper und Geist mit jenen spezifischen Qualitäten, die genauso in Qualität, Substanz und Wirkung des Planeten Mars Ausdruck finden.

In der Pflanzenwelt herrschen sie in Brennnessel, Safran, Oleander, Pfeffer, Zimt und Senf, in Stabwurz, Rauten und in vielen anderen mehr vor.

In seinem Labor findet sie der Alchymist im Feuer des Athanors, im sulfurischen Öl, in den Farben und im hellen, scharfen Salz.

Unter den Gestirnen ist diese Kraft u.a. ausgedrückt im Sternbild des Widders, und durch die Sonne wirkt sie am stärksten zu Mittag nach dem Laufe der Zeit.

Das, was uns die Dinge sichtbar und insgesamt sinnlich erfahrbar macht, ist die ‚Planetenkraft Venus'. „Ohne Venus gäbe es keine Künste," sagt *Paracelsus*. In diesem einen Satz stecken vielerlei Bedeutungen. ‚Venuskraft' heißt, dass alles in seiner natürlichen Proportion steht. Dadurch wird das ‚Webwerk Natur' erst möglich. Ohne diese Komponente kann ein ‚Ding' schlichtweg nicht existieren. Die gesamte Natur ist ein Kunstwerk. Wenn es „keine Künste gäbe", dann gäbe es keine Natur. Im Menschen ist das Empfinden dieser Harmonie vollständig angelegt. Wie er sie ausdrückt, signiert seine ‚Kultur'. Aber selbst dort, wo der Mensch meint, völlige Disharmonie zu erkennen, wirkt in Wahrheit genauso jene venerische Kraft – sonst würde der Mensch nämlich gar nichts erkennen können, auch nicht Disharmonie. Aus Sicht der Natur ist ‚Harmonie und Proportion' ungleich vielfältiger, als nach dem Empfinden des Menschen.

Deshalb ist jene Kraft an den Äußerungen aller anderen Kräfte beteiligt, eben weil sie es den Kräftemischungen ermöglicht, ‚hier in der Natur zu sein'. Der signaturkundige Iatrochymiker *J. H. Cardilucio* (16. Jhdt.) spricht vom *Motus Veneris*, einer venerischen Art, die an allen ‚natürlichen Dingen' und Vorgängen genauso beteiligt ist wie Sonne und Mond als ‚Wirken der großen Polarität'. Dieser *Motus Veneris* gilt für Kräuter und Körperteile, welche dem Mars angehören, genauso wie für solche, die dominant luft-, äther- oder wassersigniert sind und ist nur durch verstärkte solare oder lunare Aspekte bedingt.

Wo aber das, was der Mensch allgemein besonders harmonisch, wohlproportioniert und anziehend empfindet, sehr stark hervortritt, zeigt sich dann eine wesentliche Dominanz jener venerischen Kraft. Das sind im Körper Hals und Nacken, Lenden, Nieren, die Proportionen der Knochen, Muskeln und Gewebe und anderes mehr, und außerhalb des Körpers ist es das Grün der Natur, sanfte Hügel, Wasser, Apfelbäume, Rosen, insgesamt die Rosengewächse und wohlschmeckende Früchte.

Solcherart weben alle Kräfte der Natur mit an einem sich ständig wandelnden Bild, in dem es viele Ähnlichkeiten, aber keine Gleichheit gibt. Körper und Geist des Menschen sind fixe Bestandteile dieses Webwerks. Der dritte Teil des Kontinuums Körper-Geist-Seele, also die Seele, hat nur bedingt Anteil an dieser Natur, denn sie ist von göttlichem Wesen. Durch sie belebt und beseelt der Mensch das Webwerk und erfüllt es mit Aufmerksamkeit. In den Mineralen schläft diese Seele, in den Pflanzen träumt sie, und im Menschenkörper beginnt sie zu erwachen. Dieses Erwachen macht den Menschen aus, und es macht ihn zugleich zum heilenden Künstler, zum Freund aller Wesen und natürlichen Dinge. Deshalb richtet sich die Natur nach dem Menschen, und dieser zieht sie mit sich – weiter in ihrer gemeinsamen Entwicklung – zur Erfüllung der Idee, zum Ausdruck der Quintessenz.

Will nun der Mensch sein heilendes, veredelndes Potential manifestieren, dann muss er die Art der Natur und die Position des Menschen darin verstehen lernen. Er muss die Zeichen und Verwandtschaften erkennen, welche alle ‚natürlichen Dinge', die Kräuter, Bäume, Minerale und Metalle sowohl mit der ‚großen Natur' – dem Makrokosmos der Himmelskörper – als auch mit dem Mikrokosmos des menschlichen Körpers, den Organen, Geweben und Säften bis hin zu den subtilen Hüllen des menschlichen Geistes verbindet. Mit dieser Signaturenkunde erlangt er eine Grundlage zur Heilkunst und darf den hermetischen Garten betreten, indem er in spagyrischlaborantischer Praxis nachvollzieht, was er aus der Natur gelernt hat. So folgt die ‚Kunst des Menschen' der Natur, und die Natur folgt dann dem Menschen und hilft ihm, wo nötig, ‚Heilung' zu erfahren.

Dies ist das Wesen von Spagyrik und Alchymie.

Spagyrik ist …

Spagyrik beginnt, wenn man die Natur auf solche Weise wahrnehmen möchte, wie sie in ihrem eigenen Lichte steht und einem die Geringheit des eigenen ‚Wissens' bewusst wird. Das Anschauen einer Landschaft, eines Berges, eines Baches, einer Pflanze, eines Menschen erweckt eine gewisse Sehnsucht danach, die göttliche Idee darin zu begreifen. Es wird fühlbar, dass ein Gewebe unsichtbarer Kräfte zur stofflichen Ausformung der Idee führt und dieses Wirken von Idee und Kraft den überwiegend wesentlichen Anteil an diesem Schauspiel innehat, während Substanz und Material den geringsten Beitrag zur Manifestation leisten. Daraus leiten sich in der Folge die drei Prinzipien ab: Das Sulfurische, welches die Idee als erstes in sich trägt, das Merkuriale, worin das Kräfteweben geschieht und das Salinische, welches den Wert der Idee für die Erde enthüllt und bestätigt.

Spagyrik beginnt also dort, wo das gewohnte, rein stofforientierte Denken von der Erkenntnis abgelöst wird, dass in der Natur die ‚Kraft' eine dem Stoff übergeordnete Instanz darstellt, und diese Erkenntnis bewegt dazu, ein gewisses Verständnis des Kräftewebens und -wesens in der Natur zu suchen.

Spagyrik beginnt, wenn man darangeht, die aus der Naturbeobachtung gewonnenen Erkenntnisse zu ihrer Vertiefung, Bestätigung oder Korrektion in laborantische Praxis umzusetzen. Soweit ist Spagyrik noch eine Art Naturkunde. Das Feld der Heilkunst wird dann betreten, wenn das gewonnene Naturverständnis sowohl zur ganzheitlichen Befundung von Mensch und Umwelt sowie zur Erstellung von Therapiekonzepten, welche die aus der Laborpraxis erlangten Produkte miteinschließen, taugt. Damit nähert sich der Spagyriker dem Ideal des *Iatros*, des Iatrochymikers, also jenem Urbild des Heilkundigen, wie es in der Natur seit Menschenbestehen existiert.

Spagyrik ist auch, wenn man nach dem Studium alchymistischer Literatur hinausgeht, ein Kraut erntet und dann beginnt, davon durch Destillation das Sulfur-Prinzip in Form eines ätherischen Öls zu scheiden, danach die Pflanzensuppe zur Gärung stellt, um das Merkur-Prinzip darzustellen, rectificiert, separiert, calciniert, kohobiert, coaguliert oder sonstige Handgriffe übt, durch die die Ausdrucksformen der bildenden Kräfte wesentlich aus ihrer stoffgebundenen Vermischung gelöst und für sich alleine betrachtet, begriffen und bearbeitet werden können, unabhängig davon, ob der Zweck dieses Tuns ein natur- oder ein heilkundlicher ist oder es einfach aus Freude an der Sache geschieht.

Es ist auch Spagyrik, wenn man den Pflanzenrückstand im tönernen Tiegel verascht und sich dabei Gedanken macht über die Signaturen der Asche und das Wesen des Erd-Elements.

Spagyrik beginnt dort, wo der ausgepresste Pflanzenrückstand eines alkoholischen Mazerates nicht bedenkenlos fortgeworfen, sondern darin eine Ausdrucksform des Erd-Elements erkannt wird, man ihn also verascht und aus der Asche das Sal-Prinzip in Form des Erd-Salzes darstellt. Das gereinigte und von den Schlacken geschiedene Erd-Salz wird mit dem alkoholischen Auszug (einer Ausdrucksform des Merkur-Prinzips) zur spagyrischen Tinktur vereint, welche sich eben durch das darin enthaltene Erd-Salz von einer ‚gemeinen' Tinktur unterscheidet. So setzt man sein Verständnis der Ganzheitlichkeit zusammen mit seiner Achtung der Erde gegenüber in die laborantische Praxis um und lernt während der Prozesse die tiefe Stimmigkeit des ‚hermetischen' Weltbildes kennen.

Spagyrik ist, wenn man beachtet, dass sich im alkoholischen Ansatz die öligen, aromatischen Anteile der Pflanze mit dem Alkohol derart verbinden, dass diese beiden ‚Prinzipien', nämlich Sulfur und Merkur, Sonne und Mond, Seele und Geist, Individualität und Naturgeist, *Anima* und *Spiritus*, danach kaum mehr zu trennen sind. Deshalb sucht man Wege, um rechtzeitig das eine vom anderen zu scheiden, was allegorisch dem ‚Mensch, erkenne dich selbst' entspricht. Erst nach ihrer Scheidung und Reinigung werden die Prinzipien dann wieder essentiell miteinander verbunden, wodurch die Pflanze eine Art Wiedergeburt erfährt und ihre ursprüngliche Idee potenter hervortritt. Allegorisch gesehen, werden bei der Coagulation zur ‚spagyrischen Essenz' Sonne und Mond durch die Kunst des Menschen vereint und diese Vereinigung in der Erde manifestiert.

Spagyrik ist, wenn eine Rezeptur ihrem Aufbau nach der Kräftekonstellation der Problemstellung entspricht und durch Einbringung aller drei Prinzipien im rechten Verhältnis nicht einseitig wirken kann. Soweit der salinische Anteil mineralischer Art ist, wird zuvor kunstgerecht aufgeschlossen, damit dem Körper auf physischer, energetischer und geistiger Ebene das volle Heilpotential vermittelt werden kann.

Manche meinen, dass das, was ‚Alchymie' genannt wird, weit über jenes hinausgeht, was man unter ‚Spagyrik' versteht. Diese Ansicht ist nicht unbegründet und mag verschiedentlich auch richtig sein. Andererseits gibt es für alle Dinge wegen der Vielfalt der Sprachen mehrere Worte. Auch im *Ayurveda* gibt es Alchymie, die dort aber nun weder so noch ‚Spagyrik' genannt wird, sondern zumeist *Rasa Vijnan*, was soviel wie „die Wissenschaft von Merkur" bzw. „das Wissen von der Natur" bedeutet. Sollte das bei uns gebräuchliche Wort *Alchymie* tatsächlich vom chinesischen *Jin Yi* stammen (aus dem die Araber *Al-Kimiva* machten), welches eben diese Kunst und zugleich das *Aurum potabile*, als ein Meisterwerk hierin, bezeichnet, und das Wort *Spagyrik* aus dem Griechischen, dann haben wir es ganz einfach mit verschiedenen Worten für die selbe Sache zu tun, und der macht es nichts aus, ob wir sie indisch, chinesisch, arabisch oder griechisch ansprechen.

Grundsätzlich wollen wir uns also an *Paracelsus* halten, welcher sagt, es sei die „Kunst *Alchymia*, die sonst *Spagyria* genannt wird" und damit das selbe meint. Trotzdem hat sich da irgendwo in den heutigen Gebrauch der beiden Wörter der Geschmack eines Unterschiedes eingenistet, den man vielleicht nur bedingt ignorieren sollte. Gewisse grobe Trennungen werden jedoch zu Unrecht unterstellt, so etwa, dass sich die Spagyrik ausschließlich mit dem Pflanzenreich und die Alchymie mit Mineralen und Metallen befasst. Es gibt – will man hier im Grad unterscheiden – wahrhaft alchymistische Pflanzenpräparate genauso wie sehr einfache, ja bloß gerade noch spagyrisch betitelbare mineralische Mittel. Die in der Neo-Esoterik beliebte Trennung, dass ‚Alchymie' ausschließlich gewisse ‚innere' Prozesse bezeichne, während ‚Spagyrik' ein ganz bestimmter Zweig der Naturheilkunde, womöglich gar der Homöopathie sein soll, trifft ebenfalls an der Sache vorbei. Vielleicht fände man jenen Unterschied (sofern wir uns überhaupt auf einen solchen einlassen wollen) am ehesten darin, dass man die typisch quintessentiellen Themen dem Gefühl nach kaum mit *Spagyrik* ansprechen würde. Danach wäre es dann eigentlich *Alchymie*, wenn man das frisch gesammelte Kraut weder destilliert noch in Alkohol ansetzt, sondern in einem dicht verschlossenen Topf in der Erde zur *Putrefaction* vergräbt – begreifend, dass jedes natürliche Ding das Mittel zu seiner eigenen Auflösung in sich trägt. Oder: wenn man nicht aufhört, sobald die Flüssigkeiten, Geister und *Phlegmas* überdestilliert sind, sondern *legem artis* weiterheizt, bis dass mit dem Rauch die eingewurzelten Kräfte erschlossen werden können und man selbst die *Salis* nicht in ihrer physischen Ruhe lässt, sondern weiterkocht, bis auch sie ihre innersten Geister freigeben. *Alexander von Bernus* vergleicht das Wort Alchymie mit *Hal-Chymia* und übersetzt: Salz-Kochung.

Da dies alles aber nur Separationen sind, Auflösungen, Öffnungen, Abscheidungen, also Aspekte des *spao*, wäre die Kunst nicht vollkommen, wenn es keine Wiedervereinigung gäbe. Und mit dieser ist dann der Sinn erfüllt, den das Wort *Spagyrik* beschreibt: Lösen und wieder Vereinigen. Erst so erlangt man die ‚kunstvolle' Arznei und nicht ein Teilprodukt. Nun sagt *Paracelsus*: „Das höchste aber ist die Arznei". Sie ist auch der höchste Sinn dieser ganzen Arbeit. Demzufolge stünde also der Begriff ‚Spagyrik' über ‚Alchymie', und wir sind veranlasst, diese müßigen Trennungs- und Unterscheidungsversuche bleiben zu lassen, bevor wir zu der Ansicht gelangen müssten, dass *Alchymie* ganz einfach deshalb für ‚höher' als *Spagyrik* gehalten werden sollte, weil das Wort selbst klangvoller ist und sofort einen geheimnisvollen, mystischen Schleier über eine Sache wirft, welcher diese Angelegenheit nur noch unverständlicher macht, wo doch diese Sache nichts anderes ist, als klare, einfache und sehr praktische Spagyrik. Es ist die Natur an sich so geheimnisvoll, dass sie nicht noch von weiteren Schleiern verhängt werden muss.

Ließe man also die ‚Alchymie' etwa dort beginnen, wo man über die Elemente Feuer und Erde nachdenkt und zu dem Schluss kommt, dass das Erd-Salz noch weiter zu erschließen wäre, indem man es mit einem Harz oder einem ätherischen Öl imbibiert und danach zum *Sal volatile* (Bild unten) kohobiert, dann hat man den Weg betreten, welcher durch den hermetischen Satz: ‚Das Fixe flüchtig machen' bezeichnet ist. Das erfordert bisweilen viel Geschicklichkeit, Geduld und Kunstverständnis. Auch wenn dadurch bereits eine Quintessenz, ein Elixir oder ein hohes *Menstruum* erreicht wird, so ist ein solches Kunstwerk erst dann vollendet – also fertig ausgearbeitet – wenn man vermag, dieses Flüchtige wieder zu fixieren – im hermetischen Sinne zu einem ‚Stein'. Diese Fixierung erfordert Wiedervereinigung, Coagulation, und ein solches Produkt dürfte dann den Titel ‚spagyrisch' tragen. Solche Überlegungen lassen uns von einer vergleichenden Wertung der Worte *Spagyrik* und *Alchymie* Abstand nehmen und führen zu der Einsicht, dass es wohl am besten ist, sie gleichbedeutend aufzufassen – soweit wir es vermögen. Sie können aber auch dazu beitragen, die Bezeichnung ‚Spagyrik' gegenüber ‚Alchymie' aufzuwerten.

Mißverstandene ‚Goldmacherkunst'

Alchymie ist keine ‚Goldmacherkunst', deren Hauptziel es wäre, weniger edle Substanzen in Gold zu verwandeln. Die Idee der ‚Veredelung' ist im wahrsten Sinne ganzheitlich zu verstehen und entspricht der Idee der Heilkunst selbst, also etwa der ‚Veredelung' des Menschen vom kranken in den gesunden Zustand. Über die Möglichkeit der Transmutation chemischer Elemente besteht in asiatischen Traditionen kein Zweifel, und spätestens seit den Erkenntnissen aus der nanometrischen Forschung ist sie auch im Westen eine ‚erwiesene' Tatsache.

Dass es im Abendland überhaupt dazu kam, dem vermutlich jahrtausendelang höchstgeschätzen Begriff *Alchymie* (oder wie auch immer hier die Bezeichnung dafür in älteren Zeiten war) die Assoziation der ‚Goldmacherkunst' auf anrüchige Weise anzuhängen, hatte mehrere Gründe. Es besteht sogar die Ansicht, dass *Paracelsus* die Bezeichnung ‚Spagyrik' anstelle von ‚Alchymie' so sehr hervorgehoben hatte, damit es den Menschen seiner Zeit leichter fiel, hierin etwas durchaus Ernstzunehmendes und Hilfreiches zu erkennen, war doch das Wort ‚Alchymie' bereits regelrecht ‚durch den Dreck gezogen'. Wir wollen hier versuchen, in aller Kürze einen Blick auf die Gründe dieses Problems und der daraus entstehenden Missverständnisse zu werfen:

1) Ungefähr ab dem 11. Jahrhundert begann der Strudel der Wirrnisse, Unsicherheiten und des Vergessens rapide übermächtig zu werden. Es ist dies in etwa jene Zeit, welche die orientalischen Texte als ‚Beginn des dunklen Zeitalters' bezeichnen. Der Beginn dieses *Kali-Yugas* wirkte sich zwar auf der ganzen Welt aus, traf aber besonders einschneidend das Abendland. Ein altes Sufi-Gleichnis, worin ein unbemerkter Sturm über das Land braust, der alle, die davor nicht gewappnet sind, verrückt mache, ohne dass sie es selbst bemerkten, beschreibt die Situation. Eine der Auswirkungen war der rasche Fall in einen intoleranten, ja zum allgemeingültigen Dogma erklärten Materialismus, der durch die häufigen, sich weiter ausdehnenden Kriege in die Menschen hineingezwungen wurde und mit dem Überlebenskampf der in jeder Weise verarmenden Bevölkerung verschmolz. Dazu schürte die Fackel der Inquisition die Angst bis zum äußersten, erreichte jede Festung und jedes Dorf mit der Absicht, alles, was nach menschlichen Weisheiten oder altem Wissen sich anfühlte, aufzuspüren und restlos zu vertilgen. Diese Ausweglosigkeit fand in der Ausrichtung des Lebens zwei Ventile: den religiösen Fanatismus und eine unsägliche und verzweifelte Gier nach Besitz und Vermögen. Unweigerlich glaubte man, dass, wenn diese Trugbilder eingefangen werden könnten, man dadurch Lebensgefühl und Sicherheit gewinnen würde. Solcherart getrieben und verwirrt, verlor man gemeinhin die Normalität menschenwürdigen Daseins aus dem alltäglichen Empfinden.

Wenn nun die verbliebenen Kundigen alter Weisheit dort und da einen Schleier lüfteten, dann fragten Reiche wie Arme gleichermaßen: Was hilft's uns? Und die Kundigen gaben einen guten Grund an: Es dient der Heilung von Gebrechen und Krankheiten. Wenn sie diese Begründung für ihre Wissenschaft überleben und sie gar noch durch die gelungene Anwendung tatsächlicher Heilkunst beweisen konnten, dann sagte der Arme weiter: Jetzt sind wir zwar die Krankheit los, aber die Armut wird uns zerfressen. Die Reichen aber sagten: Der Schmerz ist zwar fort, aber er wird wiederkommen, wenn wir nicht genügend Gold haben, unsere Position zu halten oder zu verbessern. Und irgendwoher wussten Arme und Reiche, dass solche Kundige bisweilen auch ‚mächtige Zauberer' waren, die noch weit mehr zu vollbringen imstande sein mussten, als eine Seuche zu bannen. Sie wussten das daher, weil in ihnen das Wissen der Alten nie vollständig versiegt war, zumindest ein Gespür dafür, oder ganz einfach aus Erzählungen, die immer noch weitergegeben wurden, oder sie wussten das von orientalischen Händlern, für die solche Dinge in ihrer Heimat noch alltäglich waren. Damit war es nur mehr ein kurzer Schritt zur Kombination: die können doch auch Gold machen! Jetzt hatten sie endlich den Bezug zu diesen Kundigen und ihrem seltsamen alten Wissen gefunden, der ihnen am wesentlichsten schien. Selbst hohe Kirchenfürsten hatten offenbar nicht selten Gedanken wie: Gut, dass wir sie nicht alle haben ermorden lassen. Die einfachen Kräuterweiber können wir ruhig weiter verheizen. Wir bauen den Bauern dafür in unseren Klöstern Spitäler, damit sie endlich merken, von wem ihr Wohl jetzt abhängt. Aber die, die da mehr zu wissen scheinen über die Dinge der Natur, die müssen uns ihre Kunst lehren – freiwillig oder unter Zwang. Der Stein der Weisen wird in unseren Händen ein weiteres Instrument allumfassender Macht. Sie sollen uns das Goldmachen beibringen und nebenbei vielleicht noch ein paar Stückchen Magie.

So oder so ähnlich hat es sich zugetragen, dass das ‚Goldmachen' im Abendland in den folgenden Jahrhunderten so bedeutsam wurde – nicht nur im Abendland, dort aber ganz besonders und nachhaltig. Dazu *J.C.W.Moehsen*, 1783:

„Die europäischen Fürsten hatten von Anfang an das Glück, Adepten in ihren Ländern zu besitzen, von sich gestoßen und sich so fürchterlich gemacht, dass *Lullius* aus England und Spanien, *Flamel* und *Zachaire* aus Frankreich und *Morienus* aus Italien flüchteten. Die britischen und schottischen Philosophen wurden durch die inneren Unruhen im Land nach den orkadischen Inseln getrieben. Auch in Deutschland hatten die Feuerphilosophen gleich von den ersten Zeiten an widrige Schicksale und mussten sich verborgen halten.

Nicht der Hass der Fürsten war ihnen entgegen, sondern deren unumschränkte Liebe zur Zudringlichkeit fiel ihnen beschwerlich, weil sie mit Gefahr verknüpft war. Wenn die Fürsten auch nur zurückhaltende, finstere Chymiker für Adepten hielten, so fassten sie schon eine solche gewaltige Neigung zu ihnen, dass sie diese zeitlebens nicht von sich lassen wollten. Sie liebten die Kunstgenossen bis zum Totschlagen, (...).

Die vielen Kriege der Türken in Europa hatten aber die inneren asiatischen Provinzen von räuberischem Volke gereinigt. Die sanfte Regierung des Mogulkaisers *Akbar* gab Indien sowie den übrigen asiatischen Provinzen große Vorzüge. Das Klima war angenehm, fruchtbar und insgesamt vortrefflich. Man lebte in Freiheit und war sich des Seinigen gewiss, Häuser, Kisten und Kasten wurden nicht visitieret. (...) konnte ein jeder in seinem Hause schmelzen, destillieren, sublimieren, transmutieren und tun, was ihm gefiel, ohne dass man sich um seine Geschäfte viel kümmerte. Asien war noch im vergangenen Jahrhundert (*M.* meint das 17.), ungeachtet zuweilen vorfallender innerlicher Unruhen, die Zuflucht und auch die Schule der europäischen Adepten. Wie *Johann Agricola* im Anfange des vergangenen Jahrhunderts nach Asien reiste, um Philosophen zu suchen, da traf er einen deutschen Arzt an, der eine Tinctur besaß (...). Dieser musste nach Asien flüchten, weil er in Deutschland bekannt geworden war und man ihn um Leben oder Freiheit bringen wollte. Auch der berühmte Adept, welcher 1666 den *Helvetius* von der Verwandlungskunst überzeugte (*J. F. Helvetii, >Vitulus aureus<*), hielt sich selbst in Holland nicht sicher und nahm seinen Weg nach Palästina. *Paul Lukas* traf noch 1705 zu *Burnus Bassy* in der Gegend von *Nicea* und *Prusia* einen hundertjährigen *Dervish* an, der nicht älter als dreißig Jahre zu sein schien und zur Gesellschaft des *Flamel* gehörte. (...)

... der Berg *Athos* war den Türken und Christen heilig. In den vielen Klöstern auf diesem Berge wurden die seltensten und kostbarsten griechischen, arabischen und anderen Handschriften aufbewahrt. *Thevet*, ein Verehrer der Alchymie, welcher Mitte des 16. Jhdts. mit *Wilhelm Posteln* diesen Berg bereiste, hat viele Bildnisse der alten Philosophen nachzeichnen und zu seinem Werke in Kupfer stechen lassen (*André Thevet, >Pourtraits et Vies des hommes illustres Grecs, Latins et Payens<*, Paris 1584). Den vornehmsten Schatz von Handschriften dieser Art besaß das Hauptkloster des heiligen *Basilius* auf dem Berge *Athos*, und das auf dem Berge *Sinai*. Diese verwahrten unter vielen anderen die chymischen Schriften des *Geber, Avicenna, Adfar, Rhazes, Calid, Alfarabi* und der weit älteren griechischen Alchymisten mit Figuren und Charakteren, unter welchen die chymischen Prozesse verdeckt wurden. Selbst unter den griechischen Mönchen dieser Klöster sind der Abt *Anastasius Sinaita*, die Mönche *Isaak, Kosmas, Sergius* und andere seit dem fünften Jahrhundert als alchymistische Schriftsteller bekannt geworden. Die Alchymisten zu *Fez*, welche noch zu *Thevets* Zeiten nach den Vorschriften des *Geber* arbeiteten, (...).

Arabien selbst gehörte ehemals unter diejenigen Reiche des Morgenlandes, welche am meisten von Deutschen besucht wurden. Die Araber waren nicht alleine als Alchymisten, sondern auch als große Ärzte, Kräuterkenner und Astrologen berühmt. (...).

Außer den Morgenländern gehörten noch die orkadischen Inseln, die nordischen Reiche und Spanien zu den Ländern, nach welchen die Wanderschaften vorzüglich gerichtet waren.

England und Schottland hatten der Abt *Cremer*, der Mönch *Rogerius Baco*, *George Ripläus*, *Thomas Northon* und die privilegierten Alchymisten *Johann Cobbe*, *Thomas Trafford*, *Thomas Atzheton*, *Johann Mistelden* und andere berühmt gemacht. Deren Schüler *Chaucer*, *Charnok*, *Blomfeld*, *Gawer*, *Michael Scotus*, *Robert Fludd*, *Wilhelm Buttler*, *Alexander Sethonius*, *Eduard Kellei*, *Johann Dee* und viele mehr sind den Liebhabern der Alchymie bekannt. Manche von ihnen hatten sich wegen der vielen Unruhen in beiden Königreichen im 15. Jhdt. nach den orkadischen Inseln begeben, wo sie sich als Eremiten ungestört der Alchymie widmen konnten. Doch kamen zuweilen einige von ihnen woanders zum Vorschein, um chymische Prozesse zu zeigen.

Mauren und Araber lehrten ehemals die Künste in Spanien in öffentlichen Schulen, und obgleich sie 1491 beinahe aus ganz Spanien vertrieben worden sind, so waren doch noch in der letzten Hälfte des sechzehnten Jahrhunderts einige Schulen zu *Granada*, *Sevillien*, *Toledo*, *Cordoba* und *Salamanca* übrig geblieben, wo diese Künste heimlich und in Gewölben unter der Erde gelehrt wurden.
(Nach dem Vorigen) wird man den Grund einsehen, warum Naturforscher, Ärzte und angehende chymische Philosophen des 16. und der vorhergehenden Jahrhunderte ihre Wanderschaften so häufig nach jenen Ländern richteten. Sie verbrachten darauf viele Jahre und reisten nicht ohne Absicht und nur allein aus Neugierde, sondern scheuten keine Beschwerlichkeiten, um zu einigen gründlichen Erkenntnissen zu gelangen.

Paracelsus war unter allen Deutschen, welche durch weite Reisen Wissenschaften erlernet, der berühmteste. Schon sein Vater führte ihn zur Arzneiwissenschaft an, und nachher hielt er sich bei dem berühmten Abt *Tritheim* und sonderlich bei *Siegmund Fuggern* zu *Schwaz* auf, einem Liebhaber der Alchymie und Bergwerksverständigen. Er las die Schriften der arabischen Alchymisten und bereiste Frankreich, Spanien und Arabien, um Wissende zu suchen. Bei seiner Rückkehr soll er zu Konstantinopel den *Trismosin* gefunden haben, der ihm das Geheimnis offenbart hat. Die Araber führten ihn zu einer Kenntnis der Astrologie, die er nachgehends in seinen Schriften als die vierte Grundwissenschaft der Medizin beschrieb ..."

nach *J.C.W. Moehsen*, >Leben des *Thurneisser*<, 1783; ausführlich zitiert in
>Der Weg des Sonnenfunkens<, S.43 ff.

In dieser Epoche des übermächtig werdenden Materialismus wurde nun endgültig aus dem Wissen der Alten das herausgenommen und hochgestellt, was alleine ‚echten Nutzen' zu bringen schien. Von da an war der Alchymie der Ruf als ‚Goldmacherkunst' gewiß, und das übrige, jahrtausendelang tradierte Naturwissen war nur noch gut zum Vergessen, für intellektuelle Philosophierereien oder geriet gar unter den Beschlag dunkler Magie. Die Heilkunst war von der Apothekerkunst per ‚Medicinaledikt' bereits abgetrennt und schließlich auch noch die Alchymie aus beiden herausgejätet. Unter diesen Umständen keimte schnell, was sich dann in den folgenden Jahrhunderten zur rein stofforientierten und aus dem lebendigen Zusammenhang gelösten ‚Chemie' entwickeln sollte.

Wenn man dann also gemeinhin von ‚Alchymie' sprach, dann dachte man zuerst je nach persönlicher Einstellung dabei entweder an eine veraltete, naive Ansicht der Natur oder an ein geheimes Wissen, unedle Metalle in Gold zu verwandeln, und man dachte nur zweitrangig an die Heilkunst und noch viel weniger daran, dass die Kundigen Träger eines uralten, sich auf beinah alle Gebiete des Menschseins erstreckenden Wissens waren. Daraus ergab sich ein weiteres Problem: Der Transport dieses alten Wissens – weil es eben nicht ‚veraltet', sondern ‚zeitlos' und genaugenommen eine Lebensgrundlage war – erschien den Kundigen genauso wichtig, wie die Heilkunst selbst – manchen sogar noch weit wichtiger. Es gehört auch zur Art dieses Wissens, dass es nicht als Teilstück, sondern nur ‚ganzheitlich' überliefert werden kann, weswegen man Wege finden musste, dieses auf eine Weise zu verpacken, die für den Empfänger des Paketes – ob ‚hoch' oder ‚niedrig' – reizvoll war, es zu öffnen. Es wurde verschlüsselt, allegorisiert, in geheimnisvollen Bildern gemalt – und auch so gestaltet, als würde dieses und jenes den Stein der Weisen zum Inhalt haben (was oft ja auch stimmte). Ein solcher ‚Empfänger', welcher den Inhalt des Paketes nur soweit auszunehmen imstande war, als er es eben vermochte, war zum Beispiel *J. G. Lehmann*, der 1757 das Werk des französischen Alchymisten *Respur* >Vom Mineralgeist. Auflösung und Verwandlung derer Metallen und von der Bewegung der Welt und ihrer Teile<, (1668) – ein in diesem Zusammenhang markant formulierter Titel – dokumentierte. Wir hören die Ansichten *Lehmanns* schon im ersten Kapitel:

> „Der Autor schreibt anfänglich von den Elementen und ihren Wirkungen auf solche Art, als ob diese Wissenschaft zur Erkenntnis des Mineralgeistes oder des Samens der Metalle nötig sei. Nun will ich es zwar dahingestellt lassen, ob die, bei solchen vorausgesetzten Vor-Erfordernissen angegebenen Sätze nicht doch ihre Richtigkeit haben möchten, gleichwohl sie, wahrscheinlich aus Mangel des Beweises oder des rechten Vortrages, nicht wenig zweifelhaft sind. (1)

Aber soviel ist nicht zu glauben, dass Herr *Respur* durch solchen weit hergeholten Weg den mercurialischen Samen gefunden habe (2), und es ist ebensowenig jemandem anzuraten, dass er aus diesem unergründlichen Abgrunde oder aus diesen ersten Anfängen der sichtbaren Welt die Alchimie studieren solle (3).

Er sagt, dass die alten Weisen zur Erkenntnis der Elemente langwierigen Fleiß hätten anwenden müssen, gleichwohl ist er selbst erst 24 Jahre alt gewesen, als er schon einen Lehrer davon abgegeben hatte (4).

Entweder haben sich nun dergleichen Leute schon vorher etwas von den Anfängen der Körper eingebildet und das Ausbringen ihrer Arbeiten damit verglichen oder gar dahin gezwungen, oder (umgekehrt) aus den ihnen gelungenen Wirkungen, aus einer kleinen Schöpfung, einen Schluss auf die wirkliche Schöpfung gemacht und sich aus dem, was sich bei dem Großen Werk geäußert hat, ein System von Kräften, Hervorbringungen und Ordnungen nach eigener Macht und Gefälligkeit aufgestellt (5).

Es folget daraus aber keineswegs, dass einer, wenn er auch das größte Kunststück in der Hand hat, deswegen auch die größte Weisheit und Erleuchtung besitze. *Respur* hat selbst im 2. Buch 2.7. angemerkt, dass von denjenigen, die den Stein der Weisen besessen hatten, viele Irrtümer eingeschlichen wären. Kurz: Der Stein der Weisen ist kein *Urim* und *Thumim*. Ja es kann ein Besitzer desselben in der Erkenntnis der Natur der unwissenste und ein anderer ohne demselben viel einsehender sein (6). (...)

nach: >*P. M. von Respurs* Besondere Versuche vom Mineral-Geist zur Auflösung und Verwandlung derer Metallen auch von der Bewegung der Welt und ihrer Theile<, neu übersetzt von *D. J. G. Lehmann*, Leipzig 1772, Nachdruck bei *Daniel Hornfisher*, 1996

Nun spricht der Text zwar angesichts der obigen Ausführungen schon für sich, es drängt aber, auf die Bemerkung *Goethes* hinzuweisen, welcher das Problem zu seiner Zeit offenbar genau erkannt hatte: Wem sollte er helfen, der Stein der Weisen, wenn ihm der Weise mangelte?

Zu oben im Text (1): Der Kommentator hält sich höflich und vorsichtig und legt die Behauptung, dass zur Erkenntnis des Mineralgeistes oder des Samens der Metalle, also für Angelegenheiten der hohen Alchymie, welche der Fertigung des Großen Werkes dienen, die gesamte alte Wissenschaft nötig sei, „mangels an Beweisen" beiseite. Vielleicht ahnt *Lehmann* noch einen tieferen Zusammenhang – er wurde ihm aber „nicht recht vorgetragen". Vielleicht wurde es ihm sogar vorgetragen, nur konnte er es aufgrund seiner neuzeitlichen Verbildung nicht mehr erfassen. Und: Wer sollte wem und wozu diese Dinge ‚beweisen'?

Lehmanns Denken scheint sich ausschließlich auf das zu beziehen, was er „die praktischen Dinge" nennt – eine mitunter sehr nützliche Ausrichtung, die in den asiatischen Texten als ein wertfreies (weil verschiedenartig einsetzbares) Charakteristikum für das ‚dunkle Zeitalter' *(Kali-Yuga)* genannt wird. Dass aber die gesamte Wissenschaft „von den Elementen, ihren Hervorbringungen und Wirkungen" genauso ‚praktisch', in allen Gebieten des Lebens anwendbar und insbesondere in der Heilkunst eine Grundlage ist, war *Lehmann* offenbar nicht vertraut.

(2) Warum soll das „nicht zu glauben" sein? Hier müsste sich *Lehmann* selbst bei der Nase nehmen und ‚Beweise' für diese Behauptung liefern. Ist es nicht viel naheliegender, dass *Respur* diese uralte Kunst eben auch aufgrund uralten Wissens (insofern ist ja die Bezeichnung „weithergeholter Weg" ganz treffend) verstehen und praktizieren gelernt hat?

(3) Hier übertrifft *Lehmann* sich selbst und versucht, das Stroh für Brot zu verkaufen. Es ist nicht nur diese Aussage völlig verdreht und unbegründet – *Paracelsus* würde sagen „der Phantasie eines spintisierenden Kopfes entsprungen", sondern zudem ein Paradestück, will man die ‚Denkart in den Anfängen der modernen Chemie' dokumentieren.

(4) Das zeichnet *Respur* ganz besonders aus. Ähnliche Fälle sind nicht wenige aus den asiatischen Traditionen bekannt, wo aufgrund ihres karmischen Hintergrundes bereits Kinder oder sehr junge Menschen tiefste Weisheiten lehrten. Um bei *Respur* die Möglichkeit solch eines Hintergrundes anzunehmen, fehlte dem Herrn Bergrath *Lehmann* wohl der ganzheitliche Blick auf Kunst und Leben. Die ayurvedischen *Vidyas* würden dazu sagen, *Lehmann* hätte unter einem übermäßigen *Tamas*-Zustand gelitten.

K. Ch. Schmieder (der selbst die ‚Alchemie' beinah ausschließlich als ‚Goldmacherkunst' verstand, allerdings nicht im üblen Sinn des Wortes, weswegen er ihr mit großem Interesse zugetan war) beschreibt *Lehmann* in seiner >Geschichte der Alchemie< (1831) als typisches Kind seiner Zeit – durchwegs materialistisch orientiert im Denken, versiert in der Praxis (immerhin war er Bergrath zu Berlin) und auf naive Weise vom Gedanken an die Metallverwandlung begeistert: „Der Bergrath *Lehmann* in Berlin war nicht abgeneigt sich, (...) für die Metallveredelung zu erklären. Er glaubte fest, dass er aus dem (...) Spießglanzoxyde künstliches Silber machen könne. Er hatte es schon oft gemacht, und der Versuch gelang ihm jedes Mal, so oft er ihn mit Spießglanz von der selben Sorte wiederholte. Als er aber (...) einen anderen Spießglanz holen ließ, erhielt er gar kein Silber. Nun prüfte er das zuvor gebrauchte *Antimonium* und fand es silberhaltig."

nach: >Geschichte der Alchemie<, *Karl Christoph Schmieder*, 1832; neu: *Wolfgang Roller*, 1997

(5) Damit beschreibt *Lehmann* eigentlich aufs genaueste die Vorgehensweisen der ‚modernen Wissenschaften', insbesondere der Chemie. Warum ihm der Gedanke so fremd zu sein scheint, dass die Alten ihr Wissen aus ‚lebendiger Naturbetrachtung', aus einem ‚Schauen im Lichte der Natur' erlangt haben könnten und sich dieses nicht künstlich zusammengeflickt hatten, ist eigenartig. Wahrscheinlich gibt es dafür die einfache Erklärung, dass *Lehmann* das ‚künstliche Überstülpen' von Denkmodellen in seiner Zeit am ‚geläufigsten' war, galt dieses doch dazumal bereits als ‚moderne' Geisteshaltung, und einer direkten Naturerfahrung mit all ihren Konsequenzen war man weitestgehend entfremdet.

(6) Das ist sicherlich richtig so, zeigt aber deutlich die große Unwissenheit, ja geradezu Dumpfheit *Lehmanns* dem alten Wissen gegenüber. Unter denjenigen, welche je den Stein der Weisen besessen hatten, werden wohl die wenigsten ‚Erleuchteten' nach dem spirituellen *Terminus* gewesen sein. Wir werden später noch genauer darauf kommen, dass Alchymie keine Art der Spiritualität ist, wo hinein dieser Begriff aber ausdrücklich gehört. Es haben zwar auch große Heilige und Mystiker des Orients (also ‚Erleuchtete' im eigentlichen Sinne) die Existenz des ‚Steins der Weisen' als selbstverständlich bestätigt, nie aber dessen physische Ausarbeitung als Bedingung ihres Zweckes – nämlich der Menschenbildung und Erlösung vom Kreislauf der Geburten – auch nur andeutungsweise gefordert. *Lehmann* vermischt hier auf äußerst oberflächliche Art zwei Bereiche, die zwar naturgemäß in weiten Zügen immer wieder zusammenstehen, dennoch aber auch völlig voneinander getrennt existieren können. Das eine ist Spiritualität, das andere jene Naturkunde, jenes ‚Wissen', welches sich eben mit den ‚Dingen der Natur' befasst, sich am Höchsten in der Heilkunst äußert, von sich aus aber keine ‚Erleuchteten' hervorbringt.

Insofern muss man *Lehmann* hier natürlich im Großen und Ganzen recht geben, denn das eine bedingt das andere nicht. Dazu ist weiters die Frage geeignet, was es denn bedeutet, den Stein der Weisen ‚in der Hand zu haben'. Wurde er von einem Kundigen wirklich erarbeitet, dann würde dieser die Bemerkung *Lehmanns*, er könne „leichtlich in den Dingen der Natur völlig unwissend sein" vielleicht mit einem beruhigenden Lächeln beantworten. Dass jeder andere – auch ohne diesen Stein erarbeitet zu haben – darin noch weit wissender sein könne, liegt auf der Hand. Der meiste Unfug ist – und das hört man nicht nur in den Legenden, sondern das zeigen auch Historien – von jenen entstanden, welche den ‚Stein', die ‚Tinctur', das ‚rote Pulver' oder dergleichen geschenkt bekamen oder gar gestohlen hatten und nicht recht zu nutzen wussten. Sie hatten ihn ebenfalls ‚in der Hand', missbrauchten ihn aber, weil ihnen eben gerade das zu seinem Besitz nötige ‚Wissen' fehlte und gaben ihn nicht selten als ihr eigenes Produkt aus. Solche Geschichten führten unweigerlich zu einem verhängnisvollen Ende.

Etwas milder, aber dennoch schlecht, geht die Geschichte aus, welche von einem orientalischen Heiligen berichtet, der ein Beispiel über Geiz und Dummheit geben wollte: Er schenkte einem Mann, der sehr über seine Armut klagte, ein Stückchen vom Stein der Weisen, mit der Kraft, Eisen in Gold zu verwandeln. Die Bedingung war, dass der Mann seine Zeit nützen müsse. Nach 3 Monaten würde der Heilige wiederkommen und nach ihm sehen. Wenn er bis dahin den Stein nicht gebraucht hätte, dann würde er ihn wieder zurückgeben müssen. Der Mann ging nach einigen Tagen zum Markt, um Eisen zu kaufen, fand es aber viel zu teuer. Im folgenden Monat war wieder Eisenmarkt – das Eisen war noch teurer geworden. Im nächsten Monat das selbe – und schließlich musste er den Stein wieder zurückgeben, ohne seinen Zustand geändert zu haben.

2) Der nächste Grund für den Verruf der Alchymie als ‚Goldmacherkunst' wird augenscheinlich durch zahlreiche Berichte von ‚betrügerischen Goldmachern', die uns besonders aus dem 16., 17. und 18. Jhdt. überliefert sind, und denen man in den meisten Fällen Glauben schenken muss. Unter diesen gab es drei Kategorien, und es ist wichtig, hier ernsthaft zu differenzieren.

a) Da waren zum einen allerlei Leute im Volk, die das ihrige an Hunger, Elend und Krieg abbekommen hatten, was sie teils aus Verwegenheit, teils aus Habgier oder einfach aus Not dazu verleitete, ihr Spiel mit der Sache zu treiben, um einigen naiven Bürgern ein paar Groschen aus der Tasche zu ziehen. Oft verschwamm dabei die Grenze zur akzeptierten Gaukelei, und nach dem Motto: die Dummen wollen betrogen sein, erheischten Händler mit schlechten Waren, Reliquienfälscher und Betrugsalchymisten ihren Lebensunterhalt. Grob ausgedrückt: Mit diesen Leuten konnte man leben, und wer ihnen aufsaß, war selbst schuld. So wie die meisten Berufe ihre Schattengänger haben, so war es auch hier.
b) Weit schlimmer wirkten sich andere aus – auf die ‚Kunst' und auf die Menschen, welche ‚ehrenhafte' Berufe und Würden innehatten, aber dennoch das, den Alten ‚heilige' Wissen heranzerrten, um ihre unersättlich wichtigen Persönlichkeiten nebst ihrer Geldbeutel bis zum Platzen aufzublasen. Dazu *Schmieder:*

> „ ... Ruhmsüchtige benutzten den allgemein verbreiteten Sinn für Alchemie, um sich schnell einen Namen zu machen, Ärzte zur Förderung ihrer Praxis, Laboranten zu mühelosem Broterwerb. Alle diese Missbräuche waren damals (im 16. Jhdt.) häufiger und schreiender als je zuvor.

Wenn die Nachkommen sich mehr und mehr dahin neigten, den Alchemisten zu misstrauen und auf der anderen Seite zu weit gehend die Idee der Alchemie selbst als ein leeres Hirngespinst zu verwerfen, so ist das zum großen Teil dem Unfug beizumessen, den damals solche Leute mit weit verbreitetem Rufe mit der Alchemie trieben. Ein solches Irrlicht war *Heinrich Cornelius Agrippa von Nettesheim* (1486-1535). Er war Doktor des Rechts und der Medizin, Theolog, Philosoph, Historiograph, Soldat, Magus, ...

In verschiedenen Perioden war er ein eifriger Alchemist und laborirte am Stein der Weisen, bis er die Geduld verlor oder Geldmangel ihn nötigte, etwas Einbringliches zu treiben. Das bekennt er selbst in seinen Briefen Nr. 4 und 10. Am Ende seines Lebens hatte er den Glauben an die Möglichkeit der Transmutation aufgegeben und eiferte nun sehr gegen die Alchemie in einer Schrift: *De vanitate Scientiarum*. Man erschrickt vor solch einem Titel, aber das wollte er eben. Sein Streben war, immerfort von sich reden zu machen, und das erreichte er so freilich am besten. ...

Was und worin er gearbeitet hat, ersieht man aus seinem Briefe: *De occulta Philosophia*, welcher im *Theatrum chemicum* abgedruckt ist. Er nahm des *Basilius* Worte buchstäblich und suchte die *Prima Materia* im Vitriol. (...)

Bartholomäus Korndörffer figurirte gegen Mitte des Jahrhunderts in Deutschland als fahrender Alchemist und ward von Vielen für einen Adepten gehalten, wenn schon einige Kenner ihn besser durchschauten. In einem Brief von *Oswald Croll* (...) heißt es z.B.,Herr *Heusser* teilte mir vor einiger Zeit einige Particularprocesse aus den Korndöfferschen Handschriften mit. Alleine ich habe sie ihm zurückgegeben und nichts davon abgeschrieben.'

Korndörffer verkaufte den Gläubigen alchemistische Vorschriften für ein *Viatikum*, wie Seinesgleichen pflegte, zufrieden, wo nicht Gold, doch Geld zu machen. Die Gläubigen haben seine Recepte nicht verloren gehen lassen, sondern gesammelt, wie *Heusser*, und sogar der Ehre des Druckes gewürdigt..."

nach: >Geschichte der Alchemie< von *Karl Christoph Schmieder*, 1832, neu: Wolfgang Roller, 1997

c) Jetzt möchte man einen Anfang erwarten wie: „zur dritten Kategorie gehörten die Schlimmsten." Wir müssen unsere Gedanken aber anders ausrichten und diesen ‚Dritten', soweit wir es vermögen, zu dem ihnen gebührenden Ansehen verhelfen: Auch in der damaligen schlimmen Zeit gab es einige geschickte Leute, von denen man heute nicht genau sagen kann, ob die Beweggründe zu ihren Taten beherzt oder verwegen waren.

Von diesen ist am meisten berichtet, man kennt ihre Namen und die Daten ihrer Schurkenstreiche und schreibt darüber gern und viel, um damit endgültig zu beweisen, dass es mit der Alchymie nichts ist. Besaßen sie doch die unerhörte Frechheit, sich mit ihren Täuschungen nicht an gutgläubigen Bürgern oder armen Bauern zu versündigen, sondern an den Höchstwohlgeborenen daselbst. Sie versprachen ihnen Gold und nahmen ihnen ihr Silber ab. *Hans Sachs*, der Nürnberger Meistersinger, nahm 1566 eine Gelegenheit zum Anlass, vor solchen Gesellen zu warnen: Es soll am Hofe *Kaiser Maximilians* ein *Venediger* gewohnt und dabei gut gelebt haben, was dem Kaiser viele Mark Goldes gekostet hat. Der Kaiser ertrug es, hatte ihm der *Venediger* doch ‚künstliches Gold' versprochen. Nachdem die Zeit vergangen war, besorgte der tatsächlich eine Transmutation – jedoch mit der Bemerkung, vorerst nur eine geringe Menge probieren zu wollen. Sie brachte dem Kaiser nur 10 Mark Gold – eine Menge, die das Gewicht der kaiserlichen Schatztruhe kaum merklich veränderte und der geringste Teil dessen war, was der Künstler die ganze Zeit über vom Kaiser erhalten hatte. Jedoch – er hatte sein Versprechen eingelöst. Danach verschwand er und ward nimmer gesehen – nicht ohne eine Notiz zu hinterlassen, etwa mit folgendem Sinn:

Oh Kaiser Maximilian!
Das ganze römische Reich sieht zu Dir auf,
da sollte doch auch einer, der diese Künste kann,
Dir zu Gnaden sein.

Man ließ den flüchtigen ‚Betrüger' suchen, aber vermutlich saß der *Venediger* inzwischen mit dem damals 20jährigen *Paracelsus* in einer abgelegenen Schenke, und sie feierten zusammen den gelungenen Spott.

Ein anderer seiner Sorte war der berühmte *Leonhard Thurneysser zum Thurn*, in Basel 1530 geboren. Es wird berichtet, dass er schon mit 18 Jahren den ‚Geldverleihern' vergoldete Bleistangen als pures Gold versetzte – ein Trick, den er als Goldschmied gelernt hatte. Als der Schwindel aufflog, musste er fliehen. Er reiste viel herum, nach England, Frankreich und andernorts und nutzte Zeit und Begegnungen, um Wissen zu erwerben und die Kunst zu erlernen. Darin stand er durchwegs in paracelsischer Tradition. 1555 kam er nach Nürnberg – nicht um dem *Hans Sachs* schlaflose Nächte zu bereiten, der vehement vor ihm warnte, sondern sich sein weiteres Reisegeld dort abzuholen, wo es sein Gewissen nicht zu beunruhigen brauchte.

Er stellte dem *Erzherzog Ferdinand* in Aussicht, das ‚große Geheimnis der Adepten' aus dem Orient zu holen. Auf diese Weise in seinem Innersten bewegt (es ist nicht überliefert, ob schon zu diesem Zeitpunkt im Jahre 1560 eine größere Waage für das Schatzamt bestellt worden ist), finanzierte dieser zehn Jahre lang *Thurneissers* Reisen, die ihn nach Schottland, Portugal, Spanien, Italien, Ungarn, Griechenland, Ägypten, Arabien und Syrien führten. Natürlich brachte er seine Perlen nie zurück an den erzherzöglichen Hof, aber er war zu einem großen iatrochymischen Arzt und Kräuterkundigen geworden, weise und erfahren in den Geheimnissen der Natur.

1570 wieder in Deutschland, wurde er Leibarzt im Dienste des Kurfürsten *Johann Georg von Brandenburg* und führte das alchymistische Laboratorium der *Kurfürstin zu Halle.* In Berlin wurde er berühmt, vermögend und von den Ärzten gehasst. Er musste deshalb die Stadt 1584 verlassen, wanderte wieder umher und bewies an manchen Orten die Transmutation. Gegen Ende seines Lebens zog er sich in ein Kloster zurück. Davor aber unternahm er noch eine letzte Fahrt: Der Kardinal *Ferdinand von Medici,* späterer Großherzog der *Toscana,* fand Gefallen an der Idee, sein Gold und seinen Ruf zu vermehren und lud *Thurneisser* am 20. November 1586 nach Rom zum Essen ein. Zum Erstaunen der Tafelgesellschaft verwandelte dieser straks eine Nagel zur Hälfte in Gold. Mehr Gold gab es zwar nicht, aber der Ruhm des Nagels zusammen mit der erlauchten Tafelrunde, die dieses Kunststück mitansehen hatte dürfen, drang weit, und das Stück wurde bis um 1700 im Florentiner Schloss stolz gezeigt. Als es sich dann endgültig herumgesprochen hatte, dass das Gold bloß am Nagel angelötet war (wahrscheinlich mit Eisenfarbe bestrichen, die sich beim Eintauchen in die ‚geheime Tinctur' ablöste), ließ man den peinlichen Nagel verschwinden.

Solcherart waren die Reaktionen manch verwegener Menschen auf die widrigen Umstände damals, waren die Antworten auf Goldgier, Hochmut, Dekadenz und Protzigkeit, und man muss auch bedenken, dass die Kundigen zu jeder Zeit ihre eigenen ‚Instanzen' hatten, vor denen sie geradestehen mussten, Kollegen, Lehrer und manche Unbekannte – ‚Instanzen', welche sehr wohl über die Qualität der Zeit und die Verhältnisse in der Welt Bescheid wussten und von ihresgleichen oder von ihren Schülern auch ein dementsprechendes Verhalten erwarteten. Dabei wurde ganz anders bemessen als danach in der von den Welt-Mächtigen geschriebenen Geschichte. „Ein Mensch, der sein eigener Herr sein kann, sei niemandes Knecht", befahl *Paracelsus.* „Ein und der selbe kann nicht zugleich Philosoph und Hampelmann sein", heißt es bei *Berger.*

„Kein rechtschaffener *Philosophus* ist ein Höfling und tritt hinter dem Fürsten her, ihm zu heucheln und zu schmeicheln, sintemal er nichts auf der Welt bedarf und ihm alles zu gering ist, was ein Fürst geben und schenken kann."

Johann Baptista van Helmont (1577-1664), >Die Morgenröthe<

Wozu hätten es die Kundigen Not gehabt, die Reichsten noch mehr zu bereichern, diejenigen, die verantwortungslos ihre Macht missbrauchten, mit noch mehr Macht zu versehen und den Ruf derer zu verklären, die das Volk als Geißel empfand? Sie hätten sich vor Ihresgleichen schämen müssen. Ein wenig ‚Uhlenspiegel' stak in manchen Weisen dieser Zeit, und sie machten sich damit selbst und anderen Mut. Sie hatten die Verpflichtung, zeitloses Menschheitsgut zu bewahren – ob sie nun heilkundlich tätig waren oder erzieherisch wirkten. Und da diese Kunst ‚hermetische Kunst' genannt wurde – wer will es ihnen verargen, dass ihre Vertreter bisweilen die schalkhafte Art des *Hermes* angenommen haben?

Dies zur dritten Kategorie derer, welche die ‚historische Alchemie' in Verruf gebracht haben sollen. Den Berichten zufolge waren es nur wenige, aber sie hinterließen (vielleicht mit Absicht) die breitesten Spuren.

Wie das Vorige, so bezieht sich auch das folgende ‚Missverständnis' nur auf einen Aspekt dessen, was im Begriffsspektrum Alchymie enthalten ist, nämlich eben auf die Transmutation von Metallen. Warum gerade dieser Aspekt im Abendland so penetrant vordergründig geworden ist, darüber haben wir schon einige Überlegungen angestellt. Wir sollten aber hierbei bedenken, dass es den Kundigen und Naturweisen genau darum nicht ging, zumindest nicht so vordergründig und nicht mit jenem typisch materialistischen Hintergrund. Sie wendeten vielmehr in dem, was sie taten, jene alte Wissenschaft auf verschiedenen Ebenen an, und das vorrangigste Anwendungsgebiet war die Heilkunst mit ihrer Diagnostik, Rezeptierkunst und Arzneimittelzubereitung – also dem Erschließen der den Pflanzen, Mineralen und Metallen innewohnenden Heilkräfte. Darüberhinaus gab es wohl nie die Frage, ob denn Transmutation tatsächlich möglich sei, weil diese ‚Möglichkeit' in der alten Wissenschaft eine Selbstverständlichkeit darstellte. Die Bemerkung‚es ist leichter, Gold zu machen, als Gold zu zerstören' drückt das aus. Würde man, zB. von Silber ausgehend, eine Quantität feinsten Goldes durch die Kunst in den Tiegel bringen – wer könnte damit geheilt werden? Kann man aber das Gold kunstvoll in eine heilsame Essenz bringen (also‚zerstören'), dann hat man ein bedeutendes Heilmittel erlangt. Darin liegt weit mehr der Zweck der Kunst. Deshalb lernt der angehende Spagyriker oder Alchymist, wie man ein Kraut aufschließt und zur Essenz oder Quintessenz vereinigt und wie man das selbe mit einem Edelstein und mit einem Metall macht und nicht, wie man Particularwerke zum Zwecke der Transmutation ausführt.

Das geht dann weiter in die Anwendung der durch die Kunst gewonnenen Arzneien, und diese gründet auf der Befundung, welche wiederum abhängt von der ‚Kunst, die Zeichen zu erkennen' und vom Wissen über das Wirken der Elemente und Planeten, deren Hervorbringungen und Zusammenspiel. Das alles ist Alchymie, und wenn einer alleine Gold machen wollte und es ihm auch gelänge, weil er dazu einen Prozess nachzukochen verstünde, dann wäre er noch lange kein Alchymist, wie das die Alten und Ältesten seit jeher verstanden haben wollten. Kann er deshalb ebenso das Kupfer in seine ‚Wahre Essenz' bringen, eine Quintessenz aus dem Schöllkraut erlangen und figieren, einen Smaragd in sein ‚erstes Wesen' bringen? Was hat er von seinem Brocken Gold, wenn er daneben am Herzen erkrankt – hätte er besser die Zeit darauf verwendet, ein Elixir für sein Herz zu erlangen.

Agricola's >Chymische Medizin< gibt Beispiele davon, dass Transmutationen und Partikulare am Weg liegen können, wenn die Sache ganz natürlich und kundig angegangen wird. Er macht keinen Hehl daraus und kein großes Geschrei darum, dass ihm dort und da ein Körnchen Silber oder Gold zugefallen ist. *Kunckel* schabt die dünne, schmutzige Goldschicht vom Boden seines Kolbens, die ihm der aus der Sublimation herabfallende Silber-Merkur zusammen mit dem rückständigen metallischen Silber ergibt. Dabei bricht sein Kolben und er denkt: Schade um das teure Stück Glas. Diese natürliche Einstellung zur Sache ist in der ayurvedischen und tibetischen Alchymie heute noch eine Voraussetzung für die Zulassung zu den hohen Prozessen. *Baghwan Dash* erläutert in >*Alchemie and Metallic Medicines in Ayurveda*< ausführlich die Zubereitung des Quecksilbers zur Arznei und setzt im selben Zuge noch einige Prozesse daran, die nötig sind, wenn man sein Präparat zur Transmutation verwenden will. Ohne Geheimnisse, Versprechungen und Mystifizierungen. Die Arbeit selbst erfordert die Qualitäten vom Alchymisten.

„Wenn man auch den *Mercurium* mit dem *Oleo Vitrioli* präcipitiert und geußt dieses *Oleum Compositum* darüber (und) figiert es bis auf den höchsten Grad miteinander, so wird ein schöner roter Stein daraus. Den soll man in das Blei tragen und miteinander schmelzen, so wird man sehen, was man daraus gemacht und bekommen wird, ..."

Johann Agricola, >Chymische Medicin<, 1638 / O. Humberg 2000

„Ich will hier auch erzählen, was mir in *Transmutatione Metallorum* mit dieser Tinctur begegnet und wieweit ichs gebracht habe. Ich habe mit diesem *Oleo* also procedieret: Ich habe diesen Präcipitat ins Blei getragen und zusammengeschmelzt, und im Schmelzen hab ich bisweilen ein wenig Markasit dareingeworfen. Danach hab ichs auf dem Test abgetrieben, habe aber aus 2 Lot mehr nicht als ½ Quintlein Cronengold erlanget. Und wenn einer sonst nichts zu tun hätte, könnte sich einer zur Not davon wohl erhalten. Ich will nicht zweifeln, dass man ihm mit einem Handgriffe (noch) weiterhelfen könnte, aber ich habe darin nicht viel experimentieren mögen, habe auch wegen meiner vielfältigen Geschäfte nicht dazu kommen können. Ich habe aber diese *Tincturam* meiner guten Freunde einem gegeben, der hat es in Zinn getragen und abtreiben lassen, so hat er mehr nicht daraus bringen können als ich mit dem Blei. Ich habe aber meine Gedanken weiter darüber, doch hab ich sie im Werk nicht probiert und halte davor, wenn man den *Mercurium Saturni* mit dem besten *Oleo Mercurii* präcipitierte, wie der rechte Gebrauch erfordert, und trüge alsdenn dieses *Oleum* darein, ich wollte nicht zweifeln, es sollte einen reichern Überschuß geben. Denn dieses *Oleum Compositum* eine gewaltige Fixität hat überkommen von dem *Sulphure fixo Talci*, welcher dem natürlichen Gold fast gleich ist. Und so solches in einem Gradierfeuer ein Monat oder sechs Wochen gehalten würde und alsdenn mit dem Blei geschmelzt, ich wollte eine Wette drauf tun, es sollte in der Prob ein mehrers geben als zuvor. Aber dieses sind nur meine Gedanken, es kanns einer versuchen, ich sage drum nicht, dass ichs probiert oder einen dazu überreden wolle, dass ers eben also machen müsse. Es kann auch sonsten wohl ein ander *Modus* sein, mit diesem *Oleo* Nutzen zu schaffen, der mir unbewußt und vielleicht einem andern bekannt ist, denn nicht einer alles kann noch erfahren hat, und Gott gibt einem diese, einem andern eine andere Gabe, je nachdem sein Herz mit Gott stehe. Derwegen sind in der Natur viel gute Kräuter, welche die Menschen nicht kennen oder wissen sie nicht zuzurichten, derhalben werden sie veracht und kommen ihre Tugend nicht an den Tag. Also ist es mit vielen Stücken in der *Alchymia*, welche oft einen guten Nutzen bringen könnten, auch beschaffen, aber aus Unerfahrenheit muss ihr Nutzen zurückebleiben und nichts daraus werden. Derwegen kann ich keinem verdenken, der in solchen Sachen fleißig arbeitet und nachsuchet."

Johann Agricola, >Chymische Medicin<, 1638 / O. Humberg 2000

Schmieder meinte (1832), dass *Demokrit* (5. Jhdt. v. Chr.) kein Alchymist gewesen sein konnte, weil *Seneka* von ihm ‚nur' berichtet „... dieser *Demokrit* war ungemein kunsterfahren in der Kunst, Steine zu schmelzen, Smaragde zu färben und das Elfenbein zu erweichen" und weil *Petronius Arbiter* im *Satyricon* ‚bloß' schrieb: „Wahrlich, dieser *Demokrit* presste aller Kräuter Säfte aus, ließ keinen Stein und kein Gesträuch unversucht, was für Kräfte darin verborgen sein möchten, ..." Weil aber keiner schreibt, dass *Demokrit* auch ‚Gold gemacht' hatte, konnte dieser nach *Schmieders* Auffassung kein rechter Alchymist gewesen sein. Welcher Art aber war jenes Wissen (nach *Schmieders* Ansicht), das der Grieche in Ägypten, Persien, Indien und anderswo erworben hatte? „Nach diesen Stellen ist kein Zweifel, dass ... (er) in der eigentlichen Chemie tätig gewesen sei, nur nannte man sie damals nicht Chemie, sondern Magie, nach den persischen Philosophen (...)." Jetzt wissen wir, was die ‚eigentliche Chemie' ist ...

„Das süße Salz *Vitrioli* zu machen hat viel Wesens. Ehe ichs zum erstenmal recht verfertigen konnte, brachte ich länger denn drei Monate damit zu. Es ist so süße als kein Zucker, und ist auch ohne das an sich selber lieblich, denn das gemeine *Sal Vitrioli* ist ihm gar nicht zu vergleichen, und wer es nicht wüsste, der sollte vermeinen, es würde ganz und gar aus einem anderen *Subjecto* gemacht, so gar einer andern Natur ist es.

Diese Medicin macht auch das Eisen zu einem schönen Kupfer, welches ganz geschmeidig ist und zu allen Arbeiten gut zu gebrauchen. Welche nun begehren, aus Eisen Kupfer zu machen, die könnens versuchen (...), ob aber dieses Kupfer zu den alchymistischen Arbeiten besser sein soll als das gemeine, wie ihrer viele dafürhalten, die könnens versuchen und sehen, ob es dem Silber so nahe verwandt sei, dass es mit ihm in lötiges Silber könne gebracht werden. Ich hab es nicht versucht, weiß auch nicht viel davon aufzuschneiden, mir will es in meinen Kopf nicht, denn es gehört eine andere *Tinctur* dazu. Und gehet so schlecht (einfach) nicht her, wie die Fabelhansen in ihren Proceßbüchern vorgeben (...). Ich trage Sorge, das gute Vertrauen dürfte zu Wasser und einem Traum werden, und wenn solches so schlecht (einfach) zuginge, so würden die lausichten ‚Alchymisten' nicht so mit dem Bettelsack herumziehen, die Leute betrügen und ihnen (den ‚echten' Alchymisten) einen bösen Namen machen, sondern würden viel tausend Centner Silber zuwege bringen, ja es würde endlich an Eisen mangeln, Pferde und anderes zu beschlagen, wenn man anderes reiten wollte, oder man müsste gar silberne Hufeisen – wie Eulenspiegel seinem Pferde güldene – aufschlagen lassen. Wer die Kunst kann, wird wohl heimlich damit sein, wer es aber nicht kann, wird den Schmieden das Eisen wohl lassen."

Johann Agricola, >Chymische Medicin<, 1638 / O. Humberg 2000

„Wer nun in dem Vitriolo etwas Denkwürdiges zu machen gedenket, Zeit und Unkosten darauf wenden will, der nehme den besten, den er haben kann, vor sich. Kann er keinen gerechten finden, so nehme er die Mühe auf sich und mache ihm selber einen, (...).

Und wenn man die *Mineram* recht antrifft, so ist sie schön blaugrün und ganz mürbe. Wenn man sie zum Schlich ziehet und kocht sie, so gibt sie einen schönen Vitriol (...). Es gibt einen guten *Spiritum* und auch viel. (...) Er darf die *Mineram* gar nicht rösten, wie bei dem Schiefer sonst gebräuchlich, sondern sie gibt ihn also rohe gern von sich. Und diesen *Vitriolum* halte ich viel besser als diesen, welcher aus dem Kupfer gemacht wird, denn sein natürlicher Geist ist noch bei ihm und durch Feuersgewalt nicht von ihm getrieben wie bei dem Kupfer, denn dieses hat das *Examen Vulcani* schon überstanden, derwegen sein Geist längst davongeflogen und seine Herberge verlassen. ...

Täten also diejenigen viel besser, welche *secundum litteram* dem *Basilio* nachfolgen und aus *Venere* den Vitriol, daraus einen *Spiritum* und *Oleum* machen und mit dem *Marte* vereinigen wollen, sie nähmen diesen *Vitriolum*, und so in diesem eine Wahrheit zu finden sein sollte, so würde es besser bei dem gemachten *Vitriolo* angehen. Ich trage aber Sorge, *Basilius* rede viel von einer andern *Venere* und *Marte* als der bloße Buchstabe besagt. Ich habe auch unter allen Laboranten noch keinen gesehen, der mit Wahrheit bejahen kann, dass er durch solchen Proceß soviel erlanget, dass er zu Nürnberg beim Bitterbold eine Mahlzeit dafür essen können, nichtsdestoweniger hält ein jeder dieses für ein großes Geheimnis, und auch hochwohlgelahrte Leute.

Wer die *Animam Vitrioli* machen will, welche ein sonderliches *Arcanum in Chymicis* ist und auch ein groß Geheimnis darinnen verborgen steckt, der kann kein besser *Vitriolum* dazu erwählen denn dieses, welches aus der coburgischen *Minera* gemacht wird. Wenn es sich in dem Wasser resolviert, so gibt sich seine *Anima* wie ein Blut in das selbe und schwimmet oben (...).

Ist etwas aus dem *Vitriolo in Transmutatione Metallorum* zu bringen, so halte ich davor, es stecke in dieser *Anima*, denn diese ist in allem Feuer – es sei auch so groß es wolle – beständig ohn allen Abgang (...). Ein *Spagyrus* mag der Sache weiter nachdenken, man kann es nicht alles einem in das Maul käuen, die Erfahrung lehret alles. Wer ihr könnte den Fluss geben, dass sie in die *Metalla* einginge und sich mit ihnen wie Wasser mit Wasser vermischte, so könnte wohl die *Luna particulariter* damit transmutiert werden (...)."

Johann Agricola, „Chymische Medicin", 1638 / O. Humberg 2000

Abschließend dürfen wir nicht vergessen, zur ‚Alchymie als Goldmacherkunst' eine vielleicht ungewöhnliche, aber durchaus begründete Überlegung anzustellen: Es ist von sehr wenigen Alchymisten bekannt, dass sie durch Transmutation zu ‚Reichtum' kamen. Eines der seltenen Beispiele dafür war *Nikolaus Flamel*. Ihm wird auch nachgesagt, er hätte sein *Elixir* genutzt, um dadurch (zusammen mit seiner Frau) ein mehrere hundert Jahre währendes Leben in voller Gesundheit und ein gewisses Vermögen zu erlangen. Ähnliche Beispiele gibt es bis heute in Asien, wo alchymistische Prozesse zur Lebensverlängerung genutzt werden.

Wir sehen also, dass der ‚Stein der Weisen', die ‚Rote Tinctur', das ‚Große Elixir' zumindest zwei Dinge in Aussicht stellt: die Möglichkeit ‚Gold zu machen' und ‚ein Leben in Gesundheit bis an die tausend Jahre' zu erreichen. Beides neigt dazu, sich als Wunsch einer materialistischen Weltanschauung anzuhängen. Typische Nicht-Materialisten, Weise und Heilige, erkannten das Leben als eine Vorbereitung auf den Tod und betrachteten jeden Reichtum jenseits der Notwendigkeit als Störfaktor in ihrer Menschwerdung. Sufis besingen gottberauscht die Erwartung des Todes, um dort endlich mit ihrem Geliebten eins zu sein.

Nun ist der Rückschluss, dass manche Alchymisten ‚arm', dh. ohne auffälligen Besitz verstarben, den ‚Alchemieforschern' in der Regel das Argument, dass diese keine ‚wahren Adepten' hätten sein können, bzw. es nicht bis zu den höchsten Rängen in der Kunst (die sie aus ihrem Blickwinkel eben für das Goldmachen halten mussten) gebracht hätten. Diese Ansicht ist aus der Perspektive der wahren Adepten sehr primitiv und zeugt von großer Unwissenheit diesen Dingen, der Kunst sowie auch dem Leben an sich gegenüber. Zum einen musste doch ein Kundiger oder ein ‚Adept', wie schon gesagt, den Historikern der Nachwelt ganz und gar nichts beweisend hinterlassen – schon gar nicht die Mär vom Sinn des außergewöhnlichen Reichtums schüren. Im Gegenteil: Es zeigt weit mehr menschliche Größe, jenes *Elixir* zu besitzen und für sich selbst nicht zu nutzen. Und wenn man damit Anderen aus der Armut geholfen hätte, dann würde dies im jeweiligen Fall einen guten Grund gehabt haben und wäre wohl kaum der Nachwelt von solchen Historikern überliefert. Dass der wahre Reichtum im Gold nicht liegt, ist den meisten zumindest theoretisch bekannt. ‚Besitzlos' heißt nicht zugleich ‚arm'. Die eigentliche Armut ist die Armut im Geiste. Die Ärmsten sind die, welche ihre Seele im weltlichen Reichtum verloren haben, usf.

Der Kundige wird manchen Fürsten als sehr arm empfunden haben, den seine aus Unwissenheit verübten Handlungen in ein würgendes Netz aus Rückwirkungen zwangen, das er selbst kaum bemerkte, der ‚Philosoph' aber bemerken musste – er hatte es ja gelernt. Und so war vielleicht oft das, was ein Historiker später ‚Betrug' nannte, ganz einfach der Versuch einer Heilung aus Mitleid.

In den asiatischen Traditionen ist es heute noch selbstverständlich, dass Handlungen wie die Erhöhung eines Metalls auf eine andere planetare Ebene (Transmutation) eine sehr starke Auswirkung gemäß dem karmischen Gesetz mit sich ziehen, weil dabei in den Gang der Natur beschleunigend eingegriffen wird. Solche Handlungen bedürfen unbedingt einer entsprechenden Rechtfertigung. Eine der in Indien offiziell dokumentierte Metallverwandlung größeren Stils erfüllte z.B. den Zweck, dem *Mahatma Gandhi* Mittel zur Herstellung der Spinnräder zu verschaffen und den ‚Salzmarsch‘ mitzufinanzieren. In diesem Zusammenhang anerkannten die Kundigen den Prozess als ‚karmisch gerechtfertigt‘. Die Aktion führte schließlich auch in der Folge zur Unabhängigkeit Indiens von der britischen Tyrranei – und wurde für die ganze Welt ein Beispiel zur Bewältigung ähnlicher Probleme. Die Angelegenheit ist nicht ohne ‚hermetische‘ Allegorien. Spinnrad und Salz sind zwei archaiische Dinge von tiefstem Symbolcharakter – nicht nur in der Alchymie. Mit dem Spinnrad ist das Weben verbunden und damit der Weber, und der Begriff ‚Weber‘ bezeichnete bis ins neuzeitliche Europa eben einen ‚Kundigen‘. Darüber wurde bereits im >Weg des Sonnenfunkens< einiges geschrieben und unter anderem das Wortspiel *Gabir – Geber – Weber* bedacht. Wir wollen es hier nur andeuten und dazu erst später in die Tiefe gehen.

Fazit dieser Überlegungen sollte ganz einfach sein, dass es weder ein Qualitätsmerkmal noch ein Prüfstein des einen oder anderen Kundigen, Philosophen oder Adepten ist, ob er auch nur irgendwann im Leben nachweislich weltlichen Reichtum besessen oder offenbart, oder ob er die übliche Lebensspanne deutlich überschritten hat.

Es soll aber keinesfalls so verstanden werden, dass sowohl die Anwendung eines solchen Elixirs, welches geeignet ist, die Lebensspanne zu verlängern, als auch seine Nutzung zur stofflichen Transmutation, eine Handlung gegen die Natur wäre. Zum ersten erinnert *Paracelsus* ausdrücklich daran, dass dem Menschen zu Zeiten an die tausend Jahre Lebenszeit von Natur aus zustünden und gedenkt dabei des hohen Alters der biblischen Propheten. Dies sei nicht alleine ein Ausdruck von Zeitqualität und Lebensweise gewesen, sondern auch eine Wirkung dementsprechend ‚hoher Medizinen', welche diese benutzen konnten. Demgemäß klären orientalische Überlieferungen bezüglich der Erdzeitalter darüber auf, dass die Lebenszeit des Menschen während eines ‚Eisernen Zeitalters' am kürzesten und mit etwa 100 Jahren, meist darunter, bemessen sei. Bereits im ‚Kupferen Zeitalter' (beginnend etwa vor 8000 Jahren) war die übliche Lebenszeit schon weit länger und noch länger im ‚Silbernen Zeitalter'. Abgesehen von allegorischen Bedeutungsebenen solch urzeitlicher Überlieferungen und der Relativität der Zeit als eine sich selbst bedingende kosmische Kraft bleibt immerhin noch der Hinweis darauf, dass in den letzten tausend Jahren nicht nur die Zeit an sich ständig ‚schneller', sondern auch die für den Menschen verfügbare Lebensspanne immer kürzer wurde. Übersetzt auf die Themen der heutigen ‚Wendezeit' dürfte man annehmen, dass die folgenden Generationen eine bedeutende Verlängerung der Lebenszeit zu erwarten haben, wozu auch die Alchymie wieder das ihrige beitragen wird, ähnlich wie bei jenen ‚Alten', deren *Paracelsus* gedachte.

Agricola über seinen Besuch bei einem alten Schüler des *Paracelsus* in Oberösterreich, *Ägidius von der Wiesen*.

„Ich habe mich oft über ihn verwundert, weil er ein Mann in einem hohen Alter war. Er war aber so gerade und frisch als etwa ein Mann von dreißig Jahren, dass er also in dem Gebirge in der Einöde gar allein wohnete. Ich hatte meine Gedanken, er müsste ein heimliches Kunststücklein haben, das er laborierte, denn er lebte im Essen und Trinken trefflich wohl und war in seinem Hause alles wohl ausgeputzt und ein schöner Vorrat darinnen. War auch am Geld kein Mangel zu spüren, war kostfrei, und wer auch zu ihm kam, dem tat er alles Gutes, wie er denn auch mein sonderlicher Freund war. Und wenn ich zu ihm kam, konnte ich in drei Tagen nicht wieder von ihm kommen, und ich hörete oft wunderliche Historien von dem *Paracelso*, was er oft vorgenommen und wie er gelebet und was vor gewaltige Curen er verrichtet."

Aus: *Johann Agricola*, >Chymische Medicin<, 1638 / O. Humberg 2000

Neben den Themen der Metallveredelung und der erweiterten Lebensspanne scheint der wesentlichste Aspekt der Alchymie oft beinahe hintergründig – die Heilkunst. Und gerade das zeichnet *Paracelsus* als einen der Großen aus, dass er hierin den höchsten Wert solcher Präparate sieht und es auslässt, Projektionen in die Tausende von Blei auf Gold oder dergleichen zu versprechen, wie wir das in vielen alchymistischen Texten finden – in gewissen davon ganz besonders oft. Anderen zur Gesundheit zu verhelfen, scheint bei *Paracelsus* der einzig relevante Grund gewesen zu sein, sich der Mühen zu Verständnis und Herstellung derartiger Elixire, bzw. des ‚Steins der Weisen' selbst, zu unterziehen.

„Stein der Weisen" nennen wir ihn darum, weil er, in den menschlichen Körper genommen, diesen sogleich tingiert. (...) dass sein Eintritt den Leib durchgeht und durchdringet und alles, was im Leib ist. Vermittels dieser Durchdringung stellt er diesen wieder her und erneuert ihn. Er tut das aber nicht so, dass er das Alte hinwegschafft und stattdessen etwas Neues an seine Stelle setzte. (...) Sondern er macht das Alte so sauber und rein, gleichwie ein Salamander mitsamt seiner Haut gereinigt wird – ohne allen Schaden, ohne Fäulnis – und es bleibt doch allemal die selbe alte Haut in Wesen und Form.

So ist der Stein der Weisen derjenige, welcher das Herz und die Hauptglieder reinigt, ebenso Mark und Geäder und was darinnen eingeschlossen ist, also dass weder Makel noch Ungesundheit mehr darin gefunden werden kann. (...) Er läutert alle Unschicklichkeiten der *Humores*, sodass diese wieder der ersten Geburt gleich werden. Alles was imstande ist, die Natur zu verderben, weicht ihm. Die Krankheiten fliehen dieser Erneuerung, wie die Würmer das Feuer fliehen. (...) Von jenem Steine ist das Gleichnis zu verstehen, vom guten Baum, dem gute Samen und Zweige erwachsen, wodurch wieder gute Bäume entstehen. (...)

Würden uns nicht gewisse Wahrzeichen entgegentreten, die wir imstande sind mit den Augen zu sehen, dann wäre es nicht zu glauben, dass solche Kräfte hervorzubringen (...) möglich wäre. (...) so verlangt er doch eine langwierige Arbeit mit vielerlei Umständen genauso wie eines unverdrossenen, wohlerfahrnen Arbeiters mit regem Fleiß. (...)
Aber in unserem (...) liegt eine solche Kraft, die Menschen damit zu versorgen, dass eine höhere und gewissere Medizin in der ganzen Welt nicht gefunden wird. Darum nennt man sie eine gesegnete Medizin, welche nicht jedermann von Gott eröffnet wird, denn sie ist besser zugerichtet, als das Dreckwerk, welches Doktor Leisetritt (...) durch seine doppelte Haube oder durch seine Narrenkappe filtriert.
(...) Solches habe ich aber nicht mit Müßiggang, Sitzen und Faulenzen erworben (...), sondern mit Umherstreifen, oder wie du sagst, mit Landfahren und viel Fleiß und Forschen erfahren müssen, auf dass ich's wissen und nicht wähnen möchte. Du aber saugst deine Medizin aus (...) einem Polster (...). Will mich derohalben meines Landfahrens nicht schämen und darin wohl ein Meister bleiben."

nach: *Paracelsus*, >Handbuch über den Stein der Weisen<, Bd. 4

Basilius Valentinus fasst die arzneiliche Anwendung des ‚großen Elixirs' zusammen, wobei er differenziert zwischen der Art des auf ‚kurzem' und des auf ‚langem Wege' bereiteten Elixirs. Beide Wege führten ihn zum ‚großen Elixir', das er entweder als Arznei gebrauchen oder damit den *Lapis Philosophorum* fertigstellen kann.

„Zwar haben solches Trithemius (...) und andere wahre Weise in dieser Kunst genügend beschrieben, doch damit du auch solches hier beisammen habest, so wisse, dass du von dem Elixir, das im kurzen Wege bereitet wurde, allezeit ein Senfkorn schwer mit einem Loth altem Wein nehmen musst, denn es ist solches stark und ein lebendig machendes Feuer.

Hingegen du von dem Elixir im langen Wege schon weniger nehmen musst, nämlich ein Rübensamenkorn schwer in 1 Loth alten Wein solviert. Alsdann kannst du solches gebrauchen in allen Krankheiten sowohl bei kleinen Kindern, als auch bei Großen und Alten allezeit, wenn es die Not erfordert früh morgens gegeben, auch abends und danach schwitzen lassen, denn diese Medicin eröffnet alle Poren und treibt mit Verwunderung alles Böse heraus, dass der Mensch dadurch gestärkt und also neu geboren wird, und es ist nächst Gott kein geschwinder Hilfsmittel zu finden. Es wird gegeben nach dem Alter, besonders von 2-12 oder bis höchstens 16 Tröpflein.

Bei alten Schäden und Wunden wird solches neben dem innerlichen Gebrauch auch äußerlich gebraucht. Nämlich dass man es mit süßem Wein in die Wunden lege, dann auch schwitze. Man braucht solches nicht mehr als einmal, denn es macht alles Halbtote wieder lebendig, daher kann man Gott nicht genug dafür danken und loben. Einer schwangeren Frau soll man nur 6 Tröpflein auf einmal geben, weil sonst diese Medizin die Natur so treibend machen würde, dass dadurch das Kind vor der Zeit geboren wird und nicht lebendig bliebe.

Wie man aber mit unserem Elixir den Lapidem Philosophorum bereitet, geschieht also: Im kurzen Wege lass von einem reinen Gold 8 Loth fließen. Dann trage bei starkem Feuer von dem Elixir 1 Quintlein behutsam darauf, denn es wird einen starken Schlag tun. Alsdenn lass es zugedeckt 1 Stunde fließen und auch also erkalten (...) so hast du den Lapidem Philosophorum.

(...) Mache dir von einem reinen Regulo Antimonii ex Venere et Marte 3 Pfund. Solchen reinige mit Sal Nitri und Sal Tartari bis er werde wie ein weißes Silber und beim Ausgießen zusammenfalle mt einem schönen Stern. Solchen reibe klein zu einem zarten Pulver. Dann gib von deinem Lapide 1 Loth darunter, tue beide in einen hohen Tiegel, so dass davon 10 Teile leer bleiben, setze solchen in ein Zirkelfeuer per gradus ignis 4 Stunden lang, so wird deine Materie in der 2. und 3. Stunde steigen wie ein Teig, welcher mit Hefe angemacht, dass alles mit vielen Farben über sich gehen tut. Aber in der 4. Stunde fließt alles zusammen zu einem roten Rubin und Stein. Dieses ist nun der rechte Stein der weisen Meister. Lass den Tiegel also unbewegt stehen und erkalten, dann zerbrich denselben, so findest du denselben in der Mitte von einem sehr großen Glanz. (...)"

Nun hat man gemeinhin ja gar keinen rechten Begriff von solchen höchsten alchymistischen Prozessen und davon, welche Art von Arbeit dahintersteckt und welch künstlerisches Vermögen. Deshalb kann die Vorstellung kaum Bilder dafür finden, womit sich jene Kundigen abgaben, von denen wir bislang einiges berichtet haben. Bald ist Mancher geneigt, diese Künstler tatsächlich ,versuchend', phantasierend, ,noch so unwissend' zu nennen, sie für Traumfänger zu halten, denen der Zufall irgendeine rote Chemikalie schenkte, die sie eben aus ihrem Unverständnis für nichts anderes als den ,Großen Stein' halten mussten. Ja, es bestand unter den sogenannten wissenschaflich gebildeten Leuten der letzten zweihundert Jahre eine Meinung über die Alchymisten der vergangenen Jahrtausende, die ein Gleichnis findet auf einem ganz anderen Gebiete, nämlich in den Bemerkungen *Darwins* über die Feuerland-Indianer (1845): Es sei kaum zu glauben, dass zwischen wilden und zivilisierten Menschen ein derart großer Unterschied besteht, nämlich ein größerer als zwischen einem wilden Tier und einem Haustier. Die Sprache dieses wilden Volkes verdient es kaum, ,artikuliert' genannt zu werden. *Captain Cook* verglich sie gar mit einem Räuspern aus vollem Halse. Ganz gewiss hat sich aber noch kein Europäer so geräuspert, mit solch heiseren, gutturalen und knackenden Lauten. (...)

Besagtes Volk der *Yaghan* war eines der ältesten auf dieser Erde. *Ashley Montagues* Recherchen wiesen es als das vermutlich älteste, ethnologisch erkennbare Volk aus, das aus Asien auf den amerikanischen Kontinent eingewandert war. In nur hundert Jahren seit der Darwinschen Bemerkung konnte es vollständig ausgerottet werden...

Thomas Bridge aber verstand die Dinge anders, zog zu den *Yaghan*, und begann ihre Sprache zu lernen und die Worte und Begriffe zu notieren. 1863 hatte er bereits ein Wörterbuch mit 7000 Vokabeln verfasst. 1879 war sein Wörterbuch – die Darwinisten ignorierten die Peinlichkeit – auf 23 000 Wörter angewachsen. Dennoch verweigerte man strikt zu akzeptieren, dass diese ,Wilden' nicht nur einen äußerst reichen Sprachschatz besitzen, sondern noch dazu vor Urzeiten eine unvorstellbar reiche geistige Kultur über die halbe Erde bis an die Südspitze des amerikanischen Kontinents getragen hatten. *J. Alden Mason* nannte schließlich erst 1950 die Sprache der letzten *Yaghan*-Familien (1948 wurde die Zahl der noch lebenden *Yaghan* auf kaum 20 geschätzt) ,außerordentlich wohlklingend, weich, melodisch, angenehm, ...'

Der Vergleich mit der Ansicht manch sozusagen ,wisssenschaftlich Gebildeter' über die Alchymisten hinkt nur insofern, als dass letztere bei weitem nicht ausgestorben sind, sondern in manchen Gebieten der Welt sich immer noch ,ihrer' *Yaghan*-Sprache bedienen. Nur alleine zur Anschauung und Vorstellung ,hoher' alchymistischer Arbeiten wollen wir sehr ausführlich einen Prozess von *Monsieur d'Anvers* (17. Jhdt.) wiedergeben:

Der gemeine *Mercurius* kann vermittels mühevoller Arbeit so verändert werden, dass er dich erfreuen wird. (...) Durch diese Präparation behält er für sehr lange Zeit seine Wärme, Trockenheit und Festigkeit. Nach der Präparation ist der *Mercurius vivus* die Fruchtbare Erde der Weisen.

Die erste Arbeit besteht in einer sehr guten Reinigung, da er sich ohne diese mit den vollkommenen Metallen nicht richtig vereinigen kann, wegen seiner Kälte und Erdigkeit.

Die zweite Arbeit besteht in einer Erwärmung, wodurch er eine solche Affinität mit dem spezifischen Merkur der vollkommenen Metalle erlangt, dass er sich mit ihnen vollkommen vereinigen kann.

Die dritte Arbeit besteht darin, ihn gut zu beseelen, und zwar mit einer fixen und vollkommenen Seele und Form. Da er diese selbst nicht besitzt, empfängt er sie von anderen Metallen, ähnlich wie Wachs den Eindruck eines Stempels aufnimmt. Auf ähnliche Art, wie die Erde den Regen empfängt und mit Tugend und Kraft des Wassers angereichert wird, weswegen sie ihre Früchte hervorbringen kann, empfängt der *Mercurius vivus*, nachdem er gereinigt und erwärmt worden ist, die Seele von Gold und Silber und wird von diesen Seeelen imprägniert und fermentiert. Er vereinigt sich mit ihnen auf natürliche Weise wie Mann und Frau und empfängt deshalb wahrhaft die Qualitäten dieser zwei vollkommenen Metalle.

Das *Antimonium*, so wie es aus dem Berg kommt, ist niemals rein, sondern vermischt. Man reinigt es mittels Digestion. Diese Digestion geschieht durch ein sulphurisches Feuer, wobei ein solches von metallischer Natur das bessere ist, obwohl die Künstler sie auch mit dem vegetabilischen Feuer des *Tartarus* vollführen, obwohl eine solche Digestion nicht natürlich ist.

Der Grund also, warum der *Tartarus* den *Regulus* von den Schlacken trennt ist, weil der unreine Schwefel des *Antimoniums* mehr vegetabilisch als metallisch ist. Deshalb musst du es mit dem, aus lebendigem Schwefel gezogenen Schwefel reinigen und nicht mit Salzen, damit dieser metallische Schwefel sich von selbst zu Merkur reduziert.

Die *Luna* findet man im magischen Mars und nicht anderswo – und dies geschieht folgendermaßen:

Gib 4 Unzen Eisen oder die Spitzen von Hufnägel in einen Tiegel und lass sie im Windofen glühen. Dann gib 9 Unzen *Antimonium* hinzu, mache starkes Feuer, bis das Eisen im *Antimonium* wie ein mineralisches Wasser fließt. Um die Vereinigung zu beschleunigen, gib mit einem eisernen Löffel 2 Unzen *Sal Nitri* dazu, und lass sie zusammen im Tiegel, welcher genügend groß sein soll, um ein Überkochen der Materie zu vermeiden, fließen, bis das Rumoren im Tiegel aufhört.

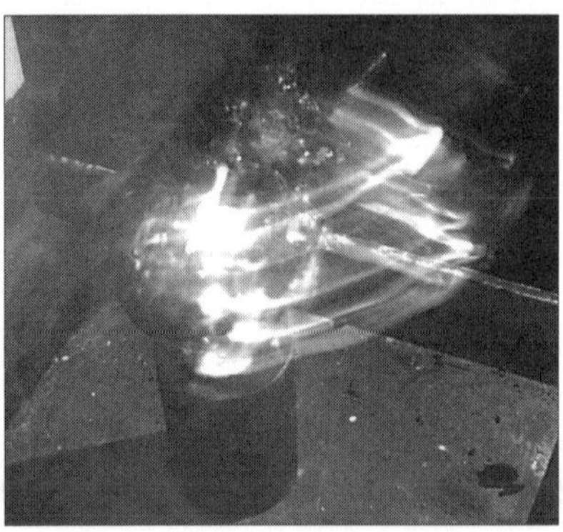

Danach wiederhole diese Operation, aber nur mit einer Unze *Sal Nitri*. Lass es auch noch ein drittes Mal fließen, indem du noch eine Unze *Sal Nitri* zufügst, und wenn das *Sal Nitri* bei starkem Feuer obenauf wie ein Öl schwimmt, sollst du den *Regulus* von dem Schlacken separieren.

Beim vierten Mal wird kein *Sal Nitri* mehr hinzugefügt. Das Zeichen eines guten Gelingens sind der Stern und goldfarbene Schlacken. Man kann diese Operation mit 4 Schmelzungen beenden und erhält dabei etwa 4 Unzen *Regulus*.

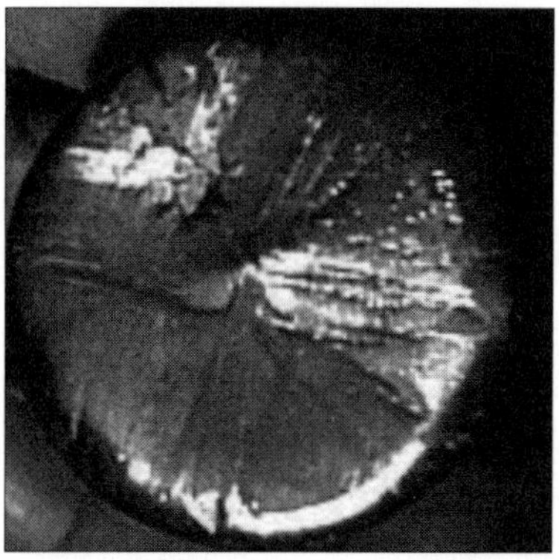

Merke, dass auf diese Weise nicht das *Sal Nitri* das *Antimonium* von seinen Schlacken trennt, sondern das *Antimonium* zieht die Seele aus dem Eisen, das heißt, es nimmt den besseren Schwefel des Eisens in sich und reduziert ihn zu Merkur. Und dieser Merkur ist nichts als Feuer, dessen Kraft und Eigenart den undigerierten, unverdauten Merkur im *Antimonium* digeriert und dadurch den Berg vom Metall trennt.

Du sollst auch wissen, dass der Merkur des Eisens in der Weiße des Merkurs dieses gereinigten *Antimoniums* verborgen liegt. Denn die Weiße der von dir gewünschten *Luna* in diesem *Regulo* kommt nicht von dessen Schwefel, sondern vom *Mercurius*, in welchem der *Mercurius Marte* versteckt liegt und der nichts anderes ist als Gold.

Der *Mercurius Marte* im Merkur des *Antimoniums* ist nichts anderes als ein Geist, oder eine lebendige Luft, bis dass er körperlich zu Gold wird. Alsdann ist er vom Merkur des *Antimoniums* separiert.

Wenn du also verstehst, welches Feuer das *Antimonium* reinigt, dann verstehst du auch, was es ist, das den *Mercurius* der Weisen reinigen und digerieren wird, denn du hast ja jetzt das *Antimonium* von den Excrementen seines Berges gereinigt. Wisse auch, dass es noch weiter Excremente und Unreinheiten im *Regulus* gibt, nämlich etwas vom verbrennlichen Schwefel. Wenn dieser völlig vom *Regulus* separiert ist, kehrt er in sein *Primum Esse*, sein erstes Wesen oder in seine *Prima Materia*, seine erste Materie zurück, die nichts anderes als ein Feuer oder ein Feuer-Wesen ist. Und das ist die *Luna*. Wenn du sie verfertigst, dann hast du das größte Geheimnis der Natur.

Die Operation, mit welcher der verbrennliche Schwefel vom *Regulus* des *Antimoniums* separiert wird, ist wunderbar:

Die *Luna* im *Regulus* des *Antimoniums* ist tot und muss daher wieder lebendig gemacht werden. Das geschieht mittels dem ihr allein eigentümlichen Leben und nicht durch das Leben von etwas anderem, fremden.

Wisse also alles darüber, wie der Merkur des *Regulus* vom *Antimonium* rectificiert werden soll. Dieses kann aber nicht ohne Vermischung geschehen. Wir sehen aber, dass sich der *Regulus* mit dem *Mercurius vivus* wegen des, im *Regulus* befindlichen verbrennlichen Schwefels nicht vermischen lässt. Denn dieser Schwefel ist nicht metallisch, folglich hat er keine Affinität mit dem *Mercurius vivus* und verhindert so die Mischung. Daher ist ein Vermittler zwischen *Regulus* und *Mercurius vivus* erforderlich:

Nimm also 2 Unzen sehr reines Silber, lass es in einem Tiegel gut glühen und gib dann 1 Unze von deinem dazu, und wenn alles zusammen in gutem Fluss steht, so gieße es in einen Inguss und du wirst eine metallische Masse von der Farbe des Bleis vorfinden, die sehr leicht zu pulverisieren ist.

Pulverisiere sie also sehr fein und gib sie in ein Glasgefäß. Darüber gieße 8 bis 10 Unzen sehr reinen Merkur und gib alles zusammen für 24 Stunden in ein *Balneum Mariae*. Während dieser Zeit wird das Pulver in den Merkur eingehen, indem du hin und wieder das Glas schüttelst. Wenn das Pulver und der Merkur gut miteinander vermischt sind, dann nimm alles aus dem Gefäß und reibe es in einem marmornem Mörser, so wird sich ein Amalgam bilden, das du zum Digerieren in das *Balneum Mariae* stellst.

Danach muss es neuerlich in einem marmornem Mörser gerieben werden, und zwar mit sehr reinem Wasser. Dieses Wasser wird schwarz. Gib dieses Wasser beiseite, stelle das Amalgam von neuem ins *B. M.*, amalgamiere und reibe solange, bis das Wasser schwarz wird, und wiederhole diese Operation, bis der Mond voll sein wird. Die Schwärze, von der das Amalgam nun endlich gereinigt sein wird, sodass es wie feines Silber erscheint, ist der bösartige Schwefel des *Regulus* vom *Antimonium*.

Um dir diese Arbeit zu erleichtern, kannst du das Amalgam zusammen mit etwas weißem *Sal gemmae* triturieren, dank dessen sich die Schwärze schneller entfernen lässt.

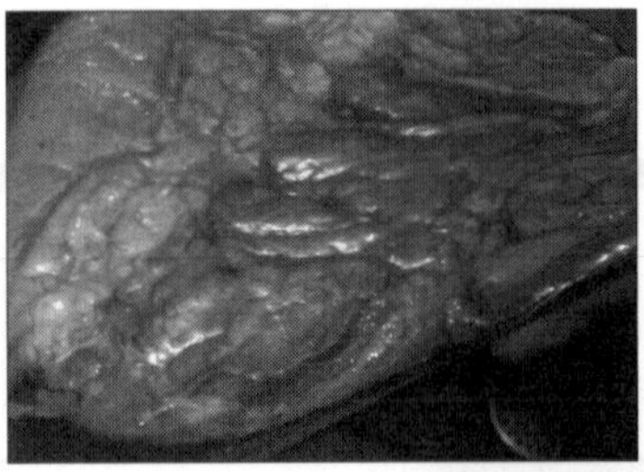

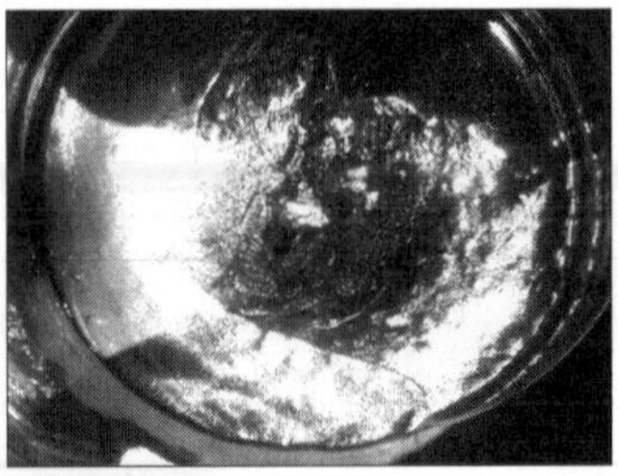

Je mehr Schwärze das Amalgam abgibt, umso mehr Merkur vom *Regulus* des *Antimoniums* schließt es auf, und der *Mercurius vivus* wird immer mehr zu dem *Mercurius*, den du begehrst. Also besteht das ganze Geheimnis darin gut zu digerieren, zu waschen und die Waschungen und alle diese Operationen zu wiederholen, um mittels einer einzigen Operation den *Mercurius vivus* zu erwärmen, zu reinigen und zu animieren (beseelen). Dann wird dieser von dreifacher Art sein: *Mercurius vivus*, Merkur des *Regulus* vom *Antimonium* und *Mercurius Marte* – alle drei in einem einzigen Merkur, welcher imstande ist Gold und Silber aufzulösen. Denn dieser Merkur, mit Silber destilliert, durchdringt die Metalle und separiert auf natürliche Weise die Elemente als *Sulphur* und *Mercurius*, was andere *Mercurii* nicht können, auch nicht der *Mercurius vivus*, außer er hat, wie du soeben erfahren hast, unsere Präparation durchschritten.

Wenn du also dein Amalgam weiß und ohne alle Schwärze siehst, dann gib es in eine Retorte und extrahiere mit starkem Feuer deinen Merkur vom Silber. Wurde die Waschung gut gemacht, so wirst du das Silber rein und weiß vorfinden, ansonsten wird es noch bleifarben sein, was ein Zeichen dafür ist, dass noch etwas vom *Regulus* vorhanden ist.

Damit nun der Merkur noch mehr erwärmt und gereinigt wird, muss man das in der Retorte zurückgebliebene Silber mit frischem *Regulus* zusammen schmelzen und neuerlich mit dem destillierten Merkur cohobieren und die Operation Punkt für Punkt bis zum 10. Mal wiederholen. Dann wird das Silber die Seele übergeben haben und du wirst wahrhaftig sagen können, den Merkur und das Solvens der Philosophen zu besitzen.

Du siehst also, dass der *Mercurius vivus* die Fruchtbare Erde der Philosophen ist, aber unvollständig, da er die Feuchte des Silbers nicht besitzt. Damit sie sich aber als vollständig erweist, muss man sie noch mit dem warmen Schwefel des Goldes animieren, was auch du auf folgende Weise machen sollst:

Nimm ebesoviel Gold, wie du Silber genommen hast, gib die Hälfte von dessen Gewicht *Regulus* in einen Tiegel, der sehr heiß sein muss, und wenn alles zusammen gut fließt, gieß es in einen Inguss, welcher die Form eines Horns oder Kegels haben sollte. Pulverisiere diese brüchige, schwarze Masse und schütte sie in ein Glasgefäß, worin du schon deinen lunaren Merkur gegeben hast. Lass es zusammen für 24 Stunden im Wasserbad digerieren, danach nimm dein Gefäß aus dem Bad, amalgamiere dein Gold zusammen mit dem Merkur gut in einem Marmormörser, indem du triturierst und immer wieder mit Wasser wäschst, bis sich keinerlei Schwärze mehr zeigt und indem du nachher alles, so wie du es zuerst gemacht hast, destillierst. Mit einem Wort: Gehe Punkt für Punkt mit dem Gold so vor, wie du es mit dem Silber gemacht hast, so wirst du den Dreifachen Merkur der Philosophen besitzen, den sie sosehr geheim gehalten haben.

Du hast nun 3 Wege, den Stein der Philosophen auszuarbeiten:

Der erste besteht darin: Nimm deinen Merkur und stelle ihn mit großer Vorsicht in einem philosophischen Ei in das geheime Feuer der Weisen. Koche und digeriere ihn bei einem Feuer, dessen Grad dem Erfordernis der Materie entspricht, welches dir, direkt und weislich von einem Philosophen regiert, den Sohn des mit seinem königlichen Purpur neu bekleideten Schwefels geben wird, den du wieder mit seinem Wasser tränken sollst, indem du wieder mit dem gleichen Regiment des Feuers wie vorher verfährst und diese Operation 7 mal wiederholst, so wirst du die Schätze der Gesundheit völlig besitzen. Beachte, dass in deinem Gefäß während der Kochung alle Farben, von denen die Philosophen sprechen, erscheinen werden.

Der zweite Weg besteht darin: Nimm 1 Teil an Gewicht vom lunaren Schwefel und inkorporiere ihn in 3 Teile seines philosophischen Merkurs. Darauf gib das ganze in das Ei der Philosophen und gib den ersten Feuergrad, das ist die Wärme einer brütenden Henne, und fahre damit fort, bis die Materie schwarz wird. Sobald sich diese Farbe zu verlieren beginnt, vermehre dein Feuer auf den 2. Grad, eine Wärme, die noch einmal so groß ist. Erhalte diese, bis die Materie grau wird. Gib nun ein Feuer des 3. Grades, bis die Materie in ein weißes Pulver verwandelt wird.

Nun hast du die *Minière Philosophique* zum Weißen, das der Anfang des Werkes ist. Die sollst du notwendigerweise 7 mal mit Deinem Merkur multiplizieren, indem du dieses Pulver jedesmal zu einem laufenden Merkur reduzierst und zwar mit Hilfe seines anfänglichen Merkurs, indem du es jedesmal mit ihm vereinigst, bis es wieder zu Pulver geworden ist. Dann wird das Werk gelungen sein.

Nachdem aber diese kostbare Materie erhalten wurde, musst du sie von neuem zu einem fließenden Merkur machen, von dem du 3 Gewichtsteile nehmen und mit 1 Teil Solarischen Schwefel vollkommen inkorporieren sollst. Gib es in ein Philosophisches Ei und gib den ersten Feuergrad, bis die Materie in ein rotes Pulver reduziert ist, das man *ad infinitum* multiplizieren kann, so wie das weiße Pulver. Aber du musst es wenigstens 7 mal für dein Werk multiplizieren, indem du es jedesmal mit seinem eigenen Merkur zuerst zu einem laufenden Merkur und dann zu einem Pulver reduzierst.

Die Alten haben dieses Pulver einen verzehrenden, verschlingenden Drachen, einen wunderbaren Löwen, eine anhaftende, klebrige Schlange genannt. Du aber sollst es Sohn des *Sulphurs* nennen, nachdem du es zu einem fixen Pulver, schmelzbar und durchdringend mittels wiederholter Sublimationen, gemacht hast. Dieser Weg ist aber viel länger als der erste und er ist auch viel länger als der dritte, den ich dir jetzt zeigen werde:

Der dritte Weg besteht darin, dass zuerst das gemeine Gold calciniert werden muss. Nimm dazu eine Unze sehr feines und mit *Antimonium* gereinigtes Gold und amalgamiere es mit 10 Gewichtsteilen Deines Merkurs. Dieses Amalgam ist gut gemacht, wenn beim Reiben zwischen den Fingern das Gold weder sichtbar noch spürbar wird. Gib acht und erwärme es beim Amalgamieren nicht zu sehr, denn es besteht die Gefahr, dass alles verdirbt.

Wenn es nun gut amalgamiert ist, dann gib es in einen mit einem anderen Gefäß bedeckten Cucurbiten und stelle es 5 oder 6 Tage ins B. M., damit durch die feuchte Wärme sich die Materien mehr konsolidieren.

Danach sammle den Merkur mit einer Feder und revifiziere, also wiederbelebe ihn wie gewöhnlich in kochendem Wasser, so wird er wieder zu einem laufenden Merkur. Sobald dies geschehen ist, vereinige ihn mit dem Gold und triturierе auf solche Weise, dass es sich amalgamiere. Stelle es wieder in das *B. M.*, sublimiere wieder, revifiziere und amalgamiere wieder von Neuem, wie beim ersten Mal. Auf solche Weise wiederhole alles bis zu 7 malen oder so oft, bis der Kalk des Goldes oder Silbers rot, laufend und so rasch schmelzbar wird wie Wachs.

Dies ist die Probe und das Zeichen, dass es der wahre, philosophische Kalk ist, den man meist schon nach der 7. Sublimation erhält.

Mit deinem auf solche Weise philosophisch calcinierten Gold musst du eine Konjunktion mit Deinem Merkur machen. Diese Konjunktion ist aber nichtsdestoweniger mysteriös. Glücklich derjenige, der dieses Ziel erreicht hat. Er schließe Gold und Silber in ihrem Gefäß ein und gebe Feuer zu Feuer und Merkur zu Merkur und fahre fort.

Nimm 1 Unze von deinem auf philosophische Weise calcinierten Gold und amalgamiere es mit 10 Unzen Deines philosophischen Merkurs. Dabei erwärme das Amalgam nicht zu sehr, denn bei zu viel und exzessiver Wärme besteht die Gefahr, dass von dieser kostbaren Materie etwas verloren geht. Das ist, weil der philosophische Merkur die *Anima Auri*, die Seele des Goldes in sich trägt. Diese aber ist sehr subtil und flüchtig.

Wenn also dein Amalgam fertig ist, passiere es durch Wildleder, um das Überflüssige auszupressen. Dann nimm was im Leder zurückbleibt und zerreibe es in einem Mörser aus Glas, bis das Amalgam zu einem subtilen Pulver geworden ist. Danach drücke es durch das Wildleder und wiederhole dies solange, bis nichts mehr durchgeht. Du kannst dann sicher sein, dass das Gewicht der Natur respektiert ist und hast also einen Teil Gold zu 4 Teilen Merkur.

Daraufhin wird es in einem Mörser aus Glas oder ähnlichem trituriert, und wenn alles gepulvert ist, dann gibt man es – nach der Konjunktion dieser Materien – in das philosophische Ei.

Diese Komposition sollst du in ein Gefäß geben um es zu kochen und zu digerieren. Es soll ein Gefäß aus Glas sein, dessen Boden nicht platt, sondern vielmehr rund sein soll, so wie das Ei einer Henne. Beachte vor allem, dass es weder zu groß, noch zu klein sein darf. Denn wenn es zu groß ist, wird sich der Merkur zerstreuen, und seine Feuchte geht verloren. Seine *Humidum Radicale*, seine Wurzelfeuchte, aber bleibt steril. Ist das Gefäß aber zu klein, dann werden unsere Blüten ersticken und keine Früchte tragen. Ich betone hier also mit Nachdruck, eine Gefäßgröße proportional zur Quantität der Materie zu wählen, und zwar darf die Materie nicht mehr als den fünften Teil des Gefäßes, also des Eivolumens, einnehmen, sodass 4 Teile leer bleiben.

Hast du also ein geeignetes philosophisches Ei von einer an die Quantität der Materie angepassten Proportion gewählt, dann versiegle es hermetisch und stelle es in den geheimen Ofen der Philosophen, wie du im Folgenden sehen wirst:

Der geheime Ofen ist ein Ofen von der gleichen Form und Bauweise wie ein solcher, der zum Brotbacken dient, aber dergestalt, dass er dir für die notwendige Bereitung deines Werkes dienlich ist.

Das Ei mit der in ihm befindlichen philosophischen Materie stellst du in eine kleine, ganz mit dem Mineralfeuer angefüllte, irdene Schale. Diese deine Schale mit dem philosophischen Ei setze auf einen kleinen Dreifuß, den du in der Mitte des Ofens platziert hast. Das philosophische Ei zusammen mit der Schale auf dem Dreifuß hängt gewissermaßen in der Luft, ist also überall von Luft umgeben. So kann auch die Wärme von allen Seiten gleichmäßig dazukommen.

Dann, nachdem du also das philosophische Ei in den geheimen Ofen gesetzt hast, gibst du ein angemessenes Feuer. Regiere das Feuer so, dass die äußere Wärme die innere nicht übersteigt, andernfalls wird sich die Vereinigung der Materien nicht vollziehen können und infolgedessen eine zu große Wärme die Materien zerstreuen und verbrennen.

Wenn aber dieses äußere Feuer schwächer ist, als es sein sollte, dann wird sich der Geist der Materien nicht bewegen und mit seiner Feuchte nicht wirken können. Er wird auch nicht trocken und fix werden, weil dann die metallischen Geister noch tot, wie betäubt und eingeschlafen sind. Also können sie weder arbeiten noch ihre Werke vollbringen, da sie ja vom lebendigen Feuer nicht erregt und zum Leben erweckt werden können.

Achte also ganz genau auf das Feuer und reguliere es so, dass es weder zu stark noch zu schwach sei. Wisse jedoch: wenn du die Kenntnis unseres mineralischen Feuers nicht besitzt, dass du in die Irre gehen wirst, wie viele andere auch, die blind, ohne den Sinn der Worte zu verstehen, darauf los arbeiten.

Verstehe also, dass wir 3 Feuer haben, ohne deren Kenntnis man in dieser Wissenschaft nichts vollbringen kann:

Das erste ist das äußere Feuer. Man bereitet es entweder mit Kohle oder mit Hilfe einer im geheimen Ofen geschickt und mit Verstand applizierten Öllampe.

Das zweite ist ein mineralisches Feuer, das einen anderen Ort als die Materie einnimmt und Anteil hat an der Natur des *Sulphurs*. Dieses digeriert, attrahiert, penetriert und fixiert alle metallischen Geister mittels seines martialischen Geistes, weil nämlich dieses zweite mineralische Feuer nichts anderes ist, als Eisenfeilspäne, die wir in eine genügend große, irdische Schale geben. In diese Eisenfeilspäne setzen wir das Ei.

Auch das dritte Feuer ist mineralisch, aber innerlich. Es ist im tiefen Zentrum von Gold, Silber und Merkur enthalten. Das sind also die 3 verschiedenen prinzipalen Feuer, deren sich die Philosophen bedienen.

Diese 3 Feuer erwirken die Corruption und Putrefaction unserer Materie bis hin zur Schwärze. Die Schwärze erscheint gewöhnlich nach 4 Tagen. Also beginnt die elementare Qualität der Körper sich zu wandeln und sich in verschiedenen Farben darzustellen. Die trockene Qualität wirkt in die Feuchte, und der Körper beginnt geistig zu werden.

Diese Schwärze und der Farbenwechsel zeigen nichts anderes als die Auflösung des Goldkörpers mittels des Merkur. Durch diese *Solution*, sobald sie abgeschlossen ist, vereinigen sich die 4 Elemente, das heißt, es verbindet sich das Wasser mit der Luft und das Feuer mit der Erde in der richtigen Mischung.

Nach all diesem, nämlich ungefähr 40 Tage, nachdem sich die Feuchte verloren hat und ausgetrocknet ist und sich die verschiedenfarbigen Blumen nicht mehr bilden und vergangen sind, beginnt sich die Materie zu weißen und sich unversehens in das schönste Weiß zu wandeln, welches ungefähr 20 Tage andauert. Dann, in den folgenden 40 Tagen, wird sie auf irgendeine Weise rot wie ein Ziegel.

Zuletzt kommt der Zeitpunkt, den die Natur ihr verordnet und verliehen hat. Während weiterer 40 Tage, mehr oder weniger, bekommt die Materie im Zustand ihrer Vollendung eine Farbe wie Blut und zwar von intensiver Röte.

Achte aber besonders darauf, die Arbeit nicht zu vernachlässigen oder das Werk zu sehr zu beschleunigen, so man es doch in 6 Monaten abschließen kann.

Ab dieser Zeit wird der König deines Glücks erwartet. Denn die ganze Zeit braucht die Materie um langsam auszutrocknen, sich zu vervollkommnen und immer mehr an Kraft und Wirksamkeit zu gewinnen. Wenn du also sehen wirst, dass der König seine hochrote Farbe annimmt und mit Glorie bekleidet ist, so zögere nicht ihn noch 40 Tage in der Wärme zu belassen, so dass das ganze Werk auf einmal seine letzte Vollendung erreicht.

Ich führte diese Operation folgendermaßen durch:

Ich stellte den Glaskolben oder das ‚Ei' am 6. März 1641 in einen *Athanor* auf Lampenfeuer mit einem Docht von 3 Baumwollfäden auf einem Messingdraht. Bis zum 17. März gab es keine sichtbare Veränderung. An diesem Tag erschien ein Rauch, eine Trübung. Danach gab es eine *Confusion* von Dämpfen vielfältiger dunkler Farben, die auf- und niederstiegen. 7 Tage lang blieb es so.

Am 24. März begann eine dunkle, gewissermaßen schwarze Farbe zu erscheinen, die sich bis zu einer Tintenschwärze steigerte, und dies während 10 Tagen.

Am 5. April erschien im Glas ein kleiner Funke, leuchtend wie ein aufgehender Stern. Das währte 8 Tage lang.

Am 13. April durchdrang das Licht die Nebel und verstärkte sich während ungefähr 15 Tagen.

Am 30. April erschien etwas wie ein Feuerregen im Glas; das dauerte 10 Tage lang.

Am 10. Mai zeigte sich eine grünliche Farbe, wie ein Regenbogen, der sich wie eine Schlange im Glase wand, und das 8 Tage hindurch.

Am 18. Mai wechselte die Farbe in ein gelbliches Grün und verblich so während 6 Tagen ohne Veränderungen.

Am 25. Mai färbte sich die Materie am Grunde des Gefäßes rot. Es sah aus wie glühende Kohle.

Am 26. Mai fügte ich noch 2 Messingdrähte hinzu – jeder mit einem Docht von 3 Baumwollfäden ausgestattet. Am Grunde des Gefäßes erschien eine glühende Farbe, während das Übrige von blasser Farbe war. Dies setzte sich 6 Tage lang fort.

Am 6. Juni steigerte sich die Farbe zu einer Röte, dann wurde sie rot wie Blut. 14 Tage lang verblieb es so. Dann schwächte sich diese Farbe ab, so wie sie sich gesteigert hatte und zwar im Verlauf von 8 Tagen.

Am 23. Juni erschien die Materie im Glas wie Wasser, vollkommen blass. Auf- und absteigend blies sie sich auf, am Grunde war sie von dunkler Farbe.

Nun erschien etwas wie rote und gelbe Flecken. Manchmal waren sie groß, manchmal klein. Dies dauerte 15 Tage lang.

Am 19. Juli erschien eine Zitronenfarbe, welche am Grunde mehr orangenfarben war und verblieb 15 Tage lang. Darauf schlug sich am Grunde des Gefäßes ein feuriges, funkelndes Gelb nieder.

Am 8. August fügte ich noch zwei Messingdrähte hinzu, jeder mit einem Docht von 3 Baumwollfäden versehen. Auf diese Weise waren es jetzt 5 Messingdrähte mit 15 Baumwollfäden zum Ansaugen des Öls.

Am 20. August zeigte sich ein Glanz im Gefäß, der wie eine Schlange aussah. Dies dauerte 5 Tage.

Am 26. August zerfiel die ganze Masse am Grunde in Stücke, und dann in goldfarbene Schuppen. Als die Masse zersplitterte, erschien das schönste Zinnober, das ich je gesehen habe.

Gegen den 8. September war alles wie kochendes Blut, das sich gegen den 18. hin zu einer feurigen Farbe veränderte. Jetzt fügte ich noch einen mit 3 Baumwollfäden versehenen Messingdraht hinzu. Das ergab für 3 Wochen ein starkes Feuer, währenddessen man das Glas knarren hörte, als ob es zerbrechen würde: Dies zeigte die Perfektion des Steins an. Sie wurde am 18. Oktober erlangt.

Wenn also dein Werk ausgekocht worden ist, kann es auf folgende Weise an Tugendkraft erhöht und an Qualität vermehrt werden:

Nimm 1 Unze Goldkalk und amalgamiere ihn mit 2 Unzen Deines Merkurs. Wenn das Amalgam fertig ist, füge 1 Unze deines roten Projektionspulvers hinzu und reibe es so lange, bis alles zu einem Körper geworden ist.

nach: >Der Prozess des *Monsieur d'Anvers*<, 17. Jhdt.

Formulierungen wie „der *Mercurius Marte* im *Mercurius* des *Antimoniums* ist nichts anderes als ein Geist oder eine lebendige Luft, bis dass er körperlich zu Gold wird" mögen Manchem klingen wie eine chemische *Yaghan*-Sprache aus Darwinschem Blickwinkel. Ernsthaft den Weg einer ‚Seele' von Gold, Silber, Eisen und anderen Substanzen während eines laborantischen Prozesses zu beschreiben, kann doch nur dunklem Aberglauben entspringen und Hilflosigkeit in der Dokumentation chemischer Prozesse ausdrücken. Muss man nicht annehmen, dass *d'Anvers* gar nicht wusste, was er da so alles zusammenschmolz, triturierte und destillierte? Vielleicht aber wurde während solcher Prozesse Porzellan oder Schießpulver entdeckt, sodass sie uns doch noch zum Segen gerieten... Inzwischen kann man diese langwierige Arbeit wohl sicher schon weit einfacher verrichten, und alle Abläufe sind genauestens in Formeln notiert. Oder man verrichtet sie gar nicht, weil ohnehin nichts dabei ‚herauskommt' außer vielleicht ein Lebenselixir oder Transmutationspulver – Dinge, die keiner brauchen kann.

Wie die *Yaghan*-Sprache eine der ältesten der Welt, so ist die Alchymie eine der ältesten Wissenschaften. Doch bevor beide ausgestorben sind, will man den Wohlklang und die Harmonie ihrer Worte nicht bemerken, weil man gewohnt ist, sich Gründe einzubilden, ‚besser' und ‚wissender' als ‚Wilde' zu sein und sich deshalb im eigenen Aberglauben verkrampft.

Lehmann konnte nicht verstehen, dass „diese Wissenschaft von den Elementen und ihren Wirkungen zur Erkenntnis des Mineralgeistes oder des Samens der Metalle nötig sei." Aus welchem Wissen erklärt *d'Anvers*, dass das Quecksilber zu ‚kalt und erdig' zur Vereinigung mit anderen Metallen ist und folgert daraus, dass es ‚erwärmt' und durch Verbindung mit Silber von seiner übermäßigen Erdhaftung befreit werden musste? Er schreibt vor, das Quecksilber zu ‚beseelen', da es keine Seele besitze. Dieser Gedanke entspringt einem tiefen Verständnis der ‚Elemente und ihrer Wirkungen'. Das Quecksilber ist ein sehr typischer Ausdruck des Merkur-Prinzips. Dieses wiederum entspricht dem ‚Naturgeist' und damit dem ‚Geist/Gemüt' im Menschen, also dem 2. Aspekt des Körper-Geist-Seele-Kontinuums. Übersetzt auf die ‚Prinzipien' der Natur ist das: Sal-Merkur-Sulfur. Während Gold und Eisen typische Ausdrucksformen des Sulfur-Prinzips und damit ‚stark beseelt' sind, vertreten Silber und Quecksilber das merkuriale Spektrum, stehen also dem Seelenhaften in gewisser Weise direkt gegenüber. Während das Silber auf eine, in der Elementegenese ursprüngliche Weise, bereits vom Gold beseelt ist (>vgl.: das Thema der ‚Spiegelung des Sonnenfunkens' beim ‚Eintritt' der Idee/Quintessenz in die Physis), bleibt das Quecksilber in der Natur davon noch unberührt. Das ist der hermetische Grund seiner typischen Signaturen (Beweglichkeit, Sphärenbildung, Unfixheit, Zerstreuen usf.).

Sulfur- und Merkur-Prinzip in der pentagrammatischen Darstellung:

Merkur-Prinzip bzw.
Merkuriales Spektrum

Spiritus, Geist

Planetenkraft Merkur – Quecksilber
Planetenkraft Mond – Silber

Sulfur-Prinzip bzw.
Sulfurisches Spektrum

Anima, Seele

Planetenkraft Sonne – Gold
Planetenkraft Mars – Eisen

D'Anvers empfiehlt, das Quecksilber „gut zu beseelen, und zwar mit einer fixen und vollkommenen Form". Die Seele vermittelt den Körpern Zentriertheit und ermöglicht dadurch erst Formbildung. Das Merkuriale in den natürlichen Dingen besteht immer auch in einer Offenheit, Unfixiertheit und Neutralität.

Diese Neutralität ist dem Naturgeist genauso zu Eigen wie dem menschlichen Geist/Gemüt. Ersterer hält sich stets mit ‚Dingen' verbunden und vermittelt deren Botschaft. Dies ist ‚Merkur, als Grundkraft der Kommunikation'. Ohne die Verbindung mit den ‚Dingen', wäre er ein unsichtbarer und unfühlbarer Geist und würde eben seiner Aufgabe als Natur-Geist nicht entsprechen. Das Gemüt des Menschen ist ebenfalls grundsätzlich neutral. Durch Verbindung mit den Objekten von Wahrnehmung und Handeln wird es ‚beseelt', d. h. die Aufmerksamkeit (Seele) des Menschen wird hineinverwoben.

Dem Spagyriker gilt der Alkohol als prinzipiell neutrale Ausdrucksform des Merkus-Prinzips im Pflanzenreich. Je besser dieser rektifiziert ist, umso neutraler ist er in hermetischem Sinne. Durch die Verbindung dieses Alkohols mit Kräutern oder anderen ‚Dingen' wird er beseelt. So entsteht eine Vermischung von Merkur mit Sulfur, und das alkoholische Mazerat erhält (in hermetischem Sinne) einen spezifischen Charakter, den der alleinige Alkohol zuvor nicht besessen hatte.

Diese Umstände werden aus der Natur selbst erkannt und können auf jeglicher Ebene erfahren bzw. nachvollzogen werden. Sie werden in de, bei den alten Alchymisten beliebten Spruch, „unsere ganze Kunst ist nur Mann und Frau", angedeutet.

Das Spektrum des Merkur-Prinzips ist vor allem von der Position des Wasser-Elements dominiert. Das des Sulfur-Prinzips vom Feuer-Element. Das feurig/sulfurisch/Seelenhafte ist aktiv und richtet sich aus. Das wässrig/merkurial/ Geisthafte ist passiv und zieht das Sulfurische an sich. Deshalb ist die Seele/Aufmerksamkeit des Menschen stets geneigt, sich mit dem Geist/Gemüt zu verbinden und sich darin zu verlieren, aufgrund der sich ausrichtenden Dynamik der Seelenkraft und der ansaugenden Passivität des Naturgeistes. Diese unterschiedlichen Qualitäten der beiden ‚Pole' sind der Grund für die Verbindung von Mann und Frau in allen Naturreichen.

Wir werden noch weiter mit diesem Thema befasst sein. Hier dient es uns vorerst zu zeigen, dass sehr wohl ein hermetisches Naturverständnis notwendig ist, damit die Arbeiten der ‚einfachen' Spagyrik genauso wie die der ‚hohen' Alchymie richtig ausgeführt werden können.

So führt *d'Anvers* zum Beispiel weiter aus: „Da (das Quecksilber) diese (Seele) nicht besitzt, empfängt er (*Mercurius vivus/* Quecksilber) sie von anderen Metallen, ähnlich wie das Wachs den Stempeleindruck aufnimmt."

Es ist zwar das Wachs als Substanz typisch sulfurisch, vergleichbar den Fetten, Harzen und Ölen, es ist aber auch relativ merkurial in seiner ‚Neutralität', solange es noch nicht durch Formen, Prägen, Zeichnen u. dgl. ‚individualisiert' wurde. Es ist stets bereit, Informationen aufzunehmen und zu vermitteln.

Weiter (nach) *d'Anvers:* „Das Quecksilber empfängt auf eine ähnliche Weise die Seele von anderen Metallen, wie die Erde den Regen erhält und sich mit Tugend und Kraft des Wassers anreichert, um ihre Früchte hervorzubringen."

Insbesondere das Frühlingsregenwasser galt den Alten als ‚vom himmlischen Feuer geschwängert'. Es vermittelt der Erde eine Quintessenz dessen, was im Herbst ‚aufgestiegen' und an der Äther-Schwelle sublimiert ist und es dient weiters als Medium für neue Quintessenz, die sich in der Physis (in der Erde) ausdrücken will und von ‚drüben', aus dem Feinstofflichen, hereinströmt. Relativ zu diesem Wasser ist die Erde merkurial, also noch neutral und unspezifisch. Auch wenn Wasser an sich grundsätzlich eine merkuriale Ausdrucksform ist, so wurde doch jenes Regenwasser bereits vom ‚ätherischen Feuer' beseelt und ist durchdrungen von Individual-Impulsen. Es besteht in der Natur sehr wohl ein Unterschied, mit welchem Wasser die Erde befeuchtet wird und wieweit es von ‚ätherischen Ideen' geschwängert ist, welche imstande sind, das Hervorbringen von Früchten aus der Erde zu initiieren. In *d'Anvers* Gleichnis wirkt das edelste Frühlingsregenwasser auf die Erde (Quecksilber), nämlich ‚Gold und Silber'. Die Verbindung der Erde mit jenem Wasser gewährleistet das Hervorbringen der besten, schönsten und zahlreichsten Früchte. Letztendlich bemerkt *d'Anvers* einfach: „Das Quecksilber vereinigt sich mit den anderen Metallen (Gold und Silber) auf natürliche Weise wie Mann und Frau und empfängt deshalb wirklich die Qualitäten (...)".

So gibt uns der obige Text viele Beispiele dazu, dass die alchymistische Handwerkskunst stets einhergeht mit jenem Naturverständnis, das ‚aus dem Lichte der Natur' stammt und nicht ‚erdacht' ist.

Gold und Silber –
das ‚Ferment' gibt seine Seele

D'Anvers hilft dem Mitdenker, wenn er anmerkt, dass „der Merkur des Eisens im Merkur des Antimons nichts anderes ist als ein Geist (also eine sehr wesenhafte Ausdrucksform des Merkur-Prinzips) oder eine ‚lebendige Luft', solange, bis dieser Geist körperlich wird, das heißt: zu Gold wird. Alsdann (dadurch) wird er vom Merkur des Antimons separiert." Der Merkur des Eisens ist als ‚Merkur' soweit neutral, dass er zuerst eins ist mit dem (ebenfalls relativ neutralen) Merkur des Antimons. Das Eisen ist ein Metall des ‚unteren' Feuer-Elements aus dem Grunde, weil ihm ein wesentlicher Grad an Seelenhaftigkeit fehlt, der dem Gold noch zu Eigen ist, weswegen das Gold das Metall des ‚oberen' Feuers ist. Die *Mercuriis* der Metalle verbinden sich mit ihresgleichen und man erhält einen ‚dreifachen Merkur', welcher umso wirksamer wird, je weiter die Materie, woran er seiner Art gemäß haftet, gereinigt werden kann. Im Stadium der vollkommensten Reinigung bzw. Erhöhung der Materie spricht *d'Anvers* dann vom ‚philosophischen Merkur', von dem aus er auf drei Wegen zum ‚Stein der Weisen' weiterarbeitet.

Für manche alchymistische Handwerker stellt dieser Text eine der klarsten Anleitungen zu jenen Arbeiten dar. Ob damit jemals einer ans ‚höchste' Ziel gekommen ist? Manche haben Jahrzehnte damit verbracht und alle Phänomene aus *d'Anvers* Tagebuch bestätigt – bis auf das allerletzte. Es schien nicht zu gelingen, die Materie zum Stein zu figieren. Vielleicht sind hier wieder nur die Termini zu hoch gegriffen, wie dies in vielen ähnlichen Texten aus jener Zeit üblich war. *Agricola* bemerkt dazu: „(...) und dass der selbe der *Mercurius Philosophorum* sein soll, wie ihnen etliche einbilden, ist nicht, denn wie schon oft gedacht, ist der selbe viel ein ander Ding." Es will ihn auch *Paracelsus secundum litteram* nicht verstanden haben.

„Wenn man den *Mercurium vivum* (Quecksilber – Anmk.) mit der *Minera Antimonii*, oder vielmehr die *Mineram* mit dem *Mercurio* solviert, so steigt zwar eine Röte mit auf als ein Zinnober, es incorporiert sich aber nicht fein miteinander. Wenn man aber den *Sulphur Antimonii* erstlichen herausziehet und procedieret, wie man sonsten den Zinnober zu machen pfleget, so wird das Sublimat schön rot und lustig aussehen. Und je öfter es sublimiert wird, je schöner wird er, wie er denn etlichen also hell als ein Rubin wird. Und ist in dieser dunklen Description anders nichts, als das *Magisterium Antimonii et Mercurii*, ..."

Die Coagulation fasst er kürzer: „Wenn man den *Mercurium sublimatum* mit dem *Antimonii*-Erz wohl untereinanderreibet und lässets acht Tage in der Kälte stehen, so solviert es sich miteinander und wird ein fetter *Liquor* daraus. Danach tut mans in eine Retorten, hernach destilliert, so bekömmt man ein schönes gelbes Öl – und dieses ist auch *Paracelsi* Meinung –, und je öfter es rectificiert wird, je schöner wird es. Und wenn man *Laminas Lunae* dareinlegt, so werden sie schwarz.

Man muss sie eine geraume Zeit darinnen liegen lassen, bis sie mürbe werden, und solches in einer steten Wärme, alsdenn reduciert und geschieden, so bekömmt man eine *Lunam fixam*. Es wird aber nicht viel, doch es wird eine wahrhaftige *Luna fixa* oder Weißgold."

Agricola scheint also nicht der Ansicht gewesen zu sein, dass aus diesen Arbeiten mit Antimon und Quecksilber tatsächlich der ‚*Mercurius Philosophorum*' oder der ‚Stein der Weisen' erlangt werden könne und richtet sich damit als erfahrener Praktiker gegen eine diesbezügliche Deutung der Schriften *Philaletas*, *Libavius* (und in der Folge auch *Starkey's*, *Newtons* und späterer Alchymisten). Nichtsdestotrotz hatte er „viel in dieser *Materiam* gearbeitet und daraus auch einige sonderliche Stücke erlangt", was sich sowohl auf kleinere Partikulare als auch auf sehr gute Arzneien bezieht. Dass „der *Mercurius Philosophorum* ein ganz anderes Ding sei", deutet er immer wieder an. Zur Bestätigung seiner Meinung führt er neben eigener praktischer Erfahrung die Schriften von *Quercetanus*, *Beguinus*, *Croll* und *Hartmann* an, erklärt aber: „(...) denn ich nicht gesonnen, alle *Arcana* in diesem Buch zu offenbaren."

Er setzt auch dazu die Bemerkung des *Paracelsus*, die er quasi direkt aus zweiter Hand von dessen Schüler *Ägidius von der Wiesen* in Oberösterreich, mit dem er gut Freund geworden war, erfahren konnte. Dieser *Ägidius* erzählte ihm von seinem Abschied aus der Lehrzeit bei *Paracelsus*: „(Er) lehrete mich daneben das *Opodeltoch* recht zuzurichten, und sagte, damit könnte ich mich reichlich ernähren. Und befahl mir daneben, ich sollte mir diese beiden Stücke, den *Mercurium* und *Antimonium* befohlen sein lassen, denn wenn ich recht damit umgehen würde, so könnte ich Zeit meines Lebens genug daran haben."

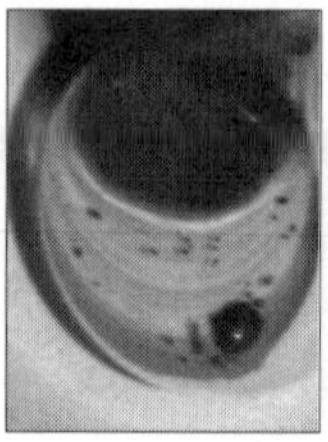
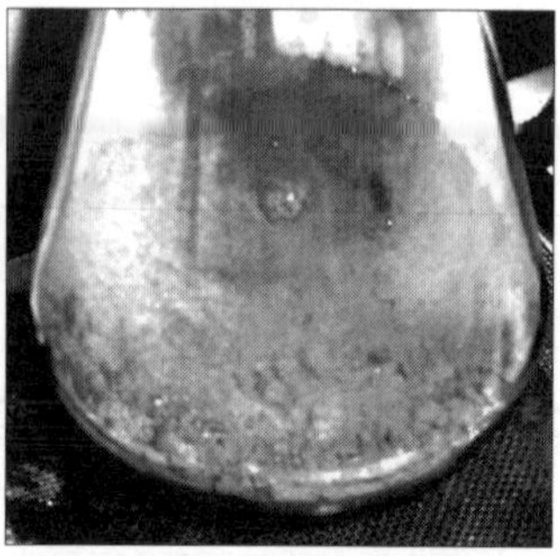

Einen Überblick über das ‚Große Werk' gibt uns der französische Alchymist *Jean d'Espagnet* und weist darin insgesamt auf das Wesen spagyrisch-alchymistischer Arbeit hin:

„Der ganze Prozess des philosophischen Werkes ist nichts anderes als Auflösen und wieder hart machen. Nämlich Auflösen den Körper und ‚hart machen', im Sinne von ‚zusammenführen', congelieren, den Geist. Und beides geschieht in einer Arbeit.

Der Stein der Weisen wird nach dem Vorbild der Schöpfung der Welt verfertigt.

Die Natur verfährt in der Bereitung ihrer Werke so, dass sie ihre Sachen von der anfangenden Generation durch unterschiedliche Stufen oder *Gradus* zur Vollkommenheit bringt. Deswegen erreicht sie ihr Ziel allgemach und nicht durch einen Sprung. Die philosophische Praktik, welche der Natur nachahmender Affe ist, soll in der Handlung ihres Werkes vom Weg und Vorbild der Natur nicht abweichen."

„Auf die Frage nun, wer denn in dieser oder jener Region oder Kultur der ‚Erste' war, der ein entsprechendes Wissen einführte, verweisen die Kundigen auf mythische Weise der Vorzeit und entrücken damit jene Frage dem formalen Intellekt. Sie antworten mit ehrfürchtigen Berichten – aber wohl auch mit einem gewissen Lächeln, so als würden sie gefragt: „Wie alt ist denn das Zeitlose?" (...)

Die taoistische Tradition beantwortet die Frage „woher?" mit dem Ur-Schamanen der Vorzeit *Fuxi*. (...) Wenn auch später der Konfuzianismus das Wesen *Fuxi* weitgehendst entmythologisierte und ihn zum ersten Kaiser machte (welcher von 2852–2737 v. Chr. das chinesische Reich regiert haben soll), erzählte man sich doch weiter von ihm, dass er ‚den Himmel fortsetzte' und ‚das Licht von Sonne und Mond nachbildete', ja er selbst wurde als eine solche Nachbildung verstanden. Seine Mutter ‚Blütenall' *(Huaxu)*, welche gleichermaßen den Sternenhimmel wie auch die irdische Materie personifizierte, soll nach den Phasen des Mondes gewandelt sein, wie in den Fußspuren eines ‚großen Menschen' *(Daren)*. Als sie dabei in *Chengji*, dem Ort des ‚Fertigstellens des Werkes' angekommen war, gebar sie *Fuxi*.

Die Idee des ‚Nachahmens der Himmelswege' und deren Vereinigung mit dem Erdgeschehen durchdringt die ‚Begründer' der Künste und Wissenschaften, wie auch deren Nachfolger, zu allen Zeiten und in den meisten Traditionen, ob wir dabei an die göttlichen *Asvins* der Veden denken, oder an die mannigfaltigen Darstellungen merkurialer Wesen der mittelalterlichen Alchymie. (...)"

nach: > Der Weg des Sonnenfunkens<

Weiter *d'Espagnet:*

> „Die Ordnung der Natur erfordert, dass eine jedwede Generation im Feuchten anfängt. Im philosophischen Werk soll man die Natur dergestalt in diese Ordnung bringen, dass die Materie des Steins, welche irdisch, compact und trocken ist, in ihr erstes Element, das Wasser, aufgelöst wird und zerfließe. Dann wird aus der Sonne der *Saturnus* geboren."

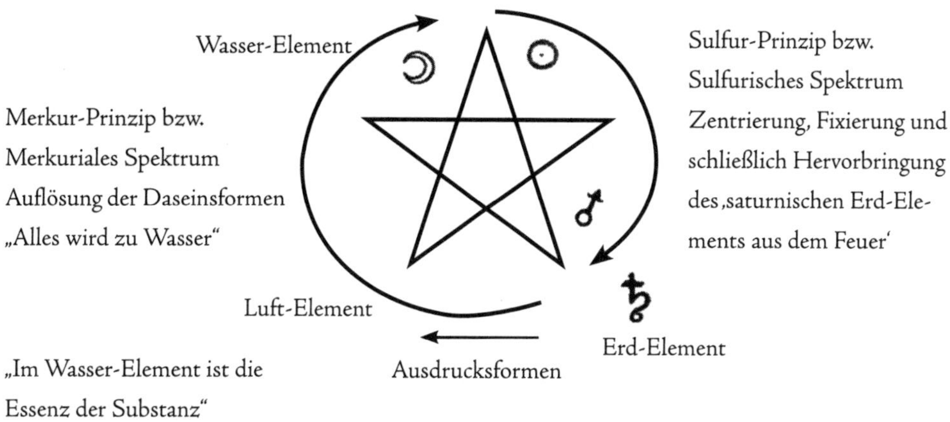

Aus dem Erd-Element heraus und ins Luft-Element hinein werden die Ausdrucks- bzw. Daseinsformen gebildet. Im Wasser-Element erfüllt sich deren Schicksal der Vergänglichkeit. Durch das ‚obere' bzw. ‚mondenhafte Wasser' öffnet sich der grobstoffliche, 4-elementare Kosmos wieder in den subtilen, ätherischen Raum und darüberhinaus dann in den feinstofflichen Kosmos. Dieser Position entsprechen die Prozesse des Welkens, Faulens und Verwesens sowie die Putrefaction. Gleichzeitig wird aber hier auch neue Stofflichkeit vorbereitet. Die Alchymisten sprechen von einer ‚Generation der Salien', welche hier stattfindet. In den Lehren der taoistisch-chinesischen Medizin heißt es: „Im Wasser-Element befindet sich die Essenz der Substanz".

Es ist also die Natur solcherart, dass direkt aus dem Prozess des Vergehens ein Werden geboren wird. Darauf bezieht sich *d'Espagnet* mit der Bemerkung, dass jedwede Generation im Feuchten anfängt, genauso wie *Thales* das Wasser als den Ursprung allen Lebens angibt. Die ‚irdische, compacte und trockene Materie des Steins' befindet sich im ‚Saturnzustand' und muss – der Natur nach – ‚in ihr erstes Element, das Wasser, aufgelöst werden', was in der alchymistischen Praxis auch ‚schmelzen' bedeuten kann (‚zerfließen').

Auflösung (oder Schmelzung) –
der ‚erste Schlüssel des Werks'

„Der Mittel, so zur Arbeit gehören, also der ‚Regierungen' oder der ‚Schlüssel des Werks' sind vier:
Der erste Schlüssel ist die Auflösung oder Schmelzung. Der zweite die Abwaschung. Der dritte die Wiederbringung oder *Reductio* und der vierte die Beständigmachung oder *Fixatio*."

Auf einer anderen Ebene – nämlich in der spagyrischen Pflanzenarbeit – folgt man demselben Weg der Natur. Der erste Schritt zur spagyrischen Essenz oder Quintessenz (im allgemeinen) ist die Auflösung. Sie geschieht aber hier tatsächlich in einem feuchten Medium und nicht vermittels Schmelzung. Die Pflanzen werden zur Gärung oder (sulfurischen) Destillation mit Wasser übergossen oder putreficieren in ihrem eigenen Saft.

Die ‚Abwaschung' (welche bei *Monsieur d'Anvers* eine tatsächliche Waschung des Amalgams und Abscheiden der Schwärze ist) entspricht der ‚philosophischen' Reinigung der Prinzipien durch Abscheiden von den festen Pflanzenteilen (bzw. in der Folge vom *Caput mortuum*) in der Regel durch Destillation und eben auch durch ‚Waschung'. Auf diese Weise wird aus der Pflanze die flüssige bzw. ölige Ausdrucksform der Prinzipien Merkur und Sulfur und dann das Erd-Salz dargestellt. ‚Die irdischen Schlacken' werden also weitestgehend von der Idee, der Kraft oder dem Prinzip ‚abgewaschen'.

Der Wiederbringung oder *Reductio* entsprechen in der Pflanzenarbeit die Rectificationen. Dadurch werden die aus dem Pflanzenkörper geschiedenen, befreiten ‚Kräfte' gewissermaßen konzentriert in die Physis ‚hereingebracht', indem man sie klärt, von *Phlegma* und *Feces* befreit und an das nunmehr subtile fluidale bzw. salinische Medium (ggf. Öl, Alkohol und Salz) bindet.

Die Beständigmachung oder Fixation entspricht in der Pflanzenarbeit der Coagulation bzw. (im Falle der Darstellung eines Pflanzensteines oder einer figierten Quintessenz) tatsächlich einer Fixation.

Weiter *d'Espagnet:*

> „Durch die Schmelzung fließen die Körper wieder zurück in die Erste Materie. Dasjenige, was zeitig, reif und gekocht ist, wird wieder roh gemacht."

Wenn die Ausdrucksformen in der Natur ‚zeitig' und ‚reif' werden, dann setzt für sie endgültig das Vergehen ein. Bis zur Reife kann man noch von Wachsen und Werden sprechen und das letztendliche Dahinsterben und Vergehen außer acht lassen. Danach folgt unweigerlich der körperliche Verfall. Schon der Geschmack der ‚überreifen' Frucht signiert die beginnende Dominanz des Wasser-Elements und den Anfang der letzten Lebensphase des Pflanzenwesens in seinem momentanen Körper.

Der Natur abgeschaut, vollzieht der Alchymist diese Phase in kurzer Zeit im Mikrokosmos seines Labors. Allegorisch bedeutet das ‚die Tötung des alten Königs'. Wir werden diesem Begriff später noch begegnen. Man muss dabei nicht annehmen, dass hier einfach ‚etwas nachgemacht' wird, was man in der Natur und im Leben der Menschen in zahlreichen Varianten beobachtet, sondern es ist viel mehr:

Der initiatische Charakter der Alchymie

Die ‚irdische, compacte Materie' ist nichts anderes als eine quintessentielle Idee, welche durch das sulfurische Spektrum hindurch verdichtet worden und schließlich Körper, also saturnische Substanz geworden ist. Vermittels dieses Körper kann sich diese Idee nun in der Stoffwelt ausdrücken, also ihren innersten Merkur offenbaren. Quintessenz, Idee und *Mercurius* sind bisweilen gleichbedeutende Begriffe. Das ist, weil ja jede einzelne Idee oder Quintessenz in der Stoffwelt einer eigenen Ausdrucksart bedarf, um sich kundzutun und so zu manifestieren.

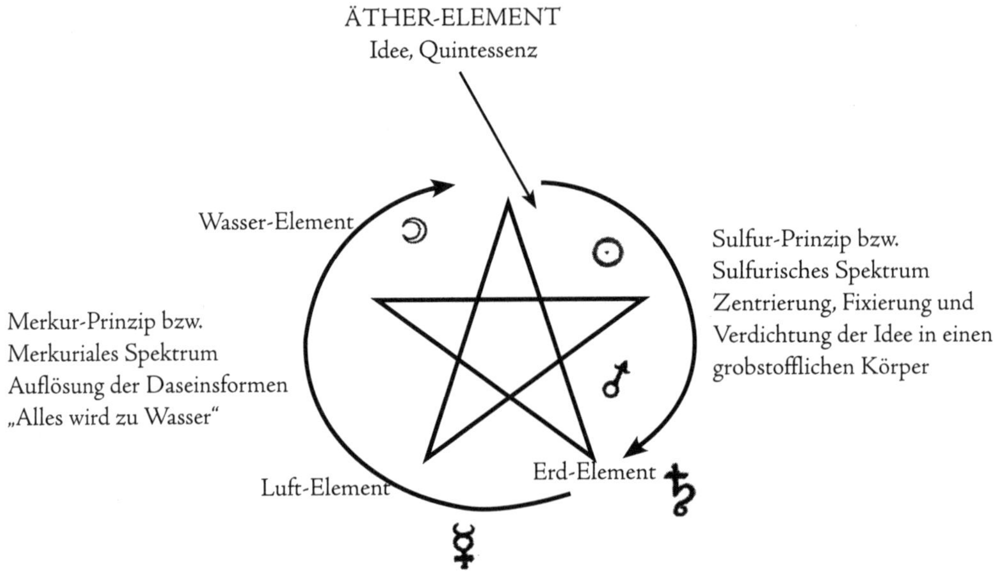

Sobald nun eine Idee in der Physis angelangt und verkörpert oder manifestiert ist, hat dieser Körper merkurialen Charakter – das heißt, er drückt etwas aus, das auf das Wesen der Idee hinweist. Die Idee einer Rose etwa wird sich als Rose zeigen und nicht als Pappel oder Quarz. Die Substanz der Rose ist insofern ‚saturnisch', als dass die Idee in dieser Substanz ‚angekommen' ist. Die körperlose Dynamik des Werdens, des Ein-Zentrierens und Verdichtens der Idee hat hier ein Ende. Zumindest auf diese Weise. Die Idee ist nicht mehr geistig, sondern ihre Geistigkeit ist nun mit grobstofflicher Substanz vermischt.

Was hier im Beispiel die Rose betrifft, so stammt die Essenz ihrer Substanz, wie alle Substanz-Essenzen aus dem Wasser-Element und ist dort noch nicht als Substanz der Rose ersichtlich. Sie ist nämlich dort nur zum allergeringsten Teil eine ‚Substanz der Rose', sondern vielmehr eine neutrale, merkuriale Substanz, aus der – bei entsprechender Imprägnierung durch die in das Äther-Element eintretende Idee – genau jene Aspekte essentieller Substanz ‚saturnisch' werden, welche die Rose zu ihrer spezifischen Ausdrucksform benötigt.

Saturn wird ‚der Hüter der Schwelle' genannt. Das saturnische Erd-Element' stellt jene Schwelle dar, welche die Idee überschreiten muss, um stofflich oder körperlich ‚anzukommen'. Beim Überschreiten dieser Schwelle ereignen sich zwei Dinge:

Erstens erreicht das Werden einen Höhepunkt und schlägt von da an ins Vergehen um. Und zweitens zieht die vordem geistige Idee materielle Substanz mit sich – gewissermaßen aus der saturnischen Schwelle heraus. So kommt es schließlich zu den Ausdrucksformen, die wir sinnlich erfahren können. In der chinesischen Medizinlehre heißt es: Erde bringt Metall (Luft-Element) hervor. Dieses Hervorkommen oder Geborenwerden aus dem Erd-Element ist das in Erscheinungtreten der Ausdrucksformen.

Beinahe allegorisch mutet die signaturkundliche Bedeutung der Gräser an: In den meisten, insbesondere in den Ried- und Sauergräsern, dominiert Merkur. Wenn man sich nun das saturnische Erd-Element als kalten, trockenen, starren, grauen Fels vorstellt, dann ist die Geburt des Luft-Elements das, wenn auf diesem Felsen Gräser zu sprießen beginnen.

Vorweg: Weit näher als in unseren Landschaften empfindet man auf vulkanischen Inseln die Geburt saturnischer Erde aus dem feurigen Spektrum. Wenn der Boden dann erst feucht und fruchtbar wird, ist – hermetisch gesehen – Mond zum Saturn getreten und hat das saturnische in venerisches Erd-Element verwandelt. Die Schwelle Saturns wurde überschritten. Jedenfalls tritt nun mit dem Grün der Pflanzenwelt aus dem Mineralhaften das Merkuriale hervor.

Wenn die Gräser aus dem kargen Boden sprießen, dann ist das wie die Geburt des Luft-Elements aus der Erde. Es ist auch wie die Bildung der Lunge im Erd-Element und die daraus folgernde Atmung als eine Äußerung des Luft-Elements. Die Lunge selbst ist fest und knorpelig, mit einer nach unten weisenden Schwere-Signatur. Sie ist wie der graue Fels. Das, was sie ermöglicht, was ihre Folge ist, das ist Luft-Element, oder mit anderen Worten: der Beginn des merkurialen Spektrums, des Austauschs, der Sprache, der Kommunikation usf.

Es heißt, die Körperbehaarung signiere die Lungenfunktionen. Das ist genauso zu verstehen. Die Körperhaare sind das Gras – der Körper der Fels. Das, was aus dem Fels heraus, aus dem Körper durch die Haut dringt, die Farben, der Schweiß, der Geruch, das teilt etwas über Zustand und Wirken der Kräfte darinnen, über die im Fels eingeschlossene Idee, mit. Es ist merkurial.

Sämtliche Ausdrucksformen sind prinzipiell von merkurialer Art. Deswegen nennt man Merkur auch den ‚Naturgeist'.

Mit dem Hervortreten dieses Merkurialen aus der Erde, mit dem Überschreiten der quintessentiellen Idee über die Schwelle des Saturn und damit mit der Stoffwerdung ist aber das vordem Geistige nunmehr mit Stoff verbunden und vermischt. Das heißt: Saturn bestimmt sein weiteres Sein. Saturn wird als ‚Knochenmann' dargestellt, oft mit Sense und Sanduhr. Als Hüter der Schwelle ist er auch Hüter der Zeit. Die körperlich geborenen ‚Ideen' unterliegen nun der Herrschaft der Zeit. Man sagt: Der quintessentielle Merkur wird von Saturn eingekerkert.

Im grobstofflichen Körper kann die Idee nicht ohne die Gesellschaft Saturns sein. Und weil es die Art der Zeit ist, ‚zu vergehen', färbt diese Art auch auf den Körper ab – der ja von der Zeit gefangen ist – und macht den Körper ebenfalls vergänglich. Vom Moment des Passierens der Schwelle an erfährt die Idee ein ständiges Vergehen der sie begleitenden und einschließenden Substanz. Der Samen vergeht in der Erde, der junge Trieb vergeht, die Frühlingsblüte vergeht, schließlich wird die Frucht reif und zeitig und – ‚alles wird zu Wasser'.

„Die Körper fließen zurück in ihre Erste Materie. Das was zeitig und reif ist, wird wieder roh gemacht." So *d'Espagnet.*

Dasselbige ereignet sich in der Natur während der jeweiligen Zeit. Manche Ideen leben lange in ihren körperlichen Ausdrucksformen, andere kürzer. Hier geht der Alchymist nun heran und vollzieht diesen Prozess des körperlichen Vergehens in relativ kurzer Zeit, (besonders wenn er mit mineralischen Substanzen arbeitet). Er ‚tötet' ihre Ausdrucksform auf solche Art, dass die Idee darin nicht beschädigt, sondern befreit wird und fasst die Idee in eine, durch die ‚Kunst' aus der Grundsubstanz herausgearbeitete, stoffliche Hülle, vom selben Körper stammend, aber essentiell, konzentriert und höchst gereinigt.

Um eine spagyrische Pflanzenessenz zu erlangen, lässt der Spagyriker die grobstoffliche, von ‚saturnischen Schlacken' begleitete Ausdrucksform der Pflanze im kochenden Wasser, im Alkohol oder im Gärgefäß ‚sterben', fängt aber ihre freiwerdenden Geister in den, ihnen entsprechenden, gereinigten, verfeinerten Substanzen auf und stellt sie vereinigt und essentiell ‚wiedergeboren' als Arznei zur Verfügung.

Zur Kunst werden durchwegs von der Natur bereits ‚fertiggekochte Dinge' verwendet – nämlich ‚Ausdrucksformen'. Die Kunst greift sich nicht zuerst die Idee des Apfels, sondern zuerst den Apfel. *D'Anvers* beginnt seine Arbeit nicht mit Geist, Kraft oder Idee des Antimons, sondern mit dem Ding selbst. Aller Werke Anfang ist: „Mache das Fixe flüchtig". Der Ausgangsstoff zu einer Veilchen-Essenz ist das Veilchen, so wie es die Natur bereits hervorgebracht hat. Und stets wird dann diese erste Ausdrucksform der Dinge auf eine Weise ‚zerstört', welche Zugang schafft zu den subtileren Ausdrucksformen der am Sein der Dinge beteiligten Kräfte bzw. zur ‚Idee'.

In der hermetischen Allegorie ist diese erste Ausdrucksform der ‚alte König'. Es ist die Individualität, welche während der Hereinbringung der Idee in die Physis zusammen mit dem Körper auftritt. Diese gewisse Individualität hat jedes Geschöpf, jede Pflanze, jeder Stein, was zu der Bemerkung veranlasst, dass ‚kein Ei dem anderen gleicht'. Es ist ein unendliches Differenzieren aus dem feinstofflichen Urgrund heraus, vermittels des sulfurischen Spektrums in den saturnischen Zustand hinein, und hindurch in die merkurialen Ausdrucksformen.

Undifferenzierter Äther – differenzierte Ausdrucksformen

Das Äther-Element ist umfassend, alles durchdringend und neutral. Es wird mit der Oberfläche eines unendlich großen Meeres verglichen. Die hawaiianischen Schamanen setzen den Begriff ‚Familie' in die Position jener fünften elementaren Naturkraft. Die tibetischen Medizinmönche sagen: „Äther hält alles zusammen."

Im Äther herrscht Undifferenziertheit vor. ‚Raum' selbst ist nicht unterschieden, er bildet nur den Grund, den ‚Platz', für alle differenzierten, unterschiedlichen Ausdrucksformen darin. ‚Äther' ist in den Formen das ‚Gefühl' des Einsseins.

Wenn sich die quintessentielle Idee im Raum verdichtet, dann entspricht das einem ‚Herausfallen aus dem Einssein' (auf seiner Ebene). Die erdhafte Manifestation der Idee ist immer ‚nur' ein Teil des Ganzen. Der Äther mag ein Meer aller Quintessenzen sein, wird an sich selbst auch Quintessenz genannt, oder ‚fünftes Element' – diejenigen Ideen, welche sich aber vermittels des sulfurischen Spektrums durch saturnische, grobstoffliche Art manifestieren, sind stets voneinander getrennte Einzelteile in diesem Meer. Unter unzähligen dieser Einzelteile gibt es nicht zwei gleiche. Unter dutzenden Pappeln oder Birken einer Allee hat jeder Baum für sich eine gewisse Individualität. Jedes Blatt, jeder Kiesel und jedes Sandkorn hat seinen eigenen Weg, seine eigene Geschichte und seine Individualität. Es kann zwar in der Natur große Ähnlichkeit geben und über lange Zeiträume Gleichsamkeit, Verbindung, Verwandtschaft – aber es gilt niemals für ein (noch so kleines) ‚natürliches Ding' genau das selbe wie für ein anderes.

Deshalb ist das Dasein ‚im Mantel des Saturn', ‚im Kerker des Saturn', also im grobstofflichen Kleide, immer mit einem Gefühl des Getrenntseins verbunden. Auf einer hohen Ebene ist eben dieses Gefühl im Menschen ein Grund der Spiritualität, welche der Sehnsucht nach dem ‚großen Einssein' folgt. Auch in den ‚natürlichen Dingen' gibt es das, was wir hier ‚Gefühl' nennen, als eine Art Sehnsucht, mit allem eins zu sein, ohne Begrenzungen.

Sobald die Idee in der Physis angekommen ist, erfährt sie dieses ‚Getrenntsein' gleichzeitig mit der Freude der körperlichen Manifestation. Das heißt: der Kieselstein birgt in sich eine unvorstellbar große Freude, dass er nicht mehr nur Idee zum Kieselstein ist, sondern diesen nun auch körperlich darstellt. Diese Freude entspricht dem inneren Feuer, der Individualität, dem Sulfur in ihm. Andererseits erfährt er durch seine Situation ein relatives Getrenntsein von allen anderen Ideen – er leidet aufgrund der körperlichen Beschränkung im ‚Gefängnis des Saturn'. Dieses Leiden entspricht Merkur in ihm, bzw.: Es ist Merkur, welcher leidet, nämlich jener Aspekt Merkurs, der zuvor noch quintessentieller Geist war. Das ist damit gemeint, wenn man sagt: Merkur ist immer stoffhaftend. Es ist nicht sein Wesen – aber seine Natur. Merkur – Naturgeist – ist stoffgebundene Idee. Anders ausgedrückt: Merkur ist sulfurischer Geist, an Substanz gebunden.

Der Sulfur, die Freude, wird zur Individualität. Im Menschen wird es zu dem, was die indischen Alchymisten *Ahankara* nennen, das Ego, der Trennungsimpuls. Daraus folgt, was die ‚Alten' den ‚Fall des Menschen in die Stoffeswelt' genannt haben. Die sulfurische Seelenkraft geht im Naturgeist auf, verliert sich selbst oder wird in ihrem freien, aktiven Wesen behindert. In den Pflanzen, Tieren und Mineralen ist es nicht eigentlich *Ahankara* im Sinne von Ego, sondern eben das, was wir hier mangels besserer Kommunikation ‚Individualität' nennen. Dort ist es der noch schlafende oder verträumte Ich-Impuls, das ‚Elemente-Ich'.

Die Freude wird dann zum Leid, wenn dieses ‚Feuer' erkennt, dass es im saturnischen Zustand nicht mehr freier Geist ist, nicht wirken kann und nicht wesenhaft eins sein kann mit allen anderen ‚Ideen'. *Philaletha* spricht in diesem Zusammenhang von einem ‚passiven *Sulphur*, welcher im *Mercurius* ist, aber eigentlich wirksam sein sollte (...).' „Das Leben liegt wie tod in seinem eigenen Zentrum verborgen." *(Philaletha, >Introitus Apertus<*, Kap. 11). *Pernety* sagt dazu: „Der Körper ist das Prinzip der Fixheit, und er entzieht den andern beiden Prinzipien (Geist und Seele) die Flüchtigkeit."

Durch diese Situation erhält der konkretisierende, verdichtende, sulfurische Impuls eine der Zentrierung, Verdichtung und Substanzwerdung entgegenwirkende Tendenz. Das bedeutet einerseits – wie wir schon angedeutet haben – einen Aspekt der Geburt des Luft-Elements aus dem Erd-Element. Die chinesische (taoistische) Medizinlehre spricht dabei von der typisch ‚zerstreuenden, das Verfestigte auflösenden' Art des Luft-Elements (Metall). Andererseits stellt dieses Entgegenwirken einen Ausdruck jener Kraft dar, welche die ‚Alten' als *ios* bezeichnet haben. Bei den griechischen Alchymisten war *ios* ein spezieller technischer Ausdruck für jene Kraft, die im Halm wirkt, also im Gras- oder Getreidehalm genauso wie im Baumstamm. Genau dieser hier beschriebenen Situation entsprechend, bewirkt das *ios* das Zerspringen des Samenkorns und das ‚die Erde verlassen', das ‚Zurückstreben in die Weite, in die Einheit des ätherischen Raums', steht also dem sulfurischen Verdichten und dem saturnischen Einschließen und Begrenzen entgegen, obwohl es eigentlich ebenfalls eine sulfurische Kraft ist. Es ist aber eine bereits verwandelte Sulfur-Kraft, verwandelt durch das Erleiden der Begrenztheit und Trennung im Saturnzustand. *Ios*, oder *vir*, ist Virulenz, welche sich zB. bei Metallen während der Oxydation äußert. Das Verrosten von Eisen ist ein Ausdruck davon. Dabei bilden sich – aus hermetischer Sicht – ‚Blumen' am ‚Halm des Metalls', analog den ‚Blüten am Halm der Pflanze', welche einen Höhepunkt und Erfolg der *ios*-Wirkung verkünden und die Pflanzen-Idee in der Blüte gleichsam wieder mit dem Äther vereinen. Die indische Tradition nennt diese Kraft *virya*. Sie wirkt dem ‚Fall in die Materie' entgegen und hebt die Dominanz der Substanz auf, indem sie den ‚philosophischen Tod' einleitet und ist gleichbedeutend mit der ‚Sehnsucht nach Initiation'.

Zuerst steht das *ios* am nächsten zur Planetenkraft Mars, aus deren Position ja auch das saturnische Erd-Element hervorgeht. Deshalb hielten die Alten den ‚Sulfur des Mars' (zB. des Eisens) für am besten geeignet, um den ‚passiven, toten Sulfur' der substanzverbundenen Idee zu aktivieren, wieder ‚aufzurichten wie den Halm'. Im Prozess des *Monsieur d'Anvers* kommen diese Zusammenhänge folgerichtig hervor. Dort wird die Seele des Antimons mittels des Eisen-Sulfurs ‚aufgerichtet'. *Jakob Böhme* bemerkt, dass man mit der Farbe (Sulfur) des Mars Gold erlangen könne, wenn man sie aus dem Eisen extrahiere. *Philaleta* meint: ‚Unser Stahl ist der wahre Schlüssel des Werkes, ohne den es völlig nutzlos ist, den Ofen anzuzünden' *(Introitus Apertus)*.

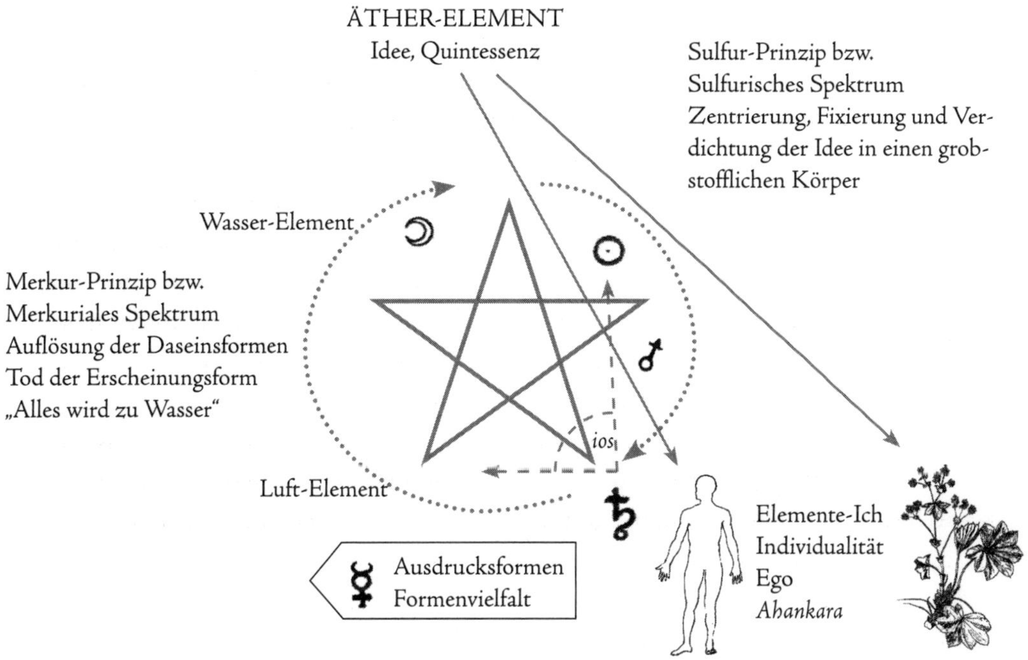

Obgleich dieses *ios* direkt aus dem Sulfur kommt, ist es schon auch in gewisser Weise ‚Merkur'. Das ist jener Merkur, den *d'Anvers* im obigen Text als ‚nichts anderes als ein reines Feuer' bezeichnet. Das ist die Schwelle vom sulfurischen ins merkuriale Spektrum. Dieses Ereignis, dieser Wandel, ist ein Aspekt der ‚Schwelle des Saturn' im Erd-Element. Wir werden uns später um dieses *ios* noch mehr bemühen.

Die gesamte Situation bis hierher tritt beinah alltäglich in Erscheinung und wird in etwa umrissen mit Formulierungen wie: Wenn man eine Idee hat, dann bringt diese eine gewisse (mehr oder weniger feurige) Dynamik mit sich. Wurde die Idee dann umgesetzt, herrscht zuerst Freude über die Umsetzung vor. Unweigerlich wird dabei aber auch jene Schwelle des Saturn überschritten, und eine relative Leere, Stagnation oder Passivität breitet sich aus.

Oder: In der Phase des Wünschens üben gewisse Dinge eine Faszination aus und erzeugen Begehren. Sobald aber der Wunsch erfüllt ist, wird die erste Freude bald von einer durch Nachlassen der Faszination entstehenden Leere abgelöst. (Was meist sofort wieder Platz für den nächsten Wunsch schafft). Daneben entsteht eine Anhaftung oder Verhaftung an das herbeigewünschte und erlangte ‚Ding', welches unmerklich zur ‚Belastung' wird und zu Unzufriedenheit führt. Diese Anhaftung ist umso stärker, je ausgeprägter das *Ahankara*, das Elemente-Ich des ‚Dinges' ist.

Je stärker wiederum diese Anhaftung ist, umso mehr empfindet man die Vergänglichkeit dieses ‚Dinges' und leidet damit. Dieses Leiden aber setzt das *ios* frei, also jene dem Ding eingewurzelte Kraft, welche geeignet ist, es zu zerstören. Mit der Freisetzung des *ios* werden Sulfur und Merkur gleichzeitig entfesselt (bzw. ‚ein Merkur, der nichts anderes als ein Feuer ist') und sie beginnen, das ‚Ding' zu töten. Und zwar jener Sulfur, jenes Feuer, aus dem heraus das ‚Ding' manifest geworden ist und jener Merkur, welcher das Wesentliche und Quintessentielle des ‚Dinges' ist und dem saturnischen Kerker entfliehen will. Deshalb fährt *d'Espagnet* im Text fort:

> „Wenn die Körper wieder in die Erste Materie zurückfließen, dann vermischen sich Mann (*Sulphur*) und Weib (*Mercurius*) miteinander, aus welcher Vermischung der Schwarze Rabe geboren wird."

Der Schwarze Rabe

Der Begriff ‚Schwarzer Rabe' hat mehrere Bedeutungen, steht aber hier zunächst einmal für ‚Tod'. Und zwar in dem Sinne, dass mit dem Überschreiten der Schwelle des Saturn kraft des ‚gegenwirkenden Feuers' und des nach Vereinigung mit dem wesensgleich Geistigen drängenden Merkurs ein bedeutender Aspekt Saturns an Einfluss verliert, nämlich der salinische bzw. die Körperlichkeit.

Von nun an – und das gilt für das ganze merkuriale Spektrum – ‚vergeht' die physische Ausdrucksform. Mit anderen Worten: Sobald man etwas hat, verliert man es auch schon wieder. Das ist ein Drama der Stoffwelt. Damit kommen wir zu dem Punkt, warum der Alchymist ein ‚natürliches Ding' nimmt, welches über kurze oder lange Zeit gewachsen ist und dieses dann ‚tötet' – aber nicht mehr in ‚Naturzeit', sondern durch die Kunst. Der erste Schlüssel ist, dass die natürlichen Körper wieder in ihre ‚Erste Materie' zurückfließen müssen.

Bei diesem ‚Tod durch die Kunst' wird die im Körper ‚geschwächt darniederliegende Seele' gestärkt, erweckt und schließlich vom Körper geschieden. Das ‚passive Feuer' wird durch das ‚aktive Feuer' des Alchymisten wiederbelebt – eben etwa durch die Handlung des Schmelzens von Eisen und Antimon in der Glut der Esse (wobei in diesem Falle drei Feuer zum Einsatz kommen: erstens das äußeres Feuer der Kohlenglut, zweitens das philosophische Feuer des Eisens und drittens die zielgerichtete Aktion des Alchymisten. Dieser übernimmt die Funktion des ‚Initiators' des ‚Dinges'. Er überlässt das ‚Ding' nicht dem durch Saturn herbeigeführten Tod der Natur, sondern tritt selbst in die Rolle des Saturn (oder besser: er nimmt dem natürlichen Saturn seine Macht aus der Hand) und ermöglicht dem ‚Ding' einen ‚philosophischen Tod' durch die ‚Kunst des Menschen'. In solchen und ähnlichen Situationen ist der Alchymist immer wieder gleichnishaft ein *Odysseus*, welcher beschließt, den Göttern nicht zu folgen, ihnen nicht freie Hand zu geben, sondern ‚den Weg des Menschen darüberzustellen'. Das spricht *Paracelsus* an, mit: Der Mensch sei niemandem untertan – auch nicht den Gestirnen.

So oder so würde das ‚Ding' sterben. Zuerst das Samenkorn in der Erde, dann der junge Trieb, danach die Blüte u.s.f. – im Jahreslauf, im Wachsen und Vergehen – der Tod ist das Schicksal alles Lebenden. Jeder weiß das, und doch liegt darin eine noch weit bedeutsamere Erkenntnis der Natur. Trotz dieses Wissens ist gerade der Mensch, je mehr er der Natur entfremdet ist, geneigt, diese Angelegenheit zu verdrängen. In initiatischen Kulturen hingegen ist der Tod das wichtigste Thema des Lebens – nicht zuletzt deshalb, weil er die einzige Realität in der materiellen Welt darstellt. Dass sich viele Menschen selbst als ‚Realisten' bezeichnen, ohne diese Dinge zu lernen, ja im Gegenteil sogar bewusst in den Bereich des Unrealistischen verwerfen, während sie die jederzeit vergänglichen, unsicheren Dinge für Realität proklamieren, ist eine typische Signatur der Paradigmenirritation und nach den Lehren alter Traditionen an sich schon ein Krankheitsbild. Dieses ‚Vorwegnehmen' und damit ‚Entschärfen' des Todes durch die Kunst entspringt ältestem Naturwissen. Es ist der Grund der Initiation. Auf der Suche nach den besten Möglichkeiten, in der Stoffwelt zu leben, hat der Mensch herausgefunden, inwieweit er selbst Natur ist und inwieweit er als Geistwesen ‚die Natur überholen kann'.

Diese Erkenntnis wendet er dann bei sich selbst – auf seiner Ebene – an und aber auch im Umgang mit den Dingen der Natur – ihrer Ebene entsprechend, wissend, dass stimmige, im Lichte der Natur gewonnene Erkenntnisse auf allen Daseinsstufen eine gültige Anwendung haben. Wenn er (zu Zeiten von ,Kultur') weiß, dass für ihn Initatiation zum essentiellen Menschsein nötig ist, dann möchte er auch die ,natürlichen Dinge' ihrer Art entsprechend gleichermaßen behandeln, um daraus eine essentielle Arznei zu erlangen. Gemeint ist hier eigentlich das ,Quintessentielle', denn dieses wirkt dann von sich aus elixirisch, also veredelnd auf das noch nicht ,Aufgeschlossene'. Aber auch alles andere auf seiner jeweiligen Ebene ist damit gemeint, was der Mensch nimmt und braucht, um sich in der Natur der Welt häuslich einzurichten. Wir erinnern uns dazu an *Paracelsus'* Worte:

„Die Natur ist so subtil und scharf in ihren Dingen, dass sie ohne große Kunst nicht gebraucht werden kann, denn sie gibt nichts an den Tag, das in seiner Art schon vollendet sei, sondern der Mensch muss es vollenden. Diese Vollendung heißt *Alchymia*. So ist der Bäck, indem er Brot macht, der Rebmann, indem er den Wein macht, der Weber indem er das Tuch macht, ein Alchymist."

nach: *Paracelsus*, *>Paragranum<*, 1530

Genaugenommen geht die ,Kunst' aber einen bedeutsamen Schritt weiter. Wir haben das eingangs bereits erwähnt. Sie nimmt nicht nur aus der Natur *(spao)*, sondern sie fügt das Genommene und Bearbeitete wieder so zusammen, dass damit eine Idee manifestiert wird, eine (neue) Quintessenz sich einkörpern kann, etwa ähnlich wie Baumstämme durch Holzfäller, Sägler und Zimmermann zum Haus werden. Die Bäume selbst werden dabei zerstört und in neuer Form wiedergeboren. Sie wären ohnehin in ihrer Eigenschaft als ,Ausdrucksformen' gestorben – und über Früchte und Wurzeln während langer Zeiträume immer wieder als Bäume inkarniert. Seit Urzeiten schon begriff der Mensch, dass es nicht einfach das ,Herausnehmen' der Bäume aus der Natur ist, was ihm sebst im Leben hilft, sondern dass er auch der Baumseele in ihrer Entwicklung hilft, wenn er ihre Hülle sinnvoll nutzt. Dazu muss das ,natürliche Ding' aber ,zeitig und reif' sein, im speziellen Fall der Baum, wie im Werk *d'Espagnets* seine Materie.

Das ist archaiisches Naturverständnis, Verständnis vom ,rechten Umgang mit den Dingen der Natur.' Es mag manchem Kind seiner Zeit heute naiv anmuten, aber es reicht weit tiefer, als dieses Kind der Zeit ermessen kann.

Es verhält sich nämlich genauso, wie *Paracelsus* andeutet: Der Ackermann, der Rebenmann, ... sind alles Alchymisten. Dann jedoch sagt er: Das höchste aber ist die Arznei! Wer aus Baumstämmen ein Haus baut, ist ein Künstler. Ebenso der, welcher ein Tuch webt, und auch der, welcher das Erz schmiedet. Wer aber aus dem Baum, aus dem Kraut, aus dem Erz eine Arznei bereitet, der ist ein ‚Philosoph'.

„Derselbe, welcher etwas, das aus der Natur dem Menschen zunutze wächst, es dahin bringt, dahin es von der Natur geordnet wird, der ist ein Alchymist. Wisset den Unterschied in der Kunst: Wenn einer eine Schafshaut nimmt und lege sie roh als Pelz oder Rock an – wie grob verhält der sich gegen einen Kürschner oder Tuchmacher! So grob und ungeschickt ist es auch, wenn einer etwas aus der Natur (zur Arznei) nimmt und das selbe nicht bereitet. Es ist noch weit mehr als grob und ungeschickt, denn es trifft an die Gesundheit, an den Leib und an das Leben. Allein in der Arznei, da das am nötigsten ist, soll dieses nicht geschehen, so das sie dergestalt die gröbste und ungeschickteste Kunst wäre. Denn darin liegt der trefflichste und letzt Punkt des ganzen Bereitens. Da muss die Philosophie, die Astronomie und all ihre Zusammenfügungen verstanden werden. (...) Gleichwie der Sommer die Birnen und Trauben zeitig macht, so soll auch deine Arznei geführt werden. (...)

Die Arznei muss aber dermaßen in die Luft gebracht werden, dass sie vom Gestirn regiert werden kann. Denn welcher Stein wird vom Gestirn hochgehoben? Keiner. Allein das *volatile*, das ist das Flüchtige. (...) Das *Quintum esse* ist nichts anderes, als wenn die vier Körper fortgenommen werden (Das Gespinst der vier grobstofflichen Elemente, welches die ‚Idee' verhüllt, aufgehoben wird), und was dann übrig ist, das ist das eigentliche *Arcanum*. (...) Es ist dem Gestirn möglich es zu führen, wie eine Feder vom Wind.

So soll nun die Bereitung der Arznei sein (...), und danach soll das Wissen da sein, was in diesem *Arcanum* das *Astrum* sei und danach, was das *Astrum* dieser Krankheit sei und was das *Astrum* in der Arznei wider die Krankheit sei.

Es ist nicht so wie sie sagen: Mache Gold, mache Silber. Hier ist das Vornehmen: Mache *Arcana*, und richte dieselbigen gegen die Krankheiten. (...) das ist der Grund, und diese Dinge alle entnimmt man aus der Anweisung der Natur und ihrer Bewährung. (...) Solches alles vollendet die Alchymie, ohne welche die Dinge nicht geschehen können.

nach: *Paracelsus*, *>Paragranum<*, 1530

Was die tieferen Gründe der Vorgehensweise betrifft, mit welcher der Alchymist seine Arznei zubereitet, sind die Erklärungen, wie man sie in der abendländischen Literatur vorfindet, sehr dunkel gehalten. Zum einen vollzog sich hier eben der Transport ältesten Wissens aus Sicherheitsgründen nur verhüllt, weil jenes Wissen weit über das kirchliche Dogma hinausging. Zum andern wussten die Praktiker des späten Mittelalters oft gar nicht mehr, wie die Bilder der – aus ihrer Sicht – ‚Alten' zu enthüllen wären, weil doch sehr viel vergessen war. Nehmen wir nur als Beispiel die Lehre von der Wiedergeburt der Seele bzw. der evolutionären Seelenwanderung durch die Schöpfungsformen. Den Druiden war sie noch eine Selbstverständlichkeit und wurde in ihren Schulen bis ins 5. und 6. Jahrhundert vermittelt. Danach scheint dieses Wissen versiegt zu sein und taucht höchstens noch allegorisch verhüllt auf. Die alchymistische Praxis hatte aber ursprünglich ihren Grund genauso darin wie im Wissen um Art und Sinn der Initiation, dem wir uns hier im Großen und Ganzen annähern wollen. Diejenigen Alchymisten und Heilkundigen, welche während ihrer weiten Reisen in anderen Regionen der Welt lernen konnten, trugen dieses Wissen mit sich, hüteten sich aber, zuviel davon zu offenbaren. Ebenso verhielt es sich mit denen, die (wie *Paracelsus*) von ‚Fahrenden' lernen konnten, welche diese Traditionen kannten. Man muss bedenken, dass jenes ‚Wissen', das im Abendland ‚erlaubt' war, nur einem winzigen, verformten Bruchteil des großen Wissens entsprach, das den Kundigen im Westen noch wenige Jahrhunderte zuvor geläufig war bzw. das in anderen Weltgegenden ohne Behinderung tradiert werde konnte. Deshalb wirkte zum Beispiel das, was *Paracelsus* von sich gegeben hatte, so mächtig, obwohl es – verglichen mit zB. dem Wissen manch ayurvedischer *Vidyas* und *Rasa Shastris* – vielleicht nicht gar so mächtig sein würde. Heute noch sagen solche *Vidyas*: „Die Menschen im Abendland haben nicht die geringste Ahnung auch nur von den Grundlagen jenes Wissens." Ein typisches Beispiel ist heute die maßlose Überbewertung der Chakrenlehre. In den indischen Traditionen eine Alltäglichkeit – im Westen eine grenzenlose Fantasie. Ähnlich verhält es sich mit dem Wissen um Seelenwanderung, Wiedergeburt und Karma. Während man im Westen heute noch mit scharfem Intellekt diskutiert, ob es denn das wirklich geben könne, meinen die *Vidyas* dazu: „Ihr wusstet es jahrzehntausendelang. Nur während der letzten tausend Jahre habt ihr es vergessen – wie so vieles andere auch. Das ist euer Problem, nicht eine Frage der Beweisführung." Nebenbei: Ganz ähnlich verhält es sich mit der Lehre von den 5 Elementen. In den asiatischen Traditionen eine Selbstverständlichkeit, im Westen eine Bildungslücke. Und wenn hier schon einer beherzt über die Natursicht der ‚Alten' spricht, dann redet er von nur vier Elementen, denn das Wissen über die Quintessenz begann schon vor bald tausend Jahren zu versiegen. In all dem liegt aber ein ursprünglicher Grund zum Verständnis der Alchymie. Deshalb kann in diesen Belangen nur eine Zusammenschau der Traditionen, ein intertraditionales Bemühen, weiterhelfen.

Im Thema des ‚philosophischen Todes' kann man besonders deutlich jene archaiische Art der ‚Kunst' erkennen, welche die Ansicht, dass die Alchymie eine mittelalterliche Zeiterscheinung gewesen wäre, *ad absurdum* führt.

Der Praktiker geht heran und nimmt die ‚lebendige' Pflanze oder das (in ganzheitlichem Sinne) lebendige Mineral heraus aus seinem ‚Leidensweg', vermittelt ihm eine Art Initiation, den ‚Tod im Leben', und lässt es auf einer höheren Daseinsebene (essentiell) wiedergeboren werden. Damit verkürzt er den langen Weg der Evolution des ‚natürlichen Dinges' und erspart ihm damit viel des (oben bezeichneten) Leidens.

„Im Blei (Saturn) herrscht ein unverschämter Dämon, der die Menschen zum Wahnsinn treibt."

Olympiodoros, 5. Jhdt.

Zugleich ermöglicht er – nach den Lehren der indischen Traditionen – der dem ‚natürlichen Ding' innewohnenden Seele eine relativ rasche Wiedergeburt auf höherer Ebene, besonders dann, wenn das ‚Ding' schließlich zur Arznei verwendet wird. Das heißt: Gemäß der Überlieferungen muss eine Seele etwa 8 400 000 Schöpfungsformen ‚durchwandern', bis sie soweit ‚reif' geworden ist, dass sie in einem menschlichen Körper inkarnieren kann. Auf diesem Weg verbringt sie sehr lange Zeiträume eingeschlossen in mineralischer Substanz und im Vergleich dazu weit kürzere (aber immer noch sehr lange) Zeit in der Pflanzenwelt. Dadurch, dass der Kundige die ‚durchlebten', beseelten Substanzen aus der Natur nimmt und sie kunstvoll aufschließt, verhilft er den innewohnenden Seelen zu weiten Entwicklungssprüngen.

Der dazu notwendige ‚philosophische Tod' ist kein ‚unnatürlicher Tod', sondern ein in der ‚Idee der Dinge' bereits angelegter ‚natürlicher Tod durch die Kunst des Menschen'. Dieser Unterschied zum ‚unnatürliche Tod' ist wesentlich und wichtig, da jener ‚ohne Kunst' und ‚gegen die Natur' geschieht. Man bedenke des *Paracelsi* Worte: „Derselbe, welcher etwas, das aus der Natur dem Menschen zunutze wächst, es dahin bringt, dahin es von der Natur geordnet wird, der ist ein Alchymist." Das ist u.a. ein Hinweis darauf, dass die quintessentielle Anlage der ‚natürlichen Dinge' bereits die Idee einer ‚kunstvollen' Begegnung mit dem Menschen enthält, weil die gesamte Natur eine ‚Natur der Erde des Menschen' ist. Auf dieses Verständnis legten die ‚Alten' großen Wert.

Wir sprechen hier also einerseits von einem ‚natürlichen Tod durch die Zeit der Natur', welchen jedes ‚Ding' erleidet, von dem Augenblick an, da es die Schwelle des Saturn in die Verkörperung hinein überschritten hat. Sterben ist langsames Vergehen und Verlassen der Form während der Ausdruckszeit (z.B. der Pflanze).

Andererseits sprechen wir vom ‚philosophischen Tod', dem ‚natürlichen Tod durch die Kunst des Menschen'. Das ist dann, wenn die Seele durch kunstvolle Auflösung und Scheidung aus dem Körper gebracht wird.

Auch wenn dieses Thema sehr diffizil erscheint, gelangt man schließlich zur Einsicht, dass man hier eine schlichte Art vor sich hat, wie sich der Mensch zur Natur als helfender Freund verhält. Es ist das Thema des ‚Sterben-Lernens im Leben' und zwar des initiatischen Sterbens und essentiellen Wiedergeboren-Werdens, welches einerseits ein zentrales Thema der hohen spirituellen, mit der Menschwerdung befassten Traditionen ist, sich aber analog auch auf den ‚rechten Umgang mit den Dingen der Natur' und hier besonders auf die Alchymie als eine Kunst ‚heilsamer Veredelung' bezieht.

> „Nimm den Grauen Wolf (...) und wirf ihm den Leib des Königs vor. Und wenn er ihn verschlungen, so mach ein großes Feuer. Da wirf den Wolf hinein, dass er ganz und gar verbrenne, so wird der König wieder erlöst werden."
>
> *Michael Maier, >Atalanta fugiens<*, 1618

Die hermetische Formulierung ‚den alten König töten' trifft auf den ‚Ackermann als Alchymisten' zu, wenn er das zeitige und reife Getreide am Feld nicht welken und verderben lässt, sondern Saturn die Sense aus der Hand nimmt und selbst erntet, um das Korn schließlich einer weiteren Veredelung zuzuführen, welche es für sich alleine auf dem Felde verkommend nicht erfahren würde. Das gereifte Korn entspricht dem ‚alten König'. In diesem Zustand empfindet sich das Pflanzenwesen nicht (mehr) quintessentiell, sondern ‚getrennt' – als Elemente-Ich, und nur im Innersten schläft das unter saturnischem Stoff verschüttete Seelenhafte, die ursprüngliche Idee. Diese würde nicht mehr erwachen, bevor der Pflanzenkörper sie über Herbst und Winter verlassen hat. Der ‚alte König' ist auch die (notwendig aus der Stoffwerdung) fixierte Individualität des Menschen, welche sein Bewusstsein der Ganzheit lähmt und seinen Blick auf sein eigenes (quintessentielles) Wesen verstellt. Analog zu den (obigen) Worten *Michael Maiers* wird das Getreidekorn in der Folge durch das Feuer des Backofens ‚erlöst' und ersteht wieder als Brot, dessen Idee von Anbeginn bereits im Getreidekorn (mit *Paracelsus'* Worten:) ‚von der Natur geordnet ist.' Dies ist ein Aspekt natürlicher Evolution, welcher auf allen Ebenen und auf zahllose Arten zu jeder Zeit stattfindet.

„Du Narr! Was du säst wird nicht lebendig, wenn es nicht stirbt."

Stolcius v. Stolcenberg, >Viridarium chymicum<, 1624

Vervollständigt durch die Ansicht der indischen Traditionen bedeutet das auch: Wenn der Mensch hier ‚veredelnd' eingreift, dann erhält die Pflanzenseele des Getreides eine höhere Wiedergeburt in kürzerer Zeit, als wenn der Mensch nicht eingegriffen hätte.

„Der Tod ist das einzige Mittel, dadurch der Geist mag in eine andere Qual und Form eingehen."

Jakob Böhme, >De Signatura<

Für die Menschen war seit der Urzeit der initiatische Tod eine gnadenvolle Möglichkeit, sich gegenüber dem ‚Lauf der Welt' zu behaupten und ein unumgängliches Ereignis auf dem Weg zur Selbsterkenntnis. Das ist, wie gesagt, Angelegenheit der Spiritualität. Wir müssen es aber erwähnen, weil in diesem zentralen Thema ein Berührungspunkt jeglicher ‚Kunst' liegt. Schließlich hat ja der Mensch seit jeher den ‚rechten Umgang mit den Dingen der Natur' aus der Natur gelernt – und aus dem menschlichen Leben. Manche Urvölker initiieren noch heute ihre Jungen, wenn sie ‚zeitig und reif' sind, sei es durch Ritual oder durch tatsächliche Erfahrung. Es ist auch eine wesentliche, vielleicht die wichtigste Signatur von ‚Kultur', ob sie initiatisch ist oder nicht, oder in welcher Weise. Noch in den christlichen Texten findet man die Aufforderung: ‚Lerne zu Sterben, damit du zu leben beginnen kannst'. Der durchaus praktische Charakter dieser Aufforderung wird heute von den ‚Nachbetern' keineswegs erfasst und die Sache bestenfalls in ein gutgemeintes ‚Erfühlen' oder in ‚so ähnlich wie' abgedrängt. In denselben Texten werden die scheinbar ‚Lebenden' aber nicht ‚im Geiste Wiedergeborenen' als ‚Tote' bezeichnet. Sie gelten als ‚tot', weil sie das Potential des Menschseins nicht nützen und vom ‚natürlichen Saturn' dahingerafft bzw. nach der tibetischen Tradition ‚vom wilden, gierigen Rachen *Mahakals* verschluckt' werden. *Mahakal* zerreißt seine Beute mit Klauen und Zähnen und schlingt sie hinunter in den tiefsten Abgrund der Unbewusstheit. Das ist das Bild Saturns als ‚Kindlifresser', als *Kronos*, welcher all seine Kinder sofort nach ihrer Geburt verschlingt – bis er durch die ‚kunstvolle Verbindung von Mensch und Natur' *(Zeus* und *Rhea)* überlistet und überwunden wird.

Den Unterschied zwischen dem ‚gemeinen' und dem ‚philosophischen Tod' erklären die Alten wie erwähnt: Im ersten Fall verlässt der von der Lebenszeit aufgebrauchte, gemarterte und welke Körper die Seele. Im anderen Fall verlässt die Seele den Körper ‚indem sie stark wird und sich von den körperlichen Fesseln befreit, ohne den Körper dabei zu verlieren, wenn sie ihn noch zu ihrer Entwicklung benötigt. *Jakob Böhme* sagt: „Das Wesen befreit sich vom Tod durch Sterben."

Im oben zitierten Text *d'Anvers* wird dieses alte Wissen analog in die alchymistische Laborpraxis umgesetzt: Das Eisen ‚initiiert' das Antimon. Die entsprechende Stelle wollen wir zur Erinnerung und kommentiert wiederholen:

> „Merke, dass auf diese Weise nicht das *Sal Nitri* das *Antimonium* von seinen Schlacken trennt (der Körper des Antimons also nicht durch eine etwaige Calcination mit Salpeter getötet wird), sondern das Antimonium zieht die Seele aus dem Eisen (es wird vom Eisen ‚initiiert'), das heißt, es nimmt den besseren Schwefel des Eisens in sich (Auch in den initiatischen Kulturen sieht man es so, dass der Initiator dem Adepten/Initianden ‚sein Bestes gibt', um seine Seele zu erwecken und zu stärken. Er gibt ihm von seiner eigenen elixirischen Essenz) und reduziert ihn zu Merkur. (Hier wird die oben beschriebene Schwelle des Saturn überschritten). Und dieser Merkur ist nichts als Feuer, ..."

Nicolaus Flamel erkärt in den >Hieroglyphischen Figuren<: „Die Auflösung heißt Tod, Zerstörung und Verderbnis, weil die Naturen dann ihre Form ändern. Da vollzieht sich eine Calcination (des Körpers) und eine Entblößung *(Denudatio)* (der beiden Geister – Sulfur und Merkur)."

Durch zahlreiche Bilder, Formulierungen und Allegorien in der abendländischen alchymistischen Literatur des späten Mittelalters und der beginnenden Neuzeit wird uns dieses urzeitliche Wissen überliefert. Es ist analog auf allen Ebenen gültig – von der hohen Spiritualität, dh. des ‚Sterbens im Leben' während Initiation und Meditation (>*Plutarch*), wie wir es heute noch bei Weisen und Mystikern des Orients finden, bis zu den schamanistischen außerkörperlichen Praktiken und darüberhinaus direkt im alchymistischen Arbeitsverständnis, einerlei ob in ‚großen Werken' oder einfachen Kräuterzubereitungen.

> „Die (initiatische) Geburt geschieht, wenn sich die Materie in einer völligen Auflösung befindet, welche ‚Fäulnis' oder ‚Schwärzestes Schwarz' genannt wird."
>
> *Pernety*

Ein Verfahren, um diesen Tod herbeizuführen, ist die Putrefaction. Nach hermetischer Sicht versinkt dabei die Ausgangsmaterie in ‚der finsteren Nacht des Chaos'. Es entsteht der ‚Schwarze Rabe'. Dieser Tod ist „schwarz, schwärzer noch als schwarz ist."

Putrefaction

Die Pflanzenmasse wird in ein gläsernes oder besser irdenes Gefäß gefüllt. Dieses Gefäß ist allegorisch der ‚Sarg' (‚Gläserner Sarg'), die Höhle, Kammer, Kiste oder ähnliches. Das Gefäß wird gut verschlossen und längere Zeit vergraben. Man stellt es (bis oben bedeckt) in ein Sand-, Aschen- oder Wasserbad, in einen beheizten ‚Brutkasten' (alleg.: Höhle, Erdhügel, Erdloch, Sumpf/Wasser/See/Meer, Turm, Baum/Baumkrone/Baumhöhle, Kammer/‚Schwitzhütte', ‚Kiste des Zauberers') oder vergräbt es in der Erde, im Misthaufen oder – was die Alten zu bevorzugen schienen – in Pferdemist. Das ‚philosophische Sterben' beginnt. Mit der Zeit wird die Materie dunkler und die Prinzipien bzw. Elemente beginnen sich zu trennen. Analog dazu wird auch der Mensch beim Vorgang des Sterbens vom Körper zurückgezogen, wobei sich die Elemente der Reihe nach aus ihren ‚Orten' lösen, im Körper von unten nach oben. Es heißt: Zuerst geht die Erde, dann das Wasser, dann das Feuer, wenn das Herz zu schlagen aufhört und dann die Luft im Kehlkopf (‚Röcheln'). Die Quintessenz in der Stirn bleibt bis zuletzt – wesensähnlich mit der Seele – und ‚geht' dann ihren eigenen Weg. Alle Elemente aber gehen zurück zu ihren ‚Orten' in der Natur. Was sich im Menschen ereignet, soll hier nur am Rande unser Thema sein, quasi als Berührungspunkt. *Paracelsus* weist darauf hin, dass man aus allen Dingen, nur nicht aus dem Menschen eine Quintessenz scheiden könne, „weil diese zu sehr mit der Seele zusammengeht."

In den Dingen der Natur, in den Pflanzen, Mineralen und Metallen, scheiden sich während der Putrefaction die Elemente, und es scheiden sich insgesamt alle ‚Kräfte' sowohl von der grobstofflichen Substanz, als auch voneinander – alle jene Kräfte, welche während der Lebenszeit der Ausdrucksform von Pflanze, Stein oder Metall diese gebildet haben und darin vermischt und verwoben waren.

In der weiten Natur würde jedes Element nach seinem Ort gehen können, wie wir das im Herbst unter der obersten Laubschicht und durch diese hindurch erleben. Kräfte des Luft-Elements riecht man – modrig, faulig, aromatisch und je nach Art. Kräfte des Wasser-Elements sieht man in unzähligen braunen Rinnsalen fortfließen. Kräfte des Feuer-Elements fühlt man als Wärme, wenn die Hand in tiefere Laubschichten wühlt. Das Erd-Element bleibt wie Asche zurück, oft erst nach mehreren Jahren der Verwesung. Diese Asche geht (wie im spagyrischen Pflanzenwerk) aus einer Calcination hervor – in diesem Fall aus der Calcination durch den Sauerstoff der Luft.

Feuer, Wasser und Luft gehen in den Äther, in den Natur-Raum zurück, woraus sie vordem zur Bildung des Pflanzenkörpers in die Materie hereingezogen wurden. Die Erde vereint sich mit dem Boden, geht also auch gewissermaßen im ,Erd-Raum' auf. Die Quintessenz geht ihren eigenen – weitesten – Weg, bis hin zu den ,Gruppenseelen' der Pflanzengattungen.

Im fest verschlossenen Gefäß des Alchymisten können diese Geister nicht zu ihren ,Orten' in der weiten Natur entfliehen, und sie (auch die Quintessenz) bleiben, wenngleich von der Materie und voneinander geschieden, darin. Ist alles fertig putreficiert, dann können sie durch Destillation abgeschieden und einzeln für sich dargestellt und rectificiert werden. Letztere Rectification ist wichtig, da es sich ja hier wiederum ,nur' um Ausdrucksformen der Elementarkräfte handelt (in der Regel *Fluida*), die mit *Phlegmata*, also ,irdischen Unreinigkeiten' vermischt – und: vom fünften Element, der Quintessenz, durchdrungen sind, da dieses das subtilste und durchdringenste und tingierendste von allen ist.

„Nimm Salbei, die zerstoße, lass es putrificieren, darnach destilliers (hier: aus einem Pferdemist-Bad), so geht zuerst das Element Feuer über, so lange bis dass sich Farbe und Dicke des ,Wassers' verändern, dann folgt die Erde nach, von der aber auch ein fixer Teil am Boden bleibt. Die beiden (destillierten) ,Wässer' stelle miteinander sechs Tage an die Sonne. Danach destilliere (aus dem Wasserbad), so geht als erstes das Element Wasser heraus, dessen nicht viel ist. Das erweist sich im Geschmack. Danach, wenn der (Geschmack) sich verändert, so folgt ihm *elementum ignis* (das Feuer-Element) nach, bis dass der Geschmack auch wieder verändert wird. Dann geht ein Teil *Terra* über, aber sehr wenig. Das andere (Erd-Element) wird am Boden gefunden. Der Aer aber (das Luft-Element) mischt sich durch die anderen. (...)"

„Der erste Weg ist: die Elemente voneinander zu scheiden, jegliches in sein eigenes Gefäß, ohne Zerstörung ihrer Kraft (...).

Der zweite Weg ist: zu scheiden das Reine vom Unreinen, nachdem die vier Elemente geschieden sind, denn wenn ein Element vom andern geschieden ist, so ist das in grober Gestalt und unförmig. Deshalb muss wieder geschieden werden, *purum ab impurum*. (...) Es ist auch so zu verstehen, dass die *Quinta Essentia* von den Dingen (Elementen bzw. deren fluidalen Ausdrucksformen) genommen werden soll. Und zwar deshalb, weil die Elemente aus ihren Körpern in der Natur des fünften Wesens, welches die Elemente mehr oder weniger tingiert, noch nicht gemeistert sind und dabei gelassen werden können. (...) Es geht auch den vier Elementen in ihrer Kraft nichts ab, wenn das prädestinierte Element, das ist die *Quinta Essentia*, fortgenommen wird."

nach: *Paracelsus, >Archidoxa<*

Auf diese Weise kann also der initiatische Tod auf der Analog-Ebene des Pflanzenreiches durch die Kunst herbeigeführt werden, und am Ende wird die selbe Pflanze als Essenz bzw. Quintessenz ‚wiedergeboren', indem man die gereinigten Ausdrucksformen der Kräfte wieder vereint (coaguliert). Als solche fällt sie dem ‚gemeinen Tod' nicht mehr anheim, ist gewissermaßen ‚unsterblich' geworden. Zur ‚Unsterblichkeit' einer Quintessenz gibt es vermutlich keine praktischen Erfahrungswerte, denn wenn ein Spagyriker eine solche Essenz gut verschlossen in einem Fläschchen ein Leben lang aufbewahren wollte, dann wird der Künstler zwar darüber vergehen, die Essenz aber bleiben, und das kann über viele Generationen so gehalten werden. Vermutlich wird das Fluid oder der figierte Stein im Glase im Laufe der Jahrhunderte immer besser und stärker. Man müsste dann ein mit Gewalt gewolltes ‚Ablaufdatum' in die Jahrhunderte setzen, würde das aber nicht fertigbringen zu Zeiten, in denen man Harzen, die tausend Jahre gut sind, nur eine willkürliche Lebenszeit von fünf Jahren anerkennt und sich darüber den Kopf zerbricht, ob Steine und Kristalle vielleicht ein wenig länger frisch bleiben würden, während man andererseits frische und lebendige Nahrung ohne jegliche ‚natürliche Kunst' zerstört und meint, sie dadurch besonders lange ‚haltbar' zu machen.

Eine eigentliche ‚Schwärze' wird sich bei der Putrefaction der Pflanzen kaum zeigen, außer bildhaft, indem die Masse dunkler wird. Am Durchdunkeln erkennt man den Fortgang der Putrefaction. Die „Erde wird faulig, stinkend und finster" *(Trismosin)*. Bei Arbeiten mit Mineralen und Metallen kann mitunter eine deutlichere Schwärze auftreten. Wie wir gehört haben, sprechen die Alten gerne von Putrefaction, auch wenn sie damit nicht den faulungsähnlichen Prozess bei Pflanzen meinen – immer aber meinen sie damit den ‚philosophischen Tod'. Die Pflanze erfährt ihre ‚Auflösung' während der Putrefaction, das Metall während der Schmelzung.

„Die Erde putreficiert mit ihrem Wasser und wird gereinigt wenn beide putridae werden. So wird vermittelst göttlicher Hilfe das ganze Magisterium verrichtet, denn es wird rein und weiß gemacht."

nach: *Rosinus*, zit. in >*Turba Philosophorum*<

Der ‚philosophische Tod' ist zuerst die Beendigung der vorigen Ausdrucksform, dann die Scheidung der Kräfte/Substanzen/Elemente etc. und führt dann also über Lösung und Reinigung direkt zur Erneuerung bzw. ‚Wiedergeburt'. Von der Urzeit bis zur Alchymie der beginnenden Neuzeit war dazu auch noch ein anderer Begriff von großer Bedeutung: die ‚Zerstückelung'.

Auflösung, Zerstückelung und Wiedergeburt

Um das Wesen von Spagyrik und Alchymie sowie deren Beitrag zur Heilkunst zu begreifen, wollen wir uns diesem Themenkreis eingehender widmen: *Michael Maier* erinnert in seiner *Atalanta fugiens* an den Tod des *Osiris*, die Zerstückelung seines Leibes und der Wiedervereinigung seiner Glieder durch *Isis*. *Osiris* wurde zwar wiederhergestellt, allein ‚es gebrach ihm das Glied des Mannes und ging im Wasser verloren. Denn der Schwefel vergeht (...).' *Maier* warnt vor ähnlichen Missgeschicken während der laborantischen Umsetzung dieser Allegorien und ermahnt (gleichsam allegorisch), dass dem König im ‚roten Meer' seine Krone nicht verloren gehen dürfte. Nun sind Stirn, Zeugungsglied und Krone allesamt sowohl ‚Bilder', wie auch tatsächliche ‚Orte' im Körper, nämlich für das Sulfur-Prinzip und darüberhinaus für die ‚Seele'. Dem *Janus Lacinus* dürfte die Arbeit besser gelungen sein. In seiner *Pretiosa Margarita* (1714) wurde der König ebenfalls getötet (vom Sohn *Azoth*), in den Sarg (und dieser in die Grube) gelegt, ersteht aber nach erfolgreichem Prozess wieder erneuert und ‚ganz geistig' auf, ohne an der Seele zu Schaden gekommen zu sein. Ähnlich bei *Trismosin*, der zuerst den König sinken sah ‚in fauliger, stinkender Erde und Finsternis', danach aber seine Auferstehung ‚im Tageslicht, gekrönt mit drei kostbaren Kronen' (was auf einer einfachen Analogebene die drei höchstgereinigten Prinzipien Sulfur, Merkur und Sal bedeuten kann) erleben durfte.

Zwar ist schon das Ernten einer Pflanze selbst bisweilen eine ‚Zerstückelung', weit mehr noch dann das Zerschneiden, Zerreißen oder Zerstampfen, aber die wesentliche ‚Zerstückelung' ereignet sich während des ‚kunstvollen Auflösens', gewissermaßen als eine ‚Zerstückelung' der Elementarteile.

Der Ethnologe und Paläonthologe *Ivar Lissner* (aus Riga) erzählt von derlei Missgeschicken (wie wir oben auf eines bei *Michael Maier* hingewiesen wurden) auf einer anderen Ebene, aber mit demselben Sinngehalt, bei den Initiationen tungusischer Schamanen. Während der außerkörperlichen Initiationserfahrung wird der ‚innere' Körper zerstückelt. Der grobstoffliche Körper liegt inzwischen einige Tage wie tot, mitunter tritt auch an manchen Stellen Blut aus. Der Initiand empfindet während diesem ‚philosophischen Tod' und der Scheidung körperlicher und feinstofflicher ‚Bestandteile' einige Qualen. Der Kopf wird ihm abgeschnitten, die Augen aus den Höhlen genommen, das Fleisch von den Knochen getrennt und vieles mehr. Schamanen mit besonders hoher Berufung müssen diesen Vorgang drei mal erdulden. Nicht selten, so heißt es, stirbt einer dabei. Mit anderen Worten: Er schafft es nicht mehr, seine Seele in den Körper zurückzumanifestieren. Der Prozess birgt für die Seele eine große Gefahr, denn wenn sie die Verbindung zum Körper verliert, dann hat sie für sehr lange Zeit in der feinstofflichen Welt ‚einen schweren Stand'.

Aus dieser Perspektive können wir nun auch die ständige, dringliche Warnung der Alchymisten verstehen, die ‚Auflösung so vorzunehmen, dass die Seele des natürlichen Dinges dabei keinen Schaden nimmt.' Die ‚Krone im roten Meer verlieren' hatte bei *Maier* auf entsprechend analoger Ebene genau die selbe Bedeutung wie Seele, *Anima*, Quintessenz oder Sulfur während der ersten Auflösung, der Schmelze, der Rückführung in die *Prima Materia*, der Putrefaction usw. zu beschädigen.

Eine einfache und verbreitete Art, die Seele zu beschädigen, ist heute das Übergießen der ‚Dinge' mit allzustarken industriellen Säuren, was entweder aus einem ‚chemischen' Alchymie-Begreifen nicht anders verstanden wird oder aus kommerziellem Denken mit einem ‚rascheren Vorwärtskommen der Arbeit' begründet wird. Schon die ‚gemeine' Übersetzung des alten Wortes ‚*Menstruum*' mit ‚Lösungsmittel' verweist auf diesen Missstand. Noch die Kundigen der beginnenden Neuzeit (wie zB. *Agricola*) sprachen davon, dass die ‚Alten' bzw. die ‚wahren Philosophen' *Menstrua* herzustellen verstanden, welche ‚gar nicht korrosivisch, sondern eher wie ein Wasser' waren. Sie aber verstünden diese Kunst in ihrer Zeit so weit nicht mehr (oder nur mehr in seltenen Fällen), und wenn dennoch einer ‚so ein Stückchen wusste, dann hielt er es insgeheim für sich.' Ein solch seltener Fall könnte jener ‚Wasser-Alkahest' gewesen sein, mit dem der berühmte, in paracelsischer Tradition stehende Alchymist *J. B. van Helmont* (1577-1664) seine Präparation aus Zedernholz bewerkstelligte. Er erwähnt es zwar dermaßen, dass viele Leser seines Werkes in den folgenden Jahrhunderten mit großer Lust nach diesem ‚Wasser' suchten, aber mehr verrät er darüber nicht.

Nach seinen allgemeinen und hermetischen Ausführungen über die Zeder, die er für den ‚Baum des Lebens' hält, erklärt er:

„Die Schwierigkeit dieser Zubereitung besteht darin, dass das Holz ohne Verlust seiner Kräfte aufgelöst werden soll, und zwar durch eine so linde Wärme, wie sie die Sonne im März hat, bis das Holz in sein Erstes Wesen tritt, in welchem Wesen alleine das selbe die urhebliche Kraft hat etwas zu erhalten, gleichsam zu würzen, weil selbiges den Eingang hat in die ersten Stücke, woraus wir bestehen und hineinzudringen weiß zu der Bekanntschaft der eingepflanzten Lebensgeister (...). Es ist in diesem aufgelösten Saft die Kraft des ganzen Zedernbaumes, nämlich die Lebenskraft mit aller sämlichen und förmlichen Eigenschaft des langen Lebens.

Die Arbeit aber, wie man es machen muss, verhält sich also: Man löse das Zedernholz stückweise, gleich an Gewicht mit dem Wasser-Alkahest, darin auf, in einem zugeschmolzenen Glase in einer gelinden Brutwärme. So wird in einer Woche das ganze Holz zuerst in einen Milchsaft verwandelt sein. Um den vierzehnten Tag aber schwimmt verschiedentlich untereinander ein doppeltes Öl, welches bis gegen Ende des Monats immer mehr wird und sich klärlich absondert. Dieses Öl kann man dann vom Wasser durch bloßes Übergießen (Dekantieren) abscheiden. Wenn man danach das Wasser aus dem Wasserbad überdestilliert, dann bleibt der Alkahest im vorigen Gewicht am Boden.

Das Öl mit dem überdestillierten Wasser soll man ein viertel Jahr lang in gelinder Wärme halten, so wird das ganze Öl die Natur des Salzes annehmen und sich unter das Wasser mischen. Dieses ist alsdann das Erste Wesen des Zedernbaumes."

nach: *Johann Baptista van Helmont*, >Die Morgenröthe<

Hier beschreibt also *Helmont* die selbe Sache: ‚Das Holz ist ohne Verlust seiner Kräfte aufzulösen' – es soll den philosophischen Tod erfahren, ohne ‚die Krone zu verlieren'. Nachdem der Todesprozess eingesetzt hat und ‚das ganze Holz zunächst in einen Milchsaft verwandelt ist', beginnen die Scheidungen der ‚Bestandteile', der Elemente oder der Prinzipien. Am Ende hat man alle drei Prinzipien, Sulfur, Merkur und Sal (hier: Öl, Wasser und Salz) zusammen – und zwar (mit *Kunckels* Worten:) ‚durch ein untrennbares Band verbunden'. Die Zeder ist quintessentiell wiedergeboren, steht in ihrem ‚Ersten Wesen', als eine Essenz von regenerierender Kraft, welche für sich vermutlich beinahe ‚unsterblich' ist und überdies, wie *Helmont* betont ‚ein langes Leben vermittelt'.

Die ‚Krone im roten Meer verlieren' bedeutet nun auch das Verunglücken des Schamanen während seines initiatischen Todes. Stirbt dieser, dann geht er den vielen, welche seine Heilkunst später benötigt hätten, verloren. Das bedeutet auch für die Menschen um ihn ein großes Missgeschick.

Analog dazu ergänzt *Michael Maier*, dass es auch schade wäre, verlöre der König seine Krone,,denn mit den Steinen darin könne man viele Krankheiten heilen.' Diese Beifügung ist – zumindest aus dem üblichen Verständnis des maier'schen Traktates – rein laborpraktisch nicht klar. Es wird aber klarer aus der Zusammenschau der Traditionen und kann als Beweis genommen werden, dass *Maier* eben nicht nur über eine praktische Laborarbeit (in Bildern verhüllt) berichtet, sondern noch (wie viele vor ihm) eine umfassende urzeitliche Tradition weitervermittelt, welche den Umständen gemäß schon zu seiner Zeit bereits nur mehr von wenigen in ihrer gesamten Reichweite nachvollzogen werden konnte.

Texte und Bilder wie die von *Maier* sind uns ja vor allem in der alchymistischen Literatur der beginnenden Neuzeit überliefert. Blicken wir von da aus weiter zurück in die Vergangenheit, zu denen, die man damals schon ‚die Alten' nannte, jene ‚Alten', welche die Kunst ‚noch weit besser verstanden' haben sollen. Wir werden dabei sehen, dass dieses Wissen tatsächlich weit tiefgreifender vorhanden gewesen sein musste. Diesen Gedanken hatten wir bereits bezüglich der jahrtausendelang tradierten und lebendig erhaltenen Alchymie in Asien, insbesondere in Indien und den im Norden angrenzenden Gebirgsregionen. Es gereichen aber auch hier (im Westen) die Forschungen von Archäologie, Ethnologie, Anthropologie und anderen mit ihnen verbundenen oder zu verbindenden Disziplinen zum selben Bild, und zwar dann, wenn man einen ‚ganzheitlichen' und gewissermaßen ‚hermetischen' Blick wagt. Wenn man Tradition, Ritual und Erzählungen heute noch lebender Naturvölker vergleicht, einerseits mit den archäologischen Bodenfunden und andererseits mit den Resten jener urzeitlichen Überlieferungen, welche seit der Erfindung des Buchdrucks in der alchymistisch-hermetischen Literatur ihren Platz fanden. Ethnologie und Prähistorie zeigen heute, dass jene Menschen, die wir ‚vorgeschichtlich' nennen, in einer ungleich höheren ‚geistigen Kraft' gestanden haben musste, als wir uns das überhaupt vorstellen können. Das bedeutet aber unweigerlich auch die Wahrheit in den Worten spätmittelalterlicher Alchymisten, wenn sie zugeben, ‚die Kunst', die ‚Dinge der Natur' und den ‚Umgang' mit ihnen nicht annähernd so gut zu verstehen, wie eben diese Alten. Und je weiter man zurückgeht – immer von den Alten zu den Älteren, von Generation zu Generation, von Jahrhundert zu Jahrhundert, von Jahrtausend zu Jahrtausend und darüber hinaus – umso mehr ist anzunehmen, dass dies zutrifft.

Es mag dem westlich geprägten Menschen schwer fallen, nachzuvollziehen, dass man eine Wissenschaft umso höherstehend befinden sollte, je weiter man sie in die Vergangenheit zurückverfolgen kann. Ist man doch zu oft dem sogenannten Fortschrittsdenken verfallen, welches suggeriert, dass das Neue immer besser ist als das Alte. Die Überwindung dieser Hürde aber lohnt sich, denn sie führt in den Umkreis der innersten Lebensfrage: ‚Was ist das eigentlich – Mensch sein?'.

Es mag auch modernen ‚chemisch‘ orientierten Alchymisten schwer fallen, besonders dann, wenn sie von Neo-Esoterismen, künstlichen Erklärungsmodellen und religiösen Dogmen beeinträchtigt sind, weil die Ansicht, dass alchymistisches oder hermetisches Wissen schon in der Urzeit sehr hochstehend und danach immer mehr verfallen ist, eine Bestätigung für jenes unbehagliche Gefühl gibt, dass diese ‚Kunst‘ gleichsam ‚erschreckend‘ alt und ‚mächtig‘ und mit ‚modernem, chemischem‘ Denken gar nicht zu begreifen, geschweige denn praktisch nachzuvollziehen sei. Es verhält sich ganz und gar nicht so, dass der Mensch jahrtausende- oder auch jahrzehntausendelang ein Naturwissen gelebt und praktiziert hat, welches ihm erst jetzt ‚wissenschaftlich erklärt werden müßte‘. Diese Einstellung hat zumindest während der letzen Jahrhunderte viel Elend über die Welt gebracht, nämlich dadurch, dass man aus jener Arroganz heraus mit Gewalt bemüht war, hochstehende Kultur- bzw. Naturvölker entweder zu ‚kultivieren‘ oder zu vernichten, ohne die Möglichkeit zu erachten, von ihnen all das zu lernen, was wir heute in den letzten Zügen dieses gigantischen Irrtums zum Überleben so nötig hätten.

Ivar Lissner bemerkt in diesem Zusammenhange: „Die Bezeichung ‚Wilde‘ verleidet einem die ganze ethnologische Literatur des vorigen Jahrhunderts. Denn es gibt doch keine ‚Wilden‘ unter den Naturvölkern. (...) In ihrer Vorstellungswelt sind viele Naturvölker außerordentlich fein und vielseitig gebildet, ganz zu schweigen von ihrer Moral, die oft auf viel höherer Stufe steht als die des Westens“ (*Ivar Lissner*, >So lebten die Völker der Urzeit<, 1958).

Wer diesen Paradigmenwechsel heute noch spät, aber doch nachvollziehen will, müsste seine indoktrinierte oder aus Bequemlichkeit angewöhnte Sicht der Dinge vollständig umwandeln und würde dabei erst mit Erstaunen feststellen, wie weit sich der ‚naturwissenschaftlich gebildete Mensch‘ während weniger Jahrhunderte von der lebendigen Realität entfernen, sprichwörtlich: ‚aus dem Einklang mit der Natur fallen‘ konnte.

Ein ayurvedischer Arzt gab seinen Besuchern aus dem Westen ein Beispiel ‚steinzeitlicher Kultur‘ anhand des *Ramayana*-Epos, dessen Geschichte mit König *Ram(a)* (Widder, ‚Rammbock‘) im steinzeitlichen Europa beginnt und aus einer Zeit vor etwa 18 000 Jahren berichtet: König *Rama* durchstreifte zusammen mit seiner Frau *Sita* und seinem Freund *Lakshman* vierzehn Jahre lang Wälder und Berge. Dann wurde *Sita* entführt, und *Lakshman* erbot sich, sie zu suchen. Er fragte überall die Leute, die er traf, ob sie die Königin, die Frau seines Freundes gesehen hätten. Die Leute wollten natürlich von ihm wissen, wie sie aussieht, und er antwortete: „Ich weiß, sie ist über die Maßen lieblich und schön. Ich kann euch ihre Sohlen beschreiben und ihre Knöchel, aber mehr habe ich niemals von ihr gesehen“ (*>Ramayana<, Tulsi Sahib*). Das gab er als Beispiel dafür, mit welcher ‚Aufmerksamkeit‘ die Menschen zu jener Zeit gelebt hatten, die wir ‚Steinzeit‘ nennen und von der uns ein völlig irriges Bild vorgestellt wird.

Wo sich der Mensch ‚im Lichte der Natur' befindet, wird – entsprechend dem natürlichen Zusammenklang – das ‚Wissen der Natur und des rechten Umgangs mit ihren Dingen' enthüllt. Wenn man sich heute mit diesen Angelegenheiten befasst und dabei sagt: ‚früher' glaubte der Mensch, die Natur bestehe aus vier oder fünf Elementen (meist ist ohnehin nur von vieren die Rede) und bei sich denkt: ‚Heute wissen wir das besser', dann wurde die Grundschule der ‚Kunst' noch nicht besucht. Vielleicht besucht, aber nichts darin gelernt.

Wenn man aber sagt: ‚Alles in der Natur besteht aus den fünf Elementen' und dabei gar nichts anderes als ebendas denken kann, weil das Früher mit dem Heute in der lebendigen Tradition verschmolzen ist, dann besteht eine gute Möglichkeit, ins Innere des ‚chymischen Gärtleins' eingelassen zu werden.

Der tiefe Grund der Heilkunst

Natürlich geht das weiter und betrifft die Heilkunst an sich, bzw. das, was wir heute ‚Naturheilkunde' nennen. Das dürfen wir nicht an dem messen, was wir heute darunter verstehen. Sondern das stammt aus einem Naturwissen und einem Verhalten des Menschen in der Natur, dem wir uns heute nur mehr vom alleruntersten Ende und mit herzlicher Demut annähern können. Es ist nicht ‚fort'. Diese Dinge sind niemals wirklich ‚verschollen' und völlig vergessen. Aber wir müssen unser lächerlich aufgeblasenes Podest verlassen, alles, was wir zu wissen wähnen, für nichts halten, ein leeres Glas werden, das mit immerfrischem Wasser gefüllt werden kann. Vielleicht ein wenig bildhaft könnte man sagen: Wir müssten unser Operationsbesteck weglegen und wieder eines aus Obsidian herstellen lernen – das hatte sich Jahrzehntausende bewährt und wird nun wieder als das beste erkannt. Genauso ist es mit den Heilkräutern. Welche verwegene Unmenschlichkeit, welch abgrundtiefes Unwissen gehört dazu, Heilkräuter und den Umgang mit ihnen auch nur einem einzigen Menschen auf dieser Welt zu ‚verbieten'? Sie haben sich Jahrzehntausende bewährt. Das muss aber nicht so ausgelegt werden, dass ‚Omas Kamillentee' schon zu der Zeit, als ‚die armen Menschen kein Geld für Medizin hatten', ganz gut fürs Bauchweh herhalten konnte. Und ein Essigwickel und andere ‚Hausmittel' ‚in manchen Fällen' eine momentane Abhilfe sind, solange kein ‚echter Arzt' zur Hand ist. Dieses Wissen, diese Heilkunde, diese Kunst, gehen dermaßen weit über solche Ansichten hinaus, dass einer, der sich diesem Niveau nähern will, keine andere Möglichkeit hat, als selbst eine vollständige Transmutation zu durchleben. Er muss eine ‚Auflösung' seines ‚Elemente-Ichs' herbeisehnen, zulassen und aushalten, auch wenn er dabei das verliert, was bislang seiner Persönlichkeit Stolz und Stütze war. Er muss ‚wissenschaftlichen' Aberglauben und Selbsttäuschung beenden, wenn er das Verlangen hat, ins ‚Licht der Natur' zu treten.

„Ich kann wohl erkennen, dass keiner seinen Kolben gern fallen lässt. Ein jeglicher, dem sein Kolben die Hand erwärmt, der behält ihn gern darin. Das tun aber allein die Narren. Der weise Mann solls nicht tun. Der weise Mann soll den Kolben fallen lassen und einen andern suchen."

Paracelsus, >Paragranum<

Jedes traditionelle medizinische System ist Naturheilkunde. Befundung, Therapie und Arzneimittelbereitung bzw. Rezeptur stehen in Einklang mit den natürlichen Zusammenhängen und Entsprechungen. Die Unterschiede zwischen den traditionellen Systemen liegen natürlicherweise in verschiedensprachigen Ausdrücken, in verschiedenartigen Formulierungen, im Blickwinkel, aus dem die selbe Sache aus historischen, kulturellen, klimatischen und anderen Gründen betrachtet wird sowie in verschiedenen Graden von Wissen, Überlieferung und Erfahrung. Die vorübergehende Irritation des Begriffes ‚Heilkunst' scheint besonders im Abendland bloß ein kurzfristiges Zeit-Phänomen zu sein. Wenngleich sich nun hier die Heilkunst bereits wieder merklich auf dem Wege der Genesung befindet, dürfte aber die Wurzel dieser Irritation weit über tausend Jahre alt und vom selben Ursprung wie das, was die Alten als ‚Erkranken der Kunst *Alchymia*' bezeichnen, sein. Wenn im Mittelalter geklagt wurde: ‚Die Alten verstanden es noch, wir aber nicht mehr', dann betrifft ja diese Klage aus ganzheitlicher Sicht nicht nur die ‚hohe Alchymie', sondern alles das, was die indische Tradition ‚*Ayurveda*' – also die ‚Wissenschaft vom Leben' – nennt. Das meinte *Paracelsus*, wenn er immer wieder bemerkt, dass ‚der Arznei der wahre Grund fehle'. War er doch einer, der mit der Restauration dieses Grundes begann, bevor noch der beinahe vollständige Verfall eingesetzt hatte. Damit war er zwar sicherlich nicht der Einzige und auch nicht der Erste oder der Letzte, aber sein Beitrag ist unermesslich wertvoll. Einerseits schimpft er kapitelweise auf die alten griechischen Intellektuellen bzw. auf deren angebliche Schriften, auf *Avicenna, Galen, Hippokrates* und andere, denen ihm zufolge dieser ‚Grund' schon abging, andererseits klingt er – was dann wieder sonderbar scheinen mag – mit Vielem gleich, was aus jener Zeit tradiert ist. Nur scheint er offenbar einen anderen Zugang zum Naturverständnis der ‚Alten' gehabt zu haben als jene, welche ‚darüber bloß viel Papier beschrieben'. Wir werden darüber später noch nachdenken, wenn wir uns fragen müssen, wo denn *Paracelsus* eigentlich gelernt haben könnte. Einstweilen soll es genug damit sein, bemerkt zu haben, dass die besagte ‚Irritation' der Heilkunst gleichen Ursprungs sein dürfte wie der beginnende Verlust der ‚hohen' alchymistischen Kunst. Heilkunst und Alchymie sind untrennbar miteinander verbunden.

„Denn der König *Artus* erzählet in seinem vierten Buch, dass die edle schwarze Kunst, die Mutter *Alchymia*, lehret und vorschreibet wie man eine Arznei suchen soll die da hilft unter den Tugenden und Kräften der Kräuter. (...)

Er sagt, dass die Mutter *Alchymia* einen kranken Körper habe, der solchermaßen geschwächt ist, dass sie sich nicht aufrichten noch aufstehen kann. So hat sie Boten zu dem weitberühmten Meister aller Kräuter *Macro (Macer)* ausgesandt, dass er zu ihr käme und die natürlichsten, besten und stärksten Kräuter zusammenlese und mitbrächte, dass er sie absolvieren und reinigen möchte von der wassersüchtigen und paralytischen Schwachheit. (...)

Der Meister *Macer* (...) ist darauf zu der edlen Mutter *Alchymia* gegangen und sagte, nachdem er sie gesehen und angerühret: ‚Meine liebe *Alchymia*, ich sage, dass ich niemals einen schöneren Körper gesehen habe, denn dein Kopf ist gülden, und die Brust ist silbern. Aber was deinen unteren Teil betrifft, so habe ich noch keinen scheuslicheren gesehen. Deine Schenkel sind dick bis auf die Knie von der Wassersucht, und von da weg bist du inficiert bis auf die Fußsohlen.‘ Da hat ihm die Mutter *Alchymia* geantwortet: ‚Verwundere dich nicht, denn ich bin von der Sonne so schön geziert bis auf die Brust und vom Mond bis auf die *Femur*. Weil ich aber von der *Femore* an bis auf die Knie vergiftet und wassersüchtig bin, ist mein Körper unvollkommen. Deswegen habe ich zu dir als einem Meister und Vater aller Kräuter geschickt, dass du mich curieren und meine Gliedmaßen, welche so sehr lahm geschlagen, erquicken sollst.‘ (...)“

nach: >*Turba Philosophorum*<

Es gibt verschiedene Auslegungsmöglichkeiten dieses Textes. Interessant ist, dass zu jener Zeit den Kundigen des Abendlandes noch das Wissen um die ‚vier Zeitalter‘ geläufig war, welches wir ja heute nur mehr in asiatischen Traditionen lebendig finden. Wie alt das Gleichnis ist, ist unbekannt. Mit Sicherheit aber geht es auf jene Epoche zurück, die das ‚kupferne Zeitalter‘ genannt wurde, wo hinein auch noch jene Periode gehört, die wir als Antike und Spätantike bezeichnen und die bis zum frühen Mittelalter andauerte. Danach setzte das sogenannte ‚eiserne Zeitalter‘ ein.

Das Bild der Mutter *Alchymia* vermittelt hier also (auch) ihren Zustand während der Zeitalter. Der Kopf entspricht dem goldenen, die Brust dem silbernen, und von da an beginnt sich eine ‚venerische‘ Krankheit zu manifestieren, also eine ‚Krankheit der Venus‘ – ‚des Kupfers‘ – eben die Wassersucht. Von den Knien abwärts wird es noch viel schlimmer. Es ist alles ‚paralysiert‘ und ‚inficiert‘. Die Nähe des Begriffes ‚inficiert‘ zur ‚Virulenz‘ und zum *ios* und damit zu einer Ausdrucksweise martialer Kraft ist signifikant. Damit entsprächen die Füße ‚von den Knien bis zur Sohle‘ dem ‚eisernen Zeitalter‘, in welchem Mutter *Alchymia* wie paralysiert ist.

Ergänzend soll hier noch berichtet werden, dass *Macer* die Mutter *Alchymia* schließlich mit zwei ‚Kräutlein' heilte. Das erste heißt *Mambre*, welches ein späterer Übersetzer der *Turba* (vmtl. 1613) *radicibus mambreis* nannte. Damit könnte die Wurzel eines schwarzen Beerenobstes gemeint sein und zwar mit großer Wahrscheinlichkeit des Brombeerstrauches. (Vielleicht aber auch die der Heidelbeere). Das andere Kraut heißt *Lolium*. Dabei handelt es sich sehr wahrscheinlich um den Taumel-Lolch (*Lolium temulentum*), den von einem giftigen Pilz befallenen Samen eines Ährengrases, welches gerne das Sommergetreide und den Hafer begleitet. (Es kann aber auch das Mutterkorn gemeint sein).

Dass dieses Gleichnis ausdrücklich dem König *Artus* zugesprochen wird, also einem Schüler des Alchymisten *Ambrosius Merlinus* (bekannt als der legendäre Zauberer *Merlin* im 6. Jhdt.) mag zudem seine besondere Bewandtnis haben. Wenn wir ein Einhergehen des Verfalls der Heilkunst mit einem Vergessen oder Verlieren der hohen alchymistischen Kunst im Abendland bemerken, dann deutet das gleichfalls daraufhin, dass der Mensch – der ‚absteigenden Qualität des Zeitalters' folgend – die Verbindung zum Leben insofern verlor, als dass er das Wissen um den Tod aufgab. Mit anderen Worten: Er verließ die initiatische Kultur. Seit der Römerzeit war es im Westen kaum mehr üblich, junge Menschen in ihrer ‚Zeitigkeit und Reife' im Sinne der Alten ‚zu initiieren' und sie dadurch – im Sinne der Alten – ‚wirklich lebendig' zu machen. Schulen, welche dies weiterpraktizierten, mussten im Verborgenen wirken und ihre Angehörigen wurden spätestens zur Zeit der Inquisition beinahe restlos vernichtet.

In den asiatischen Traditionen, insbesondere den indischen, blieb diese Lebensart erhalten. Noch bis vor wenigen Generationen galt ein ‚nichtinitiierter Jüngling' als ‚vaterlos'. Erst durch die Erfahrung eines (wie auch immer gearteten und vollzogenen ‚philosophischen Todes') wurde das Tor zum verborgenen Wissen des Lebens und der Natur entriegelt. Dadurch war es nur folgerichtig, auch die alchymisitsch-laborantischen Arbeiten in diesem Zusammenhang zu begreifen und durchzuführen. So wird dort zum Beispiel bis heute ein Verfahren angewandt, welches bei den abendländischen Alchymisten der Neuzeit gar nicht mehr bekannt oder relevant zu sein scheint – die ‚Vorbereitung auf den philosophischen Tod', das *Shodana*. Dazu werden (zB.) dünne Blättchen des zur Präparation vorgesehenen Metalls erhitzt und in verschiedenen Flüssigkeiten abgelöscht – in der Regel in Buttermilch, Essig, Kuhurin, *Ghee* (Butterfett) und einem entsprechenden Pflanzensaft. Analog dazu steht das ‚Reinigen', welches der angehende Initiand vornimmt, damit er auf seinen initiatischen Tod vorbereitet ist, wie etwa das Einhalten vegetarischer Ernährung über längere Zeit, insgesamt enthaltsames Leben, auch Fasten oder körperliche Ausleitungsverfahren, Aufenthalt in der Einsamkeit / gedankliche Reinigung – je nach Schule und System.

Bei den Naturvölkern ist diese Vorbereitungszeit oft mit dem Einhalten einer ganzen Reihe von Tabus verbunden. Wird auch nur eines davon verletzt, dann besteht die Gefahr, dass der Initiand während des ‚philosophischen Todes' verunglückt und im schlimmsten Falle (mit *Maier's* Worten) seine ‚Krone' i.e. seine ‚Seele' verliert. Wird das *Shodana* laborantisch nicht richtig ausgeführt, dann ist die gesamte weitere Arbeit (Auflösen, Schmelzen, ...) im selben Sinne gefährdet.

Die Schulen der Heilkunst

Bezeichnend für die ursprüngliche Schlüsselposition des ‚philosophischen Todes' beim ‚Erlernen der Kunst' sind Initiationserfahrungen samojedischer Schamanen-Anwärter, welche von *T. Lehtisalo* und *A. Popov* nach Berichten vor Ort wiedergegeben wurden: Der Initiand *Ganykka* geriet in den außerkörperlichen Zustand und wurde dort von den ihn begleitenden Geistern in Stücke zerschnitten. Sein grobstofflicher Körper blieb währenddessen sieben Tage lang bewusstlos am Boden liegen. Seine Seele aber ging mit dem Donnergeist im Himmel spazieren und stattete sogar dem höchsten Gott *(Mikkulei)* einen Besuch ab. Nach *A. Popow* trifft sein Initiand ‚die Krankheit' selbst und wird von ihr zu ihrem Heiler ausgebildet. Dann trifft er die ‚Herrin des Wassers' und wird von ihr als ‚Kind' angenommen und gesäugt. Im Reich des ‚Herrn der Unterwelt' lernt er ‚alle Krankheiten kennen und heilen, die den Menschen quälen.' Im ‚Land der Schamaninnen' erhält er die Gabe, seine Stimme zu Gesang, Lehre und Vortrag recht zu gebrauchen. Dann traf er neun Kräuter – die Ahnen aller Pflanzen der Erde – und lernte von ihnen, sie zu verstehen und zu nutzen. Dann trifft er im Wipfel einer großen Birke Menschen verschiedener Völker und erkennt, dass sie alle quasi ‚an der selben Sache arbeiten'. Nochmals hört er eine Stimmen, die ihm die Kräfte von sieben anderen Pflanzen enthüllt und Hinweise, sein eigenes Leben betreffend, gibt. Danach findet er sieben Steine, die ihn darin unterrichten, was Schwere und Leichte ist, was die Formen sind, wie man Eisen schmilzt und insgesamt mit den Steinen, Mineralen und Metallen verfährt.

Schließlich wird er zu einem hohen Gebirge geführt und betritt darin eine leuchtende Höhle, wo er von zwei Frauen unterrichtet und mit Gaben beschenkt wird. Daraufhin muss er eine Wüste durchqueren, in der er drei Tage in Abgeschiedenheit verbringt. Noch einmal darf er ins Gebirge eintreten und begegnet dort einem Schmied am Blasebalg. Über dem Feuer der Esse hängt ein großer Kessel. Der Schmied fasst *Ganykka* mit einer großen Zange, trennt seinen Kopf ab und zerstückelt ihn (wieder). Die Körperteile kocht er im Kessel drei Jahre lang. Den Kopf aber schmiedet er auf einem Amboss, während er ihm die Geheimnisse der Heilkunst verrät – etwa, woran man erkennt, wieweit eine Krankheit fortgeschritten sei, wo keine Hilfe mehr möglich ist oder wo Aussicht auf Heilung besteht.

Danach lehrte er ihm noch Zeichen und Symbole lesen und die Sprache der Pflanzen zu verstehen. Dann setzte er den Körper wieder zusammen und trug ihn auf den Gipfel eines Berges, wo der Kanditat erwachte und sich, wieder zurück in seinem grobstofflichen Körper, zu Hause in seiner Jurte befand.

Solche Beispiele bedenkend, erweitert sich der Begriff dessen, was wir gemeinhin unter ‚aus der Natur lernen' verstehen um den in der ‚Schule der Natur' seit Urzeiten enthaltenen initiatischen Faktor.

Was ist diese ‚Schule der Natur', von der *Paracelsus* behauptet, dass nur aus ihr ein ‚wahrer' Heilkundiger hervorgehen kann? Beinahe wäre man aufgrund besagter ‚Irritation der Heilkunst' bzw. des ‚aus dem Begriff Geratens' im Abendland der letzten Jahrhunderte geneigt gewesen, zur Heilkunst (also zur Naturheilkunde) – gleichsam als Pol – den Begriff ‚Schulmedizin' anzunehmen. Als Polarität sind diese beiden Begriffe jedoch untauglich, weder sich ergänzend noch entgegenstehend, denn wenn es dabei um ‚Heilkunst' geht, dann ist es unweigerlich eben ‚Naturheilkunde' und damit auch ‚Schulmedizin' im eigentlichen Sinne des Wortes.

Anstandsgemäß dem Paradigmenwechsel folgend, den der Mensch stets vollzieht, um in der Welt zu überleben bzw. diese lebenswert zu erhalten, muss man bedenken, dass es nirgendwo je einen Heilkundigen gab noch gibt, der nicht ‚geschult' wäre. Da nun Heilkunst nichts anderes ist als Naturheilkunde, ist der Begriff ‚Naturheilkunde' gleichbedeutend mit ‚Schulmedizin'. Und je mehr eine Sache wirklich Heilkunst ist, umso weniger gibt es darin eine tatsächliche Polarität von Systemen. Es gibt zwar zahlreiche Fakultäten und Wege zum ‚Heilen', aber grundsätzlich haben die Kundigen in ihren Schulen und von ihren Lehrern gelernt, was Menschsein bedeutet, und arbeiten von Natur aus liebevoll, selbstlos und ohne Neid oder Böswillen wo nötig zusammen. Sie sind stetig ernsthaft um diese Haltung bemüht, weil sie ihre Entwicklung nicht gefährden und die Sache selbst mit großem Respekt ehren wollen. Alle heilkundlichen Systeme entstammen dem selben Hintergrund und fließen zusammen in ein und demselben Bild – dem des Menschen in der Natur.

Wer unterrichtet in den ‚Schulen' der Heilkunst? Im besten Falle heilkundige Menschen, deren Wissen auf den ‚vier Säulen' fußt, die *Paracelsus* definiert, naturkundige Menschen, welche die Natur kennen, weil sie in ihrem jeweiligen Umgang mit ihr Alchymisten sind, und die Natur selbst. Jeder Heilkundige ist bei einem oder mehreren Lehrern in die Schule gegangen. *Paracelsus, Tanckius, Thurneisser, Cardilucio* und all die Anderen, bei Kräuterfrauen, Fahrenden, Bergleuten, Schmieden, Astrologen, Alchymisten, Bauern und Winzern. Andere gehen in die Schule bei Vidyas, Schamanen, Hakims, Lamas oder Großmüttern.

Sie lernen das älteste Wissen in den ältesten Schulen – Schulen, die wahrscheinlich bestehen, seit der Mensch selbst diese Erde belebt. Die Schulungsorte sind mannigfaltig. Wiesen, Wälder, Hütten, Höhlen, Ashrams, Zelte, Hallen, Akademien und was man sich noch alles vorstellen kann – alles gleich bedeutend, wenn nur Schüler und Schülerinnen dort die ‚Kunst' erlangen, und zwar – paracelsisch ausgedrückt – ‚von Gott und durch die Natur'.

Die vier Säulen der Heilkunst

Paracelsus nennt die Alchymie die ‚dritte Säule der Heilkunst'. Die anderen drei Säulen sind:

Anatomie – die Kenntnis des physischen Körpers, seiner Säfte, Gewebe und Organe, deren kosmologische Positionen, Funktionen und Signaturen. Die Beziehungen dieser Teile untereinander sowie zur äußeren Natur, einschließlich der Lehre über Art und Entstehen der Formen, Körper, Gewebe usf.

Astronomie/Astrologie – einerseits als Kenntnis der tatsächlichen Position der Himmelskörper und deren Wirken und Einfluß, andererseits als Verständnis der allegorischen Bedeutungen des Zodiaks und der Manifestationen seiner Bildekräfte, wie etwa des Zusammenhangs zwischen Vorgängen in der Natur, Vorgängen im menschlichen Gemüt, Signaturen und Funktionen der Körperorgane, Einwirkung der Planetenkräfte und philosophischer Hintergrund der spagyrischen Prozessen.

Wesenheit/Tugend/Theosophie/ Philosophie – Selbst Quintessenz sein, mit dem Wissen über die Beschaffenheit von Mensch, Natur und Kosmos und eine entsprechende Haltung, wie es die Natur von einem Wesen erwartet, dass sie für einen Menschen, ganz besonders dann noch für einen Heilkundigen, halten soll.

Jede Säule ist mit den anderen so sehr verbunden, dass *Paracelsus* betont:

„Alles ist ein einziges Wissen, und nur
ein Teil davon für sich alleine ist nichts."

Die verschiedenen Fakultäten dieser Schule sind solcherart mit Lernenden und Praktizierenden belegt, wie es deren ‚Hintergrund' entspricht. Es werden zum einen Traditionen gepflegt, welche im Rahmen von Familien, Clans und Völkerschaften üblich sind und zahlreiche Bereiche der Heilkunst beinhalten können, sei es, dass nur ein bestimmtes Rezept über Generationen weitervererbt wird,oder alleine die Techniken, Brüche und Verrenkungen zu richten und Wunden zu heilen.

Woanders werden bestimmte ‚alchymistische Stückchen' jahrhundertelang überliefert, aber auch ganze Systeme des Umgangs mit Heilpflanzen und Systeme des Heilens insgesamt, die Zubereitung von Kräuterpulvern, Ölen, Pillen und Rezepturen sowie einfache spagyrische oder hohe alchymistische Verfahren. Die einen befunden dazu durch direktes Erkennen des Problems, andere durch verschiedentlich übliche und wieder andere durch außergewöhnliche Techniken. Anderen wird das Heilwissen nicht direkt vererbt, sondern sie werden durch Vision oder Intuition in die Schule der Natur aufgenommen.

In allen Weltgegenden wirken Heilkundige verschiedenster Fakultäten. Zum Teil besitzen sie ein ‚philosophisches' Grundwissen, sei es erlernt oder aus der Natur direkt erkannt. Zum Teil besitzen sie kein solches Wissen, entweder weil sie es für die Art ihrer Arbeit nicht brauchen oder weil es vergessen wurde. Während tibetische Medizinmönche und hawaiianische Schamanen sich gleichermaßen darüber sorgen, dass ihrer ‚jüngeren Generation' die Notwendigkeit des Wissens über das ‚fünfte Element' ausgeredet wird, sind die Kräuterleute im Abendland schon lange nicht mehr mit der Aktualität solcher ‚philosophischer' Themen konfrontiert. Hier hat die ‚Irritation der Heilkunst' schon lange vorher begonnen, wodurch der naturphilosophische Aspekt weitestgehend ausgelöscht worden ist, während dieses Phänomen – als eine letzte Woge des ‚dunklen Zeitalters' – anderswo erst die letzteren Generationen erfasst. Dementsprechend sind die Berichte der Vertreter jenes überaus reichen Schatzes abendländischer, naturheilkundlicher Traditionen allerorten zeitgemäß ‚gefärbt' – nicht durch ‚philosophische Probleme', sondern von Angst um das Fortbestehen der ‚einfachsten' Zweige der Heilkunst.

Auf einer Insel im Südosten Europas wirkte bis vor etwa 30 Jahren ein berühmter Heilkundiger namens *Babianos*. Es soll kaum jemanden gegeben haben, dem er nicht helfen konnte. Dann wurde ein Arzt in der Stadt stationiert, einer von der Art, die *Paracelsus* des öfteren erwähnt.

In dem Land aber herrschte damals schon offene Diktatur, und so wurde dem Kundigen seine Arbeit verboten und die armen Menschen dem ‚Neuen' ausgeliefert. Der zeigte bald, was er verstand, und die Leute, die ihn versucht hatten, kehrten ihm sofort wieder den Rücken und blieben ihrem *Babianos* treu. Bald darauf folgte die Anzeige und *Babianos* wurde vor den Richter gebracht. Er kam dorthin – aber nicht alleine. Eine Menge Volk begleitete ihn und – eine junge Ziege. *Babianos* sagte nicht viel, hatte sich auch in nichts zu rechtfertigen, doch er trat mit der Ziege im Arm vor den Richter, und mit einigen schnellen, zarten Griffen renkte er ihr die Glieder aus. Er setzte das arme Tier auf den Boden – es fiel sofort kläglich zusammen. Er nahm es wieder auf und reichte es dem ‚Arzt': ‚Stell sie hin', befahl er. Der versuchte es nun wirklich, bemühte sich auch sehr, verließ mit der Ziege den Raum, eilte in sein kleines Hospital, arbeitete mit Gehilfen und Gerätschaft an der Sache, während Richter und Volk warteten.

Nach gut zwei Stunden kam er zornig wieder, die lahme Ziege im Arm. *Babianos* ergriff sie sofort – ein paar schnelle, zarte Griffe, und das Tier sprang glücklich davon, nur von dem einen Wunsch getrieben, nie mehr wieder in eine ähnliche Situation zu geraten. *Babianos* wurde sehr alt, und viele konnten von ihm lernen. Sie wirken heute noch insgeheim.

Beherrschte er nicht die Kunst, wie man etwas löst und wieder zusammenfügt? Er war auch ansonsten ein guter ‚*Iatros*' und verstand sich auf allerlei Kräuter, aufs Rezeptieren und vor allem auf gute Worte. Es gibt keinen Grund anzunehmen, dass das, was er vermochte, nicht bereits vor Jahrzehntausenden zur Schule der Heilkunst gehörte.

In manchen Gegenden wirken immer noch Kräuterkundige, die imstande sind, überaus kunstvolle Rezepturen, bestehend aus Dutzenden Kräutern, Wurzeln und Gewürzen, zu erstellen. Einige von ihnen erwarben von Kindheit an ein generationenaltes Wissen. Wieviele Generationen konnte so ein Wissen alt sein? Gar nicht, denn es hat niemals wirklich einen ‚Ersten' gegeben. Jeder hat wieder bei jemandem gelernt. Das ist die ‚Schule', von der *Paracelsus* sprach.

So gibt es eine ‚Alte' im Gebirge – sie muss schon an die neunzig Jahre alt sein – die empfängt den Fremden bescheiden und scheu, erkennt bald seinen Zustand und verabschiedet ihn nach erfolgter Konsultation mit herzhaft guten Wünschen und einer Mischung von etwa 40 Kräutern handvermessen. (Auch über 70 Kräuter soll sie zu mischen verstehen). Ihre heilsame Arbeit stellt sie nicht in Rechnung, freut sich aber ehrlich über einen geringen Ausgleich – womit kann denn Gesundheit bezahlt werden? Manche bezahlen sie reichlich, besonders wenn sie ihre letzte Hilfe war, andere bringen einen Laib Käse mit. Und wenn ihre Teemischung bald darauf das schwere Leberleiden beseitigt hat, aber ein lästiger Kopfschmerz geblieben ist, dann addiert sie noch ein einziges Kraut, und auch das Kopfweh verschwindet.

Besagte Kundige ist etwa neunzig Jahre alt, hellwach und noch halbwegs tüchtig im Körper. Ihre Großvater war weit berühmter als sie. Der hatte von seinem Vater gelernt. Auch so an die neunzig Jahre alt oder darüber, das ergibt eine ungefähre Geburtszeit Ende Siebzehnhundert. Damit sind wir schon weit über ‚Omas Kamillentee' hinaus bei einer echten Tradition, dergleichen Vertreter in der Urzeit wohl genauso wie bei lebendigen Naturvölkern in höchsten Ehren gehalten wurden. Die Natur selbst bestätigt ihre Kompetenz. Das ist es, was *Paracelsus* meinte mit: Ein Heilkundiger wird man ‚von Gott durch die Natur.'

Noch vor weniger als hundert Jahren gab es unzählige Beispiele solcher Rezeptierkunst. Die Kinder und Enkelkinder der Kundigen verstehen zumeist die Bedeutung der Notizen und Manuskripte nicht mehr und verlieren dieses Wissen im Staub der Dachböden oder in den Händen solcher, denen sie die Rezepte vertrauensvoll zum ‚Prüfen auf Tauglichkeit' überlassen.

Ein Anderer aus den südlichen Hügelländern wurde auch sehr alt. Aber keiner konnte wirklich sagen, wie alt. Er lebte im siebzehnten Jahrhundert, unterrichtete die Leute, welche zu ihm kamen, in allerlei Dingen und wirkte insgesamt typisch iatrochymisch. Er war ‚Erzieher', Heilkundiger und verstand sich auf Destillationen und Schmelzungen. Wahrscheinlich hatte er ‚ein chymisches Stückchen' gelernt, welches ihm das hohe Alter (und eine große Kinderschar) vergönnte. Seine Nachfahren erzählen noch heute von ihm.

Es kam auf, dass er der Obrigkeit ‚zu' alt wurde. Die einfachste Lösung war damals: Der ist mit dem Teufel im Bunde. Letztendlich wurde das Urteil mit dem Schwert vollstreckt. Dagegen war auch sein Elixir machtlos. Einige Generationen später versucht eine späte Verwandte ‚in seine Fußstapfen zu treten'. Nur ist sie vorsichtiger, ja eigentlich ängstlich. Sie meint, dass es heute gar nicht nötig, sei in den Ruf eines Elixir-Besitzers zu geraten (was sie auch nicht ist), weil man alleine damit schon ‚gefährlich' lebe, wenn man anderen mit Kräutern hilft.

Eine Andere erfühlt den Bruch und die Verrenkung mit birkenzweigengleichen Fingern, setzt alles zurecht und streicht mit einer großen Vogelfeder eine ‚geheime Schmiere' aus Schwefel, Leinöl und Baumharzen auf. Die Schmerzen lassen nach und der Bruch verheilt, als wäre er nie gewesen. Gerade erst war ein junger Mann mit einem älteren, noch immer schmerzenden Bruch angekommen. Seine Hand wurde damals im Hospital verarztet. Sie flucht (neben ihrem Rosenkranz) und murmelt etwas wie ‚gefühllos', knackt den Bruch mit einem geschickten Griff auf und schafft es beinahe, den stechenden Schmerz zu überrumpeln, indem sie mit den nächsten Griffen alles richtig setzt, ‚beschmiert' und einbindet.

Einmal, da hat man ihr die Vogelfeder von offizieller Seite weggenommen. Eigentlich hätte man ihr ja ein festeres Haus bauen sollen, denn ihres war nicht im besten Zustand. Man hätte sie loben sollen und viele Leute zu ihr schicken, damit sie ihre Kunst erlernen könnten. Man hätte ihr jede erdenkliche Erleichterung zukommen lassen sollen, damit sie sich im hohen Alter völlig ihrer Kunst widmen kann. Und jetzt, wo sie gestorben ist, sollte man ihr ein Denkmal errichten, welches viele Menschen dazu anregen würde, ihr nachzustreben. Wir haben aber bereits über die Signaturen von ‚Kultur' gesprochen.

Man tat das alles nicht, sondern man beschlagnahmte ihre Vogelfeder. Ein Erbstück, ein Familien-Werkzeug, und vielleicht schon über 100 Jahre im Einsatz. Die Begründung für den Raub war: Da mussten doch jede Menge Bakterien und Ungeziefer daran kleben, mit denen sie die Verletzten verseuchte. Beherzte Freunde von ihr erreichten eine genaue Untersuchung der Feder – sie war völlig rein und ihr Gebrauch wurde für unbedenklich befunden. Man brachte sie ihr zurück.

Ihre ‚stinkende, schwarze Schmiere' bereitete sie mit Hilfe von ‚Pech und Schwefel'. Deshalb hielten dann auch Knochen und Gelenke ‚wie Pech und Schwefel' zusammen, meinte sie. Solcherlei Rezepte sind unsagbar alt und in den spagyrischen Traditionen bis heute bekannt und gebräuchlich. Bisweilen mischte sie ihrer ‚Schmiere' das eine oder andere Kraut bei, besonders gerne die Beinwellwurzel. Von ihr hielt sie große Stücke. Der Gebrauch des Beinwells ist durch archäologische Bodenfunde seit der Steinzeit nachgewiesen. Erst vor wenigen Jahren wurde diese Pflanze ‚verboten', und man weiß nun endlich, dass sich die Menschen vom Beinwell jahrtausendelang haben täuschen lassen, da er in Wahrheit ein sehr gefährliches Kraut ist. Es klingt lächerlich, aber es ist verwerflichster Zynismus. Wenns aber trotzdem jemand glaubt, dann ist er eben die Kunst jener Alten nicht wert und soll sehen, wie er mit seinen Brüchen zurechtkommt.

Birkenteer machen:

Einen großen Topf ein wenig über Erdniveau herausragend eingraben. In die Mitte des Topfes einen kleineren Topf stellen. Den großen Topf mit einer Platte oder einem Deckel bedecken, am besten mit einer Vertiefung nach unten. Dieser Deckel muss in der Mitte ein Loch haben. Einen anderen Topf mit klein zerrissenen Birkenrinden vollstopfen und mit einem Drahtgitter verschließen. Dann diesen Topf auf den Deckel des eingegrabenen Topfes stülpen. Wenn kein Drahtgitter vorhanden ist, um ein Durchfallen kleiner Rindenteile zu verhindern, können einige geignete Steine in der Mitte (um das Loch) des Deckels platziert werden. Danach den überstülpten Topf und den Deckel mit reichlich Lehm verlutieren. Wenn das *Lutum* trocken ist, entfacht man ein oder zwei Schritte daneben ein Feuer. Daraus kann man nun Glut und heiße Asche über die Vorrichtung schaufeln, dazwischen auch immer wieder kleinere Holzstückchen und Reisig mischen, sodass eine schnelle, starke Hitze auf die Rinden einwirken kann. Nach ca. 30 Minuten nicht mehr nachlegen und abkühlen lassen. Wenn der Haufen noch gut warm ist, Asche entfernen und *Lutum* vorsichtig brechen. Vorsicht mit den freiwerdenden Gasen. Im kleineren Topf unter der Erde befindet sich das reine, flüssige Birkenpech, im größeren ein wäßriges Kondensat. Die Rinde ist weitestgehend verascht. Ein Teil der Asche kann verrieben und mit Pech vermischt werden, sofern man dieses als Klebstoff verwenden will. Zu diesen Arbeiten vorher die Hände einfetten.

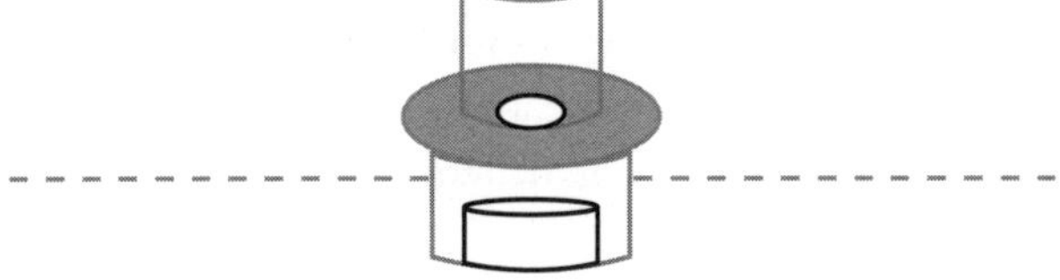

Andere besitzen aus irgenwelchen Gründen ein intuitives Naturverstehen. Vielleicht bringen sie das aus vergangenen Geburten bereits mit. Sie erkennen die Verwandtschaften zwischen Pflanzenwelt und Menschenkörper oder nehmen Bilder hinter dem Natur- oder Krankheitsgeschehen wahr und vermögen daraus überraschend effektiv an leichtere und schwere Leiden heilend heranzugehen.

Wie jene Alte, die ihren Besucher ansieht und dabei an Landschaften und Pflanzen erinnert wird. Sie hat einen völlig intuitiven Zugang zur Sache und meint – obwohl völlig ‚ungebildet' – den hätte sie aus ihrem vorigen Leben mitbekommen. Heute noch bestehen bei fast allen Naturvölkern die beiden Möglichkeiten zum Heilkundigen, Medizinmann/-frau, Zauberer oder Schamanen erwählt zu werden, nebeneinander: durch Erbe oder durch unmittelbare Bestimmung aus der Natur – das heißt: je nach Tradition ‚durch die Geister', die ‚Götter', die ‚Ahnen', durch ‚Gott' oder – in den meisten Fällen – durch mehrere dieser Instanzen zusammen. Auch in der alchymistischen Literatur finden wir stets den Hinweis: Wem Gott es nicht zuteilt, der kann die Kunst nicht erlernen.

Besagte ‚Alte' empfindet so: ‚Der sieht aus wie ein Rosenbusch'. Oder: ‚Der sieht aus wie ein kahler Berg'. Oder: ‚... wie ein Birkenhain.' Dann sucht sie nach Kräutern in ihrem Garten, die zur Vollständigkeit (oder ‚Heilung') der Szenerie passen. Wenn aber einer anklopft, der aussieht wie eine Pflanze im Wald, die eigentlich in so einem Wald gar nicht vorkommt, dann verjagt sie ihn rau von der Schwelle.

Mit den Erzählungen über und von solchen Menschen, mit ihren Erfahrungen und Begebenheiten, könnte man dicke Bücher füllen. Aber nicht, um diese Bücher dann in Museen abzulegen unter dem Titel ‚Es war einmal'. Denn darin steckt kaum verborgen das zynische Grinsen „zum Glück sind wir sie heute los."

So sind unzählige Fakultäten in der ‚Schule der Heilkunst' der Natur, und je nach ‚Kultur' befindet sich auch der Grad der Unterrichtsfächer. Das ist nun eine der Bedeutungen, welche die Alchymie in der abendländischen Heilkunst hat: Sie ist eine ‚Kultur' darin – die ‚Kultur des Naturverständnisses'. Sie ist nicht die ‚Schule der Heilkunst' selbst, sondern die ‚Hohe Schule' für jene, die in der ‚Schule der Natur' bereits aufgenommen sind – und doch ist beides auch eine einzige Sache.

Im fernen Indien hatte kürzlich ein Sohn, Schüler seines Vaters, Grund zum Feiern. Er war nun 30 Jahre alt geworden und konnte rechtzeitig zu diesem Alter alle alchymistischen Arbeiten vollenden, einschließlich der höchsten. So war es in der Tradition seit Genaerationen (für den Idealfall) vorgesehen. Nunmehr stand ihm der Weg offen die Heilkunst zu erlernen – nachdem er mit dem ‚rechten Umgang mit den Dingen der Natur' vertraut war.

Während des Festes erzählte er aus seiner ‚Schule'. Sein Vater hatte ihn auf die Art unterrichtet, indem er ihm nie eine Frage direkt beantwortete. Dadurch lernte er von Kindheit an sehr genau beobachten, selbst nachdenken und selbst probieren. Wenn er bisweilen Fragen stellte wie: ‚Wie kann ich das machen?' oder ‚Wie ist das zu verstehen?', dann empfing er darauf als Anweisung: ‚Sieh genau hin' oder ‚denk selbst nach'.

Ganz anders als sein Vater hatte der Großvater vieles genau erklärt und sogar niedergeschrieben. Die Art des Vaters aber, ihn zu lehren, wurde durch ein Ereignis in dessen Leben bewirkt:

Jener (Vater) war schon in jungen Jahren ein berühmter Heilkundiger geworden, der es verstand, mit großer Sicherheit zu befunden und die Ursache eines Leidens zu erkennen. Dazu erwarb er ein derartiges Kräuterwissen, dass er die ‚Tugend' jeder Pflanze, auch in fremden Ländern, sofort erkannte. Doch so sehr er sich bemühte, er konnte die hohen alchymistischen Prozesse nicht verstehen. Er zog sich in ein einsames Haus im Wald zurück und übte sich in Versenkung, im Lesen von Traktaten, im Nachdenken, im Schmelzen und Destillieren und blieb dort sechs oder sieben Jahre lang. Doch er begriff nicht das Wesen des Quecksilbers – und damit, seinem archaiischen Naturverständnis gemäß, auch nicht das Wesen des Naturgeistes (Merkurs), als auch das der Natur selbst. Darunter litt er so sehr, dass er meinte, die Heilkunst bald aufgeben zu wollen, denn sie sei schließlich nicht vollendet.

Traurig machte er sich auf zu einer langen, einsamen Wanderschaft in die Berge. Eines Tages begegnete ihm ein würdiger Mann, der sehr alt und zugleich jung zu sein schien. Sie unterhielten sich lange – auf seine Fragen bekam er aber keine direkten Antworten. Und was seine brennendste Frage betraf: Der Weise teilt ihm darüber direkt nichts mit. Er zog bloß ein Fläschchen mit fließendem Quecksilber hervor, goss es in eine Schale und streute ein wenig rötliches Pulver darüber. In wenigen Augenblicken wurde das Metall verwandelt. Davon aßen beide ein Krümelchen, und auf die unausgesprochene Frage kam die Antwort: ‚Sieh zu und denk nach'.

In seinem Innersten berührt, folgte bald ein Verstehen dem andern, und er führte danach alle Arbeiten mit großer Kunstfertigkeit aus. Ein einziges Mal hatte er seinem Sohn diese Begebenheit erzählt, nämlich gerade jetzt, als diesem selbst das ‚Werk' gelungen war, und er fügte der Erzählung bei, dass dort in den Bergen und auch woanders solche Menschen leben, die man aber von sich aus nicht finden konnte.

Solcherart sind die ‚Schulen der Heilkunst' – nach dem Grad der ‚Kultur' verschieden – und die Menschen wählen darin die Fächer nach Möglichkeit, Notwendigkeit oder Hintergrund – bewusst oder durch Fügung.

Paracelsus erklärt, dass Lehre und Heilung gleichermaßen von „Gott durch die Natur" kommen. (... Wer meint, von Gott und der Natur losgelöst zu heilen, der ist ein großer Künstler in der Kunst, zu betrügen – sich selbst und die anderen...) Glücklicherweise versteht man das heute, aber zu *Paracelsus*' Zeiten mussten sich manche noch solche Reden anhören:

„... das beweist ihr Werk und bezeugen ihre eigenen Kranken, desgleichen die Natur an sich selbst und aller Grund, auf dem die Arznei steht. Sie wissen nicht eine einzige Krankheit mit gewisser und tröstlicher Arznei zu heilen, während doch Gott nicht einen solchen ungewissen, sondern einen gewissen Arzt haben will. Gibt er dem Ackermann, dem Steinmetz usw. Gewißheit, noch viel mehr gibt er sie dem Arzt, (...). Doch sie machen daraus einen verzweifelten Grund und meinen dabei nach Gottes Willen zu verfahren. Und dann verlangen sie, dass die Hand Gottes solche Bescheißerei verteidigt, und sie hätten recht und Gott Unrecht. Ihre ‚Kunst' wäre gerecht – und Gott hat dieses Recht gebrochen. Sind das keine Bescheißer, so wird es keiner mehr."

nach: *Paracelsus*, >*Paragranum*<

Wenn *Paracelsus* die Natur als seine „große Lehrmeisterin' bezeichnet, dann heißt das: er ist bei ihr in die Lehre gegangen. Er wurde von ihr als Schüler angenommen. Das ist keine Selbstverständlichkeit, sondern setzt gewisse menschliche Qualitäten, entsprechende Lebensführung und Haltung voraus. Ähnlich wie die Aufnahme als Student an eine Universität verschiedenen Kriterien unterliegt, gilt das auch für Anwärter, welche eine Akzeptanz in die Hohe Schule der Natur anstreben. Auch diese Schule steht jedem offen, sofern er die dazu nötigen Voraussetzungen mit sich bringt. Ohne diese Voraussetzungen kann man zwar ‚die Natur von außen erforschen' (ihre grobstofflichen Aspekte), solange man will, wird aber nie die Verständnis-Schwelle zum Inneren, Wesentlichen hin (zu Sinn und Kräftewirken) überschreiten können.

„Es sind Etliche, die da nur die Natur für sich erforschen. Die haben die Erze, Tiere, Bäume, Kräuter, Metalle und köstliche Steine beschrieben.

Die andern aber sind viel vortrefflicher, die nicht allein die Natur, sondern auch endlich das Geheimnis der Natur und ihre inwendigen heimlichen Orte, auf das fleißigste und schärfste nachsuchen und durchgehen. Die haben mit einem wahrhaftigeren Worte den Namen eines ‚*Philosophi*' gebraucht und zugeeignet bekommen."

nach: >*Turba Philosophorum* / Kap.: Das Buch von der Kunst *Chymia* ... <

Das ‚den Dingen nachdenken' bedeutet hier ‚hermetisches Denken'. Dabei steht dem Intellekt bloß die Rolle eines Werkzeuges zu. Das Wesentliche ist das ‚Dahinterschauen' – das Geheimnis darin entdecken. Nicht der Verstand forscht alleine, sondern vornehmlich das Herz.

„Das soll viel mehr geschehen und geschieht doch zum wenigsten oder gar nicht, dass ein *Philosophus* aller Dinge Natur, soviel ihm möglich ist, erkennen lernt.

Denn die natürliche Kunst (...) ist die, so man lernet der Natur Kräfte und Werk erkennen. Der Hirsch lockt die Schlange durch seinen Atem aus der Steinritze und frißt sie. Wegen der großen Hitze ihres Giftes aber, dürstet ihn danach nach einer frischen Quelle.

Es trug sich auch einmal zu, dass ich eine große Kröte gesehen habe, über der ließ sich eine große Spinne von oben herunter und sticht sie. Die Kröte tut einen hellen Pfiff, läuft auf, schwillt an und kriecht, kaum dass sie es vermag, zu einem Kräutlein Wegebreit (lateinisch *Plantago* genannt). Von dessen Blättern ißt sie und wird alsbald gesund, und die Geschwulst ihres Leibes setzt sich, sodass nichts mehr davon zu spüren gewesen.

Da machte sich die Kröte zum andern mal auf und hüpfet wieder dorthin. Wie sie kommt, lässt sich die Spinne gar geschwinde wieder hernieder und sticht sie abermal. Die Kröte aber tut wieder einen hellen Pfiff, läuft abermals dick auf und macht sich wieder zu dem vorigen Kräutlein Wegebreit. Und wie sie abermal davon gegessen, geneset sie und wird gesund. Sie lässt auch zum drittenmal nicht ab, und wieder ist die Spinne geschwind vorhanden und sticht sie. Die Kröte tut wieder einen hellen Pfiff, läuft auf und eilt zurück zum Kraut Wegebreit (...).

Allhier mag sich ein *Philosophus* und *Medicus* üben (...) und die Ursachen anzeigen. Woher kommt dieser Streit zwischen der Spinne und der Kröte? Auf welche Weise ist der Spinne Gift kräftiger und mächtiger, als der Kröte Gift? Worin liegt die Kraft und Tugend des Wegebreit, dass es so geschwinde der Spinne Gift wegnimmt? Warum löscht und stillt der Hirsch das Gift der Schlange mit einem Trunk frischen Wassers? Warum operiert das Gift der Spinne so schnell im Leib der Kröte und weicht dann dennoch so geschwind Gift und Geschwulst nur durch des Kräutleins Anbiß? (...) Die Feindschaft zwischen Spinne und Kröte ist wohl Etlichen wissend.

In solchen Künsten haben die Alten studiert. Darin steckt große Weisheit verborgen und auch, wie der Mensch sich halten soll in seinem Leben. Aber dieser Kunst gedenkt man jetzo nicht in den hohen Schulen, und es wissen die Bauern mehr davon, als unsere hohen Magister, weswegen man diese auch unbillig ‚natürliche Narren' nennt, die mit soviel Kosten und Mühe nur hintersich lernen und des Teufels Spottvögel sind. (...)

Die Wahrheit in solchen (und unzähligen mehr) Dingen ist am Tage und kann nicht geleugnet werden, es sei denn einer spricht, dass der Tag finster und die Nacht licht wäre. Wer aber die *Magiam naturalem* weiß und darin so kundig ist, wie die Weisen, die aus dem Orient zu *Christo* kamen, der kann selbst der Lehrer und Magister von *Aristoteles* und *Galen* mitsamt deren Schülern sein. Gott aber gibt dieses Verständnis nicht allen Menschen, sondern nur den Frommen und die Ihn darum bitten. Die eine böse Art haben, dürfen sich darum gar nicht kümmern.

Dergleichen mehr unzählige Werke der Natur sind unsere Lehrmeister. Sie erklären des Menschen Natur und Wesen in vielen Dingen. (...) Der Schüler muss jederzeit in seinem Studieren und Meditieren (über die Schriften) auf die Natur sehen und ihre geheime Wirkung betrachten. Tut er solches nicht und weicht davon ab, dann gebiert er *Monstra* und bringt Mißgeburt. Daneben muss auch die Handarbeit einhergehen und mit Verstand geübt werden, auf daß, wenn geirret worden, der Irrtum aus der *Practica* billich corrigirt werden möge."

nach: *Joachim Tanckius, >Promptuarium Alchemiae<*, 1610

Die Lehrer dieser Schulen

Die alten Texte weisen vielfach darauf hin, dass die Kunst nicht jedem offenbart werde, sondern nur den ‚Gutmütigen'. Wer aber waren diese ‚Gutmütigen', und vor allem: wer die Offenbarer unter den Menschen, nämlich die Übersetzer der Zeichen und die Vermittler der Kunst, denn alles Natur-Wissen ist durch die Natur offenbar. Auf die an *Paracelsus* gerichtete Frage, warum er behaupte, dass dieser oder jener in Wahrheit kein ‚Künstler' sei, antwortete er: „Sonst würden wir uns kennen." Was bedeutet das?

Zum einen gab es innerhalb der Traditionen und Schulen immer schon eigene ‚Chroniken'. Da heißt es: ‚Dieser und jener wirkte zu der Zeit und hatte solche Qualitäten. Er gab sein Wissen an den und den, an die und die weiter,' u.s.f. Die vielen Namen sind dem ‚gemeinen Historiker' oft gänzlich unbekannt, zumindest weiß ein Außenstehender weder über Rolle und Kompetenz der einzelnen Adepten Bescheid, noch darüber, in welchem Zusammenhang sie untereinander standen. Zudem gab es zwischen den Kundigen neben fachlichen Verständnisgrundlagen auch geistige bzw. menschliche. Und es gibt ein zutiefst erfühltes ‚Wiedererkennen' quer durch Raum und Zeit, wie zum Beispiel irgendein zeitgenössischer Künstler seines Faches jeglichen Künstler der Vergangenheit in seinem Wesen und seiner Kunstfertigkeit ‚erkennen' kann. Dumpfheit, Neid, Missgunst, Dogmatismus und dergleichen waren und sind ein deutlicher Beweis dafür, dass die ‚Schule' verfehlt wurde – ‚man kennt sich nicht'.

Das sind fürs erste klare Kriterien der Kundigen und haben vermutlich auch auf die ‚Bekannten' des *Paracelsus* zugetroffen, welche er als ‚Kundige' kannte. Aber kann man gewisse Qualitäten diesbezüglich nicht auch von denen annehmen, über die sich *Paracelsus* so gerne auslässt – von *Avicenna, Hippokrates, Dioskurides, Galen* und all den anderen?

Von wem wusste *Paracelsus* denn nun wirklich solche Dinge, wie man sie mitunter ausschließlich und mit dieser besonderen Kompetenz in seinen Texten findet? Welche Lehrer nannte er nicht beim Namen? Konnten in Europa nach den kulturellen Verwüstungen, welche das römische Reich hinterlassen hatte, vielleicht Traditionen weitergeführt werden, die noch dazu weder typisch griechisch, lateinisch oder auch arabisch waren? Lag darin ein weiterer Grund für die Vehemenz der Inquisition – um etwaige Reste ‚verborgener' Traditionen, die wohl nicht bedeutungslos waren, auszutilgen? Es mangelt an Texten, aber nicht an Hinweisen. Wen von den Alten beschimpfte *Paracelsus* ständig und behauptet von ihnen, sie ‚hättens nur gelesen aber nicht gewusst' und schrieben nur nach dem Hörensagen? Eben die ‚Schreiber' wie *Avicenna, Hippokrates, Dioskurides* u. a. sowie deren Nachfolger. Waren sie alle wirklich nur Abschreiber und ‚keine in der wahren Philosophie Kundige und deshalb ein Nichts'?

Paracelsus lässt sich darüber immer wieder aus, weil ihn die Dummheit der Nachplapperer und Abschreiber schmerzte, denn er war ein Praktiker und ein in den Dingen der Natur tatsächlich Wissender. Und Wissende scheinen die Art zu haben, Ihresgleichen zu erkennen sowie andere auch. Heute wissen wir, dass die meisten, noch bis vor hundert Jahren, dem *Hippokrates* zugeordneten Texte gar nicht von diesem stammen. Die Texte sind Stückwerk, die Lehren unvollständig und unverstanden, und *Paracelsus* hatte recht. „Sie werden mir Recht geben müssen", „es wird sich erweisen", hatte er immer wieder prophezeit. Hinter *Hippokrates,* seinem Schriftwerk, seiner Art und seinem Bildnis steckt vermutlich mehr Legende, als hinter der Sagengestalt *Merlins*. Nur eben auf andere Art. Beide aber mögen in ihrem echten Leben weise Menschen und große Persönlichkeiten gewesen sein.

Hippokrates

Die einzigen zeitgenössischen Hinweise auf *Hippokrates* finden wir in zwei Dialogen des *Platon*. Im Dialog *Phaidros* erwähnt dieser, dass *Hippokrates* lehrt, man könne über den menschlichen Körper nichts erfahren, wenn man keine Kenntnis der gesamten Natur, keine Kenntnis des Ganzen überhaupt und keine Kenntnis der Natur der menschlichen Seele hätte. Er lehrt auch vom Wesen und von der Scheidung der Dinge, von Auswirkungen und Rückwirkungen u. dgl. Diese Stellen bei *Platons ‚Sokrates'* ehren den echten *Hippokrates*.

Das zweite und (bis auf eine Anmerkung in der >Politik< des *Aristoteles)* letzte bekannte Dokument über den historischen *Hippokrates* ist ein kurzer Traktat von *Meno* (einem Schüler des *Aristoteles)*, welcher niederschrieb, wie er es von *Aristoteles* gehört hatte, was *Hippokrates* über den Ursprung von Krankheiten sagte. Die Inhalte sind nicht neu, und in der arabischen und indischen Medizin *(Unani* und *Ayurveda)* damals wie heute geläufig. Dabei geht es um den Wind (die Winde) und Ernährungsfehler als Ursache von Leiden und kurz um die ‚Körpersäfte'. Der Traktat endet mit der Bemerkung *Menos*: „Man muss aber sagen, dass das, was *Aristoteles* von *Hippokrates* berichtet, nicht mit dem übereinstimmt, was *Hippokrates* selbst (...) sagt." Hatte die Galionsfigur der abendländischen ‚Schulmedizin' gar bei arabischen Hakims und indischen Weisen gelernt? Dann müsste man dem Vorbild doch unweigerlich darin folgen.

Der *Hippokrates*, gegen den sich *Paracelsus* wendet, ist nicht der historische Mensch, sondern Papier, welches seinen Namen trägt. Ähnlich ist man einige Jahrhunderte später mit einem Mann aus Palästina verfahren, einem Schüler spiritueller, asiatischer Traditionen. Es interessierte nicht, wer er war, noch was er meinte, und noch viel weniger, woher sein Wissen stammte, sondern aufgrund dessen, was man von ihm wollte, wurden Jahrhunderte nach ihm einige ausgesuchte Texte in ein Buch verpackt und dieses Buch zur einzigen Wahrheit erklärt.

Vermutlich ist *Hippokrates* durch ebendiese, ihm zugeschobenen Schriften eine verdrehte Heldenfigur geworden, in der Zukunft ein Sinnbild für den Verfall der Heilkunst im Abendland, ähnlich wie – allerdings in umgekehrter Bedeutung – *Isaak Newton*, der ‚große Physiker' im Grunde Alchymist war, wie seine (den strenggläubigen Physikern peinlichen) alchymistischen und hermetischen Texte bezeugen. Die dem *Hippokrates* zugeschriebene Textsammlung *Corpus Hippocraticum* ist bestenfalls zu einem geringen Teil wirklich sein Werk. Die verschiedenen Bücher der Sammlung entstammen nachweislich der Feder verschiedener Autoren, und diese wiederum gehörten auch verschiedenen Schulen an, sind also nicht einmal allesamt seine Schüler gewesen. Es schrieben Vertreter der Schule von *Kos*, von *Knidos*, sizilianischer Schulen und andere, jeder nach seinem Wissen oder oft nur nach seiner Meinung. Das dadurch entstandene, insgesamt im Wesentlichen irreführende Stückwerk wurde zum Dogma erklärt und auf solchem Sand ein mächtig protziges Gebäude erbaut, an dem *Paracelsus* litt. Soweit auf diese Schriften ‚angeblich' auch *Galen* (131-200), *Rhazes* (850-932) und *Avicenna* (980-1037) bauten, richteten sich *Paracelsus*' Reden auch gegen diese – nicht gegen die Personen, sondern gegen die ihren Namen tragenden Flickwerke. *Paracelsus* wollte kein Stückwerk, sondern (mit seinen Worten) „einen ganzen Arzt" – im Sinne des von ihm beschriebenen ‚Philosophen' – einen iatrochymischen Künstler.

„Wer Eines beurteilt, der weiß das Andere, und obwohl da verschiedene Namen sind, sind aber die Künste nicht geschieden oder geschieden die Wissen – das ist Wissenschaft (*scientiae*). Denn eins ist in allen. (...)

Damit ich aber den *Philosophus* beschreibe, so wißt, dass er in zwei Wegen wächst: Einer ist im Himmel, der andere in der Erde. Obwohl das hier in Bezug auf beide Seiten ein *Philosophus* genannt wird, so ist es doch dem Namen nach nicht so. Sondern der ist ein *Philosophus*, der die untere Sphäre erkennt. Derjenige der die obere weiß, ist ein *Astronomus*. Es sind aber auch beide *Astronomi*, denn der ist ein *Astronomus*, der da das Herkommen der Metalle und die Eigenschaft der Erze weiß. Der ist auch ein *Astronomus*, der die Früchte der Erde kennt (...), der den Saturn und Jupiter usw. weiß und kennt. Hingegen ist auch der ein *Philosophus*, der des Himmels Impression, seinen Einfluß und seinen Lauf weiß. Auch ist der ein *Philosophus*, der die Luft kennt, ebensowohl wie der, der die Erde kennt. Denn das, was die Natur betrifft, ist die Philosophie.

Nun ist es eine (einzige) Anatomie, eine Essenz, eine Materie in den vier (Säulen). Denn der Saturn ist nicht allein im Himmel, sondern auch im Untersten des Meeres und im Hohlsten der Erde. Die Melisse ist nicht allein im Garten, sondern auch in der Luft und auch im Himmel. Was meint ihr was Venus sei, als allein der Beifuß? Und was der Beifuß, als allein Venus? Und was sind sie beide? Matrix und Mutter, Konzeption, Empfangen und Samengefäß. Was also ist das Eisen? Nichts als Mars. Was ist Mars? Nichts als Eisen. Das heißt, sie sind beide Mars. Dasselbe ist auch die Brennnessel, auch das Harz – und es ist alles eins! Wer Mars kennt, der kennt das Eisen, und wer das Eisen kennt, der weiß was Mars ist, und wer diese kennt, der weiß was Harz ist und auch was die Brennnessel ist.

Darum so ist ein *Philosophus* einer, der eins in dem einen weiß und der auch das selbe in den andern weiß mit dem alleine die Formen betreffenden Unterschied und nichts weiter. Und das deshalb, weil nicht vier (Säulen) sind, sondern hier ist nur eins.

Das ergibt nun: Wer meint Philosoph zu sein und nur in einem bekannt ist, der ist aber nichts. Etliche sind in zweien bekannt – ist auch nichts und nichts in dreien. In allen vieren bekannt ist dann schon etwas. Wenn sie aber alle viere einschließen, so ist es ganz da. Das ist: die Arznei ist das Ganze und Letzte all dieser Dinge. Denn wozu soll das gut sein, wenn der *Astronomus* den Regen und den Schnee weiß, und er weiß aber nicht wozu sie tauglich sind? So ist er nichts. Er muss auch seines Subjekts Eigenschaft wissen – ohne das ist er nichts. Ganz sein macht den *Medicus*."

nach: *Paracelsus, >Paragranum<*

Woher aber hatte *Paracelsus* sein Wissen, wenn nicht aus den Schriften der zuvor Genannten und ihresgleichen? Von Vater, Bauern, Bergleuten, Fahrenden, vom ‚Lichte der Natur', von *Trismosin* – soweit klar. Wenn er aber ‚seine Monarchie in der Heilkunst' proklamiert und dabei die Schreiber und Abschreiber verachtet, dann muss man zumindest überlegen, ob nicht dort, wo auch er ehrfurchtsvoll von den ‚Alten' sprach, damit vor allem solche gemeint waren, die prinzipiell nicht oder kaum geschrieben haben. Wie wäre es sonst zu verstehen, dass er andererseits vielen von derlei ‚Alten', welche in den folgenden Jahrhunderten immer wieder gepriesen, zitiert und hochgestellt wurden, eben nicht jene Größe zuspricht. Wenngleich sich darunter klangvolle Namen befinden – aber wenn das, was durch sie transportiert wird, weder der Erwartung (der ‚wahren Kundigen') noch dem Bedürfnis (der Menschen) entspricht – ? Womöglich kamen auf einen bekannten, ‚großen' Namen hundert oder mehr Unbekannte, die in Wahrheit weit ‚größer' waren.

Das ‚unbekannt bleiben' jener Vielen mag einer großen Demut, einer persönlichen Eigenart oder einem besonderen Auftrag entsprungen sein, es war aber schon zur Römerzeit auch eine Überlebenschance für die im alten Wissen Kundigen und ist es in der Folgezeit geblieben.

Die Druiden

Es ist deshalb sehr wahrscheinlich, dass *Paracelsus* manch mündliche Lehren angenommen hatte, welche schon Jahrhunderte zuvor die Druiden *Cathbadh, Divitiacus, Dumnorix* und andere in abgeschiedenen Hainen vorgetragen hatte.

Für die Griechen gehörten die Druiden in denselben Rang wie alle anderen, welche sie ‚Philosophen' nannten. Wenn wir heute von den ‚alten Philosophen' sprechen, dann denken wir dabei an griechische, alexandrinische, vielleicht auch an ein paar römische, aber kaum darüber hinaus. Wenn die Alten von ‚Philosophen' sprachen, dann meinten sie damit unter all den Anderen auch jene der druidischen und ähnlicher Schulen. Diese ‚Philosophen', diese ‚Kinder des Wissens', wie man sie nannte, beschrieben kaum Papier, aber sie standen auf allen vier Säulen.

Mela bezeichnet sie als ‚Meister der Weisheit' *(Magistri sapientiae). Ammianus* betont, dass ‚sie sich zu Erforschung verborgener und hoher Dinge erhoben hätten'. Der Alexandriner *Dio Chrysostomus* stellt sie in eine Reihe ‚mit den Philosophen Persiens, Ägyptens und Indiens'. *Klemens von Alexandria* geht ihretwegen so weit, den Ursprung der Philosophie bei den Kelten zu suchen. *Cäsar* schreibt in seinem >Gallischen Krieg< (6.14): „Sie (die Druiden) stellen (...) häufig Erörterungen an über die Gestirne und ihre Bahn, über die Größe der Welt und des Erdkreises, über die Natur der Dinge, ...".

Sie wussten sogar noch über das Gesetz der Wiedergeburt Bescheid – ein Wissen, das in weiten Teilen der westlichen Welt zur Zeit der Römer bereits verlorengegangen war und später sogar per Dogma verboten wurde. *Ammianus* schrieb, dass ‚die Druiden das sterbliche Los des Körpers verachteten' und die Unsterblichkeit der Seele lehrten, was noch Jahrhunderte später in den Lehren der (von ihren Schulen inspirierten) *Katharer* weiterwirkte.

Lukan, Plinius, Mela, sie alle berichten von den astronomischen Kenntnissen der Druiden. Sie berechneten Kalender anhand des Mondes, wie andere es schon Jahrtausende vor ihnen taten, und benutzten pythagoräische Mathematik, wobei man nicht sagen kann, wer von wem gelernt hatte oder ob sich hier einfach altes Wissen ergänzte.

Sie waren als Weise die eigentlichen Führer des Volkes und standen über Königen oder Stammesfürsten. Von dieser Regelung – und davon, dass solches von Volk und Fürsten gerne und mit Vertrauen getragen wurde – konnten die griechischen Philosophen jener Zeit nur mehr träumen. Sie nannten sie ‚die Gerechtesten aller Menschen', welche ‚die Sprache der Götter sprechen'.

Rudolf Steiner bemerkt in seinem Vortrag über die Druiden, dass diese gegen alle Krankheiten Mittel gehabt hätten, bloß gegen die Schreiberei und Schriftgelehrsamkeit nicht. Menschen, die ‚einfach so' Zeichen schrieben, galten für die Druiden als krank, und obschon sie kein ausschließliches und dogmatisches ‚Schriftverbot' kannten, hielten sie eine aus gemalten Zeichen entstehende, intellektuelle Zivilisation für unheilbar dem Tode geweiht. Da klingt doch direkt *Paracelsus* durch.

Hierzu muss man Folgendes bedenken: Sowohl in der Alchymie als auch in der Heilkunst insgesamt war es stets wesentlich, dass die Kunst ‚funktionierte'. Weltanschauliche oder religiöse Hintergründe konnten bestenfalls zweitrangig sein. Wenn sich ein System in der Praxis bewährte, dann bewies es dadurch seine Stimmigkeit im Leben und im Umgang mit den Dingen der Natur. Diese durchaus natürliche Einstellung bewirkte immer wieder, dass es ernsthaften und intensiven Austausch zwischen den Kundigen verschiedener Völker und Traditionen gab. Ganz anders verhielt es sich ja mit den religiösen Modellen, welche alleine auf ‚Glauben' beruhten, und wenn dieser fehlte, per Dogma erzwungen wurden.

Die Praxisorientiertheit bewirkte auch, dass die Druiden, Filids und andere, die man eigentlich als ‚Kundige' der keltischen Traditionen kennt, die ‚Kunst' und das Wissen derjenigen Kundigen, welche vor ihrer Einwanderung nach Europa hier gewirkt hatten, ganz selbstverständlich annahmen, integrierten – oder einfach wiedererkannten. So gab es zum Beispiel durch die Einwanderung der Kelten in Irland (etwa um 300 v. Chr.) keinerlei ‚kulturellen Bruch' zur dort ansässigen jungsteinzeitlichen Bevölkerung. Deren Kundige aber verstanden sich bereits damals seit etwa 3000 Jahren auf die Arbeit mit Metallen, insbesondere Kupfer und Zinn, und es gibt keinen Grund anzunehmen, dass es sich dabei ausschließlich um das handelte, was wir heute unter ‚Schmiedekunst' verstehen, sondern dass vielmehr einhergehend mit der Herstellung von Äxten, Musikinstrumenten, Schwertern und Gefäßen aus Metallen und Legierungen auch eine hohe alchymistische Kunst bestanden hatte. Ja selbst lange Zeit, bevor Metalle zu alltäglichen Gerätschaften verarbeitet wurden, konnten sie – aufgrund des damals sehr hohen Naturverständnisses – ohne weiteres in alchymistischen Tätigkeiten und Heilmittelpräparationen verwendet worden sein. Wieder lohnt hier der Vergleich mit der noch lebendigen ayurvedischen Alchymie, wo, jahrtausendelanger mündlicher Überlieferung folgend, hochkomplexe Aufschließungsverfahren von Mineralien und Metallen zu medizinischen und anderen Zwecken durchgeführt werden. Die Vorgangsweisen deuten nicht nur auf eine höchststehende Wissenschaft der Substanzen hin, sondern auf ein ebenso hohes Wissen über Beziehungen und Zusammenhänge zwischen den ‚natürlichen Dingen' des Pflanzen- und Mineralreiches bzw. darüber hinaus der kosmischen Einflüsse und des menschlichen Körpers, was wiederum alle anderen Aspekte der Heilkunst, von der Befundung bis zur Rezeptierung, einschließt. In diesem Sinne muss man sich das Wissen und Wirken der ‚steinzeitlichen Kundigen' vorstellen. Und auf genau diese Art wurde Heilkunst und Alchymie auch tradiert bzw. intertraditional überliefert. Die Kunst musste funktionieren. Für den Künstler, für den Suchenden und Lernenden spielte es dabei keine große Rolle, welchem Volk oder welcher Religion der eine oder andere Kollege, mit dem er sich austauschte oder von dem er lernte, angehörte.

Diese Irrelevanz künstlich geschaffener oder durch Volkszugehörigkeit bedingter Grenzen brachten stets jene zum Ausdruck, deren Wissen ‚stimmig' und in der Praxis bewiesen war. So wie alle Weisen und Heiligen der spirituellen Schulen ein Wissen alleine den Menschen und Gott betreffend, in verschiedenen Sprachen zwar und verschiedenen Terminologien, aber in der Essenz ganz das selbe – wie es eben der Natur der Sache entspricht – vermittelten, genauso erwarben die Heilkundigen und Alchymisten, ja genaugenommen eben die ‚Handwerker und Künstler' der verschiedensten Völker und Kulturen das Wissen von ein und der selben Natur und lernten deswegen schon in der ‚Grundschule' zwischenmenschliche Grenzen zu überschreiten, weil diese dem Erlernen und Ausüben ihrer Kunst bloß hinderlich wären und ohnehin – insbesondere die religiösen Abgrenzungen – keinen Grund in der Natur haben. Wir erinnern uns dazu des Initiationsberichtes jenes samojedischen Schamanen, welcher in einer großen Birke Kundige vieler Völker sah und erkannte, dass ‚alle an einer Sache arbeiten'.

Es scheint allerdings diese Einstellung wiederum in verschiedenen Völkern, Religionen und zu verschiedenen Zeiten auch nur relativ vorhanden gewesen zu sein, und wir können annehmen, dass nicht jeder ‚Kundige' sie konsequent lebte. Es gibt zum Beispiel heute noch in Indien eine den Brahmanen (der hinduistischen Priesterkaste) oft typische Intoleranz Nicht-Brahmanen gegenüber, obwohl der Rang meist bloß vererbt wird. Da nun hauptsächlich Brahmanen im *Ayurveda* gelehrt sind, fällt ihnen der Austausch und die Weitergabe ihres Wissens an ‚Andere' oft sehr schwer. Ähnliches findet oder fand man auch bei anderen Völkern, wobei das Problem darin liegen dürfte, dass wohl schon seit ältesten Zeiten die Formel Wissen=Macht ein Unterbau hierophantischer Strukturen war. Ein wenig paradox – und damit vielleicht typisch indisch – ist es, dass die Schmiede seit altersher oft in der Alchymie sehr kundig, ja mancherorts schlichtweg ursprüngliche Träger dieses Wissens waren, aber im hinduistischen Religionssystem zu den Kastenlosen und Ausgestoßenen gehörten, also an das untere, dem den Brahmanen entgegengesetzte Ende der Gesellschaftsstruktur verbannt sind. In der Praxis fällt jedoch auf – ganz entsprechend unserem Thema – dass die Kundigen, je höher sie aus Sicht der Natur stehen mögen, egal ob Brahmane oder nicht, und wenn sie nicht nur im allgemeinen *Ayurveda*, sondern auch in der Alchymie kundig sind (also nicht nur *Vidyas*, sondern auch *Rasa Shastris* sind), diese religionsüblichen Trennungen für irrelevant halten und eine ganz natürliche Toleranz üben. Sie besitzen dann die natürliche Autorität und Kompetenz zu lernen und zu lehren wo und wen sie wollen. Solche Dinge scheinen ganz einfach ‚in der Natur des Menschen' zu liegen. Je weniger jemand in der einen oder anderen Weise entwickelt ist, umso sturer beharrt er auf seiner gesellschaftlichen Position und den daraus folgernden Vorrechten und Abgrenzungen. Umgekehrt werden diese Dinge immer bedeutungsloser.

Wenn der Rang eines ‚Kundigen' (im einen oder anderen Wissen) bloß vererbt wurde, ohne dass durch die Natur, durch gewisse Prüfungen oder wodurch auch immer die Fähigkeit, dieses Wissen recht zu nutzen, gesichert werden konnte, war man wieder in jener Situation angelangt, welche *Paracelsus* stets zu neuen Schimpfausbrüchen anregte. Bei den Naturvölkern war das in der Regel nicht so der Fall, weder heute noch vermutlich vor Jahrtausenden. Dort finden wir fast überall alle möglichen Zugänge zur ‚Kunst' offenstehen – ein Schamane konnte (und kann) gewählt, durch Erbe bestimmt, durch ein besonderes Ereignis ‚gezeigt' oder durch eigenen Wunsch, Eignung und Bemühen ‚hervorgebracht' werden. Die Prüfungen bleiben ihm auf keinen Fall erspart – Prüfungen welche durch ‚die Natur', durch Geister oder tatsächliche (ältere) Kundige abgenommen werden und nicht durch ‚kunstfremde' Personen, wie das in den Reglementversuchen der abendländischen, ‚offiziellen Naturheilkunde' noch üblich ist, wo allzu oft Leute prüfen und bewerten, gewähren oder ablehnen, die gar nicht imstande sind, das Potential eines Kundigen wahrzunehmen, geschweige denn aus Sicht der Natur ein geringes Wissen haben.

Doch zurück zu unserem Versuch eines Rückblicks in die Vergangenheit der ‚Kunst'. Es zeichnet die druidische Tradition besonders aus, dass sie einen fließenden Übergang von den älteren, seit vielen Jahrtausenden regional vorherrschenden bis zu den neuzeitlichen Traditionen begleitete.

Wesentliche Teile der selben Traditionen wurden dann später im christianisierten Abendland in der Alchymie weitergeführt. Es ist also nicht nur so, dass wir uns unter den ‚Alten' Griechen und Araber vorstellen müssen, sondern auch die Angehörigen der, (zumindest) bis ins 6. Jhdt. bestehenden, Druidenschulen. Das fällt nur nicht so auf, weil daraus kaum schriftliche Dokumente bekannt sind. Andererseits weiß man, dass gerade auch alchymistische Lehren sehr oft nur mündlich weitergegeben wurden. Es ist aber dann nicht alleine ein Wissen, welches wir als ein ‚Wissen der Druiden' bezeichnen könnten, sondern zudem noch ein sehr viel älteres, wofür (ebenso wie für seine Vertreter) heute kaum Bezeichnungen üblich oder tauglich sind und am ehesten noch die aus den ethnologisch-paläonthologisch vergleichenden Erkenntnissen bestätigte Nähe zu den urzeitlichen Schamanen, Zauberern (oder wie man sie eben nennen will) sinnvoll erscheint. Deshalb wollen wir noch einmal den Versuch wagen und weiter in der Zeit zurückblicken, damit nicht der ‚vorgestellte' Druide als ein Urbild des ‚alten' Alchymisten zu sehr manifest wird, wenngleich dies bis zu einem gewissen Grade sicherlich seine Richtigkeit hat. Zu solch forschendem Rückblick gibt es mehrere Herangehensweisen: Zum einen kann man die Angelegenheit ‚von außen' betrachten und dann versuchen, einen Blick auf das Innere zu erhaschen. Zum andern kann man sie ‚aus der Sache selbst', also ‚von Innen' betrachten, was allerdings für ‚Außenstehende' oft schwer nachvollziehbar ist.

'Von außen betrachtet' kann man ja die archäologische Sicht genauso zu Hilfe nehmen, wie etwa die 'chemische Sicht' dort und da eine kleine Türe zum Verständnis eines alchymistisichen Prozesses öffnen kann. Kohlereste zur Datierung erinnern den Hermetiker etwa an den Begriff 'Feuerkunst'. Wenn zum Beispiel die Datierung der ältesten bislang entdeckten menschlichen Wohnstätte Irlands (vor über 9000 Jahren) durch 'zufällig' gefundene verkohlte Haselnussschalen möglich wurde, bestünde eine erste Überlegung (im Zusammenhang) darin, wozu man wohl damals Haselnussschalen verascht oder genau damit geheizt hatte. Man mag natürlich sagen: um ein wärmendes Feuer zu nähren. Aber gewisse Dinge bergen beinahe aufdringlich einen umfassenderen, tieferen informativen Charakter. Die Haselnuss ist eine typisch merkuriale Pflanze, wächst an den Waldrand 'heraus', will gesehen werden, sich dem Menschen mitteilen.

Zu heilkundlichen Zwecken wurde immer schon 'verascht'. Die Ureinwohner Papua-Neuguineas stellen zum Beispiel heute noch ein pflanzlich-mineralisches Präparat mittels Veraschung her. Sie gehen zu einer Quelle, deren Wasser bestimmte – in ihrer Terminologie entsprechend bezeichnete – Minerale enthält. Da hinein legen sie eine gewisse Zeit lang die Kräuter der Rezeptur. Danach werden diese schnell verascht und die Asche mit frischem, zerkautem Ingwer zu Pillen geformt. Das Verfahren erinnert an eine grobe Art ayurvedischer Präparation. Über die Verwendung von Aschen und deren Salze in der neuzeitlichen Alchymie bzw. Iatrochymie und im *Ayurveda* (dort: *Ksharas*) werden wir im praktischen Teil noch einiges besprechen.

Zur Veraschung dienen Töpfe und Tiegel oder einfach nur Steine, lehmiger Boden und eine saubere Feuerstelle. Dass mit diesem reduziertem Aufwand alle Grade alchymistischer Arbeiten ausgeführt werden können, bezeugen die neuzeitlichen Alchymisten selbst (und natürlich auch die ayurvedischen *Rasa Shastris*).

> „Darum so wisse, dass die Alten keine Gläser gehabt, doch haben sie den Stein der weisen Meister wahrhaftig gemacht, und will also die ältesten Arbeiten am Anfang beschreiben, welche die kürzesten sind. Es gehen solche aus dem metallischen und mineralischen Reiche (...).
>
> Nimm von der saturnalischen *Minera Antimonii* 1 Pfund, glühe zuvor in einem guten Tiegel guten Stahl 1 Loth, mische das *Antimonium* darunter, dazu ein halbes Viertelpfund Weinsteinsalz, und wenn alles wohl fließet, so trage auch ein Viertel Pfund Salpeter nach und nach darauf, und wenn es fließt wie ein Öl, dann gieße es in einen warmen, mit Fett geschmierten Inguß. Wenn es erkaltet stehen die Schlacken oben und der *Regulus* unten. (...)"
>
> nach: *J. G. Tölde*, >*Coelum Reseratum Chymicum*< (Anfang 1600),
> neu: Edit. Hornfisher 1994

Die ganze – gar nicht so kurze – Arbeit wird nur alleine mit guten Tiegeln und ‚Feuerkunst' bewerkstelligt und führt (nach *Tölde*) in der Folge vom Regulus aus zur ‚Roten' und zur ‚Weißen Blume', zwei sehr hohen alchymistischen Präparaten und Ausdrucksformen des männlichen und weiblichen Prinzips.

Die ‚Ältesten' haben ihre Destillationen mittels Tonretorten durchgeführt, wie das heute noch, besonders in Indien der Fall ist. In der abendländischen Literatur findet man gelegentlich den Vermerk, der ‚Künstler' solle weit mehr zum Töpfer als zum Glaser gehen. Tongefäße haben in vielen Arbeiten zweifellos Vorteile gegenüber Gläsern. Natürlich könnte man auch aus ‚Kupferblasen' destillieren. Ob diese im vorgeschichtlichen Abendland verwendet wurden, ist nicht bekannt, wäre aber möglich, weiß man doch selbst aus konventionell-archäologischer Sicht (durch allgemein veröffentlichte Bodenfunde erwiesen) von der Verwendung des Kupfers seit etwa 8000 Jahren. Die hermetischen Traditionen berichten dazu ohnehin von ganz anderen Zeiträumen bzw. höheren und niedrigeren Formen von Technologie, sodass sich der wahre Sachverhalt ‚im Dunkel der Vergangenheit verliert'. Nach den vedischen Überlieferungen, welche (der Erzählung über König *Rama* folgend) ihren ‚Ursprung' in Europa vor etwa 18 000 Jahren hatten, könnte der ‚Bericht' von der ‚Entdeckung' des destillierten Alkohols eine Ahnung vom Alter dieses Verfahrens wecken. Dort sind insbesondere die Erzählungen von *Shiva* und *Parvati* von sehr archaiischem Charakter und sollen aus Zeiten vor über 60 000 Jahren berichten.

Wo war der Ursprung des Alkohols?
In den ältesten Zeiten wusste Jedermann offen bescheid, wenn der Tod nahte. Ein Schmied wollte aber noch nicht sterben und grub deshalb einen langen Tunnel in dem er hintereinander sieben Zimmer errichtete. Die Wände verkleidete er mit starken Eisenplatten und an den Türen brachte er schwere Schlösser an. Der Tod (Dharam Raj) kam, fand den Schmied im hintersten Zimmer im Tunnel und forderte ihn auf mitzukommen. Der Schmied hieß ihn kurz warten und verließ den Raum. Er verließ auch den Tunnel und verschloß einen Raum nach dem anderen, bis der Tod hinter sieben verschlossenen Türen eingesperrt war. Von nun an starben die Menschen nicht mehr, und niemand wusste wo der Tod war.

Das bereitete Shiva große Sorge. Seine Frau Parvati erbot sich das Versteck des Todes ausfindig zu machen. Sie kaufte Getreide und destillierte daraus aus einer großen kupfernen Retorte Schnaps. Dann lud sie alle Leute zum Trinken ein, warnte aber immer, dass sie sterben könnten, wenn sie zuviel trinken. Eines Tages kam auch der Schmied zu ihr. Der Alkohol löste seine Zunge und er entgegnete ihrer Warnung, dass er nichts zu fürchten brauchte, weil er den Tod hinter sieben Schlössern in eisernen Kammern gefangen hätte. Parvati befreite den Tod, welcher füderhin stets heimlich zu den Menschen kam. Die aber verstanden es nun den Schnaps zu bereiten.

Das Eisen

Das Eisen wurde aus archäologischer Sicht seit etwa 5000 Jahren verwendet. Wir wollen dies einmal so belassen, obwohl gewisse Objekte aus Eisen heute auf das Mehrhundertfache zurückdatiert werden. Eine ‚industrielle' Metallurgie in größerem Stil will man erst vor etwa 3200 Jahren in der Kaukasus-Region nachgewiesen haben. Vor dieser Zeit sammelte man das Erz der ‚vom Himmel gefallenen' Meteoriten oder nahm es aus Lagerstätten an der Erdoberfläche und goss es nicht in Formen, um daraus dann Werkzeuge und Waffen zu schmieden, sondern achtete es als ‚heiliges' Material zu ‚rituellen Zwecken'. ‚Rituell' im weitesten Sinne. Man denke nur an die eisenhältigen Erden, welche seit jeher schon als Farben für Körper- und Höhlenmalereien gedient hatten. Wenn es aber ‚rituell' auf den Körper aufgetragen worden ist, dann liegt auch eine arzneiliche Nutzung nahe. Urzeitliche, im *Ayurveda* integrierte Stammestraditionen Indiens kennen eine überraschend aufschlussreiche Art der Befundung: Sie vermischen bestimmte Erden mit Wurzeln und Kräutern und ‚beschlagen' damit den Körper.

Dann befunden sie die Signaturen der Reaktion dieser Mischung – an welchen Stellen sie früher oder später trocknet, wo sie Risse bekommt, wo sie Blasen schlägt und vieles mehr. Auflagen, Pflaster und innere Anwendung eisenhältiger Erden und verschiedener Arten von Sand sind ebenfalls üblich. Die Sublimation eines ‚Eisensulfurs' aus Ockererde mittels Sal Ammoniak hat ebenfalls urzeitlichen Charakter.

Zu dieser alchymistischen Präparation bedarf es nur Tongefäßen und der ‚Feuerkunst'. Als *Sal Ammoniak* wurde der im Sand der Karawanenstraßen versiegte und an der heißen Sonne wieder auskristallisierte Urin der Last- und Reittiere verwendet. Mitunter wurde dem auch ‚gemeines' Salz beigemischt.

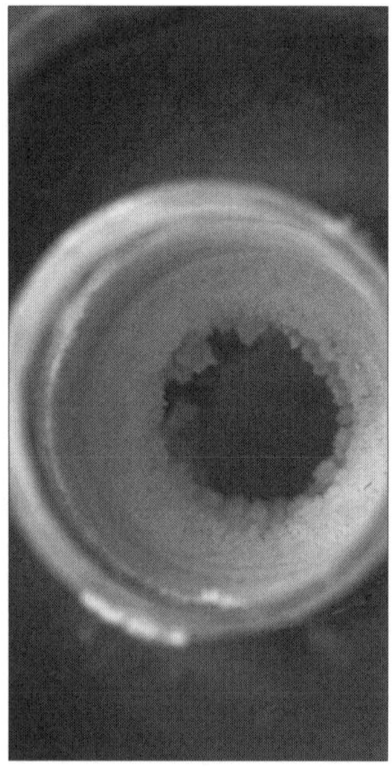

Diese Arbeit ist auch insofern interessant, als dass hier schließlich in der Folge mit Hilfe von Eisen das Kupfer aufgeschlossen wird. Es ist damit einer der wenigen Wege, das Kupfermetall (der alchymistischen Literatur zufolge) in einen ‚genießbaren (also ungiftigen) und zur Arznei tauglichen Zustand' zu transformieren, und weist darüberhinaus noch auf weit ‚höhere' alchymistische Prozesse. Nicht ganz unähnlich haben wir die Verwendung des Eisens bereits in der Arbeit zum *Regulus Antimonii* bei *Monsieur d'Anvers* kennengelernt. *Tölde* bezeichnet das Antimon als eine ‚saturnalische' Materie. Beides – Antimon und Kupfer – entsprechen dem Erd-Element. Dass Ausdrucksformen des Erd-Elements durch das Feuer-Element (Eisen) sowohl hervorgebracht als auch geöffnet werden, ist uraltes, hermetisches Wissen. Denken wir dabei nur an die Positionen von Feuer und Erde in der pentagrammatischen Darstellung der Elemente.

„Wiewohl nun dieses, was *Helmontius* hier rühmet, mit guter Wahrheit von dem *Sulphure Antimonii* gesagt werden kann, so gebührt dennoch dies herrliche Lob vielmehr der Tinctur des Kupfers, weil es von der Natur höher geadelt ist als jenes. Deshalb wird seine *Anima* von *Paracelso, Basilio* und anderen *Sulphur Philosophorum* genannt, weil dessen Kraft die Kranken in ihren Schmerzen und großen Nöten erfreuen kann. Deshalb wird dieses von Vielen mit großem Verlangen gesucht, ist aber bisher von fast keinem gefunden worden, und man hört allenthalben die Klagen bei ihnen, dass sie zu einem solchen Geheimnis nicht gelangen und diese güldische Tinctur vom aussätzigen Leibe (des Kupfers – Komm.) nicht absondern können.

Nachdem ich unsers *Philosophi* lehrreiche Bücher fleißig gelesen und doch keinen Begriff wie zur Aufknüpfung dieses Geheimnisses zu gelangen daraus formieren konnte, hab ich bei mir beschlossen solches durch Ersuchung weiser und verständiger Leute zu erfahren, hab derowegen mir keine Müh der Reise lassen zu schwer sein, sondern hin und wieder in unterschiedlichen Königreichen, auch oft mit großer Lebensgefahr, erfahrne Leute in dieser geheimen Kunst aufgesucht und mit ihnen von diesem hochwichtigen Schatze der Gesundheit conferirt. (...)

(...) Nun müssen wir, so wir Jünger der *Philosophorum* sein wollen, mit ganz andern Augen als der gemeine Mann, den Schwefel anschauen, wenn er durch Kochung mit einem gemeinen Öl dergestalt erweicht wird, dass seine innerste rote Tincur nach auswärts gekehrt wird. Der gemeine Haufen lässt es dabei gut sein wenn er solch eine Kochung (etwa zum *Oleum Lini sulphurisati* – Komm.) vollendet und solches ihm die Krätze hinwegnimmt, die Wunden heilt, gequetschte und gebrannte Glieder genesen lässt oder andere heilsame Wirkungen verrichtet. Aber ein Erforscher der geheimen Weisheit geht tiefer auf den Grund und fragt nach, wie das denn zugehe, dass das Öl unter allen Dingen alleine solche Macht habe, dass es den Schwefel so sanft und natürlich auflöst und sein Inwendiges nach außen kehrt, da doch selbst das stärkste Corrosiv solches nicht vermag. So wir nun solchen Grund recht einsehen, befinden wir, dass diese Auflösung aus Gleichheit und Einwesigkeit des Öls und des Schwefels herkommt (...).

So du den *Salmiac* mit Blut-Stein (oder Ockererde – Komm.) aufs genaueste vermengst und destillierst, so kommt alsbals ein Teil des feurigen urinosischen *Spiritus*, und soviel dessen entbunden ist, soviel Corrosiv hat sich mit dem Blut-Stein verbunden, und wenn man zuletzt etwas stärker Feuer gibt, so steigt auch ein Teil des *Salmiacs* in seiner vorigen Natur auf, aber weil er von doppelter Kraft ist, hat er die subtilste Kraft vom Blut-Stein mit in die Höhe genommen. (...) Das Corrosiv aber musst du unten im *Caput Mortuum* suchen. Dieses stoße klein, wirf es in einen *Spiritus vini*, so wird sich dieses alsbald goldgelb färben. Davon ziehe den *Spiritum* ab, so wirst du ein Salz von balsamischem Geruch finden. Daraus destilliere nach dem Corrosiv sehr schöne Blumen, ganz leicht wie eine Flaum-Feder, glitzernd in vielerlei Farben, mit großer Lust anzusehen. (...)

nach: *Joachim Polemann*, >Vom Geheimnis des *Sulphuris Philosophorum*< 1659

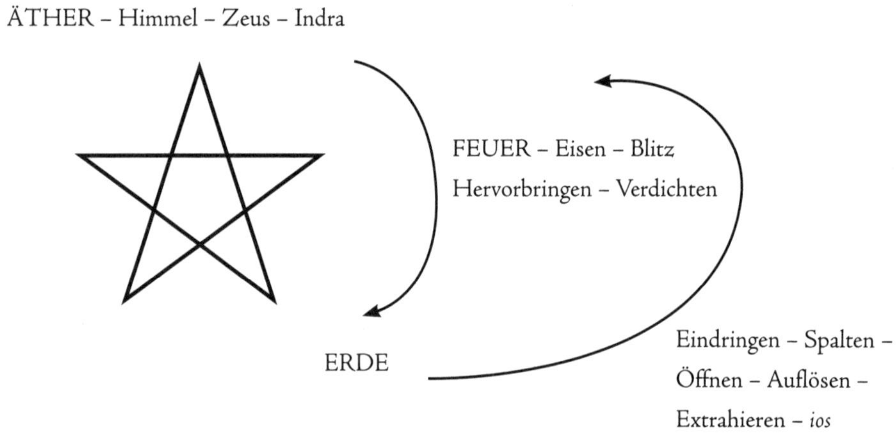

Dass das Eisen eine typische Ausdrucksform jener ‚himmlichen' Kraft ist, welche die Erde sowohl hervorbringt als auch öffnet, war den ‚Alten' bekannt. *Sulphur* ist bei den alten Griechen *theos*. Das Wirken von Eisen auf Erde ist eine Äußerung des Feurig-Göttlichen. Das geschieht etwa durch das Herabfallen von Meteoriten. Von der Urzeit bis vor etwa 3000 Jahren wurde zu Arbeiten und Ritualen mit und um das Eisen fast ausschließlich Meteor-Eisen verwendet. In den Meteoriten erkannte man die Urform der Manifestation von ‚Erd-Element' aus dem Himmel (Äther) vermittels des Feuer-Elements. Und zugleich sah man darin auch die ‚gegenwirkende Feuerkraft', das Öffnen der Erde. Diese große, dem Feuer wesenseigene auflösende und extrahierende Kraft, bezeichnet die ayurvedische Alchymie mit einem einzigen Wort, welches gleichermaßen ‚Eisen', ‚extrahieren' und ‚Flamme' bedeutet, nämlich *Loha* (oder *Luh*). Die fixierende Kraft des Eisens aber lässt den Geist, den Merkur, die Essenz während des Vorgangs des Öffnens nicht entkommen. „Nur mit der Kraft des Mars fängst du den merkurialen Hasen" heißt es in einem alchymistischen Text der frühen Neuzeit.

Der Hervorbringungs- und Nähr-Impuls Feuer-Erde war den ‚Alten' klar. Ob sie auch wussten, dass ein stofflicher Ausdruck dieses Impulses der ist, dass der ‚meteoritische Staub' (bestehend aus Teilchen in Tausendstel-Millimeter-Größe) die Erde derart ‚nährt', dass sie davon einen täglichen (!) Massezuwachs von etwa 6000 Tonnen erfährt?

Eigentlich sind es zwei ‚Dinge', welche auf diesem Wege ‚aus dem Himmel auf und in die Erde' wirken: Eisen und Quarz. Entweder die Meteoriten bestehen (zu etwa 90%) aus Eisen oder es sind (die häufiger vorkommenden) ‚Stein-Meteorite', welche dann zwar auch viel Eisen beinhalten, aber ansonsten hauptsächlich aus *Silicium* bestehen. Ansonsten gibt es natürlich auch Mischformen beider Meteoritenarten. Beides nun – Eisen und Quarz – haben seit der Urzeit genau diese entprechenden Bedeutungen.

Quarz, Kiesel, Bergkristall und Meteoriten

Einerseits gemeinhin als ‚Feuerstein' bekannt, war der Quarz (bzw. Silikate allgemein) tatsächlich ein Ausgangsstoff für die Alchymisten, um den ‚Feuerstein' zu bereiten, also ein hohes alchymistisches Präparat – wiederum tauglich in der Arznei und zur Anwendung und Weiterarbeit in der ‚hohen' Alchymie (s.u.).

Dass der Quarz seinen Namen ‚Feuerstein' davon hat, weil er dazu verwendet worden sein soll, Feuer zu entzünden, ist unwahrscheinlich, da diese Verwendungsmöglichkeit alleine durch die Funken assoziiert wird, welche entstehen, wenn man die Steine aneinander schlägt. Feuermachen mittels Reibung von Hölzern und Zunder ist um vieles praktischer und dürfte wohl eher der übliche Weg gewesen sein.

Die auf der Erde angekommenen Meteoriten stammen aber stets von einem Meteor ab, den die alten Griechen *meteoros* nannten, mit der Bedeutung von ‚in der Luft schwebend'. Neben dieser Bezeichung wurden Meteore allgemein ‚Feuerkugeln' oder eben ‚Feuersteine' genannt. Wenn der ‚Meteor' beim Durchstoßen der Erdatmosphäre nicht vollständig verdampft, schlägt er als ‚Meteorit' auf die Erde nieder. Beim Herabfallen verliert er sein Leuchten, löst aber donnerartigen Lärm aus. Dieselbe Abfolge und die selben Signaturen kennen wir vom Gewitter. Zuerst die Lichterscheinung, dann der Donner. In diesem Zusammenhang ist der ‚Donnerkeil' *Indras (Vaj)* oder *Thors* bzw. ist *Zeus/Jupiter* als Blitzeschleuderer zu verstehen.

ÄTHER – Himmel – Zeus – Indra

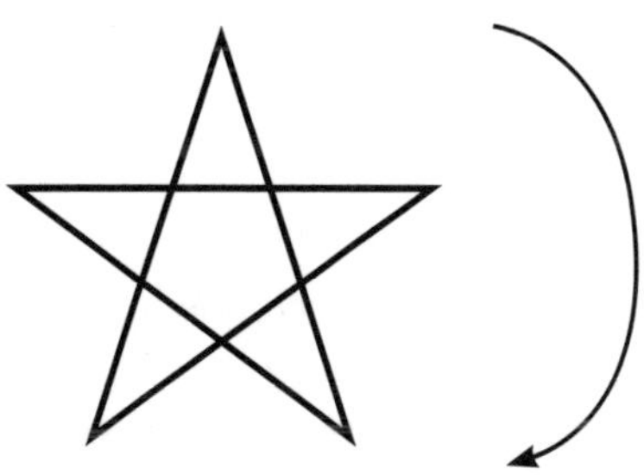

SONNE – ‚oberes Feuer' – Licht – Erdferne

Sulfurisches Spektrum – theos –
‚Göttliches Wirken'

MARS – ‚unteres Feuer' – Kraft – Erdnähe

ERDE

Aufprall und Einsetzen der
‚gegenwirkenden Feuerkraft' –
Öffnen der Erde

Über die Bedeutung des ‚Blitzmals' *(Putreal)* noch bei den Römern haben wir im >Weg des Sonnenfunkens< in Zusammenhang mit den Themen der Putrefaction schon geschrieben und wollen diese Sache deshalb hier beiseite lassen. Nicht nur Orte, an denen der Blitz in die Erde schlug, sondern naheliegend auch solche, wo Meteoriten eingeschlagen haben, wurden seit Urzeiten als besondere Plätze erkannt. Jene keltischen Stämme, in deren Gebiet besonders viele Meteoriten gefunden wurden, schmiedeten daraus in späterer Zeit die Schwerter, welche Rom zum Weltreich machten. Die Kunst dieses spezielle Eisen zu bearbeiten, wurde lange geheim gehalten. Das Geheimnis konnte wohl schließlich Verrat und Erpressung nicht länger standhalten, und die Kelten produzierten und verkauften selbst die Waffen, die ihnen den Tod bringen sollten. Damit wurden auch die als ‚heilig' empfundenen Orte, an denen Meteoriten auf die Erde getroffen sind, entweiht.

Lange Zeit, bevor das Eisen zu Werkzeug und Waffen geschmiedet wurde, machte man diese Dinge aus Quarz. Die gesamte Masse der steinernen Erdkruste besteht hauptsächlich aus Kiesel (Silikat). Weiter im Erdinneren nimmt der Anteil an Eisen und Schwefel bedeutend zu. Dort wirkt – mit den Worten der Alten – das ‚innere Feuer, welches die Metalle zu ihrer Reife kocht' – ein ‚Feuer', ausgedrückt vor allem durch Eisen und Schwefel.

Eisen und Quarz sind also in jeder Weise erdbildende Substanzen – und zugleich auch solche, die die Erde zu ‚öffnen' und – im übertragenen Sinne – zu ‚durchlichten' vermögen. Das weist aus hermetischer Sicht wieder darauf hin, dass, sobald ein ‚Ding' hervorgebracht ist, es seiner Auflösung anheim fällt. Mit dem Werkzeug aus Quarz oder Eisen zerstört man zum Beispiel die Form eines Holzes – man löst die ‚erste' Ausdrucksform auf, und in der Folge erschafft man eine neue Ausdrucksform ‚durch die Kunst des Menschen', indem man aus diesem Holz ein anderes Werkzeug oder Mobilar anfertigt.

Quarze und Eisen sind beide eng mit den Themen ‚jener Auflösung' bzw. dem ‚philosophischen Tod' und der Initiation verbunden. Die Schamanen urzeitlicher Völker benutzen Bergkristall und Quarze für ihre Initiationsrituale. Diese ‚Steine aus Licht' sind verbunden mit visionärem Schauen. Das ‚Schauen in die Kristallkugel' hat sich bis in unsere Zeit als Mittel zum ‚Wahrsagen' erhalten. Da ja die ‚Erde' ursprünglich aus dem Äther kommt, ist der ätherische Raum, der Himmel, eine geistige Vorform dessen, was wir auf der Erde vorfinden, und diese Analogie drückt sich im Kristall am vollkommensten aus. Diese Vorstellung finden wir in vielen sehr alten Traditionen. Heute noch glauben australische Ureinwohner, dass der gesamte Himmel aus Bergkristall besteht und der Thron des Himmelsgottes aus Quarz. Wenn von diesem Thron ein Teil wegbricht, dann wird er zum Meteoriten. Es heißt, dass die Quarze alles widerspiegeln, was auf Erden geschieht und auch, was sich im Kranken abspielt.

Das ist genau die Bedeutung jener ‚Himmelsschau', welcher wir an anderen Stellen noch begegnen werden – zum Beispiel bei der Pulsbefundung – der vielleicht genauesten und tiefgreifendsten Befundungsmethode. Da wir davon an anderer Stelle handeln, soll hier nur angemerkt sein, dass analog dazu die Pulsbefundung auf denselben Umständen bzw. Analogien beruht wie die ‚Himmelsschau', die ‚Vogelschau', die ‚Leberschau', die Astrologie und schließlich auch das ‚Schauen in den Kristall'. Immer werden ‚im Himmel' die ursprünglichen Zusammenhänge und Abläufe dessen erkannt, was sich im erdhaften Körper ausdrückt.

Wer einmal in China, Indien, Tibet oder durch einen arabischen *Hakim* der alten griechischen Schule *(Unani)* erfahren durfte, was die Pulsbefundung alles vermag, und vor allem dann der Sache nachgegangen ist, um ein wenig zu begreifen warum sie das vermag, der wird die analoge Verbindung dieser ‚Befundungsebenen' akzeptieren, wenngleich die ‚Leberschau' der alten Babylonier für manche nichts an ihrem abstoßenden Charakter verliert. Aber auch dort wurde die gewünschte Information aus dem ‚dem Himmel (Äther-Element) entsprechenden Körperteil (Leber) erlangt, analog zu den Informationen, welche das Blut, von der Leber kommend, an den Pulstaststellen zugänglich macht, oder der Kristall ‚widerspiegelt'.

So ist also der physische Quarz ein Ausdruck jener (jovialen/ätherischen) Himmelskraft, welche (vermittels des sulfurischen Spektrums) alles, was sich ereignen und verkörpern soll, manifestiert.

Hierzu mag *Töldes* Traktat >Wie das Feuer der Alten aus dem Silice zu verfertigen, und wie es zu gebrauchen< eine interessante Ergänzung sein. Auch das ist eine sehr archaiisch anmutende Arbeit, welche mit tönernem Geschirr und ‚Feuerkunst' noch heute bewerkstelligt werden könnte, sofern man die Sprache des Textes versteht und vor allem die hier verlangte Destillation (am besten aus einem tönernen Gefäß) schafft. Er deutet an, dass ‚den Alten' dieses Mittel wohl vertraut gewesen sein musste, zu seiner Zeit (Anfang 1600) aber schon zu den schwierigen Geheimnissen der ‚Kunst' gehöre. Das nimmt nicht Wunder, wenn man bedenkt, dass jene ‚Alten' genau wussten, was sie taten und nicht einfach ein Rezept nachkochten. Es war nicht nur das Wissen um die jeweils notwendigen laborantischen Schritte, die Zeitqualitäten und die ‚verborgenen Handgriffe', sondern auch das ‚hermetische Denken', das Naturverständnis, welches seit Urzeiten vermittelt wurde.

Tölde beschreibt dabei die Darstellung eines Mittels zur Weiterarbeit mit den vorhin schon erwähnten ‚Roten und Weißen Blumen'. Ein Auszug des Textes soll hier nur zur Anschauung und gekürzt wiedergegeben werden. Ob die von *Tölde* angegebenen *Termini* der Sache gerecht werden, sei dahingestellt:

„Ich muss dir zeigen das Feuer der Alten, worin sie die obigen Blumen gebraten und also in dieses Feuer gesetzt und solches nachgehends in 3 Stunden verfertigt haben.

Wisse, dass der Alten ihr Feuer in dem Kiesel verborgen ist, und es ist solches der selbe Feuer-Stein, so sie von den rohen Metallen reden, ja es ist der alten *Gluten* ihr Feuer und auch ihr *Sigillum Hermetis*, denn es ist außer diesem in kurzem Wege keine wahre *Fixation* zu hoffen.

Stoße die Kiesel klein wie ein Mehl, tue davon 6 Pfund in eine wohlbeschlagene Retorte, und treibe daraus einen grasgrünen *Liquor*, welcher überaus feurig und penetrant sein wird. Du wirst davon beinahe ein Pfund bekommen, musst aber alles wohl lutieren und mit großer Vorsicht die *Gradus Ignis* observiren. In diesem *Liquor* ist ein wahrhaftiges *Arcanum* für das *Podagra*, die Lähmung der Glieder und für den Stein verborgen, wenn du dazu ein Pfund guten *Spiritum vini* gießt und die beiden überdestillierst.

Ansonsten aber gieße den obigen *Liquor* wieder über ein Pfund frisch gestoßenen Kiesel, lass ein wenig miteinander digerieren, so wirst du sehen wie es sich tummelt dieweil er seinen Vater ermordet und von ihm ausgefahren sich zu seiner Mutter legt und sie ineinander agieren, und einen andern Sohn zeugen, denn sie wird zu einem glutinosischen Schleim, welchen du *per Retortam* überdestillieren musst.

Dieses liquorische Feuer biete nun dem roten Drachen an, welcher ermordet ist, aus dem *Kybric, Atrop* und *Zeubach*, und beständige Rote Blume. Dieses ist nun das Feuer der Natur ohne welches man fehlt, wenn man gedenkt im kurzen Wege etwas Fruchtbares zu erlangen. Das muss durch Feuer geschehen, denn das ist der Stein, der in der Luft wohnt und sich zugleich im Wasser ernährt. Er ist ein Salamander, den das Feuer nicht verzehren kann. Es ist der Stein, welchen der *Saturnus* in der Meinung es sei sein Sohn *Jovis* verschlungen und wieder ausgespien hat, das wahrhaftige *Gluten* und Philosophische Ei, mit seinem roten Dotter unter dem Grünen verborgen, das rechte *Corallinum* der Alten, welches im Wasser zäh und weich und in der Luft erhärtet, ja der wahre grüne rauchende Löwe.

nach: *J. G. Tölde, >Coelum Reseratum Chymicum<*, neu: Edit. Hornfisher 1994

Feuerkunst

Die fixierende und zugleich auflösende Kraft des Feuers bewirkt den Übergang der Materie von einem Zustand in den anderen. Die wesenseigene Ausdrucksart des Feuer-Elements ist das ‚physische' Feuer. Andere Arten sind ‚verborgene' oder (im Grade die höchsten) ‚philosophische' Feuer. ‚Verborgene' Feuer wohnen eigentlich in allen ‚Dingen der Natur', besonders aber in Dingen wie Gold, Eisen, Holz, Kristallen und Salzen. Wer mit dem einen oder anderen dieser Feuer umzugehen vermag, indem er damit den Zustand der Materie transformiert, ist ein ‚Feuerkünstler'. Diese Ansicht hat sich gehalten – von den Töpfern der Urzeit, zu den Schmieden und zu den Alchymisten.

Die Verwendung von Feuer ist eine typische ‚Kunst des Menschen'. Wie das ‚natürliche' Feuer (die Sonne, das Erdinnere oder der Misthaufen) gewisse Dinge gemächlich reifen lässt, bewirkt das ‚Feuer der Kunst' ebenfalls eine Reifung, aber in kurzer, durch die Feuerführung beeinflussbarer Zeit. Wenn man sich in das Empfinden des archaiischen Menschen ein wenig hineinversetzt, dann kommt man zu der Ansicht, er fand durch den Umgang mit Feuer eine Möglichkeit, etwas ‚schneller als die Natur' zu machen, die ‚Zeit der Natur' zu überholen und seine eigene ‚Idee' einzubringen. Damit sind wir beim selben Gedanken angelangt, der auch zum Grund für das Herbeiführen des ‚philosophischen Todes' führt. Zudem erwies sich das Feuer immer schon als Mittel, ‚etwas anderes' zu machen – etwas, das in der Natur noch nicht auf diese Art vorhanden war. Darin liegt auch einer der wesentlichsten Gründe der Alchymie als ‚Kunst des Menschen', welche der ‚Kunst der Natur' hinzugefügt wird, um etwas zu vervollkommnen, das die Natur noch nicht vollkommen ‚ausgebacken' hat.

„Alle Wirkungen geschehen durch den Lauf, den die Zeit gibt, und eine Zeit ist die der äußeren Welt und eine ist die des Menschen. Ehe das Korn wächst wird es in der Erde faul, danach gibt es seine Früchte – so geschieht im Feuer die Zerbrechung. (Im Feuer des Alchymisten) fermentieren sich die Arkane zu ihren Erhöhungen, und deren Zeit ist calcinieren, sublimieren, reverberieren, solvieren usw."

nach: *Paracelsus*, ›*Paragranum*‹

In der ayurvedischen Alchymie gilt die Ansicht, dass ‚Feuer bereits im Holz verborgen ist'. Man muss es nur durch Reibung hervorbringen. Das ist soweit bekannt und vor allem auch in der Hinsicht verständlich, wenn man die pentagrammatische Reihung der Elemente besieht und für ‚Äther' die in der chinesischen Medizin noch gebräuchliche allegorische Bezeichnung ‚Holz' einsetzt.

Der Grund dieser Ansicht ist aber noch ein tieferer: Wenn die Natur ein ‚Ding' (aus ihrem Vermögen) ‚vollendet' hat, also zum Beispiel die ursprüngliche Idee zu einem Stein, einem Baum oder einem Grashalm stofflich manifestiert worden ist, dann ist das Feuer darin eingeschlossen. Dazu sei nur am Rande auf die Verschachtelung der ‚platonischen Körper' hingewiesen, die, wenn sie stimmig zur natürlichen Geometrie vorgenommen wird, den Feuertetraeder (Bild u.: 4. v. li.) im Inneren des Erdwürfels (3. v. li.) zeigt. Dass der Feuer-Körper dabei ein doppelter *(ios)* sein muss, erweist hier sogar die geometrische Notwendigkeit. Verdichtende und lösende Kraft sind im ‚Sterntetraeder' (5. v. li.) verbunden.

Die Alchymie holt dieses Feuer aus der ‚Erde' hervor und vollzieht damit die Veränderung der Materie.

„Das Feuer (des Alchymisten) ist die Erde."

Paracelsus, >Paragranum<

Bei den ‚Alten' war ‚das Feuer im eigenen Körper' die Voraussetzung zum ‚philosophischen Tod' des Menschen selbst, im Sinne einer Initiation. *Yogis* sitzen im Schnee der Berggipfel und rezitieren einen Spruch, damit das innere Feuer entfacht, der Körper in der Kälte durchwärmt wird und sogar der Schnee rundum schmilzt. Die Ureinwohner Australiens ‚füllen' den Körper des Initianden rituell mit Bergkristall, um sein inneres Feuer zu erwecken. Taoistische Alchymisten hüten das innere Feuer, indem sie keine sexuelle Kraft verschwenden.

„(...) denn er hielte von dem Weibernehmen nicht viel, hatte auch mit den Weibern ganz keine Gemeinschaft – glaubte auch nicht, dass er die Zeit seines Lebens eine Frau berühret habe (...)"

Ägidius von der Wiesen über *Paracelsus*. Aus: *Johann Agricola*, >Chymische Medicin<, 1638 / O. Humberg 2000

Mann und Frau in der Kunst

Wie immer gestaltet geht die Meisterung des inneren Feuers mit der kunstvollen Handhabung des äußeren Feuers einher. Die Frau entspricht der (venerischen) Erde. In den Mythen archaiischer Völker birgt die Frau ‚von Natur aus' das Feuer in sich, während der Mann dieses ‚in sich bergen' erst lernen muss. Der Natur nach dem Sulfurischen nahestehend ist er eher veranlagt, sein Feuer ‚nach außen' zu bringen. Der Mann ist der Donnerkeil, der Meteorit, der das in der Erde verborgene Feuer hervorrufen will, um es zu seiner Kunst zu nutzen. Eines der höchsten Kunststücke der ayurvedischen Alchymie ist die Herstellung eines *Shiva-Lingam* (eines Phallus-Symbols) aus ‚fixiertem Quecksilber', also aus ‚reinem Mond' bzw. zur höchsten Stufe gebrachten *Mercurius*. Dergleichen jahrhundertealte Objekte sollen sich noch heute in verborgenen Tempeln Indiens befinden, und es mag den rationalen Materialisten verwundern, dass davon nichts zur Transmutation oder für ein Lebenselixir abgeschabt wird.

> „Jeder Mann, der einen schönen Schmelztiegel bekommen hat, wird sie natürlich von ganzem Herzen lieben."
>
> *Hsui Chen Yen I* (taoistischer Alchymist)

In den Vorstellungen wohl aller Völker und zu jeder Zeit ist das Auftreten und Agieren eines Kundigen (jeglichen Ranges) immer mit Licht und/oder Feuer verbunden. Die ‚Hüterin des Feuers' aber ist die Erde. Darin ist nicht das zum Primitivismus verkommene Bild der ‚Frau am Herd' enthalten. Es besitzt nur die Frau von Natur aus eine Einigkeit mit der erdhaften Natur, welche der Mann erst erwerben kann, wenn er ‚zum Kundigen', zum ‚Feuerkünstler' wird.

Alchymie ist keine typische Männerangelegenheit – diese Hoffnung mag mancher Leser oder mehr noch manche Leserin immer noch hegen, obwohl sie – im Text bis hierher geraten – dabei fast ausschließlich *Masculina* in den Betitelungen, Begriffen und Bezeichnungen der Kundigen vorfanden. Spätestens aber, wenn wir das Bild vom kesselrührenden Urzeit-Weisen aus der verschütteten Erinnerung zu wecken versuchen, wird Manchem neben dem Kessel aus gutem Grunde eine weibliche Gestalt erscheinen. Was tut sie dabei, wo doch er sein Süppchen ganz alleine zu kochen scheint? Sollte man nicht eher ihn, wenn überhaupt anwesend, mit seinen Kollegen im Abseits sitzen und Bier-Rezepte diskutieren sehen? Nun – sobald der Mann kochend am Kessel (oder Tiegel) steht, wird er zum ‚Künstler' transformiert, vom Prozess der Vereinigung der Pole absorbiert, wird Handlanger, reine aktive Kraft, welche in der Erde, im Kessel, wirkt.

Der legendäre *Yü der Große* verstand diese Kunst, und davon berichten viele taoistische Erzählungen. Er war ein „Bergmann, der die Erde gesünder machte, anstatt sie zu verderben." Als solcher trug er den Beinamen ‚Durchbohrer der Berge'. Weil er die männlichen und weiblichen Erze unterscheiden konnte, gelang es ihm, die ‚fünf zauberkräftigen Kessel der *Hsia*' zu gießen. Er goss sie als Paar und Abbild der kosmischen Ganzheit, indem er fünf *Yang*- und vier *Yin*-Kessel anfertigte. Beim Schmelzen halfen Knaben und jungfräuliche Mädchen. Die Mädchen gossen Wasser auf das rotglühende Metall und vollzogen damit eine Vereinigung von Feuer und Wasser. Auch das Herstellen von Legierungen glich bei den Alten einem Hochzeitsritus. Wer weiß, was *Monsieur d'Anvers* dabei empfand, als er das ‚männliche' Eisen mit dem ‚erdigen' Antimon unter Zusatz des feurig-luftigen, ‚kindlichen' Salpeters schmolz und am Ende gar Silber zusetzte.

Was immer wir hier mit der Bezeichnung ‚Kunst' verbinden wollen – die Frau war mindestens ebenso maßgeblich daran beteiligt wie der Mann, und das ist bis heute so geblieben. Seit der Mensch in den beiden Ausdrucksformen Mann-Frau in der Welt lebt (was, wie ja unter anderen, die große Theosophin *H. P. Blavatsky* in der >*Anthropogenesis*< ausführt, nicht immer und nicht ‚von Anfang an' der Fall gewesen sein soll) gilt bei den Kundigen der Grundsatz, dass alleine die Frau imstande ist, den Mann ‚zu entwickeln' und dieser kein ‚Naturverständnis' ohne sie erlangen könnte. Das wiederum liegt ‚in der Natur der Sache', weil die Frau ihrem Wesen nach der Natur näher steht und der Mann in sich die Eignung mitbringt, eine Distanz zur Natur herzustellen. Ohne Frau würde sich der Mann hilflos im großen Webwerk verlieren und nie seinen Zweck erfüllen können. Und ohne Mann würde in der Natur kein Störefried sein, der die Frau dazu veranlasste, sich im Spiegel zu betrachten.

Das ließe sich mit zahlreichen Allegorien aus den Erzählungen aller Kulturen (teils nach dem Thema ‚der Schönen und dem Biest') bebildern. Worte und Gleichnisse sind hierzu schon tausendfach gefunden worden, weshalb wir uns eine Ausweitung dieser Angelegenheit sparen dürfen. Bemerkenswert aber ist, dass die Mann-Frau-Beziehung wahrscheinlich die älteste, noch immer lebendige alchymistische, zum überwiegenden Teil mündlich tradierte Praxis geblieben ist. *Alexander von Bernus* meinte noch zu Beginn des vorigen Jahrhunderts, dass die letzten kontinuierlich lebendigen alchymistischen Praktiken (im Abendland) das Bierbrauen und das Herstellen von Whiskey seien...

Weil wir uns aber hier so insgesamt mit dem Erkennen eines Bildes – nicht nur der Alchymie im Besonderen, sondern der heilkundlichen Traditionen insgesamt – befassen, sollten wir die oft zitierte Formulierung ‚Weiberwerk und Kinderspiel' kurz betrachten:

Frauen und Kinder im Großen Werk der Alchymie

Manche Autoren setzten, wenn es um die Beteiligung der Frauen am alchymistischen Werk ging, noch einen Gedanken bzw. einen Hinweis dazu, weswegen ‚Weiberwerk und Kinderspiel' gleichsam zum geflügelten Wort wurde. Es bezeichnet aber in der Tat höchste Prozesse. So etwa bei *Trismosin* im *Splendor Solis* (16. Jhdt.). Einerseits finden wir diese Ausdrucksweise mit dem Zusatz: „Von da an ist alles (Weiberwerk und Kinderspiel)." Das ist keinesfalls abwertend gemeint, wie es uns oberflächlich betrachtet und patriarchal doktriniert erscheinen möchte. Im Gegenteil: Das ‚Kinderspiel' bezieht sich zum einen auf die sogenannte ‚Nacharbeit' zum ‚Stein der Weisen', welche dem Künstler ‚unmöglich gelingen' kann, wenn er nicht ‚im Herzen völlig rein und klar' ist. Schon Manche hatten das Ziel in greifbarer Nähe, ihren ‚Dreifachen Merkur' erlangt, Farben und Baum im Philosophischen Ei gesehen, scheiterten dann aber in dieser letzten Phase und wussten sich dabei nicht zu helfen, denn die Natur scheint hier ein letztes Siegel angebracht zu haben, welches nur mit dem ‚Seid wie die Kinderlein' zu öffnen ist. Diese Angelegenheit ist bedenkenswert. Die ‚kindliche Unschuld', das ‚reine Herz' ist ja nicht nur im großen Werk der Alchymie ein Schlüssel, sondern in der gesamten Natur- und Heilkunde und darüberhinaus selbstverständlich in den Themen der ‚Menschwerdung' und Spiritualität, wenngleich nicht einfach so oberflächlich mit gängigen moralischen Vorstellungen zu bewerten.

Das ‚Kinderspiel' ist aber auch eine Parabel auf die Art und Weise, an die ‚Dinge der Natur', von der ersten Signaturenschau bis zur Praxis der hohen Alchymie heranzugehen. Bei *Trismosin* (einem ‚fahrenden Alchymisten' aus Sachsen, welcher vermutlich der initiatische Lehrer des *Paracelsus* gewesen ist) bedeutet ‚Kinderspiel' auch, dass das, was zuvor oben war, danach unten liegt, und er meinte damit vermutlich weder seinen Kleiderschrank noch sein Bücherregal, sondern die Fixierung des *Mercurius*, des Naturgeistes, der eben zuerst noch flüchtig war, nun figiert am Boden bleibt, oder den ‚König', welcher aufgrund seiner Position stets der Oberste ist, sich aber dann im Tiegel oder Inguss unter den Schlacken befindet.

Was sich hinter dem allegorischen Bild des ‚Kinderspiels' noch alles verbergen könnte, werden wir wohl erst dann erkennen, wenn man die Kinder wieder mit ‚natürlichen Dingen' spielen lässt und dabei die musse findet, sie zu beobachten.

In der Werkstatt der zeitgenössischen Spagyriker wird das Kinderspiel wenn, dann nur allegorisch erwünscht sein – Wirkung und Arbeit der Frauen (in welcher Weise auch immer) spielt aber heute eine mindestens ebenso große Rolle wie in alten und ältesten Zeiten.

Man muss in der Geschichte nicht weit zurückblicken, um zu bemerken, dass besonders in den letzten Jahrhunderten immer wieder Frauen die Bewahrer und Übermittler alten Wissens waren. Das gilt hier ganz besonders für die Pflanzenheilkunde, und bei den Natur- und Kulturvölkern gilt es darüberhinaus auch noch für viele mit der Heilkunst verbundenen ‚Künste'.

Was die Alchymie im Besonderen anbelangt, so waren kundige Frauen im Abendland wieder relativ sicher, nachdem die erste Welle der Verfolgung vorerst abgeklungen war (und sich in dieser derben Art auch nicht mehr wiederholte). *K. Ch. Schmieder* schreibt dazu: „Um die Mitte des (18.) Jahrhunderts gewann es den Anschein, als ob (sinngemäß:) die Alchemie in der Tat zur Frauenarbeit werden sollte." Als Beispiele führt er die in der Kunst weit fortgeschrittenen *Schwestern zu Rodaun*, die *Frau von Grabau* und andere an. Für ihre Arbeiten am *Aurum Potabile* berühmt war die *Frau von Pfuel* mit ihren zwei Töchtern.

Aus den Jahrhunderten davor wird uns (verständlicher Weise) kaum von weiblichen Künstlerinnen berichtet. Die wenigen Ausnahmen bezeugen aber dennoch deren Wirken, wie zB. *Isabella Cortese*, welche um 1560 in Venedig (wo damals die Alchymie per Gesetz verboten, nichtsdestotrotz aber sehr beliebt war) eine >Italienische Abhandlung über die Alchymie< schrieb (die 1592 das erste Mal ins Deutsche übersetzt wurde).

Um 1330 wurde in Frankreich *Nicolaus Flamel* geboren, der bis heute zu den berühmtesten Alchymisten gehört. Es heißt, dass ihm das Große Werk dreimal gelungen war – und zwar zusammen mit seiner Frau *Petronell*. Die *Flamels* kamen damit auch zu einem gewissen Reichtum, von dem sie an die Kirche eine Art Schutzgeld (vermutlich überwiegend in Form von Stiftungen) bezahlten und so die Arbeiten ungehindert fortsetzen konnten. Manche behaupten, *Nicolaus* und *Petronell* leben immer noch im fernen Indien. Beider (um 1407) erbautes Haus *‚au Pignon'* kann heute in der *Rue de Montmorency* Nr. 51 besucht werden. Von *Flamel* existiert ein Portrait, welches *Rembrandt* schuf. Es herrscht aber die Meinung vor, dass dieses Bildnis kaum seinem wahren Aussehen entspricht.

Im Besonderen bezieht sich das ‚Weiberwerk' auf das Große Werk der Alchymie und darin auf die ‚Waschungen'. Es ist dies die durch Allegorien verhüllt beschriebene Phase, in der dem bleifarbenen Amalgam aus König, Hofnarr und Mond mittels Wasser seine Schwärze ausgewaschen wird, ‚bis dass der Mond voll ist', also bis es rein wie Silber erscheint. Auch die Frauen wuschen die Leinen bis gegen den Vollmond hin und hingen sie zum Trocknen auf, weil der Mond die Tücher weißer macht. Solcherart wird dem König der ‚böse Schwefel', das zerstörerische Feuer, entzogen und der ‚philosophische Merkur' kann geboren werden – kurzum eine Rolle der Frau, verbunden mit den Attributen Mond, Reinigung des Mannes von übermäßigem Feuer und Gebärung des Kindes Merkur.

„Sieh dem Weibe zu, und lerne wie sie tut ihr Leinen waschen, mit warmem Wasser aufgeschütt und mischt mit Aschen. Folge ihr nach, so wird dir alles wohl geraten, denn den Leib welcher schwarz ist wäscht das Wasser ganz rein."

Michael Maier, >Atalanta fugiens<, 1618

In der *Turba philosophorum* wird der arabische Alchymist *Morienes* angegeben, welcher ebenfalls die Rolle der Frau bzw. das Miteinander von Mann und Frau im alchymistischen Werk hervorhebt:

„*Morienes* sagt: Das Wissen über unser Meisterstück wird mit der Ordnung und Erschaffung eines Menschen verglichen. Erstlich geschieht der *Coitus*, zum andern die Empfängnis, zum dritten die Schwängerung/Schwangerschaft, zum vierten die Geburt und zum fünften die Ernährung. Diese Worte will ich erklären, damit du sie wohl verstehest:

Unser Same, welcher ist *Argentum vivum*, wird mit der Erde, das ist der vollkommene Körper, zusammengesetzt. Diese Erde nennt man Mutter, denn die Erde ist die Mutter aller Elemente. Dieses wird *Coitus* genannt.

Wenn dann die Erde das *Argentum vivum* eine Weile bei sich hält, dann wird dies *Conceptio* oder die Empfängnis geheißen. Wenn also der Mann mit der Frau wirkt, dann ist das das *Argentum vivum* in der Erde. Deshalb haben die *Philosophi* gesagt: Unser *Magisterium* ist nichts anderes als ein Mann und Weib – und ihre Zusammenfügung.

Wenn nun das Wasser, welches das *Argentum vivum* ist, dazukommt, dann wächst es in der Erde und wird gemehret, und die Erde wird weiß. Das ist die *Impraegnatio* oder Schwängerung/Schwangerschaft.

Danach wird mit diesem (noch) unvollkommenen Körper, welcher auf vorige Weise zugerichtet ist, das *Fermentum* (Sonne oder Mond) *coagulirt*, also zusammengesetzt, bis es eins wird in Farbe und Ansehen. Dieses wird dann *Ortus* genannt, und es ist die Geburt unseres Steins.

Nähret nun diesen König, bis er zu seinem rechten und vollkommenen Alter kommt. Sein Vater ist die Sonne, seine Mutter der Mond. (...) Zum fünften und letzten folget also die Ernährung, denn je mehr genährt wird, desto mehr wird er gemehret. Er wird aber genährt mit seiner Milch, das ist der Samen, aus dem sein Anfang gewesen ist. Es wird dieses *Argentum vivum* so oft imbibiert, bis er davon zwei Teile getrunken, oder bis er genug hat."

nach: *>Turba philosophorum<*

Morienes stammte aus einer griechischen Familie, wurde (um 1050) in Rom geboren und bezeichnet sich als ‚Schüler der Araber', schrieb auch seine Werke in arabischer Schrift. Er galt als Lieblingsschüler des arabischen Alchymisten *Adfar* und war einer der vielen ‚Christen', welche ihre Weisheit nicht im römischen Schafstall suchten. Vielleicht hätte er sonst keine derartige Allegorie finden wollen. Er verließ also nach einigen ersten Studien Rom, um in *Alexandria* den Weisen *Adfar* zu treffen, bei dem er vermutlich bis zu seinem Tode blieb.

Morienes kaufte sich ein Stück Land in Palästina, um sich ungestört der hermetischen Kunst widmen zu können. Als er hörte, dass *Khalid*, der Sultan Ägyptens, viele Alchymisten um sich versammelt hatte, weil er nach den Vorschriften *Adfar's* arbeiten wollte, ging er hin und bemerkte bald, dass der Sultan hauptsächlich Prahler gefunden hatte, welche keine wahren Kenntnisse besaßen. *Morienes* klärte ihn auf, indem er ihm das Große Elixir vollständig ausarbeitete. Daraufhin verschwand er, ohne um Erlaubnis zu fragen oder seine zugesicherte Belohnung anzunehmen. Er hinterließ aber eine Notiz: ‚Wer das versteht, bedarf keines Goldes.' Der Kalif geriet darüber in Zorn und ließ alle Prahler hinrichten. Als sein Zorn sich legte, sandte er aus, den wahren Kundigen zu finden. Das soll auch einem gewissen *Khaled* gelungen sein, welcher *Morienes* schließlich dazu bewegen konnte, dem Sultan das Geheimnis zu lehren.

Die Verbindung von Mann und Frau ist ein zentraler Aspekt der gesamten alchymistischen Kunst, welche in Asien ja noch heute auch als eine tantrische Wissenschaft bekannt ist.

Dieser Aspekt durchwirkt nicht nur viele Ebenen und Phasen der laborantischen Arbeit, sondern war im Abendland lange Zeit ein bedeutender Teil dessen, was zusammen mit der ‚Feuerkunst' an altem, urzeitlichem Wissen mittransportiert wurde. Er schließt in sich eine Kultur der Beziehung und der Sexualität, welche in der Minnezeit des Hohen Mittelalters eine letzte Blüte trieb, um danach sehr schnell beinahe vollständig zu verkommen. Wenn aber die Alten immer wieder betonen: „Unsere Kunst ist nichts anderes als Mann und Frau", wie wollen wir dann diese Kunst verstehen, wenn wir ‚Mann und Frau' nicht mehr verstehen können? In dieser Kunst gibt es nicht nur ‚einiges', sondern sehr viel zu lernen, und das ist für den Menschen moderner westlicher Prägung nicht mehr möglich, ohne ergänzende Zusammenschau mit noch lebendigen Kulturen.

Die Minnedichter standen in direkter Nachfolgeschaft der Barden und Filid keltischer Druidenschulen, und diese wiederum schöpften aus urzeitlichen Traditionen. Vor dem Hintergrund des iberisch-keltischen Romaniens ‚wussten' jene Kundigen des Mittelalters noch um die Mysterien von Mann und Frau, weswegen *Wolfram von Eschenbach* (nach *Guyot*, 12. Jhdt.) in seinen Vorbemerkungen zum *Parcival* schlicht bemerkt:

> „Doch sollt' ich hier euch Weib und Mann
> vollkömmlich schildern, wie ich kann,
> es würd' euch Zeit und Weile lang."

Zu jener Zeit waren die allegorischen Deutungsebenen dieses Themas noch selbstverständlich und – je nachdem man dieses Wissen einsetzte, im Alltag, in kleineren und größeren Künsten oder in hoher Mystik (worin besonders die Katharer, die ‚Reinen', eine geistige Deutung der Minnelieder verstanden) – austauschbar. Den drohenden Verfall jener Kultur vor Augen, verließen Abgesandte der Katharer immer wieder ihre Verstecke in der Einsamkeit der Pyrenäen, um in Dörfern und Burgen die uralten Mythen zu erzählen, ihre Bilder, Muster, Strukturen in Form von Märchen, Liedern oder Gedichten den Menschen einzuprägen. Heute noch werden in jenen Gegenden ‚Volksmärchen' erzählt, deren ‚hermetischer' Charakter beinahe dazu auffordern möchte, nicht nur zuzuhören, sondern ‚laborantisch' nachzuarbeiten. Fast immer werden nach geglückten Erlebnissen schließlich Mann und Frau vereinigt oder es wird darüber berichtet, was ein Kind, welches solch einer geglückten Vereinigung entsprungen ist, alles vermag. Zumindest hört man dann über die unglaublichen Heilkräfte des Kindes, über Tod und Wiedererweckung, über Verwandlungen und Veredelungen verschiedener Art und dergleichen Angelegenheiten – stets mehrdeutige Stimmigkeit versprechend.

„Nunmehr nimm dein Gold, unsere *Prima Materia*, so noch in kein Feuer gekommen und scheide den König und die Königin. Tue es in eine Retorte, die setze in Sand und lutiere eine Vorlage daran. Treibe *per gradus ignis* in 48 Stunden einen *Spiritum* und einen vielfarbigen Sublimat herüber. Letzterer legt sich im Hals der Retorte an. Zurück bleibt das fixe Salz. Den *Spiritum* verwahre wohl. Den Sublimat nimm heraus und sublimiere noch 2 mal. Solchergestalt ist der König, unser Gold, bereitet. Das zurückgebliebene Salz nimm und löse in destilliertem Rosmarienwasser auf. Solviere, evaporiere und coaguliere zu einem trockenen Salz, so hast du das Weib, die Königin, unser Silber.

Wenn diese beiden Eheleute nun copuliert werden sollen, so musst du sie in ein Behältnis aus Glas geben, damit sie sich in der Wärme miteinander vereinigen, verbinden und combinieren. So werden sie Kinder bis in das siebente Glied zeugen, denn ihr Alter und Vermehrung kann man nicht ausrechnen. Der Brunnen, welcher aus ihnen entspringt, ist unerschöpflich, sowohl auf die Metalle, als auch auf die kranken und breßhaften Menschen, da schon ein einziges Gran oder Tröpflein (...)."

nach: *Zoroaster, >Clavis Artis<*, 1236

Conjunctio (des Königs mit der Königin)

„Nimm aber von deinem geistigen Gold 1 Loth und von dem geistigen Silber 1 ½ Loth und mach diese beiden recht trocken, denn die keusche *Diana* badet sich gerne in der Luft, d.h. sie ziehet die selbe an sich und wird davon wassersüchtig, wodurch das ganze Werk verdorben wäre. Tue sie also trockengemacht in eine Phiole, sigilliere und setze sie in einen hierzu dienlichen Ofen. Gib demselben die *gradus ignis*, nämlich im ersten Monat *Termon*, den 1. Grad, im 2. Monat *Herne*, den 2. Grad, im 3. Monat *Humo*, den 3. Grad und im 4. Monat *Algir*, den 4. Grad.

Wenn du die *gradus ignis* fleißig abwartest, so wirst du Wunder sehen und erfreulich schauen, wie dein Werk so vortrefflich durch die Farben passieret. Wenn es nun zuletzt fix und feuerbeständig, auch wie ein *Al acipi*, ein roter Löwe, geworden ist, so freue dich herzlich, denn du hast hier in Händen die wahre essentialische Seele der ganzen Natur, ja die himmlische und paradiesische Erde wie solche noch vor dem Fall unserer ersten Eltern gewesen ist. Diesen geistigen Felsen nimm heraus, so ist die erste Augmentation geschehen, nämlich dass Mann und Weib eins geworden sind.

Diese beiden müssen nun noch einigemale gespeist werden, da solche auch Kinder zeugen und ihre Früchte zu deinem Genuß hervorbringen mögen, welches also geschieht:

Nimm deine *Quintam Essentiam* heraus und wäge sie. Wenn sie nun 2 ½ Loth wieget, so reibe in einer Glasschale ½ Loth von der rotgüldischen, sternflüssigen Blume darunter; dann auch ½ Loth von deiner *Diana* oder fixen *Lunaria*. Tue diese beiden miteinander in eine Phiole, sigilliere die selbe *hermetice*, setze sie wiederum in unseren Ofen und gib ihnen per gradus Feuer:

Den 1. *gradum* gebrauche in einer dämpfenden Wärme 3 Wochen lang, so wird alles in *Putrefactionem* gehen, mit dem 2. Grad bringe deine Phiole in eine trockene Wärme, wie die Brut der Henne ist, für den 3 Grad stelle deine Phiole in Asche, auch 3 Wochen lang, für den 4. Grad setze deine Phiole in den Sand, abermals 3 Wochen, so wird es wiederum fix zusammenfließen zu einer rubinroten, durchsichtigen Tinctur. Solche augmentiere auch noch ein drittes mal nach dem gleichen Gewichte, wenn du nämlich noch von dem roten und weißen Blut der Natur vorrätig hast. Jetzt wird es alle 14 Tage und Nächte durch die Farben gehen. Wenn es nun solchergestalt fix ist, so hast du hier einen unbeschreiblichen Schatz, ja die allerhöchste Tinctur der Welt in den Händen (...).

Nimm jetzt ein durch das Antimon gereinigtes Gold, lasse davon 16 Loth fließen und trage von deinem roten, feurigen Löwen 1 Quentchen darauf, so wird es von dem Löwen verschlungen und mit ihm zu lauter Tinctur werden. Wenn du nun auf diese Art den roten Löwen mit dem Gold gespeiset, so hast du den gesegneten *Acus*, den *Lapis*, in Händen. Diese geheime Arbeit haben die Alt-Väter derart hochgehalten, dass sie nicht einmal öffentlich darüber gesprochen, (...) sondern sie vor dem Tod ihren Kindern hinterlassen haben."

nach: *Zoroaster*, >*Clavis Artis*<, 1236

Das Feuer hüten

Das ‚Hüten des Feuers' muss man sich vor dem Hintergrund vorstellen, dass sich das Leben der Menschen tatsächlich viele Jahrzehntausende lang um behütete Feuerstellen bewegte und in vieler Hinsicht von ihnen abhing. In gewisser Weise ist das noch heute so. Das waren aber oft keine Feuerstellen, die dann und wann ganz einfach entzündet wurden. Es waren Feuer, die über Generationen ohne Unterlass bra nnten. Diese archaiische Situation finden wir nur mehr in der Alchymie, am wenigsten jedoch bei den westlichen Praktikern, sofern wir dort ‚echtes' Feuer suchen. Moderne Heizquellen haben in den meisten Fällen die Feuerstelle ersetzt. Bei den asiatischen Alchymisten ist es meist noch das ‚echte' Feuer, welches jahrelang unterhalten wird. Um etwa zu einem *Sahasraputi Loha Bhasma* (ein sehr hohes Eisenpräparat) zu gelangen, ist die Feuerstelle des *Rasa Vidya* bis zu 15 Jahren in Betrieb. Die Arbeit wird kaum anders, als schon viele tausend Jahre zuvor, ausgeführt.

Wenn man sich die Menschen der Urzeit sehr lange Zeit am Feuer sitzend, ja eigentlich rund ums Feuer wohnend, vorstellt, dann muss man überlegen, ob sie sich denn daran nur gewärmt und gelegentlich ein Essen gekocht haben oder ob nicht naheliegend dieses ‚heilige' Feuer auch zur Medizin genutzt wurde. Medizin in jeder Hinsicht, vor allem aber im Sinne von ‚Entwicklung'. Entwicklung zum Menschen, Entwicklung der Gemeinschaft und – Entwicklung, also Veredelung der ‚Dinge der Natur'. Die ‚philosophischen' Erkenntnisse aus diesen Entwicklungen waren wohl die selben bei den Kundigen, welche vor 30 000 Jahren in den Höhlen Südfrankreichs und Nordspaniens am Feuer saßen, bei den heute noch lebenden Schamanen der Taiga, bei den indischen *Vidyas* und am *Athanor*, an dem *Paracelsus* weilte.

Schmiede

Schamane, Zauberer, Medizinmann, Alchymist sind Feuerkünstler, ebenbürtig mit Töpfer und Schmied, wobei der Typus des Töpfers meist noch ein weit älterer, steinzeitlicher ist, als der Schmied. Noch heute besagt ein jakutisches Sprichwort: ,Der erste Schmied, der erste Schamane und der erste Töpfer waren Blutsbrüder.' In den ältesten Sagen ist der Schmied derjenige, welcher die ,Waffen' der ,Helden' anfertigt. Es ist nicht alleine die Kunstfertigkeit, sondern vor allem das ,geheime' Wissen, der Zauber, der dann den ,Helden' (zB: König *Artus – Excalibur*) zur Verfügung gestellt wird. Die Erzählungen um solche magischen Waffen sind fast immer von tieferer hermetischer Bedeutung.

Der Schmied steht in Verbindung zum Zwergenreich und hat damit direkten Zugang zum Verstehen der Minerale und Metalle. Ja die ,Zwerge' förderten die Ausgangsmaterien für das alchymistische Werk aus dem Berg und wurden deshalb immer wieder damit in allegorischen Zusammenhängen dargestellt, so wie der rote und der schwarze Knappe im Bilde der >*Aurora consurgens*< (16. Jhdt.), die beiden Knappen ,Sonne und Mond', welche das dritte Prinzip (Sal) in *Trismosins* >*Splendor Solis*< vom Berg brechen, oder die bekannten ,sieben Zwerge', als Allegorie auf die sieben coagulierten Planetenmetalle. Der Schmied erhält das Erz (sofern er es nicht selbst gebrochen hat) von diesen ,Zwergen' und ,veredelt' es – durch die Kunst des Menschen. Damit nimmt er aus Sicht der Natur einen königlichen Rang ein. Wie die eigentliche Alchymie, so ist auch die Schmiedekunst eine ,königliche Kunst'. Noch heute bedeutet das mongolische Wort für ,Schmied' *(darkhan)* gleichzeitig auch ,Held', ,freier Ritter', ,Wissender' u. dgl., und nicht selten wurden Schmiede zu Königen. So heißt es zB., dass *Dschingis Khan* ursprünglich ein Schmied war.

Der Schmied *Kavi* war der Vorfahre des iranischen Herrscherhauses, der *Kavya*-Dynastie. Das avestische Wort *Kavay* bedeutet auch ,Weiser'. Auf Java bestand zwischem dem Schmied (*kyai*=Meister) und dem Fürsten eine Beziehung wie zwischen Blutsbrüdern, und auf Bali werden noch uralte Initiationsriten mit den Schmiedelehrlingen vollzogen. Die aus Java stammende balinesische Schmiedetradition soll ursprünglich von *Brahma* initiiert worden sein, der dem Schmied die mystische Kraft (*Shakti*, die Kraft *Shiva's*) zu seinem Handwerk verleiht.

Immer ist der Schmiedeberuf mit heilkundlichen Fähigkeiten bis hin zur Alchymie verbunden. Es hat besonders in Asien kaum eine Unterbrechung der Kontinuität zwischen metallurgischer Mystik und Alchymie gegeben.

Lange Zeit, bevor es den Schmied gab, hatte der Töpfer dessen Position inne. Auch dieser Feuerkünstler ,kannte' das Pflanzen- und Mineralreich und hatte Anteil an den Geheimnissen der Natur. Betrachtet man die Beziehungen von Alchymie und Heilkunst zu vielen, insbesondere handwerklichen ,Berufen' und ,Künsten', dann scheint es, als wäre die alchymistische Handwerkskunst nie alleine gestanden, sondern immer begleitend, nährend und ergänzend zusammen mit den anderen Künsten. Da die Alchymie beinahe untrennbar mit Heilkunst verbunden ist, wird der Begriff ,Heilkunst' nun ein sehr weiter – auch die Handwerke an sich einschließender. Und diese Fächerung geht noch weit über die spezifischen Handwerke hinaus zu den Tätigkeiten, zum ,Umgang' des Menschen schlechthin. Mit den verschiedenen Tätigkeiten war auch stets ein Wissen um die tätigkeitsspezifischen Heilmittel verbunden, und mehr noch: eine Kosmologie, eine geistige Tradition und dadurch wieder ein Eingebettet-Sein in die lebendige Ganzheit. Aus diesem Hintergrund hatte *Paracelsus* betont, bei den Bauern, Winzern usf. solle man lernen.

„Ich habe, günstiger Leser, von den hohen Würden und großen Nutzbarkeiten der Alchymey klärliche Meldung getan, nämlich dass sie dem *Medico* und Kranken so hochnötig, ja viel nötiger und nützlicher als die Apotecken. Sie lehret uns, wie aus dem Gift die beste Arzney zugericht, scheidet und sondert ab das Böse vom Guten, damit das Gute eine nützliche Arzney und das Böse, Unreine, so schädlich und hinderlich ist, weggetan werde.

Über dieses ist sie vielen Künsten dienlich, so ihrer nicht wohl entraten können, als Goldschmieden, Mahlern, Bergleuten, dass ich geschweige, dass dadurch groß Reichtumb kann erlanget werden. Ob nun solches wenigen wiederfahren, so muss man es zum Teil Gott zuschreiben, der nicht will, dass wir alle gleich reich sollen sein. Zum andern Teil mangelt es an Erfahrenheit und Schicklichkeit, in dem etliche aus Geiz und Unverstand der Natur wider die Natur arbeiten, (...)."

„Denn die *Alchymia* ist ein sonderlich Geschenk und Gabe Gottes des Allmächtigen, wie solches zum Teil aus heiliger, göttlicher Schrift kann erwiesen werden, zum Teil aber bezeugen es der Philosophen Schriften und die Bücher nicht allein derer, so diese Kunst selbst beschrieben, sondern es ist auch klärlich zu vernehmen aus den Poeten, welche die philosophischen, parabolischen und allegorischen Gedichte, darinnen auch diese Geheimnisse verborgen sind, zusammen getragen und ihre Bücher damit gezieret und wie Kleinodien darein versetzet. (...)

Denn insgemein alle Künste, so mit den Metallen und Mineralien umgehen, fließen aus der Alchymey und erkennen die selbe als ihre Mutter und Gebährerin. Wie denn sonsten aus der Alchymey vielerlei Künste und herrliche Bereitung vieler Sachen entspringen."

Joachim Tanckius, >Promptuarium Alchemiae<, 1610

Zufluchts- und Pilgerorte

Die Zersetzung menschlicher Kultur während des ‚Abstiegs der Zeitalter' wurde mancherorten unerträglich, und die Kundigen suchten Zuflucht in abgelegenen Gebieten. Von dort kehrten sie immer wieder in den immer größer werdenden Scherbenhaufen zurück, um noch brauchbare Trümmer zu retten. Gerade in jenen Traditionen, welche man heute als ‚taoistische Traditionen' zusammenfasst, ist eine große Anzahl Überlieferungen von unvorstellbar hohem Alter enthalten, darunter auch Praktiken, welche sich direkt vom urzeitlichen ‚Proto-Schamanismus' der Jägervölker ableiten lassen. Es möchte wohl der Eindruck entstehen, dass die chinesischen, mongolischen, taoistischen Alchymisten, welche sich in schier unzugängliche Gebirgs- und Wüstenlandschaften zurückgezogen hatten, bereits in der Zeit, die wir ‚Antike' nennen, nur mehr ‚versuchten', jene alten Weisheiten wiederzufinden, die durch gewisse ‚Umschichtungen' zu Beginn der ‚antiken Periode' verlorenzugehen drohten. Ähnlich verhielt es sich mit den Alchymisten der alten Griechen und Araber. Es galt, während Zeiten epochalen Wandels, während des ‚Abstiegs der Zeitalter', den ‚roten Faden' zum Wissen der ‚Ältesten' zu bewahren, und diese ‚Ältesten' sind in der Zeit vermutlich (zumindest) dort zu finden, wo (nach den Traditionen) das ‚silberne Zeitlalter' sich seinem Ende zuneigte, also etwa vor 20 000 Jahren.

In Ost und West musste sich der Alchymist der ‚Antike' dann schon durch ‚Bildung' das zu erwerben suchen, was seine Vorfahren ‚erleben' durften – eingehüllt in mündliche Überlieferungen und bestätigt durch praktische Initiation.

Von Anfang an suchten die Taoisten nach den Vertretern älterer Traditionen, wandten sich mit großem Eifer den volkstümlichen Überlieferungen, sogenannten ‚Berufsgeheimnissen' und dergleichen zu und sammelten Fragmente jenes alten Wissens. Ihre Adepten verstanden mit hoher Kompetenz Authentizität zu erkennen und eine vielschichtige ‚Kunst' zu restaurieren, welche geistiges und handwerkliches Erbe, zurückreichend bis zur steinzeitlichen Jagdmagie, Ackerbau, Metallurgie oder zur ‚Erfindung' der Töpferei und vor allem ein unvorstellbar tiefes Wissen vom menschlichen Wesen und Körper, von Heilmitteln, Rezepten und iatrochymischen Prozessen enthielt. Dieses Wirken ging weit über das Erstellen eines ‚Flickwerks' hinaus. Es kann zu Recht und erwiesenermaßen als eine echte Fortsetzung uralter Traditionen bezeichnet werden. Es scheint insgesamt so, dass während der vergangenen 2000 Jahre ein gewisser Drang zur Restauration den Menschen wie eine lebenserhaltende Hand geführt hat.

„Zur Verwandlung der Metalle braucht man die Wahre Medizin. Man kann diese Verwandlung aber nicht in einem Palast vollziehen, sondern braucht dazu die Einsamkeit, fern von den weltlichen Menschen.

Was in den Büchern zu finden ist, genügt nur den Anfängern. Das übrige ist nur auf mündlichem Wege übertragbar."

nach: *Pao P'u-tsu* (254-334)

Ebenso wie die taoistischen Alchymisten schöpften auch die Kundigen des Abendlandes aus dem urzeitlichen Schatz. Bei den Naturvölkern, die noch heute in den nördlichen Gebieten der Erde leben, findet man einerseits bis in unzählige Einzelheiten eine Verbindung mit der Steinzeit und andererseits ein Spektrum des hermetisch-alchymistischen Weltbildes, ähnlich dem der beginnenden Neuzeit. Das hilft uns beim Bau einer Verständnisbrücke auf der Suche nach jenen ‚Alten', welche nicht nur in Aspekten und ‚Kunstgriffen', sondern ‚in der ganzen Kunst' wissend gewesen sein sollen. Wo könnte man diese ‚Alten' noch angetroffen haben? Wir erinnern uns an die Bemerkungen des Historikers *Moehsen* zu den Zielorten des Lernens und der Flucht abendländischer Alchymisten des späten Mittelalters und der beginnenden Neuzeit:

„Die britischen und schottischen Philosophen wurden durch die inneren Unruhen im Land nach den orkadischen Inseln getrieben. (...) Außer den Morgenländern gehörten noch die orkadischen Inseln, die nordischen Reiche und Spanien zu den Ländern, nach welchen die Wanderschaften vorzüglich gerichtet waren."

„Auch in Deutschland hatten die Feuerphilosophen gleich von den ersten Zeiten an widrige Schicksale und mussten sich verborgen halten. (...)

(...) Manche von ihnen (*Moehsen* nennt u. a.: *Johann Cobbe, Thomas Trafford, Thomas Atzheton, Johann Mistelden, Chaucer, Charnok, Blomfeld, Gawer, Michael Scotus, Robert Fludd, Wilhelm Buttler, Alexander Sethonius, Eduard Kellei* und *Johann Dee*) hatten sich wegen der vielen Unruhen in beiden Königreichen (England und Schottland – Komm.) im 15. Jhdt. nach den orkadischen Inseln begeben, wo sie sich als Eremiten ungestört der Alchymie widmen konnten. Doch kamen zuweilen einige von ihnen woanders zum Vorschein, um chymische Prozesse zu zeigen."

nach: *J.C.W. Moehsen*, >Leben des *Thurneisser*<, 1783

Die orkadischen Inseln *(Orkney Islands*, nach griech: *Cape Orcas)* sind ein Archipel nördlich Schottlands, etwa in der Höhe Südgrönlands gelegen, bestehend aus ca. 100 Inseln. Manche davon könnten schon vor etwa 8500 Jahren regelrechte Pilgerzentren gewesen sein. Zahlreiche Megalith-Anlagen sind bis heute erhalten, wie die 58 ‚*Stalled Cairns*', die ‚Galerien' *(Passage Tombs*, wie man sie in Irland, Schottland und auch im kontinentalen Europa findet), Steinkreise und baulich einmalige Anlagen wie die gemauerten Felskammern von *Crantit*. Neben ‚üblichen' archäologischen Funden, wie Keramiken und Goldschmuck, stellen andere Objekte noch gewissermaßen ‚Rätsel' dar. So etwa die sogenannten *Ancient cooking places*, also riesige ‚Kochstellen' mit Haufen verbrannter Steine, deren größte an die 30 Meter Durchmesser haben und die stets in der Nähe von Süßwasser gefunden werden (dzt. sind etwa 200 solcher Stellen bekannt). Was dort wohl jahrhundertelang so intensiv gekocht wurde?

Aus allen Gegenden haben sich dort Besucher eingefunden. Aus Griechenland ist der Reisende *Pytheas* dokumentiert, der sich im Jahre 325 v. Chr. dort aufgehalten hatte. Er nannte die Einwohner der *Cape Orcas ‚Pretani'*, was an ‚Bretonen' erinnert. Selbst die Römer bemühten sich, diese kargen, kalten Inseln unter ihre Kontrolle zu bringen und entsandten unter dem Statthalter *Tacitus Agricola* im Jahre 83 n. Chr. eine Flotte. Aber auch aus Frankreich und vor allem aus Skandinavien zog es Leute dorthin. Der Dialekt der Bevölkerung ist ein Gemisch aus ‚Insel-Schottisch' und skandinavischem *Norn*. Die geographische Lage der Orkneys bot ideale Voraussetzungen für ein Zusammentreffen circumpolarer und kontinental-europäischer Überlieferungen. Offenbar zogen sich auch die von *Moehsen* genannten Alchymisten bisweilen dorthin zurück, um ungenannte Vertreter urzeitlicher Traditionen anzutreffen.

Der ‚Jakobsweg' war ebenfalls eine beliebte Pilgerroute der Alchymisten und bot ihnen noch lange nach der katholischen Occupation die Möglichkeit, mit maurischen Schulen in Verbindung zu treten. Dass in *Cordoba, Granada* und anderen Städten solche Schulen im Verborgenen weiterwirkten, ist bekannt. In Südfrankreich und den Pyrenäen konnte man Abkömmlingen der iberisch-keltischen und katharischen Schulen begegnen. Die ersten ‚Kreuzzüge' wurden gegen diese Gebiete (und nicht nach Palästina) geführt – ein Zeichen ihrer Wichtigkeit.

„Um jene Zeit (hier: um das Jahr 1000 – Komm.) nahm der wissenschaftliche Verkehr eine andere Richtung, indem die Lateiner anfingen, sich mehr an die Araber als an die Griechen anzuschließen, weil diese von jenen weit übertroffen wurden. Die Araber, welche sich in Spanien festgesetzt hatten, wollten die Christen weder bekehren, noch bedrücken und waren milde Herren, wenn man die Steuern zahlte. Das hatte zur Folge, dass Sieger und Besiegte sich einander vertraulich näherten. Diese Araber waren weit unterrichteter als ihre christlichen Untertanen, und das regte bald Nacheiferung an."

nach: >Geschichte der Alchemie< *Karl Christoph Schmieder*, 1832, neu: W. Roller, 1997

Dass diese Umstände Mitschuld trugen an der ‚Notwendigkeit' der Kreuzzüge, ist anzunehmen, denn die ‚christlichen' Horden betrieben auch dort ihren Vernichtungsfeldzug, ganz nach dem Vorbild der alten Römer.

„Sie nahmen von ihren Lehrern mehr an, als sich mit dem Christentum gut vertragen wollte, doch wurde dem des Bedürfnisses wegen nachgesehen, und die Mode, genannt: Zeitgeist, beschönigte Vieles. Astrologie und Magie beschäftigte viele Hoch- und Ehrwürdige der Kirche. Auf diesem Wege wurde auch das Lieblingsprojekt der Araber – die Alchemie – als Gegenstand gelehrter Spekulationen in alle Länder verbreitet. (...)"

nach: >Geschichte der Alchemie< *K. Ch. Schmieder*, 1832, neu: W. Roller, 1997

Nun aber gilt die Überlegung ähnlich wie zu den Orkaden: War es nur das mittelalterliche Zusammentreffen und Verbergen Kundiger, oder reicht eine derartige Bedeutung dieser Orte vielleicht sehr viel tiefer, in Epochen, die wir aus unserem ‚historischen' Blickfeld nahezu verdrängt haben?

Es scheint, als wären jene französischen und spanischen Gebiete schon seit der Urzeit vielbesucht von ‚Pilgern' gewesen, ganz ähnlich den abgelegenen Gegenden im Norden. Sogar die ältesten archäologischen Fundstücke menschlicher Kunst wurden dort gefunden: fein gearbeitete Steinkugeln, sogenannte *Sphäroide*, mit einem Alter von etwa 300 000 Jahren.

Um die Feuer im Eingangsbereich der Höhlen des *Dèp. Seine-et-Marne*, *des Dèp. Loiret* und an zahlreichen anderen Orten saßen jahrzehntausendelang keine ‚primitiven' Affenmenschen, sondern Menschen von unvorstellbar hoher geistiger Art, von denen wohl manche im wahrsten Sinne ‚genug Zeit' damit verbrachten, die Natur in ihrem Lichte zu betrachten und die Feuerkunst zu üben.

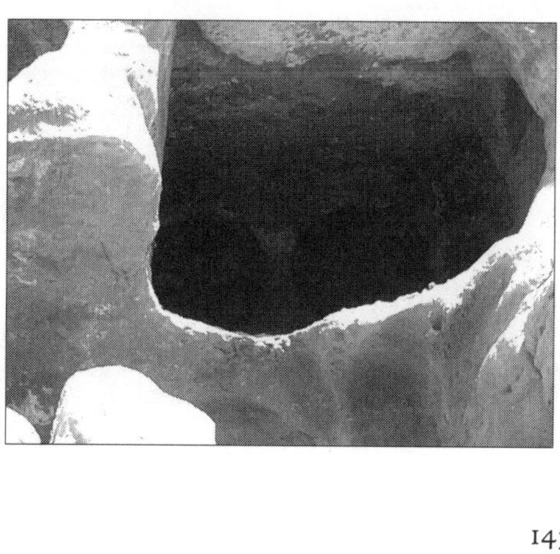

„Ähnlich kann man die vielfältigen Ausdrucksformen des Vergehens und Werdens betrachten. Wie sollte man meinen, dass die eiszeitlichen Ideogramme, wenn sie Leben, Tod und Erneuerung darstellen – wie etwa in der Höhle *Rochers des Potets* als drei parallele Striche, daneben ein Pfeil und darunter ein Linienkreuz; oder die Ritzzeichnung eines Bisons in der Höhle *Gabillou (Dordogne)* mit drei parallelen Linien und Pfeil –, dass solche Ideogramme nicht zumindest jenem Wissen entsprangen, aus dem heraus auch *Aristoteles* bemerkt: ‚Die Zerstörung eines jeglichen Dinges ist die Gebärung eines anderen' – eine Erkenntnis, die in ihrer universalen Aussage auch eine der Grundlagen der Alchymie darstellt. Dasselbe beinhaltet die Formulierung des *Menaldes*: ‚Ich gebot allen meinen Nachkommen, dass sie die Körper geistlich machen durch die Auflösung und wiederum die geistlichen Dinge körperlich, durch eine linde Kochung' – hiermit schon die praktische Anwendung eines Urwissens andeutend. Diese Anwendung würden wir heute als ‚spagyrisch' bezeichnen. Wir können sie den ‚Alten' der Vorzeit nicht einfach deswegen absprechen, weil sie darüber kein Papier beschrieben und Retortenscherben hinterlassen, sondern ‚nur' Symbole in den Stein geritzt haben."

nach: >Der ImpoWeg des Sonnenfunkens<

Es ist eigentlich eher unwahrscheinlich, dass nicht damals schon die selben Erfahrungen gemacht und das selbe Wissen gepflegt, ja in weit vollkommenerer Form geschaut, gewusst und praktiziert wurde, wie es die Alchymisten der frühen Neuzeit eben jenen Alten und noch viel mehr den ‚Älteren' vor ihnen zusprachen:

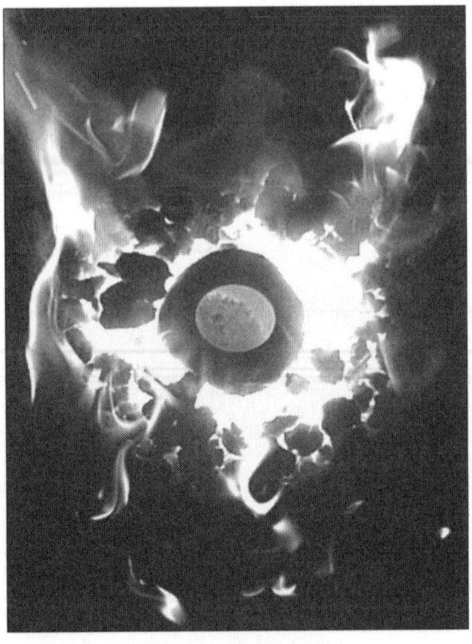

„Das eine Feuer erweckt das im andern verborgene Feuer zum Leben, indem es dieses aus dem Gefängnis befreit, worin es eingeschlossen war."

nach: *Pernety*

„Unser Feuer zerstört, löst auf, calciniert, verändert und durchdringt. Es ist subtil, luftig, nicht heftig, umschließend, beinhaltend und einzig."

nach: *Artephius*

„Macht dampfendes, digerierendes, beständiges, umhüllendes, luftiges, geschlossenes, veränderndes Feuer."

nach: *Bernhard v. Treviso*

Ursprung der Alchymie?

Eines ist vor allem klar, nämlich dass die Alchymie an sich keine typisch spätmittelalterliche Zeiterscheinung ist, wie mancherorts noch behauptet wird. Dieser Eindruck mag durch das Erscheinen zahlreicher alchymistischer und pseudo-alchymistischer Bücher seit der Erfindung des Buchdrucks entstanden sein. Dass alchymistische Traditionen jahrtausendelang überliefert wurden, zeigt uns unter anderem die lebendige ayurvedische Alchymie (*Rasa Shastra* bzw. *Rasa Vijnan*) mit ihren oft steinzeitlich anmutenden Arbeitsweisen, fundierend auf einer dem abendländischen, noch im engen mechanisch-materialistischen Weltbild verhafteten Forscher unvorstellbar hohen Kenntnis der tatsächlichen Naturgesetze, also der Vernetzung von Geist, Kraft und den Dichtegraden der Materie. Signaturkundliches und hermetisches Denken finden wir in allen Überlieferungen urzeitlicher Kulturen. Es ist anzunehmen, dass dieses Wissen in selbstverständlicher Folge immer wieder und auf die eine oder andere Art in eine handwerkliche Praxis umgesetzt wurde, die man als alchymistisch bezeichnen kann. Ausschließlich mündliche Überlieferungen sind dabei bis heute üblich, älteste Allegorien und Symbole in der alchymistischen Literatur gebräuchlich.

Dennoch sollen diese Überlegungen nicht dazu führen, einen etwaigen ‚Ursprung' der Alchymie in dieser oder jener Zeit und Region anzunehmen. Nehmen wir als Beispiel das Wort selbst: Lange Zeit hatte man vage Vermutungen (oder war auch überzeugt), dass das Wort ‚Alchymie' (mitsamt der Kunst selbst) aus Ägypten stamme. Diese Ägyptenfixiertheit mag durch den Einfluss gewisser Kreise hervorgerufen worden sein, ist aber weder historisch noch nach den hermetischen Traditionen realistisch. Im tiefsten Materialismus der letzen Jahrhunderten und der weitesten Entfernung vom eigentlichen ‚Wissen' war man stets geneigt ‚alles in einen Topf zu werfen', was nach ‚etwas mehr' als alltäglichen Gedanken aussah. Da passte das Land der Pyramiden nur zu gut ins Bild. Dazu *Schmieder:*

„Das gewaltige Geheimtun der Priester (des alten Ägyptens –Komm.) ließe wohl allenfalls vermuten, dass so etwas (die Alchymie – Komm.) dabei im Spiel gewesen sein möchte, allein wenn wir unbefangen erwägen, so streitet es gegen alle Regeln des Wahrscheinlichen, dass (dort) der erste Anfang einer Kunst mit dem größten Meisterstücke gemacht worden sei."

nach: >Geschichte der Alchemie< *K. Ch. Schmieder*, 1832, neu: W. Roller, 1997

Zwar bemerkt *Galen* (ca. 100 n. Chr.) in seinem >Buch von der Zusammensetzung der Arzneien< (4-1): „Diese Zubereitung hat ihren Namen von dem Priester *Hermon* in Ägypten. Sie ist aus dem innersten Tempel des *Vulcan* hergenommen, wo noch viele andere Bereitungen gefunden werden." *Clemens von Alexandria* übersetzt etwa hundert Jahre später den Namen *Hermon* bereits mit *Hermes*, was fürderhin ein zusätzliches Argument wurde, dass die Alchymie von einem ‚Hermes in Ägypten erfunden' worden sein soll. Derselbst *Clemens* beschreibt jedoch den Inhalt der 42 von diesem Priester erhaltenen ‚hermetischen' Schriften: Gesetze, Götterlehre, allgemeine Astrologie, Astronomie, Geographie, Schriftkunst u. dgl. – keine Alchymie. Selbst die darunter befindlichen sechs Bücher über Heilkunst hätten ‚bloß' Physiologie, Pathologie, Chirurgie, allg. Therapie, Augen- und Frauenheilkunde zum Inhalt. Nach *Galen* (bzw. ihm zugeschriebenen Texten) stammen letztere arzneikundliche Bücher vom Priester *Hermon* selbst, die anderen (obigen) von anderen Autoren. Dass darin Pharmazie oder alchymistische Zubereitungen bzw. die Präparation von Metallen enthalten wären, dafür gibt es nicht den mindesten Hinweis.

> „Die Beweise für die Alchymie der Ägypter bestehen also nicht in der Prüfung. Sie wird aber noch unwahrscheinlicher, wenn man vergleicht, was *Agatharchides von Knidos* (2. Jhdt. v. Chr.) (...) von dem Goldbergbau der Ägypter meldet. (...)"
>
> nach: >Geschichte der Alchemie< *K. Ch. Schmieder*, 1832, neu: W. Roller, 1997

Andererseits ist bekannt, dass die alten Ägypter eine sehr hohe Naturlehre kannten, deren ein zentrales Thema das Feuer war. Dies war aber auch bei anderen Völkern jener Zeit und vieler Jahrtausende davor so. Insbesondere die letzte (historische) Phase der ägyptischen Reiche glich eher einer ‚Diktatur der Priesterschaft', und da wäre es nicht verwunderlich, wenn manches Wissen zurückgehalten und weder im Bergbau noch in der Arzneikunst genutzt wurde. Bei vielen *Rasa Shastris* Indiens gilt die Ansicht, dass jenes ‚geheime Wissen' der ägyptischen Priester weit mehr mit Magie als mit Heilkunst und Alchymie befasst war, und man bezieht sich auf die Allegorie, der ‚*Pharao*' bedeute die ‚Seele' und die Priesterkaste das ‚Gemüt' *(Kal)*, wobei letztere stets bemüht ist, erstere zu kontrollieren und in ihrer Macht zu bannen. Alleine *Echnaton* habe es geschafft, für einige Zeit diesen Bann zu durchbrechen. Die Option einer ‚Anwesenheit' alchymistischer Tradition muss man jedoch zugestehen und offenlassen (wobei wiederum anzunehmen ist, dass die ‚wahren Kundigen' nicht innerhalb jener Priesterschaften zu finden waren). All das reicht aber bei weitem nicht aus, einen ‚Ursprung' der Alchymie ins alte Ägypten zu verlegen.

Eine gute Möglichkeit wäre, dass sich das Wort ‚Alchymie' über das arabische *al-kimiva* ursprünglich vom chinesischen Wort *Jin Yi*, der Bezeichnung für das *Aurum potabile* (in *Quanzhou* heute noch: *Kim Ya*) ableitet. Zwischen China und dem arabischen Raum gab es bereits in frühester Zeit regen kulturellen Austausch. Auch bezüglich der Alchymie sind hier die Verbindungen auffällig. Unter vielen Beispielen ist hierzu die starke Ähnlichkeit der Texte des arabischen Alchymisten *Jabir Ibn Hayyan* (8. Jhdt.) über die unterirdische Bildung und Wandlung der Metalle zum chinesischen Werk *Huai Nan Zi* (2. Jhdt. v. Chr.) bedeutend. Wir finden diese Vorstellung der unterirdischen Kochung der Erze aber nicht erst bei den chinesischen Alchymisten der Antike, sondern bereits lange Zeit vorher (und bis heute) bei den urzeitlichen Naturvölkern der Taiga und anderswo. Es kann das Wort also nicht einen etwaigen Ursprung der Alchymie im alten China erweisen.

Der Charakter der Kunst ist weder regions- noch kulturspezifisch, und ihre Wurzeln verschwinden schlichtweg im Dunkel der Zeit.

Manche meinen auch, sie wäre von den Vorfahren der antiken Griechen nach China gelangt, was sicherlich so gewesen sein könnte, finden wir doch in den Gegenden der Seidenstraße durch Zentralasien Spuren von Besiedelung durch alte griechische oder minoische Völker, deren Pfaden dann auch *Alexander der Große* gefolgt ist (>*Kafiristan* in *Afghanistan*, *Blue Hills* in Indien, u.a.). Deshalb aber sogleich einen ‚Ursprung' der Kunst ins alte Griechenland zu verlegen, wäre voreilig. Schon das *Veda* (übers.: *Wissen*) Indiens belehrt uns eines Besseren, indem es dazu ein urzeitliches Wissen überliefert, welches die Periode der klassischen Griechen sogar um Jahrzehntausende überragt.

Genauso wie wir keinen Ursprung der Alchymie im Mittelalter finden werden, finden wir ihn nicht im klassischen Altertum. In den von uns als ‚vorgeschichtlich' und dann ‚steinzeitlich' bezeichneten Epochen kann ebenfalls kein zeitlicher oder regionaler Ursprung der Alchymie fixiert werden, es sei denn, man erwäge eine gemeinsame Geburtszeit der Kunst, Feuer zu machen mit der ‚Feuerkunst' selbst, welche aus gängiger archäologischer Sicht vor etwa einer halben Million Jahren anzusetzen wäre, aus der Sicht unabhängiger Forscher gar vor mehreren Millionen Jahren. Die sicherste Annahme bleibt wohl die, dass die ‚Kunst' eben so alt ist, wie der Mensch selbst – was naheliegend für die Heilkunst an sich gilt und gerade deshalb einen ‚Ursprung' der Alchymie sinnvoll erscheinen lässt. Dazu muss man nicht unbedingt in den Resten jener ‚Hochkulturen' suchen, welche mächtige Ruinen hinterlassen haben und auch nicht annehmen, dass eine alchymistische Praxis umso wahrscheinlicher wäre, je mehr Schriften aus jener Zeit gefunden werden. Sondern man muss auch zu den Hinterbliebenen jener urzeitlichen Kulturen gehen, die immer noch – und das seit prähistorischen Zeiten – in vorwiegend mündlicher Tradition stehen.

„Sie (jene Urvölker – Komm.) haben nur eines nie erfaßt, dass nämlich mit dem materiellen Wechsel die eigene geistige Kraft erlahmt. Sie, wie auch wir, haben nicht begriffen, (...) dass der Mensch im festen Hause die Jahrzehntausende nicht besser überdauert, als hinter dem Windschirm. (...) Der technische Fortschritt und die ihm eigene Neigung, sich auszubreiten, das ist eine Kur, die nur wenige Völker überleben. Selbst Hochkulturen können in Fabriken zugrunde gehen. Und wo der Medizinmann durch den Arzt ersetzt wird, beginnt die Anfälligkeit des Körpers größer zu werden als in der Taigawildnis, im Staube Turkestans oder auf der *Tierra del Fuego.*"

Ivar Lissner, >So lebten die Völker der Urzeit<, Vlg. Walter, 1958

Schreiben und Lesen

Dass die ältesten schriftlichen historischen Dokumente hierzu (in Europa) fast ausschließlich von den Griechen stammen, mag den alleinigen Grund in den typischen mündlichen und symbolischen Überlieferungen anderer Völker dieser Zeit haben. Wenn aber ein Pfeilmuster nach oben und daneben eines nach unten von einem Jungpaläolithiker in eine Höhlenwand eingeritzt wurde, vielleicht noch durch Quadrate mit Diagonalen ergänzt, in Mühlebrettmustern verbunden und mit Pentagrammen unterlegt, dann sollte vermutlich ein Alchymist jeglicher anderer Zeitepoche diesen Notizen ohne weiteres erfolgreich nacharbeiten können. Nicht dass jener Jungpaläolithiker diese Arbeit ebenfalls und unbedingt mit ‚äußerer' Feuerkunst und Tontöpfen ausgeführt haben müsste (ob er das nun getan hat oder nicht) – es ist nur die ‚hermetische Stimmigkeit', die Erkenntnis im Lichte der Natur, die sich auf allen Ebenen sinnvoll nachvollziehen lässt. Wider unserer Denkgewohnheit ist das hermetische Wissen bzw. die Art der Überlieferungen (mit *Lengyels* Worten) ‚antihistorisch' konzipiert – zu jeder Zeit und auf jeder ‚Ebene' gültig.

Meist dürften auch dem Kundigen Verstehen und Praxis im Hier und Jetzt dermaßen wesentlich sein, dass sich die Frage nach dem ‚Wann ist das erfunden worden' gar nicht zu sehr erhebt. Im Moment des Verstehens, des Erkennens selbst, empfindet der Adept eine ‚Erinnerung' und erlebt im Innersten die Wahrheit des Erfahrenen. So hat der Mensch dieses zeitlose Wissen in keiner Epoche als ‚etwas Neues' empfunden. Er hat das Gelernte und Verstandene stets seiner Situation und seinem Vermögen entsprechend in die Praxis umgesetzt und sich dabei als ‚in einer endlos langen Reihe stehend' empfunden. Immer haben die Alten wieder von den Alten und diese von den noch Älteren gelernt und dabei stets weit mehr ein Erinnern und Bemerken angeregt, als verlangt, etwas völlig Neues zu erlernen.

So ist es auf diesem Gebiete grundsätzlich falsch, anzunehmen, zB. *Paracelsus* hätte das ‚erfunden' und dieser jenes. Er hat höchstens einen Aspekt einer Sache wieder verstanden, die vielleicht tausende Jahre zuvor noch Vielen vertraut war, und hat dies nach seinem Begreifen und seiner Möglichkeit praktisch umgesetzt. Selbst der Begriff ‚Spagyrik' wurde nicht erst von *Paracelsus* eingeführt, sondern war schon den alten Griechen vertraut (man denke dabei etwa an die platonischen Ansichten betreffs *spago* und *ageiro* des Trennens und Vereinigens). In dieser Spagyrik konnte die urzeitliche, noch in der Antike vorhandene Idee des heilkundlichen Adepten, dessen Heilkunst auf den (später von *Paracelsus* ‚wieder' definierten) 4 Säulen ruht, bis heute fortbestehen.

Der – ebenfalls urzeitliche – Wortstamm ATR zur Wesensbezeichnung eines solchen ‚Kundigen' war bei den Griechen im *Iatros* erhalten und wurde in den späteren abendländischen Traditionen zum ‚Iatrochymiker'. Jahrtausendealte vedische Texte nennen den ‚Urvater' der Heilkunst *Atreya*. *Friedrich Will* gibt aufgrund seiner Erkenntnisse einer jungsteinzeitlichen Kultursprache anhand minoischer ‚Texte' als Deutung für ATR an: „Einer, der das Zentrum, das Wesentliche, das Innere (A) kennt und dieses Wissen auf praktische Weise (T) seinem Umfeld (R) zur Verfügung stellt."

‚Alleine vom Lesen kommt kein Wissen' – darin waren und sind sich alle einig. Unvorstellbar lange Zeit wurde ‚Wissen' hauptsächlich mündlich und praktisch vermittelt. Aus der Abwesenheit schriftlicher Dokumente zu schließen, dass es dieses Wissen nicht gegeben hätte oder dass es ‚minderwertig' gewesen wäre, ist schlichtweg ein Irrtum. Der Irrtum der Schriftfixiertheit bewirkt eigendynamisch immer weiteren Irrtum. So ist es – und wir wollen dabei vor allem an alchymistische und heilkundliche Literatur denken – für die Stimmigkeit des Geschriebenen ausschlaggebend, ob der jeweilige Verfasser ‚aus der Sache heraus' schreibt oder ‚über' sie. Letzterer ist bestenfalls ehrlich bemüht, kann zumeist aber nur ‚viele kluge Worte machen', vielleicht seine ‚Gelehrtheit an den Tag bringen' und daraus mitunter auch spannende Erkenntnisse enthüllen, er wird aber in keinem Falle, es sei denn durch zufällige Fügung, den Kern der Sache treffen. Eher wird er Lehrmeinungen verbreiten, die nicht selten zu Verwirrungen und Streitigkeiten führen und das eigentliche Wissen der Kundigen korrumpieren. ‚Das kommt leichtlich vom Wähnen', wie *Paracelsus* es ausdrückt. Gerade zum Studium heilkundlicher und alchymistischer Literatur benötigt man diesbezüglich Feinfühligkeit. Vergleichen wir nur die Gedanken des Historikers *Schmieder* mit den Worten des *Paracelsus*. Zweifellos ist *Schmieders* Werk sehr wertvoll, interessant, aufschlussreich und man muss ihm für seine Mühe danken – aber es trifft nicht den Kern. Er schreibt ‚über' Alchymie, beansprucht aber auch nicht mehr, als eine *Historia* zu schreiben, womit die Sache sicherlich rund wäre. *Paracelsus* aber schreibt ‚aus der Sache heraus' und trifft dabei immer wieder den Kern.

Was wusste Paracelsus?

Dieser Titel ist und bleibt vorerst eine Frage; der Leser möge bitte darauf hier keine klare Antwort erwarten. Unter den oben angeführten Gesichtspunkten sollte man diese Frage aber nun noch einmal stellen und darin eine Weite und Tiefe erkennen, woraus vielleicht in der Zukunft eine Antwort zu erhoffen wäre.

Zuerst ist da die Idee beinahe aufdringlich, dass *Paracelsus* in der einen oder anderen mündlichen Tradition stand, welche schon zu seiner Zeit nur mehr verborgen existieren konnte und deshalb auch heute den ‚allgemeinen' Historien unbekannt sein muss. Wenn wir diese Tradition nicht bei den ‚schreibenden' Völkern annehmen, dann bieten sich abendländische Schulen an, welche insgesamt auf die Zeit der keltischen Druiden und darüberhinaus auf die Kundigen der Megalithzeit zurückgehen und erwiesenermaßen schon über tausend Jahre vor *Paracelsus* gute Gründe hatten, ‚insgeheim' zu bleiben. Einer mündlichen Tradition aber nachzuspüren ist schwierig. Andererseits prägen sich gewisse Bilder im Bewusstsein des Menschen während hunderter Inkarnationen ein – man bedenke dazu den Begriff der ‚Wanderseele *Hun*' in der taoistischen Medizin. Um ein wenig besser zu verstehen, wer denn diese ‚Alten' waren und was sie ‚wussten', darf man in seinem eigenen Inneren nach Bildern forschen.

Ein dem Menschen tief eingeprägtes, archaiisches Bild, welches man nicht einfach deswegen ignorieren kann, weil es oft als abgetretenes Klischee verrufen wird, ist der Alchymist in der Person des klassischen ‚Zauberers', jenes gutmütigen, aber respekteinflößenden alten Weisen mit langem weißen Bart und einer hohen, mit Sonnen, Monden und Sternen gezierten, spitzen Mütze – das Klischee einer Merlin- oder Gandalf-Persönlichkeit, die Erzählungen und Legenden verkörpernd und über dieses Klischeehafte hinaus eine tiefe Erinnerung in uns berührend, die nicht zuletzt auch in der Gestalt des ‚Weihnachtsmannes' noch manch unbewusste Assoziationen zu wecken imstande war, zumindest bevor vor einigen Jahrzehnten mit einer konsequenten Verkitschung dieses ‚Alten' begonnen wurde. Naheliegend wird dieses Urbild auch vorgestellt in seinem Labor im Turmzimmer, über brodelnde Retorten und dampfende Tiegel gebeugt.

Im Hintergrund dieses Bildes mag dann ein noch älteres Gestalt annehmen – diesem sehr ähnlich, nur dass ein märchenhafter Mischwald die Kulisse bildet, darin unser Zauberer auf einer Lichtung neben großen Felsbrocken einem Kessel unterheizt und Körben und Flaschen entnommene Zutaten hineinrührt – das Urbild eines Druiden, eines Weisen der alten Zeit, in den Tiefen unserer Erinnerung verwurzelt. Wir dürfen diesen Bildern weiter nachfühlen, denn es sind keine Traumgespinste, sondern Erbanteile, die wir unseren Vorfahren verdanken, wenngleich ihnen der Mensch westlicher Prägung heute weitestgehend entwöhnt ist. Manch alter *Hakim* des Orients und mancher Weise Indiens erweckt diese vergessenen Menschenbilder beeindruckend zur Realität.

Von solchen Persönlichkeiten sprechen die hermetischen Texte des Mittelalters, wenn sie von den ‚Alten' berichten, die ‚noch in der Kunst kundig waren, weil sie die Zeit dazu hatten' (*Turba Philosophorum*). Man nannte sie ‚Philosophen' und meinte damit ‚Weise' oder ‚Wissende', verband damit also ein völlig anderes Bild, als wir es haben, wenn wir an diejenigen denken, die das, was auch noch ‚Philosophie' genannt wird, ‚an hohen Schulen' studiert haben. Bei den Alten bedeutete ‚Philosophie' nichts anderes als ‚die natürliche Wissenschaft', und diese war direkt mit hohen menschlichen Qualitäten und einem universalen, venetzten und tatsächlichen Wissen verbunden.

Was ‚wusste' nun unser Zauberer des Silbernen oder Kupfernen Erdzeitalters? Worin war er kundig? Dem in Ahnung und Erinnerung nachfühlend, kommt uns besonders die Heilkunst in den Sinn – und darüberhinaus genau die paracelsische Formulierung: ‚der rechte Umgang mit den Dingen der Natur'. Er kannte den Körper, die Gewebe und Organe, deren Zustände und Krankheiten, er verstand es, die Heilmittel zu bereiten, wusste über die Wege der Himmelskörper und deren Wirken in Natur und Mensch Bescheid und war vertraut mit der Beschaffenheit des Makrokosmos mit all seinen grob- und feinstofflichen Regionen, der Art der Physis, des Astralen und Causalen und je nach seinem Grade darüberhinaus. Er war also (zumindest) Astronom, Anatom, Theosoph und Alchymist und stand damit auf denselben ‚vier Säulen der Arznei', die *Paracelsus* bezeichnete.

Einer jener ‚Alten', der offenbar in der Tradition jener ‚noch Älteren' stand, hat im Abendland tiefe Spuren hinterlassen und soll hier zum Thema beachtet werden:

Merlin

Von *Merlin* sind ungleich weniger Texte erhalten als von *Paracelsus*, dafür werden von ihm aber ungleich mehr Geschichten erzählt. Auch wenn er einen Teil seiner Ausbildung tatsächlich in *Alexandria* als Schüler des *Zosimos* absolviert haben sollte, waren ihm zweifellos noch andere Traditionen zugänglich als jene, die dort zusammenflossen. Die walisische Erzählung stellt ihn vor allem als einen Spross keltischer Druidenschulen dar. Ob er selbst den Rang eines Druiden innehatte oder den eines *Fili* oder einfach deren Wissen erwerben konnte, ist nicht bekannt. Dass er aber den Eindruck darin, ein Kundiger zu sein, der Nachwelt unauslöschlich hinterlassen hat, gibt ihm als Persönlichkeit seine besondere Bedeutung.

Er bildet einen Pol in der Geschichte der Alchymie, wo älteste mündliche Überlieferungen keltischer bzw. jungsteinzeitlicher Völker mit den literarischen Zeugnissen mediterraner bzw. orientalischer Alchymie zusammentreffen. Dazu kommt, dass er selbst auch geschrieben hat. Und das macht ihn so wertvoll für den Blick auf die abendländische Alchymie. Seine Legende ist wie ein Mahnmal, welches besagt, dass wir genaugenommen von keiner typisch ‚alexandrinischen Alchymie' sprechen können, sondern dass zu ihrer späteren abendländischen Form ebenso andere Kulturkreise einen vordergründig unauffälligen, weil mündlich tradierten Beitrag geleistet haben. Und dieser Beitrag ist wahrscheinlich wesentlicher, als angenommen wird.

Merlin

Auf seine Art mindestens ebenso berühmt wie *Paracelsus* wurde der in den altenglischen Volkssagen als ‚mächtiger Zauberer' bezeichnete *Merlin*, welcher um das Jahr 500 in Südwales lebte und wirkte. Das in der Stadt *Maridunum* in *Caer Merdhin* gebürtige Besatzerkind wurde mit römischem Namen *Ambrosius* genannt und hieß nach dem Volke seiner Mutter eigentlich *Merdhin*. „*Merlin*" dürfte erst später aus einem Schreibfehler entstanden sein. Genaugenommen nannten ihn seine Zeitgenossen *Merdhin Emrys*, den Eremiten von *Merdhin*, weil er bald in einem Waldgebirge bei *Maridunum* als Einsiedler lebte. In der Abgeschiedenheit der Wälder wuchs er zu einem ‚großen Zauberer' heran und fand zugleich Schutz vor den angelsächsischen Horden. In seine Ausbildung flossen die Kenntnisse der Griechen und Römer mit denen der Weisen seiner Heimat zusammen. Nach dem Zeugnis der Chroniken war er kundig in Arzneikunst und Alchymie. Es wird vermutet, dass er auch in *Alexandria* studiert hatte und dort ein Schüler des *Zosimos* war. Unter den Kundigen galt er als ein „wahrhaftiger Besitzer des Steins der Weisen". Er hat die Alchymie nicht nur praktiziert, sondern auch darüber geschrieben, wovon uns allerdings nur wenige, nichtsdestotrotz sehr interessante Fragmente überliefert sind. Berühmt darunter ist die „Allegorie von dem Geheimnisse des Steins" *(Merlini Allegoria de Arcano Lapidis)*. Daneben kennen wir von ihm noch „Weissagungen von der Burg des Vortigern" und eine Streitschrift gegen die Astrologen Vortigerns. Schon die Gelehrten des 12. Jhdts. würdigten diese alten Texte, und *Alanus* und *Galfredus* schrieben Kommentare zu den „Weissagungen". Die ältesten Drucke der „*Allegoria*" sind zum Teil nur Fragmente, die selbst nach defekten Handschriften gemacht wurden. Äußerst interessant für zukünftige Forschungen und Erkenntnisse dürfte sein, dass nach *Merlin* (mit *Schmieders* Worten) „eine Lücke von 3 Jahrhunderten" in der (dokumentierten) Nachfolge ähnlicher Alchymisten entstand, was eigentlich (auch den Historiker) verwundert.

In der Folge ist dann bemerkenswert, dass erst durch *Karl den Großen* wieder dergleichen Literatur gesammelt wird – und zwar von nun an in den Klosterbibliotheken. *Haymo*, der erste dokumentierte, diesbezüglich ‚Gelehrte' war (wie *Karl*) ein Schüler von Alkuin – zugleich aber Lektor der Theologie an der Abtei *Fulda* (866). Neben typisch theologischen Schriften sind von ihm Neufassungen handschriftlicher, alchymistischer Dokumente bekannt, welche dieser aus verschiedenen Klosterbibliotheken nahm. Bemerkenswert darunter ist die *Epistola de lapidibus philosophicis*, welche 1622 im *Theatrum chymicum* abgedruckt wurde. *Schmieder* führt die „Lücke von 3 Jahrhunderten" kurzum auf die „Ereignisse jener stürmischen Zeit" zurück, in der „die Abendländer" wohl keine Gelegenheit zu solchen Wissenschaften gehabt hätten. Dass die Zeiten in der Folge jedoch immer ‚stürmischer' wurden, bedenkt er nicht. Worin der wahre Grund des Mangels an Dokumenten aus dieser Zeit liegt, wird die Zukunft weisen.

Wahrscheinlich war es *Arisläus*, der um 1150 die >*Allegoria*< *Merlins* unter die ausgewählten Texte des ersten Teiles seines klassischen Werkes >*Turba philosophorum*< fügte. Hier einige Auszüge nach einer Druckversion aus dem Jahre 1750 (neu 1996, *Edit. Hornfisher*):

Allegoria Merlini,
die da vollkommlich begreift
das höchste Geheimnis des Philosophischen Steins
„(...) Der König spricht: ‚Das Wasser, das ich suche ist das Wasser, das ich oft begehret, und es liebet mich vor allen andern.' Und indem er (der Alchymist) es recht bedacht, ist er bald gegangen und hat ihm solches gebracht. Der König aber, wie er es genommen, hat er getrunken, und abermal wieder, bis all seine Gliedmaßen erfüllt worden und all seine Adern davon aufgeblasen sind, und danach ist er sehr verfärbet worden.

(Die Alchymisten stellen nun dem König ein Pferd zur Seite und raten ihm aufzusteigen. Dieser wehrt ab, weil er zu sehr Schwere und Schmerzen empfindet und befiehlt den Alchymisten:) ‚... dass ihr mich setzen sollt in eine helle Kammer, und diese soll an einem warmen Orte stehen und einen Tag und eine Nacht temperiert werden. So werde ich schwitzen, und das Wasser, welches ich getrunken, wird in mir austrocknen, und also werde ich frei und los gemacht werden."

(Komm: Das taten die Alchymisten. Als sie aber nach der Zeit die Kammer öffneten, fanden sie den König darin tot. Daraufhin ließen sie die ‚größten Meister' in ihrer Kunst kommen, um den König ‚zu erlösen'. Im Text werden diejenigen, welche der Kommentator ‚Alchymisten' nennt, als ‚Landsknechte' bezeichnet. Dabei handelte es sich vermutlich um Alchymisten geringeren Grades, vielleicht in der Art der ‚gemeinen Laboranten', welche die Vorarbeit leisten durften. Nachdem diese abgeschlossen und der physische Leib des Königs zerstört war, kommen nun die eigentlichen Meister der Kunst zum Zuge. Man könnte dem Text auch entnehmen, dass die ‚Auflösung' von diesen ‚Landsknechten' nicht sehr kunstvoll durchgeführt wurde.)

„Da haben die vorbemeldten Meister den König genommen, ihn in kleine Stücke zerhauen, sie gerieben, ein wenig vermischet mit ihren Arzneien die da feucht machen, und haben ihn also zugerichtet wieder eingesetzt in seine Kammer und wie zuvor einen Tag und eine Nacht temperiert. Danach haben sie ihn halb gestorben und halb lebendig herausgezogen und ihn mit süßem Wasser gewaschen, bis der Gestank der Arzneien vergangen war. Dazu haben sie wieder etwas von frischer Arznei getan und an seinen Ort gesetzt. (...)"

Wenn es um die ‚Auflösung', den ‚philosophischen Tod' der Ausgangsmaterie geht, bezieht sich *Trismosin*, vielleicht einer der bedeutendsten Lehrer des *Paracelsus* im *Splendor Solis* (16. Jhdt.), auf den griechischen Alchymisten (4. Jhdt.), bei dem sich *Merlin* längere Zeit aufgehalten haben soll. Berühmt ist sein Bildnis der gleichnishaften Zerstückelung durch das ‚zweischneidige Schwert' (durch die beiden Feuer): „Ich hab dich darum getötet, dass du ein überfließendes Leben bekommst." An anderer Stelle berichtet *Trismosin:*

> „*Ovid* schreibt von einem weisen Alten, der sich verjüngen wollte. Er solle sich zerteilen lassen und kochen bis zur vollkommenen Kochung und nicht weiter. Dann würden sich die Glieder wieder vereinigen und verjüngen in großer Kraft."
>
> *S. Trismosin*

Das ‚Einsetzen in die Kammer' (bei *Merlin*) und der Tod bzw. die Zerstückelung darin ist in beinah allen initiatischen Kulturen ein Motiv von zentraler Bedeutung. Wir haben darüber bereits gesprochen. Auch die Bedeutung von ‚zwei Feuern' taucht immer wieder auf. Bei der Initiation der Kundigen selbst sind es ‚das eigene innere Feuer' und das ‚Feuer des Lehrers bzw. des Initierenden'. Selbst bei den Ureinwohnern von *Warburton Range* in West-Australien tritt der Anwärter in eine Höhle (‚Kammer') ein und wird dort von zwei Totems getötet und zerstückelt. Der geöffnete Körper und die ausgenommenen Organe werden mit ‚magischen Substanzen' eingestrichen. („Mit süßen Wassern gewaschen" – *Merlin*). Die Ethnologen *Spencer* und *Gillen* berichten von der Initiationserzählung *Ilpailurknas*, des berühmten Zauberers des *Unmatjera*-Stammes. Er wurde von einem greisen Medizinmann (initiatisch) getötet, sein Körper zerstückelt, ausgenommen und mit *Atnongara*-Steinen und Blättern gefüllt, bis dass er schließlich völlig aufschwoll. („ ... bis all seine Gliedmaßen erfüllt worden und all seine Adern davon aufgeblasen sind" – *Merlin*).

Offensichtlich folgten die Alchymisten stets einem urzeitlichen Verfahren, welches sowohl für Initianden als auch analog dazu in der laborantischen Praxis entsprechend angewandt wurde. Der Hinweis vieler Texte, dass ‚die Auflösung der Schlüssel zum Werk‘ sei, erhält so eine besondere und erweiterte Bedeutung. An der Kunst der ‚Auflösung‘, bzw. am Wissen darüber, kann der Grad des Kundigen ermessen werden.

Wahrscheinlich ist darin auch die Ursache verborgen, warum *Paracelsus* unter all den abendländischen Alchymisten und Heilkundigen die vielleicht berühmteste und anerkennungswürdigste Persönlichkeit war. Sicherlich war er aber derjenige, welcher am nachhaltigsten wirkte. Obwohl seine Schriften, was die Praxis betrifft, weit mehr im Dunkeln lassen als die Schriften der vielen nach ihm, unter denen sich wieder aber die meisten auf ihn beziehen. Er war ihnen Jahrhunderte nach seinem Tode noch Lehrer und Vorbild – eben, weil er ‚etwas ganz besonderes an sich hatte‘, und was dieses Besondere war, das wussten einige noch. Wie *Tölde:* ‚Auf diese Weise hat *Paracelsus* sein *Elixir Proprietatis* gemacht‘. ‚Das hat *Paracelsus* unter der *Piemontischen Magnesia* verstanden‘. ‚Das war des *Paracelsi* geheimster Tincturen eine, auf Menschen und Metalle‘, usf. Und im Vorwort die Bemerkung *Töldes,* dass ‚die Unwiedergeborenen die Kunst nicht verstehen könnten‘.

Je weiter weg von diesem *Paracelsus,* umso weniger wert wurden die Anleitungen. Man bedenke die häufigen Hinweise *Agricolas,* dieses wahrhaft ernsthaften, erfahrenen und verlässlichen Praktikers, der stets Bemerkungen wie: „Ich bin aber der Ansicht, dass *Paracelsus* darunter etwas ganz anderes verstanden hat, als unser Autor meinen will,“ oder: „Des *Paracelsus Mercurius essentiale* war ein ganz anderes Ding als das gemeine, welches der Herr (so und so) vorgibt,“ wohl unter Stirnrunzeln und Aufstöhnen niederschrieb. Diese Angelegenheit setzte sich eigendynamisch bis in die heutige Zeit fort, wo man ‚paracelsische Elixire‘, ‚paracelsische Goldtincturen‘, ‚paracelsische Medizin‘ überhaupt und auch sonst viel ‚Paracelsisches‘ verkauft und zu kaufen meint.

„Er aber hätte einen verschlossenen Ofen gehabt, darinnen wären unterschiedliche Philolen gestanden, da hätte niemand dazukommen können, noch wissen mögen, was vor *Materialia* er darinnen gehabt, er könnte aber leicht erachten, was er müsste unter Händen gehabt haben. Sagte ferner, er hätte ihm einmal einen Prozess von dem *Antimonio* zu laborieren gegeben, darinnen hätte er 16 Wochen gearbeitet, und als es verfertigt worden, hätte ers ihm *(Paracelsus)* zugestellt. Da hätte *Paracelsus* diese Worte gebraucht: ‚*Ägidi,* wenn du wüßtest, wie du jetzt gearbeitet, nimmermehr würdest du es mir gegeben haben.‘“

Ägidius von der Wiesen über *Paracelsus.* Aus: *Johann Agricola,* >Chymische Medicin<, 1638 / O. Humberg 2000

Das Bild des philosophischen Todes, welches auch *Merlin* – darin in urzeitlicher Tradition stehend – verwendet, ist kein im Mitellalter neu erfundenes oder eines, das sich die ‚damals besonders vom Leben geplagten Menschen' in ihrere ‚unwissenschaftlichen' Naivität ‚zusammengedacht' hätten. Es ist kein Gleichnis, sondern eine Realität der Natur, vielleicht die älteste menschliche Wissenschaft. Die gesamte spagyrische Arzneimittelzubereitung fundiert genauso darauf wie das ‚große Werk' der Alchymie. Es beinhaltet den ‚Erlösungsfaktor', welcher zugleich mit dem Menschen selbst von Anbeginn in diese Stoffwelt gelangt ist. Es ist der Kern des ‚rechten Umgangs mit den Dingen der Natur'. Und es ist schließlich eines der deutlichsten Zeichen, wie die Kundigen des Abendlandes gleichzeitig mit ihren heilkundlichen und laborantischen Ausführungen tiefstes und ältestes Menschheitswissen tradieren konnten.

Paracelsus und Merlin

Als ein Beispiel dafür, was *Paracelsus* im Sinne unseres Themas besonders ausgezeichnet haben könnte, sei auf eine Legende hingewiesen:

Zahlreiche Volkssagen erzählen von *Paracelsus* und berichten von Begebenheiten, die sich so, oder vielleicht auch etwas anders, dort und da zugetragen haben. Eine Alpensage weiß davon, dass er einmal sein Schwert nahm und sich in einen ‚Sarg' legte, seinem Gehilfen befehlend, ihn vollständig mit ‚dem Elixir' zu übergießen. Dann sollte er den Deckel aufsetzen und ihn auf den Tag genau ein Jahr lang so liegen lassen und währenddessen seine Ruhe beschützen. Übers Jahr müsse er die Kiste wieder öffnen, worin sich der rundum erneuerte und verjüngte *Paracelsus* befinden würde. Gegen Ende der Frist plagte den Gehilfen schon zu sehr die Ungeduld. Er öffnete einige Tage zu früh und fand seinen Meister zwar tatsächlich verjüngt, gesund und wohlauf darin – nur sein Schädel war noch nicht wieder vollständig hergestellt und wies ein klaffendes Loch auf. Seinen Kopf konnte *Paracelsus* alsbald kurieren, den Gehilfen aber soll er mit Schimpf und Schande fortgejagt haben.

Nun dürfte gerade das Letzte eine volkstümliche Folgerung aus der Handlung der Erzählung sein, denn in Wahrheit hatte *Paracelsus* wohl kaum einem ‚einfachen' und in der Kunst der Geduld so unerprobten Gehilfen diese verantwortungsvolle Arbeit anvertraut. Die Geschichte erzählt über eine (mögliche) Initiation des *Paracelsus* – und zwar nicht eigentlich durch einen ‚Gehilfen' als das, was wir gemeinhin darunter verstehen, sondern vielmehr durch seinen Lehrer, bzw. aus adeptischer Sicht: durch seinen Meister und Initiator. Dass der dabei die Rolle eines ‚Gehilfen' spielt, ist nur folgerichtig und findet seine Bestätigung in ähnlichen Prozessen bei anderen Traditionen.

Einerseits steckt in dieser Erzählung eine Allegorie über den Tod des alten Königs, seine Reinigung und Erneuerung, wie sie in Bezug auf ‚Das Große Werk der Alchymie' immer wieder in Texten und Erzählungen vorkommt. Ähnliches haben wir ja auch vorhin in der *Allegoria Merlini* gefunden. Der Hinweis auf die Stirnwunde des *Paracelsus* in oben erwähnter Volkssage reicht darüberhinaus tief in ältestes Wissen hinein. Stirn und Schädel entsprechen der ersten Begegnung von Sonne und Mond. In den indischen Traditionen ist es besonders das innere Auge *(Tisra Til)*, worin sich die ‚Sonne' ausdrückt. Signaturkundlich zeigt sich an der Beschaffenheit der Zonen von Nasenwurzel und Stirn, wieweit der Funktionskreis der Leber ‚Durchgängigkeit' gewährleistet bzw. die Situation des ‚Körpersaftes Galle' – mit anderen Worten, wie mit dem Eintreten von Idee/Tugend/Quintessenz – durch das solare Äther-Element das Sulfur-Prinzip im Körper installiert oder im Falle der Initiation regeneriert wird.

Maier lässt den König ins Rote Meer tauchen und danach ins Schwitzbad setzen, *Trismosin* lässt ihn in der chaotischen, faulen Erde versinken und danach von der Sonne trocknen (die Praktiker des Großen Werkes mögen die Kurzbündigkeit solcher Zusammenfassungen verzeihen), und nun lässt *Merlin* – den wir hier als Mittler der Traditionen betrachten – den König von ‚seinem Wasser' trinken und ihn dann in die heiße Kammer setzen.

Der keltische Barde *Taliesin* wurde der Legende nach am 29. April von *Keridwen (Demeter)* im Meer ausgesetzt und am 1. Mai, genau zu *Beltaine*, dem Fest der Frühlingssonne und neuen Geburt der Natur, von *Elphin* an Land getragen. *Beltaine* wird zuweilen mit *Bel-Tene* übersetzt, was ‚mächtiges Feuer' bedeutet.

Gwydion trank vom Inhalt *Keridwens* (der Erde) Kessel, wonach einige Verwandlungen einsetzten, bis er ‚fertig ausgebacken' war.

Die erste Berührung mit jenem ‚Wasser', die Faulung, das Einsinken etc. folgt dem Tod der Ausgangsmaterie in der Alchymie in der Phase des Saturn. Saturn zwingt zur Verwandlung. Auf die Nacht, die Putrefaction, folgt das Ausschwitzen, Wiederbeleben, Trocknen usf. in der Phase des Merkur und schließlich der Sonne. Ob es sich dabei um Holzkisten, Höhlen, Kammern, Schwitzbäder, Erdgruben, Erdhügel, Krüge, Tiegel oder was auch immer handelt – der Sinn ist der gleiche. *Paracelsus* lag nach der Erzählung ein volles Jahr in der Kiste, Andere, von einem Datum zum andern, manche öfters (wie in der *Allegoria)*, *Merlin* verweilte in einer Höhle, sein Schüler *Artus* schlief unter einem Haselnussstrauch bei *Craig-y-Dinas*, bis er vom Schüler eines Weisen geweckt wurde (dabei war es ebenfalls angeblich noch zu früh)… . Alle werden zuerst ‚getötet', von Saturn zur Verwandlung gedrängt und erstehen danach (sofern der Prozess gelungen ist) wieder erneuert auf – ‚ganz geistig geworden'.

Auf einen zweiten, in der obigen *Paracelsus*-Sage verborgenen Hinweis soll hier nur kurz eingegangen werden: Schon in steinzeitliche Kulturen war es üblich, Adepten durch ein Verfahren zu initieren, das am Schädel eine Wunde hinterließ. Dadurch wurde ihre Wahrnehmung für das ‚Licht der Natur' sensibilisiert, sie wurden in gewisser Weise hellsichtig. Ähnliche Techniken werden heute noch in Tibet angewandt, wobei die einen nur so weit gehen, bis dass ein Blutstropfen am Schädel austritt, andere regelrecht die Stirn im Bereich des Inneren Auges oder die Schädeldecke an einer bestimmten Stelle öffnen. Diese Methoden sind zum Teil für heutige Verhältnisse ‚grob' und auch riskant, wurden aber während Jahrzehntausenden praktiziert. Zahlreiche solche ‚durchlöcherte Schädel' wurden aus dem *Neolithikum* gefunden. Da die Struktur der Archäologen zumeist nicht in geisteswissenschaftliche bzw. hermetische Gefilde hineinreicht, erklärt man diese Funde für vermeintliche Schädeloperationen, was sie im Grunde ja auch bisweilen waren. Aber auch ‚sanftere' Methoden, von tatsächlichen Initiationen bis hin zu rein rituellen ‚Initiationen', wurden seit jeher ausgeübt.

Dass zur Zeit *Merlins* solche Methoden noch im Westen in manchen Gegenden üblich waren, ist wahrscheinlich, und diejenigen, welche, wenngleich nicht als ‚Meister', so doch als Adepten ihrer Kunst galten, dürften dergleichen Verfahren durchgestanden haben – genauso wie dies heute noch in manchen Schulen Asiens üblich ist. Wenn dies vor allem die wirklich alten Traditionen betrifft und im jungsteinzeitlichen Europa nachweislich gehandhabt wurde, dann standen jenen Kundigen, welche aus solchen Traditionen schöpfen konnten, solche Techniken eben zur Verfügung, und sie wussten um ihre allegorische Bedeutung in der ‚Kunst des rechten Umgangs mit den Dingen der Natur'. Das ist der Grund, warum man sagt, die Alchymie sei eine initiatische Wissenschaft. Dass *Paracelsus*, und viel später auch noch andere in diesen Traditionen standen, aus diesem Geiste heraus sprachen, schrieben und handelten, niemals aber die Sache selbst oder ihre Bewahrer verrieten, ist anzunehmen.

> „Merkt, dass ihr auflösen müßt! Den Weg zurück gehen! All die Werke, welche die Natur vorangetrieben hat, von einer Staffel zur andern, die müßt ihr wieder auflösen."
>
> *Paracelsus*, >Paragranum<

All diese Dinge sind bedenkenswert, wenn man über heilkundliche und alchymistische Traditionen im Abendland spricht, und das ist zweifellos mit ein Verdienst *Merlins* – so wenig wir auch von ihm wissen, aber alleine durch seine legendäre Präsenz.

Von Columban bis Basilius Valentinus

Aus archäologischer Sicht waren mit der Etablierung der Kelten in Mitteleuropa vor etwa 3000 Jahren erwiesenermaßen alle Technologien bekannt, um das hermetische Naturwissen in alchymistisch-laborantischer Praxis zu erproben. Mit hohen Temperaturen schmolz man das Eisen, die Tongefäße wurden mit Graphit versetzt, damit sie feuerbeständiger waren, die Glasherstellung wurde intensiviert und anderes mehr. Man reiste auch gerne und konnte sich dabei schon auf einem ausgedehnten Netz aus Handelswegen, welches spätestens durch den Bronzehandel (seit etwa 4500 v. Chr.) ganz Europa durchzog, vom äußersten Norden bis Griechenland, Spanien und Arabien begeben.

Lange Zeit floss das Wissen der Alten in den Druidenschulen zusammen. Die Druiden hatten die Aufgabe, das Volk zu unterrichten, und bedienten sich dabei der jeweiligen keltischen Dialekte, welche in den Regionen verstanden wurden. Die Sprachgemeinschaft und die Erziehung und Bildung durch die Druiden waren die gemeinsamen und verbindenden Faktoren zwischen den zahlreichen keltischen Stämmen.

Diese Verbindung suchten nun die römischen Invasoren zu lösen, indem sie die Druiden verfolgten und mit der Zeit der Volkssprache die lateinische Hochsprache überstülpten. Während der späteren ‚Fortsetzung' des Römerreiches mittels der neuen zwangsverordneten ‚Religion' hüllten manche Kundige, vor allem im Norden Europas, ihre Schulen in ‚christliche Mäntelchen' und verstanden es so, noch jahrhundertelang weiterzuwirken. Die ‚Römer' waren froh, ‚Bekehrungen' verzeichnen zu können, und das Wissen der Alten blieb in weiten Teilen erhalten. Dieser Prozess setzte etwa um 400 n. Chr. ein. Was die ‚Bekehrungen' betrifft, so findet man dazu ‚historische' Feststellungen wie: ‚Die Tradition der Druiden endete mit der lokalen Verbreitung des neuen, christlichen Glaubens.' Solcherlei Formulierungen gehen natürlich weit an der Realität vorbei und täuschen eine Gegebenheit vor, welche nie der Angelegenheit entsprochen hatte. Das Mysterium jenes orientalischen Heiligen, auf das sich die neue Ismus-Doktrin bezog, war den Druiden weder neu noch unbekannt. Sie vermochten es genauso zu integrieren wie die Mysterien anderer, uns unbekannter Heiliger noch älterer Zeiten. Die zahlreichen autonomen Klöster der ‚keltischen Christen' waren Stätten des Wissens – nicht der Missionierung zu einer ‚neuen Religion'. Man sammelte Literatur, fertigte Übersetzungen an und erforschte die Natur. Mehr als in den Jahrhunderten davor wurde Wert gelegt auf das geschriebene Wort, ging es doch nun vor allem darum, dem drohenden Verlust ältester Überlieferungen entgegenzusteuern. Mit dem, was man aus sich selbst schrieb, hielt man sich allerdings immer noch sehr gering. Vor allem notierte man Historien *(Book of Kells)* und andere Angelegenheiten, welche die Tiefen der eigentlichen Lehren nicht weiter berührten.

Andererseits trachtete man danach, jeglicher Literatur habhaft zu werden, damit sie schlichtweg zur Verfügung stand. Dabei wurden nicht nur seltene griechische, chaldäische und arabische Texte gesammelt, sondern auch die neueren lateinischen beschafft und kopiert. Im 6. Jhdt. – also zur Zeit *Merlins* – wirkte zum Beispiel *Columban von Iona* in Irland und Schottland, den die ‚Historien' in die Reihe der ‚ersten christlichen Mönche und Missionare Irlands' stellen. Er beschaffte das Buch >*Cathach*< (Krieger) heimlich aus der Bibliothek des Mönches *Finnian* (einer von denen, die später den Titel ‚hl.' erhielten) und schrieb es ab. Dieses, als ‚erste Urheberrechtsverletzung in der Geschichte' bekannte Ereignis löste einen wahren ‚Bücherkrieg' aus. Nach der Schlacht (von *Cùl Dreimne*, 561) zwischen *Columbans* Sippe und König *Diarmait*, wurde der ‚Mönch' exkommuniziert. Er geht ins Exil nach *Alba* (Schottland) und gründet auf der, schon den Alten ‚heiligen' Insel *Iona* ein Kloster. Von dort aus pflegte er engen Kontakt mit den Pikten. *Iona* blieb noch viele hundert Jahre danach ein Pilgerort, an dem man so manches lernen konnte.

Obwohl es dafür keine historischen Belege gibt, wird *Columban* in altirischen Liedern als *Filid* bzw. als Barde verehrt. Als im Zuge der eigentlichen ‚Missionierung' die *Filids* aus Irland vertrieben werden sollten, gelang es ihm (beim ‚Königstreffen von *Druim Cett*', 575), diese Vertreibung zu verhindern.

Schüler *Columbans* gründeten in ganz Europa die ‚Schottenklöster' (welche in Wirklichkeit mit irischen ‚Mönchen' besetzt waren), in denen bis ins Hohe Mittelalter noch das ‚Wissen der Alten' vermittelt werden konnte.

Ein anderer Spross der Druidenschulen waren die sogenannten *Culdee (Cèli Dè)* (nicht zu verwechseln mit namensgleichen modernen Institutionen). Sie standen durchaus in sehr alter mündlicher Tradition und lehrten in den regional verständlichen keltischen Dialekten.

Erst um 1172 wurden die verschiedenartigen Äußerungen jenes sogenannten ‚keltischen Christentums' offiziell verboten.

Aus wieder einem anderen Spross, welcher allerdings nie unter einem ‚christianisierten Mäntelchen' gedieh, gingen im Mittelalter die *Katharer* hervor, in deren Tradition auch *Merlin* eine besondere Rolle spielte.

Vemutlich stark von einem alten (‚historisch' nicht bekannten) Kundigen der *Culdees* beeinflusst, verfasste *Benedikt von Nursia* im Jahre 529 (nach einer gewissen *Regula magistri*) die *Regula benedicti* – nichts anderes, als eine schlichte Erläuterung des Spruches ‚*ora et labora*'. Wieder entsprach der Auftrag durchwegs dem der Druidenschulen: Literatur sammeln, studieren, übersetzen, kopieren, lehren, unterrichten, meditieren und sich heilkundlich und kunsthandwerklich betätigen. Erst beinahe 500 Jahre später begann sich aus diesem ‚Spross' ein typisch christlicher ‚Orden', der der Benedikter, zu entwickeln.

Auch die ersten Nachfolger des *Benedikt von Nursia* trugen ihr Wissen den Menschen in ihrer eigenen Sprache vor und erfüllten ihren erzieherischen Auftrag unabhängig von großen religiösen Zentren. Sogar noch im 12. Jhdt. konnten sie nur schwer die nun neu aufkommenden lateinischen Universitäten akzeptieren, obwohl sie zu dieser Zeit selbst schon weit von ihren eigentlichen Gründen abgekommen waren.

Die Vorgehensweise war immer die selbe: Kundige ‚historisch' unbekannter, alter Schulen begannen mit Sammeln, Übersetzen, etc. interessanter Literatur. Dazu nahmen sie ihren Lehrauftrag wahr und unterrichteten Menschen jeglicher Bevölkerungsschichten in den Dingen, welche diese lernen konnten und wollten. Dabei bedienten sie sich der volkseigenen Sprachen. Trotz des Umgangs mit vielerlei Texten war ihnen ein noch weit größeres Vermächtnis an mündlichen Überlieferungen zugänglich. Welcher Art diese zum Teil waren, ersehen wir auch aus den natur- und heilkundlichen Abhandlungen der (späten) Benediktinerin *Hildegard von Bingen* (1098-1179), obgleich diesbezüglich ihre Quellen und Kontakte umstritten sind und auch die Art, lateinisch zu lehren, nicht mehr ‚traditionsgemäß' war.

Im 5., 6. und 7. Jhdt. waren Haltung und Handeln jener Kundigen noch nicht doktriniert, sondern eine aus der Zeitqualität unabhängig voneinander erkannte Notwendigkeit. Ganz ähnlich dem nur durch Sprache und der Betreuung durch die Druiden losen Zusammenhalt der keltischen Stämme arbeiteten sie selbstständig, dezentral und nicht in eigentliche ‚Orden' eingebunden, sondern eröffneten im Gegenteil ständig neue Varianten zur Erfüllung ihrer Aufgaben. Hierzu ließe sich noch vieles sagen – über manche mehr oder weniger bekannte Persönlichkeiten, was und warum sie dies und jenes taten usf. Die Art dieses Wirkens setzte sich noch bis ins Mittelalter und zwar insbesondere im Rosenkreuzertum fort, allerdings nur mehr im Ansatz dieser ‚Bewegung'. Wo möglich, wurden Kundige solcher Art immer wieder in ‚Orden' eingebunden (in ‚christliche' und andere), wobei von ihrem eigentlichen Grund sehr bald nur noch täuschende Äußerlichkeiten übrigblieben. Hier wollen wir nicht mehr weiter ins Detail gehen, sondern nun noch einen Namen erwähnen, welcher in der Alchymie einen bedeutenden Platz eingenommen hat: *Basilius Valentinus.*

Bis heute ist *Basilius* ‚historisch' nicht nachgewiesen. Dennoch haben die von ihm bzw. unter seinem Namen verfassten Werke tiefe Spuren hinterlassen und gelten heute noch als literarische Autoritäten in hohen alchymistischen Praktiken. Man spricht dabei vom ‚Benediktinermönch *B. V.*' und muss gerade hierzu überlegen, ob damit nicht eine Art Titulierung gemeint war, ähnlich der Art, wie wir sie bei der Bezeichnung ‚Geber' bzw. ‚Weber' und bei den Alten als ‚Hermes' vorfinden.

Über die beiden letzten Begriffe haben wir im >Sonnenfunken< schon gesprochen. Auch darüber, dass die Abkömmlinge der *Katharer* noch im 13. Jhdt. vom Volk die ‚*Guoten*', also die ‚guten Menschen', genannt wurden. Nun bedeutet ein *Benedictus* nichts anderes, als ‚einer, der die Wahrheit sagt' bzw. ‚der wohl spricht.' Kurz und gut: Es mag die Betitelung ‚Benediktiner' für die damaligen Kundigen eine Art Code gewesen sein, um jemanden zu bezeichnen, der tatsächlich ‚im alten Wissen stand' – unabhängig von einer ‚offiziellen' Ordenszugehörigkeit.

Ans Licht gebracht wurde der >Triumphwagen des *Antimonii*< eines Benediktinermönches namens *Basilius Valentinus* im Mai 1604 von den Alchymisten *Johann Tölde* und *Joachim Tanckius (Tancke)*. Hatte zuvor schon *Paracelsus* ‚traditionsgemäß' in deutscher Sprache geschrieben und unterrichtet – und zwar für Menschen jeglichen Standes, dann entsprachen die Werke *Basilius*' der selben Art, waren sogar noch praktischer und (in weiterem Sinne) verständlicher, wenngleich bei weitem nicht so umfassend und universal wie die des *Paracelsus*. Er beschreibt die Herstellung hoher alchymistischer Heilmittel und die sowohl zum ‚Großen Werk', als auch zu kleineren Partikular-Werken nötigen Kunstgriffe. Es scheint sogar, dass seine Anleitungen bisweilen ‚zu praktisch' gehalten bzw. aufgefasst wurden und deshalb bis heute manchen zu einer ‚chemischen Alchymie' verleitet hatten, wobei nicht bekannt ist, ob sie, so verstanden, je zu ‚höheren' Erfolgen führen konnten. Auch *Basilius*' Bücher sind eben keine ‚Kochbücher'. Der niederländische Arzt *Theodor Kerckring*, welcher um 1670 den >Triumphwagen< kommentierte (>*Microscopium Basilii Valentini*<) wies eindringlich darauf hin, dass diese Texte nicht vordergründig und oberflächlich aufgefasst, sondern ihre verborgenen Inhalte erforscht werden sollten.

Einen anderen Text des *Basilius* bzw. mit seinem Namen gezeichneten Text beginnt der Verfasser:

> „(...) Du sollst wissen, daß, nachdem der Allerhöchste mir aus lauter Gnade das allergrößte Geheimnis (...) hat offenbaren wollen, wie ich denn von solchem zu verschiedenen Zeiten auch also Verschiedenes geschrieben, denn man findet diese Perle nicht auf einmal, sondern durch die große und überschwengliche Barmherzigkeit Gottes eines nach dem andern, wie ich denn durch alle meine vorigen Schriften 2 Wege erkannt, als den trockenen und nassen Weg, hier aber in diesem großen Werk werde ich der selben nicht nur 2, auch nicht 4 Wege, wie einige der Alten gefunden, sondern deren mehr als 14 (vortragen – Komm.). Ja der Brunnen der Weisheit ist in der Natur unerschöpflich, denn wenn du erstlich weißt, was die geister in der Natur bewirken und wie solche (...) in unserem Feuer gebunden, gefesselt fix und feuerbeständig gemacht werden, so wird dir also hier mehr als ein Weg offenbart werden. (...)"

Danach folgen zahlreiche praktische Anleitungen zur Präparation ‚natürlicher Dinge' – einerseits zu ‚hohen Medizinen auf Menschen und Metalle', andererseits als ‚Zutaten zum Großen Werk'. *Basilius* beginnt mit einer diffizilen Beschreibung der Zubereitung des ‚Vöglein Hermetis', welches er auch den ‚philosophischen Salmiac' und das ‚himmlische Feuer' nennt, aus Tau und Regenwasser:

Wie man aus dem Tau und Regen unser himmlisches Salz bereiten soll, als den allergrößten Universal-Schlüssel

Willst du nun, mein Freund und Nachfolger in unserer himmlischen Kunst diesen bunten Vogel aus dem Tau und Regen bereiten, nach den Anleitungen des Königs *Xophor*, so lauten seine Worte also: „Ich mache den Anfang von den anfangenden Dingen in der Natur, welches kommt aus dem Chaos der Götter und vermischt sich hier unten mit den Elementen, denn dieser himmlische Geist wirkt aus den oberen Elementen in die untersten, daher ist das Unterste dem Obersten gleich geworden. Dies ist die größte Heimlichkeit in der Kunst. Diesen Geist findest du im Tau und Regen. Dieser ist das lebendigmachende himmlische Feuer und der Adler, der über sich wächst in Gestalt eines feurigen Glanzes. Er ist ein durchdringendes, geistiges Salz. Dieses ist das allerköstlichste Ding und auch das allerheimlichste in der Natur, aus welchem wir bereiten die Grundfeuchtigkeit zu unserem Steine."

Xophor will damit soviel sagen: Mit diesem kann man den Stein der weisen Meister bereiten, denn es ist ein Schlüssel unseres durchscheinenden Salamanders.

Darum, wenn du nun aus dem Tau und Regen diesen Geist erhalten willst, so sammle dir gleich im Frühjahr so viel vom Tau und Regen als du dessen nur haben kannst und bringe denselben in ein großes Gefäß und stelle es an einen Ort, wo die Sonne darauf scheinen kann, aber wohl zugedeckt. Lass es also stehen und faulen bis sich ein brauner Schaum oben auf begibt, dann ist es Zeit, dass du die Scheidung vornimmst.

Dann rühre es untereinander und fülle damit eine Blase an. Dieses ist ein Gefäß, woraus man Brandtwein brennen tut.

Wenn du vom Wasser 100 Pfund (1 Pfund entspricht ca. 360g – Komm.) hast, so destilliere mit lindem Feuer 20 Pfund davon, als den flüchtigsten Teil, und verwahre das Überdestillierte. Dann destilliere von dem Zurückgebliebenen noch 12 Pfund herüber (in einen anderen Recipienten – Komm.). Dieses Wasser verwahre, denn du wirst dessen nötig haben. Was aber in der Blase zurückgeblieben gieße in ein rundes Gefäß und verwahre es. Also scheide all deinen Tau und Regenwasser durch die Destillation in 3 Teile. (Komm: Die Ausgangsmenge von 100 Pfund bezieht sich auf die Größe der Destillationsblase und ist also nur ein Teil der Menge des gesammelten Wassers. Weil diese ursprüngliche Ansatzmenge sehr groß ist >"soviel als du haben kannst", werden zur Scheidung in 3 Teile mehrere Destillationsansätze nötig sein. Basilius spricht im Folgenden von 90 Pfund,allerflüchtigstem Teil'. Davon erhält er aus der ersten Destillation nur 20 Pfund. Es sind also noch soviele Destillationen zu bewerkstelligen, dass auch noch die restlichen 70 Pfund erlangt werden. Daraus kann man auf eine ursprüngliche Ansatzmenge von ca. 450 Pfund Tau und Regenwasser schließen, welche also einzusammeln wäre.)

Wenn du nun alles in 3 Teile geschieden, so fülle deine Blase mit dem flüchtigen Teil. Wenn es 90 Pfund hat, so destilliere 25 Pfund davon vom allerflüchtigsten Teil. Was hier in der Blase zurückbleibt ist ein totes Wasser. Solches schütte heraus. (Komm: Der Geist / *Spiritus* / Merkur-Anteil von 450 Pfund Ausgangsmenge an putreficiertem Wasser macht nach der ersten Destillation 90 Pfund aus und nach seiner Rectification 25 Pfund. Der rückständige Rest gilt als *Phlegma* und geht nicht ins Werk ein.)

Das Flüchtige bringe wiederum in die Blase, und wenn desselbigen 50 Pfund, so destilliere mit lindem Feuer 25 Pfund davon. (Komm: Mit dieser Anmerkung verdoppelt sich also die urssprüngliche Ansatzmenge und alle weiteren, dh. wir erhalten bei einer Ausgangsmenge von 900 Pfund putrefiziertem Wasser in der zweiten Rectification 25 Pfund *Spiritus*.) Das Zurückgebliebene schütte hinweg.

Dann destilliere also fort: Nimm von diesem allerflüchtigsten Teil 25 Pfund und destilliere 12 Pfund davon herüber. Was hier zurückbleibt wird wiederum hinweg geschüttet. Diesen Geist schütte in einen Kolben (Komm: *Basilius* empfiehlt also jetzt den Wechsel von der Kupferblase zum Glaskolben) und rectificiere denselben im Bad unserer lieben Frauen (aus dem Wasserbad) also auch 4 bis 5 mal, und wenn nun des flüchtisten Teiles 5 Pfund übrig bleibt, dann verwahre denselben in einem starken steinernen Gefäß an einem kalten Ort. Dieses ist also der lebendigmachende Universal-Geist mit welchem man fürwahr in der Kunst vieles ausrichten kann, denn seinesgleichen findest du in der Natur nicht, denn du kannst mit diesem Geist alle *Salia* in ‚unseren *Salmiac*' oder *Mercurium* verwandeln, wie du hören wirst an seinem Ort.

Dann nimm die Erde, in welcher das *Sal* und der *Sulphur* sowie auch unsere jungfräuliche Erde verborgen sind, rühre solches untereinander und lasse es auf einer linden Wärme in einem kupfernem Gefäß bis auf 2 bis 3 Pfund evaporieren. (Komm: Der Rückstand der allerersten Destillation wird sehr gelinde eingedickt.)

Dann lasse es erkalten und filtriere solches, so bleibt die Erde im *Filtrum*, solche musst du verwahren. (Komm: Man kann natürlich den Filter trocknen und dann das meiste vom darin klebenden Rückstand herausbröseln, es bleibt aber immer noch ein Verlust, was bei der sehr geringen Menge an Festmaterial abgeht. Besser scheint es, während des Evaporierens in eine kleine Kupferschale zu wechseln und darin alles – sehr gelinde – einzutrocknen und aufzubewahren. Dazu muss die Schale verschlossen werden, denn der nun folgende, nächste Arbeitsschritt, das Darstellen des Salzes aus der Flüssigkei,t kann ziemlich lange Zeit beanspruchen.)

Das Durchfiltrierte gieße in eine steinerne Schale, setze diese in die Asche und lass solches mit lindem Feuer abrauchen bis zur Honigdicke. Dann lass es kalt werden und setze das Gefäß in den Keller, damit Kristalle anschießen. Diese sammle. Den zurückgebliebenen *Liquorem* lass noch weiter auskristallisieren, wenn du ihn vorher im Aschenbad etwas evaporiert hast. Reinige alle Kristalle durch solvieren und coagulieren, wie gebräuchlich, dann mache sie in linder Wärme trocken, dass du solche zerreiben kannst. (Komm: Je nach Wasserqualität des ursprünglichen Ansatzes differiert die Menge der Kristalle. Auch können sie entweder sofort sehr rein und klar erscheinen oder schmutzig und mit vielen *Feces*. In jedem Falle wird – bei sehr gelinden Temperaturen – umkristallisiert; etwaige *Feces* kommen zum Erd-Rückstand.)

Dann mache auch die braune Erde trocken (Nachtrocknen – Komm.) und reibe sie klein. Mische das Salz darunter. Feuchte mit dem *Spiritus* an und mache Kügelchen daraus. Solche mache in linder Wärme trocken. (Komm.: Nach dieser Coagulation darf nicht mehr vollständig getrocknet werden.)

Tue die Kügelchen in eine Retorte und lege diese in ein Sandbad oder (besser) nimm statt des Sandes zarte Eisenfeilspäne und lutiere eine Vorlage an in welcher 1 Pfund von dem *Spiritus* vorgeschlagen. Dann treibe *per gradus ignis* in 48 Stunden alles herüber, so steigt zuletzt viel von dem flüchtigen, himmlischen Geist *Mercurium* mit. Sobald alles kalt, bringe den Sublimat in den übergestiegenen *Spiritum* und verwahre solches. (Komm.: Verwendet man zu dieser Arbeit anstatt der von *Basilius* empfohlenen Retorte einen niedrigen Kolben mit kurzer Brücke, dann lässt sich der merkuriale Sublimat leichter mit dem *Spiritus* auswaschen.)

Das (rückständige) *Caput mortuum* nimm und extrahiere daraus das Salz mit destilliertem Wasser. Dieses filtriere, so bleibt die Erde wiederum zurück. Solche mache trocken, denn du musst die selbe haben. Das Durchfiltrierte evaporiere bis auf ein Salz. Dieses Salz tue in eine Retorte und gieße von dem himmlischen Geist 12 mal so schwer als das Salz gewesen ist darauf und vermache die Retorte, stelle solche (Salz und *Spiritus*) in eine linde Wärme zum putreficieren 8 Tage und Nächte. Dann nimm solche und lege die selben in Eisenfeilspäne mit einer Vorlage und treibe den Geist mit den meisten *Sal fixum per gradus ignis* herüber, welches in 24 Stunden verrichtet wird. Nachdem alles kalt geworden, so nimm das wenige Salz, welches in der Retorte noch zurückgeblieben, heraus, tue solches wiederum in eine Retorte und gieße das Davondestillierte wieder darauf.

Digeriere es 2 Tage und Nächte, dann lege die Retorte in den Sand und in 24 Stunden also miteinander überdestilliert, so ist dein *Sal fixum* auch flüchtig. (Komm: Wenn nicht, dann nach einigen Kohobationen wenigstens ein guter Teil davon.)

Gieße dieses Destillat in einen hohen Kolben, und gieße auch den obigen Spiritum mit dem Sublimat dazu. Nimm deine jungfräuliche Erde, tue solche in einen Tiegel, glühe sie ein wenig aus, doch ohne Verlust, dass aber auch nichts Unreines dazukomme. Dann bringe solche auch in den Kolben. Setze alles in ein Wasserbad, gib einen Alembik darauf und schließe eine Vorlage an. Dann destilliere den Spiritum cohobando 6 bis 7 mal davon, damit der allerflüchtigste Geist sich dabei etwas figiere.

Dann lasse es erkalten, das Überdestillierte wiederum zurückgegossen, dann setze den Kolben in die Asche und destilliere den flüchtigen Geist auch 4 bis 5 mal davon. (Komm: Es folgt also auf die ‚zarte' Kohobation nun eine bei höherer Temperatur.)

In der 6. Destillation aber destilliere alles ab bis zur Trockene. Dann lass das Feuer abgehen.

Wenn alles kalt worden, so findest du im Kolben ein blutrotes Salz, welches aber nicht fix. Solches reibe klein und auch ein wenig von einem gestoßenen Glas gleich viel darunter, und bringe solche miteinander in einen hohen Kolben, setze solchen in den Sand, schließe Alembik mit Vorlage an und gib gradweise Feuer, so steigt unser bunter Vogel in die Höhe, und weil die Geister hoch steigen müssen, so fallen sie wieder herunter, wo sich dann alles in einen perlweiß glänzenden Sublimat aufsublimiert. Dieses ist nun der himmlische *Mercurius* der Weisen. Dieses wird verrichtet *per gradus ignis* in 48 Stunden. Dann lass alles erkalten und zerbrich den Kolben, nimm dein himmlisches Feuer, deinen wahren, lebendigmachenden, philosophischen *Salmiac* heraus und bringe denselben in ein Gefäß und verwahre solchen vor der Luft. Dieses ist das Vöglein *Hermetis*.

Mein Freund und Nachfolger in unserer geheimen Kunst, es können also mit diesem Schlüssel die geheimsten Arbeiten vollbracht werden.

Wie man solchen himmlischen *Salmiac* gebrauchen soll in der Kunst:

Es finden sich nun sehr viele Wege mit unserem *Mercurio* also zu Werke zu gehen. (...) das Gold aufzuschließen: Das Gold soll zuerst durch das *Antimonium* gereinigt werden, damit es feurig und also in seiner solarischen Kraft erhöht werde, denn das *Antimonium* reinigt das Gold und erhöht es auch noch in seiner Farbe, denn das *Antimonium* ist das Bad des Königs, weil es von seinem Geblüt ist. Wenn du nun ein solches reines Gold in Händen hast, so lass solches in dünne Blättlein schlagen, dann reibe es klein und nimm von solchem 1 Loth (ca. 15g – Komm.). Reibe von unserem himmlischen *Salmiac* 4 Loth darunter, tue solche miteinander in einen reinen Kolben und setze solchen in eine linde Wärme wohl vermacht 4 Tage und Nächte.

Dann nimm den Kolben heraus und setze solchen in den Sand und gib Sublimierfeuer, also dass unser himmlischer Adler sich mit Sulphur, Salz und Merkur des Goldes zusammen aufsublimiere.

Solcher Sublimat wird wie ein Gras so grün sein und mit sehr großem Glanz über sich steigen, und es wird die Erde des Goldes zurückbleiben, also hast du das Gold wiederum von seinen Banden mit unserem *Salmiac* entledigt.

Dieses himmlische grüne Salz ist jetzt eine sehr große Medicin, denn es schmilzt in der Wärme, sodass man es danach als das wahre *Aurum potabile* gebrauchen kann, denn dieses Salz ist ohne jedes Corrosiv.

Mein Freund, es wollen die meisten *Philosophi*, dass man dieses himmlische Salz vorher mit dem geheimen Feuer vereinigen und solche miteinander in ein Elixir und eine Tinctur verwandeln soll.

Und wenn du dein solarisches Salz alleine zur Fixation bringen wolltest, so würdest du eher eine Medicin als ein Elixir erhalten und würdest lange Zeit dazu vonnöten haben. Wenn du (aber) nun von diesem *Mercurium* 5 Loth hast, so nimm von dem geheimen Salz aus der hundertjährigen Hofmeisterin 10 Loth, mische solche in einer gläsernen Schale untereinander, dann (...).

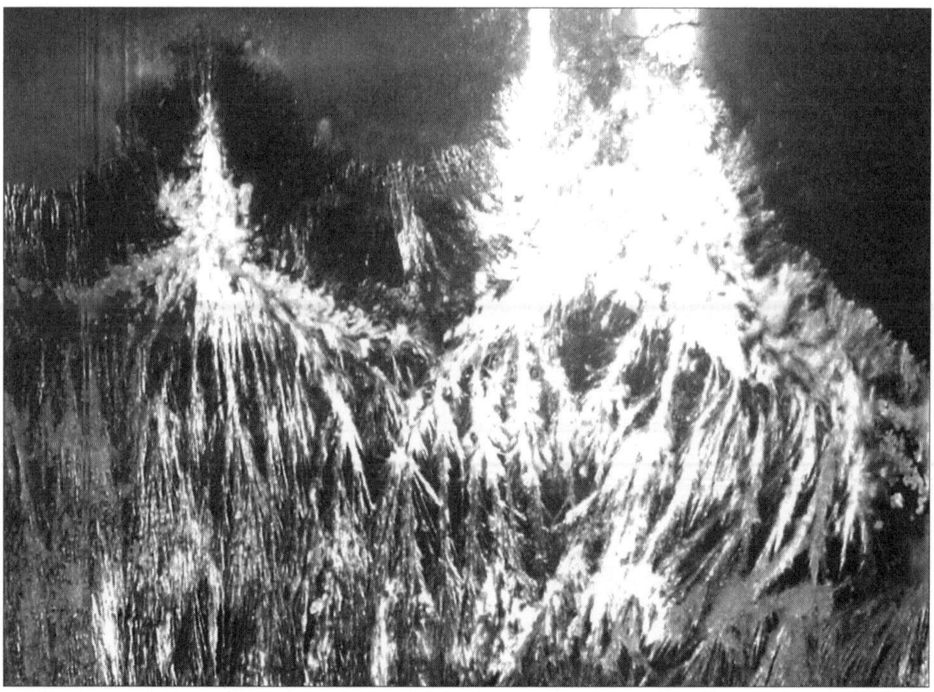

Salz

Wir haben bereits darüber gesprochen, dass die ‚Kunst' in der ältesten Zeit oft bei den Töpfern anzutreffen war, befassten sich doch diese ganz besonders mit ‚Feuerkunst' und mit dem ‚Umgang mit den natürlichen Dingen'. Nach den Töpfern folgten seit der Epoche intensiverer Metallbearbeitung die Schmiede – nicht als ausschließliche, aber doch als prädestinierte Hüter der Kunst. Außer Erden (Ton) und Metalle ist da aber noch ein anderes ‚natürliches Ding', das in diesem Zusammenhang seit je von zentraler Bedeutung ist – das Salz.

Seit je ist es ein täglicher Begleiter des Menschen auf seinem Erdengang und bei allen Völker ein ganz besonderes, oft sogar ‚heiliges' Ding. Das Stein- oder Meersalz selbst gilt als ein Vertreter des Sal-Prinzips und gilt als ein Ausdruck von Wesen und Kraft der Erde. Es ist ‚Hal', ‚Hel', ‚heil', ‚heilig', ‚Licht', das fixierte, eingebundene, greifbare Sonnenfeuer.

Aus archäologischer Sicht wird in Mitteleuropa nachweislich seit der Jungsteinzeit aus Solequellen Salz gewonnen. Seine Transportwege sind noch weit älter als die der Bronze, und es waren jene Wege, auf denen Händler, Pilger und Kundige gleichermaßen reisten. Ein Begriff für die Alchymie als ‚königliche Kunst' (nicht in weltlichem Sinne) ist der selbe wie für diese Salzwege: ‚Königspfad'. Man reiste ‚auf der königlichen Straße', der *Via regia* – Wege, welche durch das Salz gefunden und jahrtausendelang benutzt wurden. Der ‚Königsweg' verband Nord und Süd, Ost und West, und diente der Vermittlung von Wesentlichem, Essentiellem. Dieser ‚Königsweg' bezeichnet gleichbedeutend die Alchymie und die Salzpfade. Dies ist eine der Bedeutungen von Spagyrik und Alchymie in der Heilkunst.

Johann Tölde, für den die Vermutung besteht, dass er selbst den >Triumphwagen< und andere Werke unter dem Pseudonym *Basilius Valentinus* geschrieben hat, schrieb eine >*Haligraphia*<, welche 1603 erschien. Dieses (textidentisch im Leipziger Reprintverlag 1992 wieder aufgelegte) Werk ist nach *Oliver Humberg*: „Das grundlegende Buch seiner Zeit über die Salzgewinnung im mitteleuropäischen Raum, in dem *Thölde* außerdem noch eine Vielzahl alchemistischer Salzpräparate nach Herstellung und Wirkung beschreibt." *Töldes* (vermutliche) Lehrer waren berühmte Alchymisten wie *Petrus Severinus*, *Quercetanus*, *Pernot* und *Martin Ruland*. Manch andere bekannte Persönlichkeiten treffen mit *Tölde* in Sachen Salz zusammen wie *Leonhard Krehoff*, Goldschmied von Eisenach (Verfasser alchymistischer Handschriften) und *Martin Rhenanus*, Vater des alchymistischen Arztes *Johannes Rhenanus*. 1609 wurde der ‚Berghauptmann' *Tölde* mit dem ‚Salzbrunnen zu Kronach' belehnt.

Der Umgang mit Steinsalz bot selbstverständlich den Menschen stets vielerlei Möglichkeiten zu ‚philosophischen' Betrachtungen, auch zu Übungen in der Praxis mit Salzen allgemein, zur Darstellung interessanter alchymistischer Präparate und zur Anwendung in vielen Aspekten der Heilkunst. Dass aber in der älteren Literatur meist allegorisch von ‚Salz' gesprochen wird und das Steinsalz die ‚geheime Materie', der Ausgangsstoff für das Große Elixir, welchen die Alten angeblich verschwiegen hätten, nicht ist, haben ernsthafte Praktiker schon lange deutlich gemacht, unter ihnen der versierte alchymistische Arzt *Johann Agricola*, ein Iatrochymiker im wahrsten Sinne. Sein diesbezüglicher Bericht bringt uns mit genüsslicher Ausdrucksweise Erfahrungen damaliger Praktiker nahe:

„Ihrer sind viel, welche auf die Gedanken geraten, wenn sie den sublimierten *Salmiac* hätten (...) aus dem Salze, so hätten sie die *Veram Terram Foliatam* der *Philosophorum*, und ich weiß auch, dass einstmals ein großer Herr damit betöret ward, der wandte viel drauf, es wollte aber nichts draus werden. Es kam auf eine Zeit auch solcher Leimstängler einer zu mir, der wollte mich mit Gewalt überreden, dieses sublimierte Salz wäre der rechte *Mercurius Philosophorum*. Ich ließ mich auch überreden, weil er so speciose von der Sachen reden konnte. Ich brauchte allen Fleiß, dass ich dieses Sublimats eine Quantität bekam, ich schlug ihm das Gold als ein *Fermentum* zu und satzte es *in Digestionem* und hielt es in der selben eine lange Zeit. Es wollte keine Veränderung der Farben sich erzeigen. *Interim* hielt ich meinen *Artificem* mit großen Unkosten, denn er wollte nicht mit geringen Tractamenten zufrieden sein.

Als ich diese *Materiam* nun bei zehn Monaten in einem Feuer gehalten, so begunte es rot zu werden. Ich gab ihm stärker Feuer, bis dass alles blutrot worden, da nahm ich meinen Lappen aus und wollte eine *Projection* tun. Da sahe ich, was ich gemacht hatte, es wollte meine Medicin nicht fließen. Ich nahm den *Saturnum* zu Rate, satzte ihn auf den Test, trug Silber drein, und meine Tinctur, als es am besten flosse, trug ich auch darein und trieb es miteinander, hoffend, dadurch würde zum wenigsten meine *Luna* in das beste Gold transmutiert werden. Als es nun blickete, da fand ich meine *Lunam*, aber gar nicht transmutiert, sondern war weiß wie zuvor. Ich legte solche in das Scheidewasser, da resolvierte sie sich und fiel ein wenig schwarzer Kalk nieder. Den glühete ich aus, schmelzte ihn, da ward es gut Gold. Als ich dieses sahe, merkt ich wohl, es wäre nur mein zugesetzt Ferment gewesen, nahm derowegen warm Wasser, gosse es auf den andern Teil meiner Tinctur, so solvierte sich das Salz und blieb das Gold liegen. Da wurde ich gewahr, was ich so eine lange Zeit ausgebrütet. Mein Künstler ging insgeheim spazieren und soll noch wieder zu mir kommen. Da war Mühe und Unkosten verloren und unsere *Principia* wurden zu Wasser. Also gehet es, wenn man ohne Erkenntnis der Natur in dieser Kunst etwas vornehmen will, denn aus dem gemeinen Sale der *Lapis Philosophorum* nimmermehr kann gemacht werden, man greife es auch an wie man wolle.

Es sind ihrer viel auf die Gedanken kommen, das *Sal Marinum* (Meersalz; Anm. d. A.) wäre der rechte *Mercurius*, aber es ist eben wie das ander und wird ebensowenig damit ausgerichtet. Es ist zwar Sal *Maris Philosphici* das rechte Kraut, das vor den Tod dienet, aber ihrer wenig wissen, wo das selbe Meer lieget oder in welchem Teil der Welt es anzutreffen. Ich vor meine Person halte davor, es sei in *India occidentali*, denn in demselben die Wässer wunderliche Arten an sich haben. Wer nun dahin kommen könnte, der könnte ein Faß voll desselben Wassers mit herausbringen und alsdenn das Salz daraus sieden, sonst trage ich Sorge, es werde kein Meer in der ganzen Welt zu finden sein, welches dieses tun könne.

Lasse sich derwegen keiner überreden und spendiere also sein Geld vergeblich. Ich habe es mit Schaden erfahren, ich komme nicht wieder – es mag auch A. O. schreiben was er wolle – , es sind falsche *Imaginationes* und haben in der *Philosophia* keinen Grund. (...)“

Ein bemerkenswerter Schlusssatz. Um die Sache abzurunden, wollen wir auch noch das letzte Stück dieses Traktates *Agricolas* anführen:

„Den *Sublimatum* oder *Salmiac Salis* kann man auch sonsten erlangen. Wenn man das *Sal* auf das höchste purgiert und gleichsam in eine Süßigkeit bringet und mit dem *Spiritu Vini* imbibiert, digeriert und alsdenn mit starkem Feuer sublimiert, so steiget ein schöner Sublimat in die Höhe, den kann man alsdenn brauchen wie oder wozu man will. So man solchen alsdenn mit dem *Spiritu Vini* vereiniget also, dass sie nicht wieder voneinander können geschieden werden, so hat man alsdenn ein gewaltiges *Menstruum*, die *Sulphura fixa* aus den Mineralien zu ziehen und auch die *Essentias* der *Vegetabilium* zu erlangen.

Dies *Menstruum* ziehet gleichergestalt dem Golde seine Tinctur aus und lässet sein *Corpus* weiß liegen, und ist diese Solution nicht zu verachten, denn es in der Arznei große Tugenden hat und ein gewaltiges *Cordiale* wird. Ingleichen ziehet es den Granaten seine Tinctur auch aus und bringet sie in eine *Essentiam*."

Johann Agricola, >Chymische Medicin< 1638/39;

Neu bearbeitet und Herausgegeben von *Oliver Humberg*, 2000

Das hohe Lob des Steinsalzes vermischt sich in den hermetisch-alchymistischen Allegorien immer wieder mit Hinweisen auf andere Salze, insbesondere auf jenes ‚geheime Mercurialsalz', welches im ‚Großen Werk' eine bedeutende Rolle spielt.

„Was nun die innerliche Wesenheit dieses trefflichen Geschöpfes (Komm.: der Autor bezieht sich hier auf die Quintessenz an sich und insbesondere auf den *Lapis philosophorum*) anbetrifft, so ist solches nichts anderes als ein Salz. Aller Dinge Ursprung und erster Anfang liegt im Salz. Aller Dinge Leben und Erhaltung ist das Salz. Aller Dinge Reinigkeit und Klarheit bestehet im Salz. Aller Dinge Fixität und Beständigkeit ruhet im Salz. Im Salz ist das größte Geheimnis verborgen. Im Salz liegt der Grund kräftiger Wiedergebärung aller Dinge. Im Salz sind alle Elemente wesentlicherweise beschlossen, auch der Anfang aller Metalle. Im Salz liegt nur geistiges und kräftiges Wesen verborgen, welches der geistige Magnet aller Dinge ist. Das ist nichts anderes als unser *Mercurius universalis* in coagulierter, reiner Gestalt. (...) Andere nennen es das Salz der Natur und das *Salarmoniac* der Weisen, den wahren, erquickenden, balsamischen *Spiritus mundi*.

Dieses Natursalz, obgleich es in allen Dingen ist, kann es doch aus allen Dingen nicht gleich erhalten werden. Es ist sehr flüchtig. Es ist die oberste geflügelte Schlange. Der *Spiritus mundi*. Der *Mercurius Philosophorum duplicatus*. Aus diesem doppelten Salz der Natur sind nun alle Creaturen gebauet. Das Salz ist der erste Stoff gewesen zu der Welt Anfang. Salz war der Erdenkloß aus dem Gott den Menschen gemacht hat.

Dieses unser lebendigmachendes Salz wird uns überflüssig durch die Luft zugeführt. Es ist die verborgene Lebensspeise, welche wir des Nachts Tau, des Tages ein verdünntes Wasser nennen. Wenn die Luft nur kurze Zeit dieses balsamischen Salzes mangelt, was entstehen nicht da für Krankheiten? Denn diese, mit heilsamen himmlischen Ausflüssen geschwängerte Luft ist eben die Ursach, dass die Natur nicht in eine allgemeine Fäulung gerät. Darum sind die Sterne allen Geschöpfen Gottes wegen ihrer Influenz und wegen ihres Natursalzes zu Leben geordnet, denn ihre Ausflüsse bringen uns ein sehr balsamisches Salz. In dem Element der Luft sind alle Dinge ganz und gar durch Hineinbildung des Feuers. Deswegen ist es voll göttlicher Tugenden und Kraft. Dieser himmlische Geist oder *Spiritus mundi* ist nun die *Prima materia* aller Dinge der ganzen Welt, unser aller Leben und Bewegung. Er wohnet in der Luft, und durch diese genießen wir ihn durch das Atemholen.

Hieraus sehen wir nun, dass überall wo wir nur hinsehen, dieses doppelte Salz, dieser zweifache Saturn, dieser flüchtige und feuerbeständige Merkur, welcher ist Salz und kein Salz, unaufhörlich vor unseren Augen gehet, stehet, lieget, flieget, fließet, schwimmet, mit Füßen getreten und sogar auf den Mist geworfen wird. Denn das *Sal celestris*, davon alles wächst und lebet, schwebet vor jedermanns Augen, und dennoch wills niemand kennen, sondern jeder pfeift wie ihm der Schnabel gewachsen.

Also ist überall in der Natur nichts als Salz, sei es nun unsichtbar und geistig oder in den untermondigen Geschöpfen geronnen und gestaltet. Was ist's denn Wunder, dass das Salz zu jeder Zeit und bei allen Völkern so heilig gehalten worden?"

nach: >Kompaß der Weisen<, 1779

Heilkunst und Alchymie

Es gehört zum Wesen der Heilkunst, dass der Mensch nie ohne sie ist. Das gilt auf gewisse Weise auch für die Alchymie, ist sie doch eine Wurzel und ein Stamm, welche zahllose Blätter und Früchte der Heilkunst nähren. Selbst im Abendland, nach Jahrhunderten der Verfolgung, Diffamierung und abertausenden Morden an solchen, die noch im späten Mittelalter die ‚guten Menschen' genannt wurden, taucht dieses ‚Phänomen Heilkunst' immer wieder mit der ihr eigenen Hartnäckigkeit bis heute auf. Und wieder beginnen die Menschen, nach all den Verwirrungen, das Wesentliche darin zu begreifen. Sie fühlen es in ihrem Innersten und richten ihre Schritte danach.

In anderen Regionen der Welt, wo die Verwirrung (bis dato) nicht so groß war, gibt es nicht mehr nur wenige, sondern noch sehr viele solch ‚guter Menschen', und sie schaffen es mit großer Kraft, zu überleben und ihr Wissen heilsam zu benutzen, auch wenn diese Regionen noch am Ende der ‚dunklen Epoche' in Kriegselend und Diktatur geraten. Dazu gibt uns die tibetische Medizin ebenso anschauliche Beispiele wie andere in den Überlebenskampf gedrängten alte Kulturvölker. Die indischen Traditionen nennen die Epoche der letzten 1000 Jahre das ‚dunkle' oder ‚eiserne Zeitalter' *(Kali-Yuga)*, welches nun gerade vom ‚Erdfrühling', dem ‚goldenen Zeitalter' *(Satya-Yuga)* ‚abgelöst wird. Das ist in weiten Teilen Asiens ein alltägliches Wissen, man nimmt es an als eine selbstverständliche Gegebenheit, bedingt durch den Lauf der Zeit *(Kal)*. Im Abendland ist dieses Wissen um die Zeitqualitäten kaum mehr lebendig. Die bekannteren Überlieferungen dazu enden mit dem Mittelalter. Dieses Wissen steht jedoch nicht für sich alleine, sondern in einem großen Zusammenhang.

Was nützt die Alchymie? Was ist ihre Bedeutung in dem, was wir als ‚Naturheilkunde' bezeichnen? Was lehrt sie? Vor allem den philosophischen Kontext. Dies ist die ‚Kunst', welche den ‚philosophischen' oder ‚hermetischen' Über- (oder: Unter-)bau zur Heilkunst an sich beinhaltet, tradiert, anwendet und lehrt. Und zwar ganz ähnlich den heute noch lebendigen naturheilkundlichen Traditionen des Ostens: Was immer geschieht – der Wandel und die Qualität der Zeit, das Werden und Vergehen der Körper, das Entstehen von Leiden und Krankheiten, deren Therapie, die Arzneimittelsuche und -zubereitung – das alles ist in einen ‚philosophischen Kontext' eingebettet und aus diesem tiefgründig zu begreifen.

„Die Alchymie ist die einige und wahre Philosophie,
worin die ganze Natur begriffen wird."

Pierre Jean Fabres, >Chymische Schriften<, Ausg. 1713

Genauso wie der Student eines asiatischen heilkundlichen Systems heute viel lernen muss über Wesen und Wirken der Elemente, der ‚Säfte' und insgesamt über das Webwerk Natur, forderten die Kundigen des Abendlandes seit jeher das selbe Studium als Voraussetzung zum Eintritt in die ‚Kunst'. Das hat sich bis heute nicht geändert, und manche Traditionen mit ihren zahlreichen Verzweigungen sind dazu erhalten geblieben.

Ein heilkundliches System ist erst dann vollständig, wenn es in einen philosophischen Kontext eingebettet ist, welcher aus der Beobachtung der lebendigen Natur erfahren werden kann. Das finden wir in allen ‚traditionellen' Systemen so vor. Der ‚Überbau' bzw. der hermetische Hintergrund von Befundung, Therapie und Pharmakologie ist dabei kein erdachter, vermuteter oder gewollter, kein durch persönliche Einbildung eigenartig und naturfremd gestalteter oder neu erfundener, sondern ein der Natur abgeschauter und damit in allen Ebenen und Anwendungsbereichen stimmiger.

Das Studium von Alchymie und Heilkunst ist heute nicht möglich, ohne dass Traditionen zu Rate gezogen werden, welche die Jahrtausende überdauerten und die einen oder anderen, oft wesentlichen Aspekte lebendig erhielten. Nirgendwo war dies wohl schwieriger als im Abendland, welches während der letzten zweitausend Jahre mit konsequent steigender Tendenz eben jenen unvergleichlichen kulturellen und menschlichen Holocaust erfahren musste, der so großflächig, durchdringend und nachhaltig in keiner anderen Region der Welt ähnlich stattgefunden hat. Im Zuge dessen hätte die natürliche Heilkunst beinahe ihre Wurzeln verloren, wären diese nicht in der Alchymie erhalten geblieben.

Es ist kaum genug zu würdigen, dass es hier Kundigen selbst unter den widrigsten Umständen gelang, jenen ‚philosophischen Kontext' in praktischer Arbeit, heilkundlicher Anwendung und oft auch in einfacher Lebenskunst zu beweisen, zu tradieren und zu lehren. Die meisten von ihnen pflegten Austausch mit anderen Kulturkreisen, was nicht nur mit den Anstrengungen des Reisens verbunden war, sondern auch die Gefahr mit sich brachte, sich in den eigenen Landen verdächtig zu machen. Viele wanderten aus, nachdem sie hierzulande gelernt und getan hatten, was sie vermochten. Oft waren die vielzitierten ‚Reisen' einfach eine sinnvoll genutzte Flucht. Der überwiegende Teil dieser großen Menschen ist bis heute unbekannt geblieben oder sie zogen es vor, pseudonym zu wirken, wie *Basilus Valentinus* bzw. jene, die unter seinem Namen geschrieben haben (oder war es doch immer der selbe?). Nur die wenigsten gerieten persönlich zu Bekanntheit wie *Nicolaus Flamel* oder gar *Paracelsus*. Ersterer noch weniger, weil er sich vor allem im Stillen der ‚Hohen Kunst' widmete, letzterer weit mehr, weil er es verstand, diese ‚Hohe Kunst' zu integrieren in dem weiten Feld der ‚Wissenschaft vom Leben'.

... nicht nur Paracelsus

... Osthanes/Comarios/Maria Prophetissa/Johannes/Chrysomallos/Midas/Demokritos/Synesios/Philippos/Heliodoros/Kallias/Maternus/Themistios/Aincias/Gazaios/Zosimos/Archelaos/Pelagios/Olypiodoros/Pappos/Kosmas/Hierotheos/Damaskenos/Anepigraphos/Kedrenos/Psellos/Abbas/Amri/Philoponos/Geber/Abu Mussah Giabr/Abu Jussuf/Alchindi/Rhases/Farabi/Satmana/Zulnun/Avicenna/Dschasar/Adsar/Thograi/Chassem/Mohieddin/Geldeki/Alchiabdachi/Ibn Mascia/Chalid/Jazichi/Kalled/Zadith/Ben Hamuel/Al Phager/Bul Hagiz/Merlinus/Hortulanus/Pythagoras/Aristoteles/Plato/Morienes/Arisläus/Artephius/Petrus v. Zalento/Alanus/Ferrarius/Alphidius/Rhodanius/Beradianus/Skotus/Belias/Albertus Magnus/Thomas v. Aquino/Christoph v. Paris/Alphosus/Bacon/Peter v. Albano/Peter v. Toledo/Villanova/Richard v. England/Montanor/John Duns/Adolph Meutha/Clopinell/Dastyn/Ramon Lull/Cremer/Bonus/Odomarus/de Rupescissa/Ortolanus/Canches/Flamel/Blanche v. Navarra/Charles VI/Valentinus/Johann v. Tetzen/Mönch v. Oderberg/Hollandus/Heinrich IV/Jaques le Cor/Lasnienoro/Johann d. Alch./Albrecht/Nikolaus de Cusa/Makarius/Gottfried v. Stendal/Johannes Piscator/Lambspringk/Trevisanus/Aurac/Trithemius/Angelus/Poysel/Eschenreuter/Ripley/Norton/Burkhard/Krapit/Trismosin/Phillip/Alsted/Paracelsus/Korndörfer/Crinot/Wenzel/Zachaire/Vallensis/Robacher/Chrysippus/Trautmannsdorf/Thurneisser/Siebenfreund/Beyer/Tarvisinus/Bragadino/Gratarolus/Carreri/Glissenti/Fioravanti/Cortese/Penot/Barnaud/Balbian/Rudolf II/Kelley/John Dee/Franz Antony/Scotus/Beuther/Honauer/Brynner/Wittstein/Dron/Kunrath/Setonius/Libavius/Pontanus/Töpfer/Tank/Croll/Graßhof/Siebmacher/Schaubert/Rhenanus/Mayer/Furich/Neri/Birelli Brenzi/Giovanni de Padua/Angelo Sala/Jean d'Espagnet/Potier/Chataigne/Lagneau/Paumier/Besard/de l'Angèlique/Drebbel/de Boot/Thornburg/Fludd/Northon/Pieroni/Sendivogius/Müller/d'Anvers/Großschedl v. Aicha/Amthor/Batsdorf/Franke/Rist/Johann Agricola/Keßler/Liberius/Jebsenius/Polemann/Tonring/Berigard/Bureau/Val/Martini/Grimaldi/Beausoleil/Fabre/Collesson/Helmont/Philaletha/Starkey/Ashmole/Honius/Harbach/Pfenniger/Isaac Newton/Glauber/Tachen/Becher/Pfeffer/Drechsler/Tölde/Ursinus/Albineus/Cramer/Concius Morhof/Borrich/Schweitzer/Bierorth/Grill/Kerkring/Breeswyk/Johnson/Chartier/Borel/Arremont/Clos/Aubry/Conti/Mazotta/Lana/Mahomed Kiuperli/Schröder/Seyler/Wagnereck/Pantaleon/Karbiluck/Balduin/Kunckel v. Löwenstern/Wedel/Orschall/Grummet/Seger v. Weidenfeld/Kirchmayer/Hannemann/Kraus/Robert Boyle/Mundan/Dickinson/Colson/Headrich/Grandeville/St. Romain/d'Acqueville/Borri/Gualdo/Lancilotti/Severino/Vansleb/Paul Lukas/Laskaris/Bötticher/Braun/Caetano/Barchusen/Wallich/Kellner/Söldner/Brebis/Müller/Horlacher/Manget/Chymiphilus/Richter/Geoffroy/Sehfeld/Richebourg/Lenglet de Fresnoy/Thersander/Horter/Hirsching/Wenzel/de Limitibus/...

… in der Naturheilkunde

Was ist eigentlich Naturheilkunde? Der Begriff ist ein neuzeitlich erdachter für die Heilkunst an sich. Wenn ein Teil des menschlichen Körpers oder ein Aspekt seines Gemüts erkrankt, dann bedeutet das, dass die Harmonie seiner natürlichen Komposition irritiert ist. Im näheren oder weiteren Umfeld dieses Menschen befinden sich Helfer, ‚natürliche Dinge', Substanzen, Kräfte-Komplexe, welche dazu geeignet sind, die entsprechenden disharmonischen Felder wieder an ihre ursprüngliche Art zu erinnern. Diese Helfer können sein: Menschen, Objekte, Betätigungen, Berührungen, Landschaften, Konstellationen, Wässer, Erden, Pflanzen, Nahrungsmittel, Minerale u. dgl. Das ist so, wie wenn zu einem ganz schwachen Menschen jemand hinzutritt und ihn wesentlich an seine Stärke erinnert und damit einen Genesungsprozess einleitet bzw. fördert.

So ist die Natur beschaffen, in welcher der Mensch seine Lebenszeiten verbringt – wie außen so innen, wie oben so unten und umgekehrt. Für jedes Problem gibt es eine Lösung, für jede Frage eine Antwort und für jede Krankheit eine Heilung. Das ist die ‚hermetische Art' der Natur, und deswegen ist ‚Naturheilkunde' eigentlich ein zu langes Wort dafür, die eine Sache zu bezeichnen: nämlich die Heilkunst an sich.

> „Keine Krankheit – wenn es nichts anderes ist als eine Krankheit – ist so schwer, dass sie nicht eine Arznei zur Heilung habe."
>
> „Denn in soviele Teile sich die Krankheiten aufteilen, in soviele teilen sich auch die Himmelskörper, soviele Ursprünge sie haben, soviele sind der Gewächse."
>
> „Die Krankheit aber, die nicht der Arznei unterworfen ist, ist diese, welche der Himmel in seiner Gewalt behält und nicht daraus entläßt. ... Arzneiest du eine Krankheit, und diese ist noch in des Himmels Gewalt, so wirst du sie dem Himmel nicht nehmen können. Der Himmel ist Meister, Arzt und Henker."
>
> *Paracelsus: >Paragranum<*

Jegliche Heilkunst gründet im Wissen über Natur und Leben. Der Heilkundige ist vertraut mit den entsprechenden Vernetzungen – je nachdem, wie sie sich nach seiner Tradition oder seinem intuitiven Erfassen darstellen. Die innere Natur des menschlichen Körpers ist ein Mikrokosmos, verglichen mit dem Makrokosmos der Planeten und Gestirne. Sie ist auch ein Mikrokosmos im Verhältnis zum Makrokosmos des gesamten ‚Brahmischen Eies', dh. zu den astralen und kausalen Dimensionen des Kosmos. Sie ist aber auch ein Mikrokosmos im Verhältnis zu der sie umgebenden, irdischen Natur.

Jedes Organ, jedes Gewebe, jeder Knochen, jeder Saft, jeder Vorgang und jede Neigung des Gemütes hat nicht nur eine einzige Entsprechung in der körperäußeren Natur, sondern vielfältige, ineinander verwobene Entsprechungen. Es gibt Entsprechungen, gleich Verwandtschaften, zu und unter den Naturreichen der Pflanzen, Metalle und Steine, darüberhinaus Entsprechungen zur näheren Umgebung, zur weiten Landschaft, zur Art des Bodens, des Wassers, dann wieder zwischen den Umständen, Ereignissen und Begegnungen usf.

Eine weite Landschaft entspricht etwa dem Äther-Element, Jupiter, Zinn, Türkis, Lapislazuli, der Königskerze, den Augen, der Leber, der Großzügigkeit und der Qualität, dass Ideen praktisch umgesetzt werden können.

Spitze Berge entsprechen dem Feuer-Element, dem Mars, dem Eisen, der Brennnessel, dem Senf, Galgant, Ingwer und Kardamom, den Harzen allgemein und besonders dem Lärchenharz, der Zähigkeit, Fasrigkeit, Abwehrkraft und dem Durchsetzungsvermögen, dem Verdauungsfeuer, der Schärfe des Blickes und des Verstandes, Konzentration und Aktivität, der Farbe in den Dingen allgemein, insbesondere den Rot-Tönen, der Wärme und Kraft des Blutes u.v.m.

Sanfte runde Hügel entsprechen dem venerischen Wasser oder der venerischen Erde, insgesamt dem Planeten Venus, dem Kupfer, den Nieren, den Brüsten und Hüften, den Haaren und Ohren, der Aufrichtekraft des Menschen und seinem Mut, den Verhältnissen und Proportionen aller Dinge, den Rosengewächsen, Bohnen, Wasserpflanzen, Wässern, harmonischen Rhythmen, Künsten u.a.m.

Nach den Lehren der chinesischen Medizin besteht zwischen dem Äther-Element (‚Holz'), dem Wind, der Geburt, dem Frühling, Fischen und Schafen, Augen, Gallenblase und Sehnen, dem Zorn und dem Schreien eine besondere, auf elementaren Qualitäten beruhende Verwandtschaft. Die Planetenkraft ‚Jupiter' gehört genauso hierher wie der ‚Funktionskreis' Leber. Auch *Paracelsus*, welcher 1520 die selbe Natur erkennt wie der Autor des *Li Shi Zhen* um 1560 oder wie *Shang Jing Yue* um 1620 und all die anderen Kundigen vor und nach ihnen merkt an, dass die Leber zum Jupiter gehöre.

Nach den Lehren der chinesischen Medizin besteht zwischen dem Feuer-Element, der Hitze, dem Wachstum, dem Sommer, den roten Farben, Vögeln und Fohlen, Freude und Lachen, Herz, Dünndarm und Zunge eine, auf elementaren Qualitäten beruhende Verwandtschaft. Danach wirkt die feurige Kraft des Mars auch besonders in den Blutgefäßen des Körpers, vor allem des Herzens.

Der in Württemberg wirkende italienische Iatrochymiker *Cardilucci*, welcher um 1650 die selbe Natur erkennt wie *Zhang Jie Bin* um 1620, aber auch lange vor ihm schon *Hua Tuo* um 180 und all die anderen Kundigen aller Traditionen, merkt an: „Der martiale Trieb des Schützen geht gewaltig in die Arterien und Herz-Geäder."

„Wenn der Mensch in seiner Zusammensetzung in jeder Hinsicht betrachtet werden soll, so wisset erstlich zu erkennen, wie ihr die Gestirne des Firmaments im Leibe des menschlichen Mikrokosmos nach ihrer Wirkstatt verstehen sollt. Denn die Gestirne im Leibe haben Eigenschaft, Art, Wesen, Natur, Lauf, Stand und Anteil gleich den äußeren Gestirnen und sind allein in Form und Substanz unterschieden. (...) In der Natur sind beide ein Ding und ein Wesen. (...)

Ebenso wie ein Kind seines Vaters Gliedmaßen alle hat, so ist der Mensch Himmel und Erde, Luft und Wasser gleich, als ob diese sein Vater wären. Er hat all ihre Art und alle ihre Teile und mangelt nicht eines Härleins. Daraus folgt, dass der Arzt wissen soll, dass im Menschen Sonne, Mond, Saturn, Mars, Merkur (...) sind. Das muss er wissen, wenn er vom Grund der Arznei reden will. Wo das nicht so ist, so ist er nix als ein klarer Bescheißer und arzneit wie der Bauer, welcher die Koloquinten in Wein hängt und alle Menschen gleichermaßen damit heilt. (...)

Soll es denn bloß ein Kleines sein, wenn wir betrachten, dass ein Mensch nach seinem Vater angesehen und anatomiert werden soll und nicht außerhalb seines Vaters? Wie der Vater, so ist auch der Sohn mit Leber, Milz, Hirn usw.

Auf solches ermahn ich alle, die da etwas von der Arznei wissen oder lernen wollen, dass sie vor allen Dingen den Vater des Menschen erkennen, denn wenn der Vater recht erkannt wird, ist der Sohn desto leichter zu erkennen. Denn der Sohn gibt sich von sich selbst nicht zu erkennen, nur der Vater offenbart den Sohn. Und wie nun der Vater den Sohn offenbart, so offenbaren auch Himmel und Erde, Wasser und Luft den Menschen. (...)

Mitnichten soll ein Arzt etwas schreiben, lehren oder gebrauchen, das nicht in der Natur auf das höchste zur Probe gebracht und angezeigt gefunden worden ist. Doch diese Leute wollen da hindurch fahren und das verachten, was sie am Anfang wissen sollten. Was soll ich anderes von ihnen sagen, als das, was alle meine Bücher ihnen mit Schelten und Lehren und andere Wege aufzeigen offenbar machen? (...)

Wie also der Himmel an sich selbst ist, so ist der Mensch in seiner Anatomie."

nach: *Paracelsus*, >Zweites Traktat<, zu Astronomie und Anatomie als Säulen der Heilkunst

Die Suche nach Heilung und das kulturelle Umfeld

Die Natur ist so eingerichtet, dass der Mensch in ihr menschlich leben kann. Mehr noch: Hier ist der Ort, wo wir ‚Wesen im Menschenkörper' uns tatsächlich zu Menschen entwickeln können. Die Auswirkung jener Menschwerdung ist der Grund dafür, dass die Natur, alles was zum Leben nötig ist, bereitstellt. Deshalb kann ein Heilsuchender in der Natur den Grund zu seiner Heilung finden. Das ist eine der für das Leben des Menschen bedeutsamsten Gesetzmäßigkeiten. Der Volksmund nennt es: „Wer suchet, der findet."

In Zeiten tiefster geistiger Armut und Vergessenheit muss diese einfache Regel erwähnt und erklärt werden. In Zeiten von ‚menschlicher Kultur' ist dies so nicht notwendig, gehört sie doch zum ‚Kindergarten' des Menschseins. Dennoch wollen wir uns kurz um dieses Thema bemühen.

Der Erdengang des Menschen gleicht dem Tasten, Klettern und Gleiten durch ein undurchschaubares Webwerk subtilster Fäden. Mancher ist während seiner ganzen Lebenszeit nur mit dem Durchqueren eines einzigen Musters in diesem Gewebe befasst, während unzählige weitere Muster – gerade erst oder in vergangenen Leben gewebt – noch darauf harren, durchwandert zu werden.

Leiden, Verletzungen und Krankheiten können relativ kurzfristig aufgrund schlechter Lebensgewohnheiten entstehen, durch stoffliche oder mentale Vergiftungen hervorgerufen werden, aber eben auch darüber hinaus ihre Wurzeln an irgendeinem ferneren Ort des großen Webwerks haben und sich dann im aktuellen Leben aufgrund geeigneter Umstände scheinbar unvermutet äußern. Das Webwerk Natur ist jedoch stets so beschaffen, dass jedes Problem zugleich einen Ansatz zu seiner Lösung in sich birgt. Wieweit dieser Ansatz sich heilsam auswirken kann, hängt ab vom Zustand des Betroffenen (aus der Sicht der Natur) sowie von seinem kulturellen Umfeld.

Ein Umfeld mit dem Grade echter Kultur schließt die Möglichkeit zur Heilung und damit auch das Vorhanden- und Verfügbarsein naturgegebener oder durch die ‚Kunst' bereiteter Medizin mit ein. Heilkundige Menschen und solche, welche imstande sind, tatsächliche Heilmittel zu bereiten, werden hoch geachtet und in jeder Weise unterstützt. Reste dieses kulturbedingten Verhaltens findet man z. B. noch in Indien, den ayurvedischen *Vidyas* gegenüber, oder bei verschiedenen indigenen Völkern gegenüber ihren Schamanen und Medizinmönchen. Die Ehrerbietung, die man dort den Kundigen selbst heute noch entgegenbringt, scheint in den meisten Fällen ehrlich und begründet zu sein. Indien gehörte noch bis vor wenigen Jahren zu den letzten Ländern der Erde, in dem die Herstellung von Heilmitteln als ‚Menschenrecht' verankert und jedem ‚erlaubt' war.

Erst kürzlich wurde jedoch die indische Regierung mittels Kreditgeschäfte gezwungen, dieses Menschenrecht fallen zu lassen. Damit konnte sich auch dort noch ein typisches Phänomen des *Kali-Yuga* manifestieren – allerdings erst zum Ende dieses Zeitalters (wie die inidischen *Vidyas* einigermaßen beruhigt erklären), während im Westen diese unnatürliche Situation schon seit Jahrhunderten von vielen für ‚normal' gehalten wird. Geht die ‚Kultur des Menschseins' ab, dann bleibt die Medizin aus, und da hilft es nicht, wenn man ähnliche oder auch ganz andere Dinge mit den Namen ‚Kultur' und ‚Medizin' betitelt – sie halten ihr Versprechen nicht, und die Krankheit nimmt überhand. Mit dem Grad dieser Kultur steigt oder sinkt auch der Grad der Medizin. Das ist nicht nur typisch paracelsisches Denken, sondern es ist weit mehr – es ist zeitlos gültig.

Das kulturelle Umfeld trägt wesentlich dazu bei, dass der Leidende dem ihm bestimmten Heiler oder Heilmittel begegnen kann. Wird ein Naturgesetz verstanden und geachtet, dann kann es sich weit mehr zum Wohle des Menschen auswirken, als wenn es unverstanden oder missachtet bleibt. Zuzeiten mögen da abertausende Heilsuchende sein, deren potentielle Heiler oder Heilmittel aufgrund eines abwesenden oder nur vermeintlichen kulturellen Umfeldes behindert und kaum verfügbar sind. Dadurch entstehen Zustände, in denen geistige und körperliche Leiden immer mehr und vielfältiger werden, bis sie ein dermaßen erschreckendes und unübersehbares Ausmaß angenommen haben, dass selbst die größte Naivität nicht das Wagnis auf einen Blick nach den größeren Zusammenhängen zu behindern vermag.

In solchen Situationen erhebt sich hilfreich im Innersten des Menschen ein anfangs noch ungewisses Erinnern an eine natürliche, ‚ganzheitliche' Art des Menschseins. Davon ist ihm ein unauslöschliches Bewusstsein eingeboren, weil er selbst – als Seele – während zahlreicher Geburten in menschlichen Körpern bereits sehr lange Zeiträume in ‚echten' kulturellen Umfeldern erfahren konnte.

Dieses Erinnern wird nun entweder nicht genügend wahrgenommen aufgrund einer bereits stark dominierenden geistigen Krankheit, welche nach der ayurvedischen Terminologie von einem übermäßigen *Tamas*-Zustand, also von allzugroßer Dumpfheit, herrührt. Diese Krankheit wird (in wie oben beschriebenen Epochen) zur allesverschlingenden Seuche. In der tibetischen Medizin heisst es dazu: „Die eigentliche Wurzel jeglicher Krankheit ist die Unwissenheit." Daraus entsteht zuerst ein Verlust der Gesamtheit, der Ganzheitlichkeit, eine dominante Ausrichtung auf Trennung und Teilung, was dann negative Neigungen, wie Hass, Gier, Ärger, Neid, Habsucht u. dgl. hervorbringt. Diese durchdringen aus dem Gedanken- und Gefühlsbereich heraus den grobstofflichen Körper und manifestieren sich dort zu Leiden und körperlichen Krankheiten. Aus demselbsen Gedanken- und Gefühlsbereich wirken sie auch ansteckend.

Auch die Tatsache, dass kein ‚eigentlich kulturelles Umfeld' vorhanden ist (und dadurch eben allgemein ‚Unwissenheit' vorherrscht) kann bewirken, dass der Erkrankte dieses ‚Erinnern' nicht genügend wahrnimmt, damit er es als Licht bei der Suche nach Heilung nutzen könnte. Es wird dann vielmehr als lästig oder gar als jene Gefahr empfunden, welche Ignoranz und falschen Stolz (auf vermeintliches ‚Wissen') zu relativieren droht.

Hier kann der Heilkundige – vorausgesetzt, er ist in solchen Zeiten ausreichend handlungsfähig – sehr behutsam und sensibel Denkhilfe leisten und diese mit entsprechenden Therapien und Heilmitteln unterstützen, welche die körperlichen und schmerzhaften Ausdrucksformen der Krankheit lindern.

Eine spürbare Erleichterung des Schmerzes alleine wird aber oft schon als Heilung empfunden, und der (tamasische) Patient ergibt sich nicht in die für ihn unverständliche Behandlung der Ursachen der Krankheit. Dies nutzen zu solchen Zeiten jene, die *Paracelsus* als ‚Scharlatane' bezeichnete, rücksichtslos aus und täuschen ihre Patienten, indem sie alleine die körperlichen, oberflächlichen und schmerzhaften Auswirkungen ihrer Krankheit vorübergehend beruhigen, ohne dabei einen echten Heilungsprozess der eigentlichen Krankheit zu fördern. Letzteres wäre ihnen auch nicht möglich, wenn man bedenkt was die ayurvedischen Schulen diesbezüglich erklären, nämlich dass ein überwiegend tamasischer Zustand eben eine Krankheit an sich ist, der auch das ‚Befunden' des Patienten behindert, weswegen ein Heilkundiger sich nicht darin befinden solle, wenn er seine Aufgabe erfüllen will.

Dieses ‚*Tamas*-Thema' spricht *Paracelsus* an, wenn er sagt, dass dort, wo Neid, Falschheit und Habgier wohnt, keine Heilkunst gefunden wird. Damit wäre auch der Begriff des ‚Scharlatans' annähernd definiert, der zuzeiten doch oft in irrigem Zusammenhange gebräuchlich ist. Wenn ‚Heilkunst' nur vermeintlich ist oder gar vorgetäuscht wird, dann kann die Suche nach Heilung stark behindert werden. Den Traditionen zufolge besteht dieses Problem glücklicherweise nur in den kurzen ‚dunklen' Epochen der Menschheitsgeschichte, während ansonsten Jahrtausende und Jahrzehntausende damit unbekümmert sind.

„Es ist eine Schande, der du den Namen allein deines Schwätzens wegen hast, nicht der Heilung wegen. In diesem Namen steckt Betrug und Falsch, so dass der Kranke meine, du seist groß in den Künsten, wie das der Name anzeigt. Also wird er gemordet und gelähmt. (...) Manchmal gibt die Stunde und das Glück eine Probe, wodurch alle Bescheißerei, die das ganze Jahr hindurch geschehen ist, verteidigt wird. Dieses Glück gibt dir der Satan, damit er dich in seinem Regiment behalte. Es ist kläglich genug, dass in euch so gar keine Seele ist, und trotzdem plärrt ihr von eurer Frömmigkeit und Weisheit mit verdeckten Lügen."

nach: *Paracelsus*, >Elf Tractat<, 1520

Die Suche nach den Ursprüngen des Leidens

Es ist aber auch möglich, dass diese obgenannte ‚Erinnerung' ausreichend wahrgenommen wird und zu einer ernsthaften, tiefschürfenden Suche nach Heilung veranlasst, wobei die Ahnung der Art eines menschenwürdigen ‚Normalzustandes' vorerst zu einem Leitfaden der Suche wird. Ist sie konsequent, dann kann auch unter schwierigen Bedingungen die Gesetzmäßigkeit „wer suchet der findet" zum Wirken kommen. Davon gibt es zahlreiche Beispiele.

Diese Suche ist untrennbar mit einem menschlichen Entwicklungsprozess und ganz besonders mit einer zunehmenden Qualität der ‚Ehrlichkeit zu sich' selbst verbunden, welche zu einer Wurzel letztendlicher Heilung wird. Zu welcher Zeit diese Heilung erfolgen kann, hängt wieder von mehreren Umständen ab. Es gibt Probleme, die in kurzer Zeit, in Stunden und Tagen beseitigt werden können, solche die um vieles länger zu ihrer Heilung benötigen und wieder welche, die in dieser Geburt alleine nicht mehr bewältigt werden. Längerfristige *Karmas* entziehen sich dem Zugriff der Heilkunst und können ohnehin von kaum einem Menschen genau befundet werden.

In Indien wird dazu das Beispiel des blinden Königs *Dhritarashtra* erzählt: Er hatte alle verfügbaren Kundigen seiner Zeit konsultiert, jedoch ohne dass sie ihn heilen oder ihm den Grund für seine Blindheit nennen konnten. Er lebte vor etwa 6000 Jahren und war, seiner Zeit entsprechend, ein hochentwickelter Mensch, selbst in der Lage, 100 Geburten zurück zu durchforschen. Doch er fand darin weder Grund noch Heilung. Durch Fügung geriet er an *Krishna*, der der vedischen Tradition zufolge in jener Zeit mit der Kompetenz eines hohen *Avatars* wirkte, der also, wenngleich selbst dem karmischen Gesetz unterworfen, immerhin in dieses umfassend einblicken konnte. So öffnete *Krishna* dem König den inneren Blick für die hundertzehnte vergangene Geburt, und der König sah, dass er damals einen Menschen geblendet hatte. Die Auswirkung dieser Tat musste er nun in diesem Leben ertragen.

Solcherlei ‚karmische' Ursachen von Leiden entsprechen dem Naturgesetz des Ausgleichs, welches verglichen wird mit dem ‚Klebstoff im Webwerk Natur'. Es gibt *Karmas*, welche zu den Umständen einer Geburt führen, solche, die während einer Geburt neu geschaffen werden und solche, die zwar angelagert sind, aber erst zu späteren Zeiten Früchte tragen. Um die hinderlichen Rückwirkungen für die Zukunft zu verringern, empfahlen die Alten oft Gewaltlosigkeit und fleischlose Ernährung. Dieser Empfehlung folgten nicht nur Heilige und Mystiker, sondern auch viele jener Kundige, von denen wir hier handeln, Philosophen der alten Griechen, Druiden, ayurvedische *Vidyas* und andere.

Die Thematik dieses Naturgesetzes ist sehr tiefgreifend und weitreichend und kann in der Heilkunst nicht außer Acht gelassen werden. Die Angelegenheiten von Tod, Wiedergeburt und Karma sind keine ‚Glaubensfragen', sondern entsprechen der alltäglichen Realität, weswegen sie bereits in der ‚Grundschule der Natur' gelehrt werden.

Weniger tief reichen die Einprägungen, die Imprägnationen. Das sind keine Karmas im eigentlichen Sinne, sondern gewisse Lebensmuster, welche ein Mensch über mehrere Geburten mit sich nehmen kann. Die chinesische Medizin kennt dafür den Begriff der ‚Wanderseele' *Hun*, welche ‚in der Leber wohnt'. Der Funktionskreis Leber entspricht dem Äther-Element. Durch dieses hindurch ereignen sich Ein- und Entstofflichung eines Menschen in den physischen Körper.

Während die mit feinstofflichen ‚Kleidern' bedeckte Seele im Zuge ihrer Wiedergeburt den Äther passiert, nimmt sie dort jene Muster wieder an sich, die im vorhergegangenen Leben entsprechend tiefe Eindrücke hinterlassen haben, soweit diese nicht in der Zeit zwischen den Geburten (im eigentlich Feinstofflichen) erkannt und korrigiert werden konnten. Diese Muster können sich nun als ‚Hintergründe', als Eignungen, Neigungen, Talente, Fähigkeiten oder auch als Krankheitsbilder, Empfindlichkeiten u. dgl. äußern.

Da hinein reicht die Kompetenz der Heilkunst. Die Muster können erkannt und behandelt werden. Hier ist zumeist der erzieherische und lebenskonzipierende Auftrag des Heilkundigen gefordert, um förderliche Muster zu unterstützen und hinderliche auflösen zu helfen. Es heißt, ein Mensch komme mit einer großen Anzahl von förderlichen Qualitäten, von ‚Tugenden', in diese Welt, von denen er aber in seinen ersten Lebensjahren bereits die meisten verliert – bedingt durch sein Umfeld.

Gemäß diesem Blickwinkel gibt es einen sehr interessanten Befundungs- und Therapieansatz im *Ayurveda*, nämlich dass Heilung schon dadurch veranlasst wird, indem man Art und Zweck, mit denen man ursprünglich in dieses Leben kam, wieder nahekommt. Das kann beim erwachsenen Menschen mit Fragen beginnen wie: „Sind sie damit zufrieden, wie Sie leben und was Sie tun?" Oder: „Was wollten Sie immer schon machen?"

Solche Ursachen von Leiden hängen wiederum zusammen mit dem, was wir oben als ‚kulturelles Umfeld' bezeichnet haben. Hat das Kind das große Glück, in eine tatsächlich menschliche Kultur geboren zu werden, dann wird es weniger leicht erkranken, weil ihm die Möglichkeiten zur Entfaltung seines Wesens gegeben werden und es Lernprozesse erfahren darf, die sich zu seinem innersten Empfinden stimmig verhalten. So kann sein Grundzustand weitestgehend intakt bleiben und es verliert und vergisst sich nicht dermaßen selbst wie ein anderer Mensch, welcher seit seiner hellwachen Kleinkindzeit gezwungen wird mit dem, was er um ihn herum vorfindet, halbherzige Kompromisse zu schließen.

So ist die Thematik der ,Wanderseele' *Hun* direkt mit jener Qualität verbunden, die man als ,Ehrlichkeit zu sich selbst' bezeichnet. Heilsuchender und Heilkundiger bemühen sich gemeinsam darum die Situationen zu orten, in denen der Leidende ,entgleist' ist. Da es dabei sehr oft um ein anfangs unüberschaubares Gewirr von Entgleisungen und daraus folgenden Wieder-Entgleisungen geht, ohne dass sich jeder Umstand sofort körperlich geäußert hätte, ja die tiefsten Irritationen von Wesen und Art meist schon in der frühen Kindheit stattgefunden haben, erfordert diese Suche einigen Mut und die Hilfe eines einfühlsamen Menschen, der selbst ehrlich zu sich ist.

Wenn also Ursache und Verlauf eines Leidens oft Hand in Hand gehen mit einer relativen Unfähigkeit, ehrlich zu sich selbst zu sein, weil man durch die Lebensumstände korrumpiert wurde, wie sollte dann jemand helfen können, der selbst kaum mit dieser Ehrlichkeit lebt?

Hier ist dem Heilsuchenden zu wünschen, dass er in der Lage ist, mit aller Konsequenz zu entscheiden, wem und wieweit er sich jemandem anvertraut. *Paracelsus* warnt in diesem Zusammenhang vor „Solchen, deren Sinn nur nach Gut und Pfennig gerichtet ist" und die den Leidenden zur Oberflächlichkeit anstiften, weil sie selbst weder von der Art der Natur noch von der Art des Menschen ein zumindest geringes Wissen haben. Sie würden die ,Eigenheilung', welche letztendlich stets die eigentliche Heilung ist und auf besagter ,Erinnerung' gründet, verwerfen und die ,Selbstvergessenheit' fördern, was dem Heilsuchenden anstatt Heilung bloß Abhängigkeit beschert.

Die Qualitäten des Heilkundigen

Wenn *Paracelsus* sagt, „der Arzt ist durch Gott und die Natur und nicht durch den Menschen", dann meint er damit auch, dass eben der Heilkundige von Geburt an sehr hohe menschliche Qualität mit sich bringt oder aufgrund einer gewissen ,Art' oder eines ,Hintergrundes' sich ernsthaft bemüht, solche Qualitäten zu entwickeln. Er will lernen, wie die Natur beschaffen ist und erkennt, dass dieses Lernen nicht anders möglich ist, als dass die Natur ihn in ihre Schule aufnimmt. In dieser Schule gewöhnt er sich daran, jeden Augenblick mit Lernen zu verbringen, immer sorgsam darauf bedacht, nicht in grobe Vergehen gegen die Natur verwickelt zu werden und dadurch sein Anrecht auf Schülerschaft zu verlieren. Er wird selbst in gewisser Weise ,heil', indem er seiner ,Erinnerung' folgt und zu allererst danach trachtet, die Bedeutung dessen zu verstehen, was es bedeutet, wenn die Kompetenten sagen, dass der Mensch ,eine Dreiheit ist', ein Körper-Geist-Seele-Kontinuum. Bevor er das alles nicht durch großes Bemühen oder guten ,Hintergrund' so weit als möglich begriffen und die damit verbundenen Qualitäten ausreichend entwickelt hat, wird er sich den Kranken und Leidenden in keiner Weise nähern.

Solches und noch viel mehr waren ‚bei den Alten' die allerersten Voraussetzungen zur Heilkunst. Manche bringen das von sich aus bereits mit ins Leben und müssen dazu nicht viel lernen. Andere verbringen den Großteil ihres Lebens damit, es zu erlernen und erwerben vielleicht ‚bloß' die Chance, in die nächste Geburt als Heilkundige geschickt zu werden. Wenn dies aber so ‚bei den Alten' war, was sagen dann die Heilkundigen unserer Zeit dazu? Warum halten sie den *Paracelsus* so hoch, dass es bisweilen scheint, es gäbe im Abendland gar keine Heilkunst mehr, wäre der nicht gewesen? „Wie froh und sicher können wir heute sein, weil *Paracelsus* alle diese Angelegenheiten schon bezeichnet und genannt hat. Wir brauchen uns nur auf ihn berufen und sagen: ‚Das hat der *Paracelsus* geschrieben.' Wir sind dann seine Schafherde, und wenn wir irregeführt werden, dann hat er schuld. Wir müssen nicht um unser Leben fürchten, wenn wir dies oder das sagen. Denn es ist nicht von uns aus, sondern von *Paracelsus*."

Eigentlich scheint dieser *Paracelsus* durch seine (vermutliche) Ermordung in Salzburg zum Märtyrer der ‚gutmeinenden Heilkünstler' avanciert zu sein. Hatte er das gemeint, als er seinen Gegnern schrieb: „Nach meinem Tod werde ich noch weit mehr wider euch sein." War er bloß nachtragend, oder fand er einfach keinen anderen Ausdruck, weil die Lage der Heilkunst schon zu seiner Zeit dermaßen verfahren war? Man zitiert immer noch seine Worte: „Wer in die Heilkunst will, der muss in meine Monarchie." Darin wird er für einen schützenden Herrn, für einen ‚ehrlichen Menschen' gehalten. Er gehörte zweifellos zu jenen, die einen bedeutsamen Satz aussprachen oder niederschrieben, der dann auf mehreren Ebenen, ja sogar in mehreren Zeiten stimmig ist. Er war kein Heiliger und hatte auch sonst so seine Art, aber die Bedeutung seines Wesens ist für die abendländische Heilkunde unermesslich tiefgreifend.

Johann Agricola über seine Begegnungen mit *Ägidius von der Wiesen*, einem vertrauten Schüler des *Paracelsus*:

„Ich wollte es aber gleichwohl nicht um etliche hundert Gulden enbehren, dass ich dieses Mannes (des *Ägidii*) Kundschaft nicht gehabt hätte, denn ich viel und mancherlei Sachen von dem *Paracelso* erzählen gehöret. Denn es schreiben ihrer viel gar närrische Sachen und fast unglaubliche Dinge von ihm, das ich doch viel anders weiß. (...) Aber dieses ist wahr, das *Paracelsus* gern getrunken habe, hingegen aber, wenn er ihm etwas zu tun vorgenommen, habe er fast weder gessen noch getrunken, bis ers vollendet. Alsdenn, wenn er Zeit gehabt, da ist er gemeiniglich lustig gewesen etc. Und dergleichen Sachen hat er mir oft bei einem halben Tag aneinander erzählet."

Johann Agricola, >Chymische Medicin<, 1638/39; neu Hrsg. Oliver Humberg 2000

Paracelsus vermochte den gedrückten, verängstigten und vom Moloch jahrhundertelanger Verfolgung traumatisierten Heilkundigen immer wieder Mut zuzusprechen, sie an wesentliche Aspekte ihrer Kunst erinnern und gemahnen – an die Tugenden, und daran, dass Natur und Kosmos ein Webwerk sind und dass es keine vollständige Heilkunst ohne die Alchymie gibt.

Seine Hinweise und Mahnungen sind für die Kundigen lebendiger Traditionen des Ostens Selbstverständlichkeiten. Ein berühmter ayurvedischer *Vidya* und *Rasa Shastri* (Alchymist) bemerkte dazu erst vor wenigen Jahren: „Soviele ‚alte Seelen' erkennen wir unter den Heilkundigen und *Rasa Shastris* im Westen, die den geforderten ‚Hintergrund' bereits mit sich bringen, zusammen mit einem bewundernswert ehrlichem Bemühen. Viele von ihnen hatten während letztvergangener Leben schlimme Qualen erdulden und durch das Feuer gehen müssen, und doch kamen sie wieder in eure Länder, um den Menschen zu dienen. Wir behandeln sie mit Nachsicht und bevorzugt, wenn sie den Weg in unsere Schulen finden, denn zu oft erscheinen sie gebeugt und verängstigt. Beinahe haben sie ihre Wurzeln verloren und wissen meist nicht mehr das Geringste von der ‚Philosophie' der Alten."

So wies also im Abendland ganz besonders *Paracelsus* auf die Qualitäten eines Heilkundigen hin und scheint diese auch selbst besessen zu haben, wofür ihm bis heute die Menschen dankbar sind.

> „Der den Kranken treu und fromm ist und der, welcher der Natur in ihrer Kunst nachfolgen will, dieser wird mich nicht fliehen. Nicht alle sind *Christo* nachgegangen, die zu seiner Zeit da waren, und viele verachteten ihn. Warum sollte dann mir die Freiheit vergönnt sein, dass mich niemand verachtet?
>
> Ich bin ihnen zuerst ebenso auf den Leim gegangen, wie die, welche *Christo* verachtet haben. Als ich aber sah, dass sie nichts anderes als Töten, Sterben, Würgen, Erkrümmen, Erlahmen und Verderben anrichten und dass sie keinen Grund haben, da war ich gezwungen der Wahrheit anderwegs nachzugehen. (...) Je länger und je mehr ich ihr und mein Verderben besehen habe (...) fand ich, dass es eine eitle, ausgeklaubte, auserlesene Bescheißerei ist."
>
> nach: *Paracelsus*, >Paragranum<

Es muss der Heilkundige soviel ‚Ehrlichkeit' besitzen, als ihm nur möglich ist und nach dem Rest streben und das entwickeln, was asiatische Medizinsysteme (im Gegensatz zu *Tamas*) ‚geistige Klarheit' (*Satva*) nennen – ein Terminus, welcher keine ausschließlich intellektuelle, sondern vor allem eine geistige und spirituelle Qualität beschreibt. Das Verlangen danach steht ihm weit über der Heilkunst selbst, ist für ihn persönlich eine Lebensnotwendigkeit. Dann folgt ihm die Natur ganz von alleine, denn jeder ‚werdende Mensch' zieht die Natur mit sich.

> „Wer keine natürliche Veranlagung und keinen Geist hat, der in die Natur und ihre Gesetze mit scharfem Blick eindringt, wer nicht imstande ist, die Natur in ihrer Tätigkeit zu belauschen, der kann die wahre Wurzel unserer höchst wertvollen Wissenschaft nicht erlangen. Es gibt viele Hohlköpfe, ohne die nötige Vernunft für Forschungen, ja solche die kaum das einfachste Wort begreifen und die gewöhnlichsten Arbeiten nur mit Mühe erlernen. Andere wiederum haben einen phantastischen Sinn. Wenn sie glauben, sie hätten etwas Rechtes gefunden, so ist es nichts als phantastisches Zeug, ohne jede Vernunft, voll von Irrtümern und weit entfernt von allem Natürlichen. Ihr Kopf ist so voll von Phantasien und Hirngespinsten, dass sie das Wahre und Natürliche nicht begreifen. Wieder andere haben einen zu beweglichen Geist. Sie kommen von einer Ansicht zur anderen, wollen und glauben dieses und jenes, ohne feste Grundlage und Vernunft, so dass sie das meiste unfertig liegenlassen. Wieder Andere sind unfähig natürliche Dinge und Wahrheiten zu verstehen, und wieder Andere verachten die Kunst und meinen sie existiere gar nicht. Solche werden aber selbst verachtet und von den Ergebnissen ferngehalten. Andere sind Sklaven ihres Geldes und Geizes. Sie möchten zwar unsere Kunst erlangen und arbeiten auch, kommen aber aus Geiz und Habsucht doch zu keinem Ergebnis. Ihnen bleibt unsere Wissenschaft verschlossen."
>
> nach: *Geber, >Summa Perfectionis<* (Textsammlung alter alchymistischer Traktate)

In den ‚Qualitäten und Tugenden' ist das veredelnde Wirken des Menschen auf die Natur begründet, das sich dann besonders in der Alchymie ausdrückt. Für jeden drückt es sich auf seine Art aus, und das meint *Paracelsus*, wenn er sagt, dass Bäcker, Ackermann, Weber, Rebenmann usf. Alchymisten sind, aber „der Arzt sollte in dieser Kunst aufs Höchste beflissen sein."

Herz-Wahrnehmung und Signatur

Diese ‚Ehrlichkeit' ist also auch ‚Klarheit' und kommt aus der Wahrnehmung – und zwar aus der direkten Wahrnehmung des Herzens. Wir haben den ayurvedischen Terminus *Tamas* schon erwähnt. Das ist (unter anderem) Dumpfheit, übermäßiger Materialismus, unterentwickeltes Fühlen und Denken. Daraus erheben sich krankmachende Eigenschaften, und aus diesen entstehen zerstörerische Handlungen.

Es erhebt sich Neid: „Jener ist glücklicher als ich – das kann ich nicht akzeptieren." „Er hat dies oder das und ich habe davon weniger." „Er kann das besser als ich, also bin ich gegen ihn." Dazu sind nicht viele Beispiele notwendig anzuführen, denn man weiß, was Neid ist.

Dann erheben sich Hass, Angst, Gier und Habsucht – meist verbunden mit Dumpfheit in Körper und Geist. „Das kann ich nicht verstehen, deshalb verachte ich es." „Ich denke und sage das, wofür ich bezahlt werde." „Was meiner Art zuwiderläuft, das darf es nicht geben." Die Natur erkennt diese Haltung, diese Gedanken und Gefühle als gefährliche, infektiöse Krankheiten.

Dem setzt der Kundige (unter anderem) *Satva* entgegen, jene Qualität, die dem Kundigen (abgesehen von den natürlichen Vermischungen aller Qualitäten in jedem Menschen) selbst entspricht. Es bedeutet in etwa ‚Klarheit'. Körperliche, geistige und seelische Klarheit. Das ist die Voraussetzung für Heilung im Sinne von Veredelung bzw. für die Kunst, das Essentielle von den Schlacken zu scheiden. Deshalb sagte *Paracelsus*: „Drum so lern *Alchymia* (...) die lernt das Wahre scheiden vom Falschen." Darin liegt Bedeutung, Praxis und Schicksal der ‚Kunst'.

Je mehr Klarheit, umso mehr Herz-Wahrnehmung. Das ermöglicht Befundung. Das erlaubt ‚Diagnose' – Sehen und Wissen. Auf der *ars signaturae*, der Kunst, die Zeichen zu lesen, gründet jegliche Philosophie sowie die Heilkunst mitsamt ihren Säulen. *Paracelsus* drückt es so aus, dass „ohne diese Kunst alles Andere ein Loch hat." Ayurvedische *Vidyas* sagen, dass, wenn jemand nicht imstande ist, das Herz eines Menschen in einem Augenblick zu erkennen, er in der Heilkunst nirgendwo steht.

„Die Natur lässt nichts von ihr gehen, ohne das zu bezeichnen was in ihm ist. Ihr seht das Beispiel an den Menschen, die euch nicht fehlanzeigen, was für ein Herz in ihnen genaturt ist. Und es ist nichts so Geheimes im Menschen, das nicht ein auswendiges Zeichen an sich hätte. Daraus folgt, dass die Physiognomie und die Chiromantie der natürlichen Dinge zum höchsten verstanden werden sollen. Wo das nicht der Fall ist, dort wird kein Geheimnis der Natur mit sicherer Probe und rechtem Wesen gefunden werden. (...)

Die Natur zeichnet jegliches Gewächs, das von ihr ausgeht, zu dem, wozu es gut ist. An dem Zeichen muss man erkennen, welche Tugenden in ihm sind. Alle Kräfte, die in den natürlichen Dingen sind, werden durch die Zeichen erkannt. (...)

Denn aus dem *Signato* wird der Grund gefunden, und nichts wird in den Dingen so geheim sein, dass es nicht hervorgebracht werde. *Signatura* ist *Scientia*, durch die alle verborgenen Dinge gefunden werden. Und ohne die Kunst geschieht nichts Gründliches, es hat alles ein Loch."

nach *Paracelsus*: >Von den natürlichen Dingen<

‚Hermetisch versiegelt' sind dem Unkundigen die ‚Sprache der Natur' sowie die Zusammenhänge im Kräftewirken und in den stofflichen Äußerungen. Das macht die Signaturenkunde zum *Alpha* und *Omega* der Heilkunst – *Alpha* betreffs der Befundung des Menschen und des Erkennens der Wirkkräfte in Pflanzen und Steinen – *Omega* in der hermetisch ‚stimmigen' der Zubereitung von Heilmitteln und in der ‚Ganzheitlichkeit' von Therapiekonzepten. Die Signaturenkunde – das Wahrnehmen und Verstehen der Zeichen in den Dingen und Prozessen – ist der Wegweiser in der Heilkunst und in den Phasen des alchymistischen Werkes. Sie ist schlechthin die naturgegebene Orientierungshilfe im ‚rechten Umgang mit den Dingen der Natur'.

Signaturenkunde ist das Verstehen der natürlichen Ausdrucksformen von Kräften in Landschaften, Gestalten, Dingen, Substanzen, Situationen und Ereignissen, indem man sie ‚sinnlich', durch Fühlen, Hören, Sehen, Riechen, Tasten, Schmecken, kurzum durch das Benutzen der nach außen in die fünffältige elementare Welt gerichteten Aufmerksamkeit als Werkzeug zu jeglicher Befundung und Diagnose wahrnimmt. So werden Art und Werdegang einer Krankheit genauso verstanden wie die Beschaffenheit der zu ihrer Behandlung nötigen Arznei. ‚Die Sprache der Natur' spricht von der Heilkraft, welche einer Pflanze oder einem Mineral innewohnt und klärt den Kundigen darüber auf, wie diese Kraft erschlossen und angewendet werden soll.

Das Thema der Herz-Wahrnehmung gibt ein schönes Beispiel, sich in hermetischer Betrachtung zu üben und Signaturenkunde zu betreiben. Welches sind die Gründe im Natur- und Kräftegeschehen, die das Herz zu unserem ersten und direktesten Wahrnehmungsorgan machen? Wir wollen dazu einen Anfang setzen und auch Einiges vorwegnehmen sowie anderes aus den bisherigen Ausführungen zusammenfassen und vertiefen:

Alles in der Natur besteht aus den 5 Elementen. Wir ordnen sie ‚naturgemäß' pentagrammatisch an:

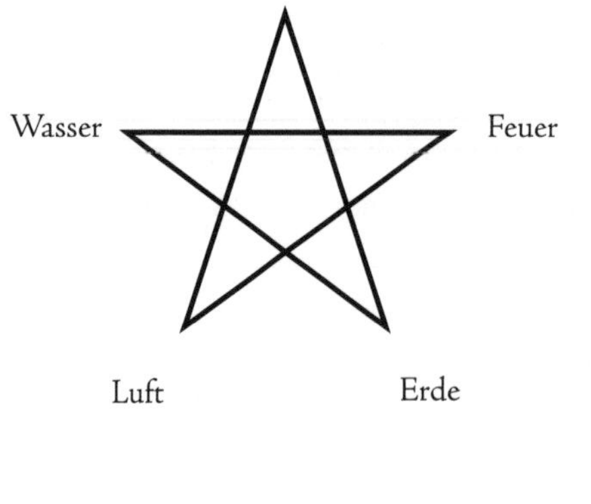

Jede Idee tritt durch den Äther in die Physis mit der Tendenz, sich zu manifestieren – die Idee zu einem Ding, einer Substanz, einer Handlung, einem Menschen. Der Begriff ‚Idee' ist in hermetischem Sinne gleichbedeutend mit dem, was Paracelsus und andere mit ihm ‚Tugend', ‚Wesenheit' und ‚Quintessenz' nennen.

Die eigentliche ‚Manifestation' vollzieht sich durch das ‚sulfurische Spektrum', beginnend im (in der Darstellung rechtsseitig positionierten) solaren Äther (der *Jyoti Akash* im *Ayurveda*). Das entspricht dem *Sheng*, der ‚Nährrichtung' oder dem ‚Hervorbringungszyklus' in der chinesischen Medizin: „Holz (Äther) gebiert Feuer. Feuer gebiert Erde." Am Ende des sulfurischen Spektrums im Erd-Element ist die Idee in der Physis ‚angekommen' – sie hat sich manifestiert bzw. durch Handlung (Feuer-Element) Materie beeinflusst.

Genauso wie in den abendländischen Traditionen gilt in der chinesischen Medizin der Funktionskreis Leber-Galle als ein Aspekt des Äther-Elementes (‚Holz'). Eine Aufgabe des Leber-Organes zeigt uns sehr praktisch, wie sich dieses Einfließen der Idee durch das sulfurische Spektrum ausdrücken kann:

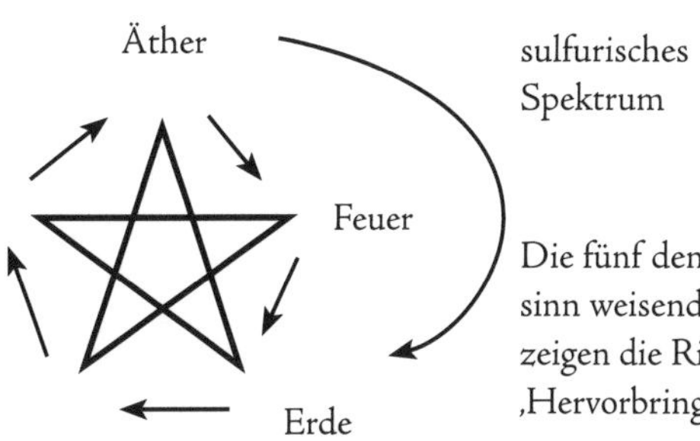

Die fünf den Uhrzeigersinn weisenden Pfeile zeigen die Richtung des ‚Hervorbringungszyklus'.

Sulfur ist eines der ‚3 philosophischen Prinzipien' (Tria principia) und kann mit Individualität, Seele und Ausdruckskraft assoziiert werden. Die anderen beiden Prinzipien sind Merkur (linksseitig in der Darstellung) und Sal (in der Position des Erd-Elements).

Eine Ausdrucksweise dieses ‚Einfließens der Idee hin zur stofflichen Manifestation' in Bezug auf den Körper ist jene Leberfunktion, welche Stärke (*Glykogen*), die in der Leber in unlöslicher Form gelagert ist, bei Bedarf zur Rückverwandlung in Traubenzucker freigibt. Der Zucker bildet dann in den Muskeln die Grundlage zur Gliedmaßenbewegung. Diese Freistellung der Kohlenhydrate setzt in der Leber bereits dann ein, wenn sich der Gedanken – also die ‚Idee' – zu einer Aktion erheben will, zu welcher dann Muskeltätigkeit notwendig ist. Was immer einem in den Sinn kommt und irgendeiner Handlung bedarf – sei es nur der Bewegung eines Fingers – beansprucht dieses Netzwerk. Somit beschreibt der Weg von der Idee zur Tat und dadurch zur Einflussnahme auf die Materie, das sulfurische Spektrum bzw. einen Verlauf, welcher im (besser: durch das) Äther-Element beginnt, das Wesen des Feuer-Elements (Aktion) annimmt und dann Erd-Element hervorbringt (oder gestaltet). Dazu heißt es in der taoistischen Medizin: „Der Funktionskreis Leber ist zuständig für den geschmeidigen Fluss der Dinge". Er ist verantwortlich für alle ‚Durchgängigkeiten'. Er lässt die Idee, die Quintessenz, ‚herein', damit sie in der stofflichen Welt ausgedrückt werden kann.

Anregung zur hermetischen Betrachtung der Energieleitbahnen

Erhaben und würdevoll gleitet der Lebermeridian, die wichtigste Energieleitbahn des Äthers im Körper, den „Königsweg“ hinauf, von der Außenseite der großen Zehe durch das breite Tal zwischen den ersten beiden Mittelfußknochen (chin: *Tai Chong*, Großer Impuls) zur oberen Front des Fußes (*Li Gou*, Muschelrinne). Oberhalb des Knöchels steigt er noch einmal galant hinab, um sich am „Kreuzungspunkt“ mit den anderen beiden *Yin*-Meridianen zu treffen.

Dort berührt er den Nieren- (Wasser) und den Milzmeridian (Erde). Es scheint, als wollte er ihnen eine letzte Audienz geben, bevor er sich anschickt, am Bein hochzusteigen, immer auf der breiten und sicheren Straße der Schienbeinfläche, danach in weit ausladenden Bögen bis zum Rumpf empor, um schließlich an der untersten Rippenspitze (*Zhang Men*, Kapitel oder „Tor am Ende“) gegen Ende seiner Bahn, den Treffpunkt aller Yin-Organe zu besuchen. Er beendet seinen Weg seitlich am unteren Brustkorb, dort wo Herz und Lunge schon den Kern des „rhythmischen Systems“ bilden, so als wolle er nun hier alles weitere dem Merkur überlassen. Hier reicht er dann auch das *Ki* an den Lungenmeridian weiter.

Solcherart würdevoll und gemessen ist die Jupiter-Signatur im Meridianverlauf. Mit seinem feurigen Gesell, dem Gallenblasenmeridian, beherrscht er am Kopf den „Ort der Krone“, wo er sich den höheren, feinstofflichen Ebenen öffnet.

gekürzt zitiert aus: >Der Weg des Sonnenfunkens<

Die Ursachen der Ideen befinden sich im Feinstofflichen, jenseits der 5-elementaren Physis. Das Äther-Element (und damit der Funktionskreis Leber) stellt eine Schleuse dar. Es ist zwar ein durchaus physisches Element, befindet sich aber an der Schwelle zum feinstofflichen Kosmos. Den Traditionen zufolge herrscht hier Zeus/Jupiter, der Repräsentant des Himmelsgewölbes. Ähnlich ‚majestätisch‘ liegt die Leber zuoberst in der durch das Zwerchfell abgegrenzten Stoffwechselregion – als Hauptstoffwechselorgan des Körpers. Aufgrund des Wissens über die analogen Verwandtschaften wurde bei alten Völkern ‚Leberschau‘ ähnlich betrieben wie Astrologie oder die Beobachtung und Auswertung des Vogelfluges. Der Eine beobachtet den ‚großen Himmel‘, der Andere erfühlt den ‚kleinen Himmel‘ im Menschen. Die Zeichen beider Himmel sind die selben. Die unübertroffene Aussagekraft der ‚Pulsbefundung‘ hat diesen Hintergrund, da hierbei die in der Leber organisierten und ausreichend stabilen Informationen des Blutes signaturkundlich erfasst werden. Das kosmische Muster des Menschen, die Art des *Hun*, seine Idee und seine Möglichkeiten sowie der Zustand der Organe, Säfte und Gewebe des Körpers können an den Pulsstellen ertastet – wahrgenommen – werden.

Der berühmte taoistische Heilkundige *Benren Yan Sandian Banxian* (um 1000 n. Chr.) soll gesagt haben, dass „selbst der sorgfältigste Astrologe denjenigen nicht an Genauigkeit übertreffen kann, welcher es versteht den Puls zu fühlen," und: „Für eine erfolgreiche Behandlung sollte es ausreichen, einmal kurz den Puls zu fühlen und eine Rezeptur zu verschreiben."

Die Leber ist ‚der Himmel im Körper', ihr Herrscher ist Zeus/Jupiter, der ‚Himmelskönig'. In ihr sind die ätherischen Komponenten des Menschen veranlagt. Leber und Herz – weiter, kalter und wässriger Mondenäther und zentrierter, heißer Sonnenäther – stellen in gewisser Weise eine Polarität dar. Diese Polarität ist auch schon in den ‚Funktionskreisen' Leber und Galle vorhanden. Auf dem Zusammenwirken dieser Polarität gründet nicht nur die Pulsbefundung, sondern die Signaturenschau insgesamt und die ‚Herzwahrnehmung'.

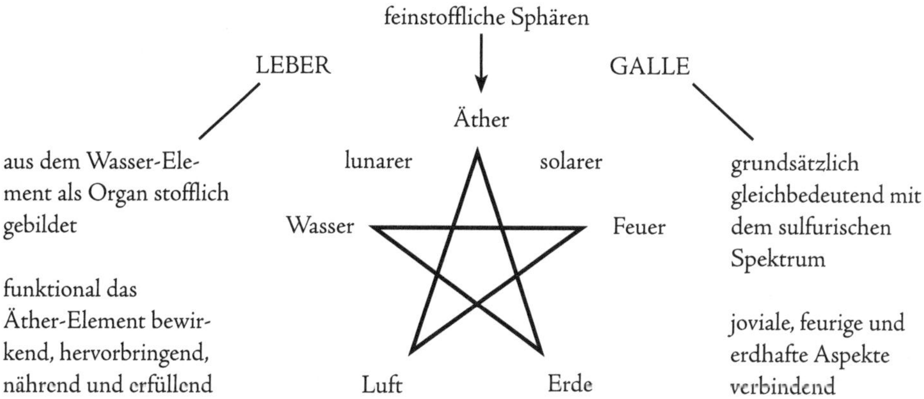

Die dabei vermittelten Zeichen sind solcherart, wie sie ‚aus der Natur erkannt' werden. Die Pulstaststelle ist keine Tastatur und der dazugehörige Körper kein Computer. Man wird nicht auf Knopfdruck eine Information erhalten, die zuvor vom menschlichen Intellekt eingespeichert wurde. Man lernt ihre ‚Bedienung' nicht in der Computerfachschule, sondern in der ‚Schule der Natur'. Die Ausdrucksweise der Natur ist eine ‚hermetische' und entspricht genau dem, wie die ‚natürlichen Dinge' – aus ‚Sicht' der Natur und nicht nach der Ansicht des menschlichen Intellektes – aufgebaut sind. Die ‚Zeichen' künden von einem sich stets wandelnden Gemälde, das die Alten als *‚Wyrd' (>wired)*, als ‚Netzwerk' oder ‚Gewebe' bezeichnet haben. Modern ausgedrückt: Die Natur, ihre Dinge, Substanzen und Kräfte, das ist das ursprüngliche ‚Internet', das eigentliche und realistische im wahrsten Sinne, und das vom Menschen technisch geschaffene gleicht dazu in seiner stolzen ‚Größe' vielleicht einem Sandkorn in der Wüste.

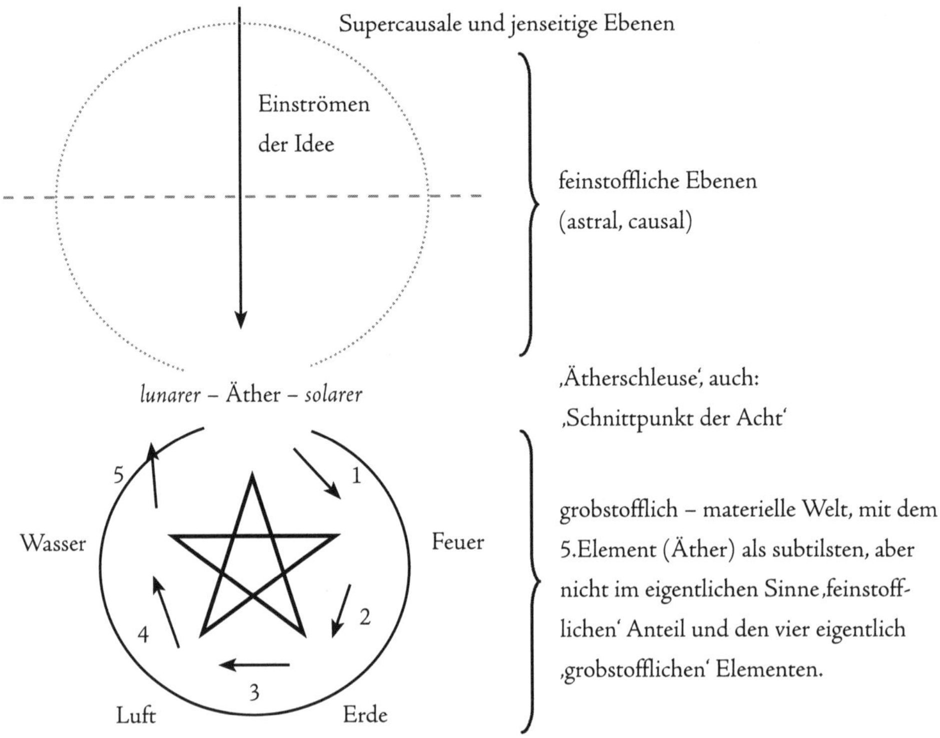

1 … Einströmen und Zentrieren der Idee (Quintessenz) / Stoffwerdungsimpuls durch Zusammenwirken von Position 5 und 1
2 … letzte Verdichtungs- bzw. Individualisationsschritte / Fixierung
3 … Struktur- und Gestaltbildung / erste Auflösungsphasen
4 … Lebensmitte der physischen Ausdrucksform / verstärkte Auflösungsphase
5 … Auflösen der Ausdrucksform (Fermentieren, Faulen, Verwesen, ...)
Scheidung der beteiligten Substanzen und Kräfte

Die Positionen der vorherrschenden Planetenkräfte (im unteren Kreis):
Beginn von Phase 1 (Äther-Element bzw. solarer Äther): JUPITER
Mitte von Phase 1 (oberes Feuer-Element): SONNE
Phase 2 (unteres Feuer-Element): MARS
Ende von Phase 2 (Erd-Element / saturnische Erde): SATURN
Beginn von Phase 3 (venerische Erde): VENUS
Mitte von Phase 3: MERKUR (darauffolgend Generation des Luft-Elements)
Ende von Phase 4 (unteres bzw. venerisches Wasser-Element): VENUS
Beginn von Phase 5 (oberes bzw. lunares Wasser-Element): MOND
Ende von Phase 5 (lunarer Äther / Klang-Äther): MOND / SATURN / JUPITER

Zurück zur im Äther sich manifestierenden, aus dem Feinstofflichen einströmenden ‚Idee': Diese Angelegenheit ist nicht ganz so einfach, wie sie bis hierher scheinen mag. Die ‚Idee' kommt nicht einer Rakete gleich aus dem jenseitigen Kosmos durch den ätherischen Raum geflogen, verringert beim Eintritt in die Physis wegen der zunehmenden Dichte (einer Qualität, die das feurige Spektrum bis ins Erd-Element hinein begleitet) ihre Dynamik und wird schließlich von der Erde abgestoppt, wo sie dann ‚ankommt'. Zu dem oben beschriebenen, beinahe geradlinig anmutenden Bild gehört noch ein ganz anderes: Die Idee tritt in den Äther ein, und sofort findet (naturgemäß) ein polares Ereignis statt: Im Lichtäther beginnt ihre Zentrierung, im Mondenäther ihre ‚Spiegelung'. Erst durch diese Spiegelung nimmt das Feuer-Element dann in der Folge seinen eigenständig-elementaren Charakter an.

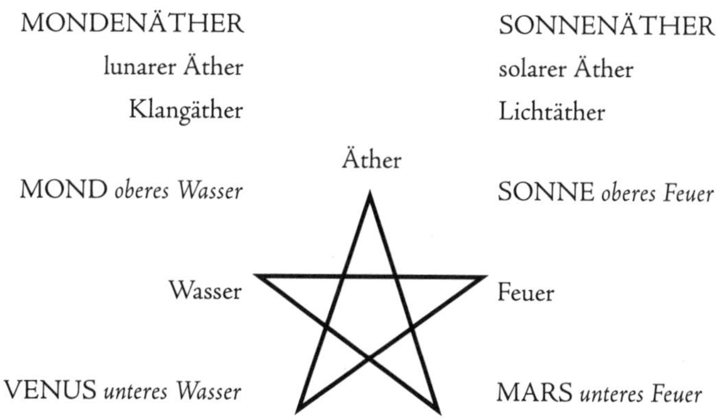

Dazu folgender Vergleich: Jemand betritt einen Raum, in dem er eine Idee, die er eben mit sich bringt, umsetzen möchte. Währenddessen ereignet sich einerseits eine Zentrierung der Idee, andererseits eine Reflektion des Raumes auf die Idee bzw. eine ‚Spiegelung der Idee im Raum'. Letzteres kann zum Beispiel ein augenblickliches Abschätzen sein, ob dieser Raum für die Umsetzung der Idee geeignet ist. Oder es kann bedeuten, dass anderen Leuten, die sich im selben Raum aufhalten, zuerst die Idee vorgetragen und ihre Reaktion empfangen wird. Ähnlich ‚betritt' die ‚Idee', von draußen kommend, vom Feinstofflichen, den Raum (Äther) der 5-elementaren Physis und wird einerseits zentriert (nach Art des solaren Äthers) und andererseits von den Grenzen bzw. von den Bedingtheiten dieses subtil-elementaren Ätherraumes reflektiert und dann erst in die Position des solaren Äthers zurückgeworfen (= ‚Geburt des Feuer-Elementes' bzw. der Aktion).

Hintergründig kann man sich vorstellen, dass das, was wir hier ‚Idee' nennen, stets seinen Ausgang in den feinstofflichen Sphären nimmt. Nicht immer liegt dort auch eine Ursache – die kann ohne weiteres vordergründig stofflicher Art sein. Alles was im stofflichen, 5-elementaren Kosmos getan und gedacht wird, wirkt in den feinstofflichen Kosmos hinaus, um dann zu seiner Zeit gleichsam, verändert, gemildert oder potenziert wieder in den grobstofflichen Kosmos zurückzuströmen, denn im Makrokosmos gibt es ebenso dieses Gesetz der Spiegelung wie in der grobstofflichen Ebene der Körper. Andererseits können aber auch in den feinstofflichen Sphären ‚neue Ideen' generiert werden, die dann quasi ‚zum ersten Mal' in die Physis strömen. Ob es grundsätzlich Neues gibt, das sich hier in den Zyklen der Äonen und Augenblicke ereignet, sei dahingestellt.

Die feinstofflichen Sphären sind unendlich ‚dünner', der Raum dort unendlich weiter, weswegen die Erfahrung der ‚Enge' und räumlichen Beschränkung für alles, was in die Ebene der Physis eindringt, typisch ist.

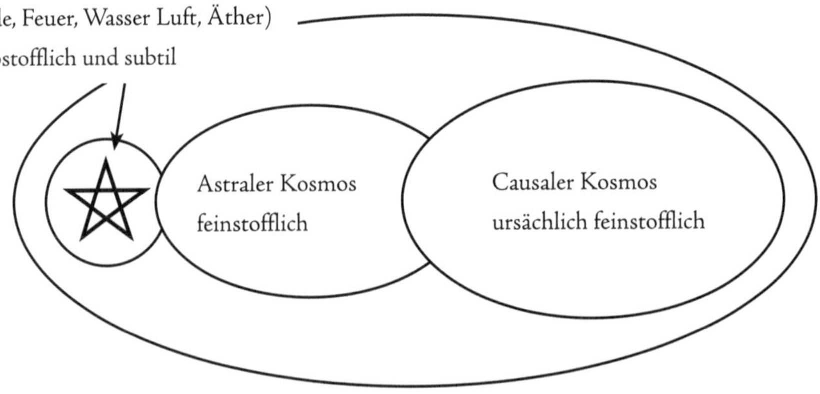

8 – elementarer Kosmos, „Brahmisches Ei"

Im menschlichen Körper entspricht dieses ‚Einströmen aus dem Feinstofflichen' zum Beispiel dem ‚sulfurischen' Seelenstrom, welcher im Augenbrennpunkt, im Bereich des ‚inneren Auges' zwischen den Augenbrauen einfließt (solarer Äther) und dann über den Zirbeldrüsen-Hypophysenbereich zerstreut und in die Schädelschale (lunarer Äther) projiziert wird. Dort am Augenbrennpunkt sind beim Menschen auch die Signaturen der ‚Äther-Organbereiche', wie Leber-Galle-Themen, ausgedrückt (zB. die sogenannten ‚Leber- oder Ärgerfalten').

Die Reflektion dieser Projektion aber trifft nun auf die Position des Herzens im Körper – ‚bildet' gewissermaßen ‚das Herz'.

Das Herz ist das erste Organ des Feuer-Elements und zugleich das erste Organ, welches die Reflektion des Umfeldes (des Raumes) auf eine Idee empfängt. Das Herz des Menschen ist so beschaffen, dass es alle Eindrücke der Außenwelt, des Körperinneren und sogar der feinstofflichen Sphären wahrnehmen kann.

So ist der Weg der ‚Idee' also kein ‚geradliniger' (vom Äther- zum Feuer-Element), sondern entspricht einem ‚ersten Pulsieren' – hinaus zu den Grenzen des stofflich-elementaren Raumes und zurück zu jener Position, an der die Zentrierung im solaren Äther stattfindet und damit gleichzeitig das eigentliche Feuer-Element ‚geboren' wird.

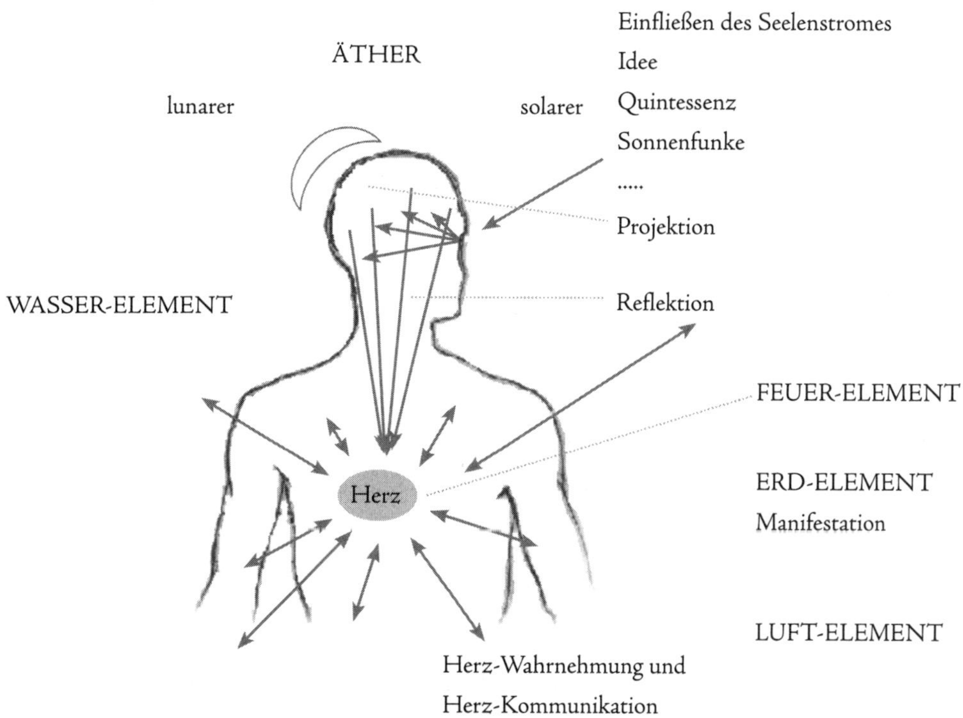

Auf der anderen Seite bezeichnet die Position des Wasser-Elementes, insbesondere das ‚obere Wasser' und natürlich der lunare Äther jene Bereiche, in denen die grobstofflichen Erscheinungsformen wieder aufgelöst werden. Hier entsprechen die Vorgänge der Fermentation, Putrefaction, Zersetzung, Verwesung u. dgl.

Im hermetischen Sinne bildet die ‚Oberfläche des Wassers' jenen Spiegel, in dem die wieder von neuem in den Elementereigen eindringende Idee und Quintessenz reflektiert und dadurch ihre stoffliche Manifestation möglich wird.

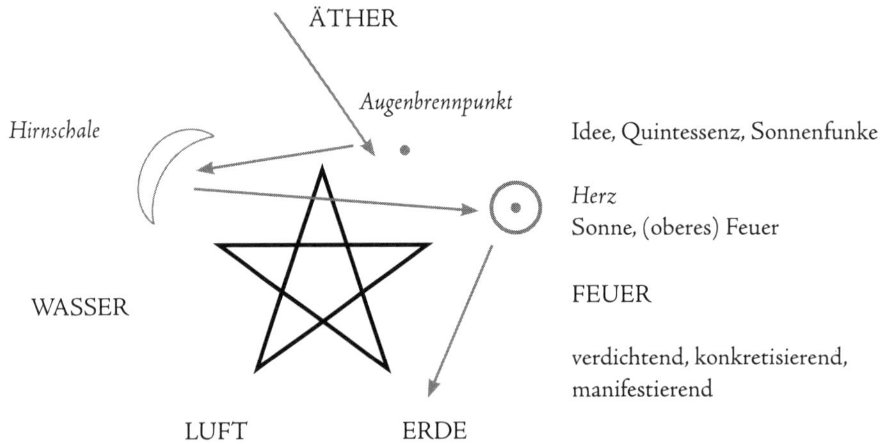

Das, was rechtsseitig in der pentagrammatischen Darstellung zuerst zentriert und dann verdichtet und manifestiert wird, gelangt mit dem Hervorbrechen des merkurialen Spektrums, besonders in der Position des Luft-Elements, zur Formenvielfalt und fällt dadurch wieder der natürlichen Vergänglichkeit anheim. Nachdem – wie es sprichwörtlich heißt – alles zu Wasser geworden ist, bleibt vom stofflichen, 5-elementaren Kosmos nur mehr ein Aspekt: der Raum. Dieser wird zu einer Grundqualität des Äther-Elements und wegen seiner qualitativen Subtilität zur Schwelle zum feinstofflichen Kosmos.

Wenn sich nun eine Idee aus dem Feinstofflichen heraus verdichtet und damit den 5-elementaren Kosmos ‚betritt', dann ist ihre erste Erfahrung die, dass der Raum der grobstofflichen Ebene ein anderer, nämlich ungleich dichterer und beschränkterer ist als der endlose Raum der feinstofflichen Ebenen. Diese ‚Erfahrung' ist gleichzeitig eben das, was wir als ‚Spiegelung' bezeichnen – oder als die ‚Spiegelung des Sonnenfunkens'.

Der ganze Vorgang ist vergleichbar dem ‚Ankommen' einer Seele, eines noch geistigen Wesens im Mutterleib. Sobald die Seele bzw. das Geistwesen beginnt, sich im Mutterleib körperlich zu manifestieren, verlässt sie auch die Sphären unendlicher Weite und erfährt nun die räumliche Bedingtheit der 5-elementaren Welt. Diese ‚Welt' wird von Seele und geistiger Essenz des Neuankömmlings bis an ihre Grenzen erfüllt, so wie jede Idee als Quintessenz stets den gesamten 5-elementaren Kosmos durchdringt. Die ‚Grenzen' sind ‚Spiegel' und ‚Mondensphäre'.

Vergleichbar dazu erfährt ein Mensch Beschränkung, räumliche Begrenzung, wenn er vor einem Spiegel steht. Die Widerspiegelung ist vom Rahmen des Spiegels bedingt. Nur ein Teil des Originals wird reflektiert – ein essentieller Teil wird ‚eingesaugt‘ (was inzwischen auch im physikalischen Experiment nachgewiesen werden konnte). Dies entspricht der einsaugenden Qualität des Wasser-Elements bzw. der lunaren Sphäre. Diese Einsaugung soll hier nicht unser Thema sein und wird an anderer Stelle eingehender behandelt werden, es entsteht aber daraus schließlich das, was die chinesische Medizin als Nieren-Essenz bzw. ‚Vorhimmels-*Jing*‘ bezeichnet – nämlich dann, wenn der Vorgang der Spiegelung analog als befruchtendes Zusammentreffen von Mann und Frau betrachtet wird.

Das, was aber reflektiert wird, gibt den ersten Anlass zur verdichtenden Tendenz des feurig-sulfurischen Spektrums. Mit anderen Worten: Die Idee wird durch Raumerfahrung, durch Reflektion zentriert. Das ist sehr klar gekennzeichnet durch das hermetische Symbol der Sonne, den Punkt (Idee, Quintessenz, Sonnenfunken) im Kreis (begrenzter Raum des grobstofflichen Kosmos).

⊙

Das, was im solaren Äther am besten einfach als Punkt symbolisiert wird, wächst durch das Zusammenwirken mit dem Raum zum Sonnensymbol – und gleichzeitig damit wird das eigentliche Feuer-Element hervorgebracht. Im menschlichen Körper geht dies einher mit der Bildung des Herzens im Embryonalstadium. Vor der Bildung des Herzens, dem ersten Pulsieren, belebt die inkarnierende Seele noch vorwiegend den feinstofflichen Bereich und hat ihre Wahrnehmung dort. Danach wird die Reflektion des nun neu belebten Raumes zu ihrer zentralen Wahrnehmung. Sie beginnt immer mehr das wahrzunehmen, was im eigenen sich bildenden Körper, im Körper der Mutter und in der stofflichen Außenwelt um die Mutter herum vor sich geht. Diese Wahrnehmung ist noch eine völlig ‚direkte‘, eben gleich einer Reflektion auf die Umgebung in nur einem einzigen Medium – dem Raum. Es gibt noch keine Streuung des Wesenspunktes durch die Sinnesorgane, geschweige denn eine Filtrierung durch den Intellekt.

Somit wird die Herz-Wahrnehmung zu unserer ersten und direktesten Wahrnehmung (aufgrund der ‚Spiegelung des Sonnenfunkens‘ aus hermetischer Sicht), und das hält dann ein Leben lang an. Die zweite Wahrnehmung (und zugleich Kommunikation) setzt später ein, wenn das Kind den Mutterleib verlässt und vermittels des Luft-Elements wahrnimmt und kommuniziert – hermetisch ausgedrückt: wenn Merkur aus dem Erd-Element geboren wird.

Wir sehen anhand dieser Darstellung auch, dass die Aussage: „Äther (‚Holz') bringt Feuer hervor" das tatsächliche Geschehen im Webwerk der Natur nur stark vereinfacht beschreibt. Genauer müsste man sagen: „Äther bringt Feuer vermittels des Wassers hervor."

Weil hier also die eigentliche Stoffbildungstendenz des Feuer-Elements eine Fortführung dessen ist, was vom Rande des ‚hohen Wassers' (aus dem Vergehen der elementaren Physis) über die Ätherschwelle wieder in den Elementereigen gelangt, liegt der Grund von Stoff, Körper und Substanz – wie schon *Thales* bemerkte – im Wasser-Element. Hierin sind auch die Themen der Quintessenz in der alchymistischen Praxis enthalten, denen wir uns später widmen wollen. Was aber an dieser Stelle bemerkenswert ist: Auch der Grund des Herz-Körpers liegt im lunaren Wasser bzw. im lunaren Äther. Das, was mit dem Herzen zusammenhängt und nicht körperlich ist, sind die ersten Äußerungen des Feuer-Elements. Das, was aber am Herzen körperlich, also in beschränkter, begrenzter Form auftritt, hat lunaren Hintergrund. Deswegen waren schon bei den Alten jene ‚natürlichen Dinge', welche höchste Mond-Signatur zeigten, auch die höchsten ‚Herzkörpermittel'. Dazu gehören Silber, Perle und Mistel. Solche unterstützen und ergänzen Herz-Aktiva wie Gold, Rubin und Safran in den Rezepturen östlicher und westlicher Traditionen.

In der chinesischen Medizin wird das Herz als ‚Herrscher und Monarch der fünf Yin- und der sechs Yang-Organe' bezeichnet, unter anderem, weil es deren stoffliche Manifestation ermöglicht und erhält. Man nennt es auch die ‚Residenz des Geistes *(Shen)*', weil es der erste Ort in der elementaren Physis ist, an dem der Geist, also auch die Quintessenz, wohnen kann. In einem Kommentar zum *Ling Shu Jing* (ca. 100 v. Chr.) wird erklärt, dass der Geist *(Shen)* eine Transformation der (oben erwähnten) Nieren-Essenz (Vor-Himmels-*Jing*) vereint mit dem *Ki* (im Sinne von ‚Nach-Himmels-*Jing*') ist. Das heißt, dass der Bewohner des Herzens *(Shen)* zuerst eben jene ‚neu' einströmende Quintessenz/Idee (der ‚Sonnenfunke') ist, vereinigt mit dem, was während ihrer eigenen ‚Reflektion im Mondenspiegel' reflektiert wird, nämlich der (im Wasser-Element generierten) ‚Essenz der Substanz'. Hierin sind auch die beiden Aspekte der Quintessenz angesprochen, welche gemäß der Formulierung ‚*de potentia in actu*' in der alchymistischen Praxis differenziert werden.

Dazu kommt aber noch ein dritter Aspekt: Im selben Text wird die Wanderseele *Hun* als ‚Ergänzung von *Shen* und *Ki*' bezeichnet. Sie ist gewissermaßen ein Aspekt der Quintessenz oder Idee – Idee vielleicht in dem Sinne, dass nicht wirklich etwas ‚Neues' den Elementereigen betritt, sondern ‚alles schon einmal dagewesen ist'. Das *Hun*, als Ein- bzw. Abdruck von etwas bereits Dagewesenen, vereint sich sowohl mit der zu ihr passenden essentiellen Substanz, als auch mit der – ebenfalls zu ihr passenden – neueinströmenden Idee.

Hun ist (der chinesisch-taoistischen Tradition zufolge) von ätherischer Natur und betritt den Körper nach der Geburt. Es heißt, ‚sie verlässt den Körper beim Tod und trägt die Erscheinung der physischen Form mit sich.' Dieses *Hun* ist der eigentlich merkuriale Anteil in diesem Geschehen, so etwas wie der subtil-ätherische Grund, eine Art ‚Vor-Gestalt' der späteren physischen Ausdrucksform. *H. P. Blavatsky* bemerkt, dass gewisse asiatische Traditionen Merkur *Pesh-Hun* nennen. Das bezeichnet jenen Aspekt eines ‚sehr hohen' Merkurs, den manche Alchymisten gleichsetzen mit den Begriffen Quintessenz und *Sal volatile*.

Wir sehen also, dass zum Verständnis des Eintritts der ‚Idee' in den 5-elementaren Kosmos mehrere Aspekte in Betracht gezogen werden müssen und bemerken darüberhinaus, wie komplex das Natur- und Lebensverständnis der ‚Alten', welches noch heute die Grundlage der praktischen Alchymie bildet, ist.

Herz-Wahrnehmung ermöglicht die direkteste Art des Verstehens – so wie das Herz-zu-Herz-Gespräch die direkteste und intimste Art der Kommunikation darstellt. Es wird dabei ‚Anwesenheit' wahrgenommen und ohne Interpretation darauf reagiert, also ‚geantwortet'.

Die chinesische Medizin stellt den ‚Funktionskreis Dünndarm' beinahe in die selbe Position wie das Herz – ins ‚obere Feuer', ins ‚königliche Feuer'. Dabei ist die Dünndarm-Qualität des unmittelbaren und direkten ‚Unterscheidens' vorrangig, sowohl stofflich als auch in der Bedeutung des Erkennens ‚aus dem Bauch heraus'. Dieses ‚Aus dem Bauch heraus' zusammen mit jener Herz-Wahrnehmung bildet nun schließlich die Grundlage zum ‚Verstehen der Sprache der Natur', zur Signaturenschau sowie zur heilkundlichen Befundung.

Im *Ling Shu Jing* heißt es: „Der Dünndarm kontrolliert Empfangen, Gefüllt-werden und Umwandeln." Das ist die Folge jenes Geschehens, das zur Herz-Bildung und Herz-Wahrnehmung führt. Mittels jener Qualitäten, die sich im Organ des Dünndarms organisch manifestieren, wird die erste Reflektion (Sonnenfunke – Mond – Sonne bzw. Idee – Raum – Herz) in gewisser Weise nachgeahmt oder ähnlich wiederholt und die körperinnere Kommunikation, besonders die stofforientierte, wird nachvollzogen. Der Dünndarm bekommt feste und flüssige Nahrung vom Magen – er empfängt sie, ähnlich wie die Reflektion der Idee (des Geistes) in der Position des ‚oberen Feuers', der Sonne, aus dem Lunaren heraus empfangen wird. Die ‚Reflektion' des Dünndarms bezieht sich dann wieder auf das Erd-Element, indem er die ‚reine' Fraktion der Nahrung zur Milz weiterreicht. Die ‚unreine' Fraktion geht an Dickdarm und Blase – alles also der Hervorbringungsrichtung folgend und in der Grobstofflichkeit wirksam. Damit ereignet sich mittels des Dünndarms bzw. auch mittels des gesamten ‚Funktionskreises Dünndarm' auf einer ‚gröberen' Ebene das selbe, wie in der Herz-Wahrnehmung und Herz-Kommunikation.

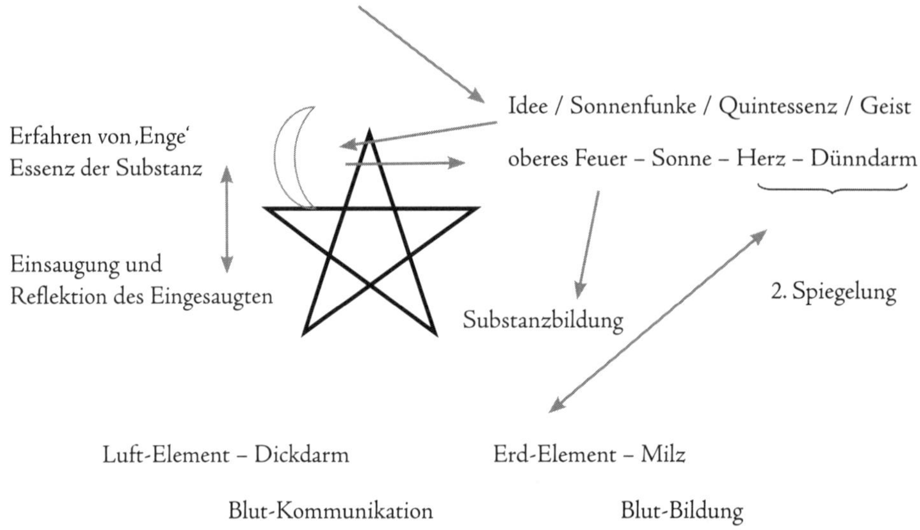

Solche Dinge lernt der Mensch aus der Natur. Er erkennt Wesen, Bedeutung und Zusammenhänge sowie ihre Stimmigkeit auf allen Ebenen. Das ist Signaturenschau, besser Signaturenkunde, denn man übt sie nicht alleine durch Schauen. Es ist nicht weniger als: die Sprache der Natur verstehen. Nicht nur die Sprache der Pflanze oder des Minerals, auch die Sprache der Gesten, der Blicke, der Gefühle, die Bedeutung dessen, ‚was in der Luft liegt' und die Ausdrucksweisen der Organe und ihrer Funktionen – kurzum das Verständnis jeglichen Ausdrucks. Ausdruck ist das, was uns kommuniziert wird – im hermetischen Terminus: ‚das Wesen Merkurs'. Wie sollte der Alchymist nicht den ‚philosophischen Merkur' erlangen, wenn er die ‚Ehrlichkeit' und Klarheit erwirbt, welche mit jener Wahrnehmung, dem ‚Schauen im Lichte der Natur', einhergeht? Merkur ist der ‚Naturgeist'. Den ‚Merkur der Weisen' zu erlangen bedeutet nichts anderes, als ‚die Natur verstehen' und dieses Verstehen in laborantische Praxis umzusetzen.

Gar manche haben ihre Heilung schon direkt im Erlangen von Ehrlichkeit zu sich selbst und damit durch das Überschreiten der Schwelle zum Naturverständnis gefunden. Und andere hat das ihrem Leben aufgebürdete Leiden zu dieser Wahrnehmung und Ehrlichkeit geführt, selbst wenn das Leiden diesmal nicht zu korrigieren war. Diese Wahrnehmung beginnt im eigenen Körper und im engsten Umfeld. Oft wachsen die zur Heilung nötigen Kräuter genau vor der Haustüre – eine der liebevollsten Äußerungen der oben angeführten Gesetzmäßigkeit, dass jedem Leidenden von Natur aus sein Heilmittel zusteht. Er muss nicht viel suchen, um zu finden. Er braucht nur wahrzunehmen. Doch selbst genau dieser Schritt vor die Haustüre bedarf bisweilen einer langjährigen Suche.

Wenn er darauf beharrt, nicht wahrzunehmen oder das unvermeidlich Wahrgenommene nicht zu nutzen, kann dem Leidenden auch ein kulturelles Umfeld kaum helfen und er muss oft noch tiefer in das Leid hineinfallen, bis er die Verantwortung sich selbst, seinem Körper, seinen Mitmenschen und der Natur gegenüber annimmt. Die Heilkräuter vor der eigenen Türe ‚wahrzunehmen' kann ein wichtiger Schritt zum Naturverständnis sein, denn in der Natur fügt sich stets das Eine zum Andern. Wegen solcher ‚Fügungen', wegen einer solch mächtigen Gesetzmäßigkeit kann man beruhigt annehmen, dass jedem, der sie sucht, auch die höchsten Arzneien zur Verfügung stehen. Es mag aber manch kleine Leiden geben, die zu großen gemacht werden, jedoch mit etwas Ehrlichkeit billig und bald geheilt werden könnten. Wenn der Mensch lernt, rechtzeitig die Reflektionen aus Umfeld und Körper wahrzunehmen, sie nach bestem Vermögen versteht und darauf angemessen reagiert, dann werden ihm in den meisten Fällen die ‚einfachen Kräuter vor seiner Türe' zum Wohle gereichen. Das erfordert Herzensbildung und ist der Beginn eines ‚Schauens im Lichte der Natur'.

„Weil nun der Alchymist mit seinem Tun hervortreibt, was in der Natur ist, so wisset bescheid über die verschiedenen Kräfte in den Blättern, wieder andere in den Blüten, andere in unreifen Früchten, andere in den reifen Früchten, und das so wunderbar, dass die letzte Frucht des Baumes ganz ungleich ist der ersten – wie in der Form, so auch in den Tugenden. Danach sollen eure Erkenntnisse stattfinden: vom ersten Hervorkommen bis zum letzten – denn so ist die Natur.

Wenn nun die Natur in ihrer Offenbarung so ist, dann ist es der Alchymist nicht weniger in den Dingen, da die Natur aufgehört hat so fortzufahren. Der Ginster erreicht nämlich den Prozess seiner Natur in der Hand des Alchymisten, auch der Thymian und alle anderen. Seht, dass ein Ding nicht nur eine Tugend hat, sondern viele. Seht es an den Blumen, die nicht nur eine Farbe haben und sind doch in einem Ding, und jede ist ein Ding und jegliche Farbe ist an sich selbst auf das Höchste gradiert. Dasselbe ist auch von den verschiedenen Tugenden zu verstehen, die in den Dingen liegen. Nun ist es die Alchymie der Farben, die Kunst und Art diese von einander zu bringen. Und wie mit den Farben, so soll auch solche Scheidung mit den Tugenden geschehen. Denn im Schwefel ist Gelbe, Weiße und Röte, dazu noch Bräune und Schwärze. Nun ist in jeder Farbe eine besondere Tugend und Kraft. Andere Dinge, die auch diese Farben haben, haben nicht die selben, sondern in diesen Farben andere Tugenden. Hierin liegt der Farben Erkenntnis, wie sie uns von den Farben zusteht. (...) Denn diese Zeichen sind in der Alchymie sehr zu beachten, damit man das wahrhaftige Ende der Wirkung und ihren Herbst wisse, damit die Zeit der reifen Tugend und die der unreifen Tugend zum Ende komme und zum rechten Verstand der Arznei. (...) Solches eröffnet und lehrt die Alchymie. Warum soll nicht der Grund der Arznei billig auf ihr stehen, und wir da Kochen lernen?"

nach: *Paracelsus, >Paragranum<*

Ansichten der Kunst und die Modernität der Inneren Alchymie

Viele Menschen befassen sich seit jeher mit der alchymistischen Kunst, um die Natur zu begreifen und ein Verständnis der Substanzen und Kräftewirksamkeiten zu erwerben bzw. zu vertiefen, ohne darüberhinaus in das Gebiet der Heilkunst eintreten zu wollen. Dieses Naturverständnis kann nicht nur in die laborantische Praxis umgesetzt, sondern zum Teil auch erst durch diese erworben, korrigiert oder bestätigt werden. Die Grundlagen dazu scheinen einfach, dem Verständnislosen beinahe naiv, im Detail entfaltet sich jedoch ein schier unüberschaubar weites Wissensgebiet, beinahe alles umfassend und auf alles anwendbar, was das menschliche Leben betrifft. Deswegen sind seit jeher zahlreiche Künste mit der Alchymie genauso verbunden wie das ‚Schauen im Lichte der Natur' mit dem hermetischen Verständnis des Gewebes Mensch, Natur und Kosmos – anders als die meisten neuzeitlichen, stofforientierten Erklärungsmodelle, auf urzeitlich-zeitlosem Wissen fundierend, das sich nicht nur in der Heilkunst bewährt hat.

In diesem Zusammenhang sollte aber über einen Begriff nachgedacht werden, der zeitgemäß sehr populär geworden ist: die ‚innere Alchymie'. Ursprünglich wurde darunter die spirituelle Entwicklung des Menschen verstanden, welche in den allermeisten Fällen ganz unabhängig von der ‚eigentlichen' Alchymie (als ‚eine Säule der Heilkunst') ist. Da aber der angehende ‚spirituelle' Mensch genauso Art und Wesen der Natur begreifen lernt wie der Alchymist oder der Heilkundige, gibt es hier dennoch auffallende Parallelen zwischen Heilkunst, alchymistischem Handwerk und Spiritualität.

Der innerste Beweggrund sowohl für die Heilkunst als auch für die Spiritualität – mit *Paracelsus*' Worten: „das Höchste aber ist die Liebe" – ist ein ähnlicher. Die Ebene der Umsetzung und Äußerung beider Gebiete ist jedoch oft wesentlich verschieden: Aus spiritueller Sicht sind Heilkunst und Alchymie ein ‚mit der stofflichen Welt befasstes Handwerk'. Dazu ist ein gewisses Wissen um die Gesetze der Polarität das geistige Werkzeug. Spiritualität selbst bedeutet den Rückzug des Menschen, bzw. der menschlichen Seele, aus der Polarität – seine Heimkehr in die Einheit. Der spirituelle Adept hat die polare Natur als reines Illusionswerk erkannt und verschwendet daran nicht mehr als die nötigste Aufmerksamkeit. Der heilkundliche Adept (Schüler) befasst sich, soweit es notwendig ist, mit bestimmten Aspekten dieses Illusionwerkes. Das ist gemeint mit der Formulierung: „der rechte Umgang mit den Dingen der Natur". Hat er das Glück, einen spirituellen Hintergrund zu besitzen, dann wird er sich in seiner Wissenschaft nicht ‚verlieren' und immer wieder imstande sein, den Blick auf die Ganzheit zu richten. Fehlt ihm dieser Hintergrund, dann kann er zwar ein bedeutender ‚Naturweiser' werden, aber er verfehlt den höchsten Zweck der menschlichen Geburt und bleibt ‚im Rad des Kommen und Gehens' gebunden.

Die beliebte Rede: „Ich arbeite nicht mit Retorten, Tiegeln, Pflanzen und Substanzen, sondern betreibe innere Alchymie“ weist heute meist darauf hin, dass es dort weder um Alchymie an sich noch um Spiritualität an sich geht und bezieht sich meist auf Probieren in asiatischen oder verwestlichten Techniken, zu deren Favoriten alles gehört, was Chakren, Kundalinis, *Pranas* und dergleichen Begriffe zum Inhalt hat oder das Gefühl, ein ‚Kabbalist' zu sein, vermittelt.

Diese Dinge schießen weit am ursprünglichen Ziel vorbei, sei dieses tatsächliche ‚spirituelle' Erkenntnis oder die ‚eigentliche' Alchymie, zu der eben auch das praktische ‚Handwerk' gehört. Aus dem Phänomen der ‚modernen inneren Alchymie' ersteht weder Weisheit noch Heilkunst, und die Lebenszeit vergeht.

Nach ‚innerer Alchymie' klingt es dem Unbedarften auch, wenn Chakren, Silben, Mantren, Hexagramme, Energiefelder und alles, was man eben so ‚kennt', mit Pflanzen, Steinen, Metallen und speziellen Präparaten in Verbindung gesetzt oder ‚innere Entwicklungen' versprochen werden, indem man sich die Pflanzen und Steine der Wochentage eintropft oder nach dem Trinken einer Tasse Johanniskrauttees sich auf einen roten Polster setzt und über das Feuer-Element ‚meditiert'. Obwohl hierin bisweilen ganz interessante Überlegungen angestellt werden und die Motivation gutmeinend, wenngleich von einem etwas hilflosen Beigeschmack ist, äußern sich solche Praktiken doch eher naiv, bleiben an der Oberfläche und täuschen den Aspiranten nicht selten durch besondere Gefühlserlebnisse oder gar durch spontane Einmischungen astralfeinstofflicher Kräfte, welche geeignet sind, ihn in die düsteren Felder der Magie zu ziehen.

Interessanterweise war weder in asiatischen noch abendländischen Traditionen ursprünglich jene ‚innere Alchymie', welche neuzeitliche Esoteriker propagieren, eine Bedingung für die eigentliche Alchymie, den ‚rechten Umgang mit den Dingen der Natur'. Diese Ansicht teilen bis heute weise Menschen – insbesondere des Ostens – obwohl sie alle Werke der Alchymie bestätigen, nicht aber ihre Abhängigkeit oder unbedingte Vereinbarkeit mit magischen Praktiken und ‚esoterischen' Systemen. Eine Ansicht, die der in geistigen Dingen meist unbewanderte oder zumindest nur gering gebildete Mensch westlicher Prägung heute erst noch vielfach überdenken und zugestehen muss, bis hier ‚alchymistisches Handwerk' und naturheilkundliche Praxis vom Anhauch eines modernen magisch-esoterischen Ballastes befreit sein kann, wird doch hier immer wieder das schlichte Erwähnen der 5 Elemente, ja sogar der Begriff ‚Naturheilkunde' selbst mit ‚Esoterik' assoziiert, und man meint nicht selten, dass man Okkultist sein müsse, um die Wirkung eines homöopathischen Heilmittels oder einer gut gesetzten Akupunkturnadel verstehen zu können, obwohl gerade das von einem simplen physikalischen Bildungsrückstand von gut 100 Jahren herrührt.

Damit verkommt aber die ganze Sache dann auch leicht zur ‚Glaubensangelegenheit' nach dem Motto: Wenn man dran glaubt, dann hilft's. Oder – auf das alchymistische Handwerk bezogen: Alleine durch seinen Glauben und seine Willenskraft schafft der Alchymist Dinge, die er durch ‚einfache Handwerkskunst' nicht vollbringen könnte.

Wenn eine hermetische – also ganzheitliche und tiefreichende – Sicht der Natur an sich schon in den Resort der ‚Esoterik' gesetzt wird, entsteht ein nicht unbedeutendes Hindernis in der Restauration der Naturheilkunde insgesamt, und es bildet sich die Meinung, dass diese Heilkunst nichts anderes sei als ein Teilgebiet jener ‚Esoterik'. Wir sind hier aber in der ‚Naturkunde' und eben in der ‚Heilkunst' und keineswegs in einer ‚Esoterik' und auch nicht in der Spiritualität.

Die Beschäftigung mit Elementen, Prinzipien, Planetenkräften und dergleichen ist Naturkunde schlechthin, während die eigentliche Esoterik von der Erforschung des feinstofflichen (astralen und kausalen) Universums handelt – ein Gebiet, welches von der Neo-Esoterik nie ernsthaft berührt wird – ja wegen der Natur der Sache gar nicht berührt werden kann, es sei denn durch – mit *Paracelsus*' Worten – ‚Wähnen und Spintisieren'. Treffend wird der Dimensionsunterschied der ‚Ebenen' von den asiatischen Weisen durch einen Vergleich dargestellt, nämlich dass der gesamte physische (5-elementare) Kosmos zum astralen (feinstofflichen) Kosmos das Verhältnis eines Sandkorns zur Wüste oder eines Tropfens zum Ozean habe. Und dabei befindet sich das Astralreich immerhin noch relativ ‚nahe' der Physis. Vom Astralen zum nächsten ‚Stockwerk', dem Causalen, gilt dann ein ähnliches Verhältnis. Genau entgegen manch propagierten ‚esoterischen' Gedanken ist es ein Bestandteil des unvorstellbaren Glücks, Mensch zu sein, dass dieser zu seiner Werdung und Entwicklung nicht in jene Ebenen einzudringen braucht, wo er unweigerlich seine menschliche Natur auf lange Zeit verspielen würde. Das ist die Ansicht der Spiritualität.

Bei aller Toleranz für verschiedenartige Termini in diesen Belangen – aber wenn jemand ‚Wasser' verlangt und ‚Feuer' erhält, dann wird ihn das nicht zufriedenstellen. Gar manche suchen sehr ernsthaft in der Heilkunst und auch nach einem praktischen Zugang zur Alchymie, müssen sich aber erst durch ein Wucherwerk an Neo-Eso-Konzepten durchkämpfen. Die elementare Natur selbst – mit ihren ‚Dingen' und Kräften – ist so unendlich mannigfaltig, subtil und weitreichend, dass uns die Beschäftigung mit ihr zeitweise wohl sehr ‚esoterisch' erscheinen mag, besonders dann, wenn wir uns über den Aberglauben ausschließlicher Stoffbezogenheit und Mechanik zu erheben beginnen und unsere Aufmerksamkeit vorzüglich auf das Weben der stoffbildenden Kräfte lenken wollen. Dennoch bleibt diese Beschäftigung stets eine ‚Naturkunde', wenngleich eine vollständigere. Die eigentliche Esoterik aber ist eine Kunde von den ‚inneren, geistig-materiellen Ebenen', die alleine den großen Mystikern schadlos zugänglich sind.

Diese Dinge müssen klar und deutlich vorgestellt werden – und wir werden an anderer Stelle noch einmal dieses Thema aufgreifen – weil sie wesentlich dazu beitragen, der Bedeutung von Spagyrik und Alchymie in der Naturheilkunde schärfere Konturen zu zeichnen. Der Grund der Alchymie ist in der Heilkunst. Sie hat keinen Grund in solch sogenannter ‚Magie'. Sie folgt der Natur und kann sie bestenfalls veredeln. Sie zwingt aber der Natur keinen Willen auf und wirkt niemals gegen sie. Zur Heilkunst braucht es nur den Menschen und ‚die Dinge der Natur'.

Seitenwege, wie die vor allem im Westen populären Arten des Geistheilens, und dergleichen neo-esoterische ‚Erfindungen' haben keinen Platz in der Naturheilkunde, denn sie wirken meist gegen die Natur – sowohl an sich, als auch der Beteiligten. Auf solche Weise ist in der Natur noch keine Heilung geschehen, sondern bloß vorübergehende Beseitigungen des einen oder anderen Leidens. Die Probleme wirken zuerst in den feinstofflichen Hüllen des Menschen weiter und ‚fallen' früher oder später potenziert auf die Beteiligten zurück. Die Worte ‚vorübergehend' oder ‚später' sollen aber hier sehr weitläufig verstanden werden, denn meist ist der Leidende damit zufrieden, auch nur eine Lösung für dieses eine Leiden oder dieses eine Leben gefunden zu haben; was danach oder viel später folgen mag, das kümmert ihn nicht. Dadurch werden diese Techniken zu einer feinstofflichen Allopathie, greifen bestenfalls Auswirkungen der eigentlichen Probleme an und zeitigen schwere Spätfolgen. Außerdem haben feinstoffliche Wesen die Art, nichts, aber auch gar nichts umsonst zu tun – weniger noch, als ein geiziger Mensch es schaffen würde. Wenn diese nun zur Hilfe herbeigezogen werden, dann stellen sie für ihre unvollendeten Taten unvorstellbar hohe Rechnungen aus, vergleichbar mit homöopathischen Potenzierungsschritten. Irgendwann müssen die Beteiligten zahlen.

Das alles ist leidig, sei aber doch angemerkt, weil es den Menschen besonders im Abendland diesbezüglich an Unterscheidungskraft mangelt. Hier ist man nur allzu geneigt, obskure Techniken mit hinein in den Topf der Naturheilkunde zu werfen und die Chakrenöffner und Engelimitatoren werden für Heilkundige gehalten. Hier strömen die ‚Meister' und Christusse nur so durch die ‚heilenden' Hände – mehr von diesen Wesen, als in der Vergangenheit von den geborgten Namen je gefunden hätten werden können, und die zahllosen Wege, die *Pranas* aufzurütteln, hat das älteste *Veda* noch nicht gekannt. Das ist nicht Heilkunst, das ist ein Unheil dieser Zeit. Damit wollen wir aber nun diese Dinge beenden, um nicht jenen auf den Zeh zu treten, die der Heilkunst an sich wohlgesonnen gegenüberstehen, aber trotzdem bereits in dem einen oder anderen solchen Felde befangen sind und wenn nicht jetzt, dann später ohnehin zusehen müssen, ob sie ihm entkommen können.

Es ist dieses Phänomen auch aus der Sicht begreiflich, dass dort, wo die Zusammenhänge von Mensch, Natur und Kosmos genauso vergessen wurden wie der ‚rechte Umgang mit den Dingen der Natur', weit eher eine Neigung besteht, sich diesen unbegründeten Dingen zuzuwenden und darin ein Heil zu suchen.

Zusammenfassung zum hermetischen Naturverständnis

Das ‚hermetische' – man könnte auch einfach sagen: das ‚ganzheitliche' – Naturverständnis erscheint auf den ersten Blick sehr simpel. Es leitet sich direkt aus der Natur ab, gewissermaßen aus dem Selbstverständnis der Natur. In den Grundzügen geht es dabei um Themen der Einheit, der Polarität und der weiteren Entfaltung und Wirkung der Naturkräfte.

Paracelsus bemerkt: „Alles kommt aus der Einheit. Sobald man beginnt, bis zwei zu zählen, fangen die Probleme schon an." Das Thema der Einheit selbst gehört seit eh und je unverrückbar in den Bereich der Spiritualität. Es betrifft nicht die mannigfaltigen Ausdrucksformen und webenden Kräfte in der Natur, sondern den Ursprung all dessen. Es betrifft vom Menschen weder Körper noch Geist, sondern ‚ihn selbst' als Seele. Alchymie ist nicht Spiritualität, sondern Naturkunde. Als solche setzt sie bei den ‚natürlichen Dingen' an, gleich wie die Heilkunst bei den lebendigen Körpern – und diese bestehen durchwegs aus polaren Kräften und polarer Substanz. Diese Art der Naturkunde im Bewusstsein der allem zugrundeliegenden Einheit zu betreiben, ist alleine schon wegen des neutralen Überblicks halber vorteilhaft. Nur im entschieden spirituellen Menschen ist die menschentypische Distanz zur Natur positiv aspektiert und insgemein förderlich. Deshalb verlangten die Weisen im alten Griechenland zu allererst das ‚Mensch erkenne dich selbst' von ihren Schülern. Daraus wächst dann unweigerlich eine echte Naturerkenntnis hervor. Die Körper und Kräfte in der Natur sind immer polar und gehören der Vielheit an, die Seelen der Einheit. Die Ebene der Seele gewährt einen Überblick, ein tatsächliches Verstehen der Natur, so als ob man aus einem dichten Wald heraustreten und auf den Gipfel des Berges steigen würde. Von da aus kann beschrieben werden, in welcher Weise der Wald in die Landschaft gebettet ist. Bleibt man im Wald, dann hat man diese Einsicht nicht.

„Der Umkreis ist im Punkt, im Samen liegt die Frucht,
Gott in der Welt – wer klug ist, der Ihn drinnen sucht."
Angelus Silesius, 1605-1660

Polarität

Überall gibt es Hitze und Kälte, Licht und Dunkelheit, Tag und Nacht u.s.f. Das stets polare Wirken in allen Belangen der Natur und des Lebens ist vordergründig und zugleich unendlich diffizil.

Es heißt: „Selbst wenn du ein ganzes Leben lang über die Polarität nachdenkst, kann es sein, dass du keinen einzigen Schritt fortschreitest."

„Alle Dinge urständen aus der Feuerwurzel als eine zweifache Gebärung."
„In dieser Welt ist Liebe und Zorn ineinander in allen Kreaturen."
„Ein jeder Mensch ist frei und ist wie ein eigener Gott – er mag sich in diesem Leben in Zorn oder ins Licht verwandeln."

Jacob Böhme, 1682

Die 3 Prinzipien

Das Wirken der Polarität bildet den ‚Hintergrund' der 3 Prinzipien *Sulphur*, *Merkur* und *Sal*, jener ‚prinzipiellen' Dreiheit in der Natur, welche anderswo auch durch die Begriffe *Anima*, *Spiritus* und *Corpus* oder Sonne, Mond und Erde bezeichnet wird.

„Diese Welt steht im vermengten Leben der Zeit zwischen Licht und Finsternis als ein wirkender Spiegel der Beiden."

Jacob Böhme, 1682

Die 3 Prinzipien begleiten das alleinige Kräftewirken zum stofflichen Ausdruck. Sie transportieren die Polarität in sich weiter auf dem Weg der quintessentiellen Idee in die vermischte Ausdrucksform.

„In jedem Holz, Stein und Kraut sind drei Dinge: Erstlich ist die Kraft daraus ein Leib wird, hernach ist darin ein Saft – das Herz des Dinges, und zum dritten ist darin eine quellende Kraft, ein Geruch oder Geschmack, das ist der Geist des Dinges davon es wächst."

Jacob Böhme, >Mörgenröthe<, 1682

Dem Verständnis des Dritten, welches zur Zweiheit hinzukommt, kann man sich von der einen Seite nähern, wenn man es als ‚Betrachter' neben Sonne und Mond stellt. Soweit dieser ‚Betrachter' stofflich manifestiert ist, entspricht er dem Sal-Prinzip, und die Dreiheit kann dann mit ‚Sonne-Mond-Erde' bezeichnet werden. Alle drei Worte benennen aber dabei eigentlich weder Körper noch Kräfte, sondern eben einen prinzipiellen Hintergrund allen Naturgeschehens, hinter dem wiederum grundsätzlich die Polarität wirkt. Ohne einen ‚Betrachter' hätte diese jedoch keinerlei Relevanz.

Eine andere Seite der Annäherung ist, wenn man anstelle von ‚Sonne' den Begriff ‚Sulfur-Prinzip' setzt. Da nun die ‚ursprüngliche Idee' sich durch das Sulfur-Prinzip hindurch zum ‚(lebendigen) Ding' oder ‚Ereignis' hin manifestiert (oder mit *Böhmes* oben zitierten Worten ‚alle Dinge aus der Feuerwurzel urständen'), stellt sich die ‚ursprüngliche Polarität' als Feuer-Erde Polarität dar. Wenn dazu ‚Wässrigkeit' kommt, dann sagt man: „Der Mond verbindet sich mit saturnischer Erde" oder: „Die durstige, aus dem Feuer geborene Erde wird befeuchtet (und damit ihr ‚Saturnzustand' relativiert)." Das ereignet sich ‚in der Erde' – bildet also die Art des ‚Sal-Prinzips'. Sal ist Salzigkeit im weitesten Sinne, zieht Feuchtigkeit an, wodurch sein fester, erdhafter Zustand relativiert bzw. zu einer Flüssigkeit aufgelöst wird. *Paracelsus* meint, dass ‚die Dinge ewig bestehen würden, wäre in ihnen nicht ein unsichtbarer Kalk, der wieder zu ihrer Auflösung beiträgt.' Dieser ‚Kalk' entspricht dem hermetischen Begriff ‚Mond', welcher eben ‚zur aus dem Feuer geborenen Erde hinzutritt'. Die ‚Erde' ist ‚Sal-Prinzip', ‚Mond' ist ‚Merkur-Prinzip'. Mond, Wasser, Luft, Merkur – das sind einander sehr ähnliche Bezeichnungen für Aspekte des ‚Merkur-Prinzips'. Es ist relativierend, kommunizierend, begleitend und auflösend. Diese 3 ‚Prinzipien' sind eine Ausdrucksweise der Dreiheit.

„Es sind drei Substanzen, die jeglichem Ding sein *Corpus* geben. Was da brennt, ist *Sulphur*, was da rauchet, ist *Mercurius*, was da Asche wird, ist *Sal*."

Paracelsus

Zusammen bilden die drei Prinzipien die ‚elementare Natur'. Jede Ausdrucksform besteht aus allen Dreien im jeweiligen Verhältnis.

„Die Drei sind Eins, nämlich Körper, Geist und Seele.
Denn alle Vollendung beruht auf der Dreizahl."

>*Aurora consurgens*<, 14. Jhdt.

Die 3 Prinzipien wirken sich schließlich durch die 5 Elemente Erde, Feuer, Wasser, Luft und Äther in der Physis aus. Im Hintergrund des Erd-Elements wirken vor allem Sulfur und Sal, des Feuer-Elements wirken Sulfur und Merkur, des Wasser-Elements wirken Merkur und Sal, des Luft-Elements Merkur und Sulfur und im Hintergrund des Äther-Elements wirken Sulfur und Sal ‚ursächlich' und Merkur ‚vollendet'. Trotz gewisser Bevorzugungen haben stets alle drei Prinzipien einen realtiven Anteil an der hintergründigen Wirksamkeit der Elementarkräfte.

Ähnlich verhält es sich mit der Wirkung der Elementarkräfte in den stofflichen Ausdrucksformen: Stets sind alle fünf vermischt, wobei dann die Dominanz von einem oder von zwei Elementen den elementaren Charakter des ‚Dinges' prägt. Vorzüglich aber verbinden sich die Elemente paarweise zu den ‚Säften' oder ‚Hüllen' (skr: *Dosha*). Diese bilden die letzte Stufe zu den Geweben der subtilen (‚energetischen') und grobstofflichen Körper.

In der Bezeichnung ‚Wind, Feuer und Schleim' für diese ‚Säfte' oder ‚Hüllen', erkennt man die Verbindung der Elemente nach dem Schema der Dreiheit. Andere Bezeichnungen dafür sind *Lung, Tripa, Peken* (tib.) oder *Vata, Pitta, Kapha* (ayu.) oder Wind, Galle und Schleim. Der Begriff ‚Galle' beschreibt das gesamte ‚sulfurische Spektrum', welches vom solaren Äther bis ins Erd-Element reicht. Deswegen wird in manchen Traditionen zwischen einer äthernahen, feurigen Galle und einer feurig-erdnahen (schwarzen) Galle unterschieden.

Elemente (mit Symbolen) und Säfte / Hüllen in pentagrammatischer Position

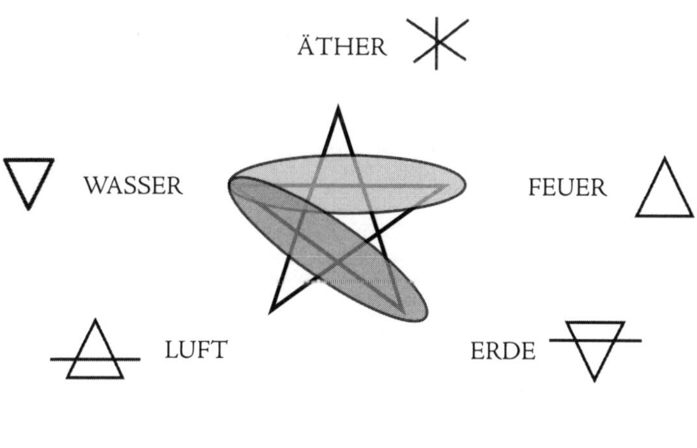

Ätherische und luftige Qualitäten verbinden sich zu WIND.
Erdhafte und wässrige Qualitäten verbinden sich zu SCHLEIM.
Feurige und wässrige Qualitäten verbinden sich zu FEUER / GALLE.

Der Begriff ‚Element' darf hier nicht im Sinne der modernen Chemie interpretiert und als ‚grobstofflicher Baustein' der Natur verstanden und mit den Atomen verglichen werden. ‚Elemente' sind spezifische Kräfteäußerungen mit mehr oder weniger stoffnaher bzw. stoffbildender Qualität. *Empedokles* nannte sie ‚Wurzeln' *(Rhizomata)*, *Plato* ‚einfache Bestandteile' *(Stoicheia)* und *Aristoteles* ‚erste dynamische Formen' *(Prota Somata)*. Älteste chinesische Texte nennen sie *Cai* (Fähigkeit, Material) oder *Fu* (Regierungssitze) und später auch *Xing* (Bewegung, Prozess, Verhalten). Der im Westen etablierte, aus der taoistischen Medizinlehre stammende Begriff ‚Wandlungsphase' beschreibt die Art der ‚Elemente' sehr treffend.

Anmerkung zum fünften Element

Ein kultur-typisch abendländisches Phänomen, vor dem schon *Paracelsus* gewarnt hatte („Vergesst die Quintessenz nicht"), erfasste weite Bereiche der Heilkunst und Spagyrik, nämlich dass der Blick für die Fünfheit der Natur verlorenging und stattdessen versucht wurde, Natur, Heilkunst und Alchymie aus der Vierheit zu begreifen, was nur lückenhaft möglich ist.

Viereck, Quader, Schachbrettmuster u. dgl. entsprechen ausschließlich dem saturnischen Erd-Element mit all seinen Attributen. Sie signieren einen bedeutenden Aspekt des Mineralreiches und reichen darüberhinaus nicht wesentlich in die höherlebendige Natur. (Zu ihrer Konstruktion ist kein Goldener Schnitt nötig).

Fünfeck und Pentagramm entsprechen der Pflanzen-, Tier- und Menschenwelt. Sie signieren die ‚lebendige Natur' und die ‚Position des Menschen in der Welt'. (Ihre Konstruktion erfolgt mittels des Goldenen Schnittes).

„*Alexander*, der Macedonier König (und Schüler des Weisen *Aristoteles*) sagt in seinem Kapitel von der Sonne, dass wohl ein gewisses Fundament ist in der Welt: Die große Welt entspricht der kleinen Welt des Menschen, und nach dem ist die Kunst gemacht. Darin ist das Erste das Begehrende, das Zweite das Verdauende, das Dritte zieht an sich oder hält, das Vierte treibt aus, und die Quintessenz zeugt und vereinigt die vier. Diese Umstände sind in der Kunst wie ein Schlüssel zum Kasten. Wenn du ihn ohne diesen Schlüssel auftust, dann wird er ganz dissolviert und ist wie ein Buch, das seine Seiten nicht öffnet.

Die *Quinta Essentia* ist das Leben, welches da ein eigenes ist, weder warm noch feucht, weder kalt noch trocken, weder Mann noch Weib. Sie erfüllt das ganze Werk."

nach: >*Turba Philosophorum*<

„*Paracelsus* unterschied in jedem Ding fünf Daseinszustände, die stufenweise ineinander übergehen können: 1. den festen (Erde), 2. den flüssigen (Wasser), 3. den flüchtigen (Luft), 4. den feurigen (Feuer) und 5. den ätherischen. Im fünften Sein besteht das eigentliche Wesen der Dinge. Erde und Wasser sind die Hauptgegenstände der Philosophie, Luft und Feuer die der Astronomie, und die Quintessenz (Äther) ist das Objekt der Alchymie."

nach: *F. Spunda*, >Das Weltbild des Paracelsus<, Wien 1941

Die Planetenkräfte

Die Elemente bzw. Elementarkräfte sind eng mit anderen Naturkräften verbunden – den Planetenkräften. Diese differenzieren die Positionen bzw. Grade der Elemente und weisen auf ihre Dynamiken und Vermischungstendenzen hin.
Die Zuordnungen der Planeten zu den Elementen in östlichen und westlichen Traditionen entsprechen einander im Wesentlichen. Dasselbe gilt auch für die mit diesen Kräften bzw. Qualitäten verbundenen Organe, Organbereiche und deren ‚Funktionskreise':

Äther – Leber, Feuer – Herz, Erde – Milz, Luft – Lunge, Wasser – Niere

In den abendländischen Traditionen werden hauptsächlich die 7 klassischen Planeten beschrieben, also Mond, Saturn, Jupiter, Merkur, Mars, Venus und Sonne. Die sonnenferneren Planeten waren den ‚Alten' nicht unbekannt, wurden allerdings, da sie für das freie Auge unsichtbar sind, entsprechend der Erkenntnis ‚wie oben so auch unten', als erdinnere Kräfte behandelt. Der Praktiker erlebt ihre Phasen unter anderem bei der Darstellung des sulfurischen Salzes aus der Pflanze, bei der er allegorisch den Weg durch den Vulkan ins Erdinnere beschreitet und dabei die uranischen und neptunischen Schichten passiert, um das plutonische Feuer zu erreichen.

Die ‚7 Planeten' sind stoff- und gestaltbildende sowie geistbegleitende Grundkräfte im Kosmos, deren typische Ausdrucksformen unter anderem die bekannten Himmelskörper sind. In der Erde entsprechen ihnen typischerweise die Metalle: Sonne – Gold, Mond – Silber, Saturn – Blei, Venus – Kupfer, Jupiter – Zinn, Mars – Eisen, Merkur – Quecksilber.

Überdies äußert sich die Wirkung der Planetenkräfte in Qualitäten, Eigenschaften, Konsistenzen, Ausdrucksformen, Farben, Prozessen und dergleichen mehr.

Mond – kalt, feucht, auflösend; Putrefaction, Verwandlung, Spiegelung, Fruchtbarkeit, Vermehrung, Generation und Regeneration. Elemente: Wasser / Äther

Saturn – hemmend, beendend, kalt, trocken, starr, dunkel, abweisend, reduzierend, zur Verwandlung zwingend. Elemente: Erde / Äther

Jupiter – reich, jovial, großzügig, majestätisch, ätherisch, weit; Stabilität, Gleichgewicht, voller Potential. Elemente: Äther / Wasser / Feuer

Merkur – immer bewegt, luftig, verspielt, schalkhaft, schillernd, vermittelnd, gestaltend, gesellig, stoffhaftend, anpassend; der ‚Naturgeist'. Elemente: Luft / Wasser / Feuer

Mars – feurig, aktiv, scharf, spitzig, reizend, ‚corrosivisch', aggressiv, konzentrierend, verdichtend, figierend / fixierend, aufbrechend. Elemente: Feuer / (Erde)

Venus – kühlend, feucht, ausgleichend, harmonisch, proportioniert, erdverbunden, rund, weich. Elemente: Wasser / Erde

Sonne – heiß, trocken, zentriert, festigt, stabilisiert, anziehend, ‚salzmachend', geraderichtend. Elemente: Feuer / Äther

Elemente, Prinzipien und Planetenkräfte in pentagrammatischer Position

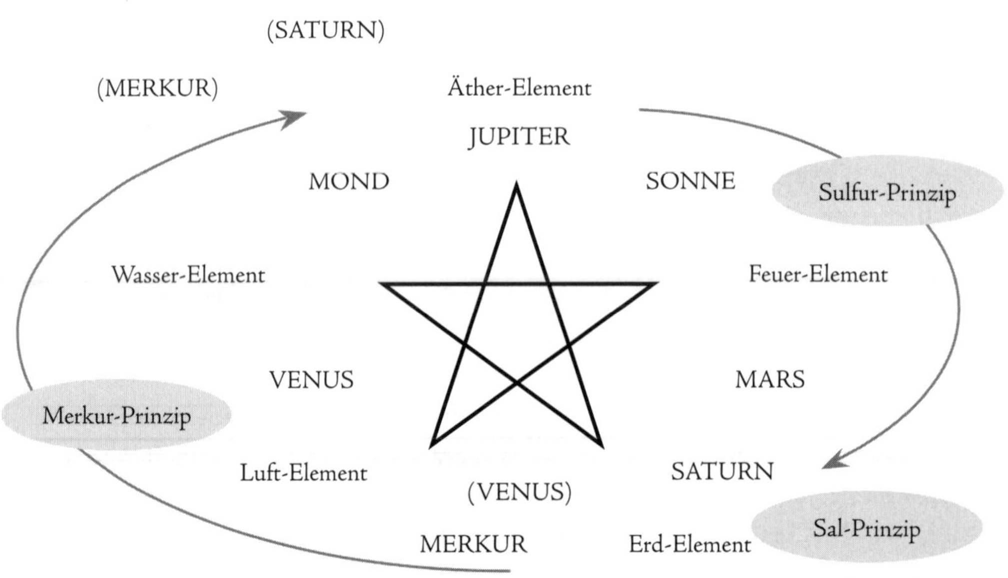

Das Planetensystem, wie wir es heute kennen, mit seinem Zentralgestirn, der Sonne, umgeben von neun Planeten, deren Monden und tausenden von Splittern und Planetoiden, bildet ein einheitliches System, welches, wirbelförmig gestaltet, wieder Bestandteil eines Riesenwirbels, der Galaxie, ist. Die zahlreichen Verhältnisse zwischen den Teilen des Planetensystems weisen überwiegend das ‚Maß der Natur' – den ‚Goldenen Schnitt' – auf.

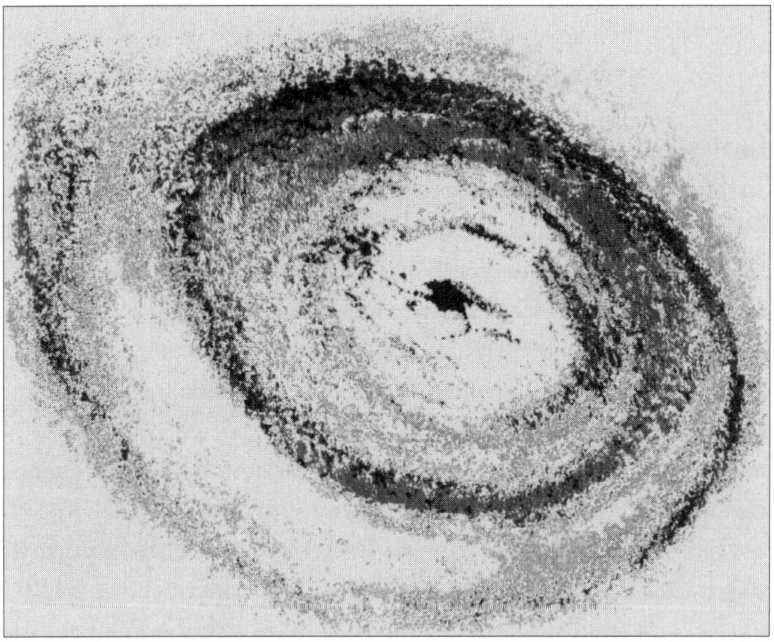

Dasselbe Gestaltungsprinzip, welches in der *Fibonacci*-Reihe dargestellt ist, finden wir im gesamten Pflanzenreich sowie in den Formen des menschlichen Körpers. Es sind also nicht vordergründig von den Himmelskörpern zur Erde fließende Kräfte, welche die ‚Dinge der Natur' und die Ereignisse hier gestalten, sondern die Himmelskörper sind Ausdrucksformen der selben Kräfte, welche das auf der Erde eben tun. Wenn es heißt: Die Sonne ist Gold, und das ist das Herz im Körper, dann bedeutet das, dass die selbe Naturkraft, das selbe Gestaltungsprinzip, welches in der Sonne vorherrscht, auch im Gold und im Herzen vorherrscht. Die Kraft selbst kann man dort oder da, je nach Gelegenheit, kennenlernen und erforschen. Man wird dabei immer auf die selbe ‚Art' kommen. Das meint *Paracelsus,* wenn er sagt: „Wer das Eisen kennt, der kennt den Mars, und wer Mars kennt, der kennt die Brennnessel und umgekehrt."

Betrachtet man etwa die,Spiraltendenz' im Pflanzenreich und nämlich im einfachsten Falle die Stellung der Blätter am Stängel, dann wird man ganz bestimmte, der betreffenden Art zugehörige Zahlenverhältnisse vorfinden. Sammelt man die verschieden gestalteten Vertreter der Pflanzenarten und legt sie nebeneinander, dann wird dabei eine mathematisch gegliederte und genau erfassbare Zahlengesetzmäßigkeit ersichtlich. Einige Pflanzen weisen pro Stengelumlauf 2, andere 3 Blätter auf. Andere zeigen während 2 Stengelumrundungen 5 Blätter, wieder andere während 3 Umrundungen 8 Blätter, usf. In Bruchform ausgedrückt lauten die Zahlenverhältnisse der verschiedenen (häufigeren) Blattzyklen:

½ - 1/3 - 2/5 - 3/8 - 5/13 - 8/21 - 13/34 - 21/55 ...

Die Zahlen des jeweils nächsten Bruches ergeben sich aus der Summe des jeweils vorangehenden Bruches. Jeder Bruch charakterisiert den betreffenden Blattzyklus in dreifacher Weise: Der Zähler gibt die Anzahl der Windungen bzw. Stengelumläufe an, der Nenner die dazugehörige Anzahl der Blätter. Der gesamte Bruch bezeichnet die genaue Größe des Winkelabstandes zweier aufeinanderfolgender Blätter (bezogen auf 360°).

Wenn zB. die Rosengewächse (Rosen, Schlehdorn, Apfelbaum, u.a.) durch 5 Blätter während 2 Stengelrunden signiert sind, dann bedeutet das, dass auf das,erste' Blatt ein zweites nach einem Winkel von 144° folgt.

Bei den meisten Kreuzblütlern findet man 8 Blätter während 3 Stengelrunden. Hier beträgt der Winkelabstand von einem Blatt zum nächsten 135°.

An den Tannenzapfen findet man (als Richtwert) 21 Blattschuppen während 8 Umrundungen. Der Winkel zwischen den Folgeschuppen beträgt damit 137° 9'. Wollte man dieses Reihung endlos fortsetzen, dann würden sich die Winkel immer mehr annähern (auf ungefähr 137° 30' 28" zu), nie aber völlig gleich werden. Dies ist die *Fibonacci*-Reihe, der ,geheime Zahlensinn' der Natur.

Die selben Verhältnisse der,stetigen Teilung' findet man nun auch in der Wirbelbildung des Wassers, in den Knochen der menschlichen und tierischen Körper sowie in den Proportionen der Glieder, Gewebe und Organe. Darüberhinaus findet man diese Verhältnisse in den Abständen, Umlaufbahnen, -zeiten und -rhytmen und diversen Eigenarten der Himmelskörper, insbesondere der Planeten. So etwa bildet die 224,7008 Erdentage dauernde Umlaufzeit der Venus um die Sonne den *Major* (der Teilung im Goldenen Schnitt) der Umlaufzeit der Erde (365, 256 Tage). Das Verhältnis des Sonnenumlaufs der Venus zu dem der Erde kann (abgerundet) durch den Bruch 8/13 genannt werden. Wenn also die Venus nach einem (siderischen) Umlauf wieder an den selben Ort zurückkehrt, dann hat die Erde erst 8/13tel ihrer Bahn hinter sich. Der Planet Merkur bewegt sich in etwa 87,9 Tagen um die Sonne. Das entspricht ziemlich genau dem *Minor* der Umlaufszeit der Venus (85,8 Tage).

Weit wichtiger als der Umlauf der Planeten um die Sonne erschien den Alten aber die Bewegung der Planeten um die Erde aus geozentrischer Sicht, da diese Ansicht für den Menschen und ‚seine' Welt relevanter ist. So beschreibt die Venus während ihres etwa 2-jährigen (584 tägigen) Umlaufes um die Erde jene 5 rückläufigen Schleifen, mit denen die Gestaltbildungskraft ‚Venus' besonders die Rosengewächse signiert. Ein schönes Abbild dieser Kraft-Äußerung zeigt der Apfelquerschnitt.

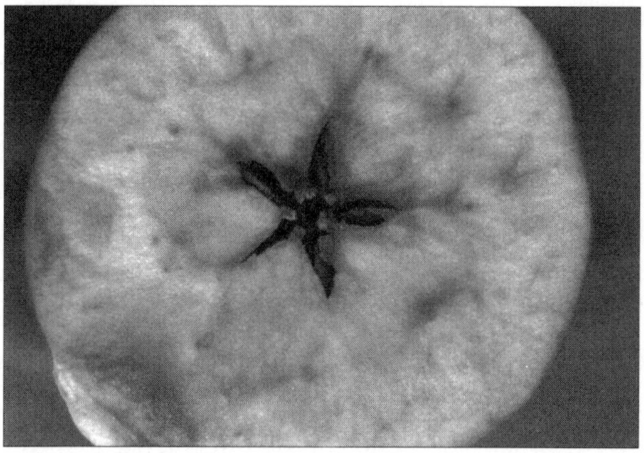

So gleicht die Bewegung der Venus um die Erde einem Pulsieren, welches sich aus Ausdehnung und Zusammenziehung der Entfernungen beider Gestirne ergibt. An den Spitzen des Pentagrammsterns – an der oberen Konjunktion (>äußerer Kreis) – ist die Venus mehr als 6x so weit von der Erde entfernt als an der unteren Konjunktion (Fünfeck innen).

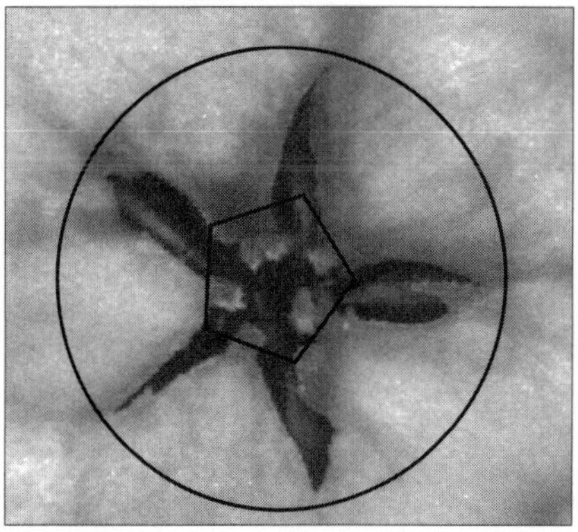

Die Zusammenziehung (zum inneren Fünfeck) begleitet das Aufleuchten der Venus in der Abendsternphase. Die Ausdehnung erfolgt spiegelbildlich in der Morgensternphase.

Während sich die Venus in der Gegend der Sternspitzen bewegt, ist sie unsichtbar.

Für die Bewegungen von Mars um die Erde kann als Näherungswert 3/8 angegeben werden. Die selbe Kraft bringt an den Stengeln, besonders der Kreuzblütler, während 3 Umläufen 8 Blätter hervor und drückt sich auch sonst in den entsprechenden Wirbeln aus. Viele Korb- und Rachenblütler entsprechen dem Umlauf-Puls des Jupiter (5/13). Dies zeigt sich zB. bei der Rosette des Breitwegerichs, beim Löwenzahn und bei der Königskerze im ersten Jahr. Das Verhältnis 8/21 ist ein Näherungswert zum Umlauf-Puls des Saturns. Die selbe Kraft findet Ausdruck in vielen Koniferengewächsen. Bei den Zapfen der Weiß- und Rottanne zählt man etwa 21 Schuppen während 8 Umläufen. Ebenso bei ihren Nadeln und denen der Lärche. Viele Zapfen weisen auch das höhere Verhältnis des nächsten Bruches (13/34) auf. Solch höhere Verhältnisse (13/34, 21/55) findet man selten, besonders aber bei ‚urzeitlichen' Gewächsen, wie etwa den Moosen. Manche Pflanzen beginnen mit einem bestimmten Verhältnis und wechseln dann im Weiterwachsen in ein anderes über. Die ‚schnellsten Planeten' sind Mond und Merkur. ‚Mondenkraft' drückt sich durch einen Zweierrhythmus aus, wie wir ihn im Pflanzenreich zB. bei gegen- und kreuzständigen Blättern finden. Merkur pulsiert mit drei rückläufigen Schleifen um die Erde und geht dabei sechs Konjunktionen mit der Sonne ein. Ihm entsprechen die Dreier- und Sechser-Signaturen zB. der Lauche, Tulpen und Riedgräser.

Solcherart sind Ausdrucksarten und Anwesenheiten der Planetenkräfte, wie sie in der Natur ‚geschaut' werden.

Die Fünf-Elemente-Lehre in chinesischer Medizin und Alchymie

Oftmals wird im Westen die Fünf-Elemente-Lehre der ‚chinesischen Medizin‘ dogmatisch übernommen, aber nicht wirklich nachvollzogen – meist aus dem einfachen Grund, weil man sie als etwas Kulturfremdes empfindet. Auch die Verbindung mit Spagyrik und Alchymie bzw. der hermetischen Naturbetrachtung an sich kann nicht erkannt werden, wenn diese als von zB. der Akupunktur oder Diätik völlig verschiedenartige Modelle empfunden werden. Oft sind die Hürden, welche eine Zusammenschau erschweren, an ‚Kleinigkeiten‘ fixiert, wie an den unterschiedlichen Begriffen für Äther und Luft (‚Holz‘ und ‚Metall‘), an den – bei oberflächlicher Betrachtung weder hier noch da verständlichen Bedeutungen und Entsprechungen von Venus und Merkur oder an der Position des Feuers oder der Erde in den Skizzen zur Darstellung des Elementereigens. Dass aber mit den ‚fünf Elementen‘ zu jeder Zeit und in jeder Region die selben Kräfte gemeint waren, muss nicht sonderlich angemerkt werden.

Organbezogene Funktionskreise (nach der Terminologie der Akupunkturlehre)

Die stofflichen Grundlagen für die jeweiligen Organe werden stets im Mutterelement geschaffen, weswegen man sagen müßte: Das Herz ist das Organ ‚aus dem Äther‘, die Milz ‚aus dem Feuer‘, die Lunge ‚aus der Erde‘, die Nieren ‚aus der Luft‘ und die Leber ‚aus dem Wasser‘. Die Signaturen beweisen den Ursprung: Das Herz als erstes Organ der Wahrnehmung und Manifestation mit seinen vier grobstofflichen Kammern und der fünften ‚verborgenen‘ Kammer, die Milz als Organ der Verwandlung mittels Feuer-Kraft, die Lunge mit ihren Schwerkraft- und Dichtesignaturen, die Nieren als Ausdruck der Bewegung voll verborgener feurig-merkurialer Essenz und die schwammige Leber mit ihrer Durchflutung und Regenerationsfähigkeit.

Das, was aber mit den stofflichen Organen verbunden ist – die ‚Organ-Bereiche', die ‚energetischen Funktionskreise' und das Zusammenwirken im Körper-Geist-Kontinuum – das gehört dann zum Wesen der entsprechenden Elemente, also Äther – Leber, Feuer – Herz, usf., ja es macht gewissermaßen den Hauptteil des ‚Elementes im Körper' aus.

Dem Uhrzeigersinn folgend bringt ein Element das nächste hervor – ‚die Mutter gebiert das Kind' und nährt und durchdringt es. Dies wird Hervorbringungszyklus, Hervorbringungs-Sequenz oder Nährrichtung genannt.

Das jeweils nächste Element überspringend (also von der Großmutter zum Enkelkind) ergibt die Verbindung der Elemente ein Pentagramm. Diese Sequenz hat ‚kontrollierende' Funktion und wird ‚Kontrollzyklus' oder ‚Überwindungssequenz' genannt.

Beide Sequenzen fließen und erfüllen dabei den gesamten Organbereich bzw. das Spektrum eines jeden Elementes. Nehmen wir als Beispiel die Verbindung des Feuer-Elements zum Luft-Element:

Das Beispiel ‚Feuer kontrolliert Luft'

Dem Luft-Element sind u. a. Lunge, Atmung, Haut, Gelenke sowie die Beweglichkeit an sich zugeordnet. Auf Heilmittel und Substanzen bezogen bedeutet das, dass jene Dinge, die dem höheren, äthernahen Bereich des feurigen Spektrums zugeordnet werden müssen (solar, solar-ätherisch / Position 1), auch auf solche Aspekte des Luft-Elements besonders wirken, welche in diesem Element eine ‚höhere', subtilere Position innehaben. Das wären etwa im Feuer die ätherischen Öle und im Luft-Element Lunge, Atmung und Haut (1'). Dem feurigen Spektrum in Richtung Erd-Element folgend werden die entsprechenden Substanzen ‚gröber' (2). Auf ebensolche gröberen Bereiche wirken diese dann im Luft-Element, weswegen Fette, Wachs, Honig und Harze eher dem Bewegungsapparat des Menschen (2') ‚kontrollierend' entsprechen.

Jede Position des sulfurischen Spektrums hat eine besondere Wirkung auf einen entsprechenden Punkt im merkurialen Spektrum

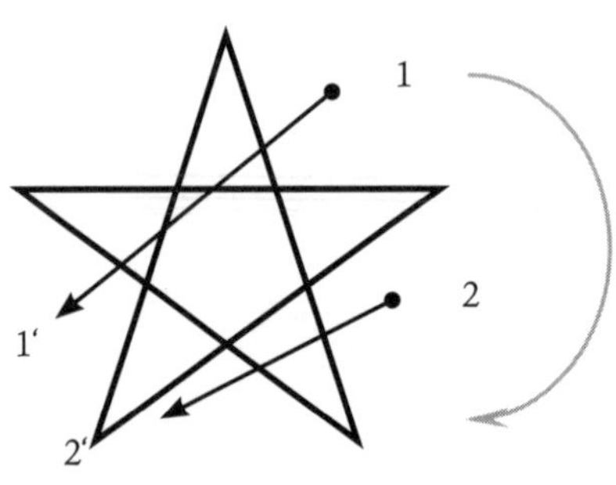

Spektrum des Feuer-Elements bzw. ‚sulfurisches Spektrum'

Von ‚ätherischer Galle' über das ‚Obere Feuer' zum ‚Secundärfeuer'.

Der Hinweis der Alchymisten „nur mit der Kraft des feurigen Mars fängst du den merkurialen Hasen“ bezeichnet die Kontrollposition Feuer – Luft. Das Luft-Element – und oft ‚Merkur‘ im Besonderen – wird mit einem Ei verglichen:

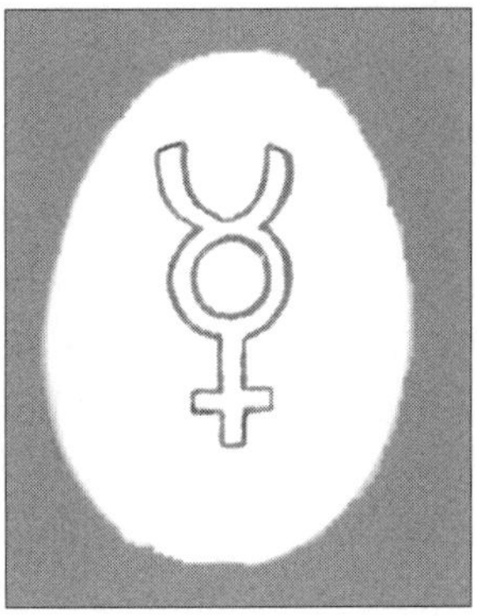

„Das Ei bewahrt Leben und Wesen.“
„Wisse, dass die Luft nichts anderes ist als ein Chaos und dieses Chaos das Klar eines Eies.“
(*Paracelsus*, >*Paragranum*<, 1530)

„Die alten Meister haben Merkur ein Ei genannt.“
(*I. Hollandus*, >Die Hand der Philosophen<, 1746)

„Das Ei der Natur man mich heißt, bekannt den Weisen allermeist, (...) *Mercurius* fein werd ich genannt (...). Ich flieg hinweg, es sei denn, dass man mich gar wohl anbindet (...).“
(*Heinrich Jamsthaler*, >*Viatorum Spagyricum*<, 1625)

Dieses ‚Anbinden‘ weist dann auf die Kontrollfunktion des Feuer-Elements hin:

„Nimm das Ei und schlage es mit einem glühenden Schwert.“
(*Michael Maier*, >*Atalanta fugiens*<, 1618)

In *Stolcenbergs Viridarium chymicum* (1624) fällt der Fuchs (Sulfur/Feuer) über die merkuriale Henne her. (Im selben Werk beteuert er auch, dass *Mercurius* erst dann Früchte tragen kann, wenn er in einem Schmelztiegel zusammen mit Gold erhitzt und fermentiert wird).

Sinngemäß in *Abraham Eleazars Uraltem chymischen Werk*, 1760:

„Der flüchtige Geist muss mit einem goldenen Nagel fixiert werden.“

In der chinesischen Medizin ist einer der wichtigsten Aspekte der Feuer-Luft-Achse die Beziehung des Herzens zur Lunge. Die Funktionen beider Organe bedingen sich in vielem gegenseitig. Sie bilden auch gemeinsam das im Thorax situierte ‚rhythmische System‘.

Besonders die Kontrollsequenz zwischen beiden Organbereichen ist auf mehreren Ebenen bedeutsam. Zum einen ist das Herz unser erstes und direktestes Organ zur Wahrnehmung und Kommunikation, also für Fähigkeiten, die dann sekundär in den Organ- und Funktionsbereichen des Luft-Elements (Lunge, Haut, Gelenke, Beweglichkeiten u. a.) weiterwirken.

Wie schon erwähnt, ist das erste Zusammentreffen von Prä-Sonne und lunarem Äther die Wurzel der ‚Wahrnehmung‘ im Menschen und führt zur Herz-Bildung. Auf ähnliche Weise wiederholt sich dieses Ereignis durch das Zusammentreffen von Mars und Venus im Erd-Element, wo es wiederum eine ‚Geburt Merkurs‘ verursacht, dh. eine (nun grobstofflicher orientierte) Kommunikationsebene ermöglicht – eben die des Erfühlens mittels der Organe des Luft-Elements, des Luftkörpers an sich und des Ausdrucks durch die Stimme. Darüberhinaus erweist sich auch die Haut durchaus informativ. Selbst die (merkurialen) Formen der Körper kommunizieren als ‚Ausdrucksformen‘ über das Luft-Element.

Herz-Kommunikation entspringt einem ‚Zusammentreffen‘, einer ‚Verbindung‘ der Pole Feuer und Wasser bzw. Sonne und Mond (im weitesten Sinne). Luft-Kommunikation entspringt einem Zusammentreffen der selben Pole – nur auf einer anderen, mehr an das grobstoffliche Geschehen gebundenen Ebene – als ‚Mars und Venus‘. Die Herz-Kommunikation steht (im Zyklus der Elemente-Genese) der Luft-Kommunikation vor und kontrolliert deren Art – „Wessen das Herz voll ist, geht der Mund über.“

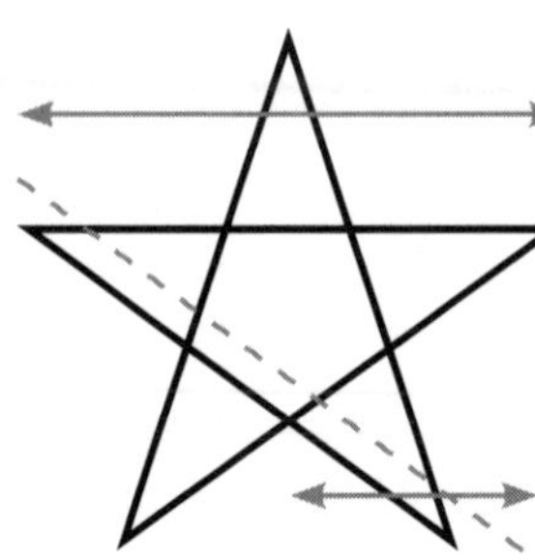

Zum andern äußert sich dieser Umstand in der Beziehung von Blut und *Ki* (ein Terminus der chinesischen Medizin, welcher eine Art der Körperenergie bezeichnet). Ein bedeutender Ort für das Zusammentreffen von Feuer- und Luftwirkung im Körper sind die Blutbahnen, weswegen ihre Art, ‚Austausch zu halten' und Informationen zu übermitteln, dem Luft-Element zugeordnet wird. Das gleiche gilt für das Blut selbst. Das *Ki*, die ‚Energie', muss sich auf die Blutgefäße konzentrieren, damit es sich durch den ganzen Körper bewegen kann. Dieses Konzentrieren ist ein Ausdruck der Kontrollsequenz Feuer-Luft in der Art einer relativen Fixierung.

Ein anderer Ausdruck ist der, dass die Lunge vom Blut des Herzens mit Nährstoffen versorgt wird. Auf diese Weise nährt das Erd-Element (dem diese Nährung aufgrund seiner ‚Mutter-Rolle' zum Luft-Element auch obliegt) indirekt über das Feuer (‚über die Großmutter') die Luft. Man kann das bildlich so verstehen: aus dem Erd-Element kommen die Nahrungsmittel. Diese werden zuerst dem Herz ‚gezeigt' (vom Herz ‚wahrgenommen'), und das, was das Herz auswählt, wird an die Lunge weitergegeben.

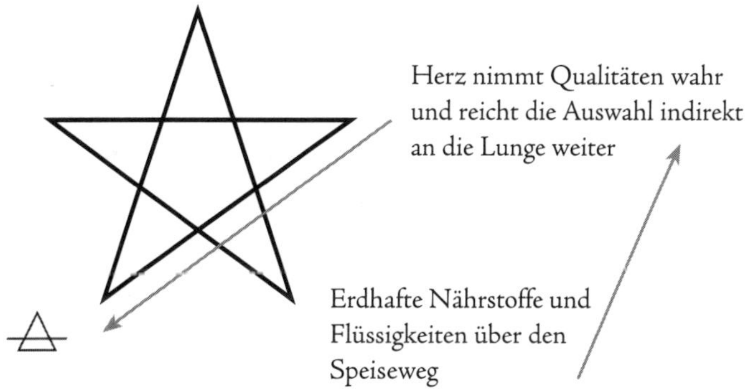

Die Beziehung Erde und Luft beinhaltet starke Gegensätze. Die Luft entflieht der Erde bzw. ‚Merkur bricht aus dem Kerker Saturns aus.' Das Luftige will nicht mehr zurück in die starre Dichte, meidet diese womöglich, wie der Wind dem Felsen ausweicht. Dennoch ist Luft immer erdgebunden – nie ‚ohne Erde'. Wie *Paracelsus* andeutet: Alle Elemente kannst du scheiden, aber die Luft wirst du niemals los. Das ist im Luft-Element eine Anwesenheit der Mutter, des Erd-Elements. ‚Merkur' (die Bezeichnung der ersten, zusammen mit dem Luft-Element aus dem Erd-Element geborenen Planetenkraft) ist der Vermittler, der Händler, der Begleiter. Er ist sich selbst genug, aber niemals wirklich alleine.

Diese Tendenz zum Entfliehen vom Erdhaften kontrolliert nun das Feuer – hier im Beispiel das Herz – durch Übernahme der Nährung. Mittels der Nährung kann der Grad der Erdhaftung bestimmt werden. Einfach ausgedrückt: Ein allzu leichter Luftikus kann sich zum Ausgleich von Wurzelgemüse ernähren, ein allzu starrer, verschlossener Griesgram, melancholisch und von Erdenschwere gebeugt, sollte leichte Nahrung zu sich nehmen und geeignete Würze, um sein Leben wieder mit Bewegung und Rhythmus zu bereichern.

Dasselbe tut das Blut mit der Lunge – ganz direkt verbunden mit der Dominanz der ersten Kommunikation (Herz) über die zweite (Luft). Das Herz fühlt immer sofort und reagiert spontan. Die Sprache kann die selbe Sache, die das Herz so erzählt, völlig anders erzählen. Die Bemerkung eines alten Weisen hierzu war: „Kaum spreche ich es aus, ist es schon verdreht." Wenn das Herz kontrollierend wirkt, findet der Mensch die richtigen Worte. Dennoch werden dabei diese ‚richtigen Worte' sofort dem ‚schalkhaften Merkur des Luft-Elements' übergeben.

Das bezieht sich nun nicht nur auf die grobstoffliche Ernährung der Lunge, sondern (nach dem Satz: ‚Herz regiert Blut – Lunge regiert *Ki*') auch auf das *Ki*. Dieses haftet aus mehreren Gründen besonders an den Blutgefäßen (obwohl kein Ort des Körpers ohne es ist):

Ein Grund ist, dass es darin ‚untertaucht' und daraus seine Nährung erhält. Dies entspricht dem doppelten Wesen der ‚Energie': Zum einen existiert sie aus einem initiatischen Impuls, welcher ‚am Anfang' gesetzt wurde, und zum andern exisitiert sie aus einer ständigen Nährung ‚aus dem Feinsten' – hier aus dem Blut. Es heißt: Das Blut kontrolliert das Untertauchen. So wird nicht nur die Nährung stofflicher Gewebe (des Luft-Elements, insbesondere von Lunge, Haut und Blutbahnen) vom Herzen kontrolliert, sondern auch die Nährung subtiler Gewebe und Körperenergien, welche dem Luft-Element entsprechen. Und immer hat dieses Ernähren auf den luftigen Merkur eine mehr oder weniger fixierende, erdbindende Auswirkung. Das ist gemeint mit dem obigen: „Der flüchtige Geist muss mit einem goldenen Nagel fixiert werden," bzw. „Feuer kontrolliert Luft." Es hat vielerlei Bedeutungen auf mehreren Ebenen – in den Geweben des Körpers sowie in der Retorte des Alchymisten.

Nährrichtung Feuer-Erde und Kontrollsequenz Äther-Erde

Stolcius v. Stolcenberg bezeichnet im *Viridarium chymicum* (1624) das Zusammenwirken der Nährrichtung Feuer-Erde mit der Kontrollsequenz Äther-Erde:

> „Der Acker benötigt Feuchtigkeit zur Faulung und Sonnenwärme zur Wachsung und Zeitigung. Solches erfordert jede Saat."

Die Befeuchtung des Erd-Elements geschieht aus hermetischer Sicht – wie bereits ausgeführt – durch die Verbindung von lunarem Äther mit saturnischer Erde. Mond und Saturn stellen dabei eine Polarität dar, die zB. in der sogenannten ‚Planetenspirale' die äußerste Sequenz bilden.

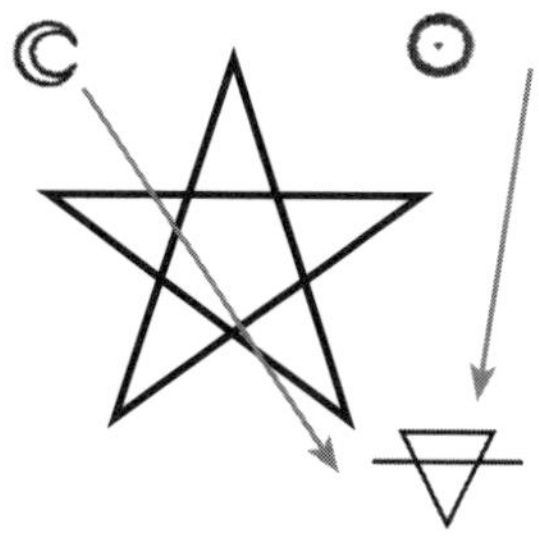

Auf die selben Elementarpositionen bezieht sich eine Anleitung zum Großen Werk der Alchymie in der *Aurora consurgens* (14. Jhdt.) :

„Säet das Gold in die weiß geblätterte Erde, ... so tingiert sie das Elixir."

Anfangs wird der Nährrichtung Feuer-Erde entsprochen, danach der Kontrollsequenz Äther-Erde. Letzteres kann so verstanden werden, dass das ‚Elixir' – also die Quintessenz – aufgrund der Kontrollposition des Äthers (ebenfalls Quintessenz) in die Erde gebracht bzw. in ihr eingeschlossen ist. Dies ist eine der bedeutsamsten Funktionen der Äther-Erde-Achse, nämlich dass (auf die entsprechenden Planetenkräfte umgedeutet) Jupiter dafür sorgt, dass nichts vollständig dem Saturnzustand anheim fällt. Äther ‚kontrolliert' Erde heißt auch: Er durchdringt sie quintessentiell. Und er reguliert einerseits das übermäßig Fixe, andererseits das, was zu wenig fix ist.

In der Alchymie entspricht die erwähnte lunar-ätherische Position den höchsten Arten des Merkur-Prinzips. Es sind also hierfür die Begriffe Äther, Mond oder auch Merkur bezeichnend. Das mag vorerst ein wenig verwirrend erscheinen, da ja der Äther selbst auch ein eigenständiges Element und nicht direkt aus den Planetenkräften Mond oder gar Merkur gebildet ist. Die Planetenkraft des Mondes entspricht in einer anderen Perspektive auch dem ‚hohen' Wasser-Element und Merkur dem Luft-Element. Insgesamt geht es dabei aber um das Merkur-Prinzip an sich, jenes merkuriale Spektrum, welches im Erd-Element einen seiner Anfänge nimmt und in den lunaren Aspekt des Äther-Elements hineinreicht.

Die aus dem Feuer geborene Erde ist von durchaus trockener Qualität, weswegen sie als ‚durstig' und ‚voller Bedürfnis' bezeichnet wird. Hier ereignet sich nun die ‚Befeuchtung', das heißt: Mondenkraft wirkt auf das saturnische Erd-Element ein. Ein praktisches Beispiel dafür ist die Anweisung, ein ‚trockenes, mittels der Kraft des Feuers dargestelltes Erdsalz' an einen feuchten Ort zu setzen, ‚bis dass es zerfließt'.

„Dann nimm dein feuriges Salz und lass es in einem feuchten Keller zu einem *Liquor* fließen. Diesen filtriere ab, coaguliere und lass es erneut fließen, solange, bis es ganz hell und rein ist wie ein Kristall."

Diese Mondenkraft ist tatsächlich als Kraft vorzustellen, wird aber ebenso mit einer äußerst subtilen Feuchtigkeit assoziiert. Das wiederum befindet sich in der Nähe des Begriffes jenes ‚hohen Merkurs', welcher zum Teil in den Wurzeln des Wasser-Elements gründet und zum Teil Anteil hat am Äther.

NB: Die aus dem Feuer geborene saturnische Erde wird durch die Verbindung mit dem Mond zur befeuchteten venerischen Erde. Die Verbindung jenes ‚hohen, äthernahen Wassers' mit Erde (im weitesten Sinne) ist auch eine Grundlage zur Bildung des eigentlichen Wasser-Elements im Sinne von venerischem Wasser.

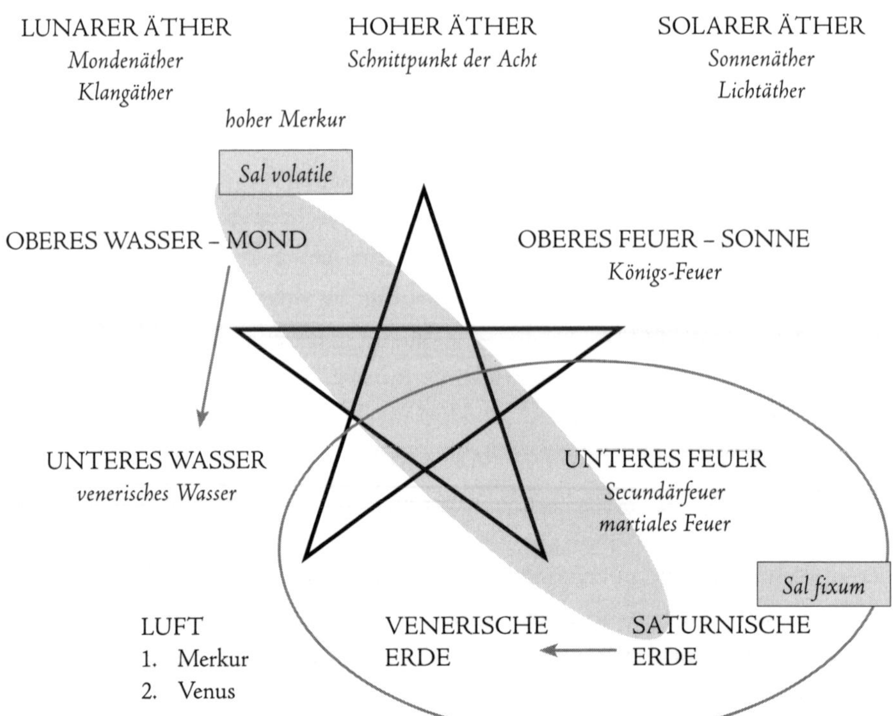

Nach diesen kurzen Übungen in ‚hermetischem Denken' wollen wir eintreten in das chymische Gärtlein und einen Blick in die Praxis von Spagyrik und Alchymie werfen.

Zur Idee der spagyrischen Präparation

Pflanzen, Minerale, Metalle, wie wir sie kennen und angreifen, sind grobstoffliche Ausdrucksformen göttlicher Idee. Die jeweilige Idee wird durch mehrere Ebenen von verschiedenem Geist-Stoff-Verhältnis in die Materialität ‚hinein' verdichtet. Die eigentlich als ‚feinstofflich' bezeichneten Ebenen sind für die spagyrische Laborpraxis nicht wesentlich relevant. Schließlich hält man ja ein ‚natürliches Ding' in den Händen, das in genau dieser Form in der grobstofflichsten Ebene, in der Physis, lebt und den Werdegang der Verdichtung zur Manifestation bereits zum größten Teil hinter sich hat. Man will dieses ‚Ding' nun weder in eine reine Kraft noch in ein völlig ‚feinstoffliches', für die groben Sinne unerfahrbares ‚Ding' zurückführen und ebensowenig ließe sich die Idee selbst in einem Glase einfangen. Die
Kräfte, welche zur Bildung des ‚Dinges' geführt haben, sind aber nicht aus dem feinstofflichen Zustand in die Grobstofflichkeit ‚gesprungen', sondern haben sich mehr oder weniger gemächlich vermischt und verdichtet.

Der Spagyriker löst nun ebenso (mehr oder weniger) gemächlich diese Vermischungen und Verdichtungen, um eine Ebene der Manifestation der Idee zu erreichen, welche zwar noch physisch fassbar ist, dennoch aber weitestgehend von ebendiesen Vermischungen und Verdichtungen befreit.

So ist also hier jene Phase der Verdichtung interessant, in welcher das ‚natürliche Ding' in der Physis ‚ankommt'. Diese Phase entspricht dem Eintritt in die Welt der 5 Elemente. Die Idee tritt durch das Äther-Element aus dem Feinstofflichen heraus, in die Physis hinein und wird deshalb dort als ‚Quintessenz' bezeichnet.

Wenn die alchymistische Zerstörung und Scheidung dermaßen kunstvoll gelingt, dass man dieses Ebene erreicht, dann bezeichnet man das Endprodukt als Quintessenz, wenngleich es, wie gesagt, nicht ‚reine Kraft' ist, sondern zumeist in fluidaler, aber auch in fester, fixer Form, sichtbar und fassbar dargestellt wird.

Bei der Herstellung spagyrischer/alchymistischer Heilmittel geht es nicht darum, sogenannte ‚Wirkstoffe' vom Ausgangsmaterial abzutrennen. Das Ziel der Scheidungen (Separationen) und Reinigungen (Rectificationen, Purificationen) ist, die Ausdrucksformen der im Ausgangsmaterial vermischt wirkenden Naturkräfte, Elemente oder Prinzipien möglichst ‚rein', dh. dem Charakter der Kraft nächststehend, darzustellen. Sei dies nun – je nach ‚Kunst' und Absicht – eine tatsächliche ‚Quintessenz' oder ‚nur' eine ‚Essenz' oder ‚einfach' eine ‚Tinctur' oder sonst ein Präparat, das durch Scheiden, Reinigen und Wiedervereinigen gewonnen wird. So liegt der Grund und Sinn dieser Kunst im Zuwegebringen eines entsprechenden ‚philosophischen Todes', mit dem die Scheidungen einhergehen, in der Reinigung des Geschiedenen und in der Vollendung des Werkes durch eine ‚stimmige' Vereinigung, welche ein auf höherer Ebene lebendiges Produkt gewährt.

Da man nun keine ‚Wirkstoffe' extrahiert, sondern prinzipielle oder elementare Ausdrucksformen der Kräfte darstellt, ist es nicht von Bedeutung, zB. Mineralstoffe in der Asche, Proteine durch schonende Vakuumdestillation oder Vitamine durch entsprechende Verfahren zu erhalten. Wenn der stofforientierte Forscher sagt, die ‚Wirkung' käme von diesen Dingen, dann ergänzt der Alchymist sein Naturverständnis mit den Worten: „Die Dinge, welche wirken können, sind an sich schon die Auswirkungen in ihrem Hintergrunde wirkender Kräfte."

Würde man sagen: „Das Kupfermetall bewirkt, dass ein Übermaß reduziert und eine Harmonie hergestellt wird", dann müsste man auch sagen, dass die Venus vom Himmel herabkommt und die Rose formt, wie ein Töpfer die Vase. In jedem Falle aber ist es die Kraft und zwar die selbe, welche die Rose (dominierend) formt, das Übermaß reduziert und das Kupfermetall hervorbringt. Die Auswirkung von Atomen, Molekülen und Verbindungen ist bereits eine Auswirkung von Auswirkungen der zu diesen stofflichen Phänomenen führenden Kräfte.

Die spagyrisch-alchymistischen Arbeitsgänge werden von tiefer Naturweisheit geleitet und nicht von rein stofforientiertem, ‚chemischem' Denken. Das sind zwei grundverschiedene Ansätze des Begreifens und Praktizierens, die nur bisweilen Berührungspunkte haben, weit weniger noch, als der Alchymist etwa (paracelsisch formuliert) die Wissenschaft des Bauern, des Bäckers, des Webers oder des Winzers berührt – Kundige, für die ‚chemisches Denken' noch nie wirklich notwendig gewesen wäre, was ein Vergleich der ‚Kunst der Alten' mit der heutigen Praxis erweist.

„Die rechte, edle *Alchymia* ist eine Kunst, die natürlichen Körper in diejenigen Teile, aus welchen sie natürlicherweise zusammengesetzt sind, zu solvieren, zu resolvieren und die resolvierten rein und stark zu machen, dass sie dem *Medico* nützliche *Remedia* gebe, oder zu den *Perfectionibus* und *Transmutationibus* der Metallen dienen. Wer da sagt, die Alchymie sei eine Kunst, das Reine vom Unreinen zu scheiden, der redet gar zu oberflächlich. Denn wenn einer die Wurzeln von der Erde reinigt, sie wäscht und den ausgepreßten Saft im Sieden abschäumt, der ist deswegen nicht gleich ein Alchymist. Obwohl ein Alchymist solche Arbeiten auch nicht ausläßt, endet doch sein Fleiß nicht in denselben gemeinen *Operationibus*. Er arbeitet weiter und scheidet nicht nur die äußerlichen Unreinigkeiten ab, sondern dringt auch zu den innerlichen vor und dissolviert die natürlichen zusammengesetzten Körper in diejenigen, daraus sie gemacht sind, scheidet ihre Teile, sondert das Nützliche vom Unnützen ab und macht alles ganz und gar kräftig. Dann komponiert er die selben wenn es nötig ist.

Deshalb scheint es, als hätte die Alchymie weder mit den alten noch mit den neuen dogmatischen Präparationen eine Gemeinschaft. Auch wenn etliche (>*Avicenna, Hippokrates, Dioskurides*) von den Griechen oder den Arabern unterwiesen wurden, dann waren das doch nicht alles Alchymisten.

(nach: *Michael Crügner*, >Chymischer Frühling<, 17. Jhdt.)

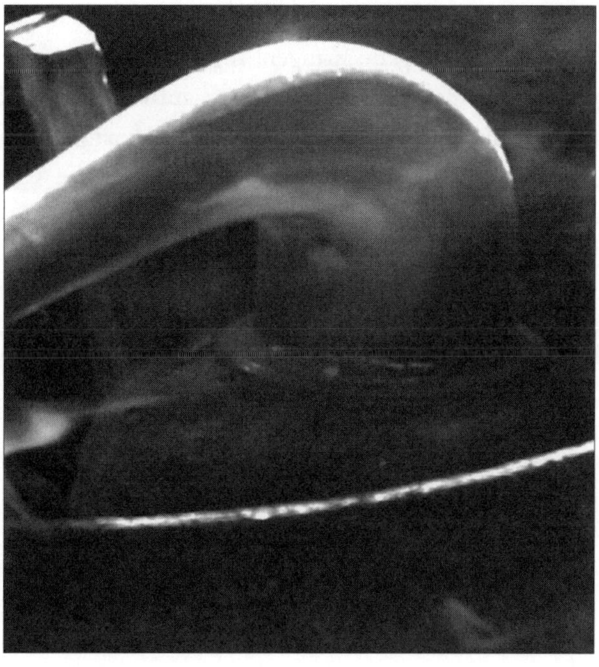

Wir haben bereits darüber gesprochen, dass Alchymie keine Vorform dessen ist, was als Chemie oder Chemo-Pharmazie bezeichnet wird. Genausowenig waren die Alchymisten und Iatrochymiker der Vergangenheit ‚Vorväter' der Chemotherapie. Das Studium der paracelsischen und anderer Schriften stellt dieses Missverständnis klar. Sowohl Denk- als auch Verfahrensweisen der Alchymie unterscheiden sich in vielem grundsätzlich von Ansicht und Arbeitsweise der heutigen Chemie. Alleine das zur laborantischen Praxis nötige ‚Kochgeschirr' mag bisweilen ein ähnliches sein, wobei aber auch hier die ‚Alten' großteils über ein weitaus gediegeneres Handwerkzeug verfügten als die heutigen Spagyriker. Die Rezepturen für gläserne und keramische Gefäße waren den Ansprüchen der Kunst angepasst, ebenso ihre Formen. Wahrscheinlich reduziert sich der Nutzen, den zeitgenössische Alchymisten diesbezüglich vom technischen Fortschritt haben, alleine auf die Verwendung elektrischer Öfen und Waagen (zugegebenermaßen ein gewaltiger Nutzen), feuerfester Handschuhe, vielleicht auch noch auf die Schliff-Verbindungen der Gläser und einige wenige solcher Dinge. Es sei denn, jemand frönt der Begeisterung für artifikale Laborglasgeräte, ist angetan von der *Industrial-Art* der Kühlschlangen, Rührmechanismen und Soxlet-Extraktoren – das ist dann eben eine persönliche Art, das Gesamtkunstwerk zu gestalten. Schon *Agricola* schrieb, er müsse lachen, was die Laboranten seiner Zeit an allerlei Gerätschaft ersonnen haben „und damit nicht annähernd das vollbringen, was die Alten mit nur einem einzigen Ofen vermochten."

Wieder hilft uns hier die ayurvedische Alchymie, ‚auf den Boden der Realität' zu kommen und zeigt uns, dass von den einfachsten Pflanzenpräparaten bis zum Großen Werk ein guter Ofen, einige Tiegeln, Zangen, Ingusse, Schalen und Retorten genug der Ausstattung abgeben. Möglichst viel teures und aufwendiges Gerät zu verwenden bedeutet nicht, zugleich auch ‚kunstvoller' zu arbeiten. Oft ist genau das Gegenteil der Fall.

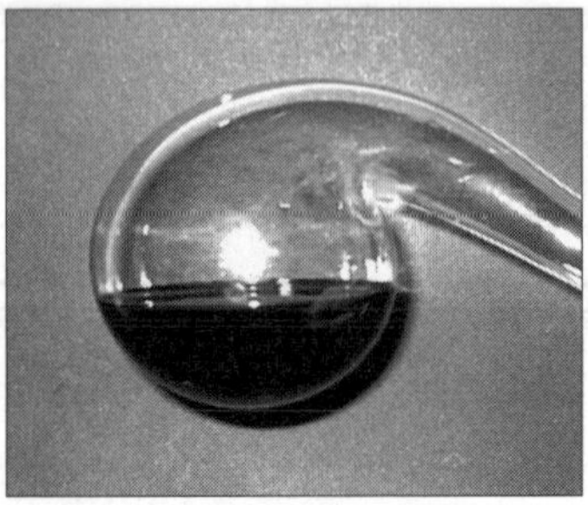

Retorte

Scheidung – Reinigung – Vereinigung

Durch diese *Separatio elementorum* sollen die Kräfte so weit als möglich von ihren materiellen (insbes. wässrigen und erdhaften) Hüllen befreit werden, damit sie dementsprechend ihre Wirkung entfalten können.

Thurneisser beschwört schon Mitte 1500 diejenigen, die sich Paracelsisten nennen, dass sie genau diese Sache nicht mehr verstehen und ihre Scheidungen und Reinigungen nicht danach richten würden. Durch zu wenige, zu ungenaue oder nur oberflächliche Rectificationen, Sublimationen u. dgl. lassen sie „das *Phlegma* beim Geist und die Schlacken bei den *Flores*." Er betont, dass derjenige, welcher nicht imstande ist, diese *Virtutes* (Kräfte, Tugenden) wahrhaft zu extrahieren, „kein rechter *Medicus*" ist, während des *Paracelsi* Kunst und Erfolg darin bestanden hatte, die Stoffe „ohne einige andre Substanz, rein, durchleuchtet, klar und ganz subtil" zu machen.

„Alles was man tut soll *in Summa* dermaßen zugerichtet werden, dass beim *Spiritus* kein *Phlegma*, bei den Ölen keine Wässrigkeit, bei *Turbit Flores* keine Irdischheit und auch bei den Tincturen nichts anderes ist."

„Weil denn die Arznei von Gott aus der Erden dem Menschen zugute erschaffen worden ist, so meine ich, dass die Alten, diese Dinge beschreibend, nicht die Materie an ihr selbst, sondern allein die Kraft daraus gemeint und verstanden haben."

nach: *Leonhard Thurneisser zum Thurn, >Quinta Essentia<*

In der Alchymie spielen die ‚stoffbildenden Kräfte' eine wesentliche Rolle, deren Kenntnis auf zeitlos gültigem Wissen fundiert, während die Chemie rein stofforientiert und deshalb eher spekulativ ist, weil sie die tieferen natürlichen Zusammenhänge nicht beachtet (obwohl ein Kräftewirken in völlig stofffreiem Raum ‚nachgewiesen' werden konnte). *Paracelsus* nennt die Alchymie „den rechten Umgang mit den Dingen der Natur", und damit ist auch schon der Unterschied zum ‚Zeitphänomen Chemie' erklärt. Gelegentlich wird noch das Bild eines angeblich ‚typischen Alchemisten' als primitiver Vor-Chemiker hingestellt, der in seinem mittelalterlichen Laboratorium verzweifelt, weil aussichtslos, den Stein der Weisen zu erringen sucht, wobei er zumindest einige nützliche Dinge wie Porzellan, Säuren oder Schießpulver entdeckte, was ihn zum Pionier einer chemischen Industrie machte, die es dann schaffte, Mensch und Natur an den Rand eines schrecklichen Abgrundes zu führen. Glücklicherweise hat dieses Bild des Alchymisten nichts mit der Realität zu tun, wiewohl es auch schon im Mittelalter ‚Sudelköche' gegeben hat, deren einziges Ziel es war, durch ihre Giftbrühen und damit verbundenen Erfolgs- und Heilsversprechen steinreich und weltmächtig zu werden. Für Ihresgleichen war aber damals die Zeit noch nicht reif genug.

„In unserer Zeit wissen die Philosophen weniger denn nichts, aber die Naturkundigen und Alchymisten haben sich darinnen geübt, damit sie der Natur Geheimnisse in dieser untersten Welt desto besser lernen möchten.

So wie die wahre *Astrologia* bei uns verloschen ist, so sind jetzt auch nur mehr wenige, die der Alchymie obliegen, die der Natur Geheimnisse erkunden, ihre Kräfte und Wirkungen suchen und ihnen nachforschen:

Es tun zwar auch zu dieser Zeit Etliche solches, doch bei denen geschieht es nicht in der Meinung, dass sie vornehmlich der Natur Geheimnisse erlernen wollen zu dem Zwecke ihrem Nächsten die Gesundheit zu erhalten und die Krankheit abzuwenden und ihm damit zu dienen, sondern ihr Trachten gehet nur darauf, dass sie mögen Reichtum dadurch erlangen und sie damit ihre Begierde und Wollust sättigen. Deshalb erlangen denn auch wenige das gewünschte Ende.

Diese Kunst und Philosophie sieht nicht allein das an, was äußerlich durch die fünf Sinne kann erforscht werden, so wie es bei fast allen anderen Künsten der Fall ist. Diese gehen nur auf die äußerliche Form und Natur, und wenn die zuwege gebracht, dann haben sie ihrer Sache genug getan. Wenn ein Schuster das Leder also richtet, dass ein Schuh oder Stiefel daraus wird, so hat er das Seine vollbracht.

Sondern die wahre Alchymey und Philosophey, welche auch die wahre Medicin einschließt, will geistige Augen haben, die in die Natur hineinsehen können und nicht allein das Irdische beschauen, sondern auch das, was in jedem Dinge verborgen ist und geistig ist.

Solche vortrefflichen Philosophen waren *Hermes, Democritus, Pythagoras* und *Plato,* (...) welche gesehen, dass die Natur allen Dingen eine verborgene und geistige Gewalt, Samen, Kraft eingepflanzt und eingegossen, so daß, wenn die selben von der irdischen Körperlichkeit und Verhinderung darin sie verwickelt und gefangen, losgesprochen und erlöset, dass sie dann von wunderbarlicher Wirkung sind, so den Unverständigen unglaublich (...) vorkommt. Denn solche Philosophen haben erkannt, dass viele Dinge in der Natur geschehen, die wir mit den körperlichen Augen nicht sehen und aus dieser Sicht nicht fassen oder ergründen können. Weil sie sich aber mit Ausflüchten nicht zufrieden geben wollten, haben sie demselben nachgedacht (...)."

nach: *Joachim Tanckius, >Promtuarium Alchemiae<*, 1610, Graz 1976

DIE SPAGYRISCHE PRAXIS

Die naturheilkundliche Praxis ist mit einfachen Worten umrissen: Befundung, manuelle und/oder substanzbezogene Therapie und Lebenskonzeption. Die spagyrisch-laborantische Praxis ist (puncto Heilkunst) am engsten mit der substanzbezogenen Therapie verbunden. In der manuellen Therapie tritt sie begleitend auf, in Form von entsprechend präparierten Salben, Pflastern, Balsamen und dergleichen. Pflanzliche und mineralische Essenzen werden sowohl innerlich als auch äußerlich (zu Einreibungen, Bädern u. ä.) verwendet.

Innerhalb der substanzbezogenen Therapie nimmt die Spagyrik als Heilmittelkunde eine zentrale Position in der Heilkunst ein. Die Ernährungskunde wird von der spagyrisch-laborantischen Praxis am Rande berührt, vom hermetischen Naturverständnis aber oft regelrecht getragen. Man denke dabei an die Heil-Ernährungssysteme, welche sich an der Elemente- oder Säftelehre orientieren. Das selbe Verständnis ist die Grundlage der Heilmittelpräparation.

„Die *Medicina* wirket nicht, so die *Alchymia* nicht bei ihr steht. Keines sich vom andern teilen lässt."

„Weil aber eine Essenz von solcher Subtilität, Kraft und Wirkung nicht ohne die Handreichung und den treuen Beistand der Kunst *Alchymia* kann ausgezogen und die Geister von den Körpern nicht können getrennt werden, genauso die Elemente nicht geschieden werden, so ist augenscheinlich, dass diese beiden Künste, *Medicina* und *Alchymia*, unzerteilt und beieinander sein müssen."

nach: *Leonhard Thurneisser, >Quinta Essentia<*, 1570

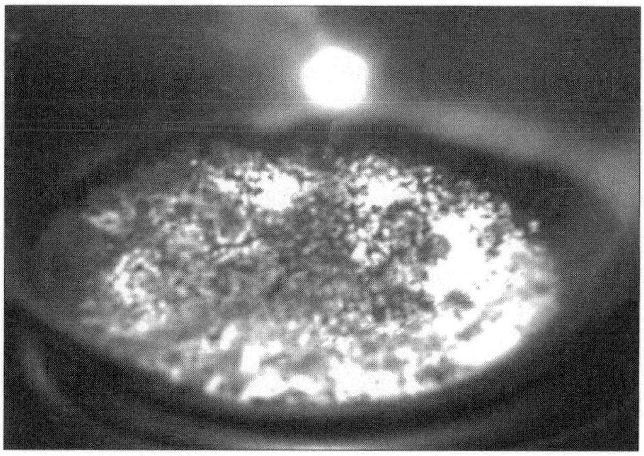

Mittels spagyrischer/alchymistischer Arbeitsgänge werden (wie bereits erwähnt) die subtileren Ausdrucksformen der Kräfte, welche in einem natürlichen Ding wirken, von ihrer grobstofflichen Erscheinungsform geschieden, gereinigt und wieder zusammengefügt. Von *Phlegma* und Schlacken befreit, erlangt man jene Teile oder Fraktionen, die gemeinhin als Sulfur, Merkur und Sal, oder (nach der Scheidung der Elemente) als Erde, Feuer, Wasser und Luft bezeichnet werden. Das sind nicht die philosophischen Prinzipien oder Elemente selbst, sondern eben deren typische, in den Dingen verborgene Ausdrucksformen wie Wässrigkeit, Dampf, Feuer, Rauch, Öl, Salz, etc. Im selben Verhältnis wie die Kräfte oder Prinzipien in der wachsenden Pflanze wirken, gehen sie in subtiler Form in die Arznei ein. Nach *Thurneisser* entstehen spagyrische Arzneimittel „durch Reduction oder Wiederbringung eines jeden Dinges in seine ‚Erste Gestalt' oder in sein ‚Erstes Wesen', ohne Mangel oder Ausbleibung einiges (von dem Ding), das vorhin dabei gewesen sei, doch nunmehr geistlich zu verstehen."

Arten der Scheidung

Es können durch die Kunst die 4 dichteren Elemente geschieden werden:

ERDE, WASSER, FEUER, LUFT

Und es kann die QUINTESSENZ, das fünfte Element (das am wenigsten dichte), generiert, geschieden und in stofflicher Form erlangt werden.
Man scheidet auch nach den philosophischen Prinzipien, nämlich:

SULFUR *(Anima)*, MERKUR *(Spiritus)*, SAL *(Corpore)*

So etwa kann zur Darstellung des Merkur-Prinzips aus einer Pflanze mittels Gärung, Destillation und Rectification der Alkohol als Vertreter des Merkur-Prinzips im Pflanzenreich erlangt werden. Aber auch andere Arten ‚merkurialer Geister' *(Spiriti)* sind bekannt.

Das Sulfur-Prinzip wird als (ätherisches) Öl, als aromatisches Hydrolat, als Farbe und/oder als Salz *(Sal sulfuris)* dargestellt. Aber auch andere ‚*Sulphuri*' sind bekannt.

Die Ausdrucksform des Sal-Prinzips wird aus der calcinierten Pflanzenasche meist kristallisch als (wasserlösliches) Salz erlangt. Salze können typisch erdhaft und fix, aber auch subtil und volatil sein. Letzteres weist über den salinischen hinaus auf einen merkurialen Charakter hin.

PRINZIPIEN	AUSDRUCKSFORMEN	ELEMENTE
Sulfur	Ätherisches Öl *Öl, Farbe, Geruch* Sal Sulphuris Hydrolat	Äther *Feuer* Erde Wasser
Merkur	Subtilität, Sal Volatile *Spiritus, Alkohol* Wässrigkeit	Äther *Luft* Wasser
	Phlegma	
Sal	*Salz* Feces, Caput Mortuum	Feuer *Erde*

DIE PRINZIPIEN IN KÖRPER UND GEIST:	DIE PRINZIPIEN IN DEN PFLANZEN:
SULFUR Individualität, Persönlichkeit, Durchsetzungsvermögen, Manifestation, Abwehrkräfte, Darstellungsdynamik, Unterscheidungskraft, Konzentration, Farben, Herz, Blut, Augen, Gesicht, Fortpflanzungsorgane	SULFUR Farbe, Geruch, Geschmack, Öligkeit, Harzigkeit, Zähigkeit, Persönlichkeit, anspruchsvoll, Blüte, Stamm, Geschlechtsteile, Dornen, Ähren, Reichhaltigkeit
MERKUR Kommunikation, Austausch, Wahrnehmung und Reaktion, Beweglichkeit, Veränderlichkeit, Lunge, Atmung, Haut, Blutgefäße, Nerven, Gelenke, Dickdarm, Haare, Stoffwechsel allg.	MERKUR Auffälligkeiten und Merkwürdigkeiten, Wachstumsschübe, Beweglichkeit, Stoffwechsel, Wässrigkeit, Durst, Veränderlichkeit, Beeindruckbarkeit, Sensibilität, Blätter, Ranken
SAL Festigkeit, Strukturkraft, Erinnern und Bewahren, Eindruck, Ernsthaftigkeit, Kargheit, Strenge, Knochen, Knorpeln, Zähne, Milz, Magen, Schädelknochen	SAL Festigkeit, Strukturkraft, Kargheit, reduzierte Formen und Vorgänge, Langlebigkeit, abweisende Art, unangenehmer Geruch oder kein Geruch, Salzigkeit, Erdnähe, Minerale, Anspruchslosigkeit, Trockenheit, Dürre, Verholzen, Schuppenbildung, Wurzel, Holz, Samenkapseln

Spagyrische und andere Präparate

In der spagyrischen Pflanzenessenz sind die (subtilen Ausdrucksformen der) drei philosophischen Prinzipien geschieden, gereinigt und wieder vereint. Die Pflanze gilt dann als ‚essentiell wiedergeboren' und wirkt nun durch die bildenden Kräfte und nicht mehr aufgrund ihrer in der Erscheinungsform vermischten Substanz. Weil zu diesem Ergebnis weder ‚Wirkstoffe' extrahiert noch wesentliche Bestandteile der Pflanze verworfen werden, stellt die spagyrische Essenz ein ‚ganzheitliches' Produkt im eigentlichen Sinne dar.

Im Mineralreich sind spagyrische Aufschließungsprozesse notwendig, um das Mineral oder Metall für den Körper absorbierbar zu machen und von Giftigkeiten (trad.: „von giftiger / schädlicher / böser Bergart") zu scheiden. Auch hier können – in ‚höheren' Arbeiten – die Prinzipien für sich dargestellt werden.

> „Denn man bringt aus dem Gold ein Öl, eine Essenz, ein *Aurum potabile*, eine *Tincturam*, einen *Sulphur*, ein *Sal*, ein *Primum Ens* (‚erstes Wesen'), ein *Magisterium*, ein *Praecipitat*, einen *Crocum*, und dergleichen. Das mag nicht alleine mit dem Gold, sondern auch mit allen anderen Metallen geschehen. Sie werden resolviert und ihre Substanz und metallisches Wesen in die Tugenden *(virtutes)* verkehrt."
>
> nach: *Leonhard Thurneisser, >Pison<*, 1572

Manche Arbeiten verbinden Pflanzen- und Mineralreich. Es werden Tinkturen, Essenzen, Quintessenzen, Elixire, *Menstrua*, Alkaheste, Arkane u. a. m. sowohl aus Pflanzen wie auch aus Mineralen und Metallen bereitet.

Die meisten, heute üblichen, naturheilkundlichen Präparate können jeweils dem einen oder anderen philosophischen Prinzip zugeordnet werden, wirken also aus hermetischer Sicht für sich alleine und nicht ‚ganzheitlich', sondern hauptsächlich kraft des dargestellten Prinzips.

Zur sulfurischen Therapie

So kann z.B. der Aromatherapeut, auch wenn er lange Zeit mit außergewöhnlichem Wissen seine Kunst betreibt, erst dann, wenn ihm der alchymistische Blickwinkel erschlossen wird, erkennen, dass das, was er vielleicht als eine ‚ganzheitliche' Methode begriffen hatte, alleine eine Arbeit mit dem sulfurischen Prinzip der Natur, also einem Teilaspekt der Heilpotenz der ganzen Pflanze ist. Diese Erkenntnis mag einerseits nichts weiter ändern, andererseits könnte er fürderhin überlegen, die anderen beiden Prinzipien (Merkur und Sal) in seine Arbeit mit einzuschließen und so z.B. anfänglich zumindest gemeinsam mit den ätherischen Ölen, Hydrolaten und Fetten auch Minerale und alkoholische Präparate zu verwenden.

Zur merkurialen Therapie

Ähnlich wird sich der Heilkundige, welcher es versteht, mit pflanzlichen Tinkturen (nach Art der ‚galenischen Tinkturen') zu behandeln (Phytotherapie, Homöopathie u. dgl.) und seine Methode bislang als ‚ganzheitlich' empfunden hat, aufgrund eines vollständigeren Verständnisses bemühen, auch die anderen beiden Prinzipien, Sulfur und Sal, zur Wirkung zu bringen, indem er etwa gleichzeitig auch ätherische Öle und Mineralstoffe bzw. Salze verwendet.

Genauso könnte es dem Homöopathen zukommen, wenn ihn seine Sehnsucht nach ‚ganzheitlicher' Heilkunst zum Verständnis der Spagyrik führt. Er wird zumindest das Sal-Prinzip in seinen Präparaten vermissen, sofern er nicht schlichtweg beschließt, diese hermetischen Gedanken fortzuweisen und hier keine Veränderungen vorzunehmen oder er aus einfacher Vorliebe bei seinem Fach bleibt – hat doch jeder Heilkundige seinen eigenen Hintergrund und Zugang zur Heilkunst und tut darin das, was ihm am meisten entspricht.

Gesetzt den Fall aber, eine Medikation mit entsprechenden Potenzen in alkoholischer oder milchzuckriger Matrix führte zu keinem weiteren Heilerfolg, dann wäre er um guten Rat nicht verlegen und könnte zu spagyrischen Präparaten greifen. Wenn er seine Urtinkturen im idealen Fall selbst anfertigt, wird er den Pflanzenrückstand nicht mehr fortwerfen, nachdem er erkannt hat, dass darin ein mindestens ebenso wertvoller Heilfaktor verborgen ist wie im alkoholischen Auszug.

Er wird dem (bislang relativ für sich alleine wirkenden) Merkur-Prinzip den Sal-Faktor hinzusetzen wollen – und sei es aus Achtung vor der Erde. So wie die Milz im Körper das Organ des Erd-Elements ist, entsprechen die Prozesse der Veraschung, Calcination und Kristallisation diesem Element. Im Geist oder Gemüt des Menschen gehört dazu die Geschmacksempfindung im weitesten Sinne. Der Therapeut wird die Zusammenhänge erkennen und guten Geschmack zeigen, indem er nicht denen folgt, welche die Milz verächtlich fortwerfen noch denen, welche den Zustand ihrer Mutter Erde ignorieren, sondern Wesen und Bedeutung des Pflanzenrückstandes (nach der alkoholischen Mazeration) respektieren. Früher oder später wird er „in die Kohlen greifen", sich geeignete Tiegel zur Veraschung richten, Freude haben am Kristallisieren des Erd-Salzes und den nunmehr tatsächlich ‚ganzheitlichen' Charakter seines Präparates hochschätzen.

Spagyrik ist keine Form der Homöopathie – weder aus historischer Sicht noch seitens des philosophischen Hintergrundes und der Verfahrensweisen. Immer mehr Homöopathen gehen aber dazu über, ihre Potenzen von spagyrischen Essenzen ausgehend zu bereiten, nicht zuletzt wegen des darin enthaltenen Sal-Aspekts, der den Mitteln die nötige erdbezogene Stabilität gibt. Aus hermetischer Sicht wirken Homöopathika alleine kraft des Merkur-Prinzips und entbehren deshalb (trotz all ihrer Vorzüge) der ‚ganzheitlichen' Qualität.

Auch solches ist eine Bedeutung von Spagyrik und Alchymie in der Heilkunst.

„Nun wisse, was von den Dreien die Form sei: Eines ist ein *Liquor*, das ist *Mercurius*, eines ist eine Öligkeit, die ist *Sulphuris*, eines ist ein *Alkali*, dieses ist vom Salz. Der *Mercurius* ist ohne *Sulphur* und Salz, der *Sulphur* ohne *Mercurio* und Salz, das Salz ohne *Sulphur* und *Mercurio*, wenn ein jegliches in seiner *potentia* steht. (...)

Darum muss in diesen Dingen der *Medicus*, wenn er ein *Simplex* nimmt, beachten, wie hoch es *in liquore* steht, wie hoch *in oleo*, wie hoch *in sale* und (sinngemäß) wie die Drei in der Krankheit zusammengesetzt sind. (...) Und in aller Kürze muss die Regel eingehalten werden (...)"

nach: *Paracelsus*, >Von den ersten drei *Principiis*<

Es wirken also galenische Tinkturen, Urtinkturen, Homöopathika, Blütenessenzen u. dgl. fast ausschließlich ‚aus dem Merkur-Prinzip'. Ebenso Mittel, deren ‚Wirk-Informationen' von Trägersubstanzen (Zucker, Erden, u.a.) transportiert werden, also auch potenzierte Mittel der sogenannten ‚Mineralstofftheraphie'. Die ‚vermittelnde', ‚informierende' Kraft ist immer Merkurialkraft.

Die in der Aromatherapie gebräuchlichen Präparate (ätherische Öle, fette Öle, Fette, Harze, u. dgl.) wirken (fast) ausschließlich ‚aus dem Sulfur-Prinzip'.

(Nicht potenzierte) Mittel der (eigentlichen) Mineralstofftherapie (Salze, Erden, Aschen u. dgl.) wirken (fast) ausschließlich ‚aus dem Sal-Prinzip' – und dies umso stärker, je mehr sie präpariert, gereinigt bzw. aufgeschlossen sind.

Alchymistische Erhöhung ‚gemeiner' alkoholischer Ansätze

Gar mancher stellt sich beim Betrachten seiner Regale voller liebevoll erstellter alkoholischer Pflanzenansätze die Frage: „Wie kann ich nun meine Tinkturen in den Rang spagyrischer Mittel erheben? Ist das überhaupt noch möglich, nachdem bereits mit dem Pflanzenrückstand die Ausdrucksform des Erd-Elements fortgeworfen wurde?"

Nicht nur wird diese Frage gestellt in Bezug auf galenische Tinkturen, sondern auch auf andere Präparate, welche kraft des Merkur-Prinzips wirken. Manche bereiten feine, energetische Pflanzenmittel etwa nach der Methode von *Dr. Bach* durch Sonnenimprägnation, durch Kochung oder auf ähnliche Weise. Von der archaischen Naturkunde beeindruckt, werfen sie aber jetzt den Pflanzenrückstand nicht mehr einfach weg, sondern veraschen ihn, ziehen daraus mit destilliertem Wasser das Erd-Salz, lassen kristallisieren und setzen es dann der Flüssigkeit zu. Auf diese Weise hat eine echte spagyrische Komponente Eingang in Präparat und Therapie gefunden.

Die ‚grobe' Ausdrucksform des Merkur-Prinzips einer Pflanze wird bei der Zubereitung einer galenischen Tinktur als gefärbter Alkohol sichtbar. ‚Grob' deshalb, weil hier diese ‚Ausdrucksform des Merkur-Prinzips' (*Spiritus;* Alkohol) nicht ‚gereinigt', sondern vermischt ist. Die Farbe weist auf eine Anwesenheit des (ebenfalls ungereinigten) Sulfur-Prinzips und die relative Dichte und Unctuosität auf das (ungereinigte) Sal-Prinzip hin.

Durch Rectification kann nun eine gereinigte, subtilere Ausdrucksform des Merkur-Prinzips erlangt werden. Der Alkohol verliert dabei die Farbe und zumindest den gröberen Teil des *Phlegmas.* Er wird auch im Wesentlichen von Sulfur und Erdhaftigkeit geschieden. Was an Sulfur noch im rectificierten Alkohol (als dem ‚Vertreter des Merkur-Prinzips im Pflanzenreich') verbleibt, sind Aroma, Schärfe und Brennbarkeit. Diese Aspekte werden bei der Zubereitung einer einfachen spagyrischen Tinktur oder Essenz in der Regel nicht geschieden – akzeptierend, dass jegliche physische Ausdrucksform – und sei es ein noch so subtiles Fluid – ohnehin immer eine ‚Vermischung' ist und jede ‚Scheidung' nur eine relative bleiben wird. Wie weit die Scheidung zum Wesenhaften der Kraft vordringt, bleibt Angelegenheit der Kunstfertigkeit.

Die rückständige, liquorische Pflanzenfarbe enthält also einen wesentlichen Aspekt des Sulfur-Prinzips, welcher nunmehr durch Veraschen, Calcinieren und Scheiden von den erdhaften Schlacken (*Caput mortuum*) in relativ subtiler Form als ‚fixer Sulfur' (*Sal sulphuris*) dargestellt werden kann. Wird dieses sulfurische Salz schließlich mit dem klaren Alkohol vereinigt (coaguliert), dann avanciert die vordem ‚gemeine' galenische Tinktur in den Rang eines ‚kunstvollen' Präparates, selbst wenn sie schon jahrelang im Regal gestanden war. Weil aber der feste Pflanzenrückstand bereits fortgeworfen wurde, kann der eigentliche Sal-Faktor des Erd-Elements (*Sal fixum*) nicht mehr ergänzt werden. Es handelt sich also trotzdem um eine ‚Notlösung', bei der ein Präparat entsteht, welches weder den Rang einer ‚spagyrischen Tinktur' noch einer ‚Essenz' beansprucht, aber immerhin durch echte alchymistische Prozesse, nämlich durch zwei Scheidungen – eine ‚zum Merkur' und eine zum sulfurischen Salz – ‚edel' geworden ist.

Eine andere ‚Notlösung' wäre, das Salz der selben Pflanze (eben in einem anderen Jahr oder in einer anderen Wachstumsperiode) frisch zu bereiten und mit dem obigen Präparat zu vereinigen. Wiederum handelt es sich dabei ‚nur' um eine Notlösung, weil ja dem genauen Mengenverhältnis der Prinzipien sowie etwaigen Qualitäten, welche von der Pflanzen-Individualität herrühren, nicht vollständig entsprochen werden kann.

Aus hermetischer Sicht noch fragwürdiger ist das spätere Hinzufügen eines ätherischen Öls – zwar der selben Pflanzenart, aber doch eines anderen Individuums. Das betrifft besonders zu diesem Zwecke zugekaufte ätherische Öle, welche kaum je genau der selben Pflanzenart entspringen, aus welcher Merkur und Salz geschieden wurde. Es ‚wohnt' aber die Pflanzen-Individualität vor allem im ätherischen Öl. Zwischen dem späteren Hinzufügen von fixem Salz oder von ätherischem Öl besteht also der wesentliche Unterschied, dass die ‚Erde' bzw. das Sal-Prinzip neutraler, umfassender und nicht dermaßen ‚individuell' ist, wie Seele/Sonne/Sulfur bzw. das ätherische Öl. Daraus folgt, dass, wenn schon eine prinzipielle Ausdrucksform einer anderen als genau der verarbeiteten Pflanze später zugesetzt werden soll, dann eher ein Salz als ein ätherisches Öl. Nebenbei: Es gibt auch Homöopathen, die – einer alten Tradition folgend – jedem ihrer Präparate ein wenig Stein- oder Meersalz zusetzen oder gar ihre Mittel nicht mit Alkohol, sondern mit Sole haltbar machen. Dadurch wird das Mittel zwar nicht ‚spagyrisch', aber der Kunstgriff entspringt hermetischem Denken. Eine moderne Erklärung für diese Methode bezieht sich auf die Verwandtschaft des Meerwassers zum Blut.

Inwieweit bei dergleichen ‚Korrektionen' bzw. ‚Erhöhungen' Zugeständnisse an die hermetische Komplexität einer Arbeit gemacht werden, bleibt dem Praktiker überlassen. Phantasie und Intuition können hier weitläufig sein, wie wir aus zahlreichen Beispielen in der alten alchymistisch-iatrochymischen Literatur ersehen. Wichtig ist, dass die Produkt-Termini dann stimmig gewählt werden.

Damit kommen wir zu jener Frage, welche die Fragestellung nach dem „wie erhebe ich meine Heilmittel in einen spagyrischen Rang?" meist direkt begleitet: „Ist es in jedem Falle, besser ein spagyrisches Präparat anzuwenden anstatt eines einfachen Pulvers, eines Tees oder einer alkoholischen Tinktur?"

Diese Frage kann nicht pauschal beantwortet werden. Grundsätzlich aber müssen wir uns darüber im Klaren sein, dass die Alten über Jahrhunderte und Jahrtausende stets sowohl spagyrisch aufgeschlossene als auch einfache Rohformen, alkoholische Auszüge, Öle, Erden, Harze usf. zusammen in ein und demselben Therapiekonzept einsetzten. Minerale und Metalle wurden dabei (zumindest in den ‚kunstvollen' Rezepturen) überwiegend mehr oder weniger ‚kunstgemäß' aufgeschlossen, während Kräuter, Gewürze und Harze sehr oft in roher Form, als Pulver, Saft, Dekokt oder ‚gemeine' Tinktur in den Rezepten enthalten sind. Diese Kombinationsweise ist heute in asiatischen Traditionen noch üblich, wo ebenfalls die Minerale oft in spagyrisch aufgeschlossener Form zusammen mit Pflanzenpulver rezeptiert werden. In der abendländischen Literatur sind viele Rezepte überliefert, die zwar einen spagyrischen oder alchymistischen ‚Grund' vermuten lassen, wenn sie aber wörtlich genommen werden, nur rohe Minerale und sogar Metalle vorschreiben. Dies scheint von einem Missstand, einer Kunstlosigkeit in der mittel- und spätmittelalterlichen ‚Apothekerkunst' herzurühren, über den ‚Kundige' wie *Paracelsus* und *Agricola* sehr klagten. Zudem wurden die Rezepte selbst sehr oft von Unkundigen abgeschrieben und überliefert. Wir kennen dazu Beispiele, wie metallisches (unaufgeschlossenes) Gold in den ‚Rezepten des *Nostradamus*', die Anwendung ‚roher' Edelsteine bei *Hildegard* und andere mehr.

Pflanzliche Mittel finden wir bei den versierteren Alchymisten oft auch in Form einer Quintessenz oder eines Elixirs solcherart bereitet, dass eben zur Erschließung der gewünschten Quintessenz zusätzliche Kräuter verwendet werden, welche zwar für sich nicht aufgeschlossen, aber zu echten, alchymistischen, verfahrensbedingten Kompositionen bzw. Rezepturen führen.

Die Wirkweise pflanzlicher Rezepturen ist dementsprechend unterschiedlich, je nachdem die Heilkraft der einzelnen Pflanzen zugänglich gemacht wird. Die Stufenleiter hört nicht damit auf, wenn man dem alkoholischen Auszug oder dem Gärprodukt noch das Erd-Salz hinzufügt, sondern reicht von den Varianten der ‚einfachen Spagyrik' bis in die hohe Alchymie. Auch werden einzelne Pflanzen zu einer Quintessenz ausgearbeitet und dann in die Mitte einer Komposition mit nicht so weit aufgeschlossenen oder rohen Pflanzen gestellt. Andere wiederum vermischen zuerst die rohen Pflanzen und bereiten dann aus dem Ganzen in einem Arbeitsgang ein spagyrisches Präparat. Als ‚kunstvoller' jedoch gilt es, die Rezepturen aus jeweils für sich aufgeschlossenen Teilen zusammenzusetzen.

Rezeptur und Komposition

Die Kombination von spagyrischen mit anderen naturheilkundlichen Präparaten ist traditionsgemäß (abgesehen von speziellen, begründeten Ausnahmen) immer möglich und oft sogar erwünscht. Die Wirkungsebenen ‚stören' sich dabei nicht, weil durch die spagyrische Präparation (je höher umso besser) nicht nur schädliche ‚Nebenwirkungen', sondern auch Behinderungsfaktoren zu anderen Präparaten aufgehoben werden.

Rezeptur oder gleichzeitige Anwendung von Spagyrika zusammen mit Rohsubstanzen, Pulvern, Harzen, Ölen, Tinkturen, Tees u. dgl. sind in der heutigen Therapie die Regel und waren noch weit umfangreicher bei den Alten üblich. Unter Einbeziehung (spag. aufgeschlossener) mineralischer und metallischer Bestandteile finden wir dazu in der alchymistischen Literatur Kompositionen höchster Rezeptierkunst, vergleichbar mit ähnlichen Präparate asiatischer Systeme, in denen die Alchymie immer noch den ihr zugehörigen Rang in der Heilkunst innehat. Schöne Beispiele dazu sind die ‚tibetischen Juwelenpillen'. Die Pille *Rinchen Ratna Samphel* enthält ungefähr 70 Bestandteile. Davon sind die Metalle und teilweise die Minerale spagyrisch aufgeschlossen, sodass sie vom Körper mühelos und vollständig absorbiert werden können und keinen Schaden verursachen. Das sind: Gold, Silber, Eisen, Kupfer, Blei, Korallen, Türkis, Perlen, Lapislazuli, Horn und andere. Die zusätzlichen Wurzeln, Kräuter und Gewürze sind zum Teil spagyrisch ‚korrigiert' oder in roher Form eingearbeitet.

Ähnlich kombiniert *Agricola* in seinem ‚Balsam zu Brüchen' Minerale, Metalle und Pflanzen – teils spagyrisch präpariert, teils roh:

Rec: Balsami Sulphuris (spag.präp.), Olei Mastichis, Croci Martis (spag. präp.), Terrae sigillatae, Terrae Vitrioli dulcis (spag. präp.), Corallorum rubrorum putr. (spag. präp.), Magnetis praeparati (spag. präp.), Extracti Consolidae Majoris und Extracti Aristolochiae rotundae.

Oder in seiner Komposition zum *Paroxysmus:*

Rec: Essentiae Antimonii (spag. präp.), Olei Auri cum Oleo Cinnamomi praeparati (spag. präp.), Spiritus corticis Citri (spag. präp.), Spiritus corticis Aurantiarum (spag. präp.), Aquae Cinnamomi, Aquae Hirundinum.

In seinem Gold-Balsam zur Apoplexie:

Rec: Olei Nucis moschatae, Destillati Rosarum, Dest. Cinnamomi, Dest. Caryophyllorum, Dest. Citri, Dest. Succini, Dest. Lavendulae, Dest. Majoranae, Dest. Asae dulcis, Dest. Rutae, Olei Camphorae (spag. präp.), Liquoris Auri (spag. präp.), Ambrae, Moschi.

In einer einfachen Komposition „zum Stillen der Flüsse':

Rec: Tincturae Smaragdi (spag. präp.), Essentiae Sulphuris (spag. präp.), Croci Martis (spag. präp.), Corallorum (spag. präp.).

Zur Stärkung des Magens:

Rec: Magisterii Corallorum rubrorum (spag. präp.), Magisterii Perlarum (spag. präp.), Olei Martis (spag. präp.), Olei Auri (spag. präp.), Essentiae Anisi und Essentiae Cinnamomi (spag. präp.), Corticis Citri conditae, Corticis Nucis Indicae, Rob Juniperi, Tincturae Croci.

(Die Rezepturen sind zum Zwecke der Anschauung und ohne Mengenangaben wiedergegeben aus: *Johann Agricola*, >Chymische Medicin. Ein Kompendium der Bereitung und Anwendung alchemistischer Heilmittel<. Nach der Erstausgabe Leipzig 1638/39 wieder herausgegeben von *Oliver Humberg*).

Die Verbindung der Prinzipien in der Präparation

Eine andere Möglichkeit, sowohl die Prinzipien als auch ‚rohe' Substanzen kunstvoll miteinander zu verbinden, welche auch die Alten hoch schätzten, ist die, Rezepturen, bestehend aus den (meist rohen) Aspekten der Prinzipien, durch eine gemeinsame spagyrische/alchymistische Präparation darzustellen. Ein berühmtes Beispiel dafür ist das *Elixir Subtilitatis* des *Paracelsus*, welches aus Olivenöl (Sulfur), Weingeist (Merkur) und Honig (Sal) zusammengesetzt wird. Die drei werden miteinander ‚kunstgemäß' zu einem echten spagyrischen Mittel vereinigt, ohne zuvor für sich selbst aufgeschlossen worden zu sein.

Auch das berühmte *Elixir Proprietatis* des *Paracelsus* ist eine Kombination von Rezeptur und Verfahren. Das Rezept ist: Safran, Aloe und Myrrhe. Vom Safran wird dazu ein einfacher alkoholischer Auszug verwendet. Das *Menstruum* für Aloe und Myrrhe ist aber von hoher alchymistischer Natur, aus Schwefel (Sulfur), Samiak (Merkur) und Kalk (Sal) bereitet (wobei die einfachste Form des alchymistischen Salmiaks wiederum eine aus den Prinzipien komponierte 3er-Formel (Russ/Sulfur, Urin/Merkur und Steinsalz/Sal) ist.

Ebenfalls in dieser Richtung befindet sich *Digby's* berühmtes Weinsteinpräparat. Hier wird gereinigtes *Sal Tartari* (Sal) mit einem beliebigen ätherischen Öl (Sulfur) imbibiert und danach mit Weingeist (Merkur) bis zur Flüchtigkeit kohobiert.

Schema der spagyrischen Pflanzenessenz

Eine ‚Essenz‘ aus der Pflanze der Wahl darzustellen, ist sozusagen der ‚Standard‘ spagyrischer Arbeit. Das gilt hauptsächlich für praktizierende Künstler. Es gilt weniger für Produzenten großer Mengen spagyrischer Pflanzenpräparate, welche meist der Einfachheit halber ‚spagyrische Tinkturen‘ erarbeiten. Der Weg zur Essenz ist aufwändiger als der zur Tinktur. Erst in der ‚hohen Alchymie‘ begegnen wir dann wieder dem Terminus ‚Tinktur‘, dort aber mit der Bedeutung ‚Quintessenz‘ oder ‚Elixir‘, also in weit höherer Position als die ‚einfache‘ Essenz.

Einige wenige, leicht verschiedene Wege zur ‚Essenz‘ folgen stets demselben Grundmuster: Die 3 Prinzipien werden aus der Pflanze geschieden, jedes für sich gereinigt und schließlich wieder vereint. In den meisten Fällen erhält man ein klares, flüssiges, farbloses Präparat, dessen Geschmack pflanzentypisch und dessen Wirkung sehr fein und durchdringend ist.

Im Folgenden wollen wir anhand des ‚Weges zur Pflanzenessenz‘ die grundlegenden Arbeiten sowie die dazugehörigen philosophischen Betrachtungen kennenlernen:

Schema einer ‚spagyrischen Pflanzenessenz‘

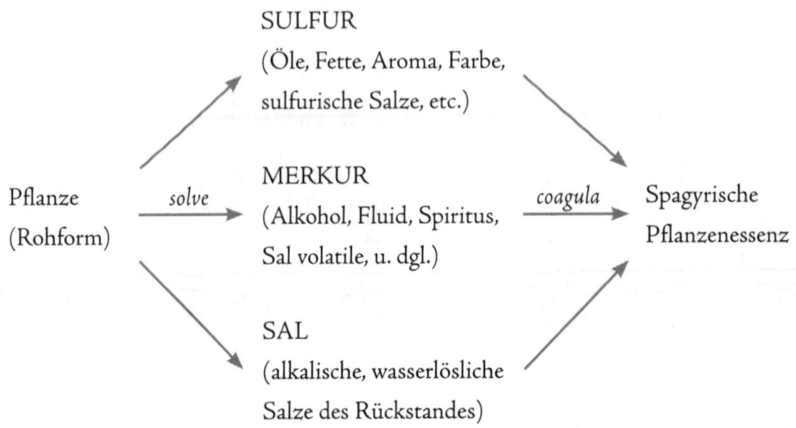

Grundriss einfacher Pflanzenpräparate im Vergleich

a) Tinktur (galenische Tinktur, Urtinktur, alkoholischer Auszug)

Mazeration der Pflanze im Alkohol

Filtrieren der fertigen Tinktur

Entsprechende philosophische Prinzipien: Merkur-Prinzip (in hermetischem Sinne ungereinigt)

b) Spagyrische Tinktur

Mazeration der Pflanze im Alkohol oder Gärung

Destillation des Alkohols im Falle des Gärungsweges oder zur Darstellung des ‚gereinigten Merkur-Prinzips'

Filtrieren des merkurialen Aspektes

Rectification des Alkohols

Veraschen des Pflanzenrückstandes

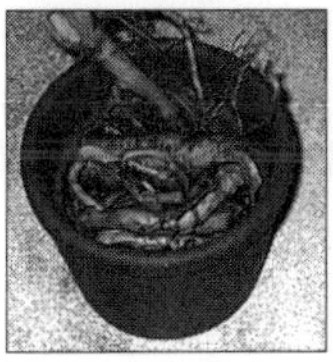

Scheidung des Erd-Salzes aus der Asche

Coagulation (Vereinigung) des Alkohols mit dem Salz (bzw. des Merkur- mit dem Sal-Prinzip der Pflanze) zur fertigen Spagyrischen Tinktur.

Ausgearbeitete philosophische Prinzipien: Merkur-Prinzip (gereinigt oder ungereinigt) und Sal-Prinzip (gereinigt).

c) Spagyrische Essenz

Anfängliche Scheidung von ätherischem Öl und Hydrolats als Ausdrucksform des Sulfur-Prinzips durch Vermittlung von Wasser

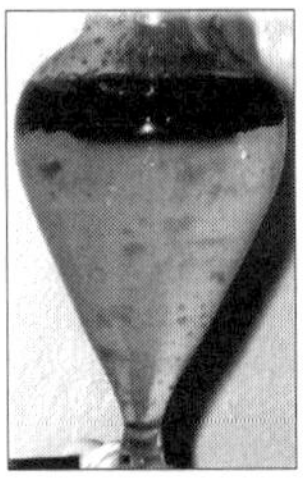

Scheiden des Öls zur Reinigung des Sulfur-Prinzips

Rectification des Hydrolats

Alkoholische Mazeration oder Gärung des Rückstandes zur Darstellung der Ausdrucksform des Merkur-Prinzips

Destillation bzw. Rectification des Alkohols zur Darstellung des gereinigten Merkur-Prinzips

Darstellung des Sal-Prinzips durch Veraschung, Calcination und Reinigung des Erd-Salzes

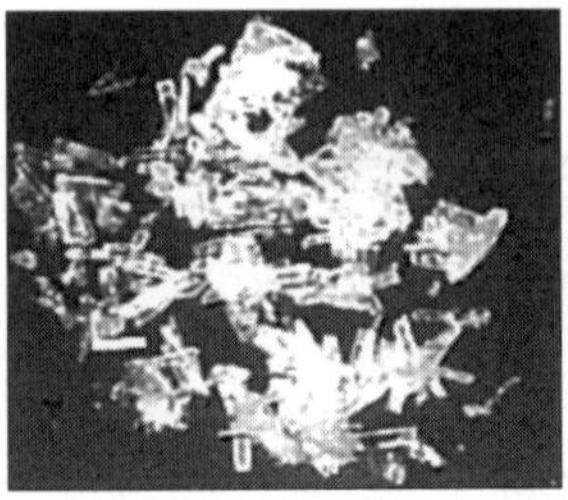

Coagulation (Vereinigung) der Ausdrucksformen der 3 Prinzipien zur fertigen Spagyrischen Essenz.

Ausgearbeitete philosophische Prinzipien: Sulfur-Prinzip (gereinigt), Merkur-Prinzip (meist gereinigt) und Sal-Prinzip (gereinigt).

Arbeitsgänge zu den Prinzipien
(für Aufschließungsverfahren von Pflanzen)

SULFUR-Prinzip:

Destillation des ätherischen Öls bzw. des Hydrolats. Scheidung des *Phlegmas* und Rectification des überwiegend sulfurischen (öligen, fettigen und/oder aromatischen) Anteils.

Darstellung des Fixen Sulfurs als Salz *(Sal sulphuris)* durch Scheidung der festen Pflanzenteile von der gefärbten Flüssigkeit des Destillationsrückstandes (‚Farbe'), Veraschung, Calcination, Scheidung vom *Caput Mortum* sowie Reinigung und Kristallisation des Salzes.

Darstellung des Inneren Sulfurs als Farbe durch Aufschmelzen des *Sal sulphuris* und Extraktion der Farbe (als eigentlicher Ausdruck des Fixen Sulfurs) durch den Alkohol aus dem Merkur-Prinzip der Pflanze.

MERKUR-Prinzip:

Techniken der Fermentation, Gärung oder Putrefaction.

Destillation, Separation und Rectification der geistigen Teile.

Mazeration von frischen oder getrockneten Pflanzen (bzw. des Rückstandes aus obiger sulfurischer Destillation) mit zugesetztem Alkohol (Weingeist, Kornbrandt u. dgl.). Nachfolgendes einfaches Filtrieren und/oder Rectificieren des Mazerats.

Techniken der Dynamisation nach erfolgter Aufschließung und Wiedervereinigung (Circulieren, Kohobieren; Rhytmisieren u. dgl.). Hierher gehören auch etwaige Potenzierungen sowie Kohobation, Circulation oder Dynamisation einzelner Aspekte (zB: Scheidung und nachfolgende Coagulation ‚nach den Elementen' als Ergänzung zur einfachen Rectification des Alkohols).

SAL-Prinzip:

Veraschen und nachfolgendes Ausglühen (Calcination) der Pflanzen bzw. des Rückstandes aus den obigen Destillationen.

Abscheidung des nicht wasserlöslichen *Caput Mortum* vom wasserlöslichen Salz *(Sal fixum)* der Asche bzw. des Calcinats.

Eindampfen der Lauge; Kristallisieren und Reinigen der Kristalle.

Etwaige weitere Calcinationen, Imbibationen bzw. Solutionen und Calcinationen und/oder Kohobation zum *Sal volatile*. In letzterem Falle geht die Arbeit am Sal-Prinzip in eine Arbeit am Merkur-Prinzip über. Gleichsam ist die Darstellung des ‚fixen Sulfurs' als Salz sowohl eine Arbeit am Sulfur- wie auch am Sal-Prinzip.

SPAGYRISCHE / ALCHYMISTISCHE PRODUKTE

A) PFLANZENMITTEL

Die Spagyrische Tinktur

Darstellungsmöglichkeiten / übliche Verfahren

1) Darstellung des Merkur-Prinzips aus der Pflanze durch

a) Gärung der frischen Pflanzen ohne Zusätze
b) Gärung der frischen Pflanzen mit Zusatz von Hefe und/oder Zucker
c) Gärung der getrockneten Pflanzen mit Zusätzen
d) Mazeration der Pflanze in Alkohol

Im ersten Fall (a) verursachen die jeder Pflanze natürlich anhaftenden Hefen die Gärung. Dieser Weg erfordert viel Geschicklichkeit und ist nicht immer gangbar. Die anhaftenden Hefen sind nach Pflanzenart und Sammelzeit verschieden. Das Endprodukt ist meist wenig, hauptsächlich bedingt durch den Zuckergehalt der Pflanze. Gut geeignet für diesen Weg sind süße Früchte wie Aprikosen, Pflaumen, Kirschen u. dgl. Ein großer Vorteil ist der geschlossene Charakter der Arbeit, bei der keine Substanzen (außer, wenn nötig, etwas Wasser) von außen zugesetzt werden.

Bei der zweiten Methode (b) ist die Gärung leichter beeinflussbar, verläuft verlässlicher und das Endprodukt ist mehr. Die Qualitäten von Hefe und Zuckerart bestimmen die ‚spagyrische' Qualität des Produktes mit. Je höher die Anforderungen dabei sind, umso weniger wird auf industriell gefertigte Massenware zurückgegriffen und stattdessen eigene Hefekulturen gezüchtet sowie naturbelassene Zuckerarten oder Honig verwendet.

Das selbe gilt auch für die Gärung getrockneter Pflanzen (c). Es ist allerdings fast in jedem Falle empfehlenswert, frische zu verwenden, da wesentliche Teile besonders des subtilsten Pflanzenmerkurs durch Trocknung und Lagerung verloren gehen (während der Anteil der sulfurischen Komponenten oft höher, aber dichter wird). Dennoch ist eine sorgfältige Trocknung die weitaus beste Art, Pflanzen, die man noch nicht verarbeiten will oder kann, einige Monate aufzubewahren. Je nach Dichte der Pflanzenteile verschwinden dann die Kräfte übers Jahr immer mehr. Harte Wurzeln, Hölzer, Rinden und Samen behalten sie länger (oft viele Jahre lang), feine Blüten und Blättchen verlieren sie bald. Die Art der Trocknung spielt dabei eine Rolle.

Die bei Massenware noch immer übliche industrielle Schnelltrocknung bei über 70° zerstört oder vertreibt den größten Teil, vor allem der feineren Wirkkräfte. Kleinere, verantwortungsvolle Betriebe trocknen heute ihre Pflanzen ebenfalls ‚in großem Stil', wobei sie aber in etwa die Grade der Körpertemperatur nicht überschreiten und dafür dem Trocknungsprozess mehr Zeit geben. Auf solche Kräuter kann der Spagyriker zurückgreifen, wenn ihm keine frischen zur Verfügung stehen.

Bei der vierten Methode (d) werden frische oder getrocknete Pflanzen in Alkohol angesetzt und einige Wochen in der Wärme mazeriert. Dazu wurde oben schon einiges gesagt. Es ist dies die einfachste Art, um die Arbeit am Merkur-Prinzip bzw. zur spagyrischen Pflanzentinktur zu beginnen – sie ist aber keinesfalls minderwertiger als die drei vorigen. Wieder spielt die Qualität des zugesetzten Alkohols eine bedeutende Rolle.

Der Alkoholgehalt kann dabei je nach Tradition, Erfahrung oder Ermessen verschieden sein. Zumeist wird man dichte, feste Pflanzen und Pflanzenteile (Wurzeln, Hölzer, Rinden, harte Samen) in höherprozentigem (Richtwert: 60-85%) und leichte, fragile Blätter in minderprozentigem (20-40%) Alkohol mazerieren. Eine besondere in der spagyischen Arbeit immer wieder geforderte Ergänzung zur Mazeration ist das Schütteln, Rühren oder sonstiges Bewegen des Ansatzes, welches öfters am Tag zu geschehen hat. Es dient der besseren ‚Lösung' und Einbindung des pflanzeneigenen Merkurs (Pflanzen-*Spiritus*) in die merkuriale Matrix (Fremdalkohol). Manche bezeichnen dieses Bewegen sogar als das ‚Hinzufügen eines zeitlichen zum stofflichen Merkur'. So wird denn auch bisweilen die Mazeration mit einer gleichzeitigen Circulation verbunden, indem man das Gefäß nur bis zu etwa 1/10 (oder weniger) befüllt und den restlichen Raum zur Ausdehnung und Kondensation freilässt.

Die Mazerations-Temperatur ist (je nach System) verschieden, von sehr gelinde (um die 20°C) bis Körpertemperatur und ‚Brutwärme', kaum jemals aber über 40°C bis zu höheren Temperaturen. Im Vakuum sind zwar Mazerationen und Circulationen bei niedrigeren Temperaturen möglich – dieses Vakuum ist aber in der Natur nicht vorhanden und es ist zu überlegen, inwieweit man hier durch technischen Aufwand Verbesserungen erzielt oder den ‚Pfad der Natur' verlässt. Die ‚schönsten spagyrischen Stücke' werden, wie schon die Alten betonten, mit einfachsten Gerätschaften hervorgebracht.

Die Dauer der Mazeration hängt wieder von den jeweiligen Systemen, Erfahrungen etc. und selbstverständlich von der Art der Pflanzenteile und dem Alkoholgehalt ab. Immer gilt aber die Regel: Besser mehr Zeit geben. Auch wenn man nach 2 Wochen den Eindruck hat, dass nichts mehr tingiert (sich der Alkohol nicht mehr weiter färbt), ist es vorteilhaft, nochmals die selbe Zeit einzusetzen.

In der klassischen Literatur findet man oft den ‚philosophischen Monat' (40 Tage) als Zeitvorgabe für derlei Mazerationen, Digestionen, etc. Insgesamt geht es hier, wie gesagt, nicht vordergründig um eine Extraktion von Wirkstoffen aus der Pflanze, sondern hauptsächlich um eine Verbindung des der Pflanze innewohnenden Natur-Prinzips mit demselben Prinzip im Alkohol.

2) Reinigung des Merkur-Prinzips der Pflanze durch

a) Destillation und Rectification des Gärproduktes
b) Filtration des Mazerats
c) Destillation und Rectification des Mazerats

Nach erfolgreicher Gärung wird auf dem Weg zur spagyrischen Tinktur bei gelinder Wärme destilliert (a), solange im Helm bzw. in der Vorlage die typischen ‚Tränen' eine Geistigkeit des Destillates anzeigen (nach der alchymistischen Terminologie der Alten erscheinen nun ‚die undeterminierten Tränen der *Diana*'). Wenn danach das wässrig-phlegmatische Kondensat erscheint, kann man die Temperatur ein wenig erhöhen und noch einige Zeit mit der Destillation fortfahren – ist doch Merkur einigermaßen tief in die Pflanzensubstanz eingewoben. Das *Phlegma*, welches jetzt mit den letzten merkurialen Anteilen übergeht, wird dann ohnehin bei den folgenden Rectificationen geschieden. Es ist auch nicht alles, was nicht typisch alkoholisch ist, bloß ‚phlegmatisch' und leer von Geist, und im weitesten Sinne gehört ja auch das *Phlegma* selbst zum Merkur-Prinzip.

Bei gut geführten, langsamen Gärungen und ebensolchen darauffolgenden Destillationen wird das Destillat sehr wohlschmeckend (bzw. von pflanzentypischen Geschmack), fein und weitestgehend frei von schädlichen Inhalten sein. Davon hängt dann die Notwendigkeit zur Abscheidung des ‚Vorlaufs' ab. ‚Wahre Künstler' erhalten keinen oder keinen nennenswerten, methylhaltigen Vorlauf. Bei zu schnellen, forcierten Gärungen und darauffolgenden Destillationen schmeckt das gesamte Destillat oft unangenehm bis jauchig – und ist deshalb in spagyrischem Sinne verwerflich.

Der naturheilkundliche Therapeut kann am Geschmack spagyrischer Pflanzenmittel, die er einsetzen möchte, viel über ihre Qualität erkennen. Schwieriger ist das bei solchen Produkten, bei denen der typische Geschmack einer Tinktur oder Essenz mittels zugesetzter aromatischen Substanzen verändert wurde. Keine Verfälschung liegt vor, wenn aromatische Substanzen (wie zB. ätherisches Zimt- oder Rosenöl) innerhalb von Rezepturen verwendet werden. Dies findet man dann auch gelegentlich so in der iatrochymischen Literatur empfohlen.

Wir haben über die Verbindung der Prinzipien in der Rezeptur bereits gesprochen. Ein Sinn solcher Zusätze kann sein, weitere Wirkkomponenten einzubringen oder unangenehme Geschmäcker zu korrigieren, auszugleichen und abzurunden. Dies ist in allen Traditionen üblich. Fraglich ist das ‚Abschmecken' ganzer Produktserien, gleichsam mit Himbeer- oder Zitronengeschmack, um das Mittel für den verwöhnten Gaumen konsumierbarer zu machen. Enzian- oder Löwenzahnwurzel schmecken nun mal bitter – Wermut und Tausendgüldenkraut noch viel mehr. Wenn diese aber ordentlich ausgearbeitet sind, dann ist jene Bitterkeit rundum in den pflanzentypischen, aromatischen Geschmack eingebettet und Jedem, in der rechten Dosierung, gut zuträglich. In Natur und im Leben hat alles seinen besonderen Geschmack und eine, zB. durch lebensfremde Ernährung erworbene Geschmacksirritation entspricht einem Krankheitsbild, welches eher längerfristig behandelt als zum Richtwert für ein Abschmecken spagyrischer Heilmittel werden sollte.

Nach der ersten Destillation des Vergorenen folgen dann in der Regel eine oder mehrere Rectificationen (a) zur weiteren Reinigung und Verfeinerung des Destillats bzw. des auf diese Weise dargestellten Merkur-Prinzips.

Wurde die Pflanze in zugesetztem Alkohol mazeriert, dann braucht nur mehr filtriert zu werden (b). Damit ist die einfachste Variante der Arbeit am Merkur vollbracht. Das Filtrat ist in den meisten Fällen klar und gefärbt (tingiert). Wenn nötig, lässt man es einige Tage lang setzen und filtriert dann nochmals. Das im Filter Verbleibende kommt zum übrigen Pflanzenrückstand. Daraus wird später das Salz gezogen. Auch so kann eine ‚spagyrische Tinktur' erlangt werden. Der Nachteil dabei ist, dass hier das Merkur-Prinzip nicht wirklich ‚rein' erhalten wird. ‚Farbe' bedeutet, wie wir schon wissen, immer auch die Anwesenheit ‚eines Sulfurs'. Je reiner, typischer und subtiler aber die Ausdrucksformen der Prinzipien dargestellt werden können, umso tiefgreifender und durchdringender wird auch das Wirkvermögen des fertigen Präparates sein.

Andererseits kann aber dennoch die Anwesenheit von Farbe selbst einen guten Wirk- oder sogar Heileffekt haben. Manche empfinden zum Beispiel eine spagyrische Tinktur oder Essenz aus dem Johanniskraut erst richtig ‚erhellend', wenn sie rot ist, und es geht ihnen diese Empfindung ab, wenn sie klar und farblos ist. Aus welchen Gründen auch immer: Der Farb-Effekt kann entweder durch einfaches Filtrieren des Mazerats erhalten oder durch späteres ‚Tingieren' des farblosen Destillates oder des fertigen Präparates mit der selben Pflanze (frisch oder getrocknet) erlangt werden. Im ersten Fall haben wir eben das Merkur-Prinzip nicht gereinigt dargestellt, im zweiten Fall setzen wir ganz im Sinne einer spagyrischen Coagulation eine sulfurische Komponente hinzu, sollten dann aber auch das zusätzliche *Sal* einbeziehen.

Eine andere ,Unreinheit' des Merkurs ist das *Phlegma*, welches im nicht rectificierten, einfach ausgepressten Mazerat enthalten ist. Es ist sowohl ein *Phlegma* des zugesetzten Alkohols, als auch ein *Phlegma* aus der Pflanzenwässrigkeit. Es werden auf diesem sehr kurzen und einfachen Wege zur Tinktur also weder das *Phlegma* noch die Farbe abgeschieden.

Das Reinigen des durch Mazeration erhaltenen Merkurs geschieht wieder durch Destillation. Nur selten wird dazu mehr als eine Rectification nötig sein, außer man ist aus bestimmten Gründen bestrebt, den Pflanzenmerkur ,noch reiner' zu erhalten, etwa wenn man zur Herstellung eines Pflanzensteines oder *Clissus* hochprozentigen Alkohol benötigt oder mit ebensolchem Alkohol auf weitere Kohobationen zur Erhöhung des Salzes abzielt).

Mittels dieser Reinigung (durch Destillation) kann man ein Präparat vom Grad einer ,spagyrischen Tinktur' erhalten, welches aber letztendlich klar und farblos, also nicht tingiert ist. Der Grad einer echten Essenz würde erst dann erreicht werden, wenn zumindest eine Ausdrucksform des Sulfur-Prinzips geschieden und gereinigt dargestellt wird.

hermetisches bzw. alchymistisches Symbol für ,Merkur' oder ,Merkur-Prinzip'

Zusammenfassende Gedanken zum Pflanzen-Merkur

Zusammenfassend sei bemerkt, dass also jede stoffliche Ausdrucksform eine Vermischung von Kräften und Substanzen ist, denn nur durch Zusammenwirken und Vermischung natürlicher Kräfte entstehen grobstoffliche Äußerungen. So hat das Pfefferkorn Farbe, Geruch, Schärfe und Geschmack vom Sulfur (Feuer), eine Feuchte und Geistigkeit von Merkur (oder: vom Luft-, Wasser- und Äther-Element) sowie Dichte, Schwere, Festigkeit und Dunkelheit vom *Sal* (vom Erd-Element). Jene Kräfte und Prinzipien, die von der ursprünglichen Idee (der Quintessenz) dazu herangezogen und bewegt wurden, das substanzielle Pfefferkorn zu manifestieren, sind darin in einem ganz bestimmten Verhältnis vermischt und haben ihre spezifischen Charaktere durch die ursprüngliche Idee oder Quintessenz angenommen. Davon ist der sulfurische Anteil der am stärksten individualisierte, also am meisten von der Quintessenz geprägt. Dem folgt das Salz und zuletzt Merkur.

Meistens stellt der Spagyriker die Ausdrucksform des Merkur-Prinzip im Pflanzenreich als Alkohol dar, was zu der (etwas oberflächlichen) Ansicht geführt hat, ‚der Merkur im Pflanzenreich sei völlig neutral und äußere sich (immer) als Alkohol'.

Dass dies bis zu einem gewissen Grade zutrifft, steht außer Zweifel – eben wegen einer oft weitgehenden, grundsätzlichen Neutralität des Merkur-Prinzips, verglichen mit den anderen beiden Prinzipien. Es kann zum Beispiel das Merkur-Prinzip einer Pflanze durch Mazeration mit den verschiedensten Arten von Alkohol prinzipiell gleich dargestellt werden. Dabei ist es beliebig bzw. bedingt durch Erfahrung, Empfehlung oder Verfügbarkeit, ob man nun die Angelikawurzel in Wein- oder Apfelgeist, Korn- oder Obstbrandt oder worin auch immer mazeriert. Hier können signaturkundliche Überlegungen oder Vorlieben maßgeblich sein. Hauptsache ist, dass die Qualität stimmt und der Alkohol ein echtes Natur- und kein denaturiertes Industrieprodukt ist. Kein ernsthaft arbeitender Spagyriker wird zur Herstellung von ‚Heilmitteln' auf zweifelhafte Flüssigkeiten zurückgreifen, die bloß durch ein Etikett als ‚Weingeist' ausgezeichnet werden, in Wahrheit aber nie mit echtem Wein in Berührung gekommen sind.

Schließlich soll sich ja ein guter Teil des Naturgeistes *Mercurius* aus der Pflanze aufgrund der Verwandtschaft und des Wiedererkennens in dieses flüssige Medium begeben, worin man mazeriert. Auch wenn sich solch denaturierte Flüssigkeiten ebenfalls mit der Zeit färben und als echte Tinktur erscheinen, sind sie doch aufgrund ihrer Herstellungsart (und ihrer Ausgangsstoffe) nur bedingt in der Lage, eine typisch merkuriale, lebendig-geistige Matrix für den Pflanzen-Merkur zu sein. Hier mag es manchem wunderlich scheinen, dass viele Homöopathika und sogenannte galenische Tinkturen, welche noch immer mit solchen Flüssigkeiten hergestellt werden, dennoch eine gewisse heilsame Wirkung entfalten können. Wieviel besser würden diese Mittel erst sein, wenn man korrekt der Natur folgte und z.B. ‚echten Weingeist' oder Kornbrand aus natürlich gewachsenem (und nicht zu letztmöglichen Wachstumsanstrengungen gezwungenem) Getreide verwendete. Ganz abgesehen von der Qualität der jeweiligen Pflanzen, die hier verarbeitet werden. Die Bezeichnung ‚technisch rein' hat in alchymistischen Belangen nicht die höchste Bedeutung. Die Qualitätsbewertung geht hier weit tiefer. Das Prädikat heißt ‚lebendig'. Technisch rein ist auch ein soeben Verstorbener. Er wird aber zu keiner Hilfeleistung mehr fähig sein, weil ihm das Wichtigste fehlt – die Lebendigkeit.

Der ‚Pflanzenmerkur' wird in den meisten Fällen entweder durch Gärung oder durch Mazeration (in zugesetztem Alkohol) dargestellt. Aufwändiger und weit tiefgreifender ist seine Darstellung als *Spiritus* durch Putrefaction, welche sehr subtile und vorsichtige Destillationen, Rectificationen und Separationen erfordert, das Endprodukt aber sehr wenig ist.

Aber auch die Gärung ist eine hohe Kunst und will dementsprechend verstanden und gehandhabt werden. Manche meinen, man könne einen ,echten' Pflanzen-Merkur durch Hitze aus dem Kraut herauszerren. Das Merkur-Prinzip ist aber insgesamt mit der Qualität der Kälte verbunden. Es kommt dabei zur ,Hasenjagd' (einem alten alchymistischen Text zufolge wird der merkuriale Hase mit der Kraft des Mars gefangen). In der Wärme verlaufen die Gärungen zwar schnell – aber möglichst viel in kurzer Zeit zu ,machen' ist so ziemlich das Gegenteil von Spagyrik und Alchymie. Das selbe gilt für die nachfolgenden Destillationen des Vergorenen. Hier wird der Geist oft nur so durch den Helm gejagt. Das Endprodukt erweist den Fehler in Geschmack, Wirkung und anderem – doch man übersieht das gerne, wenn man möglichst viel und schnell produzieren will.

Fünf oder sechs Tropfen pro Minute – das ist ein ,spagyrisches Tempo' (*Cardilucio* verlangte bei ähnlichen Arbeiten 2-3 Tropfen pro Tag!). Ein recht vergorener und destillierter Pflanzenmerkur wird nichts Schädliches enthalten (selbst der Gehalt an Methylalkohol ist minimal, wenn nicht gar abwesend), und er wird einen sehr angenehmen Geschmack haben (soweit dieser in der ,Idee' der jeweiligen Pflanze inbegriffen ist).

Der Weg der Putrefaction führt in der Regel über eine ,einfache' Tinktur oder Essenz hinaus zu einem Präparat, welches in der alchymistischen Terminologie als Quintessenz bezeichnet wird. Diese gehört zu den höchsten Formen spagyrischer (im wahrsten Sinne ,alchymistischer') Pflanzenmittel.

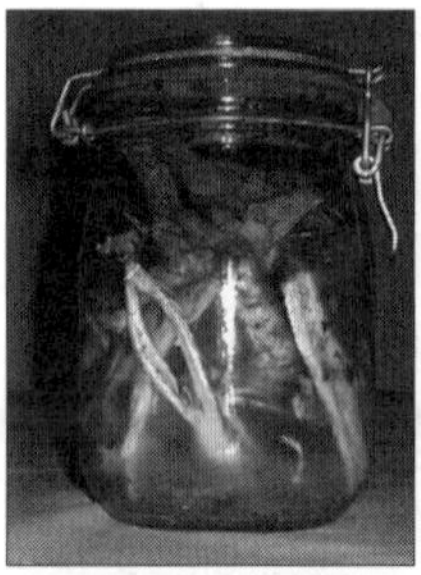

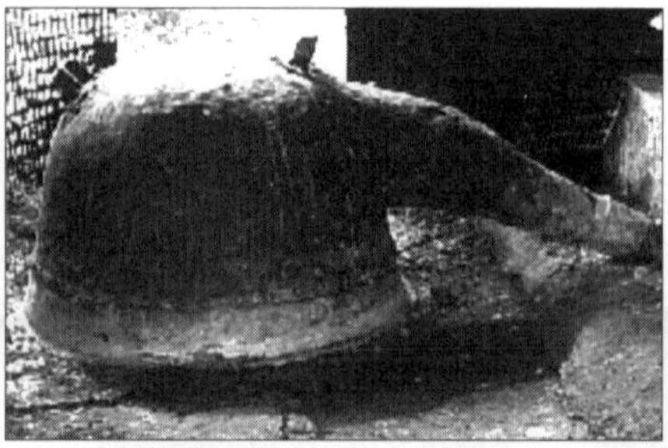

Bilder zur spagyrischen Arbeit am Pflanzen-Merkur (von oben links nach unten rechts):

1) Alkoholischer Wurzel-Ansatz zur Mazeration.
2) Destillation des alkoholischen Mazerats ‚über den Helm' zur Scheidung und Reinigung der merkurialen Ausdrucksform von Erde, Farbe und *Phlegma*.
3) Abfließende Tropfen des im Helm kondensierenden Alkohols – von den Alten ‚die undeterminierten Tränen der Diana' genannt.
4) Ein kupferner Helm bedeckt den eingemauerten, mit Gärgut befüllten Kessel, unter dem sich die sich Feuerstelle befindet. Die Fugen sind mit einem Lehmgemisch lutiert.
5) Krüge oder Fässer aus Keramik oder Holz (mit Wasser befüllt) werden seit altersher zur Kühlung benutzt.

Der dreifältige Mensch, die 5 Elemente und Merkur

Merkur ist Geist. Im dreifältigen Menschen (Körper – Geist – Seele) ist er Geist im Sinne von Gemüt – ein Geist der Natur also. Als Gemüt im Menschen hält er die Seele gefangen und führt mit ihr einen oft lebenslang dauernden Kampf, um sie zu überwinden. Zugleich existiert er aber alleine aus dieser Seelenkraft, welche in den ‚natürlichen Dingen' dem Sulfur entspricht. Es heißt: ‚Der Erwachte bedient sich Merkurs nach seinem Willen, der Schlafende verliert in ihm Sinn und Leben'.

Merkur ist Schalk und der lustige Begleiter, er ist das dunkelste Chaos und der alles verschlingende Drache in den hermetischen Traditionen der Völker.

Als Seele ist der Mensch nicht Natur, sondern ein Tropfen aus dem Meer göttlicher Bewusstheit. ‚In der Natur manifestiert' bedeutet: ‚in die Hüllen des Gemüts eingewoben' bzw. ‚Merkur zum Begleiter'. Diese Dinge sind zentrale Themen der Spiritualität, worüber wir hier aber nicht zu handeln haben, sondern uns um die Beschaffenheit der Natur und darin besonders um das Wesen von Heilkunst und Heilmitteln kümmern. Es ist aber andererseits gerade in der Heilkunst unumgänglich, immer wieder den Zusammenhang zwischen den ganz großen und den kleineren Angelegenheiten zu betrachten, denn erst dadurch erlangen wir Schritt für Schritt eine echte ‚ganzheitliche' Perspektive, welche uns in die Lage setzt, die Art der Kräfte und Dinge, von denen wir hier handeln, zu unterscheiden.

Alle Dinge der Natur, Steine, Pflanzen, Bäume, Metalle und Tiere sind direkt mit dem Spiel der 5 Elemente verbunden und haben ihre Existenz daraus. Ihre Seele befindet sich in der einen oder anderen Phase einer langen Wanderschaft, vom tiefen Schlaf im Mineralreich zum Traum in den Pflanzen bis zur Möglichkeit des Erwachens im menschlichen Körper.

Dort erst ist es der Seele gegeben, das Spiel der Elemente abzustreifen und den Zweck ihrer Wanderung zu erfüllen. Das, was Mineral, Pflanze oder Tier aus ihren Elementen ganz ‚natürlich' erhalten, wird für die menschliche Seele oft zur Entscheidungsfrage: Will ich, oder will ich nicht. Deshalb beschreiben die ältesten Texte eine Wirkung der Elemente nicht nur auf den Körper, sondern auch auf die Seele insbesondere des Menschen. In den asiatischen Traditionen bedeutet ein Aspekt der Fünf – „Lust, Ärger, Gier, Verhaftung und Egoismus" – die ‚5 Krokodile' für die Seele, welche diese (als eigentlichen Bewohner des Körpers) schwächen und das Leben des Menschen zerfressen. Dennoch sind sie für das Fahrzeug der Seele, den Körper, grundlegendes Konstrukt.

In (durch die Physis) abgeschwächter Form wirken sich diese Elementarkräfte auch auf den körperlichen und geistigen Gesundheitszustand aus. Ihre Auswirkung kann von förderlicher oder hinderlicher Art sein. Diese beiden Wirkrichtungen betreffen den geistigen Zustand (den Gemütszustand) und die Verfassung des Körpers.

Ist zum Beispiel das Wasser-Element im Menschen in Ordnung, dann ist er mutig und besitzt gute ‚Aufrichtekraft'. Er gleicht dem Halm des Strandhafers, der, viel vom Wind gebeugt, dennoch nicht knickt, sondern sich immer wieder aufrichtet und seine Früchte zur Reife trägt. Wirkt das Element gegen seinen Körper, dann ist der Mensch ängstlich, mutlos und es fehlt ihm die Kraft, sich aufzurichten. Das kann sich auf alle drei Aspekte des Körper-Geist-Seele-Kontinuums auswirken. Im Spirituell-Seelischen kann der Mut zur Selbsterkenntnis versagen, Geist bzw. Gemüt können in der Farbe der Angst gefärbt sein, und im grobstofflichen Körper kann ein Nieren- oder Knochenleiden entstehen.

Ein anderes Beispiel: Ärger und Egoismus gehören zum Wesen des Äther-Elements. Das ist zwar eine sehr allgemeine Aussage, denn genaugenommen ist Ärger ein Aspekt des aus dem Äther hervorbrechenden Feuer-Elements (Galle – Äther/Feuer), woraus dann mitunter das eigentliche Feuer als ‚Zorn' entflammt, doch wir wollen hier nur die Wirkmechanismen der Naturkräfte betrachten und, wo möglich, der Versuchung ausweichen, immer tiefer und detaillierter in die einzelnen Aspekte einzudringen. Ein ärgerlicher Mensch lebt also in dieser Variante des Gemüts und je länger, umso mehr ‚vergisst' seine Seele, wird überwunden von diesem Aspekt des ‚Naturgeistes', welcher für sich ja eine neutrale Kraft darstellt, und der Mensch verliert sich selbst. Am Ende triumphiert das Element, ohne es ‚(gut oder) böse gemeint' zu haben. Zu ihrer Existenz im Menschen benötigt diese Naturkraft einen Teil der Seelenkraft des Menschen und dadurch wird sie in gewisser Weise ‚sulfurisiert', dh. sie bekommt eine eigene, individuelle Art und Weise, in ‚ihrem' Körper zu wirken und eine Bezogenheit auf bestimmte, dem Individuum entsprechende Angelegenheiten.

Ähnlich werden wir das bei den Themen des Sulfurs kennenlernen: Wenn die Pflanze im Alkohol mazeriert, dann verbindet sich der innere Pflanzenmerkur mit dem Merkur des Alkohols und beide werden eins aufgrund ihrer Neutralität. Zugleich gehen aber auch sulfurische, seelenhafte Aspekte der Pflanze (Farbe, Harze, Öle) in den Alkohol ein, wodurch das Mazerat dann eigentlich individualisiert wird. Das eröffnet einen anderen Blickwinkel auf den Spruch: ‚Der sulfurische Mars fängt den merkurialen Hasen' bzw. auf die Kontrollposition Feuer-Luft. Dieses Wirken ist ein ‚Zusammenwirken', Merkur und Sulfur bedingen sich gegenseitig, nicht einseitig (vom Feuer zur Luft). Merkur wird zwar eingefangen und kontrolliert, aber er zieht dieses geradezu heran. So als ob die Seele das Gemüt kontrollieren möchte, indem sie sich darin vertieft und diesen Irrtum nicht bemerkt, weil sie dermaßen von ihm angezogen wird.

Wir wollen noch beim obigen Beispiel bleiben und andere Qualitäten des Äther-Elements bedenken: Großzügigkeit, Ausgeglichenheit und geistige Stabilität. Das scheinen auf den ersten Blick durchaus positive Eigenschaften zu sein, und der Mensch lebt damit besser als mit Ärger und Egoismus. Es sind aber ebenfalls Eigenschaften und damit Teilaspekte einer Naturkraft, welche sich von der Seelenkraft des Menschen ernähren. Wenn die Ätherqualitäten überhand nehmen und im Gemüt des Menschen dominieren, dann gefährden sie die Seele, indem sie diese einbinden und daran hindern ‚zu erwachen. So lebt dann der Großzügige und der Ausgeglichene dahin und lässt ebenfalls seine Zeit in solchem Wohlbefinden vergehen, ohne den Zweck des Lebens zu erkennen und die Chance des Körpers genutzt zu haben. Das ist eine Auswirkung der Elemente auf die Seele.

Das Gemüt muss unter den Qualitäten der Elemente nicht leiden. Merkur verträgt alles – nur nicht das Missachtet-werden. Im Körper – als Austragungsort des Seele-Gemüt-Dramas – kann sich im obigen Beispiel gegebenenfalls ein Leberleiden anbahnen, wenn das Äther-Element durch die (sulfurische) ‚Anwesenheit' dermaßen blockiert wird, dass die Qualität der ‚Durchgängigkeit' zu sehr darunter leidet. Bedenkend, dass Leber- bzw.Ätherthemen typischerweise mit den Augen in Zusammenhang stehen, könnte man ganz einfach sagen: Der Eine bleibt kurzsichtig – der Andere wird weitsichtig, unabhängig davon ob er selbst von ‚negativem' elementarem Ärger oder von ‚positiver' elementarer Großzügigkeit überwunden wird. Weder ein positiv noch ein negativ aspektiertes Äther-Element gewährleistet Harmonie und Gesundheit. Harmonie kann Zielvorstellung oder Ausrichtung sein, ist aber keine längerfristige Realität im Spiel der Elemente. Oder einfacher: „Keiner ist rundum gesund. Irgendwo mangelts jedem." Das kommt nicht aus den Elementen, sondern aus der ihnen zugrundeliegenden Polarität von Sulfur und Merkur und soweit es den Menschen betrifft, kommt es vom Drama zwischen Seele und Geist.

Soweit Merkur dieser Geist ist – und er ist es eigentlich nur wirklich in den natürlichen Dingen, während er im Menschen eher allegorisch mit dem Gemüt verglichen werden darf – narrt er jeden, der ihm folgt. Deswegen stellt der Mensch seit Urzeiten die Spiritualität höher als Natur- oder Heilkunde.

Das alles ist sehr sensibel abzuhandeln. Es ist der Grenzbereich zwischen Spiritualität und Heilkunst. An letzterem ist beinah jeder interessiert, zumindest soweit es mit seinem körperlichem Wohlbefinden zu tun hat. An ersterem findet kaum jemand Gefallen, weil es eine Scheidung der unbekannten Seele vom gewohnten Gemüt mit sich bringt – eben das, was die Alten mit ‚Mensch, erkenne dich selbst' meinten. Dazwischen liegt jener Bereich der Heilkunst, der den Geist des Menschen betrifft. Hier gibt es schon starke Sympathieschwankungen. Ist nicht im Allgemeinen dem Patienten ein solcher Heilkundiger lieber, der eine Pille für die Leberkrankheit verabreicht als einer, der ihn mahnt ‚sein Denken zu wandeln und dies zudem noch zum Grund einer möglichen Heilung erklärt? Nur der Tausendste wird seine Ärgerlichkeit erkennen und beseitigen wollen, die Anderen schlucken Pillen und denken weiter wie gewohnt. Was also sollte reizvoll sein an einem Versprechen von ‚Gesundheit für Körper, Geist und Seele'? Es ist zu einem ‚geflügelten Wort' geworden und bleibt doch meist unüberlegt. Gesundheit für den Körper will jeder. Gesundheit für den Geist will man nur bedingt, nämlich soweit es die gewohnte Geisteshaltung nicht in Frage stellt. Und Gesundheit für die Seele würde Erleuchtung bedeuten, Heiligkeit und dergleichen, wozu kaum einer die Notwendigkeit erkennt. Deshalb heißt es in den asiatischen Traditionen: „Die Wurzel jeder Krankheit ist Unwissenheit." Unwissenheit ist die Verstrickung der Seele im Gemüt/Geist, die Vermischung von Sulfur und Merkur. Unwissenheit ist mangelnde Scheidekunst. Sie treibt das Rad der Natur weiter und hält den Menschen darin gefangen, wirft ihn herum wie der Wind das Herbstblatt. Das ist die Verbindung von Sulfur und Merkur im rasenden Tanz der großen Polarität. Die ‚Kunst des Menschen' setzt dort an, wo die Natur nicht mehr weitermacht (bzw. ‚vielleicht' in sehr, sehr langen Zeiträumen weitermacht). Das ist die Scheidung von Sulfur und Merkur. Damit endet die Unwissenheit und die Wurzel der Krankheit.

Genau darin mag wieder eine Bedeutung von Spagyrik und Alchmie in der Naturheilkunde liegen: Durch die kunstgemäße Präparation der Pflanzen und Steine wird die heilende Wirkung tiefer geführt. Sehr hohe Präparate greifen weit über die grobstofflich-körperliche Struktur des Menschen hinein – in den Geist. Nicht das eigentlich Feinstoffliche, aber doch zumindest die subtilen und energetischen Gewebe werden tief durchwirkt und in ihrer Natur zurechtgerückt. Das beeinflusst auch Denken und Lebensart auf hilfreiche Weise und wird so im wahrsten Sinne zu einer ‚sanften Medizin', weil sie nicht nur den Körper, sondern auch den Geist bis zu einem gewissen Grad reinigen und veredeln und dadurch Auswirkungen auf den Körper lindern kann.

Anders ausgedrückt: Die Einnahme eines sehr kunstvoll präparierten alchymistischen Mittels zieht zwar nicht selbst den Karren aus dem Dreck, aber die Kraft des Mittels begleitet nicht nur den Körper, sondern auch den Geist bei Überwindung und Verwandlung und verhilft dadurch zu einer nachhaltigen Genesung.

Hier in der grobstofflichen Natur lebend, ist der Mensch mit der Ausnahme beschenkt, am Erwachen seiner Seele mitzuarbeiten. Minerale, Pflanzen und Tiere besitzen dieses Geschenk nicht. Sie sind für die gesamte Dauer ihrer körperlichen Existenz dem Gemüt unterworfen, mit anderen Worten: vom Naturgeist regiert. Das Seelenhafte hat in ihnen noch nicht jene Wachheit erlangt, die eine Erinnerung an einen Zustand jenseits der Natur, an den Ozean grenzenloser Bewusstheit, mit sich bringt. In der elementaren Natur ist Merkur der König, und nur der Mensch hängt potentiell nicht von seiner Herrschaft ab. Der Mensch hat diese ‚Erinnerung'. Deshalb kann auch nur er absichtlich gegen die Natur wirken und missbraucht seine spirituelle Kompetenz auf hilflose, verwirrte und dumpfe Art und weiß nicht, dass der Grund seiner blinden Zerstörungstaten in dem Versuch besteht, die Merkurialkraft in seine Schranken zu weisen, so als wolle er die Natur dafür strafen, dass sie ihm den Blick auf seine eigene Seele verstellt. Zuinnerst sein Unvermögen erfühlend, beginnt er, alles Natürliche zu hassen. Dadurch wird der Mantel um die Seele immer dichter, denn sie kann nur durch Natürlichkeit erkannt werden. Zu spät bemerkt er, dass er auf diese Weise alles verliert: seinen Körper, die Natur und sich selbst. Das geschieht aus Unverständnis und Verwirrung.

Paracelsus sagt: „Die Natur hat keinen Feind, außer den, der sie nicht versteht." Wie relativ gering dieses Verstehen auch sein mag – immer wird es als Voraussetzung zum Menschsein taugen. Zwischen Natur und Mensch besteht ein Verhältnis wie zwischen Gastgeber und Gast. Wenn der Gast den Gastgeber missachtet, dann wird er erfrieren und verhungern, denn er kennt die Regeln des Hauses nicht, kann selbst die Vorratskammer nicht öffnen. Es ist auch eine große Dummheit darin, denn er wird zu seiner großen Bestürzung bald merken, dass er den Gastgeber nicht ‚los wird'. Der Naturgeist kann nicht ‚sterben'. Er kann zwar in tiefster Stille verharren, wird aber irgendwann immer seine Anwesenheit kundtun, sein Spiel wieder fortsetzen, wird wieder und wieder auf die Umstände hinweisen und schließlich den Gast verjagen. Und was geschieht, wenn ein Mensch aus der Natur verjagt wird? Während des Verjagt-Werdens entsteht in ihm eine große Angst – und er schlägt immer mehr um sich – und er wird immer mehr verjagt. Er versucht zu schützen, abzuwehren, zu bekämpfen, zu kontrollieren, zu sichern, zu impfen, zu manipulieren, zu vernichten – bis er sich mit seinem unwürdigen Treiben selbst vernichtet hat.

Ein gutmütiger Gast auf lange Zeit geht dem Gastgeber zur Hand. Beide entwickeln Verständnis füreinander und richten sich wohlig ein. Sie wissen, dass sie eine Zeit miteinander verbringen werden und beschließen,‚das Beste daraus zu machen'. Sicherlich ist es mit diesem Gastgeber keine einfache Angelegenheit. Er ist launisch – doch seine Launen kündigen sich an. Der Gast sammelt Brennholz, bevor sein Gastgeber seinen schneeweißen Mantel anzieht. Er sammelt Wasser, bevor sein Gastgeber Dürre und Trockenheit um sich verbreitet. Er legt haltbare Vorräte an, bevor sein Gastgeber die Speisekammer verschließt. Der Gastgeber achtet und schätzt einen solchen Gast und weist ihn an, Arzneien zu bereiten für den Fall, dass er sich in seinem Hause anstößt und verletzt.

Die Alchymie, als eine uralte initiatische Kunst, stellt in die merkuriale Natur die ‚Kunst des Menschen'. „Der Alchymist führt das weiter, was die Natur nicht in seine Vollendung gebracht hat", sagt *Paracelsus*. Das ist das genaue Gegenteil dessen, was wir immer noch global erleben, wenngleich in seinen letzten Zügen.

Es sind zwei Seiten ein und der selben Angelegenheit. Hier befindet sich die Zerstörung der Natur, zu der der Mensch alleine fähig ist, und dort ist die Veredelung der Natur, ebenfalls alleine dem Menschen möglich. Beide Wege führen letztendlich zu Auflösung, Scheidung und Wiedervereinigung. Der eine ‚kunstlos', der andere ‚kunstvoll'. Der eine durch vollständigen Verlust, der andere durch Entfaltung und Vollendung.

Wenn wir nun hinter all dem begriffen haben, dass das merkuriale Prinzip für den Menschen eine erweiterte Bedeutung hat als für Pflanzen, Minerale u. dgl., dann betrachten wir die Ähnlichkeit mit *Paracelsus'* signaturkundlicher Betrachtung der Seerose, die wir später noch eingehender besprechen werden: Die Seerose besteht alleine aus dem Wasser-Element, und wenn man dieses aus ihr fortnimmt, dann verschwinden alle anderen Elemente, das Feuer, die Erde, usf. mit ihm. Das ist, weil das Wasser das dominierende Element in dieser Pflanze ist – unabhängig davon, dass auf einer anderen, umfassenderen Betrachtungsebene im gesamten Pflanzenreich nur alleine das Wasser-Element voll entfaltet ist. Nun transformieren wir diese Ansicht in unser obiges Thema und versuchen den Gedanken: Die gesamte Natur besteht alleine aus dem Merkur-Prinzip. Wenn dieses verschwinden würde, gäbe es weder Sulfur, noch Sal, noch irgendeine Ausdrucksform dieser Kräfte. Der Mensch aber besteht nur in Geist und Körper aus der Natur – die Seele ist nicht die Natur selbst, sondern das Leben darin. Wenn die Natur vom Menschen zurückgenommen wird, dann bleibt er dennoch als Seele – der Gast hat seinen Gastgeber verlassen. Verdelung bedeutet: sich nach Belieben zu verabschieden und immer wieder willkommen zu sein. ‚Kunstlos' bedeutet: hinausgeworfen zu werden. Wenn die Seele vom Menschen genommen wird, dann bleibt der Mensch bei der Seele und hinterlässt die Natur. Wir haben darüber bereits bei den Themen des ‚philosophischen Todes' nachgedacht.

Deshalb sagt *Paracelsus:* „Aus allen Dingen der Natur kann eine Quintessenz geschieden werden. Aus dem Menschen aber kann man die Quintessenz nicht scheiden, weil sie (sinngemäß) so eng mit der Seele zusammen ist." Sonst würde sein Wesen, seine Seele fortgehen und einen toten, kraftlosen, verwesenden Körper hinterlassen, worin diese Quintessenz nicht mehr bleiben könnte. Aus den Pflanzen und Steinen wird die Quintessenz *in forma* einer Farbe, einer Feiste, eines Fluids oder Ähnlichem gezogen und bleibt ‚lebendig' erhalten. Dort gleicht die Quintessenz der Seele und diese ist vollständig an eine, wenngleich bisweilen sehr subtile, stoffliche Ausdrucksform gebunden. Dieselbe Seele oder Quintessenz ist auch im Menschen. Aber dort hat sie bereits jene Reife erreicht und erinnert sich an ihre Herkunft von jenseits der Natur, und bei entsprechender Scheidung zieht sie dorthin zurück. Das wäre dann eben nicht Alchymie, sondern Spiritualität. Deshalb heißt es: ‚Der menschliche Körper ist eine Aufforderung, das Rad der Geburten und Tode für immer zu verlassen.'

Die anderen natürlichen (pflanzlichen, mineralischen, metallischen) Körper bieten den in ihnen wohnenden Seelen diese Möglichkeit nicht. Ihre Idee, ihre Quintessenz, ihre Seele haftet noch vollständig an der einen oder anderen Erscheinungsform, sei sie grob- oder feinstofflicher Art. Die vedischen Texte sprechen von 8 400 000 Schöpfungsformen, welche eine Seele durchwandert, bis hin zu der Chance, einen menschlichen Körper zu erhalten.

So ist also mit ‚Merkur' der Naturgeist gemeint, welcher in der gesamten Natur obligat und nur im Menschen relativ ist. Aus spiritueller Sicht hat nur der Mensch die Möglichkeit, sich seinem Einfluss und damit dem Einfluss der Elemente zu entziehen. Deshalb kann er sie auch – gewissermaßen als Außenstehender – betrachten, ihr Zusammenspiel begreifen und daraus die ‚Kunst' erlernen. Solche Betrachtungen sind zum Verständnis der Natur und damit zum Verständnis der Heilkunst notwendig. Man kann sie nicht einfach beiseite lassen und sagen: Das gehört ganz woanders hin. Diejenigen, welche diese Dinge wirklich ‚wissen', werden mit unseren kleinen Bemühungen um ein Verständnis der Natur nachsichtig sein und nicht meinen, man wolle ihnen auch nur ein einziges Wort aus dem Mund nehmen.

Es haben Analogien die selbe leidige Eigenart wie Vergleiche: sie hinken, aber sie sind unverzichtbar. Man ist geneigt, dem hermetischen Merkur-Prinzip als Gegenstück das Sulfur-Prinzip gegenüberzustellen, so als wolle man eine typische Polarität mit den Worten ‚Sonne – Mond' ausdrücken. Dazu analog die Begriffe ‚Seele – Gemüt' zu setzen, ist durchaus stimmig und wertvoll zu Anschauung und Verständnis. Hinkend daran ist, dass Seele und Gemüt als Gegensatzpaar bzw. deren Scheidung erst für den Menschen wirklich relevant ist, während die Natur alleine durch die Vermischung ‚Natur' sein kann.

Wenn man etwa vom ätherischen Öl als der ‚Seele der Pflanze' spricht, dann darf hier der Begriff ‚Seele' nicht genau mit der selben Bezeichnung, den Menschen betreffend, gleichgesetzt werden. Die Analogie ist richtig – nur die Betrachtungsebene ist anders. Von einem Naturreich zum anderen und besonders von den Naturreichen zum Menschen ist zum Analogisieren kein zwei- oder dreidimensionaler Maßstab tauglich, sondern es bedarf Wendelungen, Spiegelungen, Vervielfachungen, Verminderungen, mehrdimensionaler Versetzungen, kurzum ein ganzes harmonisches Geflecht aus Veränderungen, um eine Situation oder Position mit einer anderen zu vergleichen. Die ‚Seele' der Pflanze ist ebenfalls ‚Gemütsstoff'. Sie ähnelt der Seele des Menschen nur in einigen Aspekten mehr als seinem Gemüt, weswegen der Ausdruck ‚Seele' dann auch für das Sulfur-Prinzip bzw. seine Erscheinungsform zB. als ätherisches Öl angewandt wird. Andererseits wäre der Schluss sinnvoll: Der Sulfur der Pflanze ist eigentlich eine Art von Merkur. Darin besteht ja eben das Wesen Merkurs: er ist unfassbar, er wird mit allen Namen bezeichnet, er hält alle zum Narren, usf.

Suchte man tiefer nach der ‚Seele' einer Pflanze, dann würde man sie wohl eher in ihrer Quintessenz finden – und damit die Analogie *ad absurdum* führen, denn: diese Art der Quintessenz ist eine Art Merkur in sehr hoher Form. Hierzu die Warnung des *Helmont* fügen, dass man seine Zeit nicht mit der sinnlosen Suche nach einem ‚Sulfur' der Metalle vergeuden, sondern aus ihnen nur Merkur und Sal scheiden sollte, könnte nun zu einem tiefen Ausflug in die Alchymie anregen, eine Anregung, der wir hier keinen Platz geben werden, obwohl wir nach genauerer Betrachtung der Termini *Sulphur* und *Sal* jeweils zugeben müßten, dass sie in der Natur niemals vollständig vorhanden sind und weder vom Gemüt noch vom Verstand eigens für sich analysiert werden können. Immer ist Merkur der Vermittler jeglichen Seins.

Auf der anderen Seite unserer ‚behelfsweisen' Polarität – also ‚gegenüber' des Sulfurs – können wir Merkur schon stehen lassen, doch diese Ansicht wird hintergründig immer in dem unabdrängbaren Gedanken eingebettet sein, dass Merkur alles durchdringt. Vielleicht ist es einer seiner Schelmenstreiche, jegliches intellektuelle Verstehenwollen der Elemente, Prinzipien und sonstiger Naturkräfte zu relativieren und bisweilen regelrecht aufzulösen.

In der taoistisch-hermetischen Naturkunde heißt es: ‚Die Erde gebiert Metall', womit, auf die Elementegenese bezogen, gemeint ist, dass das Luft-Element aus dem Erd-Element hervorgebracht wird. Damit beginnt aber dann auch schon ein Auflösen unserer intellektuellen Vorstellung anhand einer zwei- oder dreidimensionalen Darstellung der Ereignisse und Positionen. Bis zum Erd-Element können wir ‚denken'. Dort herrscht Saturn. Von da an obliegt alles dem Wohlwollen Merkurs. Was sind das für Elemente, die auf das Erd-Element folgen?

Manche Texte vertauschen absichtlich die Positionen von Venus und Merkur, taoistische Texte nennen Venus den ‚Luftstern' und Merkur den ‚Wasserstern'. Immer wieder werden wir darauf hingewiesen, dass der reine Intellekt zum Verständnis der Natur nur begrenzt tauglich und brauchbar ist. Unser intellektuelles Denken ist dann ‚saturnisch', wenn es ‚aus dem Feuer' kommt, klar, einfach, reduziert und punktgenau ist. Auf die Frage aber, welchem Prinzip der Intellekt zugeordnet werden soll, antworten die Alten: „Merkur." Das erinnert an eine ähnliche Schalkhaftigkeit: die grüne Farbe ist die Farbe der Natur und in gewisser Weise ein Signat von Venus. Grün tritt dann in Erscheinung, wenn das saturnische, aus dem Feuer geborene Erd-Element vom Mond befeuchtet und dadurch zu ‚venerischer Erde' wird. Im Augenblick dieses ‚venerisch Werdens' bricht Merkur aus dem saturnischen Kerker aus. Zusammen mit Venus kann Merkur dann den Saturn überwinden und die lebendige Natur begründen (Bild unten). Damit ist alles Grüne merkurial (Gräser). Auf die Frage, was denn die ‚Farbe Merkurs' sei, soll dieser geantwortet haben: „Alle Farben sind mein, außer dem Grün."

„Aus dem *Saturno* mit dem *Mercurio vivo* und dem *Oleo Veneris* die Tinctur auszuarbeiten."
(aus: *Zoroaster*, >*Clavis Artis*<, 1236)

„Der *Mercurius* ist der Brunnen, in welchem der König und die Königin ersaufen und umkommen. Durch den Gesang der Sirenen, nämlich des *Mercurius*, wird dies verrichtet, aber durch den Trank der Göttin Venus wird aus allen dreien eine Geburt vieler Kinder.“ *(Zoroaster)*

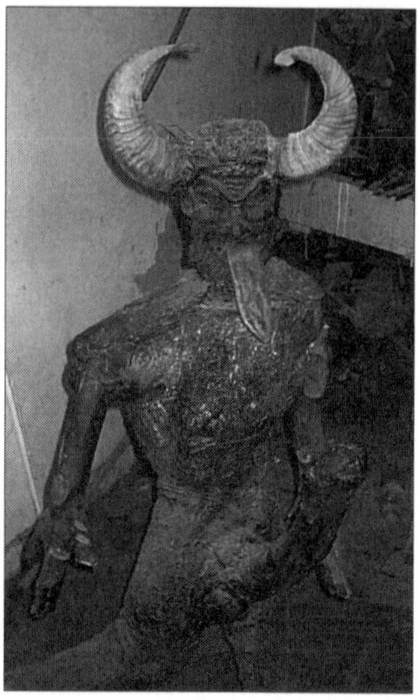

Hermes/Merkur nach einer archaischen Vorstellung (minoisch)

3) Darstellung des Sal-Prinzips aus der Pflanze durch

a) Veraschung der fixen Pflanzenreste
b) Veraschung des gesamten Destillationsrückstandes
c) Calcination
d) Scheidung und Reinigung des Salzes

Nach dem Filtrieren der oben beschriebenen merkurial-wässrigen Anteile bleiben die festen Pflanzenteile zurück. Diese sind im hermetischen Sinne eine rohe Ausdrucksform des Erd-Elements. Sie werden nun in einem geeigneten (am besten keramischen oder zumindest emaillierten) Topf zu Asche verbrannt (a) und diese, wenn nötig, zu einem lockeren Pulver zerdrückt, wodurch man die Reste eines in der Kohle als Struktur der Pflanze sichtbaren, immer noch anwesenden ‚Merkurs‘ beseitigt.

Die Asche ist grau – eine Signatur des saturnischen Erd-Elements mit der Aussage, dass hier eine Verbindung von weiß und schwarz vorliegt. Allegorisch gesehen ist nur in der grobstofflichen Materie ein Nebeneinander des prinzipiell Hellen mit dem prinzipiell Dunklen in all seinen Schattierungen möglich.

Deshalb nennt die ayurvedische Alchymie die Welt *Kam-Lok*, den Ort, an dem alle Menschen und ‚natürlichen Dinge' jedweder Art zusammentreffen können, um ihre Händel zu begleichen und in ein- und derselben Schulklasse zu lernen. In den feinstofflichen Bereichen ist diese Nebeneinander-Gräue nicht möglich – es zieht einen jeden in sein Wesensfeld.

Am einfachsten wäre es nun, die Asche mit dem filtrierten Mazerat zu vereinigen (*Coagulation*). Das wird auch bisweilen gemacht, besonders wenn Eile not tut. Nach öfterem Schütteln / Bewegen und wenigen Tagen Digestion an der Wärme hat sich ein wesentlicher salinischer Teil im Mazerat gelöst und es kann von den Schlacken abgefiltert werden. Ob ein auf diese Weise fertiggestelltes Podukt den Titel einer ‚spagyrischen' Tinktur verdient, bleibt fraglich. Es ist zwar der Sal-Faktor enthalten und damit eines der wichtigsten Kriterien erfüllt, aber es hat keinerlei spagyrische Scheidung und Reinigung stattgefunden. Weder das Merkur- noch das Sal-Prinzip wurden ‚kunstvoll' ausgearbeitet. Das Präparat wird sicherlich ganzheitlicher und wirkungsvoller sein als ein gewöhnliches, alkoholisches Extrakt. Andererseits ist die Arbeit am Sal ein offenbares Fundament der ‚natürlichen Kunst', und in der Stoffwelt ist keine vollständige Heilung ohne die unterstützende Kraft der Erde und damit des Sal-Prinzips möglich.

Diese Arbeit am Salz zu unterlassen rechtfertigt tatsächlich nur eine etwaige notwendige Eile. Klar – je weiter man das Sal-Prinzip ausarbeiten will, umso zeitaufwendiger wird der Prozess.

Andererseits wurden seit undenklichen Zeiten vermutlich in allen Medizinsystemen Aschen für sich oder in Rezepturen als spezifische Heilmittel eingesetzt. Reste oder Abkömmlinge dieser Praxis finden wir auch heute in verschiedenen Formen der Mineralstofftherapien. Aschen werden noch in vielen Zweigen alter naturheilkundlicher Traditionen verwendet. *Michael Crügner* gibt eine einfache Zubereitung zur „Kur der Mutter-Krankheiten, die aus der *Colica* oder Grimmen entspringen" im >Chymischen Frühling< an:

„(...) Denn die Matrix der Frauen wird gleicher Gestalt als eine kleine Welt erkannt sowie der Mensch für sich selbst als eine kleine Welt erkannt wird. Nun ist aber dies die allerkleinste Welt und werden doch alle Kräfte der großen Welt darin gespürt, denn der Mensch, als die edle Creatur, wird darin empfangen und geboren, deswegen sind auch die Kräfte der vier Elemente und aller Planeten unsichtbarer Weise darin verborgen. Auch der Wind und dergleichen (...). Der occidentalische Wind steigt in die Höhe, die Mutter wird perturbieret und erzeiget sich mit einer Gewalt, als wenn er das Herz erdrücken wollte. Der orientalische Wind hat seine Natur, die Kräfte der Mutter unter sich zu treiben und aus ihrer Ruhe in ein Schmerzen und Grimmen der Mutter zu verwandeln. Und das kann nicht gestillt werden, es sei denn, dass man solche Bläst und Winde stille und verzehre."

„In dieser sonderbaren Beschwerung wird vielmals ohne Frucht alles umsonst angewendet. Unter anderem sage ich einem rechten *Sale Coralliorum* und *Oleum Succini* viel gutes nach, vielmehr aber folgendem:

Sammlet den 16. Mai die Wurzel der *Bryoniae*, reibet solche, und sammlet den Saft, und lasset die *Faeculas* zu Boden setzen. Die zurückbleibende Wurzel samt dem Kraut brennet zu Aschen, mischet solche unter den Saft und bereitet mit Hilfe des Zuckers einen *Syrupum*."

nach: *Michael Crügner*, >Chymischer Frühling<, 1654, neu: Edit. *Hornfisher* 1998

Auf wunderbare Weise hat die Natur in einer einfachen Asche eine hochkomplexe Rezeptur aus mineralischen, pflanzlichen und energetischen Komponenten erstellt. Besonders Letzteres wird bei der therapeutischen Anwendung von Aschen auffällig. Auch wenn verschiedene Aschen sehr ähnliche chemische Formeln aufweisen, ist die Wirkung deutlich unterschiedlich. Als ob sich ein, durch das Verbrennen von der stofflichen Form befreiter Pflanzengeist nun erst entfalten kann.

Die ‚Veraschungen' dazu sind oft nicht einfach nur ein Verbrennen, sondern bisweilen langwährendes ‚in der Glut halten', Rühren, Reiben und gegebenenfalls prozessgenaue Beigabe von Zutaten. Die Geist-Bindung an die Asche ist auch in den magischen Traditionen bekannt (und wird dort leider nicht immer zum Guten genutzt).

Dass es selbstverständlich auch einen substantiellen Unterschied im Endprodukt ergibt, ob zum Pflanzenmerkur bloß die Asche oder ein gereinigtes Salz gegeben wird, zeigt sich unter anderem an der Färbung. Was bei eher dunklen Tinkturen kaum erkennbar ist, wird bei sehr hellen Tinkturen oder beim farblosen alkoholischen Destillat klar ersichtlich: die rohe Asche vermittelt dem Alkohol eine Farbe, während das gereinigte Salz (sofern davon während der Calcination nichts aufgeschmolzen ist) den Alkohol (weitestgehend) farblos belässt. Daraus wird ersichtlich, dass in der Asche noch ein gewisser Aspekt des sulfurischen Prinzips enthalten ist – und dies umso mehr, je dunkler sie nach dem Verbrennen bleibt bzw. je mehr Kohle noch darin ist. Es ist ja ‚alles Sulfur, was verbrennt'. In den dunklen Aschenanteilen sind noch wesentliche Ausdrucksformen der Prinzipien miteinander vermischt, davon die sulfurischen überwiegen und sich im Alkohol als Farbe ausdrücken. Von diesem ‚unreinen Schwefel' sollte eigentlich nichts ‚ins Werk mit eingehen'.

Insgesamt ist es jedenfalls ein Qualitätsmerkmal, wenn das Sal-Prinzip in einem Pflanzenpräparat sorgfältig ausgearbeitet wurde, und strenggenommen gilt der Terminus ‚spagyrisch' erst dann.

Cardilucio, ein typischer Iatrochymiker des 17. Jahrhunderts, schreibt in seinem >*Clavis*< über Zubereitung und Verwendung von Aschen und Pflanzensalzen. Er findet zur Anwendung der Salze für sich alleine die Simplicien (Kräuter, Wurzeln, Hölzer, ...) des luftigen Prinzips besonders geeignet und darunter solche mit jovialen Signaturen, die er den Zwillingen zuordnet (verschiedene Minzen, Rauten, Myrobalanen, Myrrhen, Kamillen, Muskatblüten, Gamander, Ampfer, Nesseln, Klettensamen und andere).

Nachdem er zuerst körperliche Zuordnungen und Rezepturen von rohen Kräutern beschrieben hat, wendet er sich den hierhergehörigen „spagyrischen, wohlpräparierten Sachen, so mit gewaltigem Nutz nebenst denen bisher erzählten antipathischen, jovialischen Simplicien zu denen dabei gemeldeten Affecten können gebraucht werden" zu. Solche „spagyrische, wohlpräparierten Sachen" sind unter anderen auch die *Salia:*

Sind die Salien dieser Kräuter
und Simplicien selbsten auch eine grosse
Vermehrung der Kräfften ihrer eigenen
Simplicien/sonderlich zu Heilung der Wun-
den und Schäden/ Eröffnung der verstopff-
ten Mutter / Beforderung deß Harns und
der Monatzeit / Reinigung der Nieren
und Blasen von Sand und
Stein/ rc.

„Dazu hat man nicht eben nötig, alle solche Simplicien zu Asche zu brennen, sondern nur einige, welche am häufigsten zu haben sind. (...) und dienen dazu wegen ihrer Menge sonderlich alle Arten der Camillen, brennenden und tauben Nesseln, Wollkräuter, Kletten, Gundelreben, Maßliebchen, Rainfarn, Eibsich, Malven, Lindenreiser und dergeleichen. Deren werden eine gute Quantität, nämlich viele Arm voll gesammelt wenn sie noch grün sind und nicht, wenn sie im Felde verdorrt und abgestorben sind, ausgetrocknet und hernach auf einem sauberen Herd ganz schlecht zu Aschen gebrannt. Die Asche kocht man mit gemeinem Wasser aus und macht eine klare, scharfe Lauge daraus, lässt solche in einem Glase oder steinern Geschirr abdünsten bis zur Trockenheit, solvieret es wieder im Wasser, solvieret solches abermal klar durch ein Maculatur, lässt das Wasser wieder verdünsten bis zur Trockenheit und wiederholt das Solviren, Filtriren und Coaguliren solches Salzes auch zum drittenmal, so ist es rein genug. Alsdenn calciniret man es ein wenig in einem irdenen Geschirr dergestalt, dass es nicht fließe, sondern nur die übrige Feuchtigkeit davon verdünste, tue es darauf alsofort in ein Glas mit einem engen Mundloch und verbinde das Glas, damit keine Luft aus oder ein könne, so hält sich das Salz in dem Glase lang."

„Wenn man es aber nicht dicht verbindet, ziehet das Salz aus der Luft Feuchtigkeit an sich und zerfließet in einen *Liquorem*, welches Zerfließen zwar Einige damit verwehren, dass sie die Asche der Kräuter woraus sie das Salz extrahieren wollen mit gemeinem Schwefel vermischen und den Schwefel davon verbrennen lassen und laugen hernach erst die Asche aus und machen ein Salz daraus, welches hernach nicht an der Luft zerfließet. Aber solche Leute betrügen sich selber und andere, denn wenn der Schwefel von der Asche verbrennt, so kriecht seine faule Flamme, welche eitel *Spiritus vitrioli* oder *Oleum sulphuris per campanam* ist, in die Asche, tötet die alcalische Art, und es wird aus beiden ein Mittelsalz oder *Tartarus vitriolatus*, so freilich nicht an der Luft zerfließet, hat aber auch nicht mehr die Tugend eines *Lixivial*-Salzes."

Cardilucio gehört zu den Vertretern jener Iatrochymiker, die ein ausgezeichnetes Kräuterwissen besaßen, was aber die Präparation ihrer Mittel anbelangt, so waren sie in den meisten Fällen mit einfachen, mehr oder weniger spagyrischen Verfahren zufrieden. Man sieht das auch hier, indem er die Salze nicht weiter durch ‚Umkristallisieren' zu reinigen empfiehlt, sondern alleine durch Eindampfen der Lauge fixiert. Er behauptet auch von sich selbst, dass er nicht noch weiter in die hermetische Kunst eingedrungen ist, weil er schnelle, leicht darzustellende Mittel braucht.

Das erklärt auch seine Position zur Methode, die Salze mit Schwefel zusammen zu calcinieren. Natürlich verändern diese sich dabei, aber es wird sicherlich kein eigentlicher *Tartarus vitriolatus* (nach der Terminologie der Alchymisten) daraus. *Johann Kunckel von Löwenstern*, dem der ‚hermetische Tiefgang' in der alchymistischen Praxis (nach seiner eigenen Aussage) weit mehr angelegen war als die Herstellung schnell verfügbarer Arzneien und die Behandlung von Patienten („Ich brauche von denen nicht einmal eine Hand voll.") betont, dass die Salze, welche mittels Schwefelzusatz in der Calcination erlangt werden, nicht nur weniger hygroskopisch und schöner anzusehen sind, sondern auch an Wirkvermögen die einfach dargestellten Salze weit übertreffen, „was aber viele nicht wissen, weil sie den Grund nicht verstehen" (>Chymische Werck-Schul 1 u. 2<).

Wenn man nun bedenkt, dass beim ersten Glühen der aus der Asche gelösten und eingedampften Salze aus diesen gewisse subtile, flüchtige Geister entfliehen, dann versucht man, dem ‚tiefer nachsinnend', Wege zu entdecken, diesen unsichtbaren *Spiriti* die Flucht zu verwehren und sie im Salz gebunden zu erhalten. Dadurch würde das Wirkvermögen des Salzes zweifellos tiefgreifender.

Diese bindende, zusammenziehende Kraft besitzt das Äther-Element, weswegen ihm alte Medizinsysteme die saure bzw. zusammenziehende oder herbe Geschmackssignatur zuordnen. Es heißt: ‚Äther kontrolliert Erde'. Eine dieser Kontrollfunktionen, welche König Jupiter über seinen verbannten Vater Saturn ausübt, ist, dass er diesen im Amt festhält. Er soll ‚hemmen, zusammenhalten, einkerkern'. Dafür sorgt Jupiter in oberster Instanz.

In jedem Element ist eine Kraft anwesend, die der ersten und eigentlichen Qualität jenes Elementes entgegenwirkt. Dieses Gefüge hält den Elementereigen in der Physis aufrecht. Diese Kraft stammt vom Kontrollelement, und wir können ihre Anwesenheit am Geschmack erkennen. Nehmen wir als Beispiel die bitter schmeckenden Kräuter – und zwar solche, die ganz einfach bitter, nicht aber von sehr grober, allzustarker Bitterkeit sind, Kräuter, die gerne etwas würzige, aromatische Schärfe bei sich haben wie das Tausendgüldenkraut oder die Schafgarbe.

Diese Kräuter besitzen eine ausgeprägt kühlende Wirkung. Ihre Geschmackssignatur entspricht aber dem Feuer-Element. Das ist so, weil im Feuer eine Kraft wirkt, die seiner Natur insoweit entgegensteht, als dass es elementar in der Physis anwesend (wirksam) bleiben kann und nicht als reine Geist-Qualität ins Feinstoffliche entweicht. Sonst wäre die Physis ohne Licht, Wärme und Aktivität. Die Kraft, welche nun als ‚bitter' geschmeckt wird, kommt vom Wasser-Element und wirkt sich im Feuer-Element aus. Das ist ein Aspekt der ‚Kontrollposition'.

Oder: Dem Wasser-Element entspricht der salzige Geschmack – so wie das Meer schmeckt. Das kommt von der Erde, um die Elementarkraft ‚Wasser' davor zu bewahren, dass sie in den Monden-Äther entweicht und ‚alles zerfließt'.

Geschmackssignaturen der Elementarbereiche:

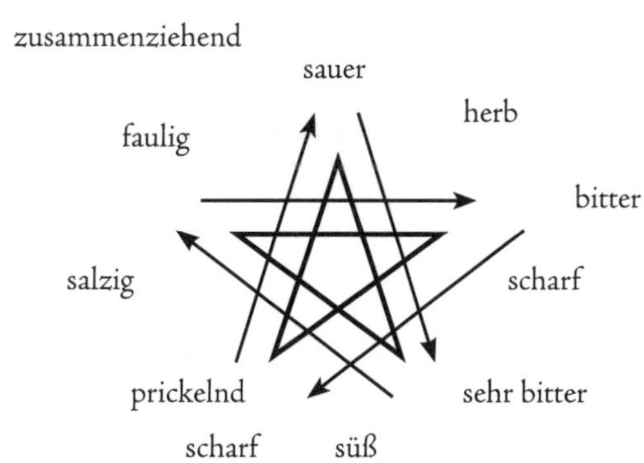

Ebenso verhält es sich nun zwischen Äther und Erde: Die Säuren sind feuriger Natur, haben aber eine wesentliche merkuriale Qualität (vom Luft-Element) in sich. Dadurch können sie (wie schon erwähnt) ‚Erde öffnen' und zugleich den darin eingeschlossenen Merkur freisetzen. Die Kontrolle über diesen Prozess hat Jupiter. Er (ebd.: der Äther) verwendet die in ihm anwesende (und vom Luft-Element stammende) Kraft zum einen, um der Zusammenziehung und Zentrierung zu entsprechen, welche für die Schwelle vom lunaren zum solaren Äther typisch ist; zum andern ‚sendet' er diese Kraft – durch das Feuer-Element hindurch – in das Erd-Element (wo sich diese Säure in Süße verwandelt). Damit gewährt er eine kontrollierte Öffnung des Erdhaften, Dichten, Festen, verhindert aber gleichzeitig ein Entfliehen dessen, das allzuschnell oder ‚zu bald' aus der Erde entweichen will.

Genau das ereignet sich im Tiegel, wenn zur Calcination des Salzes die Asche zuvor mit Schwefel verbrannt wird. Wie *Cardilucio* sagt, wird ein saurer Geist, eigentlich ein *Spiritus sulphuris*, freigesetzt, welcher ‚in das Salz kriecht'. Das entspricht einer Anerkennung des Pflanzensalzes, als bereits dem Mineralreich zugehörig. (Minerale und Metalle werden sehr oft mit Zusatz von Schwefel calciniert). Dieses Erd-Element wird dadurch besser ‚geöffnet', damit die darin verschlossenen Wirkkräfte frei und ‚zugänglich' werden. Zugleich verhindert jener joviale, zusammenziehende, festhaltende, in weiterer Folge ja auf ein Fixieren gerichtete Aspekt der schwefligen Säure das Entfliehen der so erschlossenen Kraft.

So geht die Überlegung *Kunckels* tiefer als die von *Cardilucio*. Es bleibt auch das von letzterem erwähnte *Oleum Sulphuris* nicht beim Salz, wenn man die Temperatur entsprechend heiß führt. Insgesamt soll das aber keine Aufforderung sein, spagyrische Pflanzensalze immer mit Schwefel-Zusatz zu bereiten. Es bleibt ja dennoch etwas Salinisches vom Schwefel beim Salz und ist dann auch im fertigen Präparat nicht davon geschieden. Das ist bei einfachen Pflanzenarbeiten meist unerwünscht und nur manchmal – im Rahmen einer Rezeptur bzw. Komposition – beabsichtigt.

Wir sehen aber an solchen Beispielen, wie Signaturenkunde und hermetische Betrachtung die spagyrisch-alchymistische Praxis grundlegend beeinflussen und fahren nun fort mit der spagyrischen Präparation der Pflanzen, wobei wir uns noch ein wenig dem Thema der Calcination im Allgemeinen widmen.

Bislang haben wir über die Veraschung des festen Pflanzenrückstandes gesprochen, wozu der feste Rückstand, die Pflanzenmasse, ausgepresst und für sich alleine verascht wird, was in weiterer Folge zum Fixen Salz (*Sal fixum*), dem typischen Vertreter des Sal-Prinzips führt.

Sofern der Alkohol klar abdestilliert wird, bleibt das *Phlegma* mit der ‚Farbe' zurück. Wie diese für sich gereinigt und aufgeschlossen wird, um eine Ausdrucksform des Sulfur-Prinzips zu erlangen (*Sulphur fix* oder *Sal sulphuris*), darüber haben wir bei den Themen des Sulfurs bereits gesprochen. Da diese Arbeiten aber in der Folge zur ‚spagyrischen Essenz' führen, wollen wir sie hier außer Acht lassen. Vollständigkeitshalber muss erwähnt werden, dass, wenn man zur ‚Tinktur' das Merkur-Prinzip klar und farblos darstellen will, der gesamte Destillationsrückstand zusammen eingedampft, getrocknet und verascht, also das Feste nicht vom Flüssigen durch Auspressen getrennt wird, weil man ja auf keine Scheidung des Sulfur-Prinzips abzielt. So wird also zur Darstellung einer ‚spagyrischen Tinktur' das *Sal fixum* aus der gesamten ‚rückständigen Pflanzensuppe' (Pflanzenmasse, *Phlegma* und Farbe) bereitet, das Salz wird also in der Bezeichnung nicht weiter unterschieden.

Der Prozess der Veraschung kann unmittelbar in die eigentliche Calcination (c) übergeleitet werden. Hier geht es zum einen darum, alle Kohle-Teile der Asche auszuglühen, zum anderen, durch fortgesetztes, längeres Glühen eine Öffnung des Erd-Elements mittels des Feuers zu erreichen bzw. hermetisch formuliert: durch Anregung des äußeren Feuers das innere Feuer der Pflanze aktivieren. Wir werden über diese doppelseitige Beziehung bei den Themen des Sulfurs noch sprechen. Die Position des Feuers als ‚Mutter-Element' zur Erde gleicht der einer Mutter zum Kind. Nehmen wir als Beispiel eine Vogelmutter: Sie füttert ihr Küken, damit es ‚kompakter' wird. Das entspricht dem ‚Hervorbringungszyklus', der ‚Nährrichtung' der Elemente sowie der Dichte schaffenden, fixierenden, individualisierenden Tendenz des Feuers. Andererseits öffnet das Küken weit den Schnabel, wenn die Mutter mit einem Wurm oder Korn herangeflogen kommt. Jedes Kind öffnet sich dem nährenden Einfluss der Mutter. Dieses Öffnen entspricht der Aktivität der ‚gegenwirkenden', öffnenden Feuer-Kraft, dem *ios*.

Dann soll auch durch die Calcination eine größtmögliche Helligkeit der Asche erreicht werden. Hat man diese erlangt, dann glüht man dennoch einige Stunden weiter – eben wegen der Kräftewirksamkeit. Auch hier gilt es wieder: Zeit geben. Rasche Calcinationen führen oft zu schlampigen oder minderwertigen Ergebnissen an sich, und dazu kommt, dass sich eben jene Feuer-Erde-Wirkung nicht genügend entfalten kann. Mitunter gerät auch die Temperatur zu hoch und Teile der Asche verschlacken, was bedeutet, dass „der den Salzen innewohnende Geist entflieht". Nicht selten werden mehrere Calcinationen mit einer Asche durchgeführt, dazwischen gerührt, befeuchtet, etc. Hier lehrt uns manch urzeitliche und doch lebendige Tradition zB. der ayurvedischen Alchymie, dass die Alten stets Zeit als Ingredienz reichlich zusetzten. Dort werden viele Stunden und Tage für Calcinationen verwendet, auch wenn das rein stofflich nicht nötig zu sein scheint.

Die Calcinationstemperatur bewegt sich je nach Konsistenz der Pflanzenteile meist etwa zwischen 450° und 750°C. Sehr feine Blüten erscheinen selbstverständlich schon bei geringer Hitze bald weiß. Andere Pflanzenteile erfordern hohe Temperaturen. Die Hitze wird nach erfolgter Veraschung langsam verstärkt (zur eigentlichen ‚Calcination') und dann meist gleichmäßig gehalten.

Schließlich lässt man abkühlen und reinigt die calcinierte Asche, indem man die hellen, wasserlöslichen Salze von den dunkleren, nicht löslichen ‚Schlacken', dem *Caput mortuum*, scheidet. Das Wasser wird abgedampft und die Salze durch wiederholtes Lösen und Auskristallisieren gereinigt, bis sie schöne, reine, klare, weißliche oder durchsichtige Kristalle bilden. Je schöner, desto besser. Das kann einige Tage dauern, aber auch Wochen und Monate in Anspruch nehmen. Darin liegt wieder ein Qualitätsunterschied.

Eine Idee der Reinigung ist folgende: Wenn das spagyrische Präparat im Augenblick der Einnahme oder äußerlichen Anwendung den Körper berührt, dann erkennt der Körper aus der Anwesenheit des Salzes, dass es sich hier um ‚seinesgleichen' handelt, allerdings auf sehr hoher Ebene. Die Reinheit des Salzes wird zum Passierschein für das gesamte Mittel, inwieweit es die Stofflichkeit des Körpers durchwirken und an jeden Ort und in jedes grobe oder subtile Gewebe transportiert werden kann, ähnlich einem sehr feinen, energetischen oder homöopathischen Präparats, nur mit dem Unterschied, dass die Heilkraft hier von einer essentiell-stofflichen Ebene mitgetragen wird.

Das noch weitverbreitete Ignorieren des Sal-Faktors in den naturheilkundlichen Präparaten ist eine momentane Zeiterscheinung. Es hat Analogien zu Missachtung und Unverständnis gegenüber der Milz im Körper und dem Erdwesen selbst. Pflanzensalze, Milz und Umwelt – das ist alles Erd-Element und der Boden unter unseren Füßen.

Ein typischer Effekt des integrierten Salzes ist, dass das Präparat eine innere Stabilität erhält, welche selbst dann, wenn davon Hochpotenzen bereitet werden, noch eine Erd-Bezogenheit gewährleistet. Deshalb gehen immer mehr Homöopathen dazu über, ihre Potenzen auf der Basis von spagyrischen Ur-Tinkturen oder Essenzen zu bereiten. Nimmt man etwa eine einfache galenische Tinktur (alkoholischer Auszug ohne Salz), schüttelt sie fest durch und macht gleich darauf eine Hochfrequenzfotografie eines Tropfens davon, dann bemerkt man, dass dieser sein strahlendes Kraftfeld durch das Schütteln fast vollständig verloren hat. Erst nach 10 bis 20 Minuten hat sich die Tinktur erholt und das Feld des Tropfens wird wieder sichtbar. (Von wegen: ‚Vor Gebrauch schütteln.' Dieser Spruch stammt vermutlich aus einer Zeit, in der noch die Salze zugesetzt wurden). Hingegen zeigt der Tropfen einer ‚spagyrischen Tinktur' unmittelbar nach dem Schütteln ein noch stärkeres Kraftfeld als zuvor.

Letzteres ist zum einen ein Hinweis auf die innere Stabilität des spagyrischen Präparates, andererseits auf die Wichtigkeit der abschließenden Dynamisationsschritte (Circulation u. dgl.) nach vollendeter Coagulation der Prinzipien – wovon wir noch sprechen werden.

Ein weiterer Vorteil dieser inneren Stabilität mag besonders in der jetzigen Zeit darin liegen, dass ein spagyrisches Präparat manche Belastungen besser erträgt als ‚ungeerdete' Zubereitungen, also nicht nur einfache Erschütterungen, sondern auch mannigfaltige Strahlungen und sonstige Schwingungen, welche zur Zeit insgesamt die Entfaltungen lebendiger Harmonien massiv beeinträchtigen. Nebenbei ‚erinnert' das präparierte Salz die Körpergewebe an einen reinen, gesunden Zustand, ‚ermutigt' sie gewissermaßen und regt zur ‚Nachahmung' an und vermittelt eine ‚Idee der Entschlackung'.

Ein anderes Zeichen der inneren Stabilität ist die vermutlich unbegrenzte Haltbarkeit spagyrischer Präparate (vorausgesetzt, die Prinzipien wurden sauber ausgearbeitet). Diese Mittel reifen wie die besten alten Weine und würden sie vermutlich lebendig und wirksam noch um Jahrtausende überdauern – hätten wir dazu Erfahrungswerte. Der Zeitraum einer menschlichen Generation nach der Fertigstellung einer spagyrischen Essenz entspricht der Zeit der ersten Reifung, was wiederum einen Grund gibt für eine abschließende kunstvolle, dynamische Coagulation, damit der Reifungsprozess der im alchymistischen Sinne ‚wiedergeborenen' Pflanze angeregt wird.

hermetisches bzw.
alchymistisches Symbol für
Sal, Salz und Sal-Prinzip

Das Einbeziehen des Sal-Faktors ist einer der wichtigsten Punkte, in dem sich spagyrische Präparate von anderen phytotherapeutischen oder homöopathischen Präparaten unterscheiden. Und es ist eine der Voraussetzungen für den Titel ‚spagyrisch' einer Pflanzenzubereitung. Abschließend noch eine Anregung zu Gedanken über das Wesen des Erd-Salzes nach einem Tractat von *Loen* (1729):

„*Quercetanus* gedenket eines *Medici* zu *Cracau* in Polen, welcher sehr zierlich und philosophisch die Asche ohne fremde Vermischung aus einer ganz vollkommenen Pflanze mit Wurzel, Stengel, Blättern und Blüte bereiten konnte und ihre *Spiritus* besonders zu bewahren wusste, also dass er deren über dreißig Arten in zugeschmolzenen Gläsern zeigte. Wollte nun jemand z. e. eine Rose sehen, nahm er das Glas, darinnen der selben Asche aufbehalten war, hielt solches über ein Lampenfeuer, und als es in etwas warm worden, ließ diese zarte und unbegreifliche Asche von sich die Gestalt der Rose sehen mit allen Farben, welche allmählich wuchs an Stengel und Blättern, auch endlich eine aufgegangene Rose vollkommen praesentirte, dass nichts Angenehmeres noch dem menschlichen Auge Lieblicheres hätte können gemalt werden und man meinen sollen, es sei eine rechte, natürlich, mit grobem, begreiflichem Körper aufgewachsene Rose, die nur bedürfe, samt ihren Wurzeln in die Erde gesetzt und weiter fortgepflanzt zu werden, da es doch nichts, als ein geistiges Bild dieser Blume mit einer dergleichen Essenz begabt gewesen, welches, sobald die Wärme aufgehöret, wieder in Asche als in sein Chaos zurückgefallen.

Ein gewaltig Zeugnis aus der Natur, gegen alle, welche die Unsterblichkeit der Seele und Auferstehung des Leibes nach seiner Verbrennung zu Asche leugnen wollen, denn hier muss die *Philosophia Aristotelica*, wenn sie sieht, dass bei der Verbrennung der Pflanzen ihre Form nicht verdirbt und verloren geht, sondern sich nur als etwas Geistiges in den unverbrennlichen Teilen der Asche verbirgt, den Mund zuhalten.

Dass eine Lauge von der Asche der verbrannten Brennnesseln gemacht, nachdem solche im Winter zu Eis gefroren, die ganze Gestalt des Krautes mit allen *Minutisimis* darstellt, haben die *Physici* gar vielfältig beobachtet. Der sehr berühmte *D. Hoffmann* hat in der Auflösung der Corallen mit *Spiritus Salis* fast ein Gleiches wahrgenommen. (...)

Dass der Pflanzen Kraft durch die Verbrennung nicht ersterbe und vergehe, sondern nur von den gröbern irdischen Teilen gereinigt werde, hat ein gewisser *Physicus* auch aus nachfolgendem Experiment demonstriert:

Das Holz, welches zu Kohlen und folgends zu Aschen gebrandt wird, tut man in einen gläsernen Kolben, zusammen mit seiner *Resina, Liquore* und Olität desselbigen Baumes oder Holzes, alles gleichviel untereinander und auf einer linden Wärme zerlassen, das gibt eine *Mucilaginische Materie*, und also hat man die 3 *Principia* beieinander, davon alles Holz wächst und geboren wird: *Phlegma*, Fett und Aschen. Die *Phlegma* ist Merkur, die Feiste der Sulfur, die Aschen das Salz, denn alles, was im Feuer raucht und verraucht ist *Mercurius*, was brennt und verbrennt ist *Sulphur*, und alles was Asche ist, das ist auch ein Salz.

So du nun diese 3 *Principia* beieinander hast, so setze es in Roßmist zu faulen auf seine Zeit. Wenn du nun danach diese *Materia* nimmst und in ein feistes Erdreich vergräbst oder drein schüttetst, so wird man sehen, dass selbige wiederum zu grünen und ein junger Baum oder Holz daraus hervor wachsen wird, welches viel kräftiger und edler in aller seiner Substanz."

„Dieses Holz heißt und ist ein reducitirt, renovirt und regenerirt Holz, das anfänglich auch ein Holz gewesen, hernach aber getötet, zerstöret, zu Kohlen und Aschen verbrannt und gar zu Nichte worden und doch wiederum aus demselben Nichtigen etwas gemacht und gewachsen ist. Das lass dir im Licht der Natur ein großes und hohes Mysterium sein, dass ein Ding seine erste Form und Gestalt ganz und gar verlieren und doch hernach aus dem vermeinten Nichts etwas, das viel edler in seiner Kraft und Tugend, als es anfänglich gewesen von Neuem werden kann." (Was nachzuprüfen wäre – Anmerk.)

„9 hat auch beobachtet, dass wenn die *Rote Essentia* von Wermuth aus der Aschen destilliert wird, im Helm *Idea Absynthii* erscheine. Aus dem Weinstein hat man, nach gewisser Bereitung, die Gestalt der vollkommensten Trauben dargestellt, und *Borrichius* meldet, wie aus dem, von zarten Cypressen-Schößlein bereiteten Salz, nachdem es mit dem *Phlegmate Vitrioli* zu verschiedenen malen angefeuchtet worden, in der Mitten des Glases ein Bäumlein, eines kleinen Fingers hoch erwachsen, welches von vielen zu Paris gesehen worden.

Der *Spiritus* aus dem Mayen-Tau zeigt ohnwiedersprechlich eine gar sonderliche Tugend in Wiederhervorbringung der Pflanzen aus ihrer Asche und Salz, denn es ist in ihm die Kraft des Geistes der Luft sehr mächtig zu spüren. Ich ließ einmal dessen eine ziemliche Portion faulen in einem hölzernen, dazu geschickten Gefäß, und es fügte sich, dass ich weiße Kieselsteine glühete, welche ich im Wasser ablöschen wollte. Weil aber nicht so bald ander Wasser zur hand war, so nahm ich von diesem putreficirten Tau etwas in eine Schüssel, da entstand, als ich die glühenden Steine hineinwarf, ein solcher unbeschreiblich lieblicher Geruch, als wenn alle Blumen und Balsame zugleich in einem Zimmer ausgestreut wären.

Diesem Mayen-Tau fügen andere noch den *Spiritum* aus der rohen *Minera Bismuthi* zu, um aus der verbrannten Pflanzenasche die selbe wieder darzustellen, so noch in kein Feuer kommen. Man destilliert nämlich einen *Spiritum* aus dem Mayen-Tau nach vorhergegangener Fermentation desselben und treibt gleichfalls einen aus der *Materia Bismuthi.*

Hernach nimmt man die Asche eines ganzen Krautes mit Wurzeln, Stengeln, Blättern und Blumen verbrannt, menget solche unter das *Caput Mortuum* von der *Materia Bismuthi* und gießt auch die beiden *Spiritus* 3 Finger hoch über die Asche, schmelzt die Phiole wohl zu und setzt es über gelinde Aschen-Wärme, so kommt die Pflanze mit ihrer Blüte hervor und verschwindet sogleich, wenn die Wärme wieder aufhört.

Die allerbeste und naturgemäßeste Arbeit (...) mag aber wohl folgende sein: Nimm Samen, welchen und wie vielerlei du willst, die schön zeitig und bei hellem Wetter gesammelt worden, tue die selben in eine Glaskolben, so ungefähr nach Beschaffenheit der Blumen in der Höhe eines Arms lang und das Mundloch eines Daumens weit, damit man es versiegeln kann."

„Den Samen also darinnen verwahrt und im Mai, eben wahrgenommen, wenn ein schöner Tau fallen wird, so setze den zerquetschten Samen in flache steinerne Büchsen, damit der Same nicht so dick aufeinanderliege, sodass der Tau mit Menge auf dem Samen falle, denselben befeuchte und ihm seine Nahr-Kraft zueigne.

Unterdessen sollst du reine Tücher auf Pfähle spannen, darinnen sich Tau sammle, sodass man ihn auswringen kann. Sammle dessen 8 Maß zu 4 Pfund Samen. Der Samen aber muss vor Sonnenaufgang wieder in sein Glas getan werden, sonst verdirbt er. Den, in den Tüchern gesammelten Tau musst du fleißig filtrieren und in *Balneo* destillieren, bis er keine Feces mehr hinter sich lässt. Die Feces calciniere, die geben ein schwarzes Salz, welches auch unter das destillierte Wasser gehört. Dieser gesalzene tau wird auf den angefeuchteten Samen gegossen, dass es 3 Zwerg-Finger hoch darübergeht, dann das Glas hermetice versiegelt, in einen feuchten, warmen Ort oder Roß-Mist gesetzt und vergraben 2 Fuß tief und einen Monat stehen lassen.

Dann nimm es wieder heraus, so wirst du den Samen in einer Gallerie sehen, oben auf mit einer Haut, den *Spiritum* und Tau aber zwischen der schleimigen Materie und der obern Haut, welche von mancherlei Farben ganz hochgrün aussieht.

Dieses Glas also versiegelt, den ganzen Sommer an einen solchen Ort gestellt, wo des Tages die Sonne und des Nachts der Mond ihre *Radios* hierauf anwerfen können, bei Regen und unstetem Wetter aber solches in einer Stuben oder an einem trockenen Ort behalten, dann bei schönem Wetter wieder hinaus gehängt.

Es kommt wohl, dass dieses Werk in 2 Monaten, auch wohl erst im andern Jahr, je nachdem unstet oder gut Wetter ist, kann verrichtet werden. Die Wahrzeichen der Wachsung und Perfection sind diese: Die schleimige *Materia in fundo* schwellet hoch auf, der *Spiritus* und die Haut nimmt von Tag zu Tag ab, und es wird die *Materia* fast alle dick. Es gibt auch der Same *Reflexiones* im Glase, ganz subtile Dämpfe und Nebel, welche die Gestalt des Gewächses figurieren, doch aber noch unfix, welche Gestalt dann auch oft auf und nieder steiget, je nachdem die Sonne stark operiert und der Mond in vollem Schein steht. Endlich wird alles miteinander in eine weißgraue Asche verwandelt, aus der erheben sich mit der Zeit Stengel, Kraut und Blumen in Farbe und Gestalt des Samens, welche Gestalt dann nach Hinwegnehmung der Wärme allezeit verschwindet und in seiner Erde sich verbirgt, aber sobald es über gelinde Wärme gesetzt wird, wieder erscheinet und dieses (...) währet ewig und ohne Ende."

4) Coagulation von Merkur- und Sal-Prinzip zur fertigen Tinktur

a) Coagulation – die Wiedervereinigung der Prinzipien
b) Dynamisation, Reifung

Wie auch immer Merkur und Sal dargestellt wurden – sie werden schließlich miteinander vereint (a). Das geschieht am einfachsten (und auf dem Wege der spagyrischen Tinktur fast ausschließlich) durch Lösen des Salzes im Merkur (Alkohol). Gutes Durchschütteln und einige Tage der Digestion in der Wärme fördern die Verbindung der Prinzipien. Danach werden etwaige *Feces* gefiltert, welche der Alkohol aus dem Salz noch abgeschieden hat. Andere lösen das Salz zuvor in destilliertem Wasser auf und schütten diese Lösung zum Alkohol.

Nach dieser eigentlichen Vereinigung ist der eine oder andere Dynamisationsschritt (b) zwar nicht unbedingt notwendig, es wäre aber schade, diese Möglichkeit einer weiteren Vervollkommnung bzw. ‚Abrundung' des Kunstwerkes auszulassen. Die bisweilige Bezeichnung ‚Erhöhung' ist dabei nicht direkt zutreffend, weil die Erhöhung eigentlich in der spagyrischen Präparation selbst besteht, und das, was nicht ‚durch die Kunst erhöht' wurde, kann auch durch nachfolgende Dynamisationen nicht mehr viel weiter gebracht werden.

Die beliebteste Art spagyrischer Abschlussarbeiten ist die Circulation. Man befüllt ein großes und weites Glas nur zu einem geringen Teil mit der fertigen Tinktur und lässt sie in der Wärme darin zirkulieren. Das Kondensat sammelt sich an den oberen Wänden und rinnt langsam wieder zurück – eigentlich nicht sehr verschieden von der Methode wiederholter Destillationen (Kohobationen).

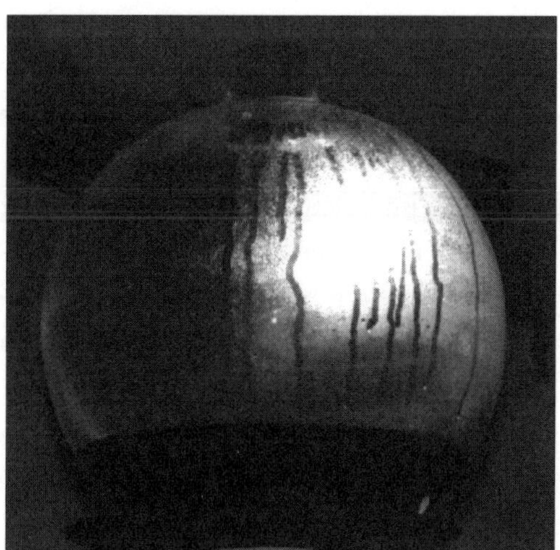

Sehr ähnlich dazu könnte man die fertige Tinktur immer wieder gelinde abdestillieren und zurückgießen (kohobieren), was allerdings weit aufwändiger wäre. Dabei wird das, von Natur aus Subtile und das durch die Kunst Verfeinerte aufsteigen und danach wieder mit dem fixen Salz am Boden vereint, welches angeregt wird, ebenfalls einen noch feineren Charakter anzunehmen. Diese Idee ist grundlegend im ganzen Naturreich und in der Menschenwelt. Das, was bereits ‚edler' geworden ist, kann aufsteigen, senkt sich dann aber wieder herab, um auf den Rest veredelnd einzuwirken.

Der Unterschied zwischen einer Circulation und einer vollständigen Kohobation ist der, dass während der Circulation immer nur die subtilste Flüssigkeit aufsteigt, kondensiert und wieder zurückrinnt. So wird der weniger subtile Teil des Präparates zur Verfeinerung angeregt. Der Rückstand am Boden bleibt aber in der Regel immer flüssig, dh. es wird nie bis zur Trockenheit abgedampft. Möglich wäre das schon (und in manchen Fällen tatsächlich interessant) – nämlich dann, wenn relativ zur Größe des Circulierglases nur eine sehr geringe Gesamtmenge an Flüssigkeit eingefüllt wird, so dass alles flüchtig in das Glas hochsteigen kann. Dabei ist bei typischen (färbigen) Tinkturen das Risiko des Anbrennens zu beachten. Bei Kohobationen ist es üblich, weitestgehend bis zur Trockenheit abzudestillieren und danach das gesamte Destillat wieder zurückzugießen. Dabei wird der eigentlich fixe Rückstand (bei farblosen Präparaten das Salz) eingehender ‚geöffnet', subtilisiert und zur eigenen Flüchtigkeit angeregt. Es findet aber kein ständiges Anregen verschieden subtiler Flüssigkeitsfraktionen statt wie bei der Circulation.

Obwohl hier nur am Rande berührt, wird deutlich, dass die Circulation mit gutem Grund zu den wesentlichen spagyrisch-alchymistischen Verfahren gehört. Eine Einreihung des Verfahrens der Circulation unter die ‚Dynamisationen' hat dabei nur zweitrangig Bedeutung.

Arbeiten, auf die der Begriff Dynamisation besser zutrifft, sind Schütteln, Rühren, Schwenken, Rhytmisieren u. dgl. Unter diesen hat wiederum das ‚Tropfen' einen hohen Rang inne: Das fertige, flüssige Präparat wird aus einer gewissen Höhe in einen Auffangbehälter getropft. Höhe und Dauer der Tropfenfolge kann je nach System oder Vorstellung verschieden sein. Ein schöner Rhythmus ist der, daß, sobald die durch den vorhergehenden Tropfen verursachten Wellen ‚ans Ufer' des Glases und wieder zur Mitte geschwungen sind, der neue Tropfen ankommt (Bild nächste Seite). Dazu gibt es wieder viele Ideen und manche Erfahrungswerte.

Insgesamt fällt auf, dass ‚erweiterte' Coagulations-Verfahren, wie eben Dynamisationen einerseits gerne zu kopfverstiegenen Klügeleien verleiten, andererseits aber auch ein erweitertes Umsetzen natürlicher Harmonien bieten.

Man dynamisiert bzw. rhythmisiert nach Tageszeiten, nach astrologischen Gesichtspunkten, nach den jeweiligen Pflanzensignaturen, nach dem Goldenen Schnitt oder nach anderen Ordnungen. Dynamisiert wird auch an bestimmten Plätzen im Labor, im Freien, in Höhlen, auf Gipfeln, in Kellern und Dachstühlen, an Kraftplätzen und dergleichen mehr.

Für all das gilt: Es ist fraglich, ob es im natürlichen Zusammenhang sinnvoll ist, für dergleichen Schritte oft mehr Aufwand zu machen als für die eigentliche Aufschließung der Pflanze. Man könnte ja dann in der selben Zeit nicht nur eine einfache Tinktur, sondern gleich eine Quintessenz, ein Elixir oder ein ähnlich höheres Präparat herstellen – wenn es die Kunstfertigkeit erlaubt. Andererseits sind diese Arbeiten verlockend ‚einfach' (sofern man entsprechende Vorgaben oder Einfälle hat), und bei einigem energetisch-feinstofflichen Vorstellungsvermögen bekommt man mitunter das Gefühl, ‚immer mehr' zu machen. Das trifft in gewisser Weise auch zu – aber eben ‚nur energetisch' und nicht im Sinne eines ganzheitlichen Naturverständnisses. Zudem kann, wer nicht tief genug im Wesen der Spagyrik gefestigt ist, hier leicht ‚vom Weg abkommen' und seinen Raum mit ‚energetischen' oder ‚dynamisierten' Wässerchen füllen, welche zweifellos eine ganz gute Wirkung entfalten mögen. Er wird jedoch dabei, anstatt in die alchymistische Kunst in einen Bereich experimenteller Homöopathie gelangen.

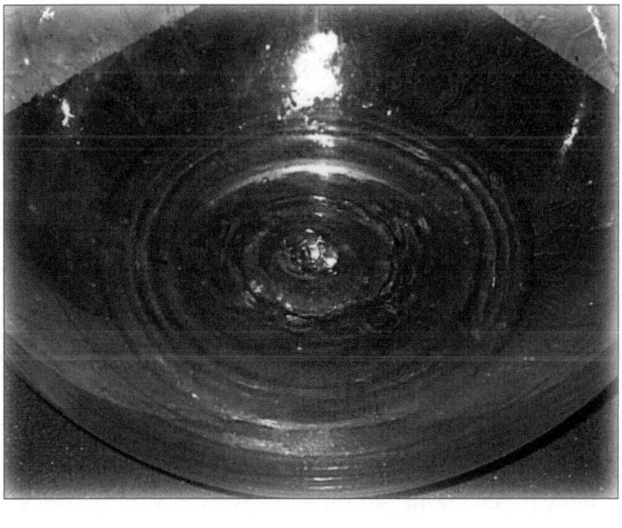

Wenn mit intellektuellen Ideenkomplexen gearbeitet wird, darf dabei nicht übersehen werden, dass diese, oft unbemerkt, starke persönliche Prägungen, Vorstellungen, enthalten. Man beginnt dann anstatt ‚im Lichte der Natur' im ‚Dämmerlicht der eigenen Projektionen' zu arbeiten. Das heißt: Man kann eine einfache spagyrische Tinktur aus Melisse herstellen und diese ‚kunstgemäß' nach der Coagulation noch 4-6 Wochen circulieren. Das ist die eine Sache. Man kann aber auch das selbe tun und nach der Circulation zum Beispiel 1000x schütteln, dann wieder zirkulieren, dann aus genau 1½ Meter Höhe in bestimmten Intervallen tropfen lassen etc. – der Charakter des Präparates wird sich auf einer bestimmten Ebene entsprechend der Vorstellung des Praktikers verändern. Die zweifellos erreichten ‚Erweiterungen' sind dabei vom spagyrischen Standpunkt nur relativ und entbehrlich.

Noch aufwändiger wird es, wenn zu solchen Verfahren bestimmte geometrische Formen verwendet werden. Natürlich bewirkt auch deren Einfluss ‚etwas' – aber was? Wenn man die tatsächliche Bedeutung der Formen – aus Sicht der Natur – verstehen kann, dann könnte man besser seine Zeit darauf verwenden, aus demselben Verständnis hohe alchymistische Präparate zu erlangen, denn ein echtes Verständnis der Körper, Maße und Formen kann u.a. zu einem Verständnis solch hoher Arbeiten führen.

Noch vor wenigen Jahrzehnten kursierte die Unart, möglichst alles unter eine Pyramide zu stellen. Man war überzeugt davon, dass es dadurch ‚hochwertiger' würde. Zweifellos bewirken solch markante Formen ‚etwas'. Inzwischen ist man auch hierzulande herangereift und hat verstanden, dass es keine energetisch-statischere Form als die Pyramide gibt und wie soll eine Lebendigkeit verbessert werden durch etwas, das lebendig-aktive Prozesse hemmt? Mit diesen Dingen mögen jene herumspielen, die meinen, dafür geschaffen zu sein – in der Spagyrik haben sie weder Platz noch Zeit, es sei denn, wir würden darin wieder zu jenem Verstehen kommen, das vermutlich die ‚Ältesten' besaßen.

Oft sind hier aber auch Grenzfälle, nämlich zwischen hermetischer Kunst und Experiment, zwischen Arbeitsvorschrift und Intuition, zwischen der ‚Kunst der Alten' und einem persönlichen Zugang zum Gesamtkunstwerk.

Strikt abzulehnen ist aber das bewusste Imprägnieren mit persönlicher Information, mit absichtlich und zusätzlich eingebrachten Gedanken- oder Gefühlsschwingungen, geschehe dies am Anfang, zwischendurch oder am Ende einer spagyrischen Arbeit. Wie kommt schließlich der Genießer eines spagyrischen Präparates dazu, dermaßen unauslotbare Aspekte des Praktikers mitgenießen zu müssen? Vermutlich ist das nur in der neuzeitlichen abendländischen Heilkunst ein Thema geworden. In alten, lebendigen Traditionen ist der Sachverhalt klar.

Ein Beispiel: Es gibt Leute, die nehmen ein Glas Wasser zwischen die Hände und laden es ‚mit ihren eigenen guten Schwingungen' auf. Einfache und ehrliche Dankbarkeit für das Glas Wasser täte es auch – aber sie meinen, damit noch mehr Gutes zu tun. Solange es nur Wollen und der Kraftstrom nicht ausgeprägt ist, bleibt das Problem vielleicht gering. Wenn es aber ‚funktioniert' und die Aufladung effektiv ist, dann wird es (ernsthaft betrachtet) zum Problem der Sondermüll-Entsorgung. Keiner ist hier so rein und klar, dass er nicht mit der einen ‚guten Schwingung' zahllose persönliche Muster mittransportieren würde.

Gesetzt den Fall, ein angehender Spagyriker stellt ein Mittel für den Verkauf her. Er ist selbst bereits ernsthaft krank, weiß aber noch nichts davon – die Krankheit hat sich noch in keiner Weise bemerkbar gemacht. Zudem steht er einigermaßen unter Stress und lebt im selben Hause mit einer zänkischen Schwiegermutter. Nach vollendeter, kunstgemäßer Produktion gedenkt er wohlmeinend, die Sache noch besser zu machen, stellt den verschlossenen, mit 3 Liter Essenz oder Tinktur befüllten Kanister auf die Anhöhe eines alten Hügelgrabes hinter seinem Haus, konzentriert sich längere Zeit darauf, dann umarmt er die Flasche und beginnt, gute Gedanken hineinzusenden. Solche Sachen werden tatsächlich praktiziert (am wenigsten allerdings von Spagyrikern). Auch mit Gebeten, Mantras, Sprüchen, Ritualen etc. Wer würde so etwas einnehmen wollen? Die Absicht mag lobenswert sein – das Ziel, weit verfehlt. Es ist nicht nötig noch weiter auf solche Dinge einzugehen – nötig aber ist es, hierzu ein Verstehen und den Mut zum Unterscheiden zu entwickeln.

Zweifellos prägt der ‚Künstler' sein Produkt. Das ist ein wichtiges Kriterium (nicht nur) spagyrischer Präparate. Sie werden nicht von Maschinen produziert. Jeder Handgriff will verstanden sein, jede Phase mitgefühlt – aber alles auf eine einfache, natürliche und in geistigem Sinne ‚losgelöste' Art. Die Kräfte der Natur sollen selbst wirken dürfen, und ohne einen ‚höheren Segen' gelangt ohnehin niemand ans Werk. Die einzelnen Schritte des Aufschließens und Vereinigens sind an sich schon ‚Dynamisation', eben ein Herausarbeiten, eine Erhöhung und bis zu einem gewissen Grade eine Befreiung der in den ‚Dingen der Natur' verborgenen, heilsamen Kräfte. Alles darüber hinaus ist mehr oder weniger nutzloses Beiwerk. Der ‚Künstler' leiht der Natur seine Hand, sein Herz, seinen Verstand, oder er leiht Gott seine Hand etc. – wie immer er die Sache versteht. Und damit sollte er bescheiden bleiben und sich selbst so weit als möglich herausnehmen und sich nicht aus einem unreifen oder missratenen Selbstverständnis absichtlich und besonders hineintun. Der Spagyriker benötigt keinen Mundschutz während der Arbeit, denn er führt ein gutes Leben, arbeitet verantwortungsvoll (und will außerdem die Dinge ungestört riechen können). Aber er trägt einen unsichtbaren Mundschutz umgebunden – er spricht keine ärgerlichen Worte während des Umgießens, projiziert keine ‚Mantras' in die Substanzen und weiß die Geheimnisse der Kunst zu wahren.

Damit wäre nun unsere spagyrische Tinktur fertiggestellt. Merkur und Sal sind vereinigt und diese Coagulation durch einige Wochen Circulation bekräftigt. Dass solcher Circulation oder ähnlichen abschließenden Arbeiten Zeit gegeben wird, ist ebenfalls ein Qualitätskriterium.

Zusammenfassung der Verfahren zur Darstellung einer spagyrischen Tinktur

Mazeration oder Gärung	Filtration Destillation Rectification	Veraschung Calcination Scheidung Reinigung Kristallisation	Coagulation Circulation

Erarbeitete philosophische Prinzipien: Merkur und Sal

Therapeutische Anwendung

Spagyrische Tinkturen können auf relativ einfache Art hergestellt werden. Wahrscheinlich gehören sie deshalb zu den beliebtesten und verfügbarsten spagyrischen Präparaten in der parapharmazeutischen und naturheilkundlich-therapeutischen Praxis. Die Materialkosten zur Produktion sind gering, der Zeitaufwand dennoch bemerkenswert groß. Trotz der allgemeinen Einfachheit der Wege zur Tinktur kann dieser Name ein äußerst kunstvolles Präparat bezeichnen, mit weitaus besserer Wirkung als andere, weniger kunstvoll bereitete ‚spagyrische' Präparate mit klangvolleren Namen.

Die innerlichen und äußerlichen Anwendungsmöglichkeiten ergeben ein schier unüberschaubar weites Spektrum. Das Wirkspektrum nur einer einzelnen Pflanze ist sehr weit und schließt (in der Regel) sowohl homöopathische als auch phytotherapeutische Indikationen mit ein.

Wie bei jedem ‚echten' spagyrischen Pflanzenpräparat ist der Sal-Faktor von größter Wichtigkeit, indem er eine Resonanz zu den Körpergeweben und eine Durchdringung der Heilkräfte gewährleistet. Da dieses Erd-Salz aber in der Tinktur nicht (oder kaum) ‚flüchtig' enthalten ist, werden bei Inhalationen fast ausschließlich Sulfur und Merkur wirksam sein.

Die Dosierungen zur Einnahme sind zum Teil bedingt durch das Verfahren bzw. durch die ‚Dichte' der Tinktur. Die Angaben dazu in der älteren und neuen Literatur schwanken beträchtlich, wobei die ‚Alten' vermutlich sehr ‚dichte' Präparate herstellten und diese dennoch für unser heutiges Empfinden oft sehr hoch dosierten.

Die Spagyrische Essenz

Darstellungsmöglichkeiten / übliche Verfahren

1) Darstellung des ersten Sulfurs aus Pflanzen

Frische oder getrocknete Pflanzen werden mit destilliertem oder anderem gutem Wasser übergossen und darin kurze Zeit digeriert (in der Regel wenige Stunden bis ein oder zwei Tage). Jedenfalls sollte während der Digestionszeit keine echte Gärung einsetzen. Eine leichte Umwandlung bzw. Öffnung, eine sogenannte ‚aromatische Gärung', ist bisweilen erwünscht. Andere lassen diesen Schritt bei Frischpflanzen aus und digerieren nur getrocknete Kräuter.

In jedem Fall wird danach – quasi als erster, eigentlicher Schritt zur Pflanzenessenz – der Anteil an ätherischem Öl und/oder aromatischen, feinriechenden, typisch schmeckendem Wasser abdestilliert. Das rectificierte Destillat behält man entweder insgesamt als Ausdrucksform des ‚flüchtigen Sulfurs' *(Sulphur volatile)* oder man scheidet davon das ätherische Öl (und gibt Hydrolat und *Phlegma* zusammen zum Rückstand). Hierzu sind mehrere Varianten üblich. Manche scheiden das Öl (sofern die Pflanze eines gibt), gießen das Wässrige zurück und destillieren von Neuem – so oft, bis alles Ölige, Fettige oder Aromatische heraußen ist.

Eine andere Möglichkeit ist, diesen ersten Schritt auszulassen und das Sulfur-Prinzip erst später (aus der ‚Farbe') darzustellen. Davon später.

Die Darstellung ‚eines Sulfurs' ist Voraussetzung für die Betitelung ‚Essenz'. Eine spagyrische Essenz herzustellen bedeutet, Sulfur, Merkur und Sal aus dem Pflanzenkörper zu scheiden, zu reinigen und wieder zu vereinen. Dies ist vielleicht der bedeutendste Unterschied zur ‚spagyrischen Tinktur' und beschreibt auch die im hermetischen Sinne ‚höhere' Form der Präparation.

Sulfur entspricht der Sonne und dem Seelenhaften. Stehen die alkoholisch-merkurialen Prozesse (Gärung, Mazeration) am Anfang der Präparation, dann bedeutet das allegorisch: Die Seele ersäuft im Alkohol / Das Feuer wird von Merkur verschluckt / Der Mensch verliert seine Aufmerksamkeit (Seele), die vollständig vom Gemüt (Mond) aufgesaugt wird / usf.

Die Scheidung von Sulfur und Merkur, von Sonne und Mond, Seele und Gemüt (allegorisch und nicht begriffsgenau) als Vorschrift zur ‚spagyrischen Essenz' weist wiederum die spagyrisch/alchymistische Kunst als eine alte initiatische Wissenschaft aus. Die initiatische Scheidung der Seele von Körper und Gemüt entspricht dem ‚Mensch, erkenne dich selbst' der alten Griechen.

Es verhält sich auch in der laborantischen Praxis so, dass, wenn das in der Pflanze enthaltene ätherische Öl (als Vertreter des Sulfur-Prinzips / Sonne / Seele) erst einmal im Alkohol (dem Vertreter des Merkur-Prinzips / Mond / Gemüt) gelöst ist, die beiden kaum mehr voneinander getrennt werden können. Setzt man etwa einen getrockneten Salbei – reich an ätherischen Ölen – zur Mazeration in Alkohol an, dann ist es danach kaum mehr möglich, das ätherische Öl auch nur annähernd vollständig aus dem Mazerat zu scheiden. (Es gibt umständliche Wege, durch die ein wenig Sulfur ‚gerettet' werden kann). Das bedeutet sinngleich auf den Menschen bezogen: Die Aufmerksamkeit, die Seelenkraft hat sich vollständig im Gemüt zerstreut – der Mensch hat sich selbst verloren. Obgleich beim Menschen die Seelenkraft nur indirekt mit dem verbunden ist, was wir unter Individualität verstehen, so gilt das bei der Pflanze voll und ganz. Der individuelle Charakter wird wesentlich beeinträchtigt, wenn man das Sulfur-Prinzip nicht (von Merkur und Sal) scheidet. Anders ausgedrückt: Erst wenn die Seele in Sicherheit gebracht worden ist, können die Dinge des Gemüts auf eine rechte Art ‚ausgegoren' werden. In der Praxis führen dabei allerdings manche Gärungen zu Kompromissen.

hermetisches bzw. alchymistisches Symbol für Sulfur oder Sulfur-Prinzip

Das Sulfur-Prinzip und die Ausdrucksformen des Feuer-Elements

Öl, Farbe, Geruch, Geschmack und Brennbarkeit gelten als typische Ausdrucksformen des Sulfur-Prinzips. Das durch die Destillation geschiedene und rectificierte ätherische Öl steht dem Wesen des Sulfur-Prinzips näher als die anderen, gröberen Ausdrucksarten dieses Prinzips, welche in der rohen Pflanze noch mit anderen Prinzipien und Kräften vermischt sind, obwohl es sich beim ätherischen Öl ebenfalls um ein Mischprodukt (aus Sulfur, Merkur und Sal) handelt, dieses aber überwiegend vom Sulfur dominiert und signiert ist – die Anteile von Merkur und Sal sind je nach Beschaffenheit und Reinheit geringfügig.

Während also in der rohen Pflanze (sei sie frisch oder getrocknet, noch wachsend oder bereits geerntet) alle drei Prinzipien nach ihrem Verhältnis vermischt und eingebunden sind, gilt das ätherische Öl bereits als ein ‚prinzipiell' sulfurischer Aspekt der Pflanze. Damit direkt verbunden sind andere sulfurische Aspekte wie Geruch, Geschmack (bedingt) und Brennbarkeit.

Genauso verhält es sich mit den anderen Produkten der Scheidung *(Separation)*, den Ausdrucksformen von Merkur und Sal bzw. der Elemente, je nach ihrer typischen Art. So werden die an der physischen Manifestation einer ‚Idee' (Quintessenz) beteiligten Kräfte entsprechend substanziell dargestellt.

Je höher die Kunst, umso weiter dringt man dabei zum Wesentlichen vor. Hierzu ist es notwendig, diese Themen tiefer zu betrachten, damit sie sich mehr und mehr eröffnen und damit das hermetische Naturverständnis zum Wegweiser in der praktischen Arbeit werden kann. Wir stellen uns also wieder die Frage: Was ist Sulfur? Und wir kommen dabei nicht umhin, weiterzufragen: Was ist das Feuer-Element, durch welches sich das ‚sulfurische Spektrum' erstreckt?

Sulfur – das ist das Auftreten, das Erscheinen, das Werden, und es ist Drängen, Treiben und Zerbrechen. Sulfur ist der hinweisende Finger, das Deutende, Zeigende, Leuchtende, Strahlende, sichtbar Machende. Es ist das, was zentriert, konkretisiert, reduziert, reinigt und fixiert, und es ist das, was die Zustände verändert. Es ist die Äußerung der Bewegung aus dem Äther heraus, welcher zuerst (nach *Aristoteles*) die ‚Bewegung an sich' ist.

Sulfur ist die Ermunterung, etwas zu tun, die Arbeit, die schaffende und die zerstörende Kraft, die Abwehr, der Angriff, das Fliehen und das Einfangen des Fliehenden. Sulfur steht in der Mitte zwischen Merkur und Sal, was seine Geistigkeit betrifft. Er ist nicht ‚subtil' wie Merkur, sondern einfach ‚Kraft'. Er ist nicht ‚fix' wie das Salinische, sondern das Fixieren und das Auflösen zugleich. Das heißt: Er ist jene Spannung, sowohl im ‚Ding', als auch im Ereignis, welche dieses im Stofflichen festhält.

Und eben deswegen und darüber hinaus ist er dem flüchtigen Merkur ein Verfolger und Gegenpart und gleichzeitig auch seine Ursache. Aber so wie Merkur sein Wesen nicht ohne Sulfur haben könnte, hätte Sulfur kein Dasein ohne Merkur. Denn Sulfur ist ein Träger des ersten und jedes neu generierten Merkurs, ein Fahrzeug , in dem der Naturgeist zu seinen Geschäften reist. Er ist das Licht auf dem Weg und seine, des Sulfurs, Natur ist vollkommen ruhig und statisch, bevor er in die Physis gewoben wird. Weil zugleich mit ihm aber auch Merkur eingewoben wird, verändert Sulfur in der Physis vollständig seine Art, stets den Merkur zur Identifikation drängend, den, dessen Namen niemand kennt, immer fragend: „Wie heißt du?" Sulfur verschlingt Merkur und speit ihn aus. Merkur lässt geschehen und imitiert: Er verschlingt Sulfur und speit ihn aus.

Sulfur ist Hitze und Flamme. Merkur ist Luft und Rauch. Stets umgibt den Sulfur der von Merkur durchtränkte Raum. Merkur erfüllt den ätherischen Raum und schließt in jedem Punkt darin Sulfur ein. Wo aber dann Dichte entsteht, umschließt Sulfur den Merkur. So ist Saturn die Schwelle einer Umkehrung der Beziehung Sulfur-Merkur.

(*Ouroboros* aus *Kirchwegers* >Goldene Kette des Homer<)

Sulfur ist Farbe – aber darin das Licht, wodurch Farben gesehen werden. Er ist nicht rot oder grün. Dieser differenzierte Ausdruck ist schon Merkur, welcher vorzüglich im Schillern und wechselndem Farbenspiel seine Art kundtut. Dazu nimmt er aber Licht und Wesen der Farbe von Sulfur.

Sulfur ist Sehen – das mittlere Schwarze des Auges. Daraus schöpft Merkur und malt aus sulfurischem Wesen die Iris rundum. Das schwarze Zentrum des Auges ist die ätherische Sonne. Im Punkt ist die Idee, die Quintessenz, noch verborgen. Dann greift Merkur gestaltend, vermittelnd ein und holt das Verborgene heraus. So werden die Farben des Auges, welche jegliche Information des menschlichen Wesens wiedergeben. Das ist merkurial. Es wäre aber nicht möglich ohne Sulfur, welcher eben dieses Verborgene ‚sichtbar' macht und dazu Farb- oder Lichtessenz zur Verfügung stellt. Sulfur spricht aus. Er enthüllt das Geheimnis, beleuchtet es, formuliert das Gefühlte und verwandelt den dunklen Punkt in einen leuchtenden Stern. Mittels der Sonne und der Augen erlebt der Mensch Sulfur.

Das edle Geschäpff die Sonne mit ihrem hell- und lieblichen Schein gibt und bringet das helle Liecht / und mit demselbigen den Tag in die gantze Welt / daß die Menschen darbey können arbeiten / handeln / wandeln / und nach Erforderung habenden Beruffes / das

das ihrige auszurichten vermögen. Über dieses gibt die Sonne auch eine angemehme natürliche Wärme von sich, dadurch der Erdboden erwärmet wird, die Früchte in der Welt wachsen, reif und zeitig werden, und das, was sonst nass und feucht gewesen, wird dürre und trocken.

Im Gegenteil findet sich oft ein schöner, lieblicher, warmer und fruchtbarer Regen und Nacht-Tau, damit das, was am Tage etwas zu trucken und dürre worden, durch den Regen und Tau wieder angefeuchtet, sich erquicke und erfrische. Wann aber der Sonnen Hitze überhand nimmt und wegen Verzögerung des Regen allzu groß und stark ist, sodass dadurch alle natürliche Feuchtigkeit der Erden ausgetrocknet, verdorret und verzehret wird, verwelken alle Früchte, fallen um und ab, dass sie ferner nicht wachsen noch aufkommen mögen.

Fället aber ein allzu kalt und nasses Wetter ein, dass es gefrieret, reifft, Schlossen wirft, unnatürlich lange regnet und wittert, so kann nichts aufkommen, wachsen, reif und zeitig werden, weil es teils frieret, wieder zu Boden geschlagen, verschwemmet, ausgerissen und zunichte gemacht wird, ja ausrostet und verfaulet, dass es nicht zu gute noch zu nütze kommen kann, obschon aller menschliche Fleiß hierbei angewendet würde.

So gehet es mit des Menschen Gesichte und Augen auch, weil die Augen ihre natürliche Gewohnheit, Wärme und Feuchtigkeit haben, so stehet es um die Augen noch wohl. Wo aber dasselbige corrumpieret und zerstört wird, so folget alsbald Schaden, Da fällt eine Hitze ein, darauf kommt großer Schmerz und Wehetage, da erhebt sich ein Fluß, es ereignet sich ein Blatter, Geschwür und Apostem, da ziehen die Felle über die Augen, es fällt, flieget oder stäubet einem in ein Auge etwas, bald wird einer drein geschlagen, geworfen, gestoßen, gestochen oder geschossen und geschiehet schnell, dass einem ein Schaden, Unrat und Unglück widerfähret, dass ein Mensch um seine gesunden Augen und fröhliches Gesichte kommen kann, ehe denn er seiner gewahr wird und sich dessen versehen hat, welches fürwahr ein armes und elendes Ding ist.

Man siehet und befindet am natürlichen hellen Tage, wenn die liebe Sonne nicht scheinet, dass es dunkel und trübe, ja alles gleichsam traurig und betrübt ist und Verlangen trägt nach dem lieben Sonnenschein. Nicht anders ist es denen Menschen auch, welche gar dunkle, trübe, blöde Augen und schwach Gesicht haben.

Wir sehen, wenn man bei Nacht wandern und reisen soll, wie ein arm und elend Ding es sei, obgleich der Mond scheinet und einer die Wege und Stege wohl weiß, noch kann er fehlen, fället oft, stößet sich und muss in großer Gefahr sein. Ist es aber gar finster und scheinet kein Mond, so ist es noch ärger, soll und muss ein Mensch doch fort, da gehets noch übler, ärger und elender daher als zuvor, da muss man Fackel und Lichter haben, eines das andere führen, da tappt und greift man um sich, noch hilft es nicht. Bald lischt das Licht aus, gehen irre und unrecht. Bald kommen sie in Wasser und Gesümpfe, in Sträuche und Hecken und wissen oft nicht wo sie sind, stehen oft in großer Gefahr Leibes und Lebens.

Gleichergestalt wie es mit Himmel und Erde, Sonne und Mond oft und manchmal veränderlich zugehet, also gehet es mit denen Menschen und deroselbe Augen und Gesichte auch zu. (...)

Bei der Formierung des menschlichen Leibes im Mutterleibe sind die Augen vor allen anderen Gliedern an dem Menschen am aller ersten deutlich zu erkennen. Sie haben ihren Ursprung von und aus dem Gehirn und sind vor allen andern die allernotwendigsten, edelsten, klärsten und subtilsten Glieder und können auch am wenigsten erleiden und ausstehen, nach dem bekannten Sprichwort: „Der Glaube, die Ehre, das Auge – das sind drei solche Stücke, die zart und edel sind und leiden keine Tücke."

Dann merkt *Georg Bartisch von Königsbrück* in seinem >Augendienst< (1568) an, dass die solarischen Kräuter und Gewürze, in denen das Sulfur-Prinzip dominiert, am besten dazu geeignet sind, das Augenlicht zu verbessern:

Eine sehr gute Artzney für dunckele und trübe Augen.

NImm Welschen Quendel/Zimmetrinden/Fenchel-Samen/ jedes drey Loth/Muscat-Blüh/Wurtz-Nägelein/jedes ein und ein halbes Loth/Cubeben/Zitwer/Wiesen-Kümmel/Cardamümlein/jedes ein Loth/Paradis-Holtz/Aloës, jedes ein halbes Loth. Alles aufs klärreste gepülvert und zusammen gemischet. Und

das mag man entweder also Pulver-Weise gebrauchen/ oder eine Latwerge oder Confect daraus bereiten und machen. Wird gebrauchet im zunehmen des Monds/Morgens und Abends/und zwar Pulver-Weise ein halbes Quintlein; im Confect aber oder in Gestalt einer Latwerge ein halbes Loth auf einmal.

Sulfur ist Herz. Das ist zuerst dann, wenn die in die elementare Physis, in den Embryo, einziehende Seele beginnt, den Raum wahrzunehmen, in den sie sich jetzt hineinlebt. Eine Seele, eine Idee, eine Quintessenz – der Art nach völlig raumlos, erfährt alleine durch ihre Ankunft im physischen Äther-Element einen Widerhall aus dem endlosen und doch beschränkten Raum der 5-elementaren Physis. Hier ereignet sich eine Art Spiegelung des ersten aufglimmenden Sulfurs und dadurch eine Beschränkung seiner Kraft und ein ‚Sichtbarwerden'. Daraus entsteht das Zentrieren und die Verdichtungstendenz im Sulfur-Prinzip und im Körper die Bildung des physischen Herzens.

Und zugleich damit entsteht die gegenwirkende, gewissermaßen zersprengende Tendenz, denn der ‚gefallene' Funke will das öffnen, worin er gefallen ist, um wieder seinen Urzustand zu erreichen. Man könnte meinen, die ewige Aktivität hat den physischen Stress satt, weil ihr Urgrund die statische Ruhe ist. Deshalb steht das Herz an der Schwelle zwischen Äther und Erde.

Durch das Herz wird die Idee, die Quintessenz in die Physis vermittelt, Erd-Element hervorgebracht, ein Körper durchgeistigt. Und durch das Herz werden die Zustände der stofflichen Welt zum Äther zurückvermittelt. Wie in der Iris des Auges, so greift auch hier Merkur ein und webt das Netz der Blutbahnen und Nerven in Körper und Natur und schöpft deren vermittelnde Kraft aus dem sulfurischen Urgrund des Herzens heraus.

Sulfur ist auch das Feuer der Verdauung – nicht nur Wärme, sondern unmittelbares Erkennen der Stoffqualitäten. Er ist die Unterscheidungskraft des Dünndarms und das spontane Entscheiden aus dem Bauch heraus, denn das liegt der feurigen Natur mehr als der phlegmatischen.

Das Sehen also, der Blick wird durch die schöne, strahlende Blüte entfacht, durch die Rose, die Magnolie, den Löwenzahn. Das ist der ätherische Sulfur der Pflanze, welcher weit in den Raum hinaus wirkt, um seinesgleichen (zB. das Schauen des Menschen) zu treffen. Ähnlich wirkt das ätherische Öl und der Geruch in den Raum. Sowohl das Leuchten der Blüte als auch der Geruch sind durchwegs sulfurischer Art. Aber die Differenzierung der Formen, die Verschiedenartigkeit der Farben und Gerüche – das ist alles kommunizierend, informativ,und, wie die Iris des Auges und die Kreisläufe in Körper und Natur, eine Reaktion Merkurs auf die Anwesenheit Sulfurs.

Was ist das erste Spiel Merkurs? Er treibt seinen Handel. Er sagt: „Sei mir gegrüßt, Sulfur! Du bist wahrhaft göttlicher Abstammung und ohne dich erkennen wir nichts. Aber was hilft das, wenn du nicht in Erscheinung trittst? Lass mich dir helfen. Du machst die Dinge sichtbar, und ich weiß, wie wir dazu die Farben und Düfte zusammensetzen – ich kann informieren, komponieren. Du behältst die Oberherrschaft – ich möchte nur beteiligt sein. ..."

Merkur weiß, wie er mit Sulfur umgehen muss, denn er kennt seine Art. Umgekehrt weiß Sulfur, dass er sich mit Merkur einlassen muss, und er weiß auch, was das für ein Gesell ist. Merkur aber kennt Sulfurs Stolz, seinen Hochmut (als Äußerung *Zeus*'), den Zorn und die helle Freude. Er kennt seine unbändige Schaffenskraft und seine Fixiertheit und weiß, dass er sich trotz alldem nicht ausdrücken kann, nicht zu kommunizieren versteht. Dabei nun tritt ihm Merkur helfend zur Seite und lässt ihm dennoch den Vortritt, wo immer möglich.

Er überläßt ihm die erste Auffälligkeit, den Blitz, die Blüte, den Geruch und sogar die erste Fraktion (zum ätherischen Öl) auf dem Weg zur Pflanzenessenz. Ja in seiner ureigensten Wohnstatt, im *Spiritus*, lässt Merkur den Königssohn wohnen und gesteht ihm Schärfe und Brennbarkeit zu. Der Naturgeist ist ein listiger Hase – er bleibt darin zufrieden, dass er alles was Sulfur darstellt, sichtbar macht, auf den Punkt bringt, eröffnet, usf. allem einen differenzierten, informativen Charakter gibt.

Je nach Blickwinkel und Sinnzusammenhang sind Sulfur, Sonne, Feuer, *Anima* (und wohl auch Mars) ähnliche und bisweilen austauschbare Termini. In Allegorien und Signaturen finden wir dazu unter anderem Sonnenscheiben, Flammen, Dornen, Stacheln, den Hahn, den Widder, Pfeile, Lanzen, Stäbe, Farben (alle aber insbes.: rot, orange, goldgelb u. dgl. warme, helle Töne), den Halm, den Stamm, Phallussymbole, das nach oben weisende Dreieck, den Kopf des Drachen, die Schnäbel mancher Vögel (signieren das ‚doppelte Feuer'), rot funkelnde Edelsteine, Gold, Eisen, Schwefel u.s.f.

Das Spektrum des Sulfur-Prinzips reicht vom solaren Äther bis ins Innerste der saturnischen Erde. Es wird initiiert durch den jovialen Funken, welcher im lunar-ätherischen, endlosen Raum erscheint – angeregt durch die aus dem Jenseitigen einströmende und zur physischen Manifestation strebende Idee. Dieser quintessentielle Funke wird in seinem eigenen Raum kurz vor der pentagrammatischen Position der Sonne zentriert und schlägt schließlich, den Blitzen des *Zeus* gleich, seiner grobstofflichen Manifestation entgegen.

Begleitet von Verdichtungsdynamik erschafft er selbst die Erde von innen heraus und führt die Formen der Natur mit sich als wachsende Individuen. Er prallt auf die Position des Saturn und nimmt dabei seine gegenwirkende Art an, wird zum *ios*, zum *viros*, bereit, alles Stoffgewordene, dem Saturn anheim Gefallene, aufzubrechen und die Formen zu zerstören. Dies ist das ‚doppelte Feuer‘ im sulfurischen Spektrum, signiert im ‚Schnabel des Hahnes‘ oder im schnabelähnlichen, spitzigen Maul des Drachen.

Deswegen zeigt auch die Ineinanderschachtelung der platonischen Körper nach der natürlichen Geometrie im Inneren des Erd-Würfels nicht nur einen einfachen, sondern einen doppelten Feuer-Tetraeder, einen Sterntetraeder (aus zweidimensionaler Sicht ein Hexagramm).

Im Inneren des Feuer-Sterntetraeders (Bild rechts) befindet sich der merkuriale Oktaeder (‚Merkur im Kerker des Saturn‘ bzw. ‚im Zentrum der Elemente‘). Außen ist der Erd-Würfel umschrieben. Diesen umhüllt der Luft-Ikosaeder und den wieder umhüllt, als äußerster Körper, der Pentagondodekaeder, das Äther-Element bezeichnend. Oktaeder, Ikosaeder und Dodekaeder (das Spektrum des Merkur-Prinzips) sind über den Goldenen Schnitt miteinander verbunden und austauschbar.

Damit wird auch Sulfur zum Kerkermeister Merkurs, während Saturn (aus diesem Blickwinkel) eher den Kerker selbst darstellt. Im Inneren des Hexagramms, des Sterntetraeders, hält er ihn verschlossen, während er selbst (Sulfur) in den Erdwürfel gezwungen, von Saturn geknechtet wird. So hält Sulfur auf Anweisung Saturns Merkur gefangen und ist dem grauen, gramen Greis für die Fixität des Naturgeistes verantwortlich – eine Verantwortung, die er niemals vollständig erfüllen kann, es sei denn, das vulcanische Feuer der alchymistischen Kunst kommt ihm zu Hilfe.

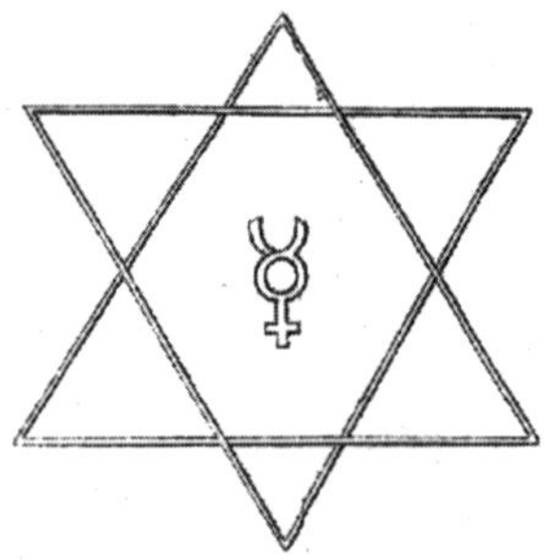

Dies sind die beiden Aspekte des Feuers: verdichtend und aufbrechend. Deshalb sagen die Alten: „Um Holz zu verbrennen, muss nur das ihm innewohnende Feuer zum Hervorbrechen angeregt werden, indem man es reibt oder mit äußerem Feuer in Verbindung bringt." Dasselbe gilt für die Nutzung der Aufbrechkraft des Feuers in der alchymistischen Praxis. Das scharfe, corrosivische *Menstruum* ‚löst' nicht das Mineral, sondern es regt das dem Mineral innewohnende Feuer an, seinen saturnischen Mantel zu zerbrechen. Damit wird dann auch gleichzeitig die Idee, die Quintessenz oder ‚Tugend', der ‚eingekerkerte Merkur' befreit, der Oktaeder im Inneren des Sterntetraeders geöffnet.

Das macht Sulfur und Merkur zu Gesellen: Beide sind in saturnischer Erde eingeschlossen, und beide werden durch deren Öffnung befreit. Deshalb geht aus den natürlichen Dingen durch Einwirkung von äußerem Feuer in der Trockendestillation ein ‚Essig' und ein Öl hervor – ein Merkur und ein Sulfur, wenngleich der merkuriale ‚Essig' (‚merkurial' nicht zuletzt aufgrund der flüchtigen Salze) vom Sulfur durchdrungen ist (Schärfe; körperöffnende Kraft; starke individuelle Art) und das sulfurische Öl einige merkuriale Charaktere (Flüssigkeit; eine gewisse Neutralität besonders im Pflanzenreich) besitzt, typisch für die Vermischung der Naturkräfte in physischen Ausdrucksformen. So durchdringt das Äußere das Innere, und so durchdringen sich oben und unten.

Deshalb wird die merkuriale Quintessenz zum sulfurischen *Specificum*, und es drängt die Quintessenz wieder zur stofflichen Verwirklichung. Vermittels des Feuers wird Dichte geschaffen, als Matrix für die ‚Tugend', und ebenso wird dadurch umgekehrt die ‚Tugend' aus ihrer hemmenden Matrix befreit. Das ist ein Aspekt des Sulfurs – das ‚Doppel'.

Sulfur ist jene Aktivität, die eine absolute Unruhe initiiert, welche allem Werden entgegenwirkt und allem dennoch Gewordenen innewohnt, nur darauf harrend, es wieder seiner Auflösung entgegenzutreiben. Die Sonne manifestiert den Stamm der Pflanze, zieht ihn zu sich wie die Flamme der Kerze, nur um ihn zur Blüte zu treiben und zum Verblühen. Und im Welken, im Putreficieren werden erneut Samen verdichtet, und diesem Verdichten steht das selbe Feuer entgegen und sprengt das Samenkorn und treibt den Spross.

Das ätherische Feuer, welches die jovial-solaren, sauren, zusammenziehenden Geschmackssignaturen hervorbringt und das Coagulat in der Phiole figiert, zertreibt und zerteilt dann alles wieder durch seine scharfe Art und zerbricht das Gefestigte, und das schlecht kontrollierte Feuer verbrennt das Coagulat in der zersprungenen Phiole, wie das Feuer eines missratenen Zorns den Halt eines friedvollen Zusammenseins zerbricht, welcher zuvor durch ein glückliches Feuer geschaffen wurde.

‚Natürliche, sulfurische Dinge'

Unter ‚natürlichen, sulfurischen Dingen' versteht man vor allem solche, in denen das Feuer in einem starken Grad verborgen ist, wie zum Beispiel in Schwefel, Öl, Fett, Holz, Rinden, Harz und Teer. Aus solchen ‚Dingen' kann es relativ leicht hervorgebracht werden. Andere ‚Dinge' sind sulfurisch wegen ihrer Farbe und Leuchtkraft (Blüten, Edelsteine, Honig). Dieses Feuer kann kaum als Flamme hervorgebracht und dargestellt werden. Je feiner die ‚Dinge' sind, umso näher stehen sie dem ätherischen Element im sulfurischem Spektrum. Ein namenstypisches Beispiel hierfür ist das ‚ätherische Öl'.

‚Natürliche Dinge', welche hauptsächlich durch Farbe, Geruch oder Vitalität wirken, gehören eher zum ‚oberen Feuer', zum solaren Feuer oder noch zum solaren Äther, wobei es kaum jemals möglich ist, nur eine einzige Position für ein ‚Ding' anzugeben, außer (in diesem Falle) das Gold bzw. zum ‚unteren Feuer' gehörend das Eisen und vielleicht auch noch den sublimierten Schwefel als typischen Vertreter der mittleren Position, also der pentagrammatischen Position des Feuer-Elements an sich.

Je ‚feister', zäher oder gröber die ‚Dinge' werden, umso weiter rücken sie in die Position des ‚unteren' bzw. des martialen Feuers. So kann man ein Spektrum der sulfurischen Dinge erstellen ‚von oben bis unten', etwa nach dem Schema: ätherisches Öl – Olivenöl – Honig – Baumharz – Teer – Erdpech. Oder aber ein Spektrum, welches die sulfurische Qualität der Salze beschreibt, wie etwa: ein helles, klares, subtiles Salz beinah ohne Geschmack, ein wenig süß – ein helles, spitziges und festes Salz von leicht bitterem und scharfem Geschmack – ein unreines, sehr bitteres und grobes Salz.

Nach ähnlichen Kriterien haben die Alten die Qualität der Pflanzensalze bewertet, werden doch diese – als Vertreter elementarer Erd-Kraft – ‚aus dem Feuer geboren'. In *Leonhard Thurneissers* >Kräuterbuch/Beschreibung influentischer, elementischer und natürlicher Wirkungen aller ... Erdgewächse< (um 1570) heißt es von der Wilden Karotte: „Die Sonne hat in der Wurzel dieses Gewächses 3 ¾ Teile, der Mars hat dabei 4 ¼ Teile und der Mond 4 Teile zu verwalten. Deshalb tut der Mond durch seine kalte und feuchte Eigenschaft den anderen beiden (weil sie fast in gleicher Proportion sind) einen großen Abbruch. Darum befindet es sich, dass diese Wurzel 6 Teile *Sulphur*, 3 Teile Salz und 3 Teile des *Mercurii* in sich hat, wodurch sie eine erhaltende, heilende und stärkende Tugend besitzt." Das Salz wird dabei mehr von martialer als von solarer Feuer-Kraft gebildet, gibt (wegen des hohen sulfurischen Anteils) in der Calcination zwar viel, muss aber (wegen des martialen Einflusses) öfters gereinigt werden.

Des Salz der Hirschwurz wird hauptsächlich von solarem Feuer gebildet. Dazu *Thurneysser*: „Die Sonne tradirt oder übergibt dieser Wurzel durch ihre influentische Kraft 7 und ¼ Teile, aber der Mars mitteilt ihr 2 Teil ¼ und der Mond 2 Teil 2/4.

Das Kraut aber concipirt oder empfängt von der Sonne 6 Teile, vom *Marte* 2 Teile ¾ und von *Luna* dem Mond 3 Teil ¼.

Der Samen ist nach der Wurzel das vornehmste und hat von der Sonne 5 ¼ Teile, vom *Marte* 4 Teil ¼ und von dem Mond 2 Teil 2/4.

Deshalb ist dieses ganze Gewächs unirt oder vereinbart von 6 ½ Teilen des alleredelsten und subtilsten *Sulphurs*, 2 ½ Teilen geläutertem und ätzendem Salz und 3 Teilen eines wohlgereinigten *Mercurii*, daher dieses ganze Gewächs nicht nur trefflich wärmet, sondern auch sanft trocknet, auszieht und renoviert. Es wird daraus kunstvoll ein Öl bereitet, welches wegen seiner Tugend ‚eine Anreizung' genannt wird. Auch wird hieraus eine Essenz bereitet sowie eine Höchste Kraft, oder wie es Etliche nennen, ein Fünftes Wesen, und auch ein wohlgereinigtes, scharfes und ätzendes Salz."

Wenn Pflanzensalze hauptsächlich aus solarer Kraft gebildet werden, sind sie sehr fein, klar und von beinahe süßem Geschmack. Das gilt zum Beispiel bis zu einem gewissen Grad für das Salz des Fenchelsamens, wo hinein jedoch auch scharfe martiale und etwas erdhafte, derbsüße Kräfte gemischt sind. So geht es weiter durch das ganze sulfurische, hervorbringende Spektrum mit allen Varianten der Vermischung bis zum ‚schmutzigen, schwefeligen Salz' (zB: Karotte) oder zum ‚unreinen, martialen, beißenden, groben Salz' (zB: Schwarze Pfefferkörner) – Salze aus der erdnahen Position des Feuers generiert. Das Salz von *Piper longum* ist dermaßen ‚hart', dass es, wenn es im coaguliertem Präparat wieder auskristallisiert, kaum mehr wasserlöslich und schmutziggelb gefärbt bleibt.

Zur Vermischung der Elemente und Prinzipien

> „Ein jegliches Gewächs, das sein Element erzeugte,
> wird in drei Dinge gesetzt, nämlich *Sal*, *Sulphur* und *Mercurium*."
>
> *Paracelsus*, >Von den ersten drei *Principiis* oder *Essentiis*<

‚Jegliches Gewächs' – das sind alle Pflanzen, Minerale, Metalle und überhaupt alles, was sich in grobstofflicher, durchlebter Form ausdrückt, Früchte, Samen, Sande, Erden, Gewebe, Organe usf. Die gesamte Natur besteht aus den fünf Elementen, die in gewisser Weise eine Detaillierung der zu ihnen ‚hintergründig' wirkenden ‚philosophischen Prinzipien' darstellen. Sowohl der prinzipielle wie auch der elementare Grund der verschiedenen Natur-Reiche ist jeweils ein besonderer, mit allem verflochten und dennoch spezifisch.

Im gesamten Pflanzenreich ist nur das Wasser-Element ‚voll aktiv', also vollständig entfaltet, während die anderen Elemente bloß in niedrigeren Entfaltungsphasen mitwirken. Der gesamte Pflanzenteppich der Erde gilt als prinzipiell venerisch-merkurial-lunar, zwischen dem noch in tiefem Seelenschlaf befindlichen Mineralreich und dem daraus erwachenden animalischen Reich. *Thurneisser* merkt an, dass die einschreibenden Kräfte in den Pflanzen zwar verschiedenartig komponiert sind, der Mond aber ‚an jeder Pflanze einen wesentlichen Anteil hat'.

Im Mineralreich ist alleine das Erd-Element voll aktiv, in den Vögeln überwiegt die Luft und danach das Feuer, in den Insekten der Äther und dann das Feuer, in anderen Insekten Äther und Wasser, in den Fischen sind es Wasser und danach das Feuer, und in den höheren Tieren gelangen dann immer mehr alle vier grobstofflichen Elemente (Erde, Feuer, Wasser und Luft) zur vollen Aktivität. Beim Menschen kommt schließlich noch der Äther als (potentiell) voll entfaltetes, fünftes Element hinzu – zumindest nach seiner natürlichen Anlage.

Dies ist eine Bedeutung der Worte: „Ein jegliches Gewächs, das sein Element erzeugte". Die elementare Rangordnung entspricht einer Art Verdichtung des ‚prinzipiellen' Hintergrundes der Geschöpfe. Im Mineralreich regiert das Sal-Prinzip. In manchen Dingen des Mineralreiches erlangt eine Entfaltung des Sulfurischen wesentliche (wie bei den Edelsteinen und Silikaten) oder dominante (wie bei Schwefel und Markasiten) Bedeutung, in anderen das Wasser-Element (wie bei den Kalken und insgesamt mehr oder weniger bei den Metallen). Die volle Entfaltung des Sulfur-Prinzips ereignet sich dann erst im Menschen.

Eine andere Bedeutung des obigen Zitates eröffnet *Paracelsus* am Beispiel der Seerose: Er bemerkt, dass sie ‚nur aus dem Wasser-Element' besteht – und zwar (nun auf einer anderen Betrachtungsebene) nicht deshalb, weil in allen Pflanzen alleine dieses Element voll aktiv ist, sondern weil unter allen in den Pflanzen mitwirkenden Elementen nur alleine im Wasser-Element die Idee, die Tugend, die Quintessenz und die Signatur insbesondere der Seerose begründet und durch dieses ausgedrückt ist. Die anderen Elemente sind auch in dieser Pflanze vorhanden, aber dann eben – wie *Paracelsus* betont – „alleine aus dem Wasser-Element wirksam". Das heißt: ‚Würde man dieses fortnehmen, dann würden auch die anderen Elemente verschwinden.' Es ist also das Feuer-Element der Seerose (auf welche Weise es hier auch ausgedrückt sein mag, als Farben, als relative Zähigkeit mancher Pflanzenkörperteile, als angedeutete Spitzigkeit der Blütenblätter, als zur Wasseroberfläche strebende Kraft der Knospe u. dgl.) ein ‚Feuer-Element des Wasser-Elements' und nicht ein Feuer-Element für sich wie beim Tausendgüldenkraut, beim Pfeffer oder beim Johanniskraut.

Das Wasser-Element des Johanniskrautes hingegen ist (auf dieser Betrachtungsebene) ein ‚Wasser-Element aus dem Feuer-Element', und in letzterem liegt die hauptsächliche Signatur und Tugend des Johanniskrautes. Würde dieses Feuer-Element aus dem Kraut herausgenommen, dann wären auch sein Wasser-, Erd- und Luft-Element fort. Diese Betrachtungsebene ist also differenzierter als die erste, oben angeführte, nämlich dass in allen Pflanzen (auch im Johanniskraut) alleine das Wasser-Element ‚voll aktiv' ist.

Oder im Beinwell – dort wirken alle Elemente aus dem Erd-Element. Würde man dieses daraus wegnehmen, dann wären auch die anderen Elemente fort. Das bedeutet zugleich, dass im Beinwell (insbesondere in der Wurzel) das salinische Prinzip dominiert, so wie im Johanniskraut oder im Tausendgüldenkraut das sulfurische und in der Seerose das merkuriale. Letzteres erstreckt sich durch das luftig-wässrige Spektrum, weswegen es zum Beispiel auch den Arnika dominiert, dessen Feuer-, Wasser-, Erd- und Äther-Element im Luft-Element gründen, durch welches sich Idee und Tugend des Arnikas ausdrückt.

Wiederum ist dies nur eine Betrachtungsebene – zwar schon detaillierter als die erste, aber immer noch gewissermaßen ‚oberflächlich'.

Tiefgründiger wird die Betrachtung erst dann, wenn die Vermischungen erkannt werden. Zwar ist der Arnika luftig-merkurial, aber unmittelbar ‚nach' dieser Dominanz ist in dieser Pflanze eine markante sonnenhaft-sulfurische Kraft anwesend. Das unterscheidet den luftig-merkurialen Arnika zum Beispiel von der ebenfalls luftig-merkurialen Quecke, welcher eine sulfurische Komponente weitestgehend fehlt, dafür aber eine erdhaft-salinische beigefügt ist.

Das sind nur wenige Beispiele davon, was die Vermischung der Prinzipien und Elemente bedeuten kann. Wir ersehen daraus nun auch, wonach der Spagyriker seine laborantischen Verfahren und seine therapeutischen Überlegungen richtet. Manche treiben das schon in der Pflanzenwelt weit, andere halten es einfach, andere führen diese Differenzierungen erst im Mineralreich zur hohen Kunst.

Man kann aus den dominant-wässrig-merkurialen Pflanzen (Seerosen, Wasserlinsen, Algen u. dgl.) besonders die Ausdrucksformen des Merkur-Prinzips (und insbesondere der Planetenkräfte von Venus, Merkur und Mond) erarbeiten, zB. durch Betonung auf Putrefactionen, Gärungen, Rectificationen und Circulationen von *Spiritus* und/oder Alkoholen, etc., um sich in seinen Produkten dann der Idee jener Pflanzen entsprechend zu nähern. Ähnlich wird man etwa das vom Feuer dominierte bittere Tausendgüldenkraut (oder den Rosmarin, die Pfefferkörner, Gewürznelken u. dgl.) wohl kaum putreficieren, wahrscheinlich nicht einmal vergären, sondern eher versuchen, das darin dominierende Sulfur-Prinzip durch ein Hauptaugenmerk auf aromatische Destillationen und Arbeiten an der Pflanzenfarbe zu achten und zu erarbeiten. Natürlich bedeutet das in keinem Fall, dass die Scheidung der anderen Kräfte unbeachtet bleiben soll.

Gedanken zum Sulfur im Pflanzenreich

Wie ist das mit dem Johanniskraut? Könnten wir sein gesamtes Feuer-Element herausnehmen, dann würden damit auch Erde, Wasser, Luft und Äther, welche in diesem dominierenden Feuer gründen, verschwinden. Es bliebe aber jenes, die gesamte Pflanzenwelt dominierende Wasser-Element prinzipiell übrig und auf der selben Ebene auch die anderen vier (eben noch nicht voll entfalteten) Elemente. Wie würde sich dann dieses ausdrücken?

Durch die alchymistische Kunst kann man zwar kaum das ‚gesamte' Feuer-Element aus dem Johanniskraut (oder aus einer anderen Pflanze) herausnehmen, man kann dies aber bis zu einem gewissen Grade tun und dabei die Ergebnisse betrachten:

(nb: Besser wir verwenden zur Anschauung Rosmarin oder Gewürznelken, da diese auch bei geringeren Ansatzmengen ein deutlicheres *ätherisches Öl* geben als Johannis- oder Tausendgüldenkraut)

a) Hydrolat und ätherisches Öl:

Als erstes würde man das Kraut einen Tag lang in destilliertem Wasser digerieren und danach mittels (zügiger) Destillation die öligen und pflanzenspezifisch-aromatischen Teile abscheiden. Das Öl (der ‚flüchtige Sulfur') aus der ersten Destillation wird (zB. durch einen Ölabscheider) abgeschieden, das Hydrolat (das aromatische Wasser) zurückgegossen, einige Stunden digeriert und wieder destilliert – so lange, bis alles Ölige heraußen ist – in der Regel 2-3x (was u. a. abhängt von der Pflanzendichte und der zugesetzten Wassermenge).

Separation des ätherischen Öls von Hydrolat und *Phlegma* mittels eines Ölabscheiders

b) die Farbe:

Der Rückstand im Kolben wird mit (je nach Pflanzenart verschieden starkem) Alkohol übergossen, einige Wochen mazeriert und damit verfahren, wie es bei der einfachen Darstellung des Merkur-Prinzips üblich ist (siehe entsprechendes Kapitel). Der Alkohol birgt nun nicht nur die Merkurialkraft des Krautes in sich, sondern auch dessen Farbe – nach dem Öl eine weitere Äußerung des Feuer-Elements bzw. des Sulfur-Prinzips.

Der klare Alkohol (als ein Ausdruck des Merkur-Prinzips und damit anteilig an Luft-, Wasser- und Äther-Element) wird durch Destillation von der mit *Phlegma* vermischten Farbe geschieden. (Vor dieser Destillation kann gegebenenfalls das obige Hydrolat dem alkoholischen Ansatz wieder zugesetzt werden).

Das im Alkohol enthaltene ‚Feuer' (hauptsächlich durch Brennbarkeit und Schärfe signiert) stammt eben aus dem Alkohol selbst und ist kein Ausdruck vom Sulfur des Krautes. Das ist wichtig zu bedenken, damit man nicht in die Meinung komme, der Alkohol sei selbst ein Sulfur. Der Alkohol – weil dominant merkurial – verbindet sich mit dem Pflanzenmerkur. Da sich Merkur aber stets gerne mit allem verbindet, würde er sich hier auch mit dem Pflanzensulfur verbinden, also das ätherische Öl und das pflanzentypisch-Aromatische in sich aufnehmen. Deswegen scheidet man zuerst den Sulfur aus dem Pflanzenkörper, damit er nun nicht von Merkur verschlungen werden kann. So geht der Alkohol dann nur mehr mit dem Pflanzenmerkur zusammen, bringt dabei aber seinen eigenen Sulfur mit, also eben jene Schärfe, Brennbarkeit und das alkoholtypische Aroma. Deshalb scheint der auf diese Weise dargestellte ‚Merkur' aus der Pflanze bis zu einem gewissen Grade ‚sulfurisch' zu sein. Es ist aber, wie gesagt, nicht mehr der Sulfur aus der Pflanze, sondern der des – im Großen und Ganzen durchwegs merkurialen – Alkohols.

Der zugesetzte Alkohol (rectificierter Schnaps oder Weingeist) hat nun zwei Ausdrucksformen der in der Pflanze anwesenden Kräfte ausgezogen: einen flüchtigen Merkur und einen fixen Sulfur – Geist und Farbe. Oberflächlich betrachtet, ist zur spagyrischen Präparation nur ersterer von Bedeutung und letzterer eher eine ‚tingierende' Begleiterscheinung. Durch die Destillation scheiden wir dieses Gemisch und erhalten so eine klare, alkoholische Ausdrucksform des Pflanzenmerkurs im Recipienten und eine wässrig-phlegmatische Farblösung als Rückstand. Wurden die festen Pflanzenteile während dieser Destillation dabeibelassen, dann müssen sie spätestens jetzt durch Auspressen und Filtrieren abgeschieden werden, vorausgesetzt, man will aus der ‚Farbe' den ‚fixen Sulfur' darstellen.

Die Farb-Flüssigkeit filtrieren wir in ein hohes, schmales Glas, lassen die *Feces* setzen, dekantieren und filtrieren danach wieder. Das wiederholen wir so oft, bis sich keine *Feces* mehr scheiden. Sie werden gesammelt und kommen zur ‚Erde' (zum ausgepressten Pflanzenrückstand). Auf diese Weise erhalten wir eine sehr intensiv gefärbte, jedoch an sich klare Flüssigkeit.

Sollten wir diese Scheidung (von den festen Pflanzenteilen) schon vor der alkoholischen Destillation vorgenommen haben, dann wird ebenfalls der Alkohol (mit dem geistig-fluidalen ‚Merkur der Pflanze') klar und farblos abdestilliert. Wenn keine ‚Tränen' mehr im Helm erscheinen, beenden wir die Destillation, lassen *Phlegma* und Farbe erkalten und klären wieder.

Aus dem Pflanzenrückstand kann nun die verbliebene, lösliche Farbe mit dem klaren, farblosen Alkohol (dem ‚Pflanzenmerkur') ausgezogen werden, solange, bis sich der Alkohol nicht mehr nennenswert färbt.

Damit haben wir die Farbe als (nach dem Öl) nächsten Ausdruck des Pflanzensulfurs aus unserer Pflanze geschieden. Sie ist noch zusammen mit *Phlegma* aus der Pflanze und *Phlegma* vom Alkohol (also mit einem Ausdruck ‚venerischen Wassers'). Dieses *Phlegma* wird als ‚Schlacke' bewertet und geht nicht ins Werk ein, wenngleich es von prinzipiell-merkurialer Natur ist. Es kann aber abdestilliert und für die späteren Arbeiten am Salz aufbewahrt werden. Das bedarf ein und desselben Schrittes, weil ohnehin die Farbe weitestgehend vom *Phlegma* befreit werden muss. Dazu zieht man das *Phlegma* durch gelinde Destillation von der Farbe ab, welche als dicker, meist rot- oder grünbrauner *Liquor* am Boden bleibt. Wegen der Gefahr des Anbrennens sollte man nicht bis zur Trockenheit destillieren. Wünscht man eine weitere Eindickung, dann kann dies in einer offenen Abdampfschale bei gelinder Wärme besser erreicht werden.

Die Farbe wird nun noch weiter gereinigt, indem man den dicken *Liquor* im klaren, alkoholischen Destillat löst und das Ganze wieder in einem hohen, schmalen Glas kaltstellt. Wenn sich keine *Feces* mehr absetzen, ist die Farbe rein und der Alkohol (mit dem darin verbliebenen *Phlegma*) kann durch Destillation wieder abgeschieden und die rückständige farbige Feiste in einer Schale getrocknet werden. Gibt es *Feces*, dann wird die Reinigung wiederholt.

Das Sulfurische an dieser ‚farbigen Feiste' ist eben die Farbe. Die Feiste selbst ist erdhaft-salinisch. Da sie aber sehr fein und gut rectificiert ist (durch wiederholtes Dekantieren und Filtrieren), kann man sie als einen Teil jener Ausdrucksform des Sulfur-Prinzips akzeptieren, ist doch jegliche physische Ausdrucksform immer Vermischung, und die Zuordnung erfolgt durch Dominanz.

c) Wurzelessig und Rotes Öl:

Will man nun aus dieser ‚farbigen Feiste' (kurz: ‚Farbe') wiederum den dem Sulfur-Prinzip nächststehenden Ausdruck scheiden, dann bietet sich der Weg der Trockendestillation an. Dabei verkohlt die farbige Feiste, wird schwarz und verliert also die Farbe. So sind alle Elemente im Feuer der Farbe. Nimmt man aber das dominierende Element (das Feuer) heraus, dann werden auch die übrigen Elemente abgeschieden,und es ist keine Farbe mehr – zumindest in der bisherigen Ausdrucksform – vorhanden. Wir wechseln also durch eine weitere Scheidung die Ebene, um wieder eine feuer-typische Ausdrucksart zu erhalten und dabei (gewissermaßen nebenbei) die anderen Elemente ebenfalls aufzuschließen und in der Folge zu scheiden:

Der Farb-*Liquor* wird bei gelinder Wärme zur zähen Extrakt-Konsistenz eingetrocknet. Dann kratzen wir ihn aus der Schale, füllen ihn in einen kleinen Kolben, legen eine kurze Brücke und einen großen Recipienten vor und beginnen bei gemächlichem Feuer zu destillieren.

(Besser man trocknet nicht zu weit ein, sondern fängt das restliche *Phlegma* in einem eigenen Recipienten auf. Die Abdampfschale kann mit destilliertem Wasser ausgewaschen und dieses zum *Liquor* gegeben werden.) Zuerst kondensiert das restliche *Phlegma*, mehr oder weniger, je nachdem, wie weit getrocknet wurde. Dann erhebt sich die Masse im Kolben und droht überzulaufen. Wir erhöhen die Temperatur sehr vorsichtig, bis alles ausgekocht ist. Dabei bilden sich Blasen, die dann platzen, die Oberfläche reißt auf und es steigen ‚schwefelige' Rauchwolken empor, was uns anzeigt, dass wir uns hier auf dem ‚Weg durch den Vulkan ins Erdinnere' befinden. Der Rauch scheidet sich in der Vorlage bald in ein rotbraunes Öl und in den ‚Wurzelessig'. Damit wird wieder eine Ausdrucksform des Feuer-Elementes sichtbar: das ‚rote Öl' der Farbe des Krautes.

Der Wurzelessig entspricht einer Äußerung des Merkur-Prinzips, ist aber von einem Aspekt Sulfurs, nämlich der ‚gegenwirkenden Feuerkraft' *(ios)* durchdrungen. Deshalb kann dieser Essig auch zusammen mit dem Roten Öl zu den Ausdrucksformen des Sulfurs gezählt werden, nämlich dann, wenn wir uns den (oben erwähnten) Feuer-Sterntetraeder als Ganzes, also mit dem Oktaeder (Merkur) als Inhalt, vorstellen. So entspricht der Wurzelessig dem ‚Inhalt' des Sterntetraeders und ist deshalb, obwohl wesenhaft merkurial, von seinem (feurigen) Wesen umschlossen und durchwirkt. Das ist sulfurisch am Wurzelessig.

Merkurial am Wurzelessig ist, dass er fließt – also Anteil hat am Wasser-Element. Und er besitzt vermittelnde Kraft – nach der Art des Luft-Elements. Dazu birgt sein durchdringendes Wesen ein flüchtiges Salz *(Sal volatile)* in sich, wie es dem Äther zukommt. Dies sind Aspekte des merkurialen Spektrums.

Seine lösende, zerteilende Schärfe ist eigentlich eine Signatur des feurigen Aspekts des Luft-Elements. Wenn Merkur und Sulfur eng verbunden sind, dann wird daraus ‚Schärfe' (wie wir es zB. von der Schärfe des Weingeistes kennen). Mars selbst hat diese Schärfe nicht, sondern er kann nur fixieren und verdichten. Wenn sich Merkur zu ihm gesellt, von martialer Kraft durchdrungen wird, dann äußert sich das (unter anderem) als ‚Schärfe'. An dieser ist das Zertrennende, Zertreibende und Zerstreuende typisch merkurial, denn ‚das, was überall hingeht, ist Merkur'. Die ‚vermittelnde' Qualität der Luft benötigt das Feuer, wie es als ‚Kontrollposition Feuer-Luft' in der taoistischen Medizinlehre beschrieben wird. Dieses Kontrollierende des Feuers zur Luft erstreckt sich auf das gesamte Spektrum des Merkur-Prinzips. Immer wird Merkur vom Feuer (Sulfur) getrieben und fixiert zugleich. Könnte man aus dieser Schärfe die Luft vom Feuer durch die Kunst scheiden, dann hätte man ein sehr subtiles Feuer – eine weitere sulfurische Ausdrucksart, und man hätte einen sehr reinen Merkur, welcher nicht korrosivisch (ohne Schärfe) ist. Dies ist der ‚hermetische' Hintergrund zur atomaren Reaktion, wie er in den vedischen Versen vor vielen Jahrtausenden beschrieben (und zB: im *Kurukshetra*-Krieg vor etwa 7000 Jahren auch in die zerstörerische Praxis umgesetzt) wurde.

Wir sehen, dass durch die Scheidung der Ausdrucksarten (der Prinzipien) mittels Trockendestillation der ‚Farbe' wieder mehrere Produkte erlangt werden, in denen ihre typischen Elemente (oder Prinzipien) vorherrschend, die anderen aber stets vorhanden sind. Damit weiter in der Praxis:

Trockendestillationen zu Wurzelessig und Rotem Öl.
Im Recipienten des rechten Bildes sind die beiden Fraktionen erkennbar.

Bevor also die Masse im Kolben verkohlt, erhalten wir aus dem Rauch Essig und Öl, davon wir ersteren als eine Art *‚Spiritus'* und letzteres vorrangig als sulfurischen Ausdruck betrachten. Die ölige Feuer-Manifestation kann jetzt durch Rectification gereinigt werden:

Die Rectification des Roten Öls:

Das Öl für sich ist schwer zu destillieren und würde zum Teil verbrennen, bevor es übersteigt. Entweder wir rectificieren es zusammen mit dem Wurzelessig oder wir scheiden die beiden zuvor im Ölabscheider und rectificieren das Öl mit destilliertem Wasser vermischt. Besser ist der erste Weg, denn das ungereinigte Öl, verklebt den Scheidetrichter und ist ohne viel Verlust kaum herauszuholen.

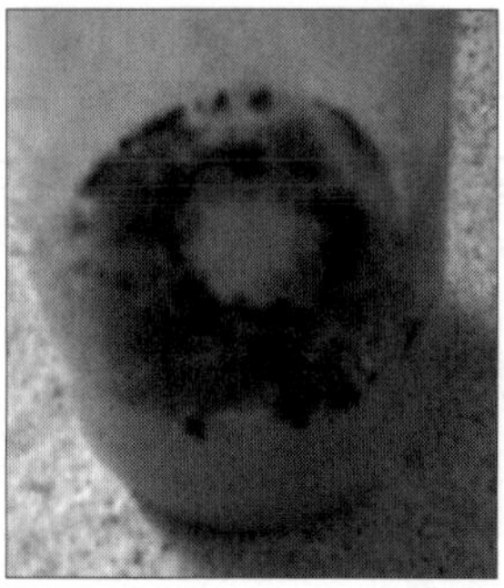

Im Recipienten setzen sich die öligen und teerigen Anteile oben ab. Unten bleibt der Wurzelessig, welcher noch mit *Phlegma* vermischt ist. Die im Bild sichtbare Trübung des Essigs kommt vom hohen Anteil an flüchtigen Salzen. Das Mengenverhältnis von Essig, Öl und *Sal volatile* ist von Pflanze zu Pflanze verschieden. Ebenso die Vermischung des eigentlichen Roten Öls mit Teer und anderen Substanzen sowie der Anteil des *Phlegmas* im Essig.

In jedem Falle aber ist darauf zu achten, dass das Öl im Kolben keine dicke, geschlossene Oberfläche bildet, da sonst ein heftiges Aufwallen des darunter befindlichen Essigs zum Zerschlagen des Gefäßes führen könnte. Dagegen lehnt man z.B. einen Glasstab in den Kolben, der mindestens einige Millimeter über die Flüssigkeitsoberfläche hinaus ragt. Oder man verwendet einen Kolben, der weit genug ist, damit die Ölschicht nur ‚zerrissen' auf dem Essig schwimmt.

Bereits bei der ersten Rectification sehen wir, dass dieses Öl zum großen Teil aus zähem, unctuosischem, anhaftendem Teer besteht und nur zum geringen Teil aus einem schönen, rötlichen Öl. Sobald es nach zwei bis drei Rectificationen halbwegs rein erscheint, scheiden wir es mittels Pipette oder Ölabscheider vom Essig, vermischen es mit destilliertem Wasser oder gemeinem destilliertem Essig und rectificieren noch einige Male. Besser noch, man vermischt den Wurzelessig schon vor der ersten Rectification mit destilliertem Obstessig oder Wasser. Die Abscheidung (so erwünscht) des eigentlichen (und nunmehr verdünnten) Wurzelessigs davon wird dadurch allerdings aufwändiger. Verwenden wir Wasser anstatt Essig, dann sollte es auf jeden Fall soweit möglich abgeschieden werden.

Auf diese Weise haben wir eine weitere Ausdrucksform des Feuer-Elementes, bzw. des Sulfur-Prinzips erlangt – das gereinigte, rote Öl.

„Die Erde mach zum Roten Öl, das ist die purpurfarb'ne Seel."

J.F. Hautnorton, >Traktat vom philosophischen Salz<, 1656

D) Themen des *Sal Sulphuris:*

Wenn wir das Rote Öl ‚aus der Farbe' nicht darstellen wollen, sondern es uns nur um die Darstellung des sulfurischen Salzes geht, dann trocknen wir die ‚Pflanzenfarbe' in einem (wegen des Schäumens ausreichend hohen) offenen Tiegel ein.

Wenn der ‚Vulkan' zu blubbern und der Rauch aufhört emporzusteigen, beginnen sich die Signaturen der ‚inneren' Planeten Uranus und Neptun zu zeigen. Die Kruste wird zuerst hart und schwarz und gleicht einem erstarrten Lava-Strom. Sie zeichnet sich durch eine gewisse Oberflächenspannung und einen tiefgründigen, schwarzen (lichtlosen) Raum darunter aus.

Dies ist die Position des *Uranos* – sowohl an der Grenze zum Erdinneren als auch an der Grenze zum jenseitigen Kosmos, jeweils als Vater-Aspekt ‚hinter' bzw. ‚in' den Positionen Saturns verborgen. Für das menschliche Auge nicht sichtbar, herrscht er in den immer heißer und dichter werdenden Gesteinsgängen unter der obersten venerischen Erdkruste und kontrolliert die weite Grenze zwischen Außen- und Innenwelt. Zugleich herrscht er am Rande der ersten erdbezogenen Planetensphäre, dort wo der eigentliche ‚Welt'-Raum endet und das physische Äther-Element sich mit astralem Äther vermischt.

Erd-Element und Innere Planeten in den pentagrammatischen Positionen:

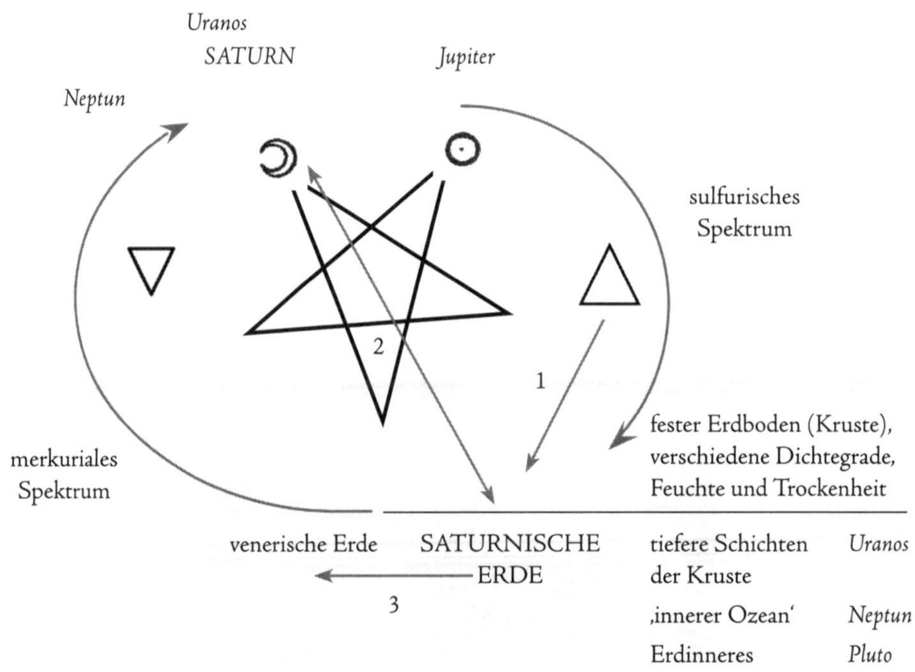

Die Pfeile 1, 2 und 3 in der Skizze beziehen sich auf die Entwicklung des Erd-Elements. Dieses wird (entsprechend der ‚Hervorbringungsrichtung') aufgrund der zuerst zentrierenden (solaren) und dann der fixierenden (martialen) Feuer-Kräfte als erstes zum ‚saturnischen Erd-Element' (1) – dicht, schwer, dunkel und kalt. Als solches verbindet es sich sogleich mit seinem Gegenpol, der mondenhaften Sphäre, dem ‚hohen Wasser' bzw. dem lunaren Äther (2). Durch diese Verbindung (siehe auch: ‚Die Planetenspirale' nach *Gichtel)* wird das Erd-Element befeuchtet und in die natürliche Harmonie eingebunden. Man nennt es dann ‚venerische Erde' (3).

Die erste saturnische Phase setzt sich in der Erdkruste als Position des *Uranos* fort. Die zweite, in der Außenwelt venerische Phase entspricht im Erdinneren *Neptun*. Deshalb verglichen die Alten das Erscheinen der ‚Weißung' im Tiegel mit dem Auftritt von *Venus/Aphrodite* ‚am Strand' *(Botticelli)*, wobei sich dieser Strand genaugenommen bereits im Erdinneren befindet, und der Ozean, aus welchem *Aphrodite* geboren wird, ist der erdinnere, neptunische Ozean. Das betrifft den einen, den erdinneren Aspekt dieser Ebenen. Der zweite Aspekt ist der ‚himmlische' in der Position des lunaren Äthers.

In der uranischen Position im lunaren Äther nähert sich auch Merkur bzw. das Merkur-Prinzip dem Ende seines Spektrums und ist dabei von äußerst subtiler und flüchtiger Art. Dem steht nun wiederum ein kontrollierender Aspekt des sulfurischen Prinzips entgegen, welcher in einer gewissen markanten Spannung und einer besonders ‚feinen', vielleicht zum Teil zähen oder elastischen Art des uranischen Ausdrucks signiert ist.

In dieser uranischen Position sind zwei Aspekte Merkurs auffallend: Das ‚der Dichte entfliehen' und das ‚die Dichte durchdringen'. Gleichsam zeigt die Haut des Körpers jene uranische Signatur. Sie empfängt gröbere und feinere Einflüsse von außen und entläßt Gröberes und Feineres vom Körperinneren an die Außenwelt. Das ist mit ein Grund, warum die Haut dem merkurialen Luft-Element zugeordnet wird. Ähnlich die Haut der Erde. An dem, was durch die Körperhaut nach außen dringt, erkennt man etwas von Zuständen und Ereignissen im Körperinneren. An der Körperbehaarung wird der Zustand verschiedener vom Luft-Element hauptsignierter Körperteile, Säfte, Organe u. dgl. befundet (also auch der Lunge und des Dickdarms), die sich im Inneren des Körpers befinden. An den (zumeist von Merkur signierten) Gräsern, den merkurialen Bäumen (Pappel) und merkurialen Sträuchern (Haselnuss) können Zustand und Ereignisse weit im Inneren der Erde befundet werden. In diesem Zusammenhang kann die Haselnussrute als klassisches Instrument zum (be)finden von Erdstrahlen, Wasseradern, Metallen u. dgl. verstanden werden. Solche merkurialen ‚Vermittlungen' geschehen aufgrund der Anwesenheit sulfurischer Kräfte.

Insgesamt ist die Position der Haut – ihre Grenzbildung, ihre Spannung, ihr Zurückhalten des Inneren und ihr Abhalten des Äußeren – im weitesten Sinne saturnisch, aber es ist genaugenommen nicht typisch Saturn, sondern eben *Uranos*, dessen Kraft und Wesen sich hier entfaltet. Was sich auf der Haut ausdrückt sind einerseits Substanzen und ‚Dinge' wie Schweiß oder Haare. Andererseits sind es Fette, Farben und Temperaturen – das ist vom Feuer-Element und es sind sulfurische ‚Dinge'. Der ‚Ausdruck' aber, einer Hitze auf der Haut oder einer roten, grünen, gelben Farbe, teilt uns etwas mit – das ist merkurial. Da entweicht Merkur der Dichte des Körpers und trägt Informationen aus dessen Inneren mit sich, welche feurige Aspekte beinhalten.

Gleichsam entfliehen Wurzelessig und -öl dem Kerker der Substanz und bringen aus dem Inneren etwas mit, etwas Individuelles, Sulfurisches. Deshalb ist ein *Spiritus Nitri* ganz anders als ein *Spiritus Salis*, und wieder anders ist der Wurzelessig aus Eichenrinden und wieder anders ‚Essig und Öl' aus der Farbe der Gewürznelken oder des Johanniskrautes. Immer ist es ‚individuell' und ölig, ‚korrosivisch', feurig, scharf – Träger der aufbrechenden Feuer-Kraft *(ios)*. Dass Essig und Öl aber überhaupt in Erscheinung treten und dadurch die Möglichkeit bieten, weitere Prozesse zu vermitteln, das ist merkurial. Merkurial vor allem beim Essig und nur wenig beim Öl.

d.i.) Die Pfauenfeder:

Wenn wir den ‚Vulkan' nicht im offenen Tiegel abrauchen, sondern aus der Farbe Wurzelessig und Rotes Öl abdestillieren, dann lässt sich die schwarze, erstarrte Kruste nun leicht aus dem Kolben nehmen, zerdrücken und zur nachfolgenden Calcination in einen Tiegel legen. Während wir langsam das Feuer unter dem Tiegel vermehren, zeigen sich zuerst vorsichtig, dann immer intensiver allerlei Farben, welche direkt aus der Schwärze hervorzutreten scheinen. Schließlich kann man das gesamte Spektrum des Regenbogens sehen, und bei stärkerem Feuer bietet eine herrliche Pfauenfeder ein eindrucksvolles Farbenspiel. Das oben Beschriebene findet jetzt im Mikrokosmos des Tiegels, des ‚hermetischen Gefäßes', statt. Hier kommen ‚aus dem Erdinneren' Farben hervor, als typischer Ausdruck eines Sulfurs, welcher immer noch in unserer Pflanzenfarbenkohle steckt. Diese Farben sind schön anzusehen und halten sich bei entsprechender Feuerführung mehrere Stunden. Sie können aber nicht eingefangen werden!

Das Irisierende, das Regenbogenspektrum, ist nur deshalb Sulfur, weil es eben Farbe ist. Es ist sehr wesentlich Farbe wegen der Intensität und Leuchtkraft und deshalb auch ein wesentlicher Ausdruck des Feuer-Elements aus der Erde. Es ist aber bereits typisches ‚inneres' Feuer, eine Äußerung der *ios*-Kraft – weit mehr noch, als dies beim Roten Öl oder gar beim Wurzelessig der Fall war. Vergleichen wir Wurzelessig und Pfauenfeder:

Der Wurzelessig ist ein überwiegend fluidales Medium für die *ios*-Kraft (weswegen er auch imstande ist, andere feste Körper aufzubrechen – und das umso mehr, je weniger Wässrigkeit/*Phlegma* er enthält). Die Kraft selbst kann man (unter üblichen Umständen) nicht sehen – man sieht nur ein ‚Wasser'. Je nach dem Stoff, woraus der Essig stammt, wird er sich mehr oder weniger verschieden äußern. Bisweilen muss er sogar in einem wässrigen Medium empfangen werden, wie wir es von der Trockendestillation des Salpeters zum *Spiritus Nitri* kennen, wo Essig und Öl als ganz rot aufsteigen, eigentlich einen öligen Eindruck machen, ansonsten aber gar nichts Fließendes an sich haben. Auch bei diesem *Spiritus* handelt es sich um einen von Sulfur durchdrungenen Merkur.

Im vorigen Beispiel tut sich die feurige Art des Essigs (zB. aus Gewürznelken oder Johanniskraut) erst in der Wirkung kund – man sieht sie ihm nicht an. Es gelangt aber ‚etwas' in den Recipienten, womit man danach weiterarbeiten kann. Der Wurzelessig für sich ist zum Beispiel gut als *Menstruum* geeignet. Der Wurzelessig aus dem inneren Bast von Rinden junger Eichen war bei den Alten zu diesem Zwecke beliebt. Er war damals leicht zu bekommen, weil die Gerber solche Rinden in großen Mengen benutzten. Dass der Honig reichlich Wurzelessig gibt, machten sich damalige und heutige Spagyriker zunutze. Er wird als *Menstruum* für Perlen empfohlen. Diese Essige werden auch gebraucht, um hochwertige Pflanzensteine, höhere *Menstrua* (im Sinne von Quintessenzen und Alkahesten) und dergleichen Arbeiten zu verrichten. Bei *Tölde* finden wir dazu zahlreiche Anleitungen. Diese Essige oder *Spiriti* sind in jedem Falle substantiell und je nach Ursprung mehr oder weniger geistig.

Verglichen mit dem teerigen, ‚roten' Öl kann der Essig nur bedingt als eine Ausdrucksart des Sulfur-Prinzips begriffen werden. Die Ausdrucksweise des Sulfurs in Essig und Öl ist weit gröber, als zB. in den Farben der Pfauenfeder. Diese sind nicht substantiell. So ist dort und da Sulfur signiert – nur auf anderen Ebenen. Die beiden Flüssigkeiten (besonders der Essig) sind auch mit Merkur verbunden – nur ist dieser dort nicht dermaßen geistig anwesend wie hier im Schillern und Farbenspiel.

Jene Pfauenfeder ist zwar schön zu betrachten, es kann aber nichts davon aufgefangen werden. Sie ist als Ausdrucksform des Sulfur-Prinzips weit elementarer als Wurzelessig oder Öl, sie steht dem Feuer-Element näher. Die Farben sind zwar tatsächlich in der Kruste vorhanden, können aber nicht stofflich herausgegriffen werden. Die Pfauenfeder teilt uns mit: Hier dringt eine Äußerung des Feuer-Elements aus der schwarzen Erde hervor, welche aufs engste verbunden ist mit einem sehr subtilen Merkur. Hier erleben wir die Geburt des Luft-Elements aus der Erde. Dieses (bzw. dieser Merkur) ist nicht die Farbe an sich, sondern die Vielfalt, das Schillern, das faszinierende Schauspiel und das (informative) Spektrum (welches ja nicht bei jeder Pflanze das selbe ist).

Und wie jede Geburt Merkurs aus der saturnischen Erde, so ist auch diese mit einem Hervorbrechen sulfurischer, feuriger Äußerungen verbunden. Auf diese Weise haben wir wieder ‚einen Sulfur' aus unserer Pflanze ‚dargestellt'.

D.2.) DAS ERDINNERE:

Wenn wir nun stärker heizen, hört das Farbenspiel schließlich auf und es setzt vom Rande der verschlackten Krusten (‚vom Ufer') aus die Weißung ein, welche wir zuerst schon beim Vergleich Venus-Neptun erwähnt haben. Im Erdinneren öffnet sich die Region Neptuns, Herrscher des inneren Ozeans – und in gewissem Sinne: des übersteigerten Mondes. Diese Region birgt Gefahren für den, der durch den Vulkan ins Erdinnere gestiegen ist. Hier ist eine Grenze zur Astralität, die man schnell durchqueren muss, will man zum Herzen der Erde, zum Reich Plutos vordringen. Die ätherische Position Neptuns weist auf die Art der Gefahr hin. Sie beschreibt die endlose Weite des Raumes, welcher sich dort öffnet, wo nicht nur das Wasser- ins Äther-Element übergeht, sondern diese Übergangsposition bereits direkt das ‚Jenseitige', das Astrale berührt. Hierher gehören Träume und Trance sowie mächtige Heil- und Giftpflanzen und Pilze. Hier bildet sich auch das Tierreich. Das Animalische wird substantiell aus Faulung und Putrefaction in der Mondposition generiert. Seine Kraft, sein Wesen wirkt ‚von drüben herein', indem das Astrale, als ‚Reich des Gemüts', alle Formen und Daseinsmuster des Tierreichs in die Physis imprimiert. Hier bilden sich aus Faulung und Astralität jene Insekten, welche dann pflanzentypisch einen Aspekt des Äther-Elements des jeweiligen Pflanzenkörpers bilden. Und hier wirken ‚von drüben' unirdische Wesen herein, die sich von den in den Putrefactionen freigewordenen Quintessenzen jeglicher physischer Körper nähren.

Dementsprechend schildern die Erzählungen, wie die Küste zum ‚inneren Ozean' von urzeitlichen Kreaturen bevölkert ist, gleichsam wilden, archaiischen Gestaltungskräften des Tierreichs. Wer diesen Kreaturen entfliehen will, muss hinaus auf das tosende Meer und wird dort von Strudeln erfasst und in die Tiefe gerissen. Die Kundigen mancher vulkanischer Inseln erwähnen in ihren Geheimlehren, dass bei gewissen Konstellationen solches Getier im heißen Krater beobachtet werden kann, wenn es durch innerirdische Gänge aus jener Region nach oben steigt, um sich an den schwefeligen Substanzen am Rande der heißen Kruste zu laben.

Wenn die Asche hier weiß (eher hellgrau) wird und der innere Ozean also erreicht ist, findet man da kein Sulfur-Prinzip und kein Feuer-Element mehr. Wir sind dort angelangt, wo der Sulfur nicht vollständig hinreicht und nur als potentielle Wurzelkraft anwesend ist. Hier herrscht tatsächlich eine Art Kraft, die dem gemeinen Feuer genau entgegensteht, etwas direkt Implosives, Einsaugendes und etwas, ganz in der Nähe des Begriffes ‚Dunkelmond'.

Es ist eine Position im ‚oberen Wasser-Element', anteilig ‚am Mond', wesentlich im lunaren Äther und darüberhinaus eine Öffnung zum Astralischen – ein chaotischer, hindurchreißender Schlund zum Jenseits.

Es ist der subtilste und oberste Rand des höchsten Merkur-Prinzips, und es ist das ‚Entfliehen des Merkurial-Geistes'. Das, was zuvor als Merkur im saturnischen Stoff eingekerkert war und selbst nach seiner Befreiung immer noch mehr oder weniger stoffhaftend bleibt, wird hier zu einer beinahe völligen Freiheit Merkurs. Ähnlich wie die Pfauenfeder (in der *Uranos*-Phase der Arbeit) den Sulfur unfassbar anzeigt, verhält es sich jetzt in der *Neptun*-Phase mit Merkur. Hier ist der ‚Brunnen' des *Berhard von Treviso* „von solch furchtbarer Macht und schrecklicher Natur, dass er, wäre er zornig, alles durchdringen würde. Entwiche er, dann wären wir alle verloren." Der Hinweis „wäre er zornig" bezeichnet die Abwesenheit Sulfurs. Im Gegensatz zu allen, vom Feuer/Sulfur durchwirkten Elementarbereichen herrscht hier völlige Nicht-Individualität.

Hier ist Merkur mit den Worten der Alten „der feuerspeiende, allesverschlingende, urzeitliche Drache", das „allestötende Gift", „das Wasser der Verderbnis und des Vergessens". Das, was er tötet und verschlingt, ist jegliche Ausdrucksart oder Ausdrucksform im Sinne von ‚Individualität' – und damit verschlingt er sich selbst, weil er sich als Meisterdieb seines eigenen Auftrags beraubt – in der ‚Hoffnung', so seiner elementaren Bindung endgültig zu entkommen.

Dieser Seite des ansonsten so durch und durch sympathischen Schelms sollten wir beginnen zu gedenken, wenn wir uns im Labor auf den Weg zum ‚Fixen Sulfur der Pflanzen' begeben, und diese Seite sollten wir schätzen, wenn wir (mit *Flamels* Worten) „den geflügelten Löwen suchen, welcher jegliche metallische Natur verschlingt und in seine eigene, wahre Substanz umwandelt".

Dass diese Phase, die wir hier die neptunische nennen, obwohl wir sie genauso als eine merkuriale oder mondenhafte bezeichnen könnten, im Tiegel allegorisch erlebt wird – und zwar auf dem Weg ins Innere der Erde – in terrestrischer genauso wie in lunar-ätherischer Position zu finden ist, verweist auf eine hermetische Ähnlichkeit von *Chaos* und *Tartaros* und enthüllt etwas vom Wesen jenes Erd-Elements, das wir so leichthin ‚saturnischen Erde' nennen und dabei zumeist an Dichte denken, welches doch ähnlich ist dem Wesen des ‚schwarzen, leeren Raumes'.

Daraus wir auch klarer, warum ‚der Mond zur Erde (zum Saturn) treten muss' und die ‚Erde', wie wir sie kennen, erst dadurch annehmbar wird. Während der Vereinigung von Mond und Saturn werden beide auf gewisse Weise zu Venus, und so bleibt es der Venus zugeschrieben, dass sie ‚die Dinge annehmbar' macht und es kommt dem Feuer zu, ihnen den Sinn zu geben.

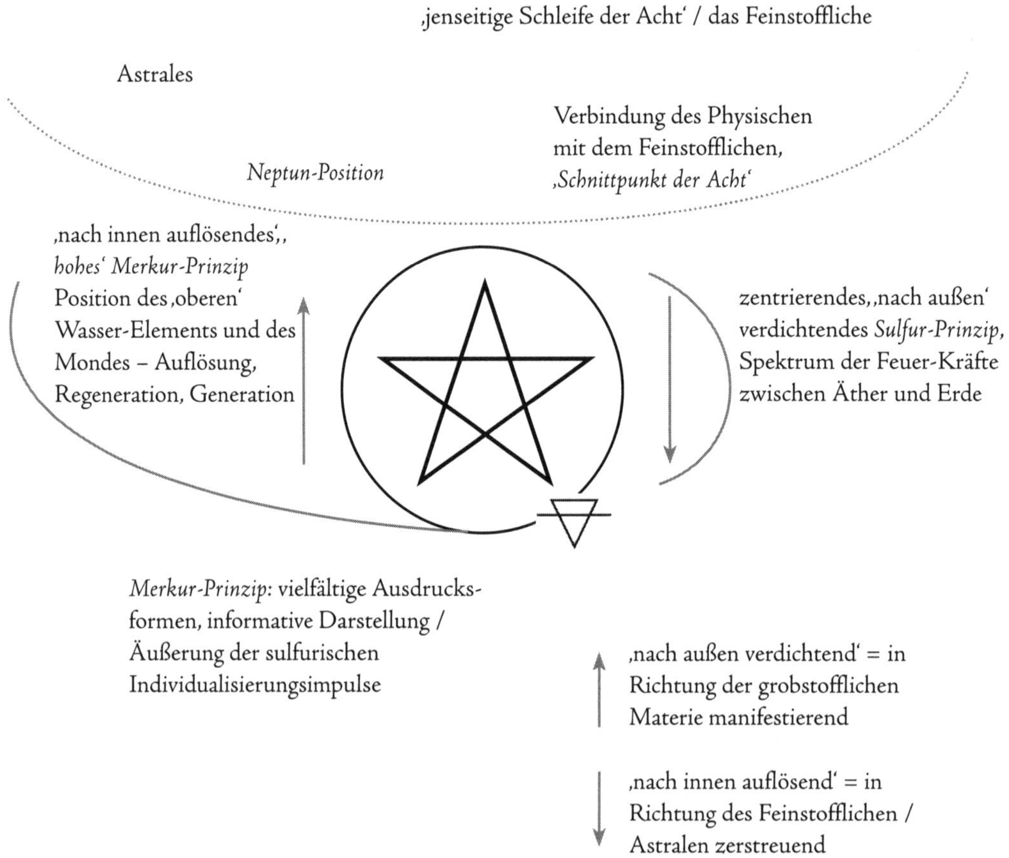

D.3.) DAS SAL SULPHURIS:

Die Signatur der hellen Asche in unserem Tiegel als venerisch/lunar/neptunisch anerkennend, lassen wir abkühlen und lösen das *Sal sulfuris* mit destilliertem Wasser heraus, wie wir es von der Darstellung des Erd-Salzes gewohnt sind. Das in der Folge durch Umkristallisieren gereinigte Salz wird gemeinhin als ‚fixer Sulfur' *(Sulphur fix)* bezeichnet, hält diesem Titel im hermetischen Sinne bei genauerer Betrachtung aber nicht wirklich stand. Es fiel ja schon bei der Feuerführung zu dieser Calcination auf, dass die Asche zur ‚Weißung' höhere Temperaturen verlangte als die Asche des gewöhnlichen Pflanzenrückstandes bei der Darstellung des Erd-Salzes *(Sal fix)*. Durch die höheren Feuergrade beginnen aber bereits Teile des Salzes aufzuschmelzen, während andere Teile der Asche noch dunkel und unaufgeschlossen bleiben.

Wir nehmen dies als Hinweis der Natur, dass das innerste, bei der Generation des Erd-Elements darin eingeschlossene Feuer zur laborantischen Darstellung seiner Ausdrucksform höhere Temperaturen braucht als das ‚oberflächlichere' Erd-Salz, welches ja auch in der Verschachtelung der platonischen Körper als Würfel den feurigen Sterntetraeder in sich einschließt.

Wir laugen also die bei ca. 800°C (bei verschiedenen Pflanzen-Sulfuris andere Temperaturen) calcinierte und zum Teil bisweilen auch verschlackte Asche mit Wasser gut aus, calcinieren die noch unaufgeschlossenen Rückstände weiter, und wenn wir alles heraußen haben, wird gereinigt und kristallisiert. Damit haben wir eine weitere Ausdrucksform des Sulfur-Prinzips, das *Sal sulphuris* erlangt und die plutonische Position im Erdinneren erreicht.

D.4.) DIE FARBE DES SULFURS:

Das reine sulfurische Salz legen wir in einen starken, glasierten Tiegel und schmelzen es (bei mindestens 1000°). Die erkaltete, erstarrte Schmelze hat nun wieder eine Farbe!

Calcination im Muffelofen

Über dieses erkaltete ‚Glas' gießen wir unseren hochrectificierten alkoholischen Pflanzenmerkur und stellen das Ganze einige Wochen in die Wärme zur Digestion. Der zuvor klare, farblose Alkohol wird sich nun mit der Zeit (oft erst nach Monaten) färben, wobei die Farbe oft eine Komplementärfarbe zur (ersten) Farbe des alkoholischen Pflanzenauszugs ist. War dieser grünlich, dann wird diese sulfurische Farbe (in den meisten Fällen) gelbbraun, orange oder rot sein. Damit sind wir bei der letzten Ausdrucksform des Sulfur-Prinzips angelangt, die wir aus den Pflanzen darstellen können. Sie entspricht am meisten dem ‚aus dem Fixen dargestellten Sulfur' und drückt sein Wesen deutlicher aus als das sulfurische Salz, welches ja farblos bzw. weiß ist.

Wollen wir diese Farbe noch ‚reiner' haben, dann scheiden wir den Alkohol durch gelinde Destillation (am besten aus dem Wasserbad) ab. Die Farbe bleibt am Boden des Kolbens mit dem restlichen *Phlegma* (aus dem Alkohol bzw. aus dem Pflanzenmerkur) und einem salinischen Anteil zurück. Das *Phlegma* kann abgedampft werden, wobei darauf zu achten ist, dass die Farbe nicht anbrennt. Genaugenommen ergibt der Rückstand im Kolben (nach der Scheidung des Alkohols bzw. des Pflanzenmerkurs) nun wieder eine Art ‚Essenz': Das merkuriale *Phlegma*, das darin gelöste Salz und die sulfurische Farbe.

Die letzten Arbeiten können mit allerlei Salzen durchgeführt werden – nicht nur mit *Salis sulphuris*. Immer wird eine sulfurische Farbe aus dem aufgeschmolzenen Salz erlangt – sehr eindrucksvoll etwa aus dem *Sal tartari*. Dabei fällt auf, dass zugleich mit der Extraktion der Farbe durch den Alkohol dieser vom *Phlegma* geschieden wird. Vermischt man etwa einen 80%igen Alkohol mit dem Salz, dann bildet sich eine obere, alkoholische und gefärbte Schicht von ungefähr 4/5 der Menge und eine untere wenig oder gar nicht gefärbte Schicht von ca. 1/5. Es bilden sich also zwei Schichten: die obere mit dem sehr hochprozentigen Alkohol und der Farbe, und die untere mit dem *Phlegma* und den salinischen Resten. Der Alkohol wird dadurch nicht vollständig vom *Phlegma* geschieden, aber doch beachtlich, und er schwimmt obenauf wie ein Öl. Darüber mag man wohl einiges nachdenken. Wir aber wollen damit zufrieden sein, uns hier nur einige wenige Gedanken zum Sulfur-Prinzip gemacht zu haben.

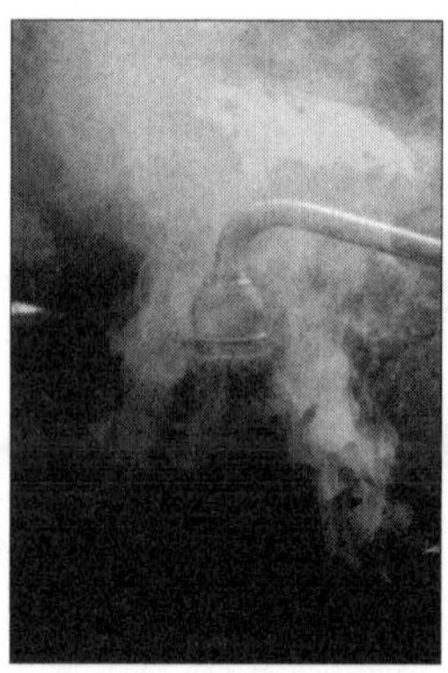
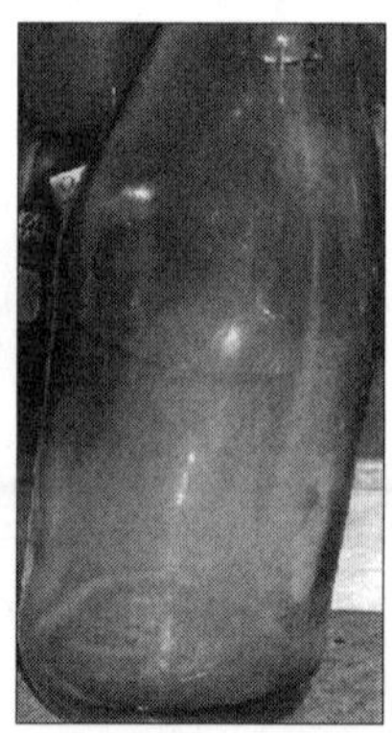

2) Darstellung des Merkur-Prinzips aus der Pflanze

Zur Darstellung des Merkur-Prinzips durch Gärung muss (nach anfänglicher sulfurischer Destillation) Hefe und (meist auch) Zucker zugesetzt werden, da die der Pflanze von Natur aus anhaftenden Hefen durch die Kochung während der Destillation für die Gärung unbrauchbar geworden sind.

Die Darstellung des Merkurs durch Mazeration in zugesetztem Alkohol kann selbstverständlich wie oben beschrieben durchgeführt werden.

Insgesamt gilt hier alles so, wie wir es bei der Darstellung des Merkurs aus Pflanzen bereits beschrieben haben.

Nach anfänglichen Gärungen und darauffolgender Destillation des merkurialen Prinzips kommt es vor, dass der Alkoholgehalt eher niedrig ist und dadurch das obgenannte Verschlingen der Seele/des Sulfurs nicht vollständig stattfindet. Das heißt: Trotz alkoholischer Gärung und Destillation kann man noch eine bedeutende Menge ätherischen Öls extra abscheiden. In diesem Falle steht der Arbeit zur ‚Essenz' nichts im Wege, wenngleich dabei das Verhältnis zu bedenken ist: Wieviel Sulfur ist im Merkur gelöst und wieviel freigeblieben? Wenn zB. ein Ansatz in sich 10ml ätherisches Öl enthalten haben würde, davon aber nur 1ml nach der Gärung separiert werden konnte, dann scheint die Betitelung ‚Essenz' etwas übertrieben. Es handelt sich eher um eine Mischform, die besser wieder mit ‚Tinktur' zu bezeichnen wäre. Allerdings ist hier immer von Bedeutung, ob zusätzlich noch andere sulfurische Ausdrucksformen, insbesondere der ‚fixe Sulfur' aus der ‚Farbe', für sich dargestellt werden und damit wieder zweifellos der Status einer ‚Essenz' erreicht wird.

3) Darstellung des Sal-Prinzips

Auch hier gilt alles, wie oben bei der ‚spagyrischen Tinktur' besprochen. Zusätzlich zum ‚flüchtigen Sulfur' wird nun aber gerne der ‚fixe Sulfur' erarbeitet, um den essentiellen Charakter zu verstärken. Im Falle, dass kein ‚flüchtiger Sulfur' erlangt werden konnte (zB: beim Ackerschachtelhalm, bei der Eibe und bei anderen Pflanzen, welche kein ätherisches Öl und nicht einmal ein nennenswertes aromatisches Destillat geben), ist dies auch unbedingt notwendig. Dieser ‚fixe Sulfur' wird, wie oben bemerkt, üblicherweise als *Sal sulphuris* – das Salz aus der ‚Farbe der Pflanzensuppe' – dargestellt – ein Prozess, der im Großen und Ganzen der Darstellung des Erd-Salzes gleichkommt.

4) Coagulation (s. o.)

Erarbeitete philosophische Prinzipien: Sulfur – Merkur – Sal

Betitelungen der Präparate

Die Herstellung einer ‚spagyrischen Essenz' ist aufwendiger als die der ‚spagyrischen Tinktur'. Sie erfordert mehr Zeit, Geduld und Können. Das mag ein Grund sein, warum heute dem Therapeuten allgemein hauptsächlich Tinkturen zur Verfügung gestellt werden und eher selten echte Essenzen.

Darüber darf eine oft irreführende Betitelung nicht hinwegtäuschen, welche ‚Essenzen' nur vorgibt oder Tinkturen als ‚Elixire' anpreist. Das verwundert nicht, und man findet dieses Phänomen der Übertreibung bei derartigen Produkt-Termini sowohl in alter Zeit wie auch heute, wo es den Anwendern an guter Aufklärung mangelt. Auch in der älteren spagyrisch/alchymistischen Literatur sind die *Termini* nicht immer und überall eindeutig oder stimmig. Viele Autoren berührten die alchymistische Kunst nur am Rande oder schrieben von den Alchymisten ab und interpretierten deren Traktate nach ihrem Verständnis. Die *Termini* und Bezeichnungen fielen dann auch dementsprechend aus. ‚Echte Künstler' stellten dann die Dinge immer wieder richtig zusammen, wie etwa *Agricola* in seiner >Chymischen Medizin<, worin er ein Werk von *Poppius* kapitelweise zitiert und dessen ‚philosophische und geheime' Präparationen in den meisten Fällen auf ‚gemeine, nicht besonders kunstvolle Präparationen' zurechtstutzen muss. Allseits bekannt, scheinen diese Umstände auch dem *Paracelsus* einigen Ärger bereitet zu haben – und würden dies heute noch viel mehr tun.

> *Poppius:* „*Oleum Vini* zu machen. (...) so man desselbigen haben will, so soll mans (...) und in einem Gläslein wohl verwahren."
>
> *Agricola:* „*Nota.* Dass das wahrhaftige *Oleum Vini* in dem *Tartaro* stecken sollte, wie der Author vorgibt, ist nicht, es stecket zwar ein *Oleum* darin, aber es ist vielmehr ein *Oleum Tartari* denn ein *Oleum Vini*. Das rechte *Oleum* muss aus (... bereitet werden)."
>
> *Poppius:* „*Salis Quintae Essentiae Praeparatio.* Nimm *Sal gemmae*, knete das unter Töpferton, (...)."
>
> *Agricola:* „*Nota.* Dieser Prozess lautet, als sollte es *Quinta Essentia Salis* sein, ist aber nicht, sondern es ist eine subtile Extraction oder Sublimation des *Spiritus Salis*, (...)."

Poppius: „Den *Salmiac* zu einem Wasser (zu) destillieren, in welchem man alle Metalle solvieren kann. (...) so tue die *Spiritus* in eine neue Retorte und rectificier dieselbigen, so ist dieses Wasser bereitet. (...)"

Agricola: „*Nota*. Unser Author sagt in diesem Prozess, wie der *Spiritus Salis Harmoniaci* soll erlanget werden, aber dieser *Spiritus* ist mehr ein Scheidewasser als ein *Spiritus Salis Harmoniaci*, wie aus der Composition zu sehen, derwegen ist er unserem in keinem Wege zu vergleichen, (...)."

Poppius: „Eine andere Art, die Corallen zu solvieren und ihre Kraft zu extrahieren. (...) In dieser *Essentia Corallorum* ist nun verborgen die ganze Kraft und Tugend der Corallen. (...)"

Agricola: „*Nota*. Unser Author setzet allhier noch einen Proceß, die Corallen zu solvieren und ihre *Essentiam* zu extrahieren, er ist aber so obscur, (...). Ich will aber fideliter erzählen, wie die Corallen auf eine rechte und sonderbare Weise sollen präpariert und ihre wahrhaftige *Essentia* hervorgebracht, und so in eine hochgelobte Medicin versetzet werden. (...)."

Poppius: „Aus den Blättern der schwarzen Niesewurzel ein *Extract* zu machen. Nimm die Blätter von schwarzer Niesewurzel, soviel du haben magst, (...). Darüber gieße guten Branntewein, der die Kraft wohl darausziehe, und den *Spiritum* hernach *in Balneo* davondestilliert, (...)."

Agricola: „Nota. Unser Author hält in allen *Extractis* nur einen Proceß und *Modum*, nämlich dass er die Species mit *Spiritu Vini* extrahiert – diese Extraction aber ist in vielem nicht zu loben, (...)."

Therapeutische Anwendung der Spagyrischen Essenz

Der Grad der Präparation einer ‚spagyrische Essenz' ist höher als der einer ‚einfachen' spagyrischen Tinktur. Genauso verhält es sich mit der Wirkung. Diese ist tiefgreifender, durchdringender und vollständiger. Ansonsten gilt auch hier alles, wie bei der Tinktur beschrieben. Die Grenze zwischen Tinktur und Essenz ist bisweilen eine fließende und die Betitelung alleine sagt nicht unbedingt etwas über eine ‚stärkere' oder ‚schwächere' Heilwirkung aus. In jedem Falle sind Arbeitsweise bzw. ‚Kunst' des Praktikers ausschlaggebend.

Die Frage, ob es immer besser ist, ein möglichst weit ausgearbeitetes Präparat dem ganz einfachen vorzuziehen, betrifft nicht nur spagyrische Mittel, sondern jegliche Heilmittel (inclusive der eigentlichen Lebensmittel). Der Heilkundige wird in der Regel beim Minimum ansetzen und zuerst versuchen, etwaige Disharmonien durch Korrektion von Ernährung und Lebensführung zu beheben. Diese Einstellung begleitet die gesamte Behandlung und führt den Patienten zu dem seinem Leiden entsprechenden Lern- bzw. Veränderungsprozess.

Andere Wege der spagyrischen Präparation von Pflanzen

Die oben beschriebenen Darstellungswege von Tinktur und Essenz sind nur Beispiele aus den vielfältigen spagyrisch-alchymistischen Schaffensmöglichkeiten und zeigen dazu die Grundzüge an. Dergleichen Produkte werden auf mannigfaltige Art in sehr guter Form hergetellt, und auch in der alchymistischen bzw. iatrochymischen Literatur findet man vielerlei Varianten. Wie bereits erwähnt, konnte sich der Titel ‚Tinktur' bei den Alten auch auf hohe alchymistische Präparate beziehen und der Titel ‚Essenz' auch eine eigentliche ‚Quintessenz' bedeuten, besonders mit der Beifügung ‚Wahre Essenz' oder ‚Fünftes Wesen', z.B: „Aus den Rosen eine Essenz zu machen damit sie in ihr Fünftes Wesen gebracht werden" *(Kunckel)*.

Richtlinien und Voraussetzungen zur Betitelung ‚spagyrisch'

Grundsätzlich gilt eine Pflanze als ‚spagyrisch aufgeschlossen', wenn mindestens zwei Prinzipien (oder Elemente) aus dem Pflanzenkörper geschieden und wieder vereinigt und davon mindestens eines gereinigt (‚die prinzipielle Ausdrucksform von den Schlacken getrennt') wird.

Des weiteren muss die Arbeitsweise einen (der Natur zufolge) stimmigen hermetischen Kontext als Grundlage haben, sich also konkret am Wesen der Prinzpien, Elemente oder Planetenkräfte orientieren und mit dem Vermögen des einzelnen Künstlers harmonieren.

Die Ausgangsqualitäten von Pflanzen und zu den Arbeiten benötigten Nebenprodukten (Wasser, Alkohol, Essig, ...) sollten möglichst hochwertig und ‚natürlich' sein. Dass viele dieser ‚natürlichen Dinge' heute extra als ‚biologisch' bezeichnet werden müssen, ist eine vorübergehende, traurige Zeiterscheinung, sofern mit diesem Begriff die Bedeutung ‚lebendig' assoziiert werden muss, denn dies würde bedeuten, dass alle ‚nicht biologischen Dinge' tot sind, und sollte dem tatsächlich so sein, dann müsste sich jeder echte Heilkundige vertausendfachen, damit dem allgemeinen Dahinsterben Einhalt geboten werden kann.

Die Ansatzmenge zu einer Arbeit darf nicht mehr ausmachen, als ein Mensch in etwa Körperflüssigkeit hat. Ansatzmengen von maximal 5 Litern bei Pflanzenarbeiten sind am besten bewährt. Besteht der Wunsch nach größeren Mengen an Endprodukten, dann werden Parallel-Ansätze gemacht. Diese Richtlinien gelten in der Regel nicht für ‚Nebenarbeiten', wie die Destillation von Wasser oder für Gäransätze zu Mazerations-Alkoholen. Hier wird auch in größeren Chargen gearbeitet. Das Verhältinis gilt aber immer für die ‚wesentliche Ausgangsmenge', zB. im Falle einer ‚spag. Essenz' für das erste Wasser-Pflanzen-Gemisch; beim Weg zur Quintessenz durch Putrefaction für die zerdrückte Pflanzenmasse; zur Darstellung einer ‚spag. Tinc.' durch Mazeration für die Menge des Pflanzen-Alkohol-Gemisches usf.

Dieses Kriterium ist unbedingt wichtig und ein typisches Qualitätsmerkmal ‚kunstvoller' Arbeit. Es schließt aber auch eine Massenproduktion von großem Umfang weitestgehend aus, was wiederum dem ‚ganzheitlichen' Konzept der Spagyrik entspricht. Dass dann gegebenenfalls erhöhte Nachfrage und Bedarf für den Handel mit solch höchstwertigen Heilmitteln nicht immer befriedigt werden kann, ist folgerichtig und trägt zu einer Korrektion des Konsumverhaltens una der Haltung ‚des Handels' gegenüber derlei Präparate bei – wieder ein ‚ganzheitlicher' Heileffekt.

Zeit muss als Ingredienz enthalten sein. Jedem Prozess sollte lieber mehr als zuwenig Zeit gegeben werden. ‚Drängen' der Prozesse vermindert den Wert und verhindert oft ein echtes spagyrisches oder alchymistisches Resultat.

Clissus (‚Pflanzenstein')

Darstellung und Gebrauch

Wie bei der spagyrischen Essenz werden auch hier die 3 Prinzipien geschieden und wiedervereinigt. Es ist dabei aber ungleich wichtiger, dass die ‚Reinheit' der prinzipiellen Ausdrucksformen sehr hoch ist. Das heißt: Das ätherische Öl ist frei von Hydrolat und *Phlegma*, der Alkohol höchstrectifiziert und das Salz aufwändig gereinigt. Zur Darstellung von Tinktur oder Essenz ist eine dermaßen weitgehende Reinigung ist nicht unbedingt notwendig (wenngleich ein außergewöhnliches Qualitätsmerkmal). Zur Coagulation zum *Clissus* ist eine solche Reinheit Voraussetzung.

Diese Coagulation findet nicht, wie bei Tinktur oder Essenz, ‚auf dem nassen Weg', sondern ‚auf dem trockenen Weg' (durch Imbibieren) statt und ergibt ein mehr oder weniger pastöses oder festes Produkt, den *Clissus*.

Diese Bezeichnung ist meist angebrachter als ,Pflanzenstein', da die Konsistenz eines Steines in der Regel nicht erreicht wird. Was mit einem ,eigentlichen Pflanzenstein' noch gemeint sein könnte, ist ein hohes alchymistisches Präparat, welches durch aufwändige und kunstvolle Arbeitsgänge tatsächlich steinartig und weitgehend feuerresistent gemacht wird. Wir finden derlei Anleitungen bisweilen in der alchymistischen Literatur (zB. bei *Tölde*). Sie sind sehr interessant, aber ebenso schwierig nachzuarbeiten (vor allem das letztendliche Figieren).

Wegen der aufwändigen Zubereitungsweise findet man den (,einfachen') *Clissus* kaum im allgemeinen Gebrauch, und er steht dem naturheilkundlichen Therapeuten äußerst selten zur Verfügung. In der Regel übertrifft er die Wirkintensität einer Tinktur und Essenz. Ein Vorteil mag auch sein, dass es sich dabei um ein (mehr oder weniger) trockenes Produkt, also um kein flüssig-alkoholisches handelt. Einige Krümelchen des *Clissus* können in Trägersubstanzen eingenommen werden. Zur äußerlichen Anwendung ist er (außer in bestimmten Pflastern) weniger geeignet.

Einfachere Abwandlungen des *Clissus* wurden von den alten Iatrochymikern gerne hergestellt und gebraucht, wie zB. die verschiedenen Arten des ,kristallinischen Steins' bei *Michael Crügner*. Man wendete dann nicht nur die gereinigten Salze alleine an, sondern imbibierte diese zuvor mit ensprechenden Pflanzensäften, *Spiriti, Mercurii* und/oder Ölen zu intensiven, meist scharfen, korrosivischen oder seifigen Präparaten zur innerlichen und äußerlichen Anwendung.

„Sammlet vom 21. bis zum 24. April *Primulas Veris, Flores et Herbae Taraxiconis* und *Consolidae Regalis ana* ein gut Teil, calciniert solche drei, und figiert das daraus bereitete *Lixivium* mit *Spiritus Principali* (? – Anmerk.), und coaguliert solches zum weißen, kristallinischen Stein. Diese Arbeit muss zum Ende gebracht werden gegen den 10. Mai."

Michael Crügner, >Chymischer Frühling<

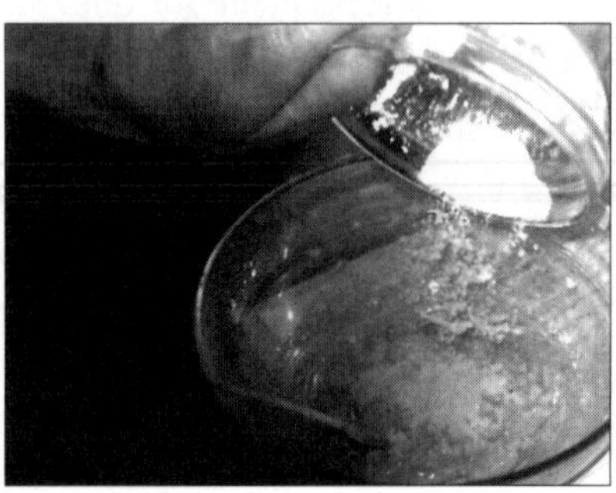

Quintessenz / Elixir

Darstellungswege, Charakter und Termini

Wenn man eine spagyrische Tinktur oder Essenz abdestilliert, dann bleibt ein nichtflüchtiger Rückstand im Kolben zurück – das Salz. Sulfur und Merkur gehen in den Recipienten über. (Vorausgesetzt, Tinktur oder Essenz sind farblos, ansonsten bleibt auch der meiste Teil der Farbe zurück). Das Erd-Salz hat einen fixen, während die sulfurischen und merkurialen Geister, Öle, etc. einen flüchtigen, volatilen Charakter haben.

Alle 3 Prinzipien zusammen bei der Destillation überzuführen bedeutet: ein Präparat vom Range einer Quintessenz zu erlangen. *Johann Kunckel von Löwenstern* nennt das „ein untrennbar verbindendes Band schaffen". Die beiden Geister lassen das Salz (den Körper) nicht mehr zurück, sondern steigen zusammen mit ihm auf.

Dieses zu erreichen gibt es mehrere Wege, welche allesamt auf eine Öffnung und Subtilisierung des Erd-Elements abzielen und geeignet sind, das fixe in ein flüchtiges Salz *(Sal volatile)* zu verwandeln. Die Bereitung des oben beschriebenen *Clissus* kann zum Beispiel als eine Art Vorübung zu einem der Wege (wobei das Salz mittels des ätherischen Öls ‚geöffnet' wird) angesehen werden. Die Wege zur Quintessenz sind vielfältiger und verschiedenartiger als die zu Tinktur oder Essenz. In jedem Fall muss hierzu eine entsprechende Aufschließung des Salzes erfolgen. In der Regel wird aber nicht das gesamte fixe Salz volatilisiert werden können, sondern nur ein mehr oder weniger großer Teil davon. Das kann ein Qualitätskriterium sein. Aus vielen alten Texten entsteht der Eindruck, dass die damaligen Scheidekünstler ‚alles, ohne einen Rückstand' zu volatisieren vermochten. Ob dem wirklich so ist, wissen wir nicht, es darf aber angenommen werden, dass sie insgesamt schönere Scheidungen vollbrachten, vielleicht einfach deshalb, weil sie ungleich mehr Sensibilität, Ruhe und Zeit in ihre Arbeiten investierten und der Empfindung des Webens der Naturkräfte näherstanden. Vergleiche mit jeglichen anderen Handwerken und Künsten führen zu der selben Ansicht. Noch vor wenigen Generationen war man fähig, Gebrauchsgegenstände herzustellen und Kunstwerke jeder Art zu schaffen auf eine Weise, die heute aus verschiedenen Gründen nicht mehr möglich sein soll. Diese Gediegenheit im Umgang mit den Dingen fand selbstverständlich auch in Alchymie und Heilmittelzubereitung ihren Ausdruck.

Die Arbeiten zur Quintessenz können weit über die Herstellung einfacher Heilmittel hinaus und hinein in die Gefilde der ‚höheren Alchymie' führen sowie unter anderem zur kunstvollen Darstellung von *Menstrua* und Alkahesten im Pflanzen- und Mineralreich.

Letzteres hat seinen Grund darin: Wann immer ein fixes Salz dermaßen aufgeschlossen wurde, nimmt es (in der Verbindung mit den anderen Prinzipien) elixirischen Charakter an – die Quintessenz ist gleichzeitig auch Elixir. Das bedeutet, dass, wenn dieses Mittel auf andere, rohe Substanzen oder Körper einwirkt, dann wird es in ihnen einen Veredelungs- bzw. Regenerationsprozess anregen und dort wieder Quintessenz hervorbringen. Das ist ein ‚elixirischer' Charakter. ‚Elixir werden' setzt (nach der alchymist-iatrochymischen Terminologie) das Wirken von etwas Quintessentiellem voraus.

Wenn wir hier von ‚Quintessenz machen' sprechen, dann ist dies beinah oberflächlich formuliert, denn eine Quintessenz wird zuallerst von der Natur selbst generiert. Der Praktiker muss nur die Voraussetzungen dazu schaffen. Dies betrifft besonders eine Kategorie von Quintessenz, die wir als ‚gewusste' bezeichnen können und die, anders als die ‚gemachte', in der Natur entsteht, generiert und freigesetzt wird. Das heißt: der Praktiker weiß, dass im ‚Mondenzustand' (in der Position des Mondes der pentagrammatischen Darstellung der Naturkräfte) jener überall beteiligte Merkur erneuert, regeneriert und zugleich ein neuer Merkur geboren (generiert) wird. Das geschieht zum Beispiel während der *Putrefaction*. Das ‚Hindurchflüchten' dieses Merkurs durch die ‚Ätherschleuse', durch den ‚Schnittpunkt der Acht', wird dort von Saturn als ‚Hüter der Schwelle' bedingt und bewirkt, dass er sich vom ‚körperlos-feuchten Geist' in einen ‚salinischen Geist' verwandelt. Dieser Aspekt Merkurs ist Quintessenz. Er ist zugleich salinisch, wässrig/luftig und zuinnerst feurig. Wenn Solches in der Natur entsteht, dann ist es eine ‚gewusste' Quintessenz – man kann sie nicht immer sehen, aber man weiß, dass sie da ist. Es ist auch eine ‚potentielle Quintessenz', weil sie zwar generiert oder freigesetzt, aber noch nicht wirkend wurde. Sobald sie nun auf eine Materie einwirkt, wird sie zum Elixir, zur ‚eigentlichen Quintessenz'. Die ‚erste Quintessenz' kommt als ‚erster Merkur' *de potentia in actu*, vom Vermögen in die Wirkung – wird zur ‚zweiten', wirkenden Quintessenz und damit zum Elixir.

Wir sehen, dass sich hier ein großes Thema der Alchymie auftut und es dem unbedarften Laien nicht so ohne weiteres möglich ist, einzudringen. Dennoch müssen wir die Dinge ein wenig skizzieren, sei es um nur einige Gedanken und Empfindungen wachzurufen, worum es sich hier bei den Begriffen und Präparaten handelt. Das ist auch deshalb notwendig, weil allerlei Mittel bei solch klangvollen Namen genannt werden, die dann aber oft mit der Sache gar nichts zu tun haben. Man findet einfache Tinkturen als Quintessenzen und einfache Kräutermischungen und Säfte in Alkohol als Elixire bezeichnet. Man findet auch obskure Darstellungswege beschrieben, so etwa, dass ein einfacher Ansatz ‚nach seiner dritten Destillation' zur Quintessenz wird und dergleichen mehr. Das ist ‚in der Kunst' nicht so gemeint.

Die ‚gewusste' Quintessenz wird während der Putrefaction erschlossen und das Wissen um diese hermetischen Prozesse während der Putrefaction erlaubt dann, diese Quintessenz elixirisch einzusetzen und damit ‚eigentliche' Quintessenzen *(in actio)* darzustellen.

Eines der bezeichnendsten Beispiele dafür ist die Anleitung des *Paracelsus*, wie man eine Quintessenz aus einem Kraut (zB. aus der Melisse) präpariert. Ausgangsmaterial sind immer frische Pflanzen. Bei getrockneten ist dieser elixirische Merkur schon verflogen. Aus ihnen kann man bestenfalls durch kunstvolle Salz-Aufschließungen, Imbibationen, Kohobationen etc. eine ‚gemachte' Quintessenz erlangen.

Paracelsus nimmt frische Pflanzen und putreficiert sie einige Wochen. Danach destilliert er die (‚erste') Quintessenz ab. Deren elixirisches Wesen ‚anwendend' übergießt er damit wieder frische Pflanzen, wodurch darin die Generation von ‚neuer' Quintessenz angeregt wird. Die ‚neue' destilliert er ab und bewahrt sie für dergleichen Arbeiten auf. Die ‚erste' hat nun bereits gewirkt und bleibt deshalb in der Retorte zurück. Durch Auspressen und Klären wird sie fertig dargestellt. Sie hat Farbe, eine gewisse Dicke und Geschmack. Soweit der Prozess in Kürze.

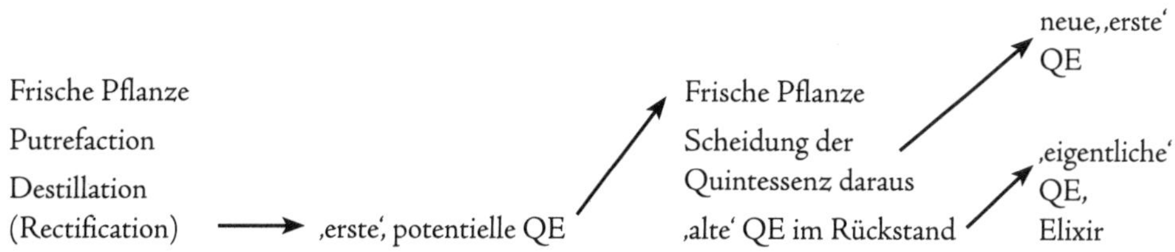

Wenn ein ‚Naturgeist' frisch generiert wird, sei dies nun durch die Natur oder mittels der Kunst, dann ist er ‚doppelt': einerseits stoffhaftend, andererseits flüchtig. Wenn er nicht vollends entflieht, sondern sein Potential wieder in den Elementereigen einbringt und ‚rohe Dinge' öffnet bzw. veredelt, dann hat er seinen Zweck erfüllt und ‚setzt sich zur Ruhe'. Er bleibt als ‚gereifte Quintessenz' im Rückstand. Dies ist die zweite Kategorie der Quintessenz, welche etwa durch Putrefaction aus frischen Pflanzen erreicht wird. Im Großen ereignet sich dieses Putreficieren unter der oberen Schicht des Herbstlaubes oder im Moor. Dort entflieht allerdings jener Geist immer wieder in den Luftraum, wenngleich er nicht ganz ins Ätherische entkommt, sondern an der Grenze zum Äther vom ‚Hüter der Schwelle', der Saturnkraft, wieder ‚zurückgesandt' wird und schließlich als Frühlingstau und -regen die nachwinterliche Natur veredelt.

Die Grundidee scheint einfach – die Ausarbeitung erfordert viel Zeit und Kunstfertigkeit. Um die lange Zeit wesentlich zu verkürzen, haben die Kundigen wohl schon seit den ältesten Zeiten Mittel ersonnen, deren Zubereitung zwar ein einmaliger, größerer Aufwand ist, die dann aber immer wieder imstande sind, aus jeglichem ‚Ding der Natur' im Nu eine Quintessenz zu ziehen. Ein solches heute noch gerne dargestelltes Mittel ist das *Circulatum Minus*, welches fast ausschließlich im Pflanzenreich wirkt. In seiner einfachen Form zieht es aus frischen Pflanzen in wenigen Minuten eine fettige oder ölige Quintessenz. Nach deren Abscheidung und folgender Klärung des *Circulatum* mittels Destillation kann es immer wieder verwendet werden. (›*Circulatum Minus* des *Baron Urbigerus*‹, wiedergegeben und kommentiert in *Manfred Junius*' ›Handbuch der Pflanzen-Alchemie‹, 1982). Andere *Circulati* oder Alkaheste greifen in der selben Weise in mineralische und metallische Substanzen in kurzer Zeit dermaßen veredelnd ein.

Gut geführte Gärungen oder Fermentationen stehen dem Wesen der Putrefaction nahe. Auch dabei können flüchtige Salze generiert werden, welche bei entsprechenden nachfolgenden Destillationen und Rectificationen bis zu einem gewissen Grade elixirisch wirken. Man wird deshalb eine sehr kunstvoll, mittels Gärung erarbeitete spagyrische Tinktur zwar nicht als Quintessenz, aber doch bis zu ihrem Grad auch als Elixir bezeichnen dürfen, wenngleich der *Terminus* auf ein höheres Verfahren hinweist. Besser wäre die Bezeichnung ‚Elixirische Tinktur'. Ähnlich verhält es sich mit sehr kunstvoll und aufwendig bereiteten spagyrischen Essenzen. Wird das Salz bei trockener Coagulation durch Imbibieren mit dem öligen und/oder harzigen Sulfur genügend geöffnet und mittels nachfolgender Kohobationen (der drei Prinzipien) zum Teil flüchtig gemacht, dann hat auch diese ‚Essenz' durchaus quintessentiellen, elixirischen Charakter.

Rezepturen alleine – seien sie noch so gut – ergeben noch kein ‚Elixir'. Dieser Titel wird erst gültig, wenn eine oder mehrere Zutaten zu einer quintessentiellen Form ausgearbeitet wurden.

Es spricht für sich, dass solche Präparate in allen Belangen anspruchsvoller sind als andere Spagyrika. Aus diesem Grunde sind sie den meisten naturheilkundlichen Therapeuten zur Zeit kaum verfügbar. Für die kommerzielle Nutzung wäre auch ihr Preis schwer zu bestimmen. Etwa 5 Liter Frischpflanzen ergeben nur einige Milliliter Endprodukt. Findet man also ‚Quintessenzen und Elixire' im Liter zu billigem Preis, dann wird es wohl das nicht sein. Welche Entwicklung die Zukunft diesbezüglich bringen wird, ist noch ungewiss. Eine zweifellos schöne Möglichkeit wäre, dass von vielen ernsthaften ‚Künstlern' viele große Alkaheste ausgearbeitet werden, wodurch dann tatsächlich größere Mengen solcher Präparate verfügbar werden. Ab einem gewissen Grad der Kunst sollen sich die Dinge ja angeblich ‚multiplizieren' lassen. Massen werden es aber wohl nie sein.

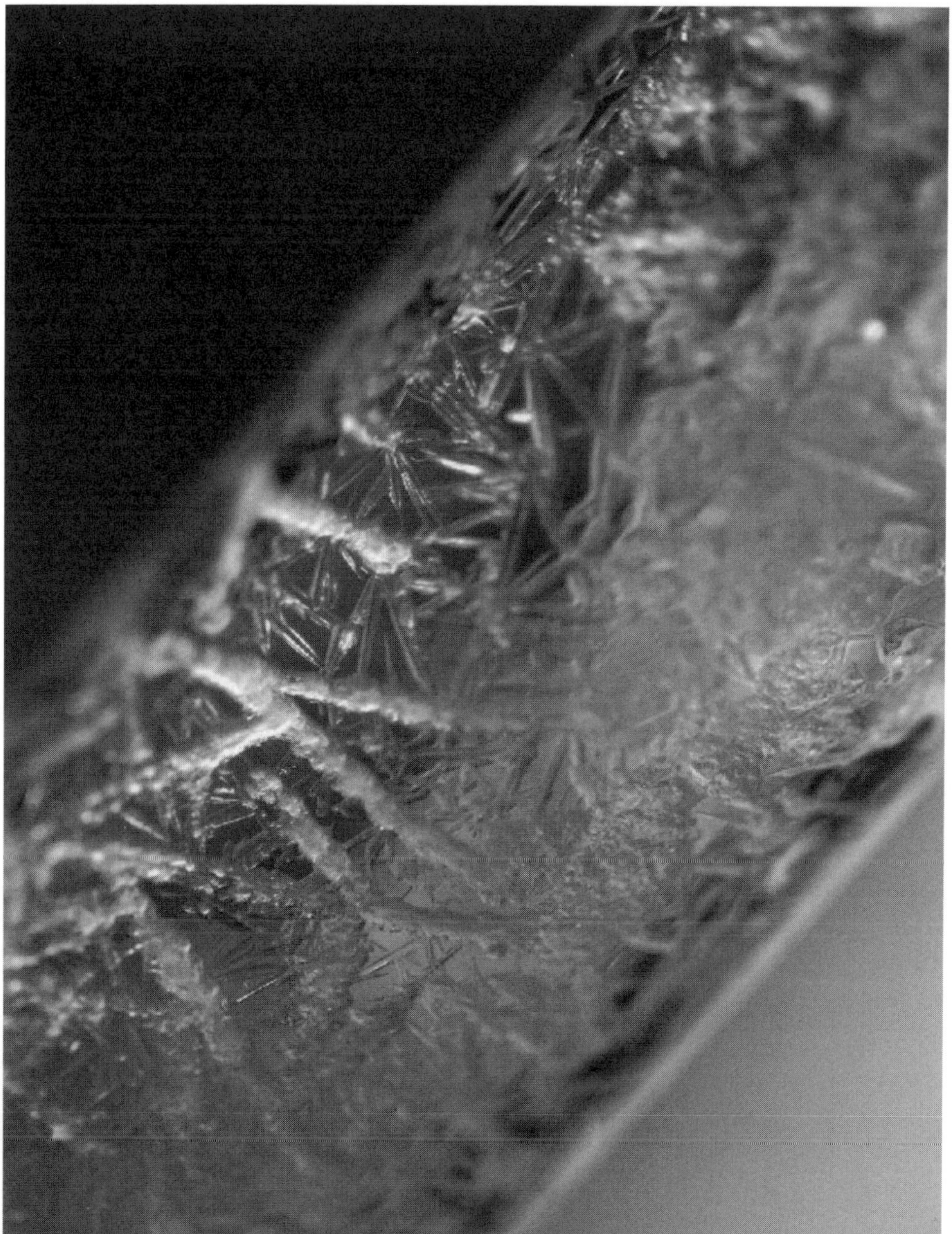

Sal volatile

Die Wirkung von Quintessenzen und (echten) Elixiren ist äußerst durchdringend und erreicht einen Grad, welcher im Allgemeinen mit dem anderer Arzneien kaum vergleichbar ist. Sowohl ‚berühmte' als auch weitestgehend unbekannte ‚hohe alchymistische Präparate' aus allen Naturreichen haben stets den Rang einer Quintessenz inne, unabhängig von ihren sonstigen Bezeichnungen und Betitelungen, nicht zuletzt deshalb, weil hier immer die typische ‚regenerierende Kraft' zur Entfaltung kommt, weil ein ‚Veredeln' im wahrsten Sinne geschieht. Bei den Alten waren dazu Vergleiche üblich wie: „Davon tut ein Tropfen mehr, als sonst die ganze Flasche", oder: „Mit dem rechten Elixir können Schäden in Stunden geheilt werden, wozu andere Tage und Wochen brauchen" u. dgl.

‚Veredelung' weist auf eine Kraft hin, welche die gesamte Natur durchdringt. Diese Kraft ist lebendig und lebensspendend und trägt zur ‚Entwicklung' jener Wesenheiten, die sich in den ‚Dingen der Natur' körperlich ausdrücken, bei. So nennen die Alten zum Beispiel das Gold ein ‚vollkommenes Metall' und die anderen Metalle ‚unvollkommen'. Dazu heißt es, die Natur bringe mit der Zeit alle Metalle zur entsprechenden ‚Reife', also ‚mache sie zu Gold', während der ‚Künstler', also der Mensch, dies in relativ kurzer Zeit vermag.

Es liegt hier eine Kraft in der Existenz der natürlichen Dinge verborgen, die nicht deren Körper direkt verändert, sondern deren einwohnende Geister und Seelen ‚erhöht', sodass diese danach aus eigener Kraft ihren Körper verändern können. Ohne Anwesenheit und Wirken dieser Kraft würde alles zu einer geistlosen Matrix stagnieren. Von außen betrachtet, muss sich die Veränderung der Körper nicht dementsprechend erzeigen. Der Körper kann altern und schrumpeln, das Kraut welken, während Geist und Seele darin wachsen und ‚edel' werden, dh. ‚über sich gehen'.

Diese Kraft ist das einzig Weitertreibende in der gesamten Natur und die einzige Kraft, welche das Lebendige ausmacht, ist eine veredelnde Kraft. Sie wird ‚das Fünfte Wesen' genannt und ist ganz eigentlich die ‚Idee'. Obwohl sehr diffizile Unterschiede bestehen zwischen ‚Merkur', ‚Quintessenz', ‚Elixir', ‚Fünftes Wesen', ‚Idee' u. dgl., auf die wir hier nicht weiter eingehen wollen, so ist doch im Großen und Ganzen ein und das selbe damit gemeint. Durch die ‚Kunst' stellt man die nächstmöglichen Ausdrucksformen dieser Kraft stofflich dar.

Unter den mannigfaltigen Wegen zur Quintessenz ist die Arbeit des in paracelsischer Tradition stehenden *Conrad Kunrath* (1555–1614) ein sehr schönes, wenngleich kein typisches Beispiel. Untypisch deshalb, weil hier nicht nur ein einziges Kraut ohne Zusätze aufgeschlossen wird, sondern am Wege dieser Aufschließung andere Kräuter beigefügt werden. Nicht unähnlich zur Idee der ayurvedischen *Bhasma*-Rezepturen. *Kunraths* Arbeit – eine ‚elixirische Rezeptur' und zugleich eine Quintessenz aus der Schwarzen Nießwurz – sei hier in ihren Grundzügen wiedergegeben:

De vera Quintae Essentiae Ellebori nigri extractione, quae est Arcanum maximum in Re Medica

Grabe *Elleborus niger* zur rechten Zeit, im rechten Planeten. Nimm gleichviel Kraut und Wurzel, reinige es, und setze es in einen Kolben zusammen mit gutem, roten Wein. Das halte einen Tag und eine Nacht im Wasserbad, dann lass abkühlen und seihe den gefärbten Alkohol ab. Den Rückstand presse fest aus und verwahre beides.

Dann wasche wieder frische Nießwurz, gib dazu Rosenblüten, Gewürznelken, Eichenmistel und Samen von Pfingstrosen, zerstoße alles im Mörser und gieße den oben abgeseihten und ausgepreßten Saft darüber. Das setze einen Tag und eine Nacht in Digestion. Dann öffne es und gieß dazu den achten Teil eines wohlrectificierten *Spiritus Vini*. Das lass 3 Tage lang digerieren. Seihe ab, presse aus, gib alle *Feces* zusammen und verwahre den Saft.

Nimm wieder frische Wurzel und Kraut, soviel du willst, mit Wein gewaschen, zerstoßen und destilliere gemächlich sein Wasser ab soviel geht. Dieses behalte. Nun destilliere mit stärkerem Feuer sein Öl ab – dieses kannst du zu anderen Dingen gebrauchen, es geht nicht in die Präparation ein. Die gesamten *Feces* calciniere bis zur Weiße 12 Stunden lang, danach abkühlen, kleinreiben und wieder calcinieren. Darauf gieße das abdestillierte Wasser, und koche beide zusammen 2 Stunden lang. Lauge alles aus, gieße ab und schütte das *Caput Mortuum* weg. Die Lauge aber lass abrauchen bis auf ein feines *Sal, Alcali* oder Salz. Trockne dieses und gieße deinen abgeseihten und ausgepreßten Saft darauf, dass es sich solviere. Dann ziehe seinen *Spiritus* mitsamt dem *Phlegma* über den Helm und bring den Rückstand wieder in ein Salz. Dieses löse im jetzigen Destillat und destilliere wieder bis zur Trockenheit ab. Danach scheide die **Quintessenz** vom *Phlegma*. Sie hat einen starken Geruch und scharfen Geschmack.

Das Salz aber reverberiere und kristallisiere mit dem obigen *Phlegma* zur höchsten Reinheit. Dazu reibe Safran und Amber. Davon einen Teil und vier Teile von deiner Quintessenz vermischt, von deinem 60. Lebensjahr an verwendet.

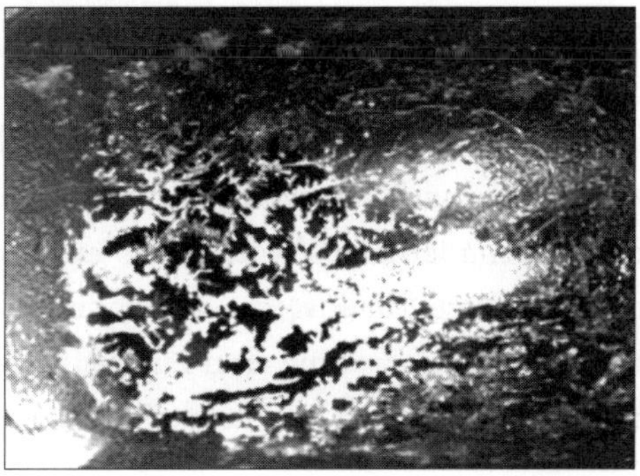

Eine spagyrische Pflanzenpräparation nach Poppius (aus Agricola's >Chymischer Medicin<, S. 1303)

(freie Wiedergabe, die typischen Ausdrücke beibehalten – nach Scheidungen geordnet)

a) Sein eigenes destilliertes Wasser:
Nimm ein Kraut, welches dir beliebt, hacke es klein und zerstoße es in einem Mörsel. Dann fülle es in einen Glaskolben oder in eine kupferne Destillierblase. Destilliere daraus sein eigenes Wasser, solange etwas gehen will (ohne dass die Pflanzenmasse anbrennt).

b) Das Öl:
Nimm ein Kraut, welches dir beliebt, hacke es klein und zerstoße es in einem Mörsel. Dann fülle es in einen Glaskolben oder in eine kupferne Destillierblase. Gieße darüber 3 Finger hoch sein eigenes destilliertes Wasser. In Mangel dessen kannst du auch destilliertes Regenwasser nehmen. Zuerst setze einen blinden Helm auf und verkleistere die Fugen. Mach darunter 5 oder 6 Tage ein sehr gelindes Feuer, sodass die Materie in der Blase zu faulen und zu circulieren anfängt.

Anmerkung: 5-6 Tage reichen in der Regel nicht für eine vollständige Faulung im Sinne einer Putrefaction aus. Außerdem wird destilliertes Wasser zugesetzt, was zur eigentlichen Putrefaction nicht üblich ist. Auch die Temperatur ist höher. Hier ist offenbar eher einer vorbereitenden Aufschließung der Kräuter mittels gelinder Wärme und Abgeschlossenheit erwünscht.

Danach nimm den blinden Helm ab und setze einen mit einem Schnabel auf. Der Schnabel soll sich in einer Röhre fortsetzen, die durch kaltes Wasser geht, damit sich das Überdestillierte abkühle. Dann fange an, mit gutem Feuer zu treiben und zu destillieren, so steigt Wasser und Öl zugleich über den Helm. Die scheide durch einen Glastrichter und hebe das Öl (welches nicht viel sein wird) auf.

c) Der subtile *Spiritus:*
Ein jedes Kraut gibt in der Destillation einen subtilen *Spiritus* von sich – je nach Hitze der Kräuter – gleich einem Branntwein. Den soll man als erstes auffangen und danach aus einem anderen Glas aus dem Wasserbade ganz linde den feurigen *Spiritus* vom *Phlegma* scheiden. Diese *Spiritus* haben die gleiche Kraft wie das eigene destillierte Wasser, sind aber in der Wirkung viel kräftiger. Was 5-6 Loth (ca. 75-90g) des Wassers verrichten, kann hier mit ½ Quintlein (knapp 2g) verrichtet werden.

Anmerkung: Dieser Spiritus wird erst dann erlangt, wenn man die Kräuter zuvor einige Wochen (tatsächlich) putreficiert. Dazu werden sie zerstoßen und in einem gut verschlossenem Gefäß an einem warmen Ort in der Erde, im Mist oder im Sand vergraben. Wenn die Putrefaction abgeschlossen ist, destilliert man bei sehr gelinder Temperatur den Spiritus über. Diese Destillation erfolgt am besten mittels einer Retorte aus dem Wasserbad und dauert mehrere Tage. Nach dem (bereits mit Phlegma vermischtem) feinsten Spiritus geht danach immer mehr Phlegma über (welches wiederum noch eine Zeit lang mit ein wenig Spiritus vermischt ist).

Der Spiritus entspricht einem Merkur aus den Kräutern, wenngleich dieser bereits von feuriger und salinischer Art sein wird, also quintessentiellen, elixirischen Charakter besitzt. Damit wäre der Weg zur Quintessenz aus den Kräutern beschritten, sofern eben eine Putrefaction vorangeht.

Eine ähliche Anleitung in Töldes >Coelum reseratum< lässt wieder diese Unklarheit bezüglich des Spiritus' bzw. Merkurs offen, wird dort allerdings von einem Kommentator nachfolgends ergänzt. Wenn es im ersten Text heißt, „nimm frische Kräuter und destilliere daraus ihren Spiritus", dann weist schließlich der Kommentator darauf hin, dass „ohne vorhergehende Putrefaction gar nichts geht."

Möge dies schon für den Beginn der vorliegenden Arbeit Poppius' gelten, so erst recht für die folgenden ‚Mercurius'.

d) Der *Mercurius:*
Es wird auch aus jedem Kraut ein *Mercurius* destilliert, das ist ein weißes, flüchtiges Salz, das sich im Helm wie ein sublimierter *Salmiac* anlegt und auf der Zunge scharf und hitzig ist. Dazu soll man die Kräuter also fein zerstoßen und den obgemeldten *Spiritum* darübergießen, hernach mit starkem Feuer aus einem Sandbad destillieren. So steigt dieser *Mercurius* mit dem Wasser als ein flüchtiges Salz über.

Anmerkung: Das entspricht nun der oben erwähnten Paracelsus-Arbeit, vorausgesetzt, genannter Spiritus ist ein ‚echter', aus der Putrefaction hervorgegangener. Dass der neu generierte Mercurius aus der frischen Pflanze auch als Sal volatile sichtbar erscheint, ist eher die Ausnahme.

e) Das fixe Salz:
Man kann auch aus jedem Kraut ein fixes, beständiges Salz machen, indem man die Kräuter dörre werden lässt und hernach im Feuer zu Asche calciniert. Über diese Asche gießt man sein eigenes destilliertes Wasser ganz warm.

(Die Lauge wird filtriert) „und die *Phlegmata* wiederum davondestillieren, so bleibet ein schwarz Pulver am Boden liegen. Dieses Pulver soll man hernach in einem Reverberierofen 24 Stunden in einer sanften Wärme reverberieren, damit die Schwärze von der Hitze des Feuers verzehret werde, allein man habe Achtung darauf, damit das Salz nicht schmelze oder zusammenfließe, sondern nur ein wenig braun glühe, so hat es genug. Dann nimm es heraus, lass kalt werden. Wie es erkaltet, so gieß sein eigen Wasser wiederum darüber und solvier es wieder zu einer Lauge, so fället ein schwarz Pulver oder Schleim zu Grunde. Diese Lauge soll man abermal durch ein Papier laufen und hernach die Feuchtigkeit davon abrauchen lassen, so bleibet am Boden des Glases ein schön weißes Salz. Auf diese Weise kann man nun aus allen Kräutern ein Salz machen und ausziehen, ..."

f) Coagulation:
Poppius unterlässt es hier, mögliche Coagulationen zu beschreiben, da er in diesem Kapitel (wie aus dem Titel „Von der Zubereitung der Oleitäten aus den Kräutern ..." hervorgeht) eigentlich bloß über die Darstellung von Ölen aus Kräutern berichten wollte und dieses Vornehmen offensichtlich ohnehin schon ausgeufert ist. Vorstellbar wäre grundsätzlich eine Vereinigung der Prinzipien Sulfur, Merkur und Sal möglichst ohne Phlegmata auf nassem oder trockenem Weg. Mit entsprechenden Kohobationen oder gar einer Fixation wäre damit eine Quintessenz erreicht (vorausgesetzt eine anfängliche Putrefaction ...).

B) DAS MINERALREICH

Minerale sind in roher Form zur Arznei untauglich – eine Ansicht, welche insgesamt tragend für die Anwendung alchymistischen Wissens in der Heilmittelherstellung ist. Sobald ihre Heilkraft aber kunstgemäß erschlossen wurde, äußert sie sich sehr oft intensiver, durchgreifender, mächtiger, jedoch universeller, nicht so spezifisch wie die von Pflanzen. Wenn das aufgeschlossene Mineral mit Pflanzenzusätzen komponiert wird, kann die Wirkung gesteuert und ins feinste Detail geführt werden. Dazu ist es nicht nötig, stets spagyrische Pflanzenpräparate mitzurezeptieren, wie wir schon besprochen haben.

Vom Sinn, mineralische Substanzen spagyrisch zu präparieren

Minerale, Edelsteine und Metalle werden nur dann vom Körper essentiell aufgenommen, wenn sie ‚durch die Kunst vorverdaut' wurden. Ihre Einnahme in roher Form ist kurz-, mittel- oder langfristig schädlich, obgleich viele davon momentan eine gewisse Wirkung zeitigen. In diesem Punkt waren sich die Heilkundigen der vergangenen Jahrhunderte einig, sofern sie zumindest ein wenig in die Gefilde der Alchymie eindringen konnten bzw. ihr Naturwissen in die Tiefe ging.

Andere (sie mögen in der Anwendung von Pflanzen sehr geschickt gewesen sein) verwendeten Minerale und sogar Metalle in ihrer rohen Form. Von ihnen sind (an sich meist sehr gute) Rezepte überliefert, in denen aber neben den Pflanzen ‚rohe' Mineralien und Metalle, wie feingeriebene Perlen, Muschelkalk, Smaragde, Amethisten und sogar Silber oder Goldblättchen vorgeschrieben sind. Weil vor allem die unpräparierten (also bloß feingeriebenen) harten Steine ihre Heilwirkung so nicht entfalten konnten, kam insbesondere die Anwendung von Edelsteinen während der letzten Jahrhunderte in Verruf und geriet schließlich in ihrerer Vielfalt und Nützlichkeit in Vergessenheit.

Iatrochymiker wie *Carrichter* und *Cardilucio* nahmen eine Zwischenposition ein: Sie empfahlen zwar einfache Arten spagyrischer Aufschließung von Metallen, Kieseln und dergleichen ‚harten Dinge', was aber die ‚weicheren Dinge' wie Perlen, Korallen, Erden und Kalke betrifft, so meinten sie, dass die Verdauungskraft des Menschen ausreiche, damit er von ihrer Heilkraft profitiere. Wie die beiden erwähnten Iatrochymiker des 17. Jahrhunderts, so gestanden auch andere in dem Zusammenhang, dass sie sich um die ‚hohe hermetische Kunst' nicht weiter bemühen wollten oder konnten, sondern nur einfache und wirksame Mittel (ohne viel Aufwand) zur Verfügung bräuchten. Nach *Cardilucio* sind …

. . . viel gebräuchliche Pulver von Edelsteinen / als Rubinen/Granaten/dermassen hart/daß wenn sie auch ein gantz Jahr im glüenden Feuer gehalten würden/ solten sie davon nicht verletzt werden/ und so man sie in den härtesten Mörseln von Stahl und Eisen zerstösset und pulverisiret / machen sie durch ihre grosse Härte so wol am Mörsel als Stämpffel viel Schrammen und Grüblein/ und so sie hernach ferner auf Reibsteinen zu einem zarten Pulver abgerieben werden / wetzen und schleiffen sie von den harten Reibsteinen so viel weg/ daß allgemach die Reibsteine tieffe Kanten und Gruben bekommen/ und werden davon die Edelsteinpulver viel schwerer im Gewicht; solten nun solche gantz unverbrennliche steinerne und beinerne Kleyen bey Krancken einigen Nutz schaffen können / und nicht viel mehr die Krancken noch kräncker machen ? denn wer weiß nicht daß die gesundeste Menschen/wenn sie harte unverdauliche Speisen essen /, die ihnen lang im Magen liegen/ davon so lang geplagt / gedruckt und endlich gantz niedergeworffen werden/ bis sie solche entweder wider von sich vomiren / oder mit grosser Müh deß Magens verdauen ? Ja nicht allein die Menschen/ sondern auch die stärckeste Thiere als Wölffe/ Hunde / Katzen / würgen die unverdauliche Dinge wider von sich/andere aber die nicht zum würgen geneigt / müssen zum öfftern daran sterben / wie man vielfältig in den Mägen der gestorbenen Hüner Steine/ und in den Kühen Stecknadeln und dergleichen findet / welche sie nicht haben verzehren können / sondern daran sterben müssen / und der arme schwache Mensch solte allein die noch weit härtere Dinge verdauen / und davon wider gesund werden ? O der unbesonnenen Püffelsköpffe / die solche Dinge erstlich außbracht / und mit was verzweiffelter Halsstarrigkeit müssen die behafftet seyn / welche solche handgreiffliche allerschädlichste Irrthümer noch verfechten / und sie zu verschreiben noch immer fortfahren ! zu deren Behelff sie nicht das allergeringste fürbringen können / als daß es nunmehr ein alter Brauch sey / welchen man nicht abschaffen dörffe ; aber wenn solches gelten solte . . .

Heute ist die Anwendung von Mineralen und Edelsteinen in der Naturheilkunde nicht die Regel, von wenigen löblichen Ausnahmen abgesehen. Damit sind natürlich nicht die Kiesel im Wasserglas oder Schmeichelsteine gemeint, sondern einige gediegene spagyrische Präparate sowie solche, bei denen es mittels anderer Verfahren gelungen ist, mehr als nur einen energetischen Abdruck ins Medium zu bringen. In den meisten Fällen ist die Wirkung überzeugend und man wird zu der Erkenntnis gelangen, dass die Pharmakologie der Heilkunst erst dann vollständig geworden ist, wenn pflanzliche Substanzen zusammen mit mineralischen und metallischen gleichermaßen kunstvoll zubereitet und angewandt werden können.

Als Vergleich zur abendländischen Situation stehen uns vor allem asiatische Traditionen zur Verfügung. Im Westen ist heute eine Zusammenfassung einiger Aspekte solcher Systeme unter der Bezeichnung TCM bekannt. Dieser bedeutende naturheilkundliche Zweig ist durchdrungen vom selben hermetischen Grundverständnis wie die abendländische Spagyrik und Alchymie. Was aber die aktuelle Heilmittelherstellung betrifft, so werden alle Substanzen, Pflanzen, Minerale, Metalle und tierische Produkte beinah ausschließlich unaufgeschlossen, aus hermetischer Sicht in ‚derber, roher' Form angewandt.

Im ‚alten China' war die Alchymie eine hochstehende Wissenschaft. Während der letzten Jahrhunderte bewirkten einige hinderliche Faktoren, dass sie (zumindest in den weiterverbreiteten Formen der Arzneimittelkunde) fast vergessen ist.

Was diese hinderlichen Faktoren betrifft, an die uns die *Historia* der Heilkunst in Indien, China und Tibet (in zeitlicher Reihenfolge) erinnert, so wollen wir sie kurz ansprechen, weil sie sich leider immer noch in manchen Gegenden zu einem Thema aufplustern. Das betrifft die Heilkunst insgesamt, und es betrifft natürlich auch die Alchymie, denn der Ort eines alchymistischen Labors ist eben relativ statisch und dadurch vor den Wirren der Zeit nicht geschützt. Die Werkstätte selbst kann nicht in Wälder oder Berge flüchten.

Die Kundigen Tibets wurden und werden noch heute manch eines würdevollen Ortes beraubt, an dem Heilmittel höchsten Ranges hergestellt wurden und werden sollten, die sowohl den Räubern als auch den Menschen dienen könnten, indem man diese Orte verwüstet und die Kundigen ermordet oder verjagt. Nicht viel anders ist es anderen Kulturvölkern ergangen. Folgt man hier dem Vorbild der Römer, die ein Volk zu entwurzeln wussten, indem sie seine ‚Kundigen' abschlachteten?

Wie hat es eigentlich *Paracelsus* geschafft, zwischen Wanderung, Reisen, Flucht und Verbergen soviele und hohe laborantische Arbeiten zu vollbringen – und daneben noch seine umfangreichen Schriften zu verfassen – in nur 48 Lebensjahren?

Er war ein Freund der Nomaden – und die verstehen es, sich zu bewegen. Vielleicht besaß er sogar ein‚Labor für unterwegs' in einem Pferdewagen oder auf einem Ochsenkarren. Gesellen der selben Art, die auch ihm keine Ruhe ließen, schlagen noch heute wild um sich, sollten aber besser die Mütze nehmen und verdunsten, wenn sie nicht als nutzlose, stinkende *Feces* der Zeitenwende erkannt werden wollen – ein wenig paracelsisch formuliert. Hierzu eignet sich die Klage *Agricolas* darüber, dass die Zeit seines Wirkens sehr unruhig war und barbarische, primitive Horden des öfteren seine Laboratorien zerstörten:

„Ich zweifle auch selber nit, wenn ich zu jetziger Zeit Gelegenheit, Zeit und Unkosten hätte, ich wollte eine solche Medicin draus machen, die sich wohl sollte sehen lassen, inmaßen ich vor der Plünderung auch schon einen Anfang dazu gehabt. Diese Sache erfordert Gelegenheit, Zeit und Unkosten, (...) und man hat länger denn ein ganzes Jahr damit zu tun. Es hat mich mein großer Verlust, welchen ich bei diesem Kriegswesen erlitten, nicht halb so sehr gekränkt als meine *Labores*, Vorrat, *Instrumenta* und dergleichen. Und obschon diese alle den Kriegsgurgeln nichts nütze gewesen, sie auch nicht wissen können, was sie bedeuten oder zu was Ende sie zugerichtet, nichtsdestoweniger hat sie der höllische Mordgeist, der dem Menschen nicht gönnet auch die Arzneien, welche Gott zur Erhaltung des Lebens erschaffen, dahin getrieben, dass sie dieses alles verwüstet. Wer solche Verwüstung hernach auch mit Augen gesehen, hat drüber weinen müssen. Und als ich auch endlichen ihren gottlosen Händen noch zuteil und gefangen wurde, trug sichs zu, dass in der selben Nacht einen solchen Plünderer und Kriegsgurgel ein Pferd schlug, dass er weder gehen noch stehen konnte. Da kam er zu mir und sagte, er hörte, dass ich ein *Medicus* wäre, ich sollte ihm an diesem Schenkel helfen. Ich führete ihn hin und zeigete ihm, wie mit meinem *Laboratorio* wär umgegangen worden, ich hätte ihm wohl helfen können, wenn dieses nicht alles so jämmerlich wäre verwüstet worden. (...) Ob nun wohl dieser Raubvogel auch ein böses Kind und aller Leichtfertigkeit voll war, nichtsdestoweniger musste er bekennen, dass um solcher Sachen willen er nicht redlich gehandelt hätte. Ich habe auch von der Zeit an zu keinem rechtschaffenen Labor mehr kommen können, weil ich nun fünf Jahr aneinander Plünderungen ausstehen hatte müssen, dass ich also wohl mit Plato sagen möge: *Ipsa Salus, si me salvare voluisset, non potuisset.* (...). Es ist mit leid, dass ich vor meinem Ende nicht sollte wieder zur Ruhe kommen und mein vertraut Gut *meditando et laborando* offenbaren. Nun wohlan, Gottes Wille muss geschehen, welchem kein Mensch wiederstreben kann. Dem sei es auch befohlen. Er weiß Zeit und Stunde dieses Wesens ein Ende zu machen. Ist es sein Wille, dass ich über dieses *scriptum* der Welt mit mehrerm dienen soll, so wird er auch Mittel schicken, ich zweifle gar nicht, es werde durch meine Wenigkeit noch etwas weiters müssen verrichtet werden, damit Gottes Ehre gesucht und gepriesen werde."

Johann Agricola, >Chymische Medicin< (*Tractatus de Vitriolo*) 1638/39;
Neu bearbeitet und herausgegeben von *Oliver Humberg*, 2000

Alchymie im alten China

Im *Zhan Guo Ce* wird berichtet, dass ein König im Staate *Chu* um 400 v. Chr. von Alchymisten ein ‚Lebenselixir‘ anfertigen ließ. *Quin Shi Huang* entsandte um 200 v. Chr. tausende Männer und Frauen in die ganze Welt, um Alchymisten zu treffen und einzuladen. Kaiser *Wu Di* (2. Jhdt. v. Chr.) betätigte sich selbst alchymistisch und hielt stets Verbindung mit Alchymisten unter seinem Volke.

Arbeiten mit Zinnober (chin.: *Dan Sha)* wurden hoch geschätzt – von der Bereitung einfacher Medizinen und Kompositionen, bis zum Großen Elixir – ganz entsprechend den Hinweisen des *Paracelsus,* dass mit diesem Mineral ein Weg zum begehrten ‚Roten Löwen‘ oder zumindest zum echten ‚*Aurum Potabile*‘ zu finden sei. Zur Aufschließung des Zinnobers war *Spiritus Vitrioli veneri* von großer Bedeutung. *Bao Pu Zi* beschreibt lange Zeit vor *Paracelsus* ein ähnliches Verfahren zur Herstellung von ‚trinkbarem Gold‘ ohne die Hilfe starker Säuren. Gleich mit den anderen Traditionen war bei den Alten Chinas das höchste, durch alchymistische Kunst zu erlangende Elixir, das höchste Heilmittel für den Menschen zugleich auch ein Transmutationsmittel für unedle Metalle. Im *Bao Zang Chang Wie Lun* (10. Jhdt.) werden 15 Arten durch ‚Kunst‘ bereitetes Gold erwähnt, wobei anzunehmen ist, dass bei den damals sehr hohen metallurgischen Kenntnissen mit Gold auch tatsächlich Gold gemeint war. Ähnliche Beschreibungen von ‚durch die Kunst bereitetem Gold‘ finden sich in jahrtausende alten vedischen Traktaten. Solches Gold von absichtlich zur Täuschung benutzen Legierungen zu unterscheiden, erklärt zB. *Li Shizhen* im *Ben Cao Gang Mu,* dem >Abriß der Arzneimittelkunde<, (16. Jhdt.) eindringlich.

Die alchymistischen Texte des alten Chinas erwähnen – wie die abendländischen – den nassen wie den trockenen Weg zur Darstellung ähnlicher Präparate. Der Alchymist *Bo Yang* (25 – 200) beschreibt im *Zhou Yi Can Tong Qi,* dass ihm etwa 600 Abhandlungen über Verfahren mittels des ‚trockenen Weges‘ zur Verfügung stünden. Ein sehr alter taoistischer Text beschreibt 54 Verfahren des ‚nassen Weges‘ zur Aufschließung von 34 Mineralen und anderen Substanzen, auch pflanzlicher Art.

Das Scheidewasser, der *Spiritus Nitri* der abendländischen Alchymisten, war als *Xiao Shi* beliebt – insbesondere zur Aufschließung von 72 Arten von Steinen.

Unter den mannigfaltigen alchymistischen Verfahren wird bereits in den ältesten Texten von ‚Aufschließen durch Schmelzen‘ *(hua),* Scheidung durch Lösen eines Teils einer Substanz *(lin),* Figieren *(feng),* Gären *(niang),* Filtirieren, Umkristallisieren und sogar von Projektion mittels eines Elixires *(dian)* berichtet. Die Auflösungen mit pflanzlichen und mineralischen *Menstrua* wurden in keramischen Gefäßen gemacht.

Beliebte Arbeiten wurden mit Blei bzw. dessen Verbindungen ausgeführt. *Qing Xu Zi* (,Meister der Einfachheit und Gemächlichkeit') berichtet im 8. Jhdt. über eine ,Methode zur Herstellung des Elixirs' aus Blei, Schwefel und Salpeter mit Hilfe von Essig. Ähnliches berichtet *Sun Simiao* (6. Jhdt.) in seinen ,Methoden der Feuerbehandlung' *(fuhuo)*. Der ,große' *Ko Hung (Ge Hong)* berichtet im 4. Jhdt. über Arbeiten mit Eisen, Kupfer, Gold, Glimmer, Zinnober, Horn u.v.m.

Für alchymistische Zwecke wurden Brennöfen, Kessel, Sublimiergefäße, Tiegel, Retorten u. dgl. hergestellt. Während der *Jijilu* ein einfacher Ofen war, hatte man dem *Weijilu* ein Wasserbad angeschlossen. Kessel und andere Gefäße wurden aus Gold, Silber, Kupfer oder Porzellan angefertigt, Vorrichtungen zur Wasserkühlung vor allem aus Silber.

Insgesamt wurde von den Alchymisten im alten China eine lange Reihe kunstvoll aufgeschlossener Präparate aus Edelsteinen, Jade, Türkis, Quarzen, Hämatit, Azurit, Magnetit, Kalk, Kreide, Salz, Perlen, Blei, Schwefel, Gold, Silber, Zinn, Zinnober, Galmei, Kupfer, Quecksilber, Alaun, Salpeter, Talk, Erden, Salmiak, Essigen, Alkoholen, Pflanzen und anderen ,natürlichen Dingen' hergestellt.

Zu den heute verwendeten mineralischen Bestandteilen der Rezepturen in der chinesischen Medizin

Heute sind diese Verfahren in der Pharmakologie der chinesischen Medizin nicht mehr bekannt, und nicht einmal die mineralischen Zusätze zu den Pflanzenmischungen werden ,kunstgemäß' aufgeschlossen. Dass in entlegenen Gegenden Chinas noch tatsächlich alchymisitsch gearbeitet wird, ist anzunehmen, stand doch die Tradition während Jahrtausenden in hohen Ehren, und die taoistsichen Lehren schließen bis heute die Alchymie als wesentlichen Aspekt mit ein.

Die Rezeptierkunst hat aber ihre Wege bis in die heutige Zeit gefunden, eng verbunden mit hochwertigsten Befundungsmethoden und jenem philosophisch-hermetischen Naturverständnis, welches die Grundlage jedes ,echten', authentischen heilkundlichen Systems darstellt. Es scheint jedoch, dass insgesamt in diesen Rezepturen mineralische und metallische Bestandteile weniger beachtet werden als dies in alten Zeiten üblich war. Nicht etwa deswegen, weil man heute besser über deren mögliche Giftigkeiten Bescheid wüsste (die subtilsten, ausgefeiltesten Gifte und Gegengifte verstanden bekanntlich die Alten zu bereiten), sondern eher, weil es den einen an der alchymistischen Kunst und den andern am Kunstverständnis mangelt. In der Regel findet man hier nur Pulver und Dekokte vor, auch dann, wenn die Rezepte Minerale enthalten, welche auf diese Art vom Körper nicht wirklich absorbiert werden können und bestenfalls nur einen kurzfristigen oder oberflächlichen Effekt zeitigen.

Im Rezept *Xuan Fu Dai Zhe Tang* (‚Dekokt von Alantblüten und Hämatit') ist zB. der (beliebte) Zusatz ‚Hämatit' in Pulverform (unaufgeschlossen) neben *Inula japonica, Panax ginseng, Pinellia ternata, Zingiber, Glycyrrhiza uralensis* und *Ziziphus jujuba* solcherart enthalten, dass sein Gewichtsanteil, verglichen mit den jeweiligen Anteilen der Pflanzen, am höchsten ist.

Das *Yu Nu Chian* (‚Dekokt des hübschen Mädchens') enthält (in einer seiner Varianten) nur unter den pflanzlichen Bestandteilen *Rehmannia glutinosa* in ‚aufgeschlossener' Form, während Gips und Hämatit, wie auch die anderen Pflanzen *(Ophiopogon japonicus, Anemarrhena asphodeloides, Achyranthes bidentata* und *Imperata cyclindrica)* unaufgeschlossen bleiben.

Im *Zi Xue Dan* (‚Purpurner Pulverschnee') sind es sogar Gips, Calcit, Talc, Salpeter, Magnetit und Zinnober sowie das Horn von Rhinozeros und Antilope, die allesamt in roher Form mit den entsprechenden Pflanzen vermischt werden.

Die zentralen Wirksubstanzen in den *Chi Chu Wan* (‚Beruhigende Pillen') sind Magnetit und Zinnober. Wieweit die in der Grundrezeptur vorgeschriebene Vermischung dieser beiden Minerale mit einer Hefe-Art *(Shen Qu)* zur Aufschließung beiträgt oder bloß als Relikt eines alten spagyrischen Prozesses erhalten blieb, ist fraglich. In den Rezept-Varianten bleiben die Pflanzen *(Lycium chinense, Ligustrum lucidum und Cuscuta chinensis)* bis auf *Rehmannia glutinosa* ebenfalls unaufgeschlossen.

Die Rezepte selbst sind zweifellos äußerst wertvoll, wirksam und basieren durchwegs auf echtem Naturwissen und Signaturenkunde. Minerale, Metalle, Horn u. dgl. werden aber in der heute üblichen Verabreichungsform nicht kunstvoll erschlossen. *Paracelsus* würde dazu auf seine Art sagen: „Sie gehen wirkungslos durch den A ...", und wenngleich hier „wirkungslos" sicherlich ein wenig überzeichnet ist, so hätten doch die Alten im fernen Asien schönere Präparate bewerkstelligt.

So zeichnet sich also im Vergleich der Traditionen eine starke Entfernung von den Blütezeiten dieser Medizinsysteme, besonders in der Arzneimittelzubereitung der heutigen ‚chinesischen Medizin', ab. Diesen Blütezeiten steht die tibetische Medizin und das *Ayurveda* noch um einiges näher. Dort ist die alchymistische Präparation von Mineralen und Metallen üblich und als hohe Kunst lebendig.

Pillen als Arzneimittelform

Die archaiische Methode, Heilmittel in Form von Pillen (‚Medizinkugeln') herzustellen, ist in der westlichen Naturheilkunde immer mehr von ‚Tropfen' verdrängt worden, sind diese doch ‚verständlicher' und ohne weitere Hemmschwelle konsumierbar. Verglichen damit muten manch unförmige, von Hand gerollte Pillen tibetischer oder ayurvedischer Herkunft herrlich natürlich an. Pillen bieten eine nahezu ideale Möglichkeit, die drei Prinzipien innerhalb einer Rezeptur miteinander zu verbinden und waren deshalb auch bei den alten Iatrochymikern des Abendlandes stets üblich und beliebt.

So wurde zB. das auf spagyrischem Wege erschlossene *Magisterium* von Kiesel mit dem Erd-Salz aus Zinnkrautasche und Myrrheharz kombiniert. Desgleichen eine Tonerde mit dem Salz von Granatapfelkernen und deren Blütensaft mit Honig und Mehl. Oder das *Magisterium* von Perlen mit einer Quintessenz der Mistel, tartarisiertem Silber und dem rohen Pulver von Ehrenpreis, Melisse, Beifuß und Safran.

Pillen-Herstellung als alte Apothekerkunst

Agricola komponiert Pillen zur Stärkung, bestehend aus: *Tincturae corticis Citri, Florum Tunicae, Granorum Kermes, Corallorum rubrorum, Salis Perlarum, Mastichis electae, Ligni Aloes, Galangae minoris, Nucis indicae, Ambrae, Croci orientalis, Moschi, Ossi de corde Cervi, Auri fulminantis, Olei destillati Rosarum, O. d. Cinnamomi, O. d. Caryophyllorum, O. d. Succini, O. d. Foeniculi, Rosarum rubrarum, Borraginis, Buglossae, Floris Rosmarini, Spiritus Vitrioli* und *Sirupo de granis Kermes quantum satis.*

Die Zutaten sind, wie ersichtlich, zum Teil roh, zum Teil ein wenig aufgeschlossen und zum Teil sehr kunstvoll präpariert, und die gesamte Mischung ergibt in sich ein geschlossenes Ganzes und eine ausgewogene Vermischung der Prinzipien. Solcherart war die Rezeptierkunst der Alten und die Darreichungsform als Pillen dazu oft die geeignetste. Ähnliche Präparate lernen wir heute in den ‚tibetischen Juwelenpillen' kennen.

Gur-Gum 8:
Carthamus tinctorius, Pisum sativum, Pterocarpus santalinus, Herpetospermum caud., Fragaria nilgeernsis, Dendrobium mon., Bärengalle, Zinnober

Mig-Nad Ide-Gu:
Terminalia chebula, Terminalia berlerica, Emblica off., Glycyrrhiza glabra, Carthamus tinctorius, Carum carvi, Butter, Honig, Mumijo, Eisen-Bhasma

Grub-Thob Ril-dKar:
Terminalia chebula, Picrorhiza kurroa, Corydalis, Prunus armeniaca, Honig, Mumijo, Calcit

Se-Bru 25:
Punica granatum, Eugenia caryophyllata, Swertia chirata, Cartamus tinctorius, Terminalia chebula, Bambusa, Emblica off., Cinnamomum, Zingiber off., Piper nigrum, Piper sp., Elettaria cardamomum, Picrorhiza kurroa, Terminalia belerica, Myristica fragrans, Cucurbita pepo, Saussurea lappa, Inula racemosa, Amomum subulatum, Salpeter, Sal ammoniak, Soda, Schwarzes Steinsalz, Kaurimuschel-Bhasma

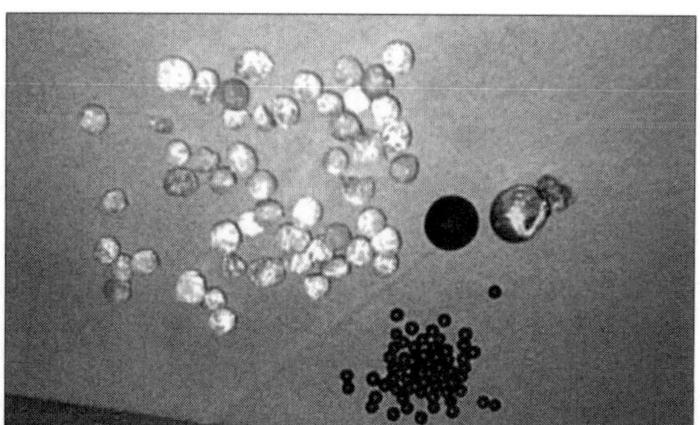

Die ayurvedischen Präparate

Im Ayurveda ist die alchymistische Erschließung der Heilkräfte aus Mineralen und Metallen seit Jahrtausenden vorgeschrieben und eine dementsprechende Zubereitung heute noch üblich. Viele Substanzen werden aber auch einfach zu einem sehr feinen Pulver *(Pishti)* verrieben und nicht weiter spagyrisch aufgeschlossen, außer dass dem Prozess sehr viel Zeit gegeben wird. Stunden- und tagelang wird unter Zusatz von Pflanzenwässern und -säften gerieben und dabei zB. die Perlen, der Muschelkalk, der Schwefel, der Smaragd u. a. enorm verfeinert. Die Pflanzenzusätze bewirken bei diesem Verfahren zwar kaum wesentlich eine ‚hermetische' Öffnung, gliedern sich jedoch in das Wirkpotential der mineralischen Substanzen ein und zeigen diesen gewissermaßen den Weg, den sie im Körper nehmen sollen, um ihre Kräfte zu entfalten. Je nach Bedarf und nach dem Grad der Kunst im Clan werden dann anstatt der *Pishtis* auch hochwertige *Bhasmas*, also eigentliche spagyrische bzw. alchymistische Zubereitungen hergestellt. Die entsprechenden Pflanzen gelangen während der Prozesse ‚ins Werk hinein', weswegen die Präparation der Minerale und Metalle selbst schließlich eine Rezeptur ergibt, in der allerdings ‚nur mehr' ein ‚Abdruck' der Pflanzen vorhanden bleibt und diese (mittels chemischer Analysen) oft gar nicht mehr nachgewiesen werden können. Im kochenden Topf, im glühenden Tiegel geben die zugesetzten Pflanzen Kraft und Wesen an die Minerale und Metalle ab, opfern sich gewissermaßen zu deren Erhöhung. Dennoch bleibt ihr spezifischer Eindruck vorhanden und sie vermitteln den Präparaten differenzierte Qualitäten. Somit wäre es etwas zu oberflächlich, würde man sagen: Das ist ein Eisen-*Bhasma*, und dafür gelten immer diese und jene Indikationen. Es werden in Indien an die 50 verschiedene Arten von Eisen-*Bhasma* hergestellt. Jedes heißt *Loha-Bhasma*, und diesem Familiennamen sind dann zusätzlich Bezeichnungen beigefügt, etwa der Name des Textes, nach dem gearbeitet wurde oder ein Begriff des Themas, wozu es auf diese und jene Art gemacht ist. Jedes dieser *Bhasmas* stellt eigentlich eine Rezeptur mit eigenen Wirkqualitäten, bedingt durch die Zusätze und das gewählte Verfahren, dar.

Der jeweilige Zeitaufwand des Verfahrens macht noch eine andere Differenzierung erforderlich. Es gibt *Bhasmas*, welche mit nur wenigen Calcinationen fertiggestellt sind, andere mit dutzenden, andere mit hunderten, mit tausend und mehr. *Loha-Bhasma* kann zB. mittels 13 Calcinationen bereitet werden, aber auch mit 1000 *(Sahasra putti Loha Bhasma)*. Da letzteres bis zu 15 Jahren Arbeit bedeutet, findet man es nur sehr selten und wenn, dann in alten Familientraditionen, wo es für einen bestimmten Zweck oder einfach nur als eine Art Meisterstück oder als meditative Freude an der Sache ausgearbeitet wurde.

Äquivalente

Selbstverständlich können *Bhasmas* auch von westlichen Alchymisten, wenn sie in diesen Verfahren kundig sind, angefertigt werden. Wo die in den indischen Vorschriften empfohlenen Pflanzenzusätze nicht verfügbar sind, verwenden sie dann Äquivalente:

Für Aufschließungsprozesse von Mineralen und Metallen kann die Rindenabkochung der *Acacia arabica* durch Eichenrindenabkochung ersetzt werden, das Guggul-Harz durch Myrrhe, das Öl der *Arachis hypogaea*-Samen durch Olivenöl, die indischen *Bethelotia lanceolata*-Blätter durch Sennesblätter, *Ajwan*, der Wilde Sellerie-Samen durch Dillsamen, Anissamen oder Thymian, Chrysanthemenblüten (insbes. *Chrysanthemum roxburghii*, bzw. *coronarium)* durch Kamillenblüten, das Dekokt aus Kokkelskörnern durch eines aus Isländisch Moos, Cumin- durch Koriandersamen, *Bhringaraj (Eclipta alba)* durch Löwenzahn, *Jatamamsi*-Wurzel *(Nardostachys jatamamsi)* durch Baldrianwurzel und die Wurzelrinde von *Arka (Calotropis gigantea und C. procera)*, dem ‚pflanzlichen Merkur' im *Ayurveda*, durch Ipecacuana, Sarsaparille und andere Pflanzen. Das sind nur wenige Äquivalenzmöglichkeiten unter sehr vielen, welche aus Überlieferung, Erfahrung oder Signatur erkannt werden, und meist spielt dabei eine wichtige Rolle, welches Ziel die Zubereitung verfolgt. Nehmen wir als Beispiel die Präparation des Hirschhorn-*Bhasmas:*

Die Präparation des Hirschhorns als Beispiel einer gegenseitig befruchtenden Verbindung östlicher und westlicher alchymistischer Tradition

Das Hirschhorn *(Marga Shringa)* wird mindestens 3x mit dem Harz von *Arka (Calotropis gigantea)* verrieben und calciniert. Anstelle von *Arka* kann man nun aber genausogut Brennnessel verwenden. Wichtig bei den Einzel-Zusätzen ist immer die *Rasayana*-Wirkung, also eine öffnende Kraft, die sich oft durch Schärfe, ‚Biss' oder Giftigkeit der Pflanze ausdrückt.

Nach diesen Verreibungen und Calcinationen ist das Horn-*Bhasma* grundsätzlich fertig. Es ist im *Ayurveda* als Herz-Mittel, insbesondere wegen seiner Wirkung auf den Herzmuskel beliebt. Üblicherweise werden zu seiner Präparation die Abwurfstangen der Hirsche gesammelt und in dünne Scheiben geschnitten. Die meiste Kraft steckt im Mark, also im dunkleren, mittigen Teil der Scheibchen, weswegen Manche den äußeren Rand verwerfen, während andere die ganzen Stücke verwenden. (Auch daraus ergibt sich wieder eine Differenzierung zwischen verschiedenen, oft gleichnamigen *Bhasmas*).

Zur Zeit ist es in Indien selbst den berühmtesten *Vidyas* nicht möglich, Hirschhorn käuflich zu erwerben. Das Verkaufsverbot gründet auf der Behauptung, dass man, um das Horn zu bekommen, die Tiere töten muss. Die *Vidyas,* welche womöglich das Gebot des Nichtverletzens *(Ahimsa)* tatsächlich beachten, meinen nun, dass die Gesetzmacher einfach nicht wissen, dass dies so nicht der Fall ist, sondern die Hörner gesammelt werden können... Irritierend dabei ist, dass Hirschhorn, zu Schmuck verarbeitet, sehr wohl gehandelt werden darf, weswegen manch ayurvedischer Arzt viel Geld für geschnitzte Objekte ausgeben muss, um sie danach im Tiegel zu verbrennen.

Vor einer ähnliche Situation stehen die Kundigen Indiens neuerdings auch bezüglich der Korallen. Da rote Edelkorallen selten und teuer sind, ging man immer mehr dazu über, billigere, rote Korallen, Schaumkorallen u. dgl. zu verarbeiten. Ihre Wirkung ist nicht sehr verschieden und dieser Notgriff wird akzeptiert. Der Handel mit Schmuckstücken aus Edelkoralle ist erlaubt, während die rohen Stücke für medizinische Zwecke nicht verfügbar sind, obwohl man sie von kundigen Tauchern sehr schonend ernten lassen könnte, ohne dass dabei der Stamm verletzt wird. Korallen sind in der Heilkunst sehr wertvoll. Auch *Paracelsus* sagt, dass er kein Leiden kenne, worin sich nicht die Koralle hilfreich auswirken würde.

Als nun aber auch billigere rote Korallen (außer als bearbeitete Schmuckstücke) nicht mehr erhältlich waren, verlegte man sich stattdessen auf den Einsatz der weißen Korallen. Deren gibt es im Meer genug und sie sind auch nicht sonderlich teuer zu erwerben. Jetzt aber können die *Vidyas* auch diese Korallen nicht mehr bekommen. Der Grund ist folgender: Es dürfen per Gesetz nur einige Tonnen Korallen jährlich geerntet werden. Und genau diese Menge kauft die Betonindustrie auf, weil sie die Korallen im Straßenbau und dergleichen Projekten, zu Kalkstaub vermahlen, ‚braucht'. Es bleiben dabei nicht einmal Abfallstücke zur Medizin.

An solchen Beispielen sehen wir, dass der destruktive Zynismus in den letzten Zügen des ‚dunklen Zeitalters' kein Land der Erde verschont. Doch zurück zur Praxis:

Während solcher Calcinationen finden keine Scheidungen statt, denn es gilt die Ansicht, dass ‚alles Natürliche', also auch die Schlacken, ‚erhöht' werden sollen. Nach der dreimaligen Calcination des Horns mit Brennnessel oder *Arka* kann noch ein vierter Durchgang hinzugefügt werden, nämlich Verreibung und darauffolgende Calcination mit Kuhmilch. Das gibt zwar ein wenig mehr Arbeit, macht das Präparat aber ‚rund' und benimmt ihm eine etwaige ‚übermäßige Hitze'. Auch solche (speziellen) Schritte, wie sie in den zahlreichen Clan- bzw. Familientraditionen mannigfaltig erscheinen, ergeben Unterscheidungskriterien der *Bhasmas.*

Ein anderes Beispiel (hierzu) ist die Addition zumindest eines weiteren Reinigungsschrittes in der ersten Phase *(Shodana)* der Metallpräparation. Üblicherweise werden die Metalle zuerst in 5 Flüssigkeiten abgelöscht, bevor die eigentliche Aufschließung beginnt. Die ,Aufschließung' entspricht dem Tod der Substanz, dem Vergehen der grobstofflichen Ausdrucksform, das fertige *Bhasma* entspricht der wiedergeborenen Idee der Substanz – immer noch körperlich, jedoch durchgeistigt bzw. ,erhöht'. Der Tod bedarf einer entsprechenden Vorbereitung, einer Reinigung der ,Gewänder', einer Loslösung von übermäßig Erdhaftem, und das eben ist *Shodana*. Dies wird nun in der Regel mittels Buttermilch, Kuhurin, *Ghee* (Butterfett), Essig und einer Abkochung der Myrobalanenmischung *Triphala* vollzogen.

Je nach Überlieferung belassen es manche aber nicht bei diesen 5 grundlegenden Reinigungen (Ablöschungen des erhitzten Metalls in den jeweiligen Flüssigkeiten), sondern setzen dem noch weitere, zum Teil auch spezifischere bzw. differenzierende Reinigungsschritte hinzu. So etwa kann eine sechste Reinigung des Goldes in der gerbstoffreichen Rindenabkochung von *Bauhinia malabarica* erfolgen, was das Gold-*Bhasma* am Ende noch subtiler und kräftiger macht. Ähnlich wie bei der oben beschriebenen zusätzlichen Horn-Milch-Calcination sieht man auch an Beispielen der Metallpräparationen, dass wieder Unterschiede bestehen zwischen den meist einfach gleichnamigen *Bhasmas* – je nach Absicht, Verstehen und Tradition.

Nun verhält es sich so, dass während der Calcination des Hirschhorns zusammen mit *Arka* oder Brennnessel im Inneren des verschlossenen Tiegels ein Prozess abläuft, den die *Vidyas* als ,Kochung' bezeichnen. ,Calcination' ist ein gültiger Überbegriff, doch eigentlich ,kochen' hier mehrere ,Dinge' während einiger Stunden zusammen und verändern sich gegenseitig. Zu den bedeutsamsten Ereignissen im Inneren des Tiegels gehört folgendes: Die Feuerung wird so gehandhabt, dass die Temperatur während 5-6 Stunden sehr langsam ansteigt, bis dass der gewünschte Grad erreicht ist, den man dann einige Zeit unterhält. Im Tiegel destilliert, sublimiert, resolviert und extrahiert der Wurzelgeist des Horns *(Spiritus Cornum Cervi)* und es destilliert, sublimiert, volatilisiert und figiert das ,rote Öl' des Horns *(Oleum Cornum Cervi)*. Selbstverständlich vermischt mit Wurzelessig und rotem Öl der Pflanzenzusätze, was wir aber hier vorerst beiseite lassen wollen. Es kommt also die Regel zur Anwendung, dass jedes natürliche Ding das Mittel zu seiner Auflösung in sich hat. Dieses eigene, eingewurzelte *Menstruum* wird während der langsamen Kochung frei und wirkt auf die feste Substanz öffnend und im Falle des Hirschhorns sogar elixirisch, weil dieses Sublimat reich an quintessentiellen, flüchtigen Salzen ist.

Im ayurvedischen Verfahren arbeiten sich diese flüchtigen Inhalte (Essig, Öl und Salz) im Tiegel vollständig auf und stehen für sich nicht mehr zur Verfügung. In der abendländischen Tradition wird Hirschhorn (zB. mit Sand vermengt) trocken destilliert, um diese flüchtigen Inhalte, seinen Geist (Wurzelessig, *Spiritus Cornum Cervi)* und sein rotes Öl zu erlangen. Beide ergeben nach mehrfachen Rectificationen hochwertige Produkte. Der Rückstand wird entweder verworfen oder noch zu einem simplen Hirschhornsalz weitercalciniert. Nun ergänzen manche ayurvedische *Vidyas* diesen ersten Teil, hierin der abendländischen Tradition folgend, und destillieren zuerst Geist und Öl aus dem Horn, anstatt dieses gleich zu ‚verbrennen'. Im Gegenzug lernen abendländische Alchymisten, den Rückstand der Destillation zu einem *Bhasma* weiter zu verarbeiten.

Dabei ist es wichtig, dass Wurzelessig (Geist) und Öl öfters über den Rückstand rectificiert werden. Das hat zwei Gründe: Erstens eben den der notwendigen, bestmöglichen Rectification der Flüssigkeiten und zweitens den der ebenso notwendigen Imbibation der Festsubstanz, welche ihrer Aufschließung und Erhöhung dient, sodass sie dann auch zum *Bhasma* taugt. Nachdem auf diese Weise destilliert, rectificiert und imbibiert wurde und man schließlich zwei wertvolle, flüchtige Produkte aus dem Horn erhalten hat, kann der Rückstand mit Brennnessel (oder *Arka)* vermischt und damit verfahren werden, wie oben beschrieben zum *Bhasma,* dem dritten wertvollen (fixen) Produkt. Solche Möglichkeiten ergeben sich aus der ergänzenden Zusammenschau der Traditionen, und auf dieses Weise haben sich seit Urzeiten jene Systeme gebildet, die wir heute als Traditionen bezeichnen. Auch das zur Calcination des Goldes zugesetzte Hirschhorn kann (wenn es anfangs noch vom Gold zu scheiden ist) zu einem *Bhasma* von ganz besonderer Qualität weiterverarbeitet werden.

Wir erkennen daraus, dass verschiedene Traditionen sich gegenseitig ergänzen und bereichern können und verstehen, wie manch abendländischer Alchymist der letzten Jahrhunderte auf seinen Reisen ihm unbekannte ‚Kunstgriffe' gelernt hat, und umgekehrt. Wir sehen bei diesen Dingen aber auch, dass die ayurvedischen *Bhasmas* in ihrer Vielfalt genausowenig oberflächlich bewertet werden dürfen wie die Präparate abendländischer Spagyriker. So wie keine Melissen-Essenz der anderen gleicht, ist *Loha-Bhasma* nicht gleich *Loha-Bhasma.* Das ist nur dann dem therapeutischen Anwender ein Hindernis und dem Kritiker ein Grund, solange sie nicht die Ganzheitlichkeit des Lebens verstehen, aus der die Heilkunst wächst. In der Natur gibt es keine Gleichheit, sondern nur Ähnlichkeit. Je näher der Gleichheit, umso näher dem Leblosen und Erstarrten. Nicht einmal Asche, Schnee und Wüstensand enthalten vollkommene Gleichheit, eben weil es ‚Dinge der Natur' sind. Dieses Gesetz betrifft auch die natürlichen Heilmittel.

Auch die Ausgangssubstanzen zu den *Bhasmas* bedingen mehr oder weniger unterschiedliche Endprodukte. Das beste *Loha-Bhasma* wird aus jenem Eisen hergestellt, welches man zuvor aus Pyrit *(Raupya Makshika)* extrahiert.

Ein anderes Beispiel: Kupfer ist das Metall der Venus und hat zudem eine ganz besondere Beziehung zur Sonne. Darüber könnte man sich wieder lange in hermetischem Denken üben. Das ayurvedische Kupfer-*Bhasma (Tamra-Bhasma)* wird deswegen auch ‚Sonnen-*Bhasma*' genannt, und man findet viele Synonyme der Sonne in seiner Namensbezeichnung gebräuchlich, zB. *Surya-Bhasma*. Weil die Sonne aber auch besonders mit der feurigen *Loha*-Kraft korrespondiert, ja diese ein Teilaspekt der Sonne ist, wird auch ein Kupfer-*Bhasma* bisweilen als *Loha-Bhasma* bezeichnet. Darin verbirgt sich eine Anspielung darauf, dass sich das feurige Spektrum mit seinen solaren und martialen Aspekten ‚noch einmal' im ‚Metall des (venerischen) Erd-Elements' ausdrückt.

Den Ausgangsstoff für ein ganz besonderes Kupfer-*Bhasma* liefern Pfauenfedern. Etwa 10 000 davon werden mit Borax *(Tankana)* verrieben. *Tankana* heißt auch *Loha-Shodana* – das *Loha*-Reinigende. Schließlich wird das Kupfer bei hoher Temperatur als Ausgangsstoff zum Kupfer-*Bhasma* aus den Pfauenfedern gewonnen. Man kann sich vorstellen, dass in diesem Falle wieder ein bedeutender Unterschied zwischen *Tamra-Bhasma* und *Tamra-Bhasma* besteht.

Diffizile Untersuchungen in modernen Laboratorien dokumentieren dazu die Unterschiede der Verfahrensweisen bezüglich grobstofflicher Veränderungen:

---

(1) Der **Hämoglobingehalt** bei Versuchstieren (Kaninchen) erhöhte sich bei der Verabreichung eines allopathischen Eisenpräparates (Eisensulfat) um 2, 345 g/dl im Blut.

(2) Nach der Gabe von *Lauha Bhasma* (Eisen-Bhasma) – durch 7 Calcinationen hergestellt – erhöhte er sich um 3, 04 g/dl, also schon beträchtlich mehr.

(3) Mit 100 x calciniertem *Bhasma (Satputti)* erhöhte er sich um 3,96 g/dl und mit 1000 x calciniertem *Bhasma (Sahasra putti)* um 5,05g/dl. (Durchschnittswerte).

(4) Durch die Calcinationen und dazwischenliegenden Aufbereitungsschritte (Mörsern mit Zusätzen) erhöhte sich die **Teilchenzahl** des Eisensulfats von 2,265 x 10 hoch 8 auf 1,130 x 10 hoch 9 / Gewichtseinheit.

(5) Ebenfalls erhöhte sich die spezifische **Oberfläche** der Teilchen von 0,460 x 10 hoch 3 cm2 / G auf 1,901 x 10 hoch3 cm2 / G und hatte sich damit also gut vervierfacht.

nach: *Ancient Science of Life, Vol. No. XV 2. Oct. 1995, pg. 140-144*

Eisen-Bhasma

Das alte Synonym für *Lauha, Loha,* erinnert an das ursprachliche *Lo* für Flamme, Licht u. dgl., wie in „lodern". Zur Darstellung des Eisen-*Bhasmas* werden gemeinhin hauptsächlich dünne Blättchen von Stahl *(tikshna)* verwendet. Diese löscht man je 7x in Sesamöl, Buttermilch, Kuhurin, Essig und zuletzt in einer Abkochung von *Kulattha (Dolichos biflorus)* ab. Diese Methode wird sowohl zur Reinigung des Eisens als auch – wie im Zusammenhang mit dem Gold schon erwähnt – für andere Metalle empfohlen. Im Speziellen verfährt man dann weiter mit der Reinigung des Eisens mittels einer Einkochung von *Triphala* (der Myrobalanenfrucht-Mischung aus *Bibhitaki, Haritaki* und *Amalaki).*

oben links: *Triphala – Bhibitaki, Haritaki, Amalaki*
oben rechts: ‚Reinigung' des Metalls durch wiederholtes Ablöschen in Flüssigkeiten
unten links: Offenes Rösten und Calcinieren mit Pflanzenzusätzen
unten rechts: *Bhasma*-Test: Eine der Prüfungen, ob das *Bhasma* richtig ausgearbeitet ist, besteht darin, zu sehen, ob es auf der Wasseroberfläche schwimmt und nicht sinkt.

Nach dem *Shodana* ist nun eine Methode, dass man solcherart gereinigte Eisenblättchen in einen irdenen Topf legt und mit Kuhurin übergießt, das ganze mit einem Eisenblech bedeckt und bis zur vollständigen Trocknung an die Sonne stellt. Danach wird wieder Urin darübergegossen und der ganze Prozess einen Monat lang ständig wiederholt.

Dieses Imprägnieren (Imbibieren) mit Urin hat auch in der abendländischen Alchymie eine ganz besondere Bedeutung. Bei der Trocknung kristallisiert eine Art des Urinsalzes aus, die mit dem Metall eigenartig zusammenwirkt – ein Thema, welches von den abendländischen Alchymisten der letzten Jahrhunderte meist nur insgeheim besprochen wurde.

Nach dieser folgen weitere Imbibationen: Zuerst mit einem Dekokt von *Triphala*, danach mit dem Saft von Aloe *(Kumari)*, jeweils einen Monat lang. Die Eisenblättchen werden dabei sehr subtil. Mit warmem Wasser sorgfältig abgewaschen, zerreibt man sie zu einem sehr feinen Pulver. So erhält man nach gut 3 Monaten ein erstes ‚Eisen-*Bhasma*'. Es wird als solches bereits medizinisch verwendet, wenngleich es noch nicht wirklich ‚kunstvoll' aufgeschlossen, sondern ‚nur gereinigt' ist. Weitere Arbeitsschritte sollten folgen:

Das Pulver wird 3 Tage lang immer wieder mit Kuhurin imprägniert und zu einer dicken Paste verrieben. Daraus formt man Plätzchen oder kleine Kugeln, die nun in einem mit sieben Lagen lehmverschmierten Tuch versiegelten Tiegel ‚verschlossen' calciniert werden.

Die Imprägnation mit Kuhurin und darauffolgende Calcination im verschlossenen Tiegel wird 3x ausgeführt. Danach imprägniert und calciniert man (ebenfalls 3x) mit *Triphala*-Dekokt und schließlich wieder mit Aloe-Saft, insgesamt also 9 Calcinationen. Dem folgen drei weitere Calcinationen mit vorhergehender Imprägnation und Verreibung zur Paste zusammen mit dem Saft von *Punarnava (Boerhaavia diff.)*.

Urzeitlich anmutende *ayurvedische Alchymie* – oft die selben Handgriffe wie in der abendländischen Praxis.

Nun wird in der Anleitung zu diesem Weg der Präparation noch ein Zusatz angegeben: Vor der 13. Calcination wird ein Zwölftel des Gewichts an ‚gereinigtem' Zinnober (s.u.) und dem gummiartigen Saft von *Arka* (s.o.) zugesetzt und wieder zur dicken Paste verrieben. Die folgenden drei Calcinationen (jeweils mit diesen Beimengungen) benötigen nur mehr geringere Hitze. Auf diese Weise wird durch weitere 15 Calcinationen aus dem Eisen nach etwa 4 Monaten ein sehr wertvolles *Bhasma*. Es ist von bittersüßem Geschmack, kühlender Energetik und schwerer Qualität. *Vaidya Bhagwan Dash* empfiehlt eine Dosierung von 125-200 mg zusammen mit Honig und Butterschmalz *(Ghee)* oder in einer entsprechenden Rezeptur.

Wir sehen also, dass, verglichen mit den anfangs beschriebenen Rezepturen der chinesischen Medizin, worin Zinnober, Gips, Hämatit und andere Minerale nur mehr in roher Form enthalten sind, die ayurvedische Tradition (genauso wie die tibetische) noch ungleich lebendiger und kunstvoller erhalten ist. Es werden zwar hier kaum Pflanzen spagyrisch aufgeschlossen, von den anderen Substanzen aber selbst die ‚weichen' einer kunstvollen Behandlung gewürdigt, etwa wie der Gips *(Godanti)*, den man eben nicht roh zur Rezeptur fügt, sondern zumindest zuerst einen Tag lang mit dem Saft der Aloe verreibt, zu Plätzchen formt, trocknet und im versiegelten Tiegel calciniert. Das wiederholt man 3x, dann gilt seine Heilkraft als für den Körper erschlossen und die ‚Ausdrucksform' als von ihrer ‚derben Erdhaftigkeit' vermittels Feuer und Pflanze befreit.

Beim oben beschriebenen Weg zum Eisen-*Bhasma* fällt der Zusatz von Zinnober in der letzten Phase auf – ein Kunstgriff, welcher in anderen Methoden der Eisen-Präparation nicht gefunden wird, sondern eben genau in diesem Verfahren (nach *Bhagwan Dash*) bemerkenswert ist. Bei anderen ‚Künstlern' fällt dieser Zusatz weg, bei wieder anderen kommen andere Zusätze dazu. Ähnlich schreibt *Bhagwan Dash* auch den Zusatz von Realgar in der letzten Calcinationsphase zum Blei-*Bhasma (Naga-Bhasma)* vor.

Der hermetische Grund zur Idee solch starker Zusätze ist die Förderung der ‚Wiedererstarkung' des Metalls nach Tod und Verwandlung. In unserem Beispiel soll der ‚gereinigte' Zinnober das durch den anstrengenden Präparationsprozess zwar erhöhte, aber auch müde gewordene Eisen wieder ‚aufrichten'. Zuerst wurde das rohe Metall durch *Shodana* auf den Tod vorbereitet. Danach hat es die Schwelle des Todes überquert (durch die Calcinationen), um dann (in den letzten Calcinationsphasen) als *Bhasma* in erhöhter Form wiedergeboren zu werden. Nach diesem langwierigen und für das Metall sehr anstrengenden Weg wird es nunmehr durch abschließende Calcinationen mit ‚kräftigen' Zusätzen gestärkt. Manche Schulen lassen, wie gesagt, diese Stärkung aus oder verwenden dazu nicht solch mächtige Substanzen wie Zinnober und Realgar.

Zum andern können verschiedentlich dermaßen starke Zusätze in gewissen Traditionen verlangt werden, um Schritte und Wege im Verfahren ‚abzukürzen'. Diese Techniken entstammen dann der sogenannten *Siddha*-Medizin, die vor allem in Südindien noch praktiziert wird. Die Sicherheit der auf diese Weise hergestellten Mittel ist umstritten – meist raten die Kundigen davon ab. *Siddha* ist nicht eigentlich *Ayurveda* und oft weniger mit ‚Wissen' als mit magischen Praktiken verwandt. Dies entspricht in manchem der auch im Abendland bekannten Mixtur von Magie und Geistheilung und jener ‚sogenannten Alchymie', die hier gerne in solche Felder gefügt wird. Arbeitsanleitungen ‚aus dem Geisterreich' gechannelt scheinen oft einfacher zu erlangen als durch ernsthaftes Studium bei guten Lehrern und jahrzehntelange Praxis. Mancher *Vidya* und *Rasa Shastri* (Alchymist) sagt dazu einfach: „Bei diesen *Siddhas* ist die ‚Unterscheidungskraft' noch nicht ausgeprägt oder irritiert." Vollständigkeitshalber sei angemerkt, dass es in Indien eine ganze Reihe echter, alchymistischer Präparate und iatrochymischer Rezepturen gibt, in deren Namensbezeichnungen das Wort ‚*Siddha*' enthalten ist, ohne dass solche aber mit diesem Zweig der Tradition und seinen fraglichen Techniken verbunden wären.

‚Verkürzte Wege' gibt es aber auch in ganz anderem Zusammenhang. Wir kennen dergleichen Berichte von abendländischen, arabischen oder chinesischen Alchymisten alter Zeiten, die das eine oder andere ‚Kunststück' erarbeitet haben und damit scheinbar ‚wunderbare' Dinge bewirken. Das sind nun nach der üblichen Terminologie keine ‚Wunder', sondern tatsächlich ‚natürliche Dinge', die durch Verstehen und Arbeit erlangt werden konnten. Beispiele dafür sind die geheimnisvollen *Arkane* und *Menstruen* mancher Alchymisten, welche dermaßen kunstvoll ausgearbeitet wurden, dass sie zu Lebenselixiren, Transmutationspulvern und dergleichen unzähligen Präparaten mehr geworden sind. Eine Ahnung davon kann der einfache Spagyriker haben, wenn ihm ein *Circulatum Minus* nach dem Weg des *Baron Urbigerus* gelingt und er dieses über frische Pflanzen gießt. Man könnte sagen, dem liegt ein einfacher ‚Trick' der Natur zugrunde, dass es sofort eine essentielle Scheidung bewirkt und vielleicht sogar ein wirkungsvolleres Präparat ergibt, als man es in wochenlanger Arbeit aus der selben Pflanze herstellen könnte. Das ist nur eine Kleinigkeit, verglichen mit wahrhaft großen Elixiren, von denen man nicht nur in der abendländischen Alchymie spricht, sondern denen man auch in der indischen Alchymie begegnen kann. Solche Elixire sind für sich selbst Arzneien und gleichzeitig Mittel, um eine Aufschließung anderer ‚Dinge' in sehr kurzer Zeit zu bewirken. Die Allegorie gilt auf allen Ebenen: Je edler etwas (oder jemand) ist, umso schneller kann es etwas anderes (oder jemanden) veredeln. Unzählige Geschichten werden in Indien darüber erzählt, dass ‚erwachte', also ‚veredelte' Menschen ‚noch schlafenden in einem Augenblick zur Erleuchtung' verhelfen konnten.

In der ayurvedischen Alchymie ist es ein hohes Ziel, ‚das Quecksilber zu stabilisieren', also ‚*Mercurius* zu figieren' bzw. ‚Merkur einzufangen'. Im Allgemeinen sind sich alle alchymistischen Traditionen über den hohen Grad dieser Arbeit einig. Merkur ist der Naturgeist, und wenn der Alchymist beginnt, das Quecksilber zu verstehen, dann hört die Natur auf, ihre Schätze und Geheimnisse vor ihm zu verbergen. Große *Vidyas* verwenden Jahre und Jahrzehnte darauf, dieses ‚Kunststück' zu vollbringen. Sie scheiden den Geist, das Öl und das Salz aus dem Quecksilber und vereinigen es zu der begehrten Substanz.

Ein berühmter alchymistischer Arzt hatte dies nach 30-jähriger Arbeit und vielen Rückschlägen erreicht und konnte es bald auf neun verschiedene Arten ausarbeiten. Er hatte gehört, dass weit im Norden ein alter Kundiger lebte, der ihn an Alter wohl um das dreifache übertraf und diese Meisterschaft ebenfalls erlangt hatte. So nahm er seine Präparate mit auf den Weg und besuchte ihn. Mit großer Ehrerbietung trug er ihm sein Anliegen vor und bat ihn, er möge seine Werke begutachten. Nachdem der Alte dies getan hatte, verbeugte dieser sich tief und berührte die Füße des Arztes. „Du bist wahrhaft mein ‚älterer' Bruder," sagte er, „ich habe in meinem ganzen Leben nur einen einzigen Weg verstanden, um dieses Werk zu vollbringen. Du aber kennst neun Wege. Schenke mir deine Gnade und erläutere mir deine Kunst." Und er blieb lange Zeit bei dem Alten, ihm seine Wege zeigend, vielleicht den einen mit Perlen, den andern mit Rubinen, mit Silber, mit Antimon und schließlich sogar ohne irgendwelche Zusätze.

Bezüglich der obgenannten Zusätze, wie Zinnober, Realgar und dergleichen muss man bedenken, dass die *Bhasmas* von Metallen in den meisten Fällen zwar ‚Metall' enthalten, allerdings eben ‚aufgeschlossen' bzw. als ‚Asche'. Durch Prozesse wie *Amriti Karana* (‚zu Nektar machen') wird jede schädliche Wirkung beseitigt. Sofern nicht eine gewisse Giftigkeit aufgrund mangelnder ‚Erhöhung' der Substanzen oder philosophisch unstimmiger Abkürzungen der Prozesse verbleibt, ist die ‚Wirk-Kraft' nicht Giftigkeit, sondern Audruck der Heilpotenz und stellt bei richtiger Anwendung und Dosierung keinerlei Gefahr für die Gesundheit dar. Die *Vidyas* betonen dabei, dass diese Heilmittel nicht ‚chemischen Experimenten' mit riesigem Aufwand, aber primitivem Naturverständnis entspringen, deren Produkte kurze Zeit an Tieren und Menschen ‚erprobt' werden, sondern sich vielmehr jahrtausendelang bewährt haben, weil sie aus Sicht der lebendigen Natur ‚stimmig' sind. Andererseits gelangen oft nur durch einen Einkaufsbummel in der Großstadt, einen Arbeitstag in mancher Fabrik oder beim Besuch einer Obstplantage schädliche, ‚tatsächlich giftige' Substanzen und Metallverbindungen in den Körper, welche keineswegs ‚rein', ‚erhöht' oder heilsam kombiniert sind. Mit primitivem Denken und rein stoffbezogenen Analysen kann dieser Sachverhalt allerdings kaum erfasst werden.

Wie weit heute ähnliche Präparate in der im Westen verbreiteten ‚ayurvedischen' Massenware tatsächlich kunstvoll ausgearbeitet wurden, ist nicht immer gewiss. Die Anleitungen zu den Prozessen wurden zwar jahrtausendelang detailliert überliefert, werden aber heute dort, wo in sehr großen Mengen produziert wird, mitunter kaum mehr genau eingehalten bzw. von den ‚Arbeitern' oft gar nicht verstanden. Während kleinere Familientraditionen seit jeher die vorgeschriebenen 100 oder 1000 Calcinationen zu dem einen oder anderen *Bhasma* ausführen, rechnen größere ‚*Bhasma*-Fabriken' nach, ob sie von den 7 ‚unbedingt nötigen' Calcinationen nicht noch ein oder zwei einsparen könnten. Aus dieser Haltung entsteht oft nicht mehr als immerhin eine Notdurft, aber keine ‚hohe Arznei'. Den Unterschied in der Haltung weist die Bemerkung eines ernsthaften Alchymisten zu der Anfrage, ob er denn in größeren Mengen das echte paracelsische *Aurum potabile* zu kosmetischen Zwecken liefern könne, es würde auch gut bezahlt, denn die Produkte seien für sehr wohlhabende Konsumenten bestimmt: „Wenn ich es hätte, würde ich es Ihnen nicht verkaufen."

Ähnlich ergeht es einem wohl überall in der Welt mit diesen Dingen. Mit dem Wissen genauso wie mit den Produkten dieses Wissens. Kein Kundiger Indiens oder Tibets würde seine höchsten Elixire in Massen herstellen und dementsprechend zum Verkauf feilbieten können oder wollen. Er würde meinen, dass dazu auch die einfacheren Mittel gut genug seien, und der Bedürftige kann im Falle aufgrund bereits erwähnter natürlicher Gesetzmäßigkeiten selbst die höchsten Arzneien finden, wenn es ihm zusteht. Ebensowenig könne das Wissen selbst ‚kommerziell' werden. Der Schüler findet seinen Lehrer durch Fügung, und auch wenn er ihm Gegenleistungen darbringt, ist der Lehrer zu keinen ‚Enthüllungen' verpflichtet. Zudem versteht ein jeder nur das, wozu er gereift ist. Alle Ereignisse und Begegnungen stehen in großen Zusammenhängen. Geld als Tauschmittel ist in Lehre und Überlieferung von geringer Bedeutung und die allerbesten Dinge sind bekanntlich nicht käuflich. Bezahlt man aber ‚hohe Präparate', dann reicht die Bezahlung knapp für das Material (Minerale, Kräuter, Brennstoffe etc.) aus, stellt aber niemals einen Gegenwert zur aufgewandten Kunst dar.

Zweifellos besteht heute eine immer stärker werdende Tendenz dazu, mit (aus hermetischer Sicht primitiven) Kopien hochwertiger Präparate große Geschäfte zu machen. Diese Entwicklung hat zwei Seiten. Zum einen werden dadurch gute Rezepturen bekannt, die selbst dann noch eine bedeutende Wirkung zeigen, wenn sie nicht nach höchster Kunst zubereitet wurden, und die Hüter jener Traditionen gewinnen respektvolle Zuneigung. Zum andern aber geht der Verfall ebendieser Traditionen typischerweise oft mit einem ‚Ausverkauf' ihrer Geheimnisse einher, und tiefes Wissen verliert sich an den Oberflächen als würzige Zutat zur Leichtlebigkeit. Was diese Angelegenheiten betrifft, so fällt im Asien der letzten hundert Jahre genau jener Prozess auf, den Europa bereits seit über tausend Jahren erdulden muss. Es bleibt zu hoffen, dass man dort für die nahe Zukunft würdigere Lösungen findet.

In den zurzeit im Westen praktizierten Formen des *Ayurveda* sind jedoch die alchymistischen Präparate noch wenig bekannt und genutzt, abgesehen von einigen Standard-Rezepturen, welche *Bhasmas* enthalten. So stammt zB. die beliebte Eisen-Wirkung des berühmten *Liv 52* von einem Präparat aus Eisenrost *(Mandoor-Bhasma)*. Die, noch weit berühmtere Latwerge *Chyavan Prash* enthält *Bhasma* aus schwarzem Glimmer *(Abhrak-Bhasma)* und Perlmutt-*Pishti*. Die, beinahe als Allzweckmittel bekannte *Boro*-Salbe enthält Zink-*Bhasma (Pushpanjana, Jasad-Bhasma)* und Talk in ‚gereinigter' Form *(Dugdha Pasana)*. Vor allem mit den beiden letzteren Präparaten hat die Alchymie unerkannt schon lange in den westlichen, naturheilkundlich ausgestatteten Hausapotheken Einzug gehalten.

Allgemeines zur ayurvedischen Alchymie

Der am meisten gebräuchliche Begriff für die indische Alchymie, *Rasa Shastra*, bedeuted ‚Wissenschaft von der absorbierbaren Form der Dinge', ‚Wissenschaft von der durchgeistigten Natur', ‚Wissenschaft von Merkur' – also ‚hermetisches Wissen'. Das bezeichnet die Kunst der Präparation pflanzlicher, tierischer und mineralischer Substanzen in vom Körper assimilierbare Mittel. Jahrtausendealte Überlieferungen ayurvedischer Pharmazie beschreiben Einzelsubstanzen, deren Kombinationen in Verfahren und Rezeptur, die zur Präparation von Mineralen und Metallen nötigen Kräuter, Aufschließungstechniken, therapeutische Eigenschaften, alchymistische Geräte wie Tiegel, Mörser und Retorten, Öfen und Feuerstellen sowie die Kunst verschiedene Feuer lange Zeit zu unterhalten.

Der am meisten diskutierte Zweig des *Rasa Shastra* – *Loha Dhatu* – ist die hohe Kunst Metalle aufzuschließen. Wir haben die Silbe *Loh* bereits im Zusammenhang mit dem Eisen kennengelernt. *Loh*, oder *luh* wird aber auch übersetzt mit ‚extrahieren', besonders in dem Vergleich, dass die Metallpräparate gleichsam die Krankheiten aus dem Körper extrahieren, genauso wie die Metalle selbst aus dem Erzgestein extrahiert werden. Und dies geschieht mittels *Loha*, der Flamme, mittels ‚Feuerkunst'.

Die 7 Metalle *(Sapta-Dhatu)*, welche seit Urzeiten einen besonderen Rang einnehmen sind: Gold *(Svarna)*, Silber *(Rajata)*, Kupfer *(Tamra)*, Eisen *(Loha)*, Zinn *(Vanga)*, Blei *(Naga)* und Zink *(Yasada)*. Das Quecksilber steht in gewisser Weise über all diesen in einer eigenen Position. Bedeutungsvoll ist sein Name *Rasa*, wie der der Kunst selbst.

Nach den *Dhatus* folgen in der Tradition die *Ratnas*. *Ratna* bedeutet ‚wertvoller Stein'. *Navaratna* sind die 9 wichtigsten Steine: Rubin *(Manikya)*, Topas *(Pushparaga)*, Smaragd *(Tarksharya)*, Perle *(Mukta)*, Diamant *(Vajra, Hiraka)*, Saphir *(Nilam)*, Tigerauge *(Vaidurya)*, Sardonyx *(Gomeda)* und Koralle *(Pravala)*.

Ihnen untergeordnet sind die *Uparatnas*, wie Bergkristall, Sonnenstein, Mondstein, Türkis, Turmalin, Lapislazuli, Jade u.a., welche günstig zu erwerben, relativ leicht zu präparieren, aber dennoch von mächtiger Wirkung sind.

Sudha-Varga sind Substanzen, welche Kalk enthalten, wie Calcit *(Khatika)*, Austernschalen und einfache Süßwasser- und Meeresmuscheln *(Shukti)*, Kaurimuschel *(Varatika)*, Muschelhorn *(Shanka)*, Eierschalen *(Kukkutandatwak)*, Schale von Tintenfischen *(Samudra phena)* sowie Talk *(Dugdhapashan)*, Selenite, Gips *(Godanti)* und Hörner von Tieren *(Mrugashrunga)*.

Darüberhinaus sind Wesen, Präparation und Wirkungen überliefert von: mineralischen Ausschwitzungen *(Shilajatu)*, Realgar *(Manaha Shila)*, Borax *(Tankana)*, Calciumsilikaten *(Badara Pashana)*, Alaun *(Sphatika)*, Auripigment *(Haritala)*, Arsenic *(Gauri Pashana)*, verschiedene Arten von Glimmer *(Abhraka)*, Ocker *(Gairika)*, Eisenvitriol *(Kashia)*, Kupfervitriol *(Tuttha)*, Eisenpyrit *(Raupya Makshika)*, Kupferpyrit *(Svarna Makshika)*, Galmei *(Rasaka, Kharpara)*, Antimon *(Srotonjana)*, verschiedene Legierungen wie Glockenmetall *(Kamsya)*, Messing *(Pittala)*, Schwefel *(Gandhaka)* und vielen anderen mehr – insgesamt also eine steinzeitliche Apotheke, die sich sehen lassen kann.

Die Beschreibung von Wesen, Präparation, Anwendung und Wirkung der Substanzen ist durchwegs hermetisch-handwerklicher Natur und gründet auf der Kenntnis der 5 Elemente, der Säfte, Hüllen, Planetenkräfte, Signaturen, entspricht also im paracelsischen Sinne dem ‚rechten Umgang mit den Dingen der Natur'.

Beliebt bei den *Rasa Vidyas* sind salinische Präparate, einige davon ganz ähnlich jenen Aschen und Salzen, welche wir bei der ‚Darstellung des Sal-Prinzips aus den Pflanzen' besprochen haben. Sie werden *Ksharas* genannt und mittels Calcination und Reinigen durch Umkristallisieren, wie wir es aus der ‚abendländischen' Tradition kennen, hergestellt. Dazu Beispiele:

Kshara Dvaya (wörtlich: ‚Zwei-Salze-Mischung') heißt die Mischung aus Soda *(Svarji Kshara)* und dem gereinigten ‚Erd-Salz' des Weizens *(Yava Kshara)*.

Fügt man noch Borax dazu, dann erhält man *Kshara Traya* (wörtlich: ‚Drei-Salze-Mischung').

Zur ‚Fünf-Salze-Mischung' *(Kshara Pancaka)* vermischt man Soda mit den Salzen der Pflanzensamen von *Musa paradisiaca, Stereospermum suaveolens, Butea monosperma* und Sesam.

Mit sieben Pflanzensalzen vermischtes Soda heißt ‚Acht-Salze-Mischung' *(Ksharastaka)*, usf.

Die, mittels Destillationen dargestellte Pflanzenessenz wird *Vanaspati Sattva* genannt.

In den indischen Formen des *Ayurveda* ist die Alchymie in der Heilkunst eine alltägliche Selbstverständlichkeit und das, was wir ‚hermetisches Wissen' nennen, bildet die Grundlage der Medizin, hineinreichend bis in das einfache und exoterische Basiswissen der Spiritualität, Lebensführung, Ernährungskunde und dergleichen betreffend. Auffallend ist, dass sich dort die Ausbildung sehr oft bloß auf Teilgebiete der Traditionen beschränkt, sodass man gelegentlich den Spruch zu hören bekommt, es gäbe in ganz Indien niemanden mehr, der das *Ayurveda* wirklich und in seiner Ganzheit versteht. Ein Grund dafür mag in den zahllosen, meist religiös bedingten, oft dogmatisch-ritualistischen, hierophantischen Strukturen verschiedener Schulen liegen. Dieser Grund wird vertieft durch die Eigenart des traditionalistisch-konservativen Menschen, dass er mit der Zeit unweigerlich eine Art Stolz im Sinne von Eitelkeit und mangelnder Einschätzung eigenen Vermögens entwickelt, was nicht nur der eigenen Reflektion, sondern eben auch der Kommunikation mit ‚Fremden' sehr hinderlich ist und eine Zusammenschau verschiedener Schulen verwehrt. Dies noch übersteigernd bemerkt man eine starke Tendenz dazu, den Mangel an ‚Gesamtwissen' durch einen fraglichen Gebrauch von Yoga-Kräften *(Riddhis, Siddhis)* zu ersetzen.

Ayurveda, Unani, Siddha und Kalpa

Wenn *Siddha* als Magie in den Resort von Alchymie und Heilkunst eingebunden wird, entstehen ähnliche Situationen wie im Abendland der beginnenden Neuzeit – nämlich dass die Grenzen zur zweckorientierten Magie verschwimmen und die eigentliche ‚Kunst' ihren gesunden Boden verliert. Therapie und Präparat verlieren ihre natürlich-heilsame Art und lösen auf körperlicher, vor allem aber auf geistiger Ebene unliebsame Rückwirkungen – sowohl auf den Therapeuten als auch auf den Patienten – aus. Dieses Phänomen in West und Ost vergleichend, kann man ebendort ein Geschehen beobachten, welches sich offenbar auf immer wieder ähnliche Art im Laufe von Epochen und im Werdegang von Kulturen ereignet. Die Struktur dieses Geschehens ist ein wenig kompliziert und alleine bei oberflächlicher oder voreingenommener Betrachtung nicht erkennbar.

Andererseits ist auch auf dieses Thema die typisch indische Art anwendbar, deren Pedant wir im Westen in Sprüche kleiden wie: „Es wird heißer gekocht als gegessen." Die Bezeichnung ‚*Siddha*' findet man immer wieder in Systemen und Produkten, welche nichts mit der eigentlichen *Siddha*-Magie zu tun haben. Auch die Ansichten der verschiedenen Schulen und Familientraditionen über Ursprung und Bedeutung von *Siddha* sind mitunter sehr unterschiedlich. Gelegentlich wird behauptet, dass es grundsätzlich zwei ‚große' Traditionen gibt: die ‚arische' und die ‚drawidische'. Die Erstere wird vor allem in Nord- und Mittelindien praktiziert, schließt das eigentliche ‚*Ayurveda*' und das *Unani* ein und enthält nur wenig ‚*Siddha*'-System.

Die *Unani*-Medizin gelangte hauptsächlich über Persien und Afghanistan im 5. Jhdt. nach Indien. Sie ist zwar altgriechischen Ursprungs, enthält aber das selbe Natur- und Heilverständnis, welches Jahrtausende zuvor wiederum aus Indien dorthin (in den Mittelmeerraum) gelangte. Noch weiter zurück in der Zeit, sollen diese Überlieferungen vor etwa 16 000 Jahren von Europa nach Indien gelangt sein. Man sieht also, dass es eher eine ständige Hin und Her, als eine deutliche Ost-West (oder umgekehrt) Bewegung des Wissens war. Nun wird aber auch erklärt, dass diese ‚arischen' Traditionen, also hier im Besonderen *Unani* und *Ayurveda*, zwar ‚*Siddha*' enthalten – allerdings nur in einem geringerem Maße. Das bedeutet für manche *Vidyas* zugleich, dass jene Traditionen sich weit mehr mit pflanzlichen Heilmitteln befassen als mit mineralischen und metallischen. So wurde die Bezeichnung ‚*Siddha*' beinahe zu einer Bezeichnung für ‚mineralische Alchymie' an sich, nämlich soweit diese besonders die Präparation der Metalle beinhaltet und weitestgehend unbekümmert von magischen Aspekten ist. Die Vertreter dieser ‚arischen' Traditionen (und ihrer zahlreichen Verzweigungen) arbeiten und behandeln nun auch überwiegend mit Pflanzen, manche vor allem mit Pulvern, andere hauptsächlich mit destillierten Ölen, wieder andere mit Kräuterweinen u.v.m. Mineralische und metallische Präparate werden zwar dargestellt und eingesetzt – aber nicht ‚vordergründig'. Die alchymistischen Techniken werden verborgen gehalten, ja man könnte fast sagen: ‚Man spricht nicht darüber.' Die Verfahren sind jedoch sehr hoch und überaus entwickelt und ‚reif'.

Anders die ‚drawidischen' Traditionen Südindiens. Hier nehmen ‚gerade noch alchymistische' Präparate, insbesondere der Metalle und manch giftiger Minerale einen vordergründigen Platz ein, während die Pflanzen eher secundär Verwendung finden. Nicht selten kommt es vor, dass zur Behandlung ‚kleiner' Leiden sehr starke mineralische und metallische Präparate (nur teilweise aufgeschlossen) eingesetzt werden, und man muss oft des Spruches gedenken: ‚Wenn du das aushältst, dann wirst du auch die Erkältung überleben.'

Manche *Vidyas* formulieren: „Das *Siddha*-System enthält auch *Ayurveda*." Es ist anzunehmen, dass das, was wir heute als *Ayurveda* (und darin als *Rasa Vijnan* bzw. *Rasa Shastra*) bezeichnen, sich vor über 20 000 Jahren in Europa formierte und dann zum Teil in den darauffolgenden Jahrtausenden durch Vermischung und Ergänzung mit noch weit älteren Systemen besonders in Süd-Indien zu *Siddha*- und *Kalpa*-Systemen entwickelte, welche vor allem von den Tamilen bis heute tradiert wurden. Manche *Vidyas* meinen: „Das *Kalpa*-System ist weder ganz *Siddha* noch ganz *Ayurveda*, hat von beiden, schließt aber auch ganz eigenständige Traditionen mit ein. Seine alchymistischen Präparate sind besser ausgearbeitet und zuträglicher als die des reinen *Siddha*."

So hat also die Bezeichnung *siddha* mehrere Bedeutungen. Wörtlich kann übersetzt werden mit: ‚Etwas erlangen.' Das bezieht sich hauptsächlich auf heil- bzw. naturkundliches Wissen und dazugehörige Fähigkeiten – insbesondere um den ‚rechten Umgang mit den natürlichen Dingen'. Eine Abart ist das ‚Erlangen übernatürlicher Kräfte' und magischer Praktiken. Solchen ‚*Siddha-Yogis*' begegnet das Volk meist mit Scheu und Furcht – oder am liebsten gar nicht (es sei denn, man trachtet nach der Erfüllung gewisser Wünsche) und die ‚echten' Kundigen warnen vor ihrer ‚Kunst'. Ein wenig erinnern diese Varianten urzeitlicher Tradition an ihr abendländisches Äquivalent der ‚guten' und ‚bösen Zauberer'.

Die Heilkundigen der *Siddha*-Systeme halten ihr Wissen für das älteste, bereits lange vor der vedischen Zeit dem Menschen übermittelte und denken (wie übrigens auch die Vertreter anderer Richtungen), dass es mit dem Menschen selbst in die Welt gekommen ist – aus dem einfachen Grund, weil eben der Mensch sich hier als erstes um den ‚Umgang mit den natürlichen Dingen' und die Heilkunst bemüht hatte.

Ebenso wie die ayurvedischen Medizinsysteme vertritt auch das *Siddha* das iatrochymische Ideal. Überwiegend in ländlichen Gegenden praktizieren heute (laut AYUSH – Dept. for Ayurveda, Yoga, Unani, Siddha and Homöopathie, Gvt. of India) etwa 17 500 (registrierte) Kundige in ganz Indien die ‚4 Säulen des *Paracelsus*'. Dazu kommt noch eine vermutlich sehr große Anzahl Kundiger, welche von dieser Organisation nicht registriert werden konnten. Sie alle verstehen zu befunden, Ernährung und Lebensart zu konzipieren, zu therapieren und sowohl ‚einfache', als auch (je nach Grad der Kunst) alchymistische Heilmittel zu bereiten. Über 270 Spitäler werden in diesem Sinne geführt, die Patienten durch iatrochymische Praktiker betreut und mit deren Arzneien behandelt. Weit mehr aber wirken diese Menschen regional, selbst ansässig in oder außerhalb der Dörfer und mit einem entsprechenden Schatz an Kenntnissen, Rezepten und Verfahren aus ihren jeweiligen Familientraditionen ausgestattet. Damit richtet sich ihr iatrochymisches Wirken zumeist nach den regionalen Gegebenheiten und Notwendigkeiten. Ihr Dasein und Schaffen gilt dem Land als ein Zeichen hoher Kultur, Stabilität und geistigen Potentials.

Mit den (laut AYUSH, Stand 2005) etwa 20 000 Praktikern des *Unani*-Systems und den 440 000 ayurvedischen *Vidyas* (ebendort registriert) sowie tausenden anerkannten Homöopathen und Vertretern zahlreicher anderer naturheilkundlicher Systeme setzt Indien ein Vorbild an Verantwortung, dem im Westen (aufgrund des dortigen Potentials heilkundlicher Traditionen) mit Eile und menschlicher Gesinnung zu folgen wäre.

Der Legende nach gelangte die alchymistische Kunst „in der Vorzeit auf den Berg *Kailash* und von dort direkt nach *Kafiristan* (heute das afghanische *Nuristan*), von wo aus sie in der ganzen Welt verbreitet wurde." Es wird erzählt, dass während der Zwangsislamisierung *Kafiristans* solch große Mengen an alten und ältesten Schriften zusammengerafft und verbrannt wurden, dass das Feuer „ein halbes Jahr lang loderte". Neben zahlreichen griechischen sollen auch zahllose Texte weit früherer Epochen vernichtet worden sein.

Von den Texten der *Siddha-* und *Kalpa-*Systeme sind heute etwa 200 ‚offiziell' (dh. zum Teil nur in diversen ‚Schulen' und Familientraditionen, aber dennoch zugänglich und einsehbar) überliefert, davon etwa ¼ nicht nur handgeschrieben, sondern auch gedruckt. Der Umfang der Texte liegt zwischen nur wenigen bis zu einigen tausend Versen, abgefasst meist in tamilischer Sprache (zB. *Balagiraga Dosham* – 8 Verse; *Virinun Muvlakaikandam Venba* – 21 000 Verse). Von einem großen Teil der Texte liegen (zumindest teilweise bzw. zumindest bemüht) englische Übersetzungen vor, welche hauptsächlich während der letzten 120 Jahre angefertigt wurden. Von einigen weiteren hundert unveröffentlichten Texten wird gesprochen.

தாதுநட்டம் பேதி சருவாசி யஞ்சிரநோய்
ஓதுசுவா சங்காசம் உட்கிரணி—வேதோ
டிலக்காய் வரும்பிணிபோம் ஏற்றமயல் பித்தங்
குலக்கா யருந்துவர்க்குக் கூறு.

Tamil-Vers (gedruckt) zu einem Myrrhe-Präparat

Vergleichbar mit anderen ayurvedischen Traditionen spielen die Beziehungen zwischen Pflanzen und Mineralen eine große Rolle. Pflanzen mit einem bedeutsamen Gehalt an bestimmten Metallen nehmen besondere Positionen ein: *Bhrin-garaj (Eclipta alba,* tamil: *Karisalai)* und das Myrobalan *Haritaki (Terminalia chebula)* wegen des Eisens; Plantanen, *Aspera, Marselia, Nymphae, Lotus* u. a. wegen des Kupfers; manche Bodenkriecher, Rinden, Gräser und Weißes Süßholz wegen des Bleis usf. Von wieder anderen Pflanzen heißt es, sie enthalten gewisse mineralische ‚Essenzen', wie zum Beispiel die ‚Essenz des Salzes', welche im Stechapfelsamen *(Datura stramonium), Gokshura (Calotropis gigantea),* Sesamsamen und in der Kokosnuß in hohem Maße anwesend und verfügbar ist.

Aufgrund solcher und anderer besonderer Eigenschaften eignen sich entsprechende Pflanzen u.a. auch als Zusätze für Aufschließungsverfahren von Mineralien und Metallen. Die Entsprechungen und Zuordnungen stimmen dabei im Wesentlichen in den klassisch ayurvedischen und in den tamilischen *Kalpa-*Texten überein. Oft beziehen die *Siddha-*Praktiker ihr Kräuterwissen ergänzend aus ayurvedischen und *Kalpa-*Texten (bzw. mündlichen Traditionen).

Diese Metallbeziehungen der Pflanzen drücken sich dann in den Texten etwa so aus:

‚Jene Pflanzen, welche dabei helfen das Quecksilber zu figieren sind: *Paliruli, Akasha-Garuda, Azhukanni, Sunankan, ...*

Devathali (Cedrus deodora) calciniert das Quecksilber. *Adhalai (Jatropha,* div. Arten) tötet das Quecksilber. *Kat-ralai* reinigt das Quecksilber. *Kal-chembu* verwandelt das Quecksilber in die Universalmedizin.

Irumbili, Kattu-karunai und *Pey-Peerkku* destillieren das Öl aus dem Schwefel. *Karu-umathai* und *Sevakanar-kilangu* figieren den Schwefel.

Serunthi und *Karunkodiveli* figieren Arsenic. Gelber Arsenic wird aber besonders durch *Purangainari* und *Surapunnai* figiert, während *Thinnai-poo* den roten Arsenic figiert.

Vetrilai, Naga-singi und *Naga-thali* figieren Zink. *Mudhalai* schmilzt Eisen. *Muthiar-kundal* und *Saranai* calcinieren schwarzes Blei, während *Attu-chevimarudhu* weißes Blei calciniert. *Pon-avarai* bringt das Blei in ein rotes Oxyd. *Nagapalai* bringt aber das Zink in ein rotes Oxyd.

Pirandai calciniert Diamant. *Maramaram* figiert Kampher. *Kala* figiert Pottasche.

Pey thuvarai destilliert das Öl aus Auripigment. *Manchevi-kalli* verändert Auripigment in ein Wachs. *Ekai* bringt Auripigment in ein rotes Oxyd. *Venpathiri* figiert Auripigment.

Vellai-kilu-kilupai (Dicotyledons polypetale discifl.) und *Chev-arali* verwandeln Gold in Kupfer. *Vanni* und *Maruthontri* verwandeln Eisen in Kupfer. *Kal-thamarai (Dic. pol. calycifl.)* verwandelt andere Metalle in Kupfer. *Chenthirai* verwandelt Quecksilber in Kupfer.

Vellai-neermulli hilft dabei, aus dem Kupfer die Universalmedizin zu bereiten. In der Präparation der Universalmedizin ist aber auch *Uroma* von hohem Nutzen. *Mu-vilaikurundhu* dient zur Bereitung des Steins der Weisen. etc.

(nach *Bogars* Notizen zur *Kalpa*-Medizin. *Bogar* wirkte als Alchymist indischer Tradition in China, 16.Jhdt.)

In den Texten aller dieser Systeme finden wir meist nur kurzgefasste Verse, die ohne zusätzlicher mündlicher bzw. praktischer Lehre nicht zum Nachkochen geeignet sind. Manche ‚Kurzfassung' wichtiger Prozesse erinnert an die Schreibweise des *Paracelsus.* Zu den urzeitlich anmutenden Wörtern passt sehr gut der ebenso urzeitliche Eindruck der alchymistischen Praxis. Auf die Frage nach seinem Labor, gestellt an einen bedeutenden ayurvedischen Alchymisten von hoher Autorität, erhält man die Antwort: „Im Garten hinter dem Haus liegen einige Ofensteine." Und bis auf wenige Gläser und Töpfe im Schrank gibt es tatsächlich nicht mehr zu sehen.

Je höher die Kundigen ‚in der Kunst stehen', umso mehr legen sie Wert darauf, dass nur mit Hölzern, Dung oder Kohlen geheizt wird, und sie meiden bewusst die ‚Einfachheit' von Elektro- oder Gasöfen. Nicht dass sie diese Geräte nie ausprobiert hätten, aber sie erkannten sie als ‚für zu viele Arbeiten' ungeeignet, besonders dort, wo konstante Hitze erforderlich ist. Sie sind glücklicherweise noch im Besitz eines Wissens über das ‚Feuermachen', über die Qualitäten verschiedenster Hölzer u. dgl., was in der abendländischen Alchymie noch keine Renaissance erfahren durfte. Ein Wissen, das in den ältesten abendländischen Traditionen, wohl auch in den jungsteinzeitlichen und druidischen sicherlich vorhanden war. Für den Beobachter nachvollziehbar erweisen sich tatsächlich gewisse ‚hohe' Arbeiten, etwa Präparationsweisen von Quecksilber, nur mithilfe einer Feuerung mit bestimmten Hölzern als durchführbar.

Calcinierofen mit Kurbel-Gebläse

Die laborantischen Gefäße sind fast ausschließlich keramisch und unglasiert. Selbst subtilste Substanzen werden darin ohne Verlust destilliert, sublimiert und calciniert. Glasgerätschaft ist oft teuer und schwer zu beschaffen, oft auch ungeeignet.

Manches, was wir für die ‚typischen ayurvedischen Calcinationen' halten, ist vielmehr ein Sublimieren und Circulieren, je nach Größe und Bauart der Gefäße und nach Temperaturführung. Bei solchen Arbeiten wird der Spruch *Flamels* verständlich, nämlich dass „alles Destillieren, Sublimieren, Kohobieren, Calcinieren usf. ein und das selbe sei". Dies betrifft insbesondere wiederum die ‚höheren' Werke.

Quecksilber, der Same Shivas

Als ein Meisterwerk wird die Fixation (,Stabilisation') des Quecksilbers (von Zinnobermineral ausgehend) betrachtet, wodurch dieses dann zur medizinischen Präparation tauglich ist und nach Wunsch und Vermögen auch zum Großen Werk geführt werden kann. Die ,einfachste' Variante ergibt ein ,feuerbeständiges' Quecksilber-*Bhasma*, meist von leicht rötlicher bis braunroter Farbe. ,Feuerbeständig' bedeutet ,Unveränderlichkeit bis mindestens zum Schmelzgrad des Goldes'.

Vom südlichsten indischen Subkontinent bis ins tibetische Hochland und in die Mongolei finden wir verschiedene Präparationsweisen des Quecksilbers. Darunter einige, den abendländischen Traditionen sehr ähnliche Verfahren (*Sindhur*, phil. Zinnober, phil. Lasur, phil. Merkur etc.).

Die ,tibetischen Juwelenpillen' *(Rinchen rilnag)* welche solche *Bhasmas* enthalten, sind nach *Orgyenpa Rinchenpal* (1229-1309) benannt, welcher der Erzählung zufolge von seinen Reisen nach Nordindien/Pakistan *(Uddyana/Swat)* einige dieser Verfahren nach Tibet brachte.

In streng geführten Traditionen führen die Adepten Stabilisation und weitere Präparationen des Quecksilbers während einiger Monate (oder Jahre) sexueller Enthaltsamkeit aus und lernen dabei den ,Samen *Shivas* zu begreifen und zu beherrschen.' Das ist nicht notwendigerweise mit religiösen Praktiken verbunden, sondern eher eine Art Disziplin zum tieferen Verstehen der Naturkräfte.

Shiva Statuette

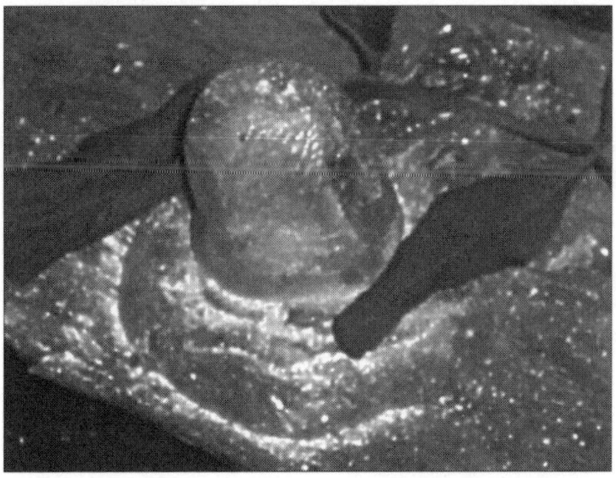

Shiva-Lingam aus figiertem Quecksilber geformt

Das Quecksilber wird aus Zinnobermineral destilliert/sublimiert und rectificiert. Die Schwärze, welches sich in der ersten Destillation/Sublimation bei hoher Temperatur bildet, wird in die Rectification mit eingeschlossen.

Danach folgt die philosophische Reinigung durch Triturationen mit Zusätzen und dann mindestens 3x7 Destillationen/Sublimationen/Calcinationen mit (oder je nach Weg auch ohne) Zusätze. Das 3x7 hat hier genau die selbe Bedeutung wie der Hinweis des *Paracelsus* „... dann fahre fort nach der Vorschrift des *Hermes*, um zum Stein zu gelangen". Damit bezieht er sich auf den Grundtext der *Tabula Smaragdina*, welcher ein Auf- und Niedersteigen der Materie fordert.

Es wird also das Quecksilber in zwei miteinander lutierten, irdenen Gefäßen zuerst 7x von unten nach oben und dann 7x von oben nach unten ‚gesandt'. Die dritte 7 bezieht sich entweder auf eine eigentliche Destillation – also seitwärts-Bewegung oder auf wiederholte vertikale Bewegungen (vergleichbar dem „Flug der Adler" in den Anleitungen *d'Anvers, Starkeys, Philaleta* u.a.). Um bei der Abwärtsbewegung ein Zerspringen des unterirdischen Gefäßes zu vermeiden, wird dieses nicht gekühlt, sondern erwärmt.

Seitliche Sublimationen (Destillationen) oder weitere Auf- und Ab-Bewegungen können mit im Quecksilber gelösten Gold, Silber, Rubinen und dergleichen fortgeführt werden. Ein Zeichen, dass das Werk gelingt ist, dass der *Mercurius* durch den Zusatz nicht mehr an Gewicht zunimmt und dass der Zusatz mit nur wenig Rückstand mitgeführt wird. Wenn also in 100g präpariertem Quecksilber 2g Gold gelöst werden, dann behält die Substanz ihr Gewicht von 100g und wird nicht um 2g schwerer. Wir haben dieses Phänomen schon gelegentlich angedeutet, wenn wir über die Position des Mondes in der pentagrammatischen Darstellung der Elemente, also über die Position des ‚hohen Merkurs' gesprochen haben. Es eröffnet sich hierin das Thema der Kolloidalität, welche die eigentliche physikalische Qualität alles Lebendigen ist. Fast alle Körpersäfte befinden sich in diesem Zustand und folgen darin nicht den ‚äußeren' Gesetzmäßigkeiten, was deutlich macht, dass ein mechanistisches Weltbild nicht zum Verständnis des Lebendigen geeignet ist und die stofforientierte Chemie in der lebendigen Realität versagen muss. Es wird auch ein Mensch nicht 1kg schwerer, wenn er einen Liter Wasser trinkt. Darin liegt auch eine Erklärung der Wirkart metallischer *Bhasmas*, in welchen metallische Teilchen zwar physisch anwesend, ihre Oberfläche aber um ein Vielfaches vergrößert wurde. Weiter wollen wir hier auf diese Thematik nicht eingehen und sie nur als Hinweis setzen. Das Bild des *Shiva* gibt einen Eindruck jener hermetischen Konstellationen, welche wir in besagter elementarer Position vorfinden.

Himmel – Äther
Eisige Berge – Kälte/Dunkelheit – ‚Hohes Wasser'
Aus dem Haarschopf entpringt der Ganges-Fluss, welcher abwärts fließt und als mondenhaftes Wasser zu venerischem Wasser wird.
(An der Seite Shivas – nicht im Bild – sitzt Parvati)
Schädelschale – Mond

Klangäther (Shabd-Akash)
Mond / ‚Hoher' Merkur /
‚Hohes' Wasser

‚Unteres Wasser' /
venerisches Wasser
Bild: in Flußrichtung sitzt
Parvati, Shiva's Frau

Shiva-Lingam (Phallus-Symbol) mit Blüte an der Spitze.

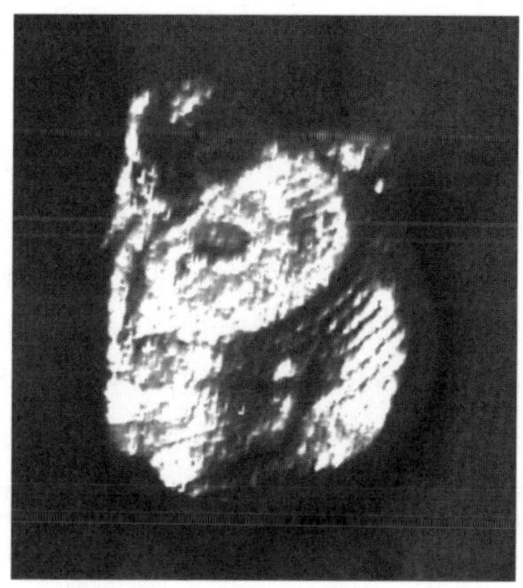

Stabilisiertes Quecksilber, knetbar – ein „Wasser, das die Hände nicht netzt" – flüssig und doch fest – ein ‚merkurialer Drache', welcher Gold, Silber, Rubin und Diamant ‚vollständig verschlingt', sodass weder Gewicht noch Rückstand verbleiben.

Zur Präparation der Minerale, Edelsteine und Metalle

Allgemeines

Wir widmen uns nun einigen typischen Präparationsweisen der ‚abendländischen' Alchymie. Manche Minerale werden calciniert und danach mit einem geeigneten *Menstruum* übergossen, welches ihre Essenz oder Tinktur extrahiert. Andere Minerale bedürfen keiner Calcination und ihre Aufschließung wird durch eine Art *Putrefaction* vorbereitet. Nachdem dann das *Menstruum* wieder abgezogen worden ist, süßt man das ‚Roh-Magisterium' ab.

Roh-Magisterium vom Smaragd

Magisterium absüßen

Darauf folgen weitere Reinigungen, Separationen, Subtilisationen und dergleichen mehr, bis man schließlich eine Tinktur oder Essenz bzw. ein ‚*Magisterium*' in trockener Form zB. aus dem Edelstein erlangt hat. Die Metalle erfordern aufwändigere Arbeiten,als Edel- und Halbedelsteine und ergeben eine Vielzahl verschiedenartiger Produkte, deren Bezeichnungen bzw. Betitelungen oft terminologisch streng gebunden sind. Aus Edelsteinen wird zumeist ein *Magisterium* dargestellt, aber auch ‚Tinkturen' und ‚Essenzen'. Höhere Arbeiten ergeben zB. eine ‚Quintessenz', ein ‚Fünftes Wesen', ein ‚Erstes Wesen' eine ‚Wahre Tinctur' usf.. Bei den Scheidungen wird, wie im Pflanzenreich, nach Prinzipien oder Elementen geschieden. Kiesel und Kalke eignen sich zu höchsten alchymistischen Präparaten.

Einen Kristall in Wasser legen und die Mischung sonnen hat mit dieser Kunst genausowenig zu tun wie den Stein, das Gold oder dergleichen 7 x zu glühen und abzulöschen und dieses Wasser dann für ein ‚alchymistisches' zu halten.

Paracelsus unterscheidet zwei Wege zum *Magisterium* aus Kalken und Edelsteinen.

Der erste betrifft Kalke, insbesondere Perlen, Perlmutt, Korallen, Muschelschalen und dergleichen. Hier ist in der Regel keine Calcination von nöten:

> „Lege gestoßene Perlen in einen korrigierten Essig und lass lösen, bis alles ein Wasser geworden ist. Dann scheide den Essig wieder davon, so hast du das *Magisterium Perlarum*."

Dazu sein Zusatz:

> „Obwohl das nur schlichte Stücke sind, so glaube einem Erfahrenen: Es sind wunderbare Wirkungen der selben. Denn die Wirkung dieser Tugenden erfolgt nicht durch die Kunst, sondern liegt in der Natur dessen, was in der groben Substanz verborgen ist. Als ‚toter' (roher) Körper kann sie nicht wirken, sondern erst in dieser *Resolution*."

Der zweite Weg betrifft die eigentlichen ‚Edelsteine' und erfordert eine vorhergehende Calcination:

> „Calciniere und Reverberiere die Steine zusammen mit dem gleichen Teil gemeinem Salpeter. Lass es auf diese Weise wohl zu einem ‚Kalk' brennen. Daraus wasche mit gebranntem Wein alles aus und siede den Wein ein zu einem Alkali. Das löse in einem Wasser auf."

Mit dem Zusatz:

> „Achte darauf nur ein Nichts davon einzugeben, weil es scharf und kalkig ist. Es ist dieses aber von solcher Art und so subtil, dass es seine Schärfe nur dem äußert, das ihm wiedersteht. Ein einziger Tropfen tingiert den ganzen Leib mit besonderen Tugenden."

In der Vorschrift zum Perlen-Magisterium lässt *Paracelsus* das ansonsten übliche Absüßen mit destilliertem Wasser und Feinscheiden mittels starkem Alkohol aus. Aus dem Edelstein beschreibt er ein vermischtes Präparat, worin der Salpeter noch enthalten ist. Folgt man diesem Weg der Calcination, dann wird danach in der Regel sorgfältig abgesüßt und wieder mit starkem Alkohol feingeschieden.

Andere beliebte Calcinationszusätze sind Schwefel, Weinsteinsalz, Steinsalz, Kräuter oder deren Vermischungen.

Aus in Wasser gelösten, alkalischen Magisterien kann mit einem sauren Fluid (zB. destilliertem Essig) eine ‚Essenz' ausgefällt werden. Das wäre dann also ein noch höheres Produkt, als das *Magisterium*, allerdings eben nur, wenn diese ‚Essenz' auf diesem Wege durch Präcipitation erstellt wurde.

Ähnlich werden auch Edelsteine, Kiesel, Sand u. dgl. mittels alkalischen Salzen, insbesondere Weinstein oder Borax, zum Schmelzfluß erhitzt, ausgegossen, abgekühlt, gemörsert und so wird mittels Extraktion und Feinscheidung durch hochprozentigen Alkohol ein (gemischtes) Präparat erlangt. Daraus kann wieder eine ‚Essenz' ausgefällt (und mit etwas Geschicklichkeit sogar zur Quintessenz sublimiert) werden.

Nach paracelsischer Tradition wird aus Perlen und Edelsteinen eine ‚Quintessenz' erlangt, wenn man diese (nach entsprechenden Vorarbeiten) in einem geeigneten *Menstruum* (z.B. Wurzelessig) löst und das eingedickte und getrocknete Extrakt an der Luft ‚fließen' lässt. Unter Einfluss der Luftfeuchtigkeit scheidet sich mit der Zeit das Quintessentielle (zumeist mit der eigentlichen Farbe des Steins oder einer ‚innewohnenden Farbe') vom erdhaften *Magisterium*. („In der Farbe ist die *Quinta Essentia*, und das *Corpus* bleibt hinten.") Diese Arbeit ist schön, spannend, zeitaufwendig und gibt äußerst wenig. *Agricola* empfiehlt dasselbse Verfahren, um ‚die Röte' aus dem Bergkristall zu erlangen.

Als *Menstrua* dienen zwar oft allerlei saure oder basische Wässer und Salze, genaugenommen darf aber der Begriff *Menstruum* nicht einfach als ‚Lösungsmittel' verstanden werden, sondern hat eine tiefreichendere Bedeutung. Wir haben darüber bei den Themen des ‚philosophischen Todes' bereits einiges angezeigt. Es soll die Heilkraft aus dem Mineral ‚erlösen', ohne sie zu beschädigen oder zu corrumpieren. Dazu sind ‚scharfe Säuren' nicht unbedingt geeignet, und das noch viel weniger, wenn diese aus großindustrieller Fertigung stammen, also nicht auf ‚kunstvollem' Wege gewonnen werden. Ein *Menstruum* sollte lebendigen Charakter haben und im Idealfall genau für das jeweilige ‚Zielobjekt' geeignet sein. Selbst starke *Menstrua* für Metalle (wie zB. das ‚Königswasser') sind nicht einfache Zusammenschüttungen konzentrierter Säuren, sondern eben für sich schon alchymistische Produkte. Begehrte und ‚geheime' *Menstrua* können sogar völlig neutral (‚wie ein einfaches Wasser'), also ohne Schärfe (‚nicht corrosivisch') sein, in ihrer Wirkung aber die starken Säuren oder Laugen weit übertreffen. Ein sehr einfaches Beispiel dazu ist die gemeinsame Präparation von Essiggeist, Weingeist und Uringeist, entsprechend der elixirischen Verbindung der drei Prinzipien Sulfur, Merkur und Sal zu einer Art ‚*Alkahest*'.

„Wer nun solche Arzneien (Edelsteine – Anmk.) gebrauchen will, der nehme sie nicht pulverisiert, sondern dass sie in eine *Essentiam*, oder doch zum wenigsten schönes weißes Salz (=*Magisterium* – Anmk.) gebracht werden, welches sich in einem jeden *Liquore* resolvieren lässet. Dieses kann sich hernach in *Succum et Sanguinem* vertieren (‚in Saft und Blut verteilen', dh. es wird vom Körper absorbiert) und der schwachen Natur zu Hilfe kommen, denn die Edelgesteine haben die Tugenden, welche die Metalle haben, denn sie kommen von einem *Principio* her, und ist ihre Tugend kein falscher Wahn, wie etliche davorhalten."

„Wenn sie nun sollen recht bereitet werden, (...)". *Agricola* beschreibt nun die oben erwähnte Calcination bzw. Reverberation mit Zusätzen. „Wenn ich sie nun in ein Salz oder auch wohl *Essentiam* bringen will, so solviere ich diese calcinierten oder reverberierten *Gemmas* in einem *Spiritus Nitri*, und wenn sich alles solviert, so ziehe ich den *Spiritum* wieder davon und procediere mit destilliertem Regenwasser mit Solvieren und Calcinieren, bis ein schönes Salz, wie eine Cristalle hell und durchsichtig, daraus werde. Alsdenn reverberier ich dasselbige in einem verschlossenen Glase, so bekömmt der Stein seine natürliche Farbe wieder, welche hernach mit dem *Spiritu Vini alcoholisato* muss extrahiert werden, aber wie gesagt, dieser (*Essentiam* – Anmk.) ist so wenig, dass man sich drüber verwundern muss."

Johann Agricola, >Chymische Medicin< *Tractatus de Gemmis*, S1227

Spritus Nitri ist eines (nicht nur) von *Agricola* gerne verwendeten, einfachen *Menstrua*, welches durch kunstvolle Destillation, Kohobation und Rectification des einer ‚Salpetergrube' entnommenen oder von mit Salpeter durchsetzten, gekalkten Wänden eines Stalles geschabten und durch Kochen und Umkristallisieren gereinigten Salpeters erlangt wird.

Ein anderes beliebtes *Menstruum* ist der *Spiritus Urinae*, dessen Präparation unter anderem in den >Alchymistischen Impressionen<, S.131 ausführlich beschrieben wird. Damit erlangt *Agricola* eine Art *Spiritus* aus den Edelsteinen, welches also der Darstellung des Merkur-Prinzips gleichkäme und deshalb verständlicherweise von ihm als ‚Tinctur' bezeichnet wird:

„So man einen *Spiritum Urinae* über die calcinierten Gesteine geußt, so extrahiert er *in Digestione* ihren *Spiritum* auch sehr schön aus, ein jeder Edelgestein nach seiner Farbe. Was aber weiche Steine sind als Smaragd und dergleichen, die soll man nur stoßen und auf einem Reibestein ganz subtil abreiben und den *Spiritum Urinae* darübergießen, so bekömmt man eine schöne grüne *Tinctur*, die selbe kann man hernach mit *Spiritu Vini* rectificieren und lieblich machen. (...) und diese Präparation ist wohl die beste."

Beide Bezeichnungen, ,Essenz' und ,Tinctur' finden wir in *Agricolas* nächstem Verfahren sinngleich vermischt:

„Etliche solvieren auch die Edelgestein in einem *Spiritu Salis* und bringen sie hernach *lege Artis* ferner in ihre *Essentiam*, welcher Modus auch nicht zu verachten, und gibt auch eine schöne *Tinctur*. Ich habe gesehen, dass zu Prag einer die Edelgesteine erstlichen zerstoßen in einem eisernen Mörsel, hernach auf einem Reibstein mit sublimierten *Salmiac*, und dessen drei Teil zu einem Teil getan, wohl gerieben und hernach mit starkem Feuer aufsublimiert, so ist in der dritten Sublimation die *Tinctur* sehr schön aufgestiegen. Von dieser hat er den *Salmiac* mit warmen Wasser gelauget und wohl abgesüßt, so hat er eine schöne *Tinctur* bekommen.

Und dergleichen Mittel hat man mehr, aber wie schon gesaget, so werden sie selten zugerichtet, weil sie so teuer zu erlangen sind, doch kann man bisweilen bei den Kaufleuten, welche in Indien segeln, solche auch um einen billigen Pfennig erlangen."

Dass in obigem Beispiel von einer Aufschließung der Edelsteine mittels *Spiritus Salis* gesprochen wird und dem dann die Beschreibung einer Arbeit mit *Salmiac* folgt, hat den Grund darin, dass die Alten ihren *Salmiac* (oder: *Sal Harmoniac*) durch Einkochung und Sublimation von Salz, Russ und Urin zubereiteten. Daraus steigt in der Sublimation zuerst der eigentliche *Salmiac* auf. Danach folgt in der Destillation bei hoher Temperatur der ,saure *Spiritus* des *Salmiacs*', als eine Art *Spiritus Salis*.

Qualitätskriterien

Die kommerzielle und damit breitgestreute therapeutische Nutzung von Edelsteinessenzen hält sich in Grenzen, zum einen wegen der oft teuren Anschaffung der rohen Steine, wegen der aufwendigen Verfahren und nicht zuletzt auch wegen vielfacher Ignoranz und Misstrauen diesen wertvollen Präparaten gegenüber – einem allgemeinen Unverständnis dieser Dinge entspringend oder durch mangelnde Unterscheidung diverser ‚Edelstein-Mittel' bedingt.

Hierin liegt sicherlich eines der bedeutsamsten Gebiete, worin die erst in jüngster Zeit umfassender bekannten asiatischen Präparate (zB. *Bhasmas*) sowohl die abendländische Alchymie als auch die naturheilkundliche Therapie positiv anregen, ja zum Teil aus einem langen Schlaf erweckt haben.

Die Qualitäten der so verwendeten ‚spagyrischen Edelstein-Präparate', wie immer diese auch letztendlich tituliert werden, sind durch mehrere Kriterien gekennzeichnet:

a) Die Qualität des Ausgangsmaterials: Bei den typischen Edelsteinen wie Rubin, Smaragd, Saphir u. dgl. gibt es nicht nur verschiedene Reinheitsgrade, sondern auch die Möglichkeit, Minerale zu verwenden, welche diesen Steinen ähnlich sind. Formal schöne Formen sind nicht notwendig, stellen aber dennoch eine eigene Qualität dar. In der Regel werden unbearbeitete Steine bevorzugt, welche nicht zB. durch Schleifen ‚in sich erschüttert bzw. traumatisiert' sind. Was die energetischen und feinstofflichen Traumata der Steine betrifft, so spielt dabei die Art ihres Abbaus eine Rolle. Sie können in kleinen Mengen gesucht, gefunden und mit Liebe und Dankbarkeit ‚geerntet' werden. Sie können aber auch durch Raubbau und großangelegte Sprengungen gerafft werden, wie dies zur Zeit etwa bei nordindischen Rubinen und ähnlichen Mineralen der Fall ist.

   Halbedelsteine, Quarze u. dgl. (Bergkristall, Granat, Rosenquarz) sind meist günstig und direkt aus den Bergen zu bekommen. Hier gelten wieder die Haltung bei der ‚Ernte', die stoffliche Reinheit der Steine, ihre Formen, Kristalle und Farben als Kriterien zur Qualitätsbewertung der Präparate.

   Perlen, Korallen, Muscheln und Schneckenhäuser gibt es in zahlreichen Varianten und Qualitäten. Iatrochymiker in Ost und West schätzten unter den Korallen die rote Edelkoralle am höchsten. Heute mag man bei ihrer Verwendung Bedenken haben, ob man dadurch wohl in die Zerstörung der Riffe eingebunden wäre. In diesem Falle kann man seine Korallen-Präparate auch in sehr guter Qualität aus Schaumkorallen zubereiten. Andererseits gehören Heilkundige und Spagyriker zweifellos zur Elite der eigentlichen ‚Naturschützer' und erhalten von Natur aus höchste Kompetenz, ihre ‚Dinge' zu verwenden.

b) Das Aufschließungsverfahren: Von einfachsten, gerade noch ‚spagyrischen' Verfahren bis hin zu hohen alchymistischen Werken gibt es sehr viele Wege die Heilkräfte des Mineralreichs zu erschließen. Aus diesen Graden und der Kunstfertigkeit ergeben sich wiederum Qualitätsunterschiede. Ebenso aus der Art der verwendeten *Menstrua*.
c) Das fertige Produkt: Eines der wesentlichsten Qualitätskriterien ist selbstverständlich die Menge von Essenz, Tinctur oder *Magisterium*, welche im ‚Produkt' enthalten ist. Da die puren Essenzen, Magisterien u. dgl. im Verhältnis zur Menge des Ausgangsmaterials nicht viel sind, werden sie kaum in purer, unverdünnter Form einen Weg in den Handel finden. Verdünnungen zu gleichen Teilen oder bis 1:5 in Wasser-Alkohol-Gemischen sind vertretbar und waren auch bei den Alten üblich. Der berühmte ‚Tropfen auf den Liter' tut zwar auch ‚etwas', gibt zumindest ein informiertes Wasser, stellt aber kein eigentlich-typisches ‚spagyrisches Präparat' mehr dar (außer es ist in einer spagyrischen Rezptur ebendieses Verhältnis so vorgeschrieben). Die Weiterverdünnung zu gleichsam homöopathischen Mitteln ist aber ein anderes Thema und korrespondiert mit einer bedeutenden zukünftigen Entwicklung der Homöopathie selbst.
   Flüssige ‚Edelsteinessenzen' sind fast ausschließlich vom Sulfur des jeweiligen Steines gefärbt. Das heißt, dass auf ein Fläschchen mit farblosem Inhalt das Etikett einer spagyrischen Edelsteinessenz nur irrtümlich geklebt worden sein kann (es sei denn, das Verdünnungsverhältnis ist angegeben). Das gilt übrigens auch für die berühmten *Aurum potabiles*, auf die wir noch zu sprechen kommen werden. Schon *Paracelsus* bemerkte dazu, dass dies ‚ein Sulfur' sei und sich durch seine Farbe auszeichnet. Und wo keine Farbe, dort kein *Potabile*.

Die Bedeutung mineralischer und metallischer Präparate in der Naturheilkunde hat nun noch einen ganz besonderen Aspekt mit unangenehmem Hintergrund: Ein Sprichwort besagt, dass ‚vor dem Sonnenaufgang die dunkelste Stunde ist.' Es kann sein, dass sich dieses Sprichwort noch in den nächsten Jahren bestätigt durch die ‚Technik der genetischen Verwirrung', welche vielleicht die hinterhältigste aller von *Paracelsus* angekündigten ‚vielen neuen Krankheiten' ist. Sie breitet sich aus durch Dummheit, Gier und Verantwortungslosigkeit oder gar durch absichtlichen Zerstörungswillen. Das kann für die Pflanzenheilkunde bedeuten, dass in fünf oder zehn Jahren etwa das Johanniskraut nur mehr dem Namen nach besteht, in Wahrheit aber eine Mixtur aus Perforata-, Kartoffelstärke-, Schlingpflanzen- und Maikäferbein-Genen ist, mal gelb, mal blau und meist gar nicht blühend. Hier geht die Entwicklung schnell, und die Seuche greift um sich.

Das Mineralreich wird bislang von Ähnlichem nicht gestört (abgesehen in strahlenverseuchten Gebieten), Edelsteine und Metalle stehen dem Menschen heute wie eh und je zur Verfügung und werden auch in kommenden Zeiten vermutlich unverändert bleiben. Dass zur Zeit eine intensive Renaissance mineralischer Heilmittel stattfindet, könnte bereits als Hilfeleistung der Natur oder sonstiger wohlmeinender Kräfte gedeutet werden, damit die Menschen der nächsten Jahrhunderte nicht ganz ohne die Heilkraft ‚natürlicher Dinge' sein müssen, abgesehen von den wenigen und schwierig zu beschaffenden, zur Arznei noch tauglichen Pflanzen. Der Gedanke ist klar – wir wollen ihm gar nicht zu sehr folgen, denn immer wieder hat die Menschheit großes Glück, und diese Dinge müßten nicht noch mehr, als ohnehin schon geschehen, eintreten. Nichtsdestotrotz kann auch auf diese besondere Bedeutung mineralischer und metallischer Heilmittel hingewiesen werden.

Die Präparation von Bernstein und Koralle mittels Destillation

Der Bernstein (Agtstein) wurde von den Alten in die Reihe der ‚Edelsteine' und ‚Gemmen' gezählt, ähnlich wie auch Perle und Koralle. Seine Affinität zur roten Koralle veranlasste dazu, die beiden mittels Destillation gemeinsam zu präparieren. Alle, die diese Verfahren beschrieben (*Agricola, Crügner,* usf.) schätzten die Produkte (Öle, Balsame, Salze, Teer, Wurzelessig) zur innerlichen oder äußerlichen Anwendung sehr hoch. Nachfolgends ein Text nach *Michael Crügner* zur Bernstein-Arbeit:

„Was nun die *Praeparationes* anlanget, so verfahre ich auf diese Weise: Ich nehme feine, reine Agtstein-Bruchstücke, tue solche in eine dazu bequeme Retorte, richte solche ins Sandbad und wende Fleiß an, dass der Retorten Hals fein hervorlange. An solchen Hals lege ich einen ziemlich dazu bequemen großen Kolben, verlutiere solchen nicht, auf dass ich ihn abnehmen kann so oft ich will und schütte um die Retorte herum soviel Sand, dass er eines Daumens hoch über den Agtstein gehe.

Alsdenn so mache ich ein Feuer im Ofen. Wenn nun in der Retorte der Agtstein zerschmelzet, so erzeiget sich alsbald im Retortenhals ein *Phlegma* in Tropfengestalt und bald darauf im Halse weiße crystallinische Fahsen, welche bald größer, bald kleiner werden. Solche nun sind nichts anderes, als das flüchtige Salz vom Agtstein. Damit mir nun solches nicht verschwinde, so habe ich zur Hand einen guten, rectificierten *Acetum Principale,* und lange mit einem *Spatul* solches crystallinisch, flüchtige Agtsteinsalz aus dem Halse der Retorte heraus und tue solches in den *Acetum* mit solcher Geschwindigkeit, als möglich sein kann."

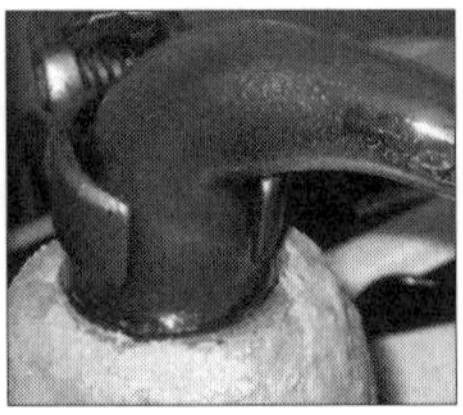

„Den Kolben lege ich wiederum an, denn es erzeigen sich alsobald die Tropfen, die des Agtstein-Öls. Weil die Retorte nicht über mit Sand bedeckt ist, so kann ich fein eben hineinsehen, wie der Agtstein zerschmelzet und ob er auch in die Höhe begehret, danach ich mich mit dem Feuer richten kann. Zu viel Feuer muss man nicht machen, denn sonst würde der Hals der Retorte zu heiß, und es verschwindet das herrliche und in der Arznei köstliche fliegende Salz.

Wenn man nun sieht, dass der Agtstein in der Mitte der Retorte gar einbricht, oder aller zerschmilzt, (denn erstlich zerschmilzt er nur am Rande herum), so kommt alsdenn das fliegende Salz haufenweise, derowegen mit dem Feuer sehr gemach muss getan werden. Und wenn ich nun sehe, dass umher im Hals der Retorte solch Salz sich angehänget, so nehme ich solches immer beizeit mit Hurtigkeit heraus und tue es zum ersten. Und fein geschwinde den Kolben wieder dran, damit das Öl nicht auf die Erde falle, sonderlich das erste, welches besser und subtiler ist, als das folgende (verstehet im Leib zu brauchen)."

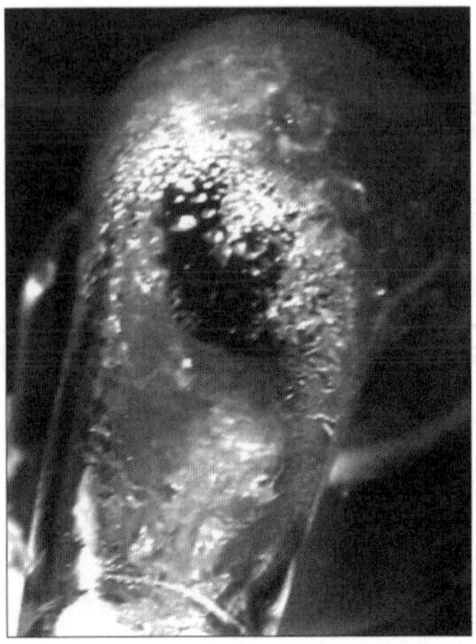

„Wenn nun aller Agtstein zerschmolzen, so hänget sich das fliegende Salz nicht mehr so weiß und crystallinisch an, sondern etwas bräunlich, und man sieht von oben hinein den Agtstein kochen wie siedend Wasser. Ich verfahre aber mit solcher Destillation und Sammlung des Salzes bis zum Ende.

Auf ein Pfund Agtstein nehme ich gemeiniglich 1 Loth *Spiritus Aceti Principalis* zur Hand. Solche zwei köstliche Arznei-Stück zerschmelzen untereinander in die schönste Röte, und es schadet nicht, wenngleich etliche Tropfen vom Agtstein-Öl, welche am Salze kleben, hinzu kommen. Die *Spiritus Principales* figieren das fliegende Agtstein-Salz, dass es nicht mehr in der Luft verschwinde, und wenn es eine Zeit miteinander steht, so crystallisiert sich solches Salz wie ein brauner Zucker-Candit. Was nun solcher *Liquor* oder solches figierte Salz nebst dem, dass es den Urin bewegt in *Podagra, Arthritica* und *Vertigine* verrichtet, wird ein jeglicher, so es recht wird anwenden, mir zu danken haben.

Aus dem rückständigen Agtstein in der Retorte kann ein fixes Salz bereitet werden, durch fernere *Calcination, Solution* und *Coagulationes*.

Das *Oleum*, welches schwarz-rot im vorgelegten Kolben erscheint, gieße ich in einen andern, dazu bequemen Kolben und tue zwei mal mehr Wasser, als des Öls ist, hinzu, setze solchen Koben ins Dampfbad und destilliere mit solcher Sorgfältigkeit, als hätte ich den subtilsten *Spiritum Vini* oder *Spiritum Astri Sulphuris Mercurii* vor mit, denn sobald nur das hinzugegossene Wasser ein wenig warm wird, so erscheinen die subtilsten Tropfen des Agtstein-Öls alsobald, und ich fahr mit solcher Destillation ferner fort, solange etwas von solch subtilen Ölen herüber steigt, und wenn es mit beliebt, so rectificiere ich solches herüber gestiegene subtile Öl noch einmal.

Was nun den Gebrauch dieses unseres herrlichen Agtstein-Öls belanget (dieweil wir den aus Indien und aus dem Fluß *Nilo* nicht haben), so ist solcher nach der Länge bei dem *Hartmanno* und *Crollio*, auch wohl anderswo zu sehen. Und soll sich keiner verwundern, dass in einem solchen, gering herkommenden Ding sollen solche herrliche und vielfältige Kräfte stecken. Aber man soll wissen, dass Gott in allen Seinen geringen Geschöpfen mächtig und kräftig ist, wenn nur wir Menschen nicht selbst übersichtige Herzen und Gedanken des Hochmuts und Verachtung geringer Dinge hätten, vielleicht gebe uns Gott auch noch bessern Verstand und Erkenntnis Seiner gering erschaffenen Wunderwerke.

Ich pflege auch in diesem Agtstein-Öl *Flores Sulphuris* oder *Lac Sulphuris* zu solvieren und solche *Solution* zum Unterschied des gemeinen Agtstein-Balsams dann *Balsamum Succini Sulphuratum* zu nennen, welcher viel kräftiger in der Operation trifft, als das Öl an sich selbsten, zur Zeit der Pestilenz, Hauptbeschwerung, Stein und Krampf, *Mensibus ciendis* und Fiebern, *item* Gelbsucht, *Colica* etc. Dessen Dosis von 8 bis auf 15 Tropfen.

Das schwarze Öl, so in der Rectification zurückbleibt, separiere ich vom hinzugetanen Wasser aufs fleißigste und verwahre es als eine Schatz in der *Chirurgia* zu gebrauchen."

„Es löscht nicht nur allerhand Brand und wenn es beizeiten geschieht, hilfts gleich, als wäre niemals ein Brand geschehen, sondern es heilt auch alle um sich fressenden Schäden am menschlichen Leibe, welche aus ihrem scharfen, resolvierenden, fliegenden Salze ihren Ursprung haben, denn dieses Öl ist als ein Magnet, oder magnetische Kraft und nähret und speiset sich von solch flüchtigen Salzen, die es begierig beraubet, wie oben in der Destillation benennet worden.

Und wenn in solchem schwarzen Agtstein-Öl *Flores Sulphuris* solvieret werden, wie insgemein bei der Präparation des Schwefelbalsams bekannt, so bekommt man gar einen vortrefflichen Balsam in der *Chirurgia* zu brauchen, welchen ich wegen seiner natürlichen Tugend halben *Balsamum Succini Magneticum Sulphuratum* nenne.

Oder auch zuzeiten, wenn ich die Destillation nicht bis zu solchem Ende bringe, dass ich das Öl gänzlich herübertreibe, sondern wenn ich merke, dass das subtilste Öl neben dem fliegenden Salz herüber ist, so lasse ich mit der Destillation nach. Was nun, nach Erkaltung, in der Retorte gleißend wie ein Spiegel zu finden ist, ist insgemein ein *Collophonium Succini*. Aber wegen köstlicher Tugend in der *Chirurgia* mag ich einen solchen tölpischen Namen auf solchem nicht sitzen lassen, sondern nenne es *Magnetem Succini*, weil es solche magnetische Natur an sich hat, über welches sich zu verwundern ist, nicht allein wie gesagt in offenen Schäden, in welchen ein fliegendes, corrosivisches Salz zugegen, sondern auch aus den entzündeten Schäden die Hitze heraus und an sich ziehen. Es erfordert aber hierzu eine *Experienz* (...).

Paracelsus setzt in *Archedoxeos Magicae lib. 1* ein *Oleum Corallorum Succini* wider das Herzzittern, wie ich denn oft gedacht, dass ich von diesen beiden, Agtstein und Corallen, gar sehr viel halte. Es lässt sich gar fein miteinander destillieren, und ich nehme mehrenteils *ana*. Alsdenn steigt das Öl gar schön rot herüber, denn das subtile, scharfe Salz im Agtstein nebst mit laufender Wärme und Hitze corrodiret die Corallen, dass sie weiß zurückbleiben. Es muss aber wie oben gesagt rectificiert werden."

nach: *Michael Crügner*, >Chymischer Frühling< 1654 (neu: Edit. *Hornfisher*)

Das Aurum Potabile

Metallische Spagyrika sind in asiatischen und westlichen traditionellen Systemen üblich und werden in vielen Formen und Qualitäten genutzt. Hierher gehören auch manche Meisterstücke der Kunst, unter denen es ein beliebtes und begehrtes ist, aus dem metallischen Gold ein *Aurum potabile* zu erlangen.

Produkte unter diesem klangvollen Namen findet man heute vermehrt im Handel. Die Qualitäten sind verschieden, und bisweilen findet man darunter auch gute, spagyrische Gold-Präparate, die auf einfachem Wege mithilfe industrieller Säuren und geeigneter Reduktionsmittel hergestellt werden.

Zum einen ergibt sich bei der Betrachtung dieses Themas die Frage: Gibt es viele verschiedene Wege verschiedenen Grades zu einem *Aurum Potabile*, oder gibt es nur einen einzigen Weg zum ‚wahren *Aurum Potabile* der Alten' bzw. des *Paracelsus?*

Daraus folgt: Was ist denn dieses ‚wahre *Aurum Potabile*', und wie stehen die anderen sogenannten Gold-Präparate im Vergleich dazu?

Nun ist hier vor allem *Agricola* (im *Tractato de Auro*) den Iatrochymikern seiner und der nachfolgenden Zeit terminologisch sehr entgegengekommen, war er doch zweifellos einer, der die *Termini* richtig, gültig und mit naturgegebener Autorität zu setzen wusste.

> „Ich habe sonder Ruhm *in Chymicis* soviel erfahren als mancher, der groß Prahlens und Aufschneidens machen kann, ich weiß gottlob auch wohl, was der Kunst möglich und wieweit sich ihre Operation erstrecket. Beschert dir Gott dieses Mittel, das *Aurum Potabile* zu erlangen, darfst du andern Prozessen nicht viel nachfragen, denn du wirst ihrer nicht viel bedürfen, sondern in kurzer Zeit und fast ohne Mühe eine *Panacaeam* zubereiten, welche den Kranken aus Nöten geschwind wird helfen können."

Mit „diesem, von Gott bescherten Mittel" meint *Agricola* hier die *Piemontische Magnesia*, ein ‚geheimer *Spiritus*', welcher aus der zum Junivollmond verfärbten Silberbergwerkserde erlangt wird. Diese Arbeit haben wir in den >Alchymistischen Impressionen< nach *Töldes* >*Coelum Reseratum*< wiedergegeben. *Agricola* sagt von diesem *Mercurius* und dem dazugehörigen Prozess:

> „Mit diesem kannst du das Gold meistern und in seine *Essentiam* bringen. Ich will nicht glauben, dass alle Scribenten soviel offenbaret als ich. Bitte Gott um glückliches Gedeihen, es ist deutsch genug, ich halte, es könnte ein Kind verstehen."

Danach aber kommt *Agricola* auf einen anderen Prozess, und zwar auf den, seiner Ansicht nach allereinfachsten ‚auch ein *Aurum Potabile*' zu erhalten.

„(...) und weil ich ihn in seiner Wirkung so gut befunden, hab ich nicht unterlassen wollen, solche hieherzusetzen, damit ein jeder sicher danach arbeiten könne. Und obwohl die Präparation schlecht scheinet, so ist daran nicht gelegen und ist oft in einem schlechten Prozess mehr Kunst als an einem, der mit vielen *Sesquipedalibus* colorieret worden, und wenn man solchen danach auf die Probe bringet, so siehet man, wie er besteht.

Nun wollen wir zur Präparation kommen: Nimm ungarisch Gold (...) solviere es in einem rechten *Oleo Salis* (...). Wenn es solviert ist, ziehe das *Oleo Salis* stark davon, so bleibt ein schöner, gelber Goldkalk am Boden des Glases liegen. Nimm ihn heraus und tue ihn in ein Kölblein, geuß soviel Zimmetöl daran, dass er wie ein Brei werde, so wird das Gold alsobald zu brausen anfangen und schwarz werden. Gieß darauf einen guten *Spiritum Vini* alcoholisatum, dass er zween Querfinger über die *Materiam* gehe, so wird er alsbald eine *Tincturam* aus dem Golde ausziehen. Gieße den *Spiritum* ab und gieße wieder andern drauf, und dieses tue, bis sich keine *Tinctura* mehr erzeigen will. Das *Caput Mortuum* tue heraus, wasche solches mit warmen Wasser und wiege solchen. So gieße alsdenn wieder eine juste Quantität *Spiritus Salis* darauf und lasse ihn wieder solvieren. Ziehe den *Spiritum* ganz trocken davon, gieße *Oleum Cinnamomi* darüber, alsdenn wenn es brauset, gieße den *Spiritum Vini* auch darauf und extrahier seine Tinctur. Und diese Arbeit wiederhole so oft, bis das Gold alles ganz und gar solviert sei. Denn nimm den klaren, tingierten *Spiritum Vini*, ziehe ab bis auf das *Oleum*, so wirst du eine herrliche Medicin oder *Aurum Potabile* erlangen. Und die Wahrheit zu sagen, hat mir unter den gemeinen *Solutionibus* eine gefallen, so ist es diese, denn es seine *Operationes* gewaltig erweiset, (...).

Johann Agricola, >Chymische Medicin<, *Tractatus de Auro*,
S. 116 ff. in der Ausgabe von *Oliver Humberg*

Dieser Prozess wird in mancherlei Variation (etwa, indem anstatt des Zimtöls ein Zuckersirup oder indem durch Destillation eines *Spiritus Nitri* über *Sal gemmae* ein *Aqua Regis* zur ersten Solution des Goldes verwendet wird) heute gerne ausgeführt, und die anwendenden Therapeuten bestätigen das hohe Lob *Agricolas* für dieses ‚auch ein *Aurum Potabile*'.

Nun beschreibt *Agricola* noch einige Wege zu anderen, kunstvolleren Zubereitungen des Goldes. Dergleichen Wege finden wir in vielen alchymistischen Texten und es bleibt der Eindruck, dass es vielleicht nicht nur ein ‚einziges, wahres *Aurum Potabile*' gegeben hat oder gibt, sondern mehrere, zwar in Graden verschieden, doch allesamt bei rechter Ausarbeitung in höchsten Graden.

Irgendwo dann am Rande, wie eben viele wesentlichen Dinge ‚so nebenbei' angemerkt werden, setzt auch *Agricola:* In Wahrheit meine er nicht, dass das ‚wahre *Aurum Potabile*' aus Gold gemacht werde, sondern dass die Alten ein ganz anderes Ding damit gemeint hätten, ja dass dieses Mittel gar nicht vom Gold stamme oder durch dieses dargestellt wird. Hier spricht er scheints ein Geheimnis an, auf das er aber in der Folge nicht mehr weiter eingeht. Vielleicht hilft uns *Zoroaster* (1236) dabei, dieses Geheimnis zu verstehen:

„Mache aus dem *Phosporos* (Kupfer – Anmk.) mit Hilfe des Schwefels und des *Spiritus Vitrioli* ein Vitriol-Salz, und aus diesem verfertige ein Öl durch ein Salz.

Diese Worte sind also zu verstehen: Mache aus Kupfer und Schwefel durch Calcinieren ein Pulver, darauf gieße einen *Spiritum* und Öl, die aus einem ungarischen Vitriol bereitet worden. Der beste *Spiritus* und Öl wird aus der *Gur* gemacht. Die Bereitung aber dieser *Gur* ist folgende:

Man nimmt und stößt die *Mineram Vitrioli Hungarici* klein und extrahieret solche mit einem warmen Regenwasser, die *Solution* filtriert man durch, lässt sie bis auf ein Häutlein abrauchen und bringt diese evaporierte Solution noch warm in verschiedene flache, steinerne Gefäße, dass diese nicht höher darin zu stehen komme als eine gute Querhand hoch. Besagte Gefäße werden so gestellt, dass die Luft beständig darüber streichen möge. Jedoch decke sie bisweilen zu, damit nichts Unreines hineinfalle. Lass es also stehen, so wird der Vitriol über sich und an den Seiten der Gefäße mit allerlei Farben auszuwachsen beginnen. Diesen ausgewachsenen Vitriol nimm immer nach und nach ab, verwahre ihn in einem reinen Gefäß und continuiere damit, bis alles über sich ausgewachsen ist, so hast du die metallische *Gur*, welche du in gelinder Wärme trocknen und einen *Spiritum* und Öl daraus treiben kannst.

Aus dem Kalk oder *Caput Mortuum* extrahiere das Salz und reinige es *solvende et coagulando*. Destilliere den *Spiritum* und das *Oleum* auch herüber, dann gieße es auf das Salz, digeriere es miteinander und destilliere es auch zusammen *cohobando* herüber. Aus dem *ultimo Capite mortuo* mache wiederum ein Salz und zwar also:

Reibe, so schwer wie das *Caput mortuum* wieget, einen reinen Schwefel darunter und calciniere es solange, bis du siehst, dass es in dem Scherben zusammenlaufen oder schmelzen will. Wenn du dieses Zeichens gewahr wirst, so wirf es noch heiß nach und nach in ein destilliertes Regenwasser, da dann das innere Salz allesamt sich in das Wasser begeben wird. Dieses filtriere durch, damit es lauter werde, lass es zu einem Salz abrauchen und solviere es noch einige Male, damit alle *Feces* davon kommen. Wenn es nun recht trocken geworden, so reibe es klein, bringe es in eine Retorte, gieße den obigen *Liquorem* auch darauf und destilliere es miteinander *cohobando* über, so erlangst du einen edlen *Liquorem*, der in allen seinen Teilen martialisch und solarisch ist. Du kannst damit alleine eine Tinctur von großer Wichtigkeit ausarbeiten."

„Willst du aber die Tinctur der Alten präparieren, so muss dieser *Liquor* (‚der einfache Alkahest' – Anm.) mit dem Kupfer combiniert werden, damit solcher eine doppelte Natur (bekommt), denn das Kupfer hat mehr Farbe als das Gold selbst, ja es liegt in ihm verborgen das wahre Feuer der Natur, wovon die Alten soviel geschrieben haben.

Nimm derowegen 5 bis 6 Pfund Kupfer, calciniere es mit gleich schwer Schwefel, bis es dahin gebracht wird, dass es zu einem sehr feinen Pulver gerieben werden kann. Dieses Pulver tue in einen Kolben und gieße von unserem einfachen Alkahest (s.o. – Anm.) nun soviel nach und nach darauf, dass es den *Liquor* nur bedecke. Wolltest du aber solchen auf einmal darübergießen, so würde es alle Gefäße zertrümmern und zerschlagen, da es sich gewaltig erhitzt. Hiernächst setze es in gelinde Wärme 4 Wochen lang und lass es also putreficieren. Alsdann gieße ein destilliertes Wasser nach und nach dazu und lass es also stehen, so wird sich alles grasgrün färben. Diese *Solution* gieße ab und wieder anderes Wasser darauf, so wird es sich abermals färben. Letztlich gieße alle gefärbten *Solutiones* zusammen und destilliere alles zusammen ab bis auf ein Häutlein. Die *Solution* stelle an einen kalten Ort, so wird ein wunderlicher Vitriol anschießen. Diesen sammle insgesamt soviel, wie der selbe anschießen will, reinige ihn noch einmal *solvendo et coagulando*, mache hernach diesen hochgrünen Vitriol, welcher lauter Tinctur ist, fein trocken, tue ihn in eine beschlagen Retorte und treibe daraus *per gradus ignis* in 48 Stunden einen *Spiritum* und *Oleum*. Dieses destilliere miteinander *per se* noch einmal herüber, so bekommst du ein feuriges, hochgelbes Öl oder *Liquorem*. Aus dem *Capite mortuo* mache das Salz, dieses reinige durch solvieren und coagulieren, bring es hernach in eine Retorte und gieße alles Überdestillierte darauf. Setze die Retorte 14 Tage und Nächte in die Putrefaction, dann lege sie in Sand mit einer daranlutierten Vorlage und destilliere es miteinander *cohobando* herüber. Das ultimum Capite mortuum calciniere noch ferner und mache das Salz daraus. Dieses musst du ebenfalls solvendo et coagulando reinigen bis es ganz purificiert und flüssig worden. Dann bringe es in eine Retorte und destilliere es auch miteinander *cohobando* herüber, so hast du den wahren *Liquorem universalem* in Händen. Mit solchem musst du nun noch weiter zu Arbeit gehen und dahin trachten, denselben noch mercurialischer zu machen, welches also geschieht.

Nimm eine reine *Mineram Antimonii*, die schön langspießig ist, tue solche kleinstgestoßen und gerieben in einen Kolben mit einem flachen Boden und gieße von unserem *Liquore* oder *Aceto* 3 Querfinger hoch darüber, setze auf den wohlverlutierten Kolben einen blinden Helm und lass es miteinander 30 Tage und Nächte, auch länger, in der Putrefaction stehen, so wird sich die *Solution* blutrot färben.

Gieß vorsichtig ab und von unserem Natur-Essig anderen darauf, bis sich nichts mehr färben will. Danach gieße die klaren Solutionen alle zusammen in ein reines Glas, stelle es abermals verschlossen und wohl zugemacht zu putreficieren hin 40 Tage und Nächte in eine dämpfende Wärme, so schließt sich das *Corpus* in der Fäulung von neuem auf und wird so schwarz werden wie ein Samt.

„Wenn man nun solche Schwärze zu Gesicht bekommt, so hat sich die Materie in den Tod begeben und realiter solviert. Danach folgt die Scheidung der Elemente:

Nimm diese schwarze *Materiam* mit dem Kolben, setze einen Helm darauf, stelle den Kolben in die Asche und destilliere herüber was gehen will, so wird nichts übersteigen als ein säuerlicher *Spiritus*, hell und klar, die Materie aber bleibt unten im Glase ganz trocken und zuckersüß liegen. Diese nimm heraus, und nachdem du solche noch etliche Male in einem destilliertem Regenwasser solvieret und wieder abdestillieret hast, so tue die *Materiam* wiederum in einen hohen Kolben oder in eine hohe Phiole. Diese sigilliere und setze sie 2 Monate lang in gelinde Wärme des 1. Grades.

Dies ist aber wohl zu merken, daß, ehe die Phiole zugemacht wird, man einen reinen und sehr starken *Spiritus Vini rectif.* drei Querfinger hoch über die *Materiam* gieße, da sich dann der Weingeist hochrot und wie ein Rubin so durchsichtig färben wird. Diesen hochroten, höchstrectificierten Weingeist gieße ab. Sollte sich etwas Trübes im Abgießen mit herüberbegeben, so muss es durch ein Papier filtriert und gereinigt werden. Nachdem nun alles dergestalt extrahieret, so wirf die rückständige *Terram* hinweg, denn man kann die selbe als eine tote Erde zu nichts gebrauchen. Die vorbesagte *Extraction* aber kann man über den Helm destillieren, so steigt die *Quinta Essentia* unter tausend Farben wie ein lauter Gold herüber, dass der Kolben und Helm sich über und über vergoldet. Dieses ist nun das wahre *Aurum Potabile* oder das flüssige Gold der Weisen."

nach: *Zoroaster, >Clavis Artis<*, 1236

LITERATURHINWEISE

- *Abraham, Georg & T. S. Kuppanna Sastry* – Relation of Babylonian and Indian Mathmethical. IJHS New Delhi 1980/82
- *Abraham, Georg* – Mean Sun and Moon in the Ancient-Greek and Indian Astronomy. New Delhi IJHS 1991
- *Abraham, Georg* – Ancient and Medical Star Catalogue. New Delhi 1997
- *Achelis, Johann Daniel* – Über den Begriff der „Alchimie" in der paracelsischen Philosophie. In: Blätter für Deutsche Philosophie, Bd. 3. Berlin 1929/30
- *Agrawal, D. P.* – Ancient Metal Technology in India, Thailand and Japan. Aryan Books Int., New Delhi 2000
- *Agricola, Johann* – Chymische Medizin, Leipzig 1638/39, Vlg. O. Humberg 2000
- *Agricola, Georg* – *De Re Metallica*, 12 Bücher vom Berg- und Hüttenwesen, 1556, Vlg. VDI, Berlin 1928
- *Allers, Rudolf* – Microcosmus. From Anaximander to Paracelsus. In: Traditio 2 (1944)
- *Archarion* – Von wahrer Alchemie, Vlg. H. Bauer 1967

- *Baader Gerhard* – Medizinisches Reformdenken und Arabismus im Deutschland des 16. Jahrhunderts. In: Sudhoffs Archiv 63 (1979), S. 261–296
- *Baumann, P.* – *praxis spagyrica*. Eine Alchemie der Heilpflanzen, Herbalchym 2002
- *Baumann, P.* – Spagyrik – Alchemie als Heilmethode, Herbalchym 1991
- *Bellinger, G. J.* – Lexikon der Mythologie, Knaur Vlg., 2002
- *Bernus, Alexander v.* – Alchemie und Heilkunst, Vlg. H. Carl, 1969/1972
- *Betschart, P. Ildefons* – Die Signaturenlehre des Paracelsus. In: *Nova Acta Paracelsica* 9 (1977), S. 164-179
- *Biedermann, Hans* – *Materia Prima*. Eine Bildersammlung zur Ideengeschichte der Alchemie, Graz 1973
- *Biedermann, Hans* – Biographisches Lexikon der hervorragenden Ärzte aller Zeiten und Völker. August Hirsch, München/Berlin 1962

- *Blavatsky, H.P.* – Die Geheimlehre, Edit. 3 Masques 1998
- *Blüher, Hans* – Traktat über die Heilkunde (insbesondere die Neurosenlehre), Vlg. O. Humberg
- *Brunschyg, Hieronymus* – Ars Destillandi (...). Straßburg 1610
- *Buchwald, Gerhard Dr. med.* – Impfen, Das Geschäft mit der Angst, Emu Vlg. 1994
- *Bühler, Walther* – Pentagramm und Goldener Schnitt, Vlg. Freies Geistesleben, Stuttgart 1996
- *Byron, Robert* – Der Weg nach Oxiana, 1937/2004

- *Cardilucio, J. Hiskia* – *Clavis* zu Carrichter. 17. Jhdt.
- *Carrichter, Bartholomäus* – Kreutterbuch (...). Straßburg 1577
- *Cerruti, Familie* – Tacuinum Sanitatis. 14. Jhdt. „Von der gesunden Lebensweise", BLV 1985
- *Cousto* – Die kosmische Oktave, Synthesis 1984
- *Croll, Oswald* – Tractatus de signaturis internis rerum (...). Frankfurt 1609
- *Crügner Michael* – Chymischer Frühling. Nürnberg 1654, Edit. Hornfisher 1998

- *Dannemann* – Die Naturwissenschaften in ihrer Entwicklung und ihrem Zusammenhang. Berlin 1910
- *Darmstaedter, Dr. E. (Hg.)* – Die Alchemie des Geber. 9. Jhdt./1922
- *Dash, Vaidya Bhagwan* – Alchemy & Metallic Medicines in Ayurveda. C.P.C. Delhi 1986
- *Dash, Vaidya Bhagwan* – Materia Medica of Ayurveda, New Delhi 1987
- *Dash, Vaidya Bhagwan* – Fundamentals of Ayurvedic medicine. New Delhi 1984
- *Debus, Allen G.* – The English Paracelsians. London 1965
- *Dieter Kuhn (Hg.)* – Chinas Goldenes Zeitalter. Die Tang-Dynastie (618-907) und das kulturelle Erbe der Seidenstraße. Heidelberg: Braus 1993
- *Dioskurides, Pedanios* – Arzneimittellehre, 1. Jhdt., Übers. v. Dr. J. Berendes, Stuttgart 1922
- *Düg, Peter* – Vom Ansehen der Arzneikunst. Historische Reflexionen in Kräuterbüchern des 16. Jhdts. In: Sudhoffs Archiv 62, S.64-79 (1978)

- *Eleazar, Abraham* – Uraltes chymisches Werk, *Crusius*, Erfurt 1735, G. Wendelholm Vlg. 1982

- *Fiedeler F.* – Yin und Yang, d. kosm. Grundmuster i. d. Kulturformen Chinas, Dumont 1993
- *Fischer, H.* – Mittelalterliche Pflanzenkunde, München 1929
- *Flamel, Nicolaus* – Chymische Werke, Paris 15. Jhdt., Vlg. AAGW, H. Frietsch 1996
- *Foucault, Michel* – Die Ordnung der Dinge. Eine Archäologie der Humanwissenschaften. Frankfurt 1971
- *Fuchs, Leonhart* – New Kreüterbuch, 1543, Vlg. Taschen, Köln 2001

- *Gimbutas, M.* – Bronze age cultures in central and eastern Europe. Den Haag 1965
- *Glaser Christoph* – *Novum Laboratorium Medico-Chymicum*, Nürnberg 1677/ VCH Weinheim 1988
- *Glauber* – *Glauberus Concentratus*, 1715/Haug Vlg. 1961
- *Goltz, Dietlinde* – Studien zur Geschichte der Mineralnamen in Pharmazie, Chemie und Medizin von den Anfängen bis Paracelsus, Wiesbaden 1972/Sudhoffs Archiv, Beiheft 14

- *Hacheney, Wilfried* – Wasser, ein Gast der Erde, Äquadukt Dingfelder Vlg., Andechs 1992
- *Hagers* – Handbuch der pharmazeutischen Praxis, Springer Vlg., Berlin 1938
- *Hahnemann, S.* – Organon der Heilkunst, Haug Vlg., 1981
- *Heinz, U. J.* – Das Handbuch der modernen Pflanzenheilkunde, Bauer Vlg., 1984
- *Helmrich, Hermann E.* – Spagyrik – Alter Wein in neuen Schläuchen, Haug Vlg. 1977
- *Herm, Gerhard* – Die Kelten. Das Volk, das aus dem Dunkel kam. Econ Vlg., 1975
- *Hiller, Johann-E.* – Die Mineralogie des Paracelsus. In: *Philosophia naturalis*. Archive für Naturphilosophie 2, S. 293-331 und 435-478 (1952/54)
- *Hochmeier Peter* – Der Weg des Sonnenfunkens. Ein Grundlagenwerk der Spagyrik, Alchymie und Signaturenkunde. Bacopa 2005
- *Hochmeier Peter* – Alchymistische Impressionen. Bilder und Texte aus der spagyrischen Praxis. Bacopa 2005

- *Holtzapfel, Walter* – Im Kraftfeld der Organe, Vlg. Goetheanum 1990

- *Jackson, K. H.* – The Oldest Irish Tradition. Cambridge 1964
- *Junius, Manfred M.* – Handbuch der Pflanzen-Alchemie, Ansata 1982

- *Kabir* – *The Bijak*, Asian P.S. India 1979
- *Kalbermatten Roger* – Wesen und Signatur der Heilpflanzen, AT Vlg., 2005
- *Karl, J.* – Phytotherapie, T. Marczell Vlg., 1978
- *Kerenyi K.* – Die Mythologie der Griechen 1 u. 2, dtv 1966
- *Kirchweger A. J.* – *Aurea Catena Homeri*, Reprint Stockholm 1972
- *Kirpal Singh* – *The Mystery of Death / Karma, The Wheel of Life*
- *Knorr v. R., Ch.* – Aufgang der Artzney-Kunst, Kösel Vlg., 1971
- *König, Marie E.P.* – Am Anfang der Kultur, Zeichensprache des frühen Menschen, Vlg.2001
- *Kunckel von Löwenstern J.* – Werckschul 1, Nürnberg, etwa 1696
- *Kunckel von Löwenstern J.* – *Laboratorium Chymicum*, Leipzig 1716
- *Kurth, Hanns* – Rezepte berühmter Ärzte aus 5000 Jahren, Keller Vlg., 1974
- *Kutumbiah, P.* – Ancient Indian medicine. Bombay: Orient Longmans, 1962

- *Lengyel, Lancelot* – Das geheime Wissen der Kelten, Vlg. H. Bauer, Freiburg 1993
- *Lenz, Hans Gerhard* (Hrsg.) – Triumphwagen des Antimons, Text-Kommentare-Studien, Vlg. O. Humberg 2004
- *Lenz, Hans Gerhard* (Hrsg.) – Der alchemistische Nachlass Friedrichs I. Vlg. O. Humberg 2005
- *Lenz, Hans Gerhard* (Hrsg.) – Der Alchemist Conrad Khunrath, Texte-Dokumente-Studien, Vlg. O. Humberg 2005
- *Leung, Albert Y.* – Chinesische Heilkräuter, Vlg. Diederichs 1995
- *Lonicerus, Adamus* – Kreuterbuch, Frankfurt 1679, Repr. Vlg. K. Kölbl 1962
- *Lorenz, Friedrich* – Vom Wesen der Elemente, Vlg. am Goetheaneum 1993

- *Lösch, Dr.* – Kräuterbuch, Hausen a. Z. 1903, Repr. Vlg. Bechtermünz, Augsburg 1997

- *Maciocia, Giovanni* – Die Grundlagen der chinesischen Medizin, VGM 1997
- *Maier, Michael* – *Atalanta Fugiens*, Oppenheim 1618
- *Mahdihassan, S.* – Indian Alchemy or Rasayana. New Delhi, 1977
- *Marzell, Heinrich* – Wörterbuch der deutschen Pflanzennamen. Bd. 1 – 4, Vlg. S. Hirzel, Leipzig 1943
- *Marzell, Heinrich* – Der Zauber der Heilkräuter in Antike und Neuzeit. In: Sudhoffs Archiv 29, S. 3-26 (1936)
- *Metzner Ralph* – Der Brunnen der Erinnerung. Von den mythologischen Wurzeln unserer Kultur, Aurum Vlg. 1994
- *Moehsen, J. C. W.* – Leben Leonhard Thurneyssers zum Thurn, Berlin 1783 Vlg. Fritsch 1976 München
- *Morton, F.* – Salzkammergut. Hallstatt 1956
- *Morys Peter* – Medizin und Pharmazie in der Kosmologie Leonhard Thurneyssers zum Thurn (1531-1596), Matthiesen Vlg. 1982
- *Mutus Liber* – Die Alchemie und ihr stummes Buch, Edit. Weber, Amsterdam 1991

- *Osman, Nabil* – Kleines Lexikon deutscher Wörter arabischer Herkunft, Beck Vlg., 1993

- *Paracelsus* – Werke, W. E. Peuckert, Vlg. Schwabe, Basel/Stuttgart 1976
- *Paracelsus* – Sämtliche Werke. Hg. Karl Sudhoff, München/Berlin 1919-1933
- *Paracelsus* – Das Bader-Büchl, Hsg. Otto Stöber, Linz 1958
- *Paracelsus* (?) – Hauptschlüssel der Paracelsischen Arcanen oder: Das zehnte Buch der Archidoxen
- *Piggott, S.* – The Druids, London 1968
- *Platon* – Werke 1 – 10, Insel Vlg., 1991
- *Plutarch* – Lebensbeschreibungen. Hg: F. Kaltwasser, München 1964
- *Pokorny, J.* – Altkeltische Dichtung. Bern 1944

- *Pomet Peter* – Der aufrichtige Materialist und Specerey-Händler 1717 neu: Leipzig 1986

- *Ray, P. & Gupta* – Caraka Samhita: A Scientific Synopsis. INSA New Delhi 1965/80
- *Reid, Daniel P.* – Chinesische Naturheilkunde. Vlg. Orac, Wien 1988
- *Reinhard, J. & Baumann, A.* – Unerhörtes aus der Medizin. Hallwag 1989
- *Respours, P. M. v.* – Besondere Versuche vom Mineral-Geist. Paris 1668/Leipzig 1772, Edit. Hornfisher 1996
- *Ring, Th.* – Astrologische Menschenkunde. Vlg. Rascher 1956
- *Roob, Alexander* – Das hermetische Museum. Alchemie & Mystik. Taschen Vlg. 1996
- *Ruska Julius* – Tabula Smaragdina. Heidelberg 1926
- *Ryff Walther* – Confect Büchlin und Hausz Apoteck. Frankfurt 1544/Leipzig 1983

- *Sankalia, H. D.* – Some Aspects of Prehistoric Technology in India. INSA New Delhi 1970
- *Scheller, E. F.* – Langlebigkeit mit Paracelsus-Arzneien (Versuch einer Geriatrie nach Paracelsus), Haug Vlg. 1977
- *Scherer, A.* – Die Urheimat der Indogermanen. Darmstadt 1968
- *Schmaltz, Dieter* – Pflanzliche Arzneimittel bei Theophrast von Hohenheim genannt Paracelsus. Stuttgart 1941
- *Schmieder K.Christoph* – Geschichte der Alchemie. Halle 1832 / W. Roller 1997
- *Schneider Wolfgang* – Lexikon alchemistisch-pharmazeutischer Symbole. Weinheim 1962
- *Seal, Brajendranath* – The Positiv sciences of ancient Hindus. New Delhi 1958
- *Sen, S. N. & Mamata Chauduri* – Ancient Glass in India. INSA New Delhi 1985
- *Sigerist, Henry E.* – Der Arzt in der griechischen Kultur. Zürich 1963
- *Spunda* – Das Weltbild des Paracelsus. W. Andermann, Wien 1941
- *Stemplinger, Eduard* – Antike und moderne Volksmedizin. Leipzig 1925
- *Sudhoff, Karl* – Iatromathematiker vornehmlich im 15. und 16. Jahrhundert. Breslau 1902 (Abhandlungen zur Geschichte der Medizin 2)

- *Tabernaemontanus, J. Th.* – Neu vollkommen Kräuterbuch. Basel 1731
- *Tacitus* – Historien. Hg. Eckstein F., München 1964
- *Tanckius Joachim* – Promptuarium Alchemiae. Leipzig 1610, Graz 1976
- *Tarn, W.* – Hellenistic civilisation. London 1959
- *Theophrast* – Naturgeschichte der Gewächse. Übers. v. E. Sprengel, Darmstadt 1971 *Thomas, Peter* – Die Astromedizin des Philosophen und Arztes *Marsilio Ficino*. Ein Beitrag zum medizinischen Denken im Zeitalter der Renaissance. Med. Diss. Münster 1970
- *Thurneisser zum Thurn, Leonhard* – *Historia* und Beschreibung (...) aller Erdgewechsen/auch irer Subtiliteten (...) aller zu der Extraction dienstlichen Instrumenten (...), Berlin 1578
- *Thurneisser zum Thurn, Leonhard* – Archidoxa. Munster 1569, Berlin 1575
- *Thurneisser zum Thurn, Leonhard* – Quinta Essentia. Munster 1570, Leipzig 1574
- *Thurneisser zum Thurn, Leonhard* – Pison. Frankfurt 1572
- *Toeltius, J. G.* – Coelum Reseratum Chymicum. Erfurt 1612/1737, Edit. Hornfisher 1994
- *Turba Philosophorum* – *Arisläus,* 11. Jhdt./Hg. Phillipum Morgenstern, Wien 1750, Edit. Hornfisher 1996

- *Valnet, Jean* – Aromatherapie. 1976
- *Vasant Lad/David Frawley* – Die Ayurveda Pflanzen-Heilkunde. Windpferd Vlg. 1995
- *Vogel, Hans Ulrich und Elisabeth Theisen-Vogel* – Pan Jixing. Übersetzung und Verbreitung von Georg Agricolas ‚De re metallica' im China der späten Ming-Zeit (1368-1644), Journal of the Economic and Social History of the Orient, 32 (1989), S. 153-202
- *Vogel, Hans Ulrich* – Naturwissenschaften und Technik im vormodernen China: Kurseinheit 2: Geschichte der chinesischen praktischen Alchemie. Studienbrief für die FernUniversität, Hagen 1993.
- *Vogel, Hans Ulrich* – Agricolas *‚De re metallica'* im China der Ming-Zeit - früher Versuch west-östlichen Technologietransfers," Spektrum der Wissenschaft, Oktober 1992, Heft 10, S. 30-34.

- *Vogel, Hans Ulrich* – Grundzüge der Geschichte der chinesischen praktischen Alchemie. China-Report, 15. Juli 1995, Nr. 23, S. 24-27.
- *Verma, P.R.P. & Prasad, C. M.* – Standardization and Bioavailability of Ayurvedic drug *Lauha Bhasma, Ancient Science of Life*, Vol.15 (1995) (S. 129-136 u. 140-144)

- *Weiss, R. F.* – Lehrbuch der Phytotherapie, Hyppokr. Vlg. 1985
- *Willfort, R.* – Gesundheit durch Heilkräuter, Vlg. R. Trauner, Linz 1977
- *Wunderli, Peter* (Hrsg.) – Reisen in reale und mythische Ferne, Droste Vlg. 1993

- *Yates, Francis A.* – Giordano Bruno and the Hermetic Tradition, London 1964
- *Yeung, Him* – Handbook of Chinese Herbs and Formulas 1 & 2, 1983
- *Yonten Pasang Arya, Dr.* – Handbuch aller Heilmittel der traditionellen tibetischen Medizin, Vlg. O. W. Barth 2001

Index

Peter Hochmeier

DER WEG DES SONNENFUNKENS

Ein Grundlagenwerk der Spagyrik, Alchymie und Signaturenkunde

Eine Zusammenschau naturkundlicher Überlieferungen. Vom Wesen der natürlichen Dinge, dem Wirken der fünf Elemente, der Planetenkräfte und der praktischen Anwendung eines zeitlosen Wissensschatzes in Spagyrik und Alchymie.

Das vorliegende Buch soll dem interessierten Leser zeitloses, hermetisches Denken nahebringen und dazu anregen, das „natürliche Gewebe" ganzheitlich zu betrachten. Wo möglich, wird es auch zur Vertiefung jener Themen beitragen. Dies mag dem Laien im alltäglichen Leben genauso dienlich sein wie dem Praktiker eines der zahlreichen natur- oder heilkundlichen Fachgebiete. Thermik, Elementar- und Planetenkräfte werden heute in hochwertigen, aus der lebendigen Natur begründeten Ernährungskonzepten beachtet, welche sich am physischen und energetischen Aufbau des Körper-Geist-Kontinuums orientieren. Sie bilden auch die Grundlage des naturgemäßen Anbaus von Lebensmitteln und Heilpflanzen. Die Wiederentdeckung asiatischer Medizinsysteme für den Westen erfordert eine tiefe Auseinandersetzung mit dem System der fünf Elemente und eine Integration der alchymistischen Themen. Mit der Signaturenkunde öffnet sich uns ein weites Gebiet, in welches viele Pfade führen – oft bedingt durch traditionelle Eigenheiten oder eine individuelle Art der Wahrnehmung. Dieses Buch soll die Grundbausteine aufzeigen und zu weiterführendem Forschen anregen. Anhand einiger praktisch nachvollziehbarer Beispiele wird die spagyrische Arbeit im Pflanzen- und Mineralreich umrissen. Hier befinden wir uns in der Grundschule jener Kunst, die Paracelsus „den rechten Umgang mit den Dingen der Natur" nennt. Dabei geht es nicht um „Philosophie" und schon gar nicht um „Esoterik" im heutigen Sinne der Worte, sondern eben um einfaches Naturbegreifen. Durch Innehalten und Wahrnehmen wird die Sprache der Natur verständlich. Manchmal stellt sich uns aber jenes uralte Wissen mit so tiefgründiger Größe dar, dass wir bloß imstande sind, seine Grenzen zu berühren. Forschergeist und Kreativität werden wachgerufen – und vor allem Freude am Dasein.

564 Seiten, Grafiken und s/w Abb.; Leinen, geb.
ISBN 3-901618-25-2
BACOPA-Verlag

Peter Hochmeier

ALCHYMISTISCHE IMPRESSIONEN

Bilder und Texte aus der spagyrischen Praxis

Dieser Bildband vermittelt einen Eindruck davon, wie die alchymistische Kunst auch heute noch (oder: wieder) praktiziert werden kann. Neben naturphilosophischen Betrachtungen und zum Teil sehr seltenen Arbeitsanleitungen aus alten Texten (photographisch dokumentiert) werden auch Verbindungen zur ayurvedischen Alchymie hergestellt. Als erstes Werk seiner Art wirft dieses Buch im wahrsten Sinne Licht auf jene zeitlose Kunst, die Paracelsus den „rechten Umgang mit den Dingen der Natur" nannte.

176 Seiten, durchg. farbige Abb.; geb.
ISBN 3-901618-38-4
BACOPA-Verlag

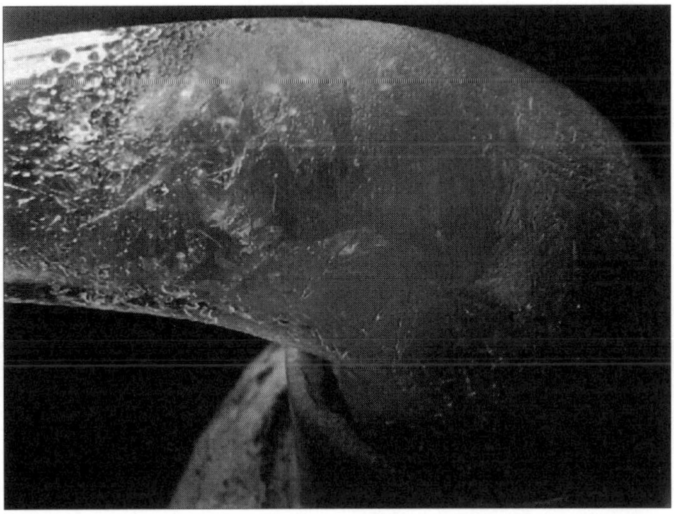